Thermodynamics and Its Applications

3rd Edition

PRENTICE HALL INTERNATIONAL SERIES
IN THE PHYSICAL AND CHEMICAL ENGINEERING SCIENCES

NEAL R. AMUNDSON, SERIES EDITOR, *University of Houston*

ADVISORY EDITORS

ANDREAS ACRIVOS, *Stanford University*
JOHN DAHLER, *University of Minnesota*
THOMAS J. HANRATTY, *University of Illinois*
JOHN M. PRAUSNITZ, *University of California*
L. E. SCRIVEN, *University of Minnesota*

BALZHISER, SAMUELS, AND ELLIASSEN *Chemical Engineering Thermodynamics*
BEQUETTE *Process Dynamics: Modeling, Analysis and Simulation*
BIEGLER, GROSSMANN, AND WESTERBERG *Systematic Methods of Chemical Process Design*
CROWL AND LOUVAR *Chemical Process Safety*
DENN *Process Fluid Mechanics*
FOGLER *Elements of Chemical Reaction Engineering, 2nd edition*
HANNA AND SANDALL *Computational Methods in Chemical Engineering*
HIMMELBLAU *Basic Principles and Calculations in Chemical Engineering, 6th edition*
HINES AND MADDOX *Mass Transfer*
KYLE *Chemical and Process Thermodynamics, 2nd edition*
NEWMAN *Electrochemical Systems, 2nd edition*
PRAUSNITZ, LICHTENTHALER, AND DE AZEVEDO *Molecular Thermodynamics of Fluid-Phase Equilibria, 2nd edition*
PRENTICE *Electrochemical Engineering Principles*
STEPHANOPOULOS *Chemical Process Control*
TESTER AND MODELL *Thermodynamics and Its Applications, 3rd edition*
TURTON, BAILIE, WHITING, AND SHAEIWITZ *Analysis, Synthesis, and Design of Chemical Processes*

Thermodynamics and Its Applications

3rd Edition

Jefferson W. Tester
Michael Modell
Chemical Engineering Department
Massachusetts Institute of Technology
Cambridge, Massachusetts

To obtain a Prentice Hall PTR mailing list, point to:
http://www.prenhall.com/register

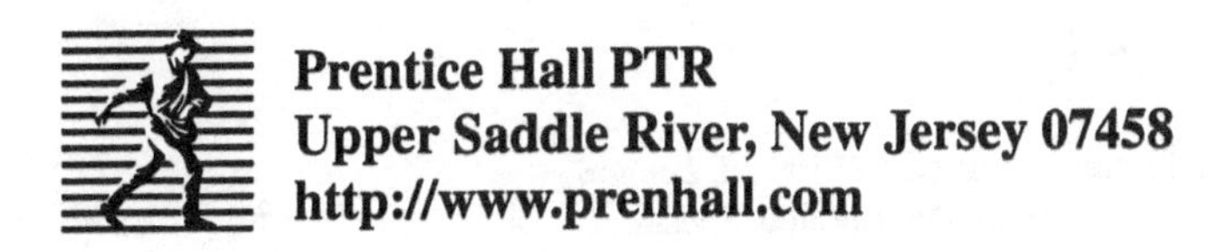

Prentice Hall PTR
Upper Saddle River, New Jersey 07458
http://www.prenhall.com

Library of Congress Cataloging-in-Publication Data

Tester, Jefferson W.

Thermodynamics and its applications / Jefferson W. Tester and Michael Modell.--3rd ed.

p. cm. -- (Prentice-Hall international series in the physical and chemical engineering sciences)

Rev. ed. of: Thermodynamics and its applications / Michael Modell, Robert C. Reid. 2nd ed. c1983.

ISBN 0-13-915356-X (cloth)

1. Thermodynamics. I. Modell, Michael Thermodynamics and its applications. II. Title. III. Series.

QD504.T37 1996

541.3'69--dc20 96-31732

CIP

Editorial/Production Supervision: Kathleen M. Caren
Acquisitions Editor: Bernard M. Goodwin
Manufacturing Manager: Alexis R. Heydt
Book Design and Layout: Susan Robson
Electronic Typesetting: Susan Robson
Figures and Artwork: Michael Kutney and Tetsuo Maejima
Cover Design Director: Jerry Votta

© 1997 by Prentice Hall PTR
Prentice-Hall, Inc.
Upper Saddle River, New Jersey 07458

The publisher offers discounts on this book when ordered in bulk quantities.
For more information, contact:

Corporate Sales Department
Prentice Hall PTR
1 Lake Street
Upper Saddle River, NJ 07458
Phone: 1-800-382-3419
FAX: 201-236-7141
email: corpsales@prenhall.com

All rights reserved. No part of this book may be
reproduced, in any form or by any means, without
permission in writing from the publisher.

Printed in the United States of America

ISBN 0-13-915356-X

Prentice-Hall International (UK) Limited, *London*
Prentice-Hall of Australia Pty. Limited, *Sydney*
Prentice-Hall Canada Inc., *Toronto*
Prentice-Hall Hispanoamericana, S.A., *Mexico*
Prentice-Hall of India Private Limited, *New Delhi*
Prentice-Hall of Japan, Inc., *Tokyo*
Pearson Education Asia Pte. Ltd., *Singapore*
Editora Prentice-Hall do Brasil, Ltda., *Rio de Janeiro*

Contents

Part II Thermodynamic Properties

Part III Applications

Appendixes

Preface

Authors' Note. Robert C. Reid was a co-author of the first and second editions of *Thermodynamics and Its Applications*. His contributions to this book and to the field of thermodynamics in general are widely known and respected. Even though Bob Reid is not listed formally as an author of this edition, his impact on us and on the contents of the third edition has been significant and is deeply appreciated.

General approach. For over 50 years, the M.I.T. Chemical Engineering Department has offered a one-semester, graduate-level subject in classical thermodynamics. Traditionally, this course has been applications oriented, with one of its primary objectives to develop problem solving skills in applying fundamentals to applications in new and sometimes unusual situations. About half of our classroom contact hours are now devoted to discussing approaches to solving problems. Today, there is an increasing need for practicing chemical engineers to connect traditional macroscopic and empirically-based property models to their roots at the molecular-level where intermolecular interactions responsible for deviations from ideal behavior are characterized in detail. Consequently, since the last edition was published in 1983, we have expanded our core graduate curriculum in thermodynamics to formally link molecular effects to constitutive property models using the principles of statistical mechanics. The text material required to achieve these enhancements evolved over time, leading to a plan for a comprehensive revision. In designing our approach, we wanted to preserve the integrity of the theoretical and conceptual treatment of classical thermodynamics contained in the earlier editions while introducing a molecular-level perspective on a comparably rigorous basis. To do so, we restructured the second edition somewhat, and added substantial new material. The third edition now is intended for a two-semester subject in graduate-level chemical thermodynamics.

About this edition. Those familiar with the first and second editions of *Thermodynamics and Its Applications*, will see that we have retained many features including the postulatory approach, the challenging homework problems, and the treatment of external fields and surfaces. Enhancements in the third edition include six new chapters that provide: (1) in-depth coverage of constitutive property models for non-ideal fluids, including a statistical mechanics basis, a discussion of van der Waals and McMillan-Mayer theories, treatments of local composition concepts and electrolyte solutions, and coverage of volumetric equations of state and activity coefficient models for non-electrolyte mixtures; (2) a unified development of combined phase and chemical equilibria using a vectorized framework; and (3) a new section on availability/exergy analysis of power cycles and other processes involving heat-to-work conversions. The book now has 19 chapters and has been divided into: Part I, Fundamental Principles; Part II, Thermodynamic Properties; and Part III, Applications.

While the subject matter has been updated to include the latest constitutive property models that practicing chemical engineers and chemists employ, we have tried to keep the coverage sufficiently general to be of lasting value both as a textbook and as a professional reference. Example problems illustrate important concepts and problem solution methods.

Part I provides the theoretical basis of classical thermodynamics. Chapters 1-6 cover the same topics as before, including the 1st and 2nd Laws, the Fundamental Equation, Legendre transformations, and general equilibrium criteria. Several enhancements were added as well, specifically more example problems, new homework problems, and revisions in notation and coverage to provide better compatibility with the traditional undergraduate chemical engineering textbooks in thermodynamics. Major changes were made in Chapter 5 to incorporate the use of enhanced graphics into the formal treatment of the Fundamental Equation to illustrate the geometric features of the *USV* Gibbs surface. In addition, Chapter 5 now contains a better derivation of Legendre transformations, and descriptions of other mathematical methods for manipulating thermodynamic properties and their derivatives. Chapter 7 covers only stability criteria, with stability applications moved to Chapter 15.

Part II contains an extensive description of how thermodynamic properties are correlated, modeled, manipulated, and estimated. Both macroscopic empirically-based and molecular-level approaches are discussed in-depth for pure components and mixtures. Formal treatments of fugacity and fugacity coefficients and activity and activity coefficients are related to appropriate $PV\underline{T}N$ equations of state and ΔG^{EX} models. Chapter 8 on pure materials and Chapter 9 on mixtures have been extensively revised. Completely new chapters include: Chapter 10 on modeling properties using statistical mechanics; Chapter 11 on non-electrolyte solution models; Chapter 12 on electrolyte models; and Chapter 13 on physical property estimation.

In Part III, applications of classical thermodynamics are treated in detail. Chapter 14 on practical heat engines and power cycles is new, emphasizing the use of availability and exergy concepts for assessing performance. Chapter 15 on phase equilibria and stability applications combines appropriate material from Chapters 9 and 10 of the second edition. Chapter 16 covers chemical equilibria with several enhancements. Chapter 17 is new, treating combined phase and chemical equilibria problems using a robust vectorized approach that is completely general. Chapter 18 covers the material on electromagnetic and other potential fields that was in Chapter 12 of the second edition. Major changes were introduced in Chapter 18 to the section on electromagnetic fields to fix some earlier problems. Chapter 19 on surface thermodynamics is essentially the same as Chapter 13 of the second edition.

In a typical two-semester subject sequence, all of Part I (Chapters 1-7) and Chapters 8, 9, 14, 15, 16, and 17 of Parts II and III are normally covered in the first semester, emphasizing the fundamentals of classical thermodynamics and their primary applications to chemical engineering, particularly to phase and chemical equilibria. In the second semester, emphasis is placed on property models and specialized applications to external field effects and surfaces, with Chapters 10, 11, 12, 13, 18, and 19 covered. To obtain a better understanding of how molecular-level effects lead to non-ideal behavior, macroscopic constitutive property models are connected to molecular theory in Chapters 11 and 12, using the concepts of statistical mechanics developed in Chapter 10. For further information please consult our web page at http://web.mit.edu/testerel/thermo/

Jefferson Tester
Michael Modell
Cambridge, Massachusetts

Acknowledgments and Credits The revisions necessary to complete this edition were done by one of us (JWT) over the past four years. Recognizing that it is absolutely impossible to thank all contributors for their support and suggestions and at the risk of offending someone we inadvertently omit, I would like to extend special thanks to several individuals who had major impact on the contents of the third edition.

Professor Kenneth Jolls of Iowa State provided major enhancements with his numerous suggestions and contributions to Part I of the book. Of particular significance were his graphical representations of the Fundamental Equation. Professors Kenneth Smith and William Deen and Howard Herzog at MIT, Professors Dudley Herschbach at Harvard and Keith Johnston at the University of Texas provided important suggestions to the sections on departure functions and constitutive equations of state. Several other colleagues including Doros Theodorou (University of Patras), Jonathan Harris (MIT), Thanassis Panagiotopoulos (Cornell), Pablo Debenedetti (Princeton), William Peters (MIT), and James Ely (Colorado School of Mines) provided substantive input to Chapter 10 on statistical thermodynamics. In addition, important course content discussions and numerous homework problems were contributed by Professor Daniel Blankschtein of MIT. Professors Greg Rutledge of MIT and Doros Theodorou also provided homework problems. Contributions from Professors Erdogan Kiran (University of Maine), Steven Penoncello (University of Idaho), and Edith Sevick (University of Colorado), and from Raymond Thorpe, Keith Gubbins, Peter Harriott, Brad Anton, and Michael Duncan (Cornell) are appreciated. Ronald Rosensweig's review of electromagnetic field effects made an important contribution to Chapter 18. I am especially grateful for the stimulating discussions on polymer thermodynamics provided by Professors Edward Merrill and Robert Cohen of MIT, by Andrew Zydney of the University of Delaware and by Ferdinand Rodriguez and Claude Cohen of Cornell.

Elizabeth Drake, Bill Peters, and other colleagues at the MIT Energy Lab unselfishly provided me with the time needed to complete the revision. In addition, during the final stages, while I was on sabbatical from MIT, colleagues at Los Alamos National Laboratory, Cornell University and the University of Colorado afforded me welcome sanctuary and encouragement. I would also like to acknowledge Professor Robert Brown, MIT's current Dean of Engineering and former Department Head of Chemical Engineering, and Professor David Litster, Vice-President for Research at MIT, for their encouragement and support of this project.

I want to acknowledge the major effort expended by Susan Robson designing and electronically typesetting the manuscript. Without her creative efforts and hard work, we would not have been able to produce such a professional document. Special thanks also go to Michael Kutney for his conscientious assistance in proofreading and preparing the figures for this edition, and to Tetsuo Maejima for his original artwork, which was electronically transferred from the earlier editions. Bonnie Caputo and Anne Carbone also provided valuable support in preparing early drafts and in the implementation of the Beta-test program.

Over the years, many students assisted us in developing problems and in proofreading and preparing charts and tables of data and parts of the appendix. For the third edition specifically, I am especially grateful to A. Cano-Ruiz, M. DiPippo, J. DiNaro,

P. Marrone, B. Phenix, R. Weinstein, M. Reagan, K. Sparks, M. Wilkinson, F. Armellini, C.O. Grigsby, P. Webley, R. Helling, R. Holgate, C. Nagel, W.G. Worley, B. Kumnick, J. Champion, and J. Cline. In addition, I would like to thank all the students who participated in the Beta-test program at MIT during the 1995-1996 academic year.

I have been blessed with the opportunity to have interacted with many great teachers of thermodynamics who have shared their enthusiasm, knowledge, and insights for this subject with me. Of these, Robert Reid, Herman P. Meissner, John Prausnitz, Raymond Thorpe, Herbert Wiegandt, Irwin Oppenheim, John Heywood, Hank Van Ness, Robert von Berg, and Jean P. Leinroth had an enormous impact on me personally and professionally.

And most of all, I am deeply thankful for the encouragement given to me by my family and friends throughout all of this. In particular, my daughter Kelsey's sensitivity and spirit and my wife Sue's steadfast love, support and good advice have enriched my life and these pages. I especially want to thank my Mother for teaching me to aim high and assuring me that "this takes care of that!"

Jefferson W. Tester

Nomenclature

Symbols [units in SI]

a, $\underline{a}$	Specific and total surface area [m^2/mol, m^2]
$\underline{A}$	Total Helmholtz free energy, extensive [J]
A	Specific Helmholtz free energy, intensive [J/mol, J/kg]
a_i	Activity of component i
a_{ij}	Activity of ionic electrolyte ij
$a_{\pm}$	Mean ionic activity of electrolyte ij
$\mathbf{a}$	Acceleration [m/s^2]
b_ℓ	Number of moles of element ℓ
$B(T)$	Second virial coefficient
$\underline{\mathcal{B}}$	Availability, extensive [J]
$\mathcal{B}$	Specific availability [J/mol, J/kg]
$\underline{B}$	General extensive property
B	General intensive property
$\mathbf{B}$	Magnetic induction [kg/s^2A or tesla]
c, C	Concentration [mol/L, etc.]
C_p	Constant pressure heat capacity [J/mol K, J/kg K]
C_v	Constant volume heat capacity [J/mol K, J/kg K]
$\mathbf{D}$	Electric displacement [sA/m^2 or coulombs/m^2]
$\underset{\approx}{\mathcal{D}}$	Elemental atom balance matrix
D_s	Static dielectric constant
d_m	Dipole moment, Debyes
d_{ij}	$\underset{\approx}{\mathcal{D}}$ matrix element, number of atoms i in component j
e	Elementary electron charge [$1.6021773 \times 10^{-19}$ coulomb (C)]
$\mathbf{E}$	Electric field strength [m^2kg/s^3A or volt/m]
$\underline{E}$	Total energy, extensive [J]
E	Specific energy, intensive [J/mol, J/kg]
F	Faraday constant [96,500 coulombs/mol or sA/mol]
f_i	Fugacity of pure i
$\hat{f}_i$	Fugacity of component i in a mixture
$\mathbf{F}$	General force vector [N]
F_i	Force component vector magnitude [N]
$\mathcal{F}$	Number of intensive degrees of freedom
$\underline{G}$	Total Gibbs free energy, extensive [J]
G	Specific Gibbs free energy, intensive [J/mol, J/kg]
$\mathbf{g}$, g	Acceleration of gravity [m/s^2]
$\underline{H}$	Total enthalpy, extensive [J]
H	Specific enthalpy, intensive [J/mol, J/kg]
$\mathbf{H}$	Magnetic field strength [A/m]
H	Hamiltonian [J]
I	Ionic strength [mol/kg]
k	Boltzmann constant [1.381×10^{-23} J/K]
K_a	Equilibrium constant
K_ϕ	Equilibrium ratio of fugacity coefficients ($\hat{\phi}_i$)
K_p	Equilibrium constant based on partial pressures (p_i)
K_y	Equilibrium ratio of mole fractions (y_i)
K_γ	Equilibrium ratio of activity coefficients (γ_i)

ℓ	Number of elements
M	Molarity [mol/liter or mol/L]
m, M	Mass [kg]
m	Molality [mol/kg of solvent]
$\dot{m}$	Mass flow rate [kg/s]
m_i	Molecular weight of component i [g/mol]
$\mathbf{M}$	Magnetization [ampere/m or A/m]
N	Number of moles [mol] or molecules
N_A	Avogadro's number [6.022137×10^{23} mol^{-1}]
n	Number of components
n_e	Moles of electrons transferred per mole
n_{in}, n_{out}	Entering or leaving mass or moles [kg or mol]
$\mathbf{n}$	Normal unit vector
P	Pressure [Pa or bar]
P_{vp}, P_{sat}	Equilibrium vapor pressure [Pa or bar]
P_o	Dead state or ambient pressure [Pa or bar]
P_i, p_i	Partial pressure of mixture component i [Pa or bar]
$\mathbf{P}$	Polarization [A s/m^2 or coulomb/m^2]
$\underset{\sim}{\mathbf{p}}^N$	Momentum vector space for N particles or molecules
$\mathbf{P}_N(\underline{E}_i)$	Probability distribution of energy states
Q	Heat [J]
$\dot{Q}$, $\dot{q}$	Heat transfer rate [J/s]
$\mathbf{q}$	Heat flux vector [J/m^2s]
$\underline{q}$	Electric charge (see Table 3.1)
R	Gas law constant [J/mol K or Pa m^3/mol K]
r	Number of independent chemical reactions and/or additional constraints on a system
r_c	Critical pressure ratio

r	Radial position coordinate
$\underset{\sim}{\mathbf{r}}^N$	Configuration space vector for N particles or molecules
$\underline{S}$	Total entropy, extensive [J/K]
S	Specific entropy [J/mol K, J/kg K]
t	Time [s]
T	Absolute temperature [K]
T_o	Dead state or ambient temperature [K]
$\underline{U}$	Total internal energy, extensive [J]
U	Specific internal energy [J/mol, J/kg]
$\underline{V}$	Total volume, extensive [m^3 or L]
V	Specific volume [m^3/mol, m^3/kg]
$\mathbf{v}$	Velocity vector [m/s]
$\lvert\mathbf{v}\rvert$, v	Velocity vector magnitude [m/s]
v_c	Sonic velocity [m/s]
W	Work [J]
$\dot{W}$	Work rate or power [J/s or W]
W_e	Electrical work [J]
W_{max}, W_{min}	Maximum or minimum work [J]
W_{rev}	Reversible work [J]
W_s	Shaft work [J]
$\dot{W}_s$	Shaft work rate or power [J/s or W]
$\mathbf{x}$, $\mathbf{z}$	General displacement vector
$x_i^{(s)}$	Mole fraction of component i in phase s
x_i	Liquid mole fraction of component i
y_i	Vapor mole fraction of component i
Z	Compressibility factor
$\mathbf{Z}^*$	Configurational integral [Eq. (10-76)]
z	Elevation [m]
z_i	General canonical variable
(z_i, ξ_i)	Conjugate coordinates
z_+, z_-	Electrical charge or valance

Greek Symbols

α_p	Isobaric thermal coefficient of expansion [K^{-1}]
γ_i	Activity coefficient of component i
$\gamma_\pm$	Mean ionic activity coefficient
Γ_{ij}	Reduced activity coefficient of electrolyte ij
$\mathcal{E}$, $\exists$	Electromotive force (emf) [volts]
ε_o	Vacuum permittivity
ε	Permittivity of a medium
η	General thermodynamic efficiency
η_c, η_{Carnot}	Carnot efficiency
η_{cycle}	Cycle efficiency
η_t	Turbine efficiency
η_p	Pump or compressor efficiency
θ	Thermometric or empirical temperature [°C, °F, ...]
κ	Heat capacity ratio $\equiv C_p/C_v$
κ_T	Isothermal compressibility [Pa^{-1} or bar^{-1}]
λ	Lagrangian multiplier
$\lambda_i(T)$	Temperature dependent term in Eq. (9-81) equal to μ_i^{ig} (T, P = 1 bar)
$\Lambda_i(T, P)$	Temperature and pressure dependent term in Eq. (9-83)
Λ	Thermal deBroglie wavelength [Eq. (10-64)]
μ_i	Chemical potential of component i
ν_i	Stoichiometric coefficient of component i
ν_+ or ν_-	Stoichiometric coefficients for an electrolyte ij
ν	Total number of ions produced per mole of electrolyte ij [$\nu = \nu_+ + \nu_-$]
π	Number of phases
Π	Osmotic pressure [Pa or bar]
ρ	Density [mol/m^3, kg/m^3]
$\rho_\pm$	Charge density [$coulombs/m^3$]
σ	Surface tension [N/m]
σ_x	Unidirectional stress ($F_x/\underline{a}$) [Pa]
ϕ_i	Fugacity coefficient of pure i
$\hat{\phi}_i$	Fugacity coefficient of component i in a mixture
ϕ_s^*	Practical (molality based) osmotic coefficient
$\phi_{s,x}^*$	Rational (mole-fraction based) osmotic coefficient
ϕ_{MM}^*	Rational (McMillan-Mayer based) osmotic coefficient
Φ_i^*	Volume fraction of component i
(ξ_i, z_i)	Conjugate coordinates in the Fundamental Equation
ξ, ξ_j	Extent of reaction [Eq. (6-51)] or ($\partial y^{(0)}/\partial z_j$) [Eq. (5-88)]
Φ	Intermolecular potential energy
ω	Acentric factor
Ω	Surface area per mole ($\underline{a}/N$) or one-dimensional strain ($\Delta x/x$)

Operators

δ	Denotes a path dependent variation
Δ	Finite change in a state property
d	Total differential
∂	Partial differential
$\int$	Integral
$\oint$	Closed path integral
ln	Natural logarithm (base e)
log or $\log_{10}$	Common logarithm (base 10)
$\prod$	Cumulative product
$\sum$	Cumulative summation

Subscripts

aq	Aqueous phase component
b	Body force component (e.g., $\mathbf{F}_b$)
c	Critical point property (e.g. P_c)
f	Formation (e.g., ΔH_f)
fus	Fusion (solid to liquid) (e.g., ΔH_{fus})
in	Incoming stream
mix	Mixing property (e.g., ΔH_{mix})
o	Ambient or dead state (e.g., T_o, P_o)
out	Exiting stream
P	Isobaric, constant pressure conditions
r	Reduced property (e.g., $P_r \equiv P/P_c$)
rx	Chemical reaction (e.g., ΔH_{rx})
ref	Reference component or conditions
s	Boundary (or surface) component or solvent
sat	Saturated condition at phase equilibrium (e.g., P_{sat})
T	Isothermal, constant temperature conditions
vap	Vaporization (liquid to vapor) (e.g., ΔH_{vap})
vp	Vapor pressure (e.g., P_{vp})
±	Mean ionic property (e.g., $\gamma_\pm$)

Special Notation

circumflex ^ (as in $\hat{f}_i$)	Denotes property of a component in a mixture
overbar ¯ (as in $\overline{H}_i$)	Denotes partial molar property
underbar _ (as in $\underline{E}$)	Denotes extensive property

Superscripts

EX	Excess property (e.g., $\overline{\Delta H}_i^{EX}$)
ID	Ideal solution or mixture (e.g., ΔG_{mix}^{ID})
ig	Ideal gas (e.g., $V^{ig} = RT/P$)
(*s*)	Phase identification (e.g., $x_i^{(s)} = x_i^{\alpha}$)
∞	Infinitely dilute condition
o	Ideal gas or single pure component state
+	General reference state
±	Mean ionic conditions
□	Hypothetical reference state at unit molality
**	Hypothetical pure component reference state
*	Attenuated condition where ideal gas behavior prevails

Part I Fundamental Principles

The Scope of Classical Thermodynamics 1

1.1 An Engineering Perspective

To the scientist, classical thermodynamics is one of a few mature fields epitomized by a rather well-defined, self-consistent body of knowledge. The essence of the theoretical structure of classical thermodynamics is a set of natural laws governing the behavior of macroscopic systems. The laws are derived from generalizations of empirical observations and are largely independent of any theory or hypothesis concerning the microscopic or molecular-level behavior of matter. From these laws, a large number of corollaries and axioms are derivable by proofs based entirely on logic.

The scientist is sometimes at a loss to understand why the engineer has so much difficulty applying the principles of thermodynamics; after all, the theoretical development is rather straightforward. From the engineer's point of view, understanding the theory as developed by the chemist or physicist is not particularly difficult; however, the neat, self-contained presentation of the subject by the scientist is not necessarily amenable to practical application. Real-world processes are usually far from reversible, adiabatic, or well mixed; very rarely are they isothermal or at equilibrium; few mixtures of industrial importance are ideal. Thus, the engineer must take a pragmatic approach to the application of thermodynamics to real systems. Frequently, a major requirement is to redefine the real problem in terms of idealizations to which thermodynamic principles and methodologies can be applied.

In the engineering context, almost all problems of thermodynamic importance can be classified into one of three types:

1. For a given process with prescribed (or idealized) internal constraints and boundary conditions, how do the properties of the system vary?
2. To cause given changes in system properties, what external interactions must be imposed? (This is the inverse of type 1.)
3. Of the many alternative processes to effect a given change in a system, what are the efficiencies of each with respect to the resources at our disposal?

Problems of the first two classes require application of the First Law, which is developed in Chapter 3:

$$\Delta \underline{E} = Q + W \tag{1-1}$$

where $\underline{E}$ is the total energy and Q and W are the heat and work interactions, respectively. The First Law may also be viewed as:

$$\text{internal changes} = \sum \text{interactions occurring at boundaries}$$

The change in energy can be related to variations of other internal properties of interest (e.g., T, P, V, etc.).

The third class of problems requires application of the Second Law, for which an idealization—the reversible process—is introduced as a standard for comparison.

There are basically only three steps required to develop a solution to any thermodynamic problem:

1. ***Problem definition and modeling***. The real-world situation must be modeled by specifying the internal constraints and boundary conditions. Idealizations must frequently be introduced to make the problem tractable. For example, is a boundary permeable, semipermeable, or impermeable? Are the contents of the system well-mixed and homogeneous? Is heat transfer fast or slow relative to the time span of interest? Which chemical reactions are known to occur under the conditions of interest?

2. ***Application of thermodynamic laws and mathematical reasoning***. As described above, these either relate effects internal to the system with external variations (the First Law) or they set limits on the extent of internal variations (the Second Law). The combined laws prescribe in part the relationships between property variations, but they do not uniquely specify the magnitude of the change in properties. For example, for a simple, non-reacting system of fixed mass (M) and moles (N) undergoing a process in which the temperature and pressure are observed to change from T_1, P_1 to T_2, P_2, we might wish to calculate the total energy change, $\Delta \underline{E}$, in order to specify the necessary heat and work interactions. We might employ the following analysis:

   (a) From *thermodynamic reasoning*, with $N=$ constant, $\Delta \underline{E}$ is a unique function of T_1, P_1 and T_2, P_2 because $\underline{E}$ is a state function. Therefore, $\Delta \underline{E}$ can be evaluated over any path between the states.

   (b) From *mathematical reasoning*, over any path for which $\underline{E}$ is defined, $d\underline{E}$ may be expressed as an exact differential, such as

   $$\Delta \underline{E} = \int d\underline{E} = \int_{T_1}^{T_2} \left(\frac{\partial \underline{E}}{\partial T}\right)_{P_1} dT + \int_{P_1}^{P_2} \left(\frac{\partial \underline{E}}{\partial P}\right)_{T_2} dP \tag{1-2}$$

   Note that $(\partial \underline{E}/\partial T)_P$ and $(\partial \underline{E}/\partial P)_T$ must be expressed as functions of T and P before Eq. (1-2) can be integrated.

   (c) *Applying thermodynamic reasoning*, $\underline{E}$ is defined as a function of T and P over a reversible path, and thus $(\partial \underline{E}/\partial T)_P$ and $(\partial \underline{E}/\partial P)_T$ can be reduced to other variable sets that are more readily quantified:

$$\Delta \underline{E} = \int_{T_1}^{T_2} \left[NC_p - P\left(\frac{\partial \underline{V}}{\partial T}\right)_P \right] dT - \int_{P_1}^{P_2} \left[T\left(\frac{\partial \underline{V}}{\partial T}\right)_P + P\left(\frac{\partial \underline{V}}{\partial P}\right)_T \right] dP \quad (1\text{-}3)$$

where C_p is the constant-pressure heat capacity. Note that Eq. (1-3) is a *general result*; it must be satisfied by any material undergoing a change from T_1, P_1 and T_2, P_2. However, the value of $\Delta \underline{E}$ is not unique; it differs from one material to the next, which leads us to the third and final step.

3. ***Evaluation of property data.*** There are property relationships that are unique characteristics of matter. For example, in Eq. (1-3), thermodynamics does not dictate the functions

$$C_p = f_1(T, P), \quad \underline{V} = NV = Nf_2(T, P) \quad (1\text{-}4)$$

required for the integration. Actual evaluation of these property data lies outside the scope of classical thermodynamics. However, they are essential to the solution of real problems and hence are within the scope of this text (see for example Chapters 8 and 10). The engineer must make recourse to a variety of methods (e.g., literature, experiments, correlations, or microscopic theories as developed with statistical mechanics) in order to determine or approximate these property relationships.

Figure 1.1 summarizes this general method for solving engineering thermodynamics problems. Before discussing the approach to classical thermodynamics used herein (Section 1.3), it is instructive to review the historical evolution of this body of knowledge.

1.2 Preclassical Thermodynamics

The origin of classical thermodynamics can be traced back to the early 1600s. The laws, as we know them today, were not formalized until the late 1800s. The interim 250 to 300 years are called the *preclassical period*, during which many of our current concepts were developed.

The chronological development is a fascinating example of the application of scientific methodology. Experimentation (e.g., thermometry) led to the development of hypotheses and concepts (e.g., the adiabatic wall) which, in turn, suggested other experiments (e.g., calorimetry) followed by new concepts (the pitfalls in scientific analysis, such as overemphasis on intuitive images, e.g., the nature of heat) which go far beyond the existing body of observations and facts. Consequently, the preclassical period was marked with pedagogical controversy and much confusion.

The beginning of the preclassical period is usually associated with Galileo's attempts to quantify thermometry (circa 1600). It is interesting to note that seventeenth-century scientists were motivated primarily by a desire to understand phenomena perceived by their senses. In contrast, scientists today require very sensitive and elaborate instrumentation to detect phenomena that are far beyond the reaches of their senses. One of Galileo's principal objectives was to quantify the subjective experiences of hot and

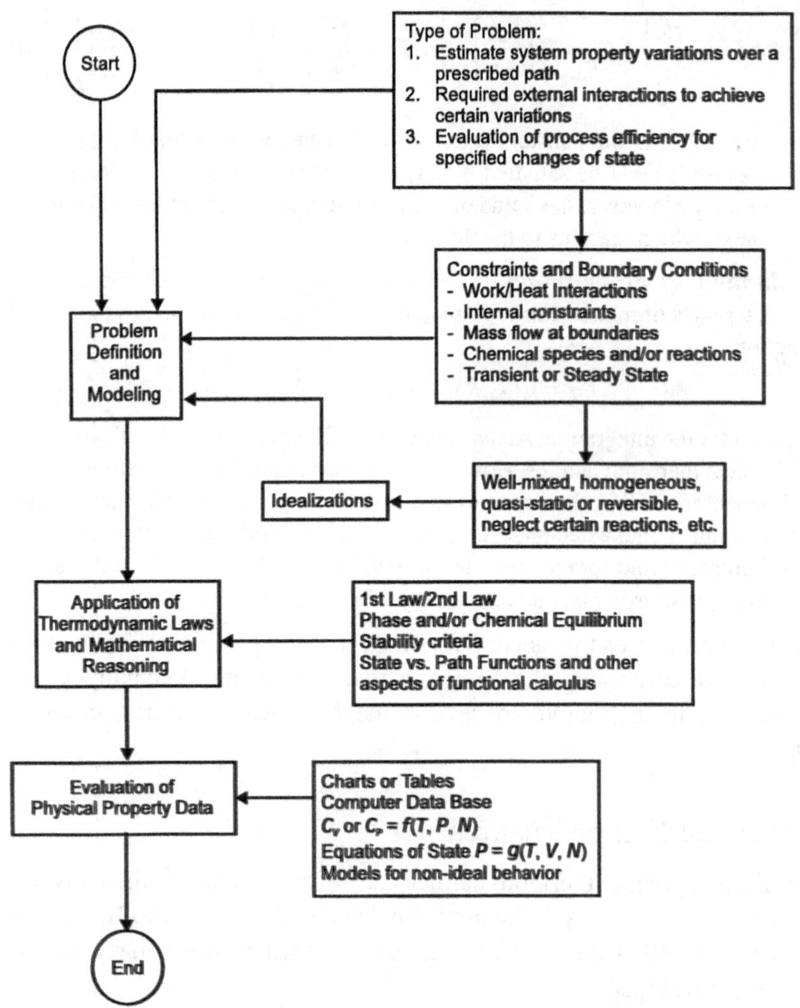

Figure 1.1 Basic thermodynamic problem solving steps.

cold. The expansion of air upon heating was appreciated in the Hellenistic era, but it was never applied. Galileo used this phenomenon in his bulb-and-stem device with the stem submerged in water. The measurements would change with time, but in the early 1600s there was no reason to assume that they should not vary. It was not until 1643, when a student of Galileo, Torricelli, developed the barometer, that it was appreciated that Galileo's device was more of a "barothermoscope" than a thermometer. As glass-blowing technology advanced, the availability of narrow capillaries led to the development of liquid thermometers in the 1630s. As might be expected, water was the first liquid used. Although difficulties that should be obvious to today's students were experienced, ten years elapsed before the sealed alcohol thermometer gained acceptance. Gas thermometers did not reappear until the 1700s, when gas properties were better understood.

When any new experimental tool is developed, invariably there is the desire to quantify it so that results among different investigators can be compared. The quantification of thermometry required the introduction of at least two fiducial or fixed points. Stop for a moment and reflect on what fixed points you might have chosen had you been a scientist in 1640. The boiling point of water? It varies from day to day. The freezing point of water? It is difficult to find ice during most of the year. Furthermore, there was no reason to believe that materials like water had unique properties that would be reproducible. Thus, it is understandable that our ancestors turned to phenomenological references, such as the "warmest water the hand could stand" or the "most severe winter cold" or the "temperature of the human body." Later, selection of the melting point of butter or the freezing point of aniseed oil were steps toward objectivity, although these transition points were not very sharp. It was not until 1694 that the freezing and boiling points of water gained acceptance.

With the advance of quantitative thermometric experimentation it soon became apparent that different types of containers had different thermal properties. Hot liquids would cool less rapidly in mica or wood vessels than in metals. These observations led to the idealized concept of the adiabatic wall, which could be approached in practice; thus, the science of calorimetry was born.

If two portions of the same fluid were mixed in a calorimeter, the final temperature θ_f could be expressed as a weighted mean of the two initial temperatures θ_1 and θ_2:

$$\theta_f = \frac{w_1\theta_1 + w_2\theta_2}{w_1 + w_2} \tag{1-5}$$

The weighting factors w_1 or w_2 could be mass or volume. Although Eq. (1-5) is of the form of a conservation law, it is not at all clear *what* is conserved. A simplistic interpretation would have temperature as the conserved quantity. The conservation law was given more structure in the mid-1700s, when Eq. (1-5), with mass or volume as weighting factors, was shown to be invalid when different liquids were mixed. In the 1760s, Joseph Black suggested a modification that was consistent with mixing data for different fluids. The constants a_i in Eq. (1-5) were subdivided into a mass-related component and an intensity parameter, the specific heat, which was a unique property of the liquid. This was the first time that it was proposed that matter had distinctive properties in the thermodynamic sense.

Black's modification indicated that something other than temperature was conserved in the mixing process. This quantity was called *heat* or *caloric*. The interpretation went far beyond the physical observations into the realm of metaphysics. The caloric theory attempted to define the microscopic nature of the conserved quantity.

Black's ingenious hypothesis led to a flurry of experimentation, during which specific heats were measured and reported in the then flourishing royal societies. The conservation law was repeatedly challenged, but by more exacting experimentation, the theory was enlarged to account for the variation of specific heat with temperature, and later, latent heats were introduced to account for phase transitions.

In the 1780s, no more than 20 years after Black's work, Count Rumford conducted his exhaustive experiments to show that mechanical work was an inexhaustible source of caloric. Hence, caloric could not be conserved and could not be of a material nature. Rumford suggested a revival of the mechanical concept of heat that had been abandoned 50 years earlier. Although we now know that Rumford's suggestion was closer to the truth, it is understandable why very few of his peers followed his lead. The statistical concepts necessary to relate micromechanical energy to the macroscopic energy of calorimetry were not to be introduced until a century later by Maxwell, Boltzmann, and Gibbs. Rumford's hypothesis failed to produce tangible phenomenological results.

Although the caloric theory remained in use for over 50 years after Rumford's work, he emphasized the dilemma between conservation and creation (or conversion) which was to perplex the best minds of the nineteenth century. The conversion crisis was firmly established by the work of Mayer and Joule in the 1840s.

In 1824, Carnot offered a partial reconciliation of the conversion and conservation phenomena with an argument based on the caloric theory (the results of which actually prolonged the life of the theory). Carnot introduced a step-change in the level of complexity and sophistication. He put forth a number of new concepts that were essential to the eventual clarification of preclassical ideas and that later led to the replacement of the caloric theory by the First and Second Laws. These firsts include the concepts of heat reservoirs and reversibility, and the requirement of a temperature difference to generate work from a heat interaction.

Carnot proposed that cyclic operation of an engine working between two heat reservoirs was analogous to water flowing over a dam. Some quantity, in being transferred from a high to a low potential, produced work, but the quantity being transferred was conserved in the process. We know today that Carnot's hypothesis is incorrect because he assumed that the conserved quantity was caloric. He carried this reasoning further to prove that there must be a limiting efficiency of heat engines. If such a limitation did not exist, then two engines could be suitably operated in a cyclic process in order to bring the heat reservoirs back to their initial states, the net effect being the production of work. This, Carnot declared, was an impossibility.

We might ask on what basis Carnot ruled out the possibility of what we today call a "perpetual motion machine of the first kind." It was not until 1847 that Helmholtz advanced the hypothesis of the conservation of energy. With few notable exceptions, scientists of the period generally agreed that the basic laws of nature existed *a priori* and were indestructible. Thus, over a period of many years, we saw the gradual acceptance of energy conservation as a basic postulate.

Carnot's engine reconciled both conversion and conservation phenomena—at least in the reversible limit. But how is this consistent with conservation in the highly irreversible mixing process of calorimetry, or conversion in the equally irreversible process of generating heat through friction? Carnot was cognizant of these difficulties and called for further experimentation and also for reconsideration of the foundations of the theory.

The preclassical period drew to a close with the quantitative work of Joule who established the equivalence of mechanical, electrical, and chemical energy to heat. We can see that at that time there were a number of concepts yet to be clarified. Caloric had to be split into heat quantity, energy, and entropy. It remained to be shown that heat and work were forms of energy transfer and that the interconvertibility was asymmetric. It is energy that is conserved in the calorimeter, in Carnot's cycle, and in frictional processes. Entropy is conserved only in the limit of the reversible process.

These further refinements occurred in relatively rapid succession beginning approximately in 1850 with the genius of Clausius, with important contributions from Thomson (Lord Kelvin), Maxwell, Rankine, Planck, Duhem, Carathéodory, and Poincaré, and terminating with the brilliance of Gibbs near the end of the century. Although we cannot cover these events in detail here, one should not underestimate their historical importance. We will attempt to reconstruct these developments, applying the insight that we have gained over the last 80 years.

1.3 The Postulatory Approach

Almost all approaches to classical thermodynamics follow one of two extremes: the *historical* approach, which parallels closely the chronological development of concepts and misconceptions, and the *postulatory* approach, in which axioms that cannot be proved from first principles are stated. There are merits and drawbacks in each extreme.

Advocates of the historical approach contend that if we are to expect our students to evolve new concepts and theories, we must expose them to the historical development of existing theories. Existing postulatory approaches make no reference to historical developments. The empirical basis for the laws of thermodynamics is impersonally stated in a small number of postulates that cannot be proved and can only be disproved by showing that consequences derived from them are in conflict with experimental facts. The postulates tend to be mathematical and abstract, but the laws of thermodynamics are derivable from them. Many students are unimpressed because little insight is provided for the necessity to define new concepts or properties. Figure 1.2 illustrates the connectivity between the postulates and their applications to engineering practice.

The approach we follow parallels the historical development in many respects. We begin by assuming the state of mind and body of knowledge available to the seventeenth-century scientist and proceed from that point to a logical development of a self-consistent set of rules that applies to the behavior of macroscopic bodies. In the process, we make use of many of the arguments put forth by our ingenious predecessors over the last 300 years, but we use hindsight to avoid the incorrect conclusions that prevailed at many points in the preclassical period and that resulted in much confusion (some of which is usually transferred to the student who studies thermodynamics by the historical approach).

At several junctures in our development we will face obstacles that cannot be obviated by invoking first principles. Our predecessors overcame these obstacles by trial-and-

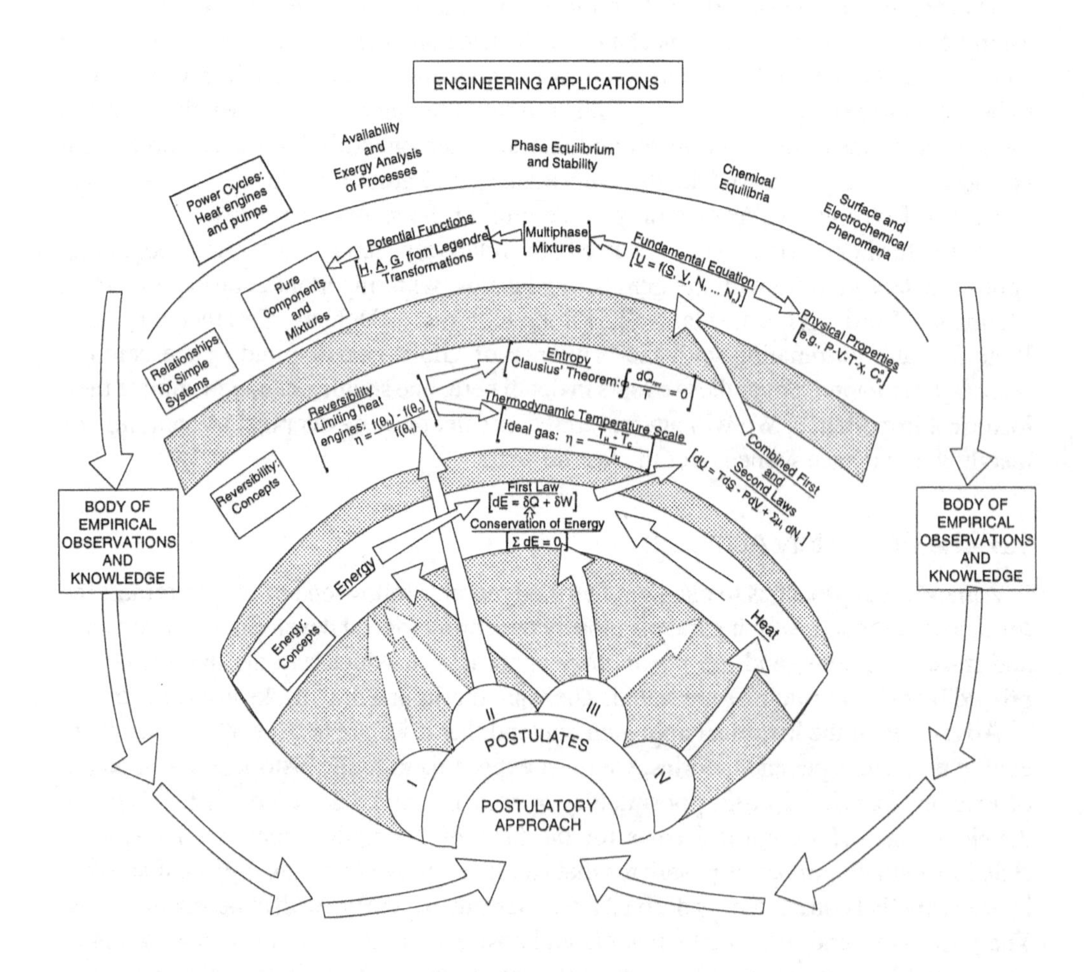

Figure 1.2 Illustration of interactions when using the postulatory approach.

error experimentation until they amassed a large body of knowledge. In this manner, generalities were stated and rules established. We clearly identify those principles that our ancestors learned to accept without proof; these are stated as postulates, but in a form that could be understood by Black, Lavoisier, Thomson, or Carnot. The ultimate verification of these postulates lies in the success of the formalism derived from them. In this vein, Schrödinger's equation is a basic postulate of quantum mechanics and Newton's laws of motion were basic postulates of classical mechanics. It is conceivable that, at some later date, new experimental information will be obtained that will necessitate revision or reformulation of the thermodynamic postulates, just as Newton's laws were found inapplicable to the motion of elementary particles of the atom.

Although the set of postulates presented in this text has the same information content as those developed in other texts, the phrasing may appear significantly different. There are obviously many different sets of postulates that provide equally valid bases for the theoretical development (see, for example, Callen (1985) and Hatsopoulos and Keenan (1964)). In developing a set of postulates, we have attempted to keep them as real as possible (as opposed to abstract) while retaining a form that will be readily acceptable to the chemical engineer.

References and Suggested Readings

Andrade, E. N. DA C. (1935), "Two historical notes—Humphry Davy's experiments on the frictional development of heat; Newton's early notebook," *Nature*, **135**, p 359.

Bejan, A. (1988), *Advanced Engineering Thermodynamics*, Wiley, New York, Chap. 1 and 2.

Bompass, C. (1817), *An Essay on the Nature of Heat, Light, and Electricity*, T & C Underwood, London.

Carathéodory, C. (1909), "Untersuchungen über die Grundlagen der Thermodynamik," *Math. Ann (Berlin)* **67**, p 355-386 (for an English translation see Kestin, J. (1976). *The Second Law of Thermodynamics*, Stroudsburg, PA: Dowden, Hutchinson and Ross part I, p 229-256)

Callen, H. B. (1985), *Thermodynamics and an Introduction to Thermostatistics*. 2nd ed, Wiley, New York.

Cardwell, D. S. L. (1971), *From Watt to Clausius*, Cornell University Press, Ithaca, NY.

Carnot, S. (1824), *Reflections on the Motive Power of Fire, and on Machines Fitted to Develop that Power*, Bachelier, Paris.

Clausius, R. (1867), *Mechanical Theory of Heat*, edited by T. A. Hirst, J. Van Voorst, London.

Gibbs, J. W. (1928), *The Collected Works of J. Willard Gibbs*, Vol. 1 and 2, Longmans and Green, New York.

Hatsopoulos, G. W. and J. H. Keenan (1964), *Principles of General Thermodynamics*. Wiley, New York.

Joule, J. P. (1963), *The Scientific Papers of James Prescott Joule*, Dawson's, London.

Joule, J. P. (1843), "On the calorific effects of magneto-electricity, and on the mechanical value of heat," *Philos. Mag.* **23** (3).

Lardner, D. (1833), *A Treatise on Heat*, Longmans, Rees, Orme, Brown, Green, and Longmans, London.

Leslie, J. (1804), *An Experimental Inquiry into the Nature and Propagation of Heat*, J. Mawman, London.

Mayer, J. R. (1842), "Remarks on the forces of inorganic nature." *Ann. Chem. Pharm.* **42**, p 233-240.

Metcalfe, S. L. (1843), *Caloric* (2 vols.), William Pickering, London.

Planck, M. (1945), *Treatise on Thermodynamics*, 3rd ed. translated by A. Ogg., Dever, New York.

Poincare, H. (1892), *Thermodynamique*. Georges Carré, Paris.

Rankine, W. J. M. (1888), *A Manual of the Steam Engine and Other Prime Movers*, 12 ed. C. Griffin, London.

Roller, D. (1950), *The Early Development of the Concepts of Temperature and Heat*, Harvard University Press, Cambridge, Mass.

Rumford, B. T. (1798), "An experimental inquiry concerning the source of the heat which is supplied by friction," *Philos. Trans.*, Royal Soc., London, **88**, p 80.

Thomson, W. (1840), *An Outline of the Sciences of Heat and Electricity*, H. Ballière, London.

Tisza, L. (1966), *Generalized Thermodynamics*, Paper 1. MIT Press, Cambridge, Mass.

Tyndall, J. (1880), *Heat, A Mode of Motion*, Longmans, Green, and Company, London.

Ure, A. (1818), *New Experimental Researches on Some of the Leading Doctrines of Caloric*, William Bulmer and Company, London.

Basic Concepts and Definitions 2

2.1 The System and Its Environment

If we are to develop a set of fundamental laws of nature without any preconceived notions, we must first develop the facility to perform some experiments. The subject of the experiment will be called the *system*, which will refer to a region that is clearly defined in terms of spatial coordinates that can be in a fixed or moving frame of reference. The surface enclosing this region will be referred to as the *boundary*. It may be an actual wall or it may be an imaginary surface whose position is defined during an experiment. The experiment itself may be a complete or partial process; it may be cyclic or non-cyclic. In most cases, initial and final conditions will be specified in some manner. The region of space external to the system and sharing a common boundary with the system is referred to as the *environment* or *surroundings*.

As will be seen shortly, work and heat effects are defined in terms of events at system boundaries; thus, the choice of a boundary is usually dictated by the kind of information desired. In any given situation, there are often many different system boundaries one can choose, each having some advantages and disadvantages. Developing a facility for choosing those system boundaries that will result in the most tractable path to the desired information is essential to the engineer. Insight into the selection process can be gained by working a given problem using several different system boundaries.

2.2 Primitive Properties

To record events that occur within a system, we must devise experimental tools that are sensitive to changes in the system. *Primitive properties* will refer to characteristics of the system that can be determined or measured by performing a standardized experiment on the system. To ensure that the measurement is a characteristic of only the system, we require that the experiment not disturb the system. The primitive property is of value because it is directly associated with the system at a particular time, and the observer need not know the history of the system to ascertain the value of the property.

A primitive property which is easily measured and is particularly useful is the *thermometric temperature* (denoted by θ). The thermometric temperature can be measured, for example, by noting the volume of a known mass of liquid in a sealed tube when the tube is brought into contact with the system. The thermal mass of the thermometer should be small in relation to that of the system so that the measurement does not alter the system. (The effect of the thermometer on the system can be determined by inserting a second thermometer and noting any change in the reading of the first thermometer.)

It should be emphasized that the value of θ is completely arbitrary and depends on the type of fluid, the materials used in construction of the thermometer tube, how the tube is notched, and the labels associated with the notches. Once the notches and labels have been made, however, the device can be used to observe *changes* occurring within a system. It could also be used to rank the θ-property of different systems. We return to a more detailed description of the thermometric temperature in Section 3.4.

Innumerable primitive properties could be defined by outlining suitable experiments. In this sense, volume, mass, pressure, index of refraction, color (given a standardized method of observation), etc., could be called primitive properties and scales for each property could be devised.

Primitive properties are useful in defining a system and in recording the occurrence of events in a system. In an *event*, at least one primitive property changes. An *interaction* is defined as events occurring simultaneously in the system and surroundings, at least one of which would not have occurred if the system were removed from the surroundings and placed in any other arbitrary environment.

2.3 Classification of Boundaries

The interactions between a system and its surroundings are governed by the nature of their common boundary. If the boundary is impermeable to mass flow, the system is called a *closed system*. An *open system* has a boundary that permits a mass flux of at least one component of the system through at least one point. In either case, the boundary may be *rigid* or *movable*.

There is one other set of conjugates needed to complete the classification of boundaries: *adiabatic* and *diathermal* walls. These boundaries represent extremes in the rate of heat exchange between the system and its surroundings, but since we have not yet defined a heat interaction, this discussion is premature. The adiabatic boundary or wall is one of the key concepts in thermodynamics, and we use it to define work and heat interactions. To avoid a circular system of definitions, we treat the adiabatic wall in detail in the next section.

In the absence of external force fields, specification of one of each of the three conjugate sets of boundaries (i.e., permeable and impermeable, rigid and movable, adiabatic and diathermal) is necessary to describe completely the external constraints placed on the system. Of the eight combinations[1] there is one that is of particular

1 It will become evident in discussing the criteria of equilibrium that only six of the eight combinations are meaningful. Of the four combinations involving permeable boundaries, there is no distinction between adiabatic and diathermal walls (see Section 6.4).

importance. This is the system enclosed by impermeable, rigid, and adiabatic walls, called an *isolated system*. It will become evident (Section 3.4) that this system can have no interactions. Any events occurring within an isolated system are independent of events in the environment.

2.4 The Adiabatic Wall

The concept of the adiabatic wall evolves from our experience; it can be illustrated by conducting a series of simple experiments. Consider a closed, rigid system in which a device to measure the thermometric temperature has been placed. Surround this system with a system having a higher thermometric temperature. If the initial system were constructed from copper, the variation of the thermometric temperature with time would look like curve 1 of Figure 2.1. Curves 2, 3, and 4 would result if the initial system were constructed from steel, glass, and asbestos, respectively. These results are, of course, based on our intuitive understanding of the effect of declining values of thermal conductivity in going from copper to asbestos. Finally, if the container were made of a double-walled, insulated Dewar flask, the variation of temperature over the time of the experiment would be quite small (curve 5). The adiabatic wall is an idealized concept representing the limiting case of curve *A* in Figure 2.1. In practice, adiabatic boundaries are approached in many situations, especially those in which events occur rapidly in relation to the time scale of the experiment.

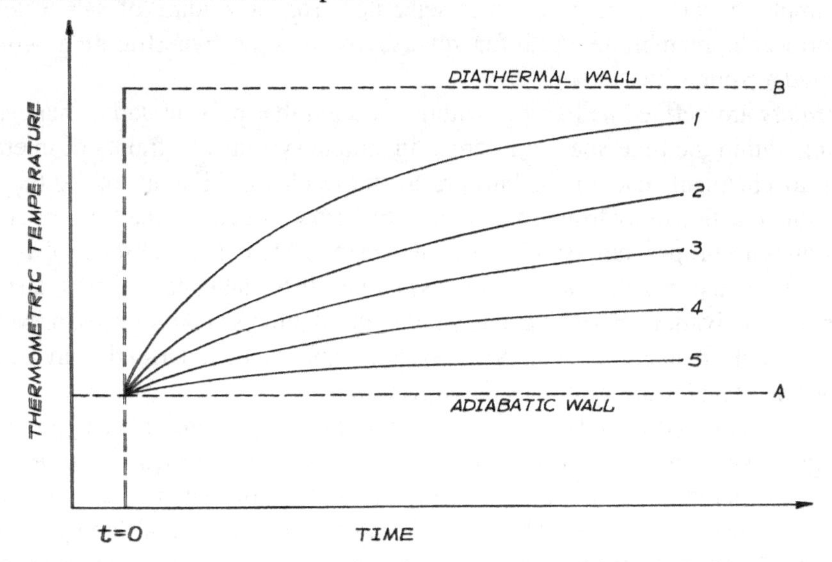

Figure 2.1 Temperature-time behavior for different materials. Limiting diathermal and adiabatic boundary behavior shown.

The case diametric to that of adiabatic is the diathermal wall (curve *B*), in which the change in thermometric temperature is very rapid in relation to the time scale of the experiment (i.e., the thermometric temperature of the two systems is always identical).

In our operational definition of an adiabatic wall, we chose the thermometric temperature to relate changes in the environment to those in the system. This selection was made for ease of visualization. Other primitive properties could also have been chosen. Some primitive properties would not be of value to define adiabatic walls. Pressure, for example, could have been varied in the surroundings with no effect on our system with rigid walls. We will specify how to select an appropriate property to define an adiabatic wall after Postulate I has been introduced.

2.5 Simple and Composite Systems

There is a special class of systems that plays a central role in the developments to follow. These systems, referred to as *simple systems*, are devoid of any internal boundaries (adiabatic, rigid, or impermeable) and are not acted upon by external force fields or inertial forces.

A stable *phase* is defined as a region within a simple system throughout which all of the properties are uniform. A single-phase, well-mixed, homogeneous system is the simplest of simple systems. A system containing multiple phases is also a simple system provided that no adiabatic, rigid, or impermeable boundary exists between any two phases.

Composite systems are systems composed of two or more simple subsystems. There are no restrictions on the kinds of boundaries separating the subsystems of the composite. For example, a volume of pure water separated from a volume of salt water by a semi-permeable membrane used for reverse osmosis or hyperfiltration would be considered a composite system.

Restraints are defined as barriers within a system that prevent some changes from occurring within the time span of interest. In simple systems, restraints of interest are barriers to chemical reaction or barriers to phase change. For example, the room-temperature reaction of hydrogen and oxygen to form water can be made to occur within milliseconds if a proper catalyst is incorporated in the system. In the absence of a catalyst, no noticeable reaction occurs within months or years. In the latter case, there is an internal restraint (the activation energy barrier) which, for all intents and purposes, prevents the occurrence of the reaction. For composite systems, internal boundaries that are adiabatic, rigid, or impermeable are also considered restraints.

Thermodynamics does not dictate the restraints that may be present in a given system. In any given situation, one must decide which restraints are present. Two factors that must be considered in making this decision are the laws of matter and the rates of the various conceivable processes. The laws of matter are comprised of three parts: (1) the continuity of matter (matter cannot move from one position to another without appearing at some time in the intervening space), (2) the conservation of electrical charge (net electrical charge must be conserved in all processes), and (3) the conservation of chemical elements (in the absence of nuclear transformations and relativistic effects, mass must be conserved). Even though process rates cannot be determined directly from thermodynamics, their magnitudes relative to time scales of interest are important in specifying tractable solution methodologies for many problems in thermodynamics. The application of some of these concepts is illustrated in the following example.

Example 2.1

A closed vessel contains water, oil, and air at room temperature. If the system is synthesized by first adding the water and then layering the oil on the water, is this system a simple or composite system? If composite, define each simple subsystem and the internal restraints. If the vessel is shaken vigorously, is the system then a simple or composite system?

Solution

There are clearly no adiabatic or rigid walls within the vessel. Thus, it is only necessary to decide if there are impermeable walls. At room temperature, a system containing water and air should have appreciable water vapor present in the air. If the system is formed by layering the oil on the water without shaking, the law of continuity of matter requires that the water pass through the oil layer. This process is slow and will not occur to any appreciable extent even after several hours. If our interest in this system did not extend beyond this time scale, the oil layer would have to be considered as an impermeable barrier to the water. In this case, the vessel would be considered a composite system of two simple subsystems: air + oil (assuming oil evaporates into the air within the time scale of the experiment), and water. If the system is shaken vigorously, water droplets will contact the air directly. Thus, the entire contents of the vessel would be considered a simple system.

2.6 States of a System

Now that we have the means for conducting experiments on a system, we would like to have a formal way of characterizing a system so that others could reproduce the experiments. For this purpose, we will identify the condition or *state* of the system by the values of those properties that are required to reproduce the system. Although this definition is functional, it is not very practical because we do not always know the number of properties that are required to specify the state of the system. We are fortunate, however, to have available a large body of experimental data, accumulated over several hundred years, which indicates that there are particular types of states that can be specified by delineating only a certain number of properties. These states, which are called *stable equilibrium states*, are defined in the next section. In general, nonequilibrium states can also be specified (for the purpose of reproducing them) from a finite number of properties; the number of such properties, however, is not given by the principles of classical thermodynamics.

2.7 Stable Equilibrium States

The aforementioned body of experimental data indicating the existence of these states is summarized in the following postulate.

Postulate I. For closed simple systems with given internal restraints, there exist stable equilibrium states which can be characterized completely by two independently variable properties in addition to the masses of the particular chemical species initially charged.

This postulate is similar to a conclusion drawn by Duhem in 1898 and is sometimes referred to as Duhem's Theorem. By "two independently variable properties" it is meant that each property could be varied (by at least a small amount) in at least one experiment during which the other property is held constant. For example, consider a closed vessel containing the simple system of a pure component liquid and its vapor in a stable equilibrium state. Given the amount of material initially charged, the thermometric temperature, and the total volume of the vessel, the system could be reproduced at will because the volume and temperature are independently variable and therefore completely specify the system. If, however, the pressure were specified instead of the total volume, the system could not be reproduced because, as we shall see later, pressure and temperature are not independently variable in this case.

We are now at an impasse. We do not know which two properties to choose to specify the state of the system because we do not yet know which properties are independently variable. For a given system with a given set of restraints, we find that the laws of thermodynamics (which we shall develop from our postulates) will result in relationships among certain variables. The relationships and the variables included therein will depend on the system and the restraints. Only from these relationships will we be able to determine which sets of properties are *not* independently variable. For an example of a closed system of liquid and vapor in a stable equilibrium state, we can show that as a result of the requirement of phase equilibrium (Chapter 15), the vapor pressure is a unique function of temperature (recall the Clausius-Clapeyron equation, for example). Thus, pressure and temperature cannot be independently variable in this case.

In another example, two independently variable primitive properties may be selected to test whether a rigid wall is adiabatic or not. Consider the following *thought experiment*. We have a simple, closed, and rigid system in a stable equilibrium state. Postulate I can be used as an aid in selecting appropriate properties. If the wall were not adiabatic, we may be fortunate to select an appropriate primitive property for the first experiment (e.g., thermometric temperature as used in Section 2.4). However, if the choice had been, for example, pressure, then variations in this variable in such a way as to keep the thermometric temperature constant (and thus independently variable) would have produced no change in pressure inside the rigid, closed system. Then, as shown above, if we keep pressure constant and vary any other independently variable property, by measuring both this new property *and* pressure inside the system we can make an unequivocal statement concerning the adiabaticity of the wall. That is, if neither change, the wall is adiabatic. If either or both change, the wall is not adiabatic.

Knowing that stable equilibrium states exist is not nearly as informative as knowing when they exist. The second postulate is directed toward establishing this fact.

Postulate II. In processes for which there is no net effect on the environment, all systems (simple and composite) with given internal restraints will change in such a way as to approach one and only one stable equilibrium state for each simple subsystem. In the limiting condition, the entire system is said to be at equilibrium.

Postulate II is specific to systems which are in effect isolated; also, for processes that consist of a series of steps, the system may interact with the environment in two or more steps, but the net effect of these steps must leave the environment unaltered.

Since the stable equilibrium state is defined as a limiting condition toward which any simple system tends to change, it follows that no property of this state varies with time. Then, from Postulate I, once a simple system has reached a stable equilibrium state, only two independently variable properties and the masses initially charged need to be specified to determine this state completely. Since all other properties are fixed in the stable equilibrium state, it follows that all other properties of the simple system are dependent variables that are determined by the two independently variable properties and the masses of the initial chemical species. Note that this conclusion is valid for each *simple* system at equilibrium, even if the simple systems are part of a composite system. The conclusion, however, does not apply to the composite system at equilibrium; that is, the state of a composite system at equilibrium cannot be specified by two independently variable properties plus the masses initially charged. The difficulty arises from the fact that all properties may vary from one subsystem to another within a composite system at equilibrium. For example, the thermometric temperature of a composite system has little significance if the subsystems of the composite are separated by adiabatic, impermeable walls. Postulate I has been restricted to simple systems in order to avoid such difficulties.

In Postulate II, it is stated that for an isolated system, there exists one and only one set of stable equilibrium states (toward which the subsystems tend) *for a given set of internal restraints*. There will be different sets of stable equilibrium states for different sets of internal restraints. For example, with reference to Example 2.1, there will be a unique equilibrium state if we assume that there is an impermeable barrier preventing water from reaching the air space; there will also be a unique equilibrium state if we assume that no such barrier exists. Although each of these states is unique, the properties of each will be different.[2] Thus, before the equilibrium state can be completely defined,

2 In some texts the former state would be referred to as metastable in the sense that *given sufficient time*, the latter state (for which the term "stable equilibrium state" would be reserved) would be reached. The problem with this set of definitions is that almost all systems of interest to chemical engineers would then have to be classified as metastable even though the final stable equilibrium state may not be obtained in any time span of interest.

the internal restraints must be identified. Specification of internal restraints is clearly an important part of specifying the system.

This connection between Postulates I and II will later be shown to provide the basis for deriving fundamental relationships for phase and chemical equilibria.

2.8 Thermodynamic Processes

A *change of state*[3] of a system is identified by a change in the value of at least one property. For systems initially in stable equilibrium states, changes of state will occur only when the system has an interaction with the environment or when internal restraints are altered. "Change of state" is usually applied to systems that are initially in one stable equilibrium state and are found after some event to be in another equilibrium state. The change of state is then fully described by the values of the properties in the two end states.

The *path* refers to the description of all the states that the system traverses during a change of state. Thus, the path is described in terms of the primitive properties that define the intermediate states. Paths for which all the intermediate states are equilibrium states are termed *quasi-static* paths.[4] From Postulate I, quasi-static paths of closed simple systems can be completely described in terms of successive values of only two independent properties.

It also follows from Postulate II that if a system progressing along a quasi-static path is isolated at some point (e.g., by temporarily altering a boundary condition), the values of all the properties will remain constant at the values observed just prior to isolation. It may, however, take more than two properties to describe a non-quasi-static path. If the system is isolated during such a path, some primitive properties will change after isolation as the system approaches a stable equilibrium state. For example, consider the system of a gas initially at two atmospheres which is contained in a cylinder fitted with a piston and stops. The stop holding the piston is removed, and the gas expands until the piston reaches a second stop. If the piston is lubricated, the expansion process will be rapid. At any instant during the process, there will be a finite and measurable pressure gradient within the gas phase. To describe such an intermediate state, it will be necessary to determine the pressure (in addition to other properties) at all points within the cylinder. The intermediate states are not stable equilibrium states; if such a state were isolated (by stopping the piston at an intermediate point), the pressure gradient would be damped out as the system approached a stable equilibrium state. Clearly, this frictionless process is not quasi-static. Alternatively, if there were external forces acting against the piston so that the expansion was very slow, no appreciable pressure gradient would be found. If

3 In common usage, a change of state often is synonymous with "change of phase." This is not the meaning we use in this book.

4 Quasi-static paths are closely related to (and sometimes confused with) reversible processes. The distinction between the two is considered in Section 4.6.

the system were then isolated at an intermediate point, no properties would change because the system was, at all times, in some stable equilibrium state. Thus, the latter process is quasi-static.

The thermodynamic *process* involved in a change of state usually refers to a description of the end states, the phenomena occurring at the system boundaries (i.e., heat and work interactions, which are discussed in Chapter 3), and the path (which is usually described only for quasi-static processes). In many instances, however, the term "process" is loosely applied to describe the path without explicitly specifying the boundary conditions. Thus, an isothermal, isobaric, or isochoric process is one in which the temperature, pressure, or volume is constant. In such cases the boundary conditions are usually implied by the nature of the process or may be immaterial to the problem in question.

2.9 Derived Properties

Primitive properties were defined in terms of an experiment or measurement made on the system at some point in time. Only experiments that do not disturb the system are allowed. By definition, primitive properties are not restricted to stable equilibrium states.

We have now established two basic postulates dealing with a particular class of states, the stable equilibrium states. These states have innumerable primitive properties associated with them. Each of these properties is measurable by definition. We might also ask ourselves if there are other *properties* of these stable equilibrium states that are not measurable by any method but could be used to characterize the system. No definite answer can be given at this point, but it will become obvious in Chapter 3 and those following that such properties do exist. We will find that we can define such properties only in terms of *changes* in the system between initial and final stable equilibrium states. (Note that the path need not be quasi-static). To distinguish these properties from the *primitive* type, we will call them *derived* properties. As defined, derived properties exist only for stable equilibrium states and, as such, may be used as variables to define a system as required in Postulate I.

Since derived properties are functions of state, and since any stable equilibrium state of simple systems can be characterized by the values of two independently variable properties plus the masses, any derived property can be expressed mathematically as a function of two other independently variable properties, derived or primitive.

It will be instructive at this point for the reader to review the mathematical relations of functions of state. A review of this topic is presented in Appendix B.

2.10 An Important Note About Nomenclature and Units

Before presenting any additional material, we need to review the conventions for nomenclature and units that have been adopted for the 3rd edition. You may have noticed that a complete listing of symbols, operators, superscripts and subscripts is found just after the Table of Contents. With few exceptions this nomenclature applies to all nineteen chapters. The rare exceptions include special functional characters that will be defined when they are introduced, for example, parameters for a particular equation of state or for a specific activity coefficient model.

For the most part, we have adopted symbols that are commonly used by chemical engineers and chemists—ones that you would find in many undergraduate textbooks in chemical thermodynamics and physical chemistry. For example, we use T for absolute temperature, P for pressure, V for volume, and U for internal energy. Our conventions closely parallel the selections endorsed by the International Union of Pure and Applied Chemistry (IUPAC). However, there are a few important conventions that will be used here that may differ somewhat from your previous experience. These conventions include:

1. ***Intensive versus extensive properties*** Earlier in this chapter both primitive and derived properties were defined. Because a primitive property like total volume or a derived property like total internal energy are extensive and depend directly on the size of the system selected while others like pressure (P) and temperature (T) are intensive and independent of system size (see Section 5.2), one needs to be able to distinguish between them in an unambiguous manner. Throughout this book, we have adopted a simple and straightforward convention for properties. All extensive properties will use an underbar, e.g. $\underline{V}$, $\underline{E}$, or $\underline{U}$, to emphasize that they depend on system size. All characteristically intensive properties, e.g. P, T, or ρ (molar density $\equiv N/\underline{V}$) or other specific properties, which are also intensive as they are scaled by dividing the total quantity by the total moles or mass of the system, will be written in upper case without the underbar, e.g. $V \equiv \underline{V} / N$ or $U \equiv \underline{U} / N$. Some other texts utilize a lower case letter for the specific quantity and an upper case letter without an underbar for the total extensive quantity, for example, upper case V for total volume in m^3 and lower case v for specific volume in m^3/mol. While we recognize that partial inconsistencies may exist between our conventions and others, we strongly feel that the use of an underbar minimizes errors by emphasizing the extensive nature of the properties.

2. ***Units*** The International System of Units (SI for Système International d'Unités) has been selected for this edition. Where appropriate in the Nomenclature section, we have given the SI units used in square brackets [ ]. In engineering practice, of course, you will need to perform unit conversions from the old English engineering system or from other hybrid metric systems such as mks or cgs. Consequently, a few selected problems in this edition will require modest unit conversions.

3. ***Mixture and partial molar properties*** Because many chemical engineering thermodynamics problems deal with multicomponent systems, a consistent nomenclature procedure is needed to distinguish between pure component and mixture properties. Although these conventions are not formally introduced until Chapter 9, some brief comments are appropriate here because different conventions may have been used in your earlier exposure to thermodynamics. Typically, subscripting will be used to designate the property for a particular component, for example V_i corresponds to the molar volume of pure i. Also, a partial molar property is defined by a differential operator. For example, the partial molar volume $\overline{V}_i$ of component i in an n-component mixture is $\overline{V}_i \equiv (\partial \underline{V}/\partial N_i)_{T,\,P,\,N_j[i]}$ where N_i = moles of component i and the differentiation is carried out with T, P, and all other moles of components $j = 1, ..., n$ except i held constant.

4. ***Scalar versus vector quantities*** From time to time reference will be made to vector quantities having both magnitude and direction, for example the body or inertial forces acting on a system. These vector quantities will be identified using boldface type for the symbol, such as **F**. In Chapters 16 and 17, one-dimensional column vectors and two-dimensional matrices will be used to describe multicomponent, multiphase behavior. In this case, a single tilde underbar ~ and a double tilde underbar ≈ will be used to indicate 1-D column vectors and 2-D matrices (see Nomenclature).

2.11 Summary

This chapter provides the framework for our postulatory development of classical thermodynamics by first defining important concepts and conventions that characterize how systems can interact with their surroundings. Some of these included:

- Simple versus composite systems
- Closed versus open versus isolated systems
- Rigid versus movable boundaries
- Permeable versus semi-permeable versus impermeable boundaries
- Adiabatic versus diathermal boundaries
- Primitive versus derived properties

Postulates I and II were introduced in the context of both defining and constraining the behavior of systems as they pass through stable equilibrium states. In a mathematical context, Postulate I describes the existence conditions for a system while Postulate II describes the uniqueness of a particular equilibrium state.

Suggested Readings

Callen, H.B. (1985), *Thermodynamics and an Introduction to Thermostatistics*, 2nd ed. Wiley, New York.

Hatsopoulos, G.W. and J.H. Keenan (1964), *Principles of General Thermodynamics,* Wiley, New York.

Energy and the First Law 3

In this chapter we adopt the definition of work from the mechanics of rigid bodies and extend this concept to thermodynamic systems. We introduce energy as a unique measure of the work required to reach one stable equilibrium state from another in certain processes. Since we cannot show *a priori* that this quantity is a function only of the end states, it will be necessary to postulate that energy is a derived property. The conservation law for energy will be a direct consequence of this new postulate. After defining work and postulating the existence of the energy property, we present an operational definition of heat and discuss the directionality of heat interactions.

3.1 Work Interactions

The mechanical work (W) associated with the movement of a rigid body is defined by the scalar product of the net force and displacement vectors as:

$$W = \int_{x_1}^{x_2} (\sum \mathbf{F}_s) \cdot d\mathbf{x} \tag{3-1}$$

or, in differential form,

$$\delta W = (\sum \mathbf{F}_s) \cdot d\mathbf{x} \tag{3-2}$$

where $\sum \mathbf{F}_s$ is the resultant force vector acting on the surface or boundary of the rigid body at a point where there is a differential displacement of the boundary, $d\mathbf{x}$. In keeping with the mechanical definition of work, *boundary forces* ($\mathbf{F}_s$) are distinguished from *body forces* or forces associated with external fields ($\mathbf{F}_b$) (i.e., centrifugal, gravitational, inertial, coulombic, etc.); in the absence of body forces, only boundary forces are used to calculate work.

As discussed in Appendix B, the symbol δ is used throughout the text to denote differentials of path-dependent functions that are not state variables. The dot $\cdot$ denotes a vector dot product to produce a scalar quantity W.

For a rigid body acted upon by both boundary and body forces, Newton's Second Law of Motion states that

$$\sum \mathbf{F}_s + \sum \mathbf{F}_b = 0 \tag{3-3}$$

In Eq. (3-3), the inertial force, $-m\mathbf{a}$, where $\mathbf{a}$ is the acceleration vector, is considered a body force and is included in the second summation. For example, consider a weight suspended by a string, as in Figure 3.1.

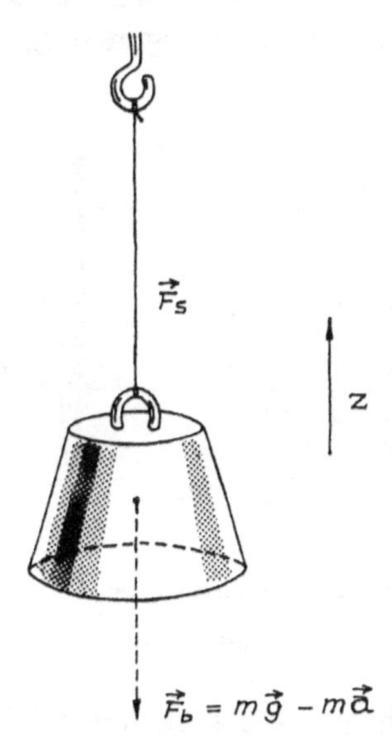

Figure 3.1 Free body diagram for a rigid body of mass m.

If the body is initially at rest, and if $\mathbf{F}_s$ and $\mathbf{g}$ are collinear, then Eq. (3-3) becomes

$$F_s - mg = 0 \tag{3-4}$$

Note that we have adopted the normal convention that collinear, unidirectional vector problems can be treated in terms of the magnitudes of the vectors only (e.g. $|\mathbf{F}_s| = F_s$ and $\mathbf{g} = g$)). If F_s is then increased so that the weight rises, at any instant during the motion, Eq. (3-3) is

$$F_s - mg - m\frac{dv}{dt} = 0 \tag{3-5}$$

where $\mathbf{a}$ has been set to dv/dt. The differential work done on the weight at any instant of time is due only to the boundary force, F_s. Thus, using Eqs. (3-2) and (3-5),

$$\delta W = \mathbf{F}_s \cdot d\mathbf{x} = \left(mg + m\frac{dv}{dt}\right)dz = mg\,dz + mv\,dv \tag{3-6}$$

The total work done on the weight between positions z_1 and z_2 is

$$W = mg(z_2 - z_1) + \frac{m}{2}(v_2^2 - v_1^2) \tag{3-7}$$

On the right-hand side of Eq. (3-7), we call the first and second terms the difference in potential and kinetic energy, respectively. Here we accept the terms for potential and kinetic energies as mgz and $mv^2/2$, but in no way have we defined the concept of energy.

We are now in a position to define the work associated with systems of interest in thermodynamics; in particular, we are concerned with systems having nonrigid as well as rigid boundaries. Consider the expansion of a gas contained in a cylinder fitted with a piston and surrounded by the atmosphere (Figure 3.2).

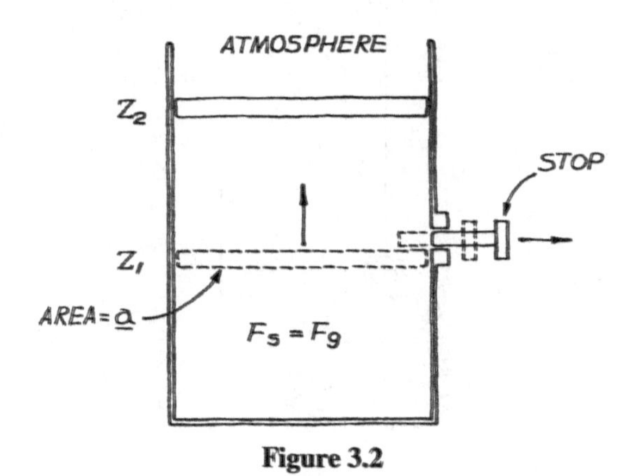

Figure 3.2

If we choose the system as the gas, the work done by the system on the environment is

$$\delta W_g = \mathbf{F}_g \cdot d\mathbf{z} = P_g \underline{a} dz = P_g d\underline{V}_g \tag{3-8}$$

where $\mathbf{F}_g$ is the force exerted by the gas on the boundary which is displaced and $\underline{a}$ is the area of the displaced boundary. (Note that this force may not be the same as that exerted on the other boundaries of the gas system.) Thus, before the work can be calculated, $\mathbf{F}_g$ must be known as a function of $\mathbf{z}$. For the gas, $\mathbf{F}_g = F_g$ which, of course, is equal to $P_g \underline{a}$, where P_g is the pressure on the boundary that is displaced.

If neither F_g nor P_g is known, the work can still be determined by measuring the effects on the environment. For example, consider a differential element of time during the expansion. To illustrate, let us suppose that there is friction between the piston and the cylinder walls. As is shown in Figure 3.3, the unknown force F_g can be evaluated by making a force balance on the object which is acted upon by this force. In this case, the object is the piston. Thus,

$$F_g - P_a\underline{a} - F_f - mg - mv\frac{dv}{dz} = 0 \tag{3-9}$$

where P_a is the pressure of the atmosphere, F_f is the magnitude of the frictional force, and m is the mass of the piston. Note that the inertial force acts in the direction opposite to that of the motion during acceleration.

Solving Eq. (3-9) for F_g and substituting into Eq. (3-8), we obtain

$$\delta W_g = P_a d\underline{V}_g + mg\, dz + mv\, dv + F_f dz \tag{3-10}$$

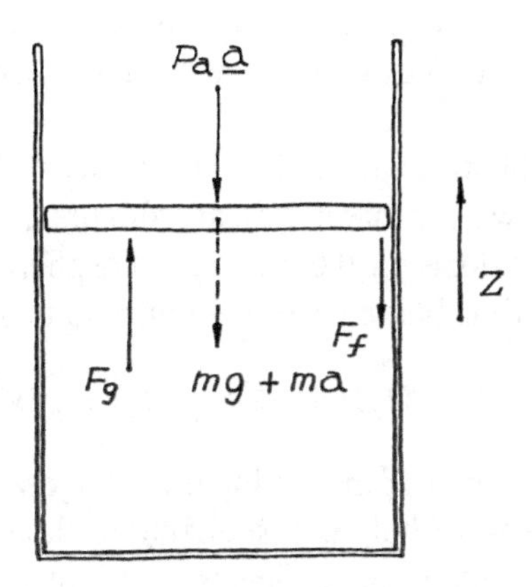

Figure 3.3

On the right-hand side of Eq. (3-10), the first term is work done by the gas in pushing back the atmosphere and increasing the volume of the gas by $d\underline{V}_g$. The second and third terms represent the work done by the system in increasing the potential and kinetic energy of the piston, and the last term is the work done on the cylinder wall to overcome friction. Note that if the frictional force were known, δW_g could be calculated from Eq. (3-10), and P_g could be calculated directly by solving Eq. (3-9) for F_g.

A different point of view is found if the piston were chosen as the system. The work done by the piston on its surroundings would then be

$$\delta W_p = \mathbf{F}_p \cdot d\mathbf{z} \tag{3-11}$$

where $\mathbf{F}_p$ is the net boundary force exerted by the piston on the environment. According to Newton's third law of motion, the net force is ***equal and opposite*** to the net boundary force exerted by the environment on the piston. Therefore,

$$\mathbf{F}_p = \frac{(-P_g\underline{a} + P_a\underline{a} + F_f)\mathbf{z}}{|z|} \tag{3-12}$$

where $|z|$ is the magnitude of the displacement vector, and

$$\delta W_p = -P_g d\underline{V}_g + P_a d\underline{V}_g + F_f dz \tag{3-13}$$

Finally, if the atmosphere were chosen as the system, the work done by the atmosphere on its surroundings would be

$$\delta W_a = \mathbf{F}_a \cdot d\mathbf{z} = P_a d\underline{V}_a = -P_a d\underline{V}_g \tag{3-14}$$

If we define $-F_f\ dz$ as the "work" done by the walls, δW_w, substitution of Eqs. (3-8) and (3-14) into Eq. (3-13) yields

$$\delta W_p = -\,\delta W_g - \delta W_a - \delta W_w \tag{3-15}$$

Thus, choosing any object in Figure 3.3 as the system, it is clear from Eq. (3-15) that the work done by the system on the environment is equal and opposite to the work done by the environment on the system. This result, which is valid for work interactions in general, is a consequence of the requirement that the sum of all body and boundary forces on a system is zero.

Returning to the problem of calculating the total work done by the gas in the expansion depicted in Figure 3.2, we see that it is clear from the discussion above that the work is equal to the total work done on the piston, atmosphere, and walls. If we neglect the frictional work for the time being (we shall return to this subject in Section 3.7), then

$$W_g = (P_a \underline{a} + mg)(z_2 - z_1) + \frac{m}{2}(v_2^2 - v_1^2) \tag{3-16}$$

There is one other method of measuring the work done by a system which, although not very practical, will be of help in visualizing a "thought" experiment which will be described shortly. We could remove the system of interest from its given environment, such as that shown in Figure 3.2, and place it in a cylinder covered with a weightless piston surrounded by a vacuum. The piston is balanced by weights placed on the top. If we then continually remove the weights so that the pressure-volume history of the gas during this process is identical to that which occurred in the original expansion, the work done by the gas during the hypothetical expansion is equal to the change in level (or potential energy) of the weights. Furthermore, this work is equal to the work done by the gas in the original expansion. Thus, the work done by a system can always be found by measuring the rise or fall of weights in the environment.

3.2 Adiabatic Work Interactions

Consider two closed systems that undergo an interaction through a common boundary. We shall call the interaction an *adiabatic work interaction* if the events occurring in each system could be repeated in such a way that the *sole* effect external to one system could be duplicated by the rise (or fall) of weights in a standard gravitational field and the *sole* effect external to the other system could be duplicated by an equivalent fall (or rise) of weights of equal magnitude. Some examples will help to illustrate the use of the definition.

Consider the situation illustrated in Figure 3.4(a). A vessel containing water initially at 0°C is bounded by adiabatic walls. A rough-surfaced disk is immersed in the water and attached to a drum by a shaft. The drum is rotated by allowing a weight to fall from the first to the second level. The first step in determining if this is an adiabatic work interaction is to designate the boundaries of the systems we wish to study. Let us choose the dashed line in Figure 3.4(a) as the boundary separating systems A and B. If we consider the events that occur external to system A, the sole effect is the fall of a weight. If the interaction is an adiabatic work interaction, we must somehow show that the events that occur external to B could be repeated solely by the raising of an identical weight by an identical amount. If, as in Figure 3.4(b), we replace system A with a drum that has a weight attached to it, and then repeat the event in system B, it is clear that the sole effect external to B is the required rise of a weight. Thus, the interaction is an adiabatic work interaction.

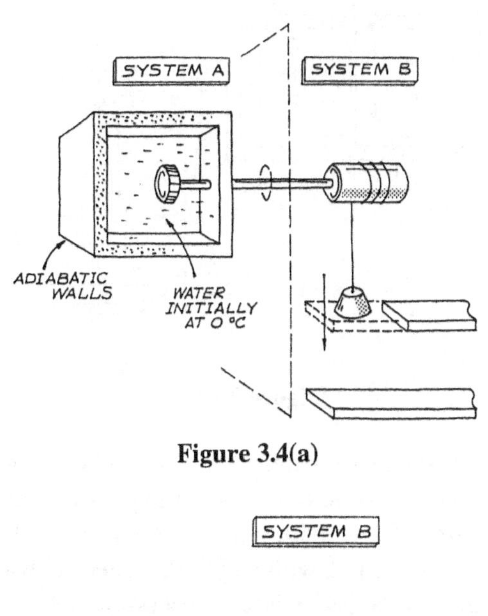

Figure 3.4(a)

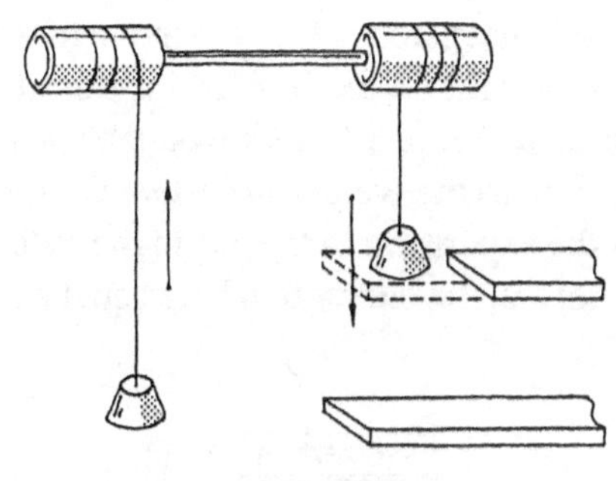

Figure 3.4(b)

Now consider the slightly more complex situation illustrated in Figure 3.5(a). If we consider the interaction of system C with the composite system $A + B$, the situation is analogous to that of the previous example and we have an adiabatic work interaction. If, however, we consider the interaction of system A with the composite system $B + C$, the analysis is quite different. External to system A, the sole effect is the lowering of a weight since the final temperature of the water in system B will be 0°C as long as some ice remains in system A. (Note that we need only consider the initial and final states; whether or not the water remains at 0°C throughout the process is immaterial.)

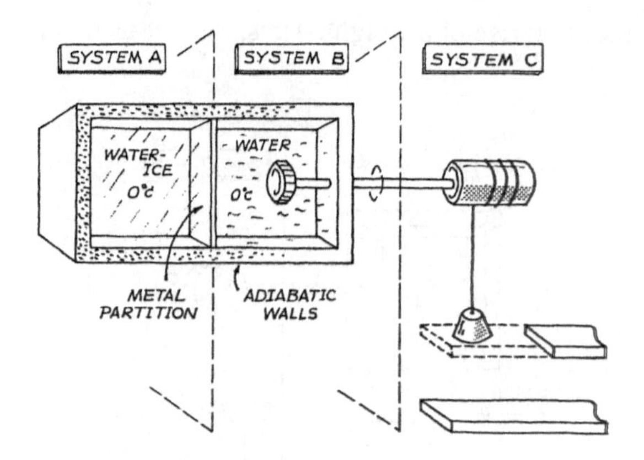

Figure 3.5(a)

Although we can devise processes external to $B + C$ which would result in the rise of a weight, it does not appear possible to devise an experiment in which the *sole* effect external to $B + C$ is the rise of a weight. (At least no such experiment is known.) For example, we could replace system A with a cylinder filled with a gas at 0°C and fitted with a piston which in turn is attached to a flywheel [system A' in Figure 3.5(b)]. As the weight in system C is lowered, the piston is freed to move to the left by removing a stop. As the piston moves out, the weight on the flywheel is raised. By judiciously choosing the components of the system in Figure 3.5(b), it would be possible to have the flywheel weight rise by the same amount as the weight in system C falls. However, the net effect external to system $B + C$ is the expansion of the gas in the cylinder in addition to the rise of the flywheel weight. Therefore, the interaction between A and $B + C$ is not an adiabatic work interaction.

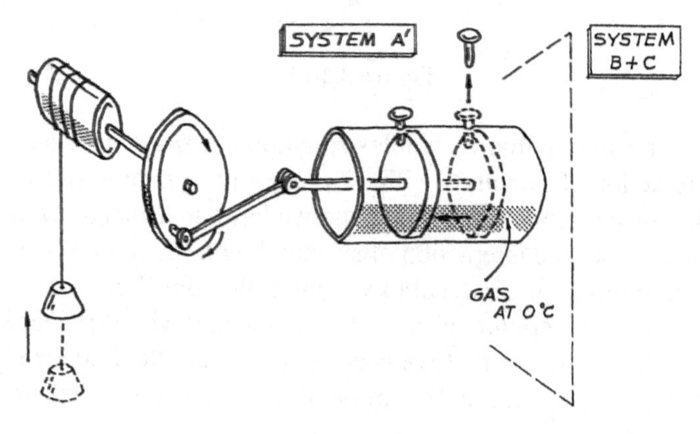

Figure 3.5(b)

Example 3.1

Consider the situation illustrated in Figure 3.6, in which an electric generator is operated by a falling weight and in which the power generated is dissipated in a resistor. Neglect any dissipative processes such as $i^2 R$ line losses, friction in bearings, etc. Is this an adiabatic work interaction?

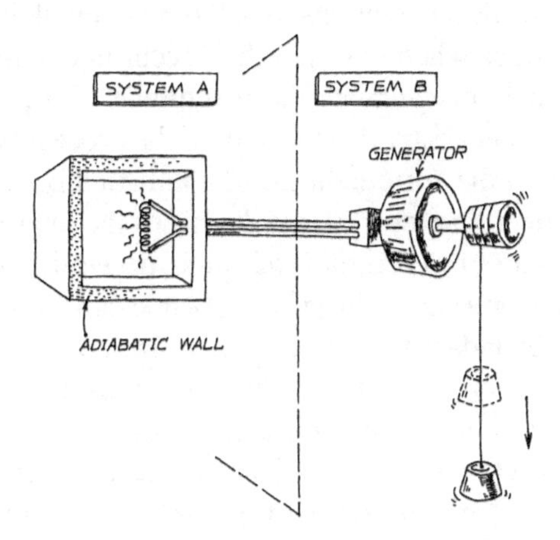

Figure 3.6

Solution

The sole effect external to system A is the fall of a weight. By replacing system A with a motor that has a weight attached to its shaft by a rope (i.e., the inverse of system B), system B could be made to execute the same process while the sole effect external to B would be an equivalent rise in the level of the weight. Hence, the electric current flowing between two systems in this manner is equivalent to an adiabatic work interaction.

Example 3.2

Return to the example shown in Figure 3.4(a). Suppose that the dashed boundary was an adiabatic wall and a bearing was placed in the wall to hold the rotating shaft. Is this process an adiabatic work interaction?

Solution

If the bearing were frictionless, the case is no different from that described in text. However, if there were friction in the bearing, it would not be possible to have *equivalent* fall and rise of weights when each system was considered separately. Thus, this interaction is *not* an adiabatic work interaction.

In summary, it should be clear that the description of a work interaction is meaningful only when systems and boundaries are carefully delineated. An adiabatic work interaction requires that all common boundaries be adiabatic walls and, if any moving shafts or electrical wires penetrate these adiabatic walls, there can be no dissipative processes occurring at the wall.

In this section we have restricted ourselves to considering what conditions are required to have a truly adiabatic work interaction which is important to further development of our postulatory approach to classical thermodynamics. Nonetheless, being able to quantify work interactions is of fundamental importance to solving problems in thermodynamics whether or not they occur under adiabatic conditions. In order to be able to estimate or calculate the magnitude of a particular type of work interaction, it is absolutely necessary that the path of the process be known either for the system or in some cases in the surroundings. Mathematically, this means that we can integrate the path dependent line integral that describes the total work.

It is possible to expand our description of adiabatic work interactions beyond the examples that have been presented. In general, work can be represented as the dot product of a generalized boundary force and a generalized displacement or $\delta W = \mathbf{F} \cdot d\underline{\mathbf{x}}$. Table 3.1 provides some examples of boundary work transfer interactions.

Operationally, the appropriate generalized force must be mapped onto the actual displacement that occurs in the real process. In essence, this is equivalent to saying that the process must proceed through a set of well-defined equilibrium states so the functional dependence of the force (e.g., P) can be quantitatively represented along the displacement path (e.g., $d\underline{V}$). This limiting condition corresponds to what is called a quasi-static process as defined earlier in Section 2.8. Although much more will be said about quasi-static processes in Chapter 4, you will be required to utilize the quasi-static approximation to solve many of the problems at the end of this chapter.

In passing, it is worth noting that Joule carried out a similar set of work interaction experiments between 1843 and 1848. His objective at that time was to verify a theory proposed independently by him and Mayer that work and "heat" are equivalent. With repeated careful experiments, Joule established that for a particular amount of work expended, the rise in the temperature of a fixed mass of liquid water was constant regardless of the type of work process involved. For the most part, the examples illustrated in Figures 3.4 through 3.6 duplicate Joule's actual experiments; thus, he was in effect carrying out a set of adiabatic work interactions with a system that consisted of a fixed mass of water. Joule viewed that the observed change in state of the water resulted from a conversion of work into an equivalent amount of "heat" with a constant conversion factor--as we know it today, 4.186 J = 1 calorie. Since we have characterized these work interactions as *adiabatic*, then no heat is transferred, so Joule was incorrect to regard the temperature increase of the water as induced by a heat interaction. Strickly speaking, these adiabatic work interactions changed the internal energy of the water as manifested by the temperature rise. In order to proceed further with our postulatory development of thermodynamics, we need to formalize the relationship between an adiabatic work interaction and the change in a system's total energy content.

Table 3.1 Types of work interactions

$$\delta W = \mathbf{F} \cdot d\underline{\mathbf{x}}$$

Specific type of work	$\mathbf{F} \cdot d\underline{x}$
Pressure - volume	$-Pd\underline{V}$
surface deformation	$\sigma d\underline{a}$
electrical charge transport	$\mathcal{E} d\underline{q}$
electric polarization	$\mathbf{E} d\underline{\mathbf{D}}$
magnetic polarization	$\mathbf{H} d\underline{\mathbf{B}}$
frictional	$F_f d\underline{x}$
stress-strain	$\underline{V}_o(F_x/\underline{a})d\Omega$ [1.]

1. $F_x/\underline{a}$ = stress; $d\Omega$ = linear strain = dx/x_o; $\underline{V}_o$ = original volume

3.3 Energy

We have gone to the trouble to define an adiabatic work interaction because the magnitude of such interactions allows us to rank stable equilibrium states. The fact that adiabatic work interactions are always possible between stable equilibrium states cannot be developed from first principles, but the truth of such a statement has been borne out by a large body of experimental evidence. Thus, it is presented in the form of a postulate.

Postulate III. For any states (1) and (2), in which a closed system is at equilibrium, the change of state represented by (1) → (2) and/or the reverse change (2) → (1) can occur by at least one adiabatic process and the adiabatic work interaction between this system and its surroundings is determined uniquely by specifying the end states (1) and (2).

As implied in the postulate, it is not always possible to go from state (1) to state (2) by an adiabatic process, but when this route is impossible, it must always be possible to find an adiabatic process from state (2) to state (1). For example, consider a system containing 1 mole of oxygen gas at a thermometric temperature θ_1 and pressure P_1. Through interactions with the environment, the temperature is increased to θ_2 but the pressure is decreased to P_2. This change of state can be effected by an adiabatic process in many ways. Suppose that the system was fitted with a movable, frictionless, and adiabatic piston. This piston moves in such a way to expand the oxygen gas and reduce the pressure from P_1 to P_2. This is an adiabatic process since the boundary walls are adiabatic and no dissipative processes occur at the (moving) boundary. During the pressure reduction, θ will also change. When we have attained the desired pressure level, we will hold this value with the use of an external pressure reservoir at P_2. We will then change the value of θ in the system with a scheme as shown in Figure 3.6 to attain the value of θ_2. This scheme is possible only if we have a thermometric scale that increases as electrical power is dissipated by an i^2R drop in the system (i.e., physiologically when the system becomes warmer). Should we have a thermometric scale that behaves in an inverse manner, we could not proceed by an adiabatic process from $1 \rightarrow 2$, but it should then be obvious that we could then go from $2 \rightarrow 1$ by similar adiabatic processes where the gas is compressed from P_2 to P_1, etc.

Since the postulate was stated for *any* states of a system at equilibrium, it follows that all stable states can be bridged by adiabatic processes originating from a given initial stable state. Thus, if state A is chosen as a reference state, any change to different states represented by $B_1, B_2, \ldots B_j, \ldots$ can be characterized by measuring experimentally the adiabatic work required for the change in state from A to B_j (or B_j to A, if the former change in state is not possible). Since the adiabatic work is only a function of the end states, the adiabatic work is a derived property of the system. We shall call this derived property the total energy, $\underline{E}$, of the system and we shall follow the convention that the energy increases when work is done on the system by the surroundings. Conversely, if work is done by the system on the surroundings, then the total energy will decrease. Thus,

$$\underline{E}_{B_j} - \underline{E}_A = + W^a_{A \rightarrow B_j} \tag{3-17}$$

where W^a is the adiabatic work and is always positive when work is done on the system by the surroundings. Although by convention we could associate a value of $\underline{E}$ with each stable state, it is clear that only differences of energy have significance.

Equation (3-17) appears as an abridged form of the *First Law of Thermodynamics* and leads directly to the *Conservation Law for Energy*: Since the adiabatic work measured in the surroundings must be equal and opposite to the adiabatic work measured in the system (see Section 3.1), the energy change of the system must be equal and opposite to the energy change of the surroundings. The approach followed here closely parallels the path taken by Carathéodory in 1909 (see Chapter 1 for the complete reference.)

It should be noted that the definitions of adiabatic work and energy are not restricted to simple systems. They are valid for both simple and composite systems even in the presence of external force fields. Throughout the text the symbol $\underline{E}$ is used to denote the total energy of such systems. When a development is limited to systems that are not acted upon by external force fields or inertial forces, the symbol $\underline{U}$ is used to denote this *internal energy*. The symbol $\underline{U}$ is used for composite and simple systems for which such forces are absent.

Total energy ($\underline{E}$), as described above using Postulate III, is a rigorous, mathematically defined quantity, requiring no further physical connotations. In practice, however, we usually dissect the total energy into three major components using the language of physics and classical mechanics: (1) kinetic energy associated with motion of the center of mass (e.g. inertial effects), (2) potential energy associated with the relative position of the system in one or more body force fields, (e.g., gravitational, electric, or magnetic) and (3) the remainder embodied in the so-called internal energy as the energy contained in the molecules comprising the system of interest. Time-averaged molecular motions (translation, rotation, vibration), intramolecular effects (nuclear spin, electronic vibrations, etc.) and intermolecular interactions due to potential energy effects between molecules all are associated with internal energy. Statistical mechanics allows us to formalize these molecular-level concepts in such a way that they can be related to macroscopic properties such as $P\underline{V}TN$ equations of state and heat capacities. Nonetheless, at this stage of our development of classical thermodynamics, we can (and should) dissociate ourselves from these molecular-level interpretations. This is true for the entire treatment of Part I, *Fundamental Principles*. In the second part of the book, *Thermodynamic Properties*, molecular-level concepts will be introduced and utilized. For example, Chapter 10 provides a basis for making connections between the statistical mechanics of interacting molecules and observed macroscopic thermodynamic properties.

For a simple system, the energy can be defined with the aid of Postulate I. If each of the end states, A and B_j, of a simple system is uniquely specified by two independently variable properties plus the masses M_i of the n components, the internal energy associated with each stable state must also be a unique function of these $n+2$ independent properties. For example, if θ and P are independently variable for a given system, then

$$\underline{U}_i = f(\theta_i, P_i, M_1, M_2, \ldots, M_n) \tag{3-18}$$

It can also be shown that as a consequence of Postulate III, $\underline{U}$ must be first order in the total mass of the system. That is, the function of Eq. (3-18) is such that

$$a\underline{U}_i(\theta_i, P_i, M_1, M_2, \ldots, M_n) = \underline{U}_i(\theta_i, P_i, aM_1, aM_2, \ldots, aM_n) \tag{3-19}$$

where a is a constant. The proof can be developed by comparing the following two processes, which have the same net effects: (1) a process in which two identical systems are acted upon simultaneously and (2) a process in which each system is acted upon separately.

Because energy is first order in mass, it can be shown that the energy of a composite of simple subsystems is equal to the sum of the energies of the subsystems of the composite. This conclusion is used repeatedly in the following discussions.

3.4 Heat Interactions

Heat is an elusive entity that is recognizable only by its effect on material substances. For our discussion of work we were fortunately able to adopt definitions and procedures from mechanics. For a discussion of heat we have no precedent to follow since the onus of developing this concept lies within the realm of thermodynamics. Thus, we must endeavor to define heat by using the definitions and concepts already presented.

The key concept needed is that the energy difference between two states *can always* be determined by measuring the work in an adiabatic process connecting the two states (Postulate III). Now, with the same initial and final states, visualize *any* process (adiabatic or non-adiabatic) as going between these states. The energy difference is the same as that found for the adiabatic process because energy is a function of state only (i.e., it is independent of the path connecting the two states). If the process is not adiabatic, the work interaction will be different from that of the adiabatic process; however, the work can always be determined by one of the methods outlined in Section 3.1. *We define heat as the difference of the total energy change and the actual work performed.* That is,

$$Q = (\underline{E}_{final} - \underline{E}_{initial}) - W \tag{3-20}$$

where by convention, W is positive if work is done on the system by the surroundings and Q is positive if heat is "added" to the system. Again, it is important to keep in mind that you may have encountered other sign conventions for work.

The definition for heat above, like the one given previously for energy, is devoid of any microscopic or molecular-level significance. Nevertheless, it is of great practical utility. We will deduce shortly under what circumstances a heat interaction is to be expected and develop a method for ranking systems with respect to the direction of heat interactions.

In Eq. (3-20), heat is defined as the difference between the actual work in the process and the adiabatic work that would be required to effect the same change in state. Since any system completely enclosed by adiabatic walls can only undergo adiabatic work interactions, it follows that for such systems $Q = 0$. Alternatively, a system must have at least one non-adiabatic or diathermal wall if it is to undergo a heat interaction. The converse however, is not necessarily true; namely, systems connected by a diathermal wall will not necessarily have a heat interaction.

The concept of a wall separating the system from its surroundings or from another system must be interpreted somewhat loosely rather than literally as a defined geometric or structural feature. For example, it is quite possible that only a portion of the surface of a real wall might be non-adiabatic. Or to solve a problem, it may be convenient to introduce an imaginary surface that is non-conducting and/or impermeable to separate two portions of a system that are undergoing such rapid changes that there is not enough time for the system's contents to mix uniformly.

In Section 2.3, an isolated system was defined as one having adiabatic, rigid, and impermeable walls. Since the walls are adiabatic, the system cannot have any heat interactions. Since the walls are rigid, the system cannot have any work interactions. The term "rigid" is used in the generic sense throughout. That is, not only are the walls immovable, but no shafts or electrical conductors pass through them. Hence, an isolated system can have no interactions with the environment, and therefore the energy of an isolated system is invariant.

We will define a ***pure heat interaction*** as one for which the actual work is zero, and therefore $\Delta\underline{E} = Q$ where Δ defines the change between final minus initial states. For a system to undergo a pure heat interaction, the system must be surrounded by rigid and impermeable walls, at least one of which is diathermal. As discussed below, the pure heat interaction is helpful in defining more specifically a thermometric temperature.

Consider two systems, A and B, which are closed, have rigid walls, and have adiabatic walls except for their common boundary, as indicated in Figure 3.7. When these two systems are first brought together, they may or may not have a heat interaction. In this particular case, the only boundary interaction will be a pure heat interaction. We can tell if any interaction occurs by observing whether or not the primitive properties of A and B change after they are brought together.

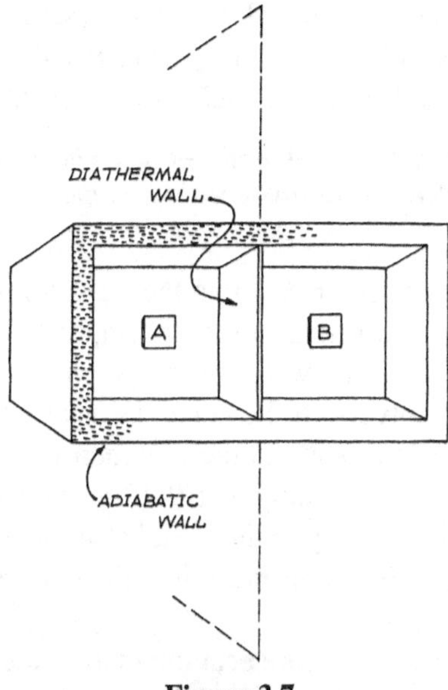

Figure 3.7

If no heat interaction occurs when the systems are placed in thermal contact, it follows from Postulate II that the composite is at equilibrium. Subsystems that are at equilibrium across a diathermal wall are said to be in *thermal equilibrium*. But if there is a heat interaction, it follows from Postulate II that the interaction must eventually cease since the subsystems of the isolated composite system will approach thermal equilibrium. If a heat interaction occurs, Q_A must be equal to $-Q_B$ because the total energy change of the isolated composite system, which is the sum of the energy changes of the subsystems, must be zero.

Let us now turn to the problem of determining the direction of a heat interaction. Let us redefine the procedure for measuring the thermometric temperature. We construct a sealed liquid-in-glass thermometer in such a way that the walls are rigid and adiabatic except for one diathermal surface which is placed in contact with the system under study. Thus, the thermometer–system interaction is equivalent to the A–B interaction of Figure 3.7. [Note: other primitive properties may be equally suitable for studying a heat interaction having no work or mass exchange. We chose the thermometric temperature as a convenient device that is easy to visualize (see Section 2.4)]. By heating or cooling the thermometer on a hot plate or in an ice bath, we can eventually find a liquid level for which no heat interaction occurs when the thermometer is placed in contact with system B, and hence the measurement of the temperature of system B can be made without disturbing system B. (Alternatively, we can make the thermometer small in relation to the system and let the two come into thermal equilibrium with a negligible change in the properties of system B.) In this manner the liquid level can be used as a measure of the stable equilibrium state of the simple system, B. In order to give more significance to the θ-property, an additional postulate is required. If we put the thermometer in contact with system B and then with system C, and if we find the same liquid level, there is still no way to prove that systems B and C are in thermal equilibrium with each other. Thus, we shall postulate such a requirement.

Postulate IV. If the sets of systems A–B and A–C each have no heat interactions when connected across nonadiabatic walls, there will be no heat interaction if systems B and C are also connected.

This postulate is sometimes referred to as the *Zeroth Law of Thermodynamics*. It requires that systems with the same thermometric temperature be in thermal equilibrium and that no heat interaction occur between these systems.

It can be shown that the converse to Postulate IV is also true. If A–B are in thermal equilibrium, and A–C result in a heat interaction, then B–C must also result in a heat interaction. The proof involves assuming that B–C does not result in a heat interaction and then showing that this is contrary to the initial statement. Thus, composite systems attain thermal equilibrium only when the temperatures are uniform throughout the subsystems.

When systems A and B undergo a pure heat interaction such that $\Delta \underline{E}_A = -\Delta \underline{E}_B < 0$, we shall use the imprecise statement that "heat is transferred from A to B" instead of the longer, more precise statement that energy is transferred from A to B as a result of a pure heat interaction. In shorthand notation, we shall say $Q_{A \to B}$ is positive, which implies that $Q_A = -Q_B < 0$.

We are now in a position to show that the thermometric temperature can be used to rank systems with respect to the direction of heat interactions. We shall prove that for the three systems A, B, and C, if $Q_{A \to B}$ is positive (i.e., if heat is transferred from A to B) and if $Q_{B \to C}$ is positive, then θ_B must lie between θ_A and θ_C. Let us choose a thermometer scale so that $\theta_A > \theta_B$ and prove that θ_C cannot be equal to or greater than θ_B.

If $\theta_C = \theta_B$, then $Q_{B \to C} = 0$, which is a contradiction to the initial statement. If $\theta_C > \theta_B$, we could have chosen C so that $\theta_A = \theta_C$. If we then allow the three systems to interact in such a way that $Q_{A \to B} = Q_{B \to C}$, so that θ_B does not change in the process (i.e., $\Delta \underline{E}_B = 0$), the net result is a heat interaction from A to C. This, however, is not allowed because $\theta_A = \theta_C$ and, therefore, θ_C cannot be greater that θ_B. Thus, the thermometric temperature can be used to rank different systems according to the directions in which heat interactions will occur.

It should be emphasized that the thermometric-temperature scale is arbitrary. We could define another scale, called the ζ-scale, for which the freezing point of water is 100°ζ and the boiling point is 0°ζ. All that we require is that $\Delta\zeta \to 0$ when systems of different ζ undergo heat interactions. Of course, once we choose a temperature scale by some convention, we must be consistent throughout.

We shall adopt the *convention* that when $\theta_A > \theta_B$, the heat interaction is such that the energy of A decreases and the energy of B increases, or equivalently:

$$\frac{dE}{d\theta} > 0 \tag{3-21}$$

3.5 The Ideal Gas

Before exploring the applications of the First Law, it is convenient to point out here another temperature scale that will be very useful in later developments.

It has been found that for a gas, such as helium, a very simple experiment may be carried out to define a particular temperature scale. Suppose that we confine a given quantity of gas in a container fitted with a piston and keep the pressure on this gas constant. For example, at 1 atm pressure immerse the gas system in boiling water and measure the volume; then repeat the experiment in an ice-water bath (the pressure above the bath should be that corresponding to water vapor, but for this discussion 1 atm pressure will not significantly affect the results). We denote the ideal-gas temperature by T instead of θ. We call the first temperature 373.15 and the second temperature 273.15 and then plot $\underline{V}$ against T as shown in Figure 3.8, drawing a straight line between the two $\underline{V}$ -T points and extrapolate to zero volume; it will be found that at $\underline{V} = 0$, our temperature scale is also zero.

This $\underline{V}$-T relationship gives us an experimental method to measure temperature. Repeating the experiments at other low pressures will give different lines but all will intersect the $\underline{V} = 0, T = 0$ singular point. Any of the various $\underline{V}$ -T lines (each at a different pressure) can be used to define a temperature scale. This particular scale is called the "ideal gas" scale. If we were clever enough, we could bring all the $\underline{V}$ -T lines (at different pressures) together by noting that if the same number of moles of gas were used in each experiment, then the product $P\underline{V}$ (rather than just $\underline{V}$) lines, which were drawn between the same two temperature points, fall on the same line and pass through the point $(P\underline{V}) = 0, T = 0$. This linearity is given analytically as $P\underline{V}/T =$ constant. Further, if we vary the moles of gas in the system, we can obtain by similar experiments the expression $P\underline{V}/NT =$ constant. This latter constant is called the *gas constant*, R, and the relation when written as

$$P\underline{V} = NRT \tag{3-22}$$

is the *ideal gas law*. If the pressure is given in pascals (1 Pa = 1 N/m^2) (note that N = newtons), the volume in m^3/mol, the temperature in Kelvins, and N in gram-moles, then R has the value 8.314 J/mol K (N-m/mol K). Appendix H gives additional values of R in different units.

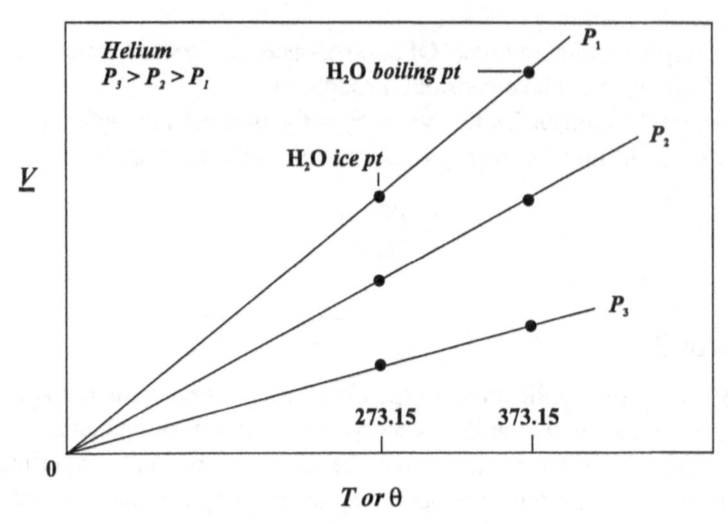

Figure 3.8 Ideal gas behavior, low pressure isobars for helium at constant N.

Now review carefully the development of this particular temperature scale. We carried out two experiments to measure the volume (or $P\underline{V}$ or $P\underline{V}/N$ products) and then we defined temperatures at these two points such that a straight line drawn between them intersects the origin.

If the ideal gas law is applied to any two experimental conditions, the expression $T_2 = T_1\ (P_2\underline{V}_2/P_1\underline{V}_1)$ results. In particular , if the two conditions are the freezing and normal boiling points of water, at 1 atm = 1.01325 bar, the ratio of $P_2\underline{V}_2/P_1\underline{V}_1$ is equal to 1.366. The ideal gas temperature scale is called the *Kelvin scale* (K) if the number 273.15 is assigned to the freezing point of water. If desired, other scales for which $T = 0$ at $\underline{V} = 0$ could be devised by assigning a different value to the freezing point of water and assigning the normal boiling point of water a value of 1.366 times the freezing point.

Alternative empirical approaches have also been used to determine the ideal gas temperature scale. For example, the pressure can be measured for a sealed gas system as a function of the thermometric temperature with the normal boiling and freezing points of water, 100.0°C and 0.0°C set as fudicial points to fix the Celsius or Centigrade scale. In the ideal gas region, a straight line will result that will intersect the temperature axis at -273.15°C. This temperature intercept will be duplicated exactly when other gases of different volumes and mass are used. The slope of each straight line will be different corresponding to different values of $N/\underline{V}$ but they all will have a common intercept.

An *ideal gas* is defined as one that obeys the ideal-gas law, Eq. (3-22), and whose total internal energy is a function only of N and T, that is,

$$\underline{U} = NU_o + N\int \frac{dU}{dT}\,dT \tag{3-23}$$

The integrand is expressed as a total derivative since, if U is a function only of T, $dU/dT = (\partial U/\partial T)_V = (\partial U/\partial T)_P$, etc., and U_o is a reference energy per mole (or mass) of material at $T = T_o$. It is obvious that in dealing with ideal gases, there is a great simplification since only the variables N and T need to be considered. For non-ideal gases (and liquids or solids), $\underline{U}$ would have to be expressed as a function of N plus *two* other independently variable properties, in accordance with Postulate I. The properties of ideal gases are summarized in Table 3.2. Further discussion of ideal-gas state heat capacities is found in Chapter 8, Section 8.4 where the importance of molecular translation, rotation, vibration and other intramolecular effects in determining the magnitude of heat capacities is outlined. In Chapter 10, a statistical mechanical interpretation of ideal gas behavior is also given.

Table 3.2 Ideal-Gas State Properties

Equation of State (EOS) $P\underline{V} = NRT$ or $PV = RT$

Internal Energy

U is only a function of temperature

$U = \int C_v^o \, dT + U_o$; $\underline{U} = N\,U$; $dU = C_v^o \, dT$

$U_o =$ reference state constant

$C_v^o =$ ideal gas state heat capacity at constant volume

$C_v^o = (\partial U/\partial T)_V$

$C_v^o = g(T) = a + bT + cT^2 + \ldots$

Enthalpy

H is only a function of temperature

$H \equiv U + PV = \int C_p^o \, dT + H_o$; $\underline{H} = N\,H = \underline{U} + P\underline{V}$; $dH = C_p^o \, dT$

$H_o =$ reference state constant

$C_p^o =$ ideal gas state heat capacity at constant pressure

$C_p^o = (\partial H/\partial T)_P$

$C_p^o = f(T) = a^* + b^*T + c^*T^2 + \ldots$

Other Useful Relationships

$C_p^o - C_v^o = R$

$dV/V = dT/T - dP/P$

$d\underline{V}/\underline{V} + dP/P = dT/T + dN/N$

1. Note: See Appendix H for values of R in different units
2. $a, b, c \ldots$ and $a^*, b^*, c^*, \ldots$ are fitted empirical constants and $a^* - a = R$, $b = b^*$, $c = c^*$, etc.

3.6 The First Law for Closed Systems

Given what has been presented in Section 3.4, we can rewrite Eq. (3.20) as:

$$\Delta \underline{E} = Q + W \tag{3-24}$$

Many authors refer to Eq. (3.24) as the *First Law*. Since Q and W are defined only in terms of interactions at boundaries for a prescribed process, Eq. (3-24) has significance only when applied to a specific system. Note that the $Q+W$ form of the first law is consistent with the convention followed by the International Union of Pure and Applied Chemistry (IUPAC), but differs from the $Q-W$ convention that has been used in earlier editions of this text. Unfortunately, the $Q-W$ convention is still followed in many other engineering texts. A major advantage of the $Q+W$ convention is that positive values of Q and W both increase the total energy of the system in an internally consistent manner.

For a closed system interacting with the surroundings, the composite of system and surroundings can always be considered as a system of constant volume surrounded by adiabatic walls. Since such an isolated composite system can have no heat or work interactions with its environment, the energy of the composite system is invariant. Therefore,

$$\Delta \underline{E}_{system} = -\Delta \underline{E}_{surroundings} \tag{3-25}$$

where again Δ is an operator that defines the change in the property ($\underline{E}$ in this case) as the final minus initial state value. Since we have already shown (see Section 3.1) that

$$W_{system} = -W_{surroundings} \tag{3-26}$$

it follows that

$$Q_{system} = -Q_{surroundings} \tag{3-27}$$

For a differential change in state, the ***First Law for closed systems*** can be written as

$$\boxed{d\underline{E} = \delta Q + \delta W} \tag{3-28}$$

where δQ signifies that heat, like work, is a path function.

It should be noted that there may be many processes for going from one stable state to another; for each process, $\Delta \underline{E}$ is the same, but Q and W may be quite different. Equations (3-24) and (3-28) apply to all processes of closed systems. The restriction of closed systems relates to the manner in which Postulate III was stated. Extension of the First Law to systems with permeable boundaries is presented in Section 3.8.

To the engineer the significance of the First Law may not be readily apparent from Eqs. (3-24) and (3-28). The fact that energy is conserved when two or more systems (forming an isolated composite) interact may have some aesthetic value, but since we do not usually measure energy directly, the statements of the First Law by themselves do not have real utility. We usually measure work effects and properties such as pressure, temperature, volume, and mass because these measurements are relatively simple. It is only when we relate energy (and other derived properties which we introduce later) to other measurable properties by using Postulate I that we obtain the full utility of the First Law. These constitutive relationships between properties (both derived and primitive) are not fixed by any principle or law of classical thermodynamics (although they place some restrictions on the form of functionality, e.g., $\underline{U}$ must be first order in mass). In general, constitutive property relationships must be determined empirically or at least validated empirically, as discussed in Chapters 8 and 9. Molecular-level concepts are related to macroscopic properties in Chapter 10 using the principles and methodology of statistical mechanics. In a few instances, statistical mechnical methods can produce an exact closed form solution. The ideal gas law given in Equation (3-22) is an example of such a relationship.

Almost all engineering applications of the First Law fall within two categories: (1) for given or measured interactions at the boundaries of a system, what are the corresponding changes in the properties of the system; and (2) for given changes in the properties, what interactions may occur at the boundaries?

3.7 Applications of the First Law for Closed Systems

Applications of the First Law are illustrated by two examples. In both, ideal gases have been chosen. Non-ideal gases (and liquids) are treated in Chapter 8.

In presenting numerical examples where only ideal gases are considered, Eq. (3-22) is used to relate the pressure-temperature-volume variables. To employ Eq. (3-23) we define a term C_v, the heat capacity at constant volume, to relate energy to temperature:

$$C_v \equiv \left(\frac{\partial U}{\partial T}\right)_V \tag{3-29}$$

This definition of C_v is *not* limited to ideal gases, but applies in all cases. Thus, using Eq. (3-23) for an ideal gas,

$$\underline{U} = NU_o + N\int C_v\, dT \tag{3-30a}$$

or, per mole (or unit mass),

$$dU = C_v\, dT \tag{3-30b}$$

The convenience of limiting the change of U with T to constant-volume cases is evident when simple systems ($E = U$) are treated by Eq. (3-28). Then, if the system has rigid walls ($\delta W = 0$),

$$dU = C_v\, dT = \delta Q \quad (N, V = \text{constant}) \tag{3-31}$$

and $C_v = dQ/dT$ for a case involving constant mass and volume. Again, it is emphasized that Eqs. (3-29) and (3-31) are general and not limited to ideal gases as are Eqs. (3-30a) and (3-30b).

Example 3.3

Two well-insulated cylinders are placed as shown in Figure 3.9. The pistons in both cylinders are of identical construction. The clearances between piston and wall are also made identical in both cylinders. The pistons and the connecting rod are metallic.

Cylinder A is filled with gaseous helium at 2 bar and cylinder B is filled with gaseous helium at 1 bar. The temperature is 300 K and the length L is 10 cm. Both pistons are only slightly lubricated.

The stops are removed. After all oscillations have ceased and the system is at rest, the pressures in both cylinders are, for all practical purposes, identical.

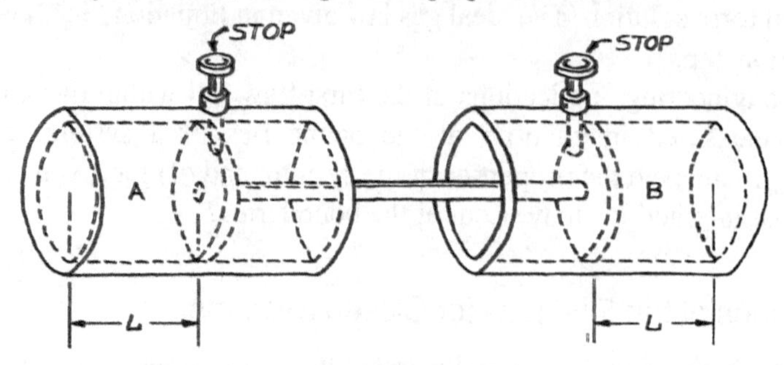

Figure 3.9

Assuming that the gases are ideal with a constant C_v and, for simplicity, assuming that the masses of cylinders and pistons are negligible (i.e., any energy changes of pistons and cylinders can be neglected), what are the final temperatures?

Solution

Let us choose the gases in compartments A and B as systems A and B, respectively. Since these are simple systems, we denote the energies by $\underline{U}$ instead of $\underline{E}$. Since the mass and initial temperature are known for the gas in each compartment, the final temperature can be calculated for each compartment, if the final energy—or energy change—of each compartment is first determined. Thus, from Eqs. (3-22) and (3-30a) as applied for each system,

$$N_A = \frac{P_{A_i} \underline{V}_{A_i}}{RT_{A_i}} \tag{3-32a}$$

$$N_B = \frac{P_{B_i} \underline{V}_{B_i}}{RT_{B_i}} \tag{3-32b}$$

$$T_{A_f} = T_{A_i} + \frac{\underline{U}_{A_f} - \underline{U}_{A_i}}{N_A C_v} \tag{3-32c}$$

and

$$T_{B_f} = T_{B_i} + \frac{\underline{U}_{B_f} - \underline{U}_{B_i}}{N_B C_v} \tag{3-32d}$$

where $\underline{U}_{A_f}$ and $\underline{U}_{B_f}$ are the only unknowns ($\underline{U}_{A_i}$ and $\underline{U}_{B_i}$ can be chosen at will because U_o in Eq. (3-30a) is an arbitrary constant).

Since the pistons and shaft have been assumed to be good conductors of heat, the final temperatures of compartments A and B will be equal:

$$T_{A_f} = T_{B_f} \tag{3-33}$$

To determine the energy change, let us apply the First Law, Eq. (3-24), to system A:

$$\underline{U}_{A_f} - \underline{U}_{A_i} = Q_A + W_A \tag{3-34}$$

The work done by system A is equal to the work done on the gas in B plus the frictional work done on the walls of A and B, or

$$W_A = -W_B - W_{W_A} - W_{W_B} \tag{3-35}$$

where W_B is work done *by* the gas in B and W_{W_A} is the work done *by* the wall of A, etc., thus a negative sign is used for all three terms on the RHS of Eq. (3-35). Now consider the wall of A as the system and apply the First Law to this system:

$$\Delta\underline{U}_{W_A} = Q_{W_A} + W_{W_A} \tag{3-36}$$

If we neglect the change in internal energy of the wall, $\Delta\underline{U}_{W_A} = 0$ and the frictional work done by the gas on the wall will be transmitted back to the gas in the form of a heat interaction; in this case W_{W_A} is negative and

$$Q_{W_A} = -W_{W_A} \tag{3-37}$$

and furthermore,

$$Q_A = -Q_{W_A} - Q_{AB} \tag{3-38}$$

where Q_{AB} is the heat conducted from A to B through the pistons. Substituting Eqs. (3-35), (3-37), and (3-38) into Eq. (3-34),

$$\Delta\underline{U}_A = -W_B - W_{W_B} - Q_{AB} \tag{3-39}$$

By analogy to Eqs. (3-37) and (3-38) as applied to the walls of B,

$$-W_{W_B} = Q_{W_B} = -Q_B + Q_{AB} \tag{3-40}$$

and by analogy to Eq. (3-36)

$$-W_B - Q_B = -\Delta\underline{U}_B \tag{3-41}$$

Equation (3-39) becomes

$$\Delta\underline{U}_A = -\Delta\underline{U}_B \tag{3-42}$$

Of course, Eq. (3-42) could have been stated at the outset because the composite system $A + B$ is equivalent to an isolated system; this result, however, may not have been immediately obvious. Combining Eq. (3-42) with Eqs. (3-32c) and (3-32d), and making use of Eq. (3-33) we obtain

$$T_f = \frac{N_A T_{A_i} + N_B T_{B_i}}{N_A + N_B} \tag{3-43}$$

or, for this special case wherein $T_{A_i} = T_{B_i} = T_i$,

$$T_f = T_i = 300 \text{ K}$$

Let us now use hindsight to reevaluate the problem. We had a qualitative feeling for the path in that we knew frictional work was involved, but we could not describe the path quantitatively because we did not know the coefficient of friction of either piston-cylinder. Nevertheless, we were able to determine the final conditions and, therefore, we did not have to describe the path to find the solution. Such a situation would be expected if the end state was independent of the path. This is obviously the case in the present example: since the composite system $A + B$ is an isolated simple system, there is only one state to which it can go, and that is the one for which $\underline{U} = \underline{U}_{A_i} + \underline{U}_{B_i}$ and $T = T_{Af} = T_{Bf}$. For an ideal gas, $\underline{U}$ is a unique function of T and N and thus the final temperature can be determined. If the gas were not ideal, $\underline{U}$ would also depend on $\underline{V}$ (or P). Thus, since the final energy, volume, and mass are known, the final temperature could still be determined.

Example 3.4

Consider the situation described in Example 3.3, but with well-insulated pistons and connecting rods of low thermal conductivity. What are the final temperatures after the oscillations have ceased and the pressures have equalized?

Solution

The composite system of $A + B$ is no longer a simple system because it contains an internal adiabatic wall. Therefore, the final composite cannot be described by a single equilibrium state; instead, the final conditions will depend on the path of the process.

With the exception of Eq. (3-33), Eqs. (3-32) through (3-42) are still valid. Combining Eqs. (3-32c) and (3-32d) with Eq. (3-42) results in

$$N_A C_v(T_{A_f} - T_{A_i}) + N_B C_v(T_{Bf} - T_{B_i}) = 0 \tag{3-44}$$

which now gives us one equation in two unknowns, T_{A_f} and T_{B_f}. We could, of course, try to juggle the other equations to find another relationship between T_{A_f} and T_{B_f}, but until we make some assumptions regarding the path, our efforts will be in vain.

If we have no information on the coefficient of friction, we are forced to use our engineering judgment to simplify the situation while obtaining a close approximation to the actual conditions. Let us assume that there is friction only in compartment B. This will give us a lower bound for T_{A_f} and the upper bound for T_{B_f}. We can then treat the case of friction only in compartment A, which will give us an upper bound for T_{A_f}. In this manner, we can bracket the true solution.

If there is no friction in compartment A and if we assume that the process is quasi-static (i.e., no pressure or density gradients within the compartment), we can write Eq. (3-28) for the simple system of the gas in A as

$$d\underline{U}_A = +\,\delta W_A = -\,P_A d\underline{V}_A \tag{3-45}$$

since

$$\delta W_{W_A} = \delta Q_A = \delta Q_{AB} = 0$$

Substituting for $\underline{U}_A$ in Eq. (3-45) from Eq. (3-32c), and for $\underline{V}_A$ from Eq. (3-22), Eq. (3-45) becomes

$$N_A C_v dT_A = -N_A R(dT_A - \frac{T_A}{P_A} dP_A) \tag{3-46}$$

or

$$\left(\frac{C_v + R}{R}\right)\frac{dT_A}{T_A} = \frac{dP_A}{P_A} \tag{3-47}$$

Integrating between initial and final conditions, we obtain

$$\frac{T_{A_f}}{T_{A_i}} = \left(\frac{P_{A_f}}{P_{A_i}}\right)^{R/(C_v + R)} \tag{3-48}$$

Equations (3-48) and (3-44) give us two equations in three unknowns, T_{A_f}, T_{B_f}, and $P_{A_f} = P_f$. The final pressure can be eliminated in the following manner. Since

$$\underline{V}_{A_f} + \underline{V}_{B_f} = \underline{V}_{A_i} + \underline{V}_{B_i} = \underline{V}_T \tag{3-49}$$

and $\underline{V}_T$ is known, substitution of Eq. (3-22) into Eq. (3-49) gives

$$\underline{V}_T = (N_A T_{A_f} + N_B T_{B_f})\frac{R}{P_f} \tag{3-50}$$

Equation (3-50) together with Eqs. (3-48) and (3-44) give us three equations in these unknowns. From Eqs. (3-50) and (3-44),

$$P_f = \frac{P_{A_i}\underline{V}_{A_i} + P_{B_i}\underline{V}_{B_i}}{\underline{V}_{A_i} + \underline{V}_{B_i}} = 1.5 \text{ bar} \tag{3-51}$$

From Eq. (3-48), with $C_v = 12.6$ J/mol K

$$T_{A_f} = \left(\frac{1.5}{2}\right)^{0.4} T_{A_i} = 267 \text{ K}$$

and from Eq. (3-44)

$$T_{B_f} = 366 \text{ K}$$

If it is assumed that there is friction only in compartment A, then we would have found $T_{B_f} = 353$ K. Since in the actual case the friction is distributed between A and B, a better approximation might be $T_{B_f} = (366 + 353)/2 = 360$ K and from Eq. (3-44), $T_{A_f} = 270$ K.

The fact that the adiabatic wall in Example 3.4 prevents a direct solution to the problem in the absence of a complete description of the path is sometimes referred to as the "adiabatic dilemma." In fact, it is no dilemma at all, but results from the difference between heat and work interactions.

3.8 The First Law for Open Systems

A system was defined as being open if it has at least one region of its boundary which is permeable to mass. The First Law, as stated in Section 3.6, was restricted to closed systems. However, since an open system can always be considered as a closed system by redefining the boundaries, the extension of the First Law to open systems requires no additional postulates.

Consider an open system bounded by the σ-surface as illustrated in Figure 3.10. Part of the σ-surface is diathermal and part is movable, so that there may be heat and work interactions, Q_σ and W_σ, with the surroundings. The σ-surface also contains a region through which mass can enter or leave. The boundary at that region may consist of an opening or a permeable membrane.

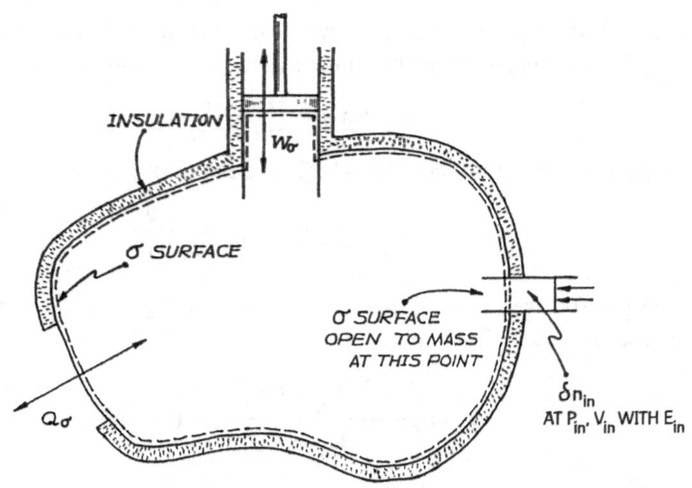

Figure 3.10 Generalized open simple system.

Consider a time, δt, during which a small quantity of mass or moles, δn_{in}, enters the system bounded by the σ-surface. The properties of the entering material are pressure, P_{in}, specific volume, V_{in} (volume per mole), and specific energy, E_{in} (energy per mole). Although the region bounded by the σ-surface is an open system, the composite system of $\sigma + \delta n_{in}$ is a closed system. Defining $\underline{E}$ as the total energy of the system bounded by the σ-surface, let us apply the First Law to the closed composite system:

$$\underline{E}_2 - (\underline{E}_1 + E_{in}\,\delta n_{in}) = Q_\sigma + W_\sigma + P_{in}V_{in}\,\delta n_{in} \tag{3-52}$$

where subscripts 1 and 2 refer to initial and final conditions, respectively, and $P_{in}V_{in}\,\delta n_{in}$ is the $P - \underline{V}$ work required to push δn_{in} into the region bounded by the σ-surface. Equation (3-52) can be rearranged to a form similar to the First Law for a closed system:

$$\Delta \underline{E} = Q_\sigma + W_\sigma + (E_{in} + P_{in}V_{in})\delta n_{in} \tag{3-53}$$

where $\Delta \underline{E} = \underline{E}_2 - \underline{E}_1$, Q_σ, and W_σ apply only to the region bounded by the σ-surface (i.e., the open system), and the last term applies to the mass flux entering the system.

The differential form of Eq. (3-53) is

$$d\underline{E} = \delta Q_\sigma + \delta W_\sigma + (E_{in} + P_{in}V_{in})\, \delta n_{in} \tag{3-54}$$

and

$$d\underline{E} = d(NE) = N\, dE + E\, dN \tag{3-55}$$

For the general case where multiple streams enter and leave the system, the First Law for the open system is

$$d\underline{E} = \delta Q_\sigma + \delta W_\sigma + \sum_{in} (E_{in} + P_{in}V_{in})\, \delta n_{in} - \sum_{out} (E_{out} + P_{out}V_{out})\, \delta n_{out} \tag{3-56}$$

or, in integrated form,

$$\Delta \underline{E} = Q_\sigma + W_\sigma + \sum_{in} \int (E_{in} + P_{in}V_{in})\, \delta n_{in} - \sum_{out} \int (E_{out} + P_{out}V_{out})\, \delta n_{out} \tag{3-57}$$

where the summations are taken over all entering and leaving streams, and where δn_{in} and δn_{out} are both taken as positive quantities. The exact differential dN for the system is given by an overall mass balance,

$$dN = \sum_{in} \delta n_{in} - \sum_{out} \delta n_{out} \tag{3-58}$$

When the system defined by the σ-surface is a simple system, E may be replaced by U. Similarly, if the entering and leaving masses are simple systems, E_{in} and E_{out} may be replaced by U_{in} and U_{out}. (Note that the composite closed system is not a simple system if the boundaries at the points where mass flux occurs are semipermeable.) We now define a term called the specific enthalpy (in units of energy per unit mass or mole) as

$$H \equiv U + PV \tag{3-59a}$$

Alternatively the specific enthalpy is sometimes expressed in lower case letters as:

$$h = u + Pv \tag{3-59b}$$

We will minimize the use of lower case h, u, and v in the treatment of this text.

Now the ***general form of the First Law for open simple systems*** is, in differential form,

$$\boxed{d\underline{U} = \delta Q_\sigma + \delta W_\sigma + \sum_{in} H_{in}\, \delta n_{in} - \sum_{out} H_{out}\, \delta n_{out}} \tag{3-60}$$

In many practical engineering situations, we do not have simple systems. In these circumstances, it is often helpful to decompose the total energy $\underline{E}$ into its components. Normal practice would subdivide $\underline{E}$ into three major parts: kinetic energy ($\underline{E}_{KE}$) due to inertial effects, potential energy ($\underline{E}_{PE}$) due to gravitational forces, and internal energy

($\underline{U}$) associated with microscopic energy storage on a molecular level. This uncoupling requires that we use constitutive equations for kinetic and potential energy which are easily obtained from our experience in physics and mechanics, namely:

$$\underline{E} = \underline{E}_{KE} + \underline{E}_{PE} + \underline{U} \tag{3-61}$$

$$d\underline{E} = d\underline{E}_{KE} + d\underline{E}_{PE} + d\underline{U} \tag{3-62}$$

where

$$\underline{E}_{KE} = \frac{m\,|\mathbf{v}|^2}{2} = \frac{mv^2}{2} \quad \text{and} \quad \underline{E}_{PE} = mgz$$

where m is the mass of the system or subsystem of interest.

By following the decomposition of $\underline{E}$ outlined above and the mathematical approach that led to Eq. (3-60), we obtain a general expression for the *First Law* for an open system with its center of mass located at a vertical distance $<z>$ from a reference plane at $z = 0$ and traveling with velocity $<\mathbf{v}>$ as shown in Figure 3.11. Note that now we will employ a mass (rather than mole) basis for the intensive enthalpy terms in the summations of Eq. (3-60): that is, H_{in} and H_{out} have SI units of J/kg. The incoming and outgoing streams also need their appropriate potential energy contributions (from $gz_{in}\delta n_{in}$ and $gz_{out}\delta n_{out}$) and kinetic energy contributions (from $v_{in}^2\,\delta n_{in}/2$ and $v_{out}^2\,\delta n_{out}/2$). For this case, the ***general form of the First Law for open, non-simple systems*** is written as:

$$d\underline{E} = \delta Q_\sigma + \delta W_\sigma + \sum_{in}\left[H_{in} + gz_{in} + \frac{v_{in}^2}{2}\right]\delta n_{in} - \sum_{out}\left[H_{out} + gz_{out} + \frac{v_{out}^2}{2}\right]\delta n_{out} \tag{3-63}$$

with
$$d\underline{E} = d\left[\underline{U} + mg<z> + \frac{m\,|<\mathbf{v}>|^2}{2}\right]$$

For any transient process we can use Eq. (3-58) to obtain the rate of change of mass or moles *(dN/dt)* for the system and differentiate Eq. (3-63) to obtain the rate of change of the system's total energy *($d\underline{E}/dt$)*. Thus,

$$\frac{d\underline{E}}{dt} = \frac{\delta Q_\sigma}{\delta t} + \frac{\delta W_\sigma}{\delta t} + \sum_{in}\left[H_{in} + gz_{in} + \frac{v_{in}^2}{2}\right]\frac{\delta n_{in}}{\delta t} - \sum_{out}\left[H_{out} + gz_{out} + \frac{v_{out}^2}{2}\right]\frac{\delta n_{out}}{\delta t} \tag{3-64}$$

To avoid ambiguities, a mass basis should be selected if kinetic and potential energy effects are to be included. The sums run over all incoming and outgoing streams. For operation at steady state with single inlet and outlet streams, Eq. (3-64) is greatly simplified because both

$$\frac{d\underline{E}}{dt} = 0 \quad \text{and} \quad \frac{dN}{dt} = 0 \tag{3-65}$$

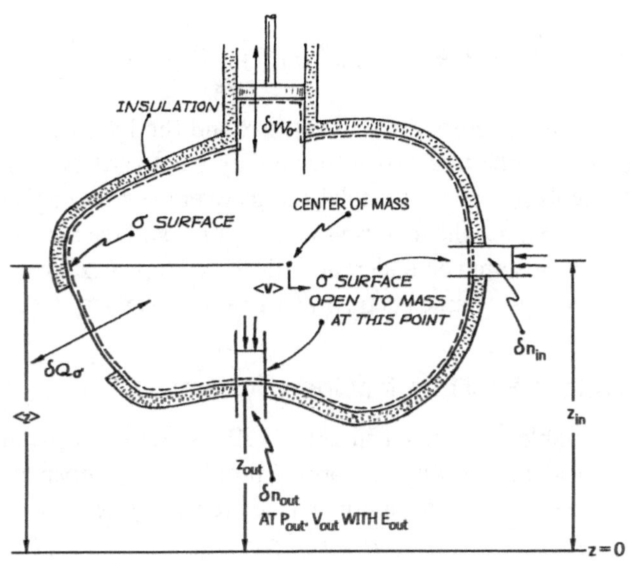

Figure 3.11 Generalized open, non-simple system moving at velocity $<\mathbf{v}>$ in a gravitational field referenced to $z = 0$.

which implies that $\delta n_{in} = \delta n_{out}$ and

$$\frac{\delta n_{in}}{\delta t} = \frac{\delta n_{out}}{\delta t} \equiv \frac{\delta n}{\delta t} = \dot{n} \tag{3-66}$$

Therefore, Eq. (3-64) is modified for *steady state conditions* to:

$$\frac{\delta Q_\sigma}{\delta t} + \frac{\delta W_\sigma}{\delta t} = \dot{Q}_\sigma + \dot{W}_\sigma = (\Delta H_{ss} + \Delta PE_{ss} + \Delta KE_{ss})\,\dot{n} \tag{3-67}$$

where

$$\Delta H_{ss} \equiv H_{out} - H_{in} \;;\quad \Delta PE_{ss} \equiv g\,(z_{out} - z_{in}) \;;\quad \Delta KE_{ss} \equiv \frac{v_{out}^2 - v_{in}^2}{2}$$

represent the time-invariant changes between the outlet and inlet stream quantities.

In some practical situations, other forms of macroscopic energy storage besides kinetic or gravitational potential energy may be important. To account for these effects, which could include storage in rotational modes, electric or magnetic fields, and elastic deformation by stress-strain effects, it is relatively straightforward to modify Eqs. (3-63) and (3-64) (see Chapter 18). Further, you may encounter problems where incoming and outgoing stream fluxes are distributed on the σ-boundary that encloses the system. In addition, the heat flux may also be distributed on the σ-surface rather than occuring at discrete locations and there may be multiple work interactions. In these cases, an integral form of Eq. (3-64) should be used:

$$\int_{\underline{V}} \frac{\partial(\rho E)}{\partial t}\, d\underline{V} = -\int_{\underline{a}_\sigma} \mathbf{q}\cdot\mathbf{n}\, d\underline{a}_\sigma + \sum_i \dot{W}_{\sigma,i} - \int_{\underline{a}_\sigma} \rho\left(H + gz + \frac{v^2}{2}\right)\mathbf{v}\cdot\mathbf{n}\, d\underline{a}_\sigma \quad (3\text{-}68)$$

where $\mathbf{q}$ and $\mathbf{v}$ represent vectors for the heat flux and fluid velocity, and $\mathbf{n}$ is the unit normal vector, perpendicular to the σ-surface pointing outward. Note that there are two types of integrals involved, one for the total energy over the volume $\underline{V}$ of the system and another for the heat and mass flux contributions over the surface area of the σ-boundary $\underline{a}_\sigma$. Note that we have inserted an underbar with $\underline{a}_\sigma$ to emphasize that as used here it is the total extensive area. The parameters $\mathbf{q}$, $\rho(H + gz + v^2/2)$ and $\mathbf{v}$ are spatially dependent.

3.9 Application of the First Law for Open Systems

The enthalpy variable introduced in Eq. (3-59) is treated in detail in Chapter 5. However, it has considerable utility in working problems with open systems. Enthalpy is a property as is clear from its definition, written for the system as

$$\underline{H} \equiv \underline{U} + P\underline{V} \quad (3\text{-}69)$$

or per mole (or unit mass)

$$H \equiv U + PV \quad (3\text{-}70)$$

Also, since U is a function only of temperature [see Eq. (3-23)] and $PV = RT$ for an *ideal gas*, then H also depends only on temperature for this special case.

Analogous to the introduction of C_v [Eq. (3-29)], we define heat capacity at constant pressure as

$$C_p \equiv \left(\frac{\partial H}{\partial T}\right)_P \quad (3\text{-}71)$$

For an *ideal gas*, the restriction to constant-pressure cases is not necessary since, as noted above, H is a function only of temperature (see also Table 3.2.). Finally, again *only* for an *ideal gas*, if the differential of Eq. (3-70) is used with Eqs. (3-22), (3-29) and (3-71), then

$$C_p = C_v + R \quad (3\text{-}72)$$

Example 3.5

A 4-m³ storage tank containing 2 m³ of liquid is about to be pressurized with air from a large, high-pressure reservoir through a valve at the top of the tank to permit rapid ejection of the liquid (see Figure 3.12). The air in the reservoir is maintained at 100 bar and 300 K.

The gas space above the liquid contains initially air at 1 bar and 280 K. When the pressure in the tank reaches 5 bar, the liquid transfer valve is opened and the liquid is ejected at the rate of 0.2 m³/min while the tank pressure is maintained at 5 bar.

What is the air temperature when the pressure reaches 5 bar and when the liquid has been drained completely?

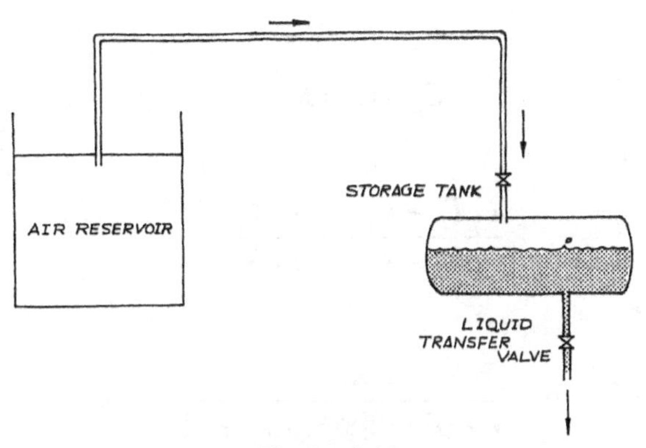

Figure 3.12

Neglect heat interactions at the gas-liquid and gas-tank boundaries. It may be assumed that the gas above the liquid is well mixed and that air is an ideal gas with a constant $C_v = 20.9$ J/mol K.

Solution

Let us treat the process in two steps: (a) the period during which the pressure rises from 1 bar to 5 bar and the volume of the gas in the tank is constant, and (b) the period during which liquid is drained.

Step (a). The most convenient system is the gas in the tank at any time; this is then an open, simple system at constant total volume. With Eq. (3-60) where

$$d\underline{U} = U\,dN + N\,dU$$

$$\delta Q_\sigma = \delta W_\sigma = 0$$

$$H_{in} = \text{constant} \quad \text{(reservoir large relative to system)}$$

$$dn_{in} = dN$$

$$\int \frac{dU}{H_{in} - U} = \int \frac{dN}{N}$$

or

$$\frac{H_{in} - U_i}{H_{in} - U} = \frac{N}{N_i} = \frac{P}{P_i}\frac{T_i}{T} \tag{3-73}$$

where the subscript i denotes the initial state and the subscript in denotes the incoming or entering stream. Since the heat capacity of air is assumed constant with temperature, and the gas behaves ideally,

$$H_{in} = C_p(T_{in} - T_o) + H_o$$

$$U = C_v(T - T_o) + U_o$$

and

$$H_o - U_o = RT_o$$
$$C_p - C_v = R$$

Then Eq. (3-73) becomes

$$\frac{C_pT_{in} - C_vT_i}{C_pT_{in} - C_vT} = \frac{P}{P_i}\frac{T_i}{T}$$

or

$$T = \frac{\kappa T_{in}}{1 + (P_i/P)[\kappa(T_{in}/T_i) - 1]} \tag{3-74}$$

where

$$\kappa \equiv \frac{C_p}{C_v} = \frac{C_v + R}{C_v} = \frac{20.9 + 8.314}{20.9} = 1.4$$

With $T_{in} = 300$ K, $T_i = 280$ K, $P_i = 1$ bar, and $P = 5$ bar,

$$T = \frac{(1.4)(300)}{1 + (1/5)[1.4(300/280) - 1]} = 382 \text{ K}$$

At this time

$$N = \frac{P\underline{V}}{RT} = \frac{(5 \times 10^5)(2)}{(8.314)(382)} = 315 \text{ mol}$$

It is interesting to note that, contrary to what might have been anticipated, the final temperature of the gas is higher than that of either the initial temperature or the temperature of the incoming gas. In the limit where $P >> P_i$, the temperature approaches κT_{in} independent of the initial conditions in the tank.

Step (b). We choose the same system as in step (a). Equation (3-60) is still applicable, but in this case, the total volume varies while the pressure remains constant. Then with

$$d\underline{U} = U\,dN + N\,dU$$
$$\delta Q_\sigma = 0$$
$$-\delta W_\sigma = P\,d\underline{V} = PN\,dV + PV\,dN$$
$$P = \text{constant}$$
$$dn_{in} = dN$$
$$H_{in} = \text{constant}$$
$$U\,dN + N\,dU = -PN\,dV - PV\,dN + H_{in}\,dN$$
$$N(dU + P\,dV) = N\,d(U + PV) = N\,dH = (H_{in} - U - PV)\,dN$$

or

$$\int \frac{dH}{H_{in} - H} = \int \frac{dN}{N}$$

$$\frac{H_{in} - H_i}{H_{in} - H} = \frac{N}{N_i} = \frac{\underline{V}}{\underline{V}_i}\frac{T_i}{T} = \frac{T_e - T_i}{T_e - T} \qquad (3\text{-}75)$$

The subscript i in this case represents the conditions existing at the end of step (a). Solving for T, we obtain

$$T = \frac{T_{in}}{(\underline{V}_i/\underline{V})[(T_{in}/T_i) - 1] + 1}$$

When the tank has been completely drained, $\underline{V} = 4\ \text{m}^3 = 2\underline{V}_i$. With $T_{in} = 300$ K and T_i [from (a)] = 382 K,

$$T = \frac{300}{(1/2)[(300/382) - 1] + 1} = 336\ \text{K}$$

Suggested Readings

Bejan, A. (1988), *Advanced Engineering Thermodynamics*, New York: Wiley, Chapter 1, p 1-42. [treatment of work interactions and First Law for open systems with kinetic and potential energy effects]

Callen, H.B. (1985), *Thermodynamics and an Introduction to Thermostatistics*, 2nd ed, New York: Wiley, Chapter 1, p 5-33. [postulatory approach]

Denbigh, K. (1981), *The Principles of Chemical Equilibrium*, 4th ed, Cambridge, UK: Cambridge University Press, Chapter 1, p 14-21. [discussion of the First Law via adiabatic work interactions]

Gyftopoulos, E.P. and G.P. Beretta (1991), *Thermodynamics: Foundations and Applications*, New York: MacMillan, Chapter 3, p 27-43. [alternative approach to the First Law using adiabatic work interactions]

Problems

3.1. A small well-insulated cylinder and piston assembly (Figure P3.1) contains an ideal gas at 10.13 bar and 294.3 K. A mechanical lock prevents the piston from moving. The length of the cylinder containing the gas is 0.305 m and the piston cross sectional area is 1.858×10^{-2} m^2.

The piston, which weighs 226 kg, is tightly fitted and when allowed to move, there are indications that considerable friction is present. When the mechanical lock is released, the piston moves in the cylinder until it impacts and is engaged by another mechanical

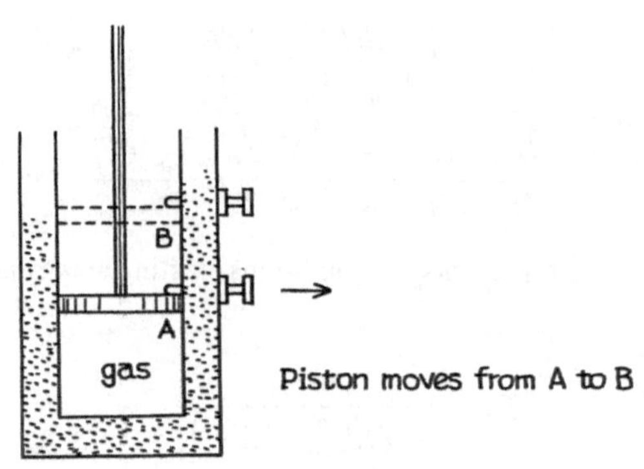

Figure P3.1

stop; at this point, the gas volume has just doubled. The heat capacity of the ideal gas is 20.93 J/mol K, independent of temperature and pressure. Consider the heat capacity of the piston and cylinder walls to be negligible.

(a) As an engineer, can you estimate the temperature and pressure of the gas after such an expansion? Clearly state any assumptions.

(b) Repeat the calculations if the cylinder were rotated 90° and 180° before tripping the mechanical lock.

3.2. Many very large liquefied natural gas (LNG) storage tanks have been built or are under construction. The LNG is predominantly liquid methane with a boiling point near 111 K at 1 bar. To avoid excessive loss, the tanks are very well insulated. In the construction of such tanks, to encourage the contractor to do the best job, there is normally a clause written into the contract which awards the builder a bonus if the heat leak into the tank is below some agreed value—but there is also a penalty clause if this specified heat leak is exceeded.

Since such penalty (or bonus) values are large, it is crucial to specify a detailed testing procedure to "prove" the heat leak after the tank has been built and filled with LNG. Normally in such a proof test, the filled tank is allowed to attain an equilibrium state with the internal pressure held constant. Ambient conditions should not vary greatly during this period. At the end of this pretest period, the tank and liquid are assumed to be in thermal equilibrium at the existing tank pressure. The actual proof test then consists of measuring the boil-off vapor over a period of several days while keeping the tank pressure equal to the pretest value.

Consider a real test. The LNG tank contains 40,000 m^3 of pure liquid methane. The tank pressure is 1.044 bar. Over the test period, the measured boil-off rate was 4.267×10^4 mol/h. Calculate as accurately as you can the heat leak into the tank (J/h). In this particular case, the contract specified a $500,000 penalty clause if the heat leak exceeded 0.35 GJ/h (0.35×10^9 J/h). A bonus of the same amount was to be awarded to the contractor if the heat leak was less than 0.35 GJ/h.

Data (from the National Bureau of Standards Report NBSIR 73-342): For methane at 112 K, 1.044 bar

$$V^V = 8.6036 \times 10^{-3}\ \text{m}^3/\text{mol} \quad V^L = 0.0380 \times 10^{-3}\ \text{m}^3/\text{mol}$$

$$dV^V/dT = -6.5015 \times 10^{-4}\ \text{m}^3/\text{mol K} \quad dV^L/dT = 1.309 \times 10^{-7}\ \text{m}^3/\text{mol K}$$

$$H^V = 1.27982 \times 10^4\ \text{J/mol} \quad H^L = 4.6052 \times 10^3\ \text{J/mol}$$

$$dH^V/dT = 26.8\ \text{J/mol K} \quad dH^L/dT = 55.96\ \text{J/mol K}$$

$$dP/dT = 0.0854\ \text{bar/K}$$

Assume that the 40,000-m^3 tank is filled with 99% liquid by volume.

(a) Would the contractor gain or lose the $500,000?

(b) The true facts for the example above were only slightly different: over the 24-h period where the boil-off vapor averaged 4.267×10^4 mol/h, despite the best intentions of the operators, the tank pressure fell from 1.044 to 1.043 bar. Would these facts change the award?

3.3. (Refer to Figure P3.3 for notation.) A piston (A) and piston rod (B) are fitted inside a cylinder of length 0.508 m and area 6.45×10^{-3} m^2. Although the piston is quite thin, it weighs 9.07 kg; the piston rod is 1.29×10^{-3} m^2 in area and weighs 4.53 kg. On top of the rod, but outside the cylinder, an 18.14-kg weight (C) is placed. Originally, gas in D is at atmospheric pressure while the piston is positioned in the middle of the cylinder. Gases in D and E are helium and under these conditions may be considered ideal with a constant $C_v = 12.6$ J/mol K. The initial temperature everywhere is 311 K.

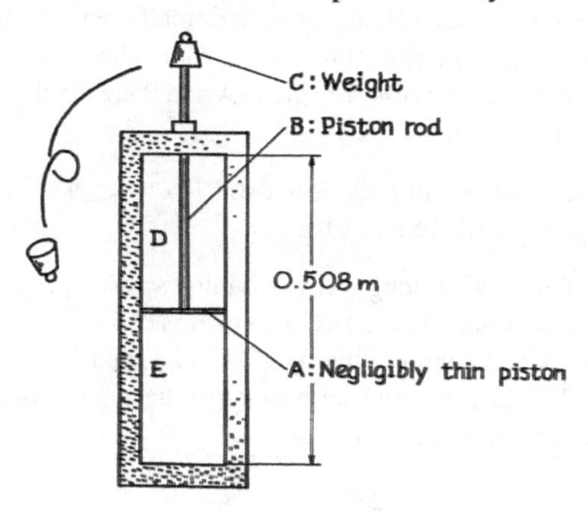

Figure P3.3

Assuming the cylinder, piston, and piston rod to be nonconducting and having a negligible heat capacity, discuss any heat or work interactions if weight (C) should fall off. What is the final state of the system when the piston has stopped and there is a balance of forces across the piston? Do not neglect the fact that during motion there may be some friction between moving parts. Consider cases in which the piston is (a) diathermal and (b) adiabatic.

3.4. Two cylinders are attached as shown in Figure P3.4. Both cylinders and pistons are adiabatic and have walls of negligible heat capacity. The connecting rod is nonconducting.

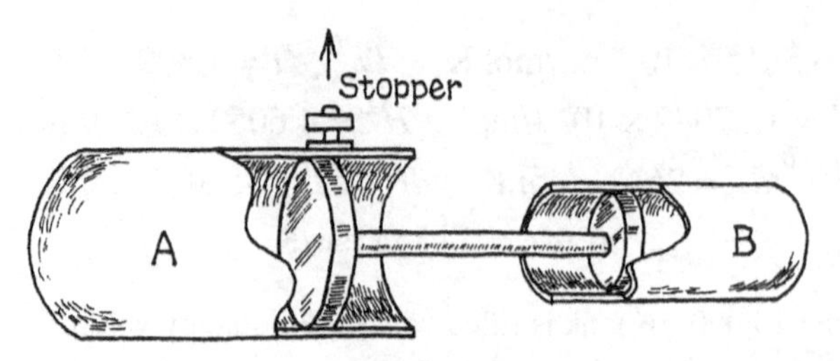

Figure P3.4

The initial conditions and pertinent dimensions are as follows:

	Cylinder A	Cylinder B
Initial pressure (bar)	10	1
Initial temperature (K)	300	300
Initial volume (m^3)	6.28×10^{-3}	1.96×10^{-3}
Piston area (m^2)	3.14×10^{-3}	1.96×10^{-3}

The pistons are, initially, prevented from moving by a stop on the outer face of piston A. When the stop is removed, the pistons move and finally reach an end state characterized by a balance of forces on the connecting rod. There is some friction in both piston-cylinders during this process. The gases A and B are ideal and have constant values of $C_v = 20.9$ J/mol K.

What are the final pressures in both A and B? Do two cases, one where the ambient pressure is 0 bar and one where it is 1 bar.

3.5. A horizontal cylinder 0.457 m long is divided into two parts, A and B (see Figure P 3.5) by a latched piston. As illustrated, the volume of A is twice that of B and contains helium at 311K and 10.13 bar. B contains hydrogen at 311 K and 1.01 bar. Both of these gases may be considered ideal with constant heat capacities as follows: C_v (helium) = 12.56 J/mol K, and C_v (hydrogen) = 20.9 J/mol K.

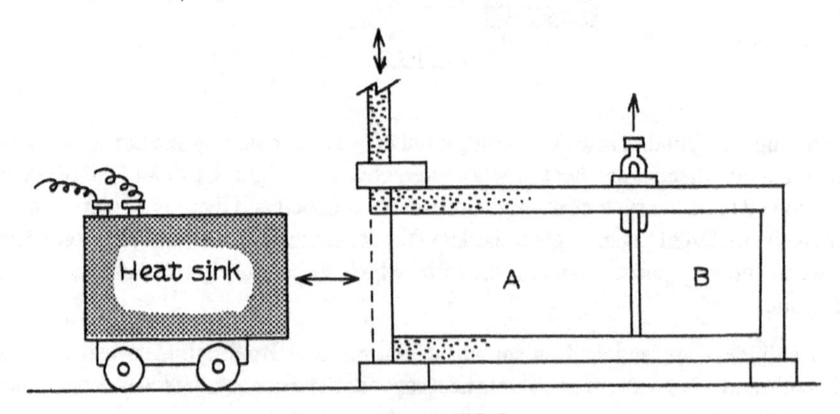

Figure P3.5

Provision has also been made to connect volume A to a constant temperature reservoir at 311 K. When the latch is removed, the piston is allowed to seek an equilibrium position so that there are equal pressures on each side.

Without neglecting friction but assuming no heat transfer to the cylinder walls, compute the final position of the piston and the temperature and pressure in both A and B for the following four cases:

No contact of volume A with the 311 K reservoir:

1. Piston is diathermal
2. Piston is adiabatic

Volume A is in diathermal contact with the 311 K reservoir:

3. Piston is diathermal
4. Piston is adiabatic

3.6. Our next research experiment is to be carried out in a vertical, cylindrical reactor 0.0929 m^2 in cross-sectional area and 0.605 m long (see Figure P3.6). A gas mixture in the reactor is at 288.8 K and may be considered to be ideal with a constant C_p = 29.3 J/mol K.

It has been suggested that we should provide a safety attachment of some kind on the reactor to prevent the pressure from exceeding 6.9 bar. One of our more creative engineers suggests that we remove the top of the reactor and weld on a pipe extension. This extension would be fitted with a heavy piston latched in place. We would also provide a pressure transducer and activation circuit in the reactor to unlatch the piston should the pressure exceed 6.9 bar.

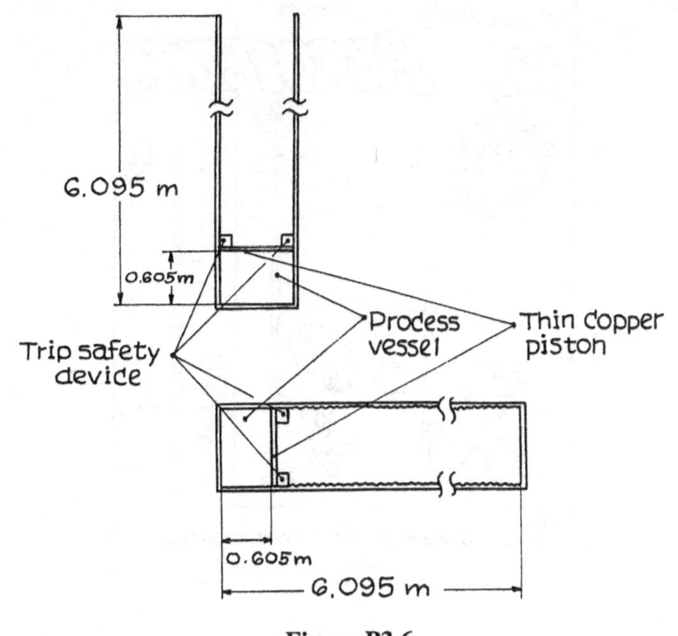

Figure P3.6

The headroom in the laboratory is only 5.49 m and, since we do not want any of our process gas escaping, we must select a piston of the correct mass so that it will not be blown out of the pipe extension. If the piston were made of copper (density = 8660 kg/m^3), what should its thickness be? Assume insignificant friction and neglect any heat transfer from the gas to the cylinder walls or piston.

Another engineer, however, has been advocating an alternative technique. He, too, wishes to remove the reactor top and weld to it a pipe extension with a latched piston. But he also wishes to put a cap on the top of this extension and rotate the cylinder on to its side. In this case, should the reactor pressure exceed 6.9 bar, the piston would move horizontally and compress the gas in the pipe extension. He also wishes to slightly roughen the walls in the pipe so that there is friction between the moving piston and pipe walls but still with no gas leakage.

(a) Again assuming negligible heat transfer, what is your best estimate of the final gas temperatures on both sides of the piston after pressures on both sides are equal? Assume initially that the gas in the closed pipe extension is air at 1 bar and 288.8 K.

(b) Which of these two techniques would you select? Why?

3.7. Advertised is a small toy that will send up a signal flare and the operation "is so simple that it is amazing" (see Figure P3.7). Our examination of this device indicates that it is a sheet metal tube 2.13 m long and 645 mm^2 in cross-sectional area. A plug shaped into the form of a piston fits into the tube and a mechanical trigger holds it in place 0.61 m above the bottom. The piston weighs 1.57 kg and contains the necessary parachute and pyro-

Figure P3.7

technics to make the show exciting. To operate the device, the volume below the piston is pumped up to a pressure of about 4.05 bar with a small hand pump, and then the trigger is depressed, allowing the piston to fly out the top. The pyrotechnic and parachute devices are actuated by the acceleration force during ejection.

When we operated this toy last summer, the ambient temperature was 305 K.

(a) Assuming no friction in the piston and no heat transfer or other irreversibilities in the operation, how high would you expect the piston to go? What would be the time required from the start to attain this height?

(b) Since you are an engineer who is never satisfied with a commercial object, suggest improvements to make the piston go even higher. What is the maximum height that could be obtained if it were limited to 4.05 bar pressure?

(c) Comment on the way you might analyze the expected performance if the restrictions in part (a) are removed.

3.8. Bottles of compressed gases are commonly found in chemistry and chemical engineering laboratories. They present a serious safety hazard unless they are properly handled and stored. Oxygen cylinders are particularly dangerous. Pressure regulators for oxygen must be kept scrupulously clean, and no oil or grease should ever be applied to any threads or on moving parts within the regulator. The rationale for this rule comes from the fact that if oil *were* present—and if it *were* to ignite in the oxygen atmosphere—this "hot" spot could lead to ignition of the metal tubing and regulator and cause a disastrous fire and failure of the pressure container. Yet it is hard to see how a trace of heavy oil or grease could become ignited even in pure, compressed oxygen since ignition points probably are over 800 K if "nonflammable" synthetic greases are employed.

Let us model the simple act of opening an oxygen cylinder that is connected to a closed regulator (see Figure P3.8). Assume that the sum of the volumes of the connecting line and the interior of the regulator is V_R. V_R is negligible compared to the bottle volume. Opening valve A pressurizes V_R from some initial pressure to full bottle pressure.

Presumably, the temperature in V_R also changes. The question we would like to raise is: Can the temperature in V_R ever rise to a sufficiently high value to ignite any traces of oil or grease in the line or regulator?

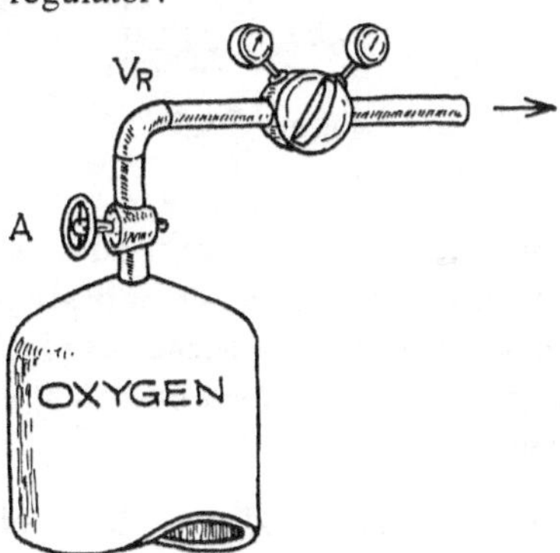

Figure P3.8

Data: The oxygen cylinder is at 15.17 MPa and 311.0 K. The connecting line to the regulator and the regulator interior (V_R) are initially at 0.101 MPa, 311 K, and contain pure oxygen.

Assume no heat transfer to the metal tubing or regulator during the operation. Oxygen is essentially an ideal gas. $C_p = 29.3$ J/mol K, $C_v = 20.9$ J/mol K, and both are independent of pressure or temperature.

(a) If the gas entering V_R mixes completely with the initial gas, what is the final temperature in V_R?

(b) An alternative model assumes that there is *no* mixing between the gas originally in V_R and that of which enters from the bottle. In this case, after the pressures are equalized, we would have two identifiable gas slugs which presumably are at different temperatures. Assuming no axial heat transfer between the gas slugs, what is the final temperature of each?

(c) Comment on your assessment of the hazard of this simple operation of bottle opening. Which of the models in (a) and (b) is most realistic? Can you suggest other improved models?

3.9. During an emergency launch operation to fill a missile with RP-4 (a kerosene-based fuel), the ullage volume of the fuel storage tank is first pressurized with air from atmospheric pressure (1.01 bar) to a pressure of 10.34 bar (see Figure P3.9).

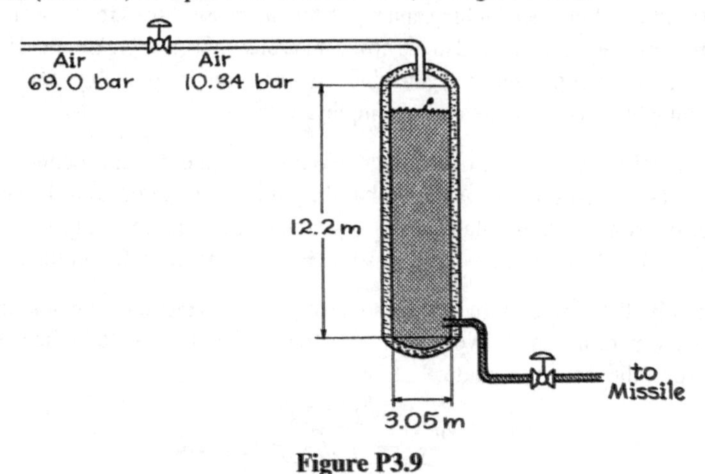

Figure P3.9

The air is available from large external storage tanks at high pressure (69.0 bar). The overall operation is to be completed as rapidly as possible. After the 10.34-bar pressure level is reached, the main transfer valve is opened and fuel flows at a steady rate until the missile is loaded. It is necessary to maintain a constant gas pressure of 10.34 bar inside the fuel tank during transfer.

The fuel storage tank can be approximated as a right circular cylinder 12.2 m tall and 3.05 m in diameter and is originally filled to 90% of capacity. Transfer of fuel to a residual volume of 10% must be completed in 18 min. Assume ideal gases and that the operation is adiabatic and all hardware have negligible heat capacity. Initial temperatures may range from 242 K (arctic sites) to 333 K (equatorial sites), but for the purposes of a first estimate, use 294 K as an initial temperature.

(a) Comment on any safety hazards that might be encountered.

(b) What problems would you anticipate if the inlet gas control valve were to malfunction and the gas space above the fuel were to reach full storage tank pressure (69.0 bar)? (The fuel tank has been hydrostatically tested to 276 bar.)

(c) What is your estimate of the time–temperature history of the gas above the fuel during the entire operation? These data are needed to size the inlet air lines.

3.10. In many installations in the chemical industry, occasions arise when compressed gas bottles are rapidly blown down. These bottles are constructed of carbon steel that becomes dangerously brittle at low temperatures. Certainly, in rapid blow down situations the gas temperature could drop to such a low value that if rapid heat transfer with the cylinder were to occur, a hazardous operation would result. It is believed that the only place where very high heat transfer rates are possible is in the cylinder neck, because velocities are highest in this region. To calculate neck wall temperatures, however, the time variation of the bulk gas temperature of the bottle must be available.

Demonstrate your ability to estimate the bulk gas temperature of the bottle gas as a function of time for the first 2 min in the following case:

Inside wall area = 9.290 m^2
Volume = 0.7788 m^3
Thickness of wall = 8.38 mm
Specific heat of wall = 0.419 J/g K
Density of wall = 8.46×10^3 kg/m^3
Initial bottle pressure = 137.9 bar
Rate of pressure decay: bottle pressure is reduced by factor of 2 every 1.6 min
Bottle gas: nitrogen (assume ideal gas behavior)
Initial bottle gas and wall temperatures = 241.5 K

It may be assumed that heat transfer from the cylinder wall to the gas occurs by a natural convection process. For purposes of computation, assume that the heat transfer coefficient is 45.4 W/m^2K for all bottle pressures in excess of 69 bar and 34.1 W/m^2K for pressures less than 69 bar.

3.11. A rigid laboratory gas cylinder of 5.66×10^{-2} m^3 volume is charged with air at 138 bar and 277.8 K. An experiment is to be carried out whereby the cylinder is rapidly vented by opening the valve on the cylinder top. The pressure in the cylinder is always so high that the flow is choked (i.e., sonic) in the valve throat. If one assumes that the gas is ideal and that there is negligible heat transfer between the gas and walls:

(a) Derive a relation to calculate the bottle pressure as a function of time.

(b) Calculate the time when the bottle pressure drops to 69 bar.

For sonic flow through a round, sharp-edged orifice (assumed to apply to the valve throat),

$$\text{mass flow} = \dot{m} = C_a A_o P \left[\frac{\kappa m}{RT} \left(\frac{2}{\kappa + 1} \right)^{(\kappa+1)/(\kappa-1)} \right]^{1/2}$$

where C_a = discharge coefficient (assume = 0.6)
A_o = orifice area = $9.29 \times 10^{-4} m^2$
P = cylinder pressure, Pa
R = gas constant, 8.314 J/mol K
κ = C_p/C_v = 1.4 for air (assume constant)
m = molecular weight = 29×10^{-3} kg/mol
$\dot{m}$ = mass flow in kg/s

3.12. A well-insulated pipe of 2.54 cm inside diameter carries air at 2 bar pressure and 366.5 K. It is connected to a 0.0283-m^3 insulated "bulge", as shown in Figure P3.12.

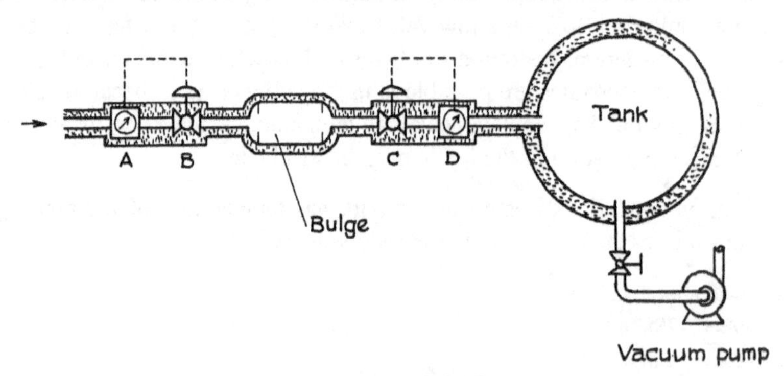

Figure P3.12

The air in the bulge is initially at one bar pressure and 311 K. A and D are flow meters which accurately measure the mass rate of airflow. Valves B and C control the airflow into and out of the bulge. Connected to the bulge is a 0.283-m^3 rigid, adiabatic tank which is initially evacuated to a very low pressure.

At the start of the operation, valve B is opened to allow 4.54 g/s of air flow into the bulge; simultaneously, valve C is operated to transfer exactly 4.54 g/s from the bulge into the tank. These flows are maintained constant as measured by the mass flow meters.

Air may be assumed to be an ideal gas with a constant C_p of 29.3 J/mol K. Assume also that the gases, both in the bulge and large tank, are completely mixed so that there are no temperature or pressure gradients present.

(a) What is the temperature and pressure of the gas in the bulge after 6 s?

(b) What is the temperature and pressure of the air in the large tank after 3 s?

3.13. A vessel containing a reactive compound is about 0.06 m^3 in volume (gas space). There is an inert atmosphere of helium maintained at 1 bar and 311K. If, however, the compound shows signs of decomposition, it is desired to increase very rapidly the helium overpressure to 10 bar. This higher pressure will then be used to dump the reactive compound to a water-soak tank. To accomplish this rapid pressurization, the vessel is connected by a short transfer line and valve to another vessel filled with high-pressure helium. This vessel is 0.18 m^3 and contains helium originally at 20 bar and 311 K.

(a) What is the pressure of the helium supply vessel after a pressurization of the reactor? Assume ideal gases and adiabatic operation.

(b) How many reactors could the supply vessel serve simultaneously? (Each reactor is 0.06 m^3 and is pressurized from 1 to 10 bar.)

3.14. To reduce gas storage costs, two companies, A and B, have built a common storage tank in the shape of a horizontal right circular cylinder 0.3 m in diameter and 30 m long (see Figure P3.14). To decide how much gas each company uses between refills, a thin piston was placed in the tank.

The piston moves freely, that is, there is essentially no friction present, and the pressure is the same on both sides. Thus, as company A uses gas, the piston moves left and as company B uses gas, the piston moves right.

When the gas company refills the tank, it must decide how much gas has been used by each company. It can easily measure the position of the piston and can, if necessary, install other instrumentation such as thermometers or pressure gauges in either or both ends of the tank.

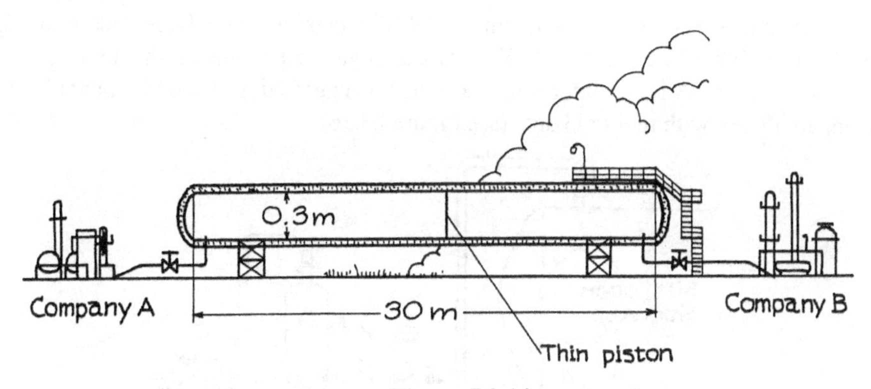

Figure P3.14

Assume that (1) the gas is ideal; (2) the piston is adiabatic; (3) the walls are well insulated and have low heat capacity; and (4) at the start of each month after filling the tank, the gas company positions the piston in the center of the tank, equalizes the temperature on both ends, and carefully meters the total amount of gas added.

List the minimum instrumentation that you would recommend, and describe how the amount of gas consumed by both companies could be determined at the time of refilling.

3.15. A thermodynamicist is attempting to model the process of balloon inflation by assuming that the elastic casing behaves like a spring opposing the expansion (see Figure P3.15). As air admitted, the spring is compressed. The pressure in the gas space is given by

$$P - P_i = k(L - L_i)$$

where k = constant = 5 bar/m

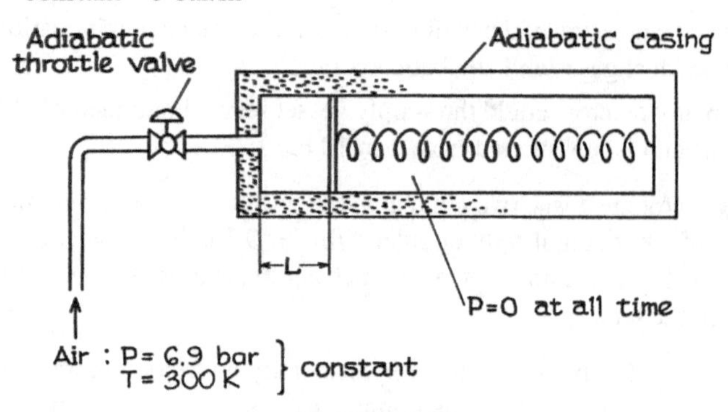

Figure P3.15

The initial conditions in the gas space are: P_i = 1 bar, T_i = 300 K, and L_i = 0.15 m. The piston area is 0.02 m^2. The air is an ideal gas and C_v = 20.9 J/mol K, independent of temperature. What is the air temperature in the gas space when L = 0.6 m?

3.16. From the memoirs of a thermodynamicist: "While relaxing near a large tank of nitrogen gas (*A*) at 687 kN/m^2 and 298 K, I began reviewing some of my knowledge of thermodynamics. A rather interesting experiment suggested itself and I thought I would compare theory with real field data (see Figure P3.16).

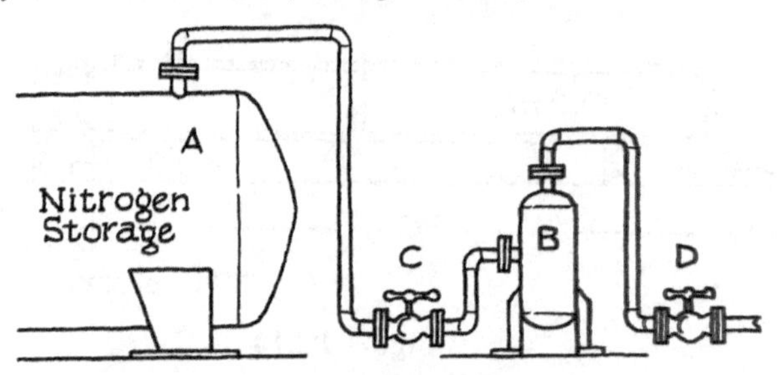

Figure P3.16

I obtained a small high pressure vessel (*B*) and two valves (*C*) and (*D*). I first filled *B* with nitrogen gas at 101 kN/m^2 and 298 K and connected it as shown. Then working quickly, I opened valve *C* (with *D* closed) and allowed the pressures in *B* and *A* to equalize. Then, I quickly closed *C* and opened *D* to blow down vessel *B* to its original pressure. I repeated this sequence a number of times. Tank *A* was so large that I did not cause any significant drop in pressure in it by my experiments. Second, I pressurized and blew down *B* so rapidly that little heat transfer probably occurred this time."

Nitrogen is an ideal gas with a value of C_p = 29.33 J/mol K.

(a) Guess the temperature of the gas in B after the second pressurization and after the second blowdown.

(b) What do you think these temperatures were after a very large number of cycles?

3.17. An all quartz Dewar flask is filled initially with liquid hydrogen at 1 bar (see Figure P3.17). The inner walls cool to the normal boiling point of hydrogen 20.6 K. The liquid is then quickly poured out and the flask evacuated to a very low pressure. Assume that at the end of the evacuation the walls are still at 20.6 K.

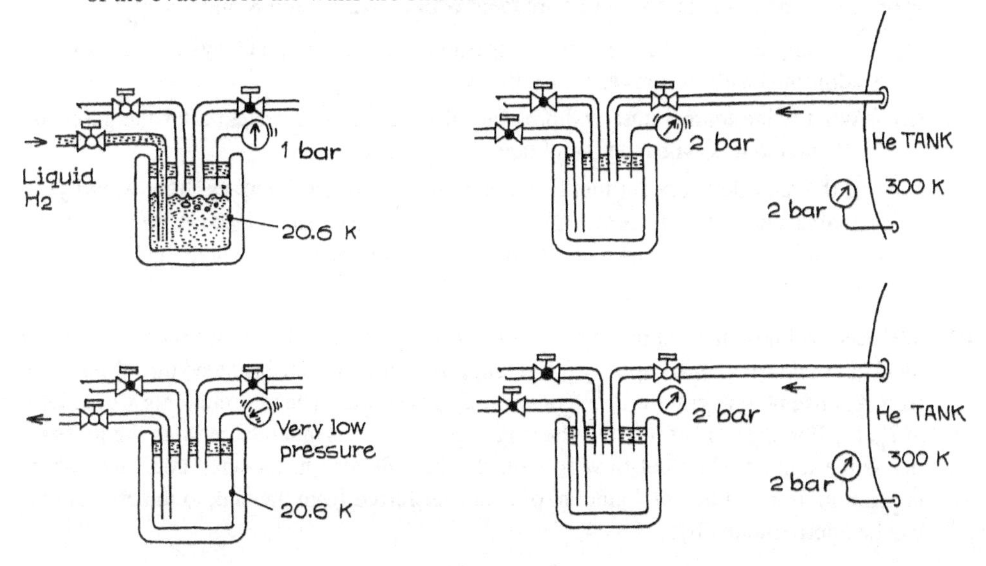

Figure P3.17

The Dewar is then connected to a large tank of helium at 2 bar and 300 K and pressurized very rapidly to 2 bar. After pressurization, the connecting line is left open to allow additional flow to occur in order to maintain a pressure of 2 bar. There is a heat transfer between the helium gas and inner Dewar walls, but assume no heat transfer by radiation, convection, or conduction across the walls of the Dewar. Under these conditions, helium behaves as an ideal gas.

Data:

C_p(helium) = 20.9 J/mol K
C_v(helium) = 12.6 J/mol K
Dewar flask volume = 8.206×10^{-3} m^3
Inner walls of Dewar = 1000 g
Temperature of the environment = 300 K

The enthalpy of quartz, H (J/g), can be approximated as a function of temperature (K) from 20 to 100 K by the relation

$$\log H = -11.43 + 10.64 \log_{10} T - 2.215(\log_{10} T)^2$$

How many moles of helium are there in the Dewar after all flow has ceased?

3.18. In problem 3.9, one model of the liquid fuel pressurization phase assumed that the gas initially in the tank was pressurized as a closed sub-system (B). It is separated from the incoming gas by an infinitesimally thin, impermeable membrane. Initially the contents of the tank are at 1 bar, 294 K and are pressurized to 10.34 bar using an air supply at 69 bar, 294 K (See Figure P3.9.) You can assume that the tank is well-insulated, the liquid fuel is incompressible and that all gases are ideal with $C_p = 29$ J/mol K and $\kappa = C_p/C_v = 1.4$.

(a) Develop an expression for the temperature of the gas in sub-system (B). List all additional assumptions and definitions.

(b) What is the temperature and pressure of the combined system after the gases are allowed to mix when the membrane is removed?

(c) If the tank developed a fuel leak below the gas-liquid interface, how would your analysis in part (a) be affected? You can assume a constant leak rate of $\dot{m}_{leak}$ (kg/s) of fuel of constant density ρ_l. Is other information needed to solve the problem? If so, state what.

3.19. We need to know the temperature-pressure history for a tank filling process where an ideal gas of constant heat capacity (C_p) is supplied across a well-insulated throttling valve from a constant temperature T_i and pressure P_i gas source. The tank contents are initially at T_o, P_o. For any case of interest, the supply pressure P_i is much larger than the pressure inside the tank, so the inlet flow is choked. The tank also has a small constant leak of $\dot{m}_{leak}$ in mol/hr and the total amount of heat transferred from the tank to the ambient air can be approximated by:

$$\langle h_o \rangle \underline{a}(T - T_o)$$

where $\langle h_o \rangle \underline{a}$ is the product of an overall average heat transfer coefficient and the surface area of the tank, T is the internal gas temperature, and T_o is the ambient air temperature. State all additional assumptions. Develop and simplify the appropriate equations and briefly describe your method of solution to obtain T and P as a function of time.

3.20. Rapid blowdown of fire extinguishers containing compressed CO_2 gas can result in enough cooling that metal embrittlement may be a problem. You are to analyze the blowdown situation for a steel cylinder containing 0.1m^3 of gas initially at 20 MPa [2900 psi] and 300 K [27°C]. For your analysis, you can assume that CO_2 is an ideal gas with $C_p = 29$ J/mol K.

The valve can be modeled as an ideal orifice with the mass flow rate $\dot{m}$ given by:

$$\dot{m}\ (\text{mol/s}) = C_o \underline{a}_o\ [f(\gamma)] \Delta P / T$$

where C_o = orifice coefficient = 0.8

$f(\gamma)$ = heat capacity function
= 0.5 for CO_2 (mol K/s Pa m^2)

$\underline{a}_o$ = orifice area = 10^{-3} m^2

The heat transfer rate from the surrounding air and container to the gas in the cylinder can be approximated by the following:

$$\dot{q} = \langle h_o \rangle \underline{a}(T_o - T)$$

where $\langle h \rangle_o$ = overall average heat transfer coefficient = 1 W/m^2 K

$\underline{a}$ = cylinder surface area = 0.2 m^2

T_o = ambient air temperature = 300 K

(a) Develop an analysis to estimate the pressure and temperature of the gas in the cylinder as a function of time. Give equations and assumptions, and explain your solution algorithm.

(b) Qualitatively sketch how the gas pressure and temperature will vary with time.

(c) Develop a criterion to determine when heat transfer to the gas can be neglected.

(d) Assuming that the heat transfer rate to the cylinder is negligible, develop an expression for gas pressure as a function of time.

(e) Using the specified heat transfer rate, solve the set of equations given in (a) to produce plots of gas pressure and temperature versus time.

Reversibility and the Second Law 4

In Postulate III we implied that only certain adiabatic processes between stable equilibrium states may be possible. We have in fact already stated all the postulates necessary for determining which processes are and are not possible. In this chapter we explore in greater depth the consequences of these postulates.

One of the underlying concepts in the developments to follow is the reversible process, which is discussed in Section 4.2. This process has not been found to occur in reality; it is in fact a limiting condition that cannot be attained, but it can be closely approached. Consequently, the reversibility concept cannot be developed fully by examining real systems in the laboratory; instead, we must utilize thought experiments. In these thought experiments, we shall use a device called a heat engine, which is described in the next section. Although many kinds of heat engines have been in use for centuries in the form of operating power cycles, we will employ almost exclusively the fictitious reversible heat engine.

The approach we follow in this chapter is similar to the historical development of thermodynamics, in that it was the creation of a reversible heat engine by Carnot in 1824 that formed the basis for the introduction of the concept of entropy by Clausius in 1850.

4.1 Heat Engines

A *heat engine* is a closed device that undergoes heat interactions with one or more systems and work interactions with a *work reservoir*. A work reservoir is a device that operates adiabatically and quasi-statically and is used for storing energy. For example, a system of weights at different levels in a gravitational field can be used as a work reservoir. The heat engine always undergoes a *cyclic* process such that any net effects appear only in the external systems and in the work reservoir. The definition above excludes various open-system power cycles, such as gas turbines and internal combustion engines (see Chapter 14).

There is great diversity in kinds of heat engines. Although we need not specify any particular one for the developments to follow, it may be instructive to illustrate the operation with a common Rankine cycle used in most stationary electric power plants (see Figure 14.1). The internal working fluid, usually water, is vaporized at a high temperature and high pressure in a boiler. Useful work is obtained by expanding the vapor to a low pressure in a turbine. The low-pressure vapor is liquefied in a condenser, and the liquid is pressurized and returned to the boiler.

The heat engine is a convenient device for evaluating processes (real and imaginary). We shall use heat engines to determine whether or not a process is consistent with the postulates and concepts that have been developed previously. A process will be considered allowable if we know of a real case in which the process occurs. (If such a process violates one of our postulates, we would have to revise that postulate.) If we have no prior experience for a given process, the process will be considered *impossible* if we can prove that the process leads to a violation of one or more of our postulates. If however, we cannot prove that the process violates a postulate, the process may be possible or impossible since we may not have been clever enough to show that a violation exists.

In the processes shown in Figure 4.1, we will choose systems A and B that are in stable equilibrium states prior to their participation in the heat engine interaction. Systems A and B are characterized by thermometric temperatures, θ_A and θ_B. We arbitrarily specify for now $\theta_A > \theta_B$ (recall that we have chosen by convention that a rise in θ corresponds to an increase in energy). Now let us consider a number of conceivable processes.

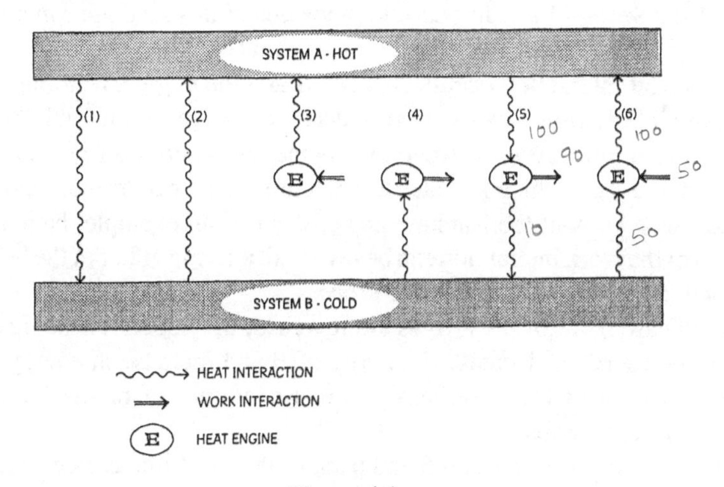

Figure 4.1

Case 1. A heat interaction occurs from A to B (i.e., the energy of A decreases and the energy of B increases) without any work being performed. Heat is being transferred from a higher to a lower temperature reservoir. Since we know of real cases in which such interactions exist, case 1 is clearly allowable.

Case 2. A heat interaction occurs from B to A without any work being performed. Since we have selected our thermometer so that $dE/d\theta > 0$, and since this process leads to an increase in energy of A and a decrease in energy of B, the net effect of this process is to increase $\Delta\theta = \theta_A - \theta_B$. This process is in violation of Postulate II since the composite system $A + B$ does not tend to a state of equilibrium (i.e., $\Delta\theta$ does not tend to zero). Since we have no prior knowledge of real cases in which such processes occur, this process is impossible. Thus, it is concluded that any process in which the net effect is the transfer of heat from a cooler to a hotter system is impossible. A similar conclusion was drawn by Clausius over a hundred years ago, and some authors refer to this result as Clausius's statement of the Second Law.

Case 3. A work interaction occurs whereby work from the reservoir passes to the engine and results in a heat interaction with A. This process is allowable since we know of real cases in which this process occurs [e.g., in Figure 3.5(a) consider B as the engine and C as the work reservoir]. Note that the heat interaction could also have been directed to system B instead of A.

Case 4. A heat interaction involving only B occurs to decrease the energy of B and all of this energy appears as work in the work reservoir. This process can be shown to lead to a violation of Postulate II. If process (4) could occur, we could use process (3) to extract the work produced in (4) and convert this to a heat interaction with A. The net result of the combined processes is equivalent to process (2), which is impossible. Since process (3) is possible, process (4) must be impossible. Thus, it is concluded that any cyclic process for which the net result is the conversion of energy of a single system to work is impossible. Many authors refer to this conclusion as the Kelvin–Planck statement of the Second Law. Impossible processes of this kind are sometimes referred to as perpetual-motion machines of the second kind (PMM2).[1]

Case 5. A heat interaction occurs between A and the engine, resulting in a decrease in the energy of A. Some work is produced in the reservoir and there is also a simultaneous heat interaction between the engine and system B to increase the energy of B. There is nothing in the postulates to prevent such an occurrence, and processes of this kind are well known; the Rankine power cycle is an example. Note that although the direction of the work interaction can be reversed, altering either of the heat interaction vector directions leads to an impossible process.

Case 6. In case 5, all of the arrows are reversed, the net effects being extraction of work from the reservoir, decrease in energy of B and increase in energy of A. Again, there is no violation of the postulates and real examples of process 6 are known as refrigerators or heat pumps.

Of all the cases discussed, cases 5 and 6 are of the most immediate interest. We note that cases 5 and 6 are opposites, and any enterprising person could immediately conjure up some interesting combined processes. As yet we have not placed any quantitative value on the heat and work interactions except to ensure that we have not violated the First Law, which necessitates the conservation of energy. Why couldn't we use case 5 and take 100 units of energy from A, put 90 of them in the work reservoir, and reject 10 to B? Then, following this, we could use case 6 to take 50 units from B, 50 units from the work reservoir, and reject 100 units to A. In this sequence, system A has undergone a cycle—but the work reservoir has gained a net 40 units while system B has lost 40 units. This particular combination of cases 5 and 6, carried out as prescribed, is in reality

---

1 Note that these processes do not necessarily violate the First Law. Perpetual-motion machines of the first kind (PMM1) refer to processes that lead to a net change in the energy of the universe.

case 4, which we found to be impossible. Thus, some combinations of cases 5 and 6 lead to impossible processes, whereas other combinations are allowable. To avoid any combinations that lead to a violation of our postulates, we must specify some limitations to the way in which the heat and work effects are split. We will find out later that the split depends on the temperatures of systems A and B, but as of now we simply recognize that there is a limitation.

A convenient way to delineate the split is to specify an *efficiency*, η, for case 5 as

$$\eta_5 \equiv \frac{\text{work done by engine}}{\text{heat transferred from hot system}} \tag{4-1}$$

$$\eta_5 = -\frac{W_E}{Q_A} \tag{4-2}$$

Following the conventions described in Chapter 3, W_E is the work for the system comprising the engine and is negative when work is done *by* the engine. Q_A is the heat interaction of the engine with system A and is positive when the interaction *decreases* the energy of A providing a positive flow of heat to the engine. In all allowable processes, only one term is negative, and therefore η is positive. Thus, the more efficient we are, the more work we can get out of a given heat interaction between system A and the engine.

We found from case 4 that the efficiency cannot equal unity; what, then does limit the efficiency? In a practical sense, we realize that factors such as friction and other resistances will decrease engine efficiency, but we have made no mention of such factors in deciding that there exists a limiting value of the efficiency. In fact, we assume in the upcoming thought experiments that we can construct an engine that is not plagued by friction and other resistances; such an engine will be referred to as a *reversible* heat engine.

The efficiency of a reversible heat engine must be less than unity:

$$\eta_5 < 1 \tag{4-3}$$

If we define the efficiency for case 6 by Eq.(4-1), then to avoid a combination of processes 5 and 6 which violates our postulates (as discussed above), we require that

$$\eta_5 \leq \eta_6 < 1 \tag{4-4}$$

Note that Eq. (4-4) must be satisfied regardless of the kind of engine used. The efficiency of refrigeration or heat pump cycles such as case 6 are more commonly measured by a coefficient of performance, COP, defined as,

$$COP \equiv + \frac{\text{heat transferred from cold system}}{\text{work done by engine}} \tag{4-5}$$

The Second Law itself describes, in mathematical terms, the physical impossibility of reversing Joule's experiments. As you should recall from earlier discussions in Chapters 1 and 3, Joule spent over a decade in the mid-1800's conducting experiments involving a series of adiabatic work interactions. At the time, Joule thought he had established the equivalence of mechanical work and "heat" (i.e., 1 calorie = 4.186 J). Even though his interpretation of heat was incorrect, the fact remained that one could convert an adiabatic work interaction into an equivalent increase in $\underline{U}$ or $\underline{E}$, or $W_{adiabatic} = \Delta\underline{U}$ or $\Delta\underline{E}$. If we tried to reverse Joule's experiments, by transferring an amount of heat Q equivalent to the original adiabatic work interaction across the system boundary, we could not reproduce that work in the surroundings. From these experimental observations, a common expression for the Second Law is developed: "*It is not possible to convert heat into an equivalent amount of work; some amount of heat must be transferred to a second body or the environment in the process of converting heat into work.*" Our task now is to quantify the relationship that limits the fractions of work produced and heat rejected per unit of heat flowing from the primary system.

4.2 Reversible Processes

It can be seen by inspection of Eq. (4-4) that the most efficient engines that could be conceived would correspond to $\eta_5 = \eta_6$. For these engines we could operate process (5) to obtain the maximum work and then use process (6) to restore systems *A* and *B and* the work reservoir to their original states. This combination is an example of a *reversible cycle*, and the engines involved are referred to as reversible engines.

In the general case, ***a process will be called reversible if a second process could be performed in at least one way so that the system and all elements of its environment can be restored to their respective initial states, except for differential changes of second order.***

It can be proved that the maximum work obtainable from, for example, an expansion process corresponds to the hypothetical case in which the boundary is moved at an infinitesimal rate for an infinite time. Thus, the difference in the maximum work obtained in an expansion and the minimum work required for the reverse compression is of the order of $(dP)(d\underline{V})$.

Because there are no dissipative or parasitic effects, each step in the path of a reversible process must be reversible. It can also be shown that in a reversible process, all systems must be in states of equilibrium at all times (i.e., all subsystems must traverse quasi-static paths). The proof follows from Postulate II:

If a system in a non-equilibrium state is isolated, it will tend toward a state of equilibrium. Since there is no way to transform a system from a state of equilibrium to a non-equilibrium state without removing it from isolation, any process involving an intermediate non-equilibrium state is irreversible. Many useful corollaries follow directly from this last conclusion. For example, simple systems involved in reversible processes can have no internal pressure or temperature gradients.

Finally, it can be shown that frictional and similar dissipative effects must not be present if a process is to be reversible. The proof follows from the fact that the work required in such processes exceeds the minimum (or the work obtained is less than the maximum) because a finite imbalance of boundary forces is required to effect the changes involved.

Example 4.1

A rigid closed vessel with adiabatic walls is divided internally by a strong diaphragm. On one side, air is present at 300 K and 1 bar; the other side is evacuated. The diaphragm is broken and air fills the entire vessel. Is this a reversible process?

Solution

It is not easy to employ the definition of a reversible process to answer the question posed above. An alternative technique to use in such cases is to assume that the given process *was* indeed reversible and then to examine whether the consequences of the assumption lead to a violation of any of the postulates.

In this case, if the expansion were reversible, then, since the heat and work interaction between the system and the environment are zero, it would be possible to propose the opposite process. That is, given a vessel filled with air at some P and T, there would be a way to have a change of state of the gas, with $Q = W = 0$, so that at the end all the gas is located at one end, while there would be a vacuum in the rest of the vessel. Then, visualize a cyclic process wherein we begin with the system as initially described. But we have a piston rather than a diaphragm. We allow the piston to move into the evacuated portion and obtain work to be stored in an external work reservoir. Since the gas will cool, we will allow a heat interaction to occur so that at the end of the expansion we have the same state of the air as in the diaphragm-breaking case, but we have had a heat interaction with the environment and obtained work. Then we invoke the assumed reversibility to go back to the initial composite case with no further Q or W interactions. It is clear that this cyclic process is identical to case 4 of Figure 4.1, which we showed violated Postulate II. Thus, the original assertion that the expansion was reversible is false. The process as described is irreversible.

In general, all real or naturally occurring processes are irreversible (at least in part). Although reversible processes represent idealizations that cannot be achieved in practice, they are quite useful in illustrating limiting behavior. The performance of a specific real process can be compared with ideal performance under reversible conditions. For example, the efficiency of a power generation system that converts the chemical energy contained in a fossil fuel to electrical work can be compared to the Second Law (Carnot) limiting efficiency of a reversible process that yields maximum work. All real or natural processes contain one or more non-reversible elements, such as frictional dissipation at boundaries or finite temperature differences between flowing streams in heat exchangers without work generation (see also Chapter 14).

All reversible processes are quasi-static, but the reverse is not necessarily true. For example, consider an adiabatic tank blowdown process where an insulated valve controls the blowdown rate. The pressure and temperature conditions inside the tank during the blowdown can closely approximate an internally reversible expansion. This quasi-static process can be modeled as a closed gas system slowly expanding against a massless piston that is frictionally damped to avoid any spatial gradients in intensive properties within the tank (P, T, ρ, or V). The pressure of the gas at the system boundary would only be differentially greater than the external pressure. Outside the tank, of course, the gas pressure is rapidly reduced to ambient conditions as it passes through the valve. This is a key irreversible step in the overall blowdown process. As we saw with several problems in Chapter 3, the governing equations for such expansion processes are identical to those for a reversible, adiabatic expansion. The *PVT* relationships for an ideal gas expansion in this closed, adiabatic system are:

$$PV^{\kappa} = \text{constant}$$

$$\left(\frac{T}{T_i}\right) = \left(\frac{P}{P_i}\right)^{\frac{\kappa-1}{\kappa}} \tag{4-6}$$

where $\kappa \equiv C_p/C_v$. One can easily derive these same equations by starting with the adiabatic, open system First Law expression and applying the assumptions of a well-mixed gas and no internal gradients in the tank. The proof is left as an exercise for the interested reader.

4.3 Thermodynamic Temperature

One of the major results of the previous section can be summarized as follows: The efficiency of all cycles involving reversible heat engines that operate between two given systems with different thermometric temperatures is a constant. It is a simple matter to extend this reasoning to show that the efficiency of any reversible engine is dependent on the thermometric temperatures of both systems with which it interacts. For example, we can prove that the efficiency must depend on the temperature of the cold system in the following manner. Let us assume that η is only a function of the temperature of the hot system and then show that this assumption cannot be valid. If we were to operate a reversible engine as in case 5 between two systems at θ_A and θ_B and also operate a reversible refrigeration cycle as in case 6 between two systems at θ_A and θ_C, the efficiencies of the two processes will be the same (if the initial assumption is correct). If we had chosen systems such that $\theta_A > \theta_B > \theta_C$, the net effect of the combined process would have been a transfer of heat from system *C* to system *B*. But since θ_B was chosen to be greater than θ_C, the net effect, which is equivalent to case 2, is in violation of our postulates. Therefore, our assumption was incorrect; the efficiency of the reversible engines must involve the temperature of the cold system. In this manner, it can be shown that the efficiency is a function of the temperatures of both systems. Thus, referring to the engine as the system in case 5:

$$\eta = \frac{-W_E}{Q_A} = f_1(\theta_A, \theta_B) \tag{4-7}$$

Since

$$W_E = -(Q_A + Q_B) \tag{4-8}$$

where W_E is negative as work is produced, Q_A is positive as it extracts heat *from* system *A*, and Q_B is negative for heat transferred *to* system *B*, Eq. (4-7) could be expressed in the equivalent form

$$\frac{Q_B}{Q_A} = f_2(\theta_A, \theta_B) = \eta - 1 \tag{4-9}$$

As a result of the foregoing deductions, the measurable primitive property, thermometric temperature, has assumed a role of prime significance. Recall the definition of a "temperature"-measuring device. It is a system, closed and rigid, which shows a variation in at least one primitive property when allowed to undergo a finite heat interaction with another system whose temperature is being measured. Of course, we could have used any number of clever techniques to define properties that have the same information content as the thermometric temperature (e.g., electrical resistance, thermocouples, thermal electron emission detectors, infrared emission analyzers, etc.), and we would have arrived at the same result expressed by Eqs. (4-7) and (4-9). In fact, until we specify the form of the function f_1 *or* f_2, any primitive temperature measurement would be equally acceptable.

Let us now look at the problem of determining the form of the function f_1 *or* f_2. We imagine our systems *A* and *B* (*with* θ_A *greater than* θ_B) operating as in case 5 to produce work with a reversible heat engine, $\mathbf{E}_1$ (see Figure 4.2(a)). Imagine that these systems are so large that they do not change in state by a significant amount during this process.

Figure 4.2(a) and (b)

Now connect to system *A* *another* heat engine, $\mathbf{E}_2$, which in this case rejects heat to a new system *C*. System *C* has a temperature intermediate between *A* and *B*. Also, any heat transferred to *C* is immediately transferred to a third reversible engine, $\mathbf{E}_3$, which rejects heat to system *B* (see Figure 4.2(b)).

If each of the processes shown in Figures 4.2(a) and (b) reduces the energy of system A by an equal amount, and if there is no accumulation of energy in system C, it is obvious that the work obtained from engine $\mathbf{E}_1$ is equal to the sum of the work from engines $\mathbf{E}_2$ and $\mathbf{E}_3$. If this were not so, the more efficient of the two procedures could be reversed, resulting in a violation of Postulate II.

Using the nomenclature shown in Figures 4.2(a) and (b), we obtain

$$Q_{A_1} = Q_{A_2} \tag{4-10a}$$

$$Q_{B_1} = Q_{B_3} \tag{4-10b}$$

$$Q_{C_2} = -Q_{C_3} \tag{4-10c}$$

$$W_{E1} = W_{E2} + W_{E3} \tag{4-11}$$

Using Eq. (4-9) for each engine yields

$$\frac{Q_{B_1}}{Q_{A_1}} = f_2(\theta_A, \theta_B) \tag{4-12}$$

$$\frac{Q_{C_2}}{Q_{A_2}} = f_2(\theta_A, \theta_C) \tag{4-13}$$

$$\frac{Q_{B_3}}{Q_{C_3}} = f_2(\theta_C, \theta_B) \tag{4-14}$$

Multiplying Eq. (4-13) by Eq. (4-14) and equating the result to Eq. (4-12), we get

$$[f_2(\theta_A, \theta_C)][f_2(\theta_C, \theta_B)] = -f_2(\theta_A, \theta_B) \tag{4-15}$$

This result places a definite restriction on the form of the function f_2; that is,

$$f_2(\theta_j, \theta_i) = -\frac{f_3(\theta_j)}{f_3(\theta_i)} \quad \text{or} \quad -\frac{f_4(\theta_i)}{f_4(\theta_j)} \tag{4-16}$$

If we adopt the former form, Eq. (4-12) reduces to

$$\frac{Q_{B_1}}{Q_{A_1}} = -\frac{f_3(\theta_A)}{f_3(\theta_B)} = \eta - 1 \tag{4-17}$$

and

$$\eta = \frac{f_3(\theta_B) - f_3(\theta_A)}{f_3(\theta_B)} \tag{4-18}$$

If we adopt the latter form,

$$\frac{Q_{B_1}}{Q_{A_1}} = -\frac{f_4(\theta_B)}{f_4(\theta_A)} = \eta - 1 \tag{4-19}$$

and

$$\eta = \frac{f_4(\theta_A) - f_4(\theta_B)}{f_4(\theta_A)} \tag{4-20}$$

This is a very profound result. The efficiency of a reversible, but otherwise arbitrary, heat engine has been related to some function of the temperatures of the systems A and B. According to our preceding arguments, since the efficiency of a reversible engine cannot be multivalued, it is apparent that f_3 and f_4 must be single-valued functions of θ.

Let us recapitulate the progress we have made thus far. We have concluded that any heat engine or refrigerator operating reversibly must have a single-valued efficiency that is a function only of the primitive property, temperature (thermometric or any other empirical temperature scale which has the same information content). The relation between efficiency and temperature must be of the form given by Eq. (4-18) or (4-20). It is clear that for given systems, A and B, the experimental technique used to define temperature is arbitrary; the efficiency, however, is not arbitrary. Systems A and B have been chosen, the efficiency is set by Nature and as such represents a fundamental principle of classical thermodynamics. If we could build a reversible engine, we could measure experimentally the value of the efficiency. Since the efficiency is not arbitrary and the temperature scale is arbitrary, the form of the functions in Eqs. (4-18) and (4-20) cannot be arbitrary. Once a temperature scale has been decided upon, there will be one and only one valid form of each of the functions f_3 and f_4.

We have said that we could measure empirically the efficiency if we could construct a reversible engine and, in this manner, we could empirically determine the functions f_3 and f_4. Nevertheless, we do not have access to any real reversible engines now, nor do we have any reason to believe that one could ever be constructed. Nonetheless, we can conduct a thought experiment for a hypothetical reversible engine in much the same manner as Carnot did over a century ago. It is only through such analysis that we are able to define unequivocally the efficiency of reversible engines and a consistent temperature scale.

Thus, let us construct our reversible heat engine cycle using an ideal gas as the working fluid. Such a cycle, involving heat input from system A at T_A and rejection of heat to system B at T_B is called a *Carnot cycle.* There are four steps to this cycle:

1. Heat flows to an ideal gas contained in a piston-cylinder device at temperature T_A.
2. The ideal gas system is then isolated from system A and allowed to expand adiabatically and reversibly to a lower pressure so that the temperature is T_B.
3. At this lower pressure, the ideal gas system is connected to system B and the gas compressed, heat being rejected to system B at a constant temperature, T_B.
4. At a particular point, the ideal gas system is isolated from system B and compressed adiabatically and reversibly to the original pressure. The point of initiation of this step is determined so that the final temperature after compression is T_A.

The heat engine (i.e., the ideal gas system) has undergone a cycle: Work has been produced in the work reservoir, the energy of system A has decreased, and the energy of system B has increased. By calculating the actual work and heat flows, and assuming that the ideal gas has the properties such that $P\underline{V} = NRT$ and the energy of the gas is not a function of pressure but only of temperature, it is possible to show that

$$\boxed{\eta = \eta_{Carnot} = \frac{T_A - T_B}{T_A}} \tag{4-21}$$

The proof is left as an exercise.

From Eq. (4-21) it is clear that the functions f_3 and f_4 are of a very simple form if the temperature is measured by the ideal gas scale. We note that Eq. (4-21) was developed without specifying *a priori* any functional form for f_3 *or* f_4. We shall have no further need for these functions since we now have a thermodynamic temperature scale that can be related to real measurements and a consistent expression for the efficiency. The functions would, of course, have been much more complicated if we had used a conventional sealed mercury thermometer. Fortunately, the ideal gas temperature scale has been adopted as *the* thermodynamic temperature scale.

We can measure approximately an ideal gas temperature since empirically we have found that simple gases at low pressures obey the ideal gas law with $U = f(T)$ only, to a high degree of approximation. From this point on we will use the T-scale instead of the θ–scale and assume that we can accurately measure T.

4.4 The Theorem of Clausius

We are now in a position to extend our analysis of heat engines to the development of an additional derived property, the entropy. There are, perhaps, as many ways to proceed logically to infer this property as there are textbooks in thermodynamics. In whichever way we choose, however, we must limit ourselves to using the stated postulates or the conclusions obtained from them. We first derive a quantitative definition of entropy and then proceed to show that it has a very significant bearing on the concept of equilibrium.

Let us first rephrase the results of the last section. Combining Eqs. (4-9) and (4-21) for a reversible heat engine cycle, we have

$$\frac{Q_B}{Q_A} = -\frac{T_B}{T_A} \tag{4-22}$$

or

$$\frac{Q_A}{T_A} + \frac{Q_B}{T_B} = 0 \tag{4-23}$$

The temperatures of A and B are assumed not to change in the heat engine cycle; if they were to change, one could still write the equations but in a differential form:

$$\frac{\delta Q_A}{T_A} + \frac{\delta Q_B}{T_B} = 0 \tag{4-24}$$

Let us now shift our attention to the interior of the closed, reversible engine. We imagine that it is charged with some material which we call the *working substance or fluid.* The only restriction we have placed thus far on the internals of the engine is that it operate reversibly. The engine may be a composite or simple system.

We assume, for simplicity, that there are no external body force fields and that only P-$\underline{V}$ work need be considered; the results, however, are valid for systems in external body force fields. Since we are considering *reversible* heat engine cycles, and since all steps within a reversible process must be reversible, it follows that we cannot have any pressure gradients across moving boundaries or any temperature gradients across diathermal boundaries. Therefore, in any $Pd\underline{V}$ work terms, P is both the external pressure and the internal pressure at the region of the moving boundary. Similarly, T_A is the temperature of external system A and also the temperature of that portion of the working substance at the diathermal boundary with system A. A similar conclusion holds for T_B. We may have many different events occurring in our system (e.g., adiabatic compressions and expansions, isothermal compressions and expansions, *etc.*). In Figure 4.3a we show a path representing an arbitrary reversible change in state of our system from i to f (i.e., a change that might occur during some part of the heat engine cycle). In Figure 4.3b the same path is shown and we have drawn through points i and f curves that represent those paths which would occur if adiabatic expansions or compressions were to take place starting from either i or f. Next, in Figure 4.3c we repeat (a) and (b) but add another path curve to represent the behavior of our system if it were at point g and were expanded isothermally (and reversibly) to h. This path, g-h, of course, represents a particular temperature level. The exact position of point g is established as shown below.

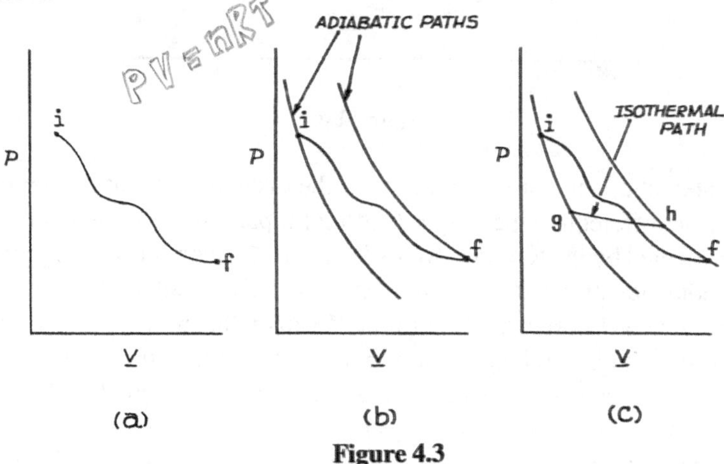

Figure 4.3

The work proceeding along the real path *i-f* is equal to the integral under the path curve. We choose point *g* so that the area under the path curve *ighf* is equal to the actual work. For these two alternative paths, $\Delta \underline{E}_{if} = \Delta \underline{E}_{ighf}$ and $W_{if} = W_{ighf}$. Therefore, $Q_{if} = Q_{ighf}$. Since it was specified that $Q_{ig} = Q_{hf} = 0$, then $Q_{if} = Q_{gh}$. This simple scheme is known as the ***Theorem of Clausius***:

Given any reversible process in which the temperature changes in any prescribed manner, it is always possible to find a reversible zigzag process consisting of adiabatic-isothermal-adiabatic steps such that the heat interaction in the isothermal step is equal to the heat interaction in the original process.

4.5 Entropy

The Theorem of Clausius is useful in analyzing the entire cycle carried out by a reversible heat engine, or for that matter, any system undergoing a reversible, cyclic process. This is shown in Figure 4.4 by the curve *jifkdc*. The unusual curve is drawn to emphasize that an actual *P*-$\underline{V}$ path is not necessarily simple.

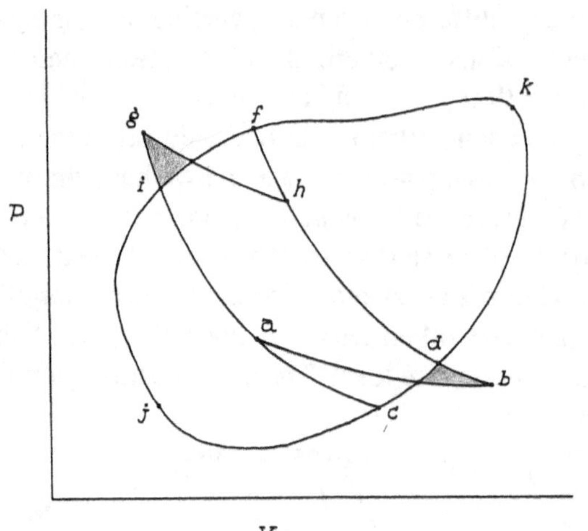

Figure 4.4

The heat interactions in various portions of the cycle may result from contacts with heat reservoirs at different temperatures. Choose for particular examination a portion of the cycle represented by the terminal points *i* and *f*. Draw path lines through these points representing adiabatic processes starting from *i* and *f*. These adiabatic path lines cut the cycle at points *c* and *d*, respectively. From Clausius's theorem we know that the path curves *i-f* and *c-d* can be broken down into a series of adiabatic and isothermal paths with the same work and heat interactions as occur in the original path. Let the heat interaction in path *i-f* be from a reservoir at T_A and in path *c-d* from a reservoir at T_B. We note that this net inner cycle *ighfdbac* is, in reality, a Carnot cycle. Thus,

$$\frac{Q_{if}}{T_{if}} + \frac{Q_{cd}}{T_{cd}} = 0 \tag{4-25}$$

If the paths *i-f* and *c-d* are made infinitesimal,

$$\frac{\delta Q_{if}}{T_{if}} + \frac{\delta Q_{cd}}{T_{cd}} = 0 \tag{4-26}$$

where T_{if} and T_{cd} become equal to the temperature of that part of the system in contact with the diathermal boundary at the respective points in the *actual* path. (The heat interactions in the differential cycle must still be equal to the heat interactions in the actual path.)

An infinite number of infinitesimal cycles, the first beginning at *j* and the last ending at *k* (see Figure 4.4), cover the entire cycle; summing, we obtain

$$\oint \left(\frac{\delta Q}{T}\right)_{rev} = 0 \tag{4-27}$$

where T and δQ are the actual values associated with the original path, and the subscript "*rev*" reminds us that this result is only valid if the original path is reversible.

Equation (4-27) could be written in the equivalent form,

$$\int_j^k \left(\frac{\delta Q}{T}\right)_{rev} = -\int_k^j \left(\frac{\delta Q}{T}\right)_{rev} \tag{4-28}$$

This result does not appear very profound. If, however, we carry our reasoning one step further, we can formalize our results in a manner that will be most useful. If we proceed from *j* to *k* via the top path shown in Figure 4.4 and then return from *k* to *j* via any other conceivable reversible path, it is readily shown that for this new cycle Eqs. (4-27) and (4-28) are still valid. Thus, we can conclude that there is a quantity, $\int (\delta Q/T)_{rev}$ which is conserved in *any* reversible cyclic process. This quantity has all the characteristics associated with a derived property of the system; we call this property the *entropy*, and we represent it by $\underline{S}$. Thus,

$$\Delta \underline{S} \equiv \int \left(\frac{\delta Q}{T}\right)_{rev} \tag{4-29}$$

or

$$d\underline{S} \equiv \left(\frac{\delta Q}{T}\right)_{rev} \tag{4-30}$$

Equations (4-29) and (4-30) were derived for systems in which only force-displacement work occurs. The derivation can be extended to include systems that are acted on by body forces as well. The result is not changed by including other forces. Clausius's theorem is still valid because we can always find a reversible adiabatic-isothermal-adiabatic path having the same work interaction as the actual path. Furthermore, Eqs. (4-29) and (4-30) follow directly from Clausius's theorem, and as defining equations for entropy, are valid for any system (simple or composite) undergoing a reversible process. In the case of a composite system consisting of two or more subsystems at different temperatures (which are separated from one another by internal adiabatic boundaries), the change in entropy of the composite is

$$\sum_i \frac{\delta Q_i}{T_i}$$

where δQ_i is the heat transferred to subsystem i at T_i. Note that in the derivation given above, only differences in entropy take on any significance, and these differences apply only between states at equilibrium.

If we follow the same line of reasoning used when energy was introduced as a property (see Section 3.3), it can be shown that total entropy must also be first order in mass. Unlike the energy, there is nothing in our previous developments or in our postulates to require that the entropy of the universe should always be conserved. On the contrary, the entropy of the universe must increase in any irreversible (or real) process and the entropy is conserved only in reversible processes. In fact, the entropy change is of great use as a means for determining how closely real processes approach reversible processes. We can summarize these concepts by the following:

$$\Delta \underline{S}_{universe} \equiv \Delta \underline{S}_{system} + \Delta \underline{S}_{surroundings}$$

$$\Delta \underline{S}_{universe} > 0 \quad \text{for all real or natural processes} \tag{4-31}$$

$$\Delta \underline{S}_{universe} = 0 \quad \text{for a completely reversible process}$$

In order to prove that $\Delta \underline{S}_{universe} \geq 0$ we need an operational method for determining the entropy difference between two stable equilibrium states of a closed, simple system. There are no entropy meters available, and there are no meters to measure directly the quantity of heat transferred. Thus, in order to determine an entropy change between two stable equilibrium states, we must visualize any convenient *reversible* path and calculate the reversible work that would be done. Then, using Eqs. (3-28) and (4-30), we have

$$T\,d\underline{S} = d\underline{E} - \delta W_{rev} \tag{4-32a}$$

which, for a closed simple system, reduces to

$$T\,d\underline{S} = d\underline{U} + P\,d\underline{V} \tag{4-32b}$$

or, over the entire reversible path,

$$\Delta \underline{S} = \int \frac{1}{T}\,d\underline{U} + \int \frac{P}{T}\,d\underline{V} \tag{4-33}$$

The units of either entropy or temperature are completely arbitrary; all that we require is that the product of the entropy and temperature units be equivalent to the units of energy. Hence, we are free to choose one or the other at will, but, once we choose one, the units of the other are fixed. The accepted convention is to specify temperature in

Kelvins (K) and total entropy $\underline{S}$ in J/K. The specific entropy S is intensive with units of J/mol K or J/kg K.

Let's now consider two very important illustrations of how entropy changes are calculated and interpreted.

Example 4.2

A bar of aluminum is placed in a large bath of ice and water (Figure 4.5). Current is passed through the bar until, at steady state, there is a power dissipation of 1000 W. A thermocouple on the surface of the aluminum reads 640 K. Film boiling is occurring at the interface with a subsequent, noisy collapse of the bubbles. What is the entropy change of the bar, water, and universe during 2 min of operation for this highly irreversible operation? Assume that there is ice remaining at the end of the 2 min.

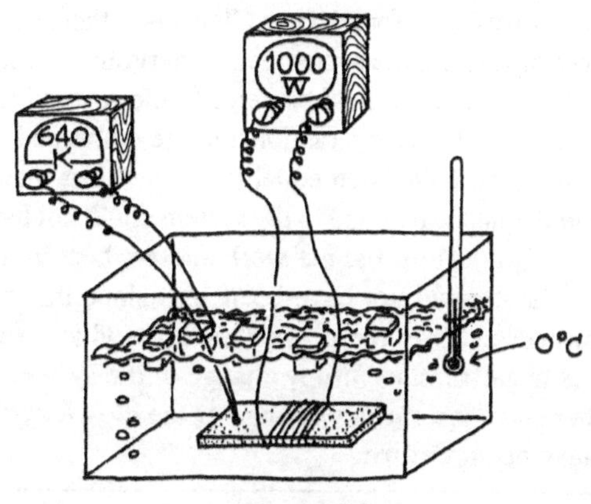

Figure 4.5

Solution

The aluminum bar undergoes no change in properties during the 2 min of operation at steady state conditions. Since entropy is a property, $\Delta \underline{S}$ (aluminum) $= 0$.

For the ice-water bath, the net result is that ice is melted but the temperature of the water does not change. (We neglect the small volume change due to melting.) To calculate the entropy change of this ice-water system, we need to devise a *reversible* process between the same initial and end states. For example, we could visualize that we contacted the system with some external heat reservoir only differentially above 0°C and allow a heat interaction to occur to melt the requisite amount of ice. This is now a reversible process so, with Eq. (4-33), assuming that $d\underline{V} \approx 0$, and $T = 273.2$ K,

$$\Delta \underline{S} \text{ (water–ice)} = \frac{\Delta \underline{U}}{T} = \frac{Q}{T}$$

But Q = power × time $= 10^3 \times 2 \times 60 = 1.2 \times 10^5$ J, and $\Delta \underline{S}$ (water–ice) $= (1.2 \times 10^5)/273.2$ $= 439$ J/K. The entropy change of the universe is then 439 J/K.

In effect, we simply calculated $\Delta\underline{S}$ (universe) as the power dissipation rate divided by the (constant) bath temperature. The reason why this is allowed is discussed further in Section 4.6.

Example 4.3

Prove that for an *adiabatic* process between defined states I and II of a closed system, the entropy change of the system must be equal to or greater than zero.

Solution

If the process was reversible, Eq. (4-29) would show that $\Delta\underline{S}_{system}$ must be zero. If the process was irreversible, we could follow this process by a second process to bring the system back to state I. Let us make this second process reversible and specify that all heat transfer between system and surroundings occurs during an isothermal step. Such a reversible process can always be found (see Clausius's theorem, Section 4.4). In the combination of the two processes, the system has undergone a complete cycle so that the overall energy and entropy changes must be zero. In the reversible portion of the cycle, there may have been work and heat interactions; in the combined processes, clearly, the net work (taken as $-W$) must have been equal to the net heat interaction. The net work done by the system and the net heat added to the system could not have been positive since this would result in a PMM2. Thus, the net work and net heat must have been negative. (The net work and net heat could not have been zero since this would correspond to a reversible adiabatic process.) Since the heat interaction (which occurs during the return reversible step only) is negative, the entropy change of the system must also be negative in the return step. Therefore, the entropy change of the closed system in the irreversible adiabatic step must have been positive.

It should be clear from Example 4.3 that the entropy change of any closed system in any adiabatic process (reversible or irreversible) cannot be negative. Since an isolated system is a special case of a closed, adiabatic system, it follows that ***the entropy change of an isolated system in any process must be equal to or greater than zero***. Furthermore, since the universe can be considered an isolated system, we are led to the same conclusion first stated by Clausius over a century ago: *"Die Entropie der Welt strebt einem Maximum zu."*

4.6 Internal Reversibility

In Section 4.2 it was shown that in a fully reversible process, all interacting systems must be in states of equilibrium at all times. Hence, a process involving a temperature gradient across a diathermal boundary separating two systems is an irreversible process because the composite system is not in thermal equilibrium. Thus, case 1 illustrated in Figure 4.1 is an irreversible process. Let us focus our attention, however, on system A in Figure 4.1. By altering the environment external to system A from that of case 1 to that of case 5, we could make the process reversible if we were to use a reversible heat engine. Since we could have effected the same changes *in system A* through either of these processes, the irreversibility in case (1) must be at or outside the system boundary. In either case, the irreversibility is considered external to system *A*. If, however, during the heat interaction a temperature gradient had been established within system *A*, the process would have been internally irreversible and at least part of the irreversibility would have occurred within system *A*.

In many engineering applications it may be necessary to minimize irreversibilities in a process in order, for example, to maximize power output or to reduce power requirements. Thus, it would be helpful to know where the irreversibilities occur in a process. In order to identify subsystems that proceed along reversible paths within a process that is irreversible *in toto, we will define a process of a system as being* ***internally reversible*** *if the process can be performed in at least one way with another environment selected such that the system and all elements of this environment can be restored to their respective initial conditions, except for differential changes of second order in external work reservoirs.*

As discussed in Section 4.2, it then follows that a system undergoing an internally reversible process must traverse a quasi-static path. Nevertheless, an entire process is not necessarily reversible even if all subsystems traverse quasi-static paths because there may be irreversibilities occurring at the *boundaries* between the subsystems.

It should be obvious that for internally reversible systems there is no need to devise artificial processes to calculate entropy changes in the system in question; the actual interactions will suffice for this purpose.

Example 4.4

Consider a heat engine operating between two systems, 1 and 2, at T_1 and T_2, respectively, with $T_1 > T_2$. If the two systems are internally reversible, and if the overall process is irreversible, show that the entropy change of the universe must be greater than zero.

Solution

If the overall process were reversible, then for a complete cycle of the engine,

$$\Delta \underline{S}_{universe} = \Delta \underline{S}_1 + \Delta \underline{S}_2 = \frac{Q_1}{T_1} + \frac{Q_2}{T_2} = 0 \qquad (4\text{-}34)$$

If we denote the work produced by a reversible and an irreversible engine as W_R and W_I, it can be shown that W_I must be less than W_R in order to avoid the possibility of a PMM2. If we run the reversible and irreversible engines in a manner such that the heat transferred from system 1 was the same in both cases, it is clear that Q_2 must be greater for the irreversible case than for the reversible case. Therefore,

$$\Delta \underline{S}_{2I} > \Delta \underline{S}_{2R}$$

while

$$\Delta \underline{S}_{1I} = \Delta \underline{S}_{1R}$$

so that for the irreversible case,

$$\Delta \underline{S}_{universe} = \Delta \underline{S}_{1I} + \Delta \underline{S}_{2I} > 0 \tag{4-35}$$

In this fashion, we have illustrated how entropy as a property can serve as a measure of the degradation of work-producing potential. (See Chapter 14 for further discussion.)

4.7 The Combined First and Second Laws

Closed, single-phase, simple systems. In this case, if the system is undergoing a reversible process with only P-$\underline{V}$ work, it was shown in Section 4.5 [Eq. (4-32b)] that

$$d\underline{S} = \frac{1}{T} d\underline{U} + \frac{P}{T} d\underline{V} \tag{4-36}$$

It should be apparent from the discussion in Section 4.6 that Eq. (4-36) is equally valid for all internally reversible or quasi-static processes of closed, simple systems. Rearranging Eq. (4-36), we obtain

$$\boxed{d\underline{U} = T\,d\underline{S} - P\,d\underline{V}} \tag{4-37}$$

Either Eq. (4-36) or (4-37) is commonly referred to as the ***combined First and Second Laws of Thermodynamics for a closed, simple system***. They represent the special case of the First Law, Eq. (3.28), applied to an infinitesimal quasi-static path of a closed, simple system.

Open, single-phase, simple systems. Consider first an internally reversible, quasi-static process with mass entering and leaving the system as shown in Figure 4.6. Under these conditions all *intensive* properties must remain the same. Thus,

$T_{in} = T_{out} = T$; $P_{in} = P_{out} = P$; $U_{in} = U_{out} = U$; $S_{in} = S_{out} = S$; and $V_{in} = V_{out} = V$

From a *mole balance*, $dN = \delta n_{in} - \delta n_{out}$. Now the First Law, which describes a differential *energy balance*, can be written with only PdV work as:

$$d\underline{E} = d\underline{U} = \delta Q_{rev} - Pd\underline{V} + (U + PV)dN \tag{4-38}$$

where the $PV\,dN$ term accounts for the net mass flow work. For this reversible process an *entropy balance* can also be formulated as:

$$d\underline{S} = \delta Q_{rev}/T + SdN \tag{4-39}$$

Now by substituting for δQ_{rev} into the First Law expression of Eq. (4-38), we get:

$$d\underline{U} = Td\underline{S} - Pd\underline{V} + (U + PV - TS)dN = Td\underline{S} - Pd\underline{V} + \mu dN \tag{4-40}$$

where μ is the molar Gibbs free energy or chemical potential defined for a single component as:

$$\mu \equiv G \equiv U + PV - TS = H - TS \tag{4-41}$$

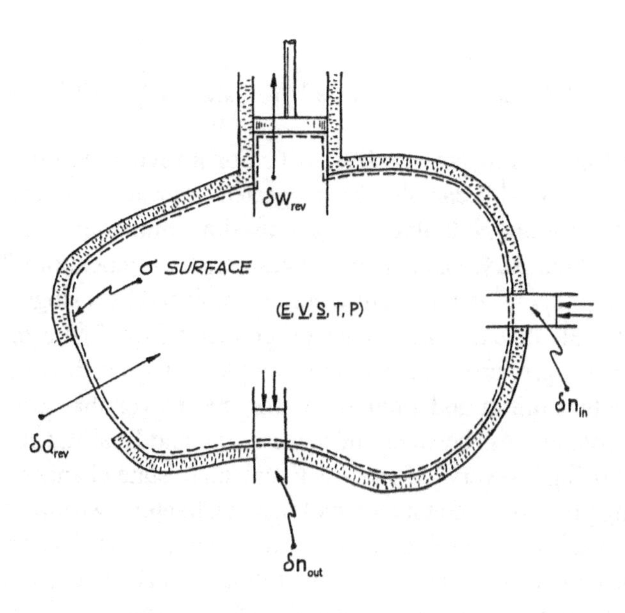

Figure 4.6

The result given in Eq. (4-40) can be generalized for a multicomponent, single phase system that is traversing a quasi-static path as,

$$d\underline{U} = Td\underline{S} - Pd\underline{V} + \sum_{i}^{n} \mu_i \, dN_i \tag{4-42}$$

where $\underline{U}$ is an explicit function of $n + 2$ variables, ($\underline{S}$, $\underline{V}$, N_1, ..., N_n), and μ_i refers to the chemical potential of component i in a multicomponent mixture (we will say more about μ_i soon).

In the derivation above leading to Eq. (4-42), we have introduced the concept of an entropy balance. What happens if irreversibilities are present? Can we still apply an entropy balance? Unlike energy, entropy is not a conserved quantity. The practical way around this is to lump the irreversibilities into a term called entropy generation or $\underline{S}_{gen}$, and to account for entropy contributions crossing the σ-boundary due to

$$\textit{mass flow:} \quad S_j \, \delta n_j$$

and (4-43)

$$\textit{heat flow:} \quad \frac{\delta Q_{\sigma,i}}{T_i}$$

Since $\underline{S}_{gen}$ is clearly a path-dependent function, the inexact differential δ should be used. Thus for a multicomponent, single-phase system, the general entropy balance is given by:

$$d\underline{S} = \sum_i \frac{\delta Q_{\sigma,i}}{T_i} + \sum_{in} S_{in}\,\delta n_{in} - \sum_{out} S_{out}\,\delta n_{out} + \delta \underline{S}_{gen} \tag{4-44}$$

For any real or natural process $\delta\underline{S}_{gen} > 0$. For a reversible process, $\delta\underline{S}_{gen} = 0$, and $\delta Q_{\sigma,i} = \delta Q_{rev,i}$. Eq. (4-44) can also be integrated to produce $\Delta\underline{S}$ for the actual process occurring in the system. Note that we have used a summation over i heat interactions that cross the σ-boundary, each with a characteristic temperature T_i. In an internally reversible system, $T_i = T$ and the sum would be replaced by a single term $\delta Q_{rev}/T$. It is very difficult to calculate the entropy generation term from first principles, but because $\underline{S}$ itself is a derived property, one can, in principle, always estimate the entropy change between the actual initial and final states for the irreversible process by devising a suitable reversible process between the same initial and final states.

The idea of using a reversible process to calculate state changes also has other uses when analyzing processes. You may recall that in Chapter 1 we introduced three general classes of thermodynamics problems. In the third class, one is concerned with comparing the efficiencies of various processes operating to effect a given change in state. Intuitively, because there are no dissipative effects, reversible processes are the most efficient--that is, ones that would yield maximum work or power or require minimum work or power to perform certain specified tasks. Later, in Chapter 14, we utilize the approach used above to derive the combined expression for the First and Second Laws to produce a maximum (or minimum) work state function which is commonly referred to as the *availability* or *exergy*, defined as $\boldsymbol{B}$. This will provide a very powerful method for evaluating the third general class of problems--particularly as they apply to practical heat engines and power cycles.

The general combined law given by Eq. (4-42) can also be derived by a completely different route. From Postulate I, for a single-phase, non-reacting system in a stable equilibrium state, we know that any property can be expressed as a function of any two independently variable properties in addition to the masses or mole numbers. Thus, we could express the internal energy as continuous function of n+2 variables

$$\underline{U} = f(\underline{S}, \underline{V}, N_1, \ldots, N_n) \tag{4-45}$$

Thus, over an infinitesimal quasi-static path for this non-reacting system, Eq. (4-45) can be written in differential form:

$$d\underline{U} = \left(\frac{\partial \underline{U}}{\partial \underline{S}}\right)_{\underline{V},N} d\underline{S} + \left(\frac{\partial \underline{U}}{\partial \underline{V}}\right)_{\underline{S},N} d\underline{V} + \sum_{i=1}^{n} \left(\frac{\partial \underline{U}}{\partial N_i}\right)_{\underline{S},\underline{V},N_j[i]} dN_i \tag{4-46}$$

By comparison with Eq. (4-37), we see that the summation term must be zero for a single-phase, *closed* system undergoing a quasi-static process since $\underline{S}$ and $\underline{V}$ are independently variable properties and the total mass is fixed. Hence, for a closed system of constant mass, Eq. (4-46) reduces to

$$d\underline{U} = \left(\frac{\partial \underline{U}}{\partial \underline{S}}\right)_{\underline{V},N} d\underline{S} + \left(\frac{\partial \underline{U}}{\partial \underline{V}}\right)_{\underline{S},N} d\underline{V} \tag{4-47}$$

It then follows that since Eqs. (4-37) and (4-47) are valid for the infinitesimal process under consideration, the coefficients of the differentials must be equal. Therefore,

$$T = \left(\frac{\partial \underline{U}}{\partial \underline{S}}\right)_{\underline{V},N} \tag{4-48}$$

and

$$-P = \left(\frac{\partial \underline{U}}{\partial \underline{V}}\right)_{\underline{S},N} \tag{4-49}$$

A similar treatment beginning with Eq. (4-36) and expressing

$$\underline{S} = f(\underline{U}, \underline{V}, N_1, \ldots, N_n) \tag{4-50}$$

would lead to

$$d\underline{S} = \left(\frac{\partial \underline{S}}{\partial \underline{U}}\right)_{\underline{V},N} d\underline{U} + \left(\frac{\partial \underline{S}}{\partial \underline{V}}\right)_{\underline{U},N} d\underline{V} \tag{4-51}$$

with

$$T^{-1} = \left(\frac{\partial \underline{S}}{\partial \underline{U}}\right)_{\underline{V},N} \tag{4-52}$$

$$\frac{P}{T} = \left(\frac{\partial \underline{S}}{\partial \underline{V}}\right)_{\underline{U},N} \tag{4-53}$$

To extend our treatment to simple, yet open, systems, we could return to Eqs. (4-45) and (4-46). Here we may have material flow across the boundary of the system (in addition to chemical reactions internal to the system). Equation (4-46) is then the general operating relationship that accounts for heat and work interactions and mass flow effects. The definitions of temperature and pressure as partial derivatives are still valid [see Eqs. (4-48) and (4-49)] and, in addition, we define a new property, the *component chemical potential*, as

$$\mu_i \equiv \left(\frac{\partial \underline{U}}{\partial N_i}\right)_{\underline{S},\underline{V},N_j[i]} \tag{4-54}$$

where the subscripting indicates the total entropy, total volume, and all mole numbers (except i) are held constant. Then, Eq. (4-46) becomes

$$\boxed{d\underline{U} = T\,d\underline{S} - P\,d\underline{V} + \sum_i \mu_i\,dN_i} \tag{4-55}$$

which is the same as Eq. (4-42). Eq. (4-55) is the commonly employed form of the ***combined First and Second Laws of Thermodynamics for a simple, open, multi-component system***. This relationship applies only to systems undergoing quasi-static processes, and no irreversibility may occur at the boundaries where mass enters or leaves. In Chapter 5 we equate Eq. (4-55) with a differential form of the ***Fundamental Equation of Thermodynamics*** and show that it has great utility in deriving relationships between system or material properties.

Had Eq. (4-50) been employed as a starting equation, the analog to Eq. (4-55) would be

$$d\underline{S} = \frac{1}{T}d\underline{U} + \frac{P}{T}d\underline{V} - \sum_i \frac{\mu_i}{T}dN_i \tag{4-56}$$

This relation could also have been obtained from Eq. (4-55) by solving for entropy.

Example 4.5

Massachusetts Innovative Tech Industries (MITI) has come up with a novel idea to conserve wasted thermal energy in chemical plants. Their concept involves using a set of "state-of-the-art" Carnoco heat engines. MITI claims it can generate enormous quantities of electric power by converting waste heat that would normally be released in cooling towers, for example. In one particular application for Cambridge, they claim that 1 MW of electric power can be generated from a 100 kg/s flow of hot process water available at a temperature of 130°C, and a pressure of 2 bar. Water is also available from the Charles River basin which has a seasonal average temperature of 17°C. Describe what you think of the proposed process. You can assume that the following properties apply to water from the Charles as well as from the plant: $C_p = 4200$ J/kg K, $\rho = 1000$ kg/m^3.

Solution

There are several approaches that can be used to analyze this problem. Initially, we can conceptualize the proposed MITI process as shown below in Figure 4.7.

First, let's try to check to see if the First Law is violated (to ensure we do not have a PMM1). For steady state operation, taking the heat engines and the process heat source as our composite system, one obtains:

$$d\underline{E} = d\underline{U} = 0 = \sum \delta Q_{ci} + \sum \delta W_i + (H_{in} - H_{out})\delta M$$

but

$$\dot{W} = \sum \delta W_i/\delta t \quad \text{and } \dot{m} = \delta M/\delta t = 100 \text{ kg/s}$$

Thus,

$$\dot{W} = -\dot{Q}_c + (H_{out} - H_{in})\dot{m} = -\dot{Q}_c + \dot{m}_H C_p(T_H^{out} - T_H^{in})$$

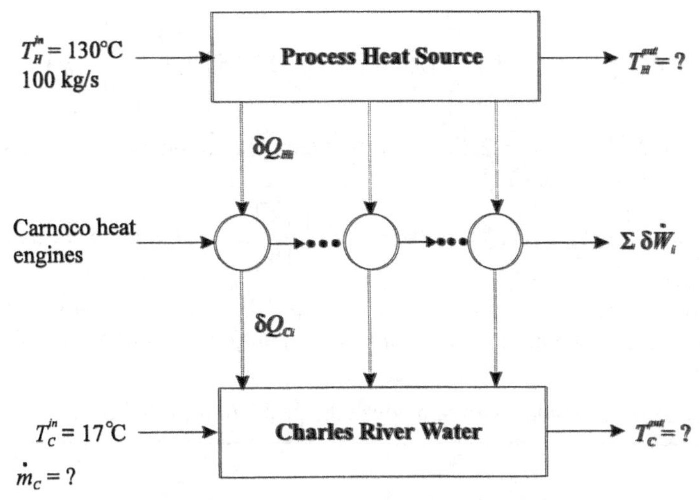

Figure 4.7 Proposed power generation process.

If $\dot{Q}_c = 0$, $\dot{W}$ is maximized, but we know this can't be true, because with Second Law limitations $\dot{Q}_c$ must be non-zero. Because both $\dot{Q}_c$ and H_{out} (or T_H^{out}) are not specified yet, we cannot do a quantitative check on the First Law. To proceed further, we need to bring in Second Law constraints. For any one of the Carnoco engines the maximum efficiency is

$$\eta_i = \frac{-\delta W_i}{\delta Q_{Hi}} = \frac{T_{Hi} - T_{Ci}}{T_{Hi}}$$

If we knew how to relate T_{Ci} in terms of T_{Hi} we could integrate the efficiency equation to determine $\dot{W}$, with $\delta Q_H/\delta t = -\dot{m}\, C_p\, dT_H$ from a First Law balance on the heat source, itself.

$$\dot{W} = \int \delta W_i/\delta t = \int \frac{T_{Hi} - T_{Ci}}{T_{Hi}} \dot{m}_H\, C_p\, dT_H$$

The best way to maximize power output would be to minimize T_{Ci} throughout. As we have assumed that the process rejects heat to the Charles River through some sort of a heat exchanger, T_{Ci} will be minimized at 17°C if $\dot{m}_C$ is infinite (or at least, if $\dot{m}_C >> \dot{m}_H = 100$ kg/s).

Integrating the equation for $\dot{W}$, and substituting values for the parameters:

$$\dot{W} \to \dot{W}_{max} = \int_{T_H^{in}}^{T_H^{out}} \frac{T_H - (273.15 + 17)}{T_H} (100)(4200)\, dT_H$$

or

$$\dot{W}_{max} = (4.2 \times 10^5)\left[T_H^{out} - T_H^{in} - (290.15)\left(\ln\frac{T_H^{out}}{T_H^{in}}\right)\right] \quad \text{in J/s}$$

With $\dot{m}_C = \infty$, $T_H^{out} = 17°C = 290.15$ K with a temperature pinch appearing at the outlet. Since the inlet temperature $T_H^{in} = 130°C = 403.15$ K

$$\dot{W}_{max} = -(4.2 \times 10^5)\left[113 - 290.15 \ln\frac{403.15}{290.15}\right] = -7.4 \times 10^6 \text{ J/s or } -7.4 \text{ MW}$$

so the proposed MITI process with a 1 MW output does not violate any thermodynamic laws subject to our assumption about $\dot{m}_C$. Note that $\dot{Q}_c = -40$ MW from the First Law expression.

As an exercise, you should try to re-analyze the problem for a finite $\dot{m}_C$.

4.8 Reversible Work of Expansion or Compression in Flow Systems

The minimum work required to pressurize a flowing fluid or the maximum work obtainable in expanding a flowing fluid is often required in process design calculations. In such an analysis, reversible engines must be employed. Since all engines have the same efficiency, we could choose any specific, reversible engine to study. We choose here, for convenience, the reciprocating engine, although the results are applicable to all other reversible engines, pumps, turbines, etc.

Consider the cyclic steady state operation of a compressor which processes δn moles of fluid in one cycle. The upstream and downstream conditions are P_1, T_1, V_1 and P_2, T_2, V_2, respectively. The compression cycle can be divided into the following four steps, which are illustrated in Figure 4.8.

Step 1. With the cylinder containing initially N_1 moles of fluid at P_1 and T_1 (condition A), δn moles are introduced at constant pressure and temperature during the upstroke of the compressor.

Step 2. The contents of the cylinder ($N_1 + \delta n$) are compressed from P_1, T_1 (condition B) to P_2, T_2 (condition C).

Step 3. The outlet check valve opens and δn moles are expelled at constant pressure and temperature as the piston completes the downstroke.

Step 4. The contents of the cylinder are expanded from P_2, T_2 (condition D) to P_1, T_1 (condition A).

Let us now analyze the work done by the fluid on the compressor in each step. We choose as an open system the compressor's fluid contents at any time, and assume that the entire compression process is reversible. The work done by the system is δW_σ, where δ represents a finite quantity corresponding to operation on the finite mass, δn. In this case, we are particularly interested in the quasi-static work interaction that occurs on the portion of the σ-boundary that is in contact with the piston.

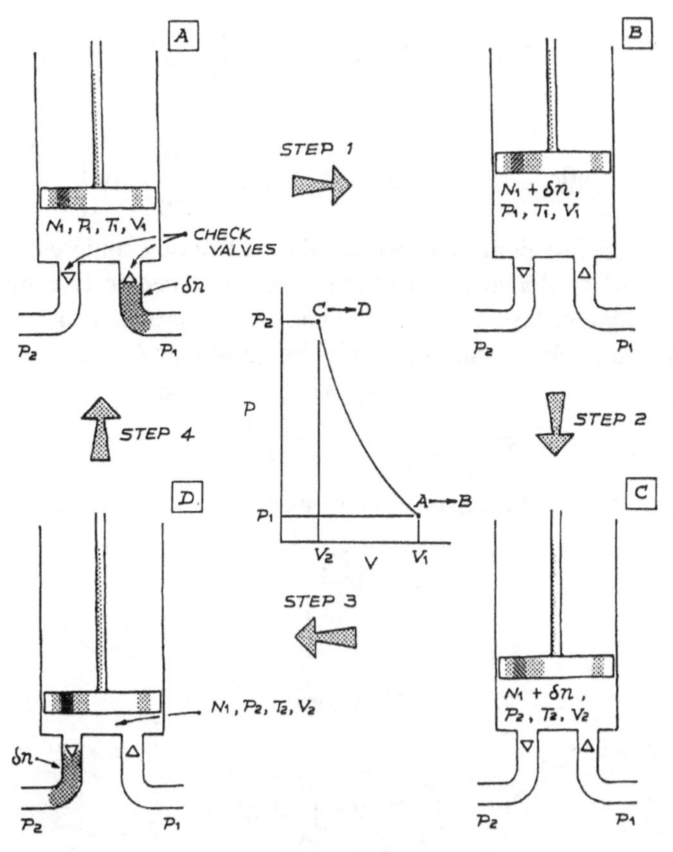

Figure 4.8

For the four steps shown in Figure 4.8 we can compute the work done:

$$\text{Step 1: } \delta W_1 = -\int_A^B P\, d\underline{V} = -P_1(\underline{V}_B - \underline{V}_A) = -P_1\, V_1\, \delta n \qquad \text{(mass flow in)}$$

$$\text{Step 2: } \delta W_2 = -\int_B^C P\, d\underline{V} = -(N_1 + \delta n)\int_B^C P\, dV \qquad \text{(compression)}$$

$$\text{Step 3: } \delta W_3 = -\int_C^D P\, d\underline{V} = -P_2(\underline{V}_D - \underline{V}_C) = +P_2 V_2\, \delta n \qquad \text{(mass flow out)}$$

$$\text{Step 4: } \delta W_4 = -\int_D^A P\, d\underline{V} = -N_1\int_D^A P\, dV \qquad \text{(expansion)}$$

The total work done by the fluid is the sum of these four steps:

$$\delta W_\sigma = \sum \delta W_i = -\delta n\left(P_1 V_1 - P_2 V_2 + \int_B^C P\, dV\right) - N_1\left(\int_B^C P\, dV + \int_D^A P\, dV\right) \qquad (4\text{-}57)$$

The first term in parantheses on the right hand side of Eq. (4-57) can be expressed using integration by parts as

$$P_1V_1 - P_2V_2 + \int_B^C P\,dV = -\int_B^C V\,dP \tag{4-58}$$

Thus,

$$\delta W_\sigma = +\delta n \int_B^C V\,dP - N_1\left(\int_B^C P\,dV + \int_D^A P\,dV\right) \tag{4-59}$$

In many cases, the functional dependence of P on V during compression (B to C) and expansion (D to A) is identical. For example, it was found in Example 3.4 that for quasi-static, frictionless (i.e., reversible) adiabatic compression of an ideal gas, the temperature-pressure relationship is given by Eq. (3-48), or

$$\frac{T_2}{T_1} = \left(\frac{P_2}{P_1}\right)^{R/(C_v+R)} = \left(\frac{P_2}{P_1}\right)^{(1-1/\kappa)} \tag{4-60}$$

Since $T_2/T_1 = P_2V_2/P_1V_1$, Eq. (4-60) becomes

$$\frac{V_2}{V_1} = \left(\frac{P_2}{P_1}\right)^{-(1/\kappa)} \tag{4-61}$$

or

$$PV^\kappa = \text{constant} \tag{4-62}$$

In cases such as this,

$$\int_B^C P\,dV = \int_A^D P\,dV \tag{4-63}$$

because $V_C = V_D = V_2$ and $V_B = V_A = V_1$. Therefore, the work done by the element of fluid on the compressor as it passes through the system as expressed in Eq. (4-59) simplifies to

$$\delta W_\sigma = \delta W_s = \delta n \int_B^C V\,dP = \int_{P_1}^{P_2} V\,dP\,\delta n \tag{4-64}$$

or

$$W_\sigma = W_s = \int_n \int_{P_1}^{P_2} V\,dP\,\delta n \tag{4-65}$$

or

$$\frac{\delta W_\sigma}{\delta n} = \frac{\delta W_s}{\delta n} = \int_{P_1}^{P_2} V\,dP \tag{4-66}$$

δW_s and W_s are sometimes referred to as the shaft work as they are directly related to the total work imparted by a reversible compressor or pump. In general, Eq. (4-66) is valid if the path traversed by the fluid is reversible and if the device itself operates without dissipative losses due to friction and other irreversible effects.

The reversible shaft work done by the compressor on the fluid is, of course, the negative of the work done by the fluid contents on the compressor. The shaft work in Eq. (4-64), however, is not equal to the *total* work done on the fluid element δn in passing from the upstream to the downstream lines. The total work is

$$\delta W_{fluid} = -\delta n \int_{V_1}^{V_2} P dV \tag{4-67}$$

where P is the same function of V as is found in step 2. This can be verified by choosing a closed system of δn moles and calculating the work through each step. The difference between the total work done on the fluid and the shaft work done by the compressor is $P_1V_1 - P_2V_2$, which is the mass flow work done on the δn moles by the fluid behind it in the upstream line and the fluid ahead of it in the downstream line. Thus, in a totally general manner, the net shaft work for a completely reversible compression or expansion from state $[P_1, V_1, T_1]$ to state $[P_2, V_2, T_2]$ is given by:

$$\delta W_s = \delta W_{fluid} - \delta W_{mass\ flow}$$

or

$$\delta W_s = -\left[\int_{V_1}^{V_2} PdV - (P_2V_2 - P_1V_1)\right]\delta n = \left(\int_{P_1}^{P_2} VdP\right)\delta n \tag{4-68}$$

where either the PdV or VdP integral defines the path of the compression or expansion process.

It should be noted that the limits of integration in Eqs. (4-64) through (4-68) will vary from one cycle to the next unless the process under consideration is operated under steady-state conditions (see Example 4.6). With steady state operation, considerable simplification results and

$$\frac{\delta W_s}{\delta t} = \dot{W}_s = \frac{\delta n}{\delta t}\int_{P_1}^{P_2} V\,dP = \dot{n}\int_{P_1}^{P_2} V\,dP \tag{4-69}$$

Rather than deal with the four-step process of Figure 4.8, we could simply consider the engine (i.e., the compressor or turbine) as a device into which a fluid flows at some defined state, and out of which fluid exits at some other state. The device has heat and work interactions with the environment. We will also assume that the device and its contents do not change with time, or, if they do, the variation in the total energy of the device is small relative to the enthalpy fluxes in and out and the work and heat interactions.

Let us focus on some finite element of the device as our system and assuming simple system behavior, write the open-system energy balance [Eq. (3-60)]:

$$d\underline{U} = 0 = \delta Q_\sigma + \delta W_\sigma + H_{in}\,\delta n_{in} - H_{out}\,\delta n_{out}$$

Since we have neglected any accumulation of mass (or energy) within the device,

$$\delta n_{in} = \delta n_{out} = \delta n \tag{4-70}$$

For our finite section, dividing by δn, we get

$$H_{out} - H_{in} = \frac{\delta Q_\sigma}{\delta n} + \frac{\delta W_\sigma}{\delta n} = \frac{\delta Q_\sigma}{\delta n} + \frac{\delta W_s}{\delta n} \tag{4-71}$$

$$= Q' + W_s' \tag{4-72}$$

Note that $W_\sigma = W_s$ (the reversible shaft work and engine work are the same); Q' is the heat interaction in our section per mole of fluid passing through the engine, and W_s' is the engine work interaction on the same basis. Now let us shrink the finite section to one of differential size; then Eq. (4-72) becomes

$$dH = \delta Q' + \delta W_s' \tag{4-73}$$

Equation (4-73) would be applicable whether the engine were or were not reversible. However, if we consider the fluid to traverse a quasi-static path and assume all heat interactions at the boundaries are reversible, then

$$dH = dU + P\,dV + V\,dP = T\,dS + V\,dP \tag{4-74}$$

and

$$\delta Q' = T\,dS \tag{4-75}$$

Substituting Eqs. (4-74) and (4-75) into Eq. (4-73) yields

$$\delta W_s' = V\,dP \tag{4-76}$$

$$W_s' = \int_{P_{in}}^{P_{out}} V\,dP = \frac{\delta W_\sigma}{\delta n} \tag{4-77}$$

or for the total work

$$W_\sigma = W_s = \int_n \int_{P_{in}}^{P_{out}} V\,dP\,\delta n \tag{4-78}$$

which is the same result we obtained before [Eq. (4-65)]. As before, the shaft work term W_s in Eq. (4-78) refers to the interaction of the chosen system (i.e., the engine) with its surroundings. If positive, the engine requires work as it compresses the fluid; if negative, the converse is true, that is, the engine produces work from the fluid expansion.

Example 4.6

A 1-m^3 tank that initially contains air at 1 bar and 300 K is to be evacuated by pumping out the contents, as illustrated in Figure 4.9. The tank contents are maintained at 300 K throughout the operation by heat transfer through the walls. The compressor discharges the air at 1 bar and is operated isothermally at 300 K. What is the total work done by the compressor? Assume that the compressor operates reversibly and that air is an ideal gas.

Solution

If it is assumed that the amount of air processed during one cycle of the compressor is small in relation to the total quantity of air expelled, the work of any one cycle can be treated as

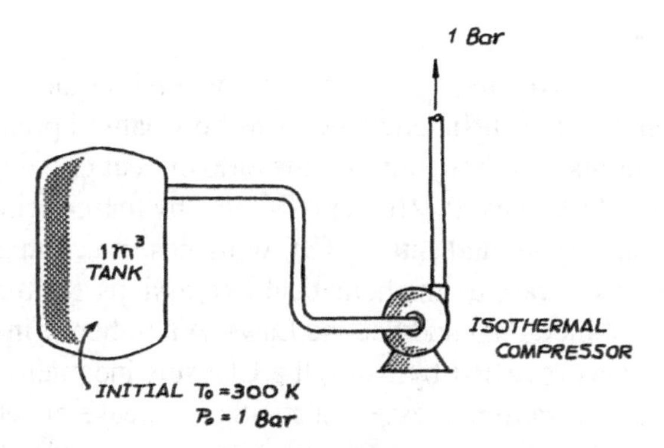

Figure 4.9

a differential and the properties within the tank can be assumed to vary smoothly with the amount of air expelled. Thus, for a differential amount $\delta n = dn$ of gas processed, the reversible shaft work done by the compressor is given by Eq. (4-77), which for this case is written as:

$$\delta W_s = dn \int_{P_t}^{P_a} V\,dP \tag{4-79}$$

where P_a is the discharge pressure (1 bar) and P_t is the tank pressure during the cycle under consideration. Introducing the ideal gas law and integrating at constant T, we obtain

$$\delta W_s = RT \ln \frac{P_a}{P_t} dn \tag{4-80}$$

Since the amount of gas processed, dn, is equal to the decrease in the gas in the tank, $-dN$, and since

$$dN = \frac{V_t}{RT} dP_t \tag{4-81}$$

Eq. (4-80) becomes

$$\delta W_s = \underline{V}_t \ln \frac{P_t}{P_a} dP_t \tag{4-82}$$

Integrating between $P_t = 1$ bar and $P_t = 0$,

$$W_\sigma = W_s = \underline{V}_t \left[P_t \ln \frac{P_t}{P_a} - P_t \right] \Bigg|_1^0 = \underline{V}_t \cdot (1\text{ bar}) \tag{4-83}$$

Thus, the reversible work done by the compressor is 10^5 J.

4.9 Summary

While the First Law gives us the thermodynamic embodiment of the principle of energy conservation, it is not sufficient to describe how natural processes will proceed. Directionality constraints and limitations on the ratio of heat to work conversions were developed by using Postulates II, III, and IV and by introducing the concepts of reversible Carnot heat engines and entropy (S). With these new fundamental principles, we were able to develop various mathematical expressions for the Second Law and ultimately for the combined First and Second Laws. A number of important constraints on system behavior was revealed by using the Clausius inequality. For example, the entropy for all isolated system processes must either increase or remain the same. All spontaneous, irreversible changes give $\Delta \underline{S} > 0$ while for reversible changes $\Delta \underline{S} = 0$. The often misunderstood concepts of mass flow and shaft work were covered in the final section to establish their relationship to the total PdV work associated with an element of fluid as it is compressed or expanded in a flowing, open system device.

Suggested Readings

Denbigh, K. (1980), *The Principles of Chemical Equilibrium*, 4th ed., Cambridge University Press, Cambridge, UK, Chapter 1, p 21-40. [Second Law concepts]

Sandler, S.I. (1989), *Chemical and Engineering Thermodynamics*, Wiley, New York, Chapter 3, p 84-118. [Entropy generation and balances]

Problems

4.1. We have four objects in our possession; the masses, heat capacities, and initial temperatures are as follows:

Object	Description	Heat capacity[a] (kJ/kg K)	Mass (kg)	T_i (K)
A	Bar of chromium	0.46	25	500
B	Oak block	2.40	8	300
C	Brick	0.84	20	400
D	Chunk of rubber	1.74	12	200

[a] *Assume that these heat capacities are independent of temperature.*

What we would like to do is to devise some process(es) that allow heat and/or work interactions between these four objects. With *no net* change in the environment, what is the minimum temperature that you could attain in any one of the four objects? In which object would you attain this temperature? Repeat if you desired the maximum possible temperature.

4.2. Most of us have seen, in novelty stores, small glass birds that appear to enjoy taking an endless series of drinks of water from a glass as illustrated in Figure P4.2. If we looked closely, we could see that these toys are simply two hollow glass bulbs separated by a tube and mounted on a swivel joint. The lower bulb is partially filled with a volatile liquid such as ethyl ether.

Ether boils at room temperature and some of the vapor condenses in the upper bulb, which is kept cold by evaporation of water on a wick placed over the bulb. In the drinking step the wick is moistened. The ether condensed in the upper bulb is prevented from returning to the lower bulb by the upward flow of vapor. Also, some of the liquid is pumped to the upper bulb by a "coffee percolator action." When the upper bulb is nearly full, the bird's center of gravity shifts, the bird swings into a horizontal position (and "drinks"), thereby allowing the ether to flow back down the tube to the lower bulb. The bird then becomes upright and the cycle is repeated.

Figure P4.2

We would like to connect this bird to some mechanism and allow it to pump water from a lower level. If we assume that *all* the water pumped is eventually evaporated from the bird's head, what is the greatest height from which it can pump water for steady-state operation? Assume normal ambient conditions and neglect any heat losses or other irreversibilities.

4.3. A Hilsh vortex tube for sale commercially is fed with air at 300 K and 5 bar into a tangential slot near the center (point *A* in Figure P4.3).

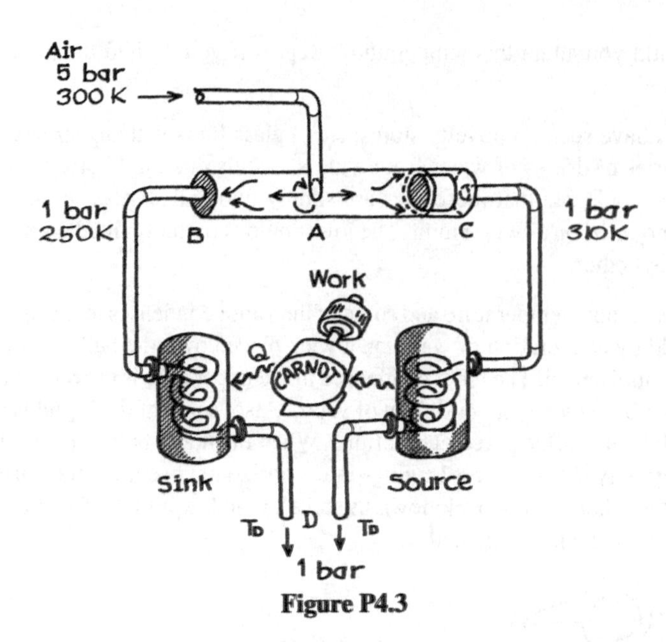

Figure P4.3

Stream B leaves from the left end at 1 bar and 250 K; stream C leaves at the right end at 1 bar and 310 K. These two streams then act as a sink and source for a Carnot engine and both streams leave the engine at 1 bar and T_D. Assume ideal gases that have a constant heat capacity $C_p = 29.3$ J/mol K.

(a) If stream A flows at 1 mol/s, what are the flow rates of streams B and C?

(b) What is T_D?

(c) What is the Carnot power output per mole of stream A?

(d) What is the entropy change of the overall process per mole of A?

(e) What is the entropy change in the Hilsh tube per mole of A?

(f) What is the maximum power that one could obtain by any process per mole of A if all heat were rejected or absorbed from an isothermal reservoir at T_D?

4.4. We have at our disposal three cylinders of pure helium gas (see Figure P4.4).

These may be designated as A, B, and C. The initial temperatures, pressures, and volumes are as follows:

Cylinder	T (K)	P (bar)	$\underline{V}$ (m^3)
A	1000	10	10×10^{-3}
B	500	1	50×10^{-3}
C	100	0.1	100×10^{-3}

We will allow interactions among these three cylinders. C_p for helium is 20.9 J/mol K. Helium may also be considered to behave as an ideal gas.

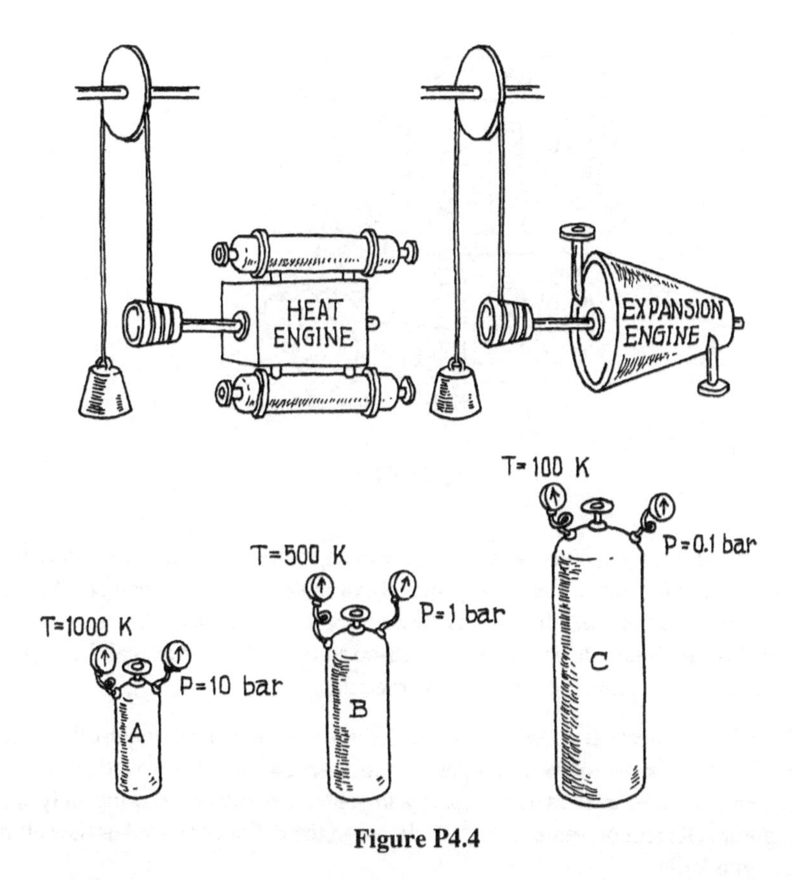

Figure P4.4

What is the *maximum* amount of work that can be produced? Heat engines, expansion engines, etc., are allowed; the only restraint is that there must be no net effect in the environment other than the rise of weights in a gravitational field (i.e., in a work reservoir). (This would then imply that the environment would not be compressed or expanded; that is, the total volume of the three cylinders initially must equal their combined volumes at the end.)

4.5. As you are no doubt aware, we in the academic world are most anxious to prepare you to solve technical problems that you may encounter in your future endeavors. Our Advanced Planning Section has been examining some unusual projected problems and requests your assistance on the particular one described below.

Decades from now, the present method of supplying energy to households (i.e., with electricity, gas, or oil) may not be possible. Instead, housepersons will shop for their energy in supermarkets (Figure P4.5). Cylinders of gas (let us assume that the cylinders contain air) may be purchased and connected to any number of Carnot engines or other such efficient devices to be stocked in the home. Work is then obtained which may be utilized by the houseperson's family. When purchased, the cylinders are packed in well-insulated bags which may be removed (if desired) when connecting to a work-producing device.

Figure P4.5

The problem we face is to devise a convenient method to allow the houseperson to compare prices for the various gas cylinders available at the supermarket. The usual size for most cylinders is 1 m^3, but the initial air pressure and temperature vary widely. One produced by R. Jones, Jr. is widely advertised to be quite economical, but our analysis indicated that the cylinder contained no air at all!

R. Nader III is expected to object to this deplorable situation and to require that we provide a simple equation to allow housepersons to calculate quickly (on their HP-1001 or equivalent) the unit cost of work energy in joules per dollar knowing only the initial temperature (K) and pressure (N/m^2) of the air in the cylinder as well as the selling price of the cylinder.

Data: Assume that the ambient temperature and pressure are 300 K and 1×10^5 N/m^2. Air has a heat capacity at constant volume of 20.7 J/mol K, and, at constant pressure, of 29.0 J/mol K. The gas constant is 8.314 J/mol K and the gravitational acceleration is 9.81 m/s^2.

Derive an equation for the unit cost of produced work and demonstrate its application for a cylinder 1 m^3 in volume initially at 8×10^5 N/m^2 and 400 K which sells for $0.32.

4.6. Under ordinary operation a steady flow of helium gas equal to 50 g/s passes from a large storage manifold through an expansion engine as shown in Figure P4.6. There are, however, certain times when the engine must be shut down for short intervals; the inlet flow cannot, however, be decreased at such times and thus the helium stream must be diverted. It is proposed to employ an adjacent system which at present is not being used. The latter system consists of a large (4 m^3) insulated tank (*C* in Figure P4.6) with a safety valve venting to the atmosphere through the plant ducting system.

The emergency diverting system is operated in the following manner. If the expansion engine must be shut down (or operates improperly), value *B* is shut and *A* is opened, letting helium into tank *C*. When the pressure in *C* rises to 2.8 bar, safety valve *D* operates, venting gas.

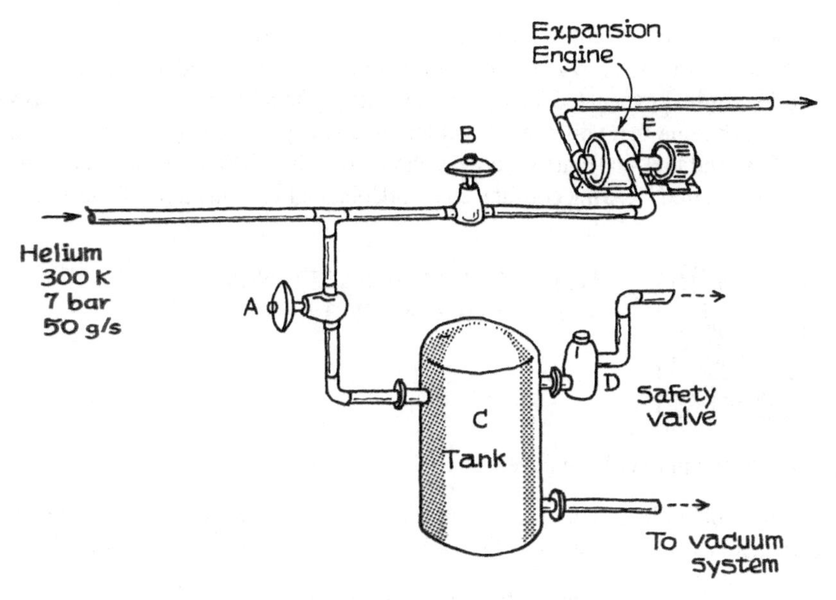

Figure P4.6

Initially, tank *C* is evacuated to a very low pressure. We shall assume that when gas enters this tank, it is well-mixed but has negligible heat transfer with the tank walls. Also assume that helium is an ideal gas with a constant $C_v = 12.6$ J/mol K.

(a) How long will flow enter tank *C* before the pressure increases to 2.8 bar and the safety valve opens?

(b) If the safety valve on tank *C* should operate, we would like to maintain a constant mass of gas in this tank equal to the mass at the time the safety tripped. The mass flow rate into the tank is, as noted above, 50 g/s; the flow rate $\dot{m}$ out of the safety valve may be expressed as

$$\dot{m}\ (\text{g/s}) = \frac{K\underline{a}P}{T}$$

where *K* is a constant, $\underline{a}$ the valve throat area, and *P* and *T* are the pressure and temperature in tank *C*. To keep the tank mass constant, the valve flow area $\underline{a}$ will be varied. What are the pressure and temperature in tank *C* 10 s after the safety valve opens? What is the variation of $\underline{a}$ with time necessary to keep the mass of gas in tank *C* constant?

(c) What is the total entropy change of the gas, the surroundings, and the universe during the time between the opening of valve *A* and just prior to the opening of relief valve *D*?

(d) What is the total entropy change of the gas, the surroundings, and the universe during the time between the opening of the relief *D* and a time 10 s later? The vented gas mixes with an infinite amount of air exterior to the tank and cools to 300 K. Leave the entropy change of mixing as an undetermined constant.

(e) Ten seconds after venting begins, the original expander comes back into operation so that valves *A* and *D* are shut and *B* is opened. It is desired to restore the tank *C* to its original evacuated state with the least possible work. You are free to select any technique that you deem feasible; all heat is to be rejected to surroundings at 300 K, and the final state of the gas should be 1 bar, 300 K. What is the minimum work required? (Do not consider any work obtainable from mixing helium with air.)

4.7.

TRANS-GALAXY-SPACELINES
— A Division of MITY, Inc.—

Welcome New Summer Employee:

As your first job with our famous old spaceline, we want you to answer a small technical problem that has arisen in our new line of boosters known to you as the Super Dodos.

We have a number of small jets in this series that will be used in attitude control in space. These jets will be powered by low-pressure nitrogen gas heated by an arc at the jet nozzle. Your problem deals with the nitrogen storage system.

The nitrogen is stored in a large, well-insulated, 0.4-m^3 sphere at 1 bar pressure. At takeoff the temperature is 280 K. The mass rate flow of N_2 will be constant and be equal to 7 g/s. Since the pressure inside the sphere must always be kept at 1 bar, a heater will be used inside the sphere (see drawing below).

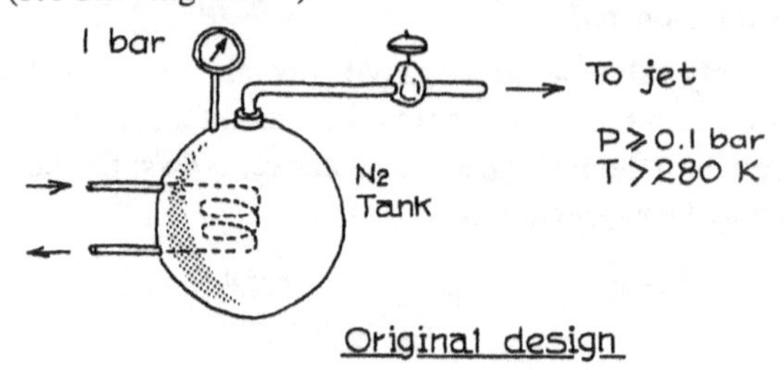

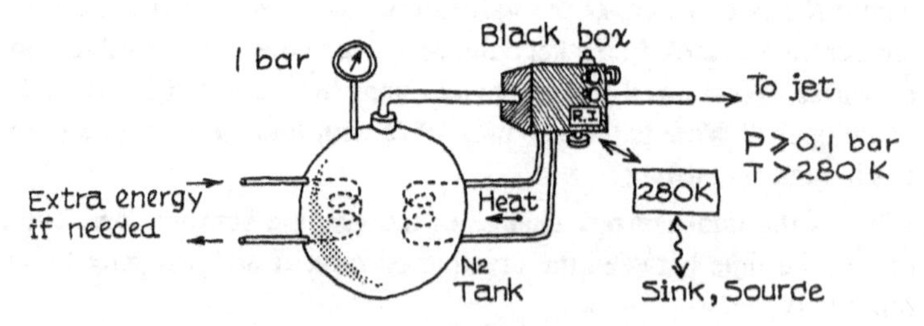

(a) Under these conditions, what will be the temperature of the N_2 in the sphere, the instantaneous rate of heat flow to the heater, and the total heat required after 10 s of operation?

(b) The energy requirement of the heater may be difficult to meet. Rocky Jones, our local genius, has made a suggestion that we would like you to evaluate. Rocky wishes to take the nitrogen from the sphere that leaves at a pressure of 1 bar and the sphere-gas temperature and put it through a ''black box'' to extract the maximum work possible before sending it to the arc jets. The interior metal of the booster may act as a heat source or sink at 280 K, and the final nitrogen pressure must be no less than 0.1 bar as fed to the jets. The work from the black box will be converted to heat and fed back to supply the sphere-heat energy. Nitrogen is an ideal gas with $C_p = 29.3$ J/mol K, independent of temperature. Work may be stored in the black box if more power is generated than is needed during some periods of the process. What is the instantaneous additional heat required at the start of the operation? Will the black box supply all the sphere heat required, and, if so, for how long? What is the temperature in the sphere at this time?

4.8. MITY, Inc. has developed another unusual device; and as usual, you are requested to aid in the evaluation.

This time, Rocky Jones has proposed that on occasion, one might desire a low-capacity high (or low)-temperature source that could be stored for indefinite times before usage. For example, long-range space flight experiments may require a short-term, high (or low)-temperature source years after launch. Obviously, one cannot hope, in such cases, to carry high-temperature (or cryogenic) fluids and expect no heat transfer to occur. Rocky says that his back-of-the-envelope calculations indicate that the way to solve this problem is simply to carry small pressurized gas tanks and, by appropriate means, convert these rapidly to low-pressure, high- or low-temperature vessels.

Our engineering department has just finished the construction of a small vessel to hold 1 mole of air at 300 K and 100 bar. This is to be the prototype unit. Upon demand, we will insert this into our conversion unit and convert the high-pressure, low-temperature gas into a low-pressure, high (or low)-temperature gas. The gas volume may vary but no gas is lost to the environment. Air is an ideal gas with a constant isobaric heat capacity of 29.3 J/mol K. However, the exact design of this conversion unit is not yet firm and we would like your help.

(a) Assume that we want the highest possible temperature from the unit described above. Suggest a sequence of processes within the conversion unit that will attain this value. You are allowed to have heat transfer to a large sink at 300 K and with the atmosphere (air) around the unit constant at 1 bar. What is the *maximum* temperature that can be obtained?

(b) Repeat part (a), but assume now that we want the lowest possible temperature. In this case, what is the minimum temperature that can be obtained?

(c) One of our younger engineers states that we should use some gas in our device other than air. Do you perceive any advantages in doing so? If you do, please elaborate.

Note: You do not have to be practical in your experimental plans. Yet it would be appreciated if you would not violate any of the laws of Nature or even of thermodynamics.

4.9. Gyro Gearloose, one of our less responsible students, neglected to close tightly a valve on his compressed air bottle last Friday. Over the weekend, the bottle pressure dropped slowly to 1 bar. Henrietta Helmholtz, his laboratory partner, berated him for his thoughtlessness. She pointed out that his action had resulted in a serious loss of available energy.

Her concern was real as it was -10°C outside. She insisted that if the bottle had had to be blown down, *she* could have devised a more efficient method to provide considerable energy to heat the laboratory. As she was a first-year student, she had not had the opportunity to benefit from our thorough thermodynamics program and thus was unable to express her feelings in a quantitative manner. Could you come to her assistance and indicate the maximum number of joules that could have been made available for heating if the bottle had been blown down in a less thoughtless manner? (Assume ideal gases.)

Data:

Laboratory temperature	20°C
Outside temperature	-10°C
Bottle volume	0.06 m^3
Initial bottle pressure	150 bar
Final bottle pressure	1 bar
C_p for air	29.3 J/mol K

(*Note:* The department is well-stocked with various sizes of Carnot engines, if needed.) Be certain that you use your most perceptive imagination to work this problem. The bottle initially is at 20°C. You may carry out the blowdown process inside or outside the laboratory, and you may employ any type of compressor, expander, and heat exchanger you desire. Neglect, however, the heat capacity of the bottle itself.

4.10. Mr. Rocky Jones, the well-known inventor and entrepreneur, visited me yesterday with a most interesting idea that we should examine carefully in this age of ever-dwindling energy sources. It appears that Mr. Jones is concerned about the inefficient use of hot water in our sinks. As he described it, one normally turns on both the hot water and the cold water and, with soap, rapidly moves the hands to-and-fro to obtain the overall sensation of warm water. The time the hands are actually in any stream of water is quite short and most of the water used is mixed in the sink and flows down the drain.

What Mr. Jones is proposing is a new kind of sink, as shown in Figure P4.10. Except in the brief intervals when the actual washing is done, the hot and cold water from the sink exit through separate drains into device *A*. Mr. Jones claims he can obtain work from this device to reheat more hot water in *B*. The work from *A* may not be quite sufficient to heat all the water, so an auxiliary electric heater in *C* is used to maintain the desired inlet hot-water temperature. We want to supply about 0.2 gal/min (757 cm^3/min) of hot water to our sinks at about 70°C (343.2 K). The average cold-water flow is twice the hot-water flow and is normally at 10°C (283.2 K).

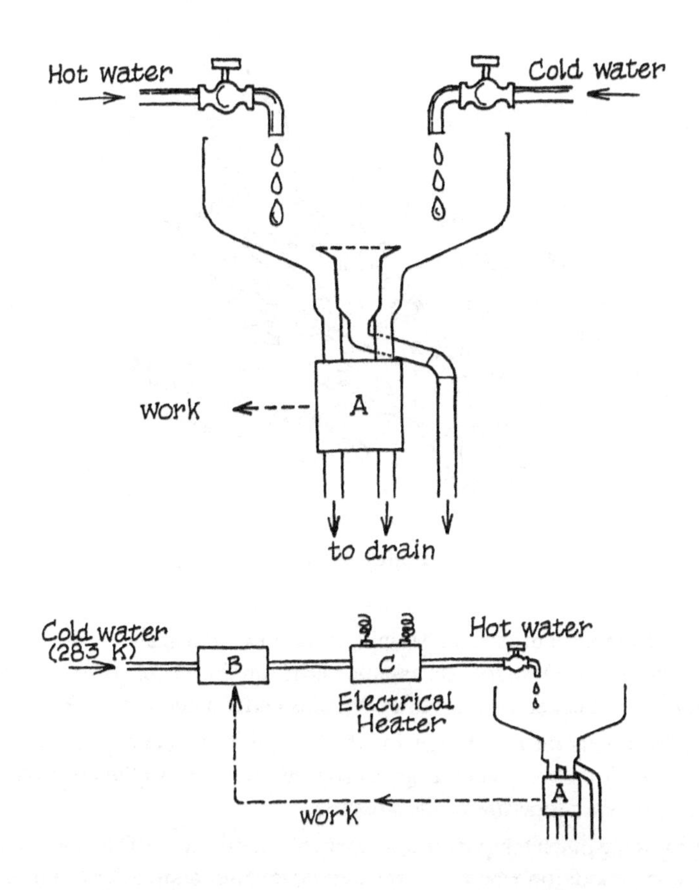

Figure P4.10

(a) If our sinks operate at steady state, what do you estimate is the necessary extra energy flow input required in unit C? Express your results in watts of electrical power. Assume no interaction with the environment in device A. You may, however, extract energy from the environment, if desirable, when considering the operation of B; the environment remains at 10°C.

(b) Can you accept the challenge to improve on Rocky's device so as to yield the desired hot water with even less electrical energy?

4.11. In several parts of the world, there exist ocean currents of differing temperatures that come into contact. An example of this takes place off the coast of southern Africa, where the warm Agulhas and the cold Benguela currents meet at Cape Point, near Cape Town (see Figure P4.11). It has been proposed that work may be obtained from these ocean currents by operating a heat engine between the warm current as a source and the cold current as a sink (shown schematically in Figure P4.11). This renewable form of energy production has been called *ocean thermal*, and the Department of Energy needs your help in evaluating the proposal.

Figure P4.11

Assume that the system may be simplified into two channels of water in contact and flowing cocurrently. Furthermore, assume that no mixing occurs between the streams and that heat transfer between the stratified streams under natural conditions is negligible.

(a) Derive a general expression for the maximum amount of power that could be obtained from the system. Express your result in terms of temperatures, flow rates, and physical properties of the streams.

(b) What is the pinch temperature, and where does it occur? The pinch is defined as the limiting condition where the stream temperatures approach each other.

(c) Repeat part (a) if the two streams flow countercurrent rather than cocurrent.

(d) It has been estimated that the Benguela current is 16×10^6 m^3/s and its initial temperature is 278 K. The Agulhas current is 20×10^6 m^3/s and its initial temperature is 300 K. Calculate and compare the power obtainable from these two currents assuming that they flow (1) cocurrently and (2) countercurrently.

4.12. Recently, we carried out an interesting experiment in our laboratory. We connected a *constant-pressure* air supply line to a small spherical vessel (*B*) as shown in Figure P4.12(a). With valves *C* and *D* we could either pressurize the sphere with high-pressure air or vent it to the atmosphere. In the test, with vent valve *D* closed, we pressurized the sphere very rapidly to the air-supply pressure. Then, quickly, we closed valve *C* and opened valve *D* to vent the sphere to atmospheric pressure. As soon as this pressure level was achieved, we closed valve *D* and repeated the cycle. This sequence was followed for many cycles. Assume that all steps are carried out rapidly and that heat transfer from the air to any piping or vessel walls is negligible. Air may be assumed to be an ideal gas with $C_p = 29.3$ J/mol K and $C_v = 21.0$ J/mol K.

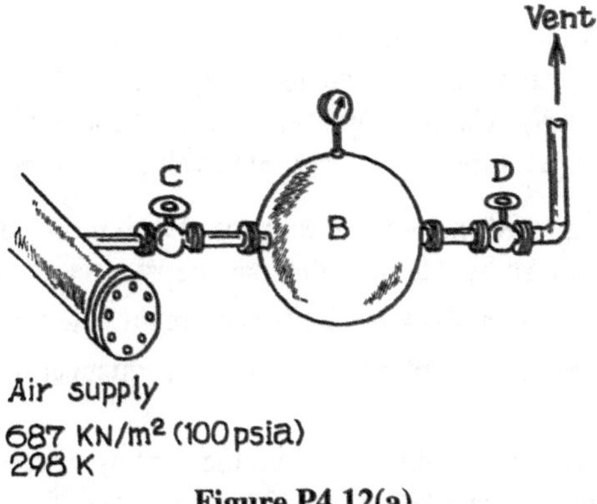

Figure P4.12(a)

(a) What was the temperature of the air in the sphere at the end of the first cycle (when the pressure was 101 kN/m^2)?

(b) After many cycles, what would be the temperature in the sphere at the end of the pressurizing step? At the end of the venting step?

As we were conducting these experiments, Rochelle Jones wandered into our laboratory. After she had watched the operation for many cycles so that the conditions at the end of the pressurization and venting steps were time invariant (see part (b)), she became excited. During the venting, she saw that the gas outlet temperature was initially high but decreases as venting proceeded. "Why not get some 'free' work from the device?" she asked. She began to explain her ideas, but I am afraid I could not understand everything she said.

She apparently wants to replace valve D with a Y tube attached to valves D_1 and D_2 [see Figure P4.12(b)]. She says she can build a sequencer to allow flow through valve D_1 for the first part of the venting step and through valve D_2 for the rest of the venting. Both flow into what she calls her Hilsch box. Both streams exit at the same temperature and at a pressure of 101 kN/m^2.

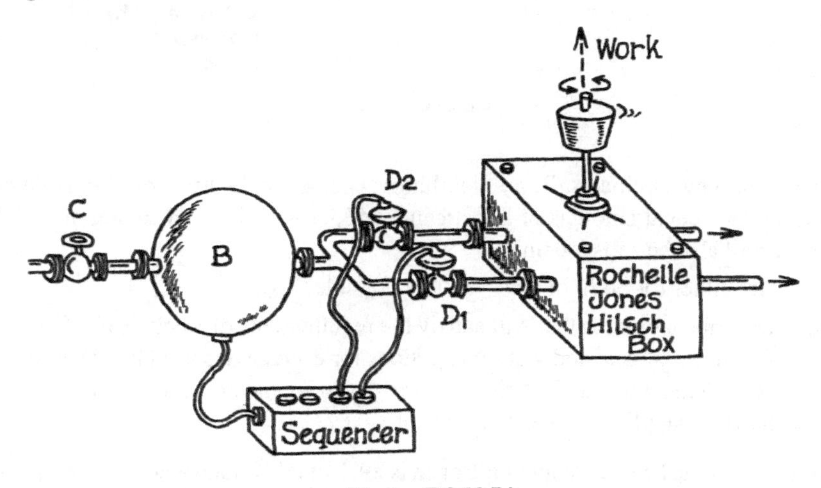

Figure P4.12(b)

No interaction with the environment is evident except for the transfer of work. We are not certain, but we believe that Rochelle's idea is to include in her Hilsch box some expansion engines and/or Carnot engines to transfer heat between the two streams since the average inlet temperature of the air through valve D_1 exceeds that through valve D_2.

Rochelle is returning on Monday to discuss her idea in more detail. Now, we don't wish to appear too ignorant, so by 9 a.m. of that day, carry out the following calculations:

(c) What is the *maximum* work Rochelle can obtain in her Hilsch box, J/kg of air?

(d) At what pressure in the sphere B should the sequencer operate to direct flow from D_1 to D_2?

(e) If you were going to obtain work from the venting of sphere B, could you suggest improvements to Rochelle's scheme? For example, if you were to allow heat transfer to the environment in her Hilsch box, what is the maximum work you could obtain per kilogram of air?

4.13. Our highly efficient plant has a nitrogen stream at 2.53 bar and 305.5 K which is presently vented to the atmosphere. Management would like to use this stream to satisfy some of the heating or cooling requirements of other processes. Our old friend, Rocky Jones, has devised a black box that will produce equal amounts of a hot stream at 500 K and a cold stream at 111 K and thus satisfy simultaneously some heating and cooling requirements (see Figure P4.13). Furthermore, Rocky claims that his device will be self-sustaining because no additional heat or work need to be supplied to the device.

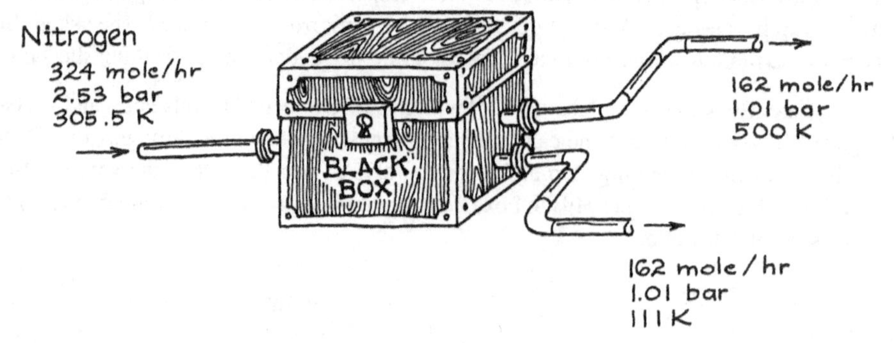

Figure P4.13

One of our new engineers, Barry Goldfinder, claims that he has a black-box device that will produce equal amounts of hot stream of 533.3 K and a cold stream of 77.7 K. His device will also be self-sustaining.

(a) Are either (or both) of these devices possible? Explain.

(b) Describe a process that will satisfy the requirements of Rocky's black box. Calculate all heat and work interactions for each device used in this process and indicate now the devices should be arranged so that no additional heat or work need be supplied by the environment.

Assume that the heat capacity of nitrogen is 29.3 J/mol K, independent of temperature or pressure.

4.14. I call your attention to the problem we now face in heating up the vapor feed to the X-13 batch synthesis unit. We would like to avoid any customary heat exchangers since hot walls may function as a catalyst and initiate decomposition of this vapor. One of our more creative consultants, R. Jones, has suggested a way to heat this vapor without even contacting hot surfaces, and I would like your opinion on her scheme.

She proposes to begin with a batch of vapor in sphere *A* (see Figure P4.14). Connected to *A* is a small piston and cylinder unit as shown. There is a check valve at *C* to allow flow only in the direction shown. As I understand the operation, the piston is drawn to the left (with valve *C* closed) until port *D* is uncovered. Gas then flows from *A* to *B* until the pressures are equalized. (Before port *D* is uncovered, you may assume a perfect vacuum in *B*.) The piston is then moved to the right, covering port *D*, and gas is pushed through valve *C* back into *A*. The cycle is repeated again and again. Jones says that the vapor in *A* becomes hotter after each cycle.

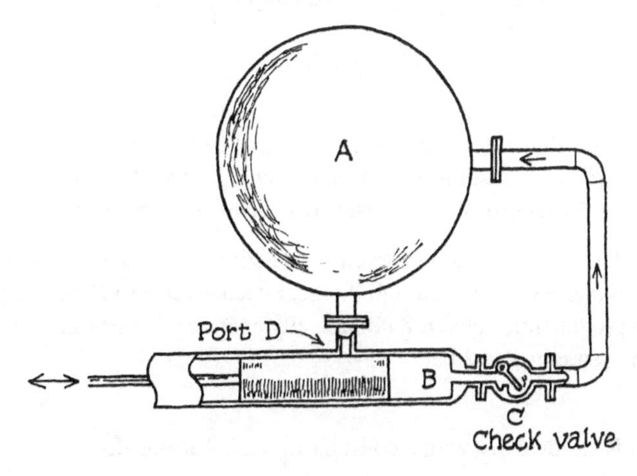

Figure P4.14

Data:

Initial pressure in *A*: 10 bar Volume of *A*: 1 m^3

Volume of cylinder *B* when the piston uncovers port *D*: 0.01 m^3

There is no significant heat transfer between the unit and the environment. The vapor being processed may be considered to be an ideal gas with a heat capacity at constant volume of 33.256 J/mol K independent of temperature. Also, when the vapor is pushed from *B* to *A*, assume that there is perfect mixing of the gas in *A*. The piston-cylinder operation is adiabatic and reversible. Can you estimate how many cycles it would take to heat the vapor from 300 to 600 K?

4.15. We have been requested to design a small system that will allow a manufacturer to test helium vacuum pumps both under transient conditions and also for long steady-state periods. The system we now propose is shown in Figure P4.15.

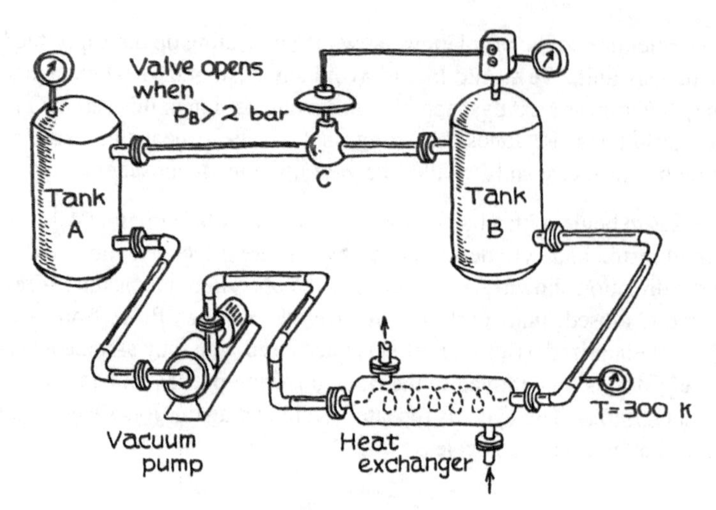

Figure P4.15

Tanks A and B are thin-walled and well-insulated, each with a volume of 0.1 m^3. The vacuum pump will take suction from A and discharge into B. Between the pump and B there is a heat exchanger to cool the gases to a constant temperature of 300 K.

Initially, both tanks A and B are charged with helium at 1 bar and 300 K. The vacuum pump is started and gas flows through the heat exchanger into B until the pressure in B reaches 2 bar. At this time valve C automatically opens and vents gas from B to A, always keeping the pressure in B at 2 bar.

Data:

- Helium is an ideal gas with a constant $C_p = 20.9$ J/mol K.
- The volumes of the connecting lines and pump are negligible compared to the volume of tanks A and B.
- The vacuum pumps to be tested are adiabatic and reversible.

We need your answers to the following:

(a) Just at the time the pressure in B reaches 2 bar, what is the temperature in both A and B, and what is the pressure in A?

(b) After the system has been running for a long time, with P_B = constant = 2 bar, what is the temperature of B and what is the pressure in A?

(c) What is the change in entropy of all the helium gas in the system from the start until the pressure in B just reaches 2 bar?

(d) Can you suggest any improvements in the system to simplify it and still test the vacuum pumps over transient and long-term periods?

(e) What is the work supplied to the vacuum pump from the start of the test until the pressure in B just reaches 2 bar?

(f) What is the heat load in the heat exchanger during the same period as described in part (e)?

4.16. It is proposed that a device be constructed to operate as follows: An evacuated tank of 1 m^3 is attached to the exhaust of an air-turbine-driven grinding wheel. Air at atmospheric pressure and 300 K will be allowed to enter the turbine inlet, while the pressure drop between the atmosphere and the tank serves to operate the turbine-driven grinder. When the pressure in the originally evacuated tank has risen to atmospheric, the turbine, of course, stops.

(a) Assuming the turbine operates in a reversible manner, how many kilowatt-hours of work may be obtained up to the time the pressure in the tank has risen to atmospheric? Consider four cases: (1) tank is adiabatic, turbine is isothermal; (2) tank is adiabatic, turbine is adiabatic; (3) tank is diathermal, turbine is adiabatic; (4) tank is diathermal, turbine is isothermal. The ideal-gas law may be assumed.

(b) Determine the air temperature in the tank at the end of the filling process. The entropy change of the universe, and the heat interaction with the environment for the four cases noted in (a).

4.17. Dear Dr. Rankine:

We received from a Ms. Jones an interesting letter that poses a somewhat unusual modification to our standard water condenser design. I won't go into the letter in any detail since, to me, parts of it were somewhat perplexing. The essence of it, however, was that Ms. Jones (who is president of MITY, Inc.) indicates that her company is diversifying and has recently bought out Carnoco, a manufacturer of diminutive reversible heat engines. Ms. Jones suggests that in all our new plants, wherein we cool and condense stream S-1 (see Figure P4.17A) from 422 K to 300 K, we eliminate the heat exchangers entirely and, instead, substitute a set of Carnoco engines. If I understand Ms. Jones correctly, her process diagram would appear as in Figure B. Ms. Jones claims that we could reduce the cooling-water requirement (and still maintain the outlet cooling water temperature at 289 K), obtain a great deal of ''free'' work, and eliminate completely the cost of our heat exchanger.

Using the following data, would you please evaluate Ms. Jones's proposal and indicate the maximum amount we could afford to pay for the purchase and installation of the Carnoco reversible engines?

Cooling water cost	*\$1/100* m^3
Value of any work output	*5.0 cents/kWh*
Present cost of heat exchanger, installed	*\$150,000*
Operational time per year	*7200 h*
Annual fixed charge rate for the capital investment	*15%*

Your attention to this request should be given priority!

Sincerely yours,

Godfrey Cross

Godfrey Cross
Boss

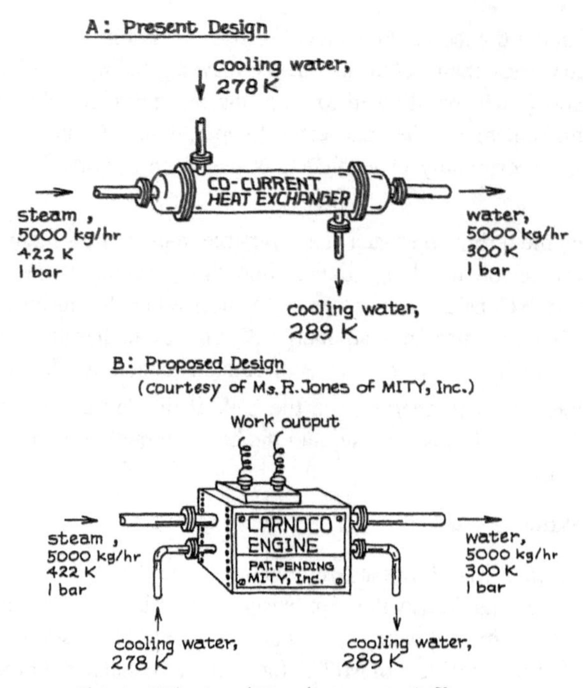

Figure P4.17

4.18. Two large gas storage spheres (0.1 m^3) each contain air at 2 bar (Figure P4.18). They are connected across a small reversible compressor. The tanks, connecting lines, and compressor are immersed in a constant temperature bath at 280 K. The compressor will take suction from one sphere, compress the gas, and discharge to the other sphere. Heat transfer between the bath and the tanks, lines and compressor is excellent. Assume that air is an ideal gas with $C_p = 29.30$ J/mol K.

(a) What is the work requirement to compress the gas in one sphere to 3 bar?

(b) What is the heat interaction with the constant-temperature bath?

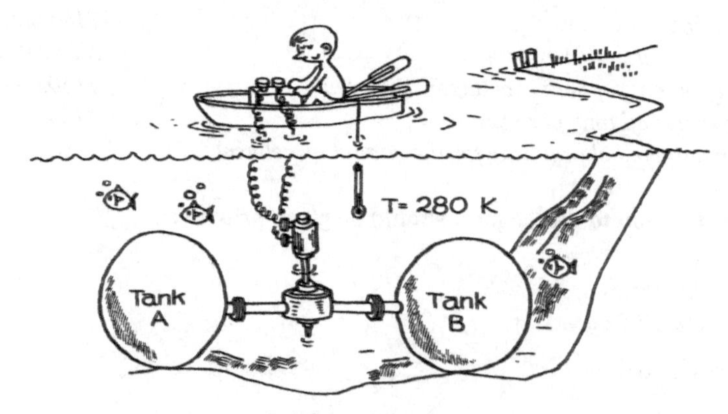

Figure P4.18

4.19. A patent on a new adiabatic device known as the "vacuum energizer" (Figure P4.19) may be filed soon. An ideal gas is stored in bulb A at some initial pressure P_{A_0}. A long, solid piston starts at the right end of the cylinder B and is pulled horizontally to the left. It uncovers the port so that gas from A can pass into cylinder B until pressures in A and B are equalized. The piston then pushes the gas out through the check valve at constant pressure. The piston is then drawn back, creating a vacuum in B until the port is again uncovered, and the cycle is repeated. The gas is air and is ideal; no heat transfer occurs from the gas to the walls or between compartments A and B.

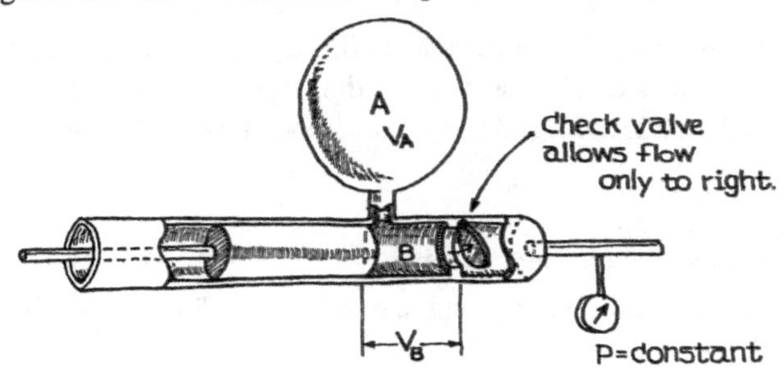

Figure P4.19

(a) If $P_{A_0} = 10$ bar, $T_{A_0} = 278$ K, $\underline{V}_A = 0.024$ m^3, and $\underline{V}_B = 0.006$ m^3, what is the temperature and pressure of the gas leaving B during the first cycle and during the nth cycle?

(b) What is the temperature and pressure in A after the first cycle and after the nth cycle?

4.20. In some experiments with a newly synthesized gaseous compound, it has been determined that it is very sensitive and will explode violently at pressures slightly above atmospheric. Other experiments have shown that the material may be unstable in a heat exchanger with hot walls. For some purposes, however, the material must be heated to a high temperature. Our project engineer has proposed the system shown in Figure P4.20.

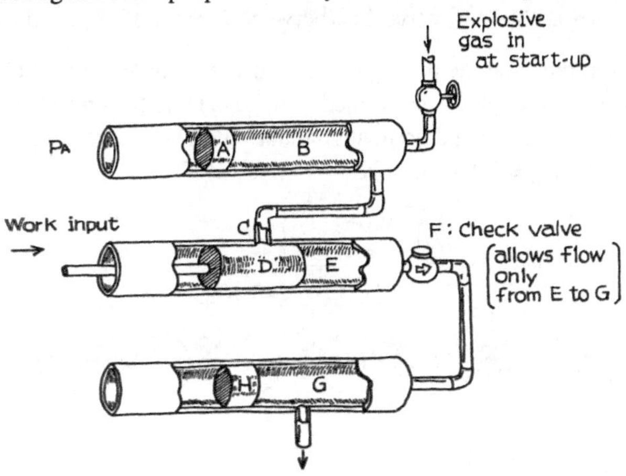

Figure P4.20

The gas initially is contained in cylinder B at 1 bar pressure. Piston A floats in a horizontal plane and prevents the pressure in B from exceeding 1 bar. When the port C in cylinder E is uncovered, gas from B enters the cylinder until pressures in the two cylinders are equalized. Piston D then moves, covers port C, and pushes the gas out slowly through check valve F into cylinder G, which in turn is fitted with a floating piston H to prevent the pressure from exceeding 1 bar. The piston D is then withdrawn to uncover port C and the cycle is repeated. Valve F only allows flow from E to G, but not the reverse. The pistons have essentially no friction.

It is claimed that the gas in G is hotter than in B. Any desired temperature may be attained by cascading such devices in series (i.e., feeding gas out from G through J into another tube, similar to the outlet of B). What is the temperature of the gas after *two* cascading steps?

Data:

- Cylinders have 7 cm inside diameters
- Assume no residual clearance volume in cylinder E when the piston is fully extended.
- Initial mass of gas in B is 1 mole and has a temperature of 311 K.
- Assume ideal gases with a constant $C_p = 41.86$ J/mol K.

4.21. Our old friends, Rochelle and Rocky Jones, have been tinkering in the laboratory and have produced an interesting device that they call an "integral pulsed shock tube." As yet, we do not see a large commercial market. In fact, we are not sure what it really does! Can you analyze the device and answer the brief questions given later?

A long, insulated tube (Figure P4.21) is divided into chambers of equal volume by rigid, adiabatic partitions. Each partition has a fast-operating valve to allow flow of gas between compartments when the valve is open. The operation of this device is as follows. Compartment A is initially filled with helium gas at 300 K and 64 bar. The remaining compartments are evacuated to zero pressure. All valves are closed. At time zero, valve AB is quickly opened and gas flows from A to B. Just when the pressures are equalized in A and B, valve AB is shut and valve BC (between B and C) is opened.

As before, just at the time the pressure in B and C are equal, valve BC is shut and valve CD is opened. This sequence is continued until gas enters the end compartment G. There is no axial heat conduction and no heat conduction across valves or partitions. Helium is an ideal gas with a $C_p = 20.9$ J/mol K.

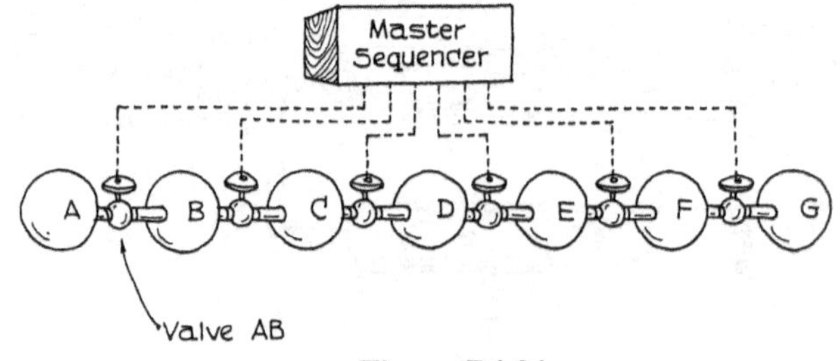

Figure P4.21

(a) When valve FG has just closed, what are the temperature and the pressure in compartment G?

(b) When the sequencing is completed, all valves are opened and the pressure allowed to equalize in all compartments. Slow axial conduction also equalizes the temperature in all compartments. What is the equilibrium temperature and pressure?

(c) The Joneses are uncertain how much work is required to prepare their integral pulse shock tube for firing. Estimate the minimum work per mole of helium initially in compartment A. In this calculation, assume that all compartments of the shock tube initially contain helium at 1 bar and 300 K. Any helium that is removed must be pumped into a large pipeline containing helium at a constant pressure of 2 bar and 300 K. Also, any helium used to charge compartment A to 64 bar is to be taken from the same pipeline. If, in your calculations, any heat transfer occurs, assume that you have a large heat sink or source at 300 K.

(d) Can you suggest any use for Rochelle and Rocky's new device?

4.22. In a laboratory experiment you must evacuate as rapidly as possible some process equipment. Initially, the equipment contains nitrogen gas at 280 K and 1 bar pressure. The volume is 0.14 m^3. You have at your disposal a typical rotary vacuum pump. It is rated to pump 0.03 m^3/min of actual volume. This pump may be considered to be a reciprocating one, with no residual volume clearance, and with an intake stroke of $\delta \underline{V}$ m^3, where $(\delta \underline{V})$ (rpm) = 0.03 m^3/min. There is one complication, however. The equipment is adiabatic except for a small heater that must be left on at all times and liberates 2 kJ/min at a constant rate. For N_2, C_p = 29.3 J/mol K, independent of temperature.

(a) What is the pressure inside the equipment after 2 min of pumping?

(b) What is the temperature of the nitrogen remaining in the process equipment after 2 min of pumping?

(c) One of your friends kindly offers to lend you his vacuum pump (which is exactly like yours) and he says that you can reach the same pressure in half the time if you connect it up parallel with yours. Do you agree?

(d) What is the lowest pressure you can pump on your system using one pump? Explain.

(e) If you wish only to pump down to 0.3 bar, what minimum horsepower motor do you recommend? The vacuum pump is neither adiabatic nor isothermal, but some tests indicate a polytropic relationship of the following form,

$$PV^{1.3} = \text{constant}$$

is applicable in the compressor. Base your minimum power estimate on the assumption that the pump is reversible.

4.23. JoAnna W. Gibbs, a long-time resident of New Haven, Conn. and distant cousin of J.W. Gibbs, has designed a new device for cooling gases--the so-called "Gibbs Chiller." The following experimental data are available for a trial run of the chiller. For an inlet gas condition of 10.1 MPa, 300 K, cooling was observed to 100 K at an outlet pressure of 0.1 MPa under steady state operating conditions. The only things we know about her device are (1) no heat interactions occur either internally or externally and (2) work interactions may be allowed to and from an internally contained "work reservoir." The fluid tested has the following properties:

$$C_p = 10 \text{ J/mol K} \quad \text{and} \quad P = RT/(V - b)$$

where b = constant = 2 x 10^{-4} m^3/mol and R = 8.314 J/mol K.What can you say about the nature of the process or processes occurring in the Gibbs Chiller?

4.24. A pumped storage power producing process that utilizes hot water at 227°C as saturated liquid (P = 25 bar) is stored on a mountaintop in New Zealand in a pressurized tank. The pumped storage system is located 3000 m above sea level. The average ambient ocean temperature is 27°C. Describe and sketch a process that would produce power and calculate the maximum work possible per kg of fluid.

Consider the case where the hot water is pumped to the mountaintop storage tank during off-peak periods from a sea-level-based geothermal fluid supply site located near the mountain. Describe how you would calculate the work associated with this periodic pumping from the geothermal site to the storage tank. The tank is a vertical right circular cylinder of constant volume with an inert ideal gas in the space above the liquid. At the start of the cycle there is no liquid in the tank. Because of the transient nature of pumping, the tank pressure increases from P = 25 bar to P = 200 bar at the end of a 10-hr pumping cycle where 10^6 kg of fluid was pumped. You can assume that the specific volume of the geothermal water is constant at 10^{-3} m^3/kg and that its initial temperature is 227°C. A multistage pump with a constant displacement stroke is also available with an efficiency of 70%. State all additional assumptions made in your analysis. You can assume that the heat capacity (C_p) of geothermal water is constant at 4.2 kJ/kg K.

4.25. The process of filling a tanker with natural gas from a high pressure storage tank is schematically shown below. Under normal operating conditions, the control valve characteristics are such that the storage tank pressure decreases exponentially with time ($P = P_o \exp[-\alpha t]$, P_o = initial pressure and α = constant). In addition, the pressure inside the tanker is never allowed to exceed 0.2 MPa (2 bar). The filling process is known to be slow, and an earlier study has suggested that some heat is exchanged with the storage tank during filling but that the valve, piping, and tanker are well insulated. Practically speaking, the tanker can be assumed to be completely evacuated at the beginning of the filling cycle. Everything is at T_o initially where T_o is the ambient air temperature.

(a) Perform a First Law analysis and describe how you would estimate how the temperature and mass vary with time in both the storage tank and tanker.

(b) If the insulated control valve was replaced with a turboexpander of constant efficiency, describe how you would calculate the work that could be recovered during the filling process.

State any and all assumptions used in answering parts (a) and (b).

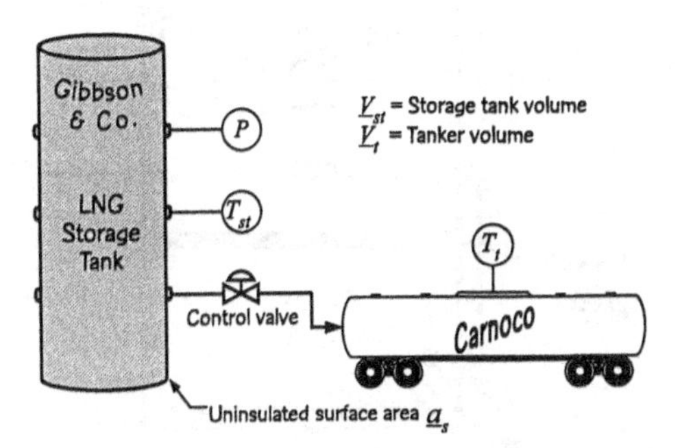

Figure P4.25

4.26. Rochelle Jones has just come up with a new scheme for underwater power based on the blowdown of a high pressure gas supply through a turbine. Basically, her design uses a gas supply available at P_i in a bare metal tank attached to a turbine expander that would rotate to turn a propeller to induce motion. The entire propulsion system is immersed in the ocean, assumed to be at a constant temperature T_o. Analyze this situation and derive an equation to estimate the maximum work-producing potential. State and justify all assumptions made. Are there any limitations to the operating depth for the system? Explain.

4.27. Rocky Jones Jr. has come up with a scheme to beat the heat this summer. To cool things off in his East Cambridge apartment, he suggests that all you have to do is open the doors of his freezer/refrigerator and close up all the windows and doors in the apartment. As the refrigerator continues to run, the apartment will cool down. What do you think of his idea? Will it work? Explain your answer.

4.28. A cylinder (200 cm in length and 50 cm^2 in cross-sectional area) is fitted with a piston which is connected to a spring that obeys Hooke's Law with a 2×10^6 dyne/cm force constant; see Figure P4.28. Hooke's Law states that the force exerted by a spring is linearly proportional to the spring elongation and acts in a direction opposite to the elongation. The chamber containing the spring is evacuated, while the other chamber contains 0.1 moles of helium initially at 25°C and 1 bar. The system is initially at equilibrium.

The helium chamber is heated slowly until the temperature reaches 200°C. Assuming that the piston is frictionless and has negligible mass and thermal conductivity and that helium can be treated as an ideal gas of constant heat capacity C_p = 20.9 J/mol K, calculate:

(a) What is the pressure of the helium when the temperature reaches 200°C?

(b) How much heat is required?

(c) What is the entropy change of the gas, the surroundings and the universe for the process described?

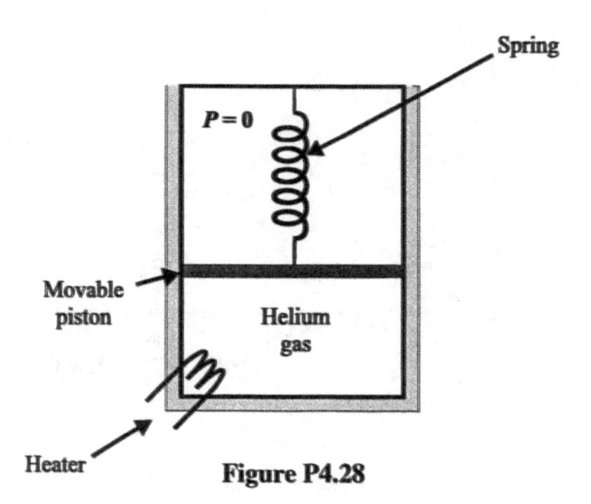

Figure P4.28

4.29. We are given a cylindrical vessel with an initial volume of 2 m^3 that is filled with helium at 1 bar, 300 K. We plan to pressurize this vessel to 10 bar using a large external source of helium gas maintained at 300 bar, 300 K.

Our model assumes that the initial helium present in the vessel does not mix with the entering helium, and that the initial gas is layered (layer A) and compressed by the entering gas (layer B) from 1 bar to 10 bar. The final system then contains two "layers" of helium, both at 10 bar, but presumably at different temperatures.

You can assume that: (1) there is no heat transfer between the gas layers, (2) there is no heat transfer to the vessel during the operation, and (3) helium behaves as an ideal gas with a constant value of C_p = 20.9 J/mol K.

(a) Calculate the final temperature in layer B.

(b) Calculate the entropy change of the universe after pressurization.

(c) If the system consisting of layers A and B was thermally isolated from the environment and no additional gas allowed to enter, what is the maximum work that one could obtain for the *reversible* mixing of A and B (in the same vessel) to some final homogeneous temperature T_f? What would be the final temperature and pressure?

4.30. The Carnoco Engine Division (CED) of MITY Industries has come up with a new concept for converting geothermal energy into electricity. They claim their design is 50% more efficient than the existing Mesa Verde geothermal power plant which employs a supercritical Rankine cycle operating at a high utilization efficiency η_u (see Figure P4.30B), where 60% of the maximum power is generated!

CED has based their design on the old Minto wheel concept depicted below (see Figure P4.30A). Because their process is proprietary, we have only sketchy details concerning its operation. According to CED's chief executive officer, Samantha Carnoto, their design utilizes a very large array of Minto-type wheels all connected to a common rotating shaft which is attached to a generator (see Figure P4.30C). The internal operating pressure of each wheel has been preset so that its contained working fluid boils at a certain

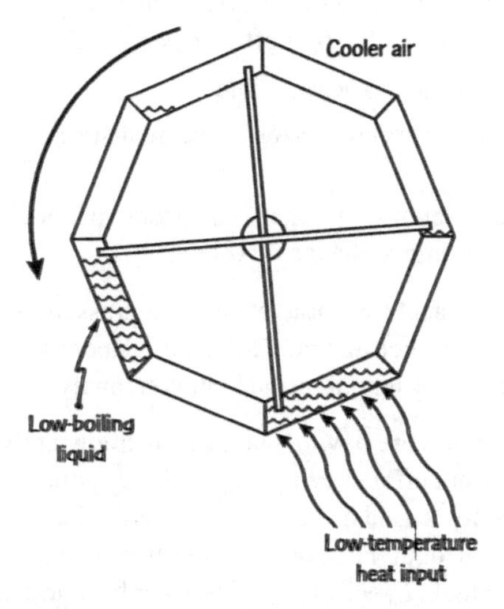

Figure P4.30A Minto wheel schematic [personal communication, W. Minto (1978)].

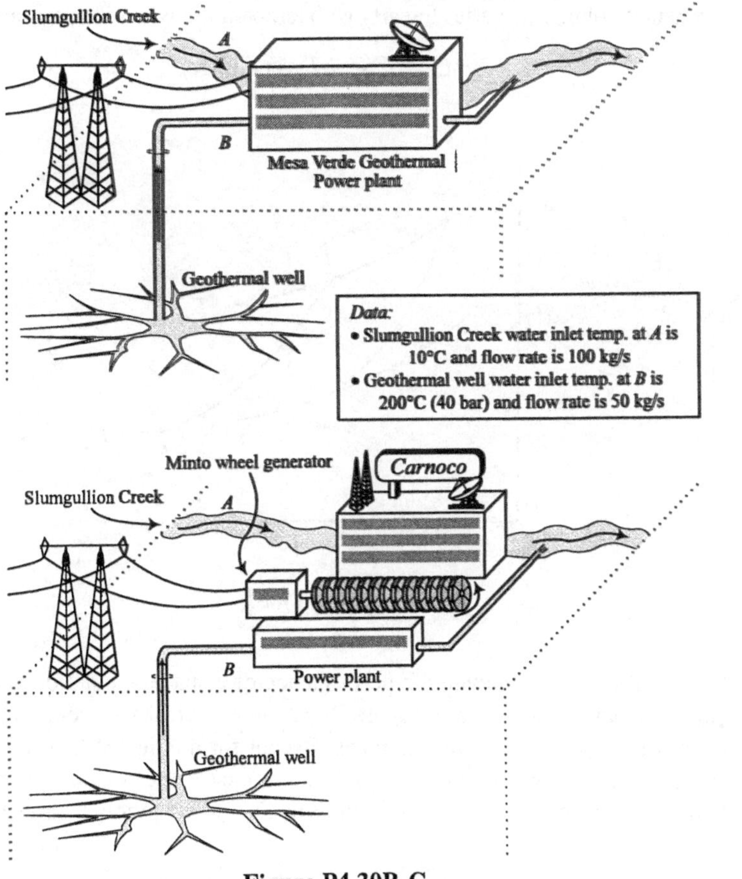

Figure P4.30B-C

pre-determined temperature. CED also claims that they need to use the full flow of Slumgullion Creek to achieve the best efficiency.

(a) What is the maximum power possible?

(b) Is CED's claim regarding performance realistic? Are there any losses? If so, describe them.

(c) If flows are countercurrent versus cocurrent for the Carnoco design, would the maximum power output change? Explain.

You can assume that the heat capacities at constant pressure (C_p) for both creek water and geothermal well water are constant at 4200 J/kg K and 4000 J/kg K, respectively. State and justify all additional assumptions made in your analysis.

4.31. Rocky and Rochelle Jones are having one of their frequent thermodynamics arguments. Rocky claims that Carnot efficiency η_{Carnot} will *only* equal $(T_A - T_B)/T_A$ if an ideal gas of constant heat capacity is used in a Carnot cycle as depicted in the adjacent *P*-*V* diagram. Rochelle claims that it does not matter if the heat capacity varies with temperature; η_{Carnot} will always equal $(T_A - T_B)/T_A$. You have been asked to settle the argument by analyzing the Carnot cycle behavior of an ideal gas (see Figure P4.31) whose heat capacity at constant volume C_v varies linearly with temperature with C_v^o and b empirical constants:

$$C_v = C_v^o + bT \quad (T \text{ in K})$$

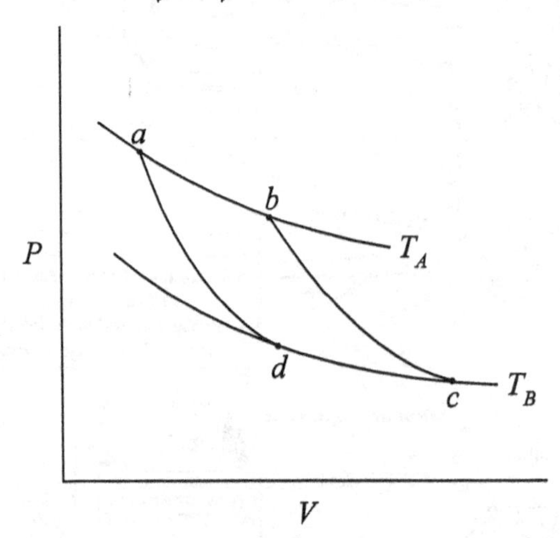

Figure P4.31

4.32. NASA is trying to evaluate the performance of a space simulation chamber and needs your assistance. As shown in Figure P4.32, an electrically driven, rotary vacuum pump is used to evacuate a 10 m^3 chamber. Except for a constant heat leak ($\dot{q}_{leak} = 10$ J/s) caused by the electronic gear kept inside the simulator, the system is adiabatic. In addition, a small air leak exists ($\dot{N}_{leak} = 10^{-7}$ mol/s at 27°C). The pump's volumetric displacement rate is 10^{-3} m^3/s at an efficiency of 80% relative to reversible adiabatic operation. The initial chamber temperature is 27°C and its initial pressure and ambient pressure are both 1.01 bar = 1.01×10^5 Pa. Air can be assumed to behave as an ideal gas of constant heat capacity (C_p = 29 J/mol K).

(a) Describe how you would calculate the chamber pressure and temperature as a function of time.

(b) Describe how you would estimate the minimum vacuum pump power rating required to pump the chamber pressure down to 0.1 bar.

(c) Estimate answers for (a) and (b) above using a suitable numerical solution procedure.

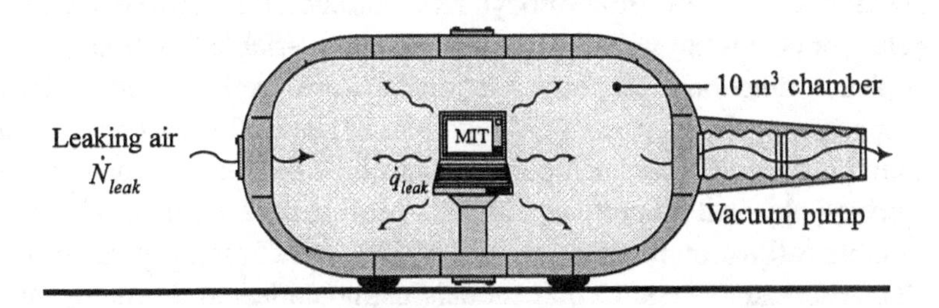

Figure P4.32 Space Simulation Chamber.

The Calculus of Thermodynamics 5

In this chapter we develop functional property relationships for simple systems. These relationships are completely general and are a direct consequence of applying the postulates and laws of thermodynamics discussed in Chapters 2 – 4. They also reflect behavior consistent with the calculus of multivariable functions.

These functional property relationships are fundamentally different than physical property models, such as $P\underline{V}TN$ equations of state, which represent practical constitutive equations specific for each pure substance or mixture composition. In order to obtain working physical property equations, experiments are performed and data correlated or suitable estimation methods are utilized to predict property values (see Chapters 8 and 13). Alternatively, molecular models using statistical thermodynamics may be used, albeit with some approximations (see Chapter 10).

A major objective of this chapter is to introduce techniques for recasting derived properties and their derivatives into forms that can be directly evaluated in terms of primitive properties which are measurable. Our starting point is to introduce a fundamental operating equation for simple systems that establishes the functional dependence of internal energy $\underline{U}$ on the variables entropy $\underline{S}$, volume $\underline{V}$, and moles (or masses) of all components present $N_1,..., N_n$. This *Fundamental Equation* has its origin from the combined first and second laws for open simple systems that was derived in Section 4.7. The $\underline{S}$, $\underline{V}$, $N_1,..., N_n$ coordinates of that equation were used extensively by J.W. Gibbs in his classic work on thermodynamics in the late 1800's, and as a result they are commonly referred to as *Gibbs coordinates*. They represent a set of $n + 2$ natural or canonical variables that provide the necessary and sufficient information required to describe the thermodynamic state of a simple system.

By employing methods from calculus and geometry, Gibbs was able to show, in a completely rigorous mathematical manner, how various thermodynamic properties were related. He clearly laid the groundwork for understanding the functional inter-relationships that exist among the so-called potential functions, $\underline{U}$, $\underline{H}$, $\underline{A}$, and $\underline{G}$. Without the computational power and graphical display methods that we have today, Gibbs had to conceptualize the structure of these property inter-relationships in his mind without being able to illustrate them graphically. He used only narratives with equations in his writings to describe his findings. In part, this restriction has limited how fast we have been able to understand the details of Gibbs' work. It has taken over a century of study to fully appreciate the significance of his contributions. Interested readers should consult

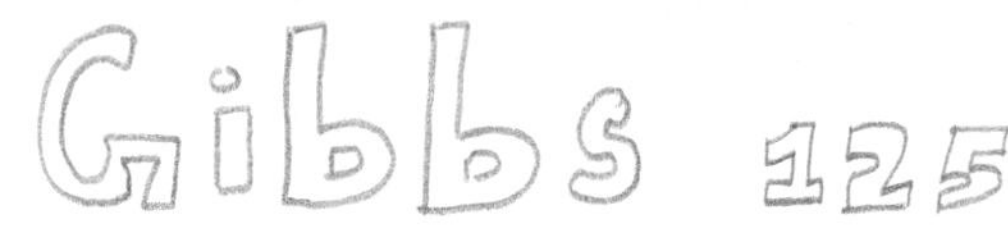

the papers by K. Jolls and his co-workers listed at the end of the chapter for a comprehensive discussion of Gibbs' contributions in this area. In the sections that follow, we will employ graphical methods to illustrate some of the geometric aspects of the Fundamental Equation and its transformations. In addition, several methods of manipulating functions and partial derivatives will be described and applied to problems frequently encountered in classical thermodynamics.

5.1 The Fundamental Equation in Gibbs Coordinates

As will be seen presently, the relationship

$$\underline{U} = f_U(\underline{S}, \underline{V}, N_1, \ldots, N_n) \tag{5-1}$$

completely describes all of the stable equilibrium states of a simple system containing n components. By solving Eq. (5-1) explicitly for $\underline{S}$, an alternative form is obtained:

$$\underline{S} = f_S(\underline{U}, \underline{V}, N_1, \ldots, N_n) \tag{5-2}$$

Either relationship is called the ***Fundamental Equation***: Eq. (5-1) is termed the *energy representation* and Eq. (5-2) the *entropy representation.*

The Fundamental Equation is represented by a (hyper) surface in $(n + 3)$ dimensional space. The points on this surface represent stable equilibrium states of the simple system. Quasi-static processes can be represented by a curve on this surface. Processes that are not quasi-static are not identified with points on this surface. (Recall that derived properties such as $\underline{U}$ and $\underline{S}$ are not defined for non-equilibrium states.)

The $n + 2$ first-order partial derivatives of the Fundamental Equation correspond to traces in the respective coordinate walls of *planes* tangent to the thermodynamic surface[1]. The significance of these tangent plane traces can be seen by expressing the Fundamental Equation in differential form. For the energy representation, Eq. (5-1),

$$d\underline{U} = \left(\frac{\partial f_U}{\partial \underline{S}}\right)_{\underline{V},N} d\underline{S} + \left(\frac{\partial f_U}{\partial \underline{V}}\right)_{\underline{S},N} d\underline{V} + \sum_{i=1}^{n} \left(\frac{\partial f_U}{\partial N_i}\right)_{\underline{S},\underline{V},N_j[i]} dN_i \tag{5-3}$$

where subscript $N_j[i]$ indicates that all N_j except $j = i$ are held constant in the differentiation. If we compare Eq. (5-3) to the combined law for a simple system (see Section 4.7), namely

$$d\underline{U} = T\,d\underline{S} - P\,d\underline{V} + \sum_{i=1}^{n} \mu_i\,dN_i \tag{5-4}$$

it is clear that

1 The term *plane* is used in the generic sense here. Even for a pure fluid there are three independent variables and three partial derivatives. Thus the tangent *plane* is really a three-dimensional hyperplane.

$$\left(\frac{\partial f_U}{\partial \underline{S}}\right)_{\underline{V},N} = T = g_T(\underline{S}, \underline{V}, N_1,\ldots, N_n) \quad (5\text{-}5)$$

$$-\left(\frac{\partial f_U}{\partial \underline{V}}\right)_{\underline{S},N} = P = g_P(\underline{S}, \underline{V}, N_1,\ldots, N_n) \quad (5\text{-}6)$$

and

$$\left(\frac{\partial f_U}{\partial N_i}\right)_{\underline{S},\underline{V}, N_j[i]} = \mu_i = g_i(\underline{S}, \underline{V}, N_1,\ldots, N_n) \quad (5\text{-}7)$$

where the functions g_T, g_P, and g_i could be obtained directly from the Fundamental Equation if it were available. Equations (5-5) through (5-7) represent a particular set of *equations of state* expressed in Gibbs $\underline{S}\,\underline{V}N$ coordinates. As shown later in Sections 5.2 and 5.3, only two of these three equations of state for a pure substance are independent.

The second-order partial derivatives of the Fundamental Equation are also related to quantities that can be measured experimentally. For example, for a *pure material*, there are four second-order partial derivatives at constant mass or moles:

$$\frac{\partial^2 \underline{U}}{\partial \underline{S}^2} = \frac{\partial}{\partial \underline{S}}\left[\left(\frac{\partial \underline{U}}{\partial \underline{S}}\right)_{\underline{V},N}\right]_{\underline{V},N} = \left(\frac{\partial T}{\partial \underline{S}}\right)_{\underline{V},N} \quad (5\text{-}8)$$

$$\frac{\partial^2 \underline{U}}{\partial \underline{V}^2} = \frac{\partial}{\partial \underline{V}}\left[\left(\frac{\partial \underline{U}}{\partial \underline{V}}\right)_{\underline{S},N}\right]_{\underline{S},N} = -\left(\frac{\partial P}{\partial \underline{V}}\right)_{\underline{S},N} \quad (5\text{-}9)$$

$$\frac{\partial^2 \underline{U}}{\partial \underline{S}\,\partial \underline{V}} = \frac{\partial}{\partial \underline{S}}\left[\left(\frac{\partial \underline{U}}{\partial \underline{V}}\right)_{\underline{S},N}\right]_{\underline{V},N} = -\left(\frac{\partial P}{\partial \underline{S}}\right)_{\underline{V},N} \quad (5\text{-}10)$$

$$\frac{\partial^2 \underline{U}}{\partial \underline{V}\,\partial \underline{S}} = \frac{\partial}{\partial \underline{V}}\left[\left(\frac{\partial \underline{U}}{\partial \underline{S}}\right)_{\underline{V},N}\right]_{\underline{S},N} = \left(\frac{\partial T}{\partial \underline{V}}\right)_{\underline{S},N} \quad (5\text{-}11)$$

Of these four derivatives, only three are independent because the last two are related by the reciprocity theorem of Maxwell (see Section 5.3, part 4).

For an *n-component mixture* there are $n + 1$ first-order and $(n + 2)(n + 1)/2$ second-order *independent* partial derivatives of the Fundamental Equation given by Eq. (5-1) or Eq. (5-2). These derivatives are particularly important because they form a basis for all other partial derivatives involving thermodynamic properties. That is, any partial derivative can be expressed in terms of an independent set of first-order and second-order derivatives of any form of the Fundamental Equation. Proof of this statement for a pure material is given in Section 5.3.

At this point, it is interesting to consider what the shape of the internal energy surface would look like in $\underline{S}\,\underline{V}N$ coordinates. Since even for a pure material this would entail a four-dimensional plot, one is forced to reduce the dimensionality of the system in order to visualize the surface in three dimensions. For a pure system, one can propose a 3-D plot of intensive U as a function of S and V. In fact, James Clerk Maxwell in the late 1800's was so intrigued with Gibbs' work that he actually made a 3-D model of the USV surface for water. He sent a plaster replica of his model to Gibbs at Yale, where it remains on display today. Figure 5.1a is a photograph of a more recent 3-D USV surface for water constructed by Clark and Katz (1939) while Figure 5.1b is a 2-D projection of the surface showing several important features. Note, for example, the point of triple tangency that corresponds to the three-phase, solid-liquid-gas, triple point equilibrium for water. We will have much more to say about such geometrical aspects and their relationship to the characterization and limits of stable equilibrium states in Section 5.7. With more accurate thermodynamic data available and computational tools for correlating and manipulating these data, it is possible to generate USV surfaces as computer-generated, 3-D plots for a wide range of pure materials.

Jolls (1990), Coy (1993) and coworkers (1991, 1992) at Iowa State have done just this for a number of pure compounds using a well-tested $P\underline{V}TN$ equation of state developed by Peng and Robinson (see Section 8.4) and a temperature-dependent correlation for the heat capacity in an ideal gas state. The actual methods for generating values of U for specified values of S and V are straightforward and described in detail in Sections 8.1 and 8.2 [see also Coy (1993)]. Figure 5.2 shows two fundamental USV surfaces—one for an ideal gas and the other generated by Jolls and co-workers for ethylene using the Peng-Robinson equation of state as a model. It is important to keep in mind that the values of S, V and U are not absolute, they depend on the specification of reference state conditions, S^o and U^o.

Example 5.1 actually derives the governing equation for U for a monatomic ideal gas that is plotted in Figure 5.2a. The USV surface is convex with respect to variations in S as the second derivative is always positive for the exponential dependence of U on S shown in Figure 5.2a. Later we will invoke even stronger arguments that this convexity must be true in general. Although the actual functional dependence of U on V will vary from substance to substance, the general power law decrease depicted in Figure 5.2a will be followed for all fluids in the dilute gas region where nearly ideal behavior is expected.

The important thing to remember is that the graphical representation of the USV surface encodes all the thermodynamic information about a pure substance as it is equivalent to the Fundamental Equation itself. The geometry of the surface represents all single phase stable equilibrium states, all two-phase coexistence points, and all triple points; it also can be used to identify limits of stability such as critical and spinodal points (see Chapter 7). A major difficulty, of course, is that a significant amount of physical property information is required to construct the Gibbs USV surface for a pure substance, e.g., the steam tables for water, or a robust $P\underline{V}TN$ equation of state that models the behavior of a compound adequately in all states of aggregation.

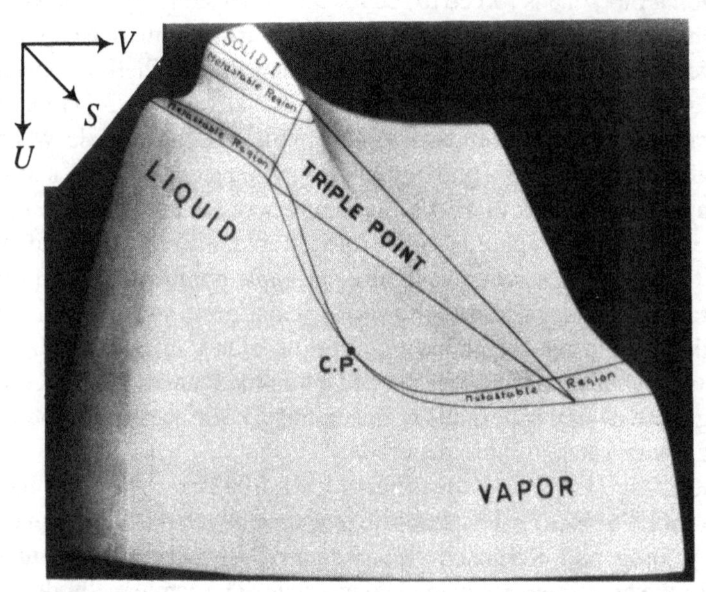

Figure 5.1a Three-dimensional model of the *USV* surface for pure water [from Clark and Katz (1939), with permission of the Royal Society of Canada].

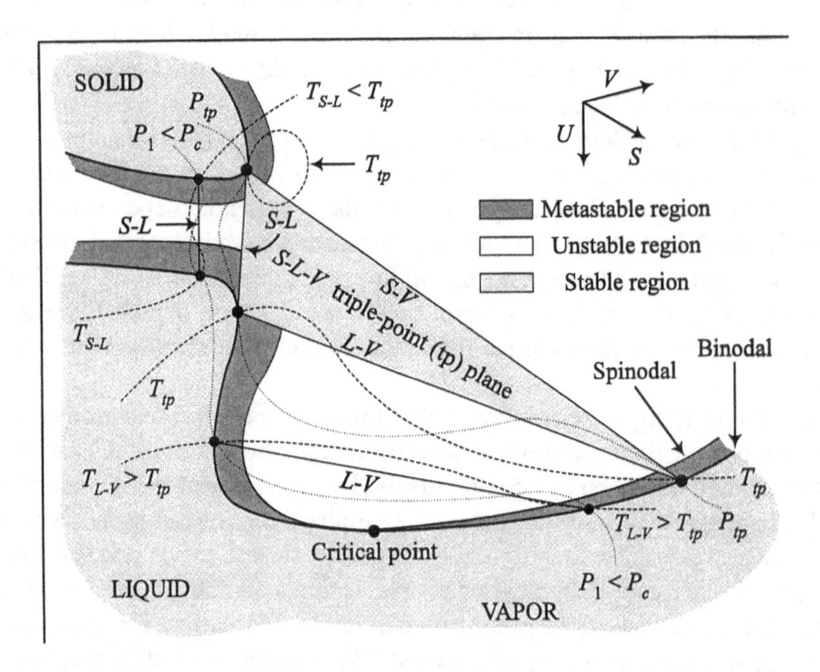

Figure 5.1b A two-dimensional projection of the *USV* surface for pure water shown in Figure 5.1a.

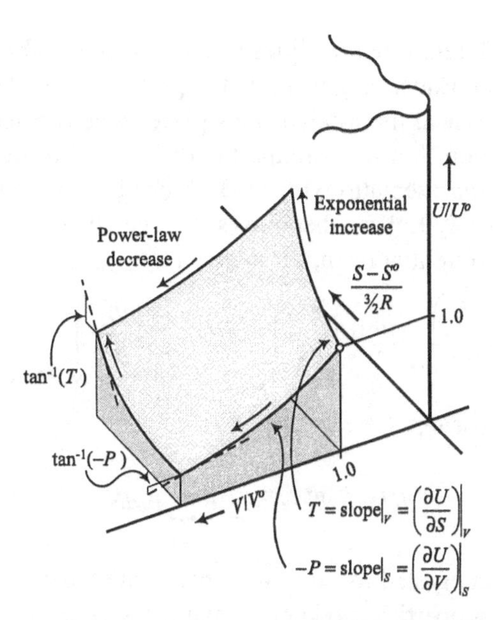

Figure 5.2a The *USV* surface from the Fundamental Equation for a pure monatomic ideal gas referenced to U^o, S^o, and V^o (see Example 5.1).

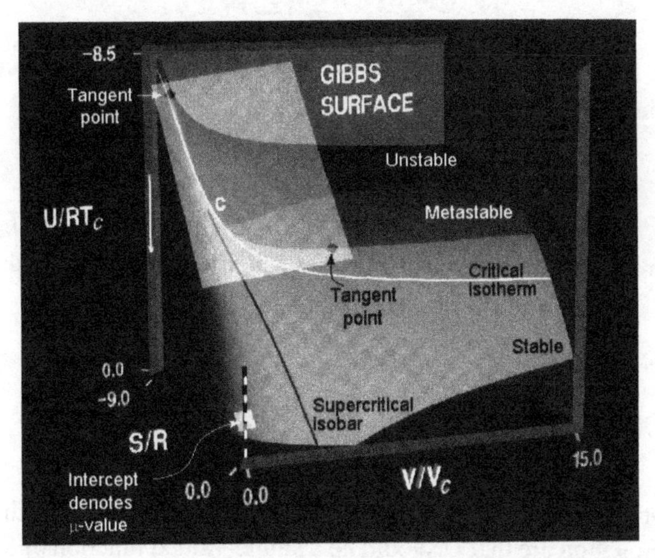

Figure 5.2b A predicted *USV* surface for the fluid phases of pure ethylene. Volumetric *PVT* properties from the Peng-Robinson equation of state and heat capacities from a fitted ideal gas state equation. ($V_{max} = 15\ V_c$) [adapted from Jolls and Coy (1992)].

It should be clear that Fundamental Equations would be of great use if they were generally available. The problem is that the complete form of the Fundamental Equation is not specified by classical thermodynamics; each substance has its own peculiarities that are reflected in different functionalities of the Fundamental Equation. Thus, there is no single Fundamental Equation governing the properties of all materials.

The postulates of classical thermodynamics place some restrictions on the form of the Fundamental Equation. Let us examine the differential form of the Fundamental Equation in the energy representation, Eq. (5-3). Since $\underline{U}$ must be first order in mass or mole number, we can apply Euler's theorem (see Appendix C) to obtain the integrated form of Eq. (5-3). In differential form, $d\underline{U}$ is given by:

$$d\underline{U} = \left(\frac{\partial f_U}{\partial \underline{S}}\right)_{\underline{V},N} d\underline{S} + \left(\frac{\partial f_U}{\partial \underline{V}}\right)_{\underline{S},N} d\underline{V} + \sum_{i=1}^{n} \left(\frac{\partial f_U}{\partial N_i}\right)_{\underline{S},\underline{V},N_j[i]} dN_i \quad (5\text{-}12)$$

or as we demonstrated earlier

$$d\underline{U} = Td\underline{S} - Pd\underline{V} + \sum_{i=1}^{n} \mu_i dN_i \quad (5\text{-}13)$$

which can be directly integrated using Euler's theorem recognizing, of course, that T, P, and μ_i are intensive properties and kept constant. The result is

$$\underline{U} = T\underline{S} - P\underline{V} + \sum_{i=1}^{n} \mu_i N_i \quad (5\text{-}14)$$

Equation (5-12) is a linear partial differential equation of the first order. Therefore, the solution must be of the form

$$\underline{U} = x\left[g\left(\frac{y}{x}, \frac{z}{x}, \ldots\right)\right] \quad (5\text{-}15)$$

where $x, y, z, \ldots$ can be $\underline{S}, \underline{V}, N_1, \ldots, N_n$ or any permutation of these variables. For a one-component system, it is often convenient to choose $x = N$, $y = \underline{S}$, and $z = \underline{V}$; we then obtain

$$\underline{U} = N\left[g\left(\frac{\underline{S}}{N}, \frac{\underline{V}}{N}\right)\right] \quad (5\text{-}16)$$

or, since $\underline{U} = NU$,

$$U = g(S, V) \quad (5\text{-}17)$$

The only other requirements our prior developments placed on the form of the Fundamental Equation are that $\underline{U}$ should be a single-valued function of $\underline{S}$, $\underline{V}$, and N (see Postulate I), and that, $(\partial f_U/\partial \underline{S})_{\underline{V},N} = T$ should be nonnegative.

5.2 Intensive and Extensive Properties

At this point, let us digress to a special case of Postulate I. In Eq. (5-17), we see that for a one-component system, only two properties, S and V, are required to obtain the specific energy of a system. This is in no way a violation of Postulate I; by delineating the *specific* properties (expressed in terms of unit mass or mole number), we have determined the "intensity" of the system but not the "extent" of the system. To specify the system completely (e.g., so that it can be reproduced by others), we must specify the mass of the system in addition to S and V.

The variables that express intensity of the system are zero order in mass and are called *intensive variables.* Variables that relate to extent of the system are first order in mass and are called *extensive variables.*

We now prove that for a single-phase system of n components, any intensive property can be defined by the values of $n + 1$ other intensive properties.[2] Let us call $b, c_1, c_2, \ldots, c_{n+1}$ *intensive* properties of a single-phase simple system containing n components. In general, we can express b as a function of $n + 2$ other properties according to Postulate I. Let us choose these $n + 2$ as $c_1, c_2, \ldots, c_{n+1}$, and the total moles (or mass) N. Thus,

$$db = \left(\frac{\partial b}{\partial c_1}\right)_{c_j[1], N} dc_1 + \ldots + \left(\frac{\partial b}{\partial c_{n+1}}\right)_{c_j[n+1], N} dc_{n+1} + \left(\frac{\partial b}{\partial N}\right)_{c_1, \ldots, c_{n+1}} dN \quad (5\text{-}18)$$

Integrating Eq. (5-18) by using Euler's theorem (see Appendix C), we have

$$\left(\frac{\partial b}{\partial N}\right)_{c_1, \ldots, c_{n+1}} N = 0 \quad (5\text{-}19)$$

Since N can be nonzero, $(\partial b/\partial N)_{c_1, \ldots, c_{n+1}}$ must be zero. Therefore, Eq. (5-18) reduces to a function of $n + 1$ intensive variables. Of course, these $n + 1$ intensive variables must be independent, so that we clearly cannot use all of the n mole fractions $x_1, \ldots, x_n$. We could, however, use $n - 1$ mole fractions in addition to two other intensive variables to obtain the required $n + 1$.

Note that this result is valid because we limited the original set of $n + 2$ variables to include only one extensive variable; if we had included two extensive variables in the original $n + 2$ set, no partial derivative would have to be zero. Thus, we could state as a corollary to Postulate I:

For a single-phase simple system, the change of any intensive variable can be expressed as a function of any $n + 1$ other independent intensive variables.

We shall use this corollary frequently in Chapter 9 in dealing with the properties of mixtures. At this point, a word of caution is in order when dealing with intensive and extensive variables in partial derivatives. For example, if for a pure material we express $\underline{U}$ as a function of $\underline{S}$ and $\underline{V}$, then using Eqs. (5-18) and (5-19), we find $(\partial \underline{U}/\partial N)_{\underline{S}, \underline{V}} = 0$. But $(\partial \underline{U}/\partial N)_{\underline{S}, \underline{V}}$ is not zero; from Eq. (5-7) applied to a pure material,

2 The proof is restricted to a single-phase system to allow us to choose any $n + 2$ variables as an independent set. The proof can be extended to specific cases of composite simple systems provided that the $n + 2$ variables chosen form an independently variable set.

$$\left(\frac{\partial \underline{U}}{\partial N}\right)_{\underline{S},\underline{V}} = \mu \tag{5-20}$$

Since $\underline{U} = NU$, we also have

$$\left(\frac{\partial \underline{U}}{\partial N}\right)_{\underline{S},\underline{V}} = U + N\left(\frac{\partial U}{\partial N}\right)_{\underline{S},\underline{V}} = \mu \tag{5-21}$$

Since U is not equal to μ, $(\partial U/\partial N)_{\underline{S},\underline{V}}$ is not equal to zero. Thus, each of the three derivatives, $(\partial U/\partial N)_{S,V}$, $(\partial \underline{U}/\partial N)_{\underline{S},\underline{V}}$, and $(\partial U/\partial N)_{\underline{S},\underline{V}}$ have different connotations. The first represents the change in the specific energy as we add more material while maintaining constant specific entropy and specific volume. Since we are holding two intensive variables constant during the process, all other intensive variables (e.g., T, P, etc.) for the pure material must remain unchanged. The only way to conduct the process is to enlarge the system in direct proportion to the added mass. The second and third cases, however, represent changes in the total and specific energy during a process in which we maintain constant total entropy and total volume. Since we are adding mass to the system, the only way to keep total entropy and total volume constant is to change the specific entropy and specific volume (e.g., by varying T and P during the addition of mass). Thus, the specific energy changes as the state of the system is varied. The total energy changes because both the specific energy and mass vary.

Example 5.1

In the entropy representation, the Fundamental Equation for a monatomic ideal gas (such as He, Ne, Kr, or Ar at low pressure) is

$$\underline{S} = N\left[S^o + R\ln\left\{\left(\frac{U}{U^o}\right)^{3/2}\frac{V}{V^o}\right\}\right] \tag{5-22}$$

where S^o, U^o, and V^o are constants representing values in a reference or base state. From the energy representation in the form of Eq. (5-1), determine the three equations of state in the form of Eqs. (5-5) through (5-7).

Solution

Solving Eq. (5-22) explicitly for U yields

$$U = U^o\left(\frac{V^o}{V}\right)^{2/3} \exp\left(\tfrac{2}{3}(S - S^o)/R\right) \tag{5-23}$$

and, thus,

$$\underline{U} = NU^o\left(\frac{V^o}{V}\right)^{2/3} \exp\left(\tfrac{2}{3}(S - S^o)/R\right) \tag{5-24}$$

Using Eq. (5-24), $U = \underline{U}/N$ was plotted in Figure 5.2a as a function of S and V. The equations of state can be found directly by partial differentiation of Eq. (5-23) or (5-24):

$$T = \left(\frac{\partial \underline{U}}{\partial \underline{S}}\right)_{\underline{V},N} = \left(\frac{\partial U}{\partial S}\right)_V = \frac{2}{3}\left(\frac{U^o}{R}\right)\left(\frac{V^o}{V}\right)^{2/3} \exp\left(\tfrac{2}{3}(S-S^o)/R\right) \tag{5-25}$$

$$-P = \left(\frac{\partial \underline{U}}{\partial \underline{V}}\right)_{\underline{S},N} = \left(\frac{\partial U}{\partial V}\right)_S = -\frac{2}{3}\frac{U^o(V^o)^{2/3}}{V^{5/3}} \exp\left(\tfrac{2}{3}(S-S^o)/R\right) \tag{5-26}$$

$$\mu = \left(\frac{\partial \underline{U}}{\partial N}\right)_{\underline{S},\underline{V}} = \frac{\partial}{\partial N}\left[NU^o\left(\frac{V^o}{V}\right)^{2/3} \exp\left(\frac{2}{3}\frac{(S-S^o)}{R}\right)\right]_{\underline{S},\underline{V}}$$

$$= U^o\left(\frac{V^o}{V}\right)^{2/3} \exp\left(\frac{2}{3}\frac{(S-S^o)}{R}\right)\left(\frac{5}{3}-\frac{2}{3}\frac{S}{R}\right) \tag{5-27}$$

The results given in Eqs. (5-25) through (5-27) may be simplified if Eq. (5-22) is used to evaluate the exponential terms. If this is done, then

$$T = \frac{2U}{3R}\ ;\ P = \frac{2U}{3V}\ ;\ \text{and}\ \mu = U\left(\frac{5}{3}-\frac{2}{3}\frac{S}{R}\right) \tag{5-28}$$

In fact, the assumptions behind the development of Eq. (5-22) are that $U = \frac{3}{2}RT$ and $PV = RT$.

5.3 Methods for Transforming Derivatives

Often one is faced with the problem of evaluating the magnitude of a particular partial derivative or integral involving non-measurable, derived thermodynamic properties. For example, suppose we needed a specific value for the entropy change ΔS of a pure fluid between two well-defined states (T_1, P_1) and (T_2, P_2). The calculus of continuous functions in two variables works to specify the intensive state for a pure component system with $n + 1 = 2$ degrees of freedom. Thus,

$$\Delta S = \int\left(\frac{\partial S}{\partial P}\right)_T dP + \int\left(\frac{\partial S}{\partial T}\right)_P dT \tag{5-29}$$

The second integral involves $(\partial S/\partial T)_P$ which equals C_p/T. Presumably, values of C_p may be tabulated for the substance of interest at particular temperatures and pressures. But what about the first integral that requires evaluation of $(\partial S/\partial P)_T$ which is not readily available? We need some convenient way to express this partial derivative in terms of measured or tabulated properties.

Again, by adhering to the calculus of continuous functions, we have several tools for manipulating these derivatives. Some of the more important ones for applications in thermodynamics are listed below. For a general function $F(x, y)$ of two variables:

1. ***Derivative inversion***

$$\left(\frac{\partial F}{\partial y}\right)_x = \frac{1}{(\partial y/\partial F)_x} \tag{5-30}$$

For example,

$$\left(\frac{\partial S}{\partial P}\right)_T = \frac{1}{(\partial P/\partial S)_T} \tag{5-31}$$

2. ***Triple product*** (xyz−1 rule)

$$\left(\frac{\partial F}{\partial x}\right)_y \left(\frac{\partial x}{\partial y}\right)_F \left(\frac{\partial y}{\partial F}\right)_x = -1 \tag{5-32}$$

For example,

$$\left(\frac{\partial H}{\partial T}\right)_P \left(\frac{\partial T}{\partial P}\right)_H \left(\frac{\partial P}{\partial H}\right)_T = -1 \tag{5-33}$$

3. ***Chain rule expansion to add another independent variable Φ***

$$\left(\frac{\partial F}{\partial y}\right)_x = \frac{(\partial F/\partial \Phi)_x}{(\partial y/\partial \Phi)_x} = \left(\frac{\partial F}{\partial \Phi}\right)_x \left(\frac{\partial \Phi}{\partial y}\right)_x \tag{5-34}$$

For example, set $\Phi = T$

$$\left(\frac{\partial S}{\partial H}\right)_P = \frac{(\partial S/\partial T)_P}{(\partial H/\partial T)_P} = \frac{C_p/T}{C_p} = \frac{1}{T} \tag{5-35}$$

4. ***Maxwell reciprocity relationship***

Maxwell's reciprocity theorem states that the value of 2nd or higher order derivatives is independent of the order of the differentiation for a smoothly varying, continuous function $F(x, y)$

$$\left(\frac{\partial(\partial F/\partial x)_y}{\partial y}\right)_x = \left(\frac{\partial(\partial F/\partial y)_x}{\partial x}\right)_y \tag{5-36}$$

or representing these 2nd order derivatives in abbreviated form:

$$F_{xy} = F_{yx} \tag{5-37}$$

The relationships cited above in (1) – (4) can easily be extended to larger functionalities, for example where $n + 2$ variables exist such as in the Fundamental Equation.

Our original problem of evaluating the integral containing $(\partial S/\partial P)_T$ can be made tractable using a Maxwell relation in the appropriate thermodynamic variables. Later in this chapter we will establish methods to select the appropriate variables for these transformations, but for now, let's employ the Gibbs free energy function, G. For a pure system, G is written in intensive form as a function of T and P, $G = f(T,P)$, and

$$dG = -SdT + VdP \tag{5-38}$$

where

$$-S = \left(\frac{\partial G}{\partial T}\right)_P \quad \text{and} \quad V = \left(\frac{\partial G}{\partial P}\right)_T$$

Thus,

$$G_{TP} = \left(\frac{\partial(\partial G/\partial T)_P}{\partial P}\right)_T = \left(\frac{\partial(\partial G/\partial P)_T}{\partial T}\right)_P = G_{PT} \tag{5-39}$$

or

$$-\left(\frac{\partial S}{\partial P}\right)_T = \left(\frac{\partial V}{\partial T}\right)_P \tag{5-40}$$

so we can use *PVT* properties or a fitted *PVT* equation of state to determine $(\partial S/\partial P)_T$ and then evaluate the integral directly. Earlier in Section 5.1, we stated that the 2nd derivatives of $\underline{U}$ with respect to $\underline{S}$ and $\underline{V}$ [see Eqs. (5-10) and (5-11)] were related by Maxwell's reciprocity theorem. Now we can show that these 2nd derivatives are indeed equal to one another.

$$\underline{U}_{\underline{V}\,\underline{S}} \equiv \frac{\partial^2 \underline{U}}{\partial \underline{S}\partial \underline{V}} = \frac{\partial^2 \underline{U}}{\partial \underline{V}\partial \underline{S}} \equiv \underline{U}_{\underline{S}\,\underline{V}} \tag{5-41}$$

where we have used the abbreviations introduced earlier for the 2nd derivative. By taking the reciprocal of Eq. (5-40), we immediately see that

$$\underline{U}_{\underline{V}\,\underline{S}} = -\left(\frac{\partial P}{\partial \underline{S}}\right)_{\underline{V}, N} = \left(\frac{\partial T}{\partial \underline{V}}\right)_{\underline{S}, N} = \underline{U}_{\underline{S}\,\underline{V}} \tag{5-42}$$

recognizing, of course, that we have returned to an extensive form for $\underline{U}$.

In addition to the derivative manipulations characterized above, it is also useful to illustrate some of the different properties of intensive and extensive variables. A partial derivative may involve intensive and extensive variables. For a pure material, only $n + 1 = 2$ intensive variables are independent; hence, a partial derivative involving only intensive variables b, c, and d can be expressed as $(\partial b/\partial c)_d$, where it is implied that N is constant. That is,

$$\left(\frac{\partial b}{\partial c}\right)_{d, N} = \left(\frac{\partial b}{\partial c}\right)_d \tag{5-43}$$

We now show that for a pure material, any partial derivative involving extensive variables can always be reduced to expressions involving partial derivatives of entirely intensive variables.

Now consider the derivative $(\partial b/\partial c)_{d,\,e}$ where one of the four variables is extensive.

1(i). If c is extensive, then

$$\left(\frac{\partial b}{\partial \underline{c}}\right)_{d,\, e} = 0 \tag{5-44}$$

The proof follows from applying Euler's theorem to $b = f(\underline{c}, d, e)$:

$$0 = \left(\frac{\partial b}{\partial \underline{c}}\right)_{d,e} \underline{c}$$

1(ii). If b is extensive, then

$$\left(\frac{\partial \underline{b}}{\partial c}\right)_{d,e} = \frac{1}{(\partial c/\partial \underline{b})_{d,e}} = \frac{1}{0} = \infty \tag{5-45}$$

1(iii). If d (or e) is extensive, then

$$\left(\frac{\partial b}{\partial c}\right)_{\underline{d},e} = \left(\frac{\partial b}{\partial c}\right)_{e} \tag{5-46}$$

Since

$$\left(\frac{\partial b}{\partial c}\right)_{\underline{d},e} = -\frac{(\partial \underline{d}/\partial c)_{b,e}}{(\partial \underline{d}/\partial b)_{c,e}} = -\frac{(\partial \underline{d}/\partial N)_{b,e}(\partial N/\partial c)_{b,e}}{(\partial \underline{d}/\partial N)_{c,e}(\partial N/\partial b)_{c,e}}$$

and

$$\left(\frac{\partial \underline{d}}{\partial N}\right)_{b,e} = \left(\frac{\partial \underline{d}}{\partial N}\right)_{c,e} = d$$

then

$$\left(\frac{\partial b}{\partial c}\right)_{\underline{d},e} = -\frac{(\partial N/\partial c)_{b,e}}{(\partial N/\partial b)_{c,e}} = \left(\frac{\partial b}{\partial c}\right)_{e,N} = \left(\frac{\partial b}{\partial c}\right)_{e}$$

Thus, if only one variable is extensive, the partial derivative is finite and nonzero only if d or e is extensive. The extensive variable may be deleted to yield a partial involving only intensive variables.

Now consider the three cases in which two of the four variables are extensive.

2(i). If b and c are extensive, then

$$\left(\frac{\partial \underline{b}}{\partial \underline{c}}\right)_{d,e} = \frac{b}{c} \tag{5-47}$$

which follows directly by applying Euler's theorem to $\underline{b} = f(\underline{c}, d, e)$:

$$\underline{b} = \left(\frac{\partial \underline{b}}{\partial \underline{c}}\right)_{d,e} \underline{c} \quad \text{and} \quad \underline{b} = Nb, \quad \underline{c} = Nc$$

2(ii). If b and d (or e) are extensive, then

$$\left(\frac{\partial \underline{b}}{\partial c}\right)_{\underline{d},e} = N\left[\left(\frac{\partial b}{\partial c}\right)_{e} - \frac{b}{d}\left(\frac{\partial d}{\partial c}\right)_{e}\right] \tag{5-48}$$

Since, expanding $\underline{b} = Nb$,

$$\left(\frac{\partial \underline{b}}{\partial c}\right)_{\underline{d},\, e} = N\left(\frac{\partial b}{\partial c}\right)_{\underline{d},\, e} + b\left(\frac{\partial N}{\partial c}\right)_{\underline{d},\, e} \tag{5-49}$$

Eq. (5-46) may be used to reduce $(\partial b/\partial c)_{\underline{d},\, e}$. The last term is reduced as follows:

$$\left(\frac{\partial N}{\partial c}\right)_{\underline{d},\, e} = \frac{-(\partial \underline{d}/\partial c)_{N,\, e}}{(\partial \underline{d}/\partial N)_{c,\, e}} = -\frac{N(\partial d/\partial c)_e}{d}$$

2(iii). If d and e are extensive, then

$$\left(\frac{\partial b}{\partial c}\right)_{\underline{d},\, \underline{e}} = -\frac{(\partial \underline{d}/\partial c)_{b,\, \underline{e}}}{(\partial \underline{d}/\partial b)_{c,\, \underline{e}}} \tag{5-50}$$

The numerator and denominator can each be reduced by applying the results of case 2(ii).

Any partial derivative involving three extensive variables can now be reduced to partials involving two extensive variables by using Eq. (5-49) followed by one or more of the steps illustrated above. Similarly, partials involving four extensive variables can be reduced to three, etc. The net result is that *any partial derivative for a pure material can be expressed in terms of partials involving only three intensive variables.*

5.4 Jacobian Transformations

A useful technique for manipulating thermodynamic properties and their derivatives involves a transformation using Jacobians or functional determinants. For the derivatives of functions of 2 variables, $f(x, y)$ and $g(x, y)$, the ***Jacobian*** is defined as:

$$\frac{\partial(f, g)}{\partial(x, y)} \equiv \begin{vmatrix} \left(\frac{\partial f}{\partial x}\right)_y & \left(\frac{\partial f}{\partial y}\right)_x \\ \left(\frac{\partial g}{\partial x}\right)_y & \left(\frac{\partial g}{\partial y}\right)_x \end{vmatrix} = \left(\frac{\partial f}{\partial x}\right)_y\left(\frac{\partial g}{\partial y}\right)_x - \left(\frac{\partial f}{\partial y}\right)_x\left(\frac{\partial g}{\partial x}\right)_y \tag{5-51}$$

Certain properties of Jacobians make them particularly useful for transforming derivatives of thermodynamic variables. These include:

1. ***Transposition***

$$\frac{\partial(f, g)}{\partial(x, y)} = -\frac{\partial(g, f)}{\partial(x, y)} \tag{5-52}$$

2. ***Inversion***

$$\frac{\partial(f, g)}{\partial(x, y)} = \frac{1}{\dfrac{\partial(x, y)}{\partial(f, g)}} \tag{5-53}$$

3. ***Chain rule expansion***

$$\frac{\partial(f, g)}{\partial(x, y)} = \frac{\partial(f, g)}{\partial(z, w)}\,\frac{\partial(z, w)}{\partial(x, y)} \tag{5-54}$$

where z and w are two additional variables. Another important property is the simplification that occurs to Eq. (5-51) if we are only interested in evaluating the first partial derivative of a function $f(z, g)$ with respect to z at constant g:

$$\left(\frac{\partial f}{\partial z}\right)_g = \frac{\partial(f, g)}{\partial(z, g)} \tag{5-55}$$

By applying the chain rule expansion and inversion property [Eqs. (5-54) and (5-53)], Eq. (5-55) becomes:

$$\left(\frac{\partial f}{\partial z}\right)_g = \frac{\dfrac{\partial(f, g)}{\partial(x, y)}}{\dfrac{\partial(z, g)}{\partial(x, y)}} \tag{5-57}$$

The properties of two-variable Jacobians are easily extended to functions in m variables $x_i(i=1,..., m), f_i(x_1, x_2,..., x_m)$. The Jacobian is

$$\frac{\partial(f_1, f_2,..., f_m)}{\partial(x_1, x_2,..., x_m)} \tag{5-58}$$

or in determinant form

$$\begin{vmatrix} \dfrac{\partial f_1}{\partial x_1} & \dfrac{\partial f_1}{\partial x_2} & \cdots & \dfrac{\partial f_1}{\partial x_m} \\ \cdot & \cdot & & \cdot \\ \cdot & \cdot & & \cdot \\ \cdot & \cdot & & \cdot \\ \dfrac{\partial f_m}{\partial x_1} & \dfrac{\partial f_m}{\partial x_2} & \cdots & \dfrac{\partial f_m}{\partial x_m} \end{vmatrix} \tag{5-59}$$

where each partial derivative holds all x_i constant except the one involved in the differentiation. Let's consider a few examples to illustrate how Jacobians are used in thermodynamics.

Example 5.2

Consider the isenthalpic (Joule-Thompson) expansion of a fluid across a well-insulated valve. Here, the derivative

$$\left(\frac{\partial T}{\partial P}\right)_H \equiv \text{Joule–Thompson coefficient} = \alpha_H$$

is of interest. Express α_H in terms of measurable properties.

Solution

For a pure fluid, the specific enthalpy, $H = f(T, P)$, so we can use Eq. (5-57) with $f = T$, $g = H$, and $z = P$. Thus,

$$\left(\frac{\partial T}{\partial P}\right)_H = \frac{\dfrac{\partial(T, H)}{\partial(x, y)}}{\dfrac{\partial(P, H)}{\partial(x, y)}} \tag{5-60}$$

Now we need to select x and y. Although we have many choices, frequently, as in this case, we want to obtain expressions in terms of measurable properties like T, P, V, etc. A good first guess for x and y would be to use T and P since they are in the derivative of interest. Thus,

$$\left(\frac{\partial T}{\partial P}\right)_H = \frac{\dfrac{\partial(T, H)}{\partial(T, P)}}{\dfrac{\partial(P, H)}{\partial(T, P)}} \tag{5-61}$$

Using the transposition property [Eq. (5-52)] and Eq. (5-55), Eq. (5-61) is quickly simplified to:

$$\left(\frac{\partial T}{\partial P}\right)_H = \frac{\dfrac{\partial(H, T)}{\partial(P, T)}}{\dfrac{-\partial(H, P)}{\partial(T, P)}} = \frac{\left(\dfrac{\partial H}{\partial P}\right)_T}{-\left(\dfrac{\partial H}{\partial T}\right)_P} = \frac{\left(\dfrac{\partial H}{\partial P}\right)_T}{-C_p} \tag{5-62}$$

In this instance, we could have used the triple product directly (Eq. (5-33)) to obtain Eq. (5-62). However, such direct transformations are not always possible. For example, in Section 5.9 we will demonstrate how Jacobian transformations are of considerable use in manipulating complex partial derivatives of multivariable functions. Before we try to evaluate $(\partial H/\partial P)_T$ in Eq. (5-62) in terms of measurable properties, let's consider another example.

Example 5.3

The isentropic (reversible, adiabatic) expansion of a pure fluid involves the derivative:

$$(\partial T/\partial P)_S = \text{isentropic coefficient} = \alpha_S$$

Express the isentropic coefficient in terms of measurable properties.

Solution

The α_s derivative can also be reconstructed using Jacobians

$$\left(\frac{\partial T}{\partial P}\right)_S = \frac{\dfrac{\partial(T, S)}{\partial(T, P)}}{\dfrac{\partial(P, S)}{\partial(T, P)}} = \frac{\left(\dfrac{\partial S}{\partial P}\right)_T}{-\left(\dfrac{\partial S}{\partial T}\right)_P} = \frac{\left(\dfrac{\partial S}{\partial P}\right)_T}{\dfrac{-C_p}{T}} = \frac{-T}{C_p}\left(\frac{\partial S}{\partial P}\right)_T \tag{5-63}$$

In order to express α_S (or α_H) in terms of measurable properties we must convert $(\partial S/\partial P)_T$ [or $(\partial H/\partial P)_T$] to another form. To start with, consider the closed system combined Law equation which is equivalent to the intensive form of the Fundamental Equation in differential form

$$dU = TdS - PdV \tag{5-64}$$

By rearranging, and differentiating with respect to P at constant T, we get

$$\left(\frac{\partial S}{\partial P}\right)_T = \frac{1}{T}\left(\frac{\partial U}{\partial P}\right)_T + \frac{P}{T}\left(\frac{\partial V}{\partial P}\right)_T \tag{5-65}$$

Again, we end up with a cumbersome derivative $(\partial U/\partial P)_T$. Some readers may recognize that we need to introduce an auxiliary potential function that has as its natural variables P and T. This is the Gibbs free energy $G = f(T, P)$ which is defined as:

$$G \equiv U + PV - TS$$

So by chain rule expansion of the differentials:

$$dG = dU + PdV + VdP - (TdS + SdT) \tag{5-66}$$

Thus, by substituting the expression for dU given in Eq. (5-64), dG becomes

$$dG = -SdT + VdP \tag{5-67}$$

And, by using Maxwell's reciprocity theorem, we obtain

$$\left(\frac{\partial S}{\partial P}\right)_T = -\left(\frac{\partial V}{\partial T}\right)_P \tag{5-68}$$

Now we can revise Eq. (5-63) for α_S to

$$\alpha_S = \left(\frac{\partial T}{\partial P}\right)_S = +\frac{T}{C_p}\left(\frac{\partial V}{\partial T}\right)_P \tag{5-69}$$

Let's see if we can use this result to help us evaluate the derivative $(\partial H/\partial P)_T$ in the α_H expression [Eq. (5-62)]. Here we begin with the definition of $H \equiv U + PV$, differentiate, and use the closed system combined law as we did before to obtain

$$dH = TdS + VdP \tag{5-70}$$

Thus, by differentiating with respect to P at constant T, we get

$$\left(\frac{\partial H}{\partial P}\right)_T = T\left(\frac{\partial S}{\partial P}\right)_T + V \tag{5-71}$$

Because we already have an expression for $(\partial S/\partial P)_T$ in terms of measurable PVT properties (Eq. (5-68)), we can express the Joule-Thompson coefficient α_H as

$$\alpha_H = \left(\frac{\partial T}{\partial P}\right)_H = \frac{-\left[T(-\partial V/\partial T)_P + V\right]}{C_p} \tag{5-72}$$

where only PVT properties and C_p values are needed to evaluate α_H.

Most vapor compression refrigeration systems in domestic use employ Joule-Thompson expansions to cool the circulating refrigerant. Thus, in these practical situations, α_H must be > 0 to obtain cooling as the pressure is lowered across the valve. What happens if the fluid being expanded is an ideal gas? In this case, $\alpha_H = 0$; thus, fluids whose *PVT* properties closely approximate ideal gas behavior would not be good choices for refrigerants when Joule-Thompson expansions are used. An alternative is to use an adiabatic turbine to expand the fluid to a lower pressure. Here, α_S is the appropriate scaling parameter as it provides the maximum cooling effect for the limiting case of a reversible, adiabatic or isentropic expansion. Now, even an ideal gas will cool upon expansion.

5.5 Reconstruction of the Fundamental Equation

The Gibbs Fundamental Equation where $\underline{U} = f(\underline{S}, \underline{V}, N_1, ..., N_n)$ provides the necessary and sufficient information needed to describe all the stable equilibrium states of any simple system. Unfortunately, $\underline{S}\ \underline{V}\ N$ coordinates are not completely amenable to direct measurement—for example, no direct-reading entropy meters exist! Thus a formal transformation is required that preserves the encoded information content of the Fundamental Equation while expressing the functional dependence in variables other than $\underline{S}$, $\underline{V}$, and N_i (i=1,..., n). Once we have this procedure, we can be sure that all other information of interest to us in classical thermodynamics can be obtained from the transformed equation. The Legendre transformation provides a rigorous mathematical route to achieve the desired reconstruction of the Fundamental Equation.

As shown in Section 5.1, if the Fundamental Equation were known, the properties T, P, and μ_i, could be determined by partial differentiation as expressed in the equations of state in Eqs. (5-5) through (5-7). Alternatively, the Fundamental Equation can be recovered, if all the equations of state are known, by substituting these equations into Eq. (5-13) or the common integrated form [Eq. (5-14)]

$$\underline{U} = T\underline{S} - P\underline{V} + \sum_{i=1}^{n} \mu_i N_i \tag{5-73}$$

As shown in Section 5.2, the $n + 2$ intensive variables, T, P, and μ_i (i=1,..., n), which are expressed explicitly by the equations of state, are not all independently variable. Any one of these variables can be expressed in differential form in terms of the other $n + 1$ variables and, upon integration, an expression connecting the $n + 2$ variables can be determined to within an arbitrary constant. It thus follows that only $n + 1$ equations of state are necessary to determine the Fundamental Equation to within an arbitrary constant (equivalent to specifying one or more reference state conditions). In Example 5.1, if Eq. (5-28) is combined with Eq. (5-22) to eliminate S, μ can be expressed in terms of reference state constants U^o, V^o and S^o; and variables T, V.

As an example, let us consider a pure material. The intensive Gibbs free energy or chemical potential, $G = \mu$, can be expressed as a function of T and P:

$$d\mu = \left(\frac{\partial \mu}{\partial T}\right)_P dT + \left(\frac{\partial \mu}{\partial P}\right)_T dP \tag{5-74}$$

In Section 5.4 [see Eq. (5-67)], it was shown that $(\partial\mu/\partial T)_P = -S$ and $(\partial\mu/\partial P)_T = V$. If Eqs. (5-5) and (5-6) were known, we could solve these simultaneously to obtain

$$S = g(T, P) \tag{5-75}$$

and

$$V = g'(T, P) \tag{5-76}$$

These equations are usually available, although not necessarily in analytical form, but the entropy is known only to within an arbitrary constant. Let us assume that we have available Eqs. (5-76) and (5-77):

$$S = S^o + g''(T, P) \tag{5-77}$$

where S^o is an arbitrary constant or reference state condition. Substitution into Eq. (5-74) and integration from an arbitrary reference state for which $T = T^o$, $P = P^o$, $V = V^o$, and $\mu = \mu^o$ leads to

$$\mu = \mu^o - S^o(T - T^o) - \int_{T^o}^{T} g''(T, P)\, dT + \int_{P^o}^{P} g'(T, P)\, dP \tag{5-78}$$

which is the desired relationship for μ as a function of T and P. Although this equation contains two arbitrary constants (S^o and μ^o), when Eqs. (5-5), (5-6), and (5-78) are substituted into Eq. (5-73) in order to obtain the Fundamental Equation, these two arbitrary constants appear as a sum; in particular, we would find

$$U = f_U(S, V) + (T^o S^o - P^o V^o + \mu^o) \tag{5-79}$$

or

$$U^o = T^o S^o - P^o V^o + \mu^o \tag{5-80}$$

Of the three arbitrary constants in Eq. (5-80) (i.e., U^o, S^o, and μ^o), it is clear that only two can be chosen independently. Thus, we can set base values of U and S at the reference state for which $T = T^o$, $P = P^o$, $V = V^o$, but having done so, the base value for μ is uniquely specified by Eq. (5-80).

5.6 Legendre Transformations

In the energy representation of the Fundamental Equation, Eq. (5-1), the properties $\underline{S}, \underline{V}, N_1, \ldots, N_n$ are treated as independent variables. This is not always an appropriate set of independent parameters. For example, since temperature can be measured much more conveniently than entropy, we might like to use $T, \underline{V}, N_1, \ldots, N_n$ as the independent variables. For a single-phase simple system, we can always express a property such as $\underline{U}$ in terms of $n + 2$ other properties such as $T, \underline{V}, N_1, \ldots, N_n$. Thus, we know that a function f exists such that

$$\underline{U} = f(T, \underline{V}, N_1, \ldots, N_n) \tag{5-81}$$

and

$$d\underline{U} = \left(\frac{\partial f}{\partial T}\right)_{\underline{V},N} dT + \left(\frac{\partial f}{\partial \underline{V}}\right)_{T,N} d\underline{V} + \sum_{i=1}^{n} \left(\frac{\partial f}{\partial N_i}\right)_{T,\underline{V},N_j[i]} dN_i \tag{5-82}$$

Given the Fundamental Equation, the function of Eq. (5-81) can be found by differentiating Eq. (5-1) to obtain Eq. (5-5),

$$T = g_T(\underline{S}, \underline{V}, N_1, \ldots, N_n)$$

and then solving Eqs. (5-1) and (5-5) simultaneously in order to eliminate $\underline{S}$. The result is an equation of the form

$$\underline{U} = f(T, \underline{V}, N_1, \ldots, N_n) = f\left[\left(\frac{\partial f_U}{\partial \underline{S}}\right)_{\underline{V},N}, \underline{V}, N_1, \ldots, N_n\right] \tag{5-83}$$

Although Eq. (5-83) is of the form desired (i.e., Eq. (5-81)), the information content of Eq. (5-83) is less than that of the Fundamental Equation. Equation (5-83) is a partial differential equation that can be integrated to yield the Fundamental Equation only to within an arbitrary function of integration.

We must now ask whether or not there are other functions with the same information content as that of the Fundamental Equation but with independent variables different than $\underline{S}, \underline{V}, N_1, \ldots, N_n$. The answer is that there are such functions if we are willing to restrict ourselves to a set of independent variables in which we choose only one from each of the following $(n+2)$ pairs: $\{\underline{S}, T\}$, $\{\underline{V}, P\}$, $\{N_i, \mu_i\}$ for $i=1, n$. These pairs of variables are usually referred to as *conjugate coordinates*. Note that there is one extensive (e.g., $\underline{S}, \underline{V}, N_i$) and one intensive (e.g., T, P, μ_i) variable in each conjugate pair. This grouping of variables is a natural consequence of the original formulation of the Fundamental Equation in Gibbs coordinates.

As mentioned in Section 5.3, conjugate pairs of additional natural variables may be added to the Fundamental Equation to account for non-PV work effects. Table 5.1 lists the pairs of conjugate coordinates that are commonly encountered in problems of classical thermodynamics.

Table 5.1 Conjugate Coordinates

Type	Extensive Parameter	Intensive Parameter
Reversible heat flow	$\underline{S}$	T
Pressure-Volume work	$\underline{V}$	$-P$
Mass flow enthalpy and entropy	N_i	μ_i
Surface deformation work	$\underline{a}$	σ
Generalized work	$\underline{x}_i$	$\mathbf{F}_i$

To formulate these functions, a *Legendre transform* is employed. Such a transformation stems from a basic theorem in line geometry, and although the rigorous proof is no simple task, the results are easy to apply. The basic principle is that a curve consisting of a locus of points can be described completely by the tangent lines that form the envelope of the curve. Let's consider a simple case to illustrate what Legendre transforms are and how they are used. We define a basis function $y^{(0)} = f(x)$ with only one independent variable. The graph of $y^{(0)}$ versus x shown in Figure 5.3 is indicative of a well-behaved, continuously differentiable function. For any value of x, a straight line tangent to the curve has a defined slope and intercept with the y-axis as shown on the figure. If we call the slope $\xi = dy^{(0)}/dx$ and the intercept $y^{(1)}$, we can construct an infinite set of line tangents to the original curve $y^{(0)} = f(x)$ such that:

$$y^{(0)} = \xi x + y^{(1)} \tag{5-84}$$

for all values of x. By knowing values of $y^{(1)}$ and ξ for every value of x we can reconstruct the original function $y^{(0)}$. The Legendre transformation is carried out by solving Eq. (5-84) for $y^{(1)}$ and inverting the functional relationship between ξ and x. Thus,

$$y^{(1)} = y^{(0)} - x\xi = f[\xi] \tag{5-85}$$

Now $y^{(1)}$ can be represented as a function of ξ with its derivative slope given by,

$$-x = \frac{dy^{(1)}}{d\xi}$$

and tangent intercept equal to $y^{(0)}$. The function $y^{(1)}$ is the Legendre transform in one dimension. It clearly contains the same information content as our original basis function $y^{(0)}$ since one can reproduce the original function plotted in Figure 5.3 from the set of line tangents.

An analogous approach can be used for functions in two variables, $y^{(0)} = f(x_1, x_2)$. Here there are two characteristic slopes corresponding to partial derivatives:

$$\xi_1 = \left(\frac{\partial y^{(0)}}{\partial x_1}\right)_{x_2} \quad \text{and} \quad \xi_2 = \left(\frac{\partial y^{(0)}}{\partial x_2}\right)_{x_1} \tag{5-86}$$

The transforms $y^{(1)} = y^{(0)} - x_1\,\xi_1$ and $y^{(2)} = y^{(0)} - x_1\,\xi_1 - x_2\,\xi_2$ are shown in Figure 5.4. In this 2-D case the transforms are obtained from planar tangents to the basis function surface rather than from line tangents to a curve.

Now we can generalize the results to m variables, where the basis function,

$$y^{(0)} = f(x_1, \ldots, x_m) \tag{5-87}$$

represents an m-dimensional surface. There are m first-order partial derivatives of $y^{(0)}$ with respect to each of the m independent variables, $x_1, \ldots, x_m$. Defining these derivatives as ξ_i,

$$\xi_i \equiv \left(\frac{\partial y^{(0)}}{\partial x_i}\right)_{x_1,\ldots,[x_i],\ldots,x_m} \tag{5-88}$$

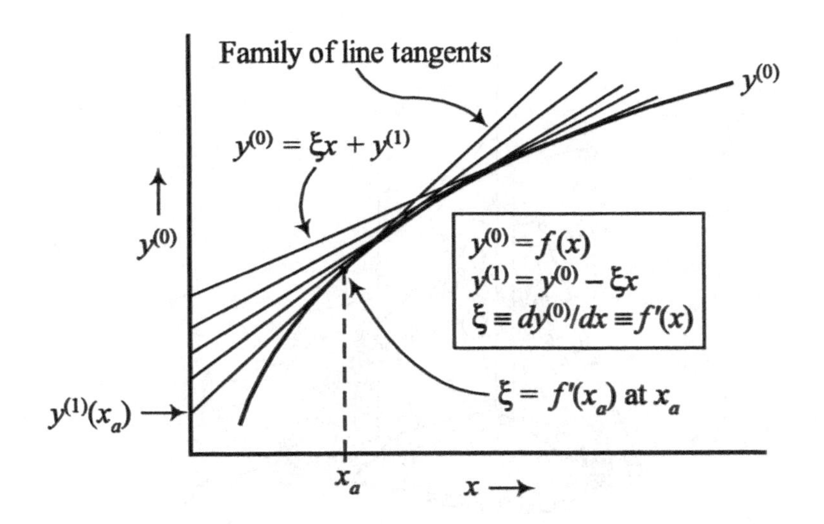

Figure 5.3 Reconstructing a function $y = f(x)$ using a family of line tangents—the one-dimensional Legendre Transform concept.

or

$$\xi_i = f(x_1, \ldots, x_m) \tag{5-89}$$

where the symbol $[x_i]$ in the subscript of the partial derivative indicates that x_i is not held constant. It follows that the variation of $y^{(0)}$ with x_1 could be described by the envelope of tangents in the $y^{(0)} - x_1$ plane. If $y^{(1)}$ is the intercept of the tangent corresponding to ξ_1,

$$y^{(1)}(\xi_1, x_2, \ldots, x_m) = y^{(0)} - \xi_1 x_1 \tag{5-90}$$

The function $y^{(1)}(\xi_1, x_2, \ldots, x_m)$ is called the *first* Legendre transform of $y^{(0)}$ with respect to x_1. In other words, a Legendre transform results in a new function in which one or more independent variables is replaced by its slope. There are obviously m different first transforms, depending on the ordering of the variables $x_1, \ldots, x_m$.

Higher-order transforms are defined in a similar manner; thus, $y^{(k)}(\xi_1, \ldots, \xi_k, x_{k+1}, \ldots, x_m)$ is the kth Legendre transform:

$$y^{(k)} = y^{(0)} - \sum_{i=1}^{k} \xi_i x_i \tag{5-91}$$

The total differential of the original basis function $y^{(0)}$ can, of course, be expressed as

$$dy^{(0)} = \sum_{i=1}^{m} \xi_i \, dx_i \tag{5-92}$$

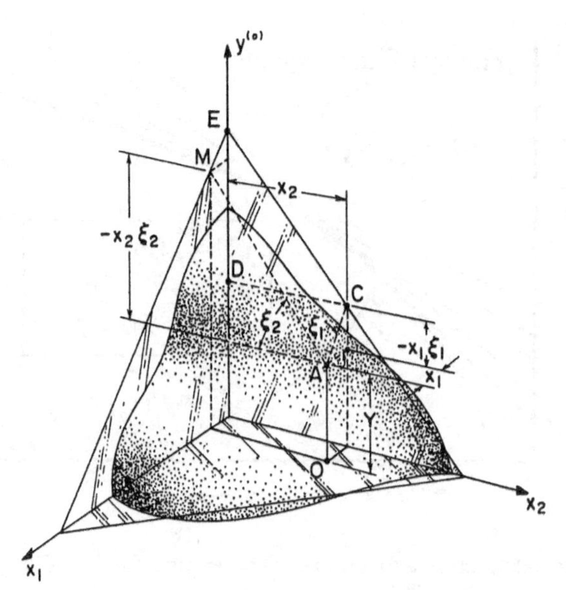

Legendre Transform for $y^{(0)} = f(x_1, x_2)$
IF: $y^{(1)} = f(\xi_1, x_2)$, $y^{(1)} =$ Ⓒ
$y^{(1)} = f(x_1, \xi_2)$, $y^{(1)} =$ Ⓜ
$y^{(2)} = f(\xi_1, \xi_2)$, $y^{(2)} =$ Ⓔ

Figure 5.4 Geometric construction for a two-dimensional Legendre transformation using $y^{(0)} = f[x_1, x_2]$. Two first order transforms $y^{(1)} = f[\xi_1, x_2]$ at Ⓒ and $y^{(1)} = f[x_1, \xi_2]$ at Ⓜ and one second order transform $y^{(2)} = f[\xi_1, \xi_2]$ at Ⓔ, are shown.

and the differential of the kth transform can be obtained by differentiating Eq. (5-91) using the chain rule and substituting Eq. (5-92) to give

$$dy^{(k)} = \sum_{i=1}^{m} \xi_i \, dx_i - \sum_{i=1}^{k} x_i d\xi_i - \sum_{i=1}^{k} \xi_i \, dx_i$$

which simplifies to

$$dy^{(k)} = -\sum_{i=1}^{k} x_i \, d\xi_i + \sum_{i=k+1}^{m} \xi_i \, dx_i \tag{5-93}$$

Since $y^{(k)}$ is a function of $\xi_1, \ldots, \xi_k, x_{k+1}, \ldots, x_m$, it follows from Eq. (5-93) that:

For transformed variables $(1 \leq i \leq k)$,

$$\left(\frac{\partial y^{(k)}}{\partial \xi_i}\right)_{\xi_1, \ldots, [\xi_i], \ldots, \xi_k, x_{k+1}, \ldots, x_m} = -x_i \tag{5-94}$$

whereas for untransformed variables $(k< i \leq m)$,

$$\left(\frac{\partial y^{(k)}}{\partial x_i}\right)_{\xi_1,\ldots,\,\xi_k,\,x_{k+1},\ldots,\,[x_i],\ldots,\,x_m} = \xi_i \tag{5-95}$$

Equation (5-94) is sometimes called the *inverse* Legendre transform. Equations (5-95) and (5-94) clearly show the canonical relationship between conjugate coordinates $\{x_i, \xi_i\}$ that is similar to forms found in classical mechanics. Eq. (5-95) is applicable for all cases where $i > k$, and thus one may generalize the result as

$$\frac{\partial y^{(i-1)}}{\partial x_i} = \frac{\partial y^{(i-2)}}{\partial x_i} = \cdots \frac{\partial y^{(0)}}{\partial x_i} = \xi_i \tag{5-96}$$

The partial derivatives in Eq. (5-96) were expressed without indicating the set of variables to be held constant. However, it is clear from the discussion above that the degree of the transform determines the set, and the only exception would be that variable used in the actual differentiation. For example, the restraints on the term $\partial y^{(i-1)}/\partial x_i$ would be that $\xi_1,\ldots, \xi_{i-1}, x_{i+1},\ldots, x_m$ would be held constant.

To illustrate the application of these relations, let $y^{(0)}$ be the total internal energy of a simple system $\underline{U}$; then the Fundamental Equation would be given in Eq. (5-1). Suppose that we desired the transform $y^{(2)}$. Using the ordering of variables given in Eq. (5-1), we can organize the transformation using the table below:

$y^{(0)} = \underline{U}$

	x_i	ξ_i
1	$\underline{S}$	T
2	$\underline{V}$	$-P$
3	N_1	μ_1
.	.	.
.	.	.
.	.	.
n+2	N_n	μ_n

$y^{(2)} = \underline{G}$

	x_i	ξ_i
1	T	$-\underline{S}$
2	$-P$	$-\underline{V}$
3	N_1	μ_1
.	.	.
.	.	.
.	.	.
n+2	N_n	μ_n

with this ordering of variables:

$$y^{(2)} = f(\xi_1, \xi_2, N_1,\ldots, N_n) \tag{5-97}$$

with

$$\xi_1 \equiv \left(\frac{\partial \underline{U}}{\partial \underline{S}}\right)_{\underline{V},N} = T \tag{5-98}$$

$$\xi_2 \equiv \left(\frac{\partial \underline{U}}{\partial \underline{V}}\right)_{\underline{S},N} = -P \tag{5-99}$$

Then, with Eq. (5-91),

$$y^{(2)} = \underline{U} - T\underline{S} - (-P\underline{V}) \equiv \underline{G} \tag{5-100}$$

where $\underline{G}$ is the total Gibbs energy. The analogs of Eqs. (5-92) and (5-93) are Eqs. (5-4) and (5-101).

$$dy^{(2)} = d\underline{G} = -\underline{S}\,dT + \underline{V}\,dP + \sum_{i=1}^{n} \mu_i\,dN_i \tag{5-101}$$

where the chemical potential μ_i can be defined in several ways as

$$\mu_i \equiv \left(\frac{\partial \underline{U}}{\partial N_i}\right)_{\underline{S},\underline{V},N_1,\ldots,[N_i],\ldots,N_n} \tag{5-102}$$

or equivalently as

$$\mu_i \equiv \left(\frac{\partial \underline{G}}{\partial N_i}\right)_{T,P,N_1,\ldots,[N_i],\ldots,N_n} \tag{5-103}$$

The significance of the Legendre transform is thus evident. The important thermodynamic property $\underline{G}$ is simply a partial Legendre transform of the energy $\underline{U}$ from $(\underline{S}, \underline{V}, N_1,\ldots, N_n)$ space to $(T, P, N_1,\ldots, N_n)$ space. Equation (5-101) is also a Fundamental Equation and no loss in information content has resulted in going from the $\underline{U}$ to the $\underline{G}$ representation.

We could have started our transformation in reverse. Referring to the table of x_i and ξ_i given above, we could have defined a new basis function as $\underline{G} = y^{(2)} \equiv y^{*(0)}$ and transformed back to $\underline{U} = y^{*(2)} \equiv y^{(0)}$, that is, to the old basis function. Again, there would be no loss of information content.

The ***total Legendre transform*** of Eq. (5-1) is

$$y^{(n+2)} = \underline{U} - T\underline{S} + P\underline{V} - \sum_{i=1}^{n} \mu_i N_i = 0 \quad \text{[by Eq. (5–73)]} \tag{5-104}$$

Thus, Eq. (5-93) becomes

$$dy^{(n+2)} = 0 = -\underline{S}\,dT + \underline{V}\,dP - \sum_{i=1}^{n} N_i\,d\mu_i \tag{5-105}$$

Equation (5-105) is known as the *Gibbs-Duhem equation.* Later in Chapter 9 on mixtures and in those chapters that follow, we will utilize the Gibbs-Duhem relationship extensively.

It is important to point out certain generalities in the use of Legendre transformations. We introduced a general functional equation [Eq. (5-87)] with arbitrary ordering of $x_1,\ldots, x_m$. We noted that for each x_i there was a conjugate coordinate variable, ξ_i [Eq. (5-88)]. We then illustrated that one could readily derive a functional relation with the same information content wherein we replaced independent variables $x_1,\ldots, x_k$ by

$\xi_1,\ldots,\xi_k$. This was the kth Legendre transform defined in Eq. (5-91) and shown in differential form in Eq. (5-93).

As we demonstrated in the example above, one may also redefine this kth Legendre transform as a new $y^{(0)}$ basis function if care is taken in defining the correct independent variables and conjugate coordinates. For example, we have shown that beginning with Eq. (5-1), we obtained the Gibbs energy potential function by a Legendre transform of $\underline{U}$ into $T, P, N_1,\ldots, N_n$ space. We could now use the Gibbs energy function as our basis function $y^{(0)}$, but the independent variable set $(x_1,\ldots,x_m)$ would be $(T, P, N_1,\ldots,N_n)$ with arbitrary ordering. The conjugate coordinate variables $\xi_1,\ldots,\xi_m$ would still be defined by Eq. (5-88); for example, for the variable T, $\xi_T=(\partial \underline{G}/\partial T)_{P,N}=-\underline{S}$; for P, $\xi_P=\underline{V}$, and for N_j, $\xi_j=\mu_j$. In fact, we may select any Legendre transform as the $y^{(0)}$ basis function by redefining the independent variable set. Examples illustrating the use of Legendre transforms for manipulating partial derivatives are found at the end of Section 5.8.

If we had started the mathematical transform development with the entropy form of the Fundamental Equation $\underline{S}=f_s(\underline{U}, \underline{V}, N_1,\ldots, N_n)$, we would have produced an equivalent set of transforms that are called Massieu-Planck functions, which are sometimes easier to use than Legendre transformations, e.g., when working in T^{-1} coordinates [see Debenedetti (1986)]. Additional treatments of Legendre transformations are given by Callen (1985), Alberty (1994), Aris and Amundson (1973), and Tisza (1966).

5.7 Graphical Representations of Thermodynamic Functions and Their Transforms

Beginning with the Fundamental Equation for $\underline{U}$ [Eq. (5-1)], we can now transform one or all of the independent variable set $\underline{S}, \underline{V}, N_1,\ldots, N_n$. Let us choose only the variable $\underline{V}$ and reorder so that $\underline{V}$ represents x_1. Then

$$y^{(1)}=y^{(0)}-\xi_1 x_1=\underline{U}-(-P)\underline{V}=\underline{U}+P\underline{V}=\underline{H} \tag{5-106}$$

where this particular Legendre transform is called the *enthalpy*. We note that

$$\underline{H}=f(P, \underline{S}, N_1,\ldots, N_n) \tag{5-107}$$

We call Eq. (5-107) a Fundamental Equation in the same way that we refer to Eq. (5-1); then P is x_1, $\underline{S}$ is x_2, etc., and $\xi_1=(\partial\underline{H}/\partial P)=\underline{V}$. We can recover Eq. (5-1) by carrying out a Legendre transform assuming that Eq. (5-107) is the $y^{(0)}$ function; that is,

$$y^{(1)}=y^{(0)}-\xi_1 x_1=\underline{H}-(\underline{V})P=\underline{U} \tag{5-108}$$

This transform can be readily shown in Figure 5.5 for a common pressure-enthalpy diagram. If a curve of constant entropy is considered, the slope is $\underline{V}$. The intercept of this tangent of the enthalpy axis is, as shown, equal to the internal energy $\underline{U}$.

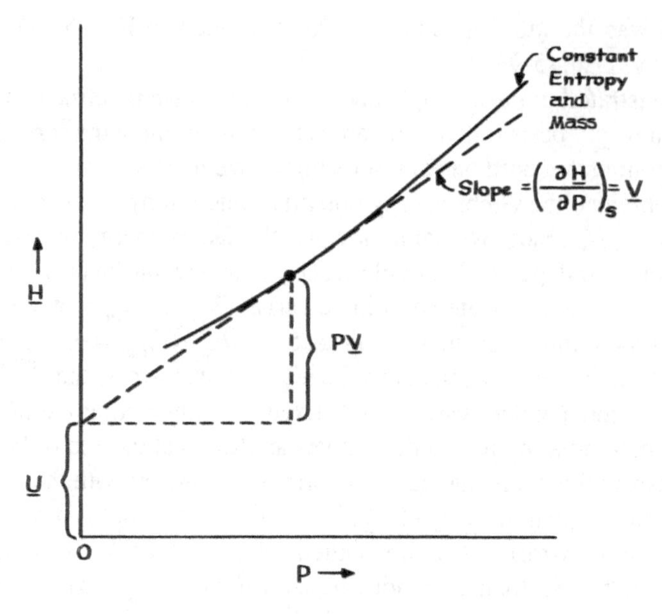

Figure 5.5

With internal energy as the basis function, there are $n + 2$ permutations of first Legendre transforms: the two common potential functions, $\underline{H}(P, \underline{S}, N_1,..., N_n)$ and $\underline{A}(T, \underline{V}, N_1,..., N_n)$, and n other functions for the independent variable set of $\underline{S}, \underline{V}, N_1,..., N_{i-1}, \mu_i, N_{i+1},..., N_n$. Since the ordering of components is arbitrary, we shall refer to the n functions as $\underline{U}'(\underline{S}, \underline{V}, \mu_1, N_2,..., N_n)$. In a similar manner, there are $(n + 2)(n + 1)/2$ second Legendre transforms: one is another potential function, the Gibbs free energy $\underline{G}(T, P, N_1,..., N_n)$; there are n of the form $\underline{A}'(T, \mu_1, \underline{V}, N_2,..., N_n)$ and n of the form $\underline{H}'(P, \mu_1, \underline{S}, N_2,..., N_n)$, and $(n)(n - 1)/2$ of the form $\underline{U}''(\mu_1, \mu_2, \underline{S}, \underline{V}, N_3,..., N_n)$. Third Legendre transforms would involve $\underline{G}'$, $\underline{A}''$, $\underline{H}''$, and $\underline{U}'''$ potential functions. Table 5.2 lists the important potential functions used in chemical thermodynamics.

Again it is instructive to re-examine the plots of the Fundamental Equation to show how various Legendre Transforms are geometrically related to the *USV* surface. Fundamental equations are plotted in Figures 5.1 and 5.2 for pure water and ethylene. Multiple phase equilibrium conditions, such as the triple point of pure water and the liquid-vapor coexistence envelope occur at specific temperatures and pressures for each particular substance. At first glance, one might expect that the thermodynamic potential that involves the natural variables of temperature T and pressure P would yield useful relationships. This function is, of course, the Gibbs free energy $\underline{G}$ which is the second Legendre transform of $\underline{U}$ and can be visualized geometrically as the intersection point of a tangent plane to the $\underline{U}\,\underline{S}\,\underline{V}$ surface with the $\underline{U}$ axis.

Table 5.2 Thermodynamic Potential Functions and the Gibbs-Duhem Relation for Simple Systems

Type	Functional Equation[2]	Symbol	Function[1]	Canonical Coordinates[3]
Internal Energy	$\underline{U} = T\underline{S} - P\underline{V} + \sum_{i=1}^{n} \mu_i N_i$	$\underline{U}$	$y^{(0)}$	$\underline{S}, \underline{V}, N_i \ (i=1,\ldots,n)$
Enthalpy	$\underline{H} = \underline{U} + P\underline{V}$	$\underline{H}$	$y^{(1)}$	$\underline{S}, P, N_i \ (i=1,\ldots,n)$
Helmholtz Free Energy	$\underline{A} = \underline{U} - T\underline{S}$	$\underline{A}$	$y^{(1)}$	$T, \underline{V}, N_i \ (i=1,\ldots,n)$
Gibbs Free Energy	$\underline{G} = \underline{U} + P\underline{V} - T\underline{S} = \underline{H} - T\underline{S}$	$\underline{G}$	$y^{(2)}$	$T, P, N_i \ (i=1,\ldots,n)$
Gibbs-Duhem Relation	$y^{(n+2)} = 0$	Total Transform	$y^{(n+2)}$	$T, P, \mu_i \ (i=1,\ldots,n)$

Type	Total Differential
Internal Energy	$d\underline{U} = Td\underline{S} - Pd\underline{V} + \sum_{i=1}^{n} \mu_i dN_i$
Enthalpy	$d\underline{H} = Td\underline{S} + \underline{V}dP + \sum_{i=1}^{n} \mu_i dN_i$
Helmholtz Free Energy	$d\underline{A} = -\underline{S}dT - Pd\underline{V} + \sum_{i=1}^{n} \mu_i dN_i$
Gibbs Free Energy	$d\underline{G} = -\underline{S}dT + \underline{V}dP + \sum_{i=1}^{n} \mu_i dN_i$
Gibbs-Duhem Relation	$dy^{(n+2)} = -\underline{S}dT + \underline{V}dP - \sum_{i=1}^{n} N_i d\mu_i = 0$

1 $y^{(0)}$ = basis function
$y^{(1)}$ = First Legendre transform, $y^{(2)}$ = Second Legendre transform, ...

2 Obtained by Euler integration

3 The canonical coordinates refer to the set of natural variables that preserves the informational content of the Fundamental Equation, $y^{(0)} = \underline{U}$.

For a pure component system, the intensive form of the Fundamental Equation can be used as we did with Figures 5.1 and 5.2 where U is plotted as a function of S and V. In this case, $y^{(0)} = U$ and $y^{(2)} = G = \mu$ and points of multiple tangency with a plane rolling over the USV surface will have the same intersection point on the U axis and thus will have identical values of $y^{(2)} = G = \mu$. Later in Chapter 6 we will show that this geometric condition rigorously describes the mathematical criteria for phase equilibrium. Thus, it is no accident that the triangular plane in Figure 5.1b corresponds to the triple point where solid ice, liquid water, and water vapor or steam coexist in equilibrium. The triple point is equivalent to a unique condition where triple planar tangency exists on the USV surface of Figure 5.1a. Similar common points for double tangency on the USV surface can be used to characterize two-phase equilibria, such as the liquid-vapor coexistence region for a pure substance.

5.8 Modifications to the Fundamental Equation for Non-simple Systems

Additional variables given in the original expression for the Fundamental Equation [Eq. (5-1)] may be needed to describe the behavior of systems in many practical situations in chemical thermodynamics. Some of these cases have already been introduced in Chapters 3 and 4. For example, there is frequently a need to account for gravitational or inertial forces—potential and kinetic energy contributions to the total energy $\underline{E}$. Systems with these effects, of course, are *not* simple systems as we have defined them in this text; but nonetheless, their behavior can be represented by modifying the Fundamental Equation given in Eqs. (5-1) or (5-2). A common approach is to introduce additional generalized work terms, such that

$$d\underline{E} \text{ or } d\underline{U} = Td\underline{S} - Pd\underline{V} + \sum_{j=1}^{m} \mathbf{F}_j \cdot d\underline{\mathbf{x}}_j + \sum_{i=1}^{n} \mu_i dN_i \tag{5-109}$$

where the last summation involving terms $\mathbf{F}_j \cdot d\underline{\mathbf{x}}_j$ represents all non-PdV work effects that may be important. A number of these are listed in Table 3.1 in Chapter 3. These effects typically include:

- Surface deformation work - $\sigma d\underline{a}$
- Electric charge transport - $\mathcal{E}\, d\underline{q}$
- Electric or magnetic polarization - $\mathbf{E}d\underline{\mathbf{D}}$ *or* $\mathbf{H}d\underline{\mathbf{B}}$
- Linear elastic deformation - $\mathbf{F}_x\, d\underline{\mathbf{x}}$

The addition of these other work or energy effects does not violate the four postulates that we have set forth to develop the laws of thermodynamics. In particular, Postulate I still holds in that two independently variable properties plus the masses of all components present are sufficient to specify the equilibrium states of a simple system. Non-simple systems with these non-PdV work effects, with potential or kinetic energy effects, or with other constraints (such as semi-permeable boundaries) can be treated.

With sufficient information we can specify the equilibrium state of these systems or the path of a particular quasistatic process.

Assuming that Eq. (5-109) can be integrated using Euler's theorem, we can write:

$$\underline{E} \text{ or } \underline{U} = f(\underline{S}, \underline{V}, N_1, \ldots, N_n, \underline{\mathbf{x}}_1, \ldots, \underline{\mathbf{x}}_m) \tag{5-110}$$

which should be a well-behaved continuous function. Then all of the relationships developed earlier in this chapter can be utilized for manipulating partial derivatives and the like. For example, Maxwell reciprocity relationships can be applied:

$$\left(\frac{\partial \mathbf{F}_k}{\partial \underline{S}}\right)_{\underline{V}, N_i, \mathbf{x}_j} = \left(\frac{\partial T}{\partial \mathbf{x}_k}\right)_{\underline{S}, \underline{V}, N_i, \mathbf{x}_j[k]} \tag{5-111}$$

Legendre transformations can also be developed to interrelate thermodynamic properties. The following example problem illustrates the use of Legendre Transforms in manipulating partial derivatives.

Example 5.4

5.4(a) Express $(\partial \underline{S}/\partial P)_{T, N_i}$ in terms of $P\underline{V}TN$ properties and/or their derivatives.

Solution

The variables held constant (T, N_i) and variable involved in the differentiation (P) suggest that $y^{(2)} = \underline{G}$ should be used

$$d\underline{G} = -\underline{S}dT + \underline{V}dP + \sum_{i=1}^{n} \mu_i dN_i$$

and we see that a Maxwell relationship for $y_{ij}^{(2)} = y_{ji}^{(2)}$ gives the desired result

$$y_{ij}^{(2)} = \underline{G}_{PT} = -\left(\frac{\partial \underline{S}}{\partial P}\right)_{T, N_i} = \left(\frac{\partial \underline{V}}{\partial T}\right)_{P, N_i} \tag{5-112}$$

5.4(b) Express $\left[\frac{\partial(\underline{G}/T)}{\partial(1/T)}\right]_{P, N_i}$ in terms of a Legendre transform of $\underline{U}$.

Solution

Because $\underline{G} = y^{(2)}$ is involved with its natural coordinates of P, T, N_i, first try to simplify by expanding the derivative, recognizing that $\underline{G}_T = (\partial \underline{G}/\partial T)_{P, N_i} = -\underline{S}$ and that $\underline{G} = y^{(2)} = \underline{U} - T\underline{S} + P\underline{V} = \underline{H} - T\underline{S}$.

$$\left[\frac{\partial(\underline{G}/T)}{\partial(1/T)}\right]_{P, N_i} = \underline{G} - T\left(\frac{\partial \underline{G}}{\partial T}\right)_{P, N_i} = \underline{G} + T\underline{S} = \underline{H} - T\underline{S} + T\underline{S} = \underline{H} \tag{5-113}$$

Eq. (5-113) is the famous *Gibbs-Helmholtz relationship* which will be used frequently to show the temperature dependencies of various derived properties.

5.4(c) Express $(\partial C_v/\partial V)_T$ in terms of PVT properties and/or their derivatives.

Solution

We start by using the definition of C_v and the basis function of $y^{(0)} = U$ in intensive form.

$$C_v \equiv \left(\frac{\partial U}{\partial T}\right)_V \quad \text{and} \quad dU = TdS - PdV = C_v dT + \left(\frac{\partial U}{\partial V}\right)_T dV$$

Therefore, with $(\partial U/\partial V)_T = T(\partial S/\partial V)_T - P$ and a Maxwell relation

$$\left(\frac{\partial C_v}{\partial V}\right)_T = \left[\frac{\partial\left[T\left(\frac{\partial S}{\partial V}\right)_T - P\right]}{\partial T}\right]_V$$

Expanding,

$$\left(\frac{\partial C_v}{\partial V}\right)_T = \left(\frac{\partial S}{\partial V}\right)_T + T\left(\frac{\partial((\partial S/\partial V)_T)}{\partial T}\right)_V - \left(\frac{\partial P}{\partial T}\right)_V$$

With the V, T variable set, use $y^{(1)} = A$ to see if a suitable Maxwell relation can be developed to express $(\partial S/\partial V)_T$ differently

$$dA = -SdT - PdV \quad \text{and} \quad A_{TV} = -\left(\frac{\partial S}{\partial V}\right)_T = -\left(\frac{\partial P}{\partial T}\right)_V = A_{VT}$$

and we can simplify the equation for $(\partial C_v/\partial V)_T$ to

$$\left(\frac{\partial C_v}{\partial V}\right)_T = +T\left(\frac{\partial^2 P}{\partial T^2}\right)_V \tag{5-114}$$

5.4(d) For systems where surface forces are important, the Fundamental Equation can be redefined by adding a term $+\sigma \underline{a}$ to account for the reversible work due to surface deformation, where σ=surface tension in J/m^2 and $\underline{a}$ = area in m^2. Develop an expression for $(\partial \underline{S}/\partial \sigma)_{T,\underline{V},N_i}$ in terms of properties that can be measured experimentally.

Solution

The modified Fundamental Equation in differential and integrated form is:

$$d\underline{U} = Td\underline{S} - Pd\underline{V} + \sigma d\underline{a} + \sum_{i=1}^{n} \mu_i \, dN_i$$

and

$$\underline{U} = T\underline{S} - P\underline{V} + \sigma\underline{a} + \sum_{i=1}^{n} \mu_i N_i$$

Again by inspecting the variable set $\sigma, T, \underline{V}, and\ N_i$ involved in the derivative, a $y^{(2)}$ transform is suggested that yields that set, so two transformations from $\underline{S}$ to T and $\underline{a}$ to σ coordinates are needed:

$$dy^{(2)} = -\underline{S}dT - \underline{a}d\sigma - Pd\underline{V} + \sum_{i=1}^{n} \mu_i \, dN_i$$

A Maxwell relation on the first two terms gives the desired result:

$$y_{T\sigma}^{(2)} = -\left(\frac{\partial \underline{S}}{\partial \sigma}\right)_{T,\,\underline{V},\,N_i} = -\left(\frac{\partial \underline{a}}{\partial T}\right)_{\sigma,\,\underline{V},\,N_i} = y_{\sigma T}^{(2)} \tag{5-115}$$

5.9 Relationships Between Partial Derivatives of Legendre Transforms

Although not obvious at this stage of our theoretical development, it is extremely helpful to be able to express derivatives of Legendre transforms. For example, later in Chapter 7 we will introduce criteria that will determine the limits of phase stability for single and multicomponent systems. Often the functional forms of various $P\underline{V}TN$ equations of state and other models of non-ideal behavior, such as activity coefficient models, require that certain variables be used to facilitate calculation of the phase stability criteria. This situation leads directly to the manipulation of partial derivatives of Legendre transformations.

The presentation below follows earlier work by Beegle, Modell, and Reid (1974) which has been updated by Kumar and Reid (1986) who applied Jacobian transformations to calculate partial derivatives of Legendre transforms.

Single variable transforms. Starting with Eq. (5-87), we wish to investigate the relations between derivatives of $y^{(0)}$ and the first transform $y^{(1)}$, where

$$y^{(1)} = f(\xi_1, x_2, \ldots, x_m) \tag{5-116}$$

$$\xi_1 = \left(\frac{\partial y^{(0)}}{\partial x_1}\right)_{x_2,\ldots,x_m} \tag{5-117}$$

There are several ways to obtain the desired results, but the most expeditious involves the use of the derivative operators

$$\left(\frac{\partial}{\partial x_i}\right)_{\xi_1,\,x_2,\ldots,\,[x_i],\ldots,\,x_m} \quad \text{(for } i > 1\text{)} \quad \text{and} \quad \left(\frac{\partial}{\partial \xi_1}\right)_{x_2,\ldots,\,x_m}$$

$$\left(\frac{\partial}{\partial x_i}\right)_{\xi_1,\,x_2,\ldots,\,[x_i],\ldots,\,x_m} = \left(\frac{\partial}{\partial x_i}\right)_{x_1,\ldots,\,[x_i],\ldots,\,x_m} - \left[\frac{y_{1i}^{(0)}}{y_{11}^{(0)}}\right]\left(\frac{\partial}{\partial x_1}\right)_{x_2,\ldots,\,x_m} \quad (i \neq 1) \tag{5-118}$$

and

$$\left(\frac{\partial}{\partial \xi_1}\right)_{x_2,\ldots,x_m} = \frac{1}{y_{11}^{(0)}}\left(\frac{\partial}{\partial x_1}\right)_{x_2,\ldots,x_m} \tag{5-119}$$

The terms $y_{1i}^{(0)}$ and $y_{11}^{(0)}$ are second-order derivatives:

$$y_{1i}^{(0)} \equiv \frac{\partial^2 y^{(0)}}{\partial x_1\, \partial x_i} \tag{5-120}$$

$$y_{11}^{(0)} = \frac{\partial^2 y^{(0)}}{\partial x_1^2} \tag{5-121}$$

Equations (5-118) and (5-119) are of little value to obtain first derivatives since, in view of Eqs. (5-94) and (5-95),

$$\left(\frac{\partial y^{(1)}}{\partial \xi_1}\right)_{x_2,\ldots,x_m} \equiv y_1^{(1)} = -x_1$$

$$\left(\frac{\partial y^{(1)}}{\partial x_i}\right)_{\xi_1, x_2,\ldots,[x_i],\ldots,x_m} = y_i^{(1)} = y_i^{(0)} = \xi_i \quad (i > 1) \tag{5-122}$$

However, to relate, for example, $y_{1i}^{(1)}$ to second-order derivatives of $y^{(0)}$, one can employ the operator equation [Eq. (5-119)] on $y_i^{(1)}$. Thus for $i \neq 1$:

$$\left(\frac{\partial y_i^{(1)}}{\partial \xi_1}\right)_{x_2,\ldots,x_m} \equiv y_{1i}^{(1)} = \frac{1}{y_{11}^{(0)}}\left(\frac{\partial y_i^{(1)}}{\partial x_1}\right)_{x_2,\ldots,x_m} = \frac{1}{y_{11}^{(0)}}\left(\frac{\partial y_i^{(0)}}{\partial x_1}\right)_{x_2,\ldots,x_m} = \frac{y_{1i}^{(0)}}{y_{11}^{(0)}} \tag{5-123}$$

where Eq. (5-96) was used to simplify the third step. In a similar manner, other second- and third-order derivatives may be readily transformed. A list of these is presented in Table 5.3.

Example 5.5

Assume that the basis function is $\underline{U} = f(\underline{S}, \underline{V}, N_1,\ldots, N_n)$. Determine $y_{112}^{(1)}$ in terms of derivatives of this basis function.

Solution

With $y^{(1)} = f(T, \underline{V}, N_1,\ldots, N_n)$ and $y^{(0)} = f(\underline{S}, \underline{V}, N_1,\ldots, N_n)$,

$$y^{(1)} = \underline{A} = \underline{U} - T\underline{S} \tag{5-124}$$

$\underline{A}$ is the Helmholtz energy. From Table 5.3,

$$y_{112}^{(1)} = \underline{A}_{TT\underline{V}} = A_{TTV} = \frac{\partial^3 \underline{A}}{\partial T^2\, \partial \underline{V}} = -\frac{\partial^2 \underline{S}}{\partial T\, \partial \underline{V}}$$

$$= \frac{y_{112}^{(0)}}{(y_{11}^{(0)})^2} - \frac{y_{111}^{(0)} y_{12}^{(0)}}{(y_{11}^{(0)})^3}$$

$$= \frac{U_{SSV}}{U_{SS}^2} - \frac{U_{SSS} U_{SV}}{U_{SS}^3}$$

where the underbars have been omitted for simplicity. Now we can write

$$-\frac{\partial^2 S}{\partial T \partial \underline{V}} = -\left(\frac{\partial^2 P}{\partial T^2}\right)_{\underline{V}, N} = \frac{\partial^2 T/\partial \underline{S} \partial \underline{V}}{(\partial T/\partial \underline{S})_{\underline{V}, N}^2} - \frac{(\partial^2 T/\partial \underline{S}^2)_{\underline{V}, N} (\partial T/\partial \underline{V})_{\underline{S}, N}}{(\partial T/\partial \underline{S})_{\underline{V}, N}^3} \qquad (5\text{-}125)$$

Table 5.3 Second- and Third-Order Derivatives of $y^{(1)}$ in Terms of $y^{(0)}$

Derivative in $y^{(1)}$ system	Derivative in $y^{(0)}$ system	Quantity operated upon:	Equation used
$y_{11}^{(1)}$	$-\frac{1}{y_{11}^{(0)}}$	$y_1^{(1)}$	(5-119)
$y_{1i}^{(1)}$	$\frac{y_{1i}^{(0)}}{y_{11}^{(0)}} \quad (i \neq 1)$	$y_i^{(1)}$	(5-119)
$y_{ij}^{(1)}$	$y_{ij}^{(0)} - \frac{y_{1i}^{(0)} y_{1j}^{(0)}}{(y_{11}^{(0)})} \quad (i, j \neq 1)$	$y_j^{(1)}$	(5-118)
$y_{111}^{(1)}$	$\frac{y_{111}^{(0)}}{(y_{11}^{(0)})^3}$	$y_{11}^{(1)}$	(5-119)
$y_{11i}^{(1)}$	$\frac{y_{11i}^{(0)}}{(y_{11}^{(0)})^2} - \frac{y_{111}^{(0)} y_{1i}^{(0)}}{(y_{11}^{(0)})^3} \quad (i > 1)$	$y_{1i}^{(1)}$	(5-119)
$y_{1ij}^{(1)}$	$\frac{y_{1ij}^{(0)}}{y_{11}^{(0)}} - \left\{\frac{y_{11i}^{(0)} y_{1j}^{(0)} + y_{11j}^{(0)} y_{1i}^{(0)}}{(y_{11}^{(0)})^2}\right\} + \frac{y_{111}^{(0)} y_{1i}^{(0)} y_{1j}^{(0)}}{(y_{11}^{(0)})^3} \quad (i, j \neq 1)$	$y_{ij}^{(1)}$	(5-119)
$y_{ijk}^{(1)}$	$y_{ijk}^{(0)} - \frac{(y_{1i}^{(0)} y_{1jk}^{(0)} + y_{1j}^{(0)} y_{1ik}^{(0)} + y_{1k}^{(0)} y_{1ij}^{(0)})}{y_{11}^{(0)}} + \frac{(y_{1i}^{(0)} y_{1j}^{(0)} y_{11k}^{(0)} + y_{1i}^{(0)} y_{1k}^{(0)} y_{11j}^{(0)} + y_{1j}^{(0)} y_{1k}^{(0)} y_{11i}^{(0)})}{(y_{11}^{(0)})^2} - \frac{y_{1i}^{(0)} y_{1j}^{(0)} y_{1k}^{(0)} y_{111}^{(0)}}{(y_{11}^{(0)})^3} \quad (i, j, k \neq 1)$	$y_{jk}^{(1)}$	(5-118)

Multiple Variable Transforms. The development shown above was limited to the case where only a single variable was transformed. Should one wish to transform more than a single variable, it is always possible to proceed a step at a time and transform each separately as was shown above. It is also possible to develop a more general technique to allow one to express the partial derivatives of a Legendre transform $y^{(j)}(\xi_1,..., \xi_j, x_{j+1},..., x_m)$ in terms of the basis function $y^{(0)}$ or, in general, some other Legendre transform $y^{(j-q)}$, where $q = 1, 2,..., j$. The equations to obtain these second-order derivatives are shown in Table 5.4 for the basis function and in Table 5.5 for other functions. In Table 5.4 there are three cases for $y^{(j)}_{ik}$: that is, $j > i, k$; $k \geq j > i$; and $j \leq i, k$. Various cases are illustrated in Examples 5.6 and 5.7. Note that we can make use of Maxwell's reciprocity relation $y^{(j)}_{ik} = y^{(j)}_{ki}$ to simplify the calculations.

Table 5.4 Relations Between Second-Order Derivatives of the *j*th Legendre Transform and the Basis Function

$$y^{(j)}_{ik} = \frac{\left|\begin{array}{c|c} D^{(0)}_j & B \\ \hline A & 0 \end{array}\right|}{D^{(0)}_j} \quad j > i, k \qquad y^{(j)}_{ik} = \frac{\left|\begin{array}{c|c} D^{(0)}_j & \begin{array}{c} D \\ \hline F \end{array} \\ \hline A & 0 \end{array}\right|}{D^{(0)}_j} \quad \begin{array}{c} j > i \\ j \leq k \end{array}$$

$$y^{(j)}_{ik} = \frac{\left|\begin{array}{c|c} D^{(0)}_j & \begin{array}{c} D \\ \hline F \end{array} \\ \hline C \mid E & G \end{array}\right|}{D^{(0)}_j} \quad \begin{array}{c} j \leq i \\ j \leq k \end{array}$$

δ_{ij}: Kronecker delta, = 1 if $i = j$, = 0 if $i \neq j$.
A: j terms; each with a value of $(-\delta_{mi})$, where $m = 1, 2,..., j$.
B: j terms; each with a value of $(-\delta_{mk})$, where $m = 1, 2,..., j$.
C: $(j - 1)$ terms; each with a value of $(1 - \delta_{ji})y^{(0)}_{mi}$, where $m = 1, 2,..., (j - 1)$.
D: $(j - 1)$ terms; each with a value of $(1 - \delta_{jk})y^{(0)}_{mk}$, where $m = 1, 2,..., (j - 1)$.
E: One term; $(1 - \delta_{ji})y^{(0)}_{ji} - \delta_{ji}$.
F: One term; $(1 - \delta_{jk})y^{(0)}_{jk} - \delta_{jk}$.
G: $(1 - \delta_{ji})(1 - \delta_{jk})y^{(0)}_{ik}$.
and

$$D^{(0)}_j = \begin{vmatrix} y^{(0)}_{11} & y^{(0)}_{12} & \cdots & y^{(0)}_{1j} \\ y^{(0)}_{21} & & & \cdot \\ \cdot & & & \cdot \\ \cdot & & & \cdot \\ \cdot & & & \cdot \\ y^{(0)}_{j1} & \cdot\ \cdot\ \cdot & \cdot & y^{(0)}_{jj} \end{vmatrix}$$

Table 5.5 Relationships Between Second-Order Derivatives of the jth Legendre Transform and the $(j-q)$ Transform[a]

$$y_{ik}^{(j)} = \frac{\begin{vmatrix} D_q^{(j-q)} & \begin{matrix} B' \\ B'' \end{matrix} \\ A' \mid A'' & H' \end{vmatrix}}{D_q^{(j-q)}} \quad j > i, k \qquad y_{ik}^{(j)} = \frac{\begin{vmatrix} D_q^{(j-q)} & \begin{matrix} D' \\ F' \end{matrix} \\ A' \mid A'' & I' \end{vmatrix}}{D_q^{(j-q)}} \quad \begin{matrix} j > i \\ j \le k \end{matrix}$$

$$y_{ik}^{(j)} = \frac{\begin{vmatrix} D_q^{(j-q)} & \begin{matrix} D' \\ F' \end{matrix} \\ J' \mid K' & L' \end{vmatrix}}{D_q^{(j-q)}} \quad \begin{matrix} j \le i \\ j \le k \end{matrix}$$

and

$$D_q^{(j-q)} = \begin{vmatrix} y_{(j-q+1)(j-q+1)}^{(j-q)} & y_{(j-q+1)(j-q+2)}^{(j-q)} & \cdots & y_{(j-q+1)j}^{(j-q)} \\ y_{(j-q+2)(j-q+1)}^{(j-q)} & y_{(j-q+2)(j-q+2)}^{(j-q)} & \cdots & \\ \cdot & & & \\ \cdot & & & \\ \cdot & & & \\ y_{j(j-q+1)}^{(j-q)} & \cdots & & y_{jj}^{(j-q)} \end{vmatrix}$$

---

[a] $Z_{rs} = 0$ if $r \le s$
$= 1$ if $r > s$

A': $(q-1)$ terms; each with a value of $[Z_{(j-q+1)i}\, y_{(j-p)i}^{(j-q)} - \delta_{(j-p)i}]$ where $p = (q-1), (q-2), \ldots, 1$

A'': $Z_{(j-q+1)i}\, y_{ji}^{(j-q)}$

B': $(q-1)$ terms, each with a value of $[Z_{(j-q+1)k}\, y_{(j-p)k}^{(j-q)} - \delta_{(j-p)k}]$ where $p = (q-1), (q-2), \ldots, 1$

B'': $Z_{(j-q+1)k}\, y_{jk}^{(j-q)}$

H': $Z_{(j-q+1)i} Z_{(j-q+1)k}\, y_{ik}^{(j-q)}$

D': $(q-1)$ terms, each with a value of $(1 - \delta_{jk})\, y_{(j-p)k}^{(j-q)}$ where $p = (q-1), (q-2), \ldots, 1$

F': $(1 - \delta_{jk})\, y_{jk}^{(j-q)} - \delta_{jk}$

I': $(1 - \delta_{jk}) Z_{(j-q+1)i}\, y_{ik}^{(j-q)}$

J': $(q-1)$ terms, each with a value of $(1 - \delta_{ji})\, y_{(j-p)i}^{(j-q)}$ where $p = (q-1), (q-2), \ldots, 1$

K': $(1 - \delta_{ji})\, y_{ji}^{(j-q)} - \delta_{ji}$

L': $(1 - \delta_{ji})(1 - \delta_{jk})\, y_{ik}^{(j-q)}$

Example 5.6

Relate the derivative $y_{ik}^{(j)}$ to second-order derivative of $y^{(0)}$ for four cases:

(a) $j = 3, i = 1, k = 2$

(b) $j = 3, i = 1, k = 3$

(c) $j=2, i=3, k=4$

(d) $j=3, i=3, k=3$

Solution

Table 5.4 is employed. The results are:

(a) $j=3, i=1, k=2$

$$y_{12}^{(3)} = \frac{\begin{vmatrix} D_3^{(0)} & \begin{vmatrix} -\delta_{12} \\ -\delta_{22} \\ -\delta_{32} \end{vmatrix} \\ \begin{matrix} -\delta_{11} & -\delta_{12} & -\delta_{13} \end{matrix} & 0 \end{vmatrix}}{D_3^{(0)}} = \frac{\begin{vmatrix} y_{12}^{(0)} & y_{13}^{(0)} \\ y_{23}^{(0)} & y_{33}^{(0)} \end{vmatrix}}{D_3^{(0)}}$$

(b) $j=3, i=1, k=3$

$$y_{13}^{(3)} = \frac{\begin{vmatrix} D_3^{(0)} & \begin{vmatrix} (1-\delta_{33})y_{13}^{(0)} \\ (1-\delta_{33})y_{23}^{(0)} \\ (1-\delta_{33})y_{33}^{(0)} - \delta_{33} \end{vmatrix} \\ \begin{matrix} -\delta_{11} & -\delta_{21} & -\delta_{31} \end{matrix} & 0 \end{vmatrix}}{D_3^{(0)}} = -\frac{\begin{vmatrix} y_{12}^{(0)} & y_{13}^{(0)} \\ y_{22}^{(0)} & y_{23}^{(0)} \end{vmatrix}}{D_3^{(0)}}$$

(c) $j=2, i=3, k=4$

$$y_{34}^{(2)} = \frac{\begin{vmatrix} D_2^{(0)} & \begin{vmatrix} (1-\delta_{24})y_{14}^{(0)} \\ (1-\delta_{24})y_{24}^{(0)} - \delta_{24} \end{vmatrix} \\ \begin{matrix} (1-\delta_{23})y_{13}^{(0)} & (1-\delta_{23})y_{23}^{(0)} - \delta_{23} \end{matrix} & (1-\delta_{23})(1-\delta_{24})y_{34}^{(0)} \end{vmatrix}}{D_2^{(0)}} = \frac{\begin{vmatrix} y_{11}^{(0)} & y_{12}^{(0)} & y_{14}^{(0)} \\ y_{21}^{(0)} & y_{22}^{(0)} & y_{24}^{(0)} \\ y_{13}^{(0)} & y_{23}^{(0)} & y_{34}^{(0)} \end{vmatrix}}{D_2^{(0)}}$$

(d) $j=3, i=3, k=3$

$$y_{33}^{(3)} = -\frac{D_2^{(0)}}{D_3^{(0)}}$$

Example 5.7

Using Table 5.5, express:

(a) $y_{23}^{(4)}$ in terms of derivatives of $y^{(1)}$.

(b) $y_{26}^{(5)}$ in terms of derivatives of $y^{(2)}$.

Solution

(a) In this case $j = 4$, $i = 2$, $k = 3$, $q = 3$. We will use the case where $j > i, k$. Notice that $Z_{(j-q+1)i} = Z_{(4-3+1)i} = Z_{2i}$. Since $2 = i$, then $Z_{2i} = 0$. Also, $Z_{(j-q+1)k} = 0$.

$$D_q^{(j-q)} = D_3^{(1)}$$

where

$$D_3^{(1)} = \begin{vmatrix} y_{22}^{(1)} & y_{23}^{(1)} & y_{24}^{(1)} \\ y_{23}^{(1)} & y_{33}^{(1)} & y_{34}^{(1)} \\ y_{24}^{(1)} & y_{34}^{(1)} & y_{44}^{(1)} \end{vmatrix}$$

Then

$$y_{23}^{(4)} = \frac{\begin{vmatrix} D_3^{(1)} & \begin{matrix} -\delta_{23} \\ -\delta_{33} \\ 0 \end{matrix} \\ -\delta_{22} \quad -\delta_{32} \quad 0 & 0 \end{vmatrix}}{D_3^{(1)}} = \frac{\begin{vmatrix} y_{23}^{(1)} & y_{24}^{(1)} \\ y_{34}^{(1)} & y_{44}^{(1)} \end{vmatrix}}{D_3^{(1)}}$$

(b) Here

$$\begin{aligned} &j > i \\ &j < k \\ &j = 5,\ i = 2,\ k = 6,\ q = 3, \qquad Z_{(j-q+1)i} = Z_{32} = 1 \end{aligned}$$

$$y_{26}^{(5)} = \frac{\begin{vmatrix} y_{33}^{(2)} & y_{34}^{(2)} & y_{35}^{(2)} & y_{36}^{(2)} \\ y_{34}^{(2)} & y_{44}^{(2)} & y_{45}^{(2)} & y_{46}^{(2)} \\ y_{35}^{(2)} & y_{45}^{(2)} & y_{55}^{(2)} & y_{56}^{(2)} \\ y_{32}^{(2)} & y_{42}^{(2)} & y_{52}^{(2)} & y_{26}^{(2)} \end{vmatrix}}{D_3^{(2)}} = -\frac{\begin{vmatrix} y_{23}^{(2)} & y_{24}^{(2)} & y_{25}^{(2)} & y_{26}^{(2)} \\ y_{33}^{(2)} & y_{34}^{(2)} & y_{35}^{(2)} & y_{36}^{(2)} \\ y_{34}^{(2)} & y_{44}^{(2)} & y_{45}^{(2)} & y_{46}^{(2)} \\ y_{35}^{(2)} & y_{45}^{(2)} & y_{55}^{(2)} & y_{56}^{(2)} \end{vmatrix}}{D_3^{(2)}}$$

Let us now consider derivatives in intensive variables only and show that they can be reduced to expressions involving only S, T, P, V. In general, a partial derivative of intensive variables may involve U, H, A, G, S, T, P, V. We have excluded N because it is extensive; μ is also excluded because μ is equal to G for a pure material.

The first step in reducing the general derivative $(\partial b/\partial c)_d$ is to eliminate any of the four potential functions U, H, A, or G, which may appear as b, c, or d. The three possibilities are treated as follows:

1. If b is U, H, A, or G, eliminate it by using the differential form of the transform. For example, with Eq. (5-101), for a closed, non-reacting system,

$$\left(\frac{\partial G}{\partial S}\right)_T = -S\left(\frac{\partial T}{\partial S}\right)_T^{\!\!\!\!\!\!\!0} + V\left(\frac{\partial P}{\partial S}\right)_T \tag{5-126}$$

2. If c is U, H, A, or G, invert the derivative using Eq. (5-30):

$$\left(\frac{\partial c}{\partial b}\right)_d = \frac{1}{(\partial b/\partial c)_d} \tag{5-127}$$

   and then proceed by step 1.

3. If d is U, H, A, or G, bring the potential function into the brackets by using the triple product relation of Eq. (5-32):

$$\left(\frac{\partial b}{\partial c}\right)_d = -\frac{(\partial d/\partial c)_b}{(\partial d/\partial b)_c} \tag{5-128}$$

   and then proceed by step 1.

Using these three steps, we can reduce any partial derivative involving only S, V, T, and P to a second-order derivative of one of the four forms of the Fundamental Equation. This is illustrated in Example 5.8 below.

Example 5.8

For a pure material there are 24 partial derivatives involving S, T, P, and V. Suggest a technique to relate any such derivative to second-order derivatives of the Gibbs energy and, therefore, to C_p, κ_T, and α_p where

$$\kappa_T \equiv -\frac{1}{V}\left(\frac{\partial V}{\partial P}\right)_T \quad \text{and} \quad \alpha_p \equiv \frac{1}{V}\left(\frac{\partial V}{\partial T}\right)_P$$

Solution

Of the 24 derivatives, 12 are simply inverses. The others are as follows: (1) $(\partial S/\partial T)_p$, (2) $(\partial S/\partial T)_V$, (3) $(\partial S/\partial P)_T$, (4) $(\partial S/\partial P)_V$, (5) $(\partial S/\partial V)_T$, (6) $(\partial S/\partial V)_p$, (7) $(\partial V/\partial P)_T$, (8) $(\partial V/\partial P)_S$, (9) $(\partial V/\partial T)_p$, (10) $(\partial V/\partial T)_S$, (11) $(\partial P/\partial T)_S$, (12) $(\partial P/\partial T)_V$. Number (3) is equal to the negative of (9) from the Maxwell reciprocity theorem; similarly for (4) and (10). Number (5) is equal to (12) and (6) to (11). Thus, there remain (1) through (8). Three of these are second-order derivatives of the Gibbs energy:

(1) $(\partial S/\partial T)_p = -G_{TT} = C_p/T$

(3) $(\partial S/\partial P)_T = -G_{TP} = -G_{PT} = -(\partial V/\partial T)_p = -V\alpha_p$

(7) $(\partial V/\partial P)_T = G_{PP} = -V\kappa_T$

In addition, derivatives (2), (5), and (10) are related by the X-Y-Z-1 rule, as are (4), (6), and (8). [The sets (1), (3), (11) and (7), (9), (12) are also so related, but these derivatives have already been eliminated.] Thus, it is only necessary to relate (2), (4), and (6) to derivatives in $\underline{G}$ to allow all derivatives to be so expressed. The reduction of (2) has been introduced as Problem 5.12. For (4) and (6), we define our basis function $\underline{G} = y^{(0)} = f(T, P, N_1, ...)$ where the x–ξ values are shown below. For (6), we define $y^{(1)} = \underline{H} = f(\underline{S}, P, N_1, ...)$ and the x–ξ values are also shown below.

$y^{(0)}$	x	ξ
1.	T	$-\underline{S}$
2.	P	$\underline{V}$
3.	N_1	μ_1, etc.

$y^{(1)}$	x	ξ
1.	$-\underline{S}$	$-T$
2.	P	$\underline{V}$
3.	N_1	μ_1, etc.

Thus, since $\underline{V} = y_2^{(1)}$,

$$\left(\frac{\partial \underline{S}}{\partial \underline{V}}\right)_{P,N} = \left(\frac{\partial S}{\partial V}\right)_P = (-y_{12}^{(1)})^{-1}$$

But, by Table 5.3,

$$y_{12}^{(1)} = \frac{y_{12}^{(0)}}{y_{11}^{(0)}} = \frac{G_{TP}}{G_{TT}} = -\frac{\alpha_p VT}{C_P}$$

and $(\partial S/\partial V)_P = C_p/\alpha_p VT$.

For derivative (4), using the same basis function, we need to use a double transform; thus, $y^{(2)} = \underline{U} = f(\underline{S}, \underline{V}, N_1, ...)$ with

$y^{(2)}$	x	ξ
1.	$-\underline{S}$	$-T$
2.	$\underline{V}$	$-P$
3.	N_1	μ_1, etc.

and $(\partial \underline{S}/\partial P)_{\underline{V},N} = N(\partial S/\partial P)_V = N(y_{12}^{(2)})^{-1}$, since $P = -y_2^{(2)}$. From Table 5.4,

$$y_{12}^{(2)} = \frac{y_{12}^{(0)}}{y_{11}^{(0)}y_{22}^{(0)} - (y_{12}^{(0)})^2} = \frac{G_{TP}}{G_{TT}G_{PP} - G_{TP}^2}$$

and

$$\left(\frac{\partial S}{\partial P}\right)_V = \frac{G_{TT}G_{PP} - G_{TP}^2}{G_{TP}} = \frac{C_p \kappa_T}{T\alpha_P} - V\alpha_P$$

Although the procedure described above is always valid and unequivocal, there are shortcut methods such as those introduced earlier in Sections 5.3 and 5.4 which, when applicable, are usually less tedious.

Example 5.9

Evaluate the following partial derivatives as functions of P, V, T, their partial derivatives, and C_p: (a) $(\partial S/\partial P)_G$; (b) $(\partial A/\partial G)_T$.

Solution

(a) Using Eqs. (5-126) and (5-128), we find that

$$\left(\frac{\partial S}{\partial P}\right)_G = -\frac{(\partial G/\partial P)_S}{(\partial G/\partial S)_P} = \frac{-S(\partial T/\partial P)_S + V}{S(\partial T/\partial S)_P} \tag{5-129}$$

Using Eq. (5-128) to eliminate $(\partial T/\partial P)_S$,

$$\left(\frac{\partial T}{\partial P}\right)_S = -\frac{(\partial S/\partial P)_T}{(\partial S/\partial T)_P} = \frac{(\partial V/\partial T)_P}{C_p/T} \tag{5-130}$$

so that

$$\left(\frac{\partial S}{\partial P}\right)_G = \frac{VC_p}{TS} - \left(\frac{\partial V}{\partial T}\right)_P \tag{5-131}$$

The entropy, S, in Eq. (5-131) can be expressed as a function of the desired variables, as demonstrated in Section 8.1 [see Eq. (8.23)].

(b) This partial derivative can be reduced by the following shortcut procedure

$$\left(\frac{\partial A}{\partial G}\right)_T = \frac{(\partial A/\partial V)_T}{(\partial G/\partial V)_T} = -\frac{P}{V(\partial P/\partial V)_T}$$

in which Eq. (5-34) was employed. The insertion of V was a convenient but not an arbitrary choice. Although P could have been used in place of V with equal simplicity, S would have been less convenient. The choice was guided by the fact that $A = f(T, V)$ and $G = f(T, P)$; since T was the constant in the differentiation, either T, V or T, P would be a convenient set of independent variables. As an alternative, Jacobian transforms could be used.

5.10 Summary

Starting with the Fundamental Equation $U = f(S, V, N)$ as a basis function, this chapter introduces a number of mathematical techniques for manipulating variables. The important methods discussed were:

1. Euler integration of the Fundamental Equation
2. Maxwell's reciprocity theorem for relating 2nd and higher order derivatives
3. Derivative inversion, triple product, and chain rule expansion for manipulating variables
4. Jacobian transformations for derivative manipulation
5. Legendre transformations to express the $\underline{U} = f(\underline{S}, \underline{V}, N)$ Fundamental Equation in terms of different variables without losing any information content
6. Modifications to the Fundamental Equation to account for non-PV work effects
7. Derivation of relationships involving partial derivatives of Legendre transforms

In addition, graphical techniques were used to illustrate the geometric relationships of the *USV* surface and various Legendre transformed functions.

References and Suggested Readings

Alberty, R.A. (1994), "Legendre transforms in chemical thermodynamics," *Chemical Reviews*, **94**(6), p 1457-1482 [Chemist's perspective on Legendre transforms]

Aris, R. and N.R. Amundson, (1973), *Mathematical Methods in Chemical Engineering*, Prentice Hall, Englewood Cliffs, NJ, Vol 2, p 197-201. [Introduction to Legendre transformations]

Beegle, B.L., M. Modell, and R.C. Reid (1974), "Legendre transforms and their application in thermodynamics," *AIChE J.* **20** (2), p 1194-1200.

Callen, H.B. (1985), *Thermodynamics and an Introduction to Thermostatistics*, 2nd ed, Wiley, New York. [Legendre transformations]

Clark, A.L. and L. Katz (1939), "Thermodynamic surfaces of H_2O," *Trans Royal Soc. Canada*, 3rd series, Section III, **33**, p. 59-72 [Figure 5.1a taken from p 72].

Coy, D.C. (1993), "Visualizing thermodynamic stability and phase equilibrium through computer graphics, PhD dissertation, Dept. of Chem. Eng., Iowa State Univ., Ames, Iowa [3-D images of variously transformed and scaled fundamental surfaces of pure, binary, and ternary systems]

Debenedetti, P.G. (1986), "Generalized Massieu-Planck functions: Geometric representation extrema and uniqueness properties," *J. Chem. Phys.* **85**(4), p 2131-2139.

Gyftopoulos, E.P. and G.P. Beretta (1991), *Thermodynamics: Foundations and Applications*, MacMillan, New York. [Mechanical engineering perspective]

Jolls, K.R. (1990), "Gibbs and the art of thermodynamics", *Proceedings of the Gibbs Symposium*, published in G.D. Mostow and D.G. Caldi (eds.) *Amer. Math. Soc.* and presented at Yale University, New Haven, CT (May 15-17, 1989) p 293-321. [computer generated fundamental surfaces for a pure fluid]

Jolls, K.R., M.C. Schmitz, and D.C. Coy (1991), "A new look at an old subject," *The Chemical Engineer*, Institution of Engineers (UK), No. 497, May 30, p 42.

Jolls, K.R. and D.C. Coy (1992), "Gibbs's models visualized," *Physics Today*, March, p 96.

Kumar, S.K. and R.C. Reid, (1986), "Derivation of the relationships between partial derivatives of Legendre transforms", *AIChE J.* **32** (7), p 1224-1226.

Tisza, L. (1966), *Generalized Thermodynamics*, MIT Press, Cambridge, MA, Ch. 2 in general and specifically, p 61, 136, 236. [Transformation methods]

Problems

5.1. (a) Demonstrate the utility of Eq. (5-96) for a system with six components and where $i = 4$. Order the Fundamental Equation as

$$\underline{U} = f(\underline{S}, \underline{V}, N_1, ..., N_6)$$

(b) If one were to write the Fundamental Equation as

$$\underline{S} = f(\underline{U}, \underline{V}, N_1, ..., N_n)$$

What would the derivative of the total Legendre transform be?

(c) Suppose that one were write the Fundamental Equation as

$$y^{(0)} = \underline{G} = f(T, P, N_1, ..., N_n)$$

Prepare a table showing the various conjugate coordinates, ξ_i, x_i. Next, write the third Legendre transform of the basis function $y^{(0)}$ shown above. Prepare a ξ_i, x_i table for this transform. What generalization can you infer from this exercise?

(d) Choose the basis function for the Fundamental Equation to be

$$y^{(0)} = \underline{A} = f(\underline{V}, N_1, ..., N_n, T)$$

Obtain an expression for $y^{(1)}_{22}$ in terms of derivatives $y^{(0)}$.

5.2. Given $y^{(0)} = f(\underline{S}, \underline{V}, N_1\ N_2, ..., N_n)$ obtain the Legendre transform and its differential if one wished to work in the following coordinate systems:

(a) $f(T, \underline{V}, \mu_1, ..., \mu_n)$

(b) $f[(1/T), \underline{V}, N_1, N_2, ..., N_n]$

5.3. Express the following partial derivatives in an equivalent form using the Fundamental Equation

$$\underline{U} = U(\underline{S}, \underline{V}, N_1, N_2, ..., N_n)$$

and Maxwell's relations. Indicate, if possible, how they might be measured experimentally.

(a) $\left(\frac{\partial y^{(2)}}{\partial \underline{V}}\right)_{U_S, N}$ (b) $\left(\frac{\partial \underline{U}_{N_j}}{\partial \underline{U}_S}\right)_{\underline{V}, N}$ (c) $\left(\frac{\partial y^{(1)}}{\partial \underline{V}}\right)_{U_S, N}$ (d) $\left(\frac{\partial \underline{S}}{\partial \underline{U}_S}\right)_{y^{(2)}, N}$

5.4. Carry out the following transformations.

(a) Express $(\partial \underline{S}/\partial \underline{V})_{T, N_1, \ldots, N_n}$ as a function of P, $\underline{V}$, T, and their derivatives.

(b) Express $(\partial \underline{H}/\partial P)_{T, N_1, \ldots, N_n}$ as a function of $\underline{G}$ and its derivatives and show how these may be given in terms of P, $\underline{V}$, T, N, and C_p.

(c) Express $[\partial(\underline{A}/T)/\partial(1/T)]_{\underline{V}, N_1, \ldots, N_n}$ as a function of $\underline{U}$ and its derivatives.

(d) Express $(\partial \underline{H}/\partial \underline{V})_{T, N_1, \ldots, N_n}$ as a function of $\underline{G}$ and its derivatives and show how these may be given in terms of P, $\underline{V}$, T, N, and C_p.

(e) Express $(\partial T/\partial N_A)_{\underline{V}, \underline{S}, \mu_B, N_c, \ldots}$ as a function of $\underline{U}$ and its *independent* derivatives.

(f) Express $(\partial T/\partial N_A)_{\underline{S}, P, \mu_B, N_c, \ldots}$ as a function of $\underline{U}$ and its *independent* derivatives.

5.5. For a one-component system, show:

(a) $$\left(\frac{\partial \mu}{\partial N}\right)_{T, \underline{V}} = \left(-\frac{\underline{V}}{N}\right)\left(\frac{\partial \mu}{\partial \underline{V}}\right)_{T, N}$$

(b) $$\left(\frac{\partial P}{\partial N}\right)_{T, \underline{V}} = \left(-\frac{\underline{V}}{N}\right)\left(\frac{\partial P}{\partial \underline{V}}\right)_{T, N}$$

5.6. The following discussion is limited to one-dimensional motion along the x coordinate and for a constant-mass system. Newtonian mechanics relates the force on a particle to the mass and acceleration; that is,

$$F = ma = m\ddot{x} \qquad \left(\ddot{x} = \frac{d^2x}{dt^2}\right)$$

Another way in which to study the dynamics of motion is with Lagrangian mechanics. In this case, a function L is defined as $L = \Im - \Phi$, where $\Im$ = the kinetic energy = $m\dot{x}^2/2$ and Φ is the potential energy = $f(x)$. Newton's Law in Lagrangian mechanics is given as

$$\left(\frac{\partial L}{\partial x}\right)_{\dot{x}} = \left(\frac{\partial}{\partial t}\right)\left(\frac{\partial L}{\partial \dot{x}}\right)_x$$

$$L = f(x, \dot{x}) \qquad \dot{x} = \frac{\partial x}{\partial t} = \text{velocity}$$

Another branch of mechanics uses a function (-**H**), which is a function of x and the momentum of a particle, p; that is,

$$-\mathbf{H} = f(p, x)$$

with

$$p = \left(\frac{\partial L}{\partial \dot{x}}\right)_x = \left(\frac{\partial \mathfrak{I}}{\partial \dot{x}}\right)_x = m\dot{x}$$

(a) Using the concepts of Legendre transforms, define $-\mathbf{H}$ in terms of L, p, and $\dot{x}$.

(b) In the latter branch of mechanics, $\mathbf{H}$ is called the Hamiltonian; with your definition of $\mathbf{H}$ complete the following equations:

$$\left(\frac{\partial \mathbf{H}}{\partial p}\right)_x = \quad \text{and} \quad \left(\frac{\partial \mathbf{H}}{\partial x}\right)_p =$$

5.7. Express the following in terms of C_p, P, V, T and derivatives of these variables.

(a) $(\partial S/\partial T)_{G,N}$

(b) $(\partial A/\partial G)_{T,N}$

5.8. Prove that for an n-component mixture there are $(n+2)(n+1)/2$ independent second-order derivatives of the Fundamental Equation.

5.9. A spherical tank contains 1 mol of helium gas at 10 bar and 300 K (see Figure P5.9). We would like to carry out an experiment in which helium is released to the atmosphere, but at the same time, the remaining contents of the sphere maintain a constant total energy,

$\underline{U}$. Heating or cooling coils may be used to keep $\underline{U}$ constant during venting. Helium behaves as an ideal gas with a constant value of $C_v = 12.6$ J/mol K. Choose a base state where the specific enthalpy is zero at 300 K.

(a) When one-half of the helium has been vented, what is the temperature and pressure of the residual helium? What is the heat interaction?

(b) Determine $(\partial T/\partial P)_{\underline{U},\underline{V}}$ at the instant when venting begins.

(c) Repeat parts (a) and (b) if the base state were chosen so that the specific internal energy, U, were zero at 300 K.

(d) For part (a), what would be the residual helium temperature when 60% of the gas had been removed?

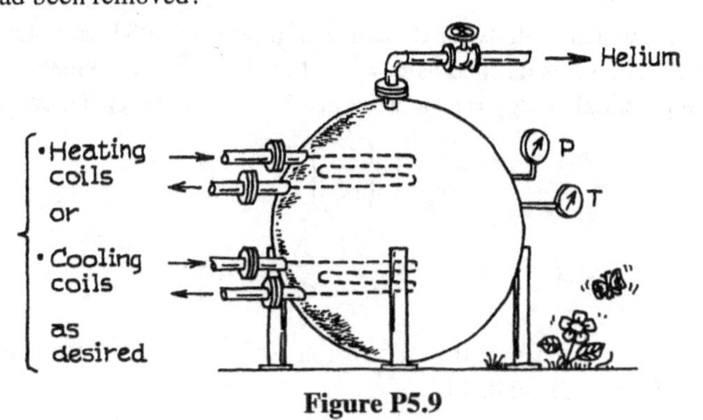

Figure P5.9

5.10. Choose as a basis function

$$y^{(0)} = f(\underline{S}, N_1, N_2, N_3, P)$$

and obtain $y_{12}^{(1)}$ in terms of derivatives of the basis function. Discuss how experiments could be designed and conducted to obtain numerical values of the $y^{(0)}$ derivatives.

Using your result, consider the following problem. We have a system containing, initially, N_i moles of a material. We wish to remove $N_i/2$ moles under conditions where the total entropy and the pressure remain constant. If the initial temperature is 400 K, what is the final temperature? Assume that the base state for entropy is such that at 400 K and the system pressure, the specific entropy is 10 J/mol K. Also assume that the heat capacity is 10 J/mol K, independent of temperature.

5.11. If the basis function is chosen as

$$y^{(0)} = \underline{U} = f(\underline{S}, \underline{V}, N_A, N_B, \ldots)$$

show that

$$y_{14}^{(2)} = G_{TN_B} = \frac{\begin{vmatrix} U_{VV} & U_{VN_B} \\ U_{SV} & U_{SN_B} \end{vmatrix}}{\begin{vmatrix} U_{SS} & U_{SV} \\ U_{SV} & U_{VV} \end{vmatrix}}$$

by performing two single-step transforms from $y^{(2)}$ to $y^{(1)}$ and then to $y^{(0)}$.

5.12. Assume that the basis Legendre transform is $y^{(0)} = f(P, T, N_1, N_2, \ldots)$. Express $y_{22}^{(1)}$ in terms of derivatives of $y^{(0)}$ and interpret the results on a physical basis.

5.13. Express $y_{44}^{(3)}$ in terms of derivatives of $y^{(1)}$ for a ternary mixture when the basis function

$$y^{(0)} = \underline{U} = f(\underline{S}, \underline{V}, N_A, N_B, N_C)$$

5.14. Express $y_{24}^{(2)}$ in terms of derivatives of the basis function

$$y^{(0)} = \underline{U} = f(\underline{S}, \underline{V}, N_A, \ldots)$$

5.15. Show that

$$y_{(n+1)(n+1)}^{(n)} = \left(\frac{\partial \xi_{n+1}}{\partial x_{n+1}}\right)_{\xi_1, \ldots, \xi_n, x_{n+2}}$$

5.16. $y^{(m-2)}$ is a Legendre transform of a basis function $y^{(0)}$, where

$$y^{(0)} = f(x_1, x_2, \ldots, x_m)$$

$$y^{(m-2)} = f(\xi_1, \xi_2, \ldots, \xi_{m-2}, x_{m-1}, x_m)$$

Derive a general equation to express the derivative $y^{(m-2)}_{(m-1)(m-1)}$ in terms of derivatives of a Legendre transform $y^{(r)}$, where $0 \le r < m-2$. Define any derivatives and show what variables are held constant.

5.17. Show that $\kappa = \dfrac{C_p}{C_v} = \left[\dfrac{\partial V}{\partial P}\right]_T \left[\dfrac{\partial P}{\partial V}\right]_S$ and check to see if the relation holds for an ideal gas.

5.18. (a) Derive Eq. (5-118), for the derivative operator:

$$\left(\frac{\partial -}{\partial x_i}\right)_{\xi_1, x_2, \ldots [x_i], \ldots, x_m} = \left(\frac{\partial -}{\partial x_i}\right)_{x_1, \ldots [x_i] \ldots, x_m} - \frac{y^{(0)}_{1i}}{y^{(0)}_{11}}\left(\frac{\partial -}{\partial x_1}\right)_{x_2, \ldots, x_m}$$

(b) Show that $y^{(1)}_{1i} = y^{(0)}_{1i}/y^{(0)}_{11}$ for $i \ne 1$ as given in Table 5.1.

(c) Express the derivative $(\partial P/\partial T)_{\underline{S}, \mu_1, N_2, \ldots, N_n}$ in terms of derivatives of the basis function $y^{(0)} = \underline{G}$.

(d) Show that the result given in part (c) above reduces to

$$\left(\frac{\partial P}{\partial T}\right)_{\underline{S}, \mu} = \left(\frac{\partial P}{\partial T}\right)_{N, \mu} \quad \text{for a pure material}$$

5.19. (a) Develop a suitable expression for $(\partial T/\partial P)_H/(\partial T/\partial P)_S$ in terms of PVT properties and their derivatives.

(b) Starting with the basis function:

$$y^{(0)} = \underline{U}^* = f(\underline{S}, \underline{V}, \underline{a}, N_1, N_2, \ldots, N_n)$$

where new conjugate coordinates $\{x_i, \xi_i\} = \{\underline{a}, \sigma\}$ have been introduced to account for surface effects, $\underline{a}$ for area and σ for surface tension, develop a suitable expression for: $(\partial \underline{S}/\partial \underline{a})_{T, \underline{V}, N}$ in terms of the appropriate 2nd derivative of a 1st order Legendre transform and its derivatives and show that it can be equated to a temperature derivative of σ.

5.20. The thermodynamics of rubber subjected to uniaxial tension is described by a fundamental equation given in differential form with all extensive quantities reduced by mass as,

$$dU = TdS + V_o \tau d\varepsilon$$

$\varepsilon = (L - L_o)/L_o$ = linear tensile strain

$\tau = F/A_o$ = uniaxial tensile stress

L = length in the direction of tension

F = tensile force exerted on the rubber

$L_o, \underline{a}_o, V_o$ = length, area, and volume in the unstressed state.

(a) Starting from this fundamental equation, express the stress τ as a linear combination of the isothermal derivatives of U and S with respect to strain.

(b) For understanding rubber elasticity, it is important to determine the relative contributions of internal energy and entropy to τ. To achieve this, we perform the following experiment: Keeping the ends of a stretched rubber band fixed, we measure the stress τ as a function of temperature T as indicated in the figure. Show how the internal energy and entropy contributions to stress can be separately evaluated on the basis of these measurements. Give a graphical interpretation on the plot.

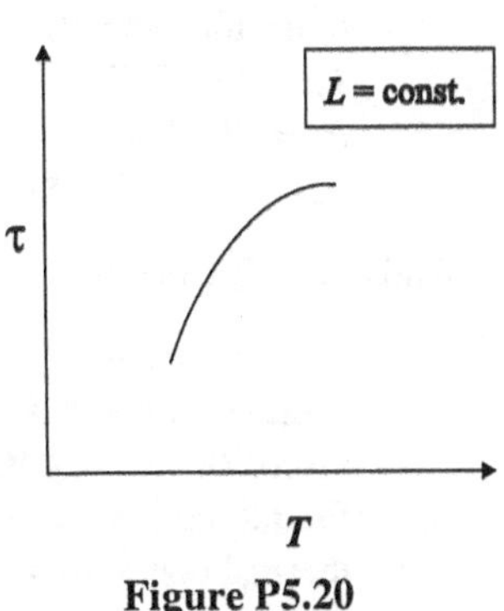

Figure P5.20

(c) A 1935 experiment by Meyer and Ferri indicated that, for a typical rubber extended to 4.5 times its original length at temperatures $T > 210$ K, the function depicted in the figure is adequately described by the equation $\tau = T/200$, with T in K and τ in *MPa*. What conclusion can you draw about entropic contributions to the elastic response?

5.21. Rocky and Rochelle Jones, while backpacking near the Presidential Range of the White Mountains of New Hampshire, were trapped in a severe thunderstorm. Rocky was very worried that they would be struck by lightning. As thunder and lightning bolts crashed along the mountain ridges, he remembered that one ought to be able to estimate the distance to a lightning strike by counting the seconds between when you first see the strike and when you first hear the thunder. Rochelle recalled her 10.40 training and remembered that the speed of sound could be related to certain thermodynamic properties. Deep in her photographic memory she recalled that:

$$v_c^2 = (\partial P/\partial \rho)_S,$$

where v_c is the sound speed, P is the pressure, ρ is the fluid density, and S is the entropy.

But neither Rochelle nor Rocky can quite figure out how to evaluate the isentropic derivative $(\partial P/\partial \rho)_S$ in terms of measurable properties. Please help them out! Rocky just timed the last strike and it was only 2 seconds away. State and justify all assumptions in your analysis.

N.B. Rocky's portable solid-state weather computer which fits into his backpack has provided some valuable meteorological data:

air temperature = 0°C (32°F)
relative humidity = 80%
barometric pressure = 0.99 bar and falling rapidly!
wind speed = 50 mph (80 kmph) from the SE

In addition, Rochelle had a copy of Perry's Handbook with her which provided some other (hopefully useful) thermophysical parameters for air.

$C_p = 29.3$ J/mol K
molecular wt = 29 g/mol
$T_c = -140.7$°C
$P_c = 37.2$ bar
$\rho_c = 1/V_c = 350$ kg/m^3

And at 25°C, 1 bar

$\rho = 1.17$ kg/m^3
Joule-Thompson coefficient = $\alpha_H = (\partial T/\partial P)_H = 0.23$°C/bar
viscosity = 0.19×10^{-4} Pa-s
Prandtl number = Pr = 0.70
thermal conductivity = $\lambda = 2.6 \times 10^{-2}$ W/m K

5.22. (a) Starting with the following alternative definition of C_p and C_v:

$$C_p = T(\partial S/\partial T)_P, \quad C_v = T(\partial S/\partial T)_V$$

show that $C_p - C_v$ can be expressed as a function of *PVT* properties only. Using the result of part (a) show that $C_p - C_v = R$ for an ideal gas.

(b) Given that $C_p - C_v = f(P, V, T)$ from part (a), is it possible to express the ratio $\kappa = C_p/C_v$ as an explicit function of *PVT* properties and their derivatives? Explain your answer.

5.23. A basis function is defined as follows:

$$\underline{\Omega} \equiv f(T, \underline{V}, N_1, N_2, \underline{Z}_1)$$

With conjugate coordinate variable pairs given as

x_i	ξ_i
T	$-\underline{S}$
$\underline{V}$	$-P$
N_1	μ_1
N_2	μ_2
$\underline{Z}_1$	F_1

where $\underline{Z}_1$ is a generalized extensive displacement (first order in mass) and F_1 is a generalized intensive force (zero order in mass) and the other terms have their usual meaning. Provide expressions for the following:

(a) $dy^{(0)}$
(b) $y^{(0)}$
(c) $dy^{(5)}$
(d) $[\partial F_1/\partial \mu_1]_{T, \underline{V}, N_2, \underline{Z}_1}$ in terms of a derivative of N_1
(e) $y^{(1)}_{SSS} = y^{(1)}_{\underline{SSS}}$ in terms of $\underline{\Omega}$ and its derivatives

You may want to use Table 5.3.

5.24. If $y^{(0)} = \underline{U} = f[\underline{S}, \underline{V}, N_1, \ldots, N_n]$, are $y^{(n+2)}$ and $dy^{(n+2)}$ always equal to zero? Answer yes or no and explain briefly.

If $y^{(1)} = f[\mu_1, P, T, N_2, N_3, \ldots, N_n]$, what are $y^{(0)}$ and $dy^{(0)}$?

What derived thermodynamic property does $y^{(0)}$ correspond to?

Give one example of $y_{ij}^{(k)} = y_{ji}^{(k)}$ for any kth Legendre transform of $y^{(0)} = \underline{U}$ and any two independent variables i and j. Express your answer as an equality of partial derivatives involving derived properties ($\underline{U}, \underline{S}, \underline{H}, \underline{G}, \underline{A}$) and/or primitive variables (T, P, V, N_i).

5.25. A basis function for a modified Fundamental Equation that includes reversible electromagnetic work effects in a two-component system is given as:

$$y^{(0)} = \underline{U}^* = f[\underline{S}, \underline{V}, N_1, N_2, \mathbf{B}] = f(x_1, x_2, \ldots, x_5)$$

where the conjugate pair of variables added for electromagnetic work effects is

$$\{x_5, \xi_5\} = \{\underline{V}\mathbf{B}, \mathbf{H}\}$$

and

$\mathbf{H}$ = magnetic field strength (amp/m)

$\underline{V}$ = total volume subject to the magnetic field, includes the system volume and any free space outside the boundaries of the system (m^3)

$\mathbf{B}$ = magnetic induction (Volt-s/m)

For all cases of interest, you can assume that the volumetric magnetic permeability is constant.

(a) What are $y^{(5)}$ and $dy^{(5)}$?

(b) Derive an expression for $(\partial\mu_1/\partial\mathbf{B})_{T, \underline{V}, M, N_2}$

(c) At constant T, P, and $\mathbf{B}$ could you calculate μ_2 as a function of mole fraction of component (x_2) from knowledge of how μ_1 varies with x_1? If so, explain your answer.

5.26. According to a recent article in the March, 1989 issue of *Scientific American*, harbor seals are believed to have sophisticated acoustic transmission and detection systems that are similar to sonar. For example, mother seals locate their pups by measuring the difference in reflected acoustic wave arrival times in air and in water.

Given that the speed of sound (v_c) can be related to an isentropic derivative of pressure with respect to density:

$$v_c^2 = (\partial P/\partial\rho)_S$$

estimate about how far a mother seal is from her pup if the time difference is 3 seconds. Air can be assumed to behave as an ideal gas with $C_p = 5/2\, R$. The properties of liquid water can be obtained from the steam tables. The air and water temperatures are about 2°C.

5.27. A non-ideal gas of constant heat capacity $C_v = 12.56$ J/mol K undergo a *reversible adiabatic* expansion. The gas is described by the van der Waals equation of state

$$(P + a/V^2)(V - b) = RT$$

where $a = 0.1362$ J m^3/mol^2 and $b = 3.215 \times 10^{-5}$ m^3/mol. Derive an expression for the temperature variation of the gas internal energy, and calculate its value when the gas volume of 400 moles is 0.1 m^3 and its temperature is 294 K.

5.28. The basic thermodynamic relationships for an axially stressed bar can be written as:

$$dQ_{rev} = Td\underline{S}, \quad dW_{rev} = -\tau d\underline{\varepsilon}, \quad \underline{\varepsilon} = N\varepsilon$$

where τ is the stress and ε is the strain. Derive the Fundamental Equation for a one-component bar and show that

$$\left(\frac{\partial \mu}{\partial T}\right)_{\tau,N} = -\left(\frac{\partial \underline{S}}{\partial N}\right)_{T,\tau}$$

5.29. A very useful property is the partial molar volume of one component in a mixture. It is defined for an n-component case as

$$\overline{V}_j \equiv (\partial \underline{V}/\partial N_j)_{T,P,N_i[j]}$$

If we have a basis function $y^{(0)}$,

$$y^{(0)} = f(\underline{S}, \underline{V}, N_A, N_B, \ldots, N_n)$$

we can see that, if we desired $\overline{V}_B$, then

$$\overline{V}_B = -y_{24}^{(2)}$$

where

$$y^{(2)} = \underline{G} = f(T, P, N_A, N_B, \ldots, N_n)$$

However, to evaluate $\overline{V}_j$, we normally only have property relations which express P as a function of the independent variable set $\underline{V}, T, N_A, N_B, \ldots, N_n$.

Use this variable set to define a new basis function $y^{(0)}$, and express $y_{24}^{(2)}$ in terms of $\underline{V}, T, N_A, N_B, \ldots, N_n$ using the following equation of state.

$$P = \frac{NRT}{\underline{V} - N_A b_A - N_B b_B - N_C b_C \cdots - N_N b_N}$$

where $b_A, b_B, \ldots, b_N$ are constants and $N = \Sigma N_i$, with $i = a, b, \ldots, N$

5.30. The extensive variable that characterizes magnetic behavior is the magnetic dipole moment $\underline{I}$. In this situation, the fundamental equation is rewritten as

$$\underline{U} = \underline{U}(\underline{S}, \underline{V}, \underline{I}, N)$$

with the added conjugate variable set $[x_i, \xi_i] = [\underline{I}, \mathbf{H}]$ where $\mathbf{H}$ is the magnetic field strength. For a pure component system develop expressions for:

(a) $\underline{U}$

(b) $y^{(4)}$ and $dy^{(4)}$

(c) $(\partial \underline{S}/\partial \mathbf{H})_{\underline{V}, \underline{I}, N}$

Equilibrium Criteria 6

In this part we will draw on the major contributions made by Clausius and Gibbs to the field of classical thermodynamics. Gibbs was a pioneer in reconstructing equivalent mathematical criteria to rigorously describe the conditions of stable equilibrium as set forth by Clausius in his famous statement that the total entropy of a complete system and its surroundings proceeds toward a maximum as equilibrium is approached.

In Postulate I it was stated that stable equilibrium states exist for all simple systems (see Section 2.7). In Postulate II it was stated that all complex systems will approach an equilibrium condition (consistent with any internal restraints) in which each simple system of the composite approaches a stable equilibrium state. It was shown in Example 4.3 that for any permissible processes occurring within a closed, adiabatic (as well as isolated) system, the change in total entropy $\Delta \underline{S}$ must be equal to or greater than zero. The total entropy is, of course, equal to the sum of the entropies of all subsystems of an isolated system. For the case of a non-isolated system, the total entropy would refer to the sum of the entropy contributions of all subsystems plus the surroundings. This, you may recall, is how we have defined $\Delta \underline{S}$ of the "universe".

In this chapter we examine the criteria that can be applied to determine if a system has reached a stable equilibrium state under various constraints. Of more importance, we investigate the consequences of the fact that a system is in a stable equilibrium state. In this way we can specify the criteria for membrane equilibrium, for phase equilibrium, and for chemical reaction equilibrium.

6.1 Classification of Equilibrium States

Up to this point we have used the term *equilibrium* only in connection with stable equilibrium states. In broader usage, any system that does not undergo a change with time is said to be in an equilibrium state. The adjective *stable* is reserved for those equilibrium states which, following a perturbation, will revert to the original equilibrium state. Alternatively, there are equilibrium states that are not stable (i.e., states that may be permanently altered as a result of even a small perturbation).

There are four classes of equilibrium states. For ease of conceptualization, they can be described with the mechanical analogy of a ball on a solid surface in a gravitational field (see Figure 6.1). If the ball were pushed to the right or the left and if it were to return to its original position, the state is stable (case (a)). If the original position were metastable (case (b)), the ball would revert to the original position after a small perturbation, but there is the possibility that a large perturbation would displace the ball to a state of lower potential energy. If the original state were unstable (case (c)), then even a minor perturbation would displace it to a position of lower potential energy. A system in a state of neutral equilibrium (case (d)) would be altered by any perturbation, but the potential energy would remain unchanged.

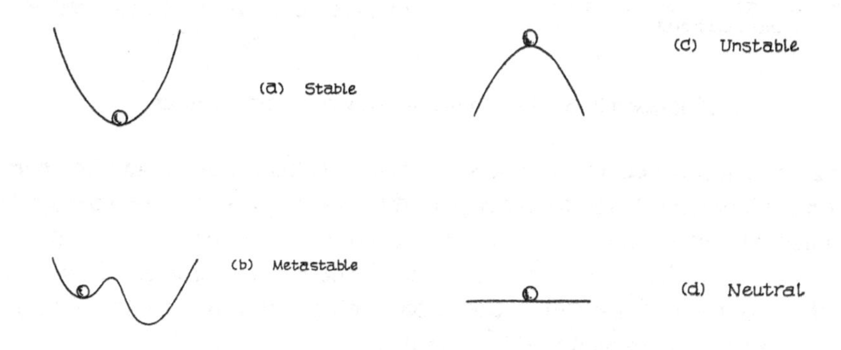

Figure 6.1 Classification of equilibrium states.

In terms of these mechanical analogies, most real systems would be classified as metastable. All organic compounds containing carbon and hydrogen atoms could attain a more stable state in the presence of oxygen by reacting to form CO_2 and H_2O; similarly, most metals are metastable in relation to their oxides. But in many such cases, the barriers to transition to the more stable state may be large enough to prevent the change from occurring within the time span of interest. Thus, practically speaking, these states can be considered stable if the barriers are large in relation to the magnitude of expected perturbations.

Since almost all real systems are metastable with respect to a perturbation of some finite magnitude yet stable with respect to minor perturbations, it is of greater utility to approach equilibrium more pragmatically than that suggested by the simple mechanical analogies of Figure 6.1. We shall consider only minor or virtual perturbations to a given system. If minor perturbations leave the system unchanged, we define the original state as a stable equilibrium state (see Figure 6.2(a)).

In real systems, the extent to which a barrier exists is dictated not by thermodynamic reasoning but by rate or kinetic considerations. If the rate of a possible transition is too slow to be significant within the time span of interest, we consider the barrier as an impenetrable internal restraint. For simple systems, such barriers may be visualized as activation energies that prevent chemical reactions or as the lack of solid nuclei needed to initiate phase transitions. Complex systems may contain additional barriers such as adiabatic, rigid, or impermeable internal walls.

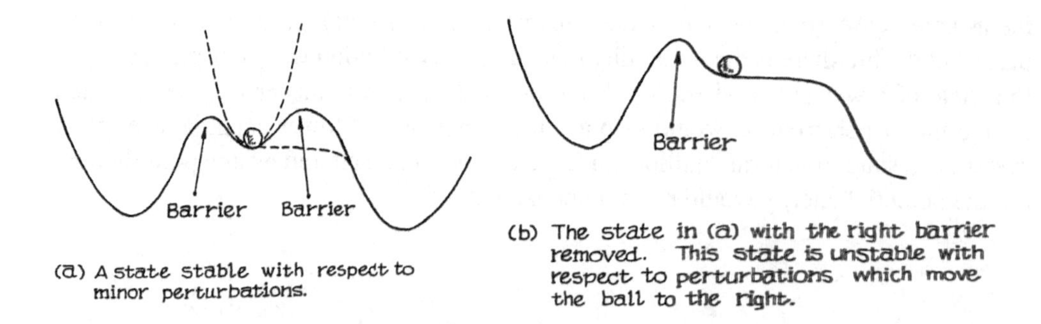

Figure 6.2 Stability limits for different perturbations.

In engineering practice, metastable systems can be treated as stable if the barriers to transition are large in relation to minor perturbations. If one of the barriers is relatively small or nonexistent, the original state is treated as unstable with respect to that particular kind of transition. It should be emphasized that the delineation of internal barriers forms part of the definition of a system. Alteration or elimination of an internal barrier can change a stable system to an unstable system.

An interesting state is depicted by the mechanical analogy in Figure 6.2(b). A minor perturbation to the right leads to an unstable state, whereas a push to the left indicates that the system is stable with respect to this kind of variation. Although the analysis may not reveal what the final state of the system might be, it will indicate that the original system is intrinsically unstable and some transition must result from a minor perturbation. Although it may be difficult experimentally to observe systems at or near their *limit of intrinsic stability*, there is sufficient evidence to suggest that such limits exist, for example the behavior observed at the critical point of a pure material.

In general, potential transitions in real systems are more complicated than the horizontal motion of a ball on a surface, which has only two kinds of perturbations (i.e., displacement to either the left or right). Instead of testing a system for all possible transitions, we usually test for only those that we suspect may occur. Hence, we classify the stability of a system with respect to particular kinds of perturbations.

The process by which we test for stability is to perform thought experiments in which we envision a transition occurring as a result of a small, finite change in one or more properties of the system. We then determine if the proposed variation leads to a more favored state. This kind of thought experiment is referred to as a *virtual displacement process* which is still consistent with any internal or external constraints placed on the system of interest.

6.2 Extrema Principles

The *stable equilibrium state* was introduced in Postulate II as that state which all isolated systems approach and eventually attain. We shall now develop criteria for testing whether or not a given system has attained a stable equilibrium state. The results of Example 4.3 provide the basis for our quantitative treatment. There, it was proved that, for an adiabatic process occurring in a closed system, the total change in entropy must be either zero if the process is reversible or positive if the process is spontaneous or irreversible. This result is also applicable to *any isolated system*, which is a special case of a closed, adiabatic system. Mathematically, we express this as,

$$\Delta \underline{S} \Big|_{\underline{E}, \underline{V}, M} = 0 \quad \textit{if reversible} \tag{6-1}$$

$$\Delta \underline{S} \Big|_{\underline{E}, \underline{V}, M} > 0 \quad \textit{if spontaneous or irreversible}$$

In this case, the conditions of isolation are: constant total energy $\underline{E}$, constant total volume $\underline{V}$, and constant total mass M.

Alternatively, one can use a totally equivalent set of criteria invoking the total energy $\underline{E}$ rather than $\underline{S}$ [see Gibbs (1876)]:

$$\Delta \underline{E} \Big|_{\underline{S}, \underline{V}, M} = 0 \quad \textit{if reversible} \tag{6-2}$$

$$\Delta \underline{E} \Big|_{\underline{S}, \underline{V}, M} < 0 \quad \textit{if spontaneous or irreversible}$$

Now the conditions that apply are: constant total entropy $\underline{S}$, constant total volume $\underline{V}$, and constant total mass M. Later we will demonstrate that the criteria of Eqs. (6-1) and (6-2) are mathematically equivalent.

From Eq. (6-1) all real processes in an isolated system occur with a zero or positive entropy change. We can therefore conclude that ***for an isolated system to be at stable equilibrium, the entropy must have a maximum value with respect to any allowed variations***. Thus, to test whether or not a given isolated system is, in fact, at equilibrium and stable, we propose virtual displacement processes to evaluate certain variations. If for such variations we can show that the total entropy decreases, then the proposed variation was impossible and the original state was an equilibrium and stable state—at least with respect to the proposed variation.

Almost all systems will be stable with respect to some variations and unstable with respect to others. Thus, the entropy maximum principle is relative and, in the last analysis, we will only be able to propose and test a finite number of variations for a given system. Therefore, we will be able to make only a qualified statement about the stability of an equilibrium state.

The method we employ to test for an entropy maximum is illustrated in Figure. 6.3 for the simple case of a single allowed variation in parameter z_1, where $\underline{S}$ is a function of variables $z_1, z_2, \ldots, z_{n+2}$. As z_1 is varied, there is a value $z_1 = z_1^e$, at which the system entropy is maximized. At this point, $(\partial \underline{S}/\partial z_1) = 0$, $(\partial^2 \underline{S}/\partial z_1^2) < 0$, and the system is in a stable equilibrium state with respect to variations of z_1. If we looked at a system originally at $z_i = z_i^o$ (i=1,..., n+2) and proposed a virtual process in which z_1 was varied by $\pm\delta z_1$, z_2 by $\pm\delta z_2$, etc., the resulting change in $\underline{S}$, or $\Delta \underline{S}$, can be calculated by expanding $\underline{S}$ in a Taylor series about *the conditions of the original state provided that* δz_i *is a small perturbation*. Thus,

$$\Delta \underline{S}^{+} \equiv \underline{S}(z_i) - \underline{S}(z_i^o) = \delta \underline{S} + \frac{1}{2!}\delta^2 \underline{S} + \frac{1}{3!}\delta^3 \underline{S} + \cdots + \frac{1}{m!}\delta^m \underline{S} + \cdots \quad (6\text{-}3)$$

where $\Delta \underline{S}^{+}$ is the resultant change in $\underline{S}$ due to the small perturbation, $\delta \underline{S}$ is the *first-order variation* of $\underline{S}$, and $\delta^m \underline{S}$ is the *mth-order variation* of $\underline{S}$. By definition,

$$\delta \underline{S} \equiv \sum_{i=1}^{n+2} \frac{\partial \underline{S}}{\partial z_i}\delta z_i = \sum_{i=1}^{n+2} \underline{S}_{z_i}\,\delta z_i \quad (6\text{-}4a)$$

$$\delta^2 \underline{S} \equiv \sum_{i=1}^{n+2}\sum_{j=1}^{n+2} \frac{\partial^2 \underline{S}}{\partial z_i \partial z_j}\delta z_i\,\delta z_j = \sum_{i=1}^{n+2}\sum_{j=1}^{n+2} \underline{S}_{z_i z_j}\,\delta z_i\,\delta z_j \quad (6\text{-}4b)$$

$$\delta^3 \underline{S} \equiv \sum_{i=1}^{n+2}\sum_{j=1}^{n+2}\sum_{k=1}^{n+2} \frac{\partial^3 \underline{S}}{\partial z_i\,\partial z_j\,\partial z_k}\delta z_i\,\delta z_j\,\delta z_k = \sum_{i=1}^{n+2}\sum_{j=1}^{n+2}\sum_{k=1}^{n+2} \underline{S}_{z_i z_j z_k}\,\delta z_i\,\delta z_j\,\delta z_k \quad (6\text{-}4c)$$

where $\delta z_i = z_i - z_i^o$ (superscript o denotes the value in the original state) and each of the partial derivatives is evaluated at the conditions prevailing in the original state. Note that the summations run over n+2 variables, as we have implicitly assumed that the masses (or moles) of n components plus 2 independently variable properties can be used to specify the state of the system (Postulate I). The shorthand notation for the partial derivatives will be used throughout this chapter.

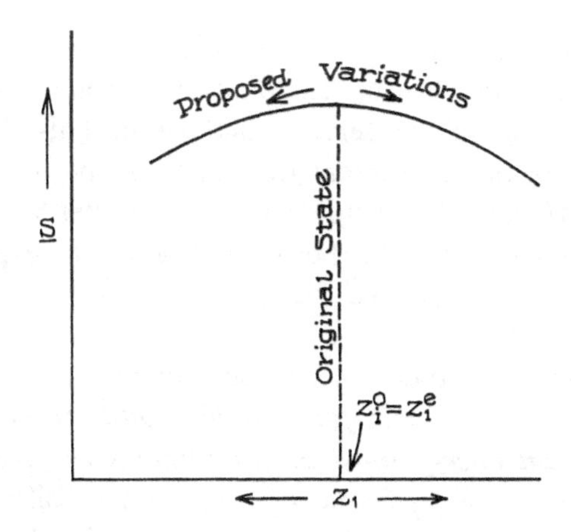

Figure 6.3 Variation in the system entropy with a single variable z_1.

If $\Delta\underline{S}^+$ represents the entropy change from the original assumed equilibrium state to the perturbed state, and if $\underline{S}$ is a maximum in the equilibrium state of an isolated system of constant $\underline{E}$, $\underline{V}$ and M, then *for all possible variations originating from a stable equilibrium state*:

$$\Delta\underline{S}^+ \equiv \underline{S}(z_i) - \underline{S}(z_i^o) < 0 \quad \text{for } i=1,\ldots, n+2 \tag{6-5}$$

If $\underline{S}$ is a smoothly varying function of z_i, the necessary and sufficient conditions for a maximum in $\underline{S}$ are that:

$$\delta\underline{S} = 0 \quad \text{and} \tag{6-6}$$

$$\begin{aligned} &\delta^2\underline{S} \le 0 \quad \text{but, if} = 0, \text{ then} \\ &\delta^3\underline{S} \le 0 \quad \text{but, if} = 0, \text{ then} \ldots \text{until } \delta^m\underline{S} < 0 \end{aligned} \tag{6-7}$$

where $\delta^m\underline{S}$ is the lowest-order, non-vanishing variation of $\underline{S}$. The equality in Eq. (6-6), $\delta\underline{S} = 0$, is the *criterion of equilibrium* in the entropy representation while the appropriate inequality in Eq. (6-7) forms the *criterion of stability*. In this chapter we are concerned primarily with Eq. (6-6); the stability of thermodynamic systems is covered in Chapter 7.

The stability and equilibrium criteria, as stated, apply only to isolated systems. Since the parameters $z_1, z_2, \ldots, z_{n+2}$ are related to $\underline{U}$, $\underline{V}$, $N_1, \ldots, N_n$ for a simple thermodynamic system, the isolation requirement places restraints upon the allowed variations of these independent parameters. In particular, if the system in question is a composite of two simple subsystems, then any proposed virtual process must be consistent with the restraining equations of isolation: constant internal energy, volume, and mass,

$$\delta\underline{U} = \delta\underline{U}^{(1)} + \delta\underline{U}^{(2)} = 0 \tag{6-8}$$

$$\delta\underline{V} = \delta\underline{V}^{(1)} + \delta\underline{V}^{(2)} = 0 \tag{6-9}$$

$$\delta M = \delta M^{(1)} + \delta M^{(2)} = 0 \tag{6-10}$$

The proposed virtual processes must also be consistent with any internal restraints (e.g., rigid walls), as discussed in more depth in Section 6.4.

We shall now consider alternative criteria for equilibrium and stability that are applicable for systems that are not isolated. We often encounter systems that interact with external heat, work, and mass reservoirs, and we must also be able to treat these cases. During such interactions, different restraints will be imposed (e.g., constant $\underline{S}, \underline{V}, M$; constant $T, \underline{V}, M$; etc.). The insight gained in Chapter 5 with alternative forms of the Fundamental Equation should, however, lead us to expect that alternative extremum principles could also be developed with the potential functions, $\underline{U}, \underline{A}, \underline{H}, \underline{G}, \underline{G}'$, etc.

The duality of the entropy and energy representation of the Fundamental Equation can readily be applied to prove that ***for a stable equilibrium system at constant*** $\underline{S}, \underline{V}, M$, ***the total internal energy must be a minimum*** as given by Eq. (6-2). To demonstrate this we shall employ a simple logical proof and follow this by an example to show that the energy minimization and entropy maximization do indeed yield identical criteria of equilibrium and intrinsic stability.

Consider a system that is supposed to be in a state of stable equilibrium. In proving this statement to be true, we isolated the system and showed that *all* allowable variations led to a decrease in the total entropy; that is, for every variation that we considered at constant $\underline{U}, \underline{V}, M$, we found $\underline{S}(z_i)$ to be less than $\underline{S}(z_i^o)$. Now, however, we wish to consider variations at constant $\underline{S}, \underline{V}, M$, and allow $\underline{U}$ to vary. We now take the system at $\underline{S}(z_i)$ and allow it to interact reversibly with an external system to return it to $\underline{S}(z_i^o)$, maintaining $\underline{V}, M$ constant. Since $\underline{S}(z_i) - \underline{S}(z_i^o) < 0$, we must transfer energy *into* the system to return to $\underline{S}(z_i^o)$. In so doing, we increase $\underline{U}$ in this two-step process. Thus, one concludes that for any variation in a system at constant $\underline{S}, \underline{V}, M$, the internal energy will increase—if the system were initially at equilibrium. The converse is also true; that is, if a variation within a system held at constant $\underline{S}, \underline{V}, M$ leads to a decrease in internal energy, the system was not initially in a state of equilibrium.

In an analogous form to Eqs. (6-3), (6-4), and (6-5), at constant $\underline{S}, \underline{V}, M$, we obtain for all possible variations originating from a stable equilibrium state

$$\Delta \underline{U}^+ \equiv \underline{U}(z_i) - \underline{U}(z_i^o) > 0 \quad \text{for } i = i, \ldots, n+2 \tag{6-11}$$

where

$$\Delta \underline{U}^+ = \delta \underline{U} + \frac{1}{2!}\delta^2 \underline{U} + \frac{1}{3!}\delta^3 \underline{U} + \cdots + \frac{1}{m!}\delta^m \underline{U} \tag{6-12}$$

$$\delta \underline{U} = \sum_{i=1}^{n+2} \left(\frac{\partial \underline{U}}{\partial z_i}\right) \delta z_i, \quad \delta^2 \underline{U} = \sum_{i=1}^{n+2} \sum_{j=1}^{n+2} \frac{\partial^2 \underline{U}}{\partial z_i \partial z_j} \delta z_i \delta z_j, \ldots, \text{ etc.}$$

Given that $\underline{U}$ is a smoothly varying, continuous function of $(n+2)$ z_i variables, the necessary and sufficient conditions for a minimum in $\underline{U}$ are that:

$$\delta \underline{U} = 0 \quad \text{and} \tag{6-13}$$

$$\begin{aligned} &\delta^2 \underline{U} \geq 0 \quad \text{but if=0 then,} \\ &\delta^3 \underline{U} \geq 0 \quad \text{but if=0 then ,... until } \delta^m \underline{U} > 0 \end{aligned} \tag{6-14}$$

where $\delta^m \underline{U}$ is the lowest-order, nonvanishing variation. The restraining equations of isolation are Eqs. (6-9), (6-10), with Eq. (6-8) replaced with:

$$\delta \underline{S} = \delta \underline{S}^{(1)} + \delta \underline{S}^{(2)} = 0 \tag{6-15}$$

The equivalence or duality of using either entropy [Eqs. (6-6) - (6-7)] or energy [Eqs. (6-13) - (6-14)] to represent stable equilibrium states is illustrated in Figure 6.4. The parameter set z_k ($k = 3, \ldots, n+2$) could represent the set of n mole numbers (N_k) or mole fractions (x_k) required to specify the state of the system in addition to $\underline{S}$ and $\underline{V}$ in energy representation or $\underline{U}$ and $\underline{V}$ in entropy representation. At equilibrium the values of these (n+2) variables are z_k^{eq} corresponding to a unique configuration as specified by Postulate II.

Example 6.1

In Figure. 6.1(a) the ball is shown at the bottom of the well to represent a case of stable equilibrium. Prove this to be so by considering a *virtual displacement* process wherein the ball moves to a point on the wall, above the bottom. Develop two proofs, one using Eq. (6-5) and the other with Eq. (6-11).

Solution

In this virtual process, the ball has gained energy (potential), but the entropy, volume, and mass have remained constant. Thus, Eq. (6-11) applies and we conclude that the ball was, initially, in a stable equilibrium state with respect to the proposed variation. If we had desired to use the equilibrium criterion given by Eq. (6-5) rather than the equivalent form, Eq. (6-11), we would have to keep the energy, volume, and mass constant during the proposed variation. Since the energy increases due to a gain in potential energy, we would have had to cool the ball sufficiently such that the energy in the higher location was the same as when it was at the bottom. Neglecting any change in volume due to the cooling, we have a variation at constant $\underline{E}$, $\underline{V}$, and M. It is obvious, however, that due to the cooling, $\Delta \underline{S} < 0$. Thus, Eq. (6-5) applies and again we conclude that the initial system was in an equilibrium state relative to the proposed variation.

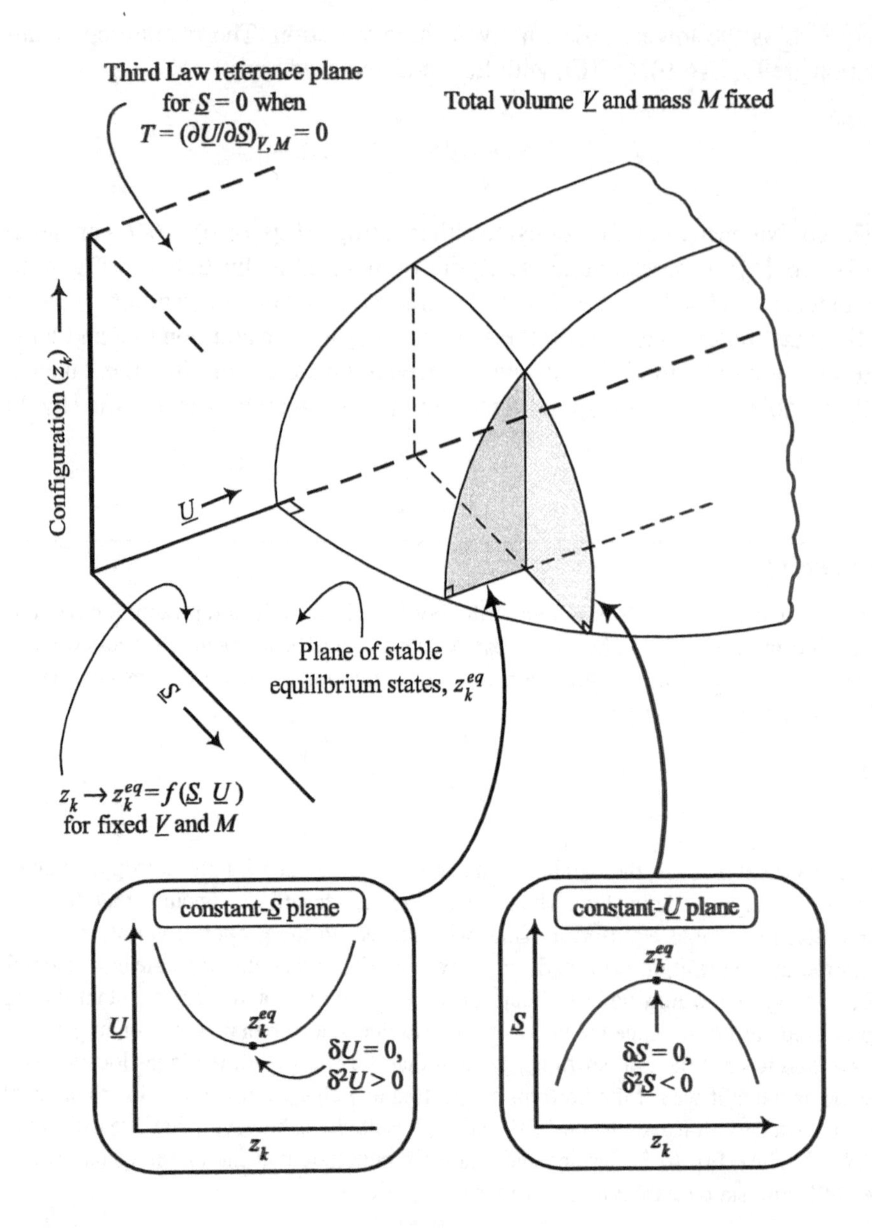

Figure 6.4 An illustration of the equivalence of the energy minimization and entropy maximization principles in defining all stable equilibrium states for a closed, constant volume system. Different states of configuration are specified by the variable set z_k ($k = 3, \ldots, n+2$) [adapted from Bejan (1988) Figure 2.7].

6.3 Use of Other Potential Functions to Define Equilibrium States

In Section 6.2 criteria of equilibrium and stability were developed for systems at constant $\underline{U}$, $\underline{V}$, and M or at constant $\underline{S}$, $\underline{V}$, and M. In the former, the total entropy was maximized and, in the latter, the total energy was minimized. It is desirable to be able to obtain alternative criteria for systems under different constraints. To accomplish this, let us consider a system that may be simple or complex, with no restrictions on the number of components or the number of phases, but with no significant body force fields. We also have available large thermal and work reservoirs (R_T and R_P) to which the system may be connected, if desired, to hold the temperature and/or pressure constant (see Figure 6.5). We still maintain our system at constant total mass during any variation, not because we could not also have used external mass reservoirs, but because the inclusion of mass variations is treated later.

We require that (1) the thermal gate, when operating, be impermeable, rigid, and diathermal, and (2) the piston between the system and work reservoir be impermeable, frictionless, adiabatic, and capable of being latched when desired. Thus, the thermal reservoir is held at constant $\underline{V}$, M while the work reservoir is held at constant $\underline{S}$, M. The system restraints may be varied depending on our desire to have interactions with one or both of the adjacent reservoirs. Finally, we assume that the reservoirs are large compared to the system so that small variations in the system as a result of movements of the piston connecting the system to R_P, or because of heat exchange with R_T, will not change the pressure level in R_P or the temperature level in R_T to any significant degree.

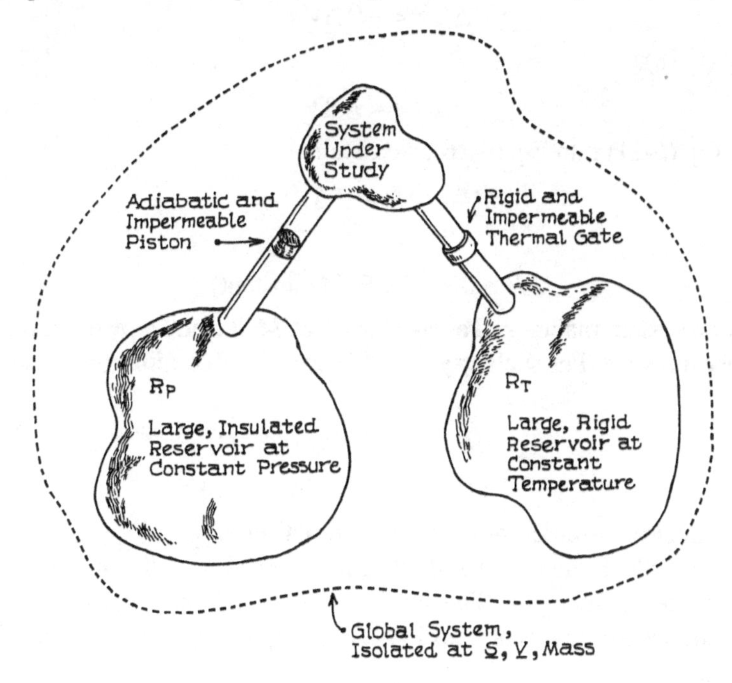

Figure 6.5 System interactions with constant pressure or temperature reservoirs.

Our global system then consists of the original system together with R_P and R_T. These three subsystems are placed in an environment such that the *total* $\underline{S}$, $\underline{V}$, and mass M are held constant. We shall investigate some possible variations of this global system and determine the consequences as we apply the energy-minimization criteria.

In general, if the global system is originally in a stable equilibrium state, then for any proposed virtual process,

$$\Delta \underline{U}^{\Sigma} = \Delta(\underline{U} + \underline{U}^{R_T} + \underline{U}^{R_P}) > 0 \tag{6-16}$$

$$\Delta \underline{S}^{\Sigma} = \Delta(\underline{S} + \underline{S}^{R_T}) = \Delta \underline{S}^{R_P} = 0 \tag{6-17}$$

$$\Delta \underline{V}^{\Sigma} = \Delta(\underline{V} + \underline{V}^{R_P}) = \Delta \underline{V}^{R_T} = 0 \tag{6-18}$$

$$\Delta M^{\Sigma} = \Delta M = \Delta M^{R_T} = \Delta M^{R_P} = 0 \tag{6-19}$$

where the superscript Σ refers to the global system, R_T and R_P to the thermal and work reservoirs, and the original system under study is not superscripted.

***Case** (a)*. The thermal gate is inoperative. The piston is unlatched in order to allow an interaction between the system and R_P in order to keep the system at constant pressure. Since the movement of the piston is also assumed to be frictionless and thus will operate in a reversible manner, its displacement will vary neither the entropy of the system nor the entropy of R_P. The results of any such variation will then apply to a system at constant $\underline{S}$, P, and M.

For the pressure reservoir, the First Law yields

$$\Delta \underline{U}^{R_P} = -P\, \Delta \underline{V}^{R_P} \tag{6-20}$$

or, using Eq. (6-18),

$$\Delta \underline{U}^{R_P} = P\, \Delta \underline{V} \tag{6-21}$$

Substituting Eq. (6-21) into Eq. (6-16), we obtain

$$\Delta \underline{U}^{\Sigma} = \Delta \underline{U} + P\, \Delta \underline{V} > 0 \tag{6-22}$$

or

$$\Delta \underline{H} > 0 \qquad (\underline{S}, P, M \text{ constant}) \tag{6-23}$$

That is, for a system maintained at constant $\underline{S}$, P, M, the enthalpy is a minimum for a stable equilibrium state. For such a system, the criteria of equilibrium and stability are, respectively,

$$\delta \underline{H} = 0 \qquad (\underline{S}, P, M \text{ constant}) \tag{6-24}$$

$$\delta^m \underline{H} > 0 \qquad (\underline{S}, P, M \text{ constant}) \tag{6-25}$$

where $\delta^m \underline{H}$ is the lowest-order, nonvanishing variation of $\underline{H}$.

***Case** (b)*. Let us lock the piston and allow no interaction of the original system with R_P. Now open the thermal gate. The small system is held at constant $\underline{V}$, T, and M. A possible variation is to allow heat transfer between R_T and the small system. Then, since $\Delta \underline{V} = \Delta \underline{V}^{R_T} = 0$,

$$\Delta \underline{U}^{R_T} = T\, \Delta \underline{S}^{R_T} = -T\, \Delta \underline{S} \tag{6-26}$$

Hence, Eq. (6-16) becomes

$$\Delta \underline{U}^{\Sigma} = \Delta \underline{U} - T\,\Delta \underline{S} > 0 \tag{6-27}$$

or

$$\Delta \underline{A} > 0 \qquad (T, \underline{V}, M \text{ constant}) \tag{6-28}$$

or

$$\delta \underline{A} = 0 \qquad (T, \underline{V}, M \text{ constant}) \tag{6-29}$$

$$\delta^m \underline{A} > 0 \qquad (T, \underline{V}, M \text{ constant}) \tag{6-30}$$

Thus, our extremum principle for systems held at constant total volume, temperature, and mass is that the Helmholtz energy should be a minimum with respect to all allowable variations.

Case **(*c*)**. If we allow simultaneous interactions for our small system with both R_P and R_T, by an approach similar to that used in cases (a) and (b),

$$\Delta \underline{G} > 0 \qquad (T, P, M \text{ constant}) \tag{6-31}$$

$$\delta \underline{G} = 0 \qquad (T, P, M \text{ constant}) \tag{6-32}$$

or

$$\delta^m \underline{G} > 0 \qquad (T, P, M \text{ constant}) \tag{6-33}$$

Thus, the Gibbs energy appears as the potential function to be minimized for systems at constant temperature, pressure, and mass. Equilibrium criteria using entropy and various energy potential functional representations are summarized in Table 6.1.

Table 6.1 Summary of Mathematical Criteria for Equilibrium and Stability

Potential	For Displacements from Equilibrium Condition	For Equilibrium	For Stability
$\underline{S}$	$\Delta \underline{S}^+ < 0 \big\rvert^{1}_{\underline{U}, \underline{V}, M}$	$\delta \underline{S} = 0$	$\delta^2 \underline{S} \le 0 \ldots \text{until } \delta^m \underline{S} < 0$ [2]
$\underline{U}$	$\Delta \underline{U}^+ > 0 \big\rvert_{\underline{S}, \underline{V}, M}$	$\delta \underline{U} = 0$	$\delta^2 \underline{U} \ge 0 \ldots \text{until } \delta^m \underline{U} > 0$
$\underline{H}$	$\Delta \underline{H}^+ > 0 \big\rvert_{\underline{S}, P, M}$	$\delta \underline{H} = 0$	$\delta^2 \underline{H} \ge 0 \ldots \text{until } \delta^m \underline{H} > 0$
$\underline{A}$	$\Delta \underline{A}^+ > 0 \big\rvert_{T, \underline{V}, M}$	$\delta \underline{A} = 0$	$\delta^2 \underline{A} \ge 0 \ldots \text{until } \delta^m \underline{A} > 0$
$\underline{G}$	$\Delta \underline{G}^+ > 0 \big\rvert_{T, P, M}$	$\delta \underline{G} = 0$	$\delta^2 \underline{G} \ge 0 \ldots \text{until } \delta^m \underline{G} > 0$

1. + denotes $\underline{S}(z_i) - \underline{S}(z_i^o)$ as the change in the potential due to a virtual displacement from the equilibrium state at z_i^o for i=1, ..., n+2 variables specified.
2. $\delta^m \underline{S} < 0, \ldots, \delta^m \underline{G} > 0$ are lowest-order, non-vanishing differentials where m is ≥ 2

Although we did not employ Legendre transformations to develop equilibrium criteria in terms of the other potential functions of $\underline{H}$, $\underline{A}$, or $\underline{G}$, we could have produced them directly using the methods developed in Chapter 5. The key attribute of the Legendre transformation that makes this route possible is that the information content of the original Fundamental Equation is fully preserved in generating these auxiliary potential functions.

6.4 Membrane Equilibrium

For an isolated system, it was shown in the preceding sections that a necessary (but not sufficient) condition of a system in a stable equilibrium state is that $\delta \underline{S} = 0$ for any virtual displacement process. In this section and the following ones we examine the consequences of this condition, first for a complex system with specific constraints and then for multiphase and chemically reacting simple systems at equilibrium.

Consider an isolated, complex system containing two subsystems, each of which contains a nonreacting binary mixture of components A and B (see Figure 6.6). The criterion of equilibrium for the isolated composite can be expanded in terms of the properties of the two subsystems:

$$\delta \underline{S} = \delta \underline{S}^{(1)} + \delta \underline{S}^{(2)} = 0$$

$$\delta \underline{S} = \frac{1}{T^{(1)}} \delta \underline{U}^{(1)} + \frac{1}{T^{(2)}} \delta \underline{U}^{(2)} + \frac{P^{(1)}}{T^{(1)}} \delta \underline{V}^{(1)} + \frac{P^{(2)}}{T^{(2)}} \delta \underline{V}^{(2)} \tag{6-34}$$

$$- \frac{\mu_A^{(1)}}{T^{(1)}} \delta N_A^{(1)} - \frac{\mu_A^{(2)}}{T^{(2)}} \delta N_A^{(2)} - \frac{\mu_B^{(1)}}{T^{(1)}} \delta N_B^{(1)} - \frac{\mu_B^{(2)}}{T^{(2)}} \delta N_B^{(2)} = 0$$

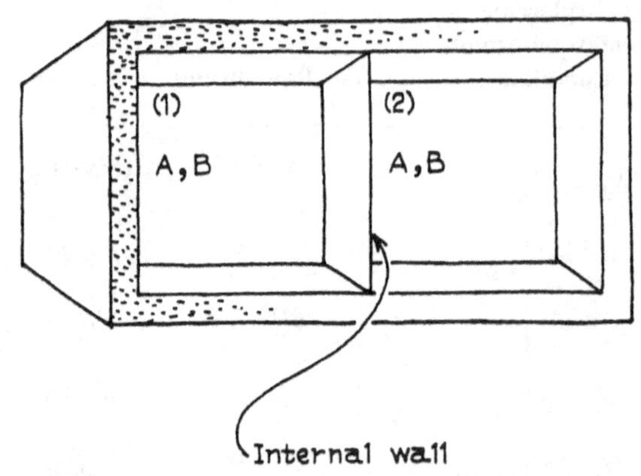

Figure 6.6 Equilibrium in a complex system.

We note that the restraining equations of isolation, Eqs. (6-8) through (6-10), place restrictions on the allowable variations in any virtual process that is considered. That is, $\delta \underline{U}^{(1)}$ and $\delta \underline{U}^{(2)}$ are not independently variable. Substituting the restraining equations of isolation into Eq. (6-34) and simplifying, one obtains

$$\delta \underline{S} = \left(\frac{1}{T^{(1)}} - \frac{1}{T^{(2)}} \right) \delta \underline{U}^{(1)} + \left(\frac{P^{(1)}}{T^{(1)}} - \frac{P^{(2)}}{T^{(2)}} \right) \delta \underline{V}^{(1)}$$
$$- \left(\frac{\mu_A^{(1)}}{T^{(1)}} - \frac{\mu_A^{(2)}}{T^{(2)}} \right) \delta N_A^{(1)} - \left(\frac{\mu_B^{(1)}}{T^{(1)}} - \frac{\mu_B^{(2)}}{T^{(2)}} \right) \delta N_B^{(1)} = 0 \tag{6-35}$$

The variations $\delta \underline{U}^{(1)}$, $\delta \underline{V}^{(1)}$, $\delta N_A^{(1)}$, and $\delta N_B^{(1)}$ are independently variable only if the composite is a simple system. By definition, a complex system contains some additional internal restraints that must be recognized in any virtual process imposed upon the system. Although there is no universal result applicable to all complex systems, we illustrate three cases that demonstrate the method of obtaining the appropriate criteria of equilibrium (see Figure 6.6).

Case (a) The internal boundary is permeable only to B, diathermal, and movable: The additional restraining equation is then

$$\delta N_A^{(1)} = \delta N_A^{(2)} = 0 \tag{6-36}$$

and Eq. (6-35) reduces to

$$\delta \underline{S} = \left(\frac{1}{T^{(1)}} - \frac{1}{T^{(2)}} \right) \delta \underline{U}^{(1)} + \left(\frac{P^{(1)}}{T^{(1)}} - \frac{P^{(2)}}{T^{(2)}} \right) \delta \underline{V}^{(1)} - \left(\frac{\mu_B^{(1)}}{T^{(1)}} - \frac{\mu_B^{(2)}}{T^{(2)}} \right) \delta N_B^{(1)} = 0 \tag{6-37}$$

For all virtual processes in which variations in $\delta \underline{U}^{(1)}$, $\delta \underline{V}^{(1)}$, and $\delta N_B^{(1)}$ are considered, if $\delta \underline{S}$ is to vanish, the coefficients of each variation must be zero. Therefore, the criteria of equilibrium are $T^{(1)} = T^{(2)}$, $P^{(1)} = P^{(2)}$, and $\mu_B^{(1)} = \mu_B^{(2)}$. Note that there is no apparent restriction on $\mu_A^{(1)}$ or $\mu_A^{(2)}$.

Case (b) The internal wall is rigid, diathermal, and permeable to both A and B: With the same approach, but with

$$\delta \underline{V}^{(1)} = \delta \underline{V}^{(2)} = 0 \tag{6-38}$$

used instead of Eq. (6-36), the temperatures and chemical potentials of both components are found to be equal in both subsystems. Note that there is no apparent restriction on $P^{(1)}$ and $P^{(2)}$. The classic case of osmotic equilibrium results when the wall is a rigid, diathermal membrane permeable only to solvent. In this instance, only the chemical potential of the solvent is constant across the membrane.

Case (c) The internal wall is adiabatic, movable, and permeable: By analogy to cases 1 and 2, it might be thought that $\delta \underline{U}^{(1)}$ and $\delta \underline{U}^{(2)}$ were zero. Mass interchange between the subsystems, however, can also vary the energy of each compartment; thus we have, in reality, no additional restraints and, at equilibrium, the temperature, pressure, and component chemical potentials are equal in each subsystem. A similar result would be found even if the boundary were rigid. Thus, the adiabatic-permeable case is very similar to the diathermal-permeable case.

An interesting dilemma may be noted in case b. At equilibrium in this binary system, μ_A, μ_B, and T were the same on both sides of the internal wall. Since each side is a simple system, the corollary to Postulate I introduced in Section 5.2 then states that only $(n+1)$ intensive variables need be specified to define completely all other intensive variables for a single-phase simple system. If each side is indeed a single phase, then $P^{(1)} = P^{(2)}$. If one or both sides contain more than one phase, then, as will be discussed in Section 15.1, fewer than $(n+1)$ variables are required to define all other intensive properties; but in any case, $P^{(1)}$ would equal $P^{(2)}$. Similar reasoning may be applied to case 1 to show that $\mu_A^{(1)} = \mu_A^{(2)}$.

It is clear that each complex system with internal membranes or walls must be examined as a special case.

6.5 Phase Equilibria

A system with more than a single phase may be considered to be a composite of simple systems with phase-separating membranes that are movable, diathermal, and permeable to all components. Thus, as will be proved below, the intensive properties T, P, and μ_j are identical in all phases.

In the proof, consider an isolated, simple multicomponent system containing π phases. The equilibrium criterion, Eq. (6-6), then becomes

$$\delta \underline{S} = \sum_{s=1}^{\pi} \delta \underline{S}^{(s)} = 0$$

Using the entropy representation of the Fundamental Equation:

$$\delta \underline{S} = \sum_{s=1}^{\pi} \underline{S}_U^{(s)} \delta \underline{U}^{(s)} + \sum_{s=1}^{\pi} \underline{S}_V^{(s)} \delta \underline{V}^{(s)} + \sum_{s=1}^{\pi} \sum_{j=1}^{n} \underline{S}_{N_j}^{(s)} \delta N_j^{(s)} \tag{6-39}$$

where the superscript (s) is a dummy variable denoting the phase. The constraints resulting from isolation, with no chemical reaction, are

$$\delta \underline{U} = \sum_{s=1}^{\pi} \delta \underline{U}^{(s)} = 0 \tag{6-40}$$

$$\delta \underline{V} = \sum_{s=1}^{\pi} \delta \underline{V}^{(s)} = 0 \tag{6-41}$$

$$\delta N_j = \sum_{s=1}^{\pi} \delta N_j^{(s)} = 0 \tag{6-42}$$

In Eq. (6-39), as indicated before, $\underline{S}_U^{(s)} = (\partial \underline{S}^{(s)}/\partial \underline{U}^{(s)})_{\underline{V},N} = (1/T^{(s)})$, $\underline{S}_{\underline{V}}^{(s)} = P^{(s)}/T^{(s)}$, and $\underline{S}_{N_j}^{(s)} = -\mu_j^{(s)}/T^{(s)}$.

To include the $(n+2)$ restraining equations in order to eliminate $(n+2)$ dependent variables, it is convenient to use the method of Lagrange undetermined multipliers. Let us define the arbitrary multipliers to Eqs. (6-40), (6-41), and (6-42) as $(1/T^{(1)})$, $(P^{(1)}/T^{(1)})$, and $(-\mu_j^{(1)}/T^{(1)})$, respectively. Then, by multiplying each constraint equation by its respective multiplier and subtracting all from Eq. (6-39), we obtain:

$$\delta \underline{S} = \sum_{s=2}^{\pi} \left(\frac{1}{T^{(s)}} - \frac{1}{T^{(1)}} \right) \delta \underline{U}^{(s)} + \sum_{s=2}^{\pi} \left(\frac{P^{(s)}}{T^{(s)}} - \frac{P^{(1)}}{T^{(1)}} \right) \delta \underline{V}^{(s)} - \sum_{s=2}^{\pi} \sum_{j=1}^{n} \left(\frac{\mu_j^{(s)}}{T^{(s)}} - \frac{\mu_j^{(1)}}{T^{(1)}} \right) \delta N_j^{(s)} = 0 \tag{6-43}$$

Since each variation in $\delta \underline{U}^{(s)}$, $\delta \underline{V}^{(s)}$, and $\delta N_j^{(s)}$ is independent, it is immediately obvious that

$$T^{(1)} = T^{(2)} = \cdots = T^{(\pi)} \tag{6-44}$$

$$P^{(1)} = P^{(2)} = \cdots = P^{(\pi)} \tag{6-45}$$

$$\mu_j^{(1)} = \mu_j^{(2)} = \cdots = \mu_j^{(\pi)} \tag{6-46}$$

These temperature, pressure, and chemical potential equalities between phases at equilibrium hold for all simple, multiphase systems, that is, those with no external force fields and with no internal barriers to restrict heat, work, or mass interactions.

6.6 Chemical Reaction Equilibria

The treatment for multiphase systems can be extended to cases in which chemical reactions occur. The procedure shown in Section 6.5 is applicable except that the constraint on mole numbers, Eq. (6-42), must be modified for those components that react chemically. If a reaction involving components 1 through i occurs, the general form may be expressed as

$$v_1 C_1 + v_2 C_2 + \cdots + v_i C_i = 0 \tag{6-47}$$

or

$$\sum_{j=1}^{i} v_j C_j = 0 \tag{6-48}$$

Equations (6-47) and (6-48) are in reality atom balances; C_j could be considered the chemical formula for component j and v_j the molar stoichiometric multiplier or coefficient. *For products, v_j is always defined as a positive number and, for reactants, a negative number.*

If the reaction is stoichiometrically balanced, there is no net change in mass, and

$$\sum_{j=1}^{i} m_j v_j = 0 \tag{6-49}$$

where m_j is the molecular weight of j. The stoichiometric coefficient defines the ratio of mole changes of reacting components. That is, for a single reaction,

$$\frac{\delta N_1}{v_1} = \frac{\delta N_2}{v_2} = \cdots = \frac{\delta N_i}{v_i} \tag{6-50}$$

Since the δN_j variations are related, only one may be varied independently. To simplify bookkeeping, a new variable, the *extent of reaction*, ξ, is introduced.

$$\delta\xi \equiv \frac{\delta N_j}{v_j}, \qquad j = 1, 2, \ldots, i \tag{6-51}$$

so that all δN_j for reacting species may be expressed in terms of $\delta\xi$. Equation (6-42) is then modified, for reacting components, to

$$\left.\begin{aligned} \delta N_j &= v_j\, \delta\xi \\ \sum_{s=1}^{\pi} \delta N_j^{(s)} &= v_j\, \delta\xi \end{aligned}\right\} \quad j = 1, 2, \ldots, i \;\; \text{(reacting)} \qquad \begin{matrix}(6\text{-}52)\\ (6\text{-}53)\end{matrix}$$

For inert species, $\delta N_j = 0$ and

$$\sum_{s=1}^{\pi} \delta N_j^{(s)} = 0 \qquad j = i+1, \ldots, n \;\; \text{(inert)} \tag{6-54}$$

Again employing the same Lagrange undetermined multipliers as used in Eq. (6-43), with Eqs. (6-53) and (6-54) in place of (6-42), Eq. (6-43) is again obtained except that there is one additional term,

$$-\frac{1}{T^{(1)}} \sum_{j=1}^{i} (v_j\, \mu_j)\, \delta\xi$$

Since ξ can be varied independently from $\underline{U}^{(s)}$, $\underline{V}^{(s)}$, and $N_j^{(s)}$, we have, at equilibrium, in addition to the equalities of temperature, pressure, and component chemical potentials between phases, the requirement that

$$\sum_{j=1}^{i} v_j \mu_j = 0 \tag{6-55}$$

The superscript on μ_j is deleted since the chemical potential of component j is equal in all phases. For cases in which multiple reactions occur, it is readily shown by an identical treatment that an equation of the form of Eq. (6-55) applies for each reaction in which there is chemical equilibrium. This general criterion will be developed further in Chapters 16 and 17.

Example 6.2

At low temperatures, a mixture of water and excess ferric chloride forms a solid phase of $FeCl_3 \cdot 6H_2O$. If equilibrium existed in an isolated system of water and ferric chloride such that there was a gas phase consisting only of water vapor, a liquid phase with only H_2O and $FeCl_3$, and a solid phase of the hexahydrate, clearly specify all equilibrium criteria that apply to this system. (Neglect any ionization of the $FeCl_3$ in solution.)

Solution

Equation (6-39) as written for this three-phase (S-L-V) system is

$$\delta \underline{S} = \frac{1}{T^V}\delta \underline{U}^V + \frac{P^V}{T^V}\delta \underline{V}^V - \frac{\mu_w^V}{T^V}\delta N_w^V + \frac{1}{T^L}\delta \underline{U}^L + \frac{P^L}{T^L}\delta \underline{V}^L - \frac{\mu_{FeCl_3}^L}{T^L}\delta N_{FeCl_3}^L$$
$$- \frac{\mu_w^L}{T^L}\delta N_w^L + \frac{1}{T^S}\delta \underline{U}^S + \frac{P^S}{T^S}\delta \underline{V}^S - \frac{\mu_{hyd}^S}{T^S}\delta N_{hyd}^S = 0$$

The constraints placed on the system are

$$\delta \underline{U}^V + \delta \underline{U}^L + \delta \underline{U}^S = 0 \tag{6-56}$$
$$\delta \underline{V}^V + \delta \underline{V}^L + \delta \underline{V}^S = 0 \tag{6-57}$$

and there is also the reaction

$$6H_2O + FeCl_3 = FeCl_3 \cdot 6H_2O$$

so that

$$\delta \xi = \frac{\delta N_w}{-6} = \frac{\delta N_{FeCl_3}}{-1} = \frac{\delta N_{hyd}}{1}$$

and

$$\delta N_w = \delta N_w^L + \delta N_w^V = -6\delta\xi \tag{6-58}$$
$$\delta N_{FeCl_3} = \delta N_{FeCl_3}^L = -\delta\xi \tag{6-59}$$
$$\delta N_{hyd} = \delta N_{hyd}^S = \delta\xi \tag{6-60}$$

Using Lagrange multipliers $(1/T^V)$ to Eq. (6-56), (P^V/T^V) to Eq. (6-57), $(-\mu_w^V/T^V)$ to Eq. (6-58), $(-\mu_{FeCl_3}^L/T^V)$ to Eq. (6-59), and $(-\mu_{hyd}^S/T^V)$ to Eq. (6-60), subtracting from the $\delta \underline{S}$ expression, we see immediately, since all variations are now independent, that

$$T = T^V = T^L = T^S \text{ and } P = P^V = P^L = P^S$$

$$\mu_w^V = \mu_w^L \text{ and } 6\mu_w^L + \mu_{FeCl_3}^L = \mu_{hyd}^S$$

These are the equilibrium criteria.

Example 6.3

Three reactors in an isolation chamber are initially charged as follows: *A* has pure methane, *B* pure hydrogen, and *C* pure ammonia. They are interconnected by rigid, semipermeable membranes as shown in Figure 6.7. All reactors have suitable catalysts so that the reactions

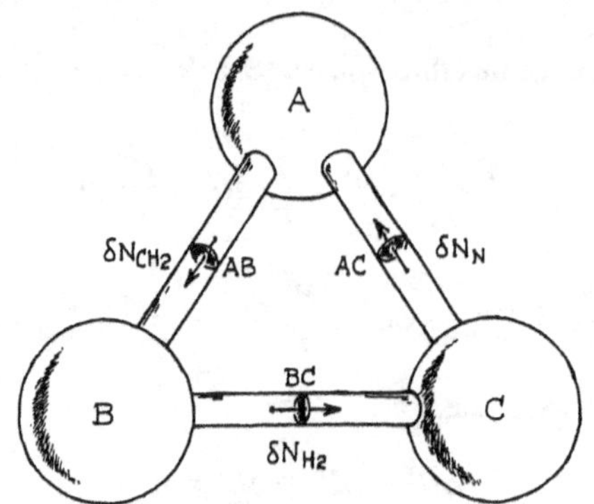

Figure 6.7 Sketch for Example 6.3.

$$CH_4 = CH_2 + H_2 \tag{R1}$$

$$NH_3 = N + \tfrac{3}{2} H_2 \tag{R2}$$

are always in equilibrium. Membrane *AB* is permeable only to CH_2, *BC* to H_2, and *AC* to nitrogen atoms. What are the equilibrium criteria for this system?

Solution

Reactions (1) and (2) can occur in *A*; call the extent of reactions for these $\delta\xi_1^A$ and $\delta\xi_2^A$, respectively. In *B* only (1) occurs with $\delta\xi_1^B$ and in *C* only (2) with $\delta\xi_2^C$. With $\delta \underline{V}^A = \delta \underline{V}^B = \delta \underline{V}^C = 0$, the entropy variation of the total system is

$$\delta \underline{S} = \delta \underline{S}^A + \delta \underline{S}^B + \delta \underline{S}^C$$

$$\delta \underline{S} = \frac{1}{T^A}\,\delta \underline{U}^A - \frac{\mu^A_{\mathrm{CH_4}}}{T^A}\,\delta N^A_{\mathrm{CH_4}} - \frac{\mu^A_{\mathrm{CH_2}}}{T^A}\,\delta N^A_{\mathrm{CH_2}} - \frac{\mu^A_{\mathrm{H_2}}}{T^A}\,\delta N^A_{\mathrm{H_2}} - \frac{\mu^A_{\mathrm{N}}}{T^A}\,\delta N^A_{\mathrm{N}} - \frac{\mu^A_{\mathrm{NH_3}}}{T^A}\,\delta N^A_{\mathrm{NH_3}}$$

$$+ \frac{1}{T^B}\,\delta \underline{U}^B - \frac{\mu^B_{\mathrm{CH_4}}}{T^B}\,\delta N^B_{\mathrm{CH_4}} - \frac{\mu^B_{\mathrm{CH_2}}}{T^B}\,\delta N^B_{\mathrm{CH_2}} - \frac{\mu^B_{\mathrm{H_2}}}{T^B}\,\delta N^B_{\mathrm{H_2}} + \frac{1}{T^C}\,\delta \underline{U}^C - \frac{\mu^C_{\mathrm{H_2}}}{T^C}\,\delta N^C_{\mathrm{H_2}}$$

$$- \frac{\mu^C_{\mathrm{N}}}{T^C}\,\delta N^C_{\mathrm{N}} - \frac{\mu^C_{\mathrm{NH_3}}}{T^C}\,\delta N^C_{\mathrm{NH_3}} = 0$$

The conditions of restraint are:

(1) $\delta \underline{U}^A + \delta \underline{U}^B + \delta \underline{U}^C = 0$

(2) $\delta N^A_{\mathrm{CH_2}} = -\delta N_{\mathrm{CH_2}} + \delta \xi^A_1$

(3) $\delta N^A_{\mathrm{H_2}} = \delta \xi^A_1 + \frac{3}{2}\,\delta \xi^A_2$

(4) $\delta N^A_{\mathrm{CH_4}} = -\delta \xi^A_1$

(5) $\delta N^A_{\mathrm{N}} = \delta N_{\mathrm{N}} + \delta \xi^A_2$

(6) $\delta N^A_{\mathrm{NH_3}} = -\,\delta \xi^A_2$

(7) $\delta N^B_{\mathrm{H_2}} = -\delta N_{\mathrm{H_2}} + \delta \xi^B_1$

(8) $\delta N^B_{\mathrm{CH_4}} = -\delta \xi^B_1$

(9) $\delta N^B_{\mathrm{CH_2}} = \delta N_{\mathrm{CH_2}} + \delta \xi^B_1$

(10) $\delta N^C_{\mathrm{H_2}} = \delta N_{\mathrm{H_2}} + \frac{3}{2}\,\delta \xi^C_2$

(11) $\delta N^C_{\mathrm{NH_3}} = -\delta \xi^C_2$

(12) $\delta N^C_{\mathrm{N}} = -\delta N_{\mathrm{N}} + \delta \xi^C_2$

Multiplying (1) by $-1/T'$, adding to $\delta \underline{S}$, and also substituting constraints (2) through (12) into the expression for $\delta \underline{S}$, there results

$$\left(\frac{1}{T^A} - \frac{1}{T'}\right)\delta \underline{U}^A + \left(\frac{1}{T^B} - \frac{1}{T'}\right)\delta \underline{U}^B + \left(\frac{1}{T^C} - \frac{1}{T'}\right)\delta \underline{U}^C + \left(\frac{\mu^A_{\mathrm{CH_2}}}{T^A} - \frac{\mu^B_{\mathrm{CH_2}}}{T^B}\right)\delta N_{\mathrm{CH_2}}$$

$$+ \left(\frac{\mu^C_{\mathrm{N}}}{T^C} - \frac{\mu^A_{\mathrm{N}}}{T^A}\right)\delta N_N + \left(\frac{\mu^B_{\mathrm{H_2}}}{T^B} - \frac{\mu^C_{\mathrm{H_2}}}{T^C}\right)\delta N_{\mathrm{H_2}} - \frac{1}{T^A}\left(\mu^A_{\mathrm{CH_2}} + \mu^A_{\mathrm{H_2}} - \mu^A_{\mathrm{CH_4}}\right)\delta \xi^A_1$$

$$- \frac{1}{T^A}\left(\mu^A_{\mathrm{N}} + \frac{3}{2}\mu^A_{\mathrm{H_2}} - \mu^A_{\mathrm{NH_3}}\right)\delta \xi^A_2 - \frac{1}{T^B}\left(\mu^B_{\mathrm{CH_2}} + \mu^B_{\mathrm{H_2}} - \mu^B_{\mathrm{CH_4}}\right)\delta \xi^B_1$$

$$- \frac{1}{T^C}\left(\mu^C_{\mathrm{N}} + \frac{3}{2}\mu^C_{\mathrm{H_2}} - \mu^C_{\mathrm{NH_3}}\right)\delta \xi^C_2 = 0$$

Therefore, the criteria of equilibrium are

$$T^A = T^B = T^C$$
$$\mu^A_{CH_2} = \mu^B_{CH_2} = \mu^A_{CH_4} - \mu^A_{H_2} = \mu^B_{CH_4} - \mu^B_{H_2}$$
$$\mu^B_{H_2} = \mu^C_{H_2}$$
$$\mu^A_N = \mu^C_N = \mu^A_{NH_3} - \tfrac{3}{2}\mu^A_{H_2} = \mu^C_{NH_3} - \tfrac{3}{2}\mu^C_{H_2}$$

Note that no statement was made on whether the semipermeable membranes were diathermal or adiabatic. Identical equilibrium criteria would result in either case.

6.7 Summary

The main purpose of this chapter was to establish rigorous criteria to define stable equilibrium states for various situations encountered in chemical thermodynamics. Starting with the principle of entropy maximization at constant total energy, volume, and mass [Eq. (6-1)], or equivalently energy minimization at constant total entropy, volume, and mass [Eq. (6-2)], we developed mathematical criteria for stable equilibrium. These criteria were expanded to other sets of constraints by using other potential functions, such as the Helmholtz energy $\underline{A}$ for constraints of constant temperature, volume, and mass or moles or the Gibbs energy $\underline{G}$ for constant temperature, pressure, and mass or moles. Table 6.1 summarizes equilibrium criteria for the important potential functions. Specific cases were treated to develop relevant constraining equations in terms of temperature, pressure, and chemical potential variables. For membrane equilibria these are summarized by three cases corresponding to the characteristics of the boundary between phases [see Section 6.4 and Eqs. (6-37) and (6-38)]; for phase equilibrium in simple systems, the equality of temperature, pressure, and chemical potential for each component between phases was established [see Eqs. (6-44) - (6-46)]; and for chemical reaction equilibrium, the constant mass constraint must be modified to account for the fact that the total moles are not necessarily conserved. This results in an additional requirement, namely that the summation of $\nu_i\ \mu_i$ consistent with the stoichiometry of the reaction is equal to zero [see Eq. (6-55)].

The equilibrium criteria developed here will be used extensively to provide fundamental understanding of phenomena in phase and chemical reaction equilibria of major importance to chemical engineers. Specific applications are explored in depth in Chapters 15, 16, and 17.

References and Suggested Readings

Bejan, A. (1988), *Advanced Engineering Thermodynamics*, Wiley, New York, p. 72-77.

Denbigh, K. (1981), *The Principles of Chemical Equilibrium*, 4th ed., Cambridge University Press, Cambridge, UK, p 82-89.

Gibbs, J.W. (1876), "On the equilibriuim of heterogeneous substances," *Transactions of the Connecticut Academy*, vol III, pp. 108-248, appearing in *The Collected Works of J. Willard Gibbs, Vol 1, Thermodynamics*, Yale University Press, New Haven, CT (1957), p 56 ff.

Gyftopoulos, E.P. and G.P. Berretta (1991), *Thermodynamics--Foundations and Applications*, Macmillan Publishing, New York, p 53-66 and 117-126.

Kyle, B.G. (1992), *Chemical and Process Thermodynamics,* 2nd ed, Prentice-Hall, Englewood Cliffs, NJ, p 158-167.

Problems

6.1. A well-insulated steel bomb initially contains liquid and vapor of compound A (see Figure P6.1). The vapor is known to decompose under the conditions of the experiment to

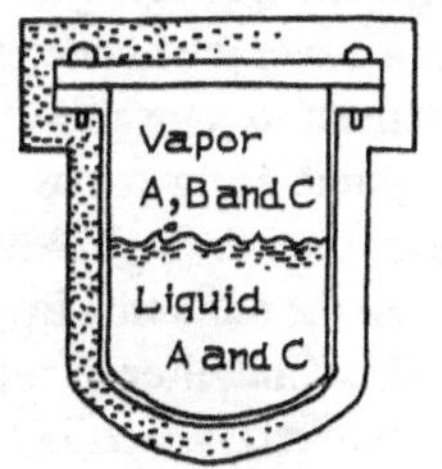

Figure P6.1

B and C by the reaction: A = 2B + C. Compound C is soluble in liquid A, but compound B does not dissolve to any measurable extent. Also, the reaction does not occur to any appreciable extent in the liquid. Derive the criteria of equilibrium for the system of liquid plus vapor.

6.2. A closed vessel of volume $\underline{V}$ contains liquid and vapor phases of a pure material. The vessel is immersed in a constant-temperature bath at temperature T^* (see Figure P6.2). Determine the criteria of equilibrium in terms of the intensive properties of the phases.

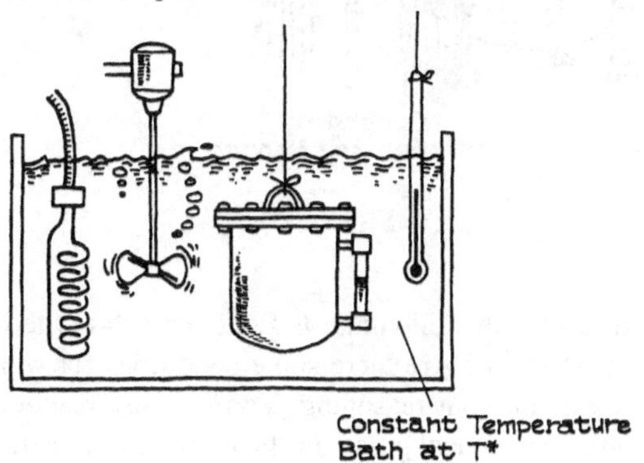

Figure P6.2

6.3. The criteria for chemical equilibrium in a multicomponent, single-phase system,

$$\Sigma \nu_i \mu_i = 0 \tag{1}$$

are derived from systems at constant ($\underline{S}$, $\underline{V}$), ($\underline{S}$, P), (T, $\underline{V}$), or (T, P), by applying the minimization principle from $\underline{U}$, $\underline{H}$, $\underline{A}$, or $\underline{G}$, respectively. There is some question about the criteria of equilibrium for a system at constant P and $\underline{V}$. For example, a constant-volume bomb may initially contain compound A and an anticatalyst to prevent the reaction

$$A = 2B \tag{R_B}$$

The anticatalyst is removed, and as the reaction proceeds, heat is added or removed in order to maintain a constant pressure (see Figure P6.3).

One student claims that the system will always reach a final state in which Eq. (1) is satisfied. Another student is not convinced and claims that under certain conditions the system may go from pure reactants to pure products, while Eq. (1) may or may not be satisfied at some intermediate point. He gives as an example the case of an exothermic reaction with the stoichiometric relation (R_B). When some A reacts to 2B, the system is cooled to maintain constant P, but the reduction in temperature results in a more favorable condition for B formation (Le Châtelier principle), so that more A reacts, and consequently more cooling is required, which in turn drives the reaction to B etc., until all of A is depleted.

Figure P6.3

You are asked to resolve this dilemma: Is Eq. (1) a valid criterion of equilibrium for a system at constant P and $\underline{V}$? Are there some conditions for which Eq. (1) need not be satisfied? Clearly explain your reasoning. You may use reaction R_B ($A = 2B$) with the assumption of a mixture of ideal gases for the purpose of illustrating your point.

6.4. A rigid, diathermal sphere contains a mixture of H_2, N_2, and NH_3. This is subsystem I. There is a tube connected to the sphere with a palladium membrane at the point of contact. This tube is closed at the other end by a free-floating, diathermal piston. The palladium membrane is permeable only to H_2. This is subsystem II.

The environment (subsystem III) consists of a large isothermal and isobaric reservoir that changes by a negligible amount for variations in subsystems I and II.

(a) What are the equilibrium criteria for the entirety of subsystems I, II, and III considered as a single system?

(b) If only subsystem I were considered, what thermodynamic potential function would be maximized (minimized) in an equilibrium state?

6.5. In Figure P6.5, the tube is well insulated and the initial conditions are:

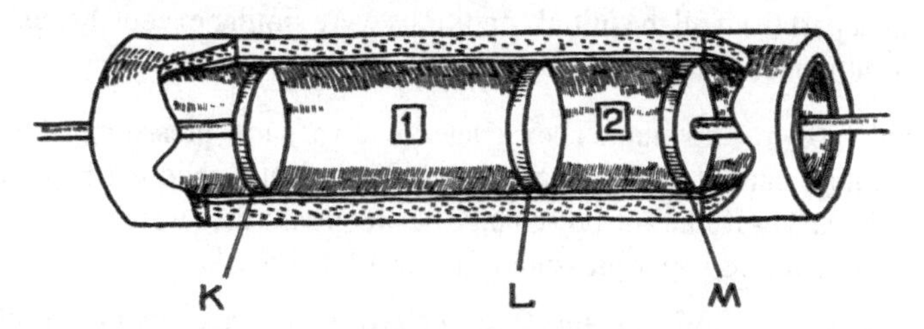

Figure P6.5

Section 1:	Section 2:
$\underline{V} = 9.086 \times 10^{-2}\ m^3$	$\underline{V} = 2.271 \times 10^{-2}\ m^3$
$T = 273.2$ K	$T = 273.2$ K
$P = 1$ bar	$P = 1$ bar
gas, pure A	gas, pure B

Piston L is rigid, diathermal, and initially impermeable. Suddenly, however, it is made permeable to gas A, but not to B. Piston K is simultaneously unlatched and connected to a large pressure reservoir at 2 bar, and piston M is unlatched and connected to a pressure reservoir at 3 bar. There is no friction in the pistons, but the pressure changes are effected so that oscillations are reduced. Assume ideal gases and $C_v = 20.9$ J/mol K.

(a) After equilibrium is attained, what are the criteria of equilibrium?

(b) Describe quantitatively the states of the gases on either side of the piston.

(c) Repeat parts (a) and (b) assuming that the piston L is permeable to both gases.

6.6. We have a cylinder whose external walls are rigid and adiabatic. There is an internal piston that is, initially, adiabatic, impermeable, and fixed in place. The two sides of the cylinder are charged with a binary mixture of ideal gases as shown below.

	Side 1	Side 2
T (K)	300	400
P (bar)	2	1
Total moles	6	5
Mole fraction A	0.4	0.8
Mole fraction B	0.6	0.2

At the start of an experiment, the dividing piston is replaced by one that is movable, diathermal, and permeable *only* to component A. Assume that $C_{v_A} = C_{v_B} = 20.9$ J/mol K.

(a) Derive the criteria of equilibria after all changes in the cylinder cease.

(b) What are the final equilibrium temperatures, pressures, and mole fractions of A in sides 1 and 2?

(c) Repeat part (b) if all the initial conditions were similar except that *no B* was initially present in side 2.

6.7. A rigid tank 0.03 m^3 in volume is evacuated to a very low pressure and then placed in a low-temperature bath at 244 K. A quantity 0.05 g of pure solid water (ice) is introduced into the tank. At 244 K, the entropy change between saturated water vapor and ice is 11.62 J/g K and the reported vapor pressure of ice is 42.8 N/m^2.

(a) Calculate $\Delta \underline{U}$, $T\Delta \underline{S}$, $\Delta \underline{A}$, and Q for the process of ice evaporation. What fraction of the ice has evaporated at equilibrium?

(b) What conclusions can you reach about the general equilibrium criterion in this case?

6.8. A rigid container has a volume of 2 liters and is divided into two chambers (A and B) by a partition. The container is maintained at 25°C in a constant temperature bath. Initially, the partition is impermeable to both components and is held by a stop such that both chambers have equal volumes and are at an equal pressure of 150 atm. Initially, chamber A contains an equimolar mixture of components 1 and 2, while chamber B contains pure component 2. There is no friction between the partition and the walls of the container.

You can assume that the pure components ***and*** the mixture obey the Clausius equation of state:

$$P(V - b_m) = RT$$

where for the mixture

$$b_m = y_1 b_1 + y_2 b_2$$

with

$$b_1 = 0.20 \text{ liter/mole}$$
$$b_2 = 0.40 \text{ liter/mole}$$

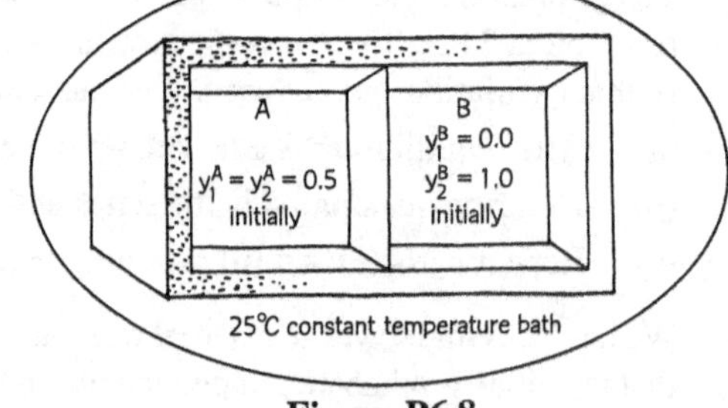

Figure P6.8

If the stop is removed, the partition is free to move. Determine the final equilibrium conditions (P_A, P_B, $\underline{V}_A$, $\underline{V}_B$, y_1^A, y_1^B, N_A, N_B) if:

(a) The partition becomes permeable only to component 1.

(b) The partition becomes permeable only to component 2.

Stability Criteria 7

In this chapter we derive the criteria of stability for thermodynamic systems and determine the conditions when such systems become unstable. This concept of stable and unstable thermodynamic states is of particular importance when one deals with *metastable* states. All phase changes can be associated with metastable states. As an example, let us select liquid water at 80°C and 1 bar as our system. If we were to heat this water isobarically to about 100°C, we would expect to see boiling begin. However, if we could suppress all active nucleation sites, it could be possible to heat ultra-pure *liquid* water well in excess of 100°C with no vaporization. In such a condition we refer to the liquid phase as superheated. It is in a metastable state. The higher the temperature we attain, the easier it becomes to initiate nucleation and rapid, explosive vaporization. Yet even in this superheated state, it is still stable in a thermodynamic sense. There is, however, a limit to the temperature that can be reached before *spontaneous nucleation* occurs. This condition occurs at the *limit of stability*; one of the important results of this chapter is the derivation of thermodynamic relations that allow one to predict these stability limits.

Superheated liquids are found in *bubble* chambers. Here a liquid exists in a state where the temperature exceeds that expected from the prevailing pressure. Nucleation in this case results from the penetration of the chamber by nuclear particles. The track of small bubbles is used to characterize the type and energy of the particle.

Another device to detect and study high-energy nuclear particles is the *cloud chamber*. In this case a vapor is *subcooled*; that is, the pressure and temperature are such that, if equilibrium existed, one would expect to find a condensed phase. But condensation requires a nucleation surface to initiate the formation of the liquid phase. (This process is described in Chapter 19.) Without an efficient nucleating surface, the vapor may be subcooled until the appearance of an energetic particle upon which condensation can take place--or, until the limit of stability of the subcooled vapor is reached. In the latter case, condensation is immediate.

Liquid mixtures at a specified temperature and pressure that contain at least one component at a concentration above its solubility limit are supersaturated and metastable. Anywhere in this metastable region, the presence of a suitable disturbance (e.g., acoustic shock) or a nucleus (e.g., a seed crystal, gas bubble, or suitable foreign body) will lead to the formation of another phase containing that component. For example, seeding a supersaturated salt solution will lead to crystallization of solid salt;

similarly an aqueous solution containing carbon dioxide above its equilibrium solubility will produce gas bubbles when appropriately disturbed.

In the metastable region, one approaches a limit of stability where spontaneous, homogeneous nucleation of a second phase will occur even in the absence of heterogeneous nuclei or external disturbances. This limit is often referred to as a spinodal point. It serves as the effective upper limiting boundary of the metastable region while the equilibrium or saturation condition, commonly called the binodal point, defines the lower boundary.

An extreme example of homogeneous nucleation can occur in polymer systems near a critical consolute point where a transition exists between a single- and two-phase region at a well-defined temperature, pressure, and composition. At the critical point itself, both the equilibrium binodal and stability-limit spinodal curves are coincident. Thus near a critical point composition, one can enter an unstable two-phase region from a stable single-phase region with a small change in temperature or pressure; because the metastable region is very small or non-existent. This leads to rapid homogeneous nucleation of the second phase in a phenomenon called spinodal decomposition. While specific applications are covered in Chapter 15, the rationale for such phenomena will be introduced here.

In another example, solvents are known to change their solvation capacity for certain solutes significantly in passing from a subcritical to a supercritical state. For example, in water, which has a critical temperature and pressure of 374°C and 220 bar, the solubility of sodium chloride (NaCl) decreases by over four orders of magnitude at 250 bar as the temperature is increased from 450 to 460°C. Such massive changes in solubility lead to rapid, instability-induced solid salt nucleation, often termed shock crystallization.

These last two examples involve very special cases where a system is both *at the limit of* stability and yet is still in a *stable state*. These states are called *critical states*. Although critical states do not occur for all phase transformations, they represent important thermodynamic phenomena. This chapter provides the basis for such processes as we develop stability and critical point criteria in a very general manner. Selection of the specific functional forms to employ in any given problem is dictated largely by the types of data or property correlations available. Legendre transforms, as developed earlier in Chapter 5, are used extensively to manipulate stability criteria into appropriate forms.

7.1 Criteria of Stability

Let us examine a simple, single-phase isolated system. Let us further suppose that the system satisfies the criteria of equilibrium; that is, T, P, and μ_j are uniform throughout the system and hence $\delta \underline{S} = 0$. To determine if $\underline{S}$ is a maximum, we must show that the lowest-order, nonvanishing variation in $\underline{S}$ is negative (see Section 6.2).

It might appear, at first glance, that no internal variations are possible for an isolated, single-phase system. Nevertheless, we can restructure our homogeneous system into one large portion α and a second small portion β (see Figure 7.1). That is, we conceptually insert a membrane enclosing some finite element β inside the system so that we may distinguish this element from the remainder of the system, α. We must, however, allow this membrane to be diathermal, nonrigid, and permeable to all components, so that the composite system is still a simple system and not restrained internally. In the original state both subsystems have properties identical to those of the α-phase.

Employing the Gibbs stability criteria of Eq. (6-7), for small variations, we should examine second-order inequalities first. Of course, if $\delta^2 \underline{S}$ should be zero, we must examine higher-order terms. As before, we write second-order derivatives in a shorthand notation such as

$$\underline{S}_{\underline{U}\,\underline{V}} = \left(\frac{\partial(\partial \underline{S}/\partial \underline{U})_{\underline{V}}}{\partial \underline{V}} \right)_{\underline{U}}$$

Then, if the original system (i.e., the α-phase) is stable,

$$\delta^2 \underline{S} = \delta^2 (\underline{S}^{\alpha} + \underline{S}^{\beta}) = \delta^2 \underline{S}^{\alpha} + \delta^2 \underline{S}^{\beta} < 0$$

or

$$0 > \delta^2 \underline{S} = \underline{S}^{\alpha}_{\underline{U}\,\underline{U}} (\delta \underline{U}^{\alpha})^2 + 2 \underline{S}^{\alpha}_{\underline{U}\,\underline{V}}\, \delta \underline{U}^{\alpha}\, \delta \underline{V}^{\alpha} + \underline{S}^{\alpha}_{\underline{V}\,\underline{V}} (\delta \underline{V}^{\alpha})^2 + \sum_{j=1}^{n} \sum_{k=1}^{n} \underline{S}^{\alpha}_{N_j N_k}\, \delta N^{\alpha}_j\, \delta N^{\alpha}_k$$

$$+ 2 \sum_{j=1}^{n} \underline{S}^{\alpha}_{\underline{U} N_j}\, \delta \underline{U}^{\alpha}\, \delta N^{\alpha}_j + 2 \sum_{j=1}^{n} \underline{S}^{\alpha}_{\underline{V} N_j}\, \delta \underline{V}^{\alpha}\, \delta N^{\alpha}_j \qquad (7\text{-}1)$$

$$+ \text{ similar terms for subsystem } \beta$$

Figure 7.1 Conceptual visualization of a subsystem β within a homogeneous system α.

Noting that in the Taylor expansions the second-order partial derivatives are evaluated at the initial conditions, and since the α-phase is identical to the β-phase at the outset of the perturbation, it follows that $N^{\alpha}\underline{S}^{\alpha}_{ij} = N^{\beta}\underline{S}^{\beta}_{ij}$, where i and j may be $\underline{U}$, $\underline{V}$, or N_j. Furthermore, the α- and β-phase variations are related by the equations of isolation:

$$(\delta \underline{U}^{\alpha})^2 = (\delta \underline{U}^{\beta})^2 \tag{7-2}$$

$$(\delta \underline{V}^{\alpha})^2 = (\delta \underline{V}^{\beta})^2 \tag{7-3}$$

$$(\delta N_j^{\alpha})^2 = (\delta N_j^{\beta})^2 \tag{7-4}$$

With these substitutions, Eq. (7-1) simplifies to

$$\delta^2 \underline{S} = \frac{N}{N^{\beta}} \left[\underline{S}^{\alpha}_{\underline{U}\,\underline{U}} (\delta \underline{U}^{\alpha})^2 + 2\underline{S}^{\alpha}_{\underline{U}\,\underline{V}}\, \delta \underline{U}^{\alpha}\, \delta \underline{V}^{\alpha} + \underline{S}^{\alpha}_{\underline{V}\,\underline{V}} (\delta \underline{V}^{\alpha})^2 \right.$$
$$\left. + 2\sum_{j=1}^{n} (\underline{S}^{\alpha}_{\underline{U}N_j}\, \delta \underline{U}^{\alpha} + \underline{S}^{\alpha}_{\underline{V}N_j}\, \delta \underline{V}^{\alpha})\, \delta N_j^{\alpha} + \sum_{j=1}^{n} \sum_{k=1}^{n} \underline{S}^{\alpha}_{N_j N_k}\, \delta N_j^{\alpha}\, \delta N_k^{\alpha} \right] < 0 \tag{7-5}$$

Note that Eq. (7-5) contains only derivatives and variations for the α-phase. Although the β-phase was introduced to allow us to vary the parameters of the α-phase, we see that the stability of the composite reduces to determining the stability of the original α-phase with the ratio N/N^{β} providing the appropriate scaling. Thus, our stability analysis, when completed, will tell us whether or not the original system is stable. If it is unstable, the analysis will not determine what other phase may form in its place, but it will show that some transformation, leading to a more stable condition, will occur.

Note that we could have eliminated the α terms instead of the β terms, in which case we would have obtained Eq. (7-5) with β superscripts instead of α superscripts. Since the properties of the α- and β-phases are identical in the initial state, and since the Taylor series expansion is for deviations from that initial state, it is immaterial which superscript we retain.

It is more convenient to explore the consequences of stability in the energy representation instead of the entropy representation, because the $\underline{U}$ criteria can be readily modified to the $\underline{H}$, $\underline{A}$, $\underline{G}$, $\underline{U}'$, etc. forms using the Legendre transform technique discussed in Section 5.6. In the internal energy representation, the analog of Eq. (7-5) is

$$\delta^2 \underline{U} = \frac{N}{N^{\beta}} \left[\underline{U}_{\underline{S}\,\underline{S}} (\delta \underline{S})^2 + 2\underline{U}_{\underline{S}\,\underline{V}}\, \delta \underline{S}\, \delta \underline{V} + \underline{U}_{\underline{V}\,\underline{V}} (\delta \underline{V})^2 \right.$$
$$\left. + 2\sum_{j=1}^{n} (\underline{U}_{\underline{S}N_j}\, \delta \underline{S} + \underline{U}_{\underline{V}N_j}\, \delta \underline{V}) \delta N_j + \sum_{j=1}^{n} \sum_{k=1}^{n} \underline{U}_{N_j N_k}\, \delta N_j\, \delta N_k \right] > 0 \tag{7-6}$$

where the superscript α has been dropped. To derive Eq. (7-6), Eqs. (7-3) and (7-4) were used. Instead of Eq. (7-2), however, one employs

$$(\delta \underline{S}^{\alpha})^2 = (\delta \underline{S}^{\beta})^2$$

The conditions where $\delta^2 \underline{U} > 0$ or $\delta^2 \underline{S} < 0$ are called the *criteria of intrinsic stability*, since they relate to a single phase. The conditions where these criteria are first violated, beginning with a stable phase, are $\delta^2 \underline{U} = 0$ or $\delta^2 \underline{S} = 0$ and are called the *limits of intrinsic stability*. Systems at the limit of intrinsic stability may or may not be stable; to answer this question, $\delta^3 \underline{U}$ or $\delta^3 \underline{S}$ (or higher-order) expansions must be examined. We return to this point in Section 7.3.

Equation (7-6) is a general quadratic and may be written in a more condensed form by defining $\underline{U}$ as a basis function $y^{(0)}$. $\underline{U}$ is a function of $n + 2$ independent variables $(\underline{S}, \underline{V}, N_1, ..., N_n)$.

$$\underline{U} = y^{(0)} = f(x_1, ..., x_m) \qquad \text{with } m \equiv n+2 \tag{7-7}$$

and the x_i's represent $\underline{S}$, $\underline{V}$, $N_1, ..., N_n$ with arbitrary ordering. Then Eq. (7-6) becomes

$$\delta^2 y^{(0)} = K \sum_{i=1}^{m} \sum_{j=1}^{m} y_{ij}^{(0)} \, \delta x_i \, \delta x_j > 0 \tag{7-8}$$

where K is a positive numerical constant. It is dropped from the subsequent treatment, as we are only interested in variations that affect the sign of $\delta^2 y^{(0)}$. For a stable system, $\delta^2 y^{(0)}$ is positive; for an unstable system, it is negative. Thus, the limit of intrinsic stability occurs when a system with an initial positive value of $\delta^2 y^{(0)}$ becomes zero.

Since the variations x_i and x_j may be either positive or negative, it is more convenient to rearrange the summations in a sum-of-squares form:

$$\sum_{i=1}^{m} \sum_{j=1}^{m} y_{ij}^{(0)} \, \delta x_i \, \delta x_j = \sum_{k=1}^{m} y_{kk}^{(k-1)} \, \delta Z_k^2 > 0 \tag{7-9}$$

with

$$\delta Z_k \equiv \delta x_k + \sum_{j=k+1}^{m} y_{kj}^{(k)} \, \delta x_j \quad \text{for } k = 1, 2, ..., m-1 \tag{7-10}$$

$$\delta Z_m \equiv \delta x_m \qquad \text{for } k = m \tag{7-11}$$

The advantage of arranging the stability criteria in a sum-of-squares form [Eq. (7-9)] is that δZ_k^2 is always positive irrespective of the sign of δZ_k. Then, since Eq. (7-9) must be positive for *all possible variations* in $\delta \underline{S}^{\alpha}$, $\delta \underline{V}^{\alpha}$, and δN_j^{α}, or for any combination of these variables as given by δZ_k, the stability of the system is dictated by the signs of $y_{kk}^{(k-1)}$. For example, we could hypothesize some variation occurring in the α–β system such that all δZ_k values were zero except one, δZ_r. In this case, the sum in Eq. (7-9) simplifies to

$$y_{rr}^{(r-1)} \, \delta Z_r^2 > 0$$

Therefore,

$$y_{rr}^{(r-1)} > 0$$

Since $y_{rr}^{(r-1)}$ is determined by the state of the original α–β system and is independent of the proposed variation, $y_{rr}^{(r-1)}$ must be positive for a stable system for *any possible variation*. It follows logically that, since the index r was chosen arbitrarily, for a stable system the necessary conditions are that

$$y_{kk}^{(k-1)} > 0, \quad k = 1, 2, \ldots, m-1 \tag{7-12}$$

Note that in Eq. (7-12), the index k did not include the case when $k = m$. For this special situation, we can show that $y_{mm}^{(m-1)}$ is identically zero. That is,

$$y_{mm}^{(m-1)} = \left(\frac{\partial^2 y^{(m-1)}}{\partial x_m^2}\right)_{\xi_1, \ldots, \xi_{m-1}} = \left(\frac{\partial \xi_m}{\partial x_m}\right)_{\xi_1, \ldots, \xi_{m-1}} = 0 \tag{7-13}$$

$y_{mm}^{(m-1)}$ must be zero since, by specifying $(m-1)$ intensive variables $(\xi_1, \ldots, \xi_{m-1})$, all other intensive variables in a system are fixed. That is, since the second derivative itself is a derivative of an intensive property with respect to an extensive property with $(m-1)$ intensive variables held constant, it must be identically zero. The limit of stability is defined as the state when any of the $y_{kk}^{(k-1)}$ derivatives in Eq. (7-12) becomes zero.

Example 7.1

What are the criteria for a stable ternary system composed of components B, C, and D?

Solution

For a ternary system, $m = n + 2 = 5$. The criteria satisfying Eq. (7-12) are that $y_{11}^{(0)}$, $y_{22}^{(1)}$, $y_{33}^{(2)}$, and $y_{44}^{(3)}$ are all positive. Note that $y_{55}^{(4)}$ is not included as, by Eq. (7-13), it is zero. If the basis function $\underline{U} = y^{(0)}$ is ordered in the manner,

$$\underline{U} = y^{(0)} = U(\underline{S}, \underline{V}, N_B, N_C, N_D)$$

then

$$y_{11}^{(0)} = \underline{U}_{\underline{S}\,\underline{S}} > 0 \qquad y_{22}^{(1)} = \underline{A}_{\underline{V}\,\underline{V}} > 0 \qquad y_{33}^{(2)} = \underline{G}_{BB} > 0$$

and

$$y_{44}^{(3)} = \underline{G}'_{CC} > 0 \qquad y_{55}^{(4)} = \underline{G}''_{DD} = 0$$

Note that we have used B, C, and D as subscripts in place of N_B, N_C, and N_D to simplify the notation. It should be clear that a reordering of variables for this basis function will allow other stability criteria to be developed. For example, if the ordering of the variables $\underline{S}$ and $\underline{V}$ were inverted, then

$$y_{11}^{(0)} = \underline{U}_{\underline{V}\,\underline{V}} > 0 \quad \text{and} \quad y_{22}^{(1)} = \underline{H}_{\underline{S}\,\underline{S}} > 0$$

but the second derivatives of $y^{(2)}$ and $y^{(3)}$ are not affected. Additional stability criteria may be formulated by using the step-down relations in Tables 5.4 and 5.5. As an illustration, for $y_{44}^{(3)}$ with the ordering of the basis function as shown originally,

$$y_{44}^{(3)} = \frac{\begin{vmatrix} y_{33}^{(2)} & y_{34}^{(2)} \\ y_{34}^{(2)} & y_{44}^{(2)} \end{vmatrix}}{y_{33}^{(2)}} = \frac{\begin{vmatrix} \underline{G}_{BB} & \underline{G}_{BC} \\ \underline{G}_{BC} & \underline{G}_{CC} \end{vmatrix}}{\underline{G}_{BB}}$$

$$= \frac{\begin{vmatrix} y_{22}^{(1)} & y_{23}^{(1)} & y_{24}^{(1)} \\ y_{23}^{(1)} & y_{33}^{(1)} & y_{34}^{(1)} \\ y_{24}^{(1)} & y_{34}^{(1)} & y_{44}^{(1)} \end{vmatrix}}{\begin{vmatrix} y_{22}^{(1)} & y_{23}^{(1)} \\ y_{23}^{(1)} & y_{33}^{(1)} \end{vmatrix}} = \frac{\begin{vmatrix} \underline{A}_{\underline{V}\,\underline{V}} & \underline{A}_{\underline{V}B} & \underline{A}_{\underline{V}C} \\ \underline{A}_{\underline{V}B} & \underline{A}_{BB} & \underline{A}_{BC} \\ \underline{A}_{\underline{V}C} & \underline{A}_{BC} & \underline{A}_{CC} \end{vmatrix}}{\begin{vmatrix} \underline{A}_{\underline{V}\,\underline{V}} & \underline{A}_{\underline{V}B} \\ \underline{A}_{\underline{V}B} & \underline{A}_{BB} \end{vmatrix}}$$

$$= \frac{\begin{vmatrix} y_{11}^{(0)} & y_{12}^{(0)} & y_{13}^{(0)} & y_{14}^{(0)} \\ y_{12}^{(0)} & \cdot & \cdot & \cdot \\ y_{13}^{(0)} & \cdot & \cdot & \cdot \\ y_{14}^{(0)} & \cdot & \cdot & y_{44}^{(0)} \end{vmatrix}}{\begin{vmatrix} y_{11}^{(0)} & y_{12}^{(0)} & y_{13}^{(0)} \\ y_{12}^{(0)} & y_{22}^{(0)} & y_{23}^{(0)} \\ y_{13}^{(0)} & y_{23}^{(0)} & y_{33}^{(0)} \end{vmatrix}} = \frac{\begin{vmatrix} \underline{U}_{\underline{S}\,\underline{S}} & \underline{U}_{\underline{S}\,\underline{V}} & \underline{U}_{\underline{S}B} & \underline{U}_{\underline{S}C} \\ \underline{U}_{\underline{S}\,\underline{V}} & \underline{U}_{\underline{V}\,\underline{V}} & \underline{U}_{\underline{V}B} & \underline{U}_{\underline{V}C} \\ \underline{U}_{\underline{S}B} & \underline{U}_{\underline{V}B} & \underline{U}_{BB} & \underline{U}_{BC} \\ \underline{U}_{\underline{S}C} & \underline{U}_{\underline{V}C} & \underline{U}_{BC} & \underline{U}_{CC} \end{vmatrix}}{\begin{vmatrix} \underline{U}_{\underline{S}\,\underline{S}} & \underline{U}_{\underline{S}\,\underline{V}} & \underline{U}_{\underline{S}B} \\ \underline{U}_{\underline{S}\,\underline{V}} & \underline{U}_{\underline{V}\,\underline{V}} & \underline{U}_{\underline{V}B} \\ \underline{U}_{\underline{S}B} & \underline{U}_{\underline{V}B} & \underline{U}_{BB} \end{vmatrix}}$$

Thus, there are four ways to express the fact that $y_{44}^{(3)} > 0$. (In fact, there are many more if the ordering of the basis function were changed.) Similar step-down relations for $y_{33}^{(2)}$ and $y_{22}^{(1)}$ are shown below. [Note that we have made extensive use of Maxwell's reciprocity relationship, e.g., $y_{32}^{(1)} = y_{23}^{(1)}$]

$$y_{33}^{(2)} = \frac{\begin{vmatrix} y_{22}^{(1)} & y_{23}^{(1)} \\ y_{23}^{(1)} & y_{33}^{(1)} \end{vmatrix}}{y_{22}^{(1)}} = \frac{\begin{vmatrix} \underline{A}_{\underline{V}\,\underline{V}} & \underline{A}_{\underline{V}B} \\ \underline{A}_{\underline{V}B} & \underline{A}_{BB} \end{vmatrix}}{\underline{A}_{\underline{V}\,\underline{V}}}$$

$$y_{33}^{(2)} = \frac{\begin{vmatrix} y_{11}^{(0)} & y_{12}^{(0)} & y_{13}^{(0)} \\ y_{12}^{(0)} & y_{22}^{(0)} & y_{23}^{(0)} \\ y_{13}^{(0)} & y_{23}^{(0)} & y_{33}^{(0)} \end{vmatrix}}{\begin{vmatrix} y_{11}^{(0)} & y_{12}^{(0)} \\ y_{12}^{(0)} & y_{22}^{(0)} \end{vmatrix}} = \frac{\begin{vmatrix} \underline{U}_{\underline{S}\,\underline{S}} & \underline{U}_{\underline{S}\,\underline{V}} & \underline{U}_{\underline{S}B} \\ \underline{U}_{\underline{S}\,\underline{V}} & \underline{U}_{\underline{V}\,\underline{V}} & \underline{U}_{\underline{V}B} \\ \underline{U}_{\underline{S}B} & \underline{U}_{\underline{V}B} & \underline{U}_{BB} \end{vmatrix}}{\begin{vmatrix} \underline{U}_{\underline{S}\,\underline{S}} & \underline{U}_{\underline{S}\,\underline{V}} \\ \underline{U}_{\underline{S}\,\underline{V}} & \underline{U}_{\underline{V}\,\underline{V}} \end{vmatrix}}$$

$$y_{22}^{(1)} = \frac{\begin{vmatrix} y_{11}^{(0)} & y_{12}^{(0)} \\ y_{12}^{(0)} & y_{22}^{(0)} \end{vmatrix}}{y_{11}^{(0)}} = \frac{\begin{vmatrix} \underline{U}_{SS} & \underline{U}_{SV} \\ \underline{U}_{SV} & \underline{U}_{VV} \end{vmatrix}}{\underline{U}_{SS}}$$

It is clear from Example 7.1 that there are many ways to express criteria for a stable thermodynamic system. The necessary conditions are, however, still given by Eqs. (7-12). If these $(m-1)$ derivatives (or any equivalent step-down form) are positive for any given ordering of the basis function, the system is stable. One does not have to test all possible forms of the criteria to answer the question of stability.

Although Eqs. (7-12) are necessary to define a stable system, it is possible to show that, if a system is initially in a stable state, as conditions change and the limit of intrinsic stability is approached, the particular derivative $y_{(m-1)(m-1)}^{(m-2)}$ goes to zero before any other. The proof of this statement depends on the use of the step-down relation shown in Table 5.5. In particular, it can be shown that

$$y_{kk}^{(k-1)} = y_{kk}^{(k-2)} - \frac{(y_{k(k-1)}^{(k-2)})^2}{y_{(k-1)(k-1)}^{(k-2)}} \tag{7-14}$$

From the stability criteria, Eqs. (7-12), both $y_{kk}^{(k-1)}$ and $y_{(k-1)(k-1)}^{(k-2)}$ must be positive. It is then clear from Eq. (7-14) that $y_{kk}^{(k-2)}$ must also be positive. It may also be noted that $y_{kk}^{(k-2)}$ would have been the coefficient of $(\delta Z_k)^2$ if the ordering of variables x_{k-1} and x_k were to have been reversed. Thus, it is reasonable to expect no unusual behavior from this term as either $y_{kk}^{(k-1)}$ or $y_{(k-1)(k-1)}^{(k-2)}$ is reduced to simulate a system that is approaching the limit of intrinsic stability. An important, logical conclusion may then be drawn. Suppose that $y_{(k-1)(k-1)}^{(k-2)}$ decreases toward zero. Equation (7-14) then indicates that as $y_{(k-1)(k-1)}^{(k-2)}$ decreases, before it can reach the value zero, $y_{kk}^{(k-1)}$ becomes negative. Thus the positive nature of $y_{kk}^{(k-1)}$ is always violated before that of $y_{(k-1)(k-1)}^{(k-2)}$.

Generalizing, one can state that the ***necessary and sufficient criterion of stability*** is

$$\boxed{y_{(m-1)(m-1)}^{(m-2)} > 0 \quad \text{with } m \equiv n+2} \tag{7-15}$$

where n = the number of components. The state of a system at a limit of stability or *spinodal* condition is then expressed as

$$y_{(m-1)(m-1)}^{(m-2)} = 0 \tag{7-16}$$

Equations (7-15) and (7-16) may also be written in terms of lower-order Legendre transforms. A particularly convenient form is obtained using the relations in Table 5.5. For a stable system,

$$\pounds_i > 0 \tag{7-17}$$

And at the limit of stability,

$$\pounds_i = 0 \tag{7-18}$$

where $0 \leq i \leq m-2$ and $\pounds_i$ is a determinant defined as

$$\pounds_i \equiv \begin{vmatrix} y^{(i)}_{(i+1)(i+1)} & y^{(i)}_{(i+1)(i+2)} & \cdots & y^{(i)}_{(i+1)(m-1)} \\ y^{(i)}_{(i+2)(i+1)} & y^{(i)}_{(i+2)(i+2)} & \cdots & y^{(i)}_{(i+2)(m-1)} \\ \vdots & \vdots & & \vdots \\ y^{(i)}_{(m-1)(i+1)} & y^{(i)}_{(m-1)(i+2)} & \cdots & y^{(i)}_{(m-1)(m-1)} \end{vmatrix} \tag{7-19}$$

Actually, with Table 5.5 one obtains the general relation

$$y^{(m-2)}_{(m-1)(m-1)} = \frac{\pounds_i}{\prod_{r=i}^{m-3} y^{(r)}_{(r+1)(r+1)}} \tag{7-20}$$

However, because $y^{(r)}_{(r+1)(r+1)}$, for all $r < m-2$, is always positive for a stable system and remains positive as $y^{(m-2)}_{(m-1)(m-1)}$ approaches zero, one can employ only the $\pounds_i$ determinant to test stability. Equation (7-20) was, in fact, illustrated in Example 7.1. In the case of a ternary system, $m = 5$, and it was shown that

$$\pounds_3 = y^{(m-2)}_{(m-1)(m-1)} = y^{(3)}_{44} = \frac{\pounds_2}{y^{(2)}_{33}} = \frac{\pounds_1}{y^{(2)}_{33}\, y^{(1)}_{22}} = \frac{\pounds_0}{y^{(2)}_{33}\, y^{(1)}_{22}\, y^{(0)}_{11}}$$

7.2 Applications to Thermodynamic Systems

7.2.1 Pure substances. In this case, with $\underline{U}$ as our basis function,

$$y^{(0)} = \underline{U} = f(\underline{S}, \underline{V}, N) = f(x_1, x_2, x_3) \tag{7-21}$$

To apply the general criterion of stability, Eq. (7-17), one must first choose the ordering of $\underline{S}$, $\underline{V}$, and N. For example, if $\underline{S} = x_1$, $\underline{V} = x_2$, $N = x_3$, then $y^{(1)} = \underline{A}$, and $m = 3$. Only two forms of Eq. (7-17) are possible, that is, when $i = 0$ and $i = 1$.

for $i = 0$:

$$\pounds_0 = \begin{vmatrix} \underline{U}_{\underline{S}\,\underline{S}} & \underline{U}_{\underline{S}\,\underline{V}} \\ \underline{U}_{\underline{S}\,\underline{V}} & \underline{U}_{\underline{V}\,\underline{V}} \end{vmatrix} > 0 \tag{7-22}$$

for $i = 1$:

$$\pounds_1 = \underline{A}_{\underline{V}\,\underline{V}} > 0 \tag{7-23}$$

Equations (7-22) and (7-23) are completely equivalent criteria. With the latter,

$$\underline{A}_{\underline{V}\,\underline{V}} = \left(\frac{\partial^2 \underline{A}}{\partial \underline{V}^2}\right)_{T,N} = -\left(\frac{\partial P}{\partial \underline{V}}\right)_{T,N} = -\frac{1}{N}\left(\frac{\partial P}{\partial V}\right)_T > 0 \tag{7-24}$$

Equation (7-24) expresses the well-known *criteria of mechanical stability* that, for stable systems of pure materials, the pressure increases as the volume decreases if the system is constrained to be isothermal and of constant mass.

Yet even Eqs. (7-22) and (7-23) are not unique ways to express the stability criterion for pure materials. Different ordering of the variables (that is, $\underline{S}$, $\underline{V}$, N) set produces equivalent criteria. In fact, there are six equivalent forms of Eq. (7-15) starting with $y^{(0)} = \underline{U}$ as a basis function, and these are listed in Table 7.1. If a pure component system were to reach the limit of intrinsic stability, all of these criteria would be violated (i.e., equal zero) simultaneously.

Other basis functions can be used as well. For example, earlier in Chapter 6, we listed general stability and equilibrium criteria using the common potential functions $\underline{U}$, $\underline{H}$, $\underline{A}$, and $\underline{G}$. (See Table 6.1). Any of these could be used to generate stability criteria.

Table 7.1 Equivalent Stability Criteria for a Pure Material with $y^{(0)} = \underline{U}$

Ordering of (x_1, x_2, x_3)	Stability criterion $y_{22}^{(1)} > 0$
$(\underline{S}, \underline{V}, N)$	$\underline{A}_{\underline{V}\underline{V}} = -(\partial P/\partial \underline{V})_{T,N} > 0$
$(\underline{S}, N, \underline{V})$	$\underline{A}_{NN} = (\partial \mu/\partial N)_{T,\underline{V}} > 0$
$(\underline{V}, \underline{S}, N)$	$\underline{H}_{\underline{S}\underline{S}} = (\partial T/\partial \underline{S})_{P,N} > 0$
$(\underline{V}, N, \underline{S})$	$\underline{H}_{NN} = (\partial \mu/\partial N)_{P,\underline{S}} > 0$
$(N, \underline{S}, \underline{V})$	$\underline{U}'_{\underline{S}\underline{S}} = (\partial T/\partial \underline{S})_{\mu,\underline{V}} > 0$[a]
$(N, \underline{V}, \underline{S})$	$\underline{U}'_{\underline{V}\underline{V}} = -(\partial P/\partial \underline{V})_{\mu,\underline{S}} > 0$

[a] Following the convention introduced in Chapter 5, potential functions in which one mole number is transformed to a chemical potential are denoted by a prime, those in which two mole numbers are transformed are denoted by a double prime, etc. Thus, $\underline{U}' = f(\mu_1, \underline{S}, \underline{V}, N_2,..., N_n)$.

Eq. (7-12) indicates that when a system is stable, all second derivatives, $y_{kk}^{(k-1)}$ for $k = 1, 2,..., m-1$, must be positive. The criteria given in Table 7.1 only represent the case of $k = m-1 = n+2-1 = 2$, thus only one second derivative is specified, $y_{22}^{(1)} > 0$. For the pure-component case, the only other positive term is $y_{11}^{(0)} = \underline{U}_{\underline{S}\underline{S}}$, $\underline{U}_{\underline{V}\underline{V}}$, or $\underline{U}_{NN}$. For example, $\underline{U}_{\underline{S}\underline{S}} = (\partial^2 \underline{U}/\partial \underline{S}^2)_{\underline{V},N} = (\partial T/\partial \underline{S})_{\underline{V},N} = T/NC_v > 0$. C_v is then always positive for a stable system. This condition is sometimes called the *criteria of thermal stability*. As noted previously, but repeated again for emphasis, $\underline{U}_{\underline{S}\underline{S}}$ never attains a zero or negative value as the system becomes unstable since higher-order criteria (Table 7.1) approach zero values before lower-order terms.

The condition of *mechanical stability* for a pure material, shown in Table 7.1, is $\underline{A}_{\underline{V}\underline{V}} > 0$ or $A_{VV} > 0$, or equivalently $(\partial P/\partial V)_T < 0$, can be visualized with the aid of Figure 7.2. This is a typical *PVT* plot for the solid, liquid, and gaseous regions of a pure substance that contracts on freezing; also shown are P-V and P-T projections. Reducing the pressure at constant temperature, one follows the isotherm to point A on Figure 7.2(d), where the liquid is called a saturated liquid and would be in equilibrium with saturated vapor at C. The isotherm for the vapor phase thus emerges normally at C. At both A and C, $(\partial P/\partial V)_T < 0$ and both phases are stable. This condition is termed coexistence and plots as a single point on the vapor pressure P-T projection curve shown in Figure 7.2(c).

Suppose, however, that we propose an experiment in which a liquid under conditions at A is further expanded isothermally to some lower pressure noted by B on Figure 7.2(d). The stability criterion $(\partial P/\partial V)_T < 0$ is still satisfied at B at least with respect to small variations in volume. We stress the fact that at B the variation must be small since it is not difficult to show that there are possible variations that could lead to a state of higher entropy; that is, with regard to some variations, B is not a stable equilibrium state. One obvious variation fitting this description would be a fluctuation large enough to perturb the volume locally to a vapor-like value and thus trigger complete vaporization of the superheated liquid. State B is, therefore, metastable. To attain this state for any prolonged period, we would need to start with a very clean, degassed liquid, free of nucleation sites to prevent vapor from forming. Superheated liquid experiments are described by Patrick-Yeboah and Reid (1981) and in a form suitable for classroom demonstration by Jolls and Prausnitz (1983).

In a similar manner, we could have begun with the saturated vapor at C and compressed isothermally to D. Again, the latter state is metastable. It is often assumed that A, B, C, and D lie on a continuous curve. If this is true, the curve would have the general shape of A-B-E-F-D-C in Figure 7.2(d). At E and F it is obvious that $(\partial P/\partial V)_T = 0$, which are the limits of intrinsic stability for the temperature in question. These points fall on the spinodal curve shown by the dot-dashed curve in Figure 7.2(d). All states on the spinodal curve are defined by the condition $(\partial P/\partial V)_T = 0$. The variations from states A to C are also shown on the P-T projection of Figure 7.2(c).

To examine more closely the stability of states on the spinodal curve, since we are interested in variations along an isotherm, we can conveniently apply the Helmholtz energy extrema principle. Equations (6-29) and (6-30) (see also Table 6.1), are written in the form

$$\delta \underline{A} = 0 \quad \text{and} \quad \delta^2 \underline{A} \geq 0, \quad \ldots \quad \delta^m \underline{A} > 0$$

where $\delta^m \underline{A}$ is the lowest-order nonvanishing variation. The significance of $(\delta \underline{A})_{T,\underline{V},N} = 0$ leads to the equality of pressures and chemical potentials throughout the system. By similar reasoning, when there is a variation in volume at constant temperature and mass, the criterion $\delta^2 \underline{A}$ reduces to

$$\delta^2 \underline{A} = -\left(\frac{\partial P}{\partial \underline{V}}\right)_{T,N} \delta \underline{V}^2$$

As we have indicated, for points E and F, $(\partial P/\partial V)_T = 0$, and thus the third-order variation of $\underline{A}$ must be considered. We must then show that

$$\delta^3\underline{A} = -\left(\frac{\partial^2 P}{\partial \underline{V}^2}\right)_{T,N} \delta\underline{V}^3 > 0$$

or, if this equals zero, then $\delta^4\underline{A} > 0$, etc.

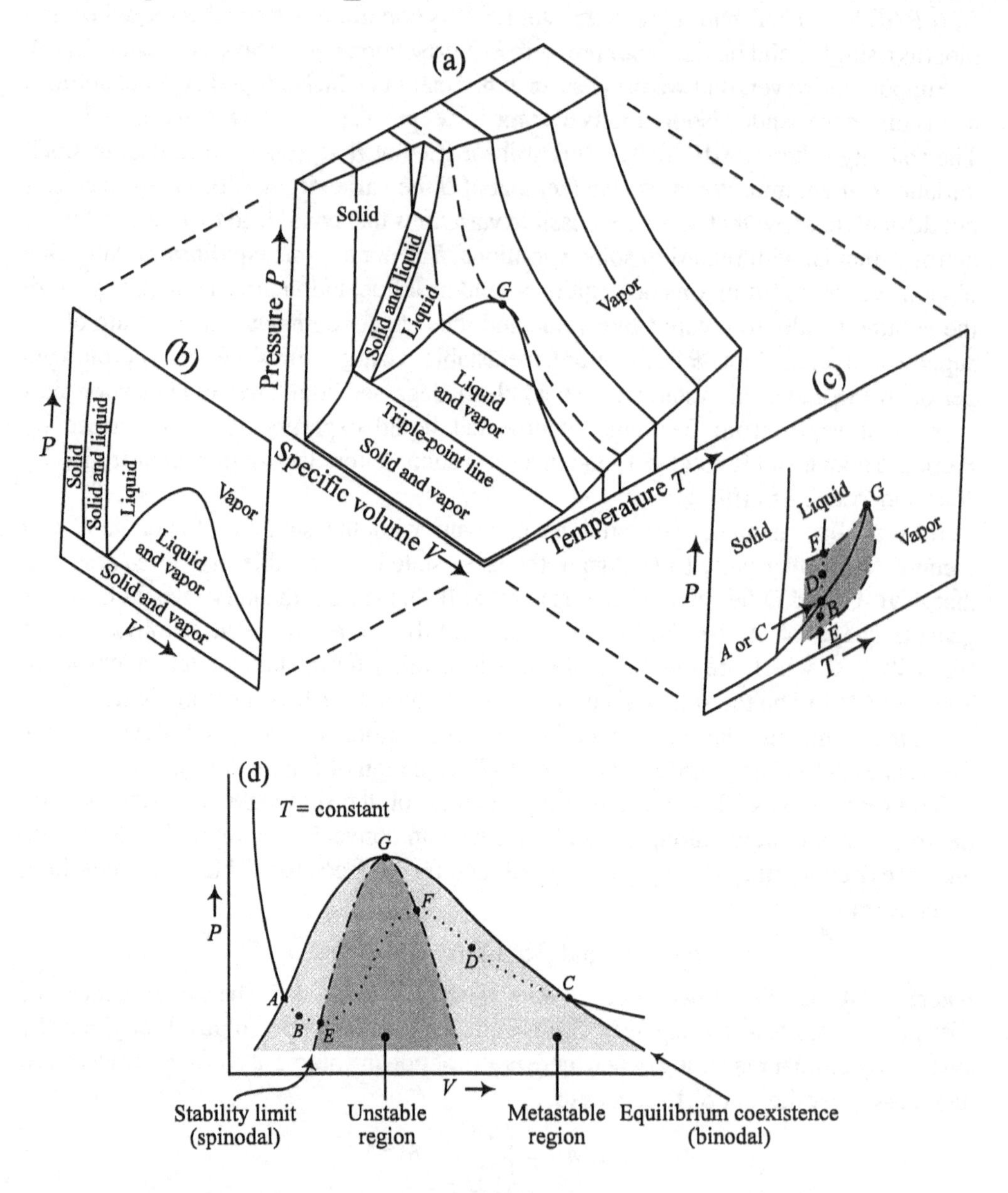

Figure 7.2 *PVT* phase diagram with *P-T* and *P-V* projections. Metastable and unstable regions shown in (d).

Now, at point E on Figure 7.2(d), $(\partial^2 P/\partial V^2)_T$ is clearly positive; thus, one would conclude that for variations that reduce the volume (i.e., $\delta \underline{V}^3 < 0$), the phase is intrinsically stable. If, however, the volume were increased, $\delta \underline{V}^3 > 0$ and the phase would be unstable. Similar reasoning applied to F would indicate that the phase is stable if $\delta \underline{V}^3 > 0$ but not if $\delta \underline{V}^3 < 0$. The arguments may be illustrated on an $\underline{A}$-$\underline{V}$ diagram as shown in Figure 7.3. Points A and C represent saturated liquid and vapor phases at the given temperature. These points have a common tangent, the slope of which is $-P$. E and F represent states in which $\partial^2 \underline{A}/\partial \underline{V}^2 = 0$, and they correspond to points E and F, respectively, in Figure 7.2. For any state between E and F, $\partial^2 \underline{A}/\partial \underline{V}^2 < 0$ and the system is unstable. Points E and F represent stable states only for particular variations (i.e., for E, $\delta \underline{V} < 0$ and for F, $\delta \underline{V} > 0$).

At higher temperatures, states comparable to E and F on the spinodal curve begin to approach each other and, at the critical point, they coincide. That is, the two phases become indistinguishable (see Figure 7.2, point G) as the intensive properties of each phase (e.g., density, specific enthalpy, etc.) become identical. For a pure substance, if a liquid in equilibrium with its vapor is heated until the meniscus disappears, we denote the vapor pressure at this critical temperature as the critical pressure. At this unique point, from the arguments shown above, it is clear that

$$\text{critical point criteria:} \quad \delta \underline{A} = 0 \qquad \delta^2 \underline{A} = 0 \qquad \delta^3 \underline{A} = 0 \qquad \delta^4 \underline{A} \geq 0 \tag{7-25}$$

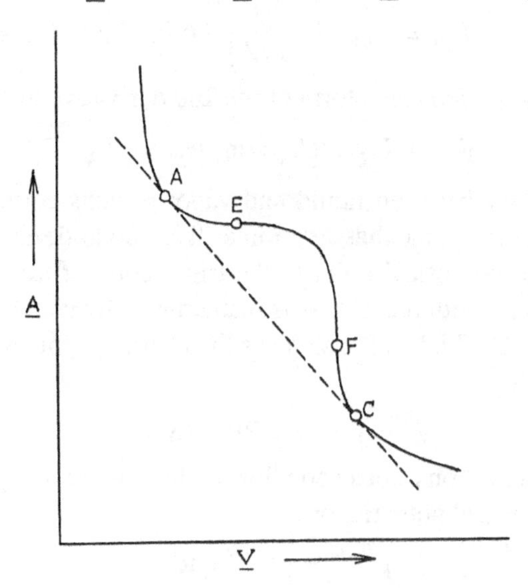

Figure 7.3 Helmholtz energy-volume diagram.

or in equivalent terms

$$\left(\frac{\partial P}{\partial \underline{V}}\right)_{T,N} = 0\,; \quad \left(\frac{\partial^2 P}{\partial \underline{V}^2}\right)_{T,N} = 0\,; \text{ and } \quad \left(\frac{\partial^3 P}{\partial \underline{V}^3}\right)_{T,N} \leq 0 \tag{7-26}$$

Above the critical point, $(\partial P/\partial V)_T$ is everywhere negative, and all such systems are intrinsically stable.

Visualization of the Gibbs *USV* surface can provide insights into how stability criteria and limits are related to certain geometric features of the surface. As before, let's consider a pure material and use the intensive form of the Fundamental Equation to define the *USV* surface as shown in Figures 5.1 and 5.2. For example, if we apply the stability criteria to the *USV* surface generated for ethylene by Jolls (1990), the boundaries between intrinsically unstable (spinodal), metastable, and equilibrium coexistence (binodal) states are shown by different shading in Figure 5.2b. The *USV* surface is convex for all regions that are stable and metastable and saddle-shaped for those regions that are unstable. The spinodal boundary between metastable and unstable states corresponds to Eq. (7-22):

$$\pounds_0 = \begin{vmatrix} U_{SS} & U_{SV} \\ U_{SV} & U_{VV} \end{vmatrix} = U_{SS}\,U_{VV} - U_{SV}^2 = 0$$

where

$$U_{SS} = \left(\frac{\partial^2 U}{\partial S^2}\right)_V \qquad U_{VV} = \left(\frac{\partial^2 U}{\partial V^2}\right)_S$$

$$U_{SV} = U_{VS} = \left(\frac{\partial^2 U}{\partial S \partial V}\right) \quad \text{(via Maxwell reciprocity)}$$

Note that we are using the *intensive* form of the 2nd derivative of $y^{(0)} = U = \underline{U}/N$:

$$y_{ij}^{(0)} = U_{SS},\ U_{VV},\ \text{or}\ U_{SV} = U_{VS}$$

The binodal boundary between liquid and vapor regions corresponds to points of double tangency where the plane that determines these conditions effectively rolls over the *USV* surface from the critical point to the triple point of the pure substance. The intersection of that plane with the U axis is mathematically equal to the 2nd Legendre transform $y^{(2)}$ of $y^{(0)} = U$ which is the same as the chemical potential μ or G for a pure material:

$$y^{(2)} = \mu = U + PV - TS = G$$

Thus, the set of conditions corresponding to liquid-vapor equilibrium will have identical values of chemical potential or

$$\mu^{liquid} = \mu^{vapor} = \mu^{sat}$$

This is what we expect from our earlier derivation of phase equilibrium criteria in Section 6.5. In addition, since the partial derivative slopes:

$$\left(\frac{\partial U}{\partial S}\right)_V = T \qquad \text{and} \qquad \left(\frac{\partial U}{\partial V}\right)_S = -P$$

are also the same for the common tangent points, then T and P must also be identical for the liquid and vapor phases, again consistent with what we expect for phase equilibrium.

In Figures 5.1a and b we also see a unique point of triple tangency corresponding to the triple point (tp) of pure water. Here,

$$\mu^{solid} = \mu^{liquid} = \mu^{vapor} = \mu_{tp_{eq}}$$

In Figure 5.2a, the Fundamental Equation for an ideal gas yields a *USV* surface that is always convex with respect to volume variations. Thus an ideal gas does not exhibit any unstable behavior; and there are no points of multiple tangency. So as we would expect, multiple phases are not possible.

Our earlier discussion of stability limits and equilibrium coexistence for single component systems using the Helmholtz energy A can be expanded to illustrate the geometry of the *ATV* surface. Because A is the first Legendre transform of U, the *ATV* surface contains the same information content as the *USV* surface of the original Fundamental Equation. Jolls (1990) and Coy (1993) have also generated an *ATV* surface for pure ethylene using the Peng-Robinson *PVT* equation of state and a fitted expression for the heat capacity. Figure 7.4 shows the complete *ATV* surface for all fluid phases with the A-axis plotted positively downward to expose the "underside" of the surface and two isothermal tangent lines bridging the two-phase region. Figure 7.5 gives a 2-dimensional projection of A versus V along a specific isotherm where $T = 0.85\ T_c$. (Note that A is plotted positively upward in Figure 7.5 so that the tangent line now lies below the curve.) If A is selected as a new basis function $y^{(0)}$ then $y^{(1)} = A + PV\ (or\ G)$ so the A-axis intercept shown in Figure 7.5 is equal to G or μ. Similar to Figure 7.3, the points of double tangency on the A versus V curve on Figure 7.5 correspond to existing liquid and vapor phases at equilibrium, and the locus of all such double tangencies on the *ATV* surface maps out the entire liquid-vapor coexistence region. Unstable regions are again identified with the condition, $A_{VV} < 0$ and the spinodal limiting points (C and D) where $A_{VV} = 0$. These conditions are particularly easy to see on Figure 7.5 where

$$A_{VV} = \left(\frac{\partial^2 A}{\partial V^2}\right)_T = -\left(\frac{\partial P}{\partial V}\right)_T$$

is negative between the spinodal points C and D and positive elsewhere. It is also interesting to see that between B, the equilibrium liquid phase point, and the maximum between C and D that $-P = A_V > 0$, that is P is negative. Thus the liquid in this metastable region is under tension. Note the difference between this isothermal A versus V curve of Figure 7.5 and the one shown in Figure 7.3 where $(\partial A/\partial V) = A_V = -P$ always yields $P \geq 0$. See the papers by Hayward (1971) and Scholander (1972) for interesting accounts of observed negative pressures.

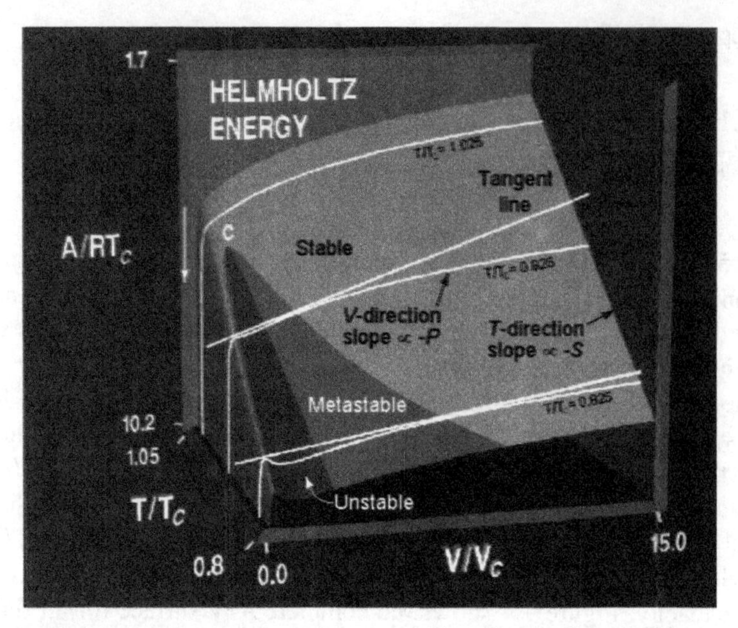

Figure 7.4 Helmholtz energy surface as predicted by the Peng-Robinson equation and a quadratic expression for $C_V^o(T)$ [from Jolls (1990) and Coy (1993)].

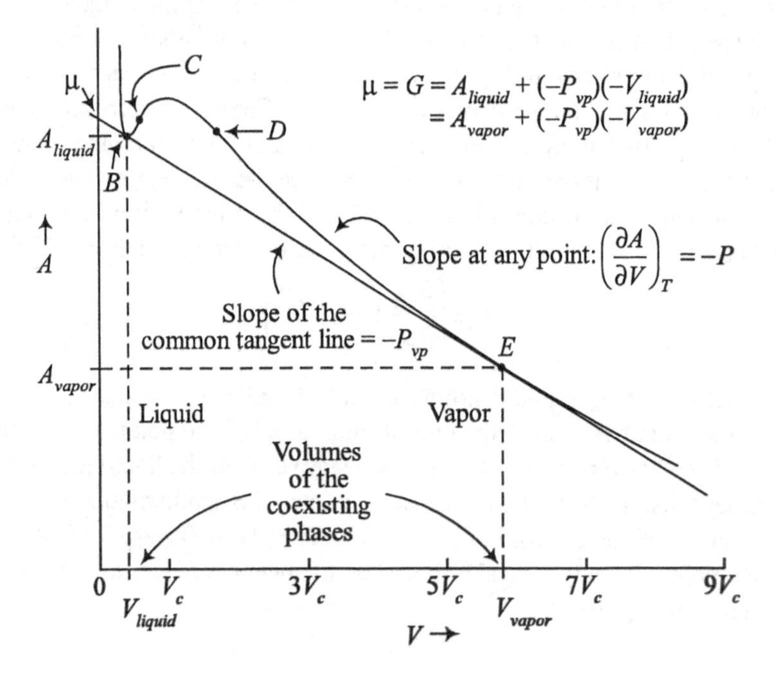

Figure 7.5 Subcritical $T_r = 0.85$ isotherm of the Helmholtz energy where P denotes the vapor pressure [adapted from Jolls and Butterbaugh (1992)].

Example 7.2

Determine the spinodal curves for both liquid and vapor phases of a van der Waals fluid.

Solution

We begin with the *PVT* van der Waals equation of state (vdW EOS) (see Section 8.3)

$$P=\frac{RT}{V-b}-\frac{a}{V^2}$$

Using limiting stability criteria to define the spinodal condition, for a pure fluid ($m=n+2=3$), we need to evaluate both

$$y_{22}^{(1)}=A_{VV}=-\left(\frac{\partial P}{\partial V}\right)_T=0 \quad \text{and} \quad y_{222}^{(1)}=A_{VVV}=-\left(\frac{\partial^2 P}{\partial V^2}\right)_T=0$$

where $A_{VV}=A_{VVV}=0$ defines the liquid-vapor critical point. The spinodal curves themselves are given by $A_{VV}=0$, and can be analytically determined by differentiating the vdW EOS directly:

$$\left(\frac{\partial P}{\partial V}\right)_T=\frac{-RT}{(V-b)^2}+\frac{2a}{V^3}=0$$

while at the critical point:

$$\left(\frac{\partial^2 P}{\partial V^2}\right)_T=\frac{+2RT}{(V-b)^3}-\frac{6a}{V^4}=0$$

as well. We can use these two criteria to obtain values of the parameters a and b in terms of V_c, T_c, and P_c. Doing this we get

$$b=\frac{V_c}{3} \quad \text{and} \quad a=\frac{27}{8}RT_c b=\frac{9}{8}RT_c V_c$$

Because V_c is frequently not known accurately, it is eliminated by using a third relationship, the vdW EOS itself. Now we obtain the conventional defining equation forms for a and b in terms of just T_c and P_c with R = gas constant:

$$a=\frac{27}{64}\frac{R^2Tc^2}{P_c} \quad \text{and} \quad b=\frac{R\,Tc}{8\,Pc}$$

To actually estimate the spinodal curves for a particular fluid, we would need to know T_c and P_c to calculate a and b which we would then use in the two equations $A_{VV}=0$ and $A_{VVV}=0$ along with the EOS. However, there is a much more straightforward and general approach. We can recast the vdW EOS in dimensionless form by scaling T, P, and V with values of these parameters at the critical point. Thus, in reduced coordinates:

$$T_r\equiv T/T_c \quad P_r\equiv P/P_c \quad \text{and} \quad V_r\equiv V/V_c$$

By making these substitutions, we rewrite the vdW EOS as,

$$P_r = \frac{8T_r}{(3V_r - 1)} - \frac{3}{V_r^2} \quad \text{with} \quad Z_c = \frac{P_c V_c}{RT_c} = \frac{3}{8}$$

We can now develop completely general spinodal and critical point criteria following the same procedures that we used before. The results are plotted in Figure 7.6. Note that we have a cubic equation in V_r so there are three roots. Below the critical temperature and pressure, three real roots are obtained; the largest volume root is taken as the vapor, the smallest volume root as the liquid, and the intermediate one is rejected because it is inside the unstable region. To see the three roots pick the $T_r = 0.9$ isotherm and draw a line at a particular pressure in the vicinity of $P_r = 0.6$. For $T_r > 1.0$, there is only one real root, the two imaginary roots are discarded.

Also shown in Figure 7.6 is the equilibrium liquid-vapor binodal curve which corresponds to the condition of phase equilibrium developed in Section 6.5 where the chemical potential of the liquid and vapor phases are equal at $T_r < 1.0$ and particular values of P_r corresponding to the reduced vapor pressure $P_{vp,r} \equiv P_{vp}/P_c$. Thus along the binodal or liquid-vapor coexistence curve:

$$\mu(T_r, P_{vp,r}, V_r^{liquid}) = \mu^{liquid} = \mu^{vapor} = \mu(T_r, P_{vp,r}, V_r^{vapor})$$

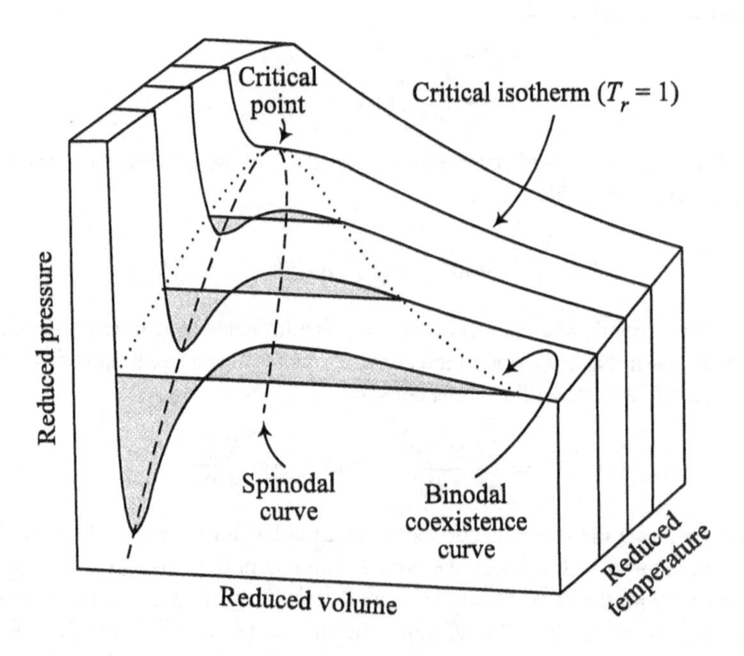

Figure 7.6 *PVT* phase diagram for the reduced form of the van der Waals equation of state. Vapor and liquid equilibrium coexistence (binodal) lines and limits of stability (spinodal) lines shown (see Example 7.2) [produced with EOS program from Jolls (1990)].

This treatment has touched only briefly on the extraordinary behavior of matter at the critical point between liquids and gases. Yet the study of matter in this region yields a number of interesting and unusual results. Critical phenomena are considered in more detail in Section 7.3.

7.2.2 Binary systems. For a simple system composed of B and C,

$$\underline{U} = f(\underline{S}, \underline{V}, N_B, N_C) \tag{7-27}$$

Choosing the $(x_1, \ldots, x_4)$ variable ordering as given in Eq. (7-27), the stability criterion is simply

$$y_{33}^{(2)} = \underline{G}_{N_B N_B} = \underline{G}_{BB} = \left(\frac{\partial^2 \underline{G}}{\partial N_B^2}\right)_{T, P, N_c} > 0 \tag{7-28}$$

A limit of stability is reached when $y_{33}^{(2)} = 0$; Eq. (7-28) is written more commonly as,

$$\left(\frac{\partial \mu_B}{\partial N_B}\right)_{T, P, N_C} > 0 \tag{7-29}$$

To visualize this inequality, examine Figure 7.7, in which, for a hypothetical binary system at constant pressure, μ_B is plotted against x_B at several different temperatures. QKR defines the locus of compositions for two phases that are in equilibrium (i.e., a phase at A is in equilibrium with a phase at C at T_1). At these points $(\partial \mu_B / \partial x_B)$ is positive and the phase is stable. A continuation of the T_1 isotherm from A to E or from C to F would define metastable equilibrium states that could be attained experimentally only if the phase transitions were inhibited. Such curves are quite similar to those in Figures 7.2 and 7.4, which were developed earlier to treat mechanical stability. States between E and F are unstable and at E and F,

$$\left(\frac{\partial \mu_B}{\partial x_B}\right)_{P, T} = 0 \tag{7-30}$$

Equation (7-30) defines states on the spinodal curve Q′KR′. Both the equilibrium phase envelope or binodal coexistence curve (QKR) and the spinodal curve become tangent at K, the critical point for this binary. The critical temperature is at T_2 and x_B^c for this particular pressure. It is obvious that Eq. (7-30) would still apply at point K.

Different stability criteria may be determined if the ordering of the independent variables in Eq. (7-27) is changed. These are shown in Table 7.2. As for pure component systems, *all* the criteria are identical, and if a system should reach the limit of intrinsic stability, *all* are violated simultaneously.

The binary stability criterion, Eq. (7-28), or any of the forms in Table 7.2, may not always be particularly convenient to use. For example, P-$\underline{V}$-T-N_i data or correlations are normally expressed in a pressure explicit equation of state; that is,

$$P = f(\underline{V}, T, N_B, N_C) \tag{7-31}$$

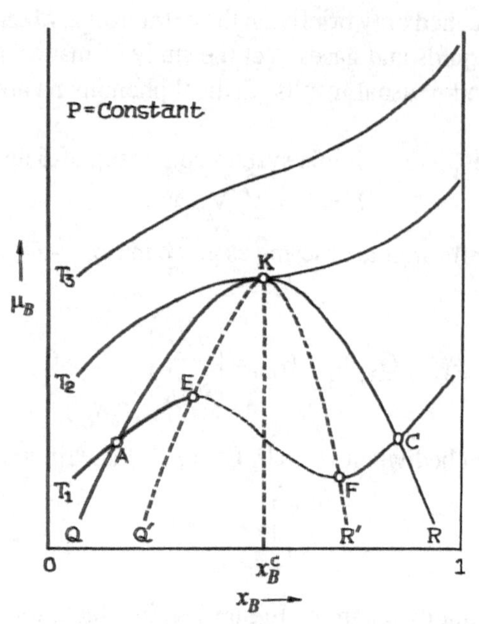

Figure 7.7 Chemical potential as a function of composition and temperature for an isobaric binary system.

In this case, criteria employing the Helmholtz energy are more desirable. With Eq. (7-17) and for $i = 1$, $y^{(1)} = \underline{A}$, $x_1 = \underline{S}$, $x_2 = \underline{V}$, $x_3 = N_B$, $x_4 = N_C$,

$$\pounds_1 = \begin{vmatrix} y_{22}^{(1)} & y_{23}^{(1)} \\ y_{23}^{(1)} & y_{33}^{(1)} \end{vmatrix} = \begin{vmatrix} \underline{A}_{\underline{V}\,\underline{V}} & \underline{A}_{\underline{V}B} \\ \underline{A}_{\underline{V}B} & \underline{A}_{BB} \end{vmatrix} > 0$$

where

$$\underline{A}_{\underline{V}} = \left(\frac{\partial \underline{A}}{\partial \underline{V}}\right)_{T,N} = -P \quad \text{and} \quad \underline{A}_{\underline{V}\,\underline{V}} = \left(\frac{\partial^2 \underline{A}}{\partial \underline{V}^2}\right)_{T,N} = -\left(\frac{\partial P}{\partial \underline{V}}\right)_{T,N}$$

$$\underline{A}_{\underline{V}N_B} = \underline{A}_{\underline{V}B} = \frac{\partial^2 \underline{A}}{\partial \underline{V}\,\partial N_B} = -\left(\frac{\partial P}{\partial N_B}\right)_{T,\underline{V},N_C}$$

and

$$\underline{A}_{N_B N_B} = \underline{A}_{BB} = \left(\frac{\partial^2 \underline{A}}{\partial N_B^2}\right)_{T,\underline{V},N_C} = \left(\frac{\partial \mu_B}{\partial N_B}\right)_{T,\underline{V},N_C} = \int_{\underline{V}}^{\infty} \left(\frac{\partial^2 P}{\partial N_B^2}\right)_{T,\underline{V},N_C} d\underline{V} + \frac{RT}{N_B} \tag{7-32}$$

Table 7.2 Equivalent Stability Criteria for Binary Systems with $y^{(0)} = \underline{U}$

Ordering of (x_1, x_2, x_3, x_4)[a]	Stability criterion[b] $y_{33}^{(2)} > 0$
$(\underline{S}, \underline{V}, N_B, N_C)$	$\underline{G}_{BB} = \left(\frac{\partial \mu_B}{\partial N_B}\right)_{T, P, N_C} > 0$
$(\underline{S}, N_B, \underline{V}, N_C)$	$\underline{A}'_{\underline{V}\underline{V}} = -\left(\frac{\partial P}{\partial \underline{V}}\right)_{T, \mu_B, N_C} > 0$
$(\underline{S}, N_B, N_C, \underline{V})$	$\underline{A}'_{CC} = \left(\frac{\partial \mu_C}{\partial N_C}\right)_{T, \mu_B, \underline{V}} > 0$
$(\underline{V}, N_B, \underline{S}, N_C)$	$\underline{H}'_{\underline{S}\underline{S}} = \left(\frac{\partial T}{\partial \underline{S}}\right)_{P, \mu_B, N_C} > 0$
$(\underline{V}, N_B, N_C, \underline{S})$	$\underline{H}'_{CC} = \left(\frac{\partial \mu_C}{\partial N_C}\right)_{P, \mu_B, \underline{S}} > 0$
$(N_B, N_C, \underline{S}, \underline{V})$	$\underline{U}''_{\underline{S}\underline{S}} = \left(\frac{\partial T}{\partial \underline{S}}\right)_{\mu_B, \mu_C, \underline{V}} > 0$
$(N_B, N_C, \underline{V}, \underline{S})$	$\underline{U}''_{\underline{V}\underline{V}} = -\left(\frac{\partial P}{\partial \underline{V}}\right)_{\mu_B, \mu_C, \underline{S}} > 0$

[a] Any ordering of (x_1, x_2, x_3, x_4) which differs only in the arrangement of the first two variables and/or in the ordering of N_B, N_C are not shown.

[b] $\underline{A}' = f(T, \underline{V}, \mu_B, N_C)$; $\underline{H}' = f(\underline{S}, P, \mu_B, N_C)$; $\underline{U}'' = f(\underline{S}, \underline{V}, \mu_B, \mu_C)$.

Thus all the $\underline{A}_{ij}$ derivatives may be found given an equation of state of the form shown in Eq. (7-31). As an example of the use of the $\pounds_i$ determinant form, consider Figure 7.8.

On this graph we have plotted the locus where $\pounds_1 = 0$ for a fixed composition (x_1 = constant) binary liquid mixture. The required derivatives $\underline{A}_{\underline{V}\underline{V}}$, $A_{\underline{V}N_1}$, and $\underline{A}_{N_1N_1}$ can be obtained directly from a pressure-explicit equation of state such as the van der Waals equation in mixture form (see Example 7.2 and Section 9.2). A particular isotherm (T=T_1) is plotted with the liquid-side of the two-phase envelope shown. The spinodal point at A defines the beginning of the unstable region at $P<P_A$ while a metastable region exists between points A and B (the equilibrium binodal point). At pressures greater than $P_B = P_{sat}$, the liquid mixture is intrinsically stable. If the mixture had been treated as a pseudo-pure component, $\underline{A}_{\underline{V}\underline{V}} = 0$ would have incorrectly defined the spinodal condition.

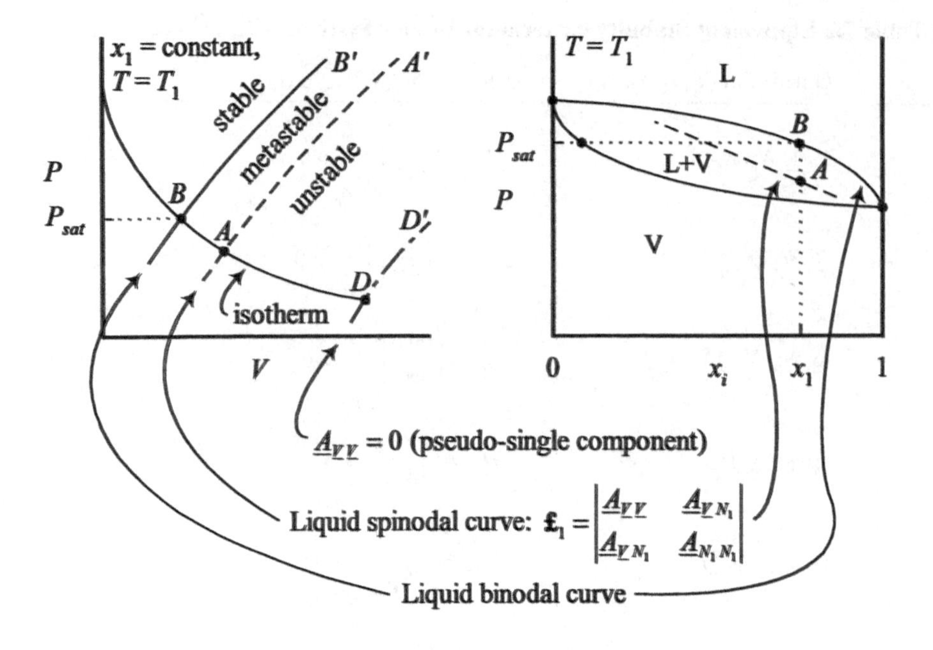

Figure 7.8 Binary spinodal line for a fixed composition liquid mixture. P_{sat} = vapor pressure of mixture at composition x_1 and temperature T_1. In general, $P_{sat} = f[T, x_i]$ for the mixture.

Example 7.3

In a binary liquid mixture of B and C at constant temperature and pressure, the excess Gibbs energy of mixing is given explicitly by an empirical equation in terms of the mole fractions of B and C

$$\frac{\Delta G^{EX}}{RT} = f[x_B, x_C] = x_B x_C [k_1 + k_2(x_B - x_C) + k_3(x_B - x_C)^2]$$

where for the particular conditions of interest, $k_1 = 2.0$, $k_2 = 0.2$, and $k_3 = -0.8$. Determine if there are regions of immiscibility and any limits of essential instability.

Solution

The total Gibbs energy of mixing is equal to the sum of the ideal and excess contributions:

$$\frac{\Delta G}{RT} = \frac{\Delta G^{ID}}{RT} + \frac{\Delta G^{EX}}{RT}$$

where

$$\frac{\Delta G^{ID}}{RT} \equiv x_B \ln x_B + x_c \ln x_c$$

and $\Delta G^{EX}/RT$ is given above. Also, we will solve this example by showing that, for a ΔG versus x_B curve, $(\partial^2 \Delta G/\partial x_B^2)_{T,P} = 0$ for any states of essential instability. Defining G_B and G_C as the Gibbs energies of pure B and C at the system temperature and pressure, ΔG can also be written as

$$\Delta G = x_B(\mu_B - G_B) + x_C(\mu_C - G_C) \equiv x_B \overline{\Delta G_B} + x_C \overline{\Delta G_C}$$

$$\left(\frac{\partial \Delta G}{\partial x_B}\right)_{T,P} = (\mu_B - G_B) - (\mu_C - G_C) + x_B\left(\frac{\partial \mu_B}{\partial x_B}\right) + x_C\left(\frac{\partial \mu_C}{\partial x_B}\right)$$

The last two terms are zero by the Gibbs-Duhem equation (see Section 9.4). Then

$$\left(\frac{\partial^2 \Delta G}{\partial x_B^2}\right)_{T,P} = \frac{\partial \mu_B}{\partial x_B} - \frac{\partial \mu_C}{\partial x_B} = \frac{\partial \mu_B}{\partial x_B}\left(1 + \frac{x_B}{x_C}\right)$$

At the limit of stability, by Eq. (7-29),

$$\left(\frac{\partial \mu_B}{\partial N_B}\right)_{T,P,N_C} = \left(\frac{\partial \mu_B}{\partial x_B}\right)\left(\frac{\partial x_B}{\partial N_B}\right) = \frac{x_C}{N}\left(\frac{\partial \mu_B}{\partial x_B}\right) = 0$$

Thus, $(\partial^2 \Delta G/\partial x_B^2)_{T,P} = 0$. The limits of stability are shown on Figure 7.9 at $x_B = 0.351$ and 0.690 (i.e., points A). The phase equilibrium limits are shown at points B, where the curve has a common tangent line indicating the equivalence of μ_i between two phases.

Point C is $\overline{\Delta G_C}/RT$ and D is $\overline{\Delta G_B}/RT$. The nomenclature used in this example is described in more detail in Chapter 9.

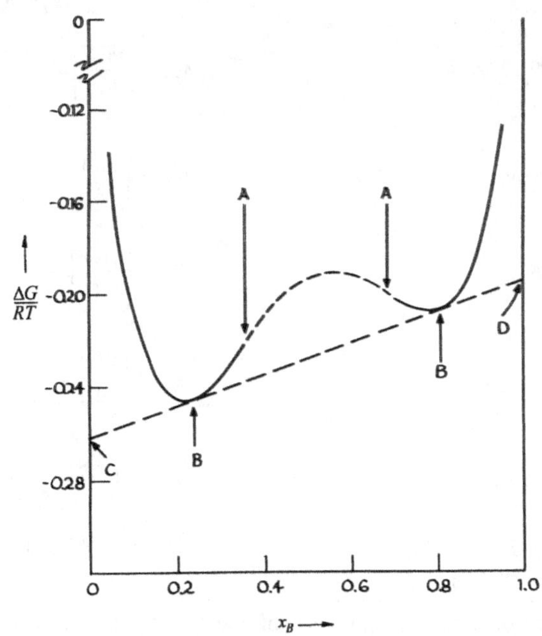

Figure 7.9 Gibbs energies of mixing for a system that exhibits phase splitting.

7.2.3 Ternary systems. The final example is a ternary system of B, C, and D. Extension to multicomponent systems of higher order ($n > 3$) is tedious but straightforward. For the system to be stable, with $m = 5$, Eq. (7-15) indicates that

$$y_{44}^{(3)} > 0 \tag{7-33}$$

or if one should desire to express this criterion using derivatives of the Helmholtz energy, then with Eq. (7-19), $i = 1$,

$$\pounds_1 = \begin{vmatrix} A_{\underline{V}\underline{V}} & A_{\underline{V}B} & A_{\underline{V}C} \\ A_{\underline{V}B} & A_{BB} & A_{BC} \\ A_{\underline{V}C} & A_{BC} & A_{CC} \end{vmatrix} > 0 \tag{7-34}$$

The second-order derivatives were defined earlier and may be readily determined from a pressure-explicit equation of state. The form of Eq. (7-33) or (7-34) changes depending on the ordering of the variables $\underline{S}$, $\underline{V}$, N_B, N_C, and N_D. We show in Table 7.3 some of the equivalent forms. To expand slightly on this point, consider the first form in this table.

Table 7.3 Stability Criteria for Ternary Systems of B, C, and D

Ordering of $(x_1, x_2, x_3, x_4, x_5)$[a]	Stability criterion $y_{44}^{(3)} > 0$
$(\underline{S}, \underline{V}, N_B, N_C, N_D)$	$\underline{G}'_{CC} = \left(\dfrac{\partial \mu_C}{\partial N_C}\right)_{T, P, \mu_B, N_D} > 0$
$(\underline{S}, N_B, N_C, \underline{V}, N_D)$	$\underline{A}''_{\underline{V}\underline{V}} = -\left(\dfrac{\partial P}{\partial \underline{V}}\right)_{T, \mu_B, \mu_C, N_D} > 0$
$(\underline{S}, N_B, N_C, N_D, \underline{V})$	$\underline{A}''_{DD} = \left(\dfrac{\partial \mu_D}{\partial N_D}\right)_{T, \mu_B, \mu_C, \underline{V}} > 0$
$(\underline{V}, N_B, N_C, \underline{S}, N_D)$	$\underline{H}''_{\underline{S}\underline{S}} = \left(\dfrac{\partial T}{\partial \underline{S}}\right)_{P, \mu_B, \mu_C, N_D} > 0$
$(\underline{V}, N_B, N_C, N_D, \underline{S})$	$\underline{H}''_{DD} = \left(\dfrac{\partial \mu_D}{\partial N_D}\right)_{P, \mu_B, \mu_C, \underline{S}} > 0$
$(N_B, N_C, N_D, \underline{S}, \underline{V})$	$\underline{U}'''_{\underline{S}\underline{S}} = \left(\dfrac{\partial T}{\partial \underline{S}}\right)_{\mu_B, \mu_C, \mu_D, \underline{V}} > 0$
$(N_B, N_C, N_D, \underline{V}, \underline{S})$	$\underline{U}'''_{\underline{V}\underline{V}} = -\left(\dfrac{\partial P}{\partial \underline{V}}\right)_{\mu_B, \mu_C, \mu_D, \underline{S}} > 0$

where $\underline{G}' = f(T, P, \mu_B, N_C, N_D)$ $\quad$ $\underline{A}'' = f(T, \underline{V}, \mu_B, \mu_C, N_D)$
$\underline{H}'' = f(\underline{S}, P, \mu_B, \mu_C, N_D)$ $\quad$ $\underline{U}''' = f(\underline{S}, \underline{V}, \mu_B, \mu_C, \mu_D)$

[a] Ordering of $(x_1, x_2, x_3, x_4, x_5)$ that differ only in the arrangement of the first three variables and/or in the ordering of N_B, N_C, N_D are not shown in this table.

The 3rd Legendre transform $y^{(3)}$ may be expressed as

$$y^{(3)} = \underline{G}' = f(T, P, \mu_B, N_C, N_D) = \underline{U} - T\underline{S} + P\underline{V} - \mu_B N_B \tag{7-35}$$

$$dy^{(3)} = d\underline{G}' = -\underline{S}\,dT + \underline{V}\,dP - N_B d\mu_B + \mu_C dN_C + \mu_D dN_D$$

$$\left(\frac{\partial y^{(3)}}{\partial x_4}\right)_{\xi_1, \xi_2, \xi_3, x_5} = \left(\frac{\partial \underline{G}'}{\partial N_C}\right)_{T, P, \mu_B, N_D} = \underline{G}'_C = \mu_C$$

and

$$y_{44}^{(3)} = \left(\frac{\partial^2 y^{(3)}}{\partial x_4^2}\right)_{\xi_1, \xi_2, \xi_3, x_5} = \left(\frac{\partial^2 \underline{G}'}{\partial N_C^2}\right)_{T, P, \mu_B, N_D} = \underline{G}'_{CC} = \left(\frac{\partial \mu_C}{\partial N_C}\right)_{T, P, \mu_B, N_D} \tag{7-36}$$

In practice, Eq. (7-36) is called the second diffusional stability limit when $y_{44}^{(3)} = 0$.

7.3 Critical States

The criterion for stable equilibrium is given by Eq. (7-15). An alternative form is

$$y_{(m-1)(m-1)}^{(m-2)} = \left(\frac{\partial \xi_{m-1}}{\partial x_{m-1}}\right)_{\xi_1, \xi_2, \ldots, \xi_{m-2}, x_m} > 0 \tag{7-37}$$

Let us assume that we have a stable equilibrium state but by changes in some state variables we are approaching a state where $(\partial \xi_{m-1}/\partial x_{m-1})_{\xi_1,\ldots,\xi_{m-2},x_m}$ is near zero. That is, we are near the limit of essential instability on the spinodal curve. Returning to the hypothetical α–β systems introduced in Section 7.1, at the very limit of essential instability, to test the system for stability, the criterion indicates that we should perturb the x_{m-1} variable and determine how ξ_{m-1} behaves. Or, in different terms, as x_{m-1} is varied, how does $y_{(m-1)(m-1)}^{(m-2)}$ respond? The conditions of restraint indicate that

$$dx_{m-1}^{\alpha} + dx_{m-1}^{\beta} = 0 \tag{7-38}$$

On the spinodal curve (approached from a stable phase), $y_{(m-1)(m-1)}^{(m-2)} = 0$ in both α and β. This must be true, since α and β are identical in the original formulation. To determine the effect of interchanging additional x_{m-1} for a system on the spinodal curve, we must consider the derivative $y_{(m-1)(m-1)(m-1)}^{(m-2)}$. If this derivative were positive, and if $dx_{m-1}^{\alpha} > 0$, then $y_{(m-1)(m-1)}^{(m-2)} > 0$ in subsystem α. That is, α is stable with respect to the transfer. However, with Eq. (7-38), $y_{(m-1)(m-1)}^{(m-2)} < 0$ and subsystem β has become unstable and must form a new phase. Similar but opposite conclusions are reached if $dx_{m-1}^{\alpha} < 0$.

A *critical state* is defined by a special set of criteria that are more restrictive than those for stability itself. Because a critical state is both a stable state and at a limit of stability on the spinodal curve, the *necessary and sufficient conditions* are as follows: A critical state is defined to be on the spinodal curve and it must satisfy both Eq. (7-16)

$$y_{(m-1)(m-1)}^{(m-2)} = 0 \tag{7-16}$$

and the condition that

$$y^{(m-2)}_{(m-1)(m-1)(m-1)} = 0 \quad \text{with } m = n+2 \tag{7-39}$$

and since the critical state is also a stable state, then

$$y^{(m-2)}_{(m-1)(m-1)(m-1)(m-1)} \geq 0 \tag{7-40}$$

If Eq. (7-40) equals zero, then the lowest even-order, non-vanishing derivative of $y^{(m-2)}$ must be positive, with all lower-order derivatives equal to zero.

A good example of a critical state satisfying these conditions occurs at the liquid-vapor critical point of a pure fluid. Here the liquid-vapor, two-phase region coalesces to a point where the liquid and vapor densities approach each other. (see Figure 7.2). Here both the stability-limiting spinodal curve and equilibrium coexistence binodal curve just touch [see Figure 7.2(d)]. Inside the liquid-vapor region we cannot simultaneously satisfy the spinodal and binodal conditions. Eqs. (7-16), (7-39), and (7-40) can only be satisfied at this unique critical point.

The second critical criterion [Eq. (7-39)] also may be written as:

$$\left(\frac{\partial^2 \xi_{m-1}}{\partial x^2_{m-1}}\right)_{\xi_1, \xi_2, \ldots, \xi_{m-2}, x_m} = 0 \tag{7-41}$$

For example, with a ternary mixture of B, C, and D. Eq. (7-33) expresses the criterion for a stable system while at the critical point with $m = 5$,

$$\begin{aligned} y^{(3)}_{44} &= \underline{G}'_{CC} = 0 \\ y^{(3)}_{444} &= \underline{G}'_{CCC} = 0 \\ y^{(3)}_{4444} &= \underline{G}'_{CCCC} \geq 0 \end{aligned} \tag{7-42}$$

For a binary system $m = 4$ and

$$y^{(2)}_{33} = \underline{G}_{N_B N_B} = \left(\frac{\partial \mu_B}{\partial N_B}\right)_{T,P,N_C} = 0 \quad \text{or equivalently} \quad \left(\frac{\partial \mu_B}{\partial x_B}\right)_{T,P} = 0$$

$$y^{(2)}_{333} = \underline{G}_{N_B N_B N_B} = \left(\frac{\partial^2 \mu_B}{\partial N_B^2}\right)_{T,P,N_C} = 0 \quad \text{or equivalently} \quad \left(\frac{\partial^2 \mu_B}{\partial x_B^2}\right)_{T,P} = 0$$

so mathematically this is a point of inflection on an isothermal, isobaric curve of μ_B versus x_B for a binary mixture. Point K on Figure 7.7 is an example of such a critical point.

The space between the equilibrium binodal and the spinodal curves on Figure 7.7 defines the metastable region for this binary. The depth of penetration into this region in moving below the binodal curve scales with the degree of supersaturation. As the metastable region widens one would expect that deeper penetrations into the region might be possible. Conversely, as the region narrows, proportionally less penetration should be allowable. At the critical point the metastable region disappears entirely; and thus it is possible to induce phase separation spontaneously as we enter the two-phase region near the critical point composition. This phenomena is called *spinodal decomposition* and can be induced experimentally by imposing a change in temperature (or pressure) near the critical point of any mixture (see also Section 15.7).

With a pure material, $m = 3$ and at the critical point,

$$y_{22}^{(1)} = \underline{A}_{\underline{V}\,\underline{V}} = -\left(\frac{\partial P}{\partial \underline{V}}\right)_{T,N} = 0 \qquad y_{222}^{(1)} = \underline{A}_{\underline{V}\,\underline{V}\,\underline{V}} = -\left(\frac{\partial^2 P}{\partial \underline{V}^2}\right)_{T,N} = 0$$

$$y_{2222}^{(1)} = \underline{A}_{\underline{V}\,\underline{V}\,\underline{V}\,\underline{V}} = -\left(\frac{\partial^3 P}{\partial \underline{V}^3}\right)_{T,N} \geq 0 \tag{7-43}$$

Again we see a point of inflection on the P-V curve at the critical point as we travel along the isotherm corresponding to the critical temperature ($T = T_c$) (see Figure 7.2).

An alternative to the criteria given above is to use lower-order Legendre transforms. The function $\pounds_i$ [Eq. (7-19)] is set equal to zero and also

$$\mathcal{M}_i = 0 \tag{7-44}$$

where $\mathcal{M}_i$ is defined as

$$\mathcal{M}_i \equiv \begin{vmatrix} y_{(i+1)(i+1)}^{(i)} & y_{(i+1)(i+2)}^{(i)} & \cdots & y_{(i+1)(m-1)} \\ y_{(i+2)(i+1)} & \cdot & \cdots & y_{(i+2)(m-1)} \\ \cdot & \cdot & & \cdot \\ \cdot & \cdot & & \cdot \\ \cdot & \cdot & & \cdot \\ y_{(m-2)(i+1)}^{(i)} & & \cdots & y_{(m-2)(m-1)}^{(i)} \\ \dfrac{\partial \pounds_i}{\partial x_{i+1}} & & \cdots & \dfrac{\partial \pounds_i}{\partial x_{m-1}} \end{vmatrix} \tag{7-45}$$

The derivation of Eq. (7-44) is presented in Appendix D of the 2nd edition of *Thermodynamics and Its Applications*. As an example, consider an n-component system and order the variables as $\underline{U} = f(\underline{S}, \underline{V}, N_B, N_C, \ldots, N_n)$. Choose $i = 2$. Then $y^{(2)} = \underline{G}$, the Gibbs energy, and with $m = n + 2$ the critical-point criteria are

$$\pounds_2 = \begin{vmatrix} \underline{G}_{BB} & \underline{G}_{BC} & \cdots & \underline{G}_{B,\,n+1} \\ \underline{G}_{CB} & \underline{G}_{CC} & \cdots & \underline{G}_{C,\,n+1} \\ \cdot & \cdot & & \cdot \\ \cdot & \cdot & & \cdot \\ \cdot & \cdot & & \cdot \\ \underline{G}_{n+1,\,B} & & \cdots & \underline{G}_{n+1,\,n+1} \end{vmatrix} = 0 \tag{7-46}$$

$$
\mathcal{M}_2 = \begin{vmatrix} \underline{G}_{BB} & \underline{G}_{BC} & \cdots & \underline{G}_{B,\,n+1} \\ \underline{G}_{CB} & \underline{G}_{CC} & \cdots & \underline{G}_{C,\,n+1} \\ \cdot & \cdot & & \cdot \\ \cdot & \cdot & & \cdot \\ \cdot & \cdot & & \cdot \\ \underline{G}_{n-2,\,B} & & \cdots & \underline{G}_{n-2,\,n+1} \\ \dfrac{\partial \pounds_2}{\partial N_B} & & \cdots & \dfrac{\partial \pounds_2}{\partial N_{n+1}} \end{vmatrix} = 0 \qquad (7\text{-}47)
$$

where $\underline{G}_{BB} \equiv \partial^2 \underline{G}/\partial N_B^2$, etc. Equations (7-46) and (7-47) are identical to the criteria given by Gibbs (1876, p 108). There are, however, many alternative but equivalent criteria, depending on the ordering of the independent variables and on the value of i.

Two examples follow to illustrate the manipulation of stability criteria in determinant formulation. Applications of stability and critical states concepts to problems in phase equilibria are discussed in Sections 15.6 and 15.7.

Example 7.4

Ordering variables as $\underline{U} = U(\underline{S}, \underline{V}, N_B, N_C)$ for a binary of B and C, write the critical-point criteria in terms of derivatives of the Helmholtz and Gibbs energies.

Solution

Here $m = 4$, $n = 2$. With $i = 1$, Eqs. (7-18) and (7-44) become

$$
\pounds_1 = \begin{vmatrix} \underline{A}_{\underline{V}\underline{V}} & \underline{A}_{\underline{V}B} \\ \underline{A}_{B\underline{V}} & \underline{A}_{BB} \end{vmatrix} = 0 \qquad \mathcal{M}_1 = \begin{vmatrix} \underline{A}_{\underline{V}\underline{V}} & \underline{A}_{\underline{V}B} \\ \dfrac{\partial \pounds_1}{\partial \underline{V}} & \dfrac{\partial \pounds_1}{\partial N_B} \end{vmatrix} = 0
$$

or

$$
\pounds_1 = \underline{A}_{\underline{V}\underline{V}} \underline{A}_{BB} - \underline{A}_{\underline{V}B}^2 = 0
$$

$$
\mathcal{M}_1 = \underline{A}_{BBB} \underline{A}_{\underline{V}\underline{V}}^2 - \underline{A}_{\underline{V}\underline{V}\underline{V}} \underline{A}_{BB} \underline{A}_{\underline{V}B} - 3\underline{A}_{\underline{V}BB} \underline{A}_{\underline{V}\underline{V}} \underline{A}_{\underline{V}B} + 3\underline{A}_{\underline{V}\underline{V}B} \underline{A}_{\underline{V}B}^2 = 0
$$

With $i = 2$,

$$
\underline{G}_{BB} = \underline{G}_{BBB} = 0
$$

where

$$
\underline{G}_{BB} = \left(\frac{\partial^2 \underline{G}}{\partial N_B^2}\right)_{T,P,N_C} = \left(\frac{\partial \mu_B}{\partial N_B}\right)_{T,P,N_C} \quad \text{and} \quad \underline{G}_{BBB} = \left(\frac{\partial^3 \underline{G}}{\partial N_B^3}\right)_{T,P,N_C} = \left(\frac{\partial^2 \mu_B}{\partial N_B^2}\right)_{T,P,N_C}
$$

Example 7.5

Ordering variables as $\underline{U} = U(\underline{S}, N_B, N_C, ..., N_n, \underline{V})$, write the stability and critical-point criteria in terms of Helmholtz energies.

Solution

Here $m = n + 2$, and $i = 1$, Eqs. (7-19) and (7-45) become

$$\pounds_1 = \begin{vmatrix} \underline{A}_{BB} & \underline{A}_{BC} & \cdots & \underline{A}_{Bn} \\ \underline{A}_{CB} & \underline{A}_{CC} & \cdots & \underline{A}_{Cn} \\ \\ \underline{A}_{nB} & \cdots\cdots\cdots & & \underline{A}_{nn} \end{vmatrix} = 0 \qquad \mathcal{M}_1 = \begin{vmatrix} \underline{A}_{BB} & \underline{A}_{BC} & \cdots & \underline{A}_{Bn} \\ \underline{A}_{CB} & \underline{A}_{CC} & \cdots & \underline{A}_{Cn} \\ \cdot & \cdot & & \cdot \\ \cdot & \cdot & & \cdot \\ \cdot & \cdot & & \cdot \\ \dfrac{\partial \pounds_1}{\partial N_B} & \cdot & \cdots & \dfrac{\partial \pounds_1}{\partial N_n} \end{vmatrix} = 0$$

By ordering $\underline{U}$ in such a manner that $\underline{V}$ is a terminal variable, the $\pounds_1$ and $\mathcal{M}_1$ determinants contain only derivatives in mole numbers. No pressure derivatives (such as A_{VV}) appear. Heidemann and Khalil (1980) followed a different approach to avoid differentiation of the $\pounds_1$ determinant in $\mathcal{M}_1$. They produced equivalent criteria and applied these to the Soave (1972) modification of the Redlich-Kwong (RKS) equation to calculate critical points of multicomponent systems (see Section 8.3 and 9.2).

7.4 Indeterminacy

A stable equilibrium state is defined by Eq. (7-12). An argument was then presented, using Eq. (7-14), to show that the derivative $y^{(m-2)}_{(m-1)(m-1)}$ in the equation set (7-12) was the first to attain a value zero as the spinodal curve was approached from a stable region. Equation (7-16) resulted and has been employed to delineate when a stable system has reached the limit of stability.

There may, however, be cases where Eq. (7-14) is indeterminate. Rewriting Eq. (7-14) for $y^{(m-2)}_{(m-1)(m-1)}$, we obtain

$$y^{(m-2)}_{(m-1)(m-1)} = \frac{\begin{vmatrix} y^{(m-3)}_{(m-2)(m-2)} & y^{(m-3)}_{(m-2)(m-1)} \\ y^{(m-3)}_{(m-1)(m-2)} & y^{(m-3)}_{(m-1)(m-1)} \end{vmatrix}}{y^{(m-3)}_{(m-2)(m-2)}} = \frac{\pounds_{m-3}}{y^{(m-3)}_{(m-2)(m-2)}} \tag{7-48}$$

If both the numerator and denominator attain a value zero simultaneously (as $y^{(m-3)}_{(m-2)(m-2)}$ is decreased to zero), the case is indeterminate. To visualize the problem in a different way, we note that in the unstable domain, there may be regions where

$$y^{(k-1)}_{kk} = 0 \qquad 1 \leq k \leq m-2 \tag{7-49}$$

For example, Heidemann (1975) shows curves where $y_{33}^{(2)} = 0$ in the unstable region (as defined by $y_{44}^{(3)} = 0$) for a ternary system at constant temperature and pressure. Teja and Kropholler (1975), show $y_{22}^{(1)} = 0$ curves in the unstable region (as defined by $y_{33}^{(2)} = 0$) for a binary system. As long as all curves defined by Eq. (7-49) are located in the unstable region [defined by Eq. (7-16)], no difficulties are encountered in numerical computations. Should, however, $y_{(m-2)(m-2)}^{(m-3)}$ become tangent to the spinodal curve at the critical (or any other) point, an indeterminacy results where $y_{(m-1)(m-1)}^{(m-2)}$ and $y_{(m-2)(m-2)}^{(m-3)}$ both become zero.

To remove the indeterminacy, the stability criterion on the spinodal curve may be modified to

$$y_{(m-1)(m-1)}^{(m-2)}\, y_{(m-2)(m-2)}^{(m-3)} = 0 \tag{7-50}$$

The only case where Eq. (7-50) would fail is when $y_{(m-3)(m-3)}^{(m-4)}$ and lower-order terms become zero simultaneously.

Equation (7-20) can be used to demonstrate this technique by writing the various products in terms of the £ determinant [see Eq. (7-19)].

$$y_{(m-1)(m-1)}^{(m-2)} = £_{m-2} = \frac{£_{m-3}}{y_{(m-2)(m-2)}^{(m-3)}} = 0$$

$$y_{(m-1)(m-1)}^{(m-2)}\, y_{(m-2)(m-2)}^{(m-3)} = £_{m-3} = \frac{£_{m-4}}{y_{(m-3)(m-3)}^{(m-4)}} = 0$$

$$\vdots \tag{7-51}$$

$$\prod_{r=i}^{m-2} y_{(r+1)(r+1)}^{(r)} = £_i = \frac{£_{i-1}}{y_{ii}^{(i-1)}} = 0$$

$$\vdots$$

$$\prod_{r=0}^{m-2} y_{(r+1)(r+1)}^{(r)} = £_0 = 0$$

The first equation of this set is Eq. (7-48). If it is indeterminate, the second form can be used—unless $£_{m-4}/y_{(m-3)(m-3)}^{(m-4)}$ is indeterminate. Continuing, if necessary, to the case where the product index r has dropped to zero, the criterion is simply $£_0 = 0$ and no indeterminacy can result.

The use of equation set (7-51) is rarely required. In essentially all cases the simple criterion for stability, Eq. (7-15), is sufficient.

Should $£_{m-3}/y_{(m-2)(m-2)}^{(m-3)}$ be indeterminate at the critical point, the criterion for criticality [Eq. (7-39)] should be multiplied by $(y_{(m-2)(m-2)}^{(m-3)})^3$ to avoid numerical difficulties.

7.5 Use of Mole Fractions in the $\mathcal{L}_i$ and $\mathcal{M}_i$ Determinants

Throughout this chapter the independent set of variables for the internal energy $\underline{U}$ was total entropy $\underline{S}$, total volume $\underline{V}$, and all mole numbers $N_1, \ldots, N_n$. This set of independent properties delineated the appropriate variables that had to be maintained constant in any derivative of $\underline{U}$ [$y^{(0)}$] or other Legendre transforms $y^{(i)}$. For example, in a ternary, with the variables ordered as $\underline{U} = U(\underline{S}, \underline{V}, N_1, N_2, N_3)$, then

$$\mathcal{L}_3 = y_{44}^{(3)} = \left(\frac{\partial^2 \underline{G}'}{\partial N_2^2}\right)_{T, P, \mu_1, N_3} \tag{7-52}$$

$$\mathcal{L}_2 = \begin{vmatrix} y_{33}^{(2)} & y_{34}^{(2)} \\ y_{43}^{(2)} & y_{44}^{(2)} \end{vmatrix} = \begin{vmatrix} \left(\frac{\partial^2 \underline{G}}{\partial N_1^2}\right)_{T, P, N_2, N_3} & \left(\frac{\partial^2 \underline{G}}{\partial N_1 \, \partial N_2}\right)_{T, P, N_3} \\ \left(\frac{\partial^2 \underline{G}}{\partial N_1 \, \partial N_2}\right)_{T, P, N_3} & \left(\frac{\partial^2 \underline{G}}{\partial N_2^2}\right)_{T, P, N_1, N_3} \end{vmatrix} \tag{7-53}$$

Suppose, however, that we selected an equally valid set of variables

$$\underline{U} = f(\underline{S}, \underline{V}, N_1, N_2, N) \tag{7-54}$$

where $N = \Sigma_j N_j$. With the same approach as used earlier, we would have reached the conclusion that $\mathcal{L}_i = \mathcal{M}_i = 0$ at the critical point. But the definitions of $\mathcal{L}_i$ and $\mathcal{M}_i$ are slightly modified. For example, Eq. (7-52) now becomes

$$\mathcal{L}_3 = y_{44}^{(3)} = \left(\frac{\partial^2 \underline{G}^+}{\partial N_2^2}\right)_{T, P, \eta_1, N} \tag{7-55}$$

where

$$\underline{G}^+ = \underline{U} - T\underline{S} + P\underline{V} - \eta_1 N_1 \tag{7-56}$$

$$\eta_1 = \left(\frac{\partial \underline{U}}{\partial N_1}\right)_{\underline{S}, \underline{V}, N_2, N} = \left(\frac{\partial \underline{G}}{\partial N_1}\right)_{T, P, N_2, N} \tag{7-57}$$

Note the difference between η_1 and the chemical potential μ_1:

$$\mu_1 = \left(\frac{\partial \underline{U}}{\partial N_1}\right)_{\underline{S}, \underline{V}, N_2, N_3} = \left(\frac{\partial \underline{G}}{\partial N_1}\right)_{T, P, N_2, N_3} \tag{7-58}$$

$$\mu_1 - \mu_n = \eta_1 \quad \text{and} \quad \mu_n = \eta_n \tag{7-59}$$

η_i is a modified chemical potential appropriate to the variable set given in Eq. (7-54). Equivalence of η_i among contacting phases is tantamount to the equivalence of μ_i itself and therefore a valid indication of phase equilibrium. Equation (7-55) could also be rewritten with $G^+ = \underline{G}^+/N$ as

$$\pounds_3 = \frac{1}{N}\left(\frac{\partial^2 G^+}{\partial z_2^2}\right)_{T, P, \eta_1, N} \tag{7-60}$$

and z is a mole fraction. With N held *constant*, it is then a simple matter to convert this derivative to one employing mole fractions.

A more useful example of this method results when $\pounds_2$ is determined using the ordering in Eq. (7-54):

$$\pounds_2 = \begin{vmatrix} y_{33}^{(2)} & y_{34}^{(2)} \\ y_{43}^{(2)} & y_{44}^{(2)} \end{vmatrix} = \begin{vmatrix} \left(\frac{\partial^2 \underline{G}}{\partial N_1^2}\right)_{T, P, N_j[1]} & \left(\frac{\partial^2 \underline{G}}{\partial N_1 \partial N_2}\right)_{T, P, N_j[1, 2]} \\ \left(\frac{\partial^2 \underline{G}}{\partial N_1 \partial N_2}\right)_{T, P, N_j[1, 2]} & \left(\frac{\partial^2 \underline{G}}{\partial N_2^2}\right)_{T, P, N_j[2]} \end{vmatrix} \tag{7-61}$$

where the Legendre transform $\underline{G}$ (Gibbs energy) is the same in Eqs. (7-53) and (7-61). Equation (7-61) can be written as

$$\pounds_2 = \frac{1}{N^2}\begin{vmatrix} \left(\frac{\partial^2 G}{\partial z_1^2}\right)_{T, P, z_2} & \left(\frac{\partial^2 G}{\partial z_1 \partial z_2}\right)_{T, P, z_j[1, 2]} \\ \left(\frac{\partial^2 G}{\partial z_1 \partial z_2}\right)_{T, P, z_j[1, 2]} & \left(\frac{\partial^2 G}{\partial z_2^2}\right)_{T, P, z_1} \end{vmatrix} \tag{7-62}$$

where, in each derivative, the mole fraction z_3 varies to maintain the requirement that $\Sigma_j\, dz_j = 0$.

The $\pounds_i$ and $\mathcal{M}_i$ determinants may always be converted to a mole fraction form [e.g., Eq. (7-60) or (7-62)], but only if the Legendre transform is $\underline{A}$, $\underline{H}$, or $\underline{G}$ does one avoid introducing modified chemical potentials η_j. If $\underline{A}$, $\underline{H}$, or $\underline{G}$ is used, the expansion shown in Eq. (7-56) may be employed. As shown in Section 9.1, the last variable in the ordering (N in this case) is always a constraint in every derivative used in stability theory. With such a constraint, for any extensive variable, it is easy to convert to an intensive form; that is, if $\underline{B}$ were an extensive variable, then *with N constant*, $d\underline{B} = N dB$, $d^2\underline{B} = N d^2 B$, etc.

Finally, it is obvious that if one should wish to employ mole fractions as working variables to specify stability and critical point criteria, one should order variables as in Eq. (7-54), although $\underline{S}$ and $\underline{V}$ may be interchanged and the particular mole number eliminated is not important.

To conclude, we summarize the convenient stability and critical-point criteria for an n-component system when mole fractions are used. In all derivatives involving mole fractions, z_n is not constant:

$$\underline{U} = f(\underline{S}, \underline{V}, N_1, \ldots, N_{n-1}, N)$$

(ordering of N_i is arbitrary) and

$$\pounds_2 = 0 = \frac{1}{N^{n-1}} \begin{vmatrix} \left(\dfrac{\partial^2 G}{\partial z_1^2}\right)_{z_2,\ldots,z_{n-1}} & \left(\dfrac{\partial^2 G}{\partial z_1\,\partial z_2}\right)_{z_3,\ldots,z_{n-1}} & \cdots & \left(\dfrac{\partial^2 G}{\partial z_1\,\partial z_{n-1}}\right)_{z_2,\ldots,z_{n-2}} \\ \left(\dfrac{\partial^2 G}{\partial z_2\,\partial z_1}\right)_{z_3,\ldots,z_{n-1}} & \left(\dfrac{\partial^2 G}{\partial z_2^2}\right)_{z_1,z_3,\ldots,z_{n-1}} & \cdots & \left(\dfrac{\partial^2 G}{\partial z_2\,\partial z_{n-1}}\right)_{z_1,z_3,z_{n-2}} \\ \vdots & \vdots & & \vdots \\ \left(\dfrac{\partial^2 G}{\partial z_{n-1}\,\partial z_1}\right)_{z_2,\ldots,z_{n-2}} & \left(\dfrac{\partial^2 G}{\partial z_{n-1}\partial z_2}\right)_{z_1,z_3,\ldots,z_{n-2}} & \cdots & \left(\dfrac{\partial^2 G}{\partial z_{n-1}^2}\right)_{z_1,\ldots,z_{n-2}} \end{vmatrix} \tag{7-63}$$

with (T, P, N) held constant in each derivative.

$\mathcal{M}_2$ is the determinant constructed from $\pounds_2$ by replacing the bottom row by $(\partial\pounds_2/\partial z_j)_{z_1,\ldots,z_{j-1},z_{j+1},\ldots,z_{n-1}}$, where j is the jth column, and $\pounds_2$ and $\mathcal{M}_2$ are $(n-1)$ by $(n-1)$ determinants. If $\pounds_1$ were desired, with the ordering as shown, then

$$\pounds_1 = 0 = \frac{1}{N^{n-1}} \begin{vmatrix} \left(\dfrac{\partial^2 A}{\partial V^2}\right)_{z} & \left(\dfrac{\partial^2 A}{\partial V\,\partial z_1}\right) & \cdots & \left(\dfrac{\partial^2 A}{\partial V\,\partial z_{n-1}}\right) \\ \left(\dfrac{\partial^2 A}{\partial z_1\,\partial V}\right) & \left(\dfrac{\partial^2 A}{\partial z_1^2}\right)_{V,z_2,\ldots,z_{n-1}} & \cdots & \left(\dfrac{\partial^2 A}{\partial z_1\,\partial z_{n-1}}\right) \\ \vdots & \vdots & & \vdots \\ \left(\dfrac{\partial^2 A}{\partial z_{n-1}\,\partial V}\right) & \left(\dfrac{\partial^2 A}{\partial z_{n-1}\,\partial z_1}\right) & \cdots & \left(\dfrac{\partial^2 A}{\partial z_{n-1}^2}\right)_{V,z_1,\ldots,z_{n-2}} \end{vmatrix} \tag{7-64}$$

with T also constant. $\mathcal{M}_1 = \pounds_1$ with the entry in the bottom row of the first column replaced by $(\partial\pounds_1/\partial V)_{z_1,\ldots,z_{n-1}}$ and each bottom entry in the other $(n-1)$ columns replaced by

$(\partial \mathcal{L}_1/\partial z_j)_{V, z_1,\ldots, z_{j-1}, z_{j+1},\ldots, z_{n-1}}$, where $1 \le j \le n-1$. In a similar manner, $\mathcal{L}_0$ and $\mathcal{M}_0$ may be written. The $\mathcal{L}_2$ and $\mathcal{M}_2$ determinants written in mole fraction form were used by Peng and Robinson (1977) to predict critical properties of multicomponent systems using the Peng-Robinson equation of state (see Sections 8.3 and 9.2 and Problems 15.29 and 15.39).

7.6 Summary

Legendre transformations were used extensively in this chapter to develop completely general stability criteria that can be applied to any simple thermodynamic system. Criteria were derived for establishing the limits of stability or spinodal conditions [see Eqs. (7-15) through (7-19)] and were applied to pure, binary, and ternary systems. Critical states criteria were also developed and discussed in the context of one- and two-component systems. Critical points and loci were treated specifically. Geometric interpretations of stable, metastable, and unstable phase regions were provided using both the Gibbs *USV* surface, the Helmholtz *ATV* surface, and other phase diagrams.

References and Suggested Readings

Beegle, B.L. (1974), "Thermodynamic stability criterion for pure substances and mixtures," *AIChE Journal*, 20 (6), p 1200.

Callen, H.B. (1985), *Thermodynamics and an Introduction to Thermostatistics*, 2nd ed., Wiley, New York.

Coy, D.C. (1993), "Visualizing thermodynamic stability and phase equilibrium through computer graphics" PhD dissertation, Dept. of Chemical Eng., Iowa State University, Ames, Iowa.

Gibbs, J.W. (1876, 1878), "On the equilibrium of heterogeneous substances," *Transactions of the Connecticut Academy*, **3**, p 108 (1876) and p 343 (1878).

Jolls, K. (1990), "Gibbs and the art of thermodynamics," *Proceedings of the Gibbs Symposium*, Yale University, New Haven, CT, May 15-17, 1989, p 293-321.

Jolls, K.R. and J.L. Butterbaugh (1992), "Confirming thermodynamic stability--a classroom example," *Chem. Eng. Education*, **26**(3), p 124.

Jolls, K.R. and J.M. Prausnitz (1983), "Laboratory demonstrations for teaching chemical thermodynamics," paper presented at the Annual Meeting of the AIChE, Washington D.C., November.

Hayward, A.T.J. (1971), "Negative pressure in liquids," *American Scientist*, **59**, p 434-443.

Heidemann, R.A. (1975), "The criteria of thermodynamic stability," *AIChE Journal*, **21**, p 824.

Heidemann, R.A. and A.M. Khalil (1980), "The calculation of critical points," *AIChE Journal*, **26**, p 769.

Patrick-Yeboah, J.R. and R.C. Reid (1981), "Superheat-limit temperatures of polar liquids," *Ind. Eng. Chem. Fundamentals*, **20**(4), p 315.

Peng, D-Y, and D.B. Robinson (1977), "A rigorous method for predicting the critical properties of multicomponent systems from an equation of state," *AIChE Journal*, **23**(2), p 137-144.

Prigogine, I. and R. Defay (1952), *Chemical Thermodynamics*, translated by D.H. Everett, Longmans, Green, and Co., New York, Chapters 15 and 16.

Rowlinson, J.S. (1969), *Liquids and Liquid Mixtures*, Butterworth, London, Chapter 5.

Soave, G. (1972), "Equilibrium constants from a modified Redlich-Kwong equation of state," *Chem. Eng. Science*, **27**, p 1197.

Scholander, P.F. (1972), "Tensile water," *American Scientist*, **60**, p 584-590.

Teja, A.S. and H.W. Kropholler (1975), "Critical states of mixtures in which azeotropic behaviour persists in the critical region," *Chem. Eng. Science*, **30**, p 435.

Tisza, L. (1966), *Generalized Thermodynamics*, MIT Press, Cambridge, MA, p 61-66.

Problems

7.1. In Table 7.1 there is a listing of equivalent stability criteria for a pure material. Consider the first two entries:

$$A_{VV} = -\left(\frac{\partial P}{\partial \underline{V}}\right)_{T,N} > 0 \quad \text{and} \quad A_{NN} = +\left(\frac{\partial \mu}{\partial N}\right)_{T,\underline{V}} > 0$$

At the limit of stability it is stated that both A_{VV} and A_{NN} go to zero simultaneously. If one had an applicable equation of state to express the $P\underline{V}TN$ relationship for a pure material, and wished to use the criterion $A_{NN} = 0$ at the limit of stability, how could this be accomplished? Compare your result with that obtained using the alternative form, $A_{VV} = 0$.

7.2. Derive Eq. (7-32).

7.3. We have liquid ethane at 1 bar pressure. What is the maximum temperature the *liquid* ethane can be heated to before it reaches the limit of stability? Use the van der Waals equation of state (see Example 7.2) with the following data:

$$T_c = 305.4\ K\ ; \quad P_c = 48.8 \text{ bar}\ ; \quad T_b = 184.5\ K \text{ (at 1.013 bar)}$$

(W. Porteous and M. Blander [(1976) *AIChE J.*, **21**, p 560] report an experimental value of 269 K at the limit of stability.)

7.4. In the storage of liquefied gases under pressure, the liquid is in equilibrium with the vapor in the ullage. In case of a severe accident in which the vessel fails and the pressure is reduced very rapidly to atmospheric, the bulk liquid is believed to undergo an isentropic expansion to the lower pressure. As the pressure drops, boiling will be initiated at the walls, but, due to the lack of nucleation sites within the bulk liquid, this liquid could become superheated. There is even the possibility that the limit-of-stability curve (the spinodal) may be reached during the pressure release. If this happens, a spontaneous generation of vapor would occur with shock waves to exacerbate the incident.

Demonstrate your ability to analyze an accident of this type by considering a cylinder containing both gas and liquid carbon dioxide. The temperature initially is at 294 K and the corresponding vapor pressure 58.4 bar. A weld seam in the vapor space fails and the pressure drops very rapidly to 1 bar. Neglecting any boiling on the cylinder walls, will the bulk liquid reach the limit of stability?

The van der Waals equation of state may be used if desired. (See Example 7.2.). Some data for saturated liquid CO_2 are also tabulated below should you find them of value.

Saturated Liquid Carbon Dioxide[a]

T (K)	*P* (bar)	*S* (J/mol K)	*V* (cm^3/mol)
270	32.034	135.93	46.502
272	33.801	136.68	47.077
274	35.638	137.44	47.687
276	37.549	138.21	48.336
278	39.533	138.98	49.029
280	41.595	139.77	49.773
282	43.737	140.56	50.576
284	45.960	141.37	51.449
286	48.269	142.21	52.404
288	50.665	143.06	53.460
290	53.152	143.94	54.641
292	55.734	144.86	55.981
294	58.415	145.83	57.531
296	61.198	146.87	59.376
298	64.090	148.00	61.661
300	67.095	149.31	64.690
302	70.220	150.95	69.297
304	73.475	154.16	81.703
304.21[b]	73.825	156.58	94.440

[a]Source: S. Angus, B. Armstrong, and K.M. deReuck, eds. (1976), *Int. Thermodynamic Tables of the Fluid State Carbon Dioxide*, (Pergamon Press, New York).
[b]Critical state.

7.5. Prove for a stable thermodynamic system that $C_p > C_V$.

7.6. A ternary liquid mixture of *B*, *C*, and *D* is at a constant pressure and temperature. The excess Gibbs energy is given as

$$\frac{\Delta G^{EX}}{RT} = \alpha(x_B x_C + x_B x_D + x_C x_D)$$

Determine the limits of essential stability (i.e., the spinodal curves) and any critical points if $\alpha = 2.5$.

7.7. A typical P-T phase diagram for a pure material is sketched below.

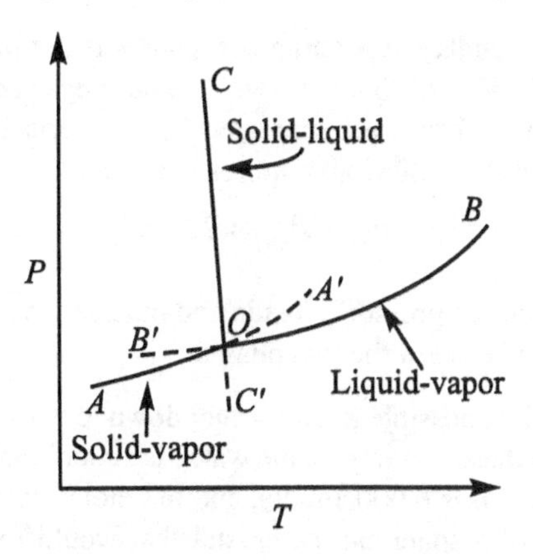

Figure P7.7

(a) Describe how you would estimate the limit of the metastable extension of the liquid-vapor curve to point B'. What criteria are needed and what property data or models are needed? Show relevant equations and state all assumptions.

(b) According to several prominent thermodynamicists, stable (solid) and metastable (dashed) monovariant lines must alternate around the triple point as shown in the figure. Show that this behavior is reasonable by considering plots of Gibbs free energy versus pressure for temperatures slightly above, slightly below, and at the triple point temperature.

7.8. The spinodal state represents the limit of stability for a thermodynamic system. For a pure material, considering the superheated liquid state, a P-T plot would be shown in Figure 7.8. Rocky Jones, our local genius, claims that the slope (dP/dT) along the spinodal is given by:

$$(dP/dT)_{spinodal} = (\partial P/\partial T)_V$$

Prove or disprove Rocky's claim.

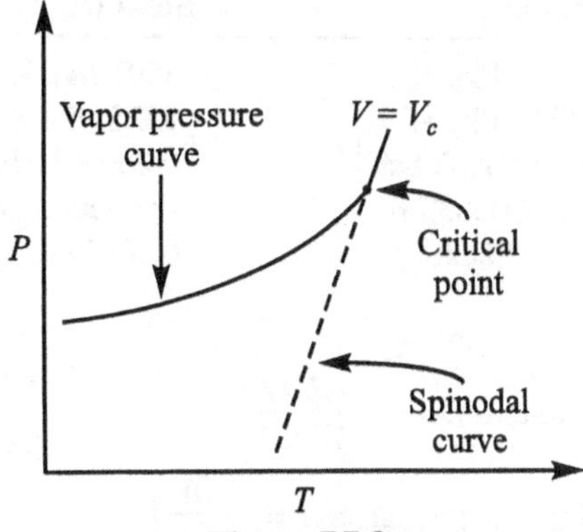

Figure P7.8

7.9. Show that C_p is always positive for a thermodynamically stable, single component system.

7.10. When the movable boundary separating two phases is *not planar*, then, at equilibrium, the pressures on both sides of that boundary are *no longer equal*. For example, Laplace showed that when two phases are separated by a spherical boundary of radius r, the condition of mechanical equilibrium requires that

$$P_{in} - P_{out} = 2\sigma/r$$

where P_{in} and P_{out} are the pressures inside and outside the spherical boundary and σ is the interfacial tension between the two phases.

In very clean water, it is possible to cool water down to – 40°C at 1 atm before crystallization occurs. Under these conditions, the water-ice interfacial tension is 22 dyne/cm, the molar enthalpy of fusion is 6,000 J/mole, and the molar volume of ice is 20 cm^3/mole. Estimate the radius of a spherical ice crystal that would form under such conditions. Clearly state and explain any approximations made in the derivation of your result.

7.11. For a pure component:

(a) Derive expressions to show what happens to the heat capacity C_p and the isothermal compressibility $\kappa_T = \frac{-1}{V}\left(\frac{\partial V}{\partial P}\right)_T$ as the critical point is approached.

(b) Sketch curves of the Gibbs free energy versus specific volume along the liquid-vapor coexistence curve at the critical ($T = T_c$) and triple ($T = T_{tp}$) points and somewhere between T_c and T_{tp}

You can assume that a suitable EOS exists to describe the *PVT* behavior in both phases.

7.12. In some metastable situations, it is important to estimate differences in derived thermodynamic properties. We are interested in knowing the magnitude of the difference between the solid and liquid phase Gibbs free energy under subcooled conditions for pure water. At –10°C and 1 bar estimate the magnitude of $\Delta G = G$ (solid) - G (liquid). Some property data for water are given below at 1 bar pressure and 0°C:

	Liquid	**Solid Ice**
C_p	4128 J/kg K	3591 J/kg K
C_v	4219 J/kg K	4225 J/kg K
κ_T	0.0000507 bar^{-1}	0.0000510 bar^{-1}
α_p	-0.0000633 K^{-1}	-0.0000802 K^{-1}
η	1.3339 cm^3/g	0.2097 cm^3/g

where:

η = refractive index

κ_T = isothermal compressibility $= \frac{-1}{V}\left(\frac{\partial V}{\partial P}\right)_T$

α_p = coefficient of thermal expansion $= \frac{1}{V}\left(\frac{\partial V}{\partial T}\right)_P$

ΔH_{fusion} = 333 kJ/kg

$\Delta H_{vaporization}$ = 2502 kJ/kg

Part II Thermodynamic Properties

Properties of Pure Materials 8

In Chapter 5 the theoretical basis for determining the information required to specify the thermodynamic properties of materials was described. In this chapter we continue this line of reasoning with reference to a specific type of system: the single-phase simple system of a pure material. In a very real sense, we will be describing the constitutive property relationships and models that are used to characterize the macroscopic behavior of pure fluids and solids. Single-phase simple systems of mixtures are treated separately in Chapter 9.

8.1 Gibbs Energy Formulation of the Fundamental Equation

For a single-phase simple system of a pure material, the independent canonical variables for the internal energy ($\underline{U}$) representation are $\underline{S}$, $\underline{V}$, and N; and for the Gibbs energy ($\underline{G}$) representation, they are T, P, and N. The latter set is usually more convenient because T, P, and N are the variables normally under the control of the experimentalist.

The Fundamental Equation for a single component in the Gibbs energy form is

$$\underline{G} = f_G(T, P, N) \tag{8-1}$$

and, in differential form,

$$d\underline{G} = -\underline{S}\, dT + \underline{V}\, dP + \mu\, dN \tag{8-2}$$

The three equations needed to define the state of the system are obtained from the first-order partial derivatives of Eq. (8-1):

$$-\underline{S} = g_1(T, P, N) = \underline{G}_T \tag{8-3}$$

$$\underline{V} = g_2(T, P, N) = \underline{G}_P \tag{8-4}$$

and

$$\mu = g_3(T, P, N) = \underline{G}_N \tag{8-5}$$

Note that these equations of state obtained from $\underline{G}$ are equivalent to ones obtained from $\underline{U}$, $\underline{H}$, or $\underline{A}$ representation. Given the three equations of state for any one representation, we could solve them simultaneously to obtain the three equations of state for any other representation. For example, compare Eqs. (8-3) to (8-5) with Eqs. (5-5) to (5-7).

The three independent second-order partial derivatives are

$$\frac{\partial^2 \underline{G}}{\partial T^2} = -\left(\frac{\partial \underline{S}}{\partial T}\right)_{P,N} = g_{11}(T, P, N) = \underline{G}_{TT} \tag{8-6}$$

$$\frac{\partial^2 \underline{G}}{\partial P^2} = \left(\frac{\partial \underline{V}}{\partial P}\right)_{T,N} = g_{22}(T, P, N) = \underline{G}_{PP} \tag{8-7}$$

$$\frac{\partial^2 \underline{G}}{\partial T\, \partial P} = -\left(\frac{\partial \underline{S}}{\partial P}\right)_{T,N} = \left(\frac{\partial \underline{V}}{\partial T}\right)_{P,N} = g_{12}(T, P, N) = \underline{G}_{TP} \tag{8-8}$$

When expressing $\underline{G}$ in extensive form, there are three more partial derivatives in addition to Eqs. (8-6) through (8-8). These derivatives, however, are either zero or are redundant with Eqs. (8-3) through (8-5). Specifically,

$$\left(\frac{\partial^2 \underline{G}}{\partial N^2}\right) = \left(\frac{\partial \mu}{\partial N}\right)_{T,P} = 0 \quad \text{(see Section 5.2)}$$

$$\left(\frac{\partial^2 \underline{G}}{\partial T \partial N}\right) = -\left(\frac{\partial \underline{S}}{\partial N}\right)_{T,P} = -S = \frac{g_1}{N} \quad \text{[see Eq. (8-3)]}$$

and

$$\left(\frac{\partial^2 \underline{G}}{\partial P \partial N}\right) = \left(\frac{\partial \underline{V}}{\partial N}\right)_{T,P} = V = \frac{g_2}{N} \quad \text{[see Eq. (8-4)]}$$

Equations (8-6) through (8-8) can be simplified because each of the g_{ij} functions are first order in mass. Thus,

$$-\left(\frac{\partial S}{\partial T}\right)_P = -\frac{1}{N}\left(\frac{\partial \underline{S}}{\partial T}\right)_{P,N} = g_{11}(T, P) \tag{8-9}$$

$$\left(\frac{\partial V}{\partial P}\right)_T = \frac{1}{N}\left(\frac{\partial \underline{V}}{\partial P}\right)_{T,N} = g_{22}(T, P) \tag{8-10}$$

and

$$\left(\frac{\partial V}{\partial T}\right)_P = -\left(\frac{\partial S}{\partial P}\right)_T = \frac{1}{N}\left(\frac{\partial \underline{V}}{\partial T}\right)_{P,N} = g_{12}(T, P) \tag{8-11}$$

In the derivatives on the left-hand sides of Eqs. (8-9) through (8-11), note that the subscript N has been omitted. Of course, n+1 variables must always be held constant in partial differentiation. It is, however, common practice to omit the mole numbers when expressing partial derivatives of intensive variables for single component systems.

These second-order partial derivatives are related to three widely used properties: *the heat capacity at constant pressure,* C_p; *the isothermal compressibility,* κ_T; and *the coefficient of thermal expansion,* α_p. By definition,

$$C_p \equiv T\left(\frac{\partial S}{\partial T}\right)_P \tag{8-12}$$

$$\kappa_T \equiv -\frac{1}{V}\left(\frac{\partial V}{\partial P}\right)_T \tag{8-13}$$

and

$$\alpha_p \equiv \frac{1}{V}\left(\frac{\partial V}{\partial T}\right)_P \tag{8-14}$$

These three properties were among the first thermodynamic properties ever reported, principally because they can be measured by relatively simple experiments.

Let us now determine which data sets have the information equivalent to Eq. (8-1). We shall approach this problem once again by determining what data are required to reconstruct the Fundamental Equation.

Applying Euler's theorem to Eq. (8-2), we have the integrated form of the Fundamental Equation in the Gibbs energy representation:

$$\underline{G} = \mu N \tag{8-15}$$

Since Eq. (8-1) can be recovered by substituting $g_3(T, P)$ from Eq. (8-5) into Eq. (8-15), we need specify only one equation of state, namely Eq. (8-5), to obtain the Fundamental Equation. This conclusion is not in contradiction to the discussion of Section 5.5, in which it was shown that two equations of state, Eqs. (5-5) and (5-6), were necessary to reconstruct the Fundamental Equation to within an arbitrary constant. As shown below, we must in fact know Eqs. (8-3) and (8-4) in order to obtain Eq. (8-5).

Let us first express Eq. (8-5) in differential form. In Section 5.4 it was shown that the Gibbs-Duhem equation for a pure material is

$$d\mu = -S\,dT + V\,dP \tag{8-16}$$

To evaluate g_3 of Eq. (8-5), we must express S and V as functions of T and P, substitute into Eq. (8-16), and integrate. The desired expressions for S and V are given by Eqs. (8-3) and (8-4) when written in intensive form. Thus,

$$d\mu = g_1(T, P)\,dT + g_2(T, P)\,dP \tag{8-17}$$

and, by integration in two steps from a reference state at $T = T^o$ and $P = P^o$, where μ is set equal to an arbitrary value of μ^o,

$$\mu = \mu^o + \int_{T^o}^{T} [g_1(T, P)]_{P^o}\,dT + \int_{P^o}^{P} [g_2(T, P)]_T\,dP \tag{8-18}$$

Note that the first integral is along an isobaric path at P^o while the second integral is along an isothermal path at T. Let us assume for the moment that Eqs. (8-3) and (8-4) are not known. In this case, we can evaluate g_1 and g_2 in the following manner.

Since S and V are state properties, we can start with the exact differentials.

$$dS = \left(\frac{\partial S}{\partial T}\right)_P dT + \left(\frac{\partial S}{\partial P}\right)_T dP \tag{8-19}$$

and

$$dV = \left(\frac{\partial V}{\partial T}\right)_P dT + \left(\frac{\partial V}{\partial P}\right)_T dP \tag{8-20}$$

Of course, the partial derivatives in Eqs. (8-19) and (8-20) are related to the second-order partial derivatives of Eq. (8-1). Therefore,

$$dS = \frac{C_p}{T}\, dT - \alpha_p V\, dP \tag{8-21}$$

and

$$dV = \alpha_p V\, dT - \kappa_T V\, dP \tag{8-22}$$

Integrating from a reference state at T^o and P^o for which $S = S^o$ and $V = V^o$, we have

$$S = S^o + \int_{T^o}^{T} \left(\frac{C_p}{T}\right)_{P^o} dT - \int_{P^o}^{P} (\alpha_p V)_T\, dP \tag{8-23}$$

$$V = V^o + \int_{T^o}^{T} (\alpha_p V)_{P^o}\, dT - \int_{P^o}^{P} (\kappa_T V)_T\, dP \tag{8-24}$$

Equations (8-23) and (8-24) are equivalent to Eqs. (8-3) and (8-4), respectively. Substitution into Eq. (8-18) yields

$$\mu = \mu^o - (T - T^o)\, S^o + (P - P^o)\, V^o - g'(T, P, T^o, P^o) + g''(T, P, T^o, P^o) \tag{8-25}$$

where:

$$g'(T, P, T^o, P^o) = \int_{T^o}^{T} \left[\int_{T^o}^{T} \left(\frac{C_p}{T}\right)_{P^o} dT - \int_{P^o}^{P} (\alpha_p V)_T\, dP \right]_{P^o} dT \tag{8-26}$$

$$g''(T, P, T^o, P^o) = \int_{P^o}^{P} \left[\int_{T^o}^{T} (\alpha_p V)_{P^o}\, dT - \int_{P^o}^{P} (\kappa_T V)_T\, dP \right]_T dP \tag{8-27}$$

Note that the second term in Eq. (8-26) vanishes because the upper limit of integration, the dummy variable P, is held constant at P^o in the second integration for dT. In Eq. (8-25), μ^o and S^o can be assigned arbitrary values; the absolute value of volume, however, has physical significance and, therefore, V^o must be the actual volume corresponding to T^o and P^o.

In summary, we have seen that the information content of the Fundamental Equation is contained in the data set consisting of the three second-order partial derivatives C_p, α_p, and κ_T in addition to one value of the specific volume in a reference state.

It should be clear that an equally valid data set consists of C_p and the PVT relationship, Eq. (8-4), since the latter contains the information content of α_p, κ_T, and the absolute value of volume. The *PVT* relationship is so commonly used that it is usually called *the* $P\underline{V}TN$ equation of state or EOS (as opposed to the equations of state, which is the general terminology for all the first-order derivatives of the Fundamental Equation). Data or equations representing C_p and the volumetric *PVT* properties of the fluid are most commonly used to solve a wide variety of problems because this information is most readily available.

Before concluding this discussion of the Gibbs energy representation, it is interesting to compare the results with those for the internal energy. From Eq. (8-25) for a change in state from T_1, P_1 to T_2, P_2,

$$\Delta G = \Delta\mu = -S^o(T_2 - T_1) + V^o(P_2 - P_1) - g'(T_2, P_2, T_1, P_1) + g''(T_2, P_2, T_1, P_1) \quad (8\text{-}28)$$

Whereas ΔU for the corresponding change in state would have a value independent of the reference values assigned to S^o and μ^o, the value of ΔG, as given by Eq. (8-28) depends on the value of S^o. Since S^o is an arbitrary constant, there can be no direct physical significance associated with ΔG for the general change in state under consideration. Note that for isothermal processes, ΔG is no longer dependent on S^o and Eq. (8-16) can be used directly to give $\Delta G|_T$

$$\Delta G|_T = \int_{P_1}^{P_2} VdP$$

In this case, we can attach physical significance to ΔG--specifically it represents the negative of the reversible flow work for an isothermal process (see Section 4.8).

8.2 *PVT* Behavior of Fluids and the Theorem of Corresponding States

As noted in Section 8.1, a knowledge of the *PVT* behavior of a pure material allows one to obtain values of the isothermal compressibility κ_T, and the coefficient of thermal expansion α_p. With these parameters, and C_p, one may then calculate all derivatives containing three of the four variables *S*, *T*, *V*, and *P*.

The volumetric *PVT* behavior of a material depicted in Figure 7.2 is often characterized by the compressibility factor Z, where

$$Z = \frac{P\underline{V}}{NRT} = \frac{PV}{RT} \quad (8\text{-}29)$$

For an ideal gas, $Z = 1.0$. For liquids, Z ranges from about 0.01 to 0.2 near the critical point; for real gases Z can be either larger or smaller than unity. At the critical point, Z_c is normally in the range 0.27 to 0.29, but values can be less for highly polar materials.

A useful concept known as the *Theorem of Corresponding States* suggests that values of thermodynamic properties of different fluids can be compared when properties are divided by their values at the critical point. Applied to Eq. (8-29), this leads directly to a reduced equation of state:

$$Z = f(Z_c, P_r, T_r) \quad (8\text{-}30)$$

where P_r is the reduced pressure, P/P_c, and T_r is the reduced temperature, T/T_c. If Z_c is now assumed to be approximately the same for all materials, then $Z=f(P_r, T_r)$. The function f is then universal in the sense that systems in the same reduced state defined by T_r and P_r will have the same reduced volume V_r (or density $\rho_r = \rho/\rho_c$). Generalized compressibility plots showing Z as a function of T_r and P_r are common in most thermodynamics textbooks.

Although experimental observations provide the most compelling evidence that the form of Eq. (8-30) is correct, there are theoretical rationalizations at the molecular level. The theory of nonideal gases is well developed on a mathematical basis, but to provide an analytical *PVT* or *P$\underline{V}$TN* relation, some assertion must be made about the intermolecular energies between the molecules comprising the system. At the present time, only approximations have been possible to delineate such energies.

If one assumes that intermolecular potential energies have the same characteristic equation for all substances, then statistical mechanics can be used to derive Eq. (8-30) (see Section 10.5).

Let's see if the *Theorem of Corresponding States* can be developed from a simple model for predicting the *PVT* properties of a fluid. Recall that in Example 7.2, we used the van der Waals EOS to relate properties at the critical point of a pure fluid to express the a and b parameters in the EOS. Namely, with

$$P = \frac{RT}{V-b} - \frac{a}{V^2} \tag{8-31}$$

and the fact that $(\partial P/\partial V)_T = 0$ and $(\partial^2 P/\partial V^2)_T = 0$ at the critical point where $T = T_c$, $P = P_c$, and $V = V_c$ (see Figure 7.6), we are able to eliminate V_c and solve for a and b in terms of T_c, P_c, and the gas constant R:

$$a = \frac{27 R^2 T_c^2}{64\, P_c} \quad \text{and} \quad b = \frac{RT_c}{8\, P_c} \tag{8-32}$$

Then we recognized that we could reformulate the vdW EOS [Eq. (8-31)] in dimensionless form by scaling P, T, and V appropriately with their critical point values. Thus in dimensionless, reduced coordinates:

$$T_r \equiv \frac{T}{T_c} \quad ; \quad P_r \equiv \frac{P}{P_c} \quad ; \quad V_r \equiv \frac{V}{V_c}$$

Equation (8-31) now can be rewritten as:

$$P_r = \frac{8T_r}{(3V_r - 1)} - \frac{3}{V_r^2} \tag{8-33}$$

which expresses the functionality given by the compressibility equation [Eq. (8-30)]. Thus, all fluids that obey the vdW EOS are implicitly following the behavior encoded by the *Theorem of Corresponding States*. These observations are, in fact, what motivated van der Waals himself to develop his famous equation of state where he proposed that at a given reduced temperature and pressure, all materials have the same reduced volume.

One can easily show that Eq. (8-33) predicts a universal value of $Z_c = 3/8 = 0.375$ which is unfortunately far from the range of values experimentally observed for most real fluids, 0.27 ± 0.04.

Because of the limited accuracy of the vdW EOS in representing the *PVT* properties of real fluids of importance to chemical engineering applications, we must resort to more extensive experimental data to further test the corresponding states concept. Assuming that the critical constants are still the proper scaling parameters, we can examine just how closely substances adhere to the Theorem of Corresponding States. In Figure 8.1, P_r is plotted as a function of T_r at various values of V_r for six aliphatic hydrocarbons from methane to heptane. And in Figure 8.2, Z is given as a function of P_r along various isotherms for ten different substances ranging from methane to water. Within experimental error the substances plotted in Figures 8.1 and 8.2 validate the corresponding states principle by following a family of curves each defined by a specific value of V_r and T_r. *PVT* data for literally hundreds of compounds have been correlated to develop the generalized compressibility correlations. Figure 8.3 is one of such correlations, which are commonly generalized to particular values of Z_c ranging from 0.23 to 0.31 to cover the behavior of most fluids. Although many have contributed to developing the uses of corresponding states, the pioneers in chemical engineering applications include Hougen, Watson, and Ragatz (1959), Meissner and Sefarian (1951), and Lydersen, Greenkorn, and Hougen (1955). Table G.1 of Appendix G provides critical constants for a number of common materials. Readers can also consult Reid *et al.* (1987) for a more extensive listing.

The seminal contribution made by van der Waals that led directly to the corresponding states principle has been validated in a macroscopic sense by the successful general correlation of the massive amounts of empirical property data in reduced form. Clearly corresponding states is a powerful concept that not only provides additional insights into how substances behave volumetrically, but it also offers a practical means of estimating *PVT* properties when only limited data are available.

If we are willing to introduce more parameters into the correlations used for corresponding states, accuracies can be improved markedly. Although this idea goes back at least to Nernst in 1907, it wasn't until the mid-1950's when Pitzer and coworkers really made progress.

Although first-order improvements are made by using Z_c as a third correlating parameter, a considerably more accurate form of Eq. (8-30) was formulated by Pitzer and coworkers (1955) who substituted a different parameter, the acentric factor, for Z_c. In their relation

$$Z = Z^{(0)}(T_r, P_r) + \omega Z^{(1)}(T_r, P_r) \tag{8-34}$$

The acentric factor, ω, was developed as a measure of the difference in structure between the material of interest and a spherically symmetric gas such as argon (where ω is essentially zero). ω is defined as

$$\omega = -\log_{10} P_{vp,r}\Big|_{T_r=0.7} - 1 \tag{8-35}$$

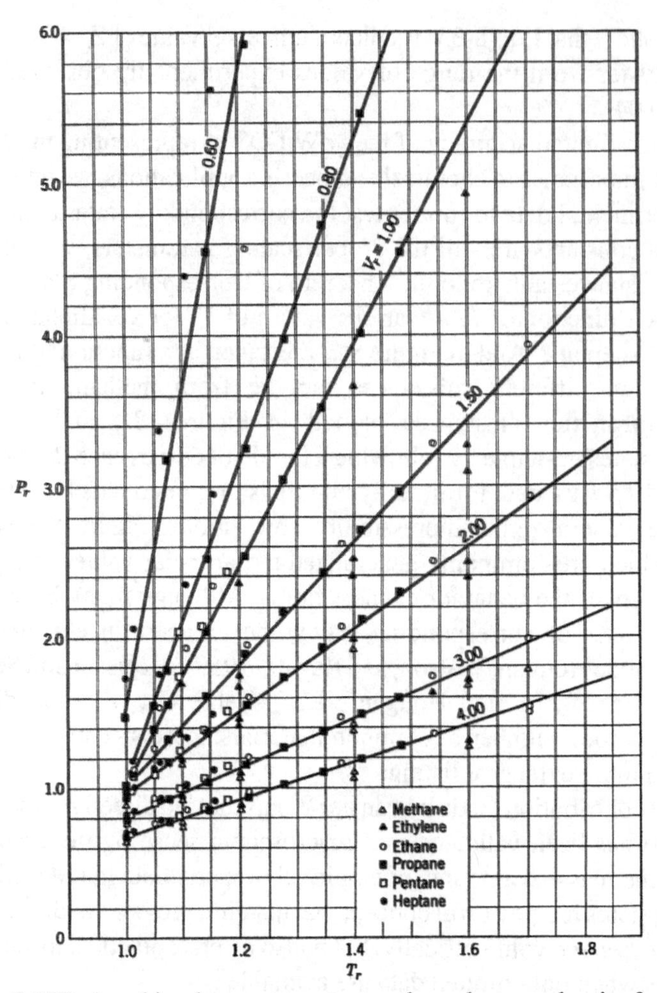

Figure 8.1 Reduced isochores at constant molar volume or density for selected hydrocarbons. [from Weber and Meissner (1957) based on data from Kay (1937) Sc.D. Thesis, Massachusetts Institute of Technology]

where $P_{vp,r}\big|_{T_r=0.7} \equiv P_{vp}(T_r = 0.7)/P_c$ is the reduced vapor pressure at $T_r = 0.7$. Values of ω have been tabulated for many materials (see Appendix G).

It is interesting to note that the reduced vapor pressure correlation for ω given by Eq. (8-35) can be justified using experimental data. Figure 8.4 plots $\log_{10} P_{vp,r}$ versus $1/T_r$ for a number of compounds. Note that a family of straight lines exist that can be correlated to a single adjustable parameter. Pitzer and coworkers selected ω and correlated it using Eq. (8-35). [see Pitzer (1995), Chapter 9 for further details] Their approach has survived in part because of the explicit dependence of ω on vapor pressure data that is readily available for most substances of interest and in part because it captures a molecular property, the sphericity of the molecule, that is theoretically reasonable.

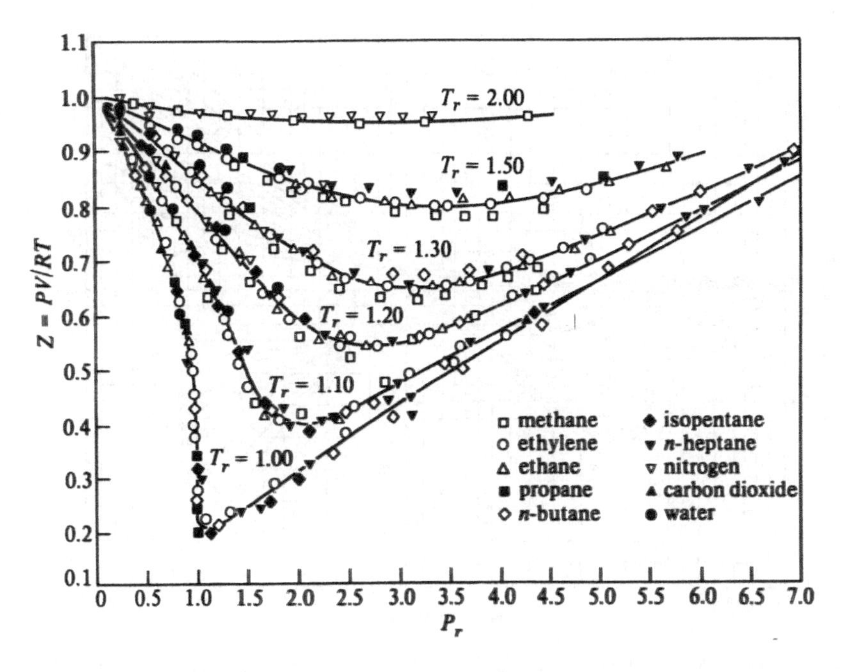

Figure 8.2 Experimentally derived compressibility factors for different fluids as a function of the reduced temperature and pressure. [Reprinted with permission from Van Ness and Abbott (1982) based on data from G.-J. Su (1946). *Ind. Engr. Chem.* 38, p 803.]

The $Z^{(0)}$ and $Z^{(1)}$ functions in Eq. (8-34) depend only on T_r and P_r as shown in Figure 8.5. The original tables of these functions given by Pitzer *et al.* were prepared from an analysis of experimental data. Most subsequent papers that present analytical forms of a $PV\underline{T}N$ equation of state were, in fact, developed to match the Pitzer *et al.* tables and not experimental data!

The generalized shape of the liquid-vapor (L-V) coexistence curve is also given by corresponding states. We can just see this region developing in Figure 8.3 along isotherms at and below the critical temperature ($T_r < 1$). This region is much clearer in the Z_0 plot of Figure 8.5a. A more comprehensive representation of this region is shown in Figure 8.6 which is a reduced temperature-density plot for methane. Note the liquid, vapor, and solid single phase regions as well as the two-phase coexistence regions. Using the Theorem of Corresponding States we can generalize the results given for methane to other substances at similar reduced conditions. As described by Herschbach and coworkers [with Ben Amotz (1990) and Xu (1992)] there are several features illustrated in Figure 8.6 that greatly extend corresponding states concepts. The so-called "law of rectilinear diameters" states that the average density in the L-V region ($\langle\rho\rangle = (\rho_L + \rho_V)/2$) varies linearly over a wide temperature range below the critical point. At the critical point, of course, $\rho_L = \rho_V$. This feature is not captured in normal corresponding states theory but nonetheless it appears to be very general.

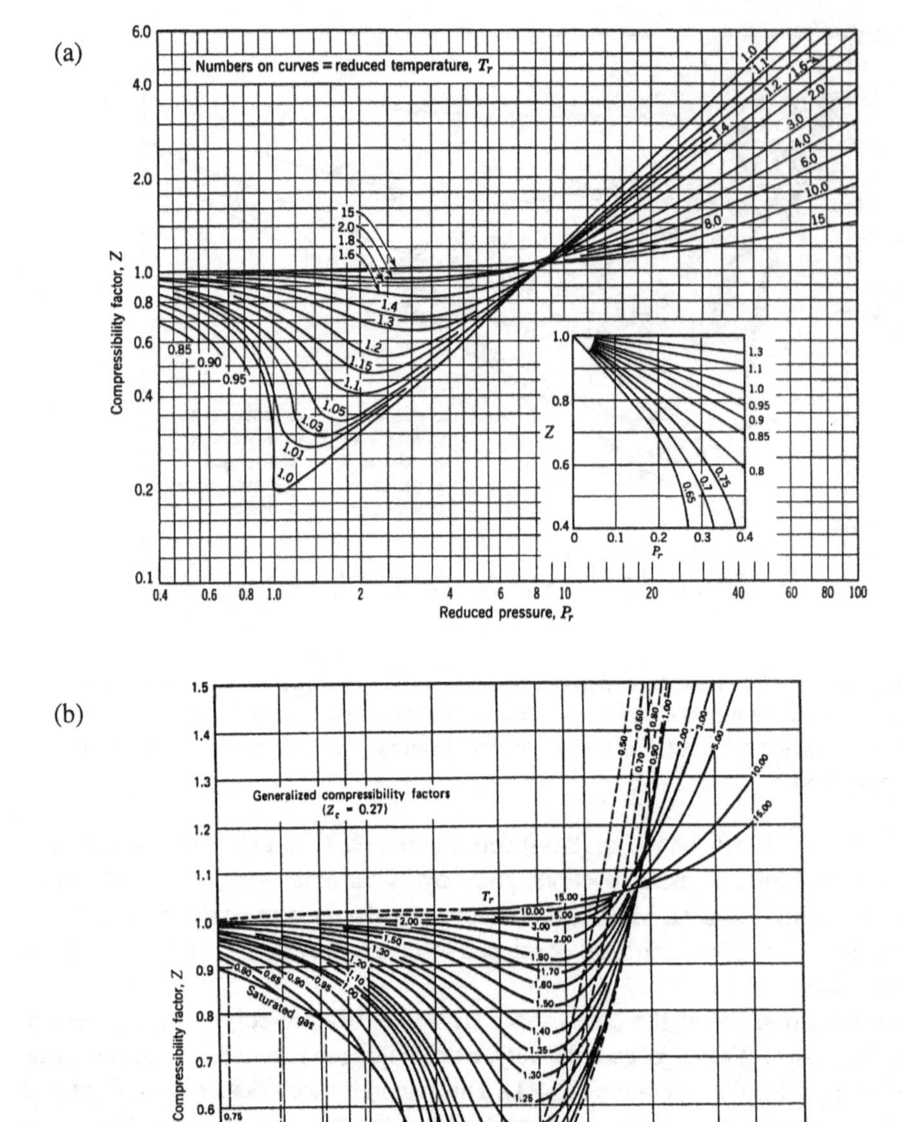

Figure 8.3 Generalized compressibility charts [(a) from Weber and Meissner (1957) Figure 2, p 132. (b) from Hougan *et al.* (1959).]

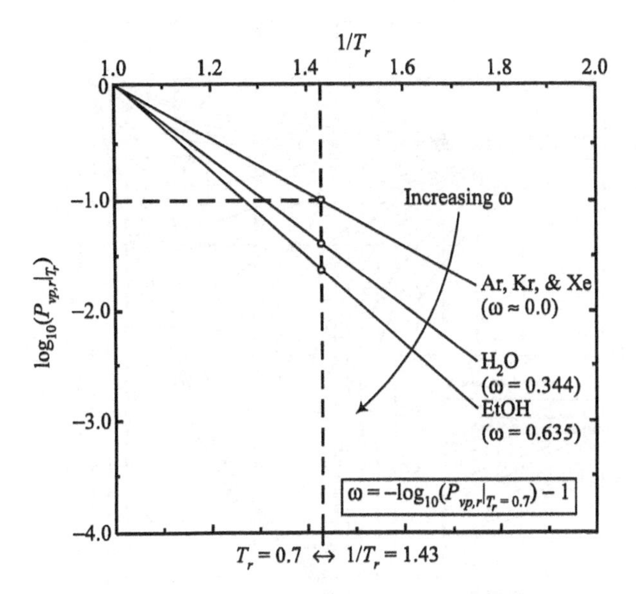

Figure 8.4 Reduced vapor pressures for several common materials (see Appendix G for other values of ω).

Corresponding states predicts that as $P_r \to 0$ or $\rho_r \to 0$ ($V_r \to \infty$) ideal gas behavior will be approached for all fluids. The Boyle point, corresponding to the condition where

$$(\partial Z/\partial P)_{T = T_{Boyle}} = 0 \quad as \quad \rho \to 0$$

is a reflection of this predicted behavior. In this attenuated state, $Z = 1$ and the fluid behaves as an ideal gas. As we can see on Figures 8.3 and 8.5a, there is another region where $Z = 1$ at much higher densities and pressures. For example, follow the $T_r = 2$ isotherm in Figure 8.5a from low pressures to where it intersects the $Z = 1$ line at $P_r \cong 6$. What is very intriguing about this feature is that the $Z = 1$ contour in the T_r versus ρ_r plane extends linearly from the Boyle point where $\rho_r \to 0$ to the dense fluid region somewhere midway between the liquid and solid density at the triple point. Herschbach and coworkers have labelled this contour, the *Zeno line*, to emphasize its "arrowlike linearity." What is particularly important is that the Zeno line linearity is observed for a wide range of substances, including the alkanes C1 - C5, CO_2, N_2, O_2, Ar, Kr, Xe, and H_2O. Further, they were able to correlate the slope and intercept of the Zeno line with the slope and intercept of the rectilinear diameters line. This correlation provides a useful means for predicting critical volumes and *PVT* behavior in the supercritical region where data are often sparse.

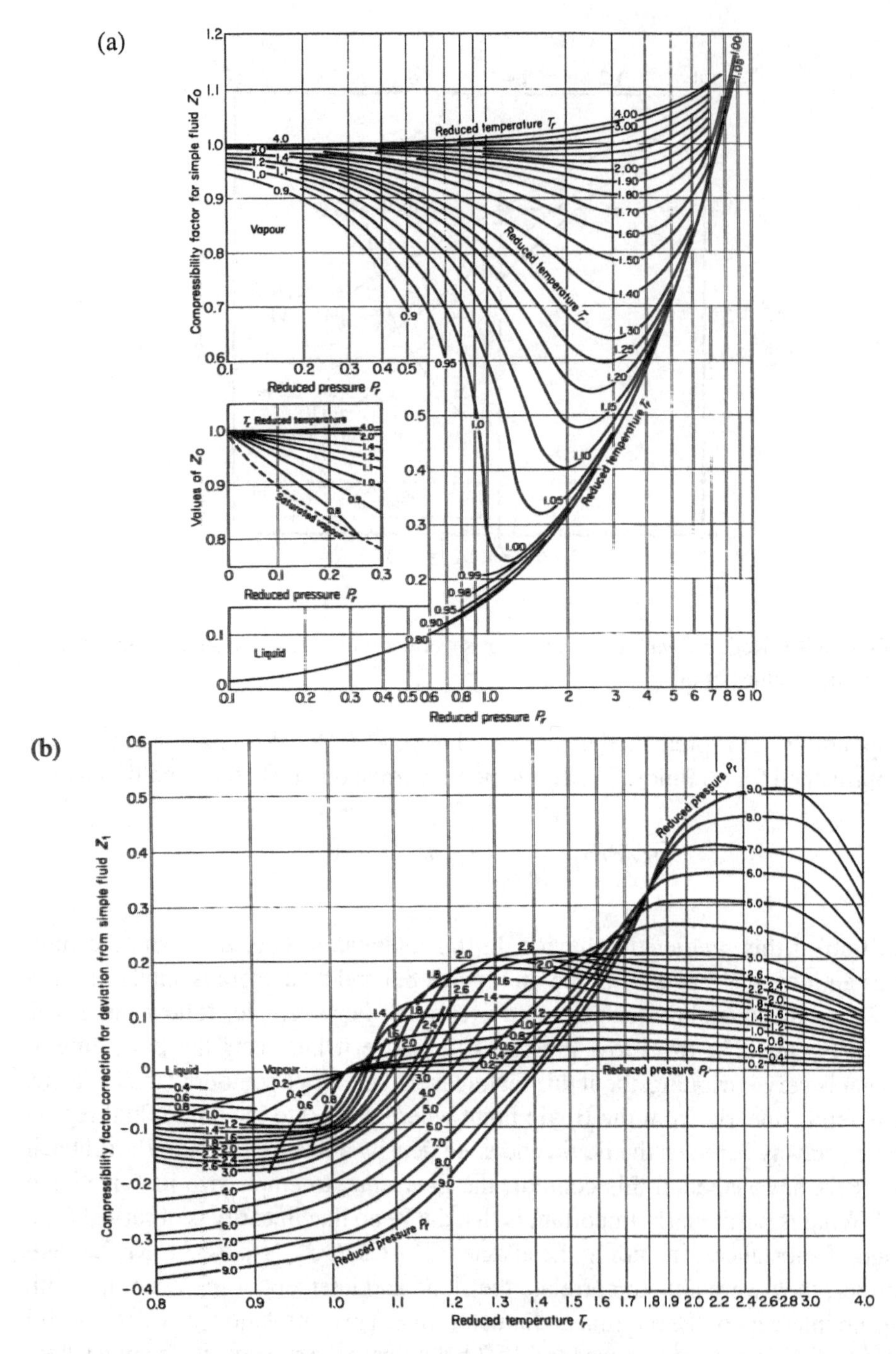

Figure 8.5 Z_0 and Z_1 contributions to generalized corresponding states correlation developed by Pitzer and coworkers (1955) [charts based on the results of Edmister (1958), reproduced with permission from *Petroleum Refiner*, April 1958, copyright 1958 by Gulf Publishing Co., all rights reserved.].

It is worthwhile to try to explain this behavior on a molecular level. The Zeno line separates the "hard" and "soft" fluid regions as shown in Figure 8.6—in principle, the Zeno line represents the boundary where attractive and repulsive intermolecular forces are in balance, and the fluid imitates an ideal gas ($Z = 1$). In the soft fluid region, attractive forces are dominant while in the hard fluid region repulsive forces are more important. Later in Chapter 10, we will re-examine these features in the context of a statistical thermodynamics treatment of hard spheres and van der Waals fluids.

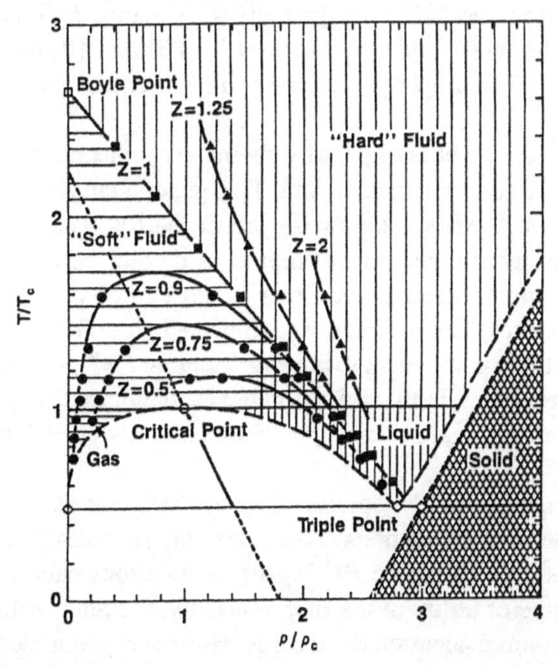

Figure 8.6 Phase diagram for methane in reduced coordinates, with experimental points for several Z-contours. Also shown is the line of rectilinear diameters in the vapor-liquid coexistence region, and its extrapolations (short-dashed lines). [From Ben-Amotz and Herschbach (1990), used with permission.]

8.3 $P\underline{V}TN$ Equations of State for Fluids

Although the corresponding states correlations described in the previous section can be made reasonably accurate by adding parameters such as the Pitzer acentric factor ω, they have a major shortcoming. Their predictions, while clearly displayed on graphs like Figure 8.3, are difficult to express in a set of general *PVT* equations of state. Thus, scientists and engineers have been motivated to develop analytic *PVT* or $P\underline{V}TN$ equations of state (EOS), which are far more convenient for the machine computations so commonly employed in engineering practice today.

In general, there have been two main approaches for formulating an EOS. One approach, pioneered by physicists and physical chemists, is to build models based on rigorous theoretical principles accounting for interactions between molecules. Model results are then related to observable macroscopic properties. Statistical mechanical techniques are used to solve this problem, frequently with deterministic molecular simulations utilizing Monte Carlo (MC) or Molecular Dynamics (MD) methods. Typically, intermolecular energy parameters are estimated by comparing predicted theoretical results with experimental data. In short, attempts to relate properties of fluids to intermolecular interactions using rigorous fundamental analysis do not yield analytical expressions except for the case of an ideal gas. The approximate methods used to solve these problems provide only partially satisfactory *PVT* predictions for real fluids, especially in the dense fluid region ($\rho_r > 0.35$). Chapter 10 provides the theoretical basis for molecular-level property modeling and describes the limitations of such approaches.

The second approach, traditionally led by chemical engineers, is to propose empirical expressions that are fit to experimental data. Two general empirical formulations have seen sustained success. The first relies on the robust characteristics of the van der Waals formulation using a cubic equation in volume to predict both volumetric *PVT* and vapor-liquid coexistence properties with two to three fitted parameters for both pure components and mixtures. The second uses a truncated virial format with modifications and numerous adjustable parameters to provide accurate *PVT* property predictions. For supercritical fluid properties, both semi-theoretical and empirical approaches have been extensively used to model real systems [see Bruno and Ely (1991) and Brennecke and Eckert (1989)].

At best, these empirical EOSs represent average behavior and are not based on the exact nature of molecular interactions. Therefore, one should not expect them to give accurate predictions over the entire *PVT* space. A judicious choice of the parameters could extend the range of utility of the EOS—usually, the greater the number of fitted parameters used, the more accurate the EOS is. However, given the limited amount of reliable experimental data available (particularly for supercritical mixtures) to fit the parameters used in the EOS, it is wise to use as few adjustable parameters as possible.

As discussed in Section 9.2, simple and straightforward mixing rules for mixtures are advantageous. Again, one wants to minimize the number of fitted parameters used to account for binary interactions, as binary data in the supercritical region for many of the hydrocarbon-gas-water mixtures of importance in SCWO are quite limited.

The empirical approach can yield reasonably accurate *PVT* predictions by the use of creatively placed, although somewhat arbitrary, parameters that correlate the observed behavior of the fluid. Typically, these empirical EOSs use fundamental phase stability criteria at the critical point in an attempt to specify pure component parameters. At this point the spinodal stability locus and the vapor-liquid equilibrium binodal coexistence locus meet, requiring that

$$\left(\frac{\partial P}{\partial V}\right)_{T_c} = 0 \tag{8-36}$$

$$\left(\frac{\partial^2 P}{\partial V^2}\right)_{T_c} = 0 \tag{8-37}$$

Evaluation of Eqs. (8-36) and (8-37) specifies two parameters contained in the EOS. Typically, these parameters are the molecular volume (or repulsive interaction) term b and attraction interaction term a. The critical temperature T_c and pressure P_c are the properties most accurately known and, therefore, are most commonly used for the evaluation of a and b. As a result, the critical volume V_c and critical compressibility Z_c are determined by the form of the EOS using specified values of T_c and P_c. Example 7.2 illustrates these methods for specifying parameters for the van der Waals equation of state. More complicated equations of state with additional parameters typically use additional experimental PVT data, in conjunction with the stability criteria, to determine the additional constants. While this approach of fitting parameters is largely empirical, it does incorporate a major result of the Theorem of Corresponding States in that pure fluid properties at the critical point are used for scaling variables and for determining empirical constants. Thus, in a completely analogous fashion we are able to predict the PVT properties of a pure fluid by knowing only its critical properties (T_c, P_c, and V_c).

The molecular volume term b used in most EOSs is qualitatively similar to the hard-sphere excluded volume b^{HS} described in Section 10.6. For a hard-sphere fluid, the values of b and b^{HS} are equal:

$$b = b^{HS} = \frac{2\pi\sigma^3 N_a}{3} = 4b_o \quad (\textit{hard–sphere fluid}) \tag{8-38}$$

where b_o = volume of a hard-sphere molecule of diameter σ. For molecules with different structures and interactions, it is difficult to estimate the value of b *a priori*. By using stability criteria instead, one obtains a fitted b from the critical data (T_c and P_c) and avoids the difficult calculation of a geometrically correct excluded volume.

Cubic Equations of State. The ideal-gas law, which has both an empirical and theoretical basis, works well for fluids at low reduced densities ($\rho_r < 0.01$):

$$P = P^{ig} = \frac{NRT}{\underline{V}} = \frac{RT}{V} \tag{8-39}$$

and was not improved upon until about 125 years ago when van der Waals proposed his two-parameter cubic equation of state:

$$P = \frac{RT}{V-b} - \frac{a}{V^2} \tag{8-40}$$

The vdW EOS can predict vapor-liquid equilibrium (VLE) for pure compounds and reduces to the ideal-gas equation at low densities. However, its PVT predictions get considerably worse as the density of the fluid increases. Since this equation has only two parameters, both of which are fit to the stability criteria operative at the critical point, it predicts a universal value of the critical compressibility factor ($Z_c = 0.375$) for all fluids. Introducing a third physical constant, analogous to the introduction of the acentric factor

Table 8.1 Commonly Used Empirical Equations of State for Pure Fluids

Cubic-type

Redlich-Kwong (RK) EOS [Redlich and Kwong (1949)] [2 parameters]

$$P = \frac{RT}{V-b} - \frac{a}{\sqrt{T}\,V(V+b)}$$

Redlich-Kwong-Soave (RKS) EOS [Soave (1972)] [2 parameters]

$$P = \frac{RT}{V-b} - \frac{a}{V(V+b)} \text{ where } a = a_c f(\omega, T_r) = a_c(1 + k(1 - \sqrt{T_r}))^2$$

Peng-Robinson (PR) EOS [Peng and Robinson (1976)] [2 parameters]

$$P = \frac{RT}{V-b} - \frac{a}{V(V+b)+b(V-b)} \text{ where } a = a_c f(\omega, T_r) = a_c(1 + \kappa(1 - \sqrt{T_r}))^2$$

Virial-type

Benedict-Webb-Rubin (BWR) EOS [Benedict *et al.* (1940)] [8 parameters]

$$P = \frac{RT}{V} + \left(B_o RT - A_o - \frac{C_o}{T^2}\right)\frac{1}{V^2} + \frac{(bRT-a)}{V^3} + \frac{a\alpha}{V^6} + \frac{c(1+\gamma/V^2)}{V^3T^2}\exp\left[-\gamma/V^2\right]$$

Modified Benedict-Webb-Rubin-Starling (BWRS) EOS [Starling and Han (1972) and Starling (1971, 1973)] [11 parameters]

$$P = \frac{RT}{V} + \left(B_o RT - A_o - \frac{C_o}{T^2} + \frac{D_o}{T^3} - \frac{E_o}{T^4}\right)\frac{1}{V^2} + \left(bRT - a - \frac{d}{T}\right)\frac{1}{V^3}$$
$$+ \alpha\left(a + \frac{d}{T}\right)\frac{1}{V^6} + \frac{c}{T^2V^3}\left(1 + \frac{\gamma}{V^2}\right)\exp\left(-\frac{\gamma}{V^2}\right)$$

Martin-Hou (in reduced form) [Martin (1967)] [21 parameters]

$$P_r = \frac{1}{Z_c}\left(\frac{T_r}{(V_r - b)} + \sum_{i=2}^{5}\frac{f_i(T_r)}{(V_r - b)^i} + f_6(T_r)\, e^{-aV_r} + f_7(T_r)\, e^{-2aV_r}\right)$$

where $f_i(T_r) = A_i + B_iT_r + C_ie^{-KT_r}$, for $i = 2, \ldots, 7$

in the application of corresponding states concepts, would alleviate the constraint of a fixed Z_c. Moreover, the vdW EOS does not give accurate estimates of derived thermodynamic properties, e.g., enthalpies and entropies, because it uses a constant value of the parameter a.

Numerous modifications of the vdW EOS have been published, and some are presented in Table 8.1 [for example, see Abbott (1973, 1979), Walas (1985) or Reid *et al.* (1987)]. Popular pressure-explicit cubic forms for chemical engineering applications are due to Redlich and Kwong (1949) (RK EOS), Soave (1972) (RKS EOS) and Peng and Robinson (1976) (PR EOS). All of these EOSs give far better predictions than the original vdW EOS over a wide range of densities.

Most of the EOSs used in engineering practice today are explicit in pressure and are of the form:

$$P = f(T, \underline{V}, N) \tag{8-41}$$

like those given in Table 8.1 and the van der Waals EOS [Eq. (8-40)]. As there are literally hundreds of different equations of state, each with its own adherents [see Walas (1985), for example], it is not possible to comment even briefly on all but the more common. We have selected several representative examples of cubic and virial type EOSs to illustrate the important features of each type.

In general, cubic equations have multiple volume or density roots and thus can be used to model both vapor and liquid phases (see Appendix E). After van der Waals, the first given is the well-known *Redlich-Kwong* (1949) (RK EOS) *relation*, as it is the precursor of many later versions. Written in intensive form, it is

$$P = \frac{RT}{V-b} - \frac{a}{\sqrt{T}\, V(V+b)} \tag{8-42}$$

where a and b are constants specific for a given material [to convert Eq. (8-42) to extensive form substitute $V = \underline{V}/N$]. These parameters may be generalized by applying the stability criteria [Eqs. (8-36) and (8-37)] at the critical point to yield

$$a = \frac{\Omega_a R^2 T_c^{2.5}}{P_c} \tag{8-43}$$

$$b = \frac{\Omega_b R T_c}{P_c} \tag{8-44}$$

where

$$\Omega_a = [(9)(2^{1/3} - 1)]^{-1} = 0.42748\ldots \quad \text{and} \quad \Omega_b = \frac{2^{1/3} - 1}{3} = 0.08664\ldots$$

Soave (1972) modified the Redlich-Kwong EOS by introducing a dependence on the Pitzer acentric factor ω and changing the temperature dependence of the attractive term (see Table 8.1):

$$a = a_c f(\omega, T_r) = a_c(1 + k(1 - \sqrt{T_r}\,))^2 \tag{8-45}$$

where a_c is equivalent to the RKEOS a but with T_c raised to 2.0 power in Eq. (8-43) and

$$k = 0.48508 + 1.55171\,\omega - 0.15613\,\omega^2$$

These modifications greatly improved the accuracy of the RK EOS for predicting liquid-vapor equilibria. Thus, the Redlich-Kwong-Soave (RKS) EOS is selected for many chemical engineering applications.

A somewhat more accurate cubic EOS that can be applied to both the liquid and vapor regions was given by Peng and Robinson (1976) (PR EOS):

$$P = \frac{RT}{V-b} - \frac{a(\omega, T_r)}{V(V+b)+b(V-b)} \tag{8-46}$$

The parameters a and b in the PR EOS, like those for the RK EOS are defined using critical point stability criteria. And like the RKS EOS, the attractive a parameter depends on both the acentric factor and reduced temperature:

$$a(\omega, T_r) = a_c \alpha(\omega, T_r) \tag{8-47}$$

with

$$a_c = \frac{0.45724 R^2 T_c^2}{P_c} \tag{8-48}$$

$$\alpha(\omega, T_r) = (1 + \kappa(1 - \sqrt{T_r}))^2 \tag{8-49}$$

$$\kappa = 0.37464 + 1.54226\omega - 0.26992\omega^2 \tag{8-50}$$

$$b = \frac{0.07780 RT_c}{P_c} \tag{8-51}$$

The Peng-Robinson relation is used as an example of an equation of state to calculate thermodynamic functions for pure materials (see Section 8.5), and also, in Chapters 9 and 15 to treat mixtures and phase equilibria.

The RK, RKS and PR EOSs predict universal values of critical compressibility like the vdW EOS (i.e., $Z_c = 0.333$ for the RK and RKS EOSs, $Z_c = 0.307$ for the PR EOS and $Z_c = 0.375$ for the vdW EOS; whereas in Table 8.1, $0.22 < Z_c < 0.29$). Furthermore, the RKS and PR EOSs treat the attractive interaction parameter a as a function of temperature and hence, predict more accurate vapor pressures as well as other derived properties such as enthalpy and entropy.

Volume translated cubic Equation of State Martin (1967 and 1979) reviewed the characteristics of 2-parameter cubic EOSs. Table 8.2 gives a general formulation for a cubic EOS. Martin also introduced the idea that density predictions near the critical point and in the compressed liquid region can be greatly improved by using an adjustable ***translation parameter*** to correct the predicted volume or density.

Table 8.2 Generalized Cubic EOS [Martin, 1979)]

$$P = \frac{RT}{V-t} - \frac{\alpha(T)}{(V+\beta-t)(V+\gamma-t)}$$

Type	$\alpha(T)$	t	β	γ	Z_c
van der Waals (vdW) (1873)	a	b	b	b	0.375
Redlich-Kwong (RK) (1949)	$a/T^{1/2}$	b	b	$2t = 2\beta = 2b$	0.333
Soave (RKS) (1972)	$a_c f(\omega, T_r)$	b	b	$2t = 2\beta = 2b$	0.333
Peng-Robinson (PR) (1976)	$a_c\alpha(\omega, T_r)$	b	$(2+\sqrt{2})b$	$(2-\sqrt{2})b$	0.307
"Best Fit" Martin[1] (1979)	$a(T_r)$	t	$t+c$	$t+c$	0.25

[1] For the "Best Fit" Martin EOS a third fitted parameter c is used:

$$P = \frac{RT}{V-t} - \frac{a}{(V+c)^2}$$

and t could have some density dependence; thus $t = f[\bar{Z}_c, \rho_r]$, $c = V_c/8\bar{Z}_c - t$, $a = 27/64\,(R^2T_c^2/P_c)\,T_r^n$ ($n = 0.5$ to ~1.0), and $\bar{Z}_c$ is the experimental critical compressibility factor.

Using Martin's generalized formulation for cubic equations of state, the 2-parameter vdW EOS can be interpreted as correcting the volume in the repulsive term $[RT/(V-t)]$ to account for the finite size of the fluid molecules. This excluded volume essentially "translates" the volume axis by subtracting $t = b$ from V where the value of b (along with a) is set by satisfying criteria at the critical point which involve isothermal derivatives [see Eqs. (8-36) and (8-37)]. In such 2-parameter EOSs, however, there is no guarantee that the predicted V_c will match the experimental value. With the vdW EOS Z_c fixed at 0.375, we observe significant errors in V_c predictions as Z_c values for real materials typically range from 0.23 to 0.31. This problem persists for the other 2-parameter cubic EOSs given in Table 8.2, although to a lesser degree as EOS-determined Z_cs are closer to reality.

Martin (1979) improved volume (density) predictions by adding a third parameter to his "best fit" cubic EOS. Along with the standard derivative stability criteria at the critical point, Martin uses the experimental value of Z_c to literally translate the volume axis (at constant T_c and P_c) until a better match between predicted and experimental densities are obtained.

Further refinements to Martin's translation approach have been implemented by Peneloux *et al.* (1982), Soave (1984), Chou and Prausnitz (1989) and Mathias *et al.* (1989). In these formulations, a two-step procedure is followed where the final predicted volume is given by a linear translation equation:

$$V = V^{UT} + t \qquad (8\text{-}52)$$

where V^{UT} is the volume estimated from the untranslated EOS and t is the volume translation which can be positive or negative. Thus, a second equation that gives V^{UT} as a function of T and P is required:

$$V^{UT} = f(P, T) \tag{8-53}$$

The volume translation equation (8-52) can be generalized:

$$V = f(V^{UT}(P, T), t) \tag{8-54}$$

The RK, RKS, and PR EOSs have been used for V^{UT} in the four studies cited above. With this two-step approach, the a and b parameters are identical to those for the original EOS, but the translation parameter t can be a complex function of several pure component properties and it is not restricted to a constant value. Thus,

$$t = f(P_c, T_c, V_c, \text{and other } PVT \text{ properties and derivatives}) \tag{8-55}$$

The value of t can be selected to minimize differences between predicted and experimental densities. Frequently, the empirical equation for t will result in a match to the experimental value of V_c at the critical point.

Hard-Sphere Equation of State Another class of EOSs incorporates knowledge of the specific molecular interactions with additional fitted parameters. A good example of these are vdW-type EOSs that model the repulsive contribution with the hard-sphere relation given in Section 10.8 and add empirical perturbation terms to model attractive interactions in real gases and fluids. There is reason to believe that the hard-sphere model, which is theoretically-based, more accurately captures the repulsive contribution due to excluded volume effects. Unfortunately, the attractive contribution has a less quantitative connection to rigorous theory and must be incorporated on a somewhat *ad hoc* empirical basis. For example, Vera and Prausnitz (1972) developed an EOS using the Carnahan-Starling result for the repulsive effects and the empirical Strobridge-Gosman EOS for attractive effects. Johnston and Eckert (1981) used a simple first-order perturbation term in their Carnahan-Starling van der Waals (CSvdW) EOS in an attempt to model the dense supercritical region:

$$P = \frac{RT}{V}\left(\frac{V^3 + b_oV^2 + b_o^2V - b_o^3}{(V - b_o)^3}\right) - \frac{a}{V^2} \tag{8-56}$$

where b_o is given in Eq. (8-38). However, the CSvdW EOS did not accurately model the highly compressible critical region. As a result, Johnston *et al.* (1982) introduced the Augmented van der Waals (AvdW) EOS which had a second-order perturbation term for more accurate modeling of the critical region.

In a recent development, the volume translation approach was applied to a Carnahan-Starling hard-sphere vdW model to produce a relatively simple EOS for high-temperature, high-pressure aqueous solutions near and above their critical conditions [Kutney *et al.* (1996)]. The resulting 6-parameter Hard-Sphere Volume-Translated vdW (HSVTvdW) EOS is defined by:

1. *Untranslated EOS to give* $P = f[V^{UT}, T]$

$$P = \frac{RT}{V^{UT}}\left(\frac{(V^{UT})^3 + b(V^{UT})^2 + b^2V^{UT} - b^3}{(V^{UT} - b)^3}\right) - \frac{a_c\alpha}{(V^{UT} + 2b)^2} \quad (8\text{-}57)$$

$$a_c = 0.4496088\frac{R^2T_c^2}{P_c} \quad (8\text{-}58)$$

$$b = 0.0245878\frac{RT_c}{P_c} \quad (8\text{-}59)$$

$$\alpha = \exp\left(\left[1 - T_r\right]\left[\frac{\alpha_A}{T_r^{0.93}} + \alpha_B T_r^{0.75}\right]\right) \quad (8\text{-}60)$$

$$Z_c^{UT} = \frac{P_c V_c^{UT}}{RT_c} = 0.3183919 \quad (8\text{-}61)$$

2. *Volume translation equation*

$$V = V^{UT} + t + (V_c - V_c^{UT} - t)\left(\frac{8\,V_r^{UT}T_r^{-9/2}}{(V_r^{UT})^3 + \frac{13}{2}\,T_r^{-13/2} + \frac{1}{2}}\right) \quad (8\text{-}62)$$

where V_r^{UT} is the untranslated reduced volume. The parameters a_c and b are determined by the critical point derivative stability criteria. The α_A and α_B coefficients for $\alpha = f(T_r)$ in Eq. (8-60) are selected to match the experimental vapor pressure $P_{vp} = f(T)$ while t is simultaneously fitted to match liquid and vapor volumes along the coexistence curve.

Virial-type Equation of State: An important EOS that has a rigorous basis in molecular theory is the virial equation which can be expressed as a polynomial in volume (or density):

$$P = \frac{RT}{V} + \frac{BRT}{V^2} + \frac{CRT}{V^3} + \ldots = RT\,[\rho + B\rho^2 + C\rho^3 + \ldots] \quad (8\text{-}63)$$

or equivalently in terms of the compressibility factor

$$Z = \frac{PV}{RT} = 1 + \frac{B}{V} + \frac{C}{V^2} + \ldots = 1 + B\rho + C\rho^2 + \ldots \quad (8\text{-}64)$$

Equation (8-64) can be recast as a polynomial expansion in pressure:

$$Z = 1 + B'P + C'P^2 + D'P^3 + \ldots \quad (8\text{-}65)$$

where

$$B' = B/RT\,; \qquad C' = \frac{C - B^2}{(RT)^2}\,; \qquad D' = \frac{D - 3BC - 2B^3}{(RT)^3}$$

Equation (8-63) may be derived from statistical mechanics (see Section 10.4), and the coefficients B, C, . . . are termed the second, third, . . . virial coefficients. If one possessed a unified model for calculating intermolecular energies as a function of molecular separation distance and density, then B, C, . . . could be calculated rigorously. Lacking this, virial coefficients are normally correlated by data or some form of the Theorem of Corresponding States noted earlier.

The virial equation of state can be truncated, for example retaining only the second virial coefficient term:

$$Z = \frac{PV}{RT} = 1 + \frac{BP}{RT} \tag{8-66}$$

$$P = \frac{RT}{V - B} \tag{8-67}$$

Equation (8-66) is obtained from Eq. (8-64) if third and higher virial coefficients C, D, . . . are set equal to zero. This is equivalent to dropping all terms in P^n where $n > 1$ in Eq. (8-65).

Equation (8-66) is quite easy to use, but it is limited to gases and its range of applicability is such that the reduced density should not exceed 0.5. Various techniques have been suggested to calculate the second virial coefficient, B. The method developed by Abbott (1987) is described here; it is similar to the form shown in Eq. (8-34).

$$B = \frac{RT_c}{P_c}(B^{(0)} + \omega B^{(1)}) \tag{8-68}$$

where

$$B^{(0)} = 0.083 - \frac{0.422}{T_r^{1.6}} \quad \text{and} \quad B^{(1)} = 0.139 - \frac{0.172}{T_r^{4.2}}$$

To achieve higher accuracy, some researchers have proposed equations with different functional forms and/or additional adjustable parameters. Among these are the ones proposed by Benedict, Webb and Rubin (1940), Starling and Han (1972), Starling (1973), and Martin and Stanford (1974). These equations typically combine a truncated virial format with exponential terms that partially compensate for the effects of higher-order virial terms for predicting high density behavior and, consequently, have many more parameters than cubic EOSs (see Table 8.1). Typically, these parameters have been evaluated only for the specific classes of compounds for which these EOSs were originally developed. In particular, the BWR and Starling (BWRS) EOSs are used for modeling the *PVT* behavior of light hydrocarbons and gases in petroleum and natural gas applications. The Martin formulation given in Table 8.1 has been extensively applied to chlorofluorocarbons.

Even higher precision is achieved in an EOS by fitting more parameters, for example to model water, the Haar, Gallagher and Kell EOS (1984) with eighty parameters is frequently used to achieve high levels of accuracy. Errors of less than 0.1% are obtained by using multiple parameter, non-linear regression methods to fit a comprehensive set of data. Such EOSs can be used to generate smoothed "data" for important pure compounds where sufficient experimental data (in both quantity and quality) exist.

Polymer fluids Equation of State: Polymer melts and liquids are more complex than simple fluids in that intramolecular chain folding can influence properties. Even with these complexities, several volumetric EOSs are available that correlate observed $P\underline{V}TN$ behavior reasonably well. In all cases, even for models that result from a rigorous statistical mechanical treatment, experimental data are needed to fit parameters.

The most important models in use today fall into two categories; purely empirical and semi-theoretical. The Tait equation is perhaps the most widely used empirical model:

$$\text{(Tait)}\quad V = [V(P \rightarrow 0, T)]\,[1 - 0.0894 \ln (1 + P/B(T))] \tag{8-69}$$

On the theoretical side, the Simka-Somcynsky (S-S), Hartmann-Haque (H-H) and Sanchez-Lacombe (S-L) models are well tested [see Hartmann and Haque (1985) and Pottiger and Laurence (1984) for further discussion]. Typically, they are written in reduced form as:

$$\text{(Simka-Somcynsky)}\quad \ln V_R = T_R^{3/2} \quad (\text{at } P = 0)$$

$$\text{(Hartmann-Haque)}\quad (P_R V_R)^5 = T_R^{3/2} - \ln V_R \tag{8-70}$$

$$\text{(Sanchez-Lacombe)}\quad V_R^{-2} + P_R + T_R[\ln (1 - V^{-1}) + (1 - 1/r)/V] = 0$$

where $V_R = V/V^*$, $T_R = T/T^*$, and $P_R = P/P^*$ and r is the chain length (number of monomer units) of the polymer. Although these are reduced parameters they are not fundamental nor as generally obtained as those that follow from the Theorem of Corresponding States. Here, the scaling parameters V^*, T^*, and P^* are not equal to V_c, T_c, and P_c, and they must be regressed to actual volumetric $P\underline{V}TN$ data over ranges of interest to investigators. The Tait, S-S, H-H and S-L EOSs for polymers are capable of matching specific volumes to accuracies of ± 0.001 cm^3/g. One advantage of the Sanchez- Lacombe model is that it has been successfully applied to polymer-solvent and polymer-polymer mixtures (see Section 15.7).

The $P\underline{V}TN$ equations of state presented in this section should be considered only as illustrative examples. There are obviously many others that represent important contributions towards predicting and correlating the volumetric behavior of pure fluids. In general, the specific requirements of the engineering application will determine the EOS that is ultimately selected. Precision, the ability to predict properties of liquids and gases, the availability of data to specify empirical constants, and the adaptability to mixtures are criteria that often influence the choice of a particular EOS. For example, a relatively simple cubic EOS, like the RKS or PR, with only a few parameters would be preferred for predicting vapor-liquid equilibrium in mixtures (see Chapters 9 and 15) while a multiple parameter, extended virial equation EOS, such as the BWR, would be more appropriate for high precision power cycle calculations involving pure working fluids (see Chapter 14). Readers interested in further discussion of EOS selection should consult Reid *et al.* (1987).

While *PVT* and *PVTN* equations of state are of value for predicting and estimating volumetric properties at given temperatures and pressures, they are most often used in chemical engineering applications to calculate *isothermal* changes in thermodynamic properties, such as enthalpy or entropy, or for phase equilibrium studies. Their application to estimating changes in thermodynamic properties is discussed in Section 8.5.

Example 8.1

A well-insulated vessel is divided into two compartments by a partition. The volume of each compartment is 0.1 m^3. One compartment initially contains 400 moles of argon at 294 K, and the other compartment is initially evacuated. The partition is then removed, and the gas is allowed to equilibrate. What is the final temperature?

Note: Under these conditions, argon is not an ideal gas. Assume that the van der Waals equation of state [Eq. (8-31)] is valid with $a = 0.1362$ J m^3/mol^2 and $b = 3.215 \times 10^{-5}$ m^3/mol. Assume that $C_v = 12.56$ J/mol K is independent of temperature and pressure and that the mass of the walls can be neglected.

Solution

If we choose the gas as the system, the process occurs at constant energy. Furthermore, we know the initial and final volume of the system. Since we are concerned with a one-component simple system, we can evaluate the change in any property if the changes of two other properties are known. Since the actual path between the initial and the final conditions is clearly irreversible we need to construct a quasi-static path. If the process were carried out reversibly, we can relate the temperature variation to the volume and energy by

$$dT = \left(\frac{\partial T}{\partial V}\right)_U dV + \left(\frac{\partial T}{\partial U}\right)_V dU \tag{8-71}$$

Equation (8-71) is, of course, an exact differential. If we choose a reversible path in which U is constant, then it can be integrated:

$$T_f - T_i = \int_{V_i}^{V_f} \left(\frac{\partial T}{\partial V}\right)_U dV \tag{8-72}$$

The problem can be solved by expressing $(\partial T/\partial V)_U$ in terms of the available data. Thus,

$$\left(\frac{\partial T}{\partial V}\right)_U = -\frac{(\partial U/\partial V)_T}{(\partial U/\partial T)_V} = \frac{-T(\partial S/\partial V)_T + P}{T(\partial S/\partial T)_V}$$

We can evaluate $(\partial S/\partial V)_T$ in terms of *PVT* derivatives by applying a Maxwell reciprocity relation to the first Legendre transform of U [i.e., $y^{(1)} = A$] in intensive form, $dy^{(1)} = dA = -SdT - PdV$. Thus,

$$\left(\frac{\partial S}{\partial V}\right)_T = \left(\frac{\partial P}{\partial T}\right)_V$$

so that $(\partial T/\partial V)_U$ becomes

$$\left(\frac{\partial T}{\partial V}\right)_U = \frac{-T(\partial P/\partial T)_V + P}{C_v} \tag{8-73}$$

From the van der Waals equation, we find that

$$\left(\frac{\partial P}{\partial T}\right)_V = \frac{R}{V-b}$$

so that Eq. (8-73) reduces to

$$\left(\frac{\partial T}{\partial V}\right)_U = \frac{-a}{C_v V^2} \tag{8-74}$$

Substituting Eq. (8-74) into Eq. (8-72), we obtain

$$T_f = T_i + \frac{a}{C_v}\left(\frac{1}{V_f} - \frac{1}{V_i}\right)$$

and with $V_f = 2V_c$

$$T_f = T_i - \frac{a}{2C_v V_i} = 294 - \frac{0.1362}{(2)(12.56)(0.1/400)} = 272 \text{ K}$$

Example 8.2

A one-component gas at 300 K and 1 bar has a heat capacity C_v = 30 J/mol K. The gas is described by the following equation of state:

$$PV = aT^2 \quad a = 5 \times 10^{-3} \text{ J/mol K}^2 \tag{8-75}$$

(a) Calculate the heat capacity C_V at 300 K and 5 and 50 bar.

(b) Calculate $(\partial P/\partial N)_{T,\underline{V}}$ for 3 moles of the above mentioned gas at 800 K and 9 bar.

(c) A stream of the above mentioned gas can be cooled either by expansion through a throttling valve or in a power turbine. To measure the relative efficiency of these two processes, the term (κ_H/κ_S) is often quoted where the isenthalpic (or Joule-Thompson) and isentropic coefficients are defined as:

$$\kappa_H \equiv (\partial T/\partial P)_H \quad \text{and} \quad \kappa_S \equiv (\partial T/\partial P)_S$$

Calculate κ_H/κ_S at 600 K and 3 bar.

Solution

(a) Since $C_V = (\partial U/\partial T)_V$

$$\left(\frac{\partial C_v}{\partial V}\right)_T = \left[\frac{\partial}{\partial V}\left(\frac{\partial U}{\partial T}\right)_V\right]_T = U_{TV} = U_{VT} = \left[\frac{\partial}{\partial T}\left(\frac{\partial U}{\partial V}\right)_T\right]_V$$

using Maxwell's relation. Using the Fundamental Equation for $U(S, V)$ and the results of Example 8.1:

$$\left(\frac{\partial U}{\partial V}\right)_T = T\left(\frac{\partial S}{\partial V}\right)_T - P = T\left(\frac{\partial P}{\partial T}\right)_V - P$$

Therefore, with $PV = aT^2$

$$\left(\frac{\partial C_v}{\partial V}\right)_T = T\left(\frac{\partial^2 P}{\partial T^2}\right)_V = \frac{2aT}{V} = \frac{2P}{T}$$

Now we can integrate isothermally from 1 bar to P,

$$C_v\,(T, P) = C_v\,(T, 1\text{ bar}) - 2a\,T \ln P$$

With $C_V = 30$ J/mol K at 300 K, 1 bar

$$C_v\,(300\text{ K}, 50\text{ bar}) = 30 - 3\ln 50 = 18.26\text{ J/mol K}$$

$$C_v\,(300\text{ K}, 5\text{ bar}) = 30 - 3\ln 5 = 25.17\text{ J/mol K}$$

(b) With $P\underline{V} = aNT^2$,

$$\left(\frac{\partial P}{\partial N}\right)_{T,\underline{V}} = \frac{aT^2}{\underline{V}} = \frac{aT^2}{\dfrac{aNT^2}{P}} = \frac{P}{N} = \frac{9\text{ bar}}{3\text{ mol}} = 3\text{ bar/mol}$$

(c) Using the triple product, the definition of $dH = TdS + VdP$, and one Maxwell relation from $y^{(2)} = G$ and $dy^{(2)} = dG = -SdT + VdP$ and $(\partial S/\partial P)_T = -(\partial V/\partial T)_P$, we get:

$$\kappa_H/\kappa_S = \frac{(\partial P/\partial S)_T\,(\partial S/\partial T)_P}{(\partial P/\partial H)_T\,(\partial H/\partial T)_P} = \frac{(\partial H/\partial P)_T}{T(\partial S/\partial P)_T} = \frac{\left[T\left(\dfrac{\partial V}{\partial T}\right)_P - V\right]}{T\left(\dfrac{\partial V}{\partial T}\right)_P}$$

$$\kappa_H/\kappa_S = \frac{\left[\dfrac{2aT^2}{P} - \dfrac{aT^2}{P}\right]}{\left[\dfrac{2aT^2}{P}\right]} = \frac{1}{2}$$

κ_H/κ_S is independent of T and P!

8.4 Ideal-Gas State Heat Capacities

In Section 8.1 we illustrated the important role that the heat capacity at constant pressure C_p has in determining derived property changes when the Gibbs free energy formulation of the Fundamental Equation is used. Unfortunately, under real gas conditions, both C_p *and* C_v are functions of T and P (or T and V) and it is frequently inconvenient to account for the changes in C_p (or C_v) that occur along the path of the real process from its initial to final state. However, if one operates at an attenuated, low density condition, that is, where $P \rightarrow 0$ or $V \rightarrow \infty$, ideal gas behavior prevails and C_p and C_v are no longer functions of P or V. Because derived properties are not dependent on path, we can create alternate paths between specified initial and final states to evaluate changes in derived properties like H or S. A portion of these alternative paths frequently can be conveniently carried out in an ideal-gas state. As we will demonstrate in the next section, the flexibility of operating under ideal-gas state conditions will greatly facilitate the calculation of non-isothermal property changes.

From our earlier introduction of ideal gas properties in Section 3.5, ideal-gas state heat capacities are defined as follows:

$$C_v^o = \left(\frac{\partial U^o}{\partial T}\right)_V = T\left(\frac{\partial S^o}{\partial T}\right)_V \tag{8-76}$$

$$C_p^o = \left(\frac{\partial H^o}{\partial T}\right)_P = T\left(\frac{\partial S^o}{\partial T}\right)_P \tag{8-77}$$

where the ^{o} superscript on U, H, and S indicates these properties are to be evaluated in an ideal-gas state. Thus, in such a state, $PV^o = RT$ and

$$H^o = U^o + PV^o = U^o + RT \tag{8-78}$$

and then

$$\frac{dH^o}{dT} = \frac{dU^o}{dT} + R = C_p^o = C_v^o + R \tag{8-79}$$

so that the restrictions of constant volume on $(\partial U^o/\partial T)$ or constant pressure on $(\partial H^o/\partial T)$ are not needed. This is true since U^o and H^o are functions only of temperature. The restriction of constant V or P on entropy in Eqs. (8-76) and (8-77) is, however, still necessary since the ideal-gas entropy is a function of temperature and volume (or pressure). Our earlier assertion that neither C_p^o nor C_v^o is dependent on volume or pressure, but only on temperature, may now be proved in a formal sense, as for example, with C_p^o,

$$\left(\frac{\partial C_p^o}{\partial P}\right)_T = \frac{\partial^2 H^o}{\partial P\,\partial T} = \frac{\partial}{\partial T}\left(\frac{\partial H^o}{\partial P}\right)_T = \frac{\partial}{\partial T}\left[V^o - T\frac{\partial V^o}{\partial T}\right]_P$$
$$= \frac{\partial}{\partial T}\left[\frac{RT}{P} - T\frac{\partial}{\partial T}\left(\frac{RT}{P}\right)\right]_P = \frac{\partial}{\partial T}(0) = 0 \tag{8-80}$$

where $(\partial H/\partial P)_T$ was related to V and T by methods shown in Chapter 5 [see Example 5.3, Eq. (5-70) ff].

While C_v^o and C_p^o refer to an ideal-gas state at (T, V^o), since they are dependent only on temperature, they could be expressed as C_v^* and C_p^* indicating a *real* state at T but at an attenuated condition where the gas density ρ and pressure P go to 0 or equivalently as the specific volume V goes to infinity. Although the physical interpretation of these states is quite different, numerical values of C_v^o and C_v^* or C_p^o and C_p^* are identical.

In most instances, experimental values of C_v or C_p have been obtained at low pressures ($P \leq 1$ bar) and, for engineering purposes, can be equated to C_v^* or C_p^*. However, most heat-capacity values available in the literature are not experimentally determined but have been obtained from theory and spectroscopic data. The essence of the method lies in expressing the internal energy of an ideal gas as a function of temperature; then, by differentiation, C_v^o is obtained.

Following common engineering practice, we will set $C_v^o = C_v^*$ and $C_p^o = C_p^*$ and use a polynominal power series in T to express the temperature dependence of C_p^o and C_v^o, for example:

$$C_v^o = f(T) = a + bT + cT^2 + dT^3 + \dots \tag{8-81}$$

$$C_p^o = C_v^o + R = a + R + bT + cT^2 + dT^3 + \dots \tag{8-82}$$

where the coefficients a, b, c, d, etc. are constants and are obtained empirically from calorimetric data, estimated using molecular group contribution techniques (see Section 13.3), or, as is most often the case, they are determined from theory utilizing spectroscopic data. The essence of this rigorous theoretical approach lies in expressing the internal energy of an ideal gas as a function of temperature; then by differentiation, C_v^o is obtained which also produces C_p^o because it is equal to $C_v^o + R$ for an ideal gas.

First, we rule out intermolecular energy since, by choosing an ideal gas, there are no interactions between molecules. Next, energies associated with electron motion, or intranuclear movement, or even energy associated with mass by relativistic considerations may be discarded since these are either constants or temperature dependent only at very high temperatures. For materials used normally in chemical engineering practice, none contribute to C_v^o.

Molecules do, however, move and have kinetic energy; in addition, these energies increase with temperature. By simple kinetic arguments it is easy to show that in each of the three possible directions of motion there is an energy of $mv^2/2$ or $kT/2$. k is Boltzmann's constant or R divided by Avogadro's number. This translational energy thus contributes $3(kT/2)$ per molecule or $3RT/2$ per mole. The translational contribution to C_v^o is then $3R/2$. Molecules having no other energy storage modes would have C_v^o values of about 12.5 J/mol K. Such is the case for monatomic gases such as He, Ar, and Ne at low pressures. Since R is 8.31 J/mol K, C_p^o is then about 12.5 + 8.3 = 20.8 J/mol K.

More complex molecules can also store energy in other ways. Most are obvious from a mechanical visualization. Rotation of the molecule may occur both with the entire molecule and, to some degree, by rotation of certain segments in relation to one another.

Individual atoms may also vibrate in relation to adjacent atoms. As the temperature increases, more energy can be stored in rotational and vibrational modes.

Estimated values of C_v^o are given classically from the equipartition principle by summing the contributions due to translation, rotation, and vibration, assuming that all modes are fully active in storing energy. Each degree of freedom provides 1/2 kT per molecule or 1/2 RT per mole to the internal energy. Thus for a general polyatomic molecule consisting of N atoms with $N > 2$ for *linear* (2-dimensional, rod-like) and *non-linear* (3-dimensional) polyatomic molecules:

	translational	rotational	vibrational		
$C_v^o =$	$3/2\,R$	$+\,2/2\,R$	$+\,(3N-5)\,R$	(*linear*)	(8-83)
$C_v^o =$	$3/2\,R$	$+\,3/2\,R$	$+\,(3N-6)\,R$	(*non–linear*)	(8-84)

Note that each vibrational mode contributes 2(½ RT) to the average energy assuming ideal harmonic behavior.

For the classical approach to be correct, the equipartition principle must be followed with a continuum of energy states accessible. These assumptions are usually satisfactory for translation and rotation, but inadequate for vibrational modes.

Observed heat capacities are always lower than predictions based on the equipartition principle and, in general, C_v^o depends on temperature. In order to capture these effects, we need to use quantum mechanics, particularly to treat the vibrational contribution correctly. We begin by assuming that translational, rotational, and vibrational stated effects can be individually treated and that no other contributions are important as stated earlier. Consequently, contributions from each mode are both separable and additive. Strictly speaking, this isn't always true as rotational-vibrational interactions can occur and there may be internal barriers to rotation; nonetheless, it is a reasonable starting point for our analysis.

Setting the translational contribution to 3/2 kT (or 3/2 RT) is rigorous for any molecule in an attenuated ideal-gas state at virtually all temperatures of interest. The rotational and vibrational contributions can be estimated by assuming that a rigid rotator, harmonic oscillator model adequately describes the energetics. The rotational part is calculated from the rotational partition function defined as follows (see Section 10.1 for a description of partition functions):

$$Q_{rotation} = Q_{rot} = \frac{\pi \,|\underset{\approx}{\mathbf{D}}|}{\sigma}\left[\frac{8\pi^2 kT}{h^2}\right]^{3/2} \tag{8-85}$$

where σ=symmetry number (for linear diatomic molecules, σ=1; for non-linear triatomics such as H_2O σ=2; for a tetrahedral molecule such as CH_4, σ=12; and for an octahedral molecule like SF_6 σ=24) and $|\underset{\approx}{\mathbf{D}}|$ = determinant containing the 9 moments of inertia for the xyz components. $|\underset{\approx}{\mathbf{D}}|$ is given by:

$$|\underset{\approx}{\mathbf{D}}| \equiv \begin{vmatrix} I_{xx} & -I_{xy} & -I_{xz} \\ -I_{xy} & I_{yy} & -I_{yz} \\ -I_{xz} & -I_{yz} & I_{zz} \end{vmatrix} \tag{8-86}$$

with $I_{xx} = \sum m_i (y_i^2 + z_i^2)$, $I_{xy} = \sum m_i x_i y_i$ and so forth where m_i=mass of the ith atom whose coordinates are x_i, y_i, z_i [see Pitzer (1995), Chapter 20 for further discussion].

The rigid rotator assumption is valid for any polyatomic molecule at room temperature or above ($T > 300$ K) and may be still apply to temperatures as low as 100 K. The heat capacity contribution is obtained directly from the rotational partition function which is related directly to the Helmholtz energy as $A_{rotation} = A_{rot} = -RT \ln Q_{rot}$. Thus,

$$C^o_{v,\,rotation} = \left(\frac{\partial U_{rotation}}{\partial T}\right)_V = \frac{\partial}{\partial T}\left(\frac{\partial (A_{rot}/T)}{\partial (1/T)}\right)_V = R\left[T^2 \frac{\partial^2 \ln Q_{rot}}{\partial T^2} + 2T \frac{\partial \ln Q_{rot}}{\partial T}\right] \tag{8-87}$$

For a non-linear, polyatomic molecule there are 3 principal rotational modes each contributing 1/2 kT (or 1/2 RT) to the rotational part of the internal energy; for a linear molecule there are only 2 modes. Thus, in a fully activated and unrestricted state the rotational contribution *per mole* yields:

$$C^o_{v,\,rotation} = 2/2\,R \qquad \textit{for a linear molecule}$$
$$C^o_{v,\,rotation} = 3/2\,R \qquad \textit{for a non-linear molecule} \tag{8-88}$$

To treat the vibrational part, we assume that the energy levels are quantized as,

$$E_s = (s + 1/2)h\nu_i \qquad \text{with } s = 1, 2, 3, \ldots \tag{8-89}$$

where h = Planck's constant and ν = characteristic vibration frequency. The contribution of this energy to the internal vibrational energy is

$$U_{vibration} = U_{vib} = \frac{1}{2} h\nu + \frac{h\nu}{\exp\,[h\nu/kT] - 1} \tag{8-90}$$

Einstein was able to simplify the treatment of vibrational effects to estimate U_{vib} as:

$$U_{vib}\,N_A = \frac{1}{2} h\nu\, N_A + \frac{R\theta}{(\exp\,[\theta/T] - 1)} \tag{8-91}$$

where N_A = Avogadro's number and $\theta = h\nu/k$ = the characteristic temperature for each oscillator. The vibrational contribution can be written in terms of the so-called Einstein function as:

$$\frac{1}{R}\left(\frac{\partial (U_{vib}\,N_A)}{\partial T}\right)_V = \frac{C^o_{v,\,vib}}{R} = \left(\frac{\theta}{T}\right)^2 \frac{\exp\,[\theta/T]}{(\exp\,[\theta/T] - 1)^2} \tag{8-92}$$

Thus, for a linear diatomic molecule in its ideal gas state at low pressure and low density ($P \to 0$, $\rho \to 0$ or $V \to \infty$)

$$\frac{C_v^o}{R} = \underbrace{3/2 + 2/2}_{\text{translation and rotation}} + \underbrace{\left(\frac{\theta}{T}\right)^2 \frac{\exp[\theta/T]}{(\exp[\theta/T]-1)^2}}_{\text{vibration}} \tag{8-93}$$

For polyatomic (non-linear) molecules in a similar attenuated state; we sum over all characteristic frequencies v_i with $\theta_i = hv_i/k$

$$\frac{C_v^o}{R} = 3/2 + 3/2 + \sum_i^{3N-6} \left(\frac{\theta_i}{T}\right)^2 \frac{\exp[\theta_i/T]}{(\exp[\theta_i/T]-1)^2} \tag{8-94}$$

where N = number of atoms and the summation runs over all $3N-6$ vibrational modes i and their characteristic temperatures θ_i at each vibrational frequency v_i.

The vibrational term (or Einstein function) can be expanded as a power series by decomposing the terms in the summation in Eq. (8-94) into two parts:

$$\frac{(\theta_i/T)^2 \exp[\theta_i/T]}{(\exp[\theta_i/T]-1)^2} = f(\theta_i/T)g(\theta_i/T) \tag{8-95}$$

First,

$$f(\theta_i/T) = \frac{(\theta_i/T)^2}{(\exp[\theta_i/T]-1)^2} = \left[1 - \frac{x}{2} + \frac{B_1x^2}{2!} - \frac{B_2x^4}{4!} + \frac{B_3x^6}{6!} - \dots\right]^2 \tag{8-96}$$

where $x = \theta_i/T$ with $x^2 < 4\pi^2$ and $B_1 = 1/6$, $B_2 = 1/30$, $B_3 = 1/42$ (Bernoulli numbers). Values of θ_i typically range from 800 to 6000 K for low molecular weight gases; thus the temperature for saturation with all vibrational modes activated will vary considerably. In addition,

$$g(\theta_i/T) = \exp[\theta_i/T] = 1 + x + \frac{x^2}{2!} + \frac{x^3}{3!} + \frac{x^4}{4!} + \dots \tag{8-97}$$

where again $x = \theta_i/T$. Thus, by combining Eqs. (8-96) and (8-97), the Einstein function is given by:

$$\frac{(\theta_i/T)^2 \exp[\theta_i/T]}{(\exp[\theta_i/T]-1)^2} = g(x)f(x)$$

$$g(x)f(x) = \left[1 + x + \frac{x^2}{2!} + \frac{x^3}{3!} + \dots\right]\left[1 - \frac{x}{2} + \frac{B_1x^3}{2!} + \dots\right]^2 \tag{8-98}$$

which can be expressed as a general power series as $a + bT + cT^2 + \dots$ etc. This suggests that the commonly employed empirical method of correlating C_p^o and C_v^o using such power series expansions in T captures the mathematical formulation encoded by a rigorous quantum statistical mechanics approach.

8.5 Evaluating Changes in Properties Using Departure Functions

In this section, we are concerned with developing methods for determining changes in the derived thermodynamics properties of a pure substance. Because changes in these properties, which include H, S, U, A, and G, are independent of the process path between initial and final states, this gives us flexibility in choosing a convenient path for calculating the change. We can utilize the constitutive relationships discussed in the previous sections to help us make this choice. Although there are a number of ways to represent the physical properties of a pure material, we must generally have access to two constitutive relationships:

1. A PVT or $P\underline{V}TN$ equation of state (EOS) (see Section 8.3)
2. An ideal-gas state heat capacity equation [$C_v^o(T)$ or $C_p^o(T)$] (see Section 8.4)

A straightforward approach to calculating property changes between two well-defined states (T_1, P_1, V_1) and (T_2, P_2, V_2) using these constitutive property representations employs a three-step process outlined in Figure 8.7 in either the P-T or V-T plane:

Step (1): *isothermal expansion at T_1 from $P = P_1$ to 0 (or $V = V_1$* to ∞)
Step (2): *isobaric (or isochoric) heating from T_1 to T_2 in an attenuated ideal gas state* [$P = 0$ (or $V = \infty$)]
Step (3): *isothermal compression at T_2 from $P = 0$ to P_2* (or from $V = \infty$ to V_2)

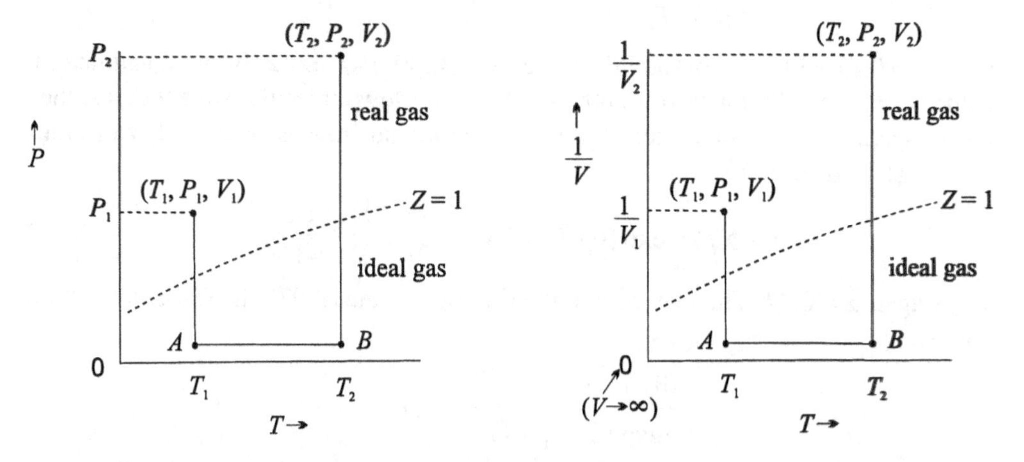

Figure 8.7 Three-step process for evaluation of derived property changes. The reciprocal of V has been plotted to represent the $V \rightarrow \infty$ ($\rho = 0$) singularity more clearly.

In general, the total property change between states 1 and 2, ΔB, is the sum over these three steps:

$$\Delta B = B(T_2, P_2, V_2) - B(T_1, P_1, V_1) = \underbrace{[B(T_1, P \to 0, V \to \infty) - B(T_1, P_1, V_1)]}_{\text{Step (1)}}$$

$$+ \underbrace{[B(T_2, P \to 0, V \to \infty) - B(T_1, P \to 0, V \to \infty)]}_{\text{Step (2)}} + \underbrace{[B(T_2, P_2, V_2) - B(T_2, P \to 0, V \to \infty)]}_{\text{Step (3)}} \quad (8\text{-}99)$$

where B is a general property (H, S, U, A, or G). The conditions in steps (1) and (3) are isothermal while in step (2) the fluid behaves as an ideal gas. Thus:

$$B_A(T_1, P \to 0, V \to \infty) = B^{ideal\ gas}(T_1, P \to 0, V \to \infty)$$
and
$$B_B(T_2, P \to 0, V \to \infty) = B^{ideal\ gas}(T_2, P \to 0, V \to \infty) \quad (8\text{-}100)$$

It is important to remember that only 2 of the 3 intensive variables T, P, and V are independent (recall Postulate I and its ramifications). Thus, the paths shown in the P-T and V-T planes of Figure 8.7 are equivalent. The choice of which variables to use depends on how the properties are represented. Because temperature changes between states 1 and 2 and ideal-gas state heat capacities (C_p^o and C_v^o) are functions of temperature only [see Eqs. (8-81) - (8-82)], we select T as one independent variable, that leaves P or V as the other. The form of the PVT EOS determines whether P or V is preferred. If we have a volume-explicit EOS, e.g., $Z = f(T_r, P_r)$ from corresponding states, then T and P are the variables to use. If we had a pressure-explicit form, such as the RK, RKS or PR EOS with $P = f(T, V)$ then T and V are preferred. Let's consider the following example:

Example 8.3

Develop expressions for ΔS between states 1 and 2 assuming you have analytic forms for (1) a volume-explicit EOS and (2) a pressure-explicit EOS and $C_p^o(T) = a + bT + cT^2 + dT^3$.

Solution

(1) *Volume-explicit formulation*—using the three-step approach of Figure 8.7 in the P-T plane:

$$\Delta S = S_2 - S_1 = S(T_2, P_2) - S(T_1, P_1) \quad (8\text{-}101)$$

$$\Delta S = \underbrace{(S_2 - S_B)}_{step\,3} + \underbrace{(S_B - S_A)}_{step\,2} + \underbrace{(S_A - S_1)}_{step\,1} \quad (8\text{-}102)$$

$$\Delta S = \int_0^{P_2} \left(\frac{\partial S}{\partial P}\right)_{T=T_2} dP + \int_{T_1}^{T_2} \left(\frac{\partial S^o}{\partial T}\right)_P \Bigg|_{ideal\ gas} dT + \int_{P_1}^{0} \left(\frac{\partial S}{\partial P}\right)_{T=T_1} dP \quad (8\text{-}103)$$

Using $y^{(2)} = G$ and a Maxwell relation

$$\left(\frac{\partial S}{\partial P}\right)_T = -\left(\frac{\partial V}{\partial T}\right)_P \quad (8\text{-}104)$$

and from Eq. (8-77) for an ideal gas

$$\left(\frac{\partial S^o}{\partial T}\right)_P = \frac{C_p^o(T)}{T} \tag{8-105}$$

Thus,

$$\Delta S = -\int_0^{P_2}\left(\frac{\partial V}{\partial T}\right)_P\Bigg|_{T=T_2} dP + \int_{T_1}^{T_2}\frac{C_p^o(T)}{T}dT + \int_0^{P_1}\left(\frac{\partial V}{\partial T}\right)_P\Bigg|_{T=T_1} dP \tag{8-106}$$

The second integral can be evaluated analytically if $C_p^o(T)$ is a simple analytic function, but a singularity exists in the integrand of the first and third integrals when $P = 0$. Because the material behaves as an ideal gas as $P \to 0$, it is possible to express this singularity analytically. In particular, adding and subtracting

$$\int_0^{P_2}\frac{R}{P}dP \quad \text{and} \quad \int_0^{P_1}\frac{R}{P}dP$$

to Eq. (8-106) to obtain:

$$\Delta S = \underbrace{-\int_0^{P_2}\left[\left(\frac{\partial V}{\partial T}\right)_P - \frac{R}{P}\right]\Bigg|_{T=T_2} dP}_{\text{ideal to real gas at } T_2} \underbrace{- R\ln\frac{P_2}{P_1} + \int_{T_1}^{T_2}\frac{C_p^o(T)}{T}dT}_{\text{net ideal gas contribution}}$$

$$\underbrace{-\int_{P_1}^{0}\left[\left(\frac{\partial V}{\partial T}\right)_P - \frac{R}{P}\right]\Bigg|_{T=T_1} dP}_{\text{real to ideal gas at } T_1} \tag{8-107}$$

With a properly constructed $V = f(T, P)$ EOS we can numerically or analytically integrate both isothermal pressure integrals which quantitatively represent a deviation from ideal gas behavior while the remaining two terms represent the ideal gas contribution to ΔS.

(2) *Pressure-explicit formulation*—again using the three-step approach shown in Figure 8.7 (V-T plane) similar to what we did before but now integrating over the path in terms of V and T changes.

$$\Delta S = \int_\infty^{V_2}\left(\frac{\partial S}{\partial V}\right)\Bigg|_{T=T_2} dV + \int_{T_1}^{T_2}\left(\frac{\partial S^o}{\partial T}\right)_V dT + \int_{V_1}^{\infty}\left(\frac{\partial S}{\partial V}\right)\Bigg|_{T=T_1} dV \tag{8-108}$$

Using $y^{(1)} = A$ and a Maxwell relation

$$\left(\frac{\partial S}{\partial V}\right)_T = \left(\frac{\partial P}{\partial T}\right)_V \tag{8-109}$$

and from Eq. (8-76)

$$\left(\frac{\partial S^o}{\partial T}\right)_V = \frac{C_v^o}{T} = \frac{C_p^o - R}{T} \tag{8-110}$$

Thus,

$$\Delta S = \int_{\infty}^{V_2} \left(\frac{\partial P}{\partial T}\right)_V \Bigg|_{T=T_2} dV + \int_{T_1}^{T_2} \frac{C_v^o}{T} dT + \int_{V_1}^{\infty} \left(\frac{\partial P}{\partial T}\right)_V \Bigg|_{T=T_1} dV \tag{8-111}$$

Again the second integral is likely to be analytic, but the two isothermal integrals have singularities at $V \to \infty$. To eliminate this problem add and subtract

$$\int_{\infty}^{V_2} \frac{R}{V} dV \text{ and } \int_{\infty}^{V_1} \frac{R}{V} dV$$

from Eq. (8-111) to obtain

$$\Delta S = \underbrace{\int_{\infty}^{V_2} \left[\left(\frac{\partial P}{\partial T}\right)_V - \frac{R}{V}\right]\Bigg|_{T=T_2} dV}_{\text{ideal to real gas at } T_2} + \underbrace{R \ln\frac{V_2}{V_1} + \int_{T_1}^{T_2} \frac{C_v^o(T)}{T} dT}_{\text{net ideal gas contribution}}$$

$$+ \underbrace{\int_{V_1}^{\infty} \left[\left(\frac{\partial P}{\partial T}\right)_V - \frac{R}{V}\right]\Bigg|_{T=T_1} dV}_{\text{real to ideal gas at } T_1} \tag{8-112}$$

Because ΔS is path independent, Eqs. (8-107) and (8-112) are equivalent. To prove this, one can transform Eq. (8-107) to Eq. (8-112) using the triple product to express

$$\left(\frac{\partial V}{\partial T}\right)_P dP\Bigg|_T = -\left(\frac{\partial P}{\partial T}\right)_V dV\Bigg|_T \tag{8-113}$$

and the relationship:

$$R\frac{dP}{P} = R\frac{d(PV)}{PV} - R\frac{dV}{V} \tag{8-114}$$

This procedure effectively performs an integration by parts. The transformation is left as an exercise.

The three-step approach we have used to calculate a property change combines two isothermal property changes from a real to an ideal-gas state and a temperature variation in the ideal-gas state. The isothermal variations can be more formally defined in terms of so-called *departure or residual functions* of a property:

A departure function is the difference between the property of interest in its real state at a specified T, P, and V and in an ideal-gas state at the same temperature T and pressure P.

There are two equivalent formulations of the ***departure function***

$$\begin{aligned} B(T, P) - B^{ideal\,gas}(T, P) &= B(T, P) - B^o(T, P) \\ B(T, V) - B^{ideal\,gas}(T, V^o) &= B(T, V) - B^o(T, V^o) \end{aligned} \tag{8-115}$$

where B is any derived property, H, S, U, A, and G and $V^o = RT/P$ is the specified ideal-gas state at T and P. Also, from Postulate I we know that $B(T, P) = B(T, V)$ for the same initial state as only 2 of 3 properties of (T, P, and V) are independent.

In passing, we note that other conventions for defining departure functions are possible. For instance, we could have selected an ideal-gas state at T and V or equivalently T and $P^o = RT/V$.

From the previous example, we see that the departure function for S is given directly by the form of the pressure integrals of Eq. (8-107). To define the departure function for S at T, first note that,

$$S(T, P) - S^o(T, P \to 0) = -\int_0^P \left(\frac{\partial V}{\partial T}\right)_P dP \tag{8-116}$$

and

$$S^o(T, P) - S^o(T, P \to 0) = -\int_0^P \frac{R}{P} dP \tag{8-117}$$

Therefore, the departure function is the difference of Eqs. (8-116) and (8-117):

$$S(T, P) - S^o(T, P) = -\int_0^P \left[\left(\frac{\partial V}{\partial T}\right)_P - \frac{R}{P}\right] dP \tag{8-118}$$

as stated above. Qualitatively, we have illustrated in Figure 8.8 how departure functions vary as a function of pressure at a specific temperature.

Departure functions can always be determined from a $P\underline{V}TN$ equation of state--or a PVT equation if the basis is 1 mole or a unit mass. Since, as shown in Section 8.3, most PVT EOS formulations are pressure-explicit with V and T as independent variables, this suggests that the intensive Helmholtz energy be used, since

$$A = U - TS = H - TS - PV = f(V, T) \quad \text{and} \quad dA = -S\,dT - P\,dV \tag{8-119}$$

Integrating at constant temperature between $A(T, V)$ and $A(T, \infty)$, we obtain

$$A(T, V) - A(T, \infty) = -\int_\infty^V P\,dV \tag{8-120}$$

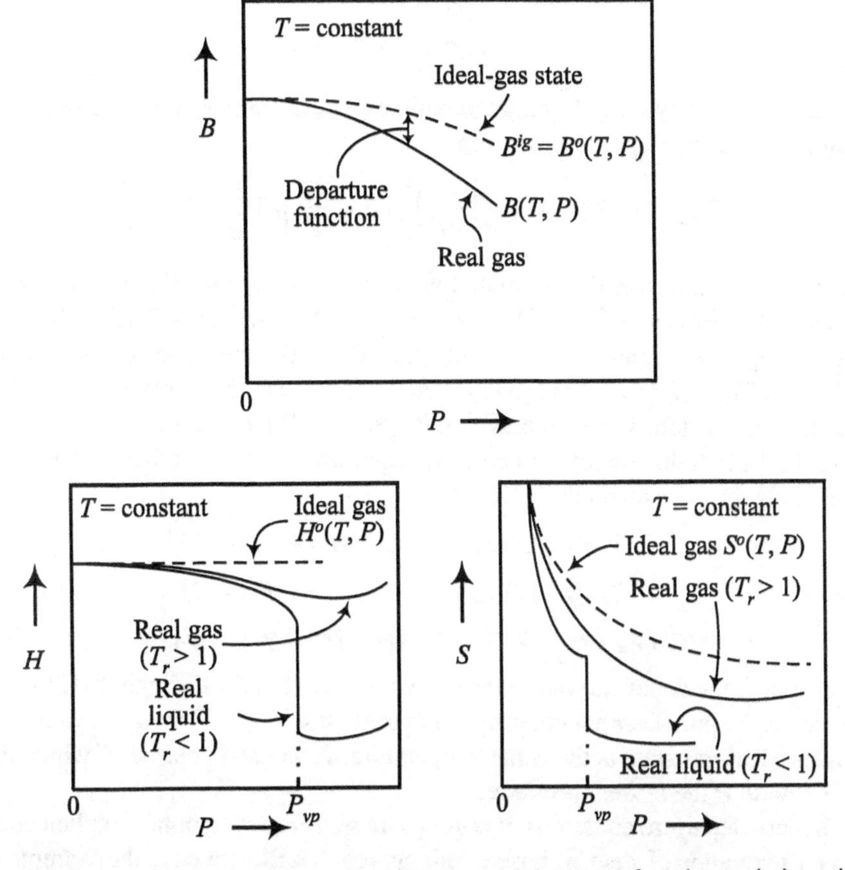

Figure 8.8 Qualitative illustration of isothermal departure function variation with pressure.

With ideal gas behavior from $V \to \infty$ to $V = V^o$, then

$$A(T, \infty) - A^o(T, V^o) = -\int_{V^o}^{\infty} P\, dV \Big|_{ideal\,gas} \tag{8-121}$$

We add Eqs. (8-120) and (8-121) and, at the same time, add and subtract

$$\int_{\infty}^{V} \frac{RT}{V}\, dV$$

to the right-hand side. The final result, when $P_{ideal\ gas} = RT/V$, is

$$A(T, V) - A^o(T, V^o) = -\int_{\infty}^{V} \left(P - \frac{RT}{V}\right) dV + RT \ln \frac{V^o}{V} \tag{8-122}$$

In Eq. (8-122) the temperature is constant. By substituting a pressure-explicit equation of state in the integral of Eq. (8-122), the Helmholtz energy departure function is obtained (see Example 8.4). Also, from Eq. (8-119),

$$\left(\frac{\partial A}{\partial T}\right)_V = -S$$

So Eq. (8-122) may be differentiated with respect to temperature at constant volume to obtain the entropy departure function,

$$S(T, V) - S^o(T, V^o) = \left[\frac{\partial}{\partial T}\int_{\infty}^{V}\left(P - \frac{RT}{V}\right)dV\right]_V - R\ln\frac{V^o}{V} \tag{8-123}$$

When Eq. (8-122) was differentiated with respect to T with V constant one should note that $(\partial A^o/\partial T)_V = -S^o - P(\partial V^o/\partial T)_V$ since $dA^o = -S^o\,dT - P\,dV^o$. However, in differentiating the right-hand side of Eq. (8-122) one also obtains the term $RT(\partial \ln V^o/\partial T)_V = (RT/V^o)\,(\partial V^o/\partial T)_V = P(\partial V^o/\partial T)_V$, so the derivatives of V^o with respect to T at constant V cancel and do not appear in Eq. (8-123).

With the Helmholtz energy and entropy departure functions, others can be obtained by simple algebra; for example,

$$U(T, V) - U^o(T, V^o) = (A - A^o) + T(S - S^o) \tag{8-124}$$

$$H(T, V) - H^o(T, V^o) = (U - U^o) + PV - RT \tag{8-125}$$

$$G(T, V) - G^o(T, V^o) = (H - H^o) - T(S - S^o) \tag{8-126}$$

To reiterate, the departure functions shown in Eqs. (8-122) through (8-126) yield the difference in the pure-component property in the real state (V, T) from the value it would have in an ideal-gas state at the same temperature, T, and at a volume V^o which is given by RT/P, with P the system pressure.

With these departure equations, it is quite straightforward to obtain explicit equations once a PVT equation of state is chosen. This approach is illustrated in the example below.

Example 8.4

Determine the departure functions shown in Eqs. (8-122) through (8-126) using the Peng-Robinson equation of state [Eq. (8-46)].

Solution

Inserting Eq. (8-46) for P in Eq. (8-122), we obtain

$$A(T, V) - A^o(T, V^o) = -\int_{\infty}^{V}\left(\frac{RT}{V-b} - \frac{RT}{V} - \frac{a}{V^2 + 2Vb - b^2}\right)dV + RT\ln\frac{V^o}{V}$$

$$= RT\ln\frac{V^o}{V-b} + \frac{a}{2\sqrt{2}b}\ln\frac{V + b(1-\sqrt{2}\,)}{V + b(1+\sqrt{2}\,)} \tag{8-127}$$

Then,

$$S(T, V) - S^o(T, V^o) = -\frac{\partial}{\partial T}(A - A^o)$$

$$= -R \ln\left(\frac{V^o}{V-b}\right) - \frac{1}{2\sqrt{2}b} \ln\left[\frac{V+b(1-\sqrt{2})}{V+b(1+\sqrt{2})}\right]\left(\frac{\partial a}{\partial T}\right)_V \tag{8-128}$$

The parameter a is given in Eqs. (8-47) through (8-49). With these,

$$\left(\frac{\partial a}{\partial T}\right)_V = \frac{-a\kappa}{(\alpha T T_c)^{1/2}} \tag{8-129}$$

In Eq. (8-129), $a = a_c\,\alpha$ is determined from Eq. (8-47), $\alpha = \alpha(\omega, T_r)$ from Eq. (8-49), and κ from Eq. (8-50). Then,

$$S(T, V) - S^o(T, V^o) = -R \ln\frac{V^o}{V-b} + \frac{a\kappa}{(\alpha T T_c)^{1/2} 2\sqrt{2}b} \ln\frac{V+b(1-\sqrt{2})}{V+b(1+\sqrt{2})} \tag{8-130}$$

The other departure functions are obtained by simple algebra; for example,

$$H(T, V) - H^o(T, V^o) = (A - A^o) + T(S - S^o) + (PV - RT)$$

$$= \left\{\frac{a}{2\sqrt{2}b} \ln\left[\frac{V+b(1-\sqrt{2})}{V+b(1+\sqrt{2})}\right]\right\}\left[1 + \frac{\kappa T}{(\alpha T T_c)^{1/2}}\right] + (PV - RT) \tag{8-131}$$

In Eq. (8-131) the pressure in the last term would be found from the original equation of state [Eq. (8-46)] at the given V and T.

Thus to determine the difference in any thermodynamic property (B) between two states 1 and 2 that differ in temperature, using a three-step process in the T–V plane, we first form the isothermal departure functions and then vary temperature in the ideal-gas state. That is,

$$B(T_2, V_2) - B(T_1, V_1) = [B(T_2, V_2) - B^o(T_2, V_2^o)]$$
$$- [B(T_1, V_1) - B^o(T_1, V_1^o)] + [B^o(T_2, V_2^o) - B^o(T_1, V_1^o)] \tag{8-132}$$

The first two brackets on the right-hand size of the identity Eq. (8-132) represent the departure functions at T_2 and T_1, respectively. They are calculated as described earlier in this section. The third bracket represents the difference in property B between two *ideal-gas states* (T_2, V_2^o) and (T_1, V_1^o). The advantage of varying temperature in an ideal-gas state is that heat capacities of ideal gases can then be used and, as shown in Section 8.4, values of C_p^o or C_v^o are known for many pure materials or, if not, they can often be estimated from molecular structure.

For example, if B were the specific internal energy U, then

$$U^o(T_2, V_2^o) - U^o(T_1, V_1^o) = \int_{T_1}^{T_2}\left(\frac{\partial U^o}{\partial T}\right)_V dT = \int_{T_1}^{T_2} C_v^o\, dT \tag{8-133}$$

and no effect of the volume difference need be considered because, in an ideal-gas state, U^o is a function of temperature only. We also note that with $V^o = RT/P$:

$$B^o(T, V^o) = B^o(T, P) \tag{8-134}$$

so that for the case where B is the enthalpy H,

$$H^o(T_2, V_2^o) - H^o(T_1, V_1^o) = H^o(T_2, P_2) - H^o(T_1, P_1)$$
$$= \int_{T_1}^{T_2} \left(\frac{\partial H^o}{\partial T}\right)_P dT = \int_{T_1}^{T_2} C_p^o \, dT \tag{8-135}$$

As with U^o, no volume (or pressure) term appears since H^o is again only a function of temperature. For B as the entropy,

$$S^o(T_2, V_2^o) - S^o(T_1, V_1^o) = \int_{T_1}^{T_2} \left(\frac{\partial S^o}{\partial T}\right)_V dT + \int_{V_1^o}^{V_2^o} \left(\frac{\partial S^o}{\partial V}\right)_T dV \tag{8-136}$$

Since $(\partial S/\partial V)_T = (\partial P/\partial T)_V$, then, for an ideal gas, $(\partial S/\partial V)_T = R/V$. Thus Eq. (8-136) becomes

$$S^o(T_2, V_2^o) - S^o(T_1, V_1^o) = \int_{T_1}^{T_2} \frac{C_v^o}{T} dT + R \ln \frac{V_2^o}{V_1^o} \tag{8-137}$$

By using Eq. (8-135) and that $C_p^o = C_v^o + R$, then an alternative form for Eq. (8-137) is

$$S^o(T_2, V_2^o) - S^o(T_1, V_1^o) = S^o(T_2, P_2) - S^o(T_1, P_1) = \int_{T_1}^{T_2} \frac{C_p^o}{T} dT - R \ln \frac{P_2}{P_1} \tag{8-138}$$

If B were the Gibbs or the Helmholtz energy, departure functions can be readily calculated if a PVT EOS is available. However, it is not possible to determine absolute changes in these properties if the temperature is changed since derivatives with respect to temperature yield terms containing the absolute entropy, S. As the absolute value of this property is only known to within an arbitrary constant, nonisothermal variations in A or G (or, in fact, any Legendre transform where $\underline{S}$ has been transformed to T) are without physical meaning. Also, if *isothermal* changes in G or A are desired, it is preferable to combine the two isothermal departure functions of Figure 8.7 to evaluate the change in G or A directly from the Fundamental Equation. For example, for the Gibbs energy, from Table 5.2, if N = constant, then in intensive form:

$$\left(\frac{\partial G}{\partial P}\right)_{T,N} = V \quad \text{and} \quad G(T, P_2) - G(T, P_1) = \int_{P_1}^{P_2} V \, dP \tag{8-139}$$

In a similar manner, for the Helmholtz energy

$$A(T_1, V_2) - A(T_1, V_1) = -\int_{V_1}^{V_2} P \, dV \tag{8-140}$$

The difference between heat capacities in a real state at T, V and an ideal-gas state at T, V^o can also be determined from an equation of state. For example, if the entropy departure function, Eq. (8-123), is differentiated with respect to temperature, at constant V, then with

$$\left(\frac{\partial S}{\partial T}\right)_V = \frac{C_v}{T}$$

one obtains

$$C_v(T, V) - C_v^o(T, V^o) = T\int_{\infty}^{V}\left(\frac{\partial^2 P}{\partial T^2}\right)_V dV \tag{8-141}$$

[See the discussion following Eq. (8-123) for the reason why there are no derivatives of V^o with respect to T.] It is less convenient to obtain $C_p - C_p^o$ from the departure functions developed above since

$$\left(\frac{\partial S}{\partial T}\right)_P = \frac{C_p}{T}$$

and a derivative at constant pressure is required, whereas the principal variables used in the present treatment use T and V. It is possible to show that

$$C_p - C_p^o = -T\int_{0}^{P}\left(\frac{\partial^2 V}{\partial T^2}\right)_P dP \tag{8-142}$$

which is similar in form to Eq. (8-141). However, since equations of state are most often explicit in pressure, the second derivative in the integral of Eq. (8-142) is not easily determined. It is probably more convenient to determine $C_v(T, V)$ by Eq. (8-141) and then use the relationship:

$$C_p = C_v + \frac{TV\alpha_p^2}{\kappa_T} \tag{8-143}$$

to relate C_p to C_v in the real state, where κ_T and α_p are defined by Eqs. (8-13) and (8-14). After rearrangement,

$$C_p = C_v - \frac{T(\partial P/\partial T)_V^2}{(\partial P/\partial V)_T} \tag{8-144}$$

The derivatives in Eq. (8-144) are readily found from a pressure-explicit equation of state. Note that for an ideal gas, Eq. (8-144) reduces to

$$C_p^o - C_v^o = R \tag{8-145}$$

Thus, combining the results of Eqs. (8-141), (8-144), and (8-145) yields

$$C_p - C_p^o = T\int_{\infty}^{V}\left(\frac{\partial^2 P}{\partial T^2}\right)_V dV - \frac{T(\partial P/\partial T)_V^2}{(\partial P/\partial V)_T} - R \tag{8-146}$$

In the absence of a suitable *PVT* EOS or volumetric experimental data, the compressibility charts from corresponding states can be used to develop generalized isothermal departure functions in reduced coordinates. In this case, we begin with a volume-explicit form $Z = PV/RT = f(T_r, P_r)$ and follow a procedure analogous to what we used for a pressure-explicit EOS, but $A(T, V)$ is replaced with $G(T, P)$. The isothermal departure functions that result are as follows:

$$G(T, P) - G^o(T, P) = RT\int_0^P (Z-1)\frac{dP}{P} \tag{8-147}$$

$$H(T, P) - H^o(T, P) = -RT^2\int_0^P \left(\frac{\partial Z}{\partial T}\right)_P \frac{dP}{P} \tag{8-148}$$

$$S(T, P) - S^o(T, P) = -RT\int_0^P \left(\frac{\partial Z}{\partial T}\right)_P \frac{dP}{P} - R\int_0^P (Z-1)\frac{dP}{P} \tag{8-149}$$

Using Eqs. (8-147) through (8-149) in non-dimensional form, and $Z(T_r, P_r)$ as given in Section 8.2, generalized correlations can be generated. Figures 8.9 and 8.10 are the corresponding states results for the isothermal enthalpy and entropy departures functions due to Hougen, Watson, and Ragatz (1959, 1960).

One final point regarding the use of departure functions needs to be made. Notice the large differences between the enthalpy and entropy departures shown in Figures 8.9 and 8.10 between gas and liquid conditions, for example look along the saturation curve. The reason why the liquid departure functions are much larger than those for vapor at similar T_r and P_r is that the phase change contributions to H and S are included in the departure function value. This has two implications. For gases, the departure or residual contribution is usually of the same order of magnitude as the ideal gas part. In the case of liquids, a robust *PVT* EOS accurate in the liquid region is usually needed to ensure accurate property predictions, as the results generated from corresponding states in Figures 8.9 and 8.10 have only limited accuracy for liquids.

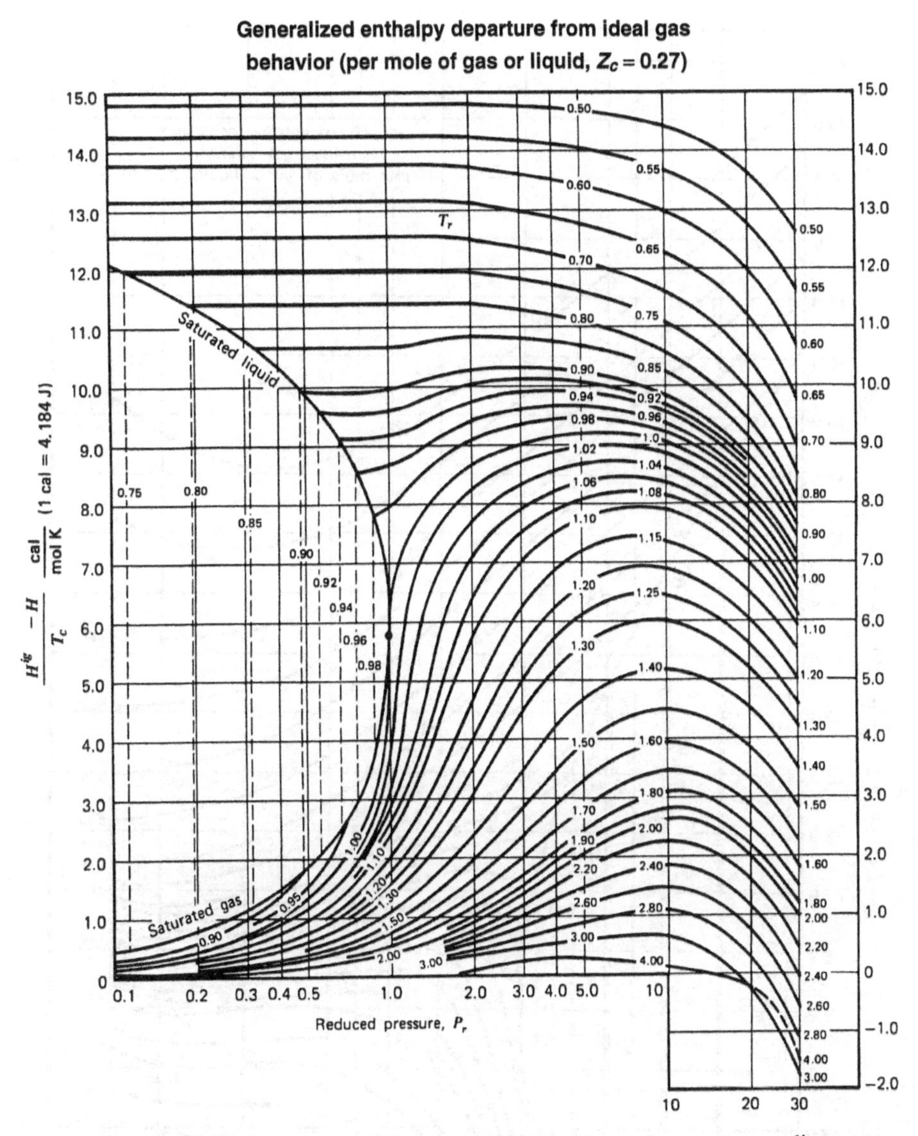

Figure 8.9 Generalized enthalpy departure function using corresponding states. [From O.A. Hougen, K.M. Watson, and R.A. Ragatz (1960), *Chemical Process Principles Charts*, 2nd ed., John Wiley & Sons, New York]

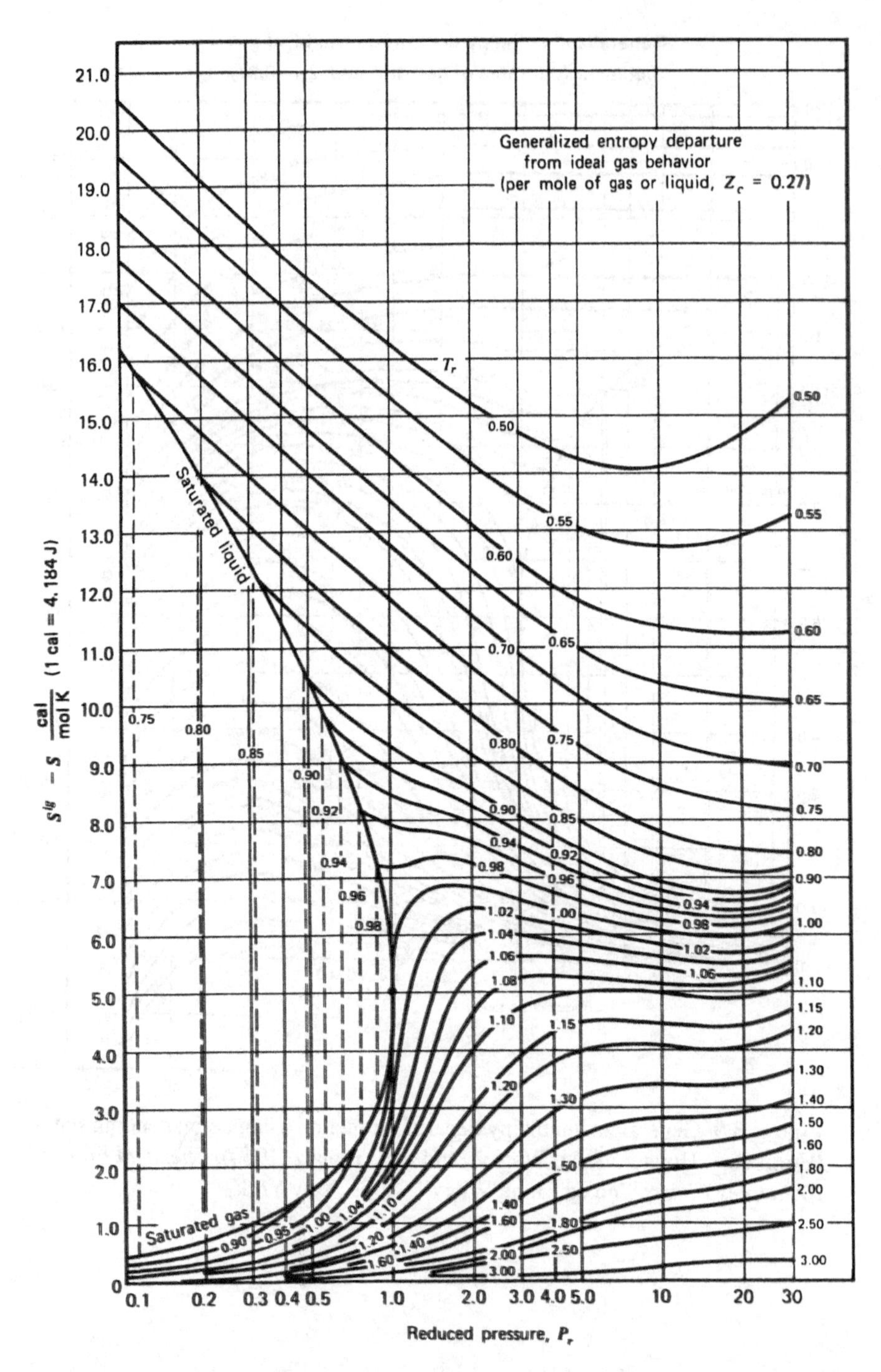

Figure 8.10 Generalized entropy departure functions using corresponding states. [From O.A. Hougen, K.M. Watson, and R.A. Ragatz (1960), *Chemical Process Principles Charts*, 2nd ed., John Wiley & Sons, New York]

8.6 Compressibility and Heat Capacities of Solids

Clearly solid phases are intrinsically different than gases, liquids, or supercritical fluids. To varying degrees, solids can be treated as elastic media (either linear or non-linear), and unlike liquids, gases, and other normal fluids, in solids, non-isotropic stresses lead to deformation (strain) but not to flow. Even with these complications, their volumetric *PVT* behavior is frequently correlated to empirical equations of state, even though it has not been possible to develop a completely general constitutive property relationship for $V = f(T, P)$ for solid materials. Recall that the isothermal compressibility κ_T and the isobaric coefficient of thermal expansion α_P are given by:

$$\kappa_T \equiv -\frac{1}{V}\left(\frac{\partial V}{\partial P}\right)_T \quad \text{and} \quad \alpha_P \equiv \frac{1}{V}\left(\frac{\partial V}{\partial T}\right)_P$$

Complete knowledge of these two quantities effectively defines the volumetric *PVT* behavior of a solid. A limiting condition is a perfectly incompressible solid, where V = constant independent of T and P. Of course, real solids have small but finite κ_T and α_P values. Table 8.3 compares these parameters for several gases, liquids, and solids at ambient conditions. In most practical situations for solids, experimental measurements are used rather than predictions from first principles or general EOSs.

Many liquids far below their critical temperatures have very low compressibilities, similar to solids. Thus, in these situations one can assume perfectly or nearly incompressible behavior as a first-order approximation.

Table 8.3 Thermal Physical Property Constants for Selected Pure Substances at 298 K, 1 bar

Substance	Phase	ρ kg/m^3	κ_T 10^{-10} Pa^{-1}	α_p 10^{-4} K^{-1}	C_p J/kg K	$\frac{T\alpha_p^2}{\rho\kappa_T C_p}$ 10^{-2}	Θ_D K
Ar	g	1.6	1.0×10^5	33.5	520	40	84
CO_2	g	1.8	1.0×10^5	33.5	661	28	140
H_2O	l	1000	4.53	2.50	4186	0.98	192
Hg	l	13,500	3.91	1.82	139	1.4	97
KBr	s	2,750	0.25	1.20	106	58.9	177
Ag	s	10,500	0.096	0.53	236	3.5	215
NaCl	s	2,160	0.268	1.20	206	36.0	281
Cu	s	8,920	0.091	0.50	386	2.4	315
Fe (pure)	s	7,860	0.05	0.35	450	2.1	453
SiO_2 (quartz)	s	2,650	0.06	0.38	1000	2.7	296
C (diamond)	s	3,510	0.01	0.35	472	22.0	1860

For heat capacities of crystalline solids, there is a sound theoretical basis for predicting the temperature dependence of C_v. Debye developed a theory that scales temperature with a characteristic constant Θ_D for each specific solid substance. In his formulation, the vibrational modes contribute to determining the magnitude of Θ_D, C_v is then correlated as follows:

$$C_v = 2R\left[12\left(\frac{T}{\Theta_D}\right)^3 \int_0^{\Theta_D/T} \frac{x^3}{e^x - 1}\,dx - \frac{3\Theta_D}{T(\exp(\Theta_D/T) - 1)}\right] \tag{8-150}$$

If $T > \Theta_D$,

$$C_v \cong 3R\left[1 - 0.05\left(\frac{\Theta_D}{T}\right)^2\right] \tag{8-151}$$

which asymptotically goes to 3R as $T >>> \Theta_D$ following the law of Dulong and Petit. At the other extreme, $T <<< \Theta_D$,

$$C_v \approx \frac{12 R \pi^4}{5}\left(\frac{T}{\Theta_D}\right)^3 \tag{8-152}$$

Selected values of Θ_D are also given in Table 8.3. Some very useful approximations are frequently used for solids that are nearly incompressible. Earlier we gave a rigorous expression for the difference between C_p and C_v [Eq. (8-143)] which when applied to most solid materials where

$$\frac{T\alpha_P^2}{\rho\kappa_T C_p} < 1 \text{ then } C_p \approx C_v$$

In addition at ambient temperatures, most solids have the following behavior:

$$dU \approx C_v dT \tag{8-153}$$

$$dH \approx C_p dT + VdP \approx C_v dT + VdP \tag{8-154}$$

$$dS \approx \frac{C_v}{T}\,dT \tag{8-155}$$

In many situations, the product of pressure and specific volume of a relatively incompressible solid or liquid is small enough to be neglected, then $H = U$. There are two key exceptions to the approximate behavior cited in the equations above. First, when operating at high pressures or stresses, even substances with low isothermal compressibilities (κ_T) exhibit deformation and specific volume (V) or density (ρ) can no longer be regarded as nearly constant. Second, there are a number of important solids, such as polymers, that have reasonably large κ_T and α_P where the nearly incompressible approximation no longer holds. In either of these situations, the contribution to the internal energy due to elastic strain energy must be accounted for. As mentioned earlier in Chapters 3 and 5, this effect can be treated in terms of a deformational work term $\sigma d\varepsilon$ where σ = stress and ε = strain ($\Delta V/V$ or $\Delta L/L$). In general, there are six directional stress and strain components, σ_{xx}, σ_{yy}, σ_{zz}, σ_{xy}, σ_{yz}, σ_{xz}, ε_{xx}, ε_{yy}, ε_{zz}, ε_{xy}, ε_{yz}, ε_{xz}, that can be represented by nine elements in a symmetric tensor, which can be summed to get the total work in a revised form of the Fundamental Equation. Roughly this is equivalent to:

$$dU = TdS + V\sum_{j=1}^{3}\sum_{i=1}^{3}\sigma_{ij}\, d\,\varepsilon_{ij} \tag{8-156}$$

Many times it is possible to simplify Eq. (8-156) by reducing the directional degrees of freedom, see for example Problem 5.20. Another key feature of the above approach is that a relatively simple linearly elastic constitutive relationship such as Hooke's law can be used to relate σ and ε, such as $\sigma_{xx} = E_x\varepsilon_{xx}$ where E_x = Young's modulus.

The Gruneisen constant γ_G is frequently introduced to characterize the behavior of crystalline or glassy solids. Here we begin with the Debye model for lattice vibration where the internal and Helmholtz energies are approximated by:

$$U = U_o(V) + U_D(T, V) \quad \text{and} \quad A = U_o(V) + A_D(T, V)$$

where U_D and A_D are the vibrational contributions and $U_o(V)$ is the ground state value. From the Helmholtz energy,

$$P = -\left(\frac{\partial A}{\partial V}\right)_T = -\left(\frac{\partial U_o}{\partial V}\right)_T - \left(\frac{\partial A_D}{\partial \Theta_D}\right)\left(\frac{\partial \Theta_D}{\partial V}\right)_T \tag{8-157}$$

Equation (8-157) assumes that the characteristic Debye temperature, Θ_D, is directly related to the volume (or density) of the solid. Non-linear, anharmonic effects are captured in the Debye model with the dependence of A_D on V. Because $A \equiv U - TS$, A_D must be of the form:

$$A_D(T, V) = Tf(\Theta_D/T) \tag{8-158}$$

Thus,

$$\left(\frac{\partial A_D}{\partial \Theta_D}\right) = \frac{U_D}{\Theta_D}$$

and

$$P = -\left(\frac{\partial U_o}{\partial V}\right)_T - \frac{U_D}{\Theta_D}\left(\frac{\partial \Theta_D}{\partial V}\right) = -\left(\frac{\partial U_o}{\partial V}\right)_T + \frac{\gamma_G U_D}{V}$$

where γ_G is the Gruneisen constant which is formally defined as:

$$\gamma_G = -\frac{\partial \ln \Theta_D}{\partial \ln V} = -\frac{V}{\Theta_D}\frac{\partial \Theta_D}{\partial V} \tag{8-159}$$

Given Eq. (8-159) it is easy to show that

$$\left(\frac{\partial P}{\partial T}\right)_V = \frac{\gamma_G C_v}{V} \tag{8-160}$$

and with the triple product:

$$\gamma_G = \frac{\alpha_p V}{\kappa_T C_v} \tag{8-161}$$

Equation (8-161) is commonly called the Gruneisen relation. Values of γ_G typically range from about 1.0 to 2.5 for most solid elements and salts.

Thermodynamics can also be applied to elasticity phenomena in polymer systems. A first-order model of a polymer on a molecular level suggests that elastic deformation above the glass transition temperature results from a quasi-static (near reversible) extension and contraction of polymer chain segments [see Eisele (1990) for details]. Entropic effects due to the structural changes related to the coiling and uncoiling of polymer molecules under tensile or compressive load tend to dominate the changes in Helmholtz and Gibbs energy with the internal energy remaining relatively constant. The differential elongation work expended during elastic deformation of a polymer is given approximately by:

$$\delta W_{rev} = \underline{a}\sigma_x\, dx \approx \underline{a}CTdx$$

For small strains,

$$\sigma_x = 3NkT\varepsilon_x \tag{8-162}$$

with

$$\varepsilon_x = \varepsilon_x^o - \frac{\alpha_p}{3}(T - T^o)$$

As we can see, Eq. (8-162) has a form analogous to the ideal gas law. For rubber systems, a thermoelastic inversion is frequently observed at strains of 10% or so. Below the inversion point, the stress (σ_x) decreases with increasing temperature while above the inversion, stress increases with temperature. This effect is related to volume changes, and we can estimate the strain at the inversion ε_x at $T_{inversion}$ by using the fact that $\partial\sigma_x/\partial T = 0$ at that point and Eq. (8-162). This yields,

$$\varepsilon_x^o = \frac{\alpha_p}{3}(2T_{inversion} - T^o)$$

For a typical rubber, $\alpha_p = 6.6 \times 10^{-4}$ K^{-1}, $T^o = 2.93$ K, $T_{inversion} = 343$ K, thus $\varepsilon_x = 0.076$ (or 7.6%) [Eisele (1990)].

8.7 Derived Property Representations

Given the necessary physical property constitutive relationships in the form of *PVT* and C_v (or C_p) correlated data or empirically-fitted equations, there are a number of ways this information can be used. Before the pervasive infusion of computers at all levels, charts or tabular data were used widely for engineering problem solving. The steam tables or Mollier (*H-S*) diagram for water (see Figure 8.11a), pressure-enthalpy (*P-H*) or temperature-entropy (*T-S*) charts (see Figures 8.11b for water and Figures 8.12a and 8.12b for *P-H* and *T-S* diagrams for carbon dioxide) are good examples of such styles

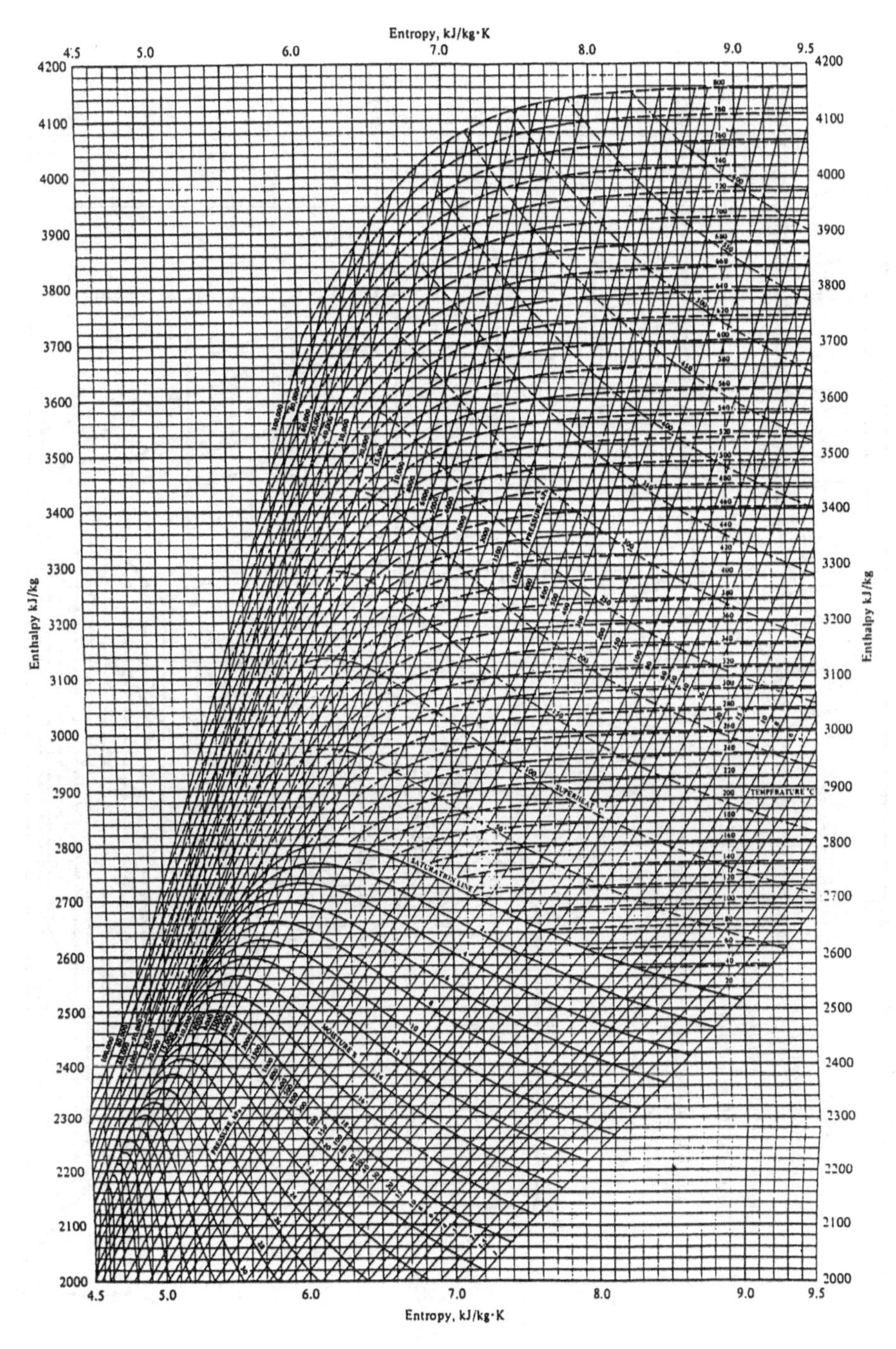

Figure 8.11a Enthalpy-entropy or Mollier diagram for water. [Source: ASME (1967), Steam Tables in SI (Metric) Units for Instructional Use, American Society of Mechanical Engineers, New York, used with permission.]

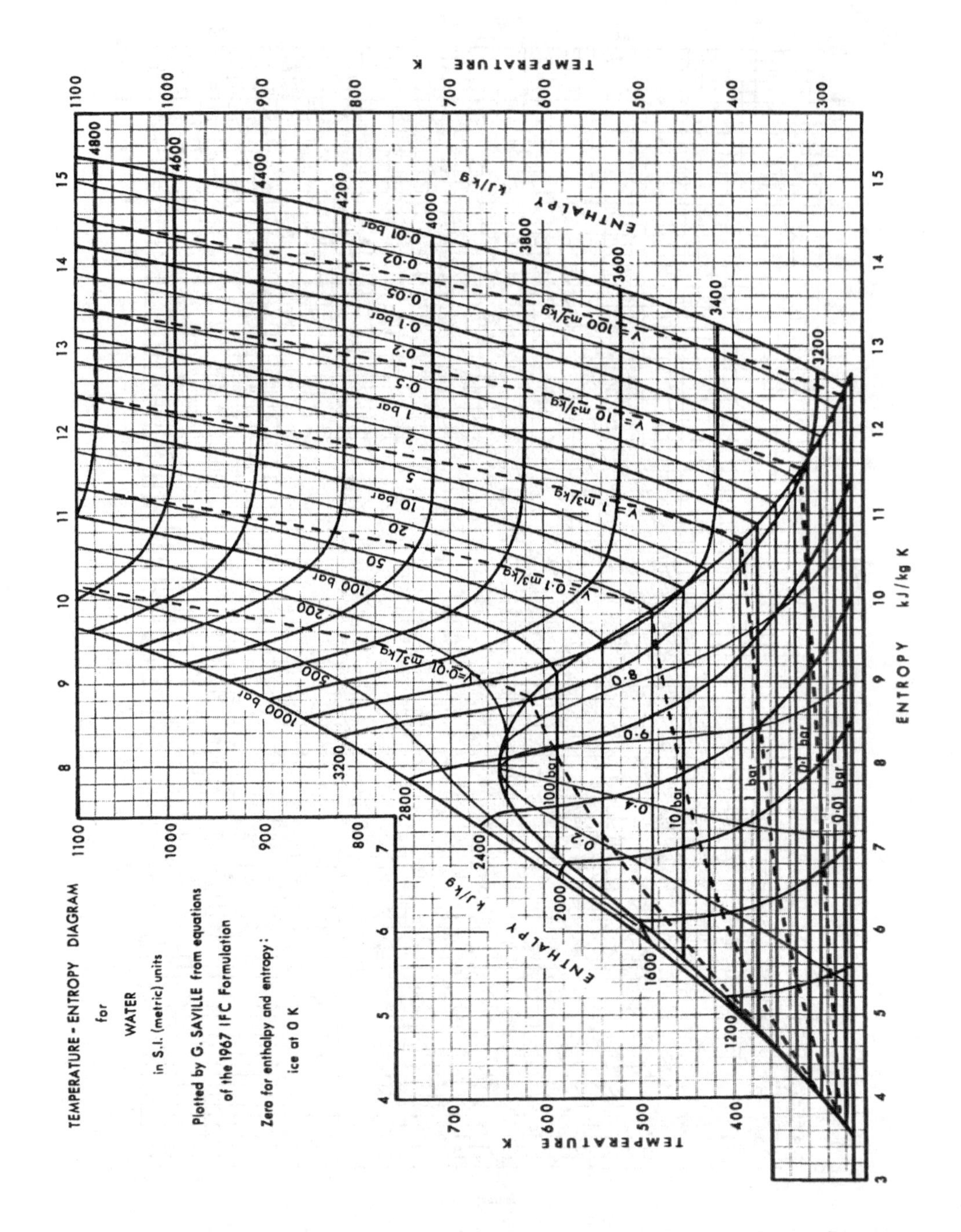

Figure 8.11b Temperature-entropy (*T-S*) diagram for pure water. [Source: Bett, Rowlinson, and Saville (©) (1975) with the permission of the Athlone Press, 1996]

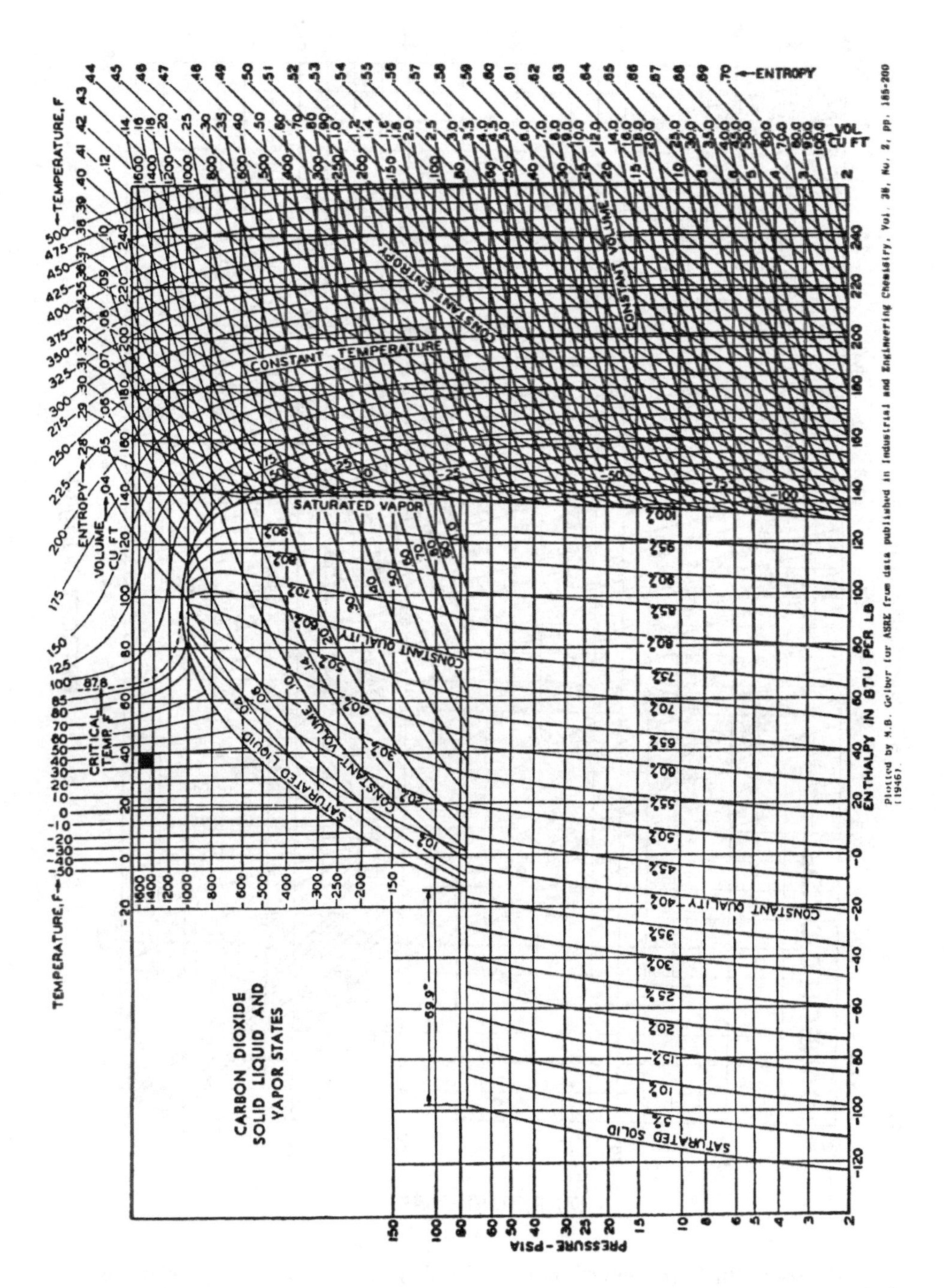

Figure 8.12a Pressure-enthalpy (P-H) diagram for pure CO_2 [used with permission of ASHRAE].

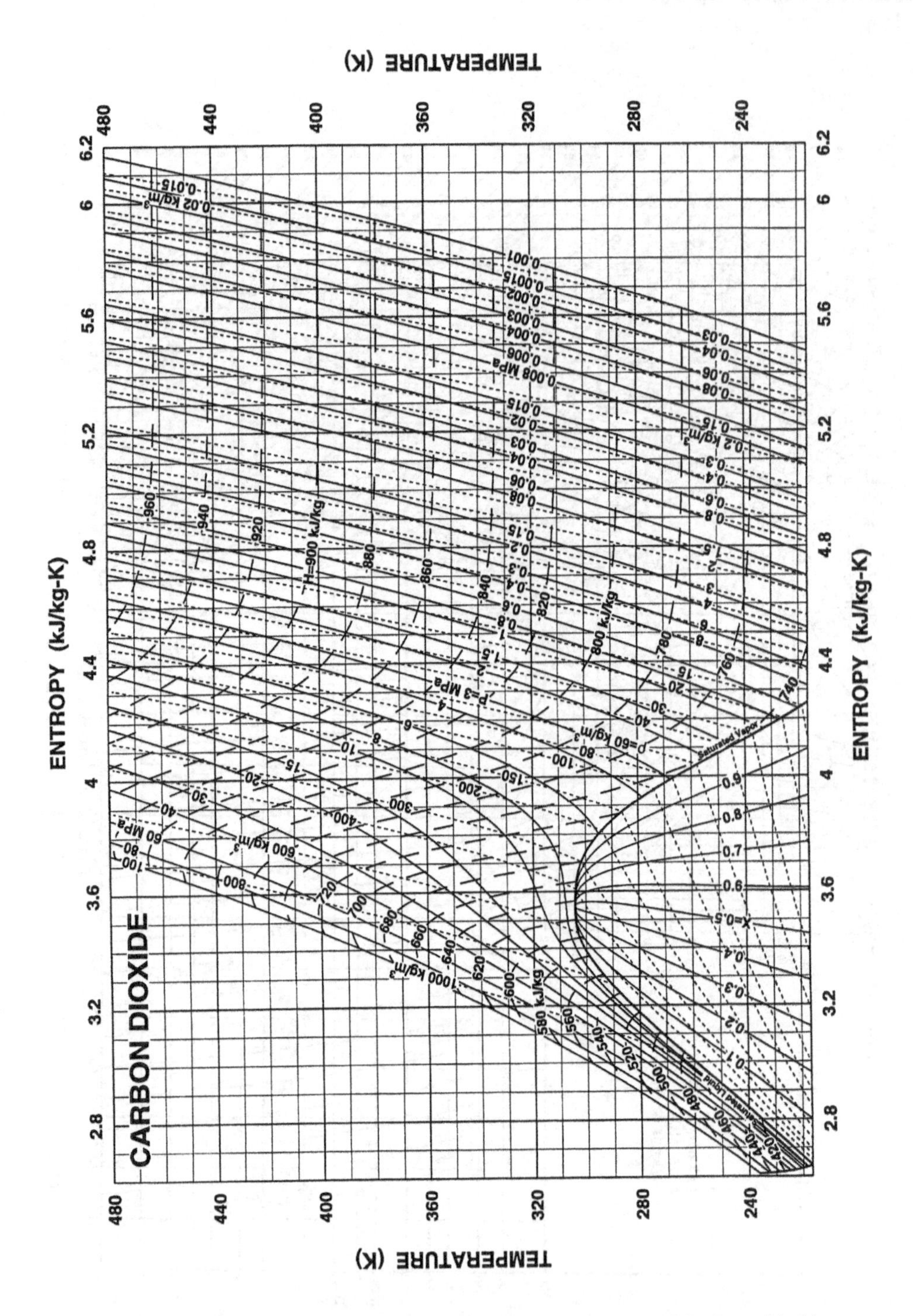

Figure 8.12b Temperature-entropy (*T-S*) diagram for pure CO_2 [provided by Professor Steven Penoncello, Center for Applied Thermodynamics Studies, University of Idaho, Moscow, Idaho].

of presenting property data. Chemical engineering sources of thermodynamic property data and charts are cited at the end of the chapter.

Today, computer-generated representations of thermodynamic properties are preferred. Although many software packages and data bases are easily accessible to practicing engineers, a central issue always is how accurately are the models and correlations used in such automated systems represent reality. Far too often, the high quality graphical output of today's computers is misinterpreted as indicating a similar high quality (and high accuracy) of predicted results. While there is no doubt that computational methods and data bases will continue to improve, considerable interpretative value is frequently obtained by using graphical thermodynamic diagrams to illustrate trends or to visually compare experimental data with predicted results.

In Chapters 5 and 7, we discussed recent work by Kenneth Jolls and coworkers who have generated three-dimensional plots of the Fundamental Equation in various canonical coordinates using a pure component Peng-Robinson equation of state and ideal-gas-state heat capacities as a basis (see Figures 5.2b and 7.4). It would be very appropriate to utilize both the high quality graphics software of such approaches along with the advanced statistical analysis packages that are available today to evaluate the quality of the models and methods you employ to represent physical properties. These computer-based techniques greatly facilitate interpretation of sensitivity and uncertainty effects in choosing parameters, for example. Section 13.4 discusses these issues further.

8.8 Standard Enthalpy and Gibbs Free Energy of Formation

In Sections 8.5 and 8.7 we were concerned with the changes in derived properties of pure substances between two well-defined state points or between a state point as a defined reference condition such as an ideal gas state at the same temperature as the state point. This defined condition might also be a specified reference state used to generate a tabular set of properties or to produce a thermodynamic diagram, for example the Mollier diagram for water given in Figure 8.11a takes the enthalpy and entropy of liquid water to be zero at the triple point. Different reference states are commonly used for different materials. Up to now this has not mattered as we are usually dealing with only differences between derived properties where the arbitrary reference condition eventually cancels out of the calculation. However, because many practical situations deal with chemical and phase transformations in mixtures of compounds, one must judiciously select reference states to avoid ambiguity.

A common approach used for dealing with reacting components is to define the reference state for each compound relative to the elements in appropriate stoichiometric ratios required to form the compound. The enthalpy of the formation reaction, defined as the difference between the enthalpy of the product compound and the sum of the enthalpies of the constituent elements that form that compound, can then be directly related to a calorimetric measurement. Very frequently for organic molecules, the calorimetric experiment conducted is one of a combustion reaction where the products would be carbon dioxide and water in varying molar ratios for any H-C organic compound. When combustion enthalpies are used to determine formation enthalpies, they must be corrected to remove the contributions from the carbon dioxide and water products. The standard reference condition usually selected for expressing enthalpies of formation is 298.15 K (25°C), 1 atm (1.013 bar) with the compound of interest placed in a defined state of aggregation, such as a pure gas, liquid, or solid, or in a solution of specified composition. Frequently, but not always, the reference state is the substance's normal stable state of aggregation at 298.15 K and 1.013 bar. When dealing with gaseous components, the ideal gas state at 1 atm is normally selected as well.

To remove further possible ambiguity, *the standard enthalpy of each element in its normal state of aggregation is set to zero.* Thus the enthalpy of the compound of interest at 298.15 K and 1.013 bar is equal to the standard enthalpy of its formation reaction from the elements. Mathematically we can show this as:

$$H^o_{product\ i} - \Sigma H^o_{element\ j} = \Delta H^o_f \text{ (298.15 K, 1.01 bar, state of aggregation)} \qquad (8\text{-}163)$$

Consider the following examples:

$$H_{2(g)} + 1/2\, O_{2(g)} \rightarrow H_2O_{(l)}$$
$$2Al_{(s)} + 3/2\, O_{2(g)} \rightarrow Al_2O_{3(s)}$$
$$1/2\, H_{2(g)} + 1/2\, I_{2(g)} \rightarrow HI_{(g)}$$

To obtain the standard reaction enthalpy $\Delta H^o_{reaction}$ from the enthalpies of formation for each reactant and product participating in the reaction, all we have to do is sum the enthalpy of formation for each component times its stoichiometric coefficient in the reaction:

$$\Delta H^o_{reaction} = \Delta H^o_{rx} = \sum_{i=1}^{n} \nu_i\, \Delta H^o_{f,i} \text{ (all at 298.15 K, 1.013 bar)} \qquad (8\text{-}164)$$

Again consider the following example

$$C_2H_{4(g)} + 3\, O_{2(g)} \rightarrow 2CO_{2(g)} + 2H_2O_{(l)}$$

$$\Delta H^o_{reaction} = 2\Delta H^o_{f,CO_{2(g)}} + 2\Delta H^o_{f,H_2O_{(l)}} - \Delta H^o_{f,C_2H_{4(g)}} - 3\Delta H^o_{f,O_{2(g)}} \quad (\text{last term} = 0)$$

Using this approach we can estimate the heat or enthalpy of any chemical reaction from the standard enthalpies of formation that are readily available for a large number of substances.

Standard Gibbs free energies of reactions are also needed for calculations involving chemical equilibria as will be shown in Chapter 16. In a completely analogous manner to the reaction enthalpies, ΔG^o_f for each component participating in the reaction multiplied by its stoichiometric coefficient is summed to generate the standard $\Delta G^o_{reaction}$, thus:

$$\Delta G^o_{reaction} = \Delta G^o_{rx} = \sum_{i=1}^{n} \nu_i\, \Delta G^o_{f,i} \quad \text{(all at 298.15 K, 1.013 bar)} \qquad (8\text{-}165)$$

As before, *the standard Gibbs free energies of all elements in their normal states of aggregation are set to zero* to generate a standard ΔG_f^o for any compound at 298.15 K and 1.013 bar in a specified state of aggregation. Because the ΔH_f^o is known, then one really determines ΔG_f^o by estimating the value of ΔS_f^o and then uses the following relationship to calculate ΔG_f^o:

$$\Delta G_f^o = \Delta H_f^o - T\,\Delta S_f^o \tag{8-166}$$

Normally, the reference state for S is defined as $S = 0$ at 0 K by the Third Law of Thermodynamics and an integration to 298.15 K is carried out. (We note that this is the consistent convention selected (see Section 11.3 for further discussion).) Heat capacity correlations derived from spectroscopic data and measured latent heats, as appropriate, are typically used for this integration.

Starting with the defined standard values at 298.15 K, 1.013 bar, we can correct ΔH_{rx}^o and ΔG_{rx}^o to the temperature and pressure conditions of interest using the techniques developed in Chapter 5. This again involves the use of heat capacities C_p and a *PVT* EOS. For example, for reactions involving only ideal gases

$$\Delta H_{rx}\,(T) = \Delta H_{rx}^o(T = 298\text{ K}) + \sum_i \int_{298}^{T} v_i C_{p_i}^o\, dT \tag{8-167}$$

$$\Delta S_{rx}\,(T) = \Delta S_{rx}^o(T = 298\text{ K}) + \sum_i \int_{298}^{T} v_i C_{p_i}^o\, d\ln T \tag{8-168}$$

where

$$\Delta S_{rx}^o\,(T = 298\text{ K}) = \sum_i v_i S_i^o\,(T = 298\text{ K}) = \sum_i v_i \left[\int_0^{298} C_{p_i}^o\, d\ln T + \sum_{j=1}^{\pi} \frac{\Delta H_{t,j}}{T_{t,j}}\right] \tag{8-169}$$

with $\Sigma \Delta H_t/T_t$ accounting for the possible occurrences of (π) phase changes occurring from 298 to 0 K and S_i^o (0 K) set to zero. Frequently S_i^o (298 K) values are tabulated. Appendix G lists the standard ΔH_f^o and ΔG_f^o values for a number of common substances. In Chapter 13, Section 13.3, we will show how molecular group contribution methods can be used to estimate ΔH_f^o and ΔG_f^o values when data are not available.

8.9 Summary

This chapter describes various constitutive property relationships that model or correlate the behavior of pure materials. All of the relationships presented are empirically-based, at least in part; and some are anchored to molecular-level theory in a direct way as well. To fully describe process changes, primitive properties that characterize the macroscopic states of systems are needed. In particular, volumetric *PVT* equations of state and heat capacity correlations usually form the basis of these constitutive relationships. Although these representations do not require a theoretical basis, they often use some element of molecular theory to structure the mathematical form of the equations selected or to test for limiting asymptotic behavior. Such is the case for certain *PVT* EOSs like the truncated virial form of the BWR EOS. In other cases, the molecular theory itself may provide a tractable route to generate the necessary macroscopic property, for example in the quantitative conversion of vibrational spectroscopic data into ideal-gas state heat capacity estimates. Thus to solve engineering thermodynamics problems to expected levels of accuracy, the physical properties of the materials involved must be quantitatively well-characterized. This requirement almost always requires a combined treatment of experimental data and relevant molecular theory.

With *PVT* and ideal-gas state relationships in hand for a pure fluid, one can unambiguously generate derived property changes for H, S, U, and V between any two states, and for A and G along any isothermal path. A most effective method of doing this involves the use of departure functions which are defined as the difference between the property of a fluid in its real state at T, P, and V and an ideal-gas state at the same T and P (or equivalently T and $V^o = RT/P$).

In this chapter we also briefly treated polymer liquids, and incompressible liquid and solid phase behavior. Various representations of thermodynamic properties, including charts, diagrams, tables, and computer-generated data bases, were discussed for different engineering applications. Finally, methods and conventions used to correlate enthalpies and Gibbs free energies of formation were covered.

References

Abbott, M.M. (1973), "Cubic equations of state (review)," *AIChE J.* 19, p 596-601.

Abbott, M.M. (1979), "Cubic equations of state: an interpretive review in *Equations of State in Engineering and Research*, ACS Adv. Series **182**, p 47-70.

Abbott, M.M. (1987) in Smith, J.M. and H.C. Van Ness, *Introduction to Chemical Engineering Thermodynamics*, 4th ed. McGraw-Hill, New York, p 91-92.

Abbott, M.M., (February, 1989), "Thirteen ways of looking at the van der Waals equation", *Chem. Eng. Prog.*, p 25-37.

Alberty, R.A. and R.J. Silbey (1992), *Physical Chemistry*, Wiley, New York, p 575 ff.

Ben-Amotz, D. and D.R. Herschbach (1990), "Correlation of Zeno (Z = 1) line for supercritical fluids with vapor-liquid rectilinear diameters," *Israel J. Chem.*, **30**, p 59-68.

Benedict, M., G.B. Webb, and L.C. Rubin (1940), "An empirical equation for thermodynamic properties of light hydrocarbons and their mixtures," *J. Chem. Physics*, **8**, p 334-345.

Bett, K.E., J.S. Rowlinson, and G. Saville (1975), *Thermodynamics for Chemical Engineers*, MIT Press, Cambridge, MA, Chapter 4.

Brennecke, J.F. and C.A. Eckert (1989), "Phase equilibria for supercritical process design," *AIChE J.*, **35**, p 1409-1427.

Bruno, T.J. and J.F. Ely (1991), *Supercritical Fluid Technology: Review of Modern Theory and Applications*, CRC Press, Boca Raton, FL.

Carnahan, N.F. and K.E. Starling (1969), "Intermolecular repulsions and the equation of state for fluids," *J. Phys. Chem.*, **51**, p 635-636.

Chou, G.F. and J.M. Prausnitz (1989), "A phenomenological correction to an equation of state for the critical region," *AIChE J.*, **35**, p 1487-1496.

Eisele, U. (1990), *Introduction to Polymer Physics*, Springer-Vetlag, New York, Chapter 10, p 107-137.

Hartmann, B. and M.A. Haque (1985), "Equation of state for polymer liquids," *J. of Polymer Science*, **30**, p 1553-1563.

Johnston, K.P. and C.A. Eckert (1981), "An analytical Carnahan-Starling-van der Waals model for solubility of hydrocarbon solids in supercritical fluids," *AIChE J.*, **27**, p 773-779.

Johnston, K.P., D.H. Ziger, and C.A. Eckert (1982), "Solubilities of hydrocarbon solids in supercritical fluids: the augmented van der Waals treatment," *Ind. Eng. Chem. Fund.*, **21**, p 191-197.

Kutney, M.C., V.S. Dodd, K.A. Smith, H.J. Herzog, and J.W. Tester (1996), "A hard-sphere volume-translated van der Waals equation of state for supercritical process modeling. Part 1. Pure components," *Fluid Phase Equilibria*.

Kyle, B.G. (1992), *Chemical and Process Thermodynamics*, 2nd ed., Prentice Hall, Englewood Cliffs, NJ, Chapter 3, section 8.5.

Lewis, G.N., M. Randall, K.S. Pitzer, and L. Brewer (1961), *Thermodynamics*, 2nd ed McGraw-Hill: New York, p 419-447 in Chapter 27 and Appendix 1.

Lydersen, A.L., R.A. Greenkorn, and O.A. Hougen (1955), *Generalized Properties of Pure Fluids*, College of Engineering, University of Wisconsin, Engineering Experimental Station, report #4, Madison, WI, Oct. 1955.

Martin, J.J. (1967), "Equations of state," *Ind. Eng. Chem*, **59**, p 34-56.

Martin, J.J. and T.G. Stanford (1974), "Development of high precision equations of state for wide ranges of density utilizing a minimum of input information: example argon," *AIChE Symp. Series*, **70**, p 1-13.

Martin, J.J. (1979), "Cubic equations of state, which?", *Ind. Eng. Chem. Fundam.*, **18**(2), p 81-97.

Mathias, P.M., Naheiri, T., and E.M. Oh (1989), "A density correction for the Peng-Robinson equation of state," *Fluid Phase Equilibria*, **47**, p 77-87.

Meissner, H.P. and R. Sefarian (1951), "*PVT* relations of gases," *Chem. Eng. Progress*, **47**, p 579-584.

Peneloux, A, E. Rauzy and R. Freze (1982), "A consistent correction for Redlich-Kwong-Soave volumes," *Fluid Phase Equilibria*, **8**, p 7-23.

Peng, D-Y. and D.B. Robinson (1976), "New two-constant equation of state," *Ind. Eng. Chem. Fundam.*, **15**, p 59-64.

Pitzer, K.S. (1995), *Thermodynamics*, McGraw-Hill, NY, Chapter 20, p 364-374 and Chapter 9, p 122-153.

Pitzer, K.S., D.Z. Lippmann, R.F. Curl, C.M. Huggins, and D.E. Peterson (1955), *J. Amer. Chem. Soc.*, **77**, p 3433.

Pottiger, M.T. and R.L. Laurence (1984), "The P-V-T behavior of polymeric liquids represented by the Sanchez-Lacombe equation of state," *J. of Polymer Science*, **22**, p 903.

Redlich, O. and J.N.S. Kwong (1949), "On the thermodynamics of solutions," *Chem. Rev.*, **44**, p 233-244 (1949).

Sandler, S.I. (1989), *Chemical and Engineering Thermodynamics*, 2nd ed., Wiley, NY, esp. p 146-150, 167-185.

Smith, J.M. and Van Ness, H.C., (1987), *Introduction to Chemical Engineering Thermodynamics*, McGraw-Hill, 4th ed., esp. p 54-63, 80-98.

Soave, G.S. (1972), "Equilibrium constants from a modified Redlich-Kwong equation of state," *Chem. Eng. Sci.*, **27**, p 1197-1203.

Soave, G.S. (1984), "Improvement of the van der Waals equation of state," *Chem. Eng. Sci.*, **39**, p 357-369.

Starling K.E. and M.S. Han (1972), "Thermo data refined for LPG, part 14: Mixtures," *Hydrocarbon Processing*, **51**, p 129-172.

Starling K.E. (1973), *Fluid Thermodynamic Properties for Lighter Hydrocarbons*, Gulf Publishing, Houston, TX.

Starling, K.E. (1971), "Thermo data refined for LPG, part 1: equation of state and computer prediction," *Hydrocarbon Processing*, **50**, p 101.

van der Waals, J.D. (1873), *Over de continuiteit van den gas-en vloeistof toestand*, Doctoral dissertation, A.W. Sijthoff, Leiden, The Netherlands (see also van der Waals, Nobel Prize lecture, 1903).

Van Ness, H.C. and M.M. Abbott (1982), *Classical Thermodynamics of Non-Electrolyte Solutions*, McGraw Hill, Figures 4-6, p 102.

Vera, J.H. and J.M. Prausnitz (1972), "Interpretative review-generalized van der Waals theory for dense fluids," *Chem. Eng. J.*, **3**, p 1-13.

Walas, S.M. (1985), *Phase Equilibria in Chemical Engineering*, Butterworth, Boston, esp. pp. 1-26, 37-78.

Weber, H.C. and H.P. Meissner (1957), *Thermodynamics for Chemical Engineers*, Wiley, New York.

Xu, J., and D.R. Herschbach (1992), "Correlation of Zeno line with acentric factor and other properties of normal fluids," *J. Phys. Chem.*, **96**, p 2307-2312.

Sources of Property Data and Correlation Methods

ASHRAE Thermodynamic Properties of Refrigerants (1969), Amer. Soc. of Heating, Refrigeration, and Air Conditioning Engineers, New York.

Benson, S.W. (1968), *Thermochemical Kinetics*, Wiley, New York.

Edmister, W.C. (1958), *Petroleum Refiner* 37(4), p 173 and (1961) *Applied Hydrocarbon Thermodynamics*, Gulf Publishing, Houston.

Haar, L., J.S. Gallagher, and G.S. Kell (1984), *NBS/NRC Steam Tables*, Hemisphere Publishing, New York.

Hougen, O.A., K.M. Watson, and R.A. Ragatz (1959), *Chemical Process Principles, Part II Thermodynamics*, Wiley, New York.

Hougen, O.A., K.M. Watson, and R.A. Ragatz (1960), *Chemical Process Principles Charts*, 2nd ed., Wiley, New York.

JANAF (1971-1975), *Thermochemical Tables*, 2nd ed and supplements, National Standards Reference Data Service, National Bureau of Standards, Washington, D.C.

Keenan, J.S., F.G. Keyes, P.G. Hill, and J.G. Moore (1969), Steam Tables-International Edition in SI Units, Wiley, New York.

Lee, B.I. and M.G. Kesler (1975), *AIChE J.*, **21**, p 510-527.

NBS Circular 500 and revisions/supplements (1952-1971), "Selected values of chemical thermodynamic properties," National Bureau of Standards, Washington, D.C.

Perry, R.H., D.W. Green, and J.O. Maloney (1984), *Perry's Chemical Engineering Handbook*, 6th edition, McGraw-Hill, New York.

Reid, R.C., J.M. Prausnitz, and B. Poling (1987), *The Properties of Gases and Liquids*, 4th ed., McGraw-Hill, New York.

Problems

8.1. A skeptical engineer questions whether the efficiency of a Carnot engine is always given by

$$\eta = \frac{-W_E}{Q_H} = \frac{T_H - T_C}{T_H}$$

where W_E = work of Carnot engine
Q_H = heat interaction with hot reservoir
T_H, T_C = temperature of hot (cold) reservoir

She agrees that the relation shown above is correct for a Carnot engine with an ideal gas as the working fluid but indicates that it may be in error if the working fluid is not an ideal gas. She suspects that the efficiency decreases as the working fluid becomes more nonideal.

How would you clarify the situation?

8.2. Our research laboratory has synthesized a new material and the properties of this substance are being studied under conditions where it is always a vapor. Two sets of experiments have been carried out and they are described below. Given the results of these experiments, the relationship $P\underline{V} = NCT$ is proposed to describe the $P\underline{V}T$ properties of the material (C is a constant). If you agree with this proposal, show a rigorous proof. If you do not agree, either prove that the relationship cannot be applicable or describe clearly what additional experiments you would recommend, and demonstrate how you would use these data to show whether or not $P\underline{V} = NCT$.

Experiment A: Vapor is contained in subsystem I of a rigid, well-insulated container (Figure P8.2). Subsystem II is initially evacuated. The partition between I and II is broken and gas fills both I and II. Over a wide range of initial temperatures, pressures, and volumes of I, the final temperature, after expansion, equals the initial temperature.

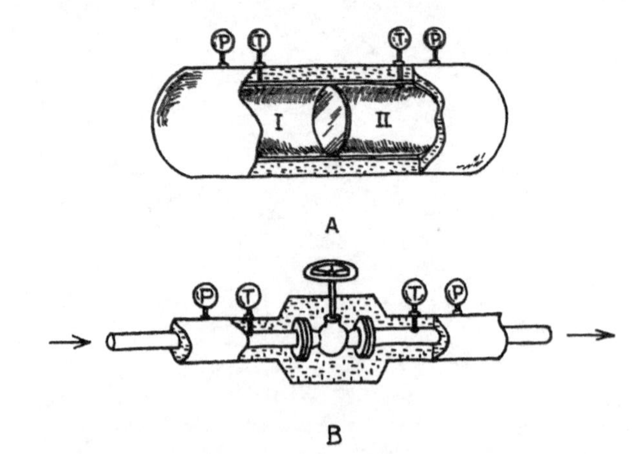

Figure P8.2

Experiment B: Vapor flows in an insulated pipe and through an insulated throttling valve wherein the pressure is reduced. Over a wide range of upstream temperatures and pressures, as well as downstream pressures, the temperature of the vapor does not change when the gas flows through the valve.

8.3. In Chapter 5 the concept of equations of state was introduced for the Fundamental Equation in energy representation. If we were, however, to use the Helmholtz energy representation of the Fundamental Equation, the three equations of state for a pure material would be

$$\underline{S} = f_1(T, \underline{V}, N) \tag{A}$$

$$P = f_2(T, \underline{V}, N) \tag{B}$$

$$\mu = f_3(T, \underline{V}, N) \tag{C}$$

Suppose that Eq. (B) were to be given by the extensive form of the Redlich-Kwong equation,

$$P = \frac{NRT}{\underline{V} - bN} - \frac{aN^2}{T^{1/2}\underline{V}(\underline{V} + bN)}$$

and the ideal-gas heat capacity were expressed as

$$C_v^o = A + BT + CT^2$$

(a) Derive an analytical expression for Eq. (A) assuming a reference state of an ideal gas at T^o, V^o.

(b) Since only two of the three equations of state are independent, derive Eq. (C) using the Redlich-Kwong equation and the result obtained for Eq. (A).

8.4. We have a constant-volume, closed vessel filled with vapor dichlorodifluoromethane. We plan to heat this vapor and would like to know how the specific entropy of the vapor varies with pressure. Derive a general relation to allow one to calculate the desired derivative, $(\partial S/\partial P)_V$ assuming that we know the total volume of the vessel, the moles of vapor, C_p as a function of T and P, and an equation of state.

Illustrate your result at the start of the heating process, where $T = 365.8$ K, $P = 16.5$ bar, and total volume = 1.51×10^{-3} m^3. Assume that at this condition, $C_p = 94.9$ J/mol K and the Redlich-Kwong equation of state is applicable with $a = 20.839$ J m^3 K$^{1/2}$/mol^2 and $b = 6.725 \times 10^{-5}$ m^3/mol

8.5. Gas streams are often cooled either by expansion through a throttling valve or in a power turbine. To measure the relative efficiency of these two processes, the term κ_H/κ_S is often quoted. The throttling (or Joule-Thompson) coefficient is defined as $\kappa_H = (\partial T/\partial P)_H$, whereas the isentropic coefficient is $\kappa_S = (\partial T/\partial P)_S$.

Calculate (κ_H/κ_S) for CO_2 at 319.4 K over a range of reduced pressures from 0.1 to 20. What conclusions can you draw? Also determine the compressibility factor over the same pressure range at 319.4 K. Use the Peng-Robinson equation of state in your calculations. For CO_2, $T_c = 304.2$ K, $P_c = 73.76$ bar, and $\omega = 0.225$.

8.6. Sulfur dioxide vapor at 520 K and 100 bar fills one-half of a rigid, adiabatic cylinder. The other half is evacuated, and the two halves are separated by a metal diaphragm. If this should rupture, what would be the final temperature and pressure? Assume that the vapor is well mixed and that expansion is sufficiently rapid so that negligible heat transfer occurs between the walls and the SO_2 vapor.

For SO_2, $T_c = 430.8$ K, $P_c = 78.8$ bar, $V_c = 1.22 \times 10^{-4}$ m^3/mol, $\omega = 0.251$, and C_p^o (J/mol K) is given by

$$C_p^o = 23.852 + 6.699 \times 10^{-2}\, T - 4.961 \times 10^{-5} T^2 + 1.328 \times 10^{-8} T^3$$

with T in Kelvins.

8.7. Saturated steam fills a rigid vessel at 472.1 K. It is desired to reduce the pressure and temperature of the steam simultaneously but to maintain always a saturated vapor. The cylinder is fitted with a piston that has negligible friction to allow for changes in volume. Heat transfer is allowed. Data for saturated steam are given below.

T(K)	P (bar)	V (m^3/kg)	H (kJ/kg)	S (kJ/kg K)
469.3	14.321	0.1378	2789.3	6.4581
472.1	15.188	0.1300	2791.0	6.4372
474.9	16.097	0.1230	2792.6	6.4163

What is the magnitude of this heat interaction at the start of the pressure reduction?

8.8. Estimate the value of C_v for propylene vapor at 125°C at a density of 92.28 kg/m^3 (48.28 bar). Perform the calculations using the experimental *PVT* data given below. Also, use the Peng-Robinson equation of state to carry out the estimation. Compare your results with the value of 1,857 J/kg K reported by N. de Nevers and J.J. Martin [(1960), *AIChE J.*, **6**, p 43].

1. C_p^o, ideal gas: The isobaric heat capacity of propylene in the ideal-gas state is given by

$$C_p^o = 88.16 + 5.574T - 2.757 \times 10^{-3} T^2 + 5.239 \times 10^{-7} T^3 \text{ (J/kg K)}$$

2. *PVT* data [(1953), *Physica*, **19**, p 287]: The data are presented in terms of amagat units as a function of pressure and temperature. An amagat unit is the ratio of the true density to the density at 1 atm pressure and 273.15 K, which for propylene is 4.544×10^{-5} g-mol/cm^3.

Density (amagat)	Pressure (bar) 100°C	125°C	150°C
1.000	1.3977	1.4932	1.5887
6.4976	8.6045	9.2626	9.9163
8.3409	10.8438	11.7045	12.5582
10.1832	12.9956	14.0657	15.1263
12.1074	15.1544	16.4500	17.7329
14.3997	17.6096	19.1839	20.7397
16.6494	19.8993	21.7574	23.5909
18.3557	21.5589	23.6388	25.6901
19.5100	22.6444	24.8775	27.0779
23.0078	25.7640	28.4762	31.1408
27.6638	29.5207	32.9048	36.2209
32.1907	32.7702	36.8431	40.8254
36.9366	35.7822	40.6151	45.3266
41.6597	38.4067	44.0276	49.4979
42.4108	38.7928	44.5451	50.1329
46.1992	40.6108	47.0201	53.2493
47.3154	41.1106	47.7175	54.1293
50.9531	42.6191	49.8806	56.9300

3. Properties for use in the Peng-Robinson equation of state: $T_c = 365.0$ K, $P_c = 46.20$ bar, $\omega = 0.148$, $m = 42.081$ g/mol

8.9. We wish to design a device to make dry ice (solid CO_2) which could be attached to the outlet tube of a cylinder of CO_2. Assume that the CO_2 in the cylinder consists of a liquid and gas in equilibrium at 300 K (67.01 bar), but that the outlet is connected only to the liquid phase by means of an eductor tube (Figure P8.9).

The enthalpy of vaporization of CO_2 at 300 K (67.01 bar) is 103.7 J/g and the enthalpy of sublimation at 1 bar is 570 J/g. The temperature of the solid-vapor mixture of CO_2 at 1 bar is 194.7 K. The heat capacity of CO_2 in the ideal-gas state may be approximated by

$$C_p \text{ (J/mol K)} = 26.62 + 0.0363T \text{ (K)}$$

between 194 and 300 K. Other properties are given in Appendix G.

Suggest what you think would be a good engineering way to design such a device. How much dry ice at 1 bar do you think you could make per kilogram of liquid CO_2 drawn from the cylinder?

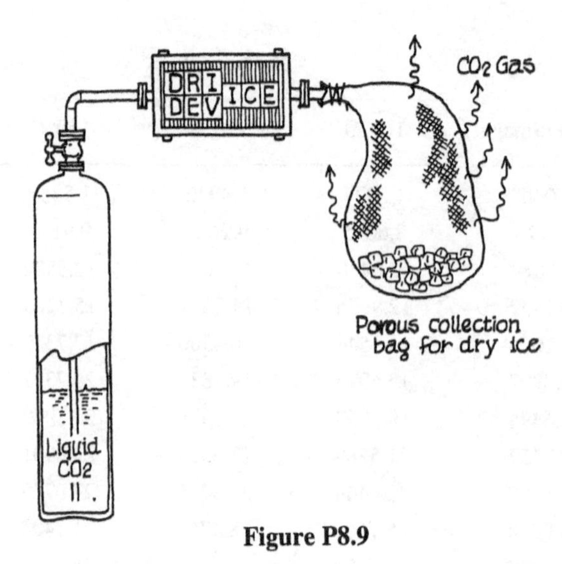

Figure P8.9

8.10. An isothermal compressor that operates essentially reversibly takes suction from a manifold containing CO_2 at 1.01 bar and 300 K. Gas is compressed and fed to a well-insulated 8.50-m^3 storage tank originally containing CO_2 also at 1.01 bar and 300 K. Assume that the CO_2 *PVT* properties may be correlated with the Redlich-Kwong equation of state with

$$a = 6.4648 \text{ J m}^3 \text{ K}^{1/2}/\text{mol}^2 \quad \text{and} \quad b = 2.9704 \times 10^{-5} \text{ m}^3/\text{mol}$$

The heat capacity of CO_2 in the ideal-gas state may be expressed as

$$C_p^o = 19.795 + 7.3436 \times 10^{-2}T - 5.6019 \times 10^{-5}T^2 + 1.7153 \times 10^{-8}T^3$$

with T in Kelvins, and C_p^o is in J/mol K.

(a) What is the final temperature of the gas in the tank when the pressure reaches 48.3 bar?

(b) Plot the tank temperature as a function of tank pressure; on the same plot show the cumulative compressor work required.

(c) Repeat part (b) assuming that CO_2 is an ideal gas with a constant $C_p \approx 36.8$ J/mol K.

8.11. A large tank contains steam at its critical point of 647.4 K and 221.2 bar. It is connected to a small well-insulated 0.1 m^3 rigid tank which, initially, has a very low pressure. Steam is allowed to enter the small tank until the pressure increases to 1 bar.

Assume that no heat transfer occurs between tanks and that the state of the steam in the large tank does not change during the process. Some data for steam are given below.

Thermodynamic Properties of Saturated Water

T(K)	P (bar)	Liquid volume (m^3/kg)	Volume change in vaporization (m^3/kg)	Liquid enthalpy (kJ/kg)	Enthalpy change in vaporization (kJ/kg)
647.4	221.2	3.17×10^{-3}	0	2.107×10^3	0
373.0	1.00	1.04×10^{-3}	1.672	4.191×10^2	2.257×10^3

(a) What is the temperature of the steam in the small tank? How many kilograms entered?

(b) Suppose that the "large" tank had contained CO_2 at 647.4 K and 221.2 bar instead of steam. What would have been the temperature in the 0.1-m^3 tank when the pressure had reached 1 bar? How many kilograms of CO_2 would have been transferred? In this case, use the Peng-Robinson equation of state to describe the properties of CO_2. For CO_2, the Peng-Robinson parameters are: T_c = 304.2 K, P_c = 73.76 bar, and $\omega = 0.225$; C_p as a function of temperature is given in Problem 8.10.

(c) If instead of either steam or CO_2, we had had helium gas in the large tank (again at 647.4 K and 221.2 bar) and had expanded this gas into the evacuated small tank, what would have been the temperature when the pressure attained 1 bar? How many kilograms would have been transferred? Assume helium to be an ideal gas at all temperatures and pressures; also let $C_p = 20.9$ J/mol K = constant.

8.12. Two moles of a gas are in a closed vessel at 10 bar and 300 K. By some series of irreversible operations, the pressure increases to 20 bar and the temperature to 320 K. Three kilojoules of work is done *on* the gas. Assume that the *PVT* properties of the gas can be modeled by an equation of state of the form $P(V - b) = RT$ where $b = 5 \times 10^{-5}$ m^3/mol. $C_v = 0.1\ T$ (K) at 10 bar over the temperature range of interest (J/mol K).

What is the net heat interaction of the system with the environment? What is the change of ΔH and ΔA for this change of state?

8.13. Memo to our thermodynamics Consultant:

As our thermo con-person, we need your help in developing a simple fire extinguisher. This device is nothing more than a small 30-liter vessel charged with liquid R-12 (dichlorodifluoromethane) at room temperature (assume 294.3 K). We are now charging so that the total R-12 mass in the extinguisher initially is 38 kg. The initial pressure reads 5.853 bar, so I presume we have an equilibrium vapor-liquid mixture.

When used, liquid R-12 is discharged so that the exit port is on the bottom. I am trying to sell the unit to a client but she refuses to negotiate until we can show her some calculations as to how the pressure in the vessel will change as liquid is expelled. She has indicated to me that she believes the pressure drop will be so great, even for a small amount of liquid expelled, that the utility of the device will be quite limited (Figure P8.13). She wants me to add an inert gas to help keep the pressure constant. I do not favor this idea, but, frankly, I am worried about the possible drop in pressure.

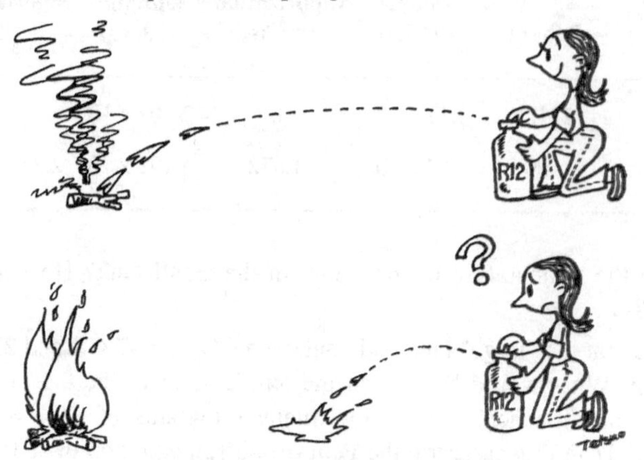

Some data for R-12 are shown below.

Some data for R-12 are shown below.

Data for Saturated Liquid and Vapor Dichlorodifluoromethane (R-12)

T (K)	P (bar)	V^L (m^3/kg x 10^4)	V^V (m^3/kg x 10^2)	H^L (kJ/kg)	H^V (kJ/kg)	S^L (kJ/kgK)	S^V (kJ/kgK)
288.8	4.994	7.437	3.486	50.628	194.009	0.193	0.690
289.9	5.158	7.458	3.378	51.686	194.456	0.197	0.690
291.0	5.325	7.480	3.274	52.744	194.900	0.201	0.689
292.1	5.497	7.502	3.174	53.807	195.342	0.204	0.689
293.2	5.673	7.524	3.078	54.873	195.782	0.208	0.688
294.3	5.853	7.547	2.985	55.940	196.219	0.211	0.688
295.4	6.037	7.570	2.895	57.013	196.870	0.215	0.688

Please write me a concise memorandum detailing how you think the pressure and temperature vary as liquid is removed. Do we have to use an inert gas?

8.14. A 1-m^3 tank is filled with supercritical CO_2 at 300 bar and 340 K. It is to be connected to a tank of identical size that contains, initially, CO_2 at 1 bar and 290 K. Both tanks are rigid and adiabatic. Assume no heat interactions between the tanks via the connecting line and neglect heat transfer to the tank walls. The valve in the interconnecting line is opened. Assume that the Peng-Robinson equation of state is applicable to CO_2. Critical data and C_p as a function of temperature are given in Appendix G.

What is the temperature and pressure in the "low-pressure" tank when the pressure in the "high-pressure" tank has dropped to 150 bar? What fraction of the CO_2 in the high-pressure tank has been transferred?

8.15. Oxygen gas at 150 K and 30 bar is to be pressurized to 100 bar in a compressor with an efficiency of 80% (based on an isentropic process). If the flow rate is 10 kg/s, what would be the power required? Carry out your calculations using:

(a) The temperature-entropy diagram in Figure P8.15.

(b) The Peng-Robinson equation of state. Data in this case are: $T_c = 154.6$ K, $P_c = 50.46$ bar, and $\omega = 0.021$. Also assume that C_p is 29.3 J/mol K, independent of temperature.

8.16. We plan to fill carbon dioxide cylinders from a supply manifold that is at 298 K and 1 bar (see Figure P8.16). Gas will be fed to the suction side of a compressor that operates in an adiabatic manner with an efficiency of 85%. Gas is discharged to a small heat exchanger and then to the cylinders.

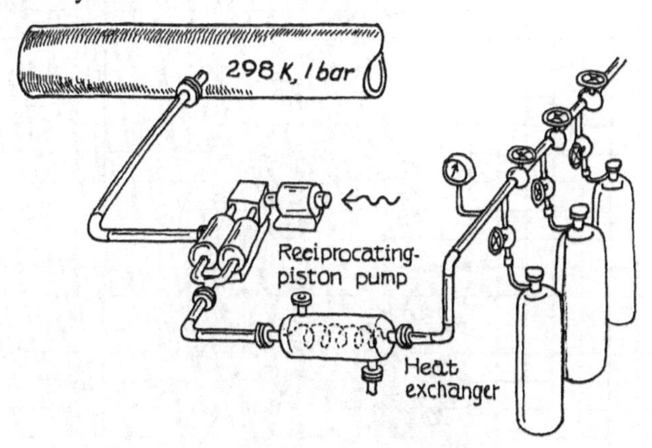

Figure P8.16

Initially, each cylinder contains CO_2 at 2 bar. The final pressure, after filling, is 140 bar. It may be assumed that the CO_2 leaves the heat exchanger at 298 K and, due to the high heat capacity of the bottle walls, the gas in the cylinder remains essentially at 298 K during filling. Each cylinder has a volume of 0.025 m^3 (25 liters).

Determine the minimum power rating of the motor if we want to fill one cylinder in 3 min. Also, what is the total work required for a filling? Use the Peng-Robinson equation of state (see Appendix G for data). the vapor pressure of liquid CO_2 at 298 K is 64.09 bar.

8.17. Differential scanning calorimeters (DSC) are now being employed to measure liquid heat capacities. Generally, liquid is placed inside a small ampule that has a locked cap to prevent losses. The energy added to effect a small temperature rise of the ampule and contents may be accurately ascertained with a DSC. In a separate experiment, the energy required to heat only the empty ampule (and cap) is also determined. The liquid and associated vapor are assumed to be in equilibrium at all times and no inert gas is present.

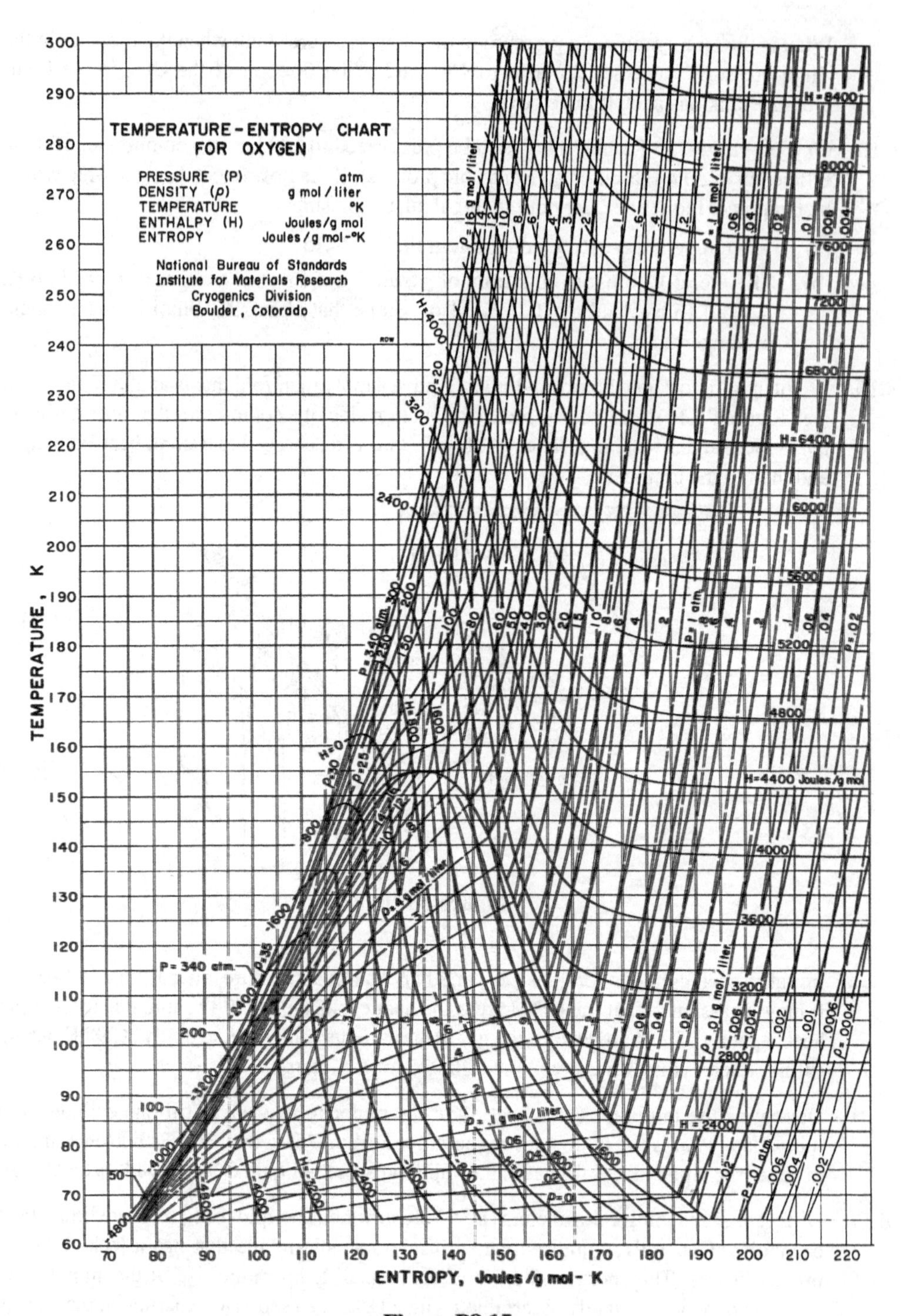

TEMPERATURE - ENTROPY CHART FOR OXYGEN
PRESSURE (P) atm
DENSITY (ρ) g mol / liter
TEMPERATURE °K
ENTHALPY (H) Joules/g mol
ENTROPY Joules / g mol-°K
National Bureau of Standards
Institute for Materials Research
Cryogenics Division
Boulder, Colorado
TEMPERATURE, K
ENTROPY, Joules /g mol- K

Figure P8.15

Prepare a rigorous analysis to indicate how one should carry out experiments to attain the highest accuracy in the measurement of the saturated liquid heat capacity, $(\partial H/\partial T)_{sat}$. In particular, comment on the effect of the initial volume fraction of vapor at the start of the experiment. Demonstrate your analysis for the case of isobutane at (a) 261.5 and (b) 316.5 K. Some property data are given below.

Sample Case: Isobutane

	(a)	(b)
T (K)	261.5	316.5
P (bar)	1.103	5.738
V^V (sat vap) (m^3/kg)	3.549×10^{-1}	6.761×10^{-2}
V^L (sat liq) (m^3/kg)	1.685×10^{-3}	1.897×10^{-3}
ΔH_v (kJ/kg)	3.643×10^{2}	3.026×10^{2}
dV^L/dT (m^3/kg K)	3.146×10^{-6}	4.944×10^{-6}
dV^V/dT (m^3/kg K)	-1.236×10^{-2}	-1.708×10^{-3}
$(\partial H^L/\partial T)_{sat}$ (kJ/kg K)	2.345	2.544
$(\partial H^V/\partial T)_{sat}$ (kJ/kg K)	1.424	1.298

8.18. Hydrogen is to be liquefied in the cycle illustrated in Figure P8.18. High-pressure gas is precooled to 40 K and passed into a heat exchanger that employs saturated hydrogen recycle gas as the coolant. The high-pressure cold vapor is then expanded to 1 atm across an insulated expansion valve. The liquid fraction is separated and the vapor recycled as shown. The heat transfer area of this final exchanger is very large.

(a) At what pressure should the compressor be operated to maximize the amount of liquid formed per kg hydrogen flow?

(b) Sketch the state of hydrogen during the process on both *T-S* and *H-P* diagrams.

(c) Determine the minimum work that would be required if the process were carried out reversibly.

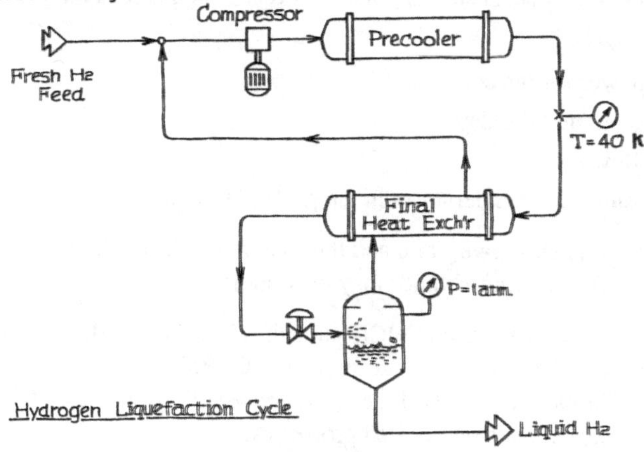

Figure P8.18

8.19.

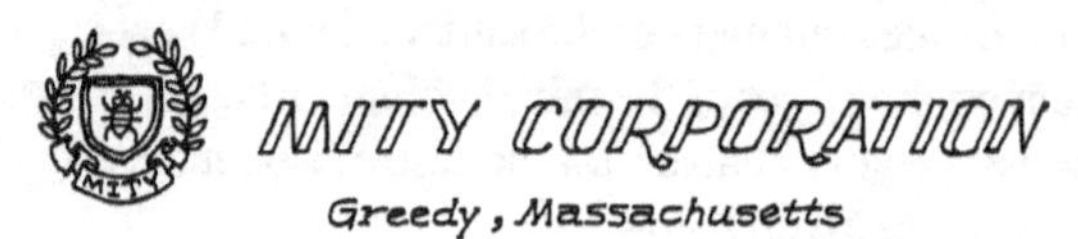

Phone : 7-11711
Cable : Eureak

Mr. James Longthorne
c/o Faroutlake, Maine

Dear Jim:

Please forgive this intrusion upon your well-deserved vacation, but a problem has arisen in the office that demands immediate attention and everyone in the design section is very busy except you and me. I am, therefore, sending this problem to you in the hope that you can handle it as soon as possible.

A client has requested us to make rough calculations on his proposed process to liquefy monochlorodifluoromethane ($CHClF_2$) as per the attached flow sheet. The process gas enters at (1) at 1 atm (1.013 bar) and 20°C and is compressed adiabatically to (2). The compressor operates at about 95% of a reversible, adiabatic device and a similar efficiency is believed applicable for the electric drive motor. The hot, high-pressure gas at (2) is cooled to 400 K by heat transfer in exchanger A and in so doing makes saturated steam at 400 K. (The feed to A is water, liquid, saturated at 400 K.)

From (3) the gas is further cooled in exchanger *B* (which is essentially infinite in area), then passes through a Joule-Thompson valve and into a liquid-vapor separator at 1 atm. Liquid is removed and saturated vapor is used as the cooling medium in exchanger *B*. The exit vapor at (*B*) is cooled and recycled, but do not worry about this part of the process.

Now, the customer demands (although we have our doubts) that there be a maximum amount of liquefaction per pass.

Please submit a flow sheet showing me your recommendations for:

(a) All line sizes
(b) Motor power for the compressor
(c) Steam generated (kg/day)
(d) Liquid flow rate
(e) All pressures and temperatures in the process steams

The flow rate at (1) is 4.54 kg/min and the customary flow velocities of 60 m/s (gas), 15 m/s (gas-liquid), and 3 m/s (liquid) may be assumed.

Also attached are all the data on $CHClF_2$ that the client has available. Sorry it is not more; I have taken the liberty to ship you the text R.C. Reid, J.M. Prausnitz, and B. Poling, Properties of Gases and Liquids, 4th ed., McGraw-Hill, New York, 1987, with this letter. I'm sure you can estimate the desired properties.

Hope to hear from you soon. Have a pleasant vacation.

With regards,

Edward Cel

Manager, Design section

EC/jh

Attachment I

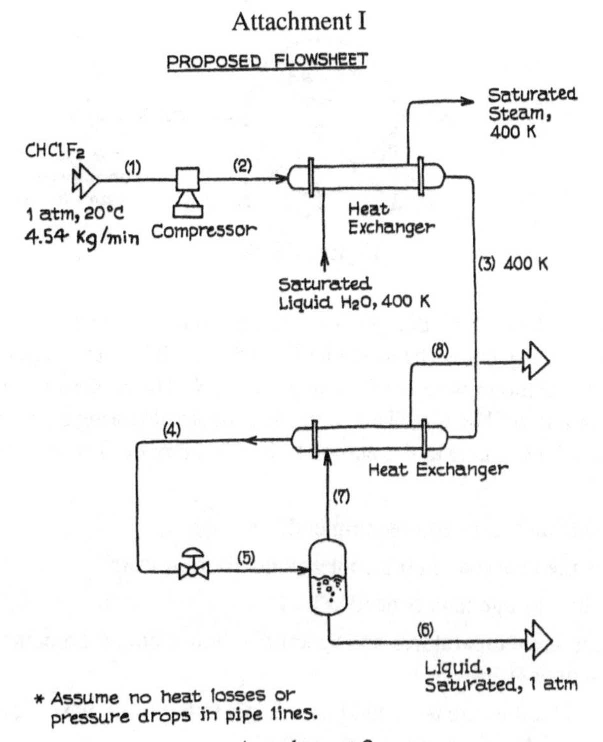

Attachment 2

Client Data on $CHClF_2$

Molecular weight = 86.48	Color: clear, water white
Boiling point at 1 atm = 232.4 K	Odor: ethereal
Melting point = 113 K	Toxicity: Group 5-A Classification of
Flammability: nonflammable	Underwriters' Laboratory Report MH-3134

8.20. While making a conceptual process design for a plant to be built in the near future, you are requested to analyze a small section of the overall design. In the section of concern to you, pure ethylidine chloride (1,1-dichloroethane) vapor is obtained from a still at about 363 K and 2.55 bar (see Figure P8.20). This vapor is to be compressed to 15.2 bar and the cooled and condensed to a saturated liquid. This hot liquid is to be stored as required for use later in the overall process. The estimated maximum flow rate is about 45.5 kg/min.

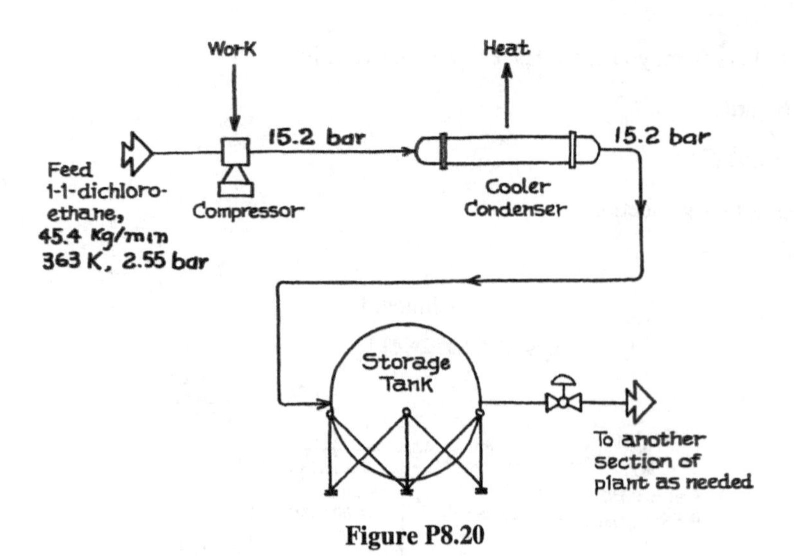

Figure P8.20

The compressor design has not yet been settled, but it is reasonable to assume that it operates adiabatically but with an efficiency of about 90% of theoretical. The pump drive unit is an electrical motor with an efficiency near 95%. The cooler-condenser is air-cooled. The storage tank is well insulated and must be capable of storing a 12-h flow of ethylidine chloride with a 10% ullage at the end of 12 h. A schematic flow sheet is given in Figure P8.20.

(a) What size motor do you recommend?

(b) What is the heat load in the cooler-condenser, in watts?

(c) What size storage tank is needed?

(d) What are the temperatures and specific volumes out of the compressor and cooler-condenser?

(e) If flow velocities are to be held near 7 m/s in the lines, what size piping do you recommend for all the connecting lines?

The *Handbook of Chemistry and Physics* lists a boiling point and liquid density at 20°C, and you may use these if desired. Use any correlation methods you desire.

8.21. We are faced with handling a stream of pure carbon monoxide at 300 atm (304 bar) and 197 K. We would like to liquefy as great a fraction as possible, but there seems to be some disagreement about the process to be used. One suggestion has been to expand this high-pressure fluid across a Joule-Thompson valve and take what liquid is formed [see Figure P8.21(a)].

Another suggestion is to expand the vapor in an adiabatic turbine to the saturation curve and then follow with a Joule-Thompson expansion [see Figure P8.21(b)]. The two-step operation has been suggested to avoid erosion of the turbine blades with a two-phase mixture.

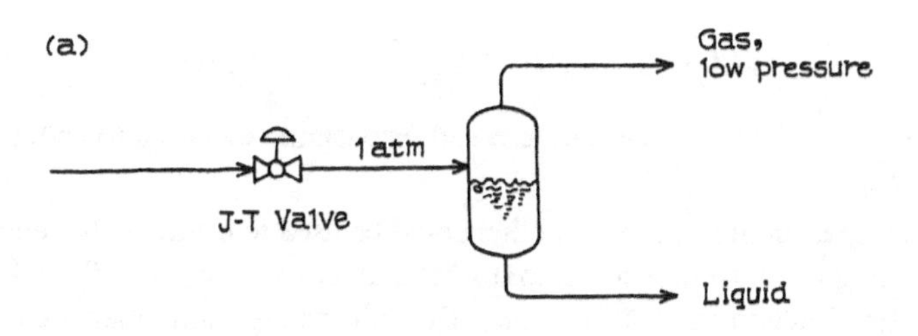

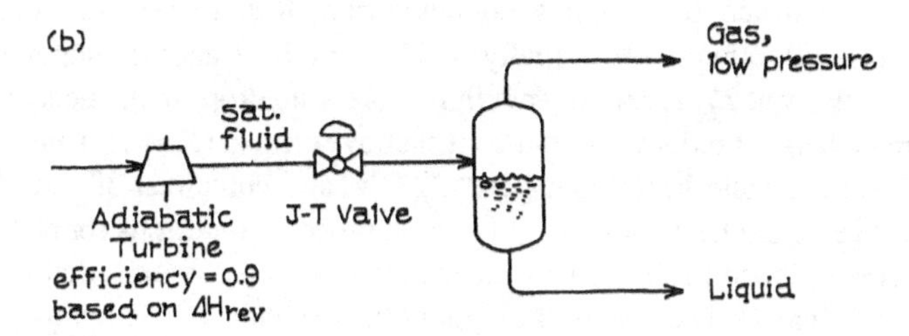

Figure P8.21

Evaluate these methods in order to indicate the fraction of initial gas one might be expected to liquefy in each. Also, suggest a better way to carry out the process to liquefy carbon monoxide that will give a larger fraction liquefied; draw a flow sheet, and calculate the fraction liquefied.

We wish we could supply you with thermodynamic property data but, unfortunately, we do not have access to any at the present time. Use any estimation methods you can find.

8.22. You have been asked to evaluate the cooling effect that would occur if a saturated vapor stream of CFC Freon R-114 at 1 MPa, 82°C is expanded across an insulated valve to 0.1 MPa (approximately 1 atm).

Your only resources for thermodynamic property data are:

(i) Critical constants for R-114 are known (P_c, V_c, T_c)

(ii) $\dfrac{C_p^o(T_r)}{R} = \alpha + \beta T_r + \gamma T_r^2$ (for ideal gas state of R-114)

where α, β, γ are specified constants.

(iii) The PVT properties of R-114 are given by the Martin-Hou equation of state in somewhat modified form as:

$$P - \frac{P_c}{Z_c}\left[\frac{T_r}{(V_r - b)} + \frac{f_1(T_r)}{(V_r - b)^2} + f_6\,(T_r)\exp\,(-2aV_r)\right]$$

where the subscripts r and c indicate reduced and critical properties, respectively. The temperature-dependent functions $f_i(T_r)$ are given by

$$f_i(T_r) = A_i + B_i T_r + C_i e^{-KT_r} \; i = 1, ..., 6$$

with A_i, B_i, C_i, and K as well as a and b constants specified to about 7 significant figures.

Develop a suitable expression that could be used to estimate the temperature resulting from the expansion to 0.1 MPa in terms P_r, V_r, and T_r, α, β, γ, and the Martin-Hou constants A_i, B_i ...etc. Describe an algorithm that could be used if machine computations were employed.

8.23. You have been asked to design a countercurrent heat recuperator that will cool a supercritical waste stream (A) initially at T_A^i to preheat another supercritical process stream (B) initially at T_B^i. Assuming negligible pressure drops in the heat exchanger, both streams are isobaric at reduced pressures just above critical ($P_r = 1.1$) and they both have identical ideal-gas-state heat capacities, $C_p^o(T)$'s, and equations of state for their PVT properties. The actual heat capacity $C_p[T, P]$ at process conditions for each stream is not constant with P_r close to unity. In your analysis you can assume that the overall heat transfer coefficient U_0 is constant throughout the exchanger. The approach temperature difference ($\Delta T = T_A - T_B$) must always be greater than or equal to 10°C at any point in the exchanger. At the waste stream inlet/process stream outlet point $\Delta T = 30$°C.

(a) On the same plot qualitatively sketch the T–$\underline{H}$ curve for each stream and label points on the graph corresponding to a schematic sketch of the heat exchanger.

(b) Develop an expression for the heat transfer area of the exchanger in terms of U_0, the temperatures and mass flow rates and specify enthalpy of each stream (A and B).

(c) Given that $C_p^o(T)$ and $P_r = f(T_r, V_r)$ are available in similar form to that presented in problem (8.22), describe a suitable computational algorithm that could be used to estimate the area. The mass flow ratio may be treated as a design variable.

8.24. Gyro Gearloose, a world-class inventor, has come up with a new scheme that he claims will revolutionize the field of thermophysical property measurements. His invention is called the "High-Temperature Drop Calorimeter" because of its mode of operation. Samples of materials to be tested are first encapsulated in an inert closed crubile, suspended in an electrically-heated tube furnace, heated to some temperature of interest, and dropped into a block of solid copper that is well-insulated from the environment. The mass of copper is much greater than combined masses of the sample and crucible by at least two orders of magnitude. Sample temperatures are determined accurately by scanning optical pyrometry while copper block temperatures are monitored with thermocouples.

Gyro claims he can measure the heat of fusion as well as the liquid and solid heat capacities of such exotic materials as gold and silver. The U.S. Patent Office wants your expert opinion on whether the inventor's claims are true. If you think Gyro's calorimeter will work, describe a set of experiments to determine C_p for liquid and solid states and ΔH_{fusion} of pure gold. Explain how you would analyze your data and sketch a plot of H

vs. T from 0°C to 2,000°C. Gold's melting point is 1064°C, while its boiling point is 2800°C. The heat capacity of copper from 0°C to 100°C is given empirically as $C_p = A + BT$ where A and B are constants.

8.25. We have a 1 kg/s stream of high pressure (40 MPa) contaminated pressurized water at 300°C that we want to feed to a turbine to produce work. Unfortunately, the salts that contaminate the water would be corrosive to the turbine, so we presently do not recover any energy from this stream. Joe Sludge has proposed a process to remove the salts from our high pressure stream. The stream is heated in a heat exchanger from 300°C to 480°C, and is then throttled through a well insulated valve from 40 MPa to 25 MPa, which is the maximum safe operating pressure in later portions of his process. An existing fuel burner adds 386.4 kJ/s to the stream at constant pressure. The stream then enters Joe Sludge's proprietary isothermal, isobaric unit which removes the contaminating salts with no change in the state of the water (many salts have very low solubilities in water at this temperature and pressure). The clean process stream now is used to preheat the incoming contaminated water in a heat exchanger. The final conditions of the clean process stream are 361.4°C and 25 MPa. Is this proposed design possible?

In your quantitative analysis, you should assume that the contaminated water has the same properties as pure water (use steam tables for thermodynamic data), and that the salt removal does not significantly alter the mass flow. You should assume that the only pressure drop in the system is that through the insulated valve.

Proposed Design Conditions:

[$T_1 = 300$°C, $H_1 = 1{,}324.6$ kJ/kg] [$T_2 = 480$°C, $H_2 = 2{,}776.0$ kJ/kg]

[$T_5 = 500$°C, $H_5 = 3{,}162.4$ kJ/kg] [$T_6 = 361.4$°C, $H_6 = 1{,}711.0$ kJ/kg]

$\dot{q} = 386.4$ kJ/s and $\Delta T_{min} = 10$°C between streams in the heat exchanger

8.26. JoAnna and Joseph Gibbs, twin cousins of the famous Connecticut thermodynamicist, J. Willard Gibbs, have designed a heat exchanger for heating supercritical organic fluids with geothermal heat. They claim that by proper design of tubes/baffles and other internal components, they can eliminate heat transfer coefficient variations and reduce the tendency of the exchanger to "pinch." According to the Gibbs twins, the pressure drop for both streams is negligible. A hot, pressurized (300°C, 14 MPa) water stream at 100 kg/s is available from a geothermal well. In their design, heat is transferred to an entering supercritical ammonia (NH_3) stream that is exiting from a Rankine cycle feed pump at $P = 12.5$ MPa and $T = 40$°C. Steam tables or charts are available for the water properties, and an equation of state is available for NH_3 with $P_r = f(T_r, V_r)$. The critical compressibility, pressure, and temperature and C_p^o for NH_3 are: $Z_c = 0.242$, $P_c = 11.28$ MPa, $T_c = 405$ K, and $C_p^o/R = 4.3 - 1.84T_r + 3.20T_r^2$.

(a) For any specified minimum ΔT condition, is it possible to set the temperatures of either the exiting water or the exiting ammonia, or both exiting streams?

(b) Incorporating your answer from part (a), sketch the temperature profiles for both streams in the heat exchanger.

(c) Following the design criteria developed in parts (a) and (b), describe how you would estimate the ammonia flow rate and area of the heat exchanger if the overall heat transfer coefficient is constant at 500 W/m^{2}K and the minimum approach temperature between streams is to be 10°C. You do not have to calculate numbers, just describe the solution procedure.

8.27. Using the Peng-Robinson equation of state generate a pressure-volume diagram for carbon dioxide at the reduced temperature isotherms, $T_r = 2.0, 1.1, 1.0, 0.9$, and 0.8. Plot your results. If you like, you can plot more isotherms. The critical properties and acentric factor (ω) for CO_2 are: $T_c = 304.2$ K, $P_c = 73.8$ bar, $V_c = 93.9$ cm^3/mol, and $\omega = 0.225$. Determine the real roots of the equation in volume for the specified conditions given below, as well.

Temperature (K)	Pressure (MPa)
220	0.1
220	0.6
220	1.0
250	1.788
280	4.0
280	4.16
280	5.0
304.2	7.376
500	1.0

8.28. Our boy-genius Rocky Jones has come up with another great idea for, as he puts it, "avoiding the inherent inaccuracy of many equations of state (EOS) in the high pressure, near critical region." If you want to know any derived thermodynamic property (H, S, U, A, G, . . ., etc.) for a pure component fluid at a specified upstream temperature and high pressure, you can use an insulated Joule-Thompson (J-T) valve to expand the fluid to a specified lower pressure. He claims that if you measure the temperature of the fluid at this lower pressure you can rigorously calculate the higher pressure properties using an EOS accurate at this expanded condition. Analyze the situation for the case where the fluid is known to follow a pressure-explicit EOS $P = f(T, V)$ for $P/P_c = P_r < 0.4$ and $T/T_c = T_r > 1.0$. In addition, the ideal-gas-state heat capacity at constant pressure $C_p^o(T_r)$ for this material is known as a polynomial function of T_r

$$C_p^o(T_r) = A_0 + A_1 T_r + A_2 T_r^2 + A_3 T_r^3 \quad \text{for } T_r < 5.0$$

where A_0, A_1, A_2, and A_3 are specified constants.

Will Rocky's scheme allow you to estimate H and S for particular values of P and T in the range $P_r > 1.0$ and $5.0 > T_r > 1.0$? Explain your answer.

8.29. Consider a closed system consisting of a single-phase pure material. Starting with the combined statement of the 1st and 2nd laws with only PdV type work:

$$dU = TdS - PdV \tag{1}$$

(a) From Postulate I, we should be able to express U as a function of any two independently variable parameters. If we choose T and V, show that equation (1) can be reformulated as:

$$dS = (1/T)(\partial U/\partial T)_V\, dT + (1/T)[(\partial U/\partial V)_T + P]dV \qquad (2)$$

which expresses a new functional dependence for $S = f(T, V)$. Derive an equation of state for P as a function of T and the derivatives $(\partial P/\partial T)_V$ and $(\partial U/\partial V)_T$.

(b) Obtain an expression for the ratio of the amount of heat removed during an isothermal compression to the work done on a material that obeys the equation of state that you derived in (a) above. You can assume that the compression occurs as a quasi-static process.

8.30. Carbon dioxide is used in enhanced oil recovery operations by flooding an oil-bearing reservoir with CO_2 gas under pressure. In one particular application sketched below, CO_2 is compressed adiabatically, then cooled at constant pressure and injected into a well encased with steel pipe.

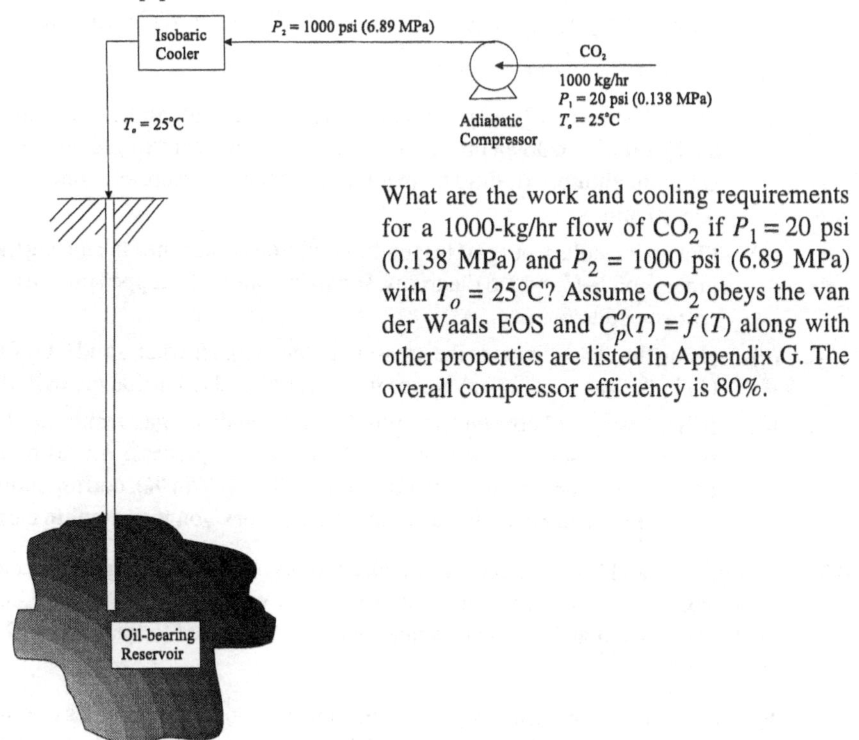

What are the work and cooling requirements for a 1000-kg/hr flow of CO_2 if $P_1 = 20$ psi (0.138 MPa) and P_2 = 1000 psi (6.89 MPa) with T_o = 25°C? Assume CO_2 obeys the van der Waals EOS and $C_p^o(T) = f(T)$ along with other properties are listed in Appendix G. The overall compressor efficiency is 80%.

8.31. Cylinders of carbon dioxide are often used as fire extinguishers. You are requested to analyze the blowdown process that produces a high-velocity, low-pressure CO_2 gas stream which may contain solid CO_2 (dry ice) particles. In your analysis, you should determine the state of the gas both inside and outside the storage cylinder as a function of time.

The gas cylinder can be approximated as a right circular cylinder about 8 in. (20.3 cm) in diameter and between 4.5 and 5.0 ft (1.37 and 1.52 m) tall. The internal volume is about 1.7 ft^3 (0.048 m^3) with an inside surface area of 10 ft^2 (0.283 m^2). The steel wall is 0.25 in. (0.64 cm) thick. The heat capacity of steel can be taken as 0.1 BTU/lb$_m$°R (418.6 J/kg K) at a density of 529 lb$_m$/ft^3 (8474 kg/m^3).

CO_2 gas cylinders are stored at an ambient temperature of 40°F (4.4°C) and are initially charged to 500 psia (3.45 MPa). For fire fighting purposes, the valve is opened quickly, resulting in a rapid discharge of fluid with cylinder pressures decaying by a factor of 2 every 120 sec.

You should assume CO_2 vapor follows the Peng-Robinson equation of state. Relevant properties for CO_2 are listed in Appendix G and below. Thermodynamic charts for *P-H* (English units) and *T-S* (SI units) are given in Figures 8.12a and b, respectively.

Sublimation temperature at 1 bar = 194.7 K (solid-gas mixture in equilibrium)

C_p^o (ideal gas state heat capacity) = 26.62 + 0.0363 T (T in K) in J/mol K

ΔH_{vap} = 103.7 J/g at 300 K, 67.01 bar

(a) For the case where heat transfer between the tank walls and the gas can be neglected, analyze the blowdown process to obtain mathematical expressions and plots of tank gas temperature, exit fluid temperature, and tank pressure as a function of time from 0 to 10 min.

(b) Plot your results from part (a) for the conditions inside the tank during the blowdown process on *P-H* and *T-S* diagrams. What does the path mapped out during blowdown correspond to?

(c) Using the charts, what is the pressure in the cylinder when solid CO_2 first can form: (1) in the gas flowing anywhere in the valve, and (2) inside the cylinder?

(d) If heat transfer between the cylinder walls and the gas inside the tank was not neglected, describe how you would analyze the process. An inside heat transfer coefficient of approximately 6 BTU/hr ft^2 °F (34 W/m^{2}K), corresponding to natural convection, can be assumed for any calculations you may want to carry out.

8.32. A well-insulated 100-gallon storage tank contains one gallon of liquid water at 90°F). The vapor space contains only water vapor (i.e., the vapor and liquid are in equilibrium at 90°F). Wet steam at 1 atm with a quality of 94% is added to the tank until the pressure rises to 1 atm.

How many pounds of steam will enter the tank, and how many pounds of liquid water are present at the completion of the process? Use the steam tables for the properties of water.

8.33. A new device has just been patented to measure temperatures between 10 and 40°C (see Figure P8.33). It consists essentially of a small rigid tube with a flexible bellows located inside. The bellows is welded to the right end and is closed from the environment except during filling. Different and nonideal gases are originally charged to *A* and *B*. The plate *C* forms a capacitor relative to bellows *B*. The bellows is quite flexible and tests indicate that $P_A \sim P_B$.

When this device is placed in a cold environment, the volumetric behavior of closed systems A and B is sufficiently different so that bellows B moves; the movement is easily related to the increase or decrease in the bellows volume. Thus, at any temperature, the volumes of both A and B can be determined because the sum is constant.

The device is originally charged to a pressure P_0 at T_0 with gas A in the tube and gas B in the bellows. The original mole ratio N_B/N_A is set at some value η. At this initial P_0, T_0, the gases A and B are essentially ideal. At lower temperatures, however, they become nonideal, but the PVT properties of both may be expressed by a simple virial equation of state:

$$Z = \left(\frac{PV}{RT}\right) = 1 + B'P$$

where $B'_A \neq B'_B$ are both functions of temperature but not of pressure. Assume $B'(T)$ values are known.

(a) Derive a relation to express $\phi_A = \underline{V}_A/(\underline{V}_A + \underline{V}_B)$ at any temperature in terms of B' values, T, and initial conditions.

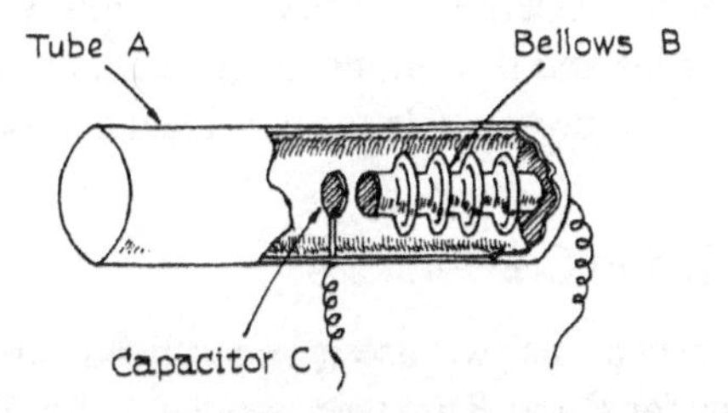

Figure P8.33

8.34. The Zeno line ($Z = 1$) condition presumably corresponds to a dynamic balance between repulsive and attractive intermolecular forces. As shown in Figure 8.6, the line is amazingly linear from the Boyle point to the triple point. Is this behavior consistent with a van der Waals EOS representation of the PVT properties. Explain, giving all relevant equations and plots to justify your answer.

Property Relationships for Mixtures 9

In Chapter 8 we developed thermodynamic relationships for pure materials. In this chapter we extend the treatment to cover single-phase, multicomponent simple mixtures. The mixtures may exist in different phase states, including solid solutions (provided that they are in a stable equilibrium state), liquids, gases, and supercritical fluids. If dissociation of any component occurs (e.g., in ionization of salts in aqueous solutions), the degree of dissociation is assumed to be known (see also Chapter 12).

As a first step, we establish conventions and then develop constitutive models for characterizing the physical properties of mixtures. Finally, methods for estimating changes in derived properties and deviations from ideal behavior are introduced. Later in Chapters 11 and 12 we explore models for non-ideal solutions in greater depth.

9.1 General Approach and Conventions

To avoid repetitive derivations, we use a generalized property $\underline{B}$ (extensive) or B (intensive). Relationships for $\underline{B}$ and B are then applicable for all derived properties such as internal energy, entropy, or any Legendre transform of the energy (or entropy), such as enthalpy, Helmholtz free energy, or Gibbs free energy. In addition, they can be applied to volume ($\underline{V}$ and V).

Two *model* mixtures are introduced: the *ideal-gas mixture* and the ***ideal solution***. Deviations of real mixtures from these models lead to the use of fugacity, fugacity coefficient, activity, and activity coefficient functions as measures of the ***non-ideality*** of the mixture.

In spite of the fact that there are few new concepts or principles, many relationships may appear unduly complex because the notation is often formidable. Yet, in the development of equations to describe mixtures, one must resort to detailed subscripting to denote the component involved or the conditions of restraint for partial differentiation. Appendix F also contains several detailed derivations relevant to mixture properties.

The state of a stable, single phase, multicomponent system can be specified as a continuous function of $n + 2$ variables as we learned from Postulate I. This concept was later utilized to formulate the Fundamental Equation (5-1) where

$$\underline{U} = f(\underline{S}, \underline{V}, N_1, ..., N_n) \tag{5-1}$$

Up to now, we have not been concerned with the dependence of $\underline{U}$ or other derived properties on N_i ($i = 1, ..., n$) except for the very trivial case of a single-component system (where i = 1). Now, we intend to develop a means of characterizing the state of a multicomponent system (where $i > 1$) in a completely general fashion. The dependence of a general extensive property $\underline{B}$ on N_i for i = 1,..., n, that is on the absolute or relative amounts of each component i present in the mixture, and two other independent properties needs to be established.

We are completely free to select these $n + 2$ independent variables from a large set of intensive and extensive properties that includes all conjugate coordinates, that is $\underline{S}$, T; P, $\underline{V}$; μ_i, N_i (i = 1,..., n). In normal engineering practice it is convenient to choose two sets.

$$(T, P, N_1, \ldots, N_n) \quad \text{or} \quad (T, P, x_1, \ldots, x_{n-1}, N) \tag{9-1}$$

where x_i is defined as the mole fraction of component i

$$x_i \equiv \frac{N_i}{\sum_{i=1}^{n} N_i} = \frac{N_i}{N} \tag{9-2}$$

where N = the total number of moles in the mixture. Note that we have only included $(n - 1)x_i$ variables as these are the only independent ones because

$$\sum_{i=1}^{n} x_i = 1 \quad \text{and} \quad x_n = 1 - \sum_{i=1}^{n-1} x_i \tag{9-3}$$

Thus the second set of variables consists of $n + 1$ intensive variables and one extensive variable (N). For a general extensive property $\underline{B}$, the convenient functional forms are

$$\underline{B} = f(T, P, N_1, \ldots, N_n) \tag{9-4}$$

or

$$\underline{B} = f(T, P, x_1, \ldots, x_{n-1}, N) \tag{9-5}$$

In many cases in dealing with *intensive* derived properties (in general B), we will find it convenient to express the functional dependence as

$$B = f(T, P, x_1, \ldots, x_n) \tag{9-6}$$

even though we know that only $n - 1$ mole fractions x_i's are independently variable. Strictly speaking we should express B as:

$$B = f(T, P, x_1, \ldots, x_{n-1}, N) \tag{9-7}$$

but since $(\partial B/\partial N)_{T,P,x} = 0$ we find Eq. (9-6) a more useful formulation for most practical situations.

Although B can be any property of interest, in most applications of mixtures, the important extensive properties are $\underline{B} = \underline{V}, \underline{H}, \underline{G}, \underline{S}, \underline{A}$, and sometimes $\underline{U}$. Among these $\underline{G}$ has special significance because the dependence of $\underline{G}$ on T, P, N,..., N_n represents the natural canonical variables for the Gibbs free energy form of the Fundamental Equation; thus preserving the total information content of Eq. (5-1).

There are several reasons for choosing these variable sets to formulate mixture relationships:

1. T, P, N_i, or x_i are measurable primitive properties.
2. The intensive set T, P, x_i of parameters is particularly convenient as all parameters are independent of the size of the system.
3. Relationships involving mole numbers $N_1 \ldots N_n$ or N are relatively easy to manipulate and visualize when determining extensive or intensive properties.

For example, to prepare an equimolar binary mixture of components A and C, $N_A = N_C$ and $x_A = N_A/(N_A + N_C) = 0.5$ and $x_C = 1 - x_A = N_C/(N_A + N_C) = 0.5$.

While Appendix F provides a completely general derivation of functional and derivative relationships for arbitrary choices of intensive or extensive parameters, it is useful to summarize some of the important results here.

First of all, a most important derivative, defined as the *partial molar property* $\overline{B}_i$, is given by

$$\boxed{\overline{B}_i \equiv \left(\frac{\partial \underline{B}}{\partial N_i}\right)_{T, P, N_j[i]}} \tag{9-8}$$

where T, P, and all mole numbers N_j, except N_i, are held constant. In Chapter 9, several equations like the one above will be boxed in to emphasize their general importance.

Since both $\underline{B}$ and N_i are first order in moles, all partial molar properties are zero order in moles or formally *intensive* properties that depend on the temperature, pressure, and composition (x_i) of the system.

By Euler integration (see Appendix F), we can easily show that

$$\underline{B} = \sum_{i=1}^{n} \overline{B}_i N_i \tag{9-9}$$

and because $\underline{B} = BN$

$$B = \sum_{i=1}^{n} \overline{B}_i x_i \tag{9-10}$$

There are several ways to physically visualize the meaning of $\overline{B}_i$. Let's consider an example where $\underline{B} = \underline{V}$, the total volume of a two-component mixture. The partial molar volume of component 2, $\overline{V}_2$, then represents how $\underline{V}$ changes when δN_2 moles of pure 2 are added at constant T and P to a specific mixture containing N_1 and N_2 moles of each component. At one extreme, we can start with a tank containing N_1 moles of pure 1, thus $x_1^o = 1.0, x_2^o = 0$ initially. As component 2 is added with T and P fixed, x_2 increases from 0 to the final state where N_2^f moles have been added. $\overline{V}_2$ then is equivalent to the slope or derivative of the curve of $\underline{V}$ versus N_2 with T, P, N_1 fixed throughout. In general, $\overline{V}_2$ (and $\overline{V}_1$) vary along the composition axis, that is their specific magnitude depends on the intensive state of the system given by T, P, and $x_2 = N_2/(N_2 + N_1)$, $x_1 = 1 - x_2$ but does not depend on the total amount of material present. Example 9.1 illustrates how the partial molar enthalpy is used to calculate the heat effect in a First-Law open system problem.

Example 9.1

A ternary mixture of 50 mole % *n*-propanol, 25 mole % *n*-pentanol, and 25 mole % *n*-heptane is prepared by a two-step process in which pentanol and heptane are mixed in vat 1 and this mixture is then added to vat 2, containing the propanol. Each vat is equipped with stirrer and heating coils; the addition is slow enough so that the process can be considered isothermal at 294 K. What is the cooling load required for each vat?

Data:

Mole fractions		Partial molar enthalpies (–J/mol)		
Heptane	**Propanol**	**Heptane**	**Propanol**	**Pentanol**
0.00	0.00	—	—	0.0
0.00	0.25	—	67.1	6.5
0.00	0.50	—	46.6	16.0
0.00	0.75	—	18.2	54.9
0.00	1.00	—	0.0	—
0.25	0.00	1153.4	—	47.5
0.25	0.25	1155.5	167.4	53.5
0.25	0.50	1165.5	136.2	74.5
0.25	0.75	1200.8	106.4	—
0.50	0.00	864.8	—	237.1
0.50	0.25	884.1	335.7	229.2
0.50	0.50	919.2	280.7	—
0.75	0.00	361.3	—	1155.8
0.75	0.25	425.6	1203.1	—
1.00	0.00	0.0	—	—

Solution

First consider vat 1; choosing the contents as the system and employing a basis of 1 mole of product, at constant pressure

$$\delta Q = d\underline{H} - H_{in}\, dN_{in}$$

$$Q = \underline{H}_{final} - \underline{H}_{initial} - H_{a5}N_{a5} - H_h N_h$$

where the subscripts $a5$ and h denote *n*-pentanol and heptane, respectively. $\underline{H}_{initial}$ is zero since the vat was initially empty. From Eq. (9-9),

$$\underline{H}_{final} = \overline{H}_{a5}N_{a5} + \overline{H}_h N_h$$

Since $N_{a5} = N_h = 0.5$ mole and the enthalpy base for the pure components is zero (i.e., $H_{a5} = H_h = 0$), then

$$-Q = (0.5)(864.8) + (0.5)(237.1) = 551 \text{ J/mol}$$

For vat 2, using the same approach,

$\underline{H}_{final} = N_{a3}\overline{H}_{a3} + N_{a5}\overline{H}_{a5} + N_h\overline{H}_h$; $\underline{H}_{initial} = 0$, pure a_3 (*n*-propanol)

$N_{in} = 0.5$; $H_{in} = -551$ J/mol (from above)

$-Q = (0.5)(136.2) + (0.25)(74.5) + (0.25)(1165.5) - (0.5)(551) = 102.6$ J/mol

---

Now that we have illustrated the usefulness of the partial molar volume and enthalpy, one final example remains before we introduce further generalizations. Consider the Gibbs energy $\underline{G}$. Here the partial molar Gibbs energy $\overline{G}_i$ has an immediate interpretation as the chemical potential μ_i

$$\mu_i = \overline{G}_i \equiv \left(\frac{\partial \underline{G}}{\partial N_i}\right)_{T, P, N_j[i]} \tag{9-11}$$

Thus, if we are dealing with situations of phase or chemical equilibrium, the equality of μ_i across phases or $\Sigma \nu_i \mu_i = 0$ for each equilibrium reaction can be related to derivative properties of $\underline{G}$. As we will see later in this chapter and in Chapters 11, 15, and 16, constitutive models for $\underline{G}$ can be used to specify conditions of phase and chemical equilibrium.

Having introduced $\overline{B}_i$ we can now formally express the differentials of $\underline{B}$ and B in terms of variations of T, P, and N_i or x_i. As derivatives are presented in Appendix F, here we summarize the important results.

$$d\underline{B} = \left(\frac{\partial \underline{B}}{\partial T}\right)_{P, N_i} dT + \left(\frac{\partial \underline{B}}{\partial P}\right)_{T, N_i} dP + \sum_{i=1}^{n} \overline{B}_i \, dN_i \tag{9-12}$$

$$dB = \left(\frac{\partial B}{\partial T}\right)_{P, x} dT + \left(\frac{\partial B}{\partial P}\right)_{T, x} dP + \sum_{i=1}^{n} \overline{B}_i \, dx_i \tag{9-13}$$

where the subscripts for the derivatives in Eq. (9-13) P, x (or x_i) and T, x (or x_i) means that the differentiation is carried out at constant P and composition or T and composition ($x_i, \ldots, x_n$ fixed). If $\underline{B} = \underline{G}$, the expected results occur in the form of the Fundamental Equation in (T, P, N_i) or (T, P, x_i) coordinates.

$$d\underline{G} = -\underline{S}dT + \underline{V}dP + \sum_{i=1}^{n} \mu_i \, dN_i \tag{9-14}$$

or

$$dG = -SdT + VdP + \sum_{i=1}^{n} \mu_i \, dx_i \tag{9-15}$$

9.2 $P\underline{V}TN$ Relations for Mixtures

A key constitutive relationship used to calculate mixture properties is a volumetric ($P\underline{V}TN$) equation of state (or EOS). Conceptually, the usefulness of an EOS is best visualized by returning to our earlier example where the partial molar volume $\overline{V}_i$ was related to a mole number derivative of the total volume $\underline{V}$ [see Eq. (9-8)]. Given a volume-explicit, constitutive EOS for $\underline{V}$ as a function of T, P, and N_i, the partial molar volume can be calculated directly. For example, with

$$\underline{V} = f[T, P, N_i] = NV = Nf[T, P, x_i] \tag{9-16}$$

$\overline{V}_i$ is determined by analytical or numerical differentiation at constant T, P, and $N_j[i]$. For a pressure-explicit EOS, more work is required.

In Section 8.3 $P\underline{V}TN$ relations for pure components were discussed in the context of cubic (van der Waals)-type and virial-type equations of state. Essentially all pure-component equations of state can also be applied to mixtures; the difficulty of application lies in delineating rules to obtain mixture parameters which now depend on the mixture composition. With the exception of the virial equation, no general theory has yet been developed to determine appropriate *mixing rules*; thus they are ordinarily formulated in a manner to yield the least error when the predictions of the equation of state are compared with experiment.

The basic approach followed is to formulate the mixture EOS parameters in terms of explicit, compositionally dependent equations that use pure component parameters and mole fractions. For example, van der Waals used combining rules for the a and b parameters in his EOS as proposed by Lorentz in 1881 and Berthelot in 1898:

$$b_m = \sum_{i=1}^{n} x_i b_i \tag{9-17}$$

$$a_m = \left[\sum_{i=1}^{n} x_i (a_i)^{1/2} \right]^2 \tag{9-18}$$

where the subscript m refers to the mixture at a specific composition (x_i, i=1,..., n) and a_i and b_i are pure component vdW parameters given by Eq. (8-32) as a function of P_c and T_c.

Because the vdW (and other cubic EOSs) can be applied to both gases and liquids, the use of y_i rather than x_i will frequently appear in Eqs. (9-17) and (9-18). [*Note: we will use x_i and y_i interchangeably to delineate mole fractions throughout the remaining chapters of the book.*]

A common alternative to the Lorentz-Berthelot formulation is to scale the critical constants directly to the specific mixture composition of interest [see for example Kay (1938) or Prausnitz and Gunn (1958)]. With this scaling, one effectively generates *pseudocritical* constants for the mixture which allow us to treat the mixture as if it was a pure component.

If these approaches are not satisfactory, which is often the case when dissimilar molecular properties between components exist, an additional parameter is introduced to account for "binary interactions" between two specific components i and j of the mixture. Commonly the symbol δ_{ij} or k_{ij} is used to designate this parameter.

For similar species i=j and the binary interactions should disappear--depending on the specific formulation of the mixing rules this usually means that $\delta_{ii} = \delta_{jj} = 1$ or 0.

Although many variations are possible, a frequently followed approach is to use Eq. (9-17) as is and to modify Eq. (9-18) to the following form

$$a_m = \sum_i \sum_j x_i x_j a_{ij} \tag{9-19}$$

where

$$a_{ij} = (1 - \delta_{ij}) \sqrt{a_i a_j} \quad i \neq j$$

$$a_{ii} = a_i \quad a_{jj} = a_j$$

$$\delta_{ii} = \delta_{jj} = 0$$

The term δ_{ij} is a ***binary interaction parameter*** that is specific for the i–j binary. δ_{ij} is often estimated by fitting experimental data, and in the most extreme (non-ideal) cases can show a dependence on T, P, and x_i. Hopefully, δ_{ij} will be constant and small, indicating that the combining rules are properly capturing the important effects.

Consider now the Redlich-Kwong equation of state given in Eq. (8-42). It is rewritten below, in intensive form in Eq. (9-20a) and in extensive form in Eq. (9-20b).

$$P = \frac{RT}{V - b_m} - \frac{a_m}{T^{1/2} V (V + b_m)} \tag{9-20a}$$

$$P = \frac{NRT}{\underline{V} - Nb_m} - \frac{N^2 a_m}{T^{1/2} \underline{V} (\underline{V} + Nb_m)} \tag{9-20b}$$

where a_m and b_m are the mixture constants. N is the total moles in the mixture; thus,

$$N = \sum_{i=1}^{n} N_i \tag{9-21}$$

The pure component parameters a_i and b_i were given in Eqs. (8-43) and (8-44) in terms of the critical constants of component i. Redlich and Kwong used the van der Waals, Lorentz-Berthelot mixing rules to determine a_m and b_m.

$$a_m^{1/2} = \sum_{i=1}^{n} y_i a_i^{1/2} \tag{9-22}$$

$$b_m = \sum_{i=1}^{n} y_i b_i \tag{9-23}$$

where y_i is the mole fraction of component i. Note that we have replaced x_i with y_i and that no binary interaction parameter is employed. With these rules, then, for Eq. (9-20b),

$$N^2 a_m = \left[\sum_{i=1}^{n} N_i a_i^{1/2}\right]^2 \tag{9-24}$$

$$Nb_m = \sum_{i=1}^{n} N_i b_i \tag{9-25}$$

Equations (9-24) and (9-25) are convenient forms should partial derivatives with respect to mole numbers be required, as illustrated in Example 9.2.

Example 9.2

Derive an expression for the partial molar volume, $\overline{V}_k$, using the Redlich-Kwong equation of state.

Solution

Using Eq. (9-8),

$$\overline{V}_k = \left(\frac{\partial \underline{V}}{\partial N_k}\right)_{T, P, N_j[k]}$$

Since the Redlich-Kwong equation of state is explicit in pressure with $\underline{V}$, T, N_1, ... as the independent variables, the defining derivative for $\overline{V}_k$ is better rewritten as

$$\overline{V}_k = \left(\frac{\partial \underline{V}}{\partial N_k}\right)_{T, P, N_j[k]} = -\frac{(\partial P/\partial N_k)_{T, \underline{V}, N_j[k]}}{(\partial P/\partial \underline{V})_{T, N}}$$

The denominator is readily found using Eq. (9-20b),

$$\left(\frac{\partial P}{\partial \underline{V}}\right)_{T, N} = \frac{-NRT}{(\underline{V} - Nb_m)^2} + \frac{N^2 a_m (2\underline{V} + Nb_m)}{T^{1/2} \underline{V}^2 (\underline{V} + Nb_m)^2}$$

To obtain the numerator from Eq. (9-20b), one must also use Eqs. (9-24) and (9-25); that is, the derivatives of $N^2 a_m$ and Nb_m are necessary.

$$\frac{\partial}{\partial N_k}(N^2 a_m)_{T, \underline{V}, N_j[k]} = \frac{\partial}{\partial N_k}\left[\left(\sum_{i=1}^{n} N_i a_i^{1/2}\right)^2\right]_{T, \underline{V}, N_j[k]}$$

$$= 2\left[\sum_{i=1}^{n} N_i a_i^{1/2}\right] a_k^{1/2}$$

$$= 2Na_m^{1/2}a_k^{1/2}$$

$$\frac{\partial}{\partial N_k}(Nb_m)_{T,\,\underline{V},\,N_j[k]} = \frac{\partial}{\partial N_k}\left(\sum_{i=1}^{n} N_i b_i\right)_{T,\,\underline{V},\,N_j[k]} = b_k$$

Therefore,

$$\left(\frac{\partial P}{\partial N_k}\right)_{T,\,\underline{V},\,N_j[k]} = RT\left[\frac{1}{\underline{V}-Nb_m} + \frac{Nb_k}{(\underline{V}-Nb_m)^2}\right] - \frac{1}{T^{1/2}\underline{V}}\left[\frac{2Na_m^{1/2}a_k^{1/2}}{\underline{V}+Nb_m} - \frac{N^2a_mb_k}{(\underline{V}+Nb_m)^2}\right]$$

To evaluate $\overline{V}_k$, one must determine $\underline{V}$, a_m and b_m for the mixture at the given state of the material; then the derivatives with respect to P can be found as shown above.

For the more complex Peng-Robinson equation of state (PR EOS), given in Eq. (8-46), the two parameters for a pure material were defined in Eqs. (8-47) through (8-51). For a mixture,

$$P = \frac{NRT}{\underline{V}_m - Nb_m} - \frac{N^2a_m}{\underline{V}_m(\underline{V}_m + Nb_m) + Nb_m(\underline{V}_m - Nb_m)} \tag{9-26}$$

with

$$a_m = \sum_i \sum_j y_i y_j a_{ij} \tag{9-27}$$

$$b_m = \sum_i y_i b_i \tag{9-28}$$

where, similar to Eq. (9-19):

$$a_{ij} = (1 - \delta_{ij})(a_i a_j)^{1/2} \quad i \neq j \tag{9-29}$$

$$a_{ii} = a_i \quad a_{jj} = a_j \quad \delta_{ii} = \delta_{jj} = 0 \tag{9-30}$$

When using the mixture form of the PR EOS, δ_{ij} is usually assumed to be independent of composition, temperature, and pressure. As it is usually obtained from experimental data with the *i–j* binary, it is frequently used in multicomponent mixtures with the implied assumption that ternary or higher-order interactions are negligible. The introduction of binary interaction parameters is characteristic of most cubic-type equations of state introduced in the past several years. While δ_{ij} is ordinarily a small number (ca. 0.1), the final result is quite sensitive to the value chosen. Note that if δ_{ij}=0, the mixing rule for a_m in the Peng-Robinson relation reduces to that for the Redlich-Kwong equation of state, Eq. (9-22).

More advanced treatments of mixing rules for cubic EOS for modeling multi-component systems with high-levels of non-ideal behavior are needed. Of particular importance is the thermodynamically consistent formulation recently developed by Wong and Sandler (1992). Here the a_m and b_m parameters for the mixture are correlated together. First we define:

$$Q_{ij} \equiv \sum_i \sum_j x_i x_j \Psi_{ij}$$

with

$$\Psi_{ij} \equiv \frac{1}{2}\left[b_i + b_j - \left(\frac{a_i + a_j}{RT}\right)\right](1 - \delta_{ij})$$

where δ_{ij} is a symmetric binary interaction parameter ($\delta_{ii} = \delta_{jj} = 0$ and $\delta_{ij} = \delta_{ji}$) that is fit to binary data and a_i, a_j, b_i, and b_j are pure component parameters. To determine a_m and b_m we need a second set of relationships

$$\frac{a_m}{RT} = \left(Q_{ij}\right)\left(\frac{D}{1-D}\right) = b_m D$$

$$b_m = \frac{Q_{ij}}{1-D} = \frac{\sum_i \sum_j x_i x_j \Psi_{ij}}{1-D}$$

$$D \equiv \sum_i x_i \frac{a_i}{b_i RT} + \frac{\Delta G^{EX}(T, x_i)}{\sigma RT}$$

where σ = EOS-dependent constant, e.g., for the PR EOS $\sigma = [\ln(\sqrt{2}-1)]/\sqrt{2}$, and any ΔG^{EX} model can be used (see Section 11.7 and Tables 11.1 and 11.2) [see Sandler (1994) for further discussion].

Finally, for the virial equation of state, Eq. (8-63), the expression for the virial coefficients in a mixture can be developed from theory; that for the second virial is

$$B_m = \sum_i \sum_j y_i y_j B_{ij} \tag{9-31}$$

while for a binary reduces to

$$B_{m\,(binary\ ij)} = y_i^2 B_{ii} + 2 y_i y_j B_{ij} + y_j^2 B_{jj}$$

Assuming that little error is introduced in using the same value of B_m for the truncated form of the virial [Eq. (8-67)], then values of B_{ii} (B_i) and B_{jj} (B_j) can be calculated as shown earlier in Eq. (8-68). To determine the *interaction* virial coefficient, B_{ij} ($i \neq j$), it would be convenient to use Eq. (8-68) but *mixture* critical temperatures, critical pressures, and acentric factors are required. Prausnitz et al. (1986) suggest that

$$T_{c_{ij}} = (1 - k_{ij})(T_{c_i} T_{c_j})^{1/2} \tag{9-32}$$

$$\omega_{ij} = \frac{\omega_i + \omega_j}{2} \tag{9-33}$$

and

$$P_{c_{ij}} = \frac{RT_{c_{ij}}(Z_{c_i} + Z_{c_j})/2}{[(V_{c_i}^{1/3} + V_{c_j}^{1/3})/2]^3} \tag{9-34}$$

where Z_{c_i} and V_{c_i} are the critical compressibility factor and critical volume of pure component i. The term k_{ij} is a binary interaction parameter similar to the δ_{ij} introduced in the Peng-Robinson equation of state. (Note that k_{ij} does not equal δ_{ij}.) Values of k_{ij} for many binaries are available, for example see Chueh and Prausnitz (1967). As with the pure-component virial equation of state, the mixture form is limited to the gas phase under conditions where the mixture density is less than one-half of the critical mixture density. An estimate of the mixture critical density can be obtained from:

$$\rho_{c_m} = V_{c_m}^{-1} \approx \left[\frac{\sum\sum y_i y_j (V_{c_i}^{1/3} + V_{c_j}^{1/3})^3}{8}\right]^{-1} \tag{9-35}$$

In this brief discussion of mixture $PVTN$ equations of state, we have seen that the basic forms of the equations do not change when treating mixtures rather than pure components. This would, of course, be expected since the mixture equations of state must reduce to the pure-component form, as all mole fractions but one approach zero. However, in any application of mixture equations of state, mixing rules to relate characteristic equation parameters to composition must be available. Most mixing rules now in use are conceptually simple, yet algebraically complex results are inevitable when operations require that derivatives be taken with respect to specific mole numbers or to compositions.

The *departure* functions for mixtures are calculated in the same manner as for pure components except that mixture parameters must be used. Thus, Eqs. (8-122) through (8-126) are still applicable, but they apply to a mixture with a specific composition. When considering variations in temperature in the ideal-gas state [e.g., Eqs. (8-133), (8-135), and (8-137)], values of C_v^o or C_p^o for the mixture are calculated as

$$C_{v_m}^o = \sum_i y_i C_{v_i}^o \tag{9-36}$$

or

$$C_{p_m}^o = \sum_i y_i C_{p_i}^o \tag{9-37}$$

That is, in the ideal-gas state, heat capacities are mole-fraction averages of the pure-component values.

9.3 Partial Molar Properties

A partial molar property was defined in Eq. (9-8) and is related to mixture properties by Eqs. (9-12) and (9-13). It follows from the definition that if $\underline{W}$ and $\underline{Z}$ are any two extensive properties, and Y is either temperature or pressure, and if $\underline{W} = Y\underline{Z}$, then $\overline{W}_i = Y\overline{Z}_i$. Thus, the conjugate sets $T\underline{S}$ and $P\underline{V}$, when differentiated as indicated in Eq. (9-8), become $T\overline{S}_i$ and $P\overline{V}_i$. It also follows that any derivative of the form $(\partial \underline{Z}/\partial Y_1)_{Y_2, N_i}$, where Y_1 and Y_2 are T and P, when operated on by $(\partial/\partial N_i)_{T, P, N_j[i]}$, becomes $(\partial \overline{Z}_i/\partial Y_1)_{Y_2, N_i}$ because the order of differentiation (Maxwell reciprocity) is immaterial when the independent variables are consistent. For example,

$$\frac{\partial}{\partial N_i}\left[\left(\frac{\partial \underline{H}}{\partial P}\right)_{T, N_i}\right]_{T, P, N_j[i]} = \frac{\partial}{\partial P}\left[\left(\frac{\partial \underline{H}}{\partial N_i}\right)_{T, P, N_j[i]}\right]_{T, N_i} = \left(\frac{\partial \overline{H}_i}{\partial P}\right)_{T, N_i} \tag{9-38}$$

Where the subscript N_i indicates that all mole numbers for i=1,..., n components are held constant in the differentiation. Using the shorthand notation for the 2nd derivative of $\underline{H}$ introduced in Chapter 5, an equivalent form to Eq. (9-38) is given by:

$$\underline{H}_{PN_i} = \underline{H}_{N_iP} = \left(\frac{\partial \overline{H}_i}{\partial P}\right)_{T, N_i} \tag{9-39}$$

Hence, from $\underline{H} = \underline{U} + P\underline{V}$, we could immediately write

$$\overline{H}_i = \overline{U}_i + P\overline{V}_i \tag{9-40}$$

Also, from $C_p = T(\partial S/\partial T)_{P, x}$, we could define $\underline{C}_p = NC_p = T(\partial \underline{S}/\partial T)_{P, N_i}$, and it then follows that

$$\overline{C}_{pi} = \frac{\partial}{\partial N_i}[NC_p]_{T, P, N_j[i]} = T\left(\frac{\partial \overline{S}_i}{\partial T}\right)_{P, N_i} \tag{9-41}$$

Transformations of potential functions require more effort. Let us apply the partial molar operator to

$$d\underline{G} = -\underline{S}\,dT + \underline{V}\,dP + \sum_{m=1}^{n} \mu_m\, dN_m$$

Then

$$d\overline{G}_i = -\overline{S}_i\,dT + \overline{V}_i\,dP + \sum_{m=1}^{n}\left(\frac{\partial \mu_m}{\partial N_i}\right)_{T, P, N_j[i]} dN_m + \sum_{m=1}^{n} \mu_m d\left[\left(\frac{\partial N_m}{\partial N_i}\right)_{T, P, N_j[i]}\right]^{\nearrow 0} \tag{9-42}$$

Since

$$\left(\frac{\partial N_m}{\partial N_i}\right)_{T,\,P,\,N_j[i]} = \begin{Bmatrix} 0 & m \neq i \\ 1 & m = i \end{Bmatrix} \tag{9-43}$$

the last term reduces $d(1) = 0$. The third term on the right-hand side can be left unchanged or transformed to a mole fraction derivative. Another route may be followed to produce Eq. (9-42), since

$$\mu_i = \overline{G}_i = \left(\frac{\partial \underline{G}}{\partial N_i}\right)_{T,\,P,\,N_j[i]} = \underline{G}_{N_i}$$

$$d\overline{G}_i = \underline{G}_{N_iT}\,dT + \underline{G}_{N_iP}\,dP + \sum_{m=1}^{n} \underline{G}_{N_iN_m}\,dN_m \tag{9-44}$$

By applying Maxwell's reciprocity relationship to all 2nd derivatives, we get:

$$\underline{G}_{N_iT} = \underline{G}_{TN_i} = -\overline{S}_i$$

$$\underline{G}_{N_iP} = \underline{G}_{PN_i} = \overline{V}_i \tag{9-45}$$

$$\underline{G}_{N_iN_m} = \underline{G}_{N_mN_i} = \left(\frac{\partial \mu_m}{\partial N_i}\right)_{T,\,P,\,N_j[i]}$$

and Eq. (9-44) reduces to Eq. (9-42):

$$d\overline{G}_i = -\overline{S}_i\,dT + \overline{V}_i dP + \sum_{m=1}^{n} \left(\frac{\partial \mu_m}{\partial N_i}\right)_{T,\,P,\,N_j[i]} dN_m \tag{9-46}$$

Partial molar quantities can be evaluated by several methods. If the property $\underline{B}$ can be conveniently measured, directly or indirectly, $\overline{B}_i$ can be found by measuring the change in $\underline{B}$ upon addition of a small amount of component i to a mixture while holding the temperature and pressure constant. From the definition, Eq. (9-8),

$$\overline{B}_i \equiv \left(\frac{\partial \underline{B}}{\partial N_i}\right)_{T,\,P,\,N_j[i]} = \lim_{\Delta N_i \to 0} \left(\frac{\Delta \underline{B}}{\Delta N_i}\right)_{T,\,P,\,N_j[i]} \tag{9-47}$$

It is essential to keep in mind the fact that $\overline{B}_i$ is a property of the mixture and not simply a property of component i. Thus, $\overline{B}_i$ will generally vary with mixture composition. To emphasize this point, we note that $\overline{B}_i$ is an intensive mixture property and, as such, we could apply any of the equations in Table F.2 in Appendix F to this property. For example, expressing $\overline{B}_i = f(T, P, x_1,\ldots, x_{n-1})$ in terms of n+1 independent variables, the total differential of $\overline{B}_i$ becomes

$$d\overline{B}_i = \left(\frac{\partial \overline{B}_i}{\partial T}\right)_{P,\,x} dT + \left(\frac{\partial \overline{B}_i}{\partial P}\right)_{T,\,x} dP + \sum_{j=1}^{n-1} \left(\frac{\partial \overline{B}_i}{\partial x_j}\right)_{T,\,P,\,x[j,\,n]} dx_j \tag{9-48}$$

where the subscript $x[j, n]$ means that all x in set $(x_1,..., x_n)$ are held constant except x_j and x_n. If experimental data or analytical expressions for $\underline{B}$ or B are available, $\overline{B}_i$ can be evaluated directly. Consider three possible cases:

(1) $\underline{B} = f(T, P, N_1,..., N_n)$
(2) $B = f(T, P, N_1,..., N_n)$
(3) $B = f(T, P, x_1,..., x_{i-1}, x_{i+1},..., x_n)$

For case (1), $\overline{B}_i$ can be evaluated by differentiation using Eq. (9-8). For case (2), we can obtain $(\partial B/\partial N_i)_{T, P, N_j[i]}$ directly from the data.

With $B = \underline{B}/N$

$$\left(\frac{\partial B}{\partial N_i}\right)_{T, P, N_j[i]} = \left(\frac{\partial(\underline{B}/N)}{\partial N_i}\right)_{T, P, N_j[i]} = \frac{\overline{B}_i}{N} - \frac{\underline{B}}{N^2}\left(\frac{\partial N}{\partial N_i}\right)_{T, P, N_j[i]} = \frac{1}{N}(\overline{B}_i - B) \quad (9\text{-}49)$$

use Eq. (9-49) to solve for $\overline{B}_i$:

$$\overline{B}_i = B + N\left(\frac{\partial B}{\partial N_i}\right)_{T, P, N_j[i]} \quad (9\text{-}50)$$

For case (3), we can obtain $(\partial B/\partial x_j)_{T, P, x[j, i]}$ directly from the data and then relate this partial to one of the forms above. For example, let us express $(\partial B/\partial N_i)_{T, P, N_j[i]}$ in Eq. (9-50) as a function of $(\partial B/\partial x_j)_{T, P, x[j, i]}$. We first express B as a differential in terms of $T, P, x_1,..., x_{i-1}, x_{i+1},..., x_n$:

$$dB = \left(\frac{\partial B}{\partial T}\right)_{P, x} dT + \left(\frac{\partial B}{\partial P}\right)_{T, x} dP + \sum_{j \neq i}\left(\frac{\partial B}{\partial x_j}\right)_{T, P, x[j, i]} dx_j \quad (9\text{-}51)$$

Dividing by dN_i and imposing the restraint of constant T, P, and $N_j[i]$, we obtain

$$\left(\frac{\partial B}{\partial N_i}\right)_{T, P, N_j[i]} = -\frac{1}{N}\sum_{j \neq i} x_j \left(\frac{\partial B}{\partial x_j}\right)_{T, P, x[j, i]} \quad (9\text{-}52)$$

Substituting into Eq. (9-50) yields

$$\boxed{\overline{B}_i = B - \sum_{j \neq i} x_j \left(\frac{\partial B}{\partial x_j}\right)_{T, P, x[j, i]}} \quad (9\text{-}53)$$

In Eq. (9-53) we have obtained $\overline{B}_i$ from a data set in which x_i was eliminated. This equation cannot be used to obtain $\overline{B}_k$ unless we transform the given data to a set in which x_k is eliminated. Alternatively, we can use Eq. (F-7) in Appendix F (with k substituted for n) to solve for $\overline{B}_k$ from the data set of case (3). Thus,

$$\overline{B}_k = \overline{B}_i + \left(\frac{\partial B}{\partial x_k}\right)_{T, P, x[k, i]} \quad (9\text{-}54)$$

or, using Eq. (9-53),

$$\overline{B}_k = B + \left(\frac{\partial B}{\partial x_k}\right)_{T,\,P,\,x[k,\,i]} - \sum_{j \neq i} x_j \left(\frac{\partial B}{\partial x_j}\right)_{T,\,P,\,x[j,\,i]} \tag{9-55}$$

For a binary system of A and C, Eq. (9-53) or Eq. (9-55) reduces to

$$\overline{B}_A = B - x_C \left(\frac{\partial B}{\partial x_C}\right)_{T,\,P}$$
$$\overline{B}_C = B - x_A \left(\frac{\partial B}{\partial x_A}\right)_{T,\,P} \tag{9-56}$$

Equation set (9-56) can be easily visualized as shown in Figure 9.1 in which B is plotted as a function of x_C at constant T and P. At any x_C, a tangent to the curve, when extrapolated, intersects the $x_C = 0$ axis at $\overline{B}_A$ and the $x_C = 1$ axis at $\overline{B}_C$.[1] The application of Eq. (9-55) to a ternary system is shown in Example 9.3. Even in this relatively simple system, the data required to calculate partial molar properties are considerable—and only rarely available.

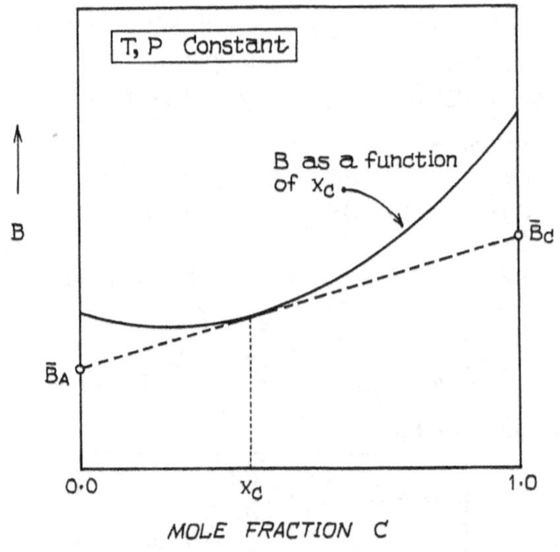

Figure 9.1 Tangent-intercept rule for the binary A-C.

---

1 There are occasions when, in determining partial molar quantities, they become either very large or very small at low concentrations (e.g., partial molar entropy or Gibbs energy), and this intercept technique is not particularly accurate. An alternative technique for handling such cases is found in H.C. van Ness and R.V. Mrazek (1959), "Treatment of thermodynamic data for homogeneous binary systems," *AIChE J.*, **5**, p 209.

Example 9.3

Apply Eq. (9-55) to a ternary system of A, D, and C to obtain $\overline{B}_A$.

Solution

For a ternary system of A, D, and C, there are several ways to express $\overline{B}_A$. Suppose that C is the component "eliminated"; then from Eq. (9-55),

$$\overline{B}_A = B + \left(\frac{\partial B}{\partial x_A}\right)_{T,P,x_D} - x_A\left(\frac{\partial B}{\partial x_A}\right)_{T,P,x_D} - x_D\left(\frac{\partial B}{\partial x_D}\right)_{T,P,x_A}$$

$$= B + (1 - x_A)\left(\frac{\partial B}{\partial x_A}\right)_{T,P,x_D} - x_D\left(\frac{\partial B}{\partial x_D}\right)_{T,P,x_A}$$

On the other hand, if D were eliminated, we would have

$$\overline{B}_A = B + (1 - x_A)\left(\frac{\partial B}{\partial x_A}\right)_{T,P,x_C} - x_C\left(\frac{\partial B}{\partial x_C}\right)_{T,P,x_A}$$

The two expressions for $\overline{B}_A$ can be shown to be equivalent by equating them, using Eq. (F-7) while noting that $1 - x_A = x_D + x_C$. The choice between the two ways of expressing $\overline{B}_A$ depends on the data available; in each case, derivatives are required with the composition of different components held constant.

9.4 Generalized Gibbs-Duhem Relation for Mixtures

In Chapter 5, we introduced the Gibbs-Duhem relation by using the total Legendre transform of a general basis function $y^{(0)}$ in n+2 extensive variables, z_i (i=1,..., n+2). With $\xi_i \equiv (\partial y^{(0)}/\partial z_i)_{z_j[i]}$

$$y^{(n+2)} = y^{(0)} - \sum_{i=1}^{n+2} \xi_i z_i = 0 \tag{9-57}$$

and

$$dy^{(n+2)} = 0 = -\sum_{i=1}^{n+2} z_i \, d\xi_i \tag{9-58}$$

Now we can apply this to the general mixture property $\underline{B}$, by setting $\underline{B} = y^{(2)}$ in n+2 variables, $(\xi_1, \xi_2, N_1, ..., N_n) = (T, P, N_1, ..., N_n)$. Here,

$$-z_1 = \left(\frac{\partial \underline{B}}{\partial T}\right)_{P,N_i}; -z_2 = \left(\frac{\partial \underline{B}}{\partial P}\right)_{T,N_i}; z_{i+2} = N_i \text{ and } \xi_{i+2} = \left(\frac{\partial \underline{B}}{\partial N_i}\right)_{T,P,N_j[i]} = \overline{B}_i \ (i = 1, ..., n)$$

Thus, by substituting these into Eq. (9-58):

$$\left(\frac{\partial \underline{B}}{\partial T}\right)_{P,N_i} dT + \left(\frac{\partial \underline{B}}{\partial P}\right)_{T,N_i} dP - \sum_{i=1}^{n} N_i\, d\overline{B}_i = 0 \tag{9-59}$$

and by rearranging

$$\boxed{\sum_{i=1}^{n} N_i\, d\overline{B}_i = \left(\frac{\partial \underline{B}}{\partial T}\right)_{P,N_i} dT + \left(\frac{\partial \underline{B}}{\partial P}\right)_{T,N_i} dP} \tag{9-60}$$

and dividing by total moles, N, we get

$$\boxed{\sum_{i=1}^{n} x_i\, d\overline{B}_i = \left(\frac{\partial B}{\partial T}\right)_{P,x} dT + \left(\frac{\partial B}{\partial P}\right)_{T,x} dP} \tag{9-61}$$

where x_i is the mole fraction of component i. Equations (9-60) and (9-61) are the most common forms of the *Gibbs-Duhem relation* for partial molar quantities. When integrated, they permit evaluation of any $\overline{B}_i$ in terms of the other $(n - 1)$ values of $\overline{B}_j$. An alternative form of the Gibbs-Duhem relation can be obtained as follows. Noting that

$$\left(\frac{\partial B}{\partial T}\right)_{P,x} = \left[\frac{\partial\left(\sum_{i=1}^{n} \overline{B}_i x_i\right)}{\partial T}\right]_{P,x} = \sum_{i=1}^{n} x_i \left(\frac{\partial \overline{B}_i}{\partial T}\right)_{P,x} \tag{9-62}$$

and

$$\left(\frac{\partial B}{\partial P}\right)_{T,x} = \sum_{i=1}^{n} x_i \left(\frac{\partial \overline{B}_i}{\partial P}\right)_{T,x} \tag{9-63}$$

Eq. (9-61) can be rewritten as

$$\sum_{i=1}^{n} x_i \left[d\overline{B}_i - \left(\frac{\partial \overline{B}_i}{\partial T}\right)_{P,x} dT - \left(\frac{\partial \overline{B}_i}{\partial P}\right)_{T,x} dP\right] = 0 \tag{9-64}$$

Substituting Eq. (9-48) for $d\overline{B}_i$ in Eq. (9-64) yields

$$\sum_{i=1}^{n} x_i \left[\sum_{j\neq k} \left(\frac{\partial \overline{B}_i}{\partial x_j}\right)_{T,P,x[j,k]} dx_j\right] = 0 \tag{9-65}$$

or, changing the order of the summations,

$$\sum_{j\neq k} \left[\sum_{i=1}^{n} x_i \left(\frac{\partial \overline{B}_i}{\partial x_j}\right)_{T,P,x[j,k]}\right] dx_j = 0 \tag{9-66}$$

Since the brackets contain the coefficients of $(n-1)$ terms in dx_j, all of which are independent (because x_k has been eliminated), it follows that each term in brackets must vanish. Thus, an equivalent Gibbs-Duhem relation is

$$\sum_{i=1}^{n} x_i \left(\frac{\partial \overline{B}_i}{\partial x_j}\right)_{T,P,x[j,k]} = 0 \tag{9-67}$$

Equation (9-67) could also have been obtained by dividing Eq. (9-61) by dx_j at constant T, P, $x[j, k]$. We shall have occasion to use this form in treating properties of phases in equilibrium, as described in Chapters 11, 15, and 17.

Example 9.4

For a binary of components 1 and 2, if the partial molar enthalpy $\overline{H}_1$ were available as a function of mole fraction x_1, show how $\overline{H}_2$ and the mixture enthalpy H can be determined. The data $\overline{H}_1 = f(x_1)$ are at constant T and P.

Solution

From Eq. (9-67),

$$x_1 \left(\frac{\partial \overline{H}_1}{\partial x_1}\right)_{T,P} + x_2 \left(\frac{\partial \overline{H}_2}{\partial x_1}\right)_{T,P} = 0$$

Separating and integrating between x_1^o and x_1, we obtain

$$(\overline{H}_2)_{x_1} - (\overline{H}_2)_{x_1^o} = -\int_{x_1^o}^{x_1} \frac{x_1}{x_2}\left(\frac{\partial \overline{H}_1}{\partial x_1}\right) dx_1$$

Let x_1^o be 0 (i.e., pure component 2). Then $(\overline{H}_2)_{x_1^o} = H_2$. If this is substituted in the above, the right-hand side may be found from the $\overline{H}_1 = f(x_1)$ data and $\overline{H}_2$ is then related to the pure component enthalpy H_2. To obtain H,

$$H = x_1\overline{H}_1 + x_2\overline{H}_2$$

Figure 9.2 shows a typical variation of $\overline{H}_1$ and $\overline{H}_2$ for a binary mixture.

We have already seen that all n $(\overline{B}_i)$s can be found from a data set of the form $B = f(T, P, x_1, \ldots, x_{n-1})$. We have also shown that any $\overline{B}_i$ can be found—to within an arbitrary constant—from the other $(n-1)$ $(\overline{B}_j)$s. It is then possible to reconstruct B from these $(\overline{B}_i)$s using Eq. (9-10). Thus, $(n-1)$ $(\overline{B}_i)$s have an information content equivalent to B.

Frequently, data are reported in the literature for all (n) $(\overline{B}_i)$s (most commonly for binary systems in which $\overline{B}_1$ and $\overline{B}_2$ are measured independently). The redundant information can be used to check the thermodynamic consistency of the data. At any given T, P, and concentration, we can calculate the partial derivative $(\partial \overline{B}_i/\partial x_j)_{T,P,x[j,k]}$

for each component, and then use Eq. (9-67) to verify that the sum does indeed vanish. As an exercise, try this on the data shown in Figure 9.2.

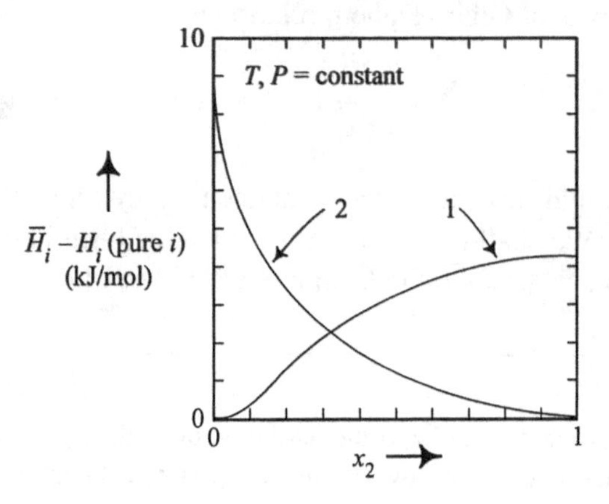

Figure 9.2 Partial molar enthalpy behavior for a binary (1-2) mixture (see Example 9.4).

In some cases, reported data will not give a satisfactory consistency check, but these may be the only available data. In that case, it is always possible to smooth the data in order to obtain a set of partial molar quantities that are consistent. The procedure is simply to reconstruct B using Eq. (9-10) and then apply an equation of the form of Eq. (9-53) or (9-55) to obtain the partial molar quantities. As shown in the following example, this set will always be thermodynamically consistent.

Example 9.5

Using Eqs. (9-53) and (9-55), prove that a set of partial molar properties obtained only from data of the form, $B = f(T, P, x_1, \ldots, x_{i-1}, x_{i+1}, \ldots, x_n)$ will always form a consistent set.

Solution

Multiply Eq. (9-55) by x_k and sum over all k except $k = i$.

$$\sum_{k \neq i} x_k \overline{B}_k = \sum_{k \neq i} x_k \left[B + \left(\frac{\partial B}{\partial x_k} \right)_{T, P, x[i, k]} - \sum_{j \neq i} x_j \left(\frac{\partial B}{\partial x_j} \right)_{T, P, x[i, j]} \right]$$

Multiply Eq. (9-53) by x_i and add to the above.

$$\sum_{k=1}^{n} x_k \overline{B}_k = B(x_i + \sum_{k \neq i} x_k) + \sum_{k \neq i} x_k \left(\frac{\partial B}{\partial x_k} \right)_{T, P, x[i, k]}$$

$$-(\sum_{k\neq i} x_k + x_i)\left[\sum_{j\neq i} x_j \left(\frac{\partial B}{\partial x_j}\right)_{T, P, x[i, j]}\right]$$

The second and third terms on the right-hand side cancel, and the result reduces to

$$\sum_{k=1}^{n} x_k \overline{B}_k = B$$

Thus, the set of $\overline{B}_k$ obtained from $B = f(T, P, x_1, \ldots, x_{i-1}, x_{i+1}, \ldots, x_n)$ will always form a consistent set.

9.5 Mixing Functions

A mixture property is often related to the properties of a reference state, which can be real or hypothetical. The difference between the value of the actual and reference state properties is denoted by the symbol Δ and is called the *mixing or solution function*. The defining equations are

$$\Delta \underline{B}_{mix} \equiv \underline{B}\,(T, P, N_1, \ldots, N_n) - \sum_{j=1}^{n} N_j \overline{B}_j^{+}\,(T^{+}, P^{+}, x_1^{+}, \ldots, x_{n-1}^{+}) \tag{9-68}$$

and

$$\Delta B_{mix} \equiv B\,(T, P, x_1, \ldots, x_{n-1}) - \sum_{j=1}^{n} x_j \overline{B}_j^{+}\,(T^{+}, P^{+}, x_1^{+}, \ldots, x_{n-1}^{+}) \tag{9-69}$$

where $\Delta \underline{B}_{mix}$ and ΔB_{mix} are the total and specific mixing (or solution) functions. For reasons that will become apparent later, the N_j or x_j (not x_j^{+}) terms within the summation are taken as the *actual mole numbers or mole fractions of the mixture*, and not that of the reference state.

It follows from the defining equations that the mixing functions are specified only when the reference states have been clearly delineated. (Note that a reference state must be defined for each component.) Reference states are chosen either for convenience (the properties are known) or for practicality (the deviations from the reference state are small).

If the mixing function is to be useful, it should depend on the properties of the mixture. That is, we desire

$$\Delta B_{mix} = f\,(T, P, x_1, \ldots, x_{n-1}) \tag{9-70}$$

This requirement limits us to either one of two choices for each reference state variable; namely, T^{+}, P^{+}, $x_1^{+}, \ldots, x_{n-1}^{+}$ are either set equal to the actual mixture properties (hence, $\overline{B}_j^{+}$ varies as the mixture conditions change) or they can be set at some fixed conditions (hence, $\overline{B}_j^{+}$ is a constant).

In practice, the most common reference state is the pure component at the same temperature, pressure, and state of aggregation of the mixture. That is, for component j, $T^+ = T$, $P^+ = P$, $x_j^+ = 1$, $x_i^+ = 0$ $(i{\neq}j)$ and, therefore, $\overline{B}_j^+ = B_j\,(T, P)$. In this case, the temperature and pressure vary as the state of the mixture changes, but the reference-state mole fractions are fixed. Note that this definition of the reference state satisfies Eq. (9-70). Whenever the pure components exist in the same state of aggregation of the mixture at T and P, this reference state actually exists as a *real* state as well. There are, however, a number of cases in which the stable phase of the pure material is different from that of the mixture and hence the reference state as defined above is *hypothetical* (e.g., an inorganic salt in an aqueous solution or a liquid mixture above the critical temperature of one of the components). In these cases, it is not uncommon to select other reference states (e.g., the saturated salt solution or an infinitely dilute solution, see Chapter 12).

The use of the mixture mole numbers in the summation of Eq. (9-68) results in a convenient definition of the partial molar mixing function. Substituting Eq. (9-9) into Eq. (9-68) yields

$$\Delta \underline{B}_{mix} = \sum_{j=1}^{n} N_j\,(\overline{B}_j - \overline{B}_j^+) \tag{9-71}$$

Applying the partial molar operator, we get

$$\overline{\Delta B}_j = \left[\frac{\partial(\Delta \underline{B})}{\partial N_j}\right]_{T,\,P,\,N_i[j]} = (\overline{B}_j - \overline{B}_j^+) \tag{9-72}$$

or, rewriting Eq. (9-71),

$$\Delta \underline{B}_{mix} = \sum_{j=1}^{n} N_j\,\overline{\Delta B}_j \tag{9-73}$$

Dividing both sides of Eq. (9-73) by N yields

$$\Delta B_{mix} = \sum_{j=1}^{n} x_j\,\overline{\Delta B}_j \tag{9-74}$$

It also follows from Eq. (9-70) that

$$d(\Delta B_{mix}) = \left(\frac{\partial(\Delta B_{mix})}{\partial T}\right)_{P,\,x} dT + \left(\frac{\partial(\Delta B_{mix})}{\partial P}\right)_{T,\,x} dP + \sum_{j \neq i} \left(\frac{\partial(\Delta B_{mix})}{\partial x_j}\right)_{T,\,P,\,x[i,\,j]} dx_j \tag{9-75}$$

Note that Eqs. (9-72), (9-74), and (9-75) are completely analogous to Eqs. (9-8), (9-10), and (9-51), with B replaced by ΔB_{mix}. Since these three equations formed the basis for all the partial molar quantity relationships developed in Sections 9.3 and 9.4, it follows that all relationships developed for B also apply to ΔB_{mix}. Thus, the analog to Eq. (9-53) is

$$\overline{\Delta B}_i = \Delta B_{mix} - \sum_{j \neq i} x_j \left(\frac{\partial(\Delta B_{mix})}{\partial x_j} \right)_{T, P, x[j, i]} \tag{9-76}$$

and, hence, for binary mixtures the slope-intercept method can be used to evaluate partial molar mixing functions from specific mixing functions. Similarly, the analogs to the Gibbs-Duhem relationships given in Eqs. (9-61) and (9-67) are

$$\sum_{i=1}^{n} x_i \, d(\overline{\Delta B})_i = \left(\frac{\partial(\Delta B_{mix})}{\partial T} \right)_{P, x} dT + \left(\frac{\partial(\Delta B_{mix})}{\partial P} \right)_{T, x} dP \tag{9-77}$$

and at constant T and P

$$\sum_{i=1}^{n} x_i \left(\frac{\partial(\overline{\Delta B}_i)}{\partial x_j} \right)_{T, P, x[j, k]} = 0 \tag{9-78}$$

and thus either form of the Gibbs-Duhem relation can be used to check consistency of data or to find one partial molar mixing function from a set of $(n - 1)$ others. The most straightforward application of the Gibbs-Duhem relation is for a 2-component, binary system illustrated in Example 9.4.

Example 9.6

If 1 mole of pure sulfuric acid is diluted with N_w moles of water, the heat evolved is about equal to

$$Q(\text{kJ}) = \frac{74.78 N_w}{N_w + 1.7983} \qquad (18°\text{C}, N_w < 20)$$

What is the differential enthalpy of solution for water and acid for a solution containing 40 mole % acid?

Solution

The heat evolved for 1 mole of acid is given in the problem statement. If we wished to obtain Q for N_A moles of acid and N_w moles of water in kilojoules,

$$Q = \frac{N_A \, (74.78 \, N_w / N_A)}{(N_w / N_A) + 1.7983}$$

Then the integral heat of solution, $\Delta \underline{H}_{mix}$, is $-Q$ and therefore

$$\overline{\Delta H}_w = \left(\frac{\partial \Delta \underline{H}}{\partial N_w} \right)_{T, P, N_A} = \frac{-134 x_A^2}{(1 + 0.798 \, x_A)^2} = -12.35 \text{ kJ/mol } H_2O$$

$$\overline{\Delta H}_A = \left(\frac{\partial \Delta \underline{H}}{\partial N_A} \right)_{T, P, N_w} = \frac{74.78 x_w^2}{(1 + 0.798 \, x_A)^2} = -15.50 \text{ kJ/mol acid}$$

9.6 Ideal Gas Mixtures and Ideal Solutions

In Chapter 3 we defined an ideal gas as one that exhibited $P\underline{V}TN$ behavior as given by

$$P^{ig} = \frac{NRT}{\underline{V}} = \frac{RT}{V} \tag{9-79}$$

and, in addition, the internal energy of an ideal gas was a function only of temperature and mass,

$$\underline{U} = f(T, N) = NU \tag{9-80}$$

Further, we indicated that this hypothetical model gas was well simulated by real gases at low pressures. As a limiting case, as pressure approaches zero, all gases are assumed to behave as ideal gases. This concept was then utilized in Section 8.5 to develop methods to calculate differences in thermodynamic properties by determining isothermal departure functions and varying temperature in an ideal-gas state. In all the previous work, we have implicitly assumed that the systems only contained a pure component.

In this section we want to examine the *ideal-gas* mixture[2] as a limiting model for a mixture of gases. Also, for mixtures, there is, in addition, a less restrictive model than the ideal-gas mixture; this is the ideal solution. The latter model can be used both for gaseous and for liquid mixtures.

Denbigh (1971) has suggested that the definitions of an ideal gas (pure), an ideal-gas mixture, and an ideal solution may all be written in a similar form:

Pure-component; ideal gas:

$$G_i = \mu_i = RT \ln P + \lambda_i (T) \tag{9-81}$$

Ideal-gas mixture; for component i:

$$\overline{G}_i = \mu_i = RT \ln p_i + \lambda_i (T) \tag{9-82}$$

Ideal solution; for component i:

$$\overline{G}_i = \mu_i = RT \ln x_i + \Lambda_i (T, P) \tag{9-83}$$

In these definitions, G_i is the Gibbs energy of pure i while $\overline{G}_i$ is the partial molar Gibbs energy of i in a mixture. P is the total system pressure, y_i is the vapor phase mole fraction of i, and p_i is the partial pressure of i,

$$p_i = y_i P \tag{9-84}$$

$\lambda_i (T)$ is a function of temperature and is specific for component i. $\lambda_i(T)$ is equivalent to the chemical potential of pure i in an ideal-gas state at T and $P = 1$ bar [μ^{ig} $(T, 1 \text{ bar})$]. $\Lambda_i (T, P)$ is a similar function for i, but it depends on both temperature and pressure.

---

2 Note that an ideal-gas mixture is not necessarily the same as a *mixture* of *ideal gases*.

$\Lambda_i(T, P)$ is equivalent to the chemical potential of pure i in an ideal-solution state at T, P, and $x_i = 1$ [μ_i^{ID} $(T, P, x_i = 1)$].

Illustrating that Eq. (9-81) satisfies the criteria stated above for a pure component, ideal gas, from Eq. (9-81)

$$\left(\frac{\partial G_i}{\partial P}\right)_T = \frac{RT}{P}$$

but, from Eq. (9-15), with $x_i = 1$,

$$\left(\frac{\partial G_i}{\partial P}\right)_T = V_i$$

Thus, Eq. (9-79) is obtained. To prove the second criterion, we utilize the definition $\underline{G} \equiv \underline{H} - T\underline{S}$ and the Gibbs-Helmholtz equation:

$$\frac{\partial}{\partial T}\left(\frac{\underline{G}}{T}\right)_{P, N_i} = \frac{\partial}{\partial T}\left(\frac{\underline{H}}{T} - \underline{S}\right)_{P, N_i} = -\frac{\underline{H}}{T^2} \tag{9-85}$$

where the identity

$$\left(\frac{1}{T}\right)\left(\frac{\partial \underline{H}}{\partial T}\right)_{P, N_i} = \left(\frac{\partial \underline{S}}{\partial T}\right)_{P, N_i} = \frac{NC_P}{T}$$

was used. Writing Eq. (9-85) in intensive form and using Eq. (9-81) for a pure ideal gas,

$$\frac{\partial}{\partial T}\left(\frac{G_i}{T}\right)_{P, N_i} = \frac{-H_i}{T^2} = \frac{d}{dT}\left(\frac{\lambda_i(T)}{T}\right)$$

Thus, H_i is only a function of temperature. Also, since as shown above, $V_i = RT/P$, then

$$U_i = H_i(T) - PV_i = H_i(T) - RT$$

so U_i depends only on temperature. Commonly for a pure ideal gas, H_i is written as H_i^o for clarity.

In treating an ideal-gas mixture, using Eqs. (9-82), (9-84), and (9-46) with constant N_i and composition, y_i, for all i=1,..., n components we get

$$\left(\frac{\partial \overline{G}_i}{\partial P}\right)_{T, N_i \text{ or } y_i} = \overline{V}_i = \frac{RT}{P} \qquad \textit{(ideal-gas mixture)}$$

Therefore,

$$V = \sum_i y_i \overline{V}_i = \frac{RT}{P}$$

As expected, an ideal-gas mixture also follows the ideal gas law [Eq. (9-79)]. Next, differentiating Eq. (9-85) with respect to N_i, keeping T, P, and all moles except i constant, we obtain the partial molar form of the Gibbs-Helmholtz relationship:

$$\frac{\partial}{\partial T}\left(\frac{\overline{G}_i}{T}\right)_{P,N_i} = -\frac{\overline{H}_i}{T^2} \tag{9-86}$$

For component i in an ideal-gas mixture $\overline{H}_i = \overline{H}_i^o$. Using Eq. (9-86) with Eq. (9-82), noting that since P and N_i (i=1,..., n) are held constant, p_i is not a variable, thus

$$-\frac{\overline{H}_i^o}{T^2} = \frac{d}{dT}\left(\frac{\lambda_i(T)}{T}\right) \tag{9-87}$$

and $\overline{H}_i^o$ depends only on temperature. Because, in general,

$$\overline{U}_i = \overline{H}_i(T) - P\overline{V}_i = \overline{H}_i(T) - RT \tag{9-88}$$

$\overline{U}_i \rightarrow \overline{U}_i^o$ also varies only with temperature. Thus, in an ideal-gas mixture, $\overline{H}_i^o$ and $\overline{U}_i^o$ are functions only of temperature. As will be seen later, the fact that neither of these properties is a function of composition will lead to the conclusion that there is no enthalpy (or internal energy) change when pure-component ideal gases are mixed to form an ideal-gas mixture.

While the consequences of the defining ideal-gas mixture equation [Eq. (9-82)] are those expected when the analogy is made with pure-component ideal gases, there is one other result that is readily obtained and is very useful in later work. In Section 6.4 we considered the equilibrium criteria for situations involving semipermeable membranes. The criteria derived were that the temperatures were equal on both sides of the membrane, as were the chemical potentials of all components for which the membrane was permeable. Therefore, for an *ideal-gas mixture*, the partial pressures of permeable components would be equal across the membrane. (see Section 9.9).

To complete the discussion of ideal-gas mixtures, we note that real-gas mixtures at low pressures simulate ideal-gas mixtures and, as the pressure is reduced toward zero, *all* gas mixtures behave in an ideal fashion. Later in Section 9.7, we will utilize this limiting ideal behavior as a reference point to define isothermal departure functions between real and ideal gas behavior.

In the case of an ideal solution (ID), as defined by Eq. (9-83), we first differentiate with respect to pressure while maintaining the temperature and composition constant.

$$\left(\frac{\partial \overline{G}_i}{\partial P}\right)_{T,y_i} = \left\{\frac{\partial[\Lambda_i(T,P)]}{\partial P}\right\}_{T,y_i} = \overline{V}_i^{ID}$$

Since Λ_i does not depend on composition, neither does $\overline{V}_i^{ID}$; therefore, in an ***ideal solution***,

$$\overline{V}_i^{ID} \neq f(\text{composition}) \quad (\textit{ideal solution}) \tag{9-89}$$

In a similar manner, using Eqs. (9-83) and (9-86), we obtain

$$\frac{\partial}{\partial T}\left(\frac{\overline{G}_i}{T}\right)_{P,y_i} = \left\{\frac{\partial[\Lambda_i(T,P)/T]}{\partial T}\right\}_{P,y_i} = -\frac{\overline{H}_i^{ID}}{T^2}$$

so that, in an ideal solution,

$$\overline{H}_i^{ID} \neq f(\text{composition}) \tag{9-90}$$

If the mixture follows *ideal-solution behavior over the entire composition range*, then Eqs. (9-89) and (9-90) imply that:

$$\overline{V}_i^{ID} = V_i \quad \text{and} \quad \overline{H}_i^{ID} = H_i \tag{9-91}$$

where V_i and H_i are pure-component specific volumes and enthalpies of i at the same temperature and pressure as the mixture and for the same phase (liquid, vapor, solid) as the mixture. Further, in such a case, it is evident from Eq. (9-83) that

$$\overline{G}_i = \overline{G}_i^{ID} = G_i + RT \ln x_i \tag{9-92}$$

where G_i is the pure-component Gibbs energy. Since, in general

$$G_i = H_i - TS_i \quad \text{and} \quad \overline{G}_i = \overline{H}_i - T\overline{S}_i \tag{9-93}$$

then

$$\overline{S}_i = \overline{S}_i^{ID} = S_i - R \ln x_i \tag{9-94}$$

Thus, the *intensive properties of the ideal solutions* may be written as:

$$G^{ID} = \sum_i^n x_i G_i + RT \sum_i^n x_i \ln x_i \tag{9-95}$$

$$S^{ID} = \sum_i^n x_i S_i - R \sum_i^n x_i \ln x_i \tag{9-96}$$

$$H^{ID} = \sum_i^n x_i H_i \tag{9-97}$$

$$V^{ID} = \sum_i^n x_i V_i \tag{9-98}$$

Mixing functions can also be defined for the ideal solution, as introduced in Section 9.5. In this case the reference state is invariably taken as *the pure components at the same temperature, pressure, and state of aggregation as that of the mixture*. Thus, the superscripts in Eq. (9-69) can be deleted:

$$\Delta B_{mix} = B - \sum_{j=1}^n x_j B_j \tag{9-99}$$

For an ideal solution,

$$\Delta B_{mix}^{ID} = B^{ID} - \sum_{j=1}^n x_j B_j \tag{9-100}$$

For the Gibbs energy, entropy, and enthalpy of an ideal solution, substituting Eqs. (9-95), (9-96), (9-97), and (9-98), respectively, into Eq. (9-100) yields

$$\Delta G_{mix}^{ID} = RT \sum_{j=1}^{n} x_j \ln x_j \tag{9-101}$$

$$\Delta S_{mix}^{ID} = -R \sum_{j=1}^{n} x_j \ln x_j \tag{9-102}$$

$$\Delta H_{mix}^{ID} = 0 \text{ and } \Delta V_{mix}^{ID} = 0 \tag{9-103}$$

$$\overline{\Delta G}_j^{ID} = RT \ln x_j \tag{9-104}$$

$$\overline{\Delta S}_j^{ID} = -R \ln x_j \tag{9-105}$$

$$\overline{\Delta H}_j^{ID} = 0 \text{ and } \overline{\Delta V}_j^{ID} = 0 \tag{9-106}$$

In Section 9.8, we will define other mixing functions using an ideal solution reference state to obtain activity and activity coefficients.

9.7 Fugacity and Fugacity Coefficients

We have seen in Chapter 6 that chemical potentials (μ_i) play an important role in defining equilibrium states: Chemical potentials are equal in all phases for multiphase systems, chemical potentials are the same for all permeable components across a membrane, and for chemical equilibrium, the sum of the products of the chemical potentials times their respective stoichiometric multipliers is zero. Thus, it is most desirable to be able to calculate such chemical potentials.

There are, however, several obstacles. First, the numerical value of any chemical potential can only be determined within an arbitrary constant, which is related to a reference state entropy. Second, chemical potentials become negatively infinite as the system pressure approaches zero. Third, the chemical potential of a component in a mixture also becomes negatively infinite as the concentration of that component approaches zero.

Unfortunately, the singularities for μ_i occur at attenuated conditions (low pressure or infinite dilution) where ideal behavior is anticipated, which make its use inconvenient for many engineering applications. Motivated by these functional characteristics of μ_i, G.N. Lewis in 1901 proposed a new property, the fugacity, that was defined and finite under attenuated ideal gas or ideal solution conditions. Furthermore, as will become apparent, the fugacity function may be employed instead of the chemical potential to define phase, membrane, or chemical equilibrium. Also, the fugacity may be numerically determined and is a well-behaved function at both low pressures and/or small concentrations.

The *fugacity of a component i in a mixture*, $\hat{f}_i$, is defined using the chemical potential as

$$\boxed{\overline{G}_i = \mu_i = RT \ln \hat{f}_i + \lambda_i(T)} \qquad (mixture\ i) \tag{9-107}$$

$\hat{f}_i$ is a function of temperature, pressure, and composition. $\hat{f}_i$ *mixture component i* is not a partial molar property. The circumflex ^ signifies only that the fugacity applies to a component in a mixture. The function $\lambda_i(T)$ is the same as introduced earlier in Eqs. (9-81) and (9-82) to define ideal gases and ideal-gas mixtures. The *fugacity of a pure component*, f_i, depends on pressure and temperature and is defined as

$$\boxed{G_i = \mu_i = RT \ln f_i + \lambda_i(T)} \qquad (pure\ i) \tag{9-108}$$

Clearly, comparing Eq. (9-81) with (9-108) or Eq. (9-82) with (9-107), the fugacity of a pure component is equal to the pressure if the gas is ideal, and the fugacity of a component in an ideal-gas mixture is equal to its partial pressure. Pressure units are used for $\hat{f}_i$ and f_i, usually in atm or bar. The units are implicitly specified by the ideal-gas (*ig*) state condition contained in the $\lambda_i(T)$ term, for example

$$\lambda_i(T) = \mu_i^{ig}\ (T, pure\ i) \text{ at } P = 1 \text{ bar or } 1 \text{ atm}$$

Since all substances (pure or mixtures) are assumed to approach an ideal-gas state as the pressure is attenuated ($P \rightarrow P^* \approx 0$) then equivalent statements would be

$$\lim_{P \rightarrow P^*} \frac{f_i}{P} = 1 \qquad (pure\ i) \tag{9-109}$$

$$\lim_{P \rightarrow P^*} \frac{\hat{f}_i}{y_i P} = 1 \qquad (mixture\ component\ i) \tag{9-110}$$

Lewis (1901) was quick to point out that many alternative definitions could be used to create a replacement for the chemical potential. But to paraphrase Lewis' own words "the fugacity is a quantity... [as defined by Eq. (9-107)]... which seems at first sight more abstruse [than other thermodynamic properties such as vapor pressure, solubility, etc] but is, in fact, simpler, more general, and easier to manipulate." Lewis went on in his 1901 paper to relate fugacity rigorously to measurable properties such as the osmotic pressure of a solution at infinite dilution and the isothermal residual volumetric properties of a pure or mixture component in a real-fluid state relative to an ideal-gas state. He demonstrated that fugacity was in all cases a property that could be calculated from experimental measurements!

The variation of fugacity with pressure is readily determined from the defining equations [Eqs. (9-107) and (9-108)] and Eqs. (9-44) and (9-45),

$$\left(\frac{\partial \ln \hat{f}_i}{\partial P}\right)_{T,N} = \frac{1}{RT}\left(\frac{\partial \overline{G}_i}{\partial P}\right)_{T,N} = \frac{\overline{V}_i}{RT} \quad (9\text{-}111)$$

$$\left(\frac{\partial \ln f_i}{\partial P}\right)_{T,N} = \frac{1}{RT}\left(\frac{\partial G_i}{\partial P}\right)_{T,N} = \frac{V_i}{RT} \quad (9\text{-}112)$$

Where the subscript N in this instance implies that all N_i (i=1,..., n) are held constant. To find the temperature effect on fugacity, it is convenient first to relate the Gibbs departure function to fugacity for a pure component. Recalling from Section 8.5 that a departure function represents the difference between the property in the real state (T, V or T, P) and in an ideal-gas state at (T, V^o), where $V^o = RT/P$, then

$$G_i - G_i^o = RT \ln \frac{f_i}{P} \quad (pure\ i) \quad (9\text{-}113)$$

where Equation (9-81) was used to represent G_i^o and Eq. (9-108) for G_i. Dividing Eq. (9-113) by RT and then differentiating with respect to temperature at constant pressure, with Eq. (9-85), we obtain

$$\frac{\partial}{\partial T}[(G_i - G_i^o)/RT]_P = \left[\frac{\partial \ln (f_i/P)}{\partial T}\right]_P = -\frac{H_i - H_i^o}{RT^2} \quad (9\text{-}114)$$

Thus, the variation in $\ln f_i$ with temperature is simply the negative of the enthalpy departure function divided by RT^2. For an ideal gas, f_i would not be a function of temperature. In a similar fashion, for a component i in a mixture, Eq. (9-113) becomes

$$\overline{G}_i - \overline{G}_i^o = RT \ln \frac{\hat{f}_i}{y_i P} \quad (mixture\ component\ i) \quad (9\text{-}115)$$

where Eqs. (9-107), (9-82), and (9-84) have been used. $\overline{G}_i^o$ is the partial molar Gibbs energy in an ideal-gas mixture at T, V^o and the same composition as the real mixture. Again dividing by RT and differentiating with respect to temperature at constant pressure and composition, with the Gibbs-Helmholtz relationship of Eq. (9-86), we get

$$\frac{\partial}{\partial T}[(\overline{G}_i - \overline{G}_i^o)/RT]_{P,y} = \left[\frac{\partial \ln (\hat{f}_i/y_i P)}{\partial T}\right]_{P,y} = -\frac{\overline{H}_i - \overline{H}_i^o}{RT^2} \quad (9\text{-}116)$$

But for an ideal-gas mixture, $\overline{H}_i^o$ is only a function of temperature and is independent of composition and pressure, so $\overline{H}_i^o = H_i^o(T, P) = H_i^* = H_i^*(T, P^*)$. Then, since the partial derivative in Eq. (9-116) is at constant P and y_i:

$$\left(\frac{\partial \ln \hat{f}_i}{\partial T}\right)_{P,y} = -\frac{\overline{H}_i - H_i^o}{RT^2} \quad (9\text{-}117)$$

We note again that $H_i^o = H_i^* =$ enthalpy of pure component i in an ideal-gas state at the same temperature and pressure as the real gas mixture. The temperature dependence of H^o can be expressed using the ideal-gas state heat capacity:

$$H^o = \sum_i y_i H_i^o \text{ and } \left(\frac{\partial H^o}{\partial T}\right)_{P,\, y_i} = \sum y_i C_P^o(T) \tag{9-118}$$

The fugacity coefficient of component i, $\hat{\phi}_i$, is given by the isothermal departure function of the chemical potential or partial molar Gibbs energy:

$$RT \ln \hat{\phi}_i = \overline{G}_i - \overline{G}_i^o = RT \ln (\hat{f}_i / y_i\, P) \tag{9-119}$$

with $\hat{\phi}_i$ defined as;

$$\boxed{\hat{\phi}_i \equiv \hat{f}_i / y_i\, P} \qquad (\textit{mixture component } i) \tag{9-120}$$

For a pure component ($y_i = 1.0$):

$$\boxed{\phi_i \equiv f_i / P} \qquad (\textit{pure } i) \tag{9-121}$$

Using Eqs. (9-109) and (9-110) both $\hat{\phi}_i$ and ϕ_i go to 1.0 as P goes to zero. In chemical engineering practice, we usually calculate the fugacity coefficient to determine $\hat{f}_i$ or f_i. To obtain numerical values, a $PVTN$ equation of state is required. Following the approach used in Section 8.5, we can define isothermal integrals for component i in a mixture or as a pure material from an ideal-gas state at P^* (where P^* approaches 0)

for component i in a mixture:

$$RT \ln (\hat{f}_i / y_i\, P^*) = \int_{P^*}^{P} \overline{V}_i \, dP \tag{9-122}$$

for pure i:

$$RT \ln (f_i / P^*) = \int_{P^*}^{P} V_i \, dP \tag{9-123}$$

By combining Eqs. (9-122) and (9-123):

$$RT \ln (\hat{f}_i / y_i f_i) = \int_{P^*}^{P} (\overline{V}_i - V_i)\, dP = \int_{P^*}^{P} \overline{\Delta V_i}\, dP \tag{9-124}$$

However, in the ideal-gas state which exists from P^* to 0, $\overline{V}_i = V_i = RT/P$, i.e. $\overline{\Delta V_i} = 0$, so Eq. (9-124) can be rewritten as:

$$RT \ln (\hat{f}_i / y_i f_i) = \int_0^P (\overline{V}_i - V_i)\, dP - \int_0^{P^*} \underbrace{(\overline{V}_i - V_i)}_{\to 0}\, dP \tag{9-125}$$

The second integral on the RHS is zero because $\overline{\Delta V_i} = 0$ over the pressure domain from 0 to P^*.

A volume-explicit formulation for evaluating the fugacity coefficient of a pure substance can be developed from its definition in Eq. (9-121):

$$RT \ln \phi_i = RT \ln f_i/P = G_i(T, P) - G_i^o(T, P) \tag{9-126}$$

Thus ϕ_i is related to an isothermal Gibbs departure function of pure i from its real state at T and P to an ideal-gas state at the same T and P, we obtain

$$\boxed{RT \ln \phi_i = \int_0^P \left(V_i - \frac{RT}{P}\right) dP} \quad \textit{(pure i)} \tag{9-127}$$

noting that $V_i^o = RT/P$.

Similarly, for component i in a mixture, we can evaluate $\hat{\phi}_i = \hat{f}_i/y_i P$ by first adding the pure component result given by Eq. (9-127) to the result given by Eq. (9-125). Thus,

$$RT \ln (\hat{f}_i/y_i f_i) + RT \ln (f_i/P) = \int_0^P (\overline{V}_i - V_i)\, dP + \int_0^P \left(V_i - \frac{RT}{P}\right) dP \tag{9-128}$$

Combining terms on both sides of Eq. (9-128) we get:

$$\boxed{RT \ln \hat{\phi}_i = RT \ln (\hat{f}_i/y_i P) = \int_0^P \left(\overline{V}_i - \frac{RT}{P}\right) dP} \quad \textit{(mixture component i)} \tag{9-129}$$

Evaluation of this integral is convenient if one has a volume-explicit EOS. Since this is rarely the case in most practical situations, an equivalent form for $\hat{\phi}_i$ is needed for a pressure-explicit EOS. In such cases, it is convenient to formulate the problem with T, $\underline{V}$, N as the independent variables. This conclusion suggests that a Legendre transform of the energy into T, $\underline{V}$, N space would be appropriate. Such a transform is the Helmholtz energy; $\underline{A}$. Recalling that

$$\underline{A} = \underline{U} - T\underline{S} \tag{9-130}$$

$$d\underline{A} = -\underline{S}\, dT - P\, d\underline{V} + \sum_i \mu_i\, dN_i \tag{9-131}$$

with

$$\left(\frac{\partial \underline{A}}{\partial \underline{V}}\right)_{T, N} = -P \quad \text{and} \quad \left(\frac{\partial \underline{A}}{\partial N_i}\right)_{T, \underline{V}, N_j[i]} = \mu_i \tag{9-132}$$

The Helmholtz energy-departure function was developed in Eq. (8-122). Rewriting it in extensive form yields

$$\underline{A}(T, \underline{V}, N) - \underline{A}^o(T, \underline{V}^o, N) = -\int_{\infty}^{\underline{V}} \left(P - \frac{NRT}{\underline{V}} \right) d\underline{V} + NRT \ln \frac{\underline{V}^o}{\underline{V}} \qquad (9\text{-}133)$$

Now if we differentiate Eq. (9-133) with respect to N_i with T, $\underline{V}$, $N_j\,[i]$ held constant, we can create an expression for μ_i which will allow us to obtain $\hat{\phi}_i \equiv \hat{f}_i/y_i\,P$. But first, consider the ideal gas reference state,

$$\underline{V}^o = NRT/P, \quad \hat{f}_i^o = y_i\,P \qquad (9\text{-}134)$$

and

$$\mu_i^o = RT \ln y_i\,P \; + \; \lambda_i(T) \; \neq \; (\partial \underline{A}^o/\partial N_i)_{T,\,\underline{V},\,N_j[i]} \qquad (9\text{-}135)$$

We can develop an expression for μ_i^o by rewriting the Fundamental Equation in T, $\underline{V}^o$, and N_i coordinates for ideal gas reference state condition as:

$$\underline{A}^o \; = f\,[T, \underline{V}^o, N_i] \;\; \text{or} \;\; d\underline{A}^o \; = -\underline{S}^o dT - Pd\underline{V}^o + \sum_i \mu_i^o dN_i \qquad (9\text{-}136)$$

Thus by differentiating Eq. (9-136) with respect to N_i at constant T, $\underline{V}$, $N_j\,[i]$, we obtain:

$$(\partial \underline{A}^o/\partial N_i)_{T,\,\underline{V},\,N_j[i]} \; = -\underline{S}^o(\partial T/\partial N_i)_{T,\,\underline{V},\,N_j[i]}$$
$$- P(\partial \underline{V}^o/\partial N_i)_{T,\,\underline{V},\,N_j[i]} + \mu_i^o\,(\partial N_i/\partial N_i)_{T,\,\underline{V},\,N_j[i]}$$

or by rearranging:

$$\mu_i^o \; = \; (\partial \underline{A}^o/\partial N_i)_{T,\,\underline{V},\,N_j[i]} + P(\partial \underline{V}^o/\partial N_i)_{T,\,\underline{V},\,N_j[i]} \qquad (9\text{-}137)$$

Now differentiating Eq. (9-133) with respect to N_i with T, $\underline{V}$, $N_j\,[i]$ held constant, with Eq. (9-137) and

$$\mu_i = (\partial \underline{A}/\partial N_i)_{T,\,\underline{V},\,N_j[i]}$$

yields:

$$(\partial \underline{A}/\partial N_i)_{T,\,\underline{V},\,N_j[i]} - (\partial \underline{A}^o/\partial N_i)_{T,\,\underline{V},\,N_j[i]} \; = \; \mu_i - \mu_i^o + P\,(\partial \underline{V}^o/\partial N_i)_{T,\,\underline{V},\,N_j[i]}$$

$$= -\int_{\infty}^{\underline{V}} \left[(\partial P/\partial N_i)_{T,\,\underline{V},\,N_j[i]} - RT/\underline{V} \right] d\underline{V} + RT \ln(\underline{V}^o/\underline{V}) + NRT \left(\frac{\partial \ln(\underline{V}^o/\underline{V})}{\partial N_i} \right)_{T,\,\underline{V},\,N_j[i]} \qquad (9\text{-}138)$$

The last term of the RHS of Eq. (9-138) simplifies to:

$$NRT \left[\left(\frac{\partial \ln \underline{V}^o}{\partial N_i} \right) - \left(\frac{\partial \ln \underline{V}}{\partial N_i} \right) \right]_{T,\,\underline{V},\,N_j[i]} = \frac{NRT}{\underline{V}^o} \left(\frac{\partial \underline{V}^o}{\partial N_i} \right)_{T,\,\underline{V},\,N_j[i]}$$

From Eq. (9-119), at the same temperature T:

$$\mu_i - \mu_i^o \; = \; \overline{G}_i - \overline{G}_i^o \; = \; RT \ln(\hat{f}_i/y_i P) \qquad (9\text{-}139)$$

Therefore, with Eq. (9-139), Eq. (9-138) becomes

$$RT \ln (\hat{f}_i/y_iP) + P(\partial \underline{V}^o/\partial N_i)_{T,\underline{V},N_j[i]} = -\int_\infty^{\underline{V}} \left[(\partial P/\partial N_i)_{T,\underline{V},N_j[i]} - RT/\underline{V}\right] d\underline{V}$$

$$+ RT \ln \left(\underline{V}^o/\underline{V}\right) + \frac{NRT}{\underline{V}^o}\left(\partial \underline{V}^o/\partial N_i\right)_{T,\underline{V},N_j[i]} \qquad (9\text{-}140)$$

But, $\underline{V}^o = NRT/P$ and $\underline{V} = ZNRT/P$ where Z is the compressibility factor for the mixture. Thus,

$$RT \ln (\hat{f}_i/y_iP) + P(\partial \underline{V}^o/\partial N_i)_{T,\underline{V},N_j[i]} = -\int_\infty^{\underline{V}} \left[(\partial P/\partial N_i)_{T,\underline{V},N_j[i]} - RT/\underline{V}\right] d\underline{V}$$

$$- RT \ln Z + P\left(\partial \underline{V}^o/\partial N_i\right)_{T,\underline{V},N_j[i]} \qquad (9\text{-}141)$$

and the two terms involving $P\left(\partial \underline{V}^o/\partial N_i\right)_{T,\underline{V},N_j[i]}$ cancel so Eq. (9-141) becomes

$$\boxed{RT \ln \frac{\hat{f}_i}{y_iP} = RT \ln \hat{\phi}_i = -\int_\infty^{\underline{V}} \left[\left(\frac{\partial P}{\partial N_i}\right)_{T,\underline{V},N_j[i]} - \frac{RT}{\underline{V}}\right] d\underline{V} - RT \ln Z} \qquad (mixture\ i) \quad (9\text{-}142)$$

Many other alternative derivations and forms for $\hat{\phi}_i$ can be developed by appropriate transformations.[3]

Equation (9-142) can also be used to determine the fugacity coefficient for a pure material, ϕ_i, by letting $y_i \rightarrow 1$, then

$$\boxed{RT \ln \phi_i = RT \ln \frac{f_i}{P} = -\int_\infty^{\underline{V}} \left[\left(\frac{\partial P}{\partial N}\right)_{T,\underline{V}} - \frac{RT}{\underline{V}}\right] d\underline{V} - RT \ln Z} \qquad (pure\ i) \quad (9\text{-}143)$$

In either Eq. (9-142) or (9-143), one employs a pressure-explicit equation of state to determine $(\partial P/\partial N_i)_{T,\underline{V},N_j[i]}$ or $(\partial P/\partial N)_{T,\underline{V}}$.

Example 9.7

Determine ϕ_i for a pure material whose $P\underline{V}TN$ behavior can be represented by the Peng-Robinson equation of state (PR EOS) (see Table 8.1).

Solution

---

3 For example, using the triple product and integrating by parts transform Eq. (9-129) into Eq. (9-142).

In this particular case, rather than use Eq. (9-143), we have already developed the Helmholtz energy departure function in Eq. (8-127) for the PR EOS. Then,

$$\frac{\partial}{\partial N}[(\underline{A}-\underline{A}^o)]_{T,\underline{V}}=\frac{\partial}{\partial N}[N(A-A^o)]_{T,\underline{V}}=\mu_i-\mu_i^o=RT\ln\phi_i$$

So

$$RT\ln\phi_i=(A-A^o)+N\left[\frac{\partial}{\partial N}(A-A^o)_{T,\underline{V}}\right]$$

To evaluate the derivative, from Eq. (8-127),

$$\frac{\partial}{\partial N}[(A-A^o)]_{T,\underline{V}}=\frac{\partial}{\partial N}\left[RT\ln\frac{\underline{V}^o}{\underline{V}-Nb}+\frac{a}{2\sqrt{2}\,b}\ln\frac{\underline{V}+Nb(1-\sqrt{2})}{\underline{V}+Nb(1+\sqrt{2})}\right]_{T,\underline{V}}$$

$$=\frac{bRT}{N(V-b)}-\frac{aV}{N[V(V+b)+b(V-b)]}=\frac{RT}{N}(Z-1)$$

Therefore,

$$RT\ln\phi_i=(A-A^o)+RT(Z-1)$$

where $A-A^o$ is given in Eq. (8-127) and Z is the compressibility factor for the pure material.

A similar treatment to determine the fugacity coefficient of a component in a mixture, using the Peng-Robinson equation, yields

$$\ln\hat{\phi}_i=\frac{b_i}{b_m}(Z-1)-\ln(Z-B)+\frac{A}{2\sqrt{2}B}\left(\frac{2\sum_k y_k a_{ik}}{a_m}-\frac{b_i}{b_m}\right)\ln\frac{Z+B(1-\sqrt{2})}{Z+B(1+\sqrt{2})} \quad (9\text{-}144)$$

where

$$A=\frac{a_mP}{(RT)^2} \quad \text{and} \quad B=\frac{b_mP}{RT}$$

Note that a_m and b_m are as given in Eqs. (9-27) and (9-28), and Z is for the mixture.

We have seen that, for a pure component that behaves as an ideal gas, $f_i = P$, and, from Eq. (9-143), ϕ_i (ideal gas) = 1.0. Also, for a component in an ideal-gas mixture, $\hat{f}_i = y_iP$; thus by Eq. (9-142), $\hat{\phi}_i$ (ideal-gas mixture) = 1.0. For the case where one wishes to model a mixture as an ideal solution over the entire range of composition, Eq. (9-92) relates $\overline{G}_i^{ID}$ to G_i. With Eqs. (9-107) and (9-108),

$$\overline{\Delta G}_i^{ID}=\overline{G}_i^{ID}-G_i=RT\ln\frac{\hat{f}_i^{ID}}{f_i}=RT\ln y_i \quad (9\text{-}145)$$

Thus,

$$\boxed{\hat{f}_i^{\,ID}=f_i\,y_i}$$

and the so-called *Lewis and Randall rule* is obtained. In general, if a system obeys the Lewis and Randall Rule, $\Delta \overline{V}_i = \overline{V}_i - V_i = 0$ over the total pressure domain from 0 to P in Eq. (9-125). The pure component fugacity must be evaluated at the same temperature and pressure as the mixture and in the same state of aggregation. For example, if the mixture were a gas at T, P, then f_i would be determined for pure i at T, P and as a gas--even though at this T and P, the stable state might be other than a gas (i.e., a liquid or solid). This problem most often occurs when the Lewis and Randall rule is used to estimate mixture fugacities in phase equilibrium calculations; a further discussion of this problem is given in Chapter 15.

The Gibbs-Duhem equation for chemical potentials was derived in Section 5.6 [see Eq. (5-105)]. A comparable relation may be obtained for fugacities. In this case it is more convenient to begin with the Fundamental Equation in entropy representation [see Problem 5.1(b)]:

$$\underline{S} = S(\underline{U}, \underline{V}, N_1, \ldots, N_n) \tag{9-146}$$

The conjugate coordinates in this case are $(T^{-1}, \underline{U})$, $(PT^{-1}, \underline{V})$, $(-\mu_i T^{-1}, N_i)$. The total Legendre transform of Eq. (9-146) is

$$-\underline{U}\, d\left(\frac{1}{T}\right) - \underline{V}\, d\left(\frac{P}{T}\right) + \sum_i N_i\, d\left(\frac{\mu_i}{T}\right) = 0 \tag{9-147}$$

Expanding the first two derivatives, and using the definition for enthalpy, $\underline{H} = \underline{U} + P\underline{V}$,

$$\frac{\underline{H}}{T^2}\, dT - \frac{\underline{V}}{T}\, dP + \sum_i N_i\, d\left(\frac{\mu_i}{T}\right) = 0 \tag{9-148}$$

From Eq. (9-107),

$$d\left(\frac{\mu_i}{T}\right) = R\, d\ln \hat{f}_i + d\left[\frac{\lambda_i(T)}{T}\right] \tag{9-149}$$

To simplify, we use Eq. (9-107) once again but differentiate with respect to temperature, keeping pressure and composition constant. Then, with Eqs. (9-86) and (9-117),

$$\frac{\partial}{\partial T}\left(\frac{\mu_i}{T}\right)_{P,y} = -\frac{\overline{H}_i}{T^2} = -\frac{\overline{H}_i - H_i^o}{T^2} + \frac{d}{dT}\left(\frac{\lambda_i(T)}{T}\right)$$

Thus, as we expect from Eq. (9-87) since $H_i^o = \overline{H}_i^o$:

$$d\left(\frac{\lambda_i(T)}{T}\right) = -\frac{H_i^o}{T^2}\, dT \tag{9-150}$$

Substituting Eq. (9-150) into (9-149) and this result into Eq. (9-148) yields

$$\sum_i N_i \, d \ln \hat{f}_i = -\left(\frac{\underline{H} - \sum_i N_i H_i^o}{RT^2}\right) dT + \frac{\underline{V}}{RT} dP \tag{9-151}$$

or

$$\sum_i N_i \, d \ln \hat{f}_i = -\sum_i \frac{N_i (\overline{H}_i - H_i^o)}{RT^2} dT + \sum_i \frac{N_i \overline{V}_i}{RT} dP \tag{9-152}$$

Either of these equations represents the Gibbs–Duhem equation for the fugacity; they may be written in intensive form by dividing by the total moles, N.

In closing this section on fugacity, we note that another type of fugacity function could have been defined. In an analogous fashion to Eq. (9-107) or (9-108), the *mixture* fugacity, f_m is given by:

$$G_m = RT \ln f_m + g(\text{composition}, T) \tag{9-153}$$

Although of little utility in itself, it is interesting to note that all the previous relationships for the general property B may be applied to this new fugacity function if $\ln f_m$ is used for B and $\ln (\hat{f}_i / y_i)$ for $\overline{B}_i$. Then, for example, using Eq (9-10), $\ln f_m$ becomes

$$\ln f_m = \sum_i y_i \ln \frac{\hat{f}_i}{y_i} \tag{9-154}$$

One could also calculate f_m by assuming the mixture to be a pseudo-pure component (i.e., by maintaining the composition constant). Then the equations developed earlier in this section to determine f_i could also be used to find f_m. In Section 15.6 we do introduce $N \ln f_m$ as an extensive parameter to allow us to treat the problem of supercritical components in multicomponent mixtures.

9.8 Activity, Excess Functions and Activity Coefficients

To calculate the fugacity of a component in a mixture, Eq. (9-142) may be used provided that an equation of state applicable for mixtures is available. Equation (9-144) illustrates such a relation when the Peng-Robinson equation of state is employed. Although Eq. (9-142) is normally limited to cases where the mixture is a gas, it can be used in some cases for liquid mixtures.[4] An alternative method, applicable primarily to liquid and solid phases, involves the use of a new function, the *activity*.

---

4 The Benedict-Webb-Rubin [M. Benedict, G.B. Webb, and L.C. Rubin (1951), *Chem. Eng. Prog.*, **47**, p 419], the Soave [G. Soave (1972), *Chem. Eng. Sci.*, **27**, p 1197], and the Peng-Robinson [D.-Y. Peng and D.B. Robinson (1976), *Ind. Eng. Chem. Fundam.*, **15**, p 59] equations of state have been shown to be applicable in determining vapor *and* liquid component fugacities in light hydrocarbon mixtures. This technique is discussed in more detail in Section 15.5.

To introduce this function, we must first define a standard or reference state for the mixture. In fact, this was also the case for treating fugacity and, in Eq. (9-115), the reference state chosen was an ideal-gas mixture. For activity, let us denote the reference state by the superscript + and define the activity in terms of a difference in the partial molar Gibbs energy between the real state and the reference state:

$$\overline{G}_i\,(T, P, x_1, \ldots, x_{n-1}) - \overline{G}_i^+\,(T, P^+, x_1^+, \ldots, x_{n-1}^+) = RT \ln a_i \tag{9-155}$$

where a_i is the activity of component i. Note that the reference-state temperature is equal to the system temperature, but that the other reference-state conditions, $P^+, x_1^+, \ldots, x_{n-1}^+$, can be chosen arbitrarily. x_i is the mole fraction of i in the real state and x_i^+ is the mole fraction of i in the reference state. (x_i is used in this section rather than y_i to emphasize that the activity concept is primarily of use in condensed phases.)

We may also use Eq. (9-107) to define a fugacity of i in the reference state,

$$\overline{G}_i^+\,(T, P^+, x_1^+, \ldots, x_{n-1}^+) = RT \ln \hat{f}_i^+ + \lambda_i(T) \tag{9-156}$$

Thus it follows that

$$\overline{G}_i - \overline{G}_i^+ = \mu_i - \mu_i^+ = RT \ln \frac{\hat{f}_i}{\hat{f}_i^+} = RT \ln a_i \tag{9-157}$$

where *the activity of component i* is defined as

$$\boxed{a_i \equiv \frac{\hat{f}_i}{\hat{f}_i^+}} \tag{9-158}$$

A number of reference states are in common use. Probably the most common is that of a *pure material* reference state, that is, pure i at T and P of the mixture *and* in the same state of aggregation as the mixture. Other reference states commonly employed include, *infinite dilution* ($x_i^+ = 0$) of component i, or *unit molality* of i following behavior at infinite dilution at the same T and P and state of aggregation of the mixture. These will be discussed later in this section. For the pure i reference state $x_i^+ = 1.0$ and

$$\hat{f}_i^+ = f_i\,(T, P) \tag{9-159}$$

and

$$RT \ln a_i = \overline{G}_i - G_i = \overline{\Delta G}_i \tag{9-160}$$

If the solution were ideal, then

$$\overline{\Delta G}_i = \overline{\Delta G}_i^{ID} = RT \ln x_i \tag{9-161}$$

or

$$a_i = x_i \quad (\textit{ideal solution}) \tag{9-162}$$

For most solutions, a_i/x_i is not unity; the difference of the ratio from unity is a measure of the nonideality. This ratio is called the *activity coefficient*, γ_i, defined as,

$$\gamma_i \equiv \frac{a_i}{x_i} \tag{9-163}$$

or

$$\boxed{\gamma_i \equiv \frac{\hat{f}_i}{f_i x_i}} \tag{9-164}$$

As written, Eq. (9-164) indicates that γ_i represents the deviation of $\hat{f}_i$ from that predicted by the Lewis and Randall rule for ideal solutions. When the pure-component reference state is employed, the activity coefficient can be related directly to the excess Gibbs energy of mixing.

In general, an excess property $\underline{B}^{EX}$ or B^{EX} *is defined as a residual of the property in its real state* [$\underline{B}(T, P, N_i, \ldots, N_n)$ or $B(T, P, x_i, \ldots, x_{n-1})$] *relative to an ideal* ***solution state at the*** *same T, P, and composition as the original mixture.* The deviation of a mixture from ideal solution behavior is commonly denoted by the *excess function*, $\underline{B}^{EX}$ or B^{EX}:

$$\underline{B}^{EX} \equiv \underline{B} - \underline{B}^{ID} \quad \text{or} \quad B^{EX} = B - B^{ID} \tag{9-165}$$

From Eq. (9-9) for $\underline{B}$ and $\underline{B}^{ID}$,

$$\underline{B} = \sum_{j=1}^{n} N_j \overline{B}_j \quad \text{and} \quad \underline{B}^{ID} = \sum_{j=1}^{n} N_j \overline{B}_j^{ID} \tag{9-166}$$

it follows that

$$\underline{B}^{EX} = \sum_{j=1}^{n} N_j (\overline{B}_j - \overline{B}_j^{ID}) \tag{9-167}$$

Applying the partial molar operator to Eq. (9-165) yields

$$\overline{B}_j^{EX} = \overline{B}_j - \overline{B}_j^{ID} \tag{9-168}$$

$$\underline{B}^{EX} = \sum_{j=1}^{n} N_j \overline{B}_j^{EX} \tag{9-169}$$

Thus B^{EX} and $\overline{B}_j^{EX}$ are completely analogous to B and $\overline{B}_j$, and we could calculate $\overline{B}_j^{EX}$ from B^{EX} using the equations developed in Section 9.3 or we could use the Duhem equation as developed in Section 9.4

By analogy to Eq. (9-165), excess mixing functions are commonly defined:

$$\Delta \underline{B}_{mix}^{EX} = \Delta \underline{B} - \Delta \underline{B}^{ID} \tag{9-170}$$

or

$$\overline{\Delta B}_j^{EX} = \overline{\Delta B}_j - \overline{\Delta B}_j^{ID} \tag{9-171}$$

But, unlike $\Delta \underline{B}$ or $\Delta \underline{B}^{ID}$, an excess function does not depend directly on the choice of the reference state. This can be readily appreciated by expanding $\overline{\Delta B}_j$ and $\overline{\Delta B}_j^{ID}$ in Eq. (9-171):

$$\overline{\Delta B}_j = \overline{B}_j - \overline{B}_j^{+} \tag{9-172}$$

$$\overline{\Delta B}_j^{ID} = \overline{B}_j^{ID} - \overline{B}_j^{+} \tag{9-173}$$

Thus,

$$\overline{\Delta B}_j^{EX} = \overline{B}_j - \overline{B}_j^{ID} = \overline{B}_j^{EX} \tag{9-174}$$

Therefore, ΔB_{mix}^{EX} and B^{EX} are identical, provided that the same reference state is used for ΔB_{mix} and ΔB^{ID}. As mentioned above, the conventional reference state for ΔB^{ID} is taken as pure components at the same temperature, pressure, and state of aggregation of the mixture. Unless otherwise specified, the same reference state is used for ΔB_{mix} whenever the excess function is applied.

Because the temperature and pressure of the ideal mixture are taken equal to those of the real mixture, it follows that

$$A^{EX} = U^{EX} - TS^{EX} \tag{9-175}$$

$$H^{EX} = U^{EX} + PV^{EX} \tag{9-176}$$

and

$$G^{EX} = H^{EX} - TS^{EX} \tag{9-177}$$

For an ideal solution, the excess functions are zero, by definition. In general, if the components of a mixture have similar force fields and are not significantly different in size and symmetry, ideal solution behavior is a good first approximation. For mixtures that are free of strong associative forces such as hydrogen bonding, solvation, or complexes, *regular solution behavior* is sometimes a more accurate approximation than ideal behavior. A regular solution is defined as one for which $\overline{S}_j^{EX}$ is zero for all components. Thus,

$$G^{EX} = H^{EX} = \Delta H_{mix} = \Delta H_{mix}^{EX} \tag{9-178}$$

or

$$\overline{G}_j^{EX} = \overline{H}_j^{EX} = \overline{\Delta H}_j^{EX} = \overline{\Delta H}_j \tag{9-179}$$

That is, the Gibbs energy of mixing of a regular solution can be synthesized from knowledge of only the enthalpy of mixing.

We mention in passing a third model that is less commonly employed; an *athermal* solution is defined as one for which $\overline{H}_j^{EX} \equiv 0$ for all j. In this case, $G^{EX} = -TS^{EX}$, and estimations of G^{EX} are made from liquid models that allow an estimate of the excess entropy of mixing. (see Chapter 11 for further discussion).

Now, we can proceed to express the activity coefficient in terms of the partial molar excess Gibbs energy. Substituting Eq. (9-163) into Eq. (9-160) and using Eq. (9-171) for $\overline{\Delta G}_i$ and Eq. (9-104) for $\overline{\Delta G}_i^{ID}$, we obtain for a pure component reference state,

$$RT \ln (\gamma_i x_i) = \overline{\Delta G}_i = \overline{\Delta G}_i^{EX} + \overline{\Delta G}_i^{ID} = \Delta \overline{G}_i^{EX} + RT \ln x_i$$

or

$$\boxed{\overline{\Delta G}_i^{EX} = RT \ln \gamma_i} \tag{9-180}$$

This simple relationship between $\overline{\Delta G}_i^{EX}$ and $\ln \gamma_i$ allows us to obtain the effects of temperature and pressure on $\ln \gamma_i$ from the derivatives of $\overline{\Delta G}_i^{EX}$:

$$\left(\frac{\partial \ln \gamma_i}{\partial T}\right)_{P,x} = \frac{1}{R}\left[\frac{\partial(\overline{\Delta G}_i^{EX}/T)}{\partial T}\right]_{P,x} = -\frac{\overline{\Delta H}_i^{EX}}{RT^2} = -\frac{\overline{\Delta H}_i}{RT^2} \tag{9-181}$$

$$\left(\frac{\partial \ln \gamma_i}{\partial P}\right)_{T,x} = \frac{1}{RT}\left[\frac{\partial(\overline{\Delta G}_i^{EX})}{\partial P}\right]_{T,x} = \frac{\overline{\Delta V}_i^{EX}}{RT} = \frac{\overline{\Delta V}_i}{RT} \tag{9-182}$$

where we used the fact that $\overline{\Delta H}_i^{ID}$ and $\overline{\Delta V}_i^{ID} = 0$ to obtain the final expressions. Differential enthalpies or volumes of mixing ($\overline{\Delta H}_i$ or $\overline{\Delta V}_i$) are normally quite small, so that activity coefficients are weak functions of temperature or pressure.

Many other relations for $\ln \gamma_i$ may be derived by utilizing the equality between $\ln \gamma_i$ and $\overline{\Delta G}_i^{EX}/RT$; that is, all the previous generalized relations for $\overline{\Delta B}_i^{EX}$ are applicable. One important result is the activity coefficient form of the *Gibbs-Duhem equation*, which is obtained from the generalized relation, Eq. (9-61), with $\overline{B}_i = \overline{\Delta G}_i^{EX}/RT$ and $B = \Delta G^{EX}/RT$:

$$\sum_i x_i\, d\left(\frac{\overline{\Delta G}_i^{EX}}{RT}\right) = \sum_i x_i\, d \ln \gamma_i = -\frac{\Delta H_{mix}}{RT^2}\, dT + \frac{\Delta V_{mix}}{RT}\, dP \tag{9-183}$$

When activity coefficients for all components in a mixture are calculated or reported in the literature, the data can be examined for consistency. That is, the Gibbs-Duhem equation, Eq. (9-183), is a relationship involving all n activity coefficients, and hence any one can be found if the other $n - 1$ activity coefficients are known. Hence all n activity coefficients are not independent; they must conform to Eq. (9-183) or to equivalent mathematical formulations. Figure 9.3 illustrates several commonly used consistency tests for binary mixtures. The method described for obtaining $\overline{B}_i$ from B as a function of composition [Eq. (9-55)] can also be used to relate $\ln \gamma_i$ to $\overline{\Delta G}_i^{EX}$. However, a more common method is to employ Eq. (9-180). If this equation is multiplied by N_i and summed over all $i = 1, \ldots, n$, then

$$\Delta \underline{G}^{EX} = \sum_i N_i \overline{\Delta G}_i^{EX} = RT \sum_i N_i \ln \gamma_i \tag{9-184}$$

Now, we differentiate with respect to N_k, keeping T, P, and all moles constant except k; then, since

$$RT \sum_i N_i \left(\frac{\partial \ln \gamma_i}{\partial N_k}\right)_{T,P,N_i[k]} = 0$$

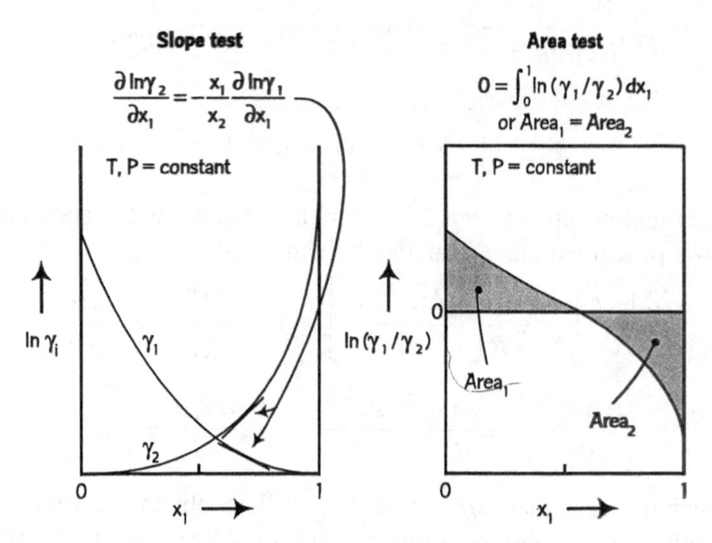

Figure 9.3 Thermodynamic consistency tests for activity coefficients in binary mixtures.

by virtue of Eq. (9-183), we have

$$RT \ln \gamma_k = \left(\frac{\partial \Delta \underline{G}^{EX}}{\partial N_k}\right)_{T, P, N_i[k]} \tag{9-185}$$

To use Eq. (9-185), one usually proposes a model for the solution's behavior that relates $\Delta \underline{G}^{EX}$ to T, P, and N_i. By simple differentiation, the activity coefficients may be determined. For example, the simplest, nontrivial relation for $\Delta \underline{G}^{EX}$ in a binary mixture of 1 and 2, which meets the obvious criteria that $\Delta \underline{G}^{EX} = 0$ as either x_1 or $x_2 \rightarrow 1.0$, is given by the two-suffix Margules equation

$$\frac{\Delta \underline{G}^{EX}}{RT} = NC\, x_1 x_2 \tag{9-186}$$

where C is not a function of composition. With Eq. (9-185) or Eq. (9-76), keeping in mind that x_i is a function of N_k:

$$\ln \gamma_1 = Cx_2^2 \quad \text{and} \quad \ln \gamma_2 = Cx_1^2 \tag{9-187}$$

These equations relating γ_1, γ_2 to composition are suitable only for very simple liquid solutions. Prausnitz *et al.* (1986) illustrates their use for the liquid binaries argon-oxygen and benzene-cyclohexane. Identical results are obtained if one assumes the solution to be regular (i.e., $\Delta \underline{S}^{EX} = 0$) and that $\Delta \underline{H} = \Delta \underline{H}^{EX}$ is symmetrical in composition. C must then be proportional to $(1/T)$.

$$\frac{\Delta \underline{H}}{RT} = NC\, x_1 x_2 = \frac{\Delta \underline{G}^{EX}}{RT} \tag{9-188}$$

For this simple case [Eq. (9-186) or (9-188)], the important properties are shown as a function of x_1 in Figure 9.4. In Figure 9.4(a), $C = 0$ (i.e., the solution is ideal).

In Figure 9.4(b), $C = 1$ and in Figure 9.4(c), $C = 2$. Note the increasing degree of nonideality with increasing C, particularly as reflected either in the activity or activity coefficients. These cases are examples of *positive deviation* from ideal solution behavior because $\gamma_1 > 1$ or $\ln \gamma_1 > 0$. Negative deviations would refer to cases where $\gamma_1 < 1$ or $\ln \gamma_1 < 0$. In general, solutions that exhibit positive deviations tend to be partially miscible over a certain range of temperatures. The case shown in Figure 9.4(c) is, in fact, in a state that is at the limit of stability, and any further increase in C (or T) would lead to immiscibility. This behavior is discussed later in Chapter 15.

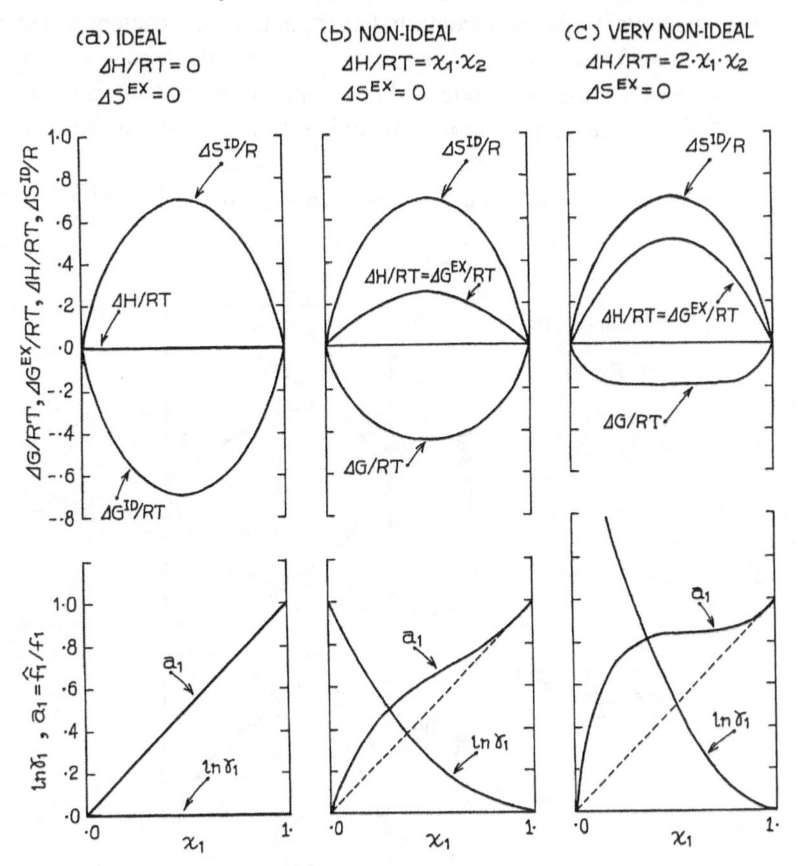

Figure 9.4 Thermodynamic functions of miscible, binary regular solutions from ideal behavior to the limit of solubility (stability).

Many liquid mixtures show very complex functionalities when enthalpies and excess entropies of mixing are plotted versus composition. This point is illustrated in Figure 9.5 for the two binary liquid solutions, ethanol-isooctane and ethanol-benzene. It is interesting to note, however, that in most cases there appear to be compensating factors in ΔH and $T\,\Delta S^{EX}$ such that ΔG^{EX} is only slightly asymmetrical. Thus, as discussed later in Chapter 11, ΔG^{EX} can often be correlated with composition by expressions containing only two (or, at most, three) constants. Chapter 11, Section 11.5, develops the functional form for ΔG^{EX} $(= \Delta \underline{G}^{EX}/N)$ for various solution models that have been proposed. By using Eq. (9-185) or Eq. (9-76), the resultant activity coefficient expressions are obtained.

Normally, one needs experimental data to be able to obtain the parameters in the ΔG^{EX} (or $\ln \gamma_i$) expressions, or, in some cases, group contribution estimation methods such as the UNIFAC technique developed by Prausnitz and his colleagues can be employed (see Section 13.4). With values of $\ln \gamma_i$ as a function of composition, fugacities of components in the solution can be found from Eq. (9-164).

In the discussion above we have concentrated on the activity coefficient, γ_i, referenced to a pure component standard state (see Figure 9.6) which is the form most commonly employed. As mentioned at the beginning of this section, there are other reference states for activity coefficients that may be more convenient to use if one of the components does not exist as a pure liquid at the temperature and pressure in question.

In general, an activity coefficient can be defined for any standard state by the equation

$$\hat{f}_i = \hat{f}_i^+ \gamma_i^+ x_i \tag{9-189}$$

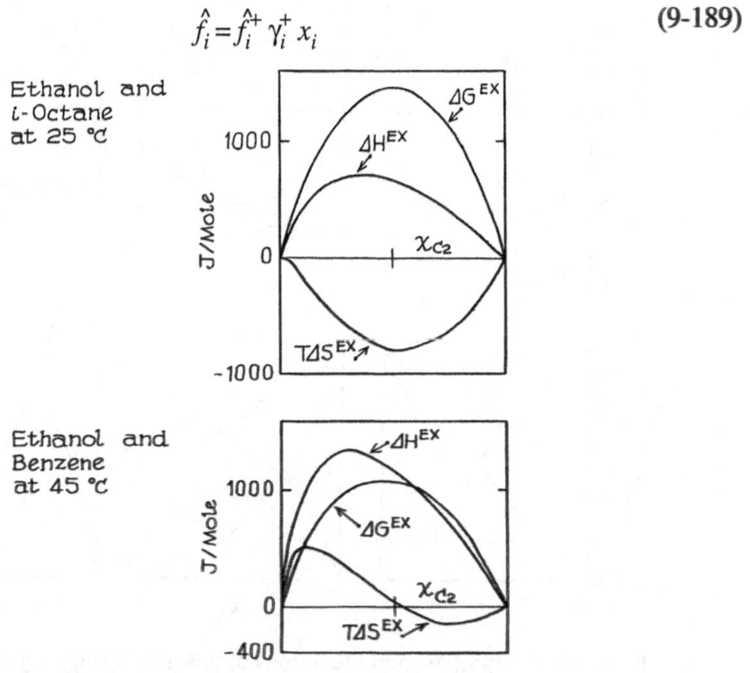

Figure 9.5 Excess functions for some binary mixtures.

where $\hat{f}_i^+$ is the fugacity of i in the standard state. By substituting Eq. (9-163) into Eq. (9-155),

$$RT \ln \gamma_i^+ + RT \ln x_i = \overline{G}_i (T, P, x_1, \ldots, x_{n-1}) - \overline{G}_i^+ (T, P^+, x_1^+, \ldots, x_{n-1}^+) \tag{9-190}$$

Note again the requirement that the standard-state temperature be equal to that of the system, but the pressure, composition, and state of aggregation may be different. Clearly the magnitude of γ_i^+ depends directly on the choice of reference state. For comparison, Eq. (9-190), for the pure solvent standard state, is

$$RT \ln \gamma_i^+ + RT \ln x_i = \overline{G}_i (T, P, x_1, \ldots, x_{n-1}^+) - G_i(T, P) \tag{9-191}$$

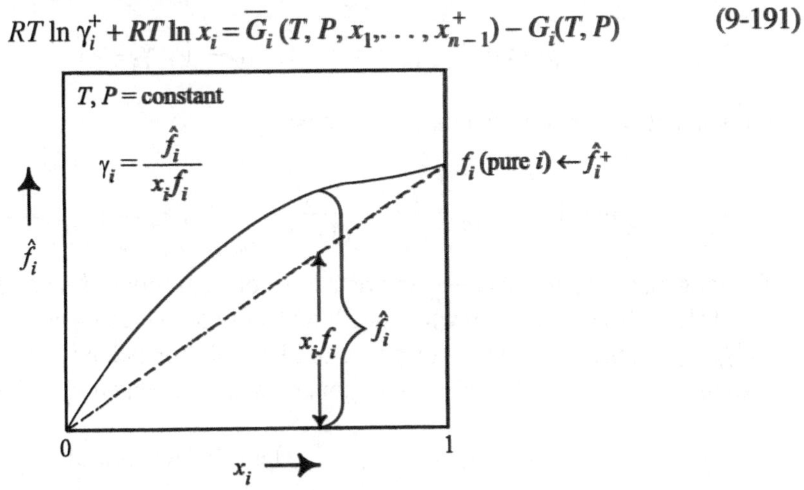

Figure 9.6 Pure component standard state, rational basis at T and P of mixture.

Let's consider how the infinite (∞) dilution and pure component standard states are related to each other. For a binary mixture of solute A in solvent B, Figure 9.7 illustrates the dependence of $\hat{f}_i$ on x_A for both components. Assume for the moment that the solubility of A is not limited, thus single phase mixtures of any composition from x_A = 0 to 1 can exist.

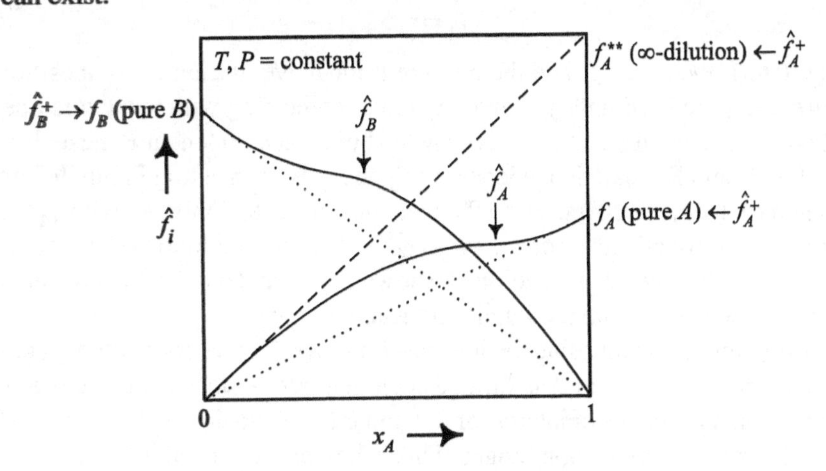

Figure 9.7 Typical $\hat{f}_i$ variation for a miscible A-B binary solution.

For solute A, there are two different reference states, that yield two different activity coefficients:

$$\textit{infinite } (\infty)\textit{-dilution of } A: \qquad \gamma_A^{**} = \frac{\hat{f}_A}{x_A f_A^{**}} \quad \text{as } x_A \to 0\,,\ \gamma_A^{**} \to 1 \tag{9-192}$$

$$\textit{pure } A: \qquad \gamma_A = \frac{\hat{f}_A}{x_A f_A} \quad \text{as } x_A \to 1\,,\ \gamma_A \to 1 \tag{9-193}$$

For solvent B, only one standard state exists:

$$\textit{pure } B: \qquad \gamma_B = \frac{\hat{f}_B}{x_B f_B} = \frac{\hat{f}_B}{(1 - x_A) f_B} \quad \text{as } x_B \to 1\ \gamma_B \to 1 \tag{9-194}$$

The tangent drawn to the $\hat{f}_A - x_A$ curve at low concentrations of A corresponds to what is called Henry's law behavior while the dotted straight line connecting the origin to the point $[\hat{f}_A = f_A$ (pure), $x_A = 1]$, corresponds to ideal solution behavior given by the Lewis and Randall rule. In the Henry's law region, data are correlated by a linear equation:

$$\hat{f}_A = K_A\, x_A = f_A^{**}\, x_A \qquad \text{(Henry's law region)} \tag{9-195}$$

Where K_A = Henry's Law constant which, in general, is a function of the temperature, pressure, and the type of solvent selected as the second component (in this case B). Strictly speaking, the f_A^{**} reference state depicted at $x_A = 1$ is only *hypothetical*, representing a linear extrapolation of behavior at infinite dilution outside the Henry's Law region. The real behavior is given by $\hat{f}_A = f_A$ (pure).

If the fluid followed the Lewis and Randall rule,

$$\hat{f}_A = f_A\, x_A \tag{9-196}$$

from $x_A = 0$ to 1, then $f_A^{**} = f_A$ and the mixture is ideal over the entire composition range from pure A to pure B with the $\hat{f}_A$ versus x_A curve coinciding with the dotted line. Also, using the Gibbs-Duhem equation, it is easy to show that component B must also follow a linear, Lewis and Randall dependence for $\hat{f}_B$ on x_B. By inspecting Figure 9.7 and Eqs. (9-192) and (9-193), we characterize the behavior of the real mixture with $\gamma_A > 1$ (pure A reference state) for all values of x_A, while $\gamma_A^{**} < 1$ (infinite-dilution reference state) for all values of x_A. In other words, solute A shows negative deviations from Henry's law and positive deviations from the Lewis and Randall rule.

A realistic setting for choosing the infinite-dilution reference state activity coefficient occurs when the solubility of A is limited such that pure A does not exist as a liquid at temperatures and pressures of interest or if A and B form two liquid phases due to limited solubility in certain composition ranges. This behavior is depicted in Figure 9.8. Again, we have selected the pure liquid standard state for solvent B. For solute A, since data for $\hat{f}_A$ only exist for low concentrations of A (i.e., pure liquid A does not exist at T and P), we define the infinite-dilution standard state for A by Henry's Law with a reference state

$(\hat{f}_A^{+} \rightarrow f_A^{**})$. It is important to recognize that to determine f_A^{**}, which allows us to calculate γ_A^{**}, only data at low concentration (approaching ∞-dilution) are required.

As before, one can express the activity coefficient in terms of the reference state fugacity for each component:

$$\gamma_B = \frac{\hat{f}_B}{x_B f_B} \qquad (\textit{pure B reference}) \tag{9-197}$$

and

$$\gamma_A^{**} = \frac{\hat{f}_A}{x_A f_A^{**}} \qquad (\textit{infinite-dilution reference for A}) \tag{9-198}$$

A pictorial representation of the activity coefficient at a particular composition x_A may also be noted on Figure 9.8 as the ratios

$$\gamma_B = \frac{\hat{f}_B^{I}}{\hat{f}_B^{II}}, \quad \text{and} \quad \gamma_A^{**} = \frac{\hat{f}_A^{IV}}{\hat{f}_A^{III}} \tag{9-199}$$

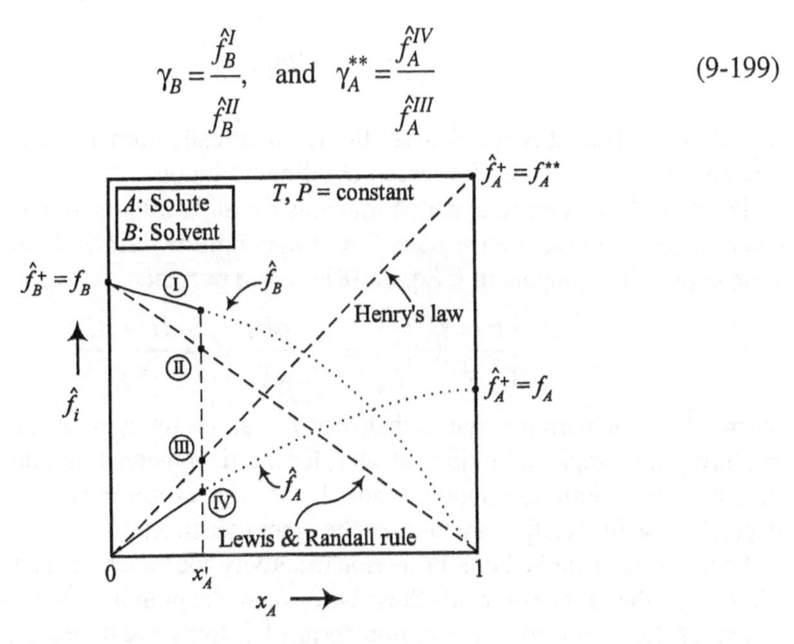

$\hat{f}_A^{+} \rightarrow \hat{f}_A^{**}$ = ∞ - dilution reference state fugacity of A in a hypothetical state (**) at $x_A = 1$ corresponding to Henry's law behavior extrapolated from an infinitely dilute state.

$\hat{f}_B^{+} \rightarrow f_B$ = pure component reference state fugacity for B in a real state at $x_B = 1$ and for A in a hypothetical state at $x_A = 1$ corresponding to
$\hat{f}_A^{+} \rightarrow f_A$ an extrapolated value of $\hat{f}_A$ to its pure state.

Figure 9.8 Rational standard states for activity coefficients with symmetrical and unsymmetrical normalization for a component of limited solubility.

In these two cases, the use of a standard state as a pure solvent is often called *symmetrical normalization*, but if the infinite dilution standard state is used for the solute, we say that this component is *unsymmetrically normalized.* It is obvious that

$$\begin{aligned} &\textit{symmetrical:} \quad \gamma_i \to 1, x_i \to 1 \\ &\textit{unsymmetrical:} \quad \gamma_i^{**} \to 1, x_i \to 0 \end{aligned} \tag{9-200}$$

Other choices of standard states may be used, but in all cases the absolute value of $\hat{f}_i$ (or μ_i) must be the same for each state of fixed T, P, and x_i ($i = 1, \ldots, n$). Thus the value of γ_i must change depending on the reference state fugacity $\hat{f}_i^+$. For the two cases considered so far, the symmetrically and unsymmetrically normalized activity coefficients are related by,

$$\hat{f}_A = \hat{f}_A(T, P, x_i) = \gamma_A x_A f_A = \gamma_A^{**} x_A f_A^{**} \tag{9-201}$$

Thus,

$$\frac{\gamma_A}{\gamma_A^{**}} = \frac{f_A^{**}}{f_A} = \underset{x_A \to 0}{\text{limit}} \ \gamma_A = \gamma_A^{\infty} \tag{9-202}$$

where the final equality for γ_A^{∞} gives the symmetrically normalized activity coefficient, referenced to a pure A standard state, at infinite-dilution.

The other basic thermodynamic relations for all cases are similar, but care must be taken to denote the reference state. For example, if the infinite dilution reference state were chosen for component j, Eq. (9-181) would become

$$\left[\frac{\partial \ln \gamma_j^{**}}{\partial T}\right]_{P,x} = -\frac{\Delta \overline{H}_j}{RT^2} = -\frac{\overline{H}_j - \overline{H}_j^{\infty}}{RT^2} \tag{9-203}$$

where $\overline{H}_j^{\infty}$ is the partial molar enthalpy of j in an infinitely dilute solution. For binary mixtures, this simple definition suffices; for multicomponent mixtures, a more explicit definition of the infinite-dilution standard state is necessary since as $x_j \to 0$, $\overline{H}_j^{\infty}$ can vary depending on the relative amounts of the other constituents.

So far we have limited our discussion to activity coefficients based on mole fractions. These are called *rational*, while those based on molal or molar concentrations are called *practical*. Let's consider a common form of activity coefficients based on molality (moles of solute/kg of solvent). This formulation is analogous to that used for rational, unsymmetrically normalized activity coefficients with an ∞-dilution reference state for solute A and a pure reference state for solvent B. However, in place of mole fractions we now introduce molality (m) as the composition parameter:

$$\hat{f}_A = \gamma_A^* m_A \hat{f}_i^+ (T, P, m^+ = 1) \tag{9-204}$$

where the reference state condition $\hat{f}_i^+$ corresponds to a hypothetical state at unit molality at the same T, P, and state of aggregation as the mixture. The limiting condition for γ_A^* is expressed as,

$$\gamma_A^* = \frac{\hat{f}_A}{m_A f_i^*} \rightarrow 1.0 \text{ as } m_A \rightarrow 0 \tag{9-205}$$

where the standard state fugacity $\hat{f}_i^+ = f_A^*$ is obtained from dilute solution behavior. This new reference state at unit molality is illustrated in Figure 9.9.

One can easily see that when $m_A \rightarrow 0$, $\gamma_A^* \rightarrow 1.0$ as the distance between the real $\hat{f}_A$ versus m_A curve and the dotted line is proportional to the magnitude of γ_A^*. For the case shown in Figure 9.9, $\gamma_A^* < 1$ for all values of $m_A > 0$. A similar set of reference states is used for dissociating electrolytes. These are discussed in Chapter 12. An alternative to the unit molality standard state is to use a saturated state with respect to a pure solid or liquid phase at a specified T and P. For example, a saturated aqueous salt solution in equilibrium with pure solid salt, might be a convenient standard state for some applications.

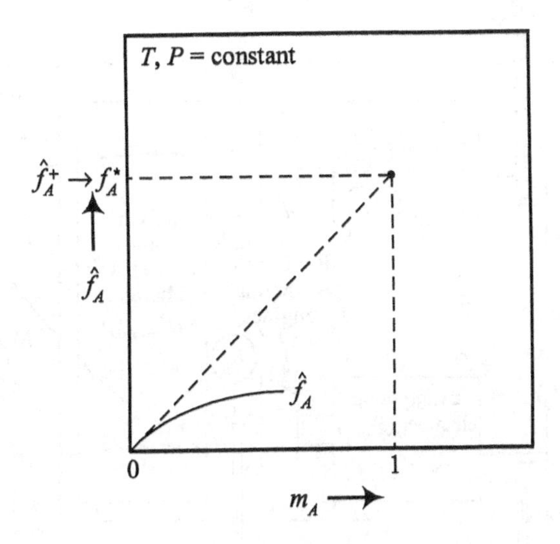

Figure 9.9 Reference state at unit molality for practical (molality-based) activity coefficients.

9.9 Reversible Work of Mixing and Separation

Another important concept for mixtures is the work effects associated with mixing or separating components. We restrict the development here to reversible processes that are isothermal and without chemical reactions between specified initial and final states. With no gradients or dissipative effects, reversible processes would yield the maximum work output for mixing components or would require the minimum work input for separating components. With such information, one could compare actual and ideal, minimum work requirements for multicomponent distillation or for liquid-liquid extraction.

Consider first the reversible, isothermal ($T = T_o$) and isobaric ($P = P_o$) mixing process depicted in Figure 9.10. Here streams of pure A and B are expanded reversibly to their respective vapor pressures at T_o, P_{vpA} and P_{vpB}. Reversible evaporation with ($\Delta T = 0, \Delta\mu = 0, \Delta P = 0$) is carried out using heat supplied by a large thermal reservoir at T_o. The pure gases leave the evaporating chambers and are expanded to lower pressures P_A, P_B such that the chemical potential (or fugacity) is equivalent across the semi-permeable membrane for each component.

$$\mu_i(T_o, P_i, \text{pure } i) = \mu_i(T_o, P_{vp,\,mix}, \text{ mixture } y_i) \tag{9-206}$$

where the vapor pressure of the mixture ($P_{vp,\,mix}$) corresponds to the saturation pressure at T_o and the mixture composition y_A, y_B. Then the gas mixture is condensed isothermally at T_o to a liquid mixture of the same composition, which is compressed reversibly to P_o to complete the process.

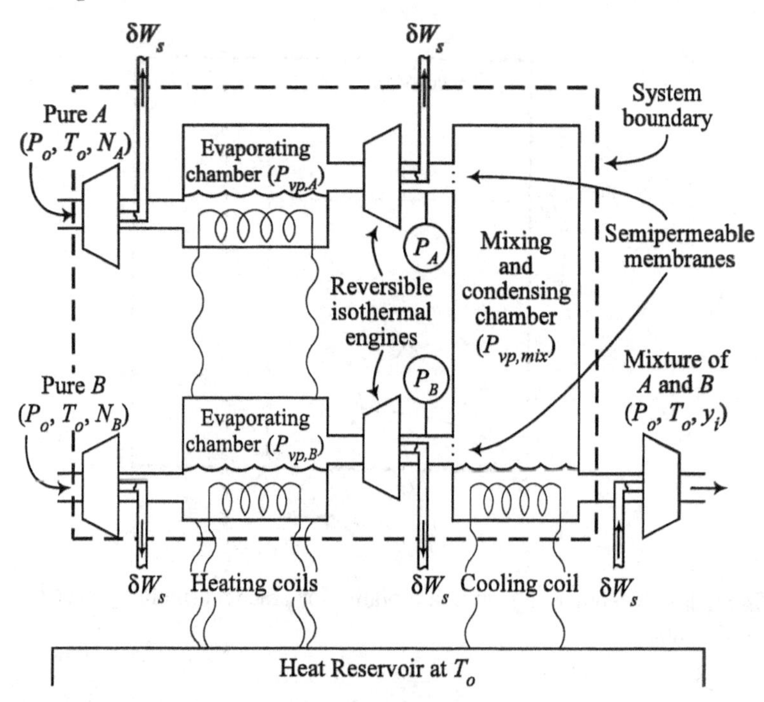

Figure 9.10 Reversible preparation of a liquid solution at constant temperature T_o and constant pressure $P = P_o$ (at the terminals of the system). All heating and cooling done reversibly with a large heat sink at T_o.

The total work would be obtained by summing over all steps involved in the mixing process.

$$W_{net} = \sum_{j\ steps} W_{s,\,rev}\Big]_{step_j} \tag{9-207}$$

The work contribution for each step can be computed using the integral expression developed in Chapter 4 [Eq. (4-68)] for reversible flow or shaft work for a steady state process. For constant inlet and outlet conditions, on a mole basis, the work component for stream *i* in step *j* is

$$W_{s,\,rev}\Big]_{\substack{stream\ i,\\ step\ j}} = N_i \int_{P_{in}}^{P_{out}} VdP\Big]_{step\ j} \tag{9-208}$$

For an isothermal path, the integration is simplified. Furthermore, the minimum work required to isothermally separate the components of a liquid mixture to their pure states is obtained by reversing flow directions in Figure 9.10. Examples 9.8 and 9.9 illustrate how the minimum work of separation can be estimated.

Example 9.8

Calculate the minimum work required to separate air into pure oxygen and nitrogen. Assume a steady-flow process operating at 300 K. The inlet air is at 1 bar and contains 80 mole % nitrogen. Air may be considered to be an ideal-gas mixture and pure oxygen and nitrogen as ideal gases.

Solution

Visualize that the inlet air contacts two semipermeable membranes. (Note that the evaporating and condensing steps can be neglected because we are dealing with gases.) One membrane is permeable only to nitrogen and the other only to oxygen. Consider the nitrogen permeable membrane first. The chemical potentials of nitrogen are equal across the membrane. Since the mixture is an ideal-gas mixture, the chemical potential equality may be replaced by an equality of partial pressures. As $p_{N_2} = Py_{N_2} = (1)\,(0.8) = 0.8$ bar, the pure nitrogen downstream of the membrane is 0.8 bar. If this stream is compressed back to 1 bar, the reversible, isothermal flow work is, per mole of nitrogen flowing,

$$\delta W/\delta n_{N_2} = \int V\,dP = RT\int d\ln P = (8.314)(300)\ln\frac{1}{0.8} = 557 \text{ J/mol } N_2$$

Similarly, for oxygen,

$$\delta W/\delta n_{O_2} = (8.314)(300)\ln\frac{1}{0.2} = 4010 \text{ J/mol } O_2$$

On a basis of 1 mole of air as feed, the net work is the sum of the two components:

$$W_{net} = (0.8)(557) + (0.2)(4010) = 1248 \text{ J/mol air}$$

Thus, 1248 J is the minimum work needed to separate 1 mole of air at 300 K, 1 bar into pure nitrogen and pure oxygen streams at 300 K, 1 bar. Note that the net minimum work is also equivalent to $-\Delta \underline{G}_{mix}^{ID}$ for an ideal gas mixture [Eq. (9-101), with $y_{N_2} = 0.8$ and $y_{O_2} = 0.2$].

Example 9.9

Repeat the calculation of Example 9.8, but treat the gases as non-ideal.

Solution

To use the solution method of the previous example, we need to specify the end states for the VdP integrals to calculate the net work. This was relatively straightforward for the ideal gas case, because the equality of partial pressures replaced the equality of component chemical potentials. However, now the situation is more complex because of the non-idealities. A convenient approach is to recognize that $W_{net} = -\Delta \underline{G}_{mix}$; that is, for an isothermal ($T = T_o$) and isobaric ($P = P_o$), reversible flow process, the net shaft work is the negative of the Gibbs free energy of mixing. On a per mole of air basis:

$$\delta W_{net}/\delta n_{air} = -\sum_{i=1}^{n} \left[\overline{G}_i(T_o, P_o, y_i) - G_i(T_o, P_o) \right] y_i \tag{9-209}$$

where $\overline{G}_i = \mu_i\,(T_o, P, y_i)$ and $G_i = \mu_i\,(T_o, P, pure)$. Substituting fugacity for chemical potential using Eq. (9-107), we obtain:

$$\delta W_{net}/\delta n_{air} = -\sum_{i=1}^{n} y_i\, RT_o \ln \left[\frac{\hat{f}_i(T_o, P_o, y_i)}{f_i(T_o, P_o, pure)} \right] \tag{9-210}$$

with $\hat{f}_i = \hat{\phi}_i\, y_i\, P_o$ the problem reduces to one of evaluating $\hat{\phi}_i$s at known conditions where $\hat{\phi}_i$ is given by Eq. (9-129) or Eq. (9-142). To perform this calculation an equation of state is needed. Note that if the gas mixture is ideal $\hat{\phi}_i = 1.0$, $\hat{f}_i = y_i\, P_o = p_i$ and Eq. (9-210) reduces to the form given in Example 9.8.

9.10 Summary

Although complex notation and tedious algebra are inherently part of mixture thermodynamics, fortunately there are only a few basic concepts and procedures. Important definitions and working equations have been highlighted so they are easy to locate. Fundamental concepts are separated by sections with the first section introducing the general approach for mixtures where partial molar properties ($\overline{B}_i$) are defined to deal with the effects of compositional variations in N_i or x_i. Mixture relationships for $P\underline{V}TN$ equations of state are described in Section 9.2 where the need for specific mixing rules is emphasized. Important operational equations for evaluating and interrelating mixture properties using differential partial molar quantities, integral mixing functions (ΔB_{mix}) and Gibbs-Duhem relationships are discussed in Sections 9.2-9.5. Ideal gas mixtures and ideal liquid solutions are defined in Section 9.6 to provide reference points for the fugacity ($\hat{f}_i$) and fugacity coefficient ($\hat{\phi}_i$) (Section 9.7) and activity (a_i) and activity coefficient (γ_i) (Section 9.8) for modeling, non-ideal behavior. Standard state conventions are discussed in detail as they can be confusing and ambiguous if not carefully specified. Minimum work requirements for separating components from a mixture are treated in the final section.

References

Chueh, P.L. and Prausnitz, J.M., (1967) *Ind. Eng. Chem. Fundam.*, **6**, p 492.

Denbigh, K., (1981) *The Principles of Chemical Equilibrium*, 4th ed. Cambridge: Cambridge University Press, Chapter 3, p 111-120.

Kay, W.B., (1938) "Liquid-vapor phase equilibrium reactions in the ethane-n-heptane systems", *Ind. Eng. Chem.*, **30**, p 459-465.

Lewis, G.N., (1901), "The law of physico-chemical change," *Proceedings of the American Academy*, **37**(4), p 49-69.

Prausnitz, J.M., Lichtenthaler, R.N., Gomes de Azevedo, E., (1986) *Molecular Thermodynamics of Fluid Phase Equilibria*, 2nd ed. Prentice Hall, Englewood Cliffs, NJ., p 132.

Prausnitz, J.M., and Gunn, R.D., (1958) "Pseudocritical constants from volumetric data for gas mixtures", *AIChE J.* **4**, p 494

Sage, B.H., (1965) *Thermodynamics of Multicomponent Systems*, Reinhold Publishing, New York, NY., Chapter 10 on fugacity.

Sandler, S.L. (1994), *Models for Thermodynamic and Phase Equilibria Calculations*, Ch. 2, "Equations of State," by Sandler, S.L., H. Orbey, and B.-I. Lee, p 87-186.

Walas, S.M., (1985) *Phase Equilibria in Chemical Engineering*, Butterworth, Boston, p 16-19.

Wong, D.S.H. and S.L. Sandler (1992), *AIChE J.*, **38**, p 671-680 and *Ind. Eng. Chem. Research*, **31**, p 2033-2039.

Problems

9.1. A small-scale experiment requires the preparation of a 40 mole % sulfuric acid solution from pure water and 80 mole % sulfuric acid. The mixture is to be prepared by metering one fluid at a constant molar rate into a tank containing the second fluid. Cooling is to be provided to maintain the bath at a constant temperature.

At the temperature of operation, 298 K, partial molar enthalpies are listed below [*International Critical Tables* (1930), Vol. 7, McGraw-Hill, New York, p 237]. The term $\overline{H}^o$ (H_2SO_4) refers to a state of infinite dilution (i.e., where the mole fraction acid approaches zero). $H^o(H_2O)$ is the enthalpy of pure water at 298 K. x is mole fraction and enthalpies are expressed in J/mol.

Data:

$x_{H_2SO_4}$	$(\bar{H}-H^o)_{H_2O}$	$(\bar{H}-\bar{H}^o)_{H_2SO_4}$
0	0	0
0.05	-183	17,290
0.10	-1,228	32,360
0.15	-2,428	38,980
0.20	-4,187	46,850
0.25	-6,071	53,090
0.30	-7,997	58,490
0.35	-10,340	63,350
0.40	-12,810	67,660
0.45	-16,250	72,180
0.50	-20,310	76,660
0.55	-23,990	79,720
0.60	-26,380	81,770
0.65	-28,010	82,650
0.70	-29,350	83,360
0.75	-30,480	83,850
0.80	-31,360	84,150
0.85	-32,240	84,300
0.90	-32,950	84,380
0.95	-33,700	84,460
1.00	-34,440	84,570

(a) Which fluid should be charged to the reactor first in order to minimize the rate of heat release?

(b) What is the peak differential heat load per mole of fluid added?

(c) What is the total heat of mixing per mole of final solution for both methods?

9.2. We are faced with the problem of diluting a 90 wt % H_2SO_4 solution with water in the following manner. A tank contains 500 kg of pure water at 298 K; it is equipped with a cooling device to remove any heat of mixing. This cooling device operates with a boiling refrigerant reflux condenser system to maintain the temperature at 298 K. Because of the peculiarities of the system, the rate of heat transfer (W/m^2) must be constant. We wish to add 1500 kg of acid solution (at a variable rate) in 1 hour. The acid is initially at 298 K. Enthalpy data are given in Problem 9.1.

(a) Plot the heat of solution (kJ/kg solution) versus weight fraction H_2SO_4 with the reference states as pure water and pure H_2SO_4, liquid, at 298 K.

(b) What is the total heat transferred in the dilution process described?

(c) Derive a differential equation to express the mass flow of 90 wt % acid, kg/min, as a function of the acid concentration in the solution.

(d) Using the result from part (c), determine the mass flow of 90 wt % acid when the overall tank liquid is 64.5 wt % acid.

9.3. In an experiment a mixture of helium and ammonia was prepared as follows. As shown in Figure P9.3, there are separate supply manifolds for the helium and ammonia.

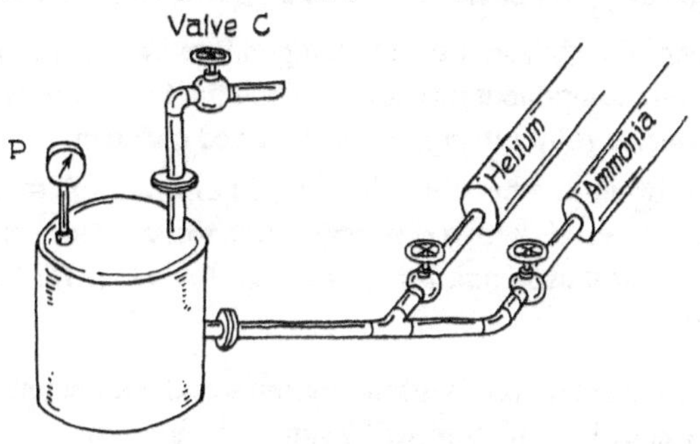

Figure P9.3

The aluminum mixing tank is first evacuated to a very low pressure. Helium gas is then admitted very rapidly until the tank pressure is at 2 bar. The helium supply valve is then closed. Ten minutes later, the ammonia supply valve is opened to allow ammonia to flow rapidly into the tank. The valve is closed when the tank pressure reaches 3 bar. A day later, after diffusive mixing, the gas mixture is drawn off through valve *C*.

Data:

Assume ideal gases. The heat capacities of helium and ammonia may be considered to be constants with the following values:

$$C_p\text{ (He)} = 20.9\text{ J/mol K} \quad \text{and} \quad C_p\text{ (NH}_3) = 35.6\text{ J/mol K}$$

The tank dimensions are: 0.3 m (inside diameter), 0.3 m tall. The wall thickness is 1.2 cm and the initial wall temperature is 310 K. C_p (aluminum) = 1 J/g K and the density of aluminum is about 2700 kg/m^3. The helium in the manifold is at 310 K and 10 bar, but the ammonia manifold is at 310 K, 5 bar.

With only this description and the given data, what is your best engineering estimate of the composition of the mixture removed from valve *C*?

9.4. For a single-phase mixture of *n* components, it is often stated that

$$\sum_{k=1}^{n} N_k \left(\frac{\partial P}{\partial N_k}\right)_{T, \underline{V}, N_i[k]} = \kappa_T^{-1}$$

where κ_T is the isothermal compressibility. Derive this relation. What is your physical interpretation of the terms in the summation for an ideal-gas mixture?

9.5. In a tank of steam at 2.068 bar and 477.6 K, the Keenan and Keyes steam tables show that the entropy is 7.510 J/g K. The reference entropy is 0.0 for saturated liquid water at 273.2 K. Suppose that you had a tank of steam at 2.068 bar and 477.6 K that contained 10 g of steam. You now injected a small quantity of additional steam (at 2.068 bar and

477.6 K) into the tank under conditions that the total entropy and pressure remained constant. The value of C_p for steam at 2.068 bar and 477.6 K is about 2.01 J/g K.

(a) Derive a general expression for the temperature change per mole of component j added to a multicomponent mixture in which the total entropy, pressure, and moles of all components (other than j) are maintained constant.

(b) Estimate the initial temperature change (K/per gram of steam). What would this derivative had been if, with a different reference state for entropy, we had chosen $S = 0$ for superheated steam, $P = 2.068$ bar, $T = 477.6$ K? How do you reconcile your answers?

9.6. Two simple systems are contained within a cylinder and are separated by a piston (Figure P9.6). Each subsystem is a mixture of ½ mole of nitrogen and ½ mole of hydrogen (consider as ideal gases).

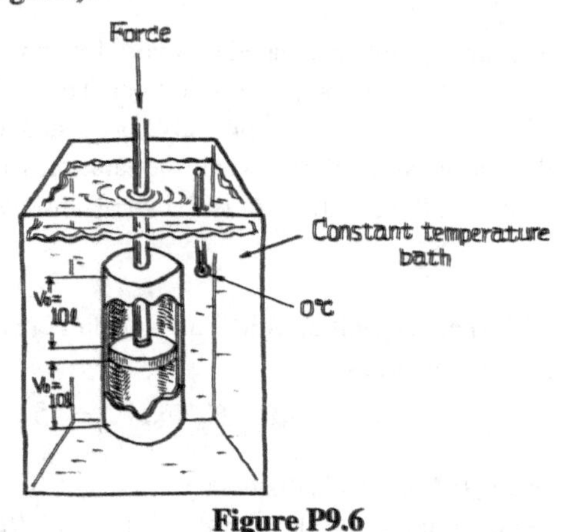

Figure P9.6

The piston is in the center of the cylinder, each subsystem occupying a volume of 10 liters. The walls of the cylinder are diathermal and the system is in contact with a heat reservoir at a temperature of 0°C. The piston is permeable to H_2 but impermeable to N_2. How much work is required to push the piston to such a position that the volumes of the subsystems are 5 and 15 liters?

9.7. A constant-volume crystallizer vessel containing a saturated salt solution at 1 bar is being fed with an unsaturated 5 wt % sodium chloride solution at the same temperature as inside the pot. The water is vaporized and removed at 1 bar, and simultaneously solid salt is removed from the bottom. The heat of vaporization and crystallization is provided by an electric heater. The system is in steady state. How would you calculate:

(a) The entropy production rate of the universe?

(b) The rate of entropy production for the system, consisting of the liquid, solid, and vapor in the crystallizer?

9.8. We have a stream on n-butane and carbon dioxide at 15 MPa and 393.2 K flowing from a well. The composition is 50 mole % carbon dioxide. We would like to make a separation of this stream to end with one stream of 2 mole % CO_2 and one with 90 % CO_2—both streams still at 15 MPa and 393.2 K.

	CO_2		*n*-butane
T_c (K)	304.2		425.2
P_c (MPa)	7.376		3.80
ω	0.225		0.193
δ_{12}		0.13	

(a) What is the minimum work required per mole of inlet feed? Use the Peng-Robinson equation of state.

(b) How does this value compare if all streams were assumed to be ideal-gas mixtures?

9.9. I am thinking about purchasing a cottage by the ocean. The location is, however, remote from most amenities of civilization and I would like a reliable power source for lights, electric blankets, etc.

There is a stream near the cottage which flows into the ocean and one of my friends, Rocky Jones, has told me I could obtain the power I need if I would only mix the fresh water in my stream with the ocean water in an appropriate manner. Simply allowing this clear, bubbly stream to cascade merrily into the ocean does seem wasteful in this day and age (Figure P9.9). But how to harness this source?

Rocky is willing to manufacture just what I need (so he says) but, friends though we are, his price is outrageous and I would be helpless to repair it since he will only sell sealed units. Before I seriously consider his offer, in any case, could you show me with appropriate detail just how much power I could expect if the unit worked perfectly? (Data are given below.)

Also, I would like some good ideas from an engineer as to just how I might build my own unit to obtain power from this mixing process.

Data:

The stream on my property is essentially at mean sea level; the average flow is 1 m^3/s. The average stream and ocean temperatures are 10°C.

There is a rapid ocean current near the shore to aid the mixing step, if desired.

The ocean averages ~ 3.5 % NaCl and the partial pressure of water over this saline solution at (10°C) is 1200 N/m^2, whereas for fresh water at 10°C, the vapor pressure is 1228 N/m^2. The densities of the ocean water and the fresh water are, respectively, 1025.5 and 1000 kg/m^3.

$\overline{\Delta H_w}$ and $\overline{\Delta V_w}$ for these dilute saline solutions are essentially zero, and both fresh water and ocean water may be assumed incompressible.

Figure P9.9

(a) Suggest at least one device (with dimensions, if relevant) and show that the power from this unit will give the same power as you calculated above.

(b) If I need about 10 kW of power, would I have to be very efficient, or is my hope an impossible dream?

9.10. We have a mixture of n components in a vessel that is maintained at constant temperature and pressure. Attached to this vessel is a small compartment that connects to the larger vessel by an ideal semipermeable membrane that is permeable only to one component, j. This small compartment is at the same temperature as the large one, but it is at a different pressure, P^+; at equilibrium it contains only component j.
Obtain relations for

$$\overline{S}_j(T, P, y_1,...) - S_j(T, P^+)$$

when the mixture is: (1) An ideal solution, (2) A regular solution, (3) An athermal solution, and (4) An ideal-gas mixture

9.11. Calculate the osmotic pressure across an ideal semipermeable membrane (permeable to water) for the system NaCl-H_2O at 25°C at pressures of 1 and 500 bar. Consider the concentration range of 0 to 25 wt % NaCl.

Compare your results with the case wherein the solution of NaCl and water is assumed to form an ideal solution.

Osmotic pressures are often calculated by the van't Hoff equation, where

$$\text{Osmotic pressure} = \Pi = CRT$$

where C is equal to the molar concentration of the solute. How is this expression derived? Data are provided in Figures P9.11(a) through (d).

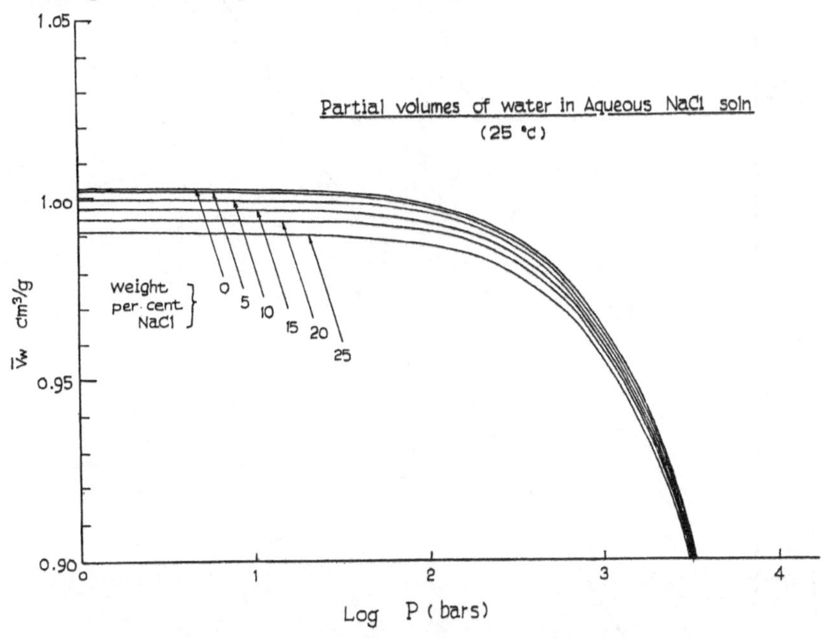

Figure P9.11(a)

9.12. Many industries face a common problem of concentrating dilute aqueous solutions with the minimum expenditure of energy. In the sugar industry, for example, dilute sucrose solutions must be concentrated before purification by crystallization may be attempted.

Numerous water removal schemes have been suggested as alternatives to the evaporative methods now used, and we wish to evaluate these new schemes.

Data:

Temperature of separation 20°C

Sucrose molecular weight = 342.30 g/mol ($C_{10}H_{22}O_{11}$)

The only reliable thermodynamic data for sucrose-water solutions show the activity of water as a function of molality (moles sucrose per kilogram of water). We show these data below and have added, for your convenience, the equivalent mole fraction water and weight fraction sucrose that correspond to the given molality. The activity of water is defined as $p_w/P_{vp,w}$ where p_w is the partial pressure of water over the sucrose solution and $P_{vp,w}$ is the vapor pressure of pure water at 20°C.

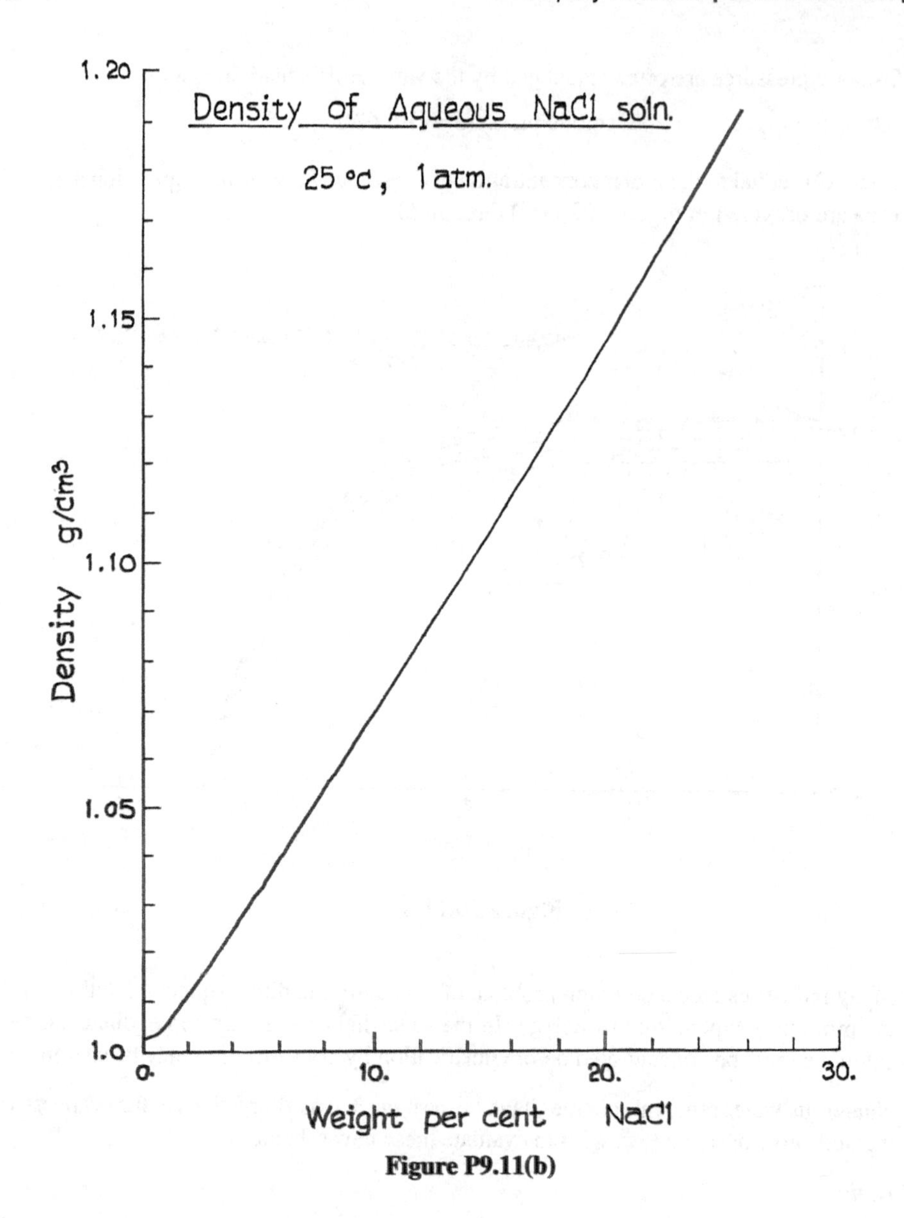

Figure P9.11(b)

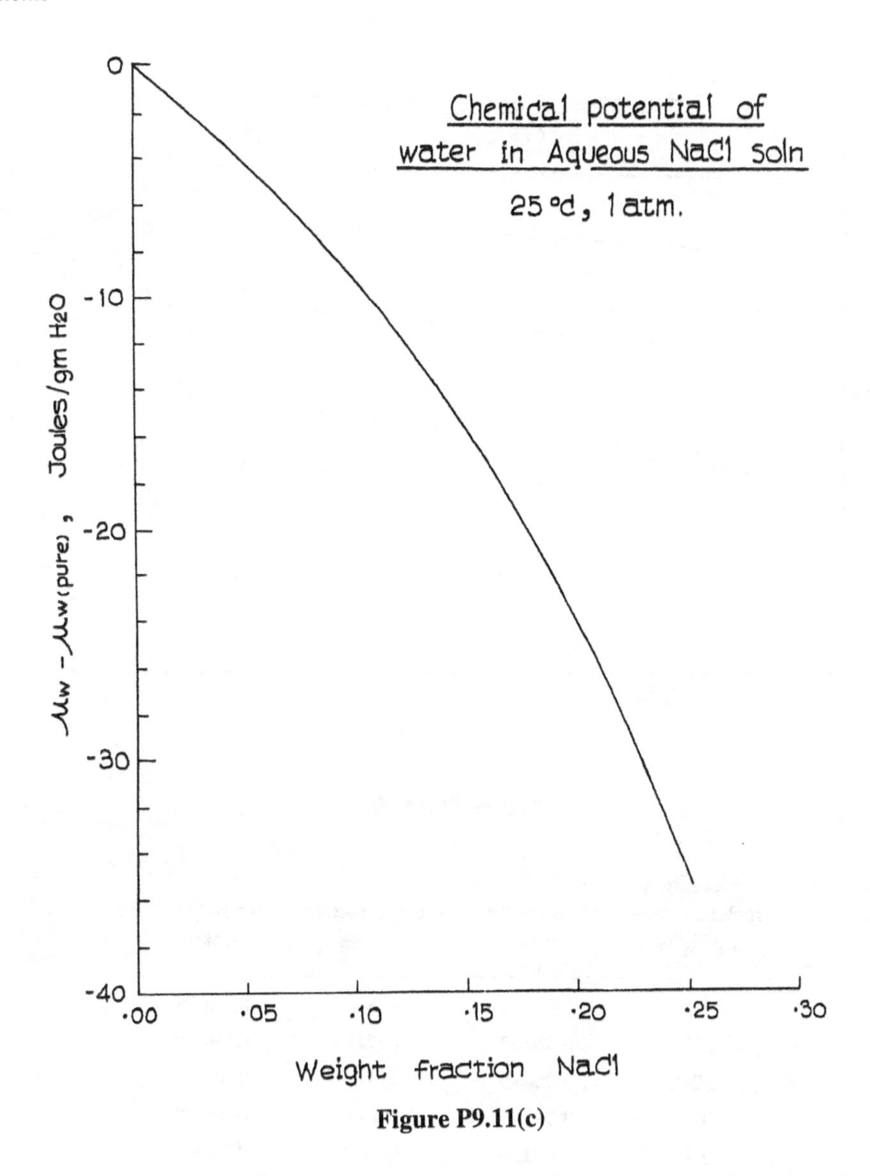

Figure P9.11(c)

(a) Calculate the *minimum* work required to concentrate a dilute sucrose solution. Assume that we begin with a 10 wt % solution of sucrose in water and we want to end with a 30 wt % sucrose solution. Pick a basis of 1 kg of original 10 % solution and express your answer in joules.

(b) Compare your answer to that which would be obtained if you assumed the sucrose-water solution to be ideal (i.e., where the activities are set equal to the mole fraction).

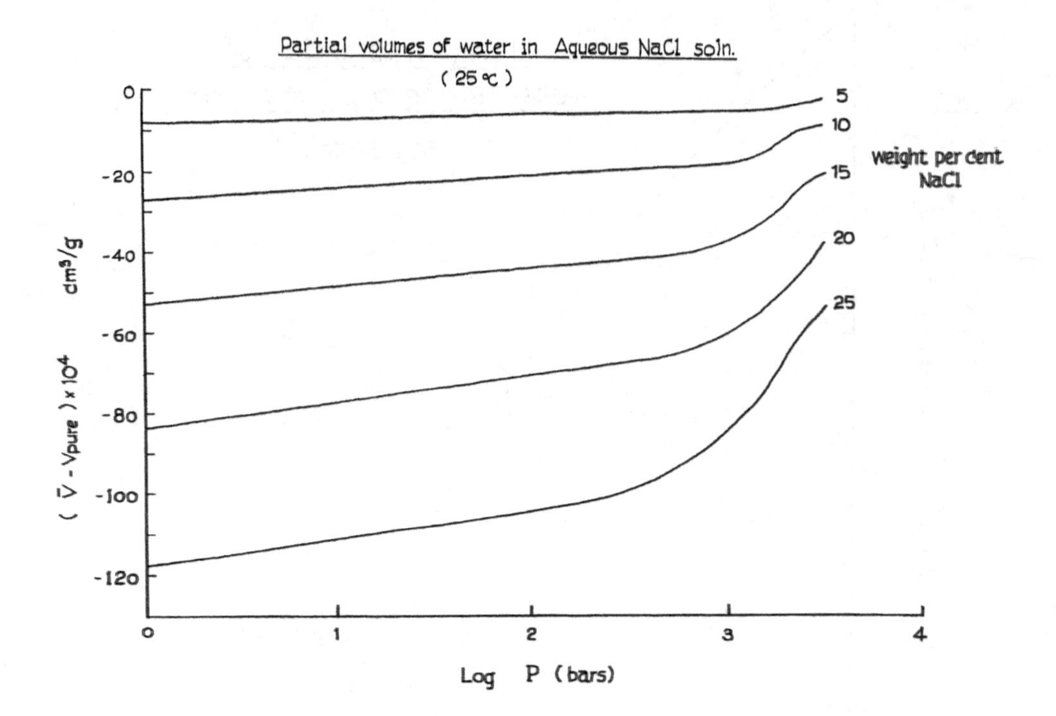

Figure P9.11(d)

Molality (moles sucrose /kg H_2O)	Mole fraction Water	Weight fraction sucrose	Activity of water
0.2	0.99641	0.064	0.99635
0.3	0.99463	0.093	0.99448
0.3249	0.99419	0.10	0.99393
0.4	0.99285	0.120	0.99259
0.5	0.99108	0.146	0.99067
0.6	0.98932	0.170	0.98871
0.8	0.98580	0.215	0.98470
1.0	0.98232	0.255	0.98057
1.2	0.97886	0.291	0.97635
1.253	0.97794	0.30	0.97596
1.5	0.97371	0.339	0.96972

Sources:

Robinson, R.A., and D.A. Sinclair (1934), *J. Am. Chem. Soc.*, **56**, p 1830.
Sinclair, D.A. (1933), *J. Phys. Chem.*, **37**, p 495.

9.13. For the liquid mixture of benzene and cyclohexane, experimental data have shown that the activity coefficient of benzene may be expressed as

$$RT \ln \gamma_B = (3800 - 8T)(1 - x_B)^2$$

where R= 8.314 J/mol K and T in K. The subscript B represents benzene. Calculate the entropy and enthalpy of dilution when 1 mole of pure benzene is added to 2 moles of a solution containing 80 mole % cyclohexane at 300 K, 1 bar. The mixing process is isothermal and isobaric.

9.14. Experimental data for the binary liquid mixture of ethanol (E) and methylcyclohexane (M) are shown below.

(1) Enthalpy of mixing and *excess* Gibbs energy of mixing at 35°C.

Mole fraction (E)	ΔH (J/mol)	ΔG^{EX} (J/mol)
0	0	0
0.0742	452.9	544.2
0.1979	625.4	1036.9
0.3456	670.6	1336.6
0.5324	641.3	1387.2
0.8004	399.3	905.4
1.0	0	0

(2) Excess liquid heat capacities as a function of composition and temperature. Excess heat capacities are defined as

$$C_p^{EX} \equiv C_{p_{mix}} - x_E C_{p_E} - x_M C_{p_M} = \sum_{j=1}^{4} b_j T^{j-1} \text{ (in J/mol K)}$$

with $b_j = f$(composition), T in K

(3) Heat capacities for pure liquid ethanol and methylcyclohexane as a function of temperature,

$$C_p = \sum_{j=1}^{5} b_j t^{j-1} \quad \text{(in J/mol K, t in °C)}$$

	Mole fraction ethanol		
	0.0742	**0.1979**	**0.3456**
b_1	3.505 019 85 E+01	6.448 759 04 E+00	2.570 176 4 E+01
b_2	-3.449 617 30 E-01	3.470 872 13 E-02	-1.710 333 5 E-01
b_3	8.597 602 38 E-04	-8.545 534 82 E-04	-1.438 276 5 E-04
b_4	0	2.635 205 92 E-06	1.844 259 9 E-06

	Mole fraction ethanol	
	0.5324	**0.8004**
b_1	1.457 230 3 E–01	-1.607 368 3 E+01
b_2	1.256 900 5 E–01	2.477 038 3 E–01
b_3	-1.262 525 8 E–03	-1.359 608 9 E–03
b_4	3.155 998 4 E–06	2.543 389 4 E–06

	Pure ethanol	**Pure methylcyclohexane**
b_1	1.036 929 3 E+02	1.749 883 1 E+02
b_2	2.986 102 6 E–01	3.870 387 7 E–01
b_3	1.878 244 3 E–03	8.778 553 5 E–04
b_4	5.558 858 1 E–06	0
b_5	1.200 221 1 E–08	0

We would like to use these data to aid in the design of a low-temperature separation process. Neglect any effect of pressure on the liquid-phase properties.

(a) Estimate the activity coefficients of both ethanol and methlcyclohexane at –85°C for a liquid mixture containing 80 mole % ethanol.

(b) Comment on the phase stability of the liquid mixture at this temperature of –85°C; do you see any evidence that the liquid mixture may split into two liquid phases? If so, indicate your best estimate of the phase compositions; if not, explain clearly why you feel there is no phase split at this temperature.

9.15. Given only the differential heat of solution of cyclohexane in methyl ethyl ketone at 18°C as determined by M.B. Donald and K. Ridgway (1956), ["The binary system cyclohexane-methylethyl ketone," *Chem. Eng. Sci.*, **5**, p 188], what is:

(a) The integral heat of solution as a function of the mole fraction cyclohexane?

(b) The differential heat of solution of methyl ethyl ketone in cyclohexane as a function of the mole fraction cyclohexane?

Mole fraction cyclohexane	Differential heat of solution of cyclohexane in methyl ethyl ketone (J/mol)
0.025	4831
0.095	4243
0.193	3541
0.400	2410
0.485	1935
0.614	1276
0.783	564
0.875	381

(c) Compare the result in part (b) with the following data. What conclusions do you reach?

Mole fraction cyclohexane	Differential heat of solution of methyl ethyl ketone in cyclohexane (J/mol)
0.124	124
0.238	263
0.381	438
0.491	894
0.598	1497
0.797	3034
0.895	4305
0.958	6651

9.16. Liquid anhydrous ammonia is now shipped in well-insulated tankers. A potential hazard exists if an accident should occur with loss of the liquid into the sea. The liquid ammonia exists at a temperature of about 239 K at 1 bar. In contacting ambient water, presumably some would dissolve and some would vaporize. A few small and medium-sized test spills have indicated, surprisingly, that the fraction of the spilled ammonia which ends up dissolved (as NH_4OH) is constant at about 0.70, independent of the quantity spilled or the exact conditions of the spill. It would be very useful to have an analytical model to describe an ammonia spill.

Develop such a model and show that it does lead to the prediction that a constant fraction of the liquid ammonia dissolves while the remaining fraction vaporizes to be dispersed downwind. Apply this model to a situation where 1000 kg of liquid ammonia contacts 293 K water at 1 bar.

Data:

At 1 bar, the enthalpy of a saturated solution of ammonia in water may be expressed empirically as:

$$H^L_{sat} = 789.04 - 905.65 \sin(2.7129 - 2.045\, x_N)$$

where x_N is the weight fraction NH_3 and H^L_{sat} is in J/g.

where $x_N = 1$, H^L (pure liquid NH_3 at 239 K, 1 bar) = 228.1 J/g.

where $x_N = 0$, H^L (pure water 1 bar) = 412.6 J/g at 373 K and 83.7 J/g at 293 K.

It may also be assumed that the vapor in equilibrium with an ammonia-water solution is essentially pure NH_3. (This is true except for very dilute solutions.) The enthalpy of ammonia vapor at 1 bar is not a strong function of temperature and may be taken as a constant equal to 1570 J/g.

9.17. Show that if a component i in a mixture obeys the Lewis and Randall Rule, its fugacity is given by an isothermal integral in pressure:

$$\hat{f}_i = y_i P \exp\left[\int_0^P \left[\frac{Z-1}{P}\right] dP\right]_{T_{fixed}}$$

9.18. Given a gas mixture of methane and *n*-pentane at 100°C and 1000 bar containing 35 percent methane, estimate the minimum work required to separate pure CH_4 from this mixture at 100°C. You should assume that the $P\underline{V}TN$ properties of pure CH_4 and the mixture are given by a volume-explicit simplified equation of state of the form:

$$V = RT/P + B_{mix}(T)$$

where

$B_{mix}(T)$ = 2nd virial coefficient for the mixture = $\sum_i B_i y_i$

B_i = pure component 2nd virial coefficient

$B_{CH_4} = -25$ cm^3/mol

$B_{n\text{-pentane}} = -800$ cm^3/mol

9.19. Two practicing engineers, M. Wilkinson, Jr. and K. Sparks, Sr. were recently examining some integral heat-of-mixing data for a liquid ternary system. The ΔH_{mix} values were obtained by adding various amounts of pure liquids *A*, *B*, and *C* together in an isothermal, isobaric calorimeter and measuring the evolution or absorption of heat. The data appear rather suspicious in that over some composition ranges, the heat of mixing is positive and over other regions, it is quite negative. The senior engineer suggests that they could use the Gibbs-Duhem equation to verify that at least the data were consistent. The junior member pooh-poohs the suggestion as being utterly worthless. They have selected you as the referee to decide who is correct.

You first suspect that much of the objection of the junior engineer stems from the fact that the onerous task of obtaining partial quantities from the data will fall entirely upon his shoulders. Yet you would also like to show clearly whether or not the Gibbs-Duhem equation is of real value in this situation.

9.20. Calculate the heating rate required to evaporate a 77°F feed stream of 10 lb/hr of 65 wt % H_2SO_4 to produce a steady flow of 80 wt % H_2SO_4 and pure saturated steam vapor at 77°F. What is your estimate of the minimum amount of work required to carry out this separation process? ΔH_{mix} data for the H_2O - H_2SO_4 system are given graphically in Figure P9.27. ΔH_{vap} (for pure water) = 1054 BTU/lb.

9.21. Consider the compression of a real gas mixture consisting of carbon dioxide and water.

(a) Estimate the power required for an isothermal compression of a 10 moles/sec stream consisting of a 40 mol % CO_2/60 mol % H_2O gas mixture from 200 to 800 bar at 700 K. For this system, a mixture fugacity coefficient $\hat{\phi}_m$ defined as

$$\ln \hat{\phi}_m = \Sigma\ y_i \ln \hat{\phi}_i$$

can be approximated as a linear function of pressure over the pressure range of interest by:

$$\hat{\phi}_m = 0.92 + 0.001\ P\ \ (P \text{ in bar})$$

You can assume that the overall compressor efficiency is 80%.

(b) Frequently in practice to achieve isothermal operation, each stage of compression is done adiabatically followed by an isobaric heat removal step. This process is commonly called adiabatic compression with intercooling. If we assume that the compressor is to be designed for optimum performance at specified values of the Balje parameters for specific speed N_S or n_S and specific diameter D_S or d_S, describe briefly how you would estimate the pitch diameter D_P of the largest or entrance stage of the compressor. If the information given in part (a) is not sufficient to make the calculation, state what additional property information is needed.

9.22. Assuming that a certain n-component gas mixture's $P\underline{V}TN$ properties are adequately described by the van der Waals equation of state

$$P = RT/(V - b_m) - a_m/V^2,$$

derive an analytical expression for the fugacity coefficient $\hat{\phi}_i$ of component i as a function of T, V, Z, a_i, b_i, and with the mixing rules given by Eqs. (9-17) and (9-19) with $\delta_{ij} = 0$.

9.23. A stream composed of a binary mixture of benzene and cyclohexane that contains 0.5 mole fraction benzene is separated into two product streams. One stream contains a 0.98 mole fraction of benzene and the second contains a 0.90 mole fraction of cyclohexane. The separation process is performed at a constant temperature of 300 K and a constant pressure of 1 bar under steady state conditions.

You can assume that the activity coefficient for benzene, γ_B, is given by

$$\gamma_B = \exp[0.56(1 - x_B)^2],$$

where x_B is the mole fraction of benzene in the mixture. What is the minimum work required to perform the separation?

9.24. The enthalpy of mixing of a ternary solution containing components 1, 2, and 3 is:

$$\Delta H_{123} = [100x_1x_2 + 50(x_1x_3 + x_2x_3)]/[x_1 + 2x_2 + 0.5x_3]$$

where ΔH_{123} is measured in J/(mole of solution), and x_i is the mole fraction of component i for $i = 1, 2$, or 3.

(a) Determine the partial molar enthalpy of mixing of component 2 at $x_1 = 0.2$, $x_2 = 0.5$. Express your result in J/(mole of component 2).

(b) A blender is used to mix two binary solutions to make an equimolar ternary solution of composition $x_1 = x_2 = x_3$. The compositions of the two binary solutions are ($x_1 = 0.333$, $x_2 = 0.667$) and ($x_1 = 0.333$, $x_3 = 0.667$), respectively. If the mixing is carried out isothermally and isobarically as a continuous process operating at steady state, determine the cooling load required. Express your result in J/(mole of ternary product).

9.25. A binary liquid mixture of components A and B has a specific volume V_m at temperature T. Isothermal measurements on this mixture indicate that over the entire composition range:

$$\Delta \overline{S}_A^{EX} = \Delta \overline{S}_B^{EX} = 0$$

$$\Delta \overline{V}_A^{EX} = \Delta \overline{V}_B^{EX} = 0$$

You can assume that the *PVT* properties of the mixture *and* both pure components can be described by the van der Waals equation of state:

$$(P + a_m/V^2)(V - b_m) = RT$$

With constant pure component parameters a_A, b_A, for component A, a_B, b_B, for component B [see Eq. (8-32)].

$$a_m = x_A^2 a_A + x_B^2 a_B + 2x_A x_B a_{AB}, \quad b_m = x_A b_A + x_B b_B, \quad \text{and} \quad a_{AB} = \sqrt{a_A a_B}$$

Derive an expression for ΔG^{EX}. Express your result in terms of x_A, V_A, V_B, a_A, a_B, b_A, b_B, and a_{AB}.

9.26. If $\hat{f}_i = \hat{\phi}_i y_i P$, derive an expression for $[\partial(\ln \hat{\phi}_i)/\partial T]_{P, y_i}$.

Describe how you would evaluate the derivative at specific temperatures over a range from 200 to 300 K at a fixed pressure 100 bar and a fixed mixture composition (y_i for $i = 1$ to n known). What property information is needed in order to obtain a numerical result for the derivative?

9.27. Gyro Gearloose Jones, one of Rocky's distant cousins, has come up with a new idea for producing electric power by taking advantage of "mixing non-idealities." In one specific application, Gyro is proposing to utilize the exothermic heat produced by mixing sulfuric acid (H_2SO_4) and water (H_2O) as a heat source for a totally self-contained portable Rankine cycle. He claims that a stream of pure acid at 25°C (77°F) flowing at 1 kg/s (2.2 lb/s), when mixed with a pure water stream at 25°C also flowing at 1 kg/s, can produce about 400 kW of electricity. His Rankine cycle is designed for a proprietary working fluid,

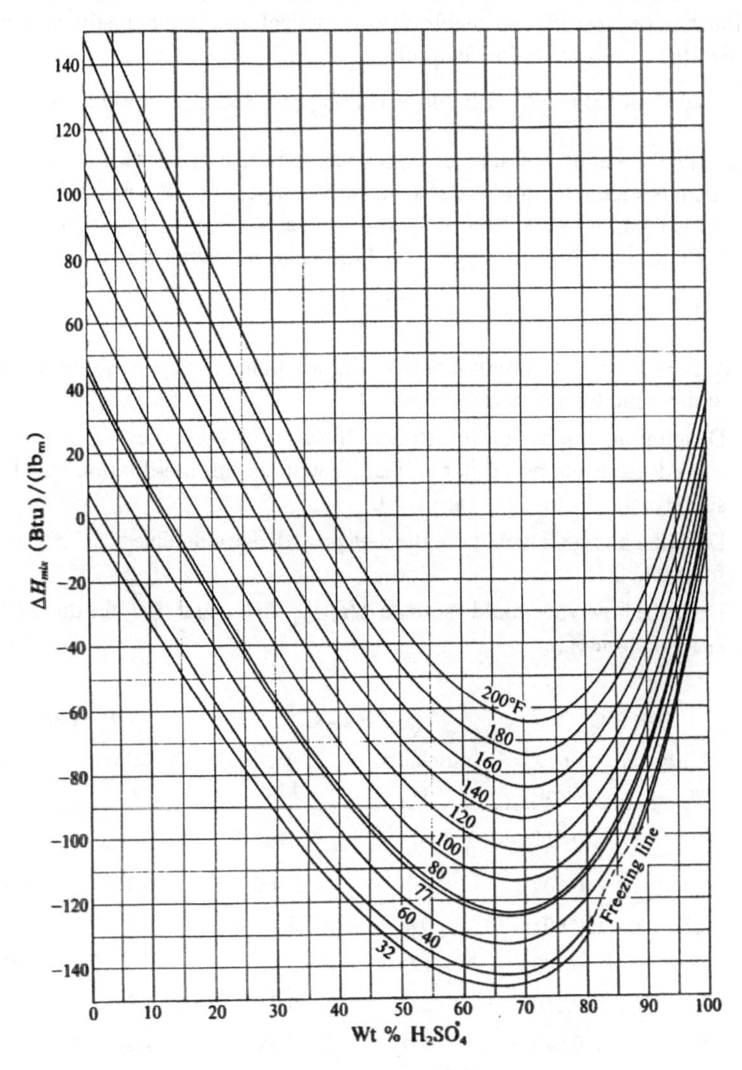

Figure P9.27 Enthalpy/concentration diagram for H_2SO_4/H_2O. [Original data from Ross, W.D. (1952), *Chem. Eng. Prog.*, **48**, p 314, Smith and Van Ness, *Chemical Engineering Thermodynamics*, 4th ed. (1987)]

"Cryptonium," to produce electricity with an η_u of 0.95 when operating under supercritical conditions. Gyro also claims that the only other thing required to make his device work is a small fan that circulates air from the room into the Rankine unit. The air temperature in this case is about 25°C.

MITY Industries has been asked to evaluate this proposal and we are naturally seeking your help. Can Gyro's design possibly work? Explain your answer, state all assumptions, and show all equations used in your analysis. Is Gyro's Rankine cycle performance specification ($\eta_u = 0.95$) reasonable from a practical standpoint? Justify your answer. You may find the data listed below helpful:

Cp_w (pure H_2O) = 4200 J/kgK = 1 Btu/lb°F and Cp_a (pure H_2SO_4) = 1400 J/kgK = 0.33 Btu/lb°F

9.28. Frequently the vapor pressure of water vapor is measured as a second non-volatile component is added to liquid water at constant temperature. For example, vapor pressure data corresponding to addition of sucrose to water are given below at 0°C, 25°C, and 100°C. The activity of water (a_w) in the liquid phase can be approximated by:

$$a_w = p_w/P_{vp,w}$$

where p_w = water partial pressure over the solution and P_{vp} = vapor pressure of pure water at the same temperature T.

(a) Develop an expression to estimate the activity coefficient γ_s of sucrose x_s. You should use asymmetric normalization with an infinite-dilution standard state for sucrose, that is, $\gamma_s = 1.0$ at $x_s = 0$

(b) Describe how you would use the data provided to calculate γ_s at 25°C for x_s varying from 0 to 0.1.

(c) Describe how you would estimate ΔH_{mixing}, $\Delta\overline{H}_w$, and $\Delta\overline{H}_s$ for this system at 50°C as a function of x_s.

$T = 0°C$		$T = 25°C$		$T = 100°C$	
molality m	p_w **kPa**	**molality** m	p_w **kPa**	**molality** m	p_w **kPa**
0	0.611	0	3.166	0	$101.33 = P_{vp,w}$
0.2	0.609	0.1	3.160	0.3	100.78
0.5	0.605	0.4	3.14	0.6	100.21
1.0	0.599	1.0	3.11	0.8	99.81
3.5	0.560	2.0	3.03	1.2	99.01
4.5	0.542	3.0	2.95	1.5	98.35
5.0	0.533	4.0	2.87		
6.0	0.516	5.0	2.78		
7.0	0.499				

Note that $\boldsymbol{m}$ in the table above is the molality ($\boldsymbol{m}$) which is equal to the number of gram formula weights per 1000 g of water.

Molecular weights: water = 18 g/mol sucrose = 342.3 g/mol

9.29. The osmotic pressure ($\Pi = P_2 - P_1$) is known at 0°C as a function of the concentration of sucrose in an aqueous system as shown below. Describe how you would use this information to calculate the activity coefficient of sucrose in the water-sucrose liquid phase as a function of composition. State any assumptions made and/or additional data needed to make your estimate.

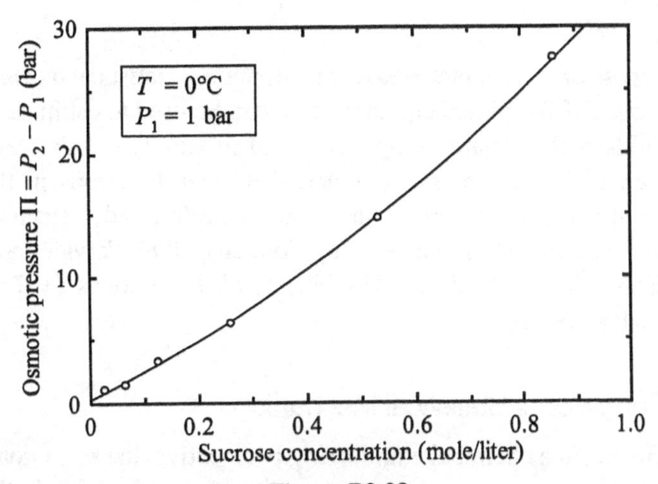

Figure P9.29

Statistical Mechanical Approach for Property Models* 10

The main purpose of this chapter is to review important statistical mechanics concepts and to develop a set of functional equations that can be used to estimate the properties of non-ideal fluids both as pure components and mixtures. In this way, we provide linkages between molecular interactions and their manifestations in the observable properties of macroscopic systems. Other more detailed and elegant treatments of statistical mechanics can be found in texts by Tolman (1938), Fowler and Guggenheim (1956), Davidson (1962), Reed and Gubbins (1973), Andrews (1975), McQuarrie (1976), and Rowley (1994).

10.1 Basic Concepts of Statistical Mechanics

10.1.1 Scope. From a chemical engineering perspective, the main goal of statistical mechanics is to estimate the macroscopic properties of systems using both rigorous and approximate theory based on the laws that control the behavior of molecules. The area of statistical mechanics that focuses on the properties of large systems of many molecules in states of thermodynamic equilibrium is frequently referred to formally as statistical thermodynamics. In general, one can use the principles and formalisms of statistical thermodynamics to derive the laws of thermodynamics based on a set of fundamental postulates that describe the microscopic states of a system on a molecular level. [NB: These postulates are different than the ones introduced in Chapters 2-4 to describe the classical macroscopic states of a system.] In a completely independent manner, one can develop rigorous theoretical expressions for thermodynamic properties: *primitive* (e.g., density, volume, temperature, pressure, heat capacity (C_v and C_p) and vapor pressure) and *derived* (e.g., internal energy, entropy, Helmholtz and Gibbs free energies and fugacity). In addition, macroscopically observable phenomena such as critical points and phase coexistence can be described at a fundamental molecular level.

---

* The authors are particularly grateful to Professors Doros Theodorou, Jonathan Harris, Pablo Debenedetti, Thanassis Panagiotopoulos, James Ely, and Dr. William Peters for their input and review of this chapter.

Linking the macroscopic world and its molecular-level roots will be new territory for some engineering students, but once mastered, these linkages will pay substantial dividends. For many systems primitive properties like density, pressure, and volume, are easily visualized and easy to measure. Connecting them to molecular phenomena, like collisions and forces between atoms and molecules has sometimes been avoided by practitioners, perhaps because the mathematics and other formalisms for doing so, initially seem foreign or unduly complicated and more suited to the interests of physicists and theoretical physical chemists. However, today's engineers must deal with increasingly complex and non-ideal pure substances and mixtures, and with operating conditions involving extremes of temperature, pressure, concentration, ionic strength, etc. For many of these substances reliable experimental measurements at such conditions are time consuming and costly. Statistical mechanics and related molecular simulation methodologies discussed in this chapter (e.g., Monte Carlo and Molecular Dynamics), enable the engineer to estimate physical and chemical properties for these and otherwise inaccessible substances or operating regimes, and shed light on how to more reliably interpolate and extrapolate data to new conditions. Familiarity with these methods also facilitates the implementation of the latest advances in molecular understanding as they emerge from research in the pure sciences and mathematics. Figure 10.1 illustrates how statistical mechanical models provide alternative routes to computing practical engineering properties, while improving the molecular underpinnings for classical thermodynamic methods for estimating these properties. A reasonable understanding of molecular-level behavior and applications thereof should be as much a part of the modern engineer's toolbox as are the microprocessor and chemical process simulation software.

10.1.2 Phase Space for Many Particle Systems. For the moment, consider a system composed of a very large number of structureless point particles, N, for example, of the order of Avogadro's number (6×10^{23}) corresponding to 1 mole of monatomic molecules. The instantaneous state of such a system in classical mechanics is defined in terms of the *position vectors* $[\mathbf{r}_i = f(x_i, y_i, z_i)]$ and the *momentum vectors* $[\mathbf{p}_i = m_i\,\mathbf{v}_i]$ of all N particles.

Each position vector $\mathbf{r}_i$ and momentum vector $\mathbf{p}_i$ has 3 components in Cartesian coordinates that can be represented by

$$\mathbf{r}_i = [x_i, y_i, z_i] \quad \text{and} \quad \mathbf{p}_i = [p_{xi}, p_{yi}, p_{zi}] = [m_i v_{xi}, m_i v_{yi}, m_i v_{zi}] \tag{10-1}$$

where each of the vectors corresponds to an orthogonal direction. Thus the composite of $N\,\mathbf{r}_i$ vectors and $N\,\mathbf{p}_i$ vectors can be written in shorthand notation as:

$$\underset{\sim}{\mathbf{r}}^N = \begin{bmatrix} \mathbf{r}_1 \\ \vdots \\ \mathbf{r}_N \end{bmatrix} = \begin{bmatrix} x_1 & y_1 & z_1 \\ \vdots & \vdots & \vdots \\ x_N & y_N & z_N \end{bmatrix} \begin{bmatrix} \mathbf{i} \\ \mathbf{j} \\ \mathbf{k} \end{bmatrix} \tag{10-2a}$$

and

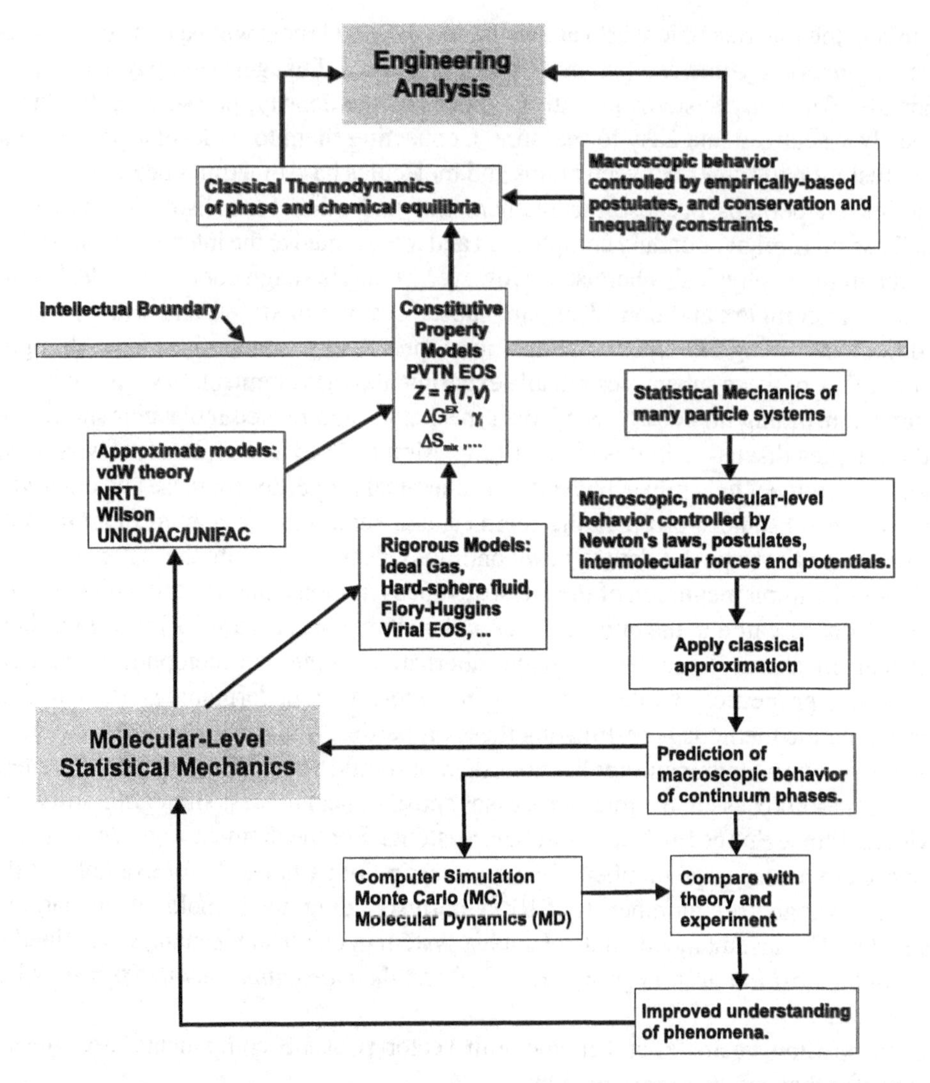

Figure 10.1 Interconnections between macroscopic engineering thermodynamic property models and molecular-level statistical mechanics.

$$\underset{\sim}{\mathbf{p}}^N = \begin{bmatrix} \mathbf{p}_1 \\ \vdots \\ \mathbf{p}_N \end{bmatrix} = \begin{bmatrix} p_{x1} & p_{y1} & p_{z1} \\ \vdots & \vdots & \vdots \\ p_{xN} & p_{yN} & p_{zN} \end{bmatrix} \begin{bmatrix} \mathbf{i} \\ \mathbf{j} \\ \mathbf{k} \end{bmatrix} \qquad (10\text{-}2b)$$

where $\mathbf{i}, \mathbf{j}$, and $\mathbf{k}$ are orthogonal unit vectors. Effectively, the dimensionality of the entire $(\underset{\sim}{\mathbf{r}}^N, \underset{\sim}{\mathbf{p}}^N)$ system is now $3 \times 2N$ or $6N$.

This 6-N dimensional vector mathematically defines a state for the N particle system in what is known as *phase space*. In essence, phase space is equivalent to a set of reference coordinates for the position $\mathbf{r}_i$ and momentum $\mathbf{p}_i$ vectors of each particle or molecule i in the system. Subsets of phase space have also been conveniently defined as: *configuration space*—the 3N-dimensional vector of spatial coordinates ($\mathbf{r}_1$, ..., $\mathbf{r}_N$) and *momentum space*—the 3N-dimensional vector of momenta ($\mathbf{p}_1$, ..., $\mathbf{p}_N$).

10.1.3 From Quantum Mechanics to the Classical Approximation. Inherent to the presentation above is the assumption that both the positions and momenta of N particles can be specified with absolute certainty to define the state of the system. Effectively, this treatment assumes a limiting behavior where the laws of classical mechanics can be applied. At sufficiently high temperatures, the classical approximation may be used in place of a rigorous quantum mechanical treatment. With quantum mechanics, the Heisenberg uncertainty principle applies and the positions and momenta of the molecules cannot be specified exactly at the same time. Quantum theory teaches us that the energy of molecules is not distributed continuously, but rather divided up into discrete packets called quantum states. Schrodinger's equation is employed to describe the behavior of the system. Specific quantum numbers can be used to label allowable states. Each state i has a particular energy level $\underline{E}_i$. If more than one quantum state has the same energy then a degeneracy Ψ_j is defined such that Ψ_j = the number of different quantum states that exist with energy $\underline{E}_j$.

Energy levels $\underline{E}_i$ correspond to eigenvalues resulting from the solution to Schrodinger's equation. The probability $P_i(\underline{E}_i)$ that a system of N particles is in a particular state with $\underline{E}_i$ energy is defined by:

$$P_i(\underline{E}_i) = \frac{\exp(-\underline{E}_i/kT)}{\sum_j \exp(-\underline{E}_j/kT)} \tag{10-3}$$

where the sum runs over all allowable quantum states j. Equation (10-3) represents, in fact, the fundamental ***Postulate of Quantum Partitioning*** in statistical mechanics: namely, that for ***an isolated, closed system of constant volume, the probability of occupying a quantum state i depends only on its energy*** $\underline{E}_i$ ***relative to the total energy of the system.***

By choosing the mathematical form of Eq. (10-3), we have effectively adopted Boltzmann statistics in that the probability that a molecule at equilibrium is in a state of energy $\underline{E}_i$ is proportional to exp $(-\underline{E}_i/kT)$. Although our postulatory approach allows us to make this assertion without further proof, eventually its validity is tested by seeing if predicted behavior matches experimental results. There are alternative theoretical routes to demonstrate that the Boltzmann formulation is indeed correct for most cases of interest to us. One popular approach is to start with a set of K systems *divided into* $n_1, n_2, ..., n_i$ *parts* in quantum states $1, 2, ..., i$ corresponding to energy levels $\underline{E}_1, \underline{E}_2, ..., \underline{E}_i$. Combinatorial analysis is used to account for states that are not distinguishable, leading to a combinatorial correction factor:

$$\Omega(n) = \frac{(n_1 + n_2 + \ldots + n_i)!}{n_1!\, n_2! \ldots n_i!} = \frac{K!}{\prod_i n_i!}$$

where $\Sigma n_i = K$ and $\Omega(n)$ is the number of different arrangements for K total systems of which $n_1, n_2, ..., n_i$ elements are indistinguishable. This analysis is also subject to the constraint that $\Sigma n_i \underline{E}_i = \underline{E}$ (total). The combinatorial relationship is used to express the statistically averaged probability of occupying quantum state i with energy $\underline{E}_i$ as,

$$< p_i(\underline{E}_i) > = P_i(\underline{E}_i) = \frac{\sum_n \Omega(n)\, n_i(n)}{K \sum_n \Omega(n)} = \frac{\exp(-\beta \underline{E}_i)}{\sum_i \exp(-\beta \underline{E}_i)}$$

The summations are replaced with a single most probable distribution using the so-called maximum term method, which is applicable for large-sized systems ($K \to \infty$). Applying this approximation leads immediately to the form of Eq. (10-3) with the term $1/kT$ replaced by β. Later in Section 10.1.5 we show that β is exactly given by $1/kT$ to complete the proof.

In a classical mechanical description of a system of N particles or molecules, we replace the discrete set of quantum states, as expressed in terms of specific energy levels $\underline{E}_i$ in Eq. (10-3), with a continuous distribution of states and energies. The classical approximation can be applied safely when the energy difference between adjacent quantum states, $\Delta E = (E_{j+1} - E_j)$, is small relative to kT (expressed on the basis of one molecule in J), or

$$\Delta E/kT << 1 \tag{10-4}$$

At normal temperatures of interest in chemical engineering applications, say $T > 50$ K, the translational modes satisfy Eq. (10-4) and for rotational modes somewhat higher temperatures, $T > 100$ K are needed. Vibrational quantum states, however, typically do not meet this criteria even at ambient temperatures of 300 K where $\underline{\Delta E}_{vib} \approx kT$, so a quantum mechanical treatment is required for correct assessment of vibrational mode contributions to thermodynamic properties. This is particularly relevant to heat capacity (C_p or C_v) calculations. Similar quantum restrictions apply to electronic energy modes as well. Later we will illustrate how these quantum limitations are dealt with.

With classical behavior, the laws of classical mechanics apply and the Hamiltonian, **H**, of the system, which is equal to the total energy of the system, can be written in terms of the phase space position and momentum coordinates for a given configuration as:

$$\mathbf{H}(\underset{\sim}{\mathbf{r}}^N, \underset{\sim}{\mathbf{p}}^N) = KE\ (kinetic\ energy) + PE\ (potential\ energy)$$

and

$$\mathbf{H}(\underset{\sim}{\mathbf{r}}^N, \underset{\sim}{\mathbf{p}}^N) = f_{KE}(\underset{\sim}{\mathbf{p}}^N) + f_{PE}(\underset{\sim}{\mathbf{r}}^N) = \sum_{i=1}^{N} \frac{\mathbf{p}_i^2}{2m_i} + \Phi(\mathbf{r}_1, ..., \mathbf{r}_N) \tag{10-5}$$

where $\Phi(\mathbf{r}_1, ..., \mathbf{r}_N)$ is the total potential energy of the N particle system. The classical equations for individual particle motion are expressed in terms of time derivatives of $\mathbf{r}_i$ and $\mathbf{p}_i$, *i.e.* $\dot{\mathbf{r}}_i \equiv \partial \mathbf{r}_i/\partial t$ and $\dot{\mathbf{p}}_i \equiv \partial \mathbf{p}_i/\partial t$:

$$\left(\frac{\partial \mathrm{H}}{\partial \mathbf{r}_i}\right)_{\mathbf{r}_j[i],\,\mathbf{p}_j} = -\dot{\mathbf{p}}_i \tag{10-6a}$$

$$\left(\frac{\partial \mathrm{H}}{\partial \mathbf{p}_i}\right)_{\mathbf{p}_j[i],\,\mathbf{r}_j} = \dot{\mathbf{r}}_i \tag{10-6b}$$

Note the canonical relationship of $\dot{\mathbf{r}}_i$ and $\dot{\mathbf{p}}_i$; this is similar to the conjugate coordinates x_i and ξ_i used in our earlier treatment of Legendre transforms in Chapter 5. In that situation, for a basis function $y^{(0)} = f[x_1, ..., x_N]$ the Legendre transform $y^{(1)}$ replaces one variable, x_i, in the set $x_1, ..., x_N$ with ξ_i. Thus,

$$y^{(1)} = f[x_1, x_2, ..., x_{i-1}, \xi_i, x_{i+1}, ..., x_N]$$

$$y^{(1)} = y^{(0)} - x_i\xi_i \ ; \quad \left(\frac{\partial y^{(0)}}{\partial x_i}\right)_{x_j[i]} = \xi_i \quad \text{and} \quad \left(\frac{\partial y^{(1)}}{\partial \xi_i}\right)_{x_j[i]} = -x_i \tag{10-7}$$

Equations (10-6a) and (10-6b) are equivalent to Newton's second law where

$$\mathbf{F}_i = m_i \frac{d^2\mathbf{r}_i}{dt^2} = m_i \frac{d\mathbf{v}_i}{dt} \tag{10-8}$$

In addition, the separation of kinetic [$f(\mathbf{p}^N)$ only] and potential [$\Phi = f(\mathbf{r}^N)$ only] energy contributions in the classical Hamiltonian is inherent to a conservative Newtonian system of particles. Points defined by particular sets of values for $(\mathbf{r}^N, \mathbf{p}^N)$ trace a path in phase space that follows the laws of Newtonian mechanics and evolve in time, consistent with all external constraints. By Liouville's theorem, the rate of change with time of the particle distribution function (also called the "density" of the particles) in phase space is mathematically analogous to the continuity equation for flow of an incompressible fluid [see, for example, Appendix C of Reed and Gubbins, (1973)]. An important analogous physical interpretation is also possible. For incompressible fluid flow there is no change in fluid density with time, for an observer moving with the flow. Similarly, as recognized by Gibbs, one of the consequences of Liouville's theorem is that particle density is conserved in phase space, i.e., as any point moves through phase space there is no change with time in the density of points in the immediate neighborhood of that point [Tolman, (1938) and Hirschfelder *et al.*, (1954)].

10.1.4 The Gibbs Ensembles. Given sufficient time, a particle (or molecule) defined by position and momentum vectors $(\mathbf{r}_i, \mathbf{p}_i)$ will occupy all positions in phase space with a particular probability distribution. In order to deduce the functional form for this probability distribution in statistical mechanics, it is convenient to introduce the concept of an ensemble. Rather than envisioning a single system evolving in time through a set of states in phase space, consider an assemblage of an infinite number of "exact" replicas of the system each with its own distribution of allowable microstates $\mathbf{r}_i, \mathbf{p}_i$ for i=1, ..., N, subject to the macroscopic thermodynamic restrictions or constraints imposed on the original N-particle system of interest. The states of all these "exact" copies define the probability distribution. This set of allowable replicated systems is called an ***ensemble***.

One can imagine different types of ensembles corresponding to different macroscopic thermodynamic constraints. The most important ones for chemical thermodynamics are:

(1) ***microcanonical***—isolated, closed system of constant $\underline{E}$, $\underline{V}$, and N
(2) ***canonical***—isothermal, closed system of constant T, $\underline{V}$, and N
(3) ***grand canonical***—open system of constant T, $\underline{V}$, and μ_i
(4) ***isobaric/isothermal***—closed system of constant P, T, and N

In a sense, one can see an immediate connection between these four ensembles using the Legendre transformation method developed in Chapter 5. If the microcanonical system is selected as the basis function then the canonical ensemble basis function is a 1st order Legendre transform and the grand canonical and the isobaric/isothermal ensembles are 2nd order transforms.

The probability density function P_N for a given ensemble of N particles at equilibrium is independent of time. By convention, we define the probability of finding the system in a particular region of $6N$-dimensional phase space in the vicinity of $(\mathbf{r}_1, \mathbf{r}_2, ..., \mathbf{r}_N, \mathbf{p}_1, \mathbf{p}_2, ..., \mathbf{p}_N)$ *as:*

$$P_N(\underset{\sim}{\mathbf{r}}^N, \underset{\sim}{\mathbf{p}}^N)\, d\underset{\sim}{\mathbf{r}}^N\, d\underset{\sim}{\mathbf{p}}^N \tag{10-9}$$

We can therefore create an ensemble average summing over all states k to estimate any observable property B using the probability density function, namely:

$$B_{observed} = \langle B \rangle = \sum_k P_N(\underset{\sim}{\mathbf{r}}^N, \underset{\sim}{\mathbf{p}}^N)\, B_k \tag{10-10}$$

Given our construction of the ensemble concept as a set of allowable microstates that represent the time evolution of the system, we can introduce the most fundamental hypothesis of statistical mechanics--the ***ergodic hypothesis which equates the time average with the ensemble average***.

Microcanonical ensemble—($\underline{E}$, $\underline{V}$, N) fixed: With $\underline{E}$, $\underline{V}$, and N constant, the microcanonical ensemble represents an isolated, closed system of constant volume. Furthermore, with classical statistics applicable, $\underline{E} = \mathsf{H}(\underset{\sim}{\mathbf{r}}^N, \underset{\sim}{\mathbf{p}}^N)$ = fixed; thus, we can assume that all variations in phase space for this ensemble occur on a constant energy surface.

At this point we need to introduce another basic postulate of statistical mechanics, namely that ***all microscopic states at a molecular level are equally likely to occur in an isolated, closed system of constant $\underline{E}$, $\underline{V}$, and N at thermodynamic equilibrium.*** Sometimes this is called the ***Postulate of Equal a Priori Probability***. Effectively, we are saying that the macroscopic state at equilibrium corresponds to the most random distribution of the N-particle system. This postulate applies equally well when mixtures of distinguishable particles exist, i.e. N_1, ..., N_n, where n = number of components. With this postulate, we can proceed to define the probability distribution P_N for states within the microcanonical ensemble as a function only of the total energy $\underline{E}$:

$$P_N = P_N(\underset{\sim}{\mathbf{r}}^N, \underset{\sim}{\mathbf{p}}^N) = \begin{bmatrix} \dfrac{C}{\Psi(\underline{E})} & \text{for } \underline{E} < \mathsf{H}(\underset{\sim}{\mathbf{r}}^N, \underset{\sim}{\mathbf{p}}^N) < \underline{E} + \delta\underline{E} \\ & \\ 0 & \text{otherwise} \end{bmatrix} \tag{10-11}$$

where again we invoke the continuum assumption to define $P_N(\underline{E})$ in an infinitely thin shell on a constant energy surface between $\underline{E}$ and $\underline{E}+\delta\underline{E}$. $\Psi(\underline{E})$ corresponds to a differential element of phase space volume given by:

$$\Psi(\underline{E}) = \int \ldots \int (d\underset{\sim}{\mathbf{r}})^N (d\underset{\sim}{\mathbf{p}})^N \text{ such that } \underline{E} < \mathsf{H}(\underset{\sim}{\mathbf{r}}^N, \underset{\sim}{\mathbf{p}}^N) < \underline{E} + \delta\underline{E} \tag{10-12}$$

The constant C in Eq. (10-11) is set equal to $h^{3N} N!$, where h = Planck's constant, to account for the phase space volume of each quantum state. The $N!$ term corrects for the indistinguishability of the particles or molecules. Because the microcanonical ensemble is defined as one of fixed $\underline{E}$, the probability distribution function given by Eq. (10-11) is rather trivial, having non-vanishing probabilities only for degenerate states with energy $\underline{E}$.

Canonical ensemble—(T, $\underline{V}$, N) fixed: The canonical ensemble corresponds to a commonly encountered set of conditions where the system is at constant $\underline{V}$ and N and is kept isothermal by having a heat interaction with a very large heat reservoir or sink. Effectively, the combination of the canonical ensemble system and the heat reservoir constitute the equivalent of a microcanonical ensemble at constant $\underline{E}$, $\underline{V}$, and N. The probability distribution of states in the canonical ensemble is given by:

$$P_N(\underline{E}_i) = a \exp(-\beta \underline{E}_i) \tag{10-13}$$

where a and β are parameters that need to be specified. The term a can be regarded as a normalization factor that represents the sum over all allowable states of the distribution given in Eq. (10-13). Thus,

$$a = \frac{1}{\sum_i \exp(-\beta \underline{E}_i)} \tag{10-14}$$

The reciprocal of this normalization factor plays an important role in statistical mechanics as it will give us a convenient way to express all classical thermodynamic variables. The accepted term for this reciprocal quantity is the *partition function*, commonly labeled Q_N. Much earlier, Tolman (1938) used the term *sum-over-states* while Planck introduced its Germanic equivalent *Zustandssumme*. We will use current conventional terminology with the ***canonical partition function*** defined by,

$$Q_N \equiv \sum_i \exp(-\beta \underline{E}_i) \tag{10-15}$$

where the summation extends over i quantum states. Using this definition of the partition function, the canonical probability distribution function given in Eq. (10-13) becomes:

$$P_N(\underline{E}_i) = \frac{\exp(-\beta \underline{E}_i)}{Q_N} \tag{10-16}$$

Now, we can easily construct the ensemble averaged energy $\langle \underline{E} \rangle$ over i as

$$\langle \underline{E} \rangle = \sum_i (P_N(\underline{E}_i))\underline{E}_i = \frac{\sum_i \underline{E}_i \exp(-\beta \underline{E}_i)}{Q_N} \tag{10-17}$$

Noting that Q_N is only a function of β and $\underline{E}_i$, we can reformulate Eq. (10-17) from the canonical ensemble at fixed $\underline{V}$ and N as:

$$\langle \underline{E} \rangle = -\left(\frac{\partial (\ln \sum_i \exp(-\beta \underline{E}_i))}{\partial \beta} \right)_{\underline{V},N} = -\left(\frac{\partial \ln Q_N}{\partial \beta} \right)_{\underline{V},N} \tag{10-18}$$

As mentioned earlier, most chemical engineering applications involve situations where the classical approximation is valid; thus, we can construct the classical equivalent of $P_N(\underline{E}_i)$ and Q_N in terms of the phase space continuum and the classical Hamiltonian $\mathbf{H}$ as:

$$P_N = P_N^c(\underset{\sim}{\mathbf{r}}^N, \underset{\sim}{\mathbf{p}}^N) = \frac{1}{Q_N^c(\underline{V}, T)} \left[\frac{\exp[-\beta \mathbf{H}(\underset{\sim}{\mathbf{r}}^N, \underset{\sim}{\mathbf{p}}^N)]}{h^{3N} N!} \right] \tag{10-19}$$

where the ***classical canonical partition function*** is expressed as an integral:

$$\boxed{Q_N = Q_N^c(\underline{V}, T) = \frac{1}{h^{3N} N!} \int \dots \int \exp[-\beta \mathbf{H}(\underset{\sim}{\mathbf{r}}^N, \underset{\sim}{\mathbf{p}}^N)]\, d\underset{\sim}{\mathbf{r}}^N\, d\underset{\sim}{\mathbf{p}}^N} \tag{10-20}$$

Note that the factor $h^{3N} N!$ has been added to account for the phase space volume of each quantum state and the indistinguishability of particles or molecules. Because the integration is carried out over all phase space $(\underset{\sim}{\mathbf{r}}^N, \underset{\sim}{\mathbf{p}}^N)$, we can interpret the source of the quantum factor h^{3N} using the Heisenberg uncertainty principle, which can be expressed for each conjugate pair of phase space coordinates $(\mathbf{p}_i, \mathbf{r}_i)$ in terms of uncertainties for $\mathbf{p}_i$ and $\mathbf{r}_i$ that scale as,

$$\Delta \mathbf{p}_i \, \Delta \mathbf{r}_i \geq h$$

Hence the complete set of $3N$ conjugate phase space coordinates leads to a normalization of the partition function by a factor of h^{3N} to account for the *effective volume* of phase space. Interestingly, in the late 1800s before quantum mechanics had been formally invented, J.W. Gibbs viewed the h^{3N} factor somewhat intuitively as a way to reconcile the classical approximation with experimental data!

Using Eq. (10-20) as a basis, the classical definition of the ensemble average of any dynamical property $\langle \underline{B} \rangle$ is given by:

$$\langle \underline{B} \rangle \equiv \frac{\int \ldots \int B(\mathbf{r}^N, \mathbf{p}^N)\, P_N(\mathbf{r}^N, \mathbf{p}^N)\, d\mathbf{r}^N\, d\mathbf{p}^N}{\int \ldots \int P_N(\mathbf{r}^N, \mathbf{p}^N)\, d\mathbf{r}^N\, d\mathbf{p}^N} \tag{10-21}$$

Later in Sections 10.1.7 and 10.1.8 we will use this classical approximation to develop $P\underline{V}TN$ equations of state for ideal and real gases.

In classical mechanics, a dynamical property is a function of only phase space variables, $\mathbf{r}^N$ and $\mathbf{p}^N$, and time. The ergodic hypothesis effectively converts the time average to an ensemble average by means of an appropriate probability distribution. Consequently, each dynamical property has a well-defined value for a given state of the system. Total energy $\underline{E}$ and pressure P are dynamical properties while entropy $\underline{S}$ and temperature T are not.

The pressure fluctuates in a manner analogous to the energy and can also be represented by an ensemble average as:

$$\langle P \rangle = \sum_i (P_N(\underline{E}_i))\, P_i \tag{10-22}$$

where P_i is the instantaneous value of the pressure of the system of N particles in their respective energy states $\underline{E}_i$. By considering a reversible, adiabatic (isentropic) expansion or compression we can develop an expression for P_i in terms for the $P_i\, d\underline{V}$ work as:

$$P_i = -\left(\frac{\partial \underline{E}_i}{\partial \underline{V}}\right)_N \tag{10-23}$$

Incorporating the concept that the ensemble average pressure $\langle P \rangle$ is equivalent to the macroscopic, non-fluctuating observable pressure P, we obtain an explicit equation for P using Eqs. (10-22) and (10-16) with Eq. (10-23):

$$P = \langle P \rangle = \frac{1}{Q_N} \sum_i -\left(\frac{\partial \underline{E}_i}{\partial \underline{V}}\right)_N \exp(-\beta \underline{E}_i) \tag{10-24}$$

and by chain rule expansion of $\ln Q_N$ from Eq. (10-15):

$$\left(\frac{\partial \ln Q_N}{\partial \underline{V}}\right)_{\beta, N} = \frac{\beta}{Q_N} \sum_i -\left(\frac{\partial \underline{E}_i}{\partial \underline{V}}\right)_N \exp(-\beta \underline{E}_i) \tag{10-25}$$

Thus,

$$P = \frac{1}{\beta}\left(\frac{\partial \ln Q_N}{\partial \underline{V}}\right)_{\beta, N} \tag{10-26}$$

which is an equation of state for statistical thermodynamics. Additional connections to macroscopic thermodynamics are introduced in Sections 10.1.5 and 10.1.6 that follow.

Grand canonical ensemble—(T, $\underline{V}$, μ_j) fixed: The grand canonical ensemble corresponds to an isothermal system of constant $\underline{V}$ where both heat and mass flow interactions occur. Consequently, this ensemble is equivalent to an *open* macroscopic system in thermodynamics connected to a much larger sink or reservoir that supplies or absorbs heat and can transfer mass. The composite system is equivalent to a canonical ensemble at constant T, $\underline{V}$, and N ($N = N_{system} + N_{reservoir}$). The probability density function for the grand canonical ensemble in a single component system is:

$$P_N(\underline{E}_i) = \frac{\exp[-\beta \underline{E}_i(\mathbf{\underset{\sim}{r}}^N, \mathbf{\underset{\sim}{p}}^N)]}{\sum_N \sum_i \exp[\beta\mu_j]^N \exp[-\beta \underline{E}_i(\mathbf{\underset{\sim}{r}}^N, \mathbf{\underset{\sim}{p}}^N)]} \tag{10-27}$$

or by introducing the Hamiltonian for $\underline{E}_i$ in a classical system:

$$P_N(\underline{E}_i) = P_N(\mathbf{H}(\mathbf{\underset{\sim}{r}}^N, \mathbf{\underset{\sim}{p}}^N, N) = \frac{\exp[\beta\mu_j]^N \exp[-\beta \mathbf{H}(\mathbf{\underset{\sim}{r}}^N, \mathbf{\underset{\sim}{p}}^N, N)]}{N! h^{3N} \sum_N \exp[\beta\mu_j]^N Q_N^C} \tag{10-28}$$

First, the absolute activity λ is defined by

$$\lambda \equiv \exp(\beta \mu_j) \tag{10-29}$$

Now $P_N(\underline{E}_i)$ for this system of N similar molecules with energy $\underline{E}_i$ becomes

$$P_N(\underline{E}_i) = \frac{\lambda^N \exp[-\beta \mathbf{H}(\mathbf{\underset{\sim}{r}}^N, \mathbf{\underset{\sim}{p}}^N, N)]}{N! h^{3N} \Xi(T, \underline{V}, \mu_j)} \tag{10-30}$$

where the normalizing factor Ξ is called the ***grand canonical partition function***:

$$\boxed{\Xi(T, \underline{V}, \mu_j) \equiv \sum_{N=0}^{\infty} \lambda^N Q_N(\underline{V}, T)} \tag{10-31}$$

where Q_N is given by Eq. (10-15) for quantum states $\underline{E}_i$ or classically by Eq. (10-20). Eq. (10-31) is easily extended to multicomponent systems by introducing separate $\lambda_j^{N_j}$ terms for each component j.

Isobaric/isothermal ensemble—(P, T, N) fixed: This ensemble provides a means for relating properties at a molecular level directly to functions in fully primitive variables such as P, T, and N. This ensemble is commonly referred to as the *NPT* ensemble. With these variables, we can relate the Gibbs free energy $\underline{G} = f(T, P, N)$ to the ***NPT partition function*** which in this situation is given by

$$\Delta(P, T, N) = \sum_{\underline{E}} \sum_{\underline{V}} \Omega(\underline{E}, \underline{V}, N) \exp[-\beta(\underline{E} + P\underline{V})] \tag{10-32}$$

where the summations are carried out over all quantum states for $\underline{E}$ and $\underline{V}$.

Now we need to link these ensembles and their partition functions (microcanonical, canonical, isobaric-isothermal and grand canonical) to thermodynamics. These connections are made by interpreting entropy statistically and by using the Fundamental Equation from Gibbs.

10.1.5 Entropy and the Boltzmann Factor β. To develop an interpretation of entropy in statistical mechanics, we first refer back to its original definition in terms of a reversible heat interaction from Chapter 4:

$$d\underline{S} \equiv \frac{\delta q_{rev}}{T} \tag{4-29}$$

and from the Fundamental Equation of thermodynamics from Gibbs (see Section 5.1) we have for a simple system that

$$\underline{E} = \underline{U} = f(\underline{S}, \underline{V}, N) \tag{10-33}$$

Now we can apply the ergodic hypothesis and the expression that relates the ensemble average $<\underline{E}>$ to the canonical partition function defined by Eq. (10-17). Note that Q_N is a function of $\underline{E}_i$ and β from its defining Eq. (10-15) thus,

$$d\ln Q_N = \left(\frac{\partial \ln Q_N}{\partial \beta}\right)_{\underline{V}, N} d\beta + \sum_i \left(\frac{\partial \ln Q_N}{\partial \underline{E}_i}\right)_{\beta, N} d\underline{E}_i \tag{10-34}$$

with further simplification, using Eqs. (10-15) and (10-18), Eq. (10-34) becomes:

$$d\ln Q_N = -<\underline{E}>d\beta - \beta \sum_i \frac{\exp(-\beta \underline{E}_i)\, d\underline{E}_i}{Q_N} \tag{10-35}$$

By chain rule expansion,

$$d(<\underline{E}>\beta) = <\underline{E}>d\beta + \beta d<\underline{E}> \tag{10-36}$$

and Eq. (10-35) can be rewritten as,

$$d(\ln Q_N + <\underline{E}>\beta) = \beta d<\underline{E}> - \beta \sum_i \frac{\exp(-\beta \underline{E}_i)\, d\underline{E}_i}{Q_N} \tag{10-37}$$

The summation term on the right hand side of Eq. (10-37) is equivalent to the reversible work done on the system as a consequence of a reversible change in volume. In other words, doing work on the system amounts to changing the energy levels of the states, while keeping the probability distribution of states constant. Thus, Eq. (10-24), which relates the pressure P to the canonical partition function Q_N, can be used to interpret work in a statistical mechanics context:

$$-\delta W_{rev} = +Pd\underline{V} = \frac{-1}{Q_N} \sum_i \exp(-\beta \underline{E}_i) \left(\frac{\partial \underline{E}_i}{\partial \underline{V}}\right)_N d\underline{V} \tag{10-38}$$

and with Eq. (10-37)

$$d(\ln Q_N + <\underline{E}>\beta) = \beta[d<\underline{E}> - \delta W_{rev}] \tag{10-39}$$

From thermodynamics, the First Law can be written for a reversible, closed system as:

$$\delta q_{rev} = d\underline{E} - \delta W_{rev} \tag{10-40}$$

For simple systems $<\underline{E}> = \underline{E} = \underline{U}$ *and* $\delta W_{rev} = -Pd\underline{V}$ and Eq. (10-40) becomes:

$$\delta q_{rev} = d\underline{U} + Pd\underline{V} \tag{10-41}$$

Therefore, the reversible heat flow is equal to the change in internal energy when the volume is constant. In general, using the definition of entropy from thermodynamics as given in Eq. (4-29)

$$\delta q_{rev} = Td\underline{S} = d\underline{U} + Pd\underline{V} = d<\underline{E}> - \delta W_{rev} \tag{10-42}$$

Thus from Eq. (10-39); $\beta\delta q_{rev} = \beta Td\underline{S} = d(\ln Q_N + <\underline{E}>\beta)$ or

$$d\underline{S} = \frac{d(\ln Q_N + <\underline{E}>\beta)}{\beta T} \tag{10-43}$$

The only way Eq. (10-43) will provide an exact differential for $d\underline{S}$ is if β is proportional to $1/T$. For now, let's set $\beta = 1/kT$ where $k \equiv$ Boltzmann's constant $= 1.38048 \times 10^{-23}$ J/K. Thus, by integrating Eq. (10-43)

$$\underline{S} = k(\ln Q_N) + \frac{<\underline{E}>}{T} \tag{10-44}$$

It is relatively straightforward to show that k is a universal constant and does not vary from substance to substance [see McQuarrie (1976), pp. 42-44]. Note that we have implicitly assumed that the entropy from classical thermodynamics is identical to the ensemble averaged quantity, that is $\underline{S} = <\underline{S}>$.

A useful interpretation of entropy is obtained by recasting the approach above using probability concepts. We can express:

$$Td\underline{S} = \delta q_{rev} = d\underline{U} - \delta W_{rev} = \sum_i \underline{E}_i \, d(P_N(\underline{E}_i)) \tag{10-45}$$

using Eqs. (10-16), (10-18), and (10-43). Thus, a differential or virtual heat interaction ($|\delta q_{rev}| > 0$) amounts to altering the probability distribution of states while keeping the energy states $\underline{E}_i$'s of the system fixed. Conversely, work can be regarded as changing the energy levels while keeping the population distribution among states constant.

10.1.6 Connection to Thermodynamics. Earlier, we introduced the concept of an ensemble average energy $<\underline{E}>$ and showed that it could be expressed using the canonical partition function Q_N. Now that $\beta = 1/kT$, Eq. (10-18) becomes:

$$<\underline{E}> = kT^2\left(\frac{\partial \ln Q_N}{\partial T}\right)_{\underline{V},N} \tag{10-46}$$

Furthermore, by manipulating the Gibbs Fundamental Equation from ($\underline{S}$, $\underline{V}$, N) coordinates to canonical ensemble (T, $\underline{V}$, N) coordinates; we get the Helmholtz free energy $\underline{A}$:

$$\underline{A} \equiv \underline{U} - T\underline{S} \tag{10-47}$$

but for simple systems $\underline{U} = \underline{E} = <\underline{E}>$ and when combined with Eq. (10-44) gives:

$$\underline{A} = -kT \ln Q_N \tag{10-48}$$

Hence, the Helmholtz free energy provides a linkage between classical thermodynamics and statistical mechanics. In addition, the pressure equation (Eq. (10-26)) provides similar linkage by

$$P = -\left(\frac{\partial \underline{A}}{\partial \underline{V}}\right)_{T,N} = +kT\left(\frac{\partial \ln Q_N}{\partial \underline{V}}\right)_{T,N} \tag{10-49}$$

which utilizes the fact that

$$d\underline{A} = -\underline{S}dT - Pd\underline{V} + \sum_i \mu_i \, dN_i \tag{10-50}$$

Consequently, by considering a N_1, N_2, ..., N_n particle system of distinguishable chemical components, the chemical potential of each component i is given by Eqs. (10-48) and (10-50) as,

$$\mu_i = \left(\frac{\partial \underline{A}}{\partial N_i}\right)_{T,\underline{V},N_j[i]} = -kT\left(\frac{\partial \ln Q_N}{\partial N_i}\right)_{T,\underline{V},N_j[i]} \tag{10-51}$$

Using the classical approximation we can show that

$$\underline{U} = \underline{E} = <\underline{E}> = <\mathbf{H}> \tag{10-52}$$

and by considering variations in energy using a fluctuation concept, we can develop an expression for the observable macroscopic heat capacity C_v. Effectively, we are calculating energy fluctuations using the probability distribution of the canonical ensemble. The variance of the distribution σ^2 in terms of energy $\underline{E}$ is defined as,

$$\sigma_E^2 \equiv \overline{(\underline{E} - <\underline{E}>)^2} = <\underline{E}^2> - (<\underline{E}>)^2 \tag{10-53}$$

Using Eq. (10-10) to define the ensemble averages, we can show that

$$\sigma_E^2 = kT^2\left(\frac{\partial <\underline{E}>}{\partial T}\right)_{\underline{V},N} = kT^2\underline{C}_v = kT^2 N C_v \tag{10-54}$$

because $<\underline{E}> = \underline{U}$. Thus, the heat capacity at constant volume is related directly to fluctuations in the total internal energy of the system. Eq. (10-54) also has important implications regarding thermodynamically stable systems, for example, what can you say about the sign of C_v? Note that σ_E, the standard deviation, itself varies like $N^{1/2}$ which for any practically sized system ($N \sim 10^{23}$) is small compared to the ensemble average energy which varies with N. Thus, the energy distribution is sharply peaked at the average energy of the system.

The above discussion illustrates another valuable application of the theory of equilibrium fluctuations. Namely, we can determine the probability of observing values for practically important physical quantities, that are different than the values we compute as ensemble averages [Davidson (1962)]. Thus statistical mechanics not only provides us with a means to use molecular understanding to calculate macroscopic quantities of engineering importance, it also provides us with the tools to check, quantitatively, the statistical reliability of these computations. In the above example, we saw that, for a closed system, except when we are concerned with a very small number of molecules or are at a critical point, energy fluctuations about a mean (ensemble average) value, are totally negligible for engineering calculations (see Problem 10.9 for another example).

The grand canonical partition function also provides some useful connections to classical thermodynamics. Equation (10-44) can be modified using the definition of Ξ from Eq. (10-31) for a single component (i) open system to give

$$\underline{S} = k\ln\Xi + \frac{<\underline{E}>}{T} - \frac{\mu_i <N>}{T} \tag{10-55}$$

where $\mu_i = \mu$ in this case. Equation (10-55) can be rearranged with

$$G \equiv \mu_i <N> = \mu_i N = \underline{U} - T\underline{S} + P\underline{V} \tag{10-56}$$

to yield an equation of state:

$$P\underline{V} = kT\ln\Xi \tag{10-57}$$

where Ξ is a function of T, $\underline{V}$, and μ_i. Other useful relationships include:

$$<N> = kT\left(\frac{\partial\ln\Xi}{\partial\mu_i}\right)_{T,\underline{V}} \tag{10-58}$$

and

$$\underline{A} = -kT\ln\Xi + <N>\mu_i \tag{10-59}$$

The grand canonical ensemble is frequently used to characterize density fluctuations in an equilibrium system. For example, the variance in particle (molecule) number σ_N^2 can be related to the isothermal compressibility κ_T, where

$$\kappa_T = \frac{-1}{V}\left(\frac{\partial V}{\partial P}\right)_T$$

with

$$\sigma_N^2 = \overline{(N - <N>)^2} = \frac{kT<N>}{V}\kappa_T = \frac{kT(<N>)^2}{\underline{V}}\kappa_T \tag{10-60}$$

For a constant volume system, variations in particle number are equivalent to density variations. Equation (10-60) is the mathematical statement of the *compressibility theorem*. As before with our discussion of C_v, as N becomes large for practical systems, the standard deviation σ_N, which varies as $<N>^{1/2}$, becomes small relative to N itself. In the thermodynamic limit, the distribution of particle densities becomes sharp near $<N>$ and the isothermal compressibility is a measure of its width. From stability arguments, $\kappa_T \geq 0$ always and as a critical point is approached $\kappa_T \to \infty$. Consequently, near the critical point, density fluctuations grow large. Experimentally, we observe the phenomenon of critical opalescence where visible light is reflected as the liquid-vapor interface or meniscus disappears indicating that the length scale of density fluctuations is comparable to the wavelength of light.

10.1.7 The Classical Monatomic Ideal Gas. In an ideal gas there are no interaction forces between molecules, the potential energy contribution ($\Phi = PE = 0$) to the classical Hamiltonian is zero and only kinetic energy effects need be considered from Eq. (10-5).

$$\mathbf{H}^{ig}(\underset{\sim}{\mathbf{r}}^N, \underset{\sim}{\mathbf{p}}^N) = \sum_{i=1}^{N} \frac{p_{x1}^2 + p_{y1}^2 + p_{z1}^2 + \ldots + p_{xN}^2 + p_{yN}^2 + p_{zN}^2}{2m} \tag{10-61}$$

where p_{x1}, p_{y1}, p_{z1}, etc. are scalar momentum components and m is the mass of a single structureless particle. Thus using Eq. (10-20) defined over a set of $3N$ identical domains for each momentum component from $-\infty$ to ∞, we get

$$Q_N^{ig}(T, \underline{V}) = \frac{1}{N!h^{3N}} \left[\int_{-\infty}^{\infty} \exp\left[\frac{-\beta p^2}{2m}\right] dp\right]^{3N} \left[\int \ldots \int_{\underline{V}} dx\, dy\, dz\right]^N \tag{10-62}$$

where $p^2 = |\mathbf{p}^2| = p_x^2 + p_y^2 + p_z^2$. Eq. (10-62) is easily separated into $3N$, single dimensional integrals of identical Gaussian integral form. With the exponential integral given by:

$$\int_{-\infty}^{\infty} \exp\,[-x^2\,]dx = \pi^{1/2}$$

We can express Eq. (10-62) as,

$$Q_N^{ig}(\underline{V}, T) = \frac{1}{N!}\left(\frac{2\pi m}{\beta h^2}\right)^{3N/2} \underline{V}^N = \frac{1}{N!}\left(\frac{\underline{V}}{\Lambda^3}\right)^N, \tag{10-63}$$

where

$$\Lambda \equiv \left(\frac{h^2}{2\pi mkT}\right)^{1/2} = \text{thermal deBroglie wavelength} \tag{10-64}$$

The factor $1/\Lambda^{3N}$ appearing in Eq. (10-63) includes the three equivalent contributions to the translational partition function for each dimension and the quantum correction term $1/h^{3N}$.

The $P\underline{V}TN$ equation of state for this ideal gas is obtained directly from Eq. (10-26) and Eq. (10-63):

$$P = kT\left(\frac{\partial N \ln \underline{V}}{\partial \underline{V}}\right)_{T,N} = \frac{NkT}{\underline{V}} \tag{10-65}$$

or

$$P\underline{V} = NkT \tag{10-66}$$

where $kN_A = R$ and the internal energy from Eq. (10-46), using Eq. (10-63), is

$$\underline{U} = kT^2 \left(\frac{\partial \ln Q_N^{ig}}{\partial T} \right)_{\underline{V}, N} = \frac{3}{2} NkT \tag{10-67}$$

Now we can express the classical equipartition result for the translational contribution to the heat capacity,

$$\underline{C}_v^{ig} = \underline{C}_v^o = \left(\frac{\partial \underline{U}}{\partial T} \right)_{\underline{V}, N} = \frac{3}{2} Nk$$

which gives the expected result of $C_v^o = 3/2\, R$ for an ideal monatomic gas (e.g., argon at moderate pressures). The absolute entropy is given by

$$\underline{S} = \frac{\underline{U} - \underline{A}}{T} = \frac{3}{2} Nk + Nk \ln\left(\frac{\underline{V}}{\Lambda^3} \right) - k \ln \underline{N}! \tag{10-68}$$

Using Stirling's approximation[1], $\ln N! \cong N \ln N - N$, thus $\underline{S}$ becomes:

$$\underline{S} = \frac{5}{2} Nk + Nk \ln\left(\frac{\underline{V}}{N\Lambda^3} \right) \tag{10-69}$$

The ideal gas EOS can also be derived directly from the grand partition function Ξ; which for a pure component (i) ideal gas is:

$$\Xi(T, \underline{V}, \mu_i) = \sum_{N=0}^{\infty} \frac{\lambda_i^N}{N!} \left(\frac{\underline{V}}{\Lambda^3} \right)^N = \exp\left(\frac{\lambda_i \underline{V}}{\Lambda^3} \right) \tag{10-70}$$

where we have employed the infinite series definition of the exponential function. Note that the sum runs from $N = 0$ to ∞ because we consider that the number of molecules in the grand canonical ensemble can be arbitrarily large. With Eq. (10-57),

$$P\underline{V} = kT \ln\left(\exp\left(\frac{\lambda_i \underline{V}}{\Lambda^3} \right) \right) = \frac{kT\lambda_i \underline{V}}{\Lambda^3} \tag{10-71}$$

$$\langle N \rangle = \frac{\lambda_i \underline{V}}{\Lambda^3} \tag{10-72}$$

Thus,

$$P\underline{V} = \langle N \rangle kT = NkT \tag{10-73}$$

---

1 Stirling's approximation can easily be derived by noting that

$$\ln N! = \sum_{i=1}^{N} \ln i$$

and then replacing the summation with the integral of $\ln x\, dx$ from $x = 1$ to N. For $N > 100$, the approximation is accurate with less than 1% error.

10.1.8 Factoring the Partition Function with the Semi-Classical Approximation. The treatment in Section 10.1.7 for an ideal gas assumed no intermolecular forces and only three translational degrees of freedom for kinetic energy effects. Strictly speaking, these restrictions require structureless, non-interacting particles or monatomic species. More complex molecules (diatomic, triatomic, etc.) have internal structure and additional degrees of freedom, e.g. rotational, vibrational, electronic, etc. that affect the total energy and thus the Hamiltonian and the magnitude of the partition function. In order to make problems in statistical mechanics tractable for polyatomic species, the partition function or Hamiltonian is first decomposed into separate contributions, one involving only the center-of-mass translational degrees of freedom and the other the internal or intramolecular degrees of freedom (rotation, vibration, etc.). Thus, for the canonical partition function

$$Q_N = Q_{int}\,(T, N, structure)\; Q_{cm}\,(T, \underline{V}, N) \tag{10-74}$$

Q_{cm} includes the translational contribution, the quantum and indistinguishability factors, and the contribution due to intermolecular forces that cause the potential energy to be finite. Note that only Q_{cm} depends on the volume or density of the system. Q_{int} depends on internal motions and is *assumed* to be independent of density--an approximation that will collapse at very high densities such as found in liquids or solids.

The next step is to assume that Q_{cm} is classical. As discussed earlier in Section 10.1.3, this is reasonable for $T > 50$ K. With these approximations we can factor the partition function using Eq. (10-20) with a further separation of the center-of-mass Hamiltonian into two components: a KE term ($\Sigma\, \mathbf{p}_i^2/2m_i$) and a PE term ($\Phi(\mathbf{r}^N)$) to give:

$$Q_N = Q_{int}\,(T, N, structure)\left[\frac{1}{N!\Lambda^{3N}}\int\ldots\int \exp\left[-\frac{\Phi(\underset{\sim}{\mathbf{r}}^N)}{kT}\right] d\underset{\sim}{\mathbf{r}}^N\right] \tag{10-75}$$

where the 3-N dimensional integral is often referred to as the ***configuration integral*** ($\mathbf{Z}^*$) as it integrates over configuration space:

$$\boxed{\mathbf{Z}^* \equiv \int\ldots\int \exp\left[-\frac{\Phi(\underset{\sim}{\mathbf{r}}^N)}{kT}\right] d\underset{\sim}{\mathbf{r}}^N} \tag{10-76}$$

Because all the density dependence of Q_N is contained in $\mathbf{Z}^*$, the pressure EOS follows from Eq. (10-26) as:

$$P = kT\left(\frac{\partial \ln \mathbf{Z}^*}{\partial \underline{V}}\right)_{T,N} \tag{10-77}$$

Therefore, the problem of obtaining an EOS for a non-ideal gas comes down to evaluating the configuration integral $\mathbf{Z}^*$ which depends explicitly on characterizing the total potential energy Φ. In general, Φ is a function of the intermolecular (or interparticle) spacing and the relative orientation of molecules with finite geometric structure. In very dense fluids, internal rotational and vibrational effects may also be important but, as stated earlier, these have been neglected so far. To be more rigorous we need to include orientation effects. Assuming each molecule rotates as a rigid body, we can modify Eq. (10-76) to,

$$\mathbf{Z}^* = \frac{1}{4\pi}\int \ldots \int \exp\left[-\frac{\Phi(\mathbf{r}^N, \underset{\sim}{\omega}_i^N)}{kT} \right] d\mathbf{r}_1 \ldots d\mathbf{r}_N \, d\omega_1 \ldots d\omega_N \tag{10-78}$$

where ω_i is the Euler orientation angle vector of molecule i. Normally, three Euler angles (α, β, γ) are sufficient to define ω_i for a non-linear rigid polyatomic molecule. The internal partition function for N molecules can easily be related to the internal partition function for a single molecule. For a single component system, $Q_{int} = (q_{int})^N$ and q_{int} receives contributions from:

$$q_{int} = q_r \, q_v \, q_{ir} \, q_e \, q_{ns} \, q_{corr} \tag{10-79}$$

where:

q_r = free rotation of the entire molecule
q_v = internal vibrations
q_{ir} = internal rotation of molecular groups
q_e = electronic motions
q_{ns} = nuclear spin

Note that q_{corr} corrects the earlier approximation for vibration and rotation coupling. The number of vibrational modes of a polyatomic molecule, consisting of n atoms $(n \geq 2)$, can be found from the fact that $3n$ degrees of freedom (coordinates) are needed to fully specify the molecule's location, orientation and structure. Within these,

- 3 coordinates are required to define translation of mass,
- 2 angle coordinates (Euler angles) are needed to define orientation for linear molecules, and 3 angle coordinates are needed to define orientation for non-linear molecules,
- r internal rotational modes can be specified

Thus, we are left with $(3n - r - 6)$ vibrational modes for non-linear molecules or $(3n - r - 5)$ vibrational modes for linear molecules.

As stated earlier, the translational modes are treated classically for $T > 50$ K. The classical or quantum mechanical contribution of the internal modes are typically treated as follows:

Free rotation: Energy separation is small. In almost all applications, the classical limit (equipartition) assumption is satisfactory.

Vibration: With large frequencies, a significant separation between vibrational energy levels exists:

$$\left[\frac{h\omega}{k} = \theta v \approx (of\ order)\ 10^3\ K\right]$$

and vibrations must be treated quantum mechanically (see for example, the Einstein-Debye treatment of heat capacity given in Sections 8.4 and 8.6).

Internal rotation: The free rotation contribution only accounts for rotation of the bulk molecule, and as such is treated essentially as a rigid body. Polyatomic molecules can have internal bond rotations that need to be accounted for as well. This is particularly important when asymmetric configurations exist around single bonds, such as around C-C bonds in long-chain polymers.

Electronic contributions: With an energy gap $\Delta E_e \approx 10^{-17}$ J, almost all molecules are in the ground state at normal temperatures $T < 10^3$ K, thus q_e is of order 1, and does not affect properties. In exceptional cases, a low-lying energy level may contribute to q_e (e.g., for NO_2).

Nuclear spin contributions: Separation of levels is extremely large ($\Delta E_{ns} = 10^{-11}$ J). Practically all molecules are in the nuclear ground state, thus q_{ns} is of order 1 and there is no effect on properties. The isotopes of molecular hydrogen are an important exception.

10.2 Intermolecular Forces

In general, intermolecular forces are caused by electrostatic or electrodynamic interactions of one type or another. Of course, charged ions separated by a fixed distance give rise to electrostatic forces, less apparent is the fact that electron clouds of neutral molecules themselves dynamically induce dipoles leading to so-called London dispersion forces. Frequently these effects are captured macroscopically by empirical parameters, for example, by the a and b terms in the van der Waals equation or they are treated on a molecular level using an empirical intermolecular potential energy function. In general, an intermolecular force $\mathbf{F}_{ij}$ is related to its potential energy as a gradient derivative:

$$\mathbf{F}_{ij} = -\nabla\Phi(\mathbf{r}_i, \mathbf{r}_j, \alpha_i, \alpha_j, \gamma_i, \gamma_j, \ldots) \tag{10-80}$$

where $\mathbf{r}_i$, $\mathbf{r}_j$ represent position vectors and α_i, α_j, γ_i, γ_j, ... orientations. If the force is purely electrostatic in nature, then between two charged particles or ions i and j, the intermolecular potential energy contribution is,

$$\Phi_{ij}^{electrostatic} = \frac{z_i z_j e^2}{4\pi\varepsilon_o D_s r} = \frac{z_i z_j e^2}{4\pi\varepsilon r}$$

where z_i, z_j are the charges, e is the charge of a single electron, $r = r_{ij} = |\mathbf{r}_i - \mathbf{r}_j|$ is the separation distance, D_s is the dielectric constant (≥ 1; for a vacuum $D_s = 1$), and ε_o = 4.802×10^{-10} (erg-cm)$^{1/2}$ is the permittivity in a vacuum. In general, $\varepsilon = \varepsilon_o D_s$. Electrostatic forces from Eq. (10-80) are proportional to r^{-2} and as such they have a relatively long distance range where they are significant. Electrostatic forces can also exist in molecules with no net electric charge. For example, the presence of neutral molecules with permanent dipoles caused by internal charge separation leads to net electrostatic effects when averaging over all orientations within a Boltzmann distribution. These dipole-dipole forces are expressed in terms of the dipole moment of each molecule $d_{m,i} = el_i$, the center of charge separation distance r between two molecules, i and j, and four orientation angles, α_i, α_j, γ_i, and γ_i. The geometric arrangement is as follows:

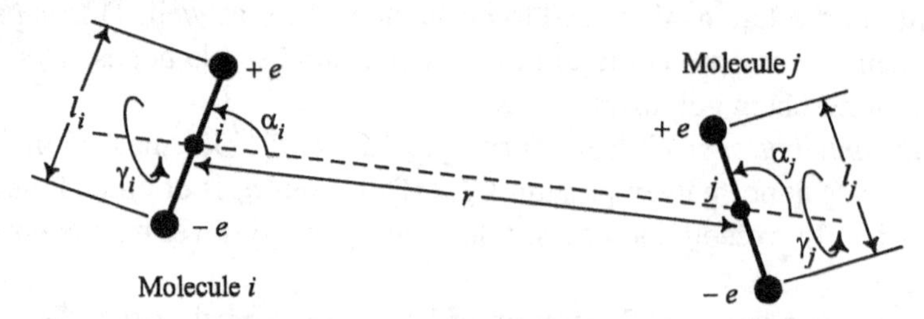

with the potential energy induced by the dipole-dipole interaction given by:

$$\Phi_{ij}^{dipole-dipole} = \frac{-d_{m,i}d_{m,j}}{4\pi\varepsilon_o r^3}\left[2\cos\alpha_i\cos\alpha_j - \sin\alpha_i\sin\alpha_j\cos(\gamma_i-\gamma_j)\right] \quad (10\text{-}81)$$

For a pair of dipolar molecules there are two opposing simultaneous effects occurring: (1) polarity causes an electric field that tends to align the dipoles and (2) thermal kinetic energy tends to randomize orientations. A Boltzmann average can be used to estimate the net effect of the dipole moments in a state of thermal equilibrium at a fixed separation distance r. Integrating over all orientations represented by solid angles $d\omega_i\, d\omega_j = \sin\alpha_i \sin\alpha_j\, d\alpha_i\, d\alpha_j\, d\gamma_i\, d\gamma_j$.

$$<\Phi_{ij}^{dipole-dipole}> = \frac{\int\ldots\int\Phi_{ij}\exp[-\Phi_{ij}/kT]d\omega_i d\omega_j}{\int\ldots\int\exp[-\Phi_{ij}/kT]d\omega_i d\omega_j} \quad (10\text{-}82)$$

$$<\Phi_{ij}^{dipole-dipole}> = -2/3\frac{d_{m,i}^2 d_{m,j}^2}{(4\pi\varepsilon_o)^2 r^6 kT} + \ldots \quad (10\text{-}83)$$

Note that the net effect is an attractive (negative) potential energy with an inverse 6th power dependence on separation distance r and a 4th power dependence on the dipole moments $d_{m,i}$ and $d_{m,j}$. Similar calculations can be made for molecules having quadrupole moments, for example CO_2. The quadrupole moment, Q_i, is usually represented as a product between the charge separation q_i and the distance l_i squared. Thus, $Q_i = q_i l_i^2$.

Because the classical contribution to the partition function is embodied in the configuration integral $\mathbf{Z}^*$ which depends directly on $\Phi(\mathbf{r}^N)$, we need to have a potential energy function that has explicit dependencies on position coordinates, e.g., separation distance r. Figure 10.2 shows how the intermolecular potential energy between a pair of neutral molecules may depend on r. Note there is a strongly repulsive $\Phi_{ij}(r) >> 0$ region as $r < \sigma$ and a region of attraction for $\sigma < r < 3r_o$ that asymptotically decreases to $\Phi_{ij}(r) = 0$ at large r.

The pair potential energy function $\Phi_{ij}(r)$ contains contributions from three major types of intermolecular forces (see Reed and Gubbins (1973) for details):

Short range **($r \leq \sigma$): "Valence or chemical" energies resulting from repulsive interactions between overlapping electron clouds.**

Intermediate range **($\sigma < r \leq r_o$): "Residual valence" interactions (hydrogen bonding, charge-transfer complexes, and forces characteristic of associated molecules).**

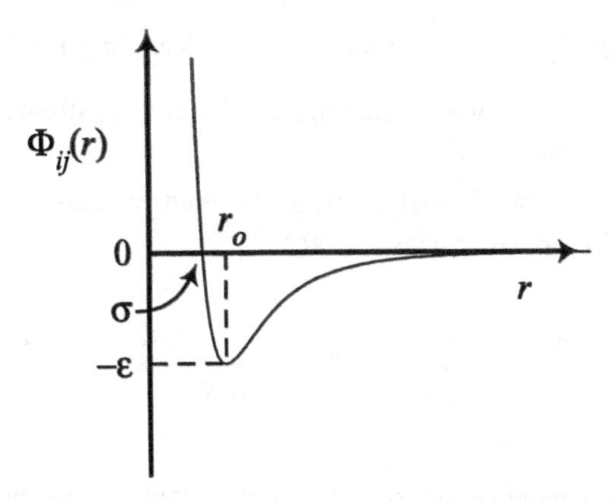

Figure 10.2 General characteristics of intermolecular potential energy function.

Long range **($r \geq r_o$):** There are four main long range effects:

(1) ***Fluctuation or dispersion energies*** (London forces): Attractive forces of quantum mechanical origin, caused by rapid fluctuations in the electronic charge distributions around atoms.

(2) ***Polarization or induction energies***: Resulting from polarizability, or the tendency to distort an electric charge distribution in an electric field of one molecule with the multiple (dipole, quadrupole, ...) moments of another molecule.

(3) ***Direct electrostatic energies:*** Caused by electrostatic forces associated with dipole moments, quadrupole moments and the higher electric moments characterizing the charge distributions within each molecule. They are estimated directly from classical electrostatic theory.

(4) ***Coulombic energies***: These forces are associated with permanently charged ionic species.

Potential energy contributions are categorized according to the type of interaction in Table 10.1 with the Boltzmann-averaged constitutive equation for $\Phi_{ij}(r)$ given. Table 10.2 provides estimates of the percent of total attractive potential energy that can be assigned to various interactions between like pairs of molecules. For only those polar molecules where the dipole moment is greater than about 1.0 do electrostatic and induction effects contribute significantly to the total attractive potential energy.

10.3 Intermolecular Potential Energy Functions

To apply statistical mechanics to model real fluid systems, one must estimate, either directly or indirectly, the configuration integral. To do this, the total potential energy Φ for all possible configurations of the system must be determined. This means we have to understand how to formulate the physics of intermolecular or interparticle interactions. Our knowledge of these interactions comes from:

(1) ***quantum theory*** (*ab initio* calculations using the Schrödinger wave equation).

(2) ***measurements*** (second virial coefficient, viscosity, diffusivity, spectroscopy, molecular-beam scattering).

The total potential (or "configurational") energy can be decomposed into contributions from binaries, ternaries, ..., etc.

$$\Phi(\underset{\sim}{\mathbf{r}}^N) = \sum_{i<j} \Phi_{ij} + \sum_{i<j<k} \Delta\Phi_{ijk} + \sum_{i<j<k<l} \Delta\Phi_{ijkl} + \ldots + \Delta\Phi_{123\ldots N} \quad (10\text{-}84)$$

where

Φ_{ij} = mutual potential energy of a pair of molecules (or, in general, centers of mass) located at positions defined by $\mathbf{r}_i$ and $\mathbf{r}_j$ and isolated from the influence of all other molecules.

$\Delta\Phi_{ijk}$ = additional mutual potential energy of the ternary group of molecules located at $\mathbf{r}_i$, $\mathbf{r}_j$, and $\mathbf{r}_k$ that arises from the changes in the electron charge distributions (wave functions) of the isolated pair located at i and j when a third molecule is in close proximity at location k.

$\Delta\Phi_{ijkl}$ = additional contribution to Φ for a quaternary group of molecules, and

$\Delta\Phi_{123\ldots N}$ = potential energy residual, characteristic of the whole system, that is not included in the previous terms.

The successive terms in Eq. (10-84) decrease in magnitude:

$$\sum \Phi_{ij} > \sum \Delta\Phi_{ijk} > \sum \Delta\Phi_{ijkl} > \ldots > \Delta\Phi_{123\ldots N} \quad (10\text{-}85)$$

The initial *pairwise additivity* assumption amounts to retaining only the first term in Eq. (10-84). Essentially, we are saying that:

$$\Phi(\underset{\sim}{\mathbf{r}}^N) = \Phi \cong \sum_{pairs} \Phi_{ij} = \sum_{i<j} \Phi_{ij} = \tfrac{1}{2} \sum_{i=1}^{N} \sum_{j=1}^{N} \Phi_{ij} \quad (10\text{-}86)$$

Implicitly, pairwise additivity assumes that $\Phi_{ij} = \Phi_{ji}$ and that the total potential energy associated with the ith particle is $2\Phi_{ij}$.

Table 10.1 Summary of Attractive Intermolecular Potential Contributions

Interaction Type	$\Phi_{ij}(r)$ Constitutive Equation (Boltzmann-averaged)
Electrostatic	
ion-ion (coulombic $i = +, j = -$)	$\dfrac{z_i z_j e^2}{4\pi\varepsilon r}$
dipole-dipole	$\dfrac{-2}{3}\dfrac{d_{m,i}^2 d_{m,j}^2}{(4\pi\varepsilon_o)^2 kTr^6} + \cdots$
quadrupole-dipole	$-\dfrac{\left(Q_i^2 d_{m,j}^2 + Q_j^2 d_{m,i}^2\right)}{(4\pi\varepsilon_o)^2 kTr^8} + \ldots$
quadrupole-quadrupole	$\dfrac{-7}{40}\dfrac{Q_i^2 Q_j^2}{(4\pi\varepsilon_o)^2 kTr^{10}}$
Electrostatic-Electrodynamic (polarization)	
dipole-induced dipole	$-\dfrac{(d_{m,i}^2\alpha_j + d_{m,j}^2\alpha_i)}{4\pi\varepsilon_o r^6}$
quadrupole-induced dipole	$-\dfrac{(Q_i^2\alpha_j + Q_j^2\alpha_i)}{4\pi\varepsilon_o r^8}$ (small and usually neglected)
Electrodynamic (fluctuation or London forces)	
induced dipole-induced dipole	$\dfrac{-3}{2}\dfrac{I_i I_j \alpha_i \alpha_j}{(I_i + I_j)r^6}$
induced dipole-induced quadrupole	$\dfrac{-45}{8}\dfrac{4\pi\varepsilon_o}{e^2}\dfrac{I_i I_j \alpha_i \alpha_j}{r^8}\left[\dfrac{\alpha_i I_i}{2I_i + I_j} + \dfrac{\alpha_j I_j}{I_i + 2I_j}\right]$

Note: all Φ_{ij} contributions are in medium of permittivity ε_o except ion-ion

Q_i = quadrupole moment, (coulomb) m^2
α_i = polarizability, m^3
$d_{m,i}$ = dipole moment, (coulomb) m
I_i = ionization potential, J
$r = r_{ij} = |\mathbf{r}_i - \mathbf{r}_j|$ = separation distance, m

$\varepsilon = \varepsilon_o D_s$ = permittivity, (coulomb)^2/Jm
$\varepsilon_o = 8.854 \times 10^{-12}$ (coulomb)^2/Jm
D_s = Dielectric strength
e = electron charge, coulomb

Table 10.2 Relative Contributions to the Total Attractive Intermolecular Potential Energy at 0°C [adapted from Syrkin and Dyatkina (1950) and London (1937)]

Molecule	$d_{m,i}$-dipole moment (debyes)	Percent Contribution to Attractive PE $d_{m,i}-d_{m,i}$ (electrostatic)	$d_{m,i}-ind\ d_{m,i}$ (induction)	$ind\ d_{m,i}-ind\ d_{m,i}$ (dispersion)
CCl_4	0	0	0	100
CO	0.10	<1	<1	>99
HI	0.42	0.1	0.5	99.4
HBr	0.80	3.3	2.2	94.5
HCl	1.08	17.6	4.4	78.0
NH_3	1.47	50.7	6.0	43.3
H_2O	1.84	80.6	4.3	15.1
CH_3OH	1.70	63.4	14.4	22.2
C_2H_5OH	1.70	55.4	12.6	32.4
$(CH_3)_2CO$	2.87	67.0	5.8	27.2

$d_{m,i}-d_{m,i}$ = dipole-dipole   1 debye = 1D = 3.336×10^{-30} coulomb – m
$d_{m,i}-ind\ d_{m,i}$ = dipole-induced dipole
$ind\ d_{m,i}-ind\ d_{m,i}$ = induced dipole-induced dipole (London dispersion)

Pairwise additivity is very satisfactory in dilute (low density) systems, because the probability that three or more molecules are close enough together to interact is negligibly small compared to the magnitude of the binary interaction between two adjacent molecules. However, this assumption leads to errors at higher densities. For liquid argon at the triple point, non-additive contributions to the internal energy are 5 to 10%, according to the calculations of Margenau and Kestner (1969). Readers can refer to Reed and Gubbins (1973) and McQuarrie (1976) for further discussion of pairwise additivity.

There are many possible choices for intermolecular potential energy functions. Although there is some theoretical justification for selecting a r^{-6} dependence for the attractive contribution, the repulsive contribution is not as readily characterized. Qualitatively, as molecules approach each other, strong repulsions will occur when negatively charged electron clouds overlap. Typically, these repulsions are characterized in terms of an empirical function with a high order dependence on r, for example:

$$\Phi_{ij}^{repulsion}(r) = B\exp[-Ar] \tag{10-87}$$

or

$$\Phi_{ij}^{repulsion}(r) = B/r^n \quad n = 9 \text{ to } 100 \tag{10-88}$$

Repulsive contributions become very important when $r \leq \sigma$ as seen in Figure 10.2; frequently, the magnitude of σ is approximated by the so-called van der Waals radii of the molecular pair (ij).

The hard-sphere, square-well, Lennard-Jones 6-12 and Kihara potentials are somewhat simplistic involving only one to three adjustable parameters as shown in Table 10.3; but nonetheless they have stood the test of time and are used widely. In many instances, these simple forms seem to capture important physical interactions by giving rather accurate predictions of specific properties. However, we must be cautious not to give too much physical significance to the values of the parameters such as σ and ε as they are frequently fit to actual data, e.g., second virial coefficient or viscosity, rather than obtained directly from molecular beam scattering measurements or from *ab initio* calculations. Thus, values of σ and ε may not be independently determined causing them to be non-unique and correlated.

In polyatomic molecules, the structural distribution of atoms gives rise to assymmetries in the potential so that orientation and angular effects need to be considered. One approach commonly used to capture these effects is to assign potential parameters for each atom i in molecule k and atom j in molecule m and to sum the potential contributions over all atom-to-atom interactions. This site-to-site approach is particularly convenient for computerized molecular simulation methods which are discussed later in Section 10.9. The basic idea is illustrated below for a 2-constant intermolecular potential Φ_{ij} for all binary pairs i-j

$$\Phi(\mathbf{r}^N) = \tfrac{1}{2} \sum_j \sum_i \Phi_{ij}(r_{ij}, \sigma_{ij}, \varepsilon_{ij})$$

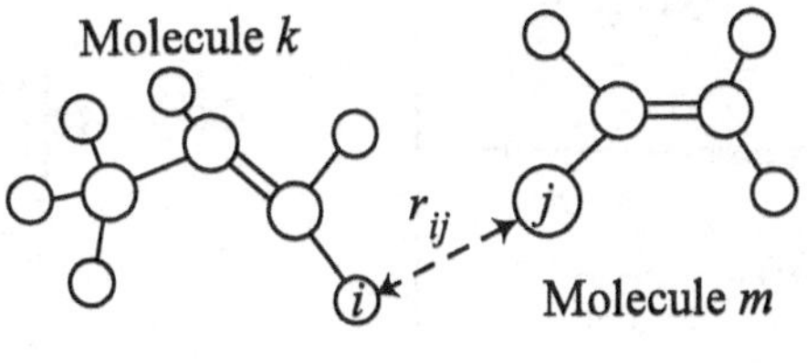

Molecule k

r_{ij} i j

Molecule m

where each summation refers to atoms in each molecule (k and m) that are interacting as a set of binary pairs of constituent atoms. If atoms i and j are different, then mixing rules are needed to calculate σ_{ij} and ε_{ij}, if they are the same then pure component σ_{ii} and ε_{ii} values can be used. Note that r_{ij} defines the center-of-mass distance between atom i and j.

Table 10.3 Commonly Used Intermolecular Potential Functions

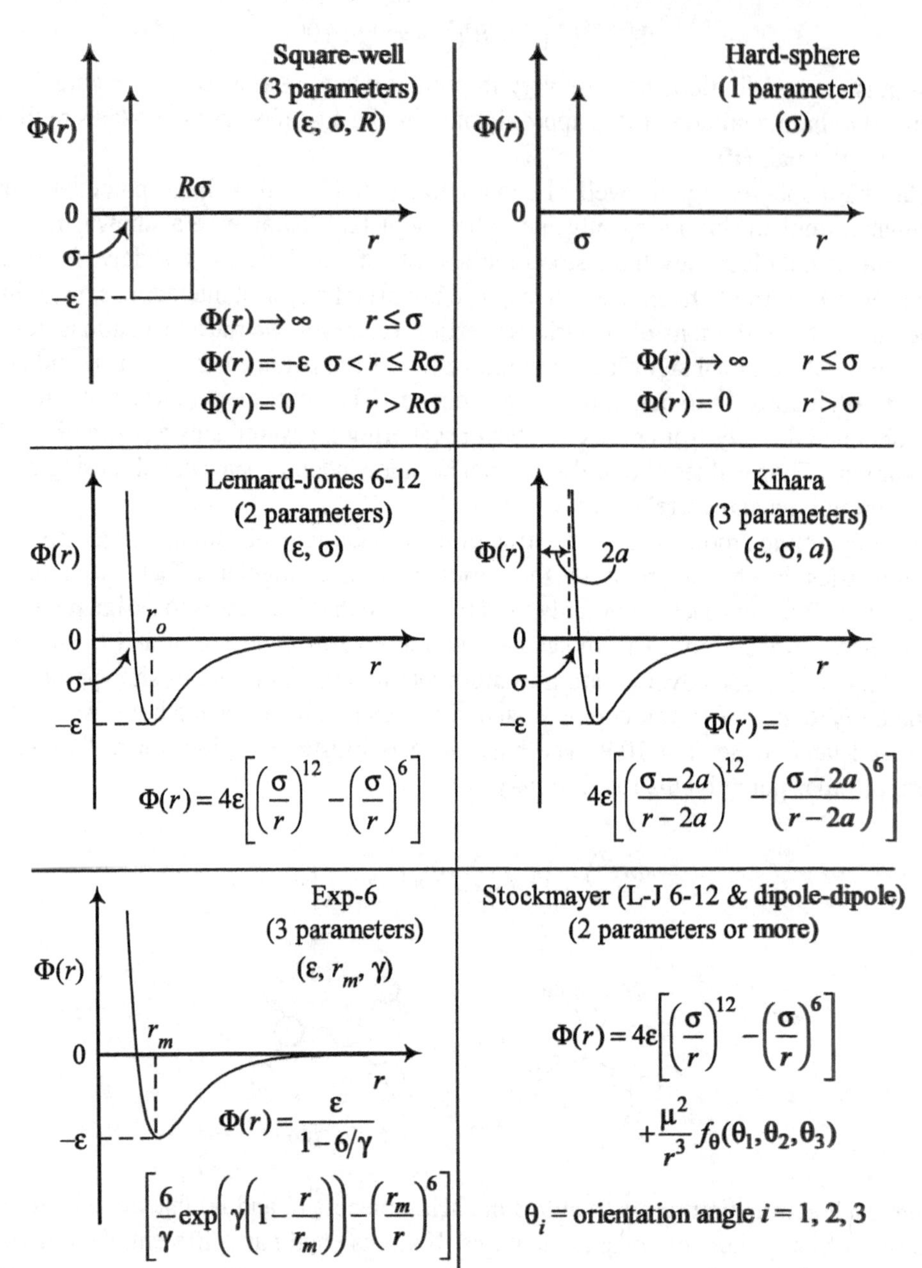

10.4 The Virial Equation of State

Recalling the classical form of the factorized canonical partition function from Section 10.1.8 [see Eq. (10-75)ff]

$$Q_N = Q(N, \underline{V}, T) = \left[\frac{1}{N!}\left(\frac{2\pi\, m\, k\, T}{h^2}\right)^{\frac{3N}{2}}\right](q_{int})^N \mathbf{Z}^*$$

or

$$Q_N = \frac{1}{N!\,\Lambda^{3N}}(q_{int})^N \mathbf{Z}^* \tag{10-89}$$

As the density goes to zero ($V \to \infty$, $\rho \to 0$), $\Phi = 0$ and $\mathbf{Z}^* \to \underline{V}^N$ and the ideal gas equation results directly from

$$P = -\left(\frac{\partial A}{\partial \underline{V}}\right)_{T,N} = kT\left(\frac{\partial \ln \mathbf{Z}^*}{\partial \underline{V}}\right)_{T,N} \tag{10-90}$$

As the density increases, intermolecular interactions become important, $\Phi \neq 0$; $\mathbf{Z}^*$ no longer equals $\underline{V}^N$; and deviations from ideal gas behavior are observed. Many equations of state (EOSs) have been formulated to describe non-ideal behavior. The most theoretically based of these is the virial equation which in density form is written as

$$\frac{P}{kT} = \rho + B(T)\,\rho^2 + C(T)\,\rho^3 + D(T)\,\rho^4 + \ldots \tag{10-91}$$

where $\rho = N/\underline{V} = 1/V$. The second virial coefficient $B(T)$ will only depend on *i-j* binary interactions, the third $C(T)$ also depends on ternary *i-j-k* interactions, and so forth. This is proved most directly using the grand canonical partition function [for example, see McQuarrie (1976) or Davidson (1962)]. Alternative approaches usually employ complex formulations involving cluster integrals [see Mayer and Mayer (1940) and Hirschfelder *et al.* (1954)].

The grand canonical partition function is given in quantum mechanical form as

$$\Xi = \Xi(\underline{V}, T, \mu) = \sum_{N=0}^{\infty} Q(N, \underline{V}, T)\lambda^N \tag{10-92}$$

with the canonical partition function Q in quantum mechanical form:

$$\Xi(\underline{V}, T, \mu) = \sum_N \sum_j \exp\left[-E_{N_j}/kT\right] \exp\left[\mu N/kT\right] \tag{10-93}$$

where λ is the absolute activity, $\lambda \equiv \exp[\mu/kT]$, and the sums run over energy states j and particles N. Note also that $\mu = \mu_i$ for a single component system. Because

$$Q(N = 0, \underline{V}, T) = 1 \tag{10-94}$$

$\Xi(\underline{V}, T, \mu)$ can be rewritten with the following abbreviations, $Q(N, \underline{V}, T) = Q_N(\underline{V}, T) = Q_N$, as

$$\Xi\,(\underline{V}, T, \mu) = 1 + \sum_{N=1}^{\infty} Q_N(\underline{V}, T)\, \lambda^N = 1 + \sum_{N=1}^{\infty} Q_N \lambda^N \tag{10-95}$$

Because from Eqs. (10-57) and (10-58), using Eq. (10-29),

$$P\underline{V} = kT \ln \Xi \quad \text{and} \quad N = kT(\partial \ln \Xi/\partial\mu)_{\underline{V},T} = \lambda(\partial \ln \Xi/\partial\lambda)_{\underline{V},T} \tag{10-96}$$

we can express P and $\rho = N/\underline{V}$ in terms of T and $ln\ \Xi$ to eliminate Ξ and get a more useful EOS form. The most obvious choice for expanding the expression for $ln\ \Xi$ is in terms of λ since we already have Ξ expressed as a polynomial sum in terms of the absolute activity λ. Thus, from Eq. (10-95)

$$\ln \Xi \;=\; \ln\left(1 + \sum_{N=1}^{\infty} Q_N \lambda^N\right) \tag{10-97}$$

Before proceeding further, it is convenient to define a new activity ρ_o that is proportional to λ and equals the density as $\rho \to 0$ (or $\lambda \to 0$). Thus, $\rho_o = \rho$ in the domain $\lambda \to 0$. We can take the limit of Eq. (10-96) as $\lambda \to 0$ to represent the average number of particles in the ensemble:

$$\lim_{\lambda \to 0} N = \lambda \left(\frac{\partial \ln \Xi}{\partial \lambda}\right)_{\underline{V},T} \tag{10-98}$$

where $\ln \Xi$ is given by Eq. (10-97) and can be approximated for small values of λ as

$$\ln \Xi \approx \sum_{N=1}^{\infty} Q_N \lambda^N - \frac{\left(\sum_{N=1}^{\infty} Q_N \lambda^N\right)^2}{2} + \ldots \tag{10-99}$$

As $\lambda \to 0$,

$$\ln \Xi \approx \lambda\, Q_1$$

Thus as $\lambda \to 0$, $\rho \to \lambda Q_1/\underline{V} = N/\underline{V}$, the particle number density. Now we can define

$$\rho_o \equiv \rho_{\lambda \to 0} \equiv \lambda Q_1/\underline{V} \tag{10-100}$$

and we can rewrite Eq. (10-95):

$$\Xi\,(\underline{V}, T, \mu) = 1 + \sum_{N=1}^{\infty} \frac{Q_N \underline{V}^N \rho_o^N}{Q_1^N} \tag{10-101}$$

Equation (10-101) expresses Ξ as a power series in ρ_o (or density as $\lambda \to 0$). We can now define another term Z_N^* analogous to the configurational integral as

$$Z_N^* \equiv N! \left(\frac{\underline{V}}{Q_1} \right)^N Q_N \tag{10-102}$$

Thus,

$$\Xi = 1 + \sum_{N=1}^{\infty} \frac{Z_N^*}{N!} \rho_o^N \tag{10-103}$$

The next key assumption requires expressing the pressure P as a power series in ρ_o:

$$P = kT \sum_{j=1}^{\infty} b_j \rho_o^j \tag{10-104}$$

and using the relationship between pressure and ln Ξ (Eq. (10-96)) to get:

$$\Xi = \exp\,[P\underline{V}/kT] = \exp\left[\underline{V} \sum_{j=1}^{\infty} b_j \rho_o^j \right] \tag{10-105}$$

where the exponential in Eq. (10-105) can be expanded in a power series as $\exp\,[x] = 1 + x/1! + x^2/2! + x^3/3! + \ldots$. By collecting terms in like powers of ρ_o and equating them to corresponding $Z_N^*/N!$ terms from Eq. (10-103), we can easily show that:

$$b_1 = \frac{1}{1!\,\underline{V}} Z_1^* = 1$$

$$b_2 = \frac{1}{2!\,\underline{V}} (Z_2^* - Z_1^{*2}) \tag{10-106}$$

$$b_3 = \frac{1}{3!\,\underline{V}} (Z_3^* - 3Z_2^* Z_1^* + 2Z_1^{*3})$$

Now we need to relate ρ_o to $\rho = N/\underline{V}$, the true particle or molecular density. From Eqs. (10-96) and (10-100):

$$\rho = N/\underline{V} = \frac{\lambda}{\underline{V}} \left(\frac{\partial \ln \Xi}{\partial \lambda} \right)_{\underline{V},T} = \frac{\rho_o}{kT} \left(\frac{\partial P}{\partial \rho_o} \right)_{\underline{V},T} \tag{10-107}$$

Using Eq. (10-104) for P we get:

$$\rho = \sum_{j=1}^{\infty} j\, b_j \rho_o^j \tag{10-108}$$

We can eliminate ρ_o via the expansion:

$$\rho_o = a_1 \rho + a_2 \rho^2 + a_3 \rho^3 + \ldots$$

by substituting this expression into Eq. (10-108) for ρ_o and equating like powers of ρ to obtain: $a_1 = 1$, $a_2 = -2\, b_2$, $a_3 = -3\, b_3 + 8\, b_2^2$, and so forth. Thus,

$$\frac{P}{kT} = \rho + B(T)\rho^2 + C(T)\rho^3 + \ldots \tag{10-109}$$

where

$$B(T) = -b_2 = -\frac{1}{2!\,\underline{V}}[Z_2^* - Z_1^{*2}]$$

$$C(T) = -4b_2^2 - 2b_3 = \frac{-1}{3!\,\underline{V}^2}\left[\underline{V}(Z_3^* - 3Z_2^* Z_1^* + 2Z_1^{*3}) - 3(Z_2^* - Z_1^{*2})^2\right]$$

If we use the classical limit to express the partition functions Q_i:

$$\text{for } N = 1 \quad : Q_1 = Q_1(\underline{V}, T) = \left(\frac{2\pi mkT}{h^2}\right)^{3/2} \underline{V} q_{int} = \frac{\underline{V}}{\Lambda^3} q_{int}$$

$$\text{for } N > 1 \quad : Q_N = Q(N, \underline{V}, T) = \frac{Z_N^*}{N!\,\Lambda^{3N}} q_{int}^N \tag{10-110}$$

But according to our earlier analyses

$$Q_N = \frac{Z_N^*}{N!}(Q_1/\underline{V})^N = \frac{1}{N!}\frac{1}{\Lambda^{3N}} Z_N^* q_{int}^N \tag{10-111}$$

Thus, in the classical limit $Z_N^* = \mathbf{Z}^*$ = configurational integral, and

$$Z_1^* = \int d\mathbf{r}_1 = \underline{V}$$

$$Z_2^* = \int\int e^{-\Phi_2/kT} d\mathbf{r}_1\, d\mathbf{r}_2 \tag{10-112}$$

$$Z_3^* = \int\int\int e^{-\Phi_3/kT} d\mathbf{r}_1\, d\mathbf{r}_2\, d\mathbf{r}_3$$

etc.

where

$$\Phi_2 = \Phi_2(\mathbf{r}_1, \mathbf{r}_2)$$

$$\Phi_3 = \Phi_3(\mathbf{r}_1, \mathbf{r}_2, \mathbf{r}_3) = \Phi_2(\mathbf{r}_1, \mathbf{r}_2) + \Phi_2(\mathbf{r}_2, \mathbf{r}_3) + \Phi_2(\mathbf{r}_3, \mathbf{r}_1) + \Delta\Phi_3(\mathbf{r}_1, \mathbf{r}_2, \mathbf{r}_3)$$

Now we can rewrite $B(T)$ with $\beta \equiv 1/kT$ as

$$B(T) = -\frac{1}{2\underline{V}}\left(Z_2^* - Z_1^{*2}\right) = -\frac{1}{2\underline{V}}\int\int\left[\exp[-\beta\Phi_2] - 1\right] d\mathbf{r}_1\, d\mathbf{r}_2 \tag{10-113}$$

which for a spherically symmetric case $[\Phi_2 = \Phi(|\mathbf{r}_2 - \mathbf{r}_1|) = \Phi(r_{12})]$ simplifies to

$$B(T) = -\frac{1}{2\underline{V}}\underbrace{\int d\mathbf{r}_1}_{\underline{V}} \int\left[\exp[-\beta\Phi(r_{12})] - 1\right] \underbrace{dr_{12}}_{4\pi r^2 dr} = -2\pi\int_o^\infty [\exp[-\beta\Phi(r)] - 1]\, r^2 dr$$

or in its more typically presented molar form

$$B(T) = -2\pi N_A \int_0^{\infty} [\exp[-\beta\Phi(r)] - 1] r^2 dr \qquad (10\text{-}114)$$

where N_A = Avogadro's number to put the virial equation on a per mole basis.

In a similar fashion, with some manipulation using the so-called cluster integrals, and restricting ourselves to two-body contributions where pairwise additivity applies and $\Delta\Phi_3 = 0$ [see Eq. (10-86)], $C(T)$ is given by,

$$C(T) = -\frac{1}{3\underline{V}} \int\int\int f_{12} f_{13} f_{23}\, d\mathbf{r}_1\, d\mathbf{r}_2\, d\mathbf{r}_3 \qquad (10\text{-}115)$$

where $f_{ij} \equiv \exp[-\Phi(r_{ij})/kT] - 1$ is the *Mayer f function.*

To summarize, the key points to remember in this section are:

(1) The virial equation of state can be derived rigorously from statistical mechanics and is expressed as

$$P = \rho kT(1 + B(T)\,\rho + C(T)\rho^2 + ...) \qquad (10\text{-}109a)$$

or equivalently in compressibility form as:

$$Z = \frac{P}{\rho kT} = \frac{P\underline{V}}{kT} = 1 + B(T)\rho + C(T)\rho^2 + \dots \qquad (10\text{-}109b)$$

(2) The virial coefficients $B(T)$, $C(T)$, etc. can be calculated with suitable models for intermolecular potential energy as a function of position and orientation.

(3) The second virial coefficient $B(T)$ representing binary ij interactions is rigorously expressed in terms of a pair potential $\Phi_{ij}(r)$.

(4) The third $C(T)$ and higher virial coefficients can be estimated assuming pairwise additivity and by applying specialized mathematical techniques such as cluster integral methods.

In practice, the second and third virial coefficients can be determined from low-density experimental PVT data with,

$$B(T) = \lim_{\rho\to 0}\left(\frac{\partial Z}{\partial \rho}\right)_T \qquad (10\text{-}116a)$$

$$C(T) = \frac{1}{2}\lim_{\rho\to 0}\left(\frac{\partial^2 Z}{\partial \rho^2}\right)_T \qquad (10\text{-}116b)$$

We can rearrange Eq. (10-109b) to give,

$$(Z - 1)/\rho = B(T) + C(T)\rho + \dots \qquad (10\text{-}117)$$

so by plotting $(Z - 1)/\rho$ versus ρ we should obtain a straight line as $\rho \to 0$. Data for pure water taken from the steam tables are presented in Figure 10.3a. Here one sees that the intercept is equal to $B(T)$ and the slope is $C(T)$. It is also possible to estimate the second and higher virial coefficients assuming pairwise additivity and using a suitable intermolecular potential model, for example see Eq. (10-114). Figure 10.3b represents the temperature dependence of the second virial coefficient for the Kihara and Lennard-Jones 6-12 potential models in terms of dimensionless parameters $B^*(T) = B(T)/(2/3\ \pi N_A \sigma^3)$ and $T^* = kT/\varepsilon$ (see Table 10.3). The term $a^* = 2a/(\sigma - 2a)$ represents the hard core component of the Kihara model. With $a^* = 0$ the Lennard-Jones 6-12 potential results. With known values of the potential parameters we can generate $B(T)$ at any temperature. However, this is usually not the situation in that universal values for the parameters, σ, ε, a, etc., do not exist. With experimental values of $B(T)$ as a function of temperature one can fit the potential parameters by a suitable regression to match the data. Unfortunately, the parameter values are correlated to one another as well. For example, for the Lennard-Jones 6-12 potential it is not possible to get a unique set of σ and ε parameters for a particular material. This non-uniqueness property becomes even more evident when other models and sets of data are used to fit potential parameters; for example, molecular level models for viscosity and diffusivity can also be used to generate potential parameters (see Reid *et al.* (1987), Section 9-4, p. 392-397 and Section 11-3, p. 581-586).

More recently, *ab initio* (from first principles) calculations are used to estimate parameters. Although the theoretical basis of these calculations is solid, namely that one is using a full quantum mechanical treatment, there are a number of approximations used to arrive at tractable solutions. So even *ab initio* potential parameters values should be viewed with appropriate skepticism. What happens if the predicted value for a property based on a statistical mechanical model doesn't match the experimental value? Is the theory wrong or are the experiments in error? If the discrepancy is larger than experimental error, most likely the problem is with the intermolecular potential model or with some of the assumptions used, for example the assumption of pairwise additivity may not be appropriate for complex dense fluids.

10.5 Molecular Theory of Corresponding States

$P\underline{V}TN$ equations of state are essential for a great number of chemical engineering calculations. Unfortunately the accuracy of such equations as discussed in Section 8.3, like the van der Waals, Peng-Robinson, or Benedict-Webb-Rubin equations are limited. Over a wide range of densities, from gas-like to liquid-like conditions, large errors can occur. Although most of the practical EOSs used today contain empirical, adjustable parameters, they still can only represent higher-order, non-ideal behavior to various degrees of approximation.

It is of interest to explore further why we are so handicapped. In fact, we do have a rigorous basis from which to start. Earlier in Eq. (10-77) we showed that pressure is related to a volume derivative of the configurational integral $\mathbf{Z}^*$ which is defined in Eq. (10-76).

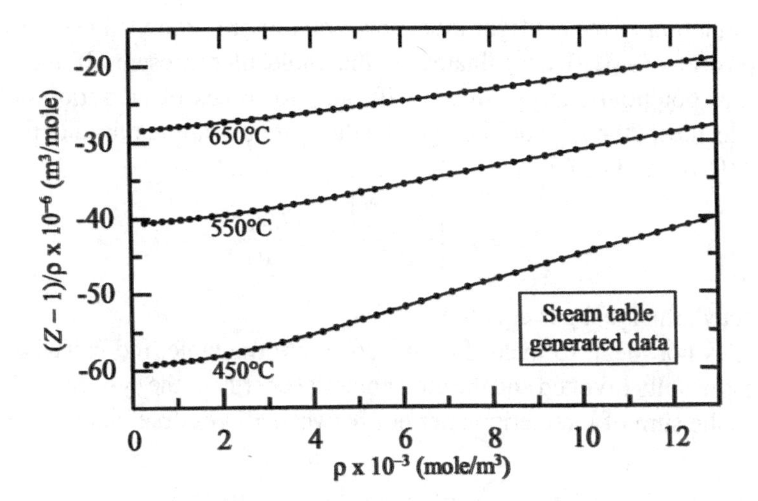

Figure 10.3a Reduced plot to determine the second [$B(T)$] and the third [$C(T)$] virial coefficients. $B(T)$ = intercept and $C(T)$ = slope of straight line as ρ approaches zero [from Kutney (1996)].

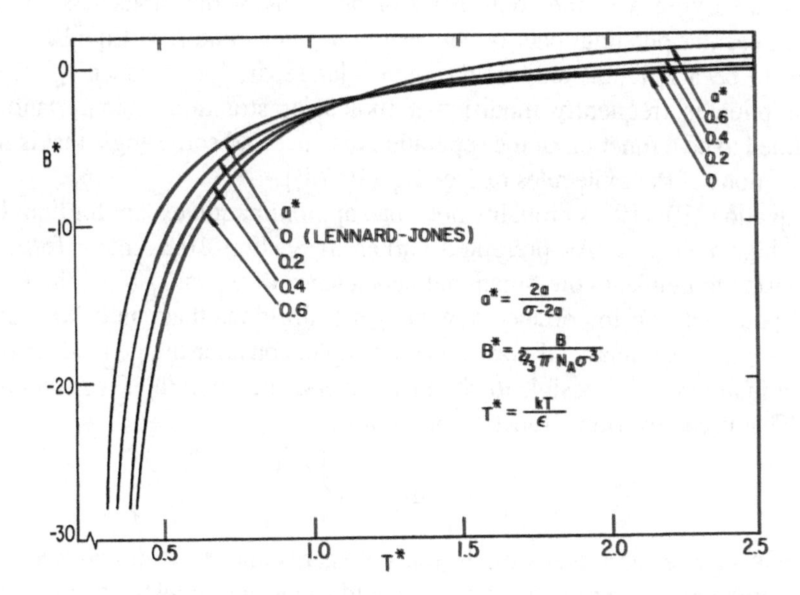

Figure 10.3b Reduced second virial coefficients calculated for the Lennard-Jones and Kihara potentials with a spherical core of radius a. Note that for $a = 0$, the Lennard-Jones 6-12 potential is recovered [adapted from Prausnitz *et al.* (1986), with permission].

The spatial coordinates of each molecule j are given by $\mathbf{r}_j = f(x_j, y_j, z_j)$, and the integration in Eq. (10-76) is over the system volume $\underline{V}$. As discussed in Sections 10.1.7

and 10.3, evaluation of the configurational integral requires knowledge of how to relate Φ to the position (x, y, z) coordinates of the molecules present. If there were no intermolecular potential energy $[\Phi(\mathbf{r}^N) = 0]$, i.e., no forces of attraction or repulsion between molecules, then $\mathbf{Z}^*$ would simply be the system volume raised to the power N. From Eqs. (10-76) and (10-77),

$$P = kT\left[\frac{\partial \ln \underline{V}^N}{\partial \underline{V}}\right]_{T,N} = \frac{NkT}{\underline{V}} \tag{10-118}$$

and as expected, the ideal gas equation results.

If the gas is not ideal, to make Eq. (10-76) more tractable, the pairwise additivity assumption is usually invoked and the total potential energy of the system of N molecules is equated to the sum of interaction energies between all possible binaries.

$$\Phi(\underset{\sim}{\mathbf{r}}^N) = \sum_{i<j} \Phi_{ij}(r_{ij}) = \tfrac{1}{2}\sum_{i=1}^{N}\sum_{j=1}^{N'} \Phi_{ij}(r_{ij}) \tag{10-119}$$

Thus, if one were examining a system of, say, three molecules, the total intermolecular potential energy for this system energy would be taken to be $\Phi_{12} + \Phi_{13} + \Phi_{23}$, with the energies varying with the coordinates or positions of the molecules. Here, Φ_{ij} is the intermolecular potential energy between molecules i and j. In Eq. (10-119), $\Phi_{ij}(r_{ij})$ is taken to be a function only of the molecular separation distance r_{ij}. This particular assumption is frequently modified if molecular structure is important; then Φ_{ij} is assumed to be a function of the separation distance and some angle that is related to the orientations of the molecules ω_i [see Eq. (10-78)].

Equation (10-119) is probably not a bad approximation except for liquids or gases at very high densities. As presented earlier in Section 10.3, many forms have been suggested to delineate the functional dependence of Φ_{ij} on r_{ij} (e.g., the Lennard-Jones 6-12 potential). Of importance now are generalizations that can be extracted from the theory briefly summarized above. For example, consider the general term $\Phi_{ij}(r_{ij})$. We might inquire if it is possible to obtain a universal function Φ' if we nondimensionalize Φ_{ij}. That is can we find a function of the form

$$\Phi_{ij} = \varepsilon \Phi'\left[\frac{r_{ij}}{\sigma}\right] \tag{10-120}$$

where ε is some parameter with the units of energy and σ is a similar term with the units of length? For a pure material, ε and σ could be taken as constants. For a mixture, there would presumably be different ε and σ values for each type of binary interaction. The significant result of Eq. (10-120) is that if it were used in Eq. (10-119) which would, in turn, be substituted into Eq. (10-76), then

$$\mathbf{Z}^* = \sigma^{3N}\int \ldots \int \exp\left[-\sum_i \sum_j \left[\frac{\varepsilon}{kT}\right] \Phi'\left[\frac{(r_{ij})}{\sigma}\right]\right] d\left(\frac{x_1}{\sigma}\right) \ldots d\left(\frac{z_N}{\sigma}\right) \tag{10-121}$$

or in functional form

$$\mathbf{Z}^* = \sigma^{3N} g\left(\frac{kT}{\varepsilon}, \frac{V}{\sigma^3}\right)$$

Equation (10-121) is both interesting and valuable. With the assumptions made, one concludes that Eq. (10-77) might be written in nondimensional form as,

$$\frac{P\sigma^3}{\varepsilon} = f\left(\frac{kT}{\varepsilon}, \frac{V}{\sigma^3}\right) \tag{10-122}$$

or

$$P^* = f(T^*, V^*) \tag{10-123}$$

where $P^* \equiv P\sigma^3/\varepsilon$, $T^* \equiv kT/\varepsilon$, and $V^* \equiv V/\sigma^3$. Note that with only two independent scaling parameters (σ and ε) we end up with a fixed compressibility factor at the critical point. Thus, for example, if the Lennard-Jones 6-12 potential were chosen for Eq. (10-120)

$$\Phi_{ij} = 4\varepsilon\left[\left(\frac{\sigma}{r_{ij}}\right)^{12} - \left(\frac{\sigma}{r_{ij}}\right)^{6}\right] \tag{10-124}$$

then the characteristic energy and distance parameters of this relation may be used to non-dimensionalize P, T, and V in Eqs. (10-122) and (10-123) for a pure material. If one further assumes linear correlations between ε/σ^3 and P_c, ε/k and T_c, and σ^3 and V_c $(= Z_c RT_c/P_c)$, then it is obvious that with $Z \equiv PV/RT$

$$Z = f(Z_c, P_r, T_r) \tag{10-125}$$

This result is the familiar relation expressing the compressibility factor as a function of two scaled parameters, $P/P_c = P_r$ and $T/T_c = T_r$, and Z_c. In many cases, Z_c is eliminated for simplicity and the famous *two-parameter Law of Corresponding States* results (see also Section 8.2).

The key step in this very rapid development was Eq. (10-120); presumably, different universal Φ_{ij} functions that included other parameters might have been introduced. The dipole moment may, for example, be important when delineating intermolecular energies. Likewise, the polarizability, the quadrupole moment, a shape factor or other parameters could be introduced. If so, a similar approach would lead to Eq. (10-125) where besides Z_c, P_r, and T_r, there would be other dimensionless groups characteristic of the material or the mixture.

Using statistical mechanics, one can even infer the functionality of Eq. (10-125). As introduced in Section 10.4, the compressibility factor Z can be closely approximated by a virial expansion in inverse volume or density:

$$Z = 1 + \frac{B}{V} + \frac{C}{V^2} + \ldots = 1 + B\rho + C\rho^2 + \ldots \tag{10-126}$$

where the coefficients B, C, etc. are expressible in terms of a specific intermolecular potential function Φ_{ij}. For example, the second virial coefficient in molar units is given by Eq. (10-114),

$$B = B(T) = 2\pi N_A \int_0^\infty \left[1 - \exp\left[\frac{-\Phi_{ij}}{kT}\right]\right] r_{ij}^2 \, dr_{ij} \tag{10-127}$$

Thus, with Φ_{ij} given say by Eq. (10-124), it is clear that $(B(T)/V)$ is a function only of (V/σ^3) and (kT/ε). It is only a short but obvious jump to obtain Eq. (10-123). A similar treatment may be applied for the third and higher virials, although the mathematics is considerably more complicated.

We now have a choice of direction if we wish to pursue further the development of an equation of state. We may elect to concentrate on the theory-based virial equation or we may wish to correlate empirical data to produce analytical or graphical forms of the relation shown functionally by Eq. (10-125). Both paths have been extensively developed, and although the latter has been of considerable value to us in other sections of the book (see Section 8.3, for example), the former approach needs at least a few words to emphasize some of its more salient advantages and disadvantages. Let us, in our further discussion of the virial equation, truncate it after the second term. This still will limit the usefulness of the equation to a region of moderate pressures or low densities, but inclusion of higher terms would merely overcomplicate matters and, in most cases, estimations of the third virial coefficient C can only be made in a very approximate manner. In the truncated virial form,

$$Z \approx 1 + \frac{B}{V} \tag{10-128}$$

The first point to be made is that B is a function only of temperature, as is obvious from Eq. (10-127). $B(T)$ is readily determinable if the binary intermolecular potential energy function $\Phi_{ij}(r_{ij})$ is accurately known. Though such functions are not available from theory for real fluids, we can at least approximate them by, for example, Eq. (10-124). Also $B(T)$ turns out to be somewhat insensitive to the exact functional form of $\Phi_{ij}(r_{ij})$. For any pure substance, we need to know the characteristic parameters (such as ε and σ in Eq. (10-124)) in order to perform any detailed calculations. Many investigators have inserted particular $\Phi_{ij}(r_{ij})$ functions, determined ε and σ from comparison of experimental B values with calculated ones, and tabulated ε, σ for a number of pure substances. Thus, even the theoretical, truncated form of the virial equation requires some specific constants in order to estimate thermodynamic properties.

Equation (10-128) has a simple analytical form; in fact, if the substitution $V = ZRT/P$ is made, one obtains

$$Z \approx 1 + \frac{B'P}{RT} \tag{10-129}$$

Z is predicted to be a linear function of P at constant T, which is certainly true at low pressures or densities where ideal gas behavior is approached.

Example 10.1

A residual function α is often employed in thermodynamic analyses.

$$\alpha \equiv \frac{RT}{P} - V$$

Determine the limiting value of α as the pressure is reduced to a very low value.

Solution

At low pressures, using Eq. (10-129)

$$\underset{P \to 0}{\text{limit}} \left[\alpha = \frac{RT}{P} - V = \frac{RT}{P} - \frac{ZRT}{P} \right] = -B$$

Should we wish to obtain a numerical value of B for a specific material, the solution of Eq. (10-127) for a reasonable potential energy function is sought for particular σ and ε values. The Lennard-Jones 6-12 is probably the most widely used potential energy function and many tabulations of σ and ε are available. A word of caution is necessary, however. When using tabulations of σ and ε, even for the same potential function, never use ε from one tabulation and σ from another. These parameters are correlated and thus there exist multiple sets of different ε and σ values which are almost equally satisfactory in calculating virial coefficients.

Leaving the virial equation, let us return to the more general functional form of Eq. (10-123). Since some liberties were taken in relating (ε/k) to T_c, (ε/σ^3) to P_c, and σ^3 to V_c, this relation is certainly not based on rigorous theory. The general concept, however, has been of immense importance to chemical engineers. The *two-parameter Corresponding States equation*, $Z = f(T_r, P_r)$ has been utilized in many so-called reduced compressibility factor plots (see Section 8.2). Only in the last 25-40 years has there been much effort expended to improve these simple and easy-to-use diagrams. The major modifications came when it was realized that a *third parameter*, such as Z_c in Eq. (10-125) should be introduced to set apart different classes of compounds. The detailed tables of Z as a function of T_r, P_r, and Z_c found in Hougen, Watson, and Ragatz (1959) attest to the effort expended in developing this correlation (see Section 8.2 for further discussion).

An alternate approach that has been equally successful involves substituting for Z_c a parameter characteristic of the vapor pressure of a material. The most commonly used parameter, originally suggested by Pitzer and co-workers (1955) is the acentric factor, ω, defined as

$$\omega \equiv -\log_{10} P_{vp,r} \ (\text{at } T_r = 0.7) - 1.000 \qquad (10\text{-}130)$$

where $P_{vp,r}$ is the reduced vapor pressure $\equiv P_{vp}/P_c$ (see Figure 8.4). As defined, ω for spherically symmetric molecules is essentially zero. For nonsymmetric molecules $\omega > 0$. Equation (10-125) is modified to include ω as the third parameter:

$$Z = f(T_r, P_r, \omega) \tag{10-131}$$

For computational ease, the functionality is further expressed as

$$Z = Z^{(0)}(T_r, P_r) + \omega Z^{(1)}(T_r, P_r) \tag{10-132}$$

where $Z^{(0)}$ and $Z^{(1)}$ are functions simply of T_r and P_r. These functions are illustrated in Figure 8.5.

Figure 8.5 (or its tabular counterpart) provides reasonable representations of many pure component *PVT* properties as shown by the data presented in Figure 8.2. Some recent analytical equations of state are but approximations to these figures. As such, Corresponding States provides a powerful method to generalize the correlation of the $P\underline{V}TN$ properties of fluids. For accurate predictions, however, the two-parameter model is limited to fairly simple molecules such as Ar, Kr, Xe, or CH_4. By introducing Pitzer's acentric factor, ω, as a third parameter, we improve the correlations for more complex, non-spherical molecules such as CH_3OH, H_2O, C_3H_8, etc.

In this section, we have established a theoretical basis for the scaling used to non-dimensionalize variables. Based on classical statistical mechanics, we established the connection between the configurational integral and the Helmholtz free energy. This means that when the functional form of the intermolecular potential energy function is identical for different fluids, we expect them to rigorously follow the Law of Corresponding States. Fluids with this characteristic are called *conformal*. Basically, we are saying that fluids are conformal when their reduced, residual Helmholtz energies due to configurational, density-dependent effects are identical. Hard-sphere or Lennard-Jones fluids are good examples of conformal fluids, and as the properties of many real molecules are reasonably modeled by these approximate potentials, we expect many compounds to follow Corresponding States as well.

As described in the review by Ely and Marrucho (1995), Corresponding States' correlations have been extended in a general manner by a number of research groups. Two basic approaches are used. The first expands the compressibility factor (Z) in a Taylor series using a set of characterization parameters (λ_i) as

$$Z(T_r, V_r, \lambda_1, \ldots, \lambda_n) = Z^o(T_r, V_r) + \sum_i \left(\frac{\partial Z}{\partial \lambda_i}\right)\lambda_i + \cdots$$

where the derivative coefficients are evaluated using reference fluids. Pitzer's model given in Eq. (10-132) is an example of a one-term expansion with $\lambda_1 = \omega$. The second approach extends the two-parameter form of Eq. (10-123) at the molecular level by making the intermolecular potential dependent on additional molecular shape factors and temperature. In general this can be cast as,

$$Z = \frac{PV}{RT} = Z(T_r^*, V_r^*)\ ; \quad V_r^* = \frac{V}{\pi\sigma^3(T, \lambda_i)} = \frac{V}{h_i}; \quad T_r^* = \frac{kT}{\varepsilon(T, \lambda_i)} = \frac{T}{f_i}$$

Reference fluids (o) are introduced to establish a basis for the shape factor correlating equations by fitting experimental $P\underline{V}TN$ data. This results in:

$$f_i = \frac{T_{c,i}}{T_{c,o}} \Theta\,(T, d_{m,i}, Q_i, \textit{geometric factors}\ldots)$$

$$h_i = \frac{V_{c,i}}{V_{c,o}} \Gamma(T, d_{m,i}, Q_i, \textit{geometric factors}\ldots)$$

where f_i and h_i are functions of T, dipole moment ($d_{m,i}$), quadrupole moment (Q_i), and other geometric factors that describe the molecular architecture. Then correlations are developed for Θ and Γ from data to establish their dependencies on well-known pure component physical properties such as T_c, P_c, V_c, Z_c, and ω.

10.6 Generalized van der Waals Theory

The objective here is to introduce the essential features of van der Waals theory for representing fluid properties over wide ranges of density from low (gas-like) to high (liquid-like). The discussion in this section follows that of Prausnitz, *et al.*, (1986) and Vera and Prausnitz (1972).

In rigorous quantum mechanical form the canonical partition Q_N function is:[2]

$$Q_N = \sum_{\substack{quantum \\ states\ i}} \exp[-\underline{E}_i\,(\underline{V}, N)/kT] = \sum_{\substack{energy \\ levels\ k}} \Omega\,(\underline{E}_k) \exp\,[-\underline{E}_k/kT] \qquad (10\text{-}133)$$

where $\Omega\,(\underline{E}_k)$ = degeneracy of kth energy level. In the classical continuum limit, Q_N can be represented by:

$$Q_N = \int_0^{\infty} \overline{\Omega}\,(\underline{E}) \exp\,[-\underline{E}/kT]\, d\,\underline{E} \qquad (10\text{-}134)$$

where $\overline{\Omega}\,(\underline{E})$ = density of states of energy $\underline{E}$.

In the semi-classical limit, the canonical partition function can be separated into internal (*int*) and center of mass (*cm*) or translational (*t*), and potential energy contributions as:

$$Q_N = Q_{int}\,(T, N)\, Q_{cm}\,(T, \underline{V}, N) = q_{int}^N \frac{q_t^N}{N!} \mathbf{Z}^* \qquad (10\text{-}135)$$

where $\mathbf{Z}^*$ is the configurational integral (see Eq. (10-76)) and $Q_{int} = q_{int}^N$ is the internal contribution due to internal rotational and vibrational effects of N particles. For simplicity, we have omitted other internal effects such as electronic or nuclear spin contributions.

From Eq. (10-63) the ideal gas translational contribution, including the h^{3N} quantum correction, can be expressed analytically as:

2 Note that we continue to adopt the $\underline{V}$ notation convention for total volume. In most treatments of statistical mechanics, such as those cited at the end of the chapter, the total volume is given by V.

$$q_t^N = \left(\frac{2\pi mkT}{h^2}\right)^{\frac{3}{2}N} = \frac{1}{\Lambda^{3N}} \tag{10-136}$$

Now Q_N can be written as:

$$Q_N = \frac{1}{N!} q_{int}^N \left(\frac{2\pi mkT}{h^2}\right)^{\frac{3}{2}N} \mathbf{Z}^* = \frac{1}{N!} q_{int}^N \frac{\mathbf{Z}^*}{\Lambda^{3N}} \tag{10-137}$$

For a so-called "van der Waals fluid," with pairwise additivity assumed, $\mathbf{Z}^*$ from Eq. (10-76) is approximated by:

$$\mathbf{Z}^* \rightarrow \mathbf{Z}^*_{vdW} = \int \dots \int \exp\left[\frac{-\Phi(\underset{\sim}{\mathbf{r}}^N)}{kT}\right] d\underset{\sim}{\mathbf{r}}^N \approx \exp\left[\frac{-|\Phi_{ij}|}{2kT}\right]^N \left[\underline{V_f}\right]^N \tag{10-138}$$

Thus $Q_N \rightarrow Q_{vdW}$ with,

$$Q_{vdW} = \frac{1}{N!} \frac{q_{int}^N}{\Lambda^{3N}} \left[\exp\left[\frac{-|\Phi_{ij}|}{2kT}\right]\right]^N \left[\underline{V_f}\right]^N \tag{10-139}$$

where

$\underline{V_f} \equiv$ free volume

$|\Phi_{ij}|$ = mean pair potential of all 2-body interactions $= \frac{N}{\underline{V}} \int_0^\infty \Phi_{ij}(r)\, g(r)\, 4\pi r^2 dr$

$g(r)$ = radial distribution function (see Section 10.7)

$\Phi_{ij}(r)$ = pair potential energy function

For a van der Waals (vdW) fluid, the lower limit in the $|\Phi_{ij}|$ integral is effectively σ (the hard sphere distance of closest approach) because $g(r) = 0$ for $r < \sigma$, and $\Phi_{ij}(r)$ is always negative (attractive) and independent of T for $r \geq \sigma$.

An alternate approach to approximating $\mathbf{Z}^*$ in Eq. (10-137) is given by Abbott and Prausnitz (1987) and Abbott (1989) where they define the configurational integral/free volume relationship as:

$$\mathbf{Z}^* = \frac{\mathbf{Z}^*}{\mathbf{Z}_o}(\underline{V}_f^N) \qquad \underline{V}_f^N \equiv \mathbf{Z}_o = \lim_{\substack{T\to\infty \\ \tau = 0}} (\mathbf{Z}^*) \tag{10-140}$$

Here the free volume ($\underline{V}_f$) is equivalent to the infinite temperature limit $[T \rightarrow \infty\ (\tau \equiv 1/T = 0)]$ of the configurational integral ($\mathbf{Z}^* \rightarrow \mathbf{Z}_o$ at $T \rightarrow \infty$). Note that the ensemble average total potential energy $\langle\Phi\rangle$ is given by:

$$\left(\frac{\partial \ln \mathbf{Z}^*}{\partial \tau}\right)_{\underline{V},N} = \frac{\int_{\underline{V}} \dots \int -\frac{\Phi}{k} \exp[-\Phi/kT] d\underset{\sim}{\mathbf{r}}^N}{\int_{\underline{V}} \dots \int \exp[-\Phi/kT] d\underset{\sim}{\mathbf{r}}^N} = \frac{-\langle\Phi\rangle}{k} \tag{10-141}$$

Now integrate from the infinite temperature limit $\tau = 0$ to the system temperature $\tau = 1/T$:

$$\ln\frac{\mathbf{Z}^*}{\mathbf{Z}_o} = \int_0^\tau -\frac{<\Phi>}{k}\,d\tau \tag{10-142}$$

or,

$$\frac{\mathbf{Z}^*}{\mathbf{Z}_o} = \exp\left[\frac{-|\Phi_{ij}|}{2kT}\right] \quad \text{where} \quad |\Phi_{ij}| \equiv \frac{2T}{N}\int_0^\tau <\Phi>\,d\tau \tag{10-143}$$

Therefore using Eq. (10-139), we can reconstruct the vdW partition function with $\underline{V}_f = \mathbf{Z}_o$ and a slightly modified interpretation of the mean pair potential $|\Phi_{ij}|$. Now we need to approximate $|\Phi_{ij}|$ from Eq. (10-143) and compare it to the first approach following Eq. (10-139):

$$|\Phi_{ij}| = \frac{N}{\underline{V}}\int_o^\infty g(r)\,\Phi_{ij}(r)4\pi r^2 dr \approx \frac{2<\Phi>}{N} \tag{10-144}$$

The approximation used in 1873 by van der Waals for $\underline{V}_f$ was as follows:

$$\underline{V}_f = \underline{V} - \frac{N}{N_A}b = \underline{V} - \frac{N}{N_A}\left[\frac{2\pi N_A\sigma^3}{3}\right] \tag{10-145}$$

where a molar basis for b has been used. In general, $b = 4 \times$ (hard sphere volume) $= 4b_o$ $= 4(\pi\sigma^3/6)$ on a per molecule basis or $4\,(\pi N_A\sigma^3/6)$ on a per mole basis. The term b is also equivalent to the ***excluded volume per molecule*** or $(1/2)\,(4/3\pi\sigma^3)$ or $2/3\pi N_A\sigma^3$ on a per mole basis. Van der Waals further assumed that only attractions were contained in the $<\Phi>$ term, which was independent of T but linearly proportional to $N/\underline{V}$ (number density) and N. Thus, the average potential [Eq. (10-144)] was approximated by,

$$|\Phi_{ij}| \approx \frac{2}{N}<\Phi> = \frac{2}{N}\left(\frac{-a}{N_A^2}\right)\frac{N}{\underline{V}}\,(N) = \frac{-2aN}{\underline{V}N_A^2} \tag{10-146}$$

where a is the constant from the macroscopic vdW EOS and the term, N_A^2, normalizes the potential to a per molecule basis.

Therefore using Eqs. (10-145) and (10-146), the vdW partition function, Q_{vdw}, from Eq. (10-139) can be written as:

$$Q_{vdW} = \frac{1}{N!}\left[\frac{q_{int}^N}{\Lambda^{3N}}\right]\exp\left[\frac{+aN}{\underline{V}N_A^2 kT}\right]^N\left[\underline{V} - \frac{bN}{N_A}\right]^N \tag{10-147}$$

An alternative interpretation of Eq. (10-144) can also be used to arrive at Eq. (10-147). We assume that molecule i is contained in a cell or lattice defined by its nearest neighbors (molecules j). The pairwise interaction potential Φ_{ij} is summed over all binary ij interactions as molecule i wanders inside the accessible free volume of the cell. If we average the potential over its trajectory inside the cell and use the average potential to characterize the potential energy contribution to the Hamiltonian, then we are making the so-called *mean field approximation.* In this case the free volume of the cell can be regarded as the difference between the total system volume ($\underline{V}$) and the volume excluded by the finite size of the N molecules in the system. With a hard core repulsive dependence for Φ_{ij}, then $g(r \leq \sigma) = 0$, and it is rather easy to estimate the minimum excluded volume. But for $r > \sigma$, $g(r)$ will vary depending on the positions of the neighboring molecules (j) relative to molecule i. Eventually for large $r >> \sigma$, $g(r) \to 1.0$. With the mean field approximation, we effectively set $g(r)$ to 1.0 for $r > \sigma$ and integrate Φ_{ij} over the free volume, yielding

$$|\Phi_{ij}| = \frac{N}{\underline{V}} \int_{\sigma}^{\infty} \Phi_{ij}\, 4\pi r^2 dr$$

If Φ_{ij} is characterized by a hard sphere repulsive core and a $1/r^6$ dependence for the attractive term, then Q_{vdw} given by Eq. (10-147) results (see Problem 10.12). Now let's see how we can use Q_{vdw} to develop the macroscopic $P\underline{V}TN$ vdW EOS.

Using Eq. (10-49), we can develop a pressure-explicit EOS from Eq. (10-147):

$$P = P_{vdw} = kT\left[\frac{\partial \ln Q_{vdW}}{\partial \underline{V}}\right]_{T,N} = \frac{-aN^2}{\underline{V}^2 N_A^2} + \frac{NkT}{\underline{V} - b(N/N_A)} \tag{10-148}$$

or

$$P = \frac{-a}{\underline{V}^2(N_A^2/N^2)} + \frac{N_A kT}{\underline{V}(N_A/N) - b} \tag{10-149}$$

Since $N_A\, k = R$ and $V = \underline{V}/(N/N_A)$, by rearranging Eq. (10-149) we get:

$$\left(P + \frac{a}{V^2}\right)(V - b) = RT \tag{10-150}$$

which is the macroscopic form of the vdW EOS.

Van der Waals actually recognized the problem associated with using the hard sphere approximation for b in the free volume expression. He noted that as density increases, the excluded volume would decrease below the simplified hard sphere (2-body) approximation. This is crudely illustrated below.

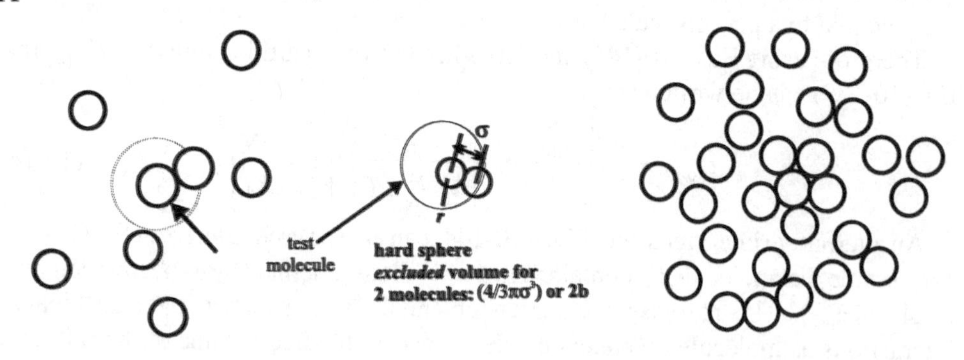

As gas density increases, the concentration of molecules around each other increases above the hard sphere excluded volume. Van der Waals expressed this effect on b using an expanded power series in density ($1/V = \rho$) as:

$$b = 4b_o\left[1 - \alpha\frac{4b_o}{V} + \beta\left(\frac{4b_o}{V}\right)^2 + \ldots\right] \tag{10-151}$$

or

$$b = 4b_o\left[1 - \alpha 4b_o\rho + \beta(4b_o\rho)^2 + \ldots\right] \tag{10-152}$$

where α, β, etc. are Taylor expansion coefficients.

An important result of the presentation in this section is the establishment of a statistical mechanical basis for the van der Waals EOS that theoretically rationalizes the use of the b and a as parameters to account for molecular size and intermolecular attractive force effects.

10.7 Radial Distribution Functions

As density increases it is difficult to use the virial equation to represent behavior accurately as the higher order terms require complex and tedious calculations with precise specification of intermolecular forces. In fact, in the functional formulation given in Section 10.4, Eq. (10-109), the density form of the virial expansion, diverges at high density. Thus, we need a theoretical approach for treating dense fluid or liquid-like systems that is fundamentally different but yet still captures the basic physics. For example, each molecule in a liquid constantly interacts with its neighbors. A common, general approach used to model such systems introduces a radial distribution function to characterize the local number density of molecules in the system.

One can examine the time-averaged variations in the local density $\rho(r)$ of molecules at various distances r from a specific molecule i in a fluid system (as shown in Figure 10.4). At very short distances, the presence of i will exclude other molecules, so $\rho(r)$ will be 0. From a point of closest contact, $\rho(r)$ will rise with increasing separation distance to reach a maximum corresponding to the nearest neighbors of atom i (first coordination shell). These nearest neighbors, in turn, will prevent other molecules from getting too close, so $\rho(r)$ will decrease again and perhaps reach a minimum, and from then on increase again to a lower maximum (second coordination shell), etc. The local density $\rho(r)$ is assumed to be an ensemble (or time) averaged quantity. Since the macroscopic density $\rho = \langle\rho\rangle$ is also ensemble-averaged, the *radial distribution function* (or pair distribution function) is defined as:

$$g(r) \equiv \frac{\rho(r)}{\rho} = \frac{\rho(r)}{\langle\rho\rangle} \tag{10-153}$$

The function $g(r)$ provides a direct measure of local structure. It is always true that as $r \to 0$, $g(r) = 0$ (excluded volume effects) and that as $r \to \infty$, $g(r) = 1$ (bulk density limit), the behavior in between is characteristic of the physical state of the system. In Figure 10.4b we show what $g(r)$ looks like for a typical dilute gas, a liquid and a solid. For a dilute gas there is only a single peak, approximately at an intermolecular separation distance at r_o corresponding to the potential energy minimum, and $g(r)$ quickly assumes its long-range limit as r increases. In a crystalline solid there are sharp local maxima, corresponding to the very well-defined atomic repeat distances imposed by the symmetry of the lattice. A liquid exhibits short-range order and long-range disorder characterized by damped oscillatory behavior with the local density $\rho(r)$ converging to $<\rho>$ at large r.

In general, the total number of particles interacting with the test particle fixed at $r = 0$ is given by

$$N - 1 = \int_0^\infty <\rho> g(r) 4\pi r^2 dr \approx N \text{ for large } N \tag{10-154}$$

where $<\rho>g(r)4\pi r^2\, dr$ is the probability of observing a second molecule in a spherical shell of thickness dr. Strictly speaking, $g(r)$ will depend on the center of mass location and orientation of each molecule relative to one another. For spherically symmetric molecules, only an r or $r_{ij} = |\mathbf{r}_i - \mathbf{r}_j|$ dependence results.

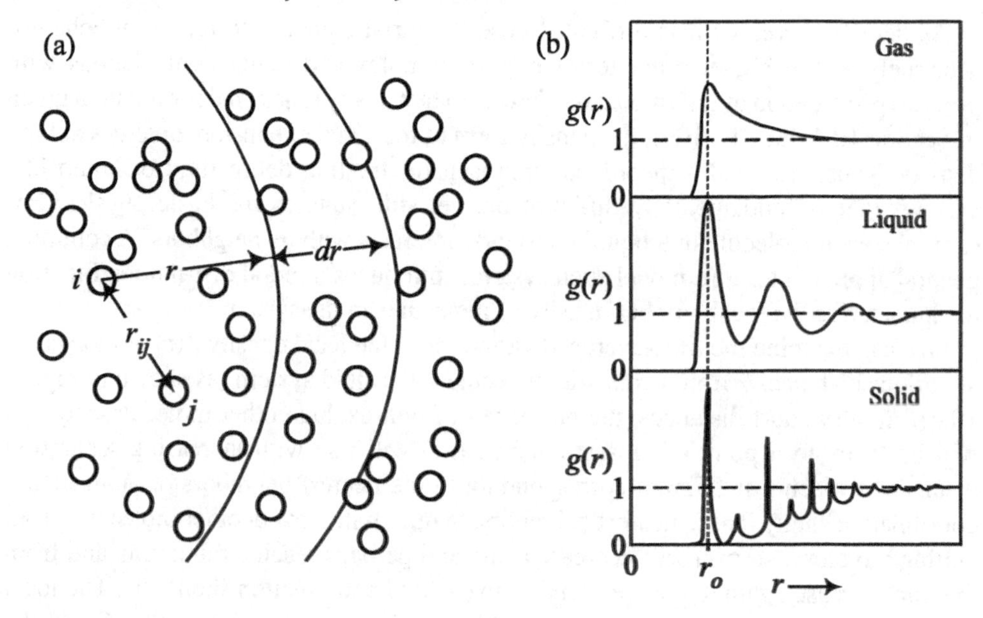

Number of molecules contained in a spherical shell of thickness dr at radius $r = \rho(r, t_1)dr$

Figure 10.4 Conceptualization of a radial distribution function about a particular molecule i at a particular time t_1 for representative gas, liquid, and solid phases.

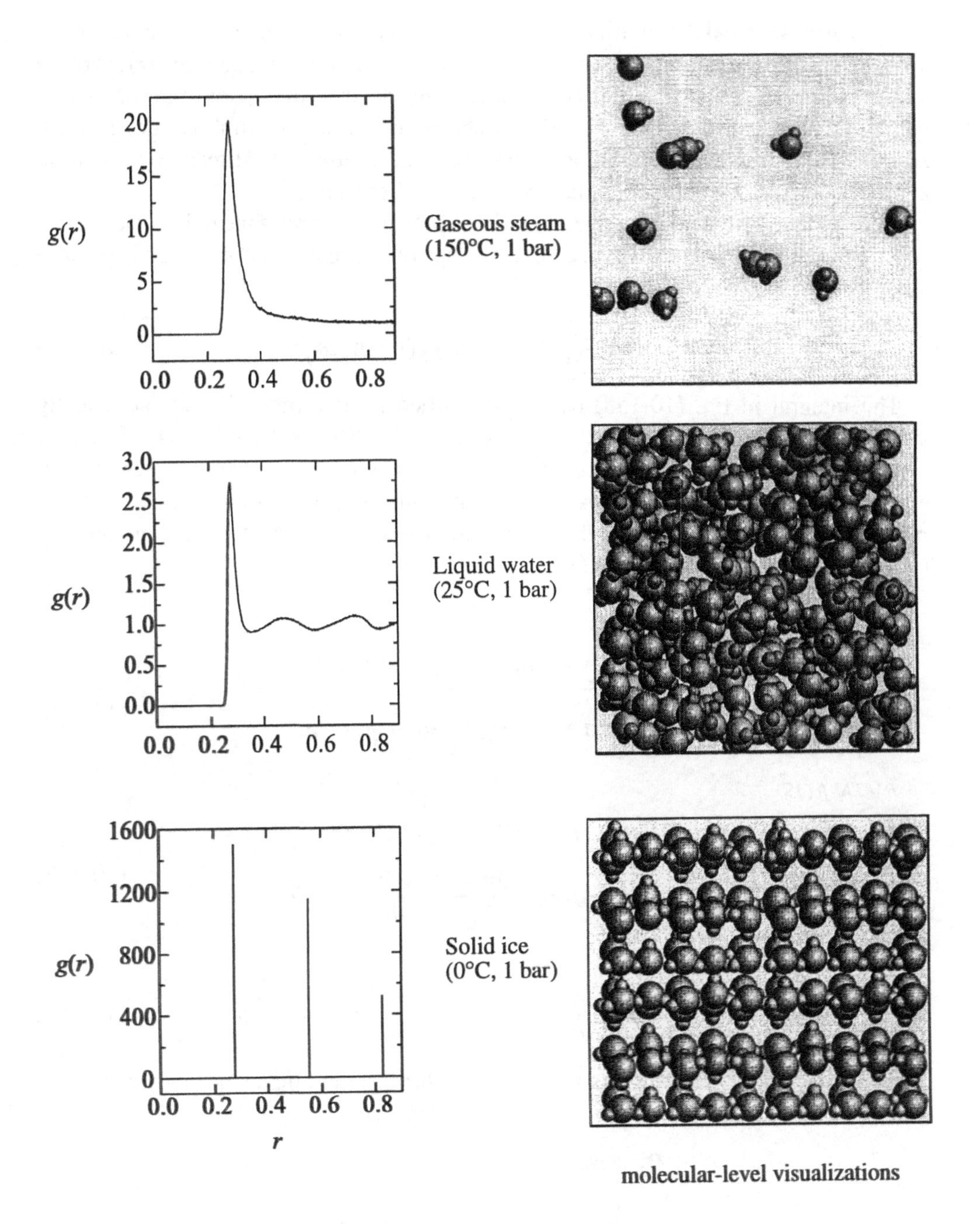

Figure 10.5 Radial distribution functions for gas, liquid, and solid phases of water and their relationship to molecular-level visualizations [provided by Reagan, Weinstein, and Harris (1996)].

The pairwise radial distribution function $g(r)$ can be experimentally determined from x-ray and neutron-scattering measurements in liquids and solids. Obtaining $g(r)$ involves taking a Fourier transformation of the scattering pattern. In liquids, the diffraction patterns are more diffuse than for solids, but short-range structure can still be determined. Alternatively, computer simulation techniques can be used to estimate $g(r)$ with an appropriate intermolecular potential specified (see Section 10.8).

Derived thermodynamic properties can be expressed in terms of the radial distribution function $g(r)$. For example, the ensemble average of any pair function can be expressed as:

$$<B> = \frac{N^2}{2\underline{V}} \int\int B(\mathbf{r}_i, \mathbf{r}_j)\, g(r_{ij})\, d\mathbf{r}_i\, d\mathbf{r}_j \tag{10-155}$$

The integral in Eq. (10-155) can be simplified to one dimension for spherically symmetric molecules with $d\mathbf{r}_i\, d\mathbf{r}_j = 4\pi r^2 dr$, $B(\mathbf{r}_i, \mathbf{r}_j) = B(r)$, and $g(r_{ij}) = g(r)$. Thus, we can develop expressions for macroscopic, ensembled-averaged quantities for internal energy $\underline{U}$, pressure P, or compressibility κ_T in terms of $g(r)$ in a rather straightforward manner for monatomic fluids where the pairwise additivity approximation (see Eq. (10-86)) is valid [see McQuarrie (1976)]:

Internal energy $\underline{U}$*:*

$$\underline{U} = \underline{U}^{\,ideal\ gas} + <\underline{U}>^{configurational}$$

$$\underline{U} = \frac{3}{2}NkT + 2\pi N\rho \int_0^\infty \Phi(r)\, g(r)\, r^2 dr \tag{10-156}$$

$P\underline{V}TN$ *EOS:*

$$P\underline{V} = (P\underline{V})^{ideal\ gas} + (P\underline{V})^{residual}$$

$$P\underline{V} = NkT - \frac{2\pi N^2}{3\underline{V}} \int_0^\infty \frac{d\Phi}{dr} g(r)\, r^3 dr \tag{10-157}$$

Compressibility:

$$kT\rho\kappa_T = kT\left(\frac{\partial\rho}{\partial P}\right)_{T,N} = 1 + 4\pi\rho \int_0^\infty [g(r) - 1] r^2 dr \tag{10-158}$$

Although the form of Eq. (10-157) is somewhat different than a normal equation of state it is easily derived from the *Clausius virial theorem*:

$$P\underline{V} = NkT + 1/3 < \sum_i \sum_j \mathbf{F}_{ij}(r_{ij}) r_{ij} > \tag{10-159}$$

where the intermolecular force $\mathbf{F}_{ij}(r_{ij}) = F(r) = -\, d\Phi(r)/dr$ for pairwise interactions. Eq. (10-155) is used to obtain the ensemble average. Eq. (10-157) can also be decomposed to produce the virial equation. This is done by expressing $g(r)$ as a perturbation expansion in density:

$$g(r) = g_o + g_1\rho + g_2\rho^2 + g_3\rho^3 + \ldots \tag{10-160}$$

By substituting Eq. (10-160) into Eq. (10-157) and comparing terms to the virial expansion, one sees that the second virial coefficient is given by,

$$B(T) = -\frac{1}{6kT}\int_0^{\infty} r\left(\frac{d\Phi(r)}{dr}\right) g_o \, 4\pi r^2 dr \tag{10-161}$$

with $g_o = \exp[-\Phi(r)/kT]$ we recover the normal expression for $B(T)$ (Eq. (10-114)).

Various models have been proposed for $g(r)$. These include:

(i) Kirkwood—using potential of mean force
(ii) Born-Green-Yvon (BGY)
(iii) Ornstein-Zernike
(iv) Percus-Yevick
(v) Hypernetted chain equation (HNC)

These models then are used to formulate so-called integral equations to describe the complete set of thermodynamic properties. (For details see McQuarrie (1976), Ch. 13, p. 254-289.) Because repulsive forces are important in determining the molecular structure and properties of liquids and dense fluids in general, hard sphere fluids represent a suitable starting point for perturbation expansions. Density-functional theories represent an extension of integral equation methods that are particularly appropriate for multicomponent, dense fluids when both attractive and repulsive effects are important in determining molecular structure and subsequent mixture properties [see Hansen and McDonald (1986), Chapter 6 for further discussion].

10.8 Hard-Sphere Fluids

For an idealized fluid consisting of hard spheres, Ree and Hoover have been able to obtain closed form solutions for the virial coefficients using the Pade approximation. They have accurately evaluated all virial coefficients up through the 6th. Thus, we can represent an "exact" hard-sphere $P\underline{V}TN$ EOS as:

$$P^{HS} = \frac{NkT}{\underline{V}}(1 + 4y + 10y^2 + 18.365y^3 + 28.24y^4 + 39.5y^5 + 56.7y^6 + \ldots) \tag{10-162}$$

where $y \equiv b/(4\underline{V})$. Again, b is related to the volume occupied by the molecules. For hard spheres b is rigorously given by $2\pi\sigma^3/3$, identical to the b introduced in Section 10.6 for a van der Waals fluid. In all cases, b scales with the excluded volume per molecule. Results for several approximate models for hard spheres are given in Table 10.4. Even though the Carnahan-Starling model is a polynomial approximation of the exact solution, it matches deterministic molecular modeling results generated by computer simulation extremely well. A polynomial expansion for the repulsive $1/(\underline{V}-b)$ part of the vdW EOS is also listed in Table 10.4. Note that the vdW EOS gives the correct 2nd virial coefficient for a hard-sphere fluid. As density or y increases, deviation between the exact and vdW expansion quickly becomes very large. Clearly, van der Waals qualitatively captured a significant portion of the correct physics with his repulsive term well before statistical mechanical methods had been invented!

The correct quantitative form for a hard-sphere fluid requires the refinements provided by classical statistical mechanics. In practice, however, we encounter difficulties in applying this rigorous approach to real systems, partly because intermolecular potential functions are not known exactly. These characterization difficulties are particularly pronounced in the case of dense liquid systems. Consequently, approximate methods have been introduced to provide tractable solutions. Many of these methods are particularly suited for *mixtures*, in which the exact problem is much more involved than for pure systems.

Perturbation and Density-Functional Theories: Analytical approaches based on the hard-sphere model with attractive interactions introduced as a perturbation. For example, the Carnahan-Starling van der Waals EOS with and without volume translation discussed in Sections 8.4 and 9.2 is an excellent example of a perturbed hard-sphere model used in chemical engineering calculations. Theories due to Barker-Henderson and Chandler-Weeks-Anderson are important examples of perturbation expansions that use a softer core for repulsion and a Lennard-Jones-like attraction that damps out at large r [see McQuarrie (1976) Chapter 14 and Hansen and McDonald (1986), Chapter 6 for further discussion]. In these models the total potential energy is separated into two parts:

$$\Phi = \Phi^{(0)} + \Phi^{(1)}$$

where the superscripts correspond (0) to the reference (e.g., hard-sphere) system and (1) to the perturbation.

Lattice-Cell Models: Simplified, periodic, "solid-like" picture of fluids. Molecules are assumed to spend most of their time at or near one location in the volume of the assemblage (in cells, centered at lattice sites). The key objective is to consider only the most probable configurations. Clearly, this is an extension of the approach followed to develop the vdW partition function. The Wilson and UNIQUAC models for non-ideal mixtures and Flory-Huggins model used for polymer solutions are based on lattice models (see Sections 11.4 and 11.5). In addition, lattice model approaches have been employed with some success in modeling liquids and solid clathrate compounds [see Section 10.9 and Hirschfelder, *et al.* (1954), p 293-311 for a description of the Lennard-Jones-Devonshire model].

Corresponding States Theory: Provides a basis for numerous correlations of physical properties (e.g., second virial coefficients, densities, viscosities, thermal conductivities (see Sections 8.2 and 10.5)). When keyed to a specific functional form for the intermolecular potential, this approach produces conformal fluid theory.

Table 10.4 Virial Expansions for a Hard-Sphere Fluid

$$Z = \frac{PV}{RT} = 1 + \sum_{i=1}^{n} b_i y^i = f(y)$$

$$y \equiv b/(4V) = (b/4)\rho = \pi\sigma^3\rho/6$$

Model	$f(y)$
Ree-Hoover exact expansion	$1 + 4y + 10y^2 + 18.365y^3 + 28.24y^4 + 39.5y^5 + 56.7y^6 + ...$
Percus-Yevick (pressure Eq.) $Z = (1 + 2y + 3y^2)/(1-y)^2$	$1 + 4y + 10y^2 + 16y^3 + 22y^4 + 28y^5 + 34y^6 + ...$
Percus-Yerick (compressibility Eq.) $Z = (1 + y + y^2)/(1-y)^3$	$1 + 4y + 10y^2 + 19y^3 + 31y^4 + 46y^5 + 64y^6 + ...$
Carnahan-Starling $Z = (1 + y + y^2 - y^3)/(1-y)^3$	$1 + 4y + 10y^2 + 18y^3 + 28y^4 + 40y^5 + 54y^6 + ...$
van der Waals (repulsion only) $Z = 1/(V-b)$	$1 + 4y + 16y^2 + 64y^3 + 256y^4 + 1024y^5 + 4096y^6 + ...$
Kirkwood (pressure Eq.)	$1 + 4y + 10y^2 + 8.96y^3 + ...$
Hypernetted Chain (HNC) model (pressure Eq.)	$1 + 4y + 10y^2 + 28.50y^3 + 29.37y^4 + ...$
Born-Green-Yvon (BGY) (pressure Eq.)	$1 + 4y + 10y^2 + 14.42y^3 + 12.16y^4 + ...$

10.9 Molecular Simulation Applications

Molecular simulation methods provide a *deterministic* means for validating and analyzing theoretical models. In essence, molecular simulation provides a framework for unambiguous testing under perfectly controlled conditions. Because we are frequently simulating molecular configurations and/or motions, information is generated that is not directly measurable in the laboratory. Processes can be studied on a very detailed, molecular level—frequently providing insight regarding such phenomena as condensation of a vapor into a liquid or density fluctuations near the critical point. The two major simulation techniques, Molecular Dynamics and Monte Carlo, are discussed in the sections that follow. Interested readers should consult Allen and Tildesley (1987) for further discussion on actual methods for implementing molecular simulations.

10.9.1 Molecular Dynamics (MD). Molecular dynamics simulates fluid structure and molecular motion by solution of Newton's equations in an isolated system of N molecules of mass m within a cell of fixed volume. Thus we can write Newton's Second Law for each molecule i as,

$$\ddot{\mathbf{r}}_i(t) = \frac{1}{m_i} \sum_{i<j}^{N} \mathbf{F}_{ij}\,(r_{ij}(t)) \tag{10-163}$$

where $\ddot{\mathbf{r}}_i(t)$ = acceleration of molecule i, m_i = mass of molecule i, and r_{ij} is the separation distance between molecule i and j. In using Eq. (10-163) we have again assumed pairwise additivity and a classical system of interacting particles.

Alder and Wainwright (1959, 1960) were the first to implement molecular dynamics (MD) methods for a hard-sphere fluid. The basic elements of MD are:

(i) Finite difference or element approximations are typically used to integrate the differential equation set over time from preset initial values for position and velocities, $\mathbf{r}_i$ and $\dot{\mathbf{r}}_i$. Time step sizes for integration are on the order of 10^{-15} s or 1 *fs*, and usually about 10^3 to 10^5 steps are required when between 100 to 10,000 particles or molecules are used in a simulation. In practice, some number of iterations are required from startup to reach a dynamic equilibration state.

(ii) Periodic boundary conditions are used (see Figure 10.6).

(iii) Translational, rotational, and intermolecular potential energy effects are included to specify forces between molecules.

(iv) Method is fully deterministic with simulation of the actual evolution of molecular configurations in time closely analogous to the microcanonical ensemble at constant N, $\underline{V}$, and $\underline{E}$. Thus, MD provides a fully dynamic simulation making it possible to obtain both thermodynamic and transport properties.

For each molecule i, the acceleration $\ddot{\mathbf{r}}_i(t)$ at a particular time t is approximated as a function of the total interaction force between all ij pairs of molecules as given in Eq. (10-163). The net force $[\mathbf{F}(\mathbf{r})^N]$ can be determined directly from the derivative of the total potential energy. Thus,

$$\ddot{\mathbf{r}}_i(t) = \frac{1}{m_i} \sum_{i<j}^{N} -\nabla\Phi_{ij}\,(r_{ij}(t)) \tag{10-164}$$

Now in order to carry out an MD simulation, Eq. (10-164) needs to be integrated for each molecule i in the system at each time step. In principle, this is straightforward as we are dealing with a conservative system of N particles that yield N ordinary differential equations. A number of algorithms are available for the integration [see Chapter 3 in Allen and Tildesley (1987) for examples]. The *Verlet algorithm* is one of the most popular with two finite difference schemes normally employed:

Ordinary Verlet

$$\mathbf{r}_i(t+\delta t) \approx 2\mathbf{r}_i(t) - \mathbf{r}_i(t-\delta t) + (\delta t)^2\,\ddot{\mathbf{r}}_i(t) \qquad (10\text{-}165)$$

where velocities $\dot{\mathbf{r}}_i(t)$ have been eliminated by adding two Taylor expansions.

$$\mathbf{r}_i(t+\delta t) = \mathbf{r}_i(t) + \delta t\,\dot{\mathbf{r}}_i(t) + \frac{(\delta t)^2}{2!}\ddot{\mathbf{r}}_i(t) + \cdots$$

$$\mathbf{r}_i(t-\delta t) = \mathbf{r}_i(t) - \delta t\,\dot{\mathbf{r}}_i(t) + \frac{(\delta t)^2}{2!}\ddot{\mathbf{r}}_i(t) - \cdots$$

The combination of these forward and backward difference equations yields Eq. (10-165). At any particular time step, the velocity can be approximated as

$$\dot{\mathbf{r}}_i(t) = \mathbf{v}_i(t) = \frac{\mathbf{r}_i(t+\delta t) - \mathbf{r}_i(t-\delta t)}{2\delta t} \qquad (10\text{-}166)$$

Verlet velocity

$$\mathbf{r}_i(t+\delta t) \approx \mathbf{r}_i(t) + \delta t\dot{\mathbf{r}}_i(t) + \frac{(\delta t)^2}{2}\ddot{\mathbf{r}}_i(t)$$
$$\mathbf{v}_i(t+\delta t) = \dot{\mathbf{r}}_i(t+\delta t) \approx \dot{\mathbf{r}}_i(t) + \frac{\delta t}{2}\left[\ddot{\mathbf{r}}_i(t) + \ddot{\mathbf{r}}_i(t+\delta t)\right] \qquad (10\text{-}167)$$

where $\ddot{\mathbf{r}}_i(t+\delta t)$ is obtained from Eq. (10-164) using the new position coordinates $\mathbf{r}_i(t+\delta t)$ for all N particles to evaluate $r_{ij}(t+\delta t)$ in the summation of Eq. (10-164). The Verlet velocity formulation only requires computer storage of N values of $\mathbf{r}_i$, $\dot{\mathbf{r}}_i$, and $\ddot{\mathbf{r}}_i$.

With either Verlet formulation, the selection of the size of the time step δt is important to produce a realistic sequence of configurations. In addition, for non-linear molecules rotational motion must be simulated as well, which adds additional computational requirements. A suitable startup period is also needed to equilibrate the system in order to generate meaningful results.

A sample calculation for computing several thermodynamic properties is shown in the example that follows.

Example 10.1

Derive relationships for estimating temperature, pressure, internal energy, and heat capacity (C_v) for a 3-D MD simulation of N molecules.

Solution

The mean temperature is obtained from the averaged of the kinetic energy:

$$\langle KE\rangle = \tfrac{3}{2}NkT = \tfrac{1}{2}\sum_{i=1}^{N} m_i \langle \mathbf{v}_i^2\rangle = \tfrac{1}{2}\sum_{i=1}^{N} m_i \left[\frac{1}{M}\sum_{j=1}^{M}[\mathbf{v}_i(j\Delta t)]^2\right]$$

where the $<\,>$ indicates a time average over the total time $\tau = M\Delta t$ of the simulation with M moves of time step Δt. The mean pressure comes directly from the virial equation (10-159), where the time-averaged virial is used

$$\frac{1}{M}\sum_{m=1}^{M}\left[\sum_{i=1}^{N}\sum_{j=1}^{N}[\mathbf{F}(r_{ij})\,r_{ij}]_m\right]$$

The total internal energy $\underline{U}$ is assumed to be equal to $\underline{E}$ for a simple system. Thus,

$$\underline{U} = <\underline{U}> = <\underline{E}> = \underline{E} = <KE> + <PE> = <KE> + <\Phi(\underset{\sim}{\mathbf{r}}^N)>$$

In a microcanonical ensemble MD simulation, $\underline{E}$, $\underline{V}$, and N are fixed, so the predicted value of $\underline{U} = \underline{E}$ should be constant; thus, we have an internal check on the consistency of the simulation. The heat capacity is given by the fluctuation equation (10-53) as:

$$C_v = \frac{\left[<\underline{E}^2> - (<\underline{E}>)^2\right]}{NkT^2}$$

where again time averages of $\underline{E}^2$ and $\underline{E}$ are calculated over the total simulation time.

The chemical potential can also be estimated using a particle insertion technique developed by Widom (1963) as illustrated in Problem 10.8.

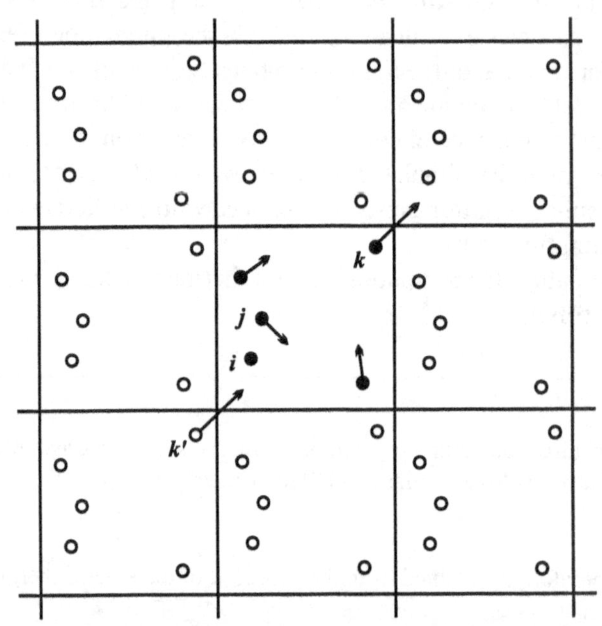

Figure 10.6 Periodic boundary condition for Monte Carlo and Molecular Dynamics. In two dimensions, the random configuration of 5 molecules given in the central square is surrounded by identical "ghost" systems. When one molecule moves out (or into) the central cell, another molecule moves in (or out) from the adjacent mirror cell (for example follow the trajectory of k and k'.)

10.9.2 Monte Carlo (MC). The central idea of Monte Carlo is to use a stochastic method to generate a sequence of configurations by successive random displacements. In this manner, we can sample configuration space of the canonical ensemble and, in principle, estimate ensemble-averaged macroscopic properties. Typically, sequences of 10^5 to 10^6 configurations of 1000 or more molecules are required to generate acceptable ensemble averages for a liquid.

Since there is no explicit time scale involved, the order in which a particular configuration occurs has no physical significance. Thus, unlike MD, the MC method is good for static (thermodynamic) properties only, and not for dynamic properties such as diffusivity and viscosity which require a simulation in the time domain.

Typically, configuration space is sampled by generating successive states that are determined by specifying random positions (and orientations if non-spherical molecules are involved). Transition probabilities have to be defined so that the sampling frequency corresponds to the Boltzmann-weighted probability distribution of the canonical ensemble at constant N, $\underline{V}$, and T.

$$P_N(\underline{E}_j) = \frac{\exp(-\beta \underline{E}_j)}{\sum_i \exp(-\beta \underline{E}_i)} \tag{10-168}$$

The MC technique was first implemented by Metropolis and co-workers at Los Alamos National Laboratory in the late 1940's and 1950's by introducing an algorithm to generate a series of configurations consistent with required transition probabilities. The Metropolis-sampling algorithm is described by the sequence of operations given in Figure 10.7. This procedure guarantees ergodic sampling with Boltzmann-weighting such that a simple arithmetic mean can be used to estimate the ensemble-averaged property, for example, the average internal energy:

$$<\underline{U}> = \frac{\sum_{j=1}^{M} \underline{U}_j}{M} \tag{10-169}$$

where M = total number of moves or configurations. The MC method with Metropolis sampling effectively generates a Markov chain of states in configurational space since the result of each trial move (j+1) depends only on the value of Φ_{j+1} and Φ_j from the preceding move (j). In using Eq. (10-169) we are assuming that the only system variations that are important to the average value of the property of interest are those associated with the configuration or arrangement of the N-particle system. For example, the kinetic energy translational contribution ($\frac{3}{2}NkT$) and the internal, intramolecular contribution from q_{int}^N to $\underline{U}$ are fixed during the MC simulation at constant values. Because of this fact, the radial distribution function [$g(r)$] is often estimated in MC simulations and used to generate thermodynamic properties of interest following the techniques illustrated in Section 10.7. Other properties that depend on fluctuations, in energy (C_v) or density (κ_T) can be computed directly from Metropolis MC averages given by the form of Eq. (10-169). In the next section, we show how MC can be used to generate a radial distribution $g(r)$ for a simple fluid.

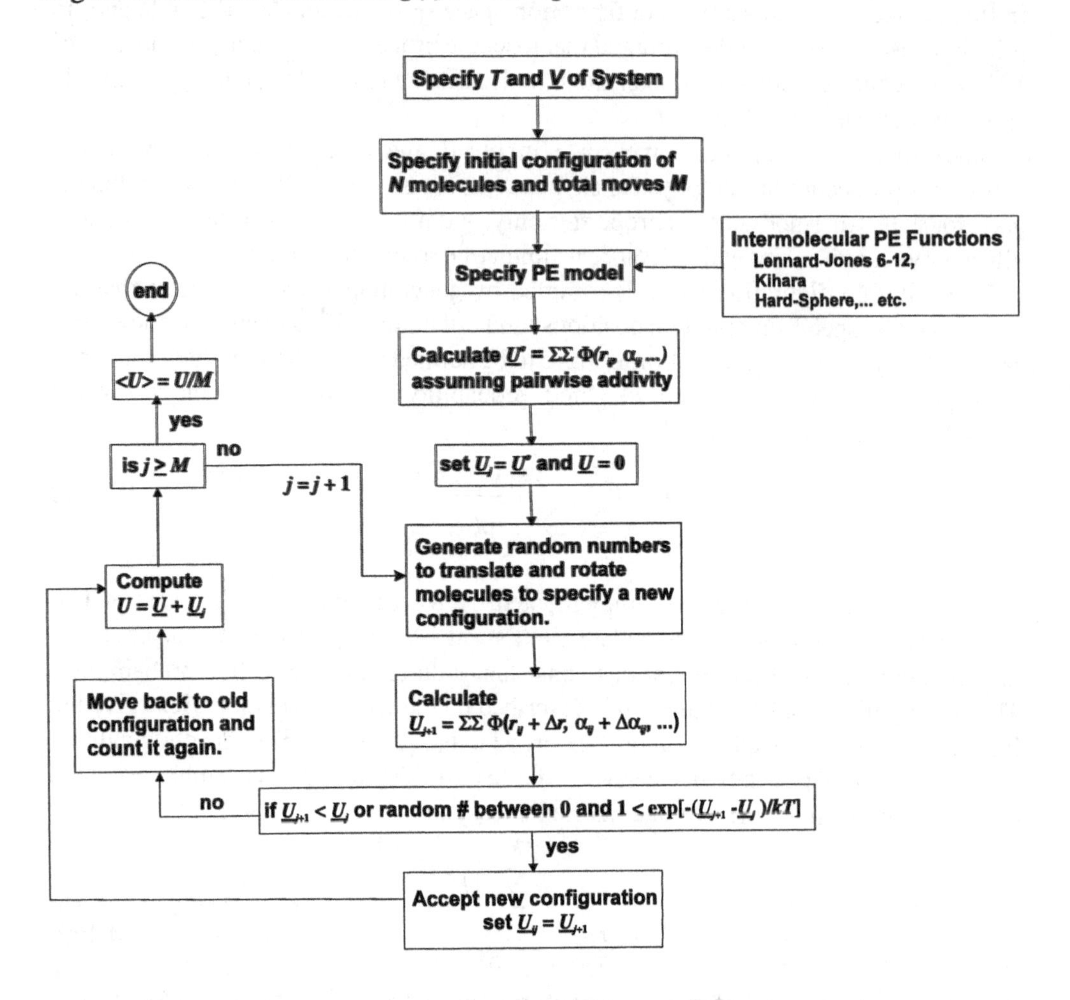

Figure 10.7 Monte Carlo using the Metropolis Sampling algorithm to estimate $\langle \underline{U} \rangle$.

Although it is possible to use the MC method to estimate the configurational integral ($\mathbf{Z}^*$) directly, there are some important limitations. For example, consider what it would take to evaluate $\mathbf{Z}^*$ from Eq. (10-76) for a 100 molecule system contained in a cubic volume $\underline{V} = L^3$. If the molecules are spherically symmetric, then 300 coordinates are needed to specify the configurational state of the system. Even a simple quadrature approach such as Simpson's rule would require multiple evaluations of $\exp[-\Phi(r^N)/kT]$ for each configuration to integrate Eq. (10-76). For example, if 10 specified configurations with $\underline{V}$ were used in the quadrature, then 10^{300} function evaluations of $\Phi_{ij}(r_{ij})$ would be needed. If the molecules are non-linear and polyatomic the calculational burden just worsens. In either situation, the quadrature approach is not feasible. Monte Carlo integration provides an alternative using the mean value theorem of calculus:

$$\mathbf{Z}^* \approx [\langle \exp(-\Phi/kT) \rangle]\, \underline{V}^N \tag{10-170}$$

where a suitable value of Φ in the domain of $\underline{V}$ must be used. If we generate configurations of the 100 molecule system randomly and average the exponential function over M configurations then with pairwise additivity assumed:

$$\mathbf{Z}^* \approx \left[\frac{1}{M} \sum_{m=1}^{M} \exp(-\tfrac{1}{2} \sum_{i=1}^{100} \sum_{j=1}^{100} \Phi(r_{ij})_m / kT) \right] L^{300} \tag{10-171}$$

If M is large enough then we should sample configurational space adequately enough to estimate $\mathbf{Z}^*$ with reasonable accuracy.

However, in a fully random generation of configurations each requiring 300 coordinates, we expect that a great many configurations will be unfavorable with values of $\exp(-\Phi/kt) << 1$ and thus will only make a minute contribution to $\mathbf{Z}^*$. For dense fluids such as liquids, direct MC integration is not practical because M would be very large. Consequently, a better algorithm for generating configurations is needed. An *importance sampling technique* that selects configurations in a biased fashion more consistent with the probability distribution given by the Boltzmann factor would greatly improve efficiency. Fortunately, the Metropolis method illustrated in Figure 10.7 provides a tractable solution to this problem. Even so, the calculation of $\mathbf{Z}^*$ using Metropolis Monte Carlo is not a trivial matter, as illustrated in the example below.

Example 10.2

Show how MC integration can be used to estimate $\mathbf{Z}^*$ for a gas hydrate clathrate compound where guest molecules such as Ar, CH_4, or CO_2 are entrapped in cages formed by host water molecules arranged on a solid lattice as shown in Figure 10.8. These entrapped molecules stabilize the host lattice via van der Waals interactions.

Solution

With the geometry of each cell known and only one mobile guest molecule per cell, the number of possible interactions between *ij* pairs is greatly reduced and MC integration now becomes feasible. For the clathrate system shown in Figure 10.8, there are two cells of type 1 (pentagonal dodecahedra) and six cells of type 2 (tetrakiadecahedra) in a unit cell containing 46 water molecules. Thus complete occupation of all cells by guest molecules (M_g) yields the following stoichiometry

$$8\, M_g \cdot 46H_2O \quad \text{or} \quad M_g \cdot 5\tfrac{3}{4}\, H_2O$$

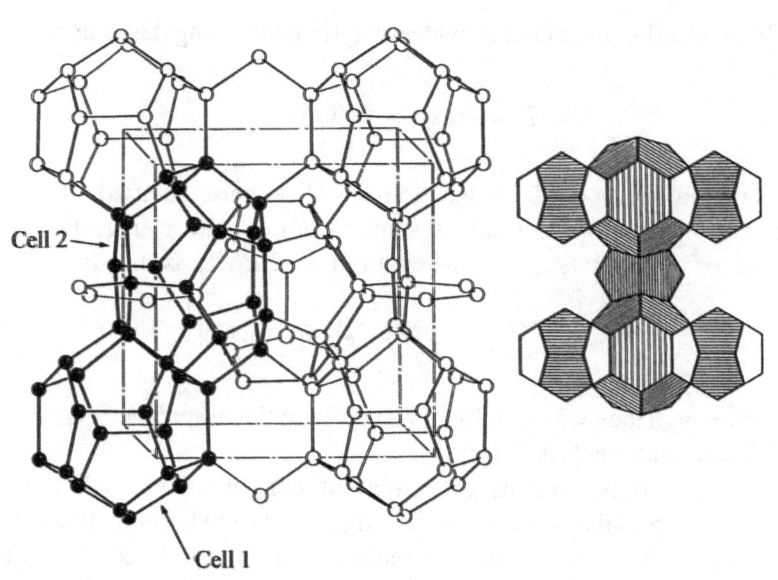

Figure 10.8 Structure I gas hydrate (water clathrate) unit cell. Guest molecules like argon, carbon dioxide, and methane are trapped inside cages (cells 1 and 2) created by a rigid lattice of water molecules.

In these systems configurational effects can be separated into contributions from each cell type. Van der Waals and Platteauw (1959) developed a model for clathrates that neglects interactions beyond those between the guest molecule and the first nearest neighbor host water molecules that comprise each cell. Their model expresses the fugacity of guest i in cell J as,

$$\hat{f}_i = \frac{1}{C_{iJ}(T)} \frac{y_{iJ}}{1 - y_{iJ}} \tag{10-172}$$

where y_{iJ} is the occupation probability of guest i in cell J that contains N_J water molecules and $C_{iJ}(T)$ is a 3-D Langmuir constant that can be related to a configurational integral $\mathbf{Z}^*_{iJ}$:

$$C_{iJ}(T) = \frac{\mathbf{Z}^*_{iJ}}{kT} \tag{10-173}$$

with

$$\mathbf{Z}^*_{iJ} \equiv \frac{1}{4\pi} \int \cdots \int \exp\left[\frac{-\Phi_{cell\,J}}{kT}\right] dr_{i1} \cdots dr_{i,N_j+1}\, d\omega_i \tag{10-174}$$

from Eq. (10-78) and with pairwise additivity assumed,

$$\Phi_{cell\,J} = \sum_{j=1}^{N_j} \Phi_{ij}(r_{ij}, \omega_i)$$

where r_{ij} = distance between guest i and the jth host molecule and ω_i = orientation angle vector for guest i

Eq. (10-174) can be rewritten using spherical coordinates (r, θ, ϕ) and three Euler angles (α, β, γ) to specify the position and orientation of the guest molecule in the cell (note that for spherical guests, such as argon, the Euler angles are not needed and for linear guests, such as carbon dioxide, only two are required):

$$\mathbf{Z}^*_{iJ} = \frac{1}{4\pi}\int_0^{R_{cellJ}}\int_0^{\pi}\int_0^{2\pi}\int_\alpha\int_\beta\int_\gamma \exp\frac{[-\Phi_{cell\,J}]}{kT} r^2 \sin\theta d\phi d\theta dr d\alpha d\beta d\gamma \tag{10-175}$$

MC integration approximates $\mathbf{Z}^*_{iJ}$ over M moves within the cell volume $\underline{V}_{cell\,J}$ as,

$$\mathbf{Z}^*_{iJ}\Big|_{MC} \approx \frac{1}{M}\sum_{m=1}^{M} \exp\left[\frac{-\Phi_{cell\,J}}{kT}\right]\underline{V}_{cell\,J} \tag{10-176}$$

For spherical guest molecules the integration is dimensionally simple enough to use quadrature methods. In addition, the Lennard-Jones-Devonshire (LJD) model provides another approximation of Z^*_{iJ} where the potential for the water cage is smeared over the surface of the cell using a mean spherical approximation. Thus

$$\mathbf{Z}^*_{iJ}\Big|_{LJD} \approx 4\pi\int_o^{R_{cellJ}} \exp\left[\frac{-\Phi_{cell\,J}}{kT}\right] r^2 dr \tag{10-177}$$

A comparison of the MC integration, quadrature approximation and the LJD model shows similar results for cell 1 of a structure 1 water clathrate containing a spherically symmetric guest with a Lennard-Jones 6-12 potential used to model binary interactions [see Sparks (1991) for details].

While Monte Carlo methods have been applied extensively to the simulation of single component, single and two phase systems, their use for multicomponent phase equilibrium problems has been restricted until recently because of computational limitations. In principle, MC methods should be able to predict non-ideal behavior in multiphase mixtures of several components as they encode the effects of intermolecular interactions using appropriate potential energy functions. By applying phase equilibrium criteria and specific constraints (for example, a specified molar feed or constant total volume), phase compositions should be predictable. Aside from the inherent difficulties of establishing a representative potential energy model with appropriate binary interaction terms and "mixing rules", the computational problems are significant.

Early approaches used the grand canonical ensemble. Unfortunately, these methods were very inefficient requiring complete configurational calculations at specified μ's to establish phase coexistence conditions. In 1987, Panagiotopoulos developed an efficient simulation algorithm, called the *Gibbs ensemble technique*, that incorporates features of the canonical ($T\underline{V}N$), grand canonical ($T\underline{V}\mu$) and the isobaric/isothermal PTN ensembles to guarantee equality of T, P, and μ among coexisting phases.

The Gibbs ensemble method is based on performing a simulation in a system with two regions, each representative of a small volume of a homogeneous phase. There are no explicit interfaces present in the system. The conditions of phase equilibrium between the two regions are satisfied by performing three distinct operations: (1) internal displacements of particles, (2) changes of the volumes of the two regions to achieve equality of pressures, and (3) particle transfers between regions to ensure chemical potential equality. The volume change steps can be performed at constant total volume or constant pressure, corresponding respectively to the physical processes of an isothermal flash into a constant-volume container or to a specified pressure. Since its inception, the method has been used extensively to obtain phase diagrams of pure fluids and mixtures.

The method in its original form is applicable without major difficulties for components that can be described with a small number of interaction centers, depending also on the strength of the interactions and the density. For components with more segments, or at high densities, the particle transfer step has a low probability of success and the statistical uncertainty of the results rapidly increases.

A major improvement in sampling efficiency for multisegment molecules results from combination of configurational-bias methods with the Gibbs ensemble, as suggested by Laso *et al.* (1992) and, independently, Mooij *et al.* (1992). Configurational-bias sampling draws upon ideas originated by Rosenbluth and Rosenbluth (1955) and relies on the "smart" insertion of one segment at a time of the molecule being transferred. A number of trial directions is selected at each growth step, and the decision as to along which direction to proceed is made based on the Boltzmann factor of the energies of interaction experienced by the growing segments. The bias introduced by the preferential sampling along favorable directions is restored through incorporation of appropriate factors (termed "Rosenbluth weights") in the acceptance criteria. Combination of configurational-bias sampling and the Gibbs ensemble has been used to predict the phase behavior of *n*-alkanes.

Readers interested in pursuing further study of Gibbs Ensemble Monte Carlo techniques should consult the reviews of Panagiotopoulos (1992, 1995) and Smit (1993).

10.9.3 Using Monte Carlo to Estimate Molecular Structure. We can easily illustrate the use of Monte Carlo for estimating the radial distribution or pair correlation function $g(r)$: our earlier definition of $g(r_{ij})$ in Section 10.7, can be reformulated for the canonical ensemble as:

$$g(r_{ij}) = g(r_1, r_2) = \frac{N(N-1)}{\rho^2 \mathbf{Z}_N^*} \underbrace{\int \cdots \int}_{N-2} \exp[-\beta\Phi(\underset{\sim}{\mathbf{r}}^N)]d\mathbf{r}_3 \ldots d\mathbf{r}_N \qquad (10\text{-}178)$$

As the choice of i=1 and j=2 is arbitrary in a system of identical molecules, $g(r_{ij})$ = $g(r)$ and

$$g(r) = \frac{\rho(r)}{<\rho>} = \frac{<\sum_i \sum_{i \neq j} \delta(r_i)\delta(r_j - r)>}{(<\rho>)^2} = \frac{V<\sum_i \sum_{i \neq j} \delta(r - r_{ij})>}{N^2} \qquad (10\text{-}179)$$

where δ( ) is the delta function of the specified argument. In practical simulations, the function is replaced with a function that is non-zero over a small but finite range of separation distances.

10.10 Summary

This chapter focused on making connections between molecular-level effects such as intermolecular forces and intramolecular modes of energy storage (vibration, internal rotation, etc.) and macroscopic thermodynamic properties such as those expressed in a $P\underline{V}TN$ equation of state or in an ideal-gas state heat capacity $[C_v^o(T)$ or $C_p^o(T)]$. Statistical mechanics provides a formal, rigorous basis for making these connections. The postulates and mathematical principles that describe the behavior of systems consisting of a large number of particles or molecules constitute the major elements of statistical mechanics. The Gibbs ensembles provided a convenient means to describe the states and interactions of these N-particle systems. Just as we have different thermodynamic functions to describe different types of systems, different ensembles are available: microcanonical with $(\underline{E}, \underline{V}, N)$ fixed, canonical with $(T, \underline{V}, N)$ fixed, grand canonical with $(T, \underline{V}, \mu)$ fixed, and isobaric/isothermal with (P, T, N) fixed. These ensembles provide alternative ways of expressing the information content of statistical mechanics. Legendre transformations among the canonical variables of the Fundamental Equation ($\underline{E}$, $\underline{S}$, $\underline{V}$, and N) and their conjugate coordinates can be used to show relationships between ensembles. Classical thermodynamics and statistical mechanics were connected by both entropy and Helmholtz free energy using the canonical partition function Q_N which specifies the distribution of allowable energy states in an N-molecule system.

In Chapter 10, we saw that for most problems of interest to chemical engineering, that a semi-classical rather than a full quantum treatment was adequate. In practical terms, this allowed us to factor the partition function into an internal, intramolecular contribution containing vibrational and internal rotational effects and an external contribution containing 3-D translation and intermolecular potential energy interactions. The internal part (q_{int}^N) depended on molecular structure and temperature but was assumed to be independent of density. Quantum effects such as the discretization of intramolecular vibrational energy levels, important to $C_v^o(T)$, are also incorporated into q_{int}^N. Other appropriate quantum and statistical corrections were added to the external part, such as the $1/N!\, h^{3N}$ factor to account for the indistinguishability of particles and the "volume" of phase space. The potential energy (Φ) contribution was expressed classically as a $3N$-fold integral over configurational space with Boltzmann-weighting $[\exp(-\Phi/kT)]$. In one important application, the resulting configurational integral could be related directly to the ensemble-averaged pressure to provide a macroscopic $P\underline{V}TN$ EOS.

The remaining sections of the chapter (10.4 - 10.8) explored various methods for approximating the configurational integral as well as the complete partition function of the system. Several important models were discussed that produced equations of state and an interpretation of their fitted parameters. These include the virial density expansion for gases, molecular-level corresponding states and conformal fluid theory, and the van der Waals partition function decomposition which introduced the concepts of free and excluded volume as well as the mean field approximation. Radial distribution functions where then used to illustrate how molecular structure and ordering would be represented for gases, liquids, and solids. The radial distribution function itself [$g(r)$] then becomes the building block for a variety of integral equation theories of dense fluids including models due to Kirkwood, Bonn-Green-Yvon, and Ornstein-Zernike. Hard sphere fluids were treated in some detail because they represent a practical reference condition for perturbation and density-functional theories for liquids.

The final section of the chapter described the basic elements of the Monte Carlo and Molecular Dynamics methods of molecular simulation. With a well-defined intermolecular potential energy function and the pairwise additivity assumption, both methods can be used to estimate the structure and configurational properties of molecular fluids ranging in density from gases to liquids.

References and Suggested Readings

Abbott, M.M. and J.M. Prausnitz (1987), "Generalized van der Waals theory: A classical perspective," *Fluid Phase Equilibria*, **37**, p 29-62.

Abbott, M.M. (1989), "Thirteen ways of looking at the van der Waals equation," *Chem. Eng. Prog.*, **85**(2), p 25-37.

Alder, B.J. and T.E. Wainwright (1959), "Studies in molecular dynamics. I. General method," *J. Chem. Phys.*, **31**, p 459-466.

Alder, B.J. and T.E. Wainwright (1960), "Studies in molecular dynamics. II. Behavior of a small number of elastic spheres," *J. Chem. Phys.*, **33**, p 1439-1451.

Allen, M.P. and D.J. Tildesley (1987), *Computer Simulation of Liquids*, Oxford Science Publications, Oxford, UK.

Andrews, F.C. (1975), *Equilibrium Statistical Mechanics*, 2nd ed., John Wiley & Sons, Inc., New York, NY.

Barker, J.A. and Henderson, D. (1976), "What is a liquid? Understanding the states of matter," *Reviews of Modern Physics*, **48**, p 587-671.

Chandler, D. (1987), *Introduction to Modern Statistical Mechanics*, Oxford University Press, Oxford, UK.

Davidson, N. (1962), *Statistical Mechanics*, McGraw-Hill, New York, NY, Ch. 15, Section 14 p 345 ff.

DePablo, J.J. and Prausnitz, J.M. (1989), "Phase equilibria for fluid mixtures from Monte-Carlo simulations," *Fluid Phase Equilibria*, **53**, p 177-189.

Ely, J.F. and Marrucho, M.F. (1995), "The corresponding states principle," in *Equations of State for Fluids and Fluid Mixtures*, IUPAC, Blackwell Scientific Publications, London, Chapter 8.

Fowler, R.H. and Guggenheim, E.A. (1956), *Statistical Thermodynamics*, Cambridge University Press, Cambridge, UK.

Hansen, J.P. and McDonald, I.R. (1986), *Theory of Simple Liquids*, 2nd ed., Academic Press, New York.

Harle, J.M. and Mansoori, G.A. (1983), *Molecule-Based Study of Fluids*, Advances in Chemistry Series, No. 204, American Chemical Society, Washington, DC.

Hirschfelder, J.O., Curtiss, C.F., Bird, R.B. (1954), *Molecular Theory of Gases and Liquids*, John Wiley & Sons, Inc., New York, NY, Chs. 1-3, p 1-173.

Hougen, O.A., Watson, K.M. and Ragatz, R.A. (1959), *Chemical Process Principles, Part II, Thermodynamics*, 2nd ed., John Wiley & Sons, Inc., New York, NY.

Kutney, M. (1996), personal communication, Massachusetts Institute of Technology, Cambridge, MA.

Laso, M., de Pablo, J.J. and Suter, U.W. (1992), "Simulation of phase equilibria for chain molecules," *J. Chem. Phys.* **97**, p 2817-2819.

Lewis, G.N. and Randall M. (1961), *Thermodynamics*, 2nd ed., rev. by L. Brewer and K.S. Pitzer, McGraw-Hill Book Company, New York.

London, F. (1937), *Trans. Faraday Society*, **33**(8).

Margenau, H. and Kestner, N.R. (1969), *Theory of Intermolecular Forces*, Pergamon Press, New York.

Mayer, J.E. and M.G. Mayer (1940), *Statistical Mechanics*, John Wiley & Sons, Inc., New York, NY, Ch. 13, p 277-294.

McQuarrie, D.A. (1976), *Statistical Mechanics*, Harper and Row, New York.

Metropolis, N., Rosenbluth, M.N., Teller, A.H. and Teller, E. (1953), "Equation of state calculations by fast computing machines," *J. Chem. Phys.*, **21**, p 1087-1092.

Mooij, G.C.A.M., Frenkel, D. and Smit, B. (1992), "Direct simulation of phase equilibria of chain molecules," *J. Phys. Condens. Matter*, **4**, p L255-L259.

Panagiotopoulos, A.Z. (1987), "Direct determination of phase coexistence properties of fluids by Monte Carlo simulation in a new ensemble," *Molec. Phys.*, **61**, p 813-826.

Panagiotopoulos, A.Z. (1992), Direct determination of fluid phase equilibria by simulation in the Gibbs ensemble: A review. *Molec. Simulation.*, **9**, p 1-23.

Panagiotopoulos, A.Z. (1995), "Gibbs ensemble simulations," In M. Baus, L.F. Rull and J.P. Ryckaert (eds.), *Observation, Prediction and Simulation of Phase Transitions in Complex Fluids*, Kluwer Academic Publishers: Dordrecht, The Netherlands, p 463-501.

Pitzer, K.S., Lippmann, D.Z., Curl, R.F., Huggins, C.M. and Petersen, D.E. (1955), *J. Am. Chem. Soc.*, **77**, p 3433.

Prausnitz, J.M., R.N. Lichtenthaler and E. Gomes de Azevedo (1986), *Molecular Thermodynamics of Fluid-Phase Equilibria*, 2nd ed., Prentice Hall, Englewood Cliffs, NJ, Section 10.5, p 474ff.

Reagan, M.T., R.D. Weinstein, and J.G. Harris (1996), personal communication, Massachusetts Institute of Technology, Cambridge, MA.

Reed, T.M. and Gubbins, K.E. (1973), *Applied Statistical Mechanics*, Butterworth-Heinemann, Boston, MA.

Reid, R.C., J.M. Prausnitz, and B. Poling (1987), *The Properties of Gases and Liquids*, 4th ed., McGraw-Hill, New York.

Rosenbluth, M.N. and Rosenbluth, A.W. (1955), "Monte Carlo calculations of the average extension of molecular chains," *J. Chem. Phys.*, **23**, p 356-359.

Rowley, R.L. (1994), *Statistical Mechanics for Thermophysical Property Calculations*, PTR Prentice Hall, Englewood Cliffs, NJ.

Smit, B. (1993), "Computer simulations in the Gibbs ensemble," in Allen, M.P.; Tildesley, D.J. (eds.), *Computer Simulation in Chemical Physics*, Kluwer Academic Publishers: Dordrecht, The Netherlands, p 173-209.

Sparks, K.A. (1991), "Configurational properties of water clathrates through molecular simulation," PhD thesis, Department of Chemical Engineering, Massachusetts Institute of Technology, Cambridge, MA.

Syrkin, Y.K. and Dyatkina, M.E. (1950), *Structure of Molecules and the Chemical Bond*, translated and revised by M.A. Partridge and D. Jordan, New York Interscience.

Tolman, R.C. (1938), *Statistical Mechanics*, Oxford University Press, London [available in unabridged and unaltered form from Dover Publications, Inc., New York (1979)].

van der Waals, J.D. (1967), "The equation of state for gases and liquids," Nobel lecture, December 12, 1910, in *Nobel Lectures - Physics 1901-1921*, Elsevier, NY.

van der Waals, J.H. and J.C. Platteeuw (1959), "Clathrate solutions," *Adv. Chem. Phys.*, **2**, p 1.

Vera, J.H. and J.M. Prausnitz (1972), "Generalized van der Waals theory for dense fluids," *Chem. Eng. J.*, **3**, p 1.

Widom, B. (1963), "Some topics in the theory of fluids," *J. Chem. Phys.*, **39**, p 2808.

Problems

10.1. For large systems, show that the canonical partition function Q_N is the Laplace transform of the inverse of the microcanonical probability density function $[C/\Phi(\underline{E})]^{-1}$ given in Eq. (10-11). What can you say about the information content of both ensembles, given the uniqueness properties of Laplace transforms.

10.2. Using the truncated form of the perturbed hard-sphere modified van der Waals equation given in part (b) of Problem 10.7, develop an expression for the enthalpy departure function from a state at (T, P, V) to a suitable ideal gas state at a volume $V^o = RT/P$. The constants a and b are independent of temperature. All integrals and/or derivatives should be evaluated.

10.3. What is the maximum value of the ratio of Φ_{ij}/kT for the dipole-dipole interaction between two water molecules with dipole moments of 1.84 Debye at 300 K separated by distances of 2, 3, 5, and 10 Å?

10.4. What is the *average* dipole interaction energy between two dipoles at a distance r apart, giving equal weight to all possible angular configurations?

10.5. Repeat 10.4, this time use the Boltzmann statistical weighing factor that favors those orientations with lower (more attractive) potential energies i.e., if $\Phi_{ij} = f[\alpha_1, \alpha_2, (\gamma_1 - \gamma_2)]$ then

$$<\Phi_{ij}> = \frac{\int_0^{2\pi}\int_0^{\pi}\int_0^{\pi} \Phi_{ij} \exp(-\Phi_{ij}/kT) \sin\alpha_1 \sin\alpha_2 \, d\alpha_1 \, d\alpha_2 \, d(\gamma_1 - \gamma_2)}{\int_0^{2\pi}\int_0^{\pi}\int_0^{\pi} \exp(-\Phi_{ij}/kT) \sin\alpha_1 \sin\alpha_2 \, d\alpha_1 \, d\alpha_2 \, d(\gamma_1 - \gamma_2)}$$

and Φ_{ij} is the dipole interaction energy at a specified separation distance r. Hint: you should expand $\exp(-\Phi_{ij}/kT)$ in a power series and truncate as appropriate.

10.6. Use the result obtained in Problem 10.5 to calculate the average dipole-dipole interaction energy between two water molecules separated by 2, 3, 5, and 10 Å. How does this compare with the dipole-dipole interaction between two carbon monoxide molecules?

10.7. (a) Assuming that a pure fluid follows the van der Waals EOS, derive an equivalent expression in virial form as:

$$PV = RT + B(T)/V + C(T)/V^2$$

that would be approached as $\rho \to 0$ where $\rho = 1/V$. Develop an expression for $B(T)$ in terms of the van der Waals constants a and b and P, V, T variables.

$$\text{Note that } (1-x)^{-1} = 1 + x + x^2 + x^3 + x^4 + \dots \quad \text{if } x^2 < 1$$

(b) If we were to modify the van der Waals EOS to more accurately account for the repulsions as "perturbed-hard-sphere" interactions, the following form for the EOS would result:

$$P = \frac{RT}{V}[1 + 4y + 10y^2 + 18.365y^3 + 28.24y^4 + \ldots] \;\; - a/V^2$$

$$\text{where } y = b/(4V) = (b/4)\rho$$

As a first order approximation of repulsions, consider only the first three terms in the square bracket (up through $10y^2$). The resulting equation is then a truncated cubic, hard-sphere EOS with van der Waals attractions. Describe how you would calculate the parameters a and b from a measured critical pressure (P_c) and critical temperature (T_c) for the fluid. (Note that the critical volume (V_c) is not known for this particular fluid.)

10.8. Frequently in the study of dense fluids in statistical thermodynamics, the chemical potential is used as a starting point for derivations:

$$\mu_i = -kT\left[\frac{\partial \ln Q_N}{\partial N_i}\right]_{T,\,\underline{V},\,N_j[i]} \qquad \text{(P10-1)}$$

(a) Show that for large $N = N_i$ with only one component i

$$\mu = \mu_i = -kT \ln \frac{Q_{N+1}}{Q_N} = -kT \ln\left[\frac{Q^*_{t,i}}{N\underline{V}} \frac{\mathbf{Z}^*_{N+1}}{\mathbf{Z}^*_N}\right] \qquad \text{(P10-2)}$$

where $Q^*_{t,i} \equiv \underline{V}(q_{r,\,v,} \ldots /\Lambda^3) = Q_1$ represents the combined translational and internal (vibrational, rotational, etc.) canonical partition function for a single ($N = 1$) molecule. Q_{N+1} and $\mathbf{Z}^*_{N+1}$ are the canonical partition function and configurational integral for a system of $N + 1$ molecules, respectively, while Q_N and $\mathbf{Z}^*_N$ correspond to those quantities for an N-molecule system. Equation (P10-2) effectively defines the basis for the Widom particle insertion method used in molecular simulation.

(b) Show that for a spherical molecule (e.g., argon or neon)

$$\frac{\mathbf{Z}^*_{N+1}}{\mathbf{Z}^*_N} = \int \exp\,[-2\Phi_{N+1}/kT]d\mathbf{r}_{N+1} \qquad \text{(P10-3)}$$

where only a single coordinate in configuration space ($\mathbf{r}_{N+1}$) is required for the integration. Note that Φ_{N+1} is the potential energy of the (N+1)th molecule as inserted into a system of N molecules.

(c) Using Eq. (P10-2) and the semiclassical approximation, show that

$$\mu = -kT \ln\left[\frac{Q^*_{t,i}}{N} < \exp\,(-2\Phi_N/kT) >\right] = -kT \ln\left[\frac{Q^*_{t,i}}{N}\,\mathrm{a}^{-1}\right]$$

where Φ_N corresponds to the total potential energy of a molecule in a system containing N other particles and $< >$ defines the normal ensemble average and a is the activity = $(\mathbf{Z}^*_{N+1}/(\underline{V}\mathbf{Z}^*_N))^{-1}$.

(d) Using the fundamental equation relationship for the single component chemical potential at constant temperature,

$$d\mu = VdP$$

and the Lewis definition of activity, a

$$\mu(T, \rho = 1/V) = \mu^o(T, \rho = 1/V) + kT \ln a$$

where μ^o refers to the standard state as an ideal gas at the same temperature and density as the fluid of interest, demonstrate that an EOS can be developed as:

$$\frac{P}{\rho kT} = 1 - \ln a^{-1} + \frac{1}{\rho}\int_0^{\rho} \ln a^{-1} d\rho$$

(e) The term a^{-1} can be equated to the probability that a point chosen at random in an N molecule system has no neighboring molecule with its center within a distance σ. For a system of hard spheres we can estimate a by the product of two terms: $a^{-1} = \boldsymbol{P}(1)\,\boldsymbol{P}(2)$.

$\boldsymbol{P}(1)$: the probability that a point r selected at random in $\underline{V}$ does not lie inside the core of one of the N molecules. For hard spheres

$$\boldsymbol{P}(1) = 1 - \frac{4}{3}\pi\left(\frac{\sigma}{2}\right)^3 \rho = 1 - \frac{1}{8}b\rho$$

$\boldsymbol{P}(2)$: the probability that about r there is also no particle within a distance σ. For large N, $\boldsymbol{P}(2)$ becomes

$$\lim_{N\to\infty} \boldsymbol{P}(2) = \lim_{N\to\infty}\left[1 - \frac{7b}{8(\underline{V} - N\omega)}\right]^N = \exp\left[\frac{-7\rho b}{8(1-\rho\omega)}\right]$$

where ω = close-packed molecular volume. Using the results of part (d), develop an EOS of the form $P/\rho kT = f(\rho)$ with b and ω as parameters.

10.9. Assuming the van der Waals equation applies to a fluid region volume $\underline{V} = L^3$, answer (a), (b), and (c).

(a) Because of molecular fluctuations, the local density at a particular point can differ from its average by an amount $\delta\rho$. Derive an expression for scaling the size of these fluctuations as a function of the following:

$$\frac{[<(\delta\rho)^2>]^{1/2}}{\rho} = f[\beta = 1/kT, \rho, a, b, L^3]$$

note that vdW EOS can be written as

$$\beta P = \frac{\rho}{1 - b\rho} - \beta a \rho^2$$

where $\rho = <N>/\underline{V}$

(b) Using appropriate criteria for the liquid-vapor critical point, determine ρ_c and β_c as a function of a and b for this vdW fluid.

(c) If $L^3 = 100b$ estimate the magnitude of the density fluctuations along the critical isochore ($\rho = \rho_c$) for $T/T_c = 1.10, 1.01, 1.001, 1.00001$. How close to the critical point must we be before light scattering (critical opalescence) phenomena are observed? Assume that $b = 5$ (Å)^3/molecule and the average wavelength of visible light $\gamma = 1000$ Å.

10.10. In its normal formulation the virial equation expresses the compressibility factor (Z) as a density expansion,

$$Z = 1 + B\rho + C\rho^2 + \ldots$$

At times, a pressure expansion is preferred, where

$$Z = 1 + B'P + C'P^2 + \ldots$$

(a) Express the virial coefficients B' and C' in terms of the constants a and b of the Redlich- Kwong-Soave (RKS) EOS (see Table 8.1).

(b) For a binary equimolar gas mixture of CO_2 and CH_4, estimate the fugacity of CO_2 at 100°C, 150 bar. Neglect virial coefficients beyond the second. For the second virial coefficient, use the relation found in (a) with assumed mixing rules. The experimental fugacity coefficient of CO_2 under these conditions is 0.72.

10.11 Compute the 2nd virial coefficient for a pure system that behaves as a hard sphere fluid and as a square well fluid (with $R\sigma = 1.5\sigma$). How do these results compare with those shown in Figure 10.3b for fluids that follow the Lennard-Jones 6-12 and Kihara potentials? What happens to $B(T)$ for the hard sphere, square well, and L-J 6-12 fluids as T gets very large, say $T^* = 1000$? Note that $B(T)$ for a L-J 6-12 fluid is given analytically from Eq. (10-127) as

$$B(T)_{L\text{-}J\,6\text{-}12} = -\frac{2\pi\sigma^3}{3}\sum_{n=0}^{\infty}\frac{2^{\frac{2n+1}{2}}}{4\,n!}\left(\frac{\varepsilon}{kT}\right)^{\frac{2n+1}{4}}\Gamma\left(\frac{2n-1}{4}\right)$$

where $\Gamma(\,)$ is the gamma function. Using the recursive relation $\Gamma(n+1) = n\Gamma(n)$ and the following tabular data one can generate all required values of Γ.

x	$\Gamma(x)$
1	1.00000
1.1	0.95135
1.2	0.91817
1.3	0.89747
1.4	0.88785
1.5	0.88623
1.6	0.89352
1.7	0.90864
1.8	0.93138
1.9	0.96177
2.0	1.00000

10.12 The Sutherland potential combines hard sphere repulsion with Lennard-Jones $1/r^6$ attraction to give:

$$\Phi_{ij}(r) = \begin{bmatrix} \infty \quad \text{for} \quad r < \sigma \\ -\varepsilon(\sigma/r)^6 \quad \text{for} \quad r > \sigma \end{bmatrix}$$

Show that the vdW canonical partition function given by Eq. (10-147) can be derived exactly from the Sutherland potential when $|\Phi_{ij}|$ is defined as the volume-averaged potential from $r = \sigma$ to ∞ with uniform weighting.

Models for Non-Ideal, Non-Electrolyte Solutions 11

The main objective of this chapter is to describe and evaluate some of the models that chemical engineers employ for characterizing non-ideal effects in condensed phase mixtures. In this context, we limit the treatment in this chapter to non-dissociated, non-ionic species in liquid solutions. In addition, we will also briefly discuss models for solid solutions. Chapters 8, 9, and 10 of *Part II - Thermodynamic Properties* provide the background material for the presentation here. In effect, we are extending those developments to establish constitutive property relationships appropriate for non-ideal mixtures subject to the constraints of classical thermodynamics, including mole/mass conservation, the First and Second Laws, adherence to Gibbs-Duhem consistency, and the required asymptotic condition that all excess properties must equal zero at the pure component limits ($x_i = 1.0$ for all $i = 1, \ldots, n$). Two general frameworks are used, the first is based on $P\underline{V}TN$ equation of state constitutive models and the second on a condensed-phase excess Gibbs free energy (ΔG^{EX}) models. Both express the mixture fugacity of each component i, the first using the fugacity coefficient $\hat{\phi}_i^L$ and the second using the activity coefficient γ_i. Using the definitions and nomenclature introduced in Sections 9.7 and 9.8, the general relationships for each are related:

$$\hat{f}_i = \hat{f}_i^{+} \exp\left[\frac{\mu_i - \mu_i^{+}}{RT}\right] = \hat{\phi}_i^L x_i P = \gamma_i x_i \hat{f}_i^{+} \tag{11-1}$$

We discuss both theoretical and empirical methods to characterize and correlate non-ideal behavior. Theoretical, molecular-based techniques using many of the statistical mechanics concepts introduced in Chapter 10 are employed when possible to provide a more fundamental basis for the constitutive equations used to model non-ideal behavior. But in every situation, no matter how general and robust the theoretical approach is, experimental data are required before any of these models can be used effectively to quantitatively represent non-idealities. Commonly, both pure component properties and adjustable binary interaction parameters, determined by fits to experimental data, are used in equations that characterize non-ideal behavior in multicomponent solutions. Consequently, one must not underestimate the value of having a few good measurements to assist the process of generating a tractable model.

In chemical engineering practice, the main sources of experimental data come from vapor-liquid, liquid-liquid, and solid-liquid phase equilibrium and solubility measurements that are of importance to distillation, extraction, adsorption, crystallization, and other separation processes. In addition, from physical chemistry sources, colligative property data are also frequently available. These are in the form of osmotic pressure, freezing point depression, vapor pressure lowering, and boiling point elevation measurements.

Readers interested in an in-depth discussion of these topics should consult the references listed at the end of the chapter. Of particular value are the monographs by Prausnitz *et al.* (1986), Walas (1985), and Reid *et al.* (1987).

11.1 $P\underline{V}TN$ EOS — Fugacity Coefficient Approach

In Section 9.7, we introduced fugacity as a parameter to express non-ideal behavior. The component fugacity in a mixture is typically evaluated through the fugacity coefficient as shown in Eq. (11-1) above. With knowledge of the volumetric properties of the mixture in terms of temperature, pressure or density, and composition we can determine the fugacity coefficient. Recalling Eqs. (9-142) and (9-129):

$$RT \ln \hat{\phi}_i^L = RT \ln \left(\frac{\gamma_i \hat{f}_i^+}{P} \right) = \int_0^P \left(\overline{V}_i - \frac{RT}{P} \right) dP$$

or

$$RT \ln \hat{\phi}_i^L = -\int_\infty^{\underline{V}} \left[\left(\frac{\partial P}{\partial N_i} \right)_{T,\underline{V},N_j[i]} - \frac{RT}{\underline{V}} \right] d\underline{V} - RT \ln Z \tag{11-2}$$

If one has experimental data or a $P\underline{V}TN$ EOS that works well for liquid phase mixtures, then $\hat{\phi}_i^L$ can be calculated directly. Example 9.7 illustrates this calculation for the Peng-Robinson EOS. In this case the constitutive property relationship is a well-tested cubic equation of state with a set of binary interaction parameters that have been regressed and tabulated for a large number of two-component mixtures of interest. With a cubic EOS, the liquid phase properties correspond to the highest density (lowest specific volume) root (see Section 8.3). A critical issue is whether the EOS accurately represents liquid phase behavior. The answer depends partly on the level of the non-idealities present. In other cases, the mathematical form of the EOS does not provide a liquid-phase root; or even if it does, the EOS may not be suited to mixture calculations.

In Section 9.2 we introduced mixing rules [Eq. (9-19)] that incorporate non-ideal effects with δ_{ij}, a binary interaction parameter, by inserting it directly into the defining equation for the attractive term a_{ij} of the EOS. The effect of δ_{ij} on $\hat{\phi}_i$ can be quite large, so having realistic values for δ_{ij} is important. Eventually some testing of the EOS against mixture data is needed. If binary interaction parameters are required then they either have been adjusted to fit an earlier set of two-component data or they must be determined from the data you have available to you. These data typically consist of some type of phase equilibrium measurement that allows you to use the equality of chemical potential or fugacity between phases to determine the liquid phase fugacity and hence the fugacity coefficient $\hat{\phi}_i$. The integrations shown in Eq. (11-2) also often introduce errors as they require a mixture $P\underline{V}TN$ EOS to represent volumetric properties from a low-density, ideal-gas-state ($P \to 0$ or $V \to \infty$) to a high-density, compressed-liquid state. For mixtures consisting of relatively non-volatile liquids or solids dissolved in a supercritical solvent (see Section 15.8) and for a number of non-polar hydrocarbon mixtures, the $P\underline{V}TN$ EOS-fugacity coefficient approach yields satisfactory results assuming, of course, that one has a reasonably accurate EOS that works for the liquid phase mixture. For other systems that exhibit strong non-idealities, for example liquid-liquid phase splitting, the EOS-fugacity coefficient is usually not acceptable.

11.2 ΔG^{EX} — Activity Coefficient Approach

Given the inherent limitations and potential inaccuracies of the liquid phase fugacity coefficient approach, an alternative means of estimating non-ideal effects in condensed phase media is needed. The activity approach provides such an alternative. Here we define the activity as the ratio of the real fugacity of component i ($\hat{f}_i$) in a multicomponent solution relative to its value in a specified standard state ($\hat{f}_i^+$). As discussed in Section 9.8, the activity is usually divided into two parts, a non-ideal contribution expressed by the activity coefficient (γ_i) and an ideal contribution. Mathematically this is given by,

$$\hat{f}_i = \gamma_i x_i \hat{f}_i^+ \tag{11-3}$$

where $\hat{f}_i^+$ equals the fugacity of component i in its reference or standard state at T^+, P^+, and x_i^+, and x_i is the mole fraction. In normal engineering practice, three reference states are used. They are all defined at the temperature and pressure of the mixture: (1) pure component i, (2) component i in a state of infinite dilution and (3) component i in a hypothetical state following Henry's law behavior at unit molality or some other composition (see Section 9.8 for further discussion). By far the most common is (1), where symmetrical normalization is used and γ_i is a direct measure of deviation from ideal solution behavior. In this case:

$$\gamma_i = \hat{f}_i / x_i f_i^{pure} = \hat{f}_i / x_i f_i$$

Here the definition of an ideal solution follows the Lewis and Randall rule with $\gamma_i = 1.0$. Thus, $\gamma_i > 1.0$ represents a positive deviation from the Lewis and Randall rule and $\gamma_i < 1.0$, a negative deviation. With the pure component standard state, γ_i is obtained from the excess Gibbs free energy of mixing in either extensive ($\Delta \underline{G}^{EX}$) or intensive ($\Delta G^{EX}$) form:

$$RT \ln \gamma_i = \overline{\Delta G_i^{EX}} = \left(\frac{\partial \Delta G^{EX}}{\partial N_i} \right)_{T,\,P,\,N_j[i]} = \Delta G^{EX} - \sum_{j \neq i} x_j \left(\frac{\partial \Delta G^{EX}}{\partial x_j} \right)_{T,\,P,\,x_k[i,\,j]} \tag{11-4}$$

So if we have an expression for

$$\Delta G^{EX} = f(T, P, x_1, \ldots, x_n) \tag{11-5}$$

and we can obtain γ_i directly. Experimental data and/or a suitable model are needed to quantitatively represent ΔG^{EX}. Important requirements for any model, whether it's theoretical or empirical, include (1) all pure component conditions at $x_i = 1$ (i=1, . . . , n) must yield $\gamma_i = 1.0$ and $\Delta G^{EX} = 0$ and (2) the Gibbs-Duhem relationship [Eq. (9-183)] must be followed. Although many mathematical functional forms for ΔG^{EX} will satisfy these criteria, only a relatively few have been employed in practice. This is partly because simple mathematical forms, for example first or second degree polynomials or exponential or logarithmic functions in composition, yield satisfactory results and partly because the underlying theories for many models are commonly expressed in similar functional forms themselves. Again we want to emphasize the key role that experimental data play in specifying γ_i. A single data point at one composition and T and P, for example a known composition of a binary azeotrope or a measured osmotic pressure or freezing point depression, is sufficient to specify the parameters needed for many ΔG^{EX} or γ_i models.

Let's consider one of the simpler models of non-ideal behavior from the theory of Van Laar, which evolved from the classic equation of state work by van der Waals in the late 1890's in The Netherlands. Van Laar was able to derive an analytic expression for ΔG^{EX} by creating a three-step process to prepare a liquid mixture at a specified T, P, and composition:

step (1) -- isothermal expansion of pure liquids to an ideal gas state at T
step (2) -- isothermal mixing in the ideal gas state
step (3) -- isothermal compression of the mixture to P at T

He used the pure and mixture forms of the van der Waals $P\underline{V}TN$ EOS to calculate the property changes in steps (1) and (3) and set $\Delta S^{EX} = 0$ for step (2). This results in,

$$\Delta G^{EX} = \frac{x_1 x_2 b_1 b_2}{x_1 b_1 + x_2 b_2} \left[\frac{\sqrt{a_1}}{b_1} - \frac{\sqrt{a_2}}{b_2} \right]^2 \tag{11-6}$$

where a_i and b_i are the van der Waals parameters for pure component i (see Eq. (8-3)). Eq. (11-4) gives the van Laar activity coefficients as,

$$\ln \gamma_1 = \frac{A^*}{\left(1 + \frac{A^* x_1}{B^* x_2} \right)^2} \qquad \ln \gamma_2 = \frac{B^*}{\left(1 + \frac{B^* x_2}{A^* x_1} \right)^2} \tag{11-7}$$

with

$$A^* = \frac{b_1}{RT}\left(\frac{\sqrt{a_1}}{b_1} - \frac{\sqrt{a_2}}{b_2}\right)^2 \quad \text{and} \quad B^* = \frac{b_2}{RT}\left(\frac{\sqrt{a_1}}{b_1} - \frac{\sqrt{a_2}}{b_2}\right)^2$$

which gives poor results for γ_i when compared to experimental values. Obviously the van der Waals EOS is not capturing the important non-ideal effects. However, if we allow the A^* and B^* parameters to be adjustable and fit them to limited experimental data (two experimental values of γ_i may be sufficient) then we obtain a much more accurate model for γ_i that works well over the entire composition range for a large number of non-ideal systems.

A further simplification occurs if we set $A^* = B^* = C$, which results in the 2-suffix Margules equation:

$$\ln \gamma_1 = C\, x_2^2 \quad \text{and} \quad \ln \gamma_2 = C\, x_1^2$$

Alternatively, if we have a low-pressure, binary azeotrope with $x_i^{az} = y_i^{az}$, γ_i for each component i appear immediately as,

$$\gamma_1 = \frac{P}{P_{vp_1}} \quad \text{and} \quad \gamma_2 = \frac{P}{P_{vp_2}}$$

where only the total pressure P and the vapor pressure of each component are needed to determine A^* and B^* in Eq. (11-7). Example 15.6 illustrates this approach for the n-butanol - water azeotropic system at 1 atm.

11.3 Ideal Entropy of Mixing and the Third Law

Using the statistical mechanical concepts developed in Chapter 10, we can provide a statistical basis for estimating the entropy of mixing in an ideal solution. Although we merely defined an ideal solution in Chapter 9, now we can show that the entropy of mixing in a binary solution is determined solely by the combinatorial effects of mixing two distinguishable species. The first step is to manipulate the expression we derived for the entropy in terms of the canonical partition function and the ensemble averaged energy $<\underline{E}>$ given in Eq. (10-44):

$$\underline{S} = k \ln Q_N + \frac{<\underline{E}>}{T} \tag{11-8}$$

to a form where the entropy can be interpreted as a statistical property. But recalling that the canonical partition function Q_N and the probability of occupying a quantum state of energy $\underline{E}_i$ are defined by,

$$Q_N \equiv \sum_i \exp\,[-\beta \underline{E}_i] \quad \text{and} \quad P_N(\underline{E}_i) \equiv \frac{\exp[-\beta \underline{E}_i]}{Q_N}$$

In addition, $<\underline{E}>$ is given by,

$$<\underline{E}> = \frac{\sum_i \underline{E}_i \exp[-\beta \underline{E}_i]}{Q_N}$$

Thus Eq. (11-8) can be rewritten as:

$$\underline{S} = k\left[\ln Q_N + \frac{\sum_i \beta\, \underline{E}_i \exp[-\beta \underline{E}_i]}{Q_N}\right] \tag{11-9}$$

Because

$$\frac{\sum_i \exp[-\beta \underline{E}_i]}{Q_N} = 1.0 \quad \text{and} \quad \ln\left(\exp[-\beta \underline{E}_i]\right) = -\beta \underline{E}_i$$

Eq. (11-9) is rearranged to give,

$$\underline{S} = -k\left[\frac{\sum_i \exp[-\beta \underline{E}_i]}{Q_N}\left(\ln\left(\exp[-\beta \underline{E}_i]\right) - \ln Q_N\right)\right] \tag{11-10}$$

and with the definition of $\boldsymbol{P}_N$:

$$\boldsymbol{P}_N(\underline{E}_i) \equiv \left(\exp[-\beta \underline{E}_i]\right)/Q_N$$

then $\underline{S}$ can be expressed just in terms of the probability distribution as:

$$\underline{S} = -k \sum_i \boldsymbol{P}_N(\underline{E}_i) \ln \boldsymbol{P}_N(\underline{E}_i) \tag{11-11}$$

Consider the special case where all quantum states are equally probable, that is one corresponding to the microcanonical ensemble ($\underline{E}$, $\underline{V}$, N) via the *Postulate of Equal a Priori Probability,*

$$\boldsymbol{P}_N(\underline{E}_i) = \boldsymbol{P}_1(\underline{E}_1) = \boldsymbol{P}_2(\underline{E}_2) = \ldots = \boldsymbol{P}_\Omega(\underline{E}_\Omega) = \frac{1}{\Omega} \tag{11-12}$$

where Ω = the total number of possible states. Because the probability distribution is normalized,

$$\sum_i \boldsymbol{P}_N(\underline{E}_i) = 1$$

and Eq. (11-11) can be modified to give the common statistical form for entropy,

$$\underline{S} = -k\left[\sum_i \boldsymbol{P}_N(\underline{E}_i)\right] \ln \frac{1}{\Omega} = k \ln \Omega \tag{11-13}$$

Ω is frequently called the thermodynamic probability. If $\Omega = 1.0$ only one configuration with a single energy is possible. Under these conditions the absolute entropy $\underline{S} = 0$. This is exactly the condition specified by the so-called *Third Law of Thermodynamics* which states that *at absolute zero* (0 K) *if we have only* **one** *stable configuration* (*a perfect crystal*), *then the entropy goes to zero.* Although this condition is, in its limit, undoubtedly hypothetical and certainly arbitrary from the perspective of classical thermodynamics, experimental measurements at extremely low temperatures (of 1 K and less) support the hypothesis that S tends to zero as T goes to 0 K. For example, low temperature heat capacity measurements show behavior consistent with the Debye model (Eq. (8-152)) that gives,

$$C_v \cong C_p \;\alpha\; T^3 \quad (\text{as } T \rightarrow 0 \text{ K}) \tag{11-14}$$

so that with $dS = C_p\, dT/T$ for isobaric variations or $dS = C_v\, dT/T$ for constant volume variations, ΔS approaches zero for all variations near 0 K as well. This behavior is embodied in the Nernst heat theorem.

The Third Law and the Nernst heat theorem provide a rational basis for specifying the standard state entropy as $S^o = 0$ at $T = 0$ K. Most modern tabulations of thermochemical properties use this zero absolute entropy state as their reference state. Strictly speaking, this choice is still arbitrary in that the calculation of properties of interest to chemical engineers involve only differences in derived properties. Even if the entropy could be specified absolutely, we still would have problems with expressing absolute energy quantities, such as the internal energy and enthalpy, and ultimately the Helmholtz and Gibbs free energies. Nonetheless, if we use this reference state for entropy, one can define an "absolute" entropy in a completely consistent manner by the following general equation:

$$S(T, P) \equiv \int_0^T \frac{C_p^o}{T}\, dT + \sum_{\text{transitions}} \frac{\Delta H^o_{tr,i}}{T_{tr,i}} + S(T, P) - S(T, P^o) \tag{11-15}$$

where the integral involving C_p^o/T and the term $\Delta H^o_{tr,i}/T_{tr,i}$ account for all phase transitions at the reference state pressure P^o and the last two terms are the isothermal departure function to compress the system to its final pressure or density. This procedure is used to generate standard entropies of formation, ΔS_f^o, values used in the calculation of standard Gibbs free energies of formation, ΔG_f^o, discussed in Section 8.8. In this case heat capacity estimates from spectroscopic data are typically integrated from the zero point entropy condition at 0 K to the standard state of 298.15 K (25°C).

The next important concept that evolves from the statistical representation of entropy is a rationalization of ΔS_{mix} for an ideal solution. In Chapter 9 we simply defined an ideal solution as one where

$$\mu_i^{ID} = \overline{G}_i^{ID} = RT \ln x_i + \Lambda_i\,(T, P) \tag{9-83}$$

which led immediately to the equation for the ideal entropy of mixing, namely,

$$\Delta S^{ID} = \Delta S^{ID}_{mix} = -R \sum_{i=1}^{n} x_i \ln x_i = -kN_A \sum_{i=1}^{n} x_i \ln x_i \quad (\text{in J/mol K}) \tag{9-102}$$

Note that the subscript "*mix*" is used here to emphasize the conceptual mixing process. As discussed earlier in Section 9.8, this subscript may be optionally omitted on excess and ideal mixing functions.

Now, using the statistical interpretation of entropy on a molecular level as expressed mathematically in Eq. (11-13), we can prove that Eq. (9-102) is correct. Let's start with a binary solution consisting of N_1 molecules of component 1 and N_2 molecules of component 2. For an ideal solution the interaction energies between like and unlike molecules (1-1, 1-2, and 2-2) are identical as are the molecular or molar volumes of both components ($V_1 = V_2$). Thus at constant total moles, constant temperature T and constant total volume $\underline{V} = \underline{V}_1 + \underline{V}_2$ we can envision a mixture of distinguishable molecular species 1 and 2 that are completely randomly distributed in a three-dimensional lattice of volume $\underline{V}$. Initially we start with an empty lattice consisting of $N_1 + N_2$ sites and add one molecule at a time in a random fashion until the lattice is filled. Simple combinatorial analysis gives the total number of distinguishable arrangements of the $N_1 + N_2$ molecules as,

$$\Omega_{mix} = \frac{(N_1 + N_2)!}{N_1!\, N_2!} \tag{11-16}$$

Keep in mind that N_i is the ***number of molecules*** and not the moles of component i. For the case of N_1 molecules of pure 1 and N_2 molecules of pure 2,

$$\Omega_1^{pure} = \frac{N_1!}{N_1!} = 1.0 \quad \text{and} \quad \Omega_2^{pure} = \frac{N_2!}{N_2!} = 1.0 \tag{11-17}$$

Then the entropy of mixing is given by applying Eq. (11-13) and the definition of a mixing property from Chapter 9 to get:

$$\Delta \underline{S}_{mix}^{ID} = k \ln \Omega_{mix} - k \ln \Omega_1^{pure} - k \ln \Omega_2^{pure} \tag{11-18}$$

Substituting Eqs. (11-16) and (11-17) into Eq. (11-18), $\Delta \underline{S}_{mix}$ for an ideal solution is expressed as,

$$\Delta \underline{S}_{mix}^{ID} = k \ln \frac{(N_1 + N_2)!}{N_1!\, N_2!} \tag{11-19}$$

But since N_1 and N_2 are very large (say of order 10^{23}), we can use Stirling's approximation (see footnote 1 in Chapter 10),

$$\ln N_i! \approx N_i \ln N_i - N_i \tag{11-20}$$

with essentially no error to simplify Eq. (11-19),

$$\Delta \underline{S}_{mix}^{ID} = k\left[(N_1 + N_2)\ln(N_1 + N_2) - N_1 \ln N_1 - N_2 \ln N_2\right] \tag{11-21}$$

which by combining terms becomes,

$$\Delta \underline{S}_{mix}^{ID} = -k\left[N_1 \ln\left(\frac{N_1}{N_1 + N_2}\right) + N_2 \ln\left(\frac{N_2}{N_1 + N_2}\right)\right] = n\, \Delta S_{mix}^{ID} \tag{11-22}$$

with the total moles $n = (N_1 + N_2)/N_A$ and $x_i = N_i/\Sigma N_i$. Thus, in general, for a n-component ideal solution on a per mole basis,

$$\Delta S_{mix}^{ID} = \Delta S^{ID} = -R \sum_i x_i \ln x_i \tag{11-23}$$

which is identical to Eq. (9-102). The statistical treatment of entropy on a molecular level is consistent with other macroscopic representations as well. These macroscopic viewpoints usually cover two conceptual frameworks: (1) that entropy is a measure of disorder (as S increases the level of disorder increases) and (2) that entropy scales inversely with the information content of a system (as S increases the knowledge about the state of the system decreases). With $S = k \ln \Omega$ as Ω increases the number of possible configurations increase and one might expect intuitively that the level of disorder would increase and the information content would simultaneously decrease. For example, imagine heating a pure substance from a solid state to a gas. Clearly S increases during heating as both C_p and $\Delta S_{tr} = \Delta H_{tr}/T_{tr}$ of all transitions are positive, and during heating we would expect an increase in the number of possible configurational states and level of disorder as well. An analogous effect appears for mixing processes. In a qualitative manner, one can rationalize that as Ω increases, the uncertainty of specifying an exact state or configuration increases, thus there is a loss of information content.

11.4 Regular and Athermal Solution Behavior

We briefly introduced regular and athermal solutions in Chapter 9 as models that capture certain types of non-idealities. A regular solution was defined following the convention introduced by Hildebrand that the excess entropy of mixing ΔS_{mix}^{EX} and volume of mixing ΔV_{mix}^{EX} are zero. Later Hildebrand at Berkeley and Scatchard at MIT, working independently, suggested a major improvement to Van Laar's theory by introducing a solubility parameter, δ_i, to correlate the excess internal energy and Gibbs free energy of mixing. The parameter δ_i was empirically related to the energy of vaporization and the molar volume of each pure component i. For a binary mixture, the activity coefficients and solubility parameters are given by the following:

$$\ln \gamma_1 = \frac{V_1 \Phi_2^{*2}}{RT}\left(\delta_1 - \delta_2\right)^2 \quad \text{and} \quad \ln \gamma_2 = \frac{V_2 \Phi_1^{*2}}{RT}\left(\delta_1 - \delta_2\right)^2 \tag{11-24}$$

where the volume fractions are defined as

$$\Phi_1^* \equiv \frac{x_1 V_1}{x_1 V_1 + x_2 V_2} \quad \text{and} \quad \Phi_2^* = \frac{x_2 V_2}{x_1 V_1 + x_2 V_2}$$

and $\delta_i \equiv (\Delta U_{vap\,i}/V_i)^{1/2}$; δ_i^2 is equal to the cohesive energy density. While the basic approach of Hildebrand and Scatchard is quite simple but not always accurate, the incorporation of solubility parameters which can be estimated from only pure component data has several advantages. In the absence of mixture data their theory provides a first-order approach for estimating γ_i that works particularly well for non-polar solvents.

One notes that regular solution theory always yields positive deviations ($\gamma_i > 1.0$) which would be incorrect when strong associative interactions were present. This is frequently the case for polar solvents. Further discussion of regular solution theory and its extensions can be found in Prausnitz *et al.* (1986) and Hildebrand *et al.* (1970).

Athermal solutions represent another general class of solutions whose enthalpy of mixing is negligible. Since the enthalpy of mixing for an ideal solution is zero, then athermal behavior corresponds to situations where the excess enthalpy of mixing, ΔH_{mix}^{EX} is zero as well. This means that γ_i is independent of temperature for an athermal mixture; a property that can be exploited in many practical situations involving vapor-liquid equilibria. In Section 11.5, we will show what conditions at a molecular level lead to athermal behavior.

In reality, mixtures do not exactly follow regular or athermal solution behavior. These are asymptotic conditions which approximate real behavior to varying degrees, and as such provide useful reference points for correlating and interpreting non-ideal effects.

11.5 Lattice Models with Configurational and Energetic Effects

One approach to modeling the behavior of liquids is to view them geometrically as more "solid-like" rather than "gas-like". Certainly the density of liquids, in general, is closer to solids than to gases (see for example, the molecular simulations depicted in Figure 10-5). Liquids, of course, do not contain the long-range geometric structural order that solids do. Nonetheless as we showed in Section 11.3, we were able to derive the rigorous expression for the entropy of mixing of an ideal solution using a lattice modeling approach. Guggenheim (1952) was one of the pioneers in utilizing lattice models for interpreting liquid solution behavior, and much of what we say in this section is due to his contributions.

The fundamental concepts of lattice theories relate first to the combinatorial aspects of arranging molecules of different types in the lattice of specified geometry and second to the nature of the energetic interactions between the molecules. As one might expect, the key elements of such an approach involve the relative molecular sizes of the constituent molecules and the magnitude of their intermolecular interactions. Like molecules do not necessarily interact with the same magnitude of intermolecular forces as unlike molecules do. After all, this difference in energetics along with differences in molecular size give rise to the observed non-idealities we are interested in modeling. Although the lattice itself is an idealization, the approximations involved that relate to evaluating the configurational part of the partition function are intuitively reasonable. In particular, they allow separation of the internal components partition function relating to vibration, rotation, etc. from those contributions that relate to the total energy of the lattice with the molecules fixed at each lattice point.

To begin our treatment of lattice models, assume that we are dealing with molecules of similar size that are coordinated on a 3-D lattice with a spatially periodic geometry and a fixed number of nearest neighbors z. Following the methods described by Guggenheim (1952) and Prausnitz *et al.* (1986), we can develop equation for the total potential energy, Φ_T and the number densities of different nearest neighbor pairs in the lattice (N_{11} for pairs of 1-1 molecules, N_{22} for pairs of 2-2 molecules, and N_{12} for pairs of 1-2 molecules):

$$\Phi_T = N_{11}\,\Phi_{11} + N_{22}\,\Phi_{22} + N_{12}\,\Phi_{12}$$

or

$$\Phi_T = \frac{z}{2} N_1\,\Phi_{11} + \frac{z}{2} N_2\,\Phi_{22} + \frac{N_{12}}{z} w \tag{11-25}$$

where w = interchange energy = $z(\Phi_{12} - 1/2(\Phi_{11} + \Phi_{22}))$, Φ_{ij} represent the potential energy contribution of an ij pair. For the lattice itself:

$$zN_1 = 2N_{11} + N_{12} \quad \text{and} \quad zN_2 = 2N_{22} + N_{12} \tag{11-26}$$

which are effectively conservation equations. From statistical arguments for a totally random mixture (the so-called *zeroth approximation* of Guggenheim):

$$N_{12} = N_{12}^{*} = \frac{z\,N_1\,N_2}{N_1 + N_2} \tag{11-27}$$

Note that we have used an * to designate the assumed complete randomness of the mixture. Strictly speaking, w must be zero for the mixture to be totally randomly distributed. We will revisit this assumption later.

The statistical mechanical treatment presented in Chapter 10 provides us means to express the canonical partition function for the lattice, $Q_{lattice}$, and develop an expression for the Helmholtz free energy of the lattice, $A_{lattice}$:

$$Q_{lattice} = \sum_{N_{12}} g(N_1, N_2, N_{12}) \exp\left[-\underline{E}_{12}/kT\right] \tag{11-28}$$

$$\underline{A}_{lattice} = -kT \ln Q_{lattice} \tag{11-29}$$

Now we need to create the mixing quantity $\Delta\underline{A}_{mix} = \underline{A}_{lattice} - \underline{A}^{pure1} - \underline{A}^{pure2}$. We can express the canonical partition function for pure components 1 and 2 by setting the combinatorial factor $g(N_i)$ to 1.0 and $\underline{E}_{12}$ to $\underline{E}_{ii} = z/2(N_i\Phi_{ii})$ for each. With these additions, we can now evaluate $\Delta\underline{A}_{mix}$ as

$$\Delta\underline{A}_{mix} = -kT\left[\ln Q_{lattice} - \ln\left[\exp\left[\frac{-z\,N_1\,\Phi_{11}}{2kT}\right]\right] - \ln\left[\exp\left[\frac{-z\,N_2\Phi_{22}}{2kT}\right]\right]\right] \tag{11-30}$$

$Q_{lattice}$ is modified using the maximum term approach because N_{12} is very large (note that if N_1 and N_2 are large, N_{12} will also be large and of order $z/2(N_1$ or $N_2)$. Summarizing, for pure 1 and 2 and the mixture (assumed random with $N_{12} = N_{12}^{*}$):

$$pure\,1: \; g(N_1) = 1.0 \quad N_2 = N_{12} = 0 \quad \underline{E}_{11}^{pure} = \frac{z}{2} N_1\,\Phi_{11}$$

$$pure\,2: \; g(N_2) = 1.0 \quad N_1 = N_{12} = 0 \quad \underline{E}_{22}^{pure} = \frac{z}{2} N_2\,\Phi_{22}$$

$$mixture \ : \ g(N_1, N_2, N_{12}) \rightarrow g(N_1, N_2, N_{12}^*) \quad \underline{E}_{12} = \frac{z}{2} N_1 \, \Phi_{11} + \frac{z}{2} N_2 \, \Phi_{22} + N_{12}^* \frac{w}{z}$$

Thus we can rewrite $\Delta\underline{A}_{mix}$ as,

$$\Delta\underline{A}_{mix} = -kT \left[\ln \left(g(N_1, N_2, N_{12}^*) \exp\left[-\frac{w \, N_{12}^*}{z \, k \, T} \right] \right) \right] \tag{11-31}$$

N_{12}^* is given by Eq. (11-27) and $g(N_1, N_2, N_{12}^*)$ comes directly from combinational analysis:

$$g\,(N_1, N_2, N_{12}^*) = \frac{(N_1 + N_2)\,!}{N_1!\, N_2!} \tag{11-32}$$

With these substitutions,

$$\Delta\underline{A}_{mix} = -kT \ln \left(\frac{(N_1 + N_2)!}{N_1!\, N_2!} \exp\left[\frac{-w \, N_1 N_2}{kT(N_1 + N_2)} \right] \right) \tag{11-33}$$

Using Stirling's approximation Eq. (11-33) can be simplified to produce an intensive form for ΔA_{mix} (per mole basis)

$$\Delta A_{mix} = kTN_A \,(x_1 \ln x_1 + x_2 \ln x_2 + \frac{w}{kT} x_1 \, x_2) \tag{11-34}$$

where again $kN_A = R$. If this solution is assumed to be regular then $\Delta V^{EX} = 0$ and $\Delta S^{EX} = 0$ and therefore

$$\Delta S_{mix} = \Delta S^{ID}$$

Now to obtain an estimate of ΔG^{EX}, we assume that $\Delta G^{EX} = \Delta A^{EX}$ which is a reasonable approximation for systems where (ΔPV) is small relative to ΔU. Thus,

$$\Delta H^{EX} \approx \Delta U^{EX} \quad \text{and} \quad \Delta G^{EX} \approx \Delta A^{EX}$$

and using Eq. (11-34), we approximate ΔG^{EX} by

$$\Delta G^{EX} = \Delta A_{mix} + T\Delta S^{ID} = w \, x_1 \, x_2 \, N_A \tag{11-35}$$

With Eq. (11-35), we can immediately obtain expressions for $\ln \gamma_i$, using Eq. (11-4), that are identical in form to the 2-suffix Margules model:

$$\ln \gamma_1 = \frac{w \, N_A \, x_2^2}{RT} \quad \text{and} \quad \ln \gamma_2 = \frac{w \, N_A \, x_1^2}{RT} \tag{11-36}$$

The next improvement introduced by Guggenheim was his *quasi-chemical approximation* to account for the existence of non-randomness in liquid solutions.[1] Certainly, this modification is a step in the right direction for mixtures with energetic

---

1 The quasi-chemical approximation was first used by Bethe (1935) in a completely different context to model order-disorder transitions in metal alloys.

non-idealities where w is non-zero. Guggenheim's approach is quite creative in that he introduces an equilibrium constant K_{eq} to relate the 1-2 interactions to 1-1 and 2-2 interactions. For canonical ensemble conditions at constant T, $\underline{V}$, and N, K_{eq} is written as,

$$K_{eq} \equiv \frac{N_{12}^2}{N_{11} N_{22}} \tag{11-37}$$

Earlier, with the zeroth-order approximation, we assumed complete randomness with $N_{12} = N_{12}^*$, now $N_{12} \neq N_{12}^*$. If $w > 0$, $N_{12} < N_{12}^*$ and if $w < 0$, $N_{12} > N_{12}^*$. In other words, if the 1-2 interaction is strongly attractive $N_{12} > N_{12}^*$, if its repulsive $N_{12} < N_{12}^*$. This is the essence of the quasichemical approximation. What follows is the incorporation of some classical thermodynamics and mathematical reformulation to obtain a working equation for ΔA^{EX}. Following our previous derivation using the zeroth approximation, we again approximate ΔG^{EX} by ΔA^{EX} to obtain γ_i. The steps are as follows:

(1) use the Gibbs-Helmholtz type relationship to express

$$\left[\partial \ln K_{eq}/\partial[1/T]\right]_{\underline{V}, N_i} = \frac{-\Delta U_{rx}}{R}$$

(2) assume $\Delta U_{rx} \cong \frac{2wR}{kz}$

(3) integrate (1) to obtain $K_{eq} = f(1/T)$

(4) the integration constant is evaluated for the limiting case of complete randomness ($w = 0$, $N_{12} \to N_{12}^*$, $N_{11} \to N_{11}^*$, and $N_{22} \to N_{22}^*$)

With all of these steps, we end up with an expression for

$$K_{eq} = \frac{N_{12}^2}{N_{11} N_{22}} = 4 \exp\left[\frac{-2w}{z\,kT}\right] = \frac{4}{\eta^2} \tag{11-38}$$

where $\eta \equiv \exp(w/zkT)$. Next another parameter β is defined to facilitate expressing K_{eq} in terms of x_1 and x_2

$$\beta \equiv (1 - 4x_1 x_2 (\eta^2 - 1))^{1/2} \tag{11-39}$$

Now when $w = 0$, complete randomness exists, $\eta = 1$ and $\beta = 1$ and we have an ideal solution. With further modification and restructuring [see Prausnitz *et al.* (1986) for details].

$$\Delta A^{EX} = \frac{zkN_A T}{2}\left[x_1 \ln \frac{(\beta - 1 + 2x_1)}{x_1(\beta + 1)} + x_2 \ln \frac{(\beta - 1 + 2x_2)}{x_2(\beta + 1)}\right] \tag{11-40}$$

Note that when $\beta = 1$, $\Delta A^{EX} = 0$ as expected, and with the quasi-chemical approximation, ΔA^{EX} now depends on z as well as w. Also note for $w \neq 0$ as $T \to \infty$, $\eta \to 1.0$, and $\beta \to 1.0$ and we induce complete randomness in the high temperature limit.

If we limit ourselves to cases where $w/zkT < 1$, we can recast Eq. (11-40) using a polynominal expansion for η^2 in Eq. (11-39) to give

$$\Delta G^{EX} \approx \Delta A^{EX} = wx_1x_2N_A\left[1 - \frac{1}{2}\left(\frac{2w}{zkT}\right)x_1x_2 + \ldots\right] \tag{11-41}$$

and obtain equations for $\ln\gamma_i$ using Eq. (11-4). These would obviously be in the form of expanded polynomials involving x_1x_2 or (or $x_1\,(1-x_1)$) terms. In this sense, the Margules and Redlich-Kister empirical models, that express $\ln\gamma_i$ in terms of polynomial functions in composition (x_i), are based theoretically on lattice theory and the quasi-chemical approximation.

11.6 McMillan-Mayer Theory

Proceeding in a completely analogous manner to the development of the virial equation which was presented in Section 10.4, it is possible to develop a polynomial expansion in concentration for the osmotic pressure and activity coefficient of a solvent in the presence of a solute. This approach originated in a theory proposed by McMillan and Mayer in 1945. The direct analogy compares a solute dissolved in a solvent in a state of infinite dilution to a gas at low pressures. McMillan-Mayer theory provides a starting point for several important theoretical activity coefficient models of electrolytes that will be discussed in Chapter 12. As only the salient features of their theory are presented in this section, readers interested in the details should consult Hill (1960) who presents derivations of the equations cited here.

Because we are dealing with an infinite dilution condition for the solute, we would like to use the unsymmetrical normalization convention for γ_i of the solute, namely as $x_i \to 0$, $\gamma_i \to 1.0$. McMillan-Mayer theory can be most easily visualized for a two-component mixture where component 1 refers to the solvent and 2 to the solute. The osmotic equilibrium condition comes directly from Chapter 6 [Eq. (6-37)] with the constraints equivalent to having an internal wall or membrane separating pure solvent from a mixture at the same temperature and a specified composition. The membrane is rigid, diathermal, and permeable to the solvent. In this case the chemical potentials of the solvent (component 1) on either side of the membrane are equal:

$$\mu_1(T, P, \text{pure}) = \mu_1(T, P+\Pi, x_i) \tag{11-42}$$

where Π is the osmotic pressure. We begin with an approach that is similar to what we used to derive the virial equation in Section 10.4, but as suggested by Hill (1960) we use the grand canonical partition function Ξ to more conveniently obtain the osmotic virial equation. From our previous discussion in Chapter 10 [see Eq.(10-57)], the $P\underline{V}TN$ EOS is given by,

$$\ln \Xi = \frac{(P+\Pi)\,\underline{V}}{kT} \tag{11-43}$$

But Ξ for the binary mixture can be written, in general, as:

$$\Xi = \sum_{N_2}\sum_{N_1} \lambda_1^{N_1}\lambda_2^{N_2} Q_{N_1,N_2}(\underline{V},T) = \sum_{N_2} \lambda_2^{N_2} \Lambda_{N_2}(\mu_1,\underline{V},T) \tag{11-44}$$

where the so-called absolute activities are defined as

$$\lambda_1 \equiv \exp(\mu_1/kT) \quad \text{and} \quad \lambda_2 \equiv \exp(\mu_2/kT)$$

$$\text{with } \Lambda_{N_2} \equiv \sum_{N_1} \lambda_1^{N_1} Q_{N_1,N_2}(\underline{V},T)$$

are employed. In this case the normal solute activity a_2 is given by,

$$a_2 = \frac{\lambda_2 \Lambda_1}{\underline{V}\Lambda_0} \quad \text{with} \quad \begin{bmatrix} \Lambda_1 = \lambda_1 Q_{N_1,N_2=1} \quad (\infty\text{–dilution}) \\ \Lambda_0 = \Xi\Big|_{\text{pure solvent}} = \exp\left[\frac{P\underline{V}}{kT}\right] \end{bmatrix} \tag{11-45}$$

With unsymmetrical normalization, $a_2 \to \rho_2$ as $\rho_2 \to 0$ with μ_1 and T constant, where ρ_2 is defined as the solute number density (number of solute molecules per unit volume of the solution). This convention is completely equivalent to saying that $\gamma_2 \to 1.0$ as $x_2 \to 0$ in the ∞-dilution limit.

Using the pure solvent condition for Λ_o, divide Eq. (11-44) by Λ_o and replace λ_2 with a_2 to get,

$$\exp\left[\frac{\Pi \underline{V}}{kT}\right] = 1 + \sum_{j=1}^{N_2} Y_j(\mu_1,\underline{V},T)[a_2]^j \tag{11-46}$$

where $Y_j = Y_j(\mu_1,\underline{V},T) \equiv \dfrac{\Lambda_j \Lambda_o^{j-1} \underline{V}^j}{\Lambda_1^j}$. The Y_j terms are completely analogous to the configurational Z_j^* terms we used in developing the virial equation in Section 10.4.

Now take the logarithm of Eq. (11-46) and divide by $\underline{V}$ to obtain an expression for the osmotic pressure from Eq. (11-43):

$$\frac{\Pi}{kT} = \sum_{j=1}^{\infty} b_j(\mu_1,T)[a_2]^j = \sum_{j=1}^{\infty} b_j [a_2]^j \tag{11-47}$$

where $b_1, b_2, b_3, \ldots$ are the osmotic virial coefficients. In a fashion analogous to Eq. (10-106), we can relate b_j to Y_j:

$$1!\,\underline{V}b_1 = Y_1 = \underline{V} \quad (\text{i.e., } b_1 = 1)$$

$$2!\,\underline{V}b_2 = Y_2 - Y_1^2$$

$$3!\,\underline{V}b_3 = Y_3 - 3Y_1Y_2 + 2Y_1^3$$

Using a standard thermodynamic transformation between activity and concentration, we find that in the dilute limit as $a_2 \rightarrow \rho_2$

$$\rho_2 = \sum_{j=1}^{N_2} j\, b_j(\mu_1, T)\, [a_2]^j$$

Eq. (11-47) can be recast into a solute concentration or number density expansion. Using the fact that $b_1 = 1.0$ and incorporating the result of the transformation of a_2 into ρ_2 in the dilute limit we produce the ***McMillan-Mayer osmotic virial equation***.

$$\boxed{\frac{\Pi}{kT} = \rho_2 + B_2\rho_2^2 + B_3\rho_2^3 + \cdots} \tag{11-48}$$

where we formally define modified osmotic virial coefficients B_i for $i = 2, \ldots, n$ and the solute number density ρ_2 as

$$B_2 \equiv \frac{-\left(Y_2 - \underline{V}^2\right)}{2\underline{V}} \quad \text{and} \quad \rho_2 \equiv \frac{N_A\, C_2}{m_2}$$

In general, a recursive relationship is used to generate the higher order terms

$$B_n = \frac{-(n-1)}{n}\beta_{n-1}; \qquad \beta_1 = 2b_2, \quad \beta_2 = 3b_3 - 6b_2^2, \ldots$$

with

C_2 = solute concentration in g/liter
m_2 = solute molecular weight (g/mol)
N_A = Avogadro's number (molecules/mol)
B_n = nth osmotic virial coefficient in (liter/molecule)$^{n-1}$

In this case, the osmotic virial coefficients are defined in terms of intermolecular forces in a completely analogous fashion to our treatment of the normal virial expansion for non-ideal gases given in Section 10.4. For example, assuming pairwise additivity, we can express the osmotic virial coefficients in terms of cluster integrals using the Mayer function definition:

$$f_{ij} \equiv \exp\left[-\langle\Phi_{ij}\rangle/kT\right] - 1 \tag{11-49}$$

where $\langle\Phi_{ij}\rangle$ is the potential of mean force between solute pairs i and j in an infinitely dilute solution. Now the second and third coefficients, B_2 and B_3, are given by forms identical to Eqs. (10-114) and (10-115) for spherically symmetric potential functions:

$$B_2 = -2\pi\int_0^\infty f_{12}\, r^2\, dr \tag{11-50}$$

$$B_3 = -\frac{1}{3\underline{V}} \int_0^{\infty} \int_0^{\infty} \int_0^{\infty} f_{12} f_{13} f_{23} \, d\mathbf{r}_1 \, d\mathbf{r}_2 \, d\mathbf{r}_3 \qquad (11\text{-}51)$$

The analogy inherent in the development of the osmotic virial equation between the dilute solute in solvent concept and the limiting condition of a gas at very low pressure is valid when the intermolecular forces are of sufficiently short range to guarantee convergence of the cluster integrals. This is the situation for many liquid solutions, for example, for dilute to even reasonably concentrated solutions of most polymers and macromolecules in low molecular weight solvents as well as for dilute solutions of dissociated electrolytes of low and high molecular weight (e.g., macroions) in aqueous media where sufficient screening of coulombic interactions exists.

Under these conditions, osmotic pressure measurements can be used to determine molecular weight, activity coefficients, and to probe our understanding of intermolecular forces. In the latter case, several approaches have been used. One common route is to develop a constitutive model for solute-solute intermolecular interactions based on a potential of mean force and use it to estimate osmotic pressure versus concentration behavior. Vilker, Colton, and Smith (1981) employed such a technique to correlate osmotic pressure variations in concentrated solutions of Bovine Serum Albumin (BSA) up to 400 g/liter in NaCl/NaOH mixtures of various pH. Representative experimental results and theoretical predictions are presented in Figure 11.1. A similar set of data for aqueous α-chymotrypsin solutions provided by Haynes *et al.* (1992) is given in Figure 11.2 where extremely accurate measurements in the dilute region (<10 g/liter) show the expected linearity of Π/C_2 with concentration C_2.

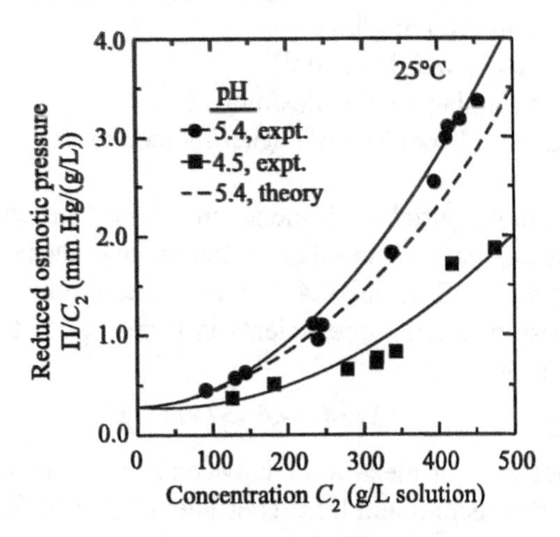

Figure 11.1 Experimental and theoretical predictions of reduced osmotic pressure as a function of BSA concentration. Eq. (11-52) used for the theoretical prediction [data from Vilker, Colton, and Smith (1981)].

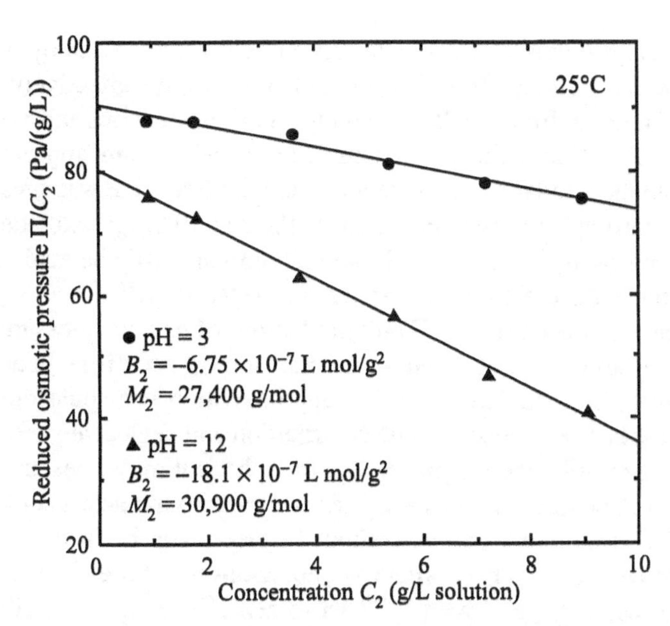

Figure 11.2 Reduced osmotic pressure versus α-chymotrysin concentration in a 0.1 M K_2SO_4 solution for two different pH's. Linearity allows estimation of second osmotic virial coefficients B_2 and solute molecular weights by regression [data from Haynes *et al.* (1992)].

The theoretical predictions for these protein solutions are based on a model that superimposes a hard-sphere, excluded-volume model of intermolecular interactions on top of estimated contributions from electrostatic, induction and dispersion force interactions. Although the hard sphere contribution can be expressed rigorously for the second, third, fourth, and fifth coefficient terms, the remaining calculations are not simple and require a number of simplifying assumptions. The details of the calculations for the electrostatic, induction, and dispersion contributions are beyond the scope of this chapter, interested readers should refer to Vilker *et al.* (1981) or Haynes *et al.* (1992). The final form of the osmotic virial expansion, obtained by superimposing hard-sphere/excluded-volume and coulombic interactions for the second and third coefficient terms and using only the excluded volume contribution for terms involving the fourth and fifth coefficients, is given by,

$$\frac{\Pi}{RT} = \rho_2 + B_2\rho_2^2 + B_3\rho_2^3 + B_4\rho_2^4 + B_5\rho_2^5 + \dots \qquad (11\text{-}52)$$

where

$$B_2 = B_2^{HS} + B_2^{coul} \quad (B_2^{HS} = 16/3\,\pi\, b_o^3 = 4\, v_m)$$

$$B_3 = B_3^{HS} + B_3^{coul} \quad (B_3^{HS} = 5/8\,(B_2^{HS})^2)$$

$$B_4 = B_4^{HS} = 0.2869\,(B_2^{HS})^3$$

$$B_5 = B_5^{HS} = 0.115\,(B_2^{HS})^4$$

with b_o = molecular radius = $\sigma/2$, σ = distance of closest approach by two molecules, and v_m = molecular volume. To obtain Eq. (11-52) we have effectively evaluated the configurational integral from $r = 2b_o$ to ∞ for the hard sphere terms and have determined the integral from $r = 0$ to ∞ for the coulombic terms using many approximations. One also should note the similarity of the development here to what was presented for van der Waals and hard-sphere fluids in Sections 10.6 and 10.8. The excluded volume per molecule is given by $B_2^{HS} = 2/3\,\pi\,\sigma^3$. Modifications can easily be made to modify the excluded volume contribution for non-spherical shapes as well.

In order to actually use Eq. (11-52) for predictions of osmotic pressure as illustrated in Figures 11.1 and 11.2, the intermolecular forces occurring between polymer molecules need to be modeled. As a first approximation, we could only include the repulsive, hard sphere/excluded volume contribution and neglect any attractive effects. This would require only one parameter (b_o) for spherical molecules, and two to three parameters for non-spherical molecules. These geometric parameters could be estimated using molecular group contribution methods or they could be fit to experiental data. Practically speaking, for this first-order case, equations like those given above would be used to specify each $B_i = B_i^{HS}$ with B_i^{coul} set to zero. To include contributions due to electrostatic, induction, and dispersion forces, a much more comprehensive intermolecular potential model is needed to capture these effects correctly. As alluded to earlier, this refinement requires much more work and many additional assumptions [see Sections 10.2 and 10.3, Vilker *et al.* (1981) and Haynes *et al.* (1992)]. Such a sophisticated calculation has been done, but the central issue once again is the accuracy of the constitutive models used for representing these complex intermolecular interactions.

11.7 Activity Coefficient Models for Condensed Fluid Phases

Several extensions to the theoretical approaches outlined in Sections 11.5 and 11.6 have led to models that are used extensively today to account for non-ideal behavior. Some of the most representative models are discussed in this section. The important extensions include:

(1) extended polynominal expansion models with multiple, empirically-fitted, adjustable parameters

(2) the Flory-Huggins extension of lattice theory to account for different molecular sizes

(3) the incorporation of local composition effects by the Wilson, Non-Random Two-Liquid (NRTL) and similar models when the concentration of mixture components at a molecular level are non-randomly distributed.

(4) modifications to the quasi-chemical approximation to account for molecular sizes (areas and volumes) by the UNIQUAC and UNIFAC models

(5) the Debye-Hückel and Pitzer extensions to McMillan-Mayer theory and the Chen extension to local composition models for dissociating electrolyte solutions (see Chapter 12).

While it is certainly satisfying to see the theoretical roots of these important models, one must keep in mind that far reaching approximations are often introduced, and empirical fitting of model parameters with real experimental data is the rule rather than the exception. These limitations are not meant to detract from the importance of having tractable models for non-ideal solutions, but only to keep them in the proper context as semi-theoretical approximations at best.

As our knowledge of solution behavior increases by advances in experimental diagnostic methods and by advances in molecular simulation, undoubtedly many of these *ad hoc* approximations will be replaced by more rigorous theory. For now we have to be content with the creativity of enlightened empiricism that pervades the modeling of these complex systems.

11.7.1 First-order polynomial models. Given that γ_i variations in composition x_i are continuous functions for homogeneous phase mixtures, one might expect that certain polynomial expansions would be well suited to correlate γ_i or ΔG^{EX} data. Any one of them would be subject to the normal thermodynamic boundary conditions and Gibbs-Duhem constraints. Thus, a general polynomial representation could be given as,

$$\Delta G^{EX}/RT = \Sigma A_j\, [f(x_i)]^j \quad \text{for } i=1 \text{ to } n \tag{11-53}$$

with $\Delta G^{EX} = 0$ when $x_i = 1.0$ and $\Sigma x_i\, d\,(\overline{\Delta G_i^{EX}}) = 0$ at constant T and P

Furthermore, our treatment of lattice theory and the quasi-chemical approximation suggested several simple forms for these polynomials, for example with terms grouped as $x_1 x_2 f(x_1, x_2)$ [see Eqs. (11-35) and (11-41)].

We have already introduced one of the simplest expansions as the 2-suffix Margules equation with $\Delta G^{EX} = Ax_1x_2$, which yields completely symmetrical γ_i behavior. Higher-order Margules and Redlich-Kister expansions and the van Laar equation are tabulated in Tables 11.1 and 11.2.

Wohl (1942) has suggested a general method for expressing ΔG^{EX} as a polynomial expansion. For a 1-2 binary,

$$\frac{\Delta G^{EX}}{RT(x_1q_1 + x_2q_2)} = 2a_{12}\,\Phi_1^*\,\Phi_2^* + 3a_{112}\,\Phi_1^{*2}\,\Phi_2^* + 3a_{122}\,\Phi_1^*\,\Phi_2^{*2}$$

$$+ 4a_{1112}\,\Phi_1^{*3}\,\Phi_2^* + 4a_{1222}\,\Phi_1^*\,\Phi_2^{*3} + \ldots \tag{11-54a}$$

where volume fractions for each component are defined as

$$\Phi_1^* \equiv \frac{x_1q_1}{x_1q_1 + x_2q_2} \quad \text{and} \quad \Phi_2^* = \frac{x_2q_2}{x_1q_1 + x_2q_2}$$

and the q_is are a direct measure of the volumetric size of the constituent molecules. Using Eq. (11-54) as a basis, it is possible to generate all forms of the Margules and Redlich-Kister expansions as well as the van Laar equation (see Table 11.1) by appropriately specifying the a_{ij}, a_{ijk}, a_{ijkl} and q_i parameters, and truncating the expansion at an appropriate order in Φ_i^*. In using Eq. (11-54) we are implicitly using symmetrical normalization for γ_i with ΔG^{EX} referenced to pure components. Thus, only terms that contain interactions between unlike molecules to some order are retained. In this setting coefficients a_{ii}, a_{iii}, or a_{iiii} are set to zero to remove any forms of the type: $a_{ii}\,\Phi_i^{*2}$, $a_{iii}\,\Phi_i^{*3}$, or $a_{iiii}\,\Phi_i^{*4}$. This is a direct result of the requirement that $\Delta G^{EX} = 0$ for x_i or $\Phi_i^* = 1$. We can reformulate ΔG^{EX} with the infinite dilution reference condition of unsymmetrical normalization. Now $\Delta G^{EX} \rightarrow \Delta G^{EX(\infty)}$ and the expansion in Φ_i^* retains *only* those similar terms. Thus, with component 2 taken at infinite dilution.

$$\frac{\Delta G^{EX(\infty)}}{RT(x_1 q_1 + x_2 q_2)} = a_{22}\,\Phi_2^{*2} + a_{222}\,\Phi_2^{*3} + a_{2222}\,\Phi_2^{*4} + \ldots \tag{11-54b}$$

Keep in mind that experimental γ_i data are needed to actually use any of these operating equations for interpolating or extrapolating $\gamma_i = f(x_i)$ behavior. The most commonly used forms involve one to three adjustable constants that will require a minimum of one to three values of γ_i to specify the constants. For example, the van Laar needs two values of γ_i to fix the A and B parameters. Of course, if additional γ_i data are available, then a more complete regression can be used to specify optimal values of the parameters subject to some objective function Ψ. For example, for M data points we may want to minimize the sums of squares of the deviation:

$$\min \Psi = \sum_{i=1}^{M} (\gamma_i^{exp} - \gamma_i^{predicted})^2$$

11.7.2 Configurational effects of molecular size (Flory-Huggins model). In the theoretical models described in Sections 11.5-11.6, we assumed that the solute and solvent molecules were similar in size. This constraint can be relaxed in the lattice model if we are careful about the combinatorial treatment. In 1941, Flory and Huggins, working independently, basically followed this path when they developed their theory to account for entropy of mixing effects in systems consisting of large polymer macromolecules in the presence of smaller solvent or monomer molecules. They initially assumed that these mixtures were otherwise ideal, that is $\Delta H_{mix}^{EX} = 0$. These entropic configurational effects are properly captured if *volume* rather than *mole* fractions are used.

To begin, we consider the two-dimensional lattice given in Figure 11.3. To illustrate how the combinatorial analysis is carried out, we assume a mixture of N_1 molecules of solvent and N_2 molecules of polymer solute where the molecular volume of the pure polymer is r times that of the pure solvent. In other words, if we treat the polymer as a connected chain of r-mers then each repeat unit of the polymer occupies the same volume on the lattice as a single solvent molecule. The empty lattice in Figure 11.3 has N available sites which will ultimately equal $N_1 + rN_2$ when the final mixture is prepared. The combinatorics and algebra are somewhat tedious in trying to reach a tractable statistical model for the entropy of the mixture, where

$$\underline{S}_{mix} = k \ln \Omega_{mix}$$

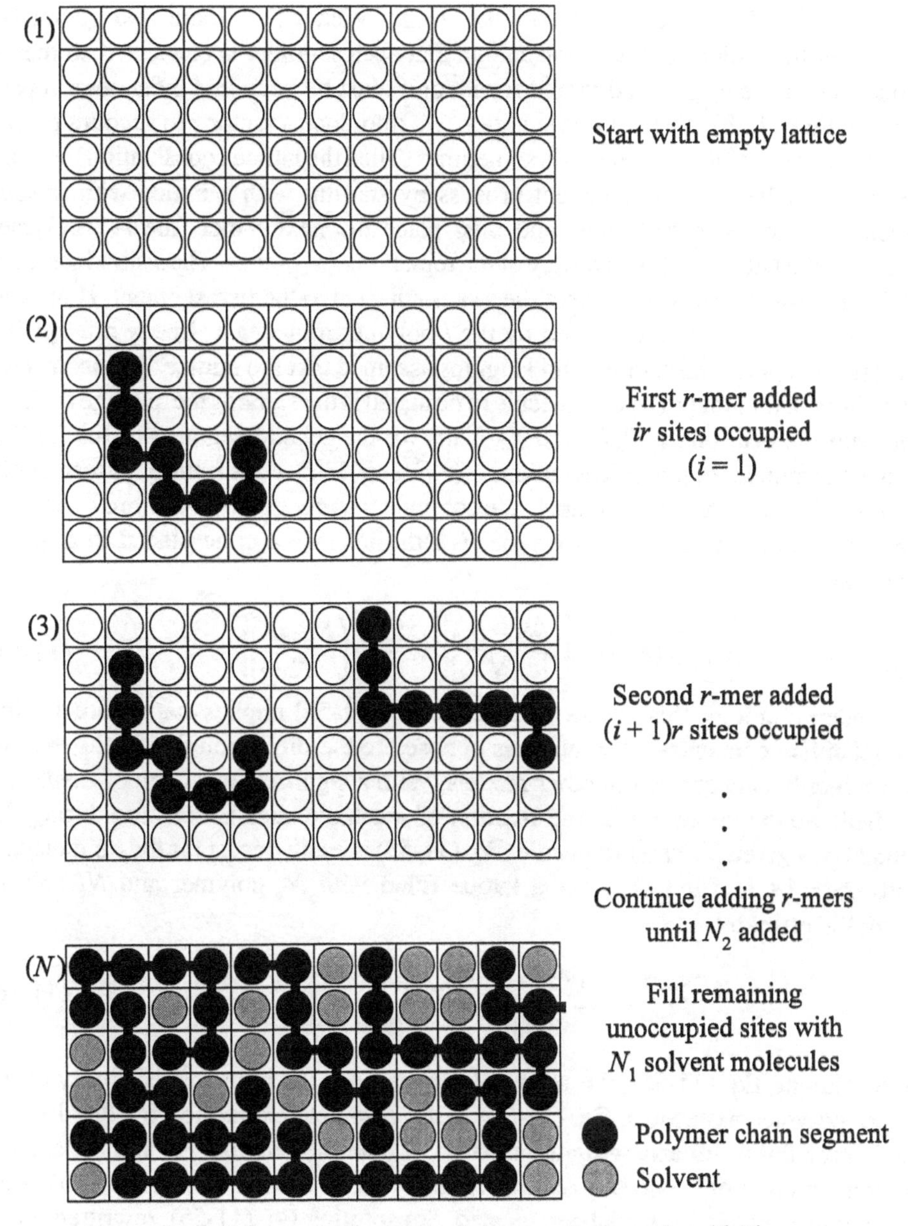

Figure 11.3 Polymer and solvent mixing on a 2-dimensional lattice.

which can then be used to evaluate the entropy of mixing as,

$$\Delta \underline{S}_{mix} = \underline{S}_{mix} - N_1 S_1 - N_2 S_2$$

As Hiemenz (1984) presents such a clear and lucid derivation of the Flory-Huggins model, we outline his approach here. The objective is to calculate Ω which corresponds to the number of ways of placing N_2 molecules in the lattice. Unlike the N_1 solvent molecules which can be placed randomly on any vacant site, there is a geometric constraint for the r-mer polymer molecule in that each segment is connected so that at least one site adjacent to an individual repeat unit must be occupied by another repeat unit (see Figure 11.3b). To treat this situation, we introduce a factor z to account for the number of nearest neighbors where z is sometimes called the lattice coordination number.

We can visualize the arrangement process by starting with a lattice with ir sites occupied by i polymer molecules and then place the next r-mer (the i+1 polymer molecule) in the lattice. It can go on any of the remaining $N - ir$ sites. The second segment could, in principle, occupy any one of the z sites adjacent to the first segment. However, some of the z sites may be occupied by the i polymer molecules already present. To account for this practically, Flory and Huggins assumed that the number of vacant sites in the vicinity of (i+1)th polymer molecule to be added is the same as the average fraction for the entire lattice, that is $[(N - ir)/N]$. Thus for the second segment of the (i+1)th molecule, the number of possible locations is $[z\,(N - ir))/N)$. For the third segment, only $[(z - 1)(N - ir)/N)]$ sites are available. We continue the process for the entire r-mer and obtain ω_{i+1} the total number of ways the (i+1)th molecule can be placed in a lattice already containing i molecules:

$$\omega_{i+1} = (N - ir)\left[z\left(\frac{N - ir}{N}\right)\right]\left[(z-1)\left(\frac{N - ir}{N}\right)\right]^{r-2} \qquad (11\text{-}55)$$

The underlying logic that allows us to derive Eq. (11-55) implies that we are dealing with a relatively concentrated solution as opposed to the dilute solution limits that we have commonly encountered in other models. With a significant fraction of the lattice sites filled, we can assume that the fraction of available sites is constant during the placement of a given polymer molecule. Eq. (11-55) is easily recast for the ith molecule by setting $i = i + 1$. Then Ω_{mix} for a lattice filled with N_2 polymer and N_1 solvent molecules is given by

$$\Omega_{mix} = \frac{\omega_1\omega_2\omega_3 \cdots \omega_{N_2}}{N_2!} = \frac{1}{N_2!}\prod_{i=1}^{N_2}\omega_i \qquad (11\text{-}56)$$

In developing Eq. (11-56), the term $N_2!$ accounts for the indistinguishability of the N_2 r-mer polymer molecules. Eq. (11-56) is the complete expression for the thermodynamic probability for arranging the mixture of both the polymer (2) and the solvent (1), because there is only one way of placing the N_1 solvent molecules on the lattice after all the polymer molecules have been located. Substituting Eq. (11-55), rewritten for ω_i into Eq. (11-56) we obtain:

$$\Omega_{mix} = \frac{z^{N_2}(z-1)^{N_2(r-2)}}{N_2!\,(N)^{N_2(r-1)}}\prod_{i=1}^{N_2}\left(\frac{N}{r} + 1 - i\right)^r \qquad (11\text{-}57)$$

where the continued product can be modified to:

$$\prod_{i=1}^{N_2}\left(\frac{N}{r}+1-i\right)^r = \left[\frac{(N/r)!}{(N/r-N_2)!}\right]^r \tag{11-58}$$

Because N is large, Stirling's approximation ($\ln N! = N \ln N - N$) can be used and with the statistical representation for the entropy of the mixture $\underline{S}_{mix} = k \ln \Omega_{mix}$ given by Eq. (11-13), we get,

$$\underline{S}_{mix} = k\left[-N_2 \ln\left(\frac{r\,N_2}{N}\right) - N_1 \ln\left(\frac{N_1}{N}\right)\right] + kN_2\,[\ln z + (r\text{–}2)\ln(z\text{–}1) + (1\text{–}r) + \ln r] \tag{11-59}$$

The pure component contributions are for pure solvent, $N_2 = 0$ and $N = N_1$ so, $S_{mix} \to S_1(pure) = 0$ and for pure polymer solute, $N_1 = 0$ and $N = rN_2$ so from Eq. (11-59) we obtain,

$$\lim_{N_1 \to 0} \underline{S}_{mix} = \underline{S}_2\,(pure) = kN_2\,[\ln z + (r\text{–}2)\ln(z\text{–}1) + (1\text{–}r) + \ln r] \tag{11-60}$$

ΔS_{mix} is then just the difference between the mixture entropy from Eq. (11-59) and Eq. (11-60). Thus,

$$\Delta \underline{S}_{mix} = -k\left[N_1 \ln\left(\frac{N_1}{N}\right) + N_2 \ln\left(\frac{rN_2}{N}\right)\right] \tag{11-61}$$

Note that the terms containing z have vanished. In the Flory-Huggins model, the terms in Eq. (11-61) with $N = N_1 + rN_2$ are immediately equated to volume fractions Φ_i^* for each component:

$$\Phi_1^* = \frac{N_1}{N_1 + rN_2} \quad \text{and} \quad \Phi_2^* = \frac{rN_2}{N_1 + rN_2} \tag{11-62}$$

With $kN_A = R$, Eq. (11-62) is converted to a per mole basis:

$$\Delta S_{mix} = -R\left[x_1 \ln \Phi_1^* + x_2 \ln \Phi_2^*\right] \tag{11-63}$$

If $r = 1$, $\Phi_i^* = x_i$ and the ideal solution result is obtained. [Eq. (11-23)]. Our derivation initially assumed that the solution was athermal ($\Delta H^{EX} = 0$). We can now relax that restriction and give the final form of the ***Flory-Huggins model*** as,

$$\Delta G_{mix} = \Delta H_{mix} - T\Delta S_{mix} = RT\left[\left[x_1 \ln \Phi_1^* + x_2 \ln \Phi_2^*\right] + \frac{\chi\,\Phi_1^*\,\Phi_2^*\,(N_1 + rN_2)}{N_1 + N_2}\right] \tag{11-64}$$

where the term containing the Flory parameter χ has been added to capture the non-athermal effects. The parameter χ is usually fit to data of one type or another. For example, we can relate χ to the Hildebrand-Scatchard solubility parameter using an empirical phenomenological approach:

$$\chi = \chi_o + \frac{V_1}{RT}[\delta_1 - \delta_2]$$

where χ_o is a constant, V_1 = molar volume of pure solvent (1), and δ_i is given by Eq. (11-24). For athermal mixtures, $\chi = 0$. If the molecules in the mixture are chemically similar then $\chi < 0.1$. For typical polymer-solvent systems $\chi = 0.1$ to 0.9.

Given an expression for ΔG_{mix} we can express the excess Gibbs free energy of mixing for a Flory-Huggins mixture as:

$$\Delta G_{mix}^{EX} = \Delta G_{mix} - \Delta G_{mix}^{ID} = RT\left[x_1 \ln\left(\frac{\Phi_1^*}{x_1}\right) + x_2 \ln\left(\frac{\Phi_2^*}{x_2}\right) + \frac{\chi\,\Phi_1^*\,\Phi_2^*\,(N_1 + rN_2)}{(N_1 + N_2)}\right] \quad (11\text{-}65)$$

and easily estimate γ_i for the solvent and solute (see Problem 15.33). Also the chemical potential or activity of the solvent is given by,

$$\mu_1 - \mu_1^o = RT \ln a_1 = RT \ln\left(\frac{\hat{f}_1}{f_1^o}\right) = RT\left[\ln(1 - \Phi_2^*) + (1 - \frac{1}{r})\,\Phi_2^* + \chi\,\Phi_2^{*2}\right] \quad (11\text{-}66)$$

where a pure solvent (o) standard state at T and P of the mixture was selected.

The Flory-Huggins model is the current model of choice for characterizing polymer solution thermodynamics. Several improvements have been implemented to more correctly account for the energetics of hindered rotation of polymer chains and for the existence of polydisperse systems of varying molecular weights (that is with *chain length* r not constant). The fundamental Flory-Huggins approach is on firm enough ground to provide a basis for characterizing polymer solutions for a long time. In addition, to being able to predict the dependence of osmotic pressure on concentration much more accurately for mixtures containing molecules of dissimilar size, the general robustness of the Flory-Huggins approach is quite striking. For example, it has been successfully applied to many systems where polymer-solvent, polymer-monomer, or polymer-polymer solubility limits are encountered and phase separation occurs (for example, see Section 15.7 on spinodal decomposition).

11.7.3 Local Composition Models (Wilson, NRTL). An interesting empirical extension of the Flory-Huggins model incorporating concepts from lattice theory and the quasi-chemical approximation was proposed by Wilson in 1964. Wilson's approach was structured to account for the effects of both differing molecular size and intermolecular forces.

Table 11.1 Models for the Excess Gibbs Energy for Condensed Phase Binary Systems

Type and Name	Operating Equations
First-order polynomial	
Two-suffix[a] Margules	$\Delta G^{EX} = Ax_1x_2$ Binary parameters = A $RT \ln \gamma_1 = Ax_2^2$ $RT \ln \gamma_2 = Ax_1^2$
Three-suffix[a] Margules	$\Delta G^{EX} = x_1x_2\,[A + B(x_1 - x_2)]$ Binary parameters = A, B $RT \ln \gamma_1 = (A + 3B)x_2^2 - 4Bx_2^3$ $RT \ln \gamma_2 = (A - 3B)x_1^2 + 4Bx_1^3$
van Laar	$\Delta G^{EX} = \dfrac{Ax_1x_2}{x_1(A/B) + x_2}$ Binary parameters = A, B $RT \ln \gamma_1 = A\left(1 + \dfrac{Ax_1}{Bx_2}\right)^{-2}$ $RT \ln \gamma_2 = B\left(1 + \dfrac{Bx_2}{Ax_1}\right)^{-2}$
Four suffix[a] Margules	$\Delta G^{EX} = x_1x_2\,[A + B(x_1 - x_2) + C(x_1 - x_2)^2]$ Binary parameters = A, B, C $RT \ln \gamma_1 = (A + 3B + 5C)x_2^2 - 4(B + 4C)x_2^3 + 12Cx_2^4$ $RT \ln \gamma_2 = (A - 3B + 5C)x_1^2 + 4(B - 4C)x_1^3 + 12Cx_1^4$
General Redlich-Kister	$\dfrac{\Delta G^{EX}}{RT} = x_1\,x_2\,[B + C(x_1 - x_2) + D(x_1 - x_2)^2 + \ldots]$ Binary parameters = B, C, D, ...
Volume-fraction based	
Hildebrand-Scatchard	$\Delta G^{EX} = \dfrac{(\delta_1 - \delta_2)^2}{\left(\dfrac{1}{V_1x_1} + \dfrac{1}{V_2x_2}\right)}$ V_1, V_2 = molar volumes of pure 1 and 2 Pure component parameters = δ_1, δ_2 $RT \ln \gamma_1 = V_1\,\Phi_2^{*2}\,(\delta_1 - \delta_2)^2$ $RT \ln \gamma_2 = V_2\,\Phi_1^{*2}\,(\delta_1 - \delta_2)^2$ $\Phi_i^* = \dfrac{V_ix_i}{V_1x_1 + V_2x_2}\; i = 1, 2$

Flory-Huggins	$\frac{\Delta G^{EX}}{RT} = x_1 \ln \frac{\Phi_1^*}{x_1} + x_2 \ln \frac{\Phi_2^*}{x_2} + \frac{\chi\, \Phi_1^*\, \Phi_2^*(N_1 + rN_2)}{(N_1 + N_2)}$ $\Phi_1^* = \frac{N_1}{N_1 + rN_2} \quad \Phi_2^* = \frac{rN_2}{N_1 + rN_2}$ r = chain length of polymer = number of monomer units Binary parameters = χ; 2(1 – *solvent/monomer* ; 2 – *polymer*)
General Wohl expansion	$\frac{\Delta G^{EX}}{RT} = (x_1q_1 + x_2q_2)\left[2a_{12}z_1z_2 + 3a_{112}z_1^2z_2 + 3a_{122}z_1z_2^2 + 4a_{1112}z_1^3z_2 + 4a_{1222}z_1z_2^3 + 6a_{1122}z_1^2z_2^2 + \ldots\right]$ $z_1 \equiv \frac{x_1q_1}{x_1q_1 + x_2q_2}$, $z_2 \equiv \frac{x_2q_2}{x_1q_1 + x_2q_2}$ } effective volume fraction q_i = volume fraction size of pure i (i = 1, 2) Binary parameters = $a_{12}, a_{112}, a_{122}, \ldots$
Local-composition based	
Wilson	$\frac{\Delta G^{EX}}{RT} = -x_1 \ln(x_1 + \Lambda_{12}x_2) - x_2 \ln(x_2 + \Lambda_{21}x_1)$ Binary parameters = $\Lambda_{12}, \Lambda_{21}$ $\ln \gamma_1 = -\ln(x_1 + \Lambda_{12}x_2) + \beta x_2$ $\ln \gamma_2 = -\ln(x_2 + \Lambda_{21}x_1) - \beta x_1$ $\beta \equiv \left(\frac{\Lambda_{12}}{x_1 + \Lambda_{12}x_2} - \frac{\Lambda_{21}}{\Lambda_{21}x_1 + x_2}\right)$
TK-Wilson[f]	$\frac{\Delta G^{EX}}{RT} = x_1 \ln \frac{(x_1 + V_2x_2/V_1)}{(x_1 + \Lambda_{12}x_2)} + x_2 \ln \frac{(V_1x_1/V_2 + x_2)}{(\Lambda_{21}x_1 + x_2)}$ V_1, V_2 = molar volumes of pure 1 and 2 β same as for Wilson model Binary parameters = $\Lambda_{12}, \Lambda_{21}$ $\ln \gamma_1 = \ln \frac{(x_1 + V_2x_2/V_1)}{(x_1 + \Lambda_{12}x_2)} + (\beta - \beta^*)\, x_2$ $\beta^* = \frac{V_2/V_1}{(x_1 + V_2x_2/x_1)} - \frac{V_1/V_2}{(V_1x_1/V_2 + x_2)}$

NRTL[b]	$\frac{\Delta G^{EX}}{RT} = x_1 x_2 \left(\frac{\tau_{21} G_{21}}{x_1 + x_2 G_{21}} + \frac{\tau_{12} G_{12}}{x_2 + x_1 G_{12}} \right)$ where $\tau_{12} = \frac{\Delta g_{12}}{RT}, \tau_{21} = \frac{\Delta g_{21}}{RT}$ $\ln G_{12} = -\alpha_{12}\tau_{12}, \ln G_{21} = -\alpha_{12}\tau_{21}$ Binary parameters = $\Delta g_{12}, \Delta g_{21}, \alpha_{12}$[c] $\ln \gamma_1 = x_2^2 \left[\tau_{21} \left(\frac{G_{21}}{x_1 + x_2 G_{21}} \right)^2 + \frac{\tau_{12} G_{12}}{(x_2 + x_1 G_{12})^2} \right]$ $\ln \gamma_2 = x_1^2 \left[\tau_{12} \left(\frac{G_{12}}{x_2 + x_1 G_{12}} \right)^2 + \frac{\tau_{21} G_{21}}{(x_1 + x_2 G_{21})^2} \right]$
UNIQUAC[d]	$\Delta G^{EX} = \Delta G^{EX}(\text{combinatorial}) + \Delta G^{EX}(\text{residual})$ $\frac{\Delta G^{EX}(\text{combinatorial})}{RT} = x_1 \ln \frac{\Phi_1^*}{x_1} + x_2 \ln \frac{\Phi_2^*}{x_2} + \frac{z}{2} \left(q_1 x_1 \ln \frac{\theta_1}{\Phi_1^*} + q_2 x_2 \ln \frac{\theta_2}{\Phi_2^*} \right)$ $\frac{\Delta G^{EX}(\text{residual})}{RT} = -q_1' x_1 \ln [\theta_1' + \theta_2' \tau_{21}] - q_2' x_2 \ln [\theta_2' + \theta_1' \tau_{12}]$ $\Phi_1^* \equiv \frac{x_1 r_1}{x_1 r_1 + x_2 r_2}, \theta_1 \equiv \frac{x_1 q_1}{x_1 q_1 + x_2 q_2} \theta_1' \equiv \frac{x_1 q_1'}{x_1 q_1' + x_2 q_2'}$ $\ln \tau_{21} \equiv -\frac{\Delta u_{21}}{RT}, \ln \tau_{12} \equiv -\frac{\Delta u_{12}}{RT}$ r, q, and q' are pure-component parameters and coordination number $z = 10$ Binary parameters = Δu_{12} and Δu_{21}[e] $\ln \gamma_i = \ln \frac{\Phi_i^*}{x_i} + \frac{z}{2} q_i \ln \frac{\theta_i}{\Phi_i^*} + \Phi_j^* \left(l_i - \frac{r_i}{r_j} l_j \right) - q_i' \ln (\theta_i' + \theta_j' \tau_{ji}) + \theta_j' q_i' \left(\frac{\tau_{ji}}{\theta_i' + \theta_j' \tau_{ji}} - \frac{\tau_{ij}}{\theta_j' + \theta_i' \tau_{ij}} \right)$ where $i = 1, j = 2$ or $i = 2, j = 1$ $l_i \equiv \frac{z}{2}(r_i - q_i) - (r_i - 1); l_j \equiv \frac{z}{2}(r_j - q_j) - (r_j - 1)$

[a]Two-suffix signifies that the expansion for ΔG^{EX} is quadratic in mole fraction. Three-suffix signifies a third-order, and four-suffix signifies a fourth-order equation.

[b]NRTL = non-random two-liquid model.

[c]$\Delta g_{12} = g_{12} - g_{22}; \Delta g_{21} = g_{21} - g_{11}$

[d]UNIQUAC = universal quasi-chemical activity coefficient model.

[e]$\Delta u_{12} = u_{12} - u_{22}; \Delta u_{21} = u_{21} - u_{11}$.

[f]works for liquid-liquid systems.

Sources: Prausnitz *et al.* (1986), 2nd ed, Chapter 6 and Walas (1985), Chapter 4 where the Margules, Redlich-Kister, van Laar, Wilson, TK-Wilson, Wohl, Hildebrand-Scatchard, NRTL, and UNIQUAC equations are discussed.

Table 11.2 Extension of ΔG^{EX} Models to Multicomponent Systems

Model	$\ln \gamma_i$	Parameters required
Two-Suffix Margules	$\sum_{k=1}^{n}\sum_{j=1}^{n}\left[A_{ki}-\frac{1}{2}A_{kj}\right]x_k x_j$ (binary interactions only)	$A_{jj}=A_{kk}=0$ $A_{kj}=A_{jk}$
Wilson	$1-\ln\left[\sum_{j=1}^{n}\Lambda_{ij}x_j\right]-\sum_{k=1}^{n}\left[\frac{\left[\Lambda_{ki}x_k\right]}{\sum_{j=1}^{n}\Lambda_{kj}x_j}\right]$	$\Lambda_{ii}=\Lambda_{jj}=1$ $\Lambda_{ij}=\frac{V_j}{V_i}\exp\left(\frac{-\lambda_{ij}}{RT}\right)$
TK-Wilson	$-\ln\left[\sum_{j=1}^{n}\Lambda_{ij}x_j\right]-\sum_{k=1}^{n}\left[\frac{\left[\Lambda_{ki}x_k\right]}{\sum_{j=1}^{n}\Lambda_{kj}x_j}\right]$ $+\ln\left(\sum_{j=1}^{n}V_j x_j/V_i\right)+\sum_{k=1}^{n}\left[\frac{(V_i/V_k)\,x_k}{\sum_{j=1}^{n}(V_j/V_k)\,x_j}\right]$	Λ_{ij} and Λ_{ji} are defined as in Wilson model $\Lambda_{ii}=\Lambda_{jj}=1$
NRTL	$\frac{\sum_{j=1}^{n}\tau_{ji}G_{ji}x_j}{\sum_{k=1}^{n}G_{ki}x_k}+\sum_{j=1}^{n}\left[\frac{G_{ij}x_j}{\sum_{k=1}^{n}G_{kj}x_k}\right]\left[\tau_{ij}-\frac{\sum_{m=1}^{n}\tau_{mj}G_{mj}x_m}{\sum_{k=1}^{n}G_{kj}x_k}\right]$	$\tau_{ji}=\frac{g_{ji}-g_{ii}}{RT}$ $\tau_{ii}=\tau_{jj}=0$ $G_{ji}=\exp(-\alpha_{ji}\tau_{ji})$ $G_{ii}=G_{jj}=1.0$
UNIQUAC	$\ln\gamma_i=\ln\gamma_i^C+\ln\gamma_i^R$ $\ln\gamma_i^C=\ln\frac{\Phi_i^*}{x_i}+\frac{z}{2}q_i\ln\frac{\theta_i}{\Phi_i^*}+l_i-\frac{\Phi_i^*}{x_i}\sum_{j=1}^{n}x_j l_j$ $\ln\gamma_i^R=q_i\left[1-\ln\left[\sum_{j=1}^{n}\theta_j\tau_{ji}\right]-\sum_{j=1}^{n}\left[\frac{\theta_j\tau_{ij}}{\sum_{k=1}^{n}\theta_k\tau_{kj}}\right]\right]$ $\Phi_i^*=\frac{r_i x_i}{\sum_{k=1}^{n}r_k x_k}$ and $\theta_i=\frac{q_i x_i}{\sum_{k=1}^{n}q_k x_k}$	$\tau_{ji}=\exp\left[-\frac{(u_{ji}-u_{ii})}{RT}\right]$ $\tau_{ii}=\tau_{jj}=1.0$ $l_i=\frac{z}{2}(r_i-q_i)-(r_i-1)$ $z=10$ (usually)

Although his approach is not based on rigorous theory, he tried to capture important effects by defining local mole fractions (x_{ij}) that scaled with the bulk mole fraction (x_i) and a Boltzmann factor that was proportional to the probability of finding a molecule of type i in the vicinity of a molecule of type j ($\boldsymbol{P}_{ij}$). This idea is easily illustrated by considering a binary (1-2) mixture. Shown in Figure 11.4 are two cases: one with a central molecule of type 1 and the other with a central molecule of type 2.

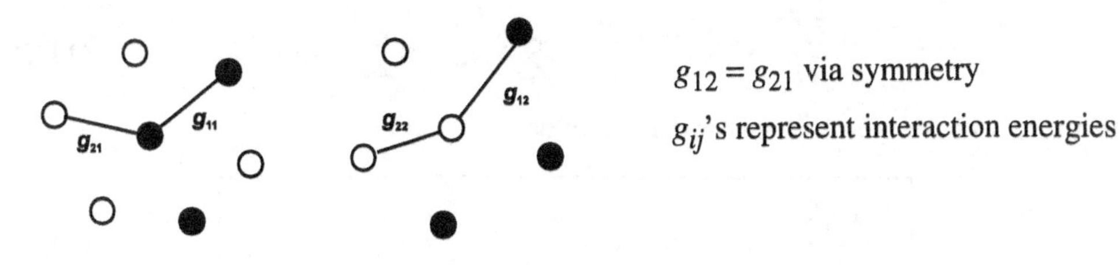

Figure 11.4 Interaction energies in the Wilson model.

With Wilson's definition of probability:

$\boldsymbol{P}_{11}$ = probability of finding a molecule of type 1 in vicinity of 1
$\boldsymbol{P}_{21}$ = probability of finding a molecule of type 2 in the vicinity of 1
$\boldsymbol{P}_{12}$ = probability of finding a molecule of type 1 in vicinity of 2
$\boldsymbol{P}_{22}$ = probability of finding a molecule of type 2 in the vicinity of 2

Ratios of $\boldsymbol{P}_{ij}$s were viewed as ratios of "local mole fractions" (x_{ij}) expressed as a product of the bulk mole fractions x_i and Boltzmann factors in terms of g_{ij}'s. For a binary solution, this results in,

$$\frac{x_{11}}{x_{21}} = \frac{x_1 \exp\left[\frac{-g_{11}}{RT}\right]}{x_2 \exp\left[\frac{-g_{21}}{RT}\right]} \qquad \frac{x_{12}}{x_{22}} = \frac{x_1 \exp\left[\frac{-g_{12}}{RT}\right]}{x_2 \exp\left[\frac{-g_{22}}{RT}\right]} \tag{11-67}$$

with conservative equations: $x_{21} + x_{11} = 1$ and $x_{12} + x_{22} = 1$.

Following the Flory-Huggins approach, we can define local volume fractions (z_i) in terms of the local mole fractions (x_{ij})

$$z_1 \equiv \frac{x_{11} V_1}{x_{11} V_1 + x_{21} V_2} \qquad z_2 \equiv \frac{x_{22} V_2}{x_{22} V_2 + x_{12} V_1} \tag{11-68}$$

which can be rewritten as

$$z_1 = \frac{x_1}{x_1 + \Lambda_{12} x_2} \qquad z_2 = \frac{x_2}{x_2 + \Lambda_{21} x_1} \tag{11-69}$$

where:

$$\Lambda_{12} \equiv \frac{V_2}{V_1} \exp\left[-\lambda_{12}/RT\right] \qquad \lambda_{12} \equiv g_{21} - g_{11}$$

$$\Lambda_{21} \equiv \frac{V_1}{V_2} \exp\left[-\lambda_{21}/RT\right] \qquad \lambda_{21} \equiv g_{12} - g_{22} \qquad g_{12} = g_{21}$$

The excess Gibbs free energy is given by assuming that these local volume fractions can be used directly to evaluate ΔS_{mix} in the Flory-Huggins model [Eq. (11-63)]. Thus,

$$\frac{\Delta G^{EX}}{RT} = \frac{\Delta G - \Delta G^{ID}}{RT} = \sum x_i \ln z_i - \sum x_i \ln x_i \tag{11-70}$$

With Eq. (11-69), the ***Wilson model*** becomes:

$$\boxed{\frac{\Delta G^{EX}}{RT} = -x_1 \ln(x_1 + \Lambda_{12} x_2) - x_2 \ln(\Lambda_{21} x_1 + x_2)} \tag{11-71}$$

Using the fact that $RT \ln\gamma_i = \overline{\Delta G_i^{EX}}$ and Eq. (9-53), we get

$$\ln \gamma_1 = \beta x_2 - \ln(x_1 + \Lambda_{12} x_2) \quad \text{and} \quad \ln \gamma_2 = -\beta x_1 - \ln(\Lambda_{21} x_1 + x_2) \tag{11-72}$$

with $\beta \equiv \Lambda_{12}/(x_1 + \Lambda_{12} x_2) - \Lambda_{21}/(\Lambda_{21} x_1 + x_2)$.

The Wilson model empirically extends the Flory-Huggins configurational entropy representation in a non-rigorous fashion. Nonetheless, Eqs. (11-71) and (11-72) give qualitatively correct interpretations of molecular-level effects and quantitatively do an excellent job of correlating activity coefficient data for many polar and non-polar systems. For example, the Wilson equation works well for very non-ideal vapor-liquid equilibrium systems without further approximations (see Table 11.2). Only two parameters (Λ_{12} and Λ_{21}) are required per binary pair, no higher-order parameters are needed.

One disadvantage of the Wilson equation is that it can only be used for miscible liquid systems (see Problem 11.6). Other potential problems are the presence of multiple roots if $\gamma_i < 1.0$ and the fact that negative values of Λ_{ij} are not permitted when representing γ_i variations over the full composition interval from $x_i = 0$ to 1.0.

An important modification to the Wilson model was made by Tsuboka and Katayama (1975) to permit its use for liquid-liquid systems. In their model (the ***TK-Wilson equation***):

$$\frac{\Delta G^{EX}}{RT} = x_1 \ln\left[\frac{x_1 + V_{12} x_2}{x_1 + \Lambda_{12} x_2}\right] + x_2 \ln\left[\frac{x_2 + V_{21} x_1}{x_2 + \Lambda_{21} x_1}\right] \quad \text{with } V_{ij} = \frac{V_i}{V_j} \tag{11-73}$$

Expressions for $\ln \gamma_i$ are given in Table 11.1. For further discussion of the Wilson model and the TK-Wilson extension, see Prausnitz *et al.* (1986), Section 6.11 and 7.8 and Walas (1985), Section 4.9.

Renon and Prausnitz (1968) developed the ***Non-Random Two-Liquid equation*** or ***NRTL*** equation, which is based on a local mole fraction concept and can be applied to both vapor-liquid and liquid-liquid equilibrium (VLE and LLE) problems. Renon and Prausnitz define modified local mole fractions similar to Eq. (11-67) as:

$$\frac{x_{21}}{x_{11}} = \frac{x_2 \exp[-\alpha_{12} g_{21}/RT]}{x_1 \exp[-\alpha_{12} g_{11}/RT]}$$
$$\frac{x_{12}}{x_{22}} = \frac{x_1 \exp[-\alpha_{12} g_{12}/RT]}{x_2 \exp[-\alpha_{12} g_{22}/RT]} \tag{11-74}$$

again with $x_{21} + x_{11} = 1$, $x_{12} + x_{22} = 1$, and $g_{12} = g_{21}$. But now an additional parameter α_{12} has been introduced to represent binary interactions characteristic of the non-randomness of the mixture. Again this is a creative extension of the local composition approach without a rigorous theoretical basis.

Now we can rewrite the equations to solve for x_{21} and x_{12}:

$$x_{21} = \frac{x_2 \exp[-\alpha_{12}(g_{21} - g_{11})/RT]}{x_1 + x_2 \exp[-\alpha_{12}(g_{21} - g_{11})/RT]} \qquad x_{12} = \frac{x_1 \exp[-\alpha_{12}(g_{12} - g_{22})/RT]}{x_2 + x_1 \exp[-\alpha_{12}(g_{12} - g_{22})/RT]} \tag{11-75}$$

By using Scott's two-liquid theory for binaries,

$$\frac{\Delta G^{EX}}{RT} = \sum x_i \frac{\overline{G}_i - G_i^o}{RT} = x_1 x_{21}(g_{21} - g_{11})/RT + x_2 x_{12}(g_{12} - g_{22})/RT \tag{11-76}$$

and defining three parameters per binary,

$$\tau_{12} \equiv \frac{g_{12} - g_{22}}{RT} = \frac{\Delta g_{12}}{RT} \; ; \; \tau_{21} \equiv \frac{g_{21} - g_{11}}{RT} = \frac{\Delta g_{21}}{RT} \; ; \; \alpha_{12}$$

the ***NRTL equation*** for ΔG^{EX} becomes

$$\boxed{\frac{\Delta G^{EX}}{RT} = x_1 x_2 \left[\frac{\tau_{21} G_{21}}{x_1 + x_2 G_{21}} + \frac{\tau_{12} G_{12}}{x_2 + x_1 G_{12}} \right]} \tag{11-77}$$

with $G_{12} \equiv \exp[-\alpha_{12}\tau_{12}]$ and $G_{21} \equiv \exp[-\alpha_{12}\tau_{21}]$. For ternaries and higher-order mixtures, one can use an expansion of binary interactions:

$$\frac{\Delta G^{EX}}{RT} = \sum_{i=1}^{n} \left[\frac{x_i \sum_{j=1}^{n} \tau_{ji} G_{ji} x_j}{\sum_{k=1}^{n} G_{ki} x_k} \right] \tag{11-78}$$

where

$\tau_{ii} = 0$; $G_{ii} = 1.0$; $g_{ij} = g_{ji}$

$\tau_{ji} = (g_{ji} - g_{ii})/RT$; $G_{ji} = \exp[-\alpha_{ji} \tau_{ji}]$ with $\alpha_{ji} = \alpha_{ij}$.

Table 11.2 gives a complete listing of multicomponent forms for five models. One should note that appropriate temperature dependence of the τ_{ij} parameters is often incorporated from data regressions.

Although the NRTL equation requires three parameters per binary, Renon and Prausnitz provide some guidelines for specifying the magnitude of α_{12} based on comparisons to classes of mixtures, typically α_{12} is about 0.2 to 0.4 on average (see Walas (1985), Section 4.10, for further discussion). Like the Wilson equation, the NRTL model has been well-tested with binary τ_{ij} and α_{ij} parameters for many mixtures tabulated in several databases such as DECHEMA and DIPPR. Furthermore, the NRTL equation can be used for highly non-ideal liquid systems where liquid-liquid phase splitting occurs, but in these cases care must be exercised in selecting parameters.

11.7.4. *Quasi-chemical models (UNIQUAC).* Abrams and Prausnitz (1975) introduced the so-called *Universal Quasi-Chemical Activity Coefficient* model or UNIQUAC as an extension of Guggenheim's quasi-chemical model that accounts for non-randomness to include the effects of molecules of different size. The UNIQUAC model like the Wilson and NRTL (with α_{ij} fixed) models has two adjustable parameters per binary, is applicable to ternaries and will work for both VLE and LLE problems.

The excess Gibbs free energy is divided into two parts:

$$\Delta G^{EX} = \Delta G^{EX\,(combinatorial)} + \Delta G^{EX\,(residual)} \tag{11-79}$$

The combinatorial part captures the major entropic contribution as determined by the sizes and shapes of the constituent molecules and their mole fractions. The residual part depends on intermolecular forces that lead to finite enthalpies of mixing. While the combinatorial contribution only requires pure component properties, the residual part requires two binary interaction parameters.

UNIQUAC follows Guggenheim's statistical mechanics approach extending his quasi-chemical theory. For example, for a binary:

$$\Delta G^{EX} \cong \Delta A^{EX} = \Delta A - \underbrace{RT\,(x_1 \ln x_1 + x_2 \ln x_2)}_{T\Delta S^{ID}} \tag{11-80}$$

Using the statistical mechanics formalism introduced in Chapter 10 we can use the configurational canonical partition function to express ΔA as

$$\Delta A = -kT \ln\left[\frac{Q_{lattice}\,(N_1, N_2)}{Q_{lattice}\,(O, N_2)\; Q_{lattice}\,(N_1, O)}\right] \tag{11-81}$$

where

$$Q_{lattice} = \sum_{all\ \theta_{ij}} \omega(\theta_{ij}) \exp\left[-U_o/kT\right]$$

$U_o = f(\theta_{ij})$ = potential energy of lattice, related to ΔU_{vap} from liquid to ideal gas

$\omega(\theta_{ij})$ = combinatorial factor

θ_{ij} = area fractions consistent with stoichiometry of mixture = fraction of external sites around molecule i occupied by molecule j.

We can express U_o in terms of local area fractions θ_{ij} as,

$$-U_o = \frac{z}{2} q_1 N_1 (\theta_{11} U_{11} + \theta_{21} U_{21}) + \frac{z}{2} q_2 N_2 (\theta_{22} U_{22} + \theta_{12} U_{12}) \tag{11-82}$$

where

q_i = structural factor proportional to external surface area of molecule i

U_{ij} = potential energy of interaction between molecules of type i and j

z = coordination number of the lattice = 10 (typically).

Note that the following conservation equations apply: $\theta_{11} + \theta_{21} = 1$ and $\theta_{12} + \theta_{22} = 1$ and $\theta_{12} = \theta_{21}$ (via symmetry).

After considerable mathematical restructuring and the introduction of several approximations, the ***UNIQUAC model*** is given by,

$$\frac{\Delta G^{EX\,(combinatorial)}}{RT} = x_1 \ln \frac{\Phi_1^*}{x_1} + x_2 \ln \frac{\Phi_2^*}{x_2} + \frac{z}{2}\left(q_1 x_1 \ln \frac{\Phi_1^*}{x_1} + q_2 x_2 \ln \frac{\Phi_2^*}{x_2} \right) \tag{11-83}$$

and

$$\frac{\Delta G^{EX\,(residual)}}{RT} = -q_1 x_1 \ln (\theta_1 + \theta_2 \tau_{21}) - q_2 x_2 \ln (\theta_2 + \theta_1 \tau_{12}) \tag{11-84}$$

where

$\tau_{21} \equiv \exp(-[u_{21} - u_{11}]/RT) \quad \tau_{12} \equiv \exp(-[u_{12} - u_{22}]/RT) \quad z = 10$

$u_{12} = u_{21} \qquad \left.\begin{array}{l}\Delta u_{12} = u_{12} - u_{22} \\ \Delta u_{21} = u_{21} - u_{11}\end{array}\right\}$ two adjustable parameters for mixture

$$\Phi_1^* = \frac{x_1 r_1}{x_1 r_1 + x_2 r_2} \qquad \Phi_2^* = \frac{x_2 r_2}{x_1 r_1 + x_2 r_2}$$

$$\theta_1 = \frac{x_1 q_1}{x_1 q_1 + x_2 q_2} \qquad \theta_2 = \frac{x_2 q_2}{x_1 q_1 + x_2 q_2}$$

The r_is and q_is can be predicted by molecular group contribution methods such as UNIFAC (see Section 13.4) or can be estimated from experimental data

$$r_i = \text{relative volume fraction} \equiv V_{wi}/V_{ws}$$
$$q_i = \text{relative area fraction} \equiv A_{wi}/A_{ws} \tag{11-85}$$

where V_{ws} and A_{ws} refer to sizes of a $-CH_2-$ group in a very high molecular weight paraffin such that $V_{ws} = \frac{4}{3}\pi R_{ws}^3$ and $A_{ws} = 4\pi R_{ws}^2$ with $R_{ws} = 10.95 \times 10^{15}$ cm/mole.

Anderson and Prausnitz (1978) modified the original UNIQUAC formulation to improve agreement for systems containing water and/or lower alcohols. In the original equation $q' = q$. The modified equation only alters the residual contribution to:

$$\frac{\Delta G^{EX\,(residual)}}{RT} = -q_1' x_1 \ln(\theta_1' + \theta_2' \tau_{21}) - q_2' x_2 \ln(\theta_2' + \theta_1' \tau_{12}) \tag{11-86}$$

where

$$\theta_1' = \frac{x_1 q_1'}{x_1 q_1' + x_2 q_2'} \qquad \theta_2' = \frac{x_2 q_2'}{x_1 q_1' + x_2 q_2'}$$

$$\tau_{12} = \exp[-\Delta u_{12}/RT] = \exp[-a_{12}/T]$$

$$\tau_{21} = \exp[-\Delta u_{21}/RT] = \exp[-a_{21}/T]$$

With the modified form, γ_1 and γ_2 are given by appropriate differentiation of ΔG^{EX} [Eq. (11-83) and Eq. (11-86)] using Eq. (9-53):

$$\ln \gamma_1 = \ln \frac{\Phi_1^*}{x_1} + \frac{z}{2} q_1 \ln \frac{\theta_1}{\Phi_1^*} + \Phi_2^* \left(l_1 - \frac{r_1}{r_2} l_2 \right) - q_1' \ln(\theta_1' + \theta_2' \tau_{21}) + \theta_2' q_1' \left(\frac{\tau_{21}}{\theta_1' + \theta_2' \tau_{21}} - \frac{\tau_{12}}{\theta_2' + \theta_1' \tau_{12}} \right) \tag{11-87}$$

$$\ln \gamma_2 = \ln \frac{\Phi_2^*}{x_2} + \frac{z}{2} q_2 \ln \frac{\theta_2}{\Phi_2^*} + \Phi_1^* \left(l_2 - \frac{r_2}{r_1} l_1 \right) - q_2' \ln(\theta_2' + \theta_1' \tau_{12}) + \theta_1' q_2' \left(\frac{\tau_{12}}{\theta_2' + \theta_1' \tau_{12}} - \frac{\tau_{21}}{\theta_1' + \theta_2' \tau_{21}} \right) \tag{11-88}$$

with $\quad l_1 = \frac{z}{2}(r_1 - q_1) - (r_1 - 1) \qquad l_2 = \frac{z}{2}(r_2 - q_2) - (r_2 - 1)$

Table 11.3 gives values for the parameters r_i, q_i and q_i' for a number of compounds. Table 11.4 provides values for a_{12} and a_{21} which can be used to determine τ_{ij} binary interaction parameters for specific binary mixtures. The UNIQUAC model can be extended to multicomponent mixtures (see Table 11.2). Note also that the binary parameters (τ_{ij}) are expressed as functions of temperature. It has the further advantage that in the absence of any experimental γ_i data, τ_{ij} parameters can also be estimated using the UNIFAC molecular group contribution method (see Section 13.4). Although the UNIQUAC model is fairly complex in its mathematical form, the total number of adjustable parameters is still only two (τ_{12} and τ_{21}) per binary as the structural parameters r_i, q_i and q_i' are specified from a pure component database. It, like the Wilson and NRTL models, has been used extensively by chemical engineers to represent VLE and LLE and consequently comprehensive tabulations of fitted binary parameters are available.

11.8 Activity Coefficient Models for Solid Phases

A good starting point for mixtures that are solids is to utilize the lattice theory that was introduced in Section 11.5. In this context, effectively we are saying that a solid phase mixture behaves as a substitutional solid solution where the molecules of either component can occupy equivalent lattice sites with minimal distortion to the lattice geometry. As we saw earlier, the two-component treatment yields a result identical to a regular solution where,

Table 11.3 Some Structural Parameters for UNIQUAC Equation [from Table 6.9 in Prausnitz *et al.* (1986)]

Component	r	q
Carbon tetrachloride	3.33	2.82
Chloroform	2.70	2.34
Formic acid	1.54	1.48
Methanol	1.43	1.43
Acetonitrile	1.87	1.72
Acetic acid	1.90	1.80
Nitroethane	2.68	2.41
Ethanol	2.11	1.97
Acetone	2.57	2.34
Ethyl acetate	3.48	3.12
Methyl ethyl ketone	3.25	2.88
Diethylamine	3.68	3.17
Benzene	3.19	2.40
Methylcyclopentane	3.97	3.01
Methyl isobutyl ketone	4.60	4.03
n-Hexane	4.50	3.86
Toluene	3.92	2.97
n-Heptane	5.17	4.40
n-Octane	5.85	4.94
Water	0.92	1.40

Component	q'	Component	q'
Water	1.00	C_4-alcohols	0.88
CH_3OH	0.96	C_5-alcohols	1.15
C_2H_5OH	0.92	C_6-alcohols	1.78
C_3-alcohols	0.89	C_7-alcohols	2.71

Note: the r, q, and q' parameters are dimensionless because they are (arbitrarily) taken relative to the size and surface area of a $-CH_2-$ unit in a high-molecular weight paraffin.

Table 11.4 Some Binary Parameters for UNIQUAC Equation [from Table 6.10, Prausnitz *et al.* (1986)]

Binary (1)/(2) system	Temperature (K)	Energy parameters (K) a_{12}	a_{21}
Acetonitrile/benzene	318	-40.70	229.79
n-Hexane/nitroethane	318	230.64	-5.86
Acetone/chloroform	323	-171.71	93.96
Ethanol/*n*-octane	348	-123.57	1354.92
Formic acid/acetic acid	374-387	-144.58	241.64
Propionic acid/methyl isobutyl ketone	390-411	-78.49	136.46
Acetone/water	331-368	530.99	-100.71
Acetonitrile/water	350-364	294.10	61.92
Acetic acid/water	373-389	530.94	-299.90
Formic acid/water	374-380	924.01	-525.85
Methylcyclopentane/ethanol	333-349	1383.93	-118.27
Methylcyclopentane/benzene	344-352	56.47	-6.47
Ethanol/benzene	340-351	-138.90	947.20
Methyl ethyl ketone/*n*-heptane	350-369	-75.13	242.53
Methanol/carbon tetrachloride	328	-29.64	1127.95
Methanol/benzene	528	-56.35	972.09
Chloroform/ethanol	323	934.23	-208.50
Chloroform/*n*-heptane	323	-19.26	88.40
Ethanol/*n*-heptane	323	-105.23	1380.30
Acetone/methanol	323	379.31	-108.42
Methanol/ethyl acetate	335-347	-107.54	579.61

$$\Delta G^{EX} = A\,x_1\,x_2;\quad \ln\gamma_1 = \frac{A\,x_2^2}{RT};\quad \ln\gamma_2 = \frac{A\,x_1^2}{RT} \tag{11-89}$$

The term A is related to the interchange energy w which expresses the potential energy difference between unlike (1-2) and like (1-1 and 2-2) molecules as

$$w = [\Phi_{12} - 1/2\,(\Phi_{11} + \Phi_{22})]$$

Regular solution behavior as indicated above in Eq. (11-89) is completely symmetrical about the $x_i = 0.5$ point. For many mixtures of solids including a range of binary metallic and ceramic solutions, this model represents thermodynamic non-idealities adequately. In other cases, behavior is almost regular, and consequently several simple modifications have been proposed. For example, by adding an additional parameter or two to account for asymmetric trends, several subregular models have been formulated (for example, see Lupis (1983)). For a binary (1-2) mixture,

$$\Delta G^{EX} = x_1\,x_2\,[A_{11} + A_{21}\,x_1 + A_{12}\,x_2 + A_{22}\,x_1\,x_2 + \ldots]$$

or

$$\Delta G^{EX} = \sum_{i=1}^{m}\sum_{j=1}^{n} A_{ij}\,x_1^i\,x_2^j \tag{11-90}$$

where various options exist to specify and relate the coefficients A_{ij} (for example, $A_{ij} = A_{ji}$ or $A_{22} = 0$ represent two possible cases). Unless strong non-idealities exist, usually three parameters are sufficient to adequately represent the data. Regular and sub-regular models are explored in Chapter 15 and 17 where phase diagrams for such condensed phase systems are discussed. Regular solution theory and its extensions although deceptively simple, in fact, actually predict quite non-ideal behavior, for example, phase separations, eutectic points, and maxima and minima in solidus and liquidus curves.

Another important approach for modeling metallic solutions in both condensed liquid and solid phases has been proposed by Lupis and Elliott in 1967. Their model is based on expanding the partial molar excess Gibbs free energy of a component of interest from a dilute state where Henry's law is followed [see Chapter 9 in Lupis (1983) for details]. Basically, they use a Taylor series expansion of $\ln \gamma_i$ from the point of infinite dilution where x_i approaches zero. For example, for a single solute (2) dissolved in a single metallic component (1) with T and P constant:

$$\ln \gamma_2 = \ln \gamma_2^\infty + \left(\frac{\partial \ln \gamma_2}{\partial x_2}\right)_{x_2 \to 0} x_2 + \frac{1}{2!}\left(\frac{\partial^2 \ln \gamma_2}{\partial x_2^2}\right)_{x_2 \to 0} x_2^2 + \ldots + \frac{1}{n!}\left(\frac{\partial^n \ln \gamma_2}{\partial x_2^n}\right)_{x_2 \to 0} x_2^n \tag{11-91}$$

For a general solute component i in a mixture of n components, the Lupis-Elliott model expresses $\ln \gamma_i$ as,

$$\ln \gamma_i = \ln \gamma_i^\infty + \sum_{j=2}^{n} \varepsilon_{ij} x_j + \sum_{j=2}^{n} \rho_{ij} x_j^2 + \sum_{j=2}^{n-1} \sum_{k>j}^{n} \rho_{ijk}\, x_j x_k + \cdots \tag{11-92}$$

where the solvent component corresponds to $i = 1$ and γ_i^∞ refers to the condition where component i is in an infinitely dilute mixture of $m = n-1$ solute components [that is one that approaches pure solvent (1)]. Thus, for $i, j \geq 2$, we are dealing only with the solute components. If $i = j$ then we generate coefficients for binary ii (ε_{ii}, ρ_{ii}) and ternary $i\,i\,k$ (ρ_{iik}, $k \neq i$) interactions; for $i \neq j$ additional binary and ternary mixture interaction coefficients (ε_{ij}, ρ_{ij}, and ρ_{ijk}) are required. The ε_{ii} and ρ_{ii} coefficients in the series have been called first- and second-order self interaction coefficients following the earlier work by Wagner and Chipman. Geometrically, if one plots $\ln \gamma_i$ versus x_i, the slope of the curve is ε_{ii} and the curvature is ρ_{ii}. Basically, these self interaction coefficients determine how the concentration of solute i (x_i) affects its activity coefficient in a mixture consisting only of that solute and the solvent. In multi-component systems with more than one solute ($m \geq 2$), both self and mixture interaction coefficients are needed. The order of the polynomial retained in the expansion for practical purposes depends on the errors involved. The form of Eq. (11-92) can be generated from the general Wohl expansion given in Eq. (11-54) with $\Delta G^{EX(\infty)}$ referenced to an infinite-dilution state for solute ($x_2 \to 0$). Thus,

$$\Delta G^{EX(\infty)} = G_{mix} - G_1 \to 0 \text{ as } x_2 \to 0$$

Eq. (11-92) is also similar to the osmotic virial expansion of McMillan and Mayer [Eq. (11-48)] but the connection here is not based on rigorous theory.

One direct application of the Lupis-Elliott model is correlating and predicting the solubility of dissolved gases, such as nitrogen and oxygen, in liquid iron containing specified amounts of dissolved carbon and other alloying elements such as sulfur, chromium, nickel, etc. This example has obvious relevance to steelmaking processes. As shown in Figure 11.5, the presence of relatively small amount of alloying component j can significantly affect the amount of nitrogen that is soluble. In this instance, the activity coefficient of dissolved nitrogen is given by Eq. (11-92) in slightly modified form, with $i = 2 = N_2$:

$$\ln\left(\frac{\gamma_{N_2}}{\gamma_{N_2}^{\infty}}\right) = \ln\frac{\gamma_2}{\gamma_2^{\infty}} = \sum_{j=2}^{n} \varepsilon_{2j}\, x_j + \sum_{j=2}^{n} \rho_{2j}\, x_j^2 \tag{11-93}$$

where $j = 1 = \text{Fe}$ (liquid iron), $j = 2 = N_2$ (nitrogen), and j (or k) $= 3, 4, \ldots, n$ represent the alloying components (S, C, V, Cr, Ni, etc.). Note that the binary mixture interaction coefficients must be known for each alloying component as well as the self interaction coefficients for nitrogen in liquid iron in order to complete the prediction of $\ln \gamma_i$. Appendix 5 of Lupis (1983) provides an extensive listing of interaction coefficients for molten iron solutions.

11.9 Summary and Recommendations

This chapter illustrates several appoaches for modeling and correlating non-idealities in condensed phase mixtures of non-electrolytes. In a thermodynamically consistent fashion, these models typically provide an expression for the excess Gibbs free energy of a mixture as a function of composition, temperature, and possibly pressure. Solution models used in practice today range from those based on rigorous theory to those that are purely empirical. Although one can successfully rely on models that are only empirical, employing molecular-based theory improves our understanding of non-ideal systems and our ability to create better models for representing their behavior.

Most importantly, thinking on a molecular-level gives us insight into both the sources of potentially dominant effects as well as to the inherent limitations of the theory itself. For example, we were able to rigorously make the connections between lattice theory, the configurational entropy of mixing, and regular solution theory. In order to quantitatively make predictions using these models, we need experimental data to specify parameters. Even for the simplest lattice model at least one data point is required to specify the constant in the $\Delta G^{EX} = Ax_1x_2$ equation.

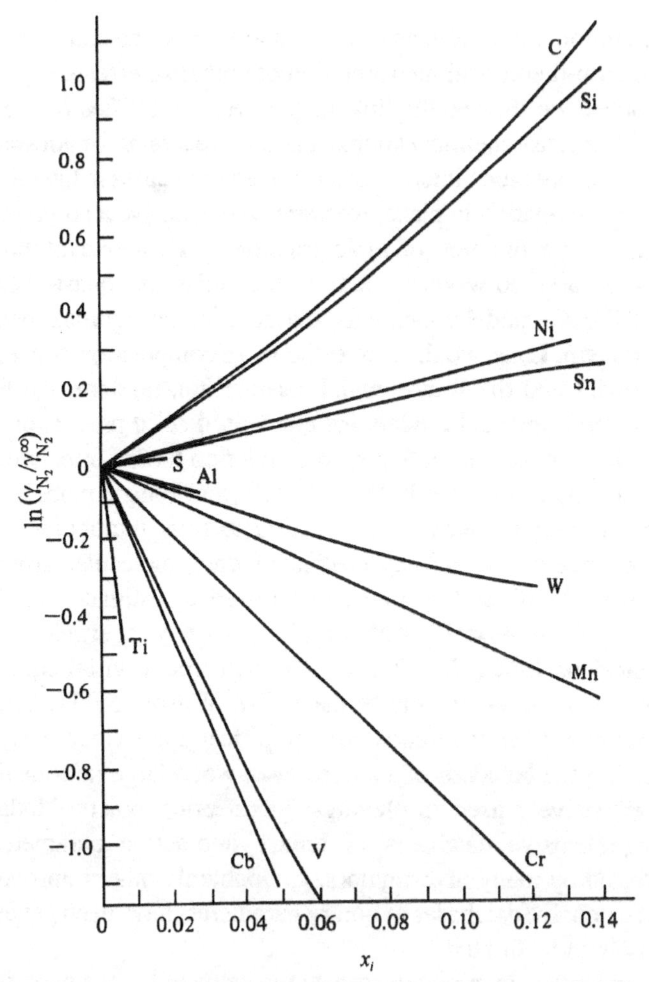

Figure 11.5 Effect of alloying components j on the activity coefficient of dissolved nitrogen (N_2) in liquid iron at 1600°C [adapted from Lupis (1983), with permission]. Unsymmetrical normalization for $\gamma^*_{N_2}$ (as $x_{N_2} \rightarrow 0$, $\gamma^*_{N_2} \rightarrow 1.0$). In this case, because of the very low solubility of N_2 in iron, we can neglect the self interaction coefficient terms, $\varepsilon_{ii}x_i$ and $\rho_{ii}x_i^2$, for nitrogen (i=2) and only include the mixture interaction coefficient terms (e.g., $\varepsilon_{23}x_3$ and $\rho_{23}x_3^2$ where 3 = S, C, V, Cr, or Ni, etc) for the alloying component.

We introduced McMillan-Mayer theory for dilute solutions to illustrate its relationship to the virial equation for gases in an attenuated state at low pressure and to show how simple models of intermolecular interactions and size effects could yield accurate predictions of osmotic pressure and activity coefficients for polymers, proteins and other macromolecules. The Flory-Huggins model for mixtures containing molecules of different size was also presented as the basis for models that capture important effects for polymer solutions. The Flory-Huggins model with its lattice framework correctly accounts for the combinatorial effects of molecular size on the configurational entropy and introduces a semi-theoretical interpretation of enthalpic effects.

In other examples, we showed the linkage between a modified lattice theory model using Guggenheim's quasi-chemical approximation to account for non-randomness due to non-uniformities in energetic interactions between molecules in the mixture. With the quasi-chemical approximation in place, we were able to derive a polynomial expansion for expressing $\ln \gamma_i$ as a function of mole fraction x_i. Further evolutionary work by Prausnitz, Abrams, and co-workers, led to the universal quasi-chemical activity coefficient (UNIQUAC) model which was in part based on rigorous theory and in part on straight empiricism. Later we discussed the local composition concept as the basis of the Wilson model and the Renon and Prausnitz non-random two-liquid (NRTL) model. Finally several practical models for condensed solid phases and liquid metals were discussed where we observed that regular solution theory often worked well.

A most useful attribute of the UNIQUAC, NRTL, and Wilson models is that they can be easily extended from a binary representation to treat multicomponent solutions. Further, in the absence of any activity coefficient data, molecular group contribution methods such as UNIFAC and ASOG can be used to estimate parameters for the UNIQUAC and Wilson models. Although the activity coefficient models using polynomial expansions (Margules, Redlich-Kister, osmotic virial equation, etc.) and other models (such as Flory-Huggins, Wilson, T-K Wilson, NRTL, and UNIQUAC) employ a number of *ad hoc* theoretical and empirical approximations, they are quite good at representing the behavior of systems even when large non-idealities exist. As such they are extensively used in chemical engineering practice today. With such popularity, comprehensive databases of binary interaction parameters have been generated and tested by many investigators. Undoubtedly other improved models will appear, but until extensive databases of binary parameters have been generated for them, their practical use will be limited.

In all of this, one must never underestimate the value of having experimental data on the system of interest. Measurements may be in the form of colligative properties (like osmotic pressure) or phase equilibria for the full set of composition, temperature, and pressure variations or for complex multicomponent mixtures, they may consist of a more limited set such as phase equilibrium data for binary pairs. Experimental data provide a starting point for developing and ultimately for validating any model.

References and Suggested Readings

Abrams, D.S. and J.M. Prausnitz (1975)," Statistical thermodynamics of liquid mixtures: a new expression for the excess Gibbs energy of partly or completely miscible systems," *AIChE J.*, **21**, p 116-128.

Anderson, T.F. and J.M. Prausnitz (1978)," Application of the UNIQUAC equation to calculation of multicomponent phase equilibria, 1: vapor-liquid equilibria; 2: liquid-liquid equilibria," *Ind. Eng. Chem. Process Design and Development*, **17**, p 552-567.

Bethe, H. (1935), *Proceedings of the Royal Society (London)* A, p 552.

Flory, P.C. (1941), *J. Chem. Phys.* **9**, p 660.

Guggenheim, E.G. (1952), *Mixtures -- The Theory of Equilibrium Properties of Some Classes of Mixtures , Solutions and Alloys,* Oxford, London, Chapter 3 and 4, p 23-46.

Haynes, C.B., K. Tamura, H.R. Korfer, H.W. Blanch, and J.M. Prausnitz (1992), "Thermodynamic properties of aqueous α-chymotrypsin solutions from membrane osmometry measurements," *J. Phys. Chem.* **96**, p 905-912.

Hiemenz, P.C. (1984), *Polymer Chemistry - The Basic Concepts,* Marcel Dekker, New York, p 513-517.

Hildebrand, J.H., J.M. Prausnitz, and R.L. Scott (1970), *Regular and Related Solutions,* Van Nostrand Reinhold, New York.

Hill, T.L. (1960), *An Introduction to Statistical Thermodynamics*, Addison-Wesley, Reading, MA, Chapter 19, p 340-370.

Huggins, M..L. (1941), *J. Phys. Chem.,***9**, p 440.

Lupis, C.H.P. (1983), *Chemical Thermodynamics of Materials,* Prentice Hall PTR, Englewood Cliffs, NJ, part VII, p 169-186 and part IX, p 235-256.

McMillan, W.G. and J.E. Mayer (1945), *J. Chem. Phys.*, **13**, p 276 ff.

Prausnitz, J.M., R.N. Lichtenthaler, E. Gomez de Azevedo (1986), *Molecular Thermodynamics of Fluid Phase Equilibria*, 2nd ed., Prentice-Hall, Englewood Cliffs, NJ: Sections 6.10-6.15 , p 226-268, and Chapter 7, p 274-370.

Reid, R.C., J.M. Prausnitz, and B.E. Poling (1987), *The Properties of Gases and Liquids,* 4th ed., McGraw-Hill, New York, Section 8.10, p 283-332.

Renon, H. and J.M. Prausnitz (1968), "Local compositions in thermodynamic excess functions for liquid mixtures," *AIChE J.,* **14**, p 135-144.

Tsuboka, T. and T. Katayama (1975), "Correlations based on local mole fraction model between new excess Gibbs energy equations," *J. Chem. Eng. Japan*, **12**, p 404-406.

Vilker, V.L., C.K. Colton, and K.A. Smith (1981), "The osmotic pressure of concentrated protein solutions: effect of concentration and pH in saline solutions of bovine serum albumin," *J. of Colloid and Interface Science,* **79**, p 548-566.

Walas, S.N. (1985), *Phase Equilibria in Chemical Engineering,* Butterworth, New York, Chapter 4, p 165-238.

Wilson, G.M. (1964), "Vapor-liquid equilibrium. XI: a new expression for the excess free energy of mixing," *J. Amer. Chem. Soc.*, **86**, p 127-130.

Problems

11.1. The polyethylene oxide (PEO)-water system can be characterized with the Flory-Huggins model (see Eq. (11-66)).

Describe how you would use the osmotic pressure data shown at 295 K to estimate the average molecular weight $<m_2>$ of the polymer in solution and the Flory parameter χ. Are there any conditions where you would expect ideal solution behavior for the PEO-water binary system? You can assume that $rN_2 << N_1$ such that

$$\Phi_2^* = N_2 \overline{V}_2 / N_1 \overline{V}_1 << 1.0$$

and that $\overline{V}_1 = 18$ cm^3/mole and $\overline{V}_2 \approx r\overline{V}_1$

where:

N_1 = moles of solvent (water)

N_2 = moles of PEO polymer

r = degree of polymerization

Π = osmotic pressure, atm

C_2 = grams of PEO per cm^3 of solution

Hint: develop an expression for $\Pi\overline{V}_1/(RTC_2)$ as a function of C_2 where

$$C_2 = N_2 <m_2>/(N_1\overline{V}_1 + N_2\overline{V}_2)$$

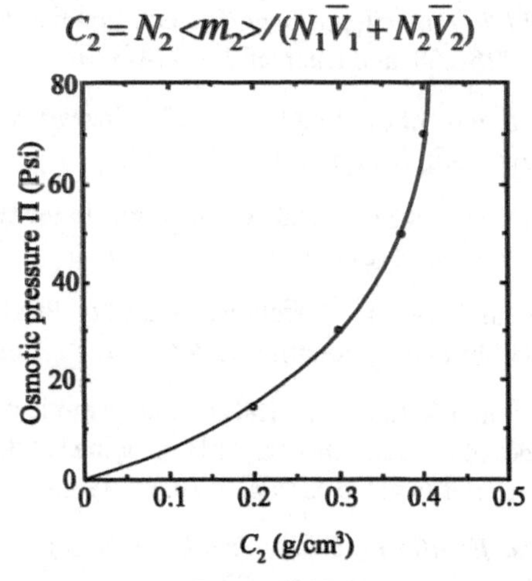

Figure P11.1

11.2. Japas and Franck [(1985), *Ber. Bunsenges Phys. Chem.* **89**, p. 1268-1275] present excess Gibbs energy as a function of water mole fraction of a supercritical water-oxygen mixture at 673 K for different pressures as shown in the graph below.

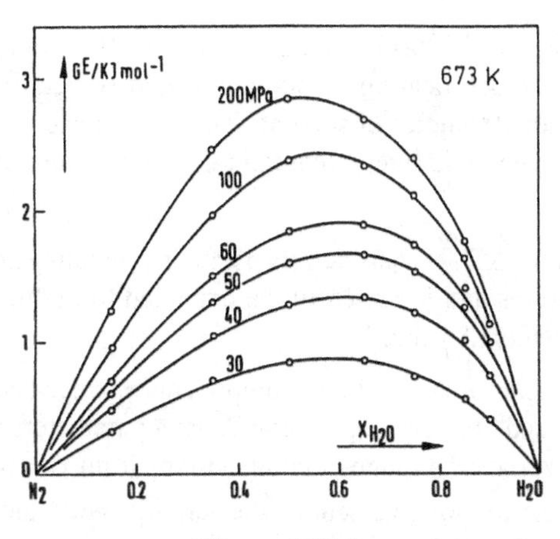

Figure P11.2

Describe how you would use these data to estimate the activity coefficients of H_2O and O_2 at 30 MPa, 673 K, and at 100 MPa, 673 K. Give all relevant equations and describe all mathematical procedures you would follow to get a numerical result.

11.3. An important step in the manufacture of many polymers is the batch mixing of monomer and solvent prior to initiation of the polymerization reaction. In some cases, very non-ideal solutions result and enthalpies of mixing can be large. In order to avoid premature initiation of the polymerization reaction, it is necessary to maintain isothermal conditions during the mixing process when monomer is added to an insulated batch kettle reactor containing the solvent. Cooling and heating coils are placed inside the reactor for this purpose. Monomer would normally be added at a constant rate using a positive displacement transfer pump.

Not much is known about the thermodynamic properties of the monomer-solvent system, except that the non-idealities are adequately accounted for by the Wilson equation (see Table 11.1):

$$\Delta G^{EX} = RT\,[-x_1 \ln (x_1 + \Lambda_{12}x_2) - x_2 \ln (x_2 + \Lambda_{21}x_1)]$$

where the subscript 1 refers to monomer and 2 refers to solvent and Λ_{12} and Λ_{21} are functions of temperature only with

$$\Lambda_{12} = \exp\,[-\lambda_{12}/RT] \text{ and } \Lambda_{21} = \exp\,[-\lambda_{21}/RT]$$

where λ_{12} and λ_{21} are constants.

(a) Develop a relationship for the heat of mixing as a function of composition in terms of λ_{12}, Λ_{12}, λ_{21}, and Λ_{21}.

(b) Assuming your result in part (a) is correct, derive an expression for determining the appropriate heating or cooling rate to maintain isothermality for the addition of N_1 moles of monomer at a feed rate of $\dot{N}_f$ (mole/sec) to a batch of N_2 moles of solvent. Both monomer and solvent are initially at the same temperature.

(c) Given that $\lambda_{12} = -12000$ J/mol and $\lambda_{21} = -90000$ J/mol for a specific system of interest, how much heating or cooling would be required to mix 100 moles of monomer with 400 moles of solvent? The solvent and monomer are both at 300 K initially and we would like to maintain the temperature of the mixture at 300 K as well.

11.4. The Wilson model for ΔG^{EX} is particularly useful in predicting non-idealities in both polar and nonpolar mixtures and it can easily be extended to multicomponent systems using only binary parameters (Λ_{ij} or λ_{ij}).

(1) For an equimolar mixture of two components $\gamma_1 = 1.2$ and $\gamma_2 = 1.5$ at a particular temperature and pressure. Determine Wilson parameters (Λ_{12} and Λ_{21}) and plot $\ln \gamma_1$ and $\ln \gamma_2$ over the range of compositions from $x_1 = 0$ to 1.0.

(2) Develop an expression to calculate the activity coefficients at infinite dilution (γ_i^∞) and evaluate γ_1^∞ and γ_2^∞.

(3) Determine values of Wilson parameters that give extrema in variations of $\ln \gamma_1$ as a function of x_1 at constant T and P.

11.5. The NRTL model for the excess Gibbs energy of mixing in a binary solution [H. Renon and J.M. Prausnitz (1968), *AIChE J.*, **14**, p 135] is

$$\frac{\Delta G^{EX}}{RT} = x_1 x_2 \left(\frac{\tau_{21}\, G_{21}}{x_1 + x_2\, G_{21}} + \frac{\tau_{12}\, G_{12}}{x_2 + x_1\, G_{12}} \right)$$

where x_1 and x_2 are the liquid mole fractions of 1 and 2, τ_{12} and τ_{21} are binary parameters, and $G_{12} = \exp(-\alpha\tau_{12})$, $G_{21} = \exp(-\alpha\tau_{21})$. α is a third parameter for the mixture. Let $x_1 = x_2$ and choose a symmetric mixture where $\tau_{21} = \tau_{12}$. What is the largest value of α that can be used for this correlation if it is to show phase splitting? What happens if the mixture is not equimolar?

11.6. G.M. Wilson (1964), *J. Am. Chem. Soc.*, **86**, 127, 133 developed an excess Gibbs energy relation for nonideal liquid mixtures that has been shown to be widely applicable. For a binary system of components B and C it is written as

$$\frac{\Delta G^{EX}}{RT} = -x_B \ln(x_B + \Lambda_{BC} x_C) - x_C \ln(x_C + \Lambda_{CB} x_B)$$

Show that irrespective of the parameters Λ_{BC} and Λ_{CB}, this relation cannot be used for systems that can phase split into two liquid fractions.

11.7. Manufacturers of plastic parts often employ solvents to aid in the processing of the neat polymer into a final product. However, the commercial polymers commonly in such use are actually mixtures of chemically similar molecules of many different molecular weights, or degrees of polymerization or chain lengths r_i (where $r_i = m_i/m_{ru}$, the ratio of the molecular weight of species i over the molecular weight of the constituent repeat unit of the polymer). The actual distribution of species of different molecular weight in any one sample of polymer may span several orders of magnitude as shown in Figure P11.7 below.

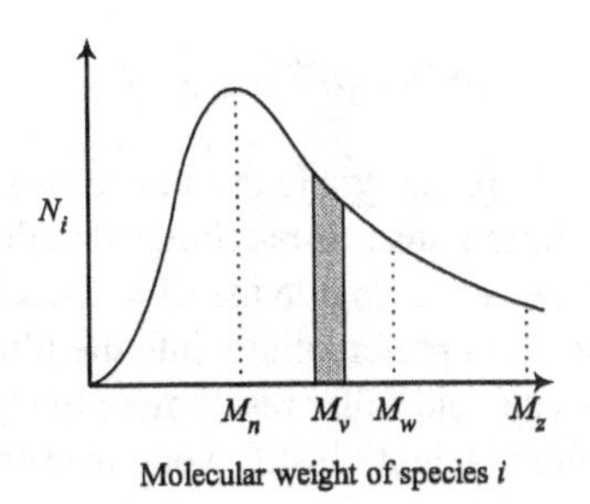

Figure P11.7 Distribution of molecular weight in a polymer mixture.

One film manufacturer proposes to use a solvent-casting technique to make large area plastic films in a continuous process—his method is to dissolve the as-supplied polymer into a large quantity of solvent and then feed the solution over a large cold roller on which a liquid film forms. The supernatent is recycled and the remaining solvent remaining in the film is evaporated on a second, heated roller. Using the Flory-Huggins model, the heat of mixing is:

$$\Delta \underline{H}_{mix} = RT\chi\, n_o\, \Phi_p^*$$

and the entropy of mixing is:

$$\Delta \underline{S}_{mix} = -R\left[n_o \ln(\Phi_o^*) + \sum_i n_i \ln(\Phi_i^*)\right]$$

where

n_o = number of moles of solvent

n_i = number of moles of polymer species i

Φ_o^* = volume fraction of solvent

Φ_i^* = volume fraction of polymer species i

Φ_p^* = the total volume fraction of polymer

χ = Flory parameter, which is independent of molecular weight and inversely proportional to temperature.

(a) What is the equation for the free energy of mixing of polymer and solvent upstream of the cold roller?

(b) Write a set of equations which may be used to determine the partitioning (i.e., volume fractions) of each species between the film (f) and the solution (s) at the cold roller. Clearly state your assumptions. How many equations are required? How would these be determined from the information supplied above? Justify your answers.

(c) The ratio of the volume fractions of polymer species i in each phase at the cold roller can be expressed simply as:

$$\ln\left(\Phi_i^{*(f)}/\Phi_i^{*(s)}\right) = \sigma\, r_i$$

where σ is a function of Φ_p^* and χ and $\sigma > 0$. If σ is of the order 10^{-3}, what fraction (by volume) of the solution must be precipitated at the cold roller in order to get half of the $r_i = 5000$ species to go into the film. Under these conditions, will the $r_i = 10000$ species partition preferentially into the film or the supernatent? Will increasing the area of the cold roller result more likely in a film of higher or lower average molecular weight? Justify your answers.

(d) At the hot roller, fractionation is not an issue. In this case, it is suitable to treat the polymer component as a single species. The activity of the solvent then may be approximated using Eq. (11-66) as:

$$\ln(a_o) = \left[\ln(1-\Phi_p^*) + \left(1 - \frac{1}{\langle r_p \rangle}\right)\Phi_p^* + \chi\,(\Phi_p^*)^2\right]$$

where $\langle r_p \rangle$ is a representative aggregate degree of polymerization. Find an expression for the boiling point elevation of the solvent at the hot roller in terms of the polymer volume fraction in the film. Again, state any assumptions you make.

11.8. Compare the two-suffix Margules expression for ΔG^{EX} [see Table 11.1 and Eq. (11-35)] with the quasi-chemical approximation result [Eq. (11-41)] in terms of their predictions of the critical temperature for phase stability of a binary mixture.

11.9. Infinite dilution activity coefficients are known for a binary mixture of acetonitrile (C_2H_3N) (1) and water (H_2O) (2) at 25°C:

$$\gamma_1^\infty = 8.22 \quad \gamma_2^\infty = 14.39$$

In addition, the ratio of molar liquid volumes, $V_1/V_2 = 2.89$. Using the two-suffix Margules, van Laar, Wilson, and NRTL (with $\alpha_{12} = 0.535$) models, tabulate and plot $\ln\gamma_i$ $(i = 1, 2)$ from $x_1 = 0$ to 1.0.

11.10. Given the data of Problem 11.9, determine UNIQUAC binary interaction parameters τ_{12} and τ_{21} with $z = 10$ if values for r_i, q_i, and q_i' are given below. Tabulate and plot predictions of $\ln\gamma_i$ $(i = 1, 2)$ from $x_1 = 0$ to 1.0 using the UNIQUAC model

	acetonitrile (1)	**water (2)**
r_i	1.87	0.92
q_i	1.72	1.40
q_i'	1.72	1.00

Models for Electrolyte Solutions 12

This chapter is devoted to introducing the models used for liquid solutions containing mixtures of ionic species that dissociate when dissolved. Most commonly these are aqueous solutions of salts, acids or bases. The nomenclature that deals with electrolytes is somewhat specialized, as it must account for the fact that positive and negative ions are produced and that electrical neutrality constrains the system. In addition, different conventions are needed to specify standard states and limiting behavior at infinite dilution or in ideal solutions requires special consideration. Following the discussion of standard states and notation, we introduce the fundamental theoretical approach due to Debye and Hückel for strong electrolytes that are completely dissociated. The Debye-Hückel model permits us to analytically describe the limiting behavior of the activity coefficient in dilute solutions.

After this, we introduce several representative models used to represent non-ideal effects in electrolytes at higher concentrations. First is Pitzer's extension of the Debye-Hückel model which incorporates McMillan-Mayer concepts to capture the effects of non-ionic intermolecular interactions. Second, we discuss the approach developed by Meissner based on an empirical extension of corresponding states theory. Then, the Chen model is treated where Pitzer's extended Debye-Hückel model for long-range interactions has been superimposed onto a local composition model based on the NRTL equation to account for short-range inter- molecular interactions. Models based on the mean spherical approximation in integral equation theory and other advanced statistical mechanical approaches are covered briefly to complete the discussion of electrolyte models. Although our treatment of electrolyte models is not intended to be comprehensive, you will develop an appreciation for the nature of the assumptions used and when empirical fitting of parameters is needed to provide an operational equation for either ΔG^{EX} or $\gamma_{\pm}$.

The final sections of the chapter compare model predictions and cover several applications of electrolyte solution thermodynamics.

12.1 Conventions and Standard States

Electrolytes represent a special class of non-ideal mixtures that are different both at the macroscopic level of classical thermodynamics and at the molecular level of statistical thermodynamics. The fact that electrolytes dissociate in solution--strong electrolytes completely and weak electrolytes to some degree--introduces changes in observable behavior that has its roots in the long-range coulombic forces between ions. These coulombic forces cause the intermolecular potential contribution to scale with $1/r$ where r is the separation distance between ions as opposed to the relatively shorter range, induced dipole-induced dipole or similar forces, which lead to an attractive Boltzmann-averaged potential of attraction which scales with $1/r^6$ (see Sections 10.2 and 10.3).

Earlier in Section 9.8, various standard state conventions were introduced. At that time we mentioned that a unit molality reference condition is commonly used where the standard state is in a hypothetical state representing ideal, Henry's law type behavior extrapolated to unit molality ($\boldsymbol{m}$ = 1 mole of solute per kg of solvent)[1]. At first glance it would seem quite straightforward to specify this definition for both electrolytes and non-electrolytes without ambiguity by simply determining the limiting slope of $\hat{f}_{ij}$ as a function of $\boldsymbol{m}$ and as $\boldsymbol{m} \to 0$ extrapolating a straight line tangent to the fugacity versus molality curve out to $\boldsymbol{m}$ = 1.0. While this works quite well for non-dissociated, non-electrolytes (for example see Figure 9.7), it does not work for electrolytes. Initially let's confine our attention to strong single (ij) electrolytes such as NaCl, HCl, or Na_2SO_4. Lewis and Randall (1923) plotted the fugacity of HCl as a function of molality as illustrated in Figure 12.1a. Here the limiting slope as $\boldsymbol{m} \to 0$ is zero so one cannot easily define the reference state at $\boldsymbol{m}$ = 1 other than to say that $\hat{f}_{ij}^{+} = 0$. Whereas if one plots the same data as $\hat{f}_{ij}$ versus $\boldsymbol{m}^2$ a straight line results that is described by a simple parabolic equation:

$$\lim_{\boldsymbol{m}\to 0} \hat{f}_{ij} = k\boldsymbol{m}^2 \tag{12-1}$$

The exponent of two in Eq. (12-1) is not a fortuitous accident. It explicitly depends on the fact that 2 ions, H^+ and Cl^-, are released for every molecule of HCl that dissolves. We can generalize the result of Eq. (12-1) to any strong electrolyte that is completely dissociated but first we need to establish a notation to define the dissociation and other conditions in a generally consistent manner.

For a fully dissociated single electrolyte (ij) consisting of positive cations (C) and negative anions (A), we have

$$C_{\nu_i} A_{\nu_j} = \nu_i C^{z_i} + \nu_j A^{z_j} = \nu_+ C^{z_+} + \nu_- A^{z_-} \tag{12-2}$$

In general, to establish a convention for mixtures of strong electrolytes we set $i = 1, 3, 5, \ldots$ *odd* to designate specific cations, e.g., H^+, K^+, Na^+, Al^{+3}, Fe^{+2}, . . . etc. and $j = 2, 4, 6, \ldots$ *even* for specific anions, e.g., Cl^-, F^-, NO_3^-, SO_4^{2-}, . . . etc. The total ionic strength of the solution is defined by

---

1 Readers should note that a lower case, bolded, and italicized $\boldsymbol{m}$ is being used as the symbol for molality not mass which is given by m.

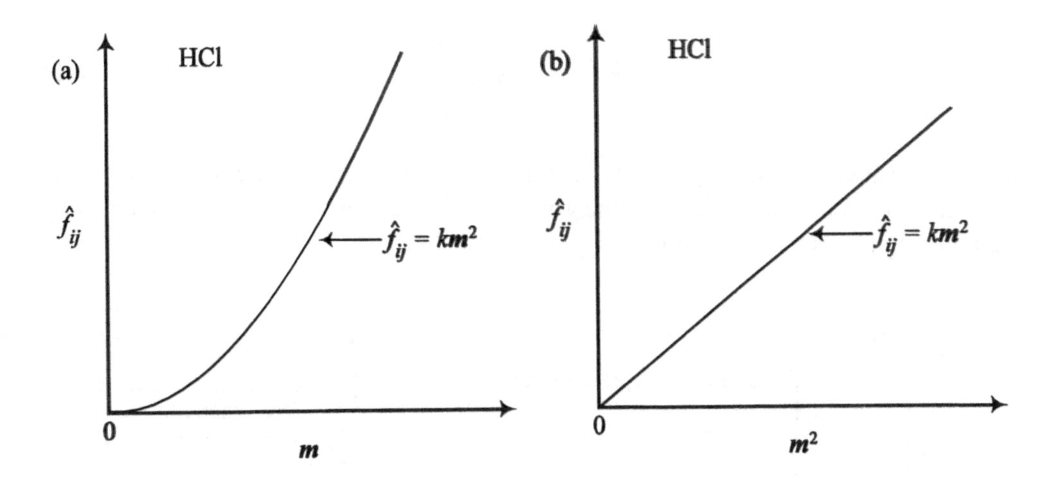

Figure 12.1 Experimentally determined fugacity of HCl as a function of molality m in water in the dilute region.

$$I \equiv \tfrac{1}{2} \sum_{k=1}^{\text{all ions}} m_k z_k^2 = \tfrac{1}{2} \sum_{i=odd}^{\text{cations}} m_i z_i^2 + \tfrac{1}{2} \sum_{j=even}^{\text{anions}} m_j z_j^2 = I_c + I_a \tag{12-3}$$

where m = molality in moles of solute/kg of solvent. For a single ij electrolyte of molality m, Eq. (12-3) simplifies to:

$$I = \tfrac{1}{2} |z_+ z_-| (\nu_+ + \nu_-) m \tag{12-4}$$

using the electroneutrality condition that $\nu_i z_i + \nu_j z_j = 0$ or $\nu_+ z_+ + \nu_- z_- = 0$.[2]

To return to our discussion of standard states and fugacity, Eq. (12-1) can be generalized for dilute solutions as $m \to 0$ to:

$$\lim_{m \to 0} \hat{f}_{ij} = \hat{f}_{ij}^{\infty} = km^{\nu} \tag{12-5}$$

where $\nu \equiv \nu_+ + \nu_- = \nu_i + \nu_j$ represents the total number of moles of ions produced when 1 mole of electrolyte ij fully dissociates.

Figure 12.2 illustrates the behavior of $\hat{f}_{ij}$ as a function of m at fixed temperature and pressure. Note that the reference line corresponds to Eq. (12-5) and would be identical to the experimental lines shown in Figure 12.1 for HCl with $\nu = 2$. The activity of the electrolyte in a real solution is defined as,

---

2 Note that we adopted the specific convention that the charges z_i or z_+ for cations are positive and z_j or z_- for anions are negative.

$$a_{ij}^{(solution)} = a_{ij} \equiv \frac{\hat{f}_{ij}}{\hat{f}_{ij}^{\square}} \tag{12-6}$$

where the standard state fugacity $\hat{f}_{ij}^{\square}$ in this case is at unit molality and follows the behavior given along the reference line given by Eq. (12-5) and which is approached by the real solution in the dilute limit:

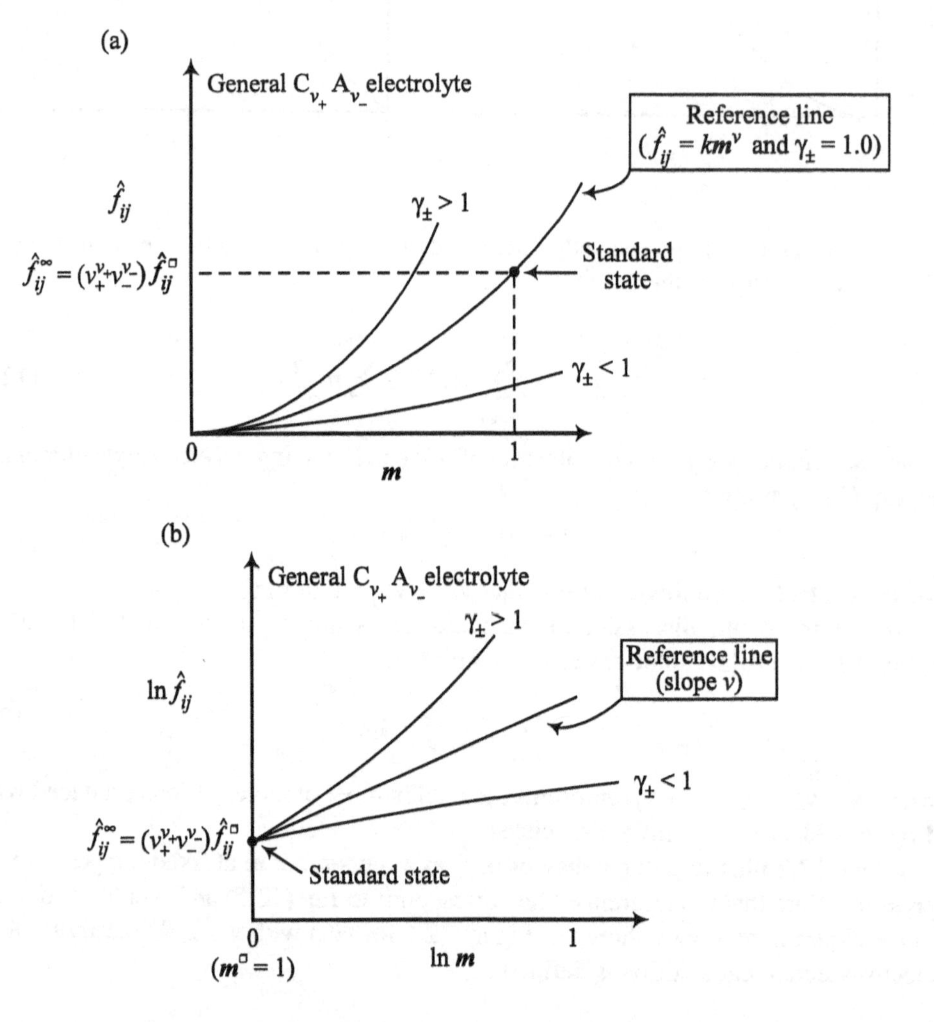

Figure 12.2 Practical (molal-based) scale standard state convention for a single strong (ij) electrolyte [i=cation (C), j=anion (A)].

$$\hat{f}_{ij}^{\square} \equiv \frac{\hat{f}_{ij}^{\infty}(T, P, m^{\square} = 1.0)}{(\nu_+^{\nu_+} \nu_-^{\nu_-})} = \frac{k m^{\nu}}{(\nu_+^{\nu_+} \nu_-^{\nu_-})} = \frac{k}{(\nu_+^{\nu_+} \nu_-^{\nu_-})} \tag{12-7}$$

where the factor $(\nu_+^{\nu_+} \nu_-^{\nu_-})$ has been introduced for convenience to ensure that the activity coefficient (γ_+) goes to unity as $m \to 0$. Thus, the practical (molal scale) based standard reference state $(\square)$ corresponds to a *hypothetical* state at unit molality and T and P of the mixture whose behavior exactly corresponds to that of an infinitely dilute solution. While these "ideal" electrolyte conditions are analogous to the Henry's law limit used to establish the standard state for unsymmetrically normalized, non-electrolytes, as discussed in Section 9.8, obvious differences exist largely as a result of the ionic dissociation process.

The fugacity of the electrolyte ij (the solute) in the solution is defined using the mean ionic activity coefficient to show a deviation from "ideal", infinite-dilution behavior along the reference line of Figure 12.2. Mathematically this is given by,

$$\hat{f}_{ij} \equiv k(\gamma_{\pm} m)^{\nu} \tag{12-8}$$

Now we return to our definition of activity in Eq. (12-6) and use Eqs. (12-7) and (12-8):

$$a_{ij} = \frac{\hat{f}_{ij}(T, P, m)}{\hat{f}_{ij}^{\square}(T, P, m^{\square} = 1)} = \frac{k \gamma_{\pm}^{\nu} m^{\nu}}{(k/(\nu_+^{\nu_+} \nu_-^{\nu_-}))} \tag{12-9a}$$

This brings us to the point where the formal definitions of mean ionic activity $(a_{\pm})$, mean activity coefficient $(\gamma_{\pm})$, and mean molality $(m_{\pm})$ can be introduced:

$$a_{\pm} \equiv (a_{ij})^{1/\nu} \qquad \nu \equiv \nu_+ + \nu_- = \nu_i + \nu_j \tag{12-10}$$

$$\gamma_{\pm} \equiv \frac{a_{\pm}}{m_{\pm}} = (\gamma_+^{\nu_+} \gamma_-^{\nu_-})^{1/\nu} \quad \text{or} \quad (\gamma_i^{\nu_i} \gamma_j^{\nu_j})^{1/\nu} \tag{12-11}$$

$$m_{\pm} \equiv m\left[(\nu_+)^{\nu_+} (\nu_-)^{\nu_-}\right]^{1/\nu} \quad \text{or} \quad m\left[(\nu_i)^{\nu_i} (\nu_j)^{\nu_j}\right]^{1/\nu} \tag{12-12a}$$

With Eq. (12-12a), a_{ij} in Eq. (12-9a) becomes

$$a_{ij} = \gamma_{\pm}^{\nu} m_{\pm}^{\nu} = a_i^{\nu_i} a_j^{\nu_j} \tag{12-9b}$$

By comparing Eqs. (12-5) and (12-8) we see that $\gamma_{\pm} \to 1.0$ as $m \to 0$. This is equivalent to saying that $a_{ij} \to m_{\pm}^{\nu}$ or $a_{\pm} \to m_{\pm}$ as $m \to 0$. For mixtures of electrolytes, complications arise because of the presence of common ions. For example, a mixture of 1 mole of NaCl and 2 moles of KCl in 1000 g of H_2O would yield 3 moles of Cl^- ions for every 1 mole of Na^+ ions. In this situation and for mixtures, in general, Eq. (12-12a) cannot be used and $\boldsymbol{m}_\pm$ is rewritten as:

$$\boldsymbol{m}_\pm = \left[(\boldsymbol{m}_i)^{\nu_i} (\boldsymbol{m}_j)^{\nu_j} \right]^{1/\nu} \tag{12-12b}$$

where $\boldsymbol{m}_i$ and $\boldsymbol{m}_j$ refer to the total molality of type *i* positive ions and type *j* negative ions, respectively.

Table 12.1 provides a summary of the conventions and formulae for various electrolyte types using the practical (molal) scale. The type of electrolyte, 1:1, 1:2, 2:1, 2:2, etc. is defined in terms of the $z_+ = z_i$ and $z_- = z_j$ charges on its dissociated ions.

Table 12.1 Practical (Molal) Based Conventions for Single Strong Electrolytes

$$C_{\nu_+}A_{\nu_-} = \nu_+ C^{z_+} + \nu_- A^{z_-}$$

Property	Electrolyte Type general $z_+:z_-$	1:1	1:2	2:1	2:2	3:1	Non-electrolyte 0:0
Representative salt	$C_{\nu_+}A_{\nu_-}$	NaCl	Na_2SO_4	$CaCl_2$	$CaSO_4$	$AlCl_3$	glucose
Dilute behavior for $\hat{f}_{ij}$	km^{ν}	km^2	km^3	km^3	km^2	km^4	km
Standard state fugacity at unit molality $\hat{f}_{ij}^{\square}$	$k/(\nu_+^{\nu_+}\nu_-^{\nu_-})$	k	$k/4$	$k/4$	k	$k/27$	k
Electrolyte activity a_{ij}	$a_+^{\nu_+}a_-^{\nu_-}$	a_+a_-	$a_+^2a_-$	$a_+a_-^2$	a_+a_-	$a_+a_-^3$	a
Mean ionic activity $a_\pm$	$(a_{ij})^{1/\nu}$	$(a_+a_-)^{1/2}$	$(a_+^2a_-)^{1/3}$	$(a_+a_-^2)^{1/3}$	$(a_+a_-)^{1/2}$	$(a_+a_-^3)^{1/4}$	a
Mean molality $\boldsymbol{m}_\pm$	$(\nu_+^{\nu_+}\nu_-^{\nu_-})^{1/\nu}\boldsymbol{m}$	$\boldsymbol{m}$	$4^{1/3}\boldsymbol{m}$	$4^{1/3}\boldsymbol{m}$	$\boldsymbol{m}$	$3^{3/4}\boldsymbol{m}$	$\boldsymbol{m}$
Ionic strength I	$\frac{1}{2}\sum_{k=1}^{2} z_k^2 \boldsymbol{m}_k$	$\boldsymbol{m}$	$3\boldsymbol{m}$	$3\boldsymbol{m}$	$4\boldsymbol{m}$	$6\boldsymbol{m}$	$\boldsymbol{m}$
Mean ionic activity coefficient $\gamma_\pm$	$(\gamma_+^{\nu_+}\gamma_-^{\nu_-})^{1/\nu}$	$(\gamma_+\gamma_-)^{1/2}$	$(\gamma_+^2\gamma_-)^{1/3}$	$(\gamma_+\gamma_-^2)^{1/3}$	$(\gamma_+\gamma_-)^{1/2}$	$(\gamma_+\gamma_-^3)^{1/4}$	γ

Notes $\nu = \nu_+ + \nu_-$; $a_{ij} = (a_\pm)^{\nu} = a_+^{\nu_+}a_-^{\nu_-}$; $a_+ = \gamma_+ \boldsymbol{m}_+$ and $a_- = \gamma_- \boldsymbol{m}_-$; $\gamma_\pm = a_\pm / \boldsymbol{m}_\pm$

Occasionally for electrolyte solutions, we will need to return to a rational or mole fraction basis for activity coefficients. Under these conditions, we still use "ideal" infinite-dilution behavior to define the standard state. But now, the standard state is a hypothetical state at unit mole fraction ($x_{ij} = 1$) rather than unit molality. The fugacity of the electrolyte at the same $T, P, x_{ij} = f(\boldsymbol{m})$ is the same as it was before, but the reference standard state has shifted, so the activity a_{ij}^{**} defined on this new basis is different than a_{ij}:

$$\hat{f}_{ij} = \gamma_{ij}^{\nu} x_{ij}^{\nu} \hat{f}_{ij}^{**} \quad \text{or} \quad a_{ij}^{**} = (\gamma_{ij} x_{ij})^{\nu}$$

where γ_{ij} and x_{ij} are defined using equations similar to Eqs. (12-11) and (12-12a).

A straightforward transformation exists between $\gamma_{\pm}$ based on molality $\boldsymbol{m}_{ij} = \boldsymbol{m}$ and γ_{ij} based on mole fraction x_{ij}:

$$\gamma_{ij} = \gamma_{\pm} (1 + 0.001 m_s \nu \boldsymbol{m}) \tag{12-13}$$

where m_s is the molecular weight of the solvent in g/mol. Later in this chapter, we will find it convenient to define a reduced form of $\gamma_{\pm}$ that is scaled to the magnitude of the charge product as follows.

$$\Gamma_{\pm} \equiv (\gamma_{\pm})^{1/|z_+ z_-|} \tag{12-14}$$

For a 1:1 electrolyte like NaCl, $\Gamma_{\pm} = \gamma_{\pm}$ while for a 2:1 electrolyte like $CaCl_2$, $\Gamma_{\pm} = \gamma_{\pm}^{1/2}$. Because of our definitions of $a_{\pm}$ and $\gamma_{\pm}$, the values of $\gamma_{\pm}$ are very close to 1.0 in dilute solutions regardless of the magnitude of non-ideal effects. [This behavior contrasts sharply with symmetrically normalized systems where γ_i (∞-dilution) is normally quite different than 1.0]. Thus, for experimental determination of $\gamma_{\pm}$ it is frequently convenient to employ the *practical osmotic coefficient* ϕ_s^* which is defined in terms of the solvent activity a_s as:

$$\boxed{\phi_s^* \equiv \frac{-1000 \ln a_s}{m_s \sum_{k=1} \nu_k \boldsymbol{m}_k}} \tag{12-15}$$

$\sum \nu_k \boldsymbol{m}_k$ is the total concentration of dissociated ionic species in solution. It turns out that ϕ_s^* is a sensitive function of molality even in dilute solution. Using a modified form of the Gibbs-Duhem equation, we obtain an expression for $\ln \gamma_{\pm}$:

$$\boxed{\ln \gamma_{\pm} = \phi_s^* - 1 + 2 \int_0^{\boldsymbol{m}} (\phi_s^* - 1) \frac{d\boldsymbol{m}^{1/2}}{\boldsymbol{m}^{1/2}}} \tag{12-16}$$

where $\phi_s^* \to 1.0$ as $\boldsymbol{m} \to 0$. The 1/2 exponent was introduced in Eq. (12-16) to avoid the singularity of $(\phi_s^* - 1)/\boldsymbol{m}$ at $\boldsymbol{m} = 0$ [see Pitzer (1995, p 264-266)]. The experimental osmotic pressure Π of the solution is related directly to ϕ_s^*. For a single ij electrolyte of molality $\boldsymbol{m}$:

$$\Pi \equiv P - P^o = \frac{RT}{\overline{V}_s} \frac{\nu \boldsymbol{m} m_s}{1000} \phi_s^* \tag{12-17}$$

where $\overline{V}_s$ = partial molar volume of solvent which is assumed constant and independent of P and x_i. Thus $\overline{V}_s \approx V_s^o$, the molar volume of pure solvent.

For completeness, we should mention that other forms of the osmotic coefficient are frequently used. One such, rational or mole-fraction-based osmotic coefficient is $\phi_{s,x}^*$ defined by,

$$\phi_{s,x}^* \equiv \frac{\ln a_s}{\ln x_s} \tag{12-18}$$

For an ideal solution, $a_s = x_s$ and $\phi_{s,x}^* \to 1.0$. $\phi_{s,x}^*$ and its molality based practical counterpart ϕ_s^* are related by,

$$\phi_s^* = \phi_{s,x}^* \left(1 - \frac{m\,m_s}{2(1000)} + \frac{m^2 m_s^2}{3(1000)^2} - \ldots\right) \tag{12-19}$$

Another rational osmotic coefficient comes directly from McMillan-Mayer theory (see Section 11.6) where ϕ_{MM}^* is defined as:

$$\phi_{MM}^* \equiv \frac{\Pi}{ckT} \tag{12-20}$$

where c is the total dissociated solute concentration in units of ions per unit volume. The practical coefficient ϕ_s^* actually is based in a Lewis and Randall framework given its definition involving $\ln a_s$ in Eq. (12-15). The rational coefficient ϕ_{MM}^* is fundamentally different as it relates specifically to the osmotic pressure Π in the McMillan-Mayer framework. In order to carry out a rigorous development of electrolyte properties, we need to be able to express the difference between ϕ_s^* and ϕ_{MM}^*. This becomes particularly important when mixed solvents are encountered with electrolytes present (see Section 12.9). The difference between them is best illustrated by describing the experiment used to measure Π. A semi-permeable membrane shown below divides the osmotic cell into two chambers:

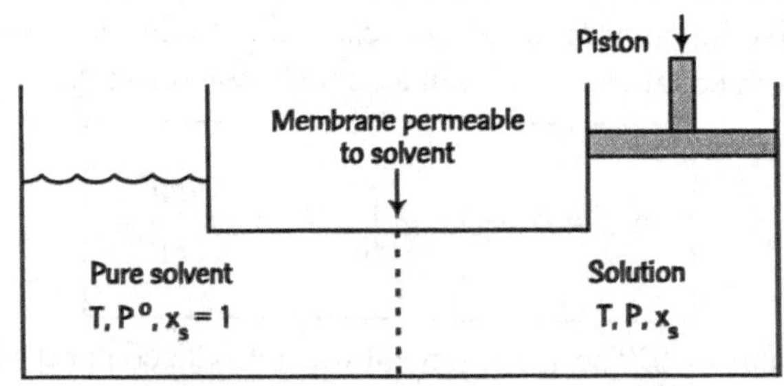

The pressure on the solution side is increased using the piston until the osmotic equilibrium condition is reached where the chemical potential on either side of the membrane are in balance

$$\mu_s^{pure}(T, P^o, x_s = 1) = \mu_s(T, P, x_s)$$

at this condition $\Pi \equiv P - P^o$. Thus in the McMillan-Mayer framework, the canonical variables are T, V, μ_s for the solvent, and N_i (or x_i) for the solute only. If we assume that the liquid solution is incompressible, then we can calculate the change in μ_s from P^o to P at fixed T and x_s:

$$\mu_s(T, P, x_s) - \mu_s(T, P^o, x_s) = \int_{P^o}^{P} \overline{V}_s \, dP \approx \overline{V}_s \Pi \qquad (12\text{-}21)$$

where P^o in this case is taken as the ambient pressure.

In the Lewis and Randall framework, the canonical variables are T, P, and N_i (or x_i) for solvent and solute. μ_s is defined at a specific T, P^o, and solvent composition x_s. If we select pure solvent at T and P^o as the standard state (+), then $\mu_s^+ = \mu_s^{pure}(T, P^o, x_s = 1)$ and the defining equation for $a_s = \hat{f}_s/\hat{f}_s^+ = \hat{f}_s/f_s^{(pure)}$ becomes

$$RT \ln a_s = \mu_s(T, P^o, x_s) - \mu_s^{(pure)}(T, P^o, x_s = 1)$$

Thus, by combining Eqs. (12-20) and (12-21) with this result, we get:

$$\ln a_s = \frac{-\overline{V}_s \Pi}{RT} = -\overline{V}_s \phi_{MM}^* C$$

where C = molar concentration of solute ions (mol/L) = c/N_A and N_A = Avogadro's number. Now we can relate ϕ_{MM}^* to ϕ_s^* since we have an expression for $\ln a_s$ which is the basis of the practical (Lewis and Randall) osmotic coefficient:

$$\phi_s^* = \left(\frac{1000\, \overline{V}_s C}{m_s \sum_k \nu_k m_k} \right) \phi_{MM}^* \qquad (12\text{-}23a)$$

for a single electrolyte ij

$$\phi_s^* = (1 - C_{ij} \overline{V}_{ij})\, \phi_{MM}^* \qquad (12\text{-}23b)$$

where C_{ij} and $\overline{V}_{ij}$ are the concentration (molarity) and partial molar volume of the salt.

For mixed solvent systems it is difficult to define ϕ_s^* unambiguously whereas ϕ_{MM}^* is rigorously given by Π, c, and T in Eq. (12-20), which is experimentally accessible. With the ϕ_{MM}^* to ϕ_s^* conversion as a starting point, we can correctly convert activity coefficients as well [see Cardoso and O'Connell (1987) and Pailthorpe *et al.* (1984) for details]. For a single electrolyte in a single solvent, the mean activity coefficient $\gamma_\pm$ based on a Lewis and Randall activity is related to a mean activity coefficient $\gamma_{\pm,MM}$ based on a McMillan-Mayer modeling approach by,

$$\ln \gamma_\pm = \ln \gamma_{\pm,MM} + \frac{\Pi \overline{V}_{ij}}{RT} \qquad (12\text{-}24)$$

For most systems of interest, this correction is small and can be neglected. For mixed solvent systems, however, the situation is different and a more careful analysis is warranted (see Section 12.9).

12.2 Experimental Measurements of Ionic Activity

In principle, any method that determines changes in equilibrium thermodynamic properties that are compositionally dependent at a specified temperature and pressure can be used to estimate the activity coefficient. For electrolytes, the convenient experimental measurements are:

(1) ***osmotic pressure, vapor pressure, and isopiestic methods***

(2) ***freezing point depression***

(3) ***solubility***

(4) ***electrochemical potential (emf)***

In general, these are properties that depend on the type and number of molecular and ionic species in solution. In very dilute solutions, purely colligative properties result where the magnitude of the measured effect depends on the properties of the pure solvent and the total concentration of solute with proper accounting for any dissociation that occurs. This is clearly the case for type (1) and (2) measurements. Solubility and emf methods depend on the properties of the solute (*ij*) as well. A brief outline of the common methods used for determining $\gamma_{\pm}$ is presented here. Interested readers should consult Pitzer (1995), Klotz (1950), and Castellan (1983) for details.

***(1) osmotic pressure, vapor pressure, and isopiestic methods*:** If the osmotic pressure is measured as a function of solution composition, that is $\Pi = f(\boldsymbol{m})$ or $f(x_{ij})$ then Eq. (12-17) can be used to determine ϕ_s^* and Eq. (12-16) to estimate $\gamma_{\pm}$. Similarly, if the equilibrium vapor pressure of a mixture containing only non-volatile solutes is known as a function of molality, the activity of the solvent can be determined as

$$a_s = \frac{\hat{f}_s}{f_s^{(pure)}} \approx \frac{P_{vp}^{(mixture)}}{P_{vp}^{(pure)}}$$

and from the Gibbs-Duhem equation, one can determine $\gamma_{\pm}$ for the solute. Isopiestic methods employ this concept using a simultaneous measurement for different electrolytes at different compositions under the same solvent activity. For an aqueous system with dissolved salts, different solutions, including a reference solution whose vapor pressure ($P_{vp}^{(ref)}$) is known as a function of molality, are kept at the same temperature and water vapor pressure long enough to reach equilibrium. Under these conditions, the water activity is the same for all solutions. Thus by measuring the composition of the liquid phases of each solution one can equate the chemical potential of water in its solution to its composition. Again, using the Gibbs-Duhem relation, $\gamma_{\pm}$ can be determined.

Frequently, ϕ_s^* is used as it is defined in terms of the solvent activity and the composition of the solution [Eq. (12-15)]. For the unknown (*ij*) and the reference salt (*ref*), a_s is the same, so with Eq. (12-15) for each molality $\boldsymbol{m}$, we get:

$$\phi_{s,ij}^* = \frac{\phi_{s,ref}^* \boldsymbol{m}_{ref} \nu_{ref}}{\nu \boldsymbol{m}} \qquad (12\text{-}25)$$

where $\nu = \nu_i + \nu_j$ and $\boldsymbol{m}$ refer to the unknown salt. With ϕ_s^* now estimated, $\gamma_\pm$ can be determined for the unknown using Eq. (12-16).

***(2) freezing point depression*:** This technique follows a similar line of analysis to the osmotic and vapor pressure measurements. Here the activity of the solvent (usually water) is related to a measured depression of the freezing point of pure solvent (for example, ice) as a function of the liquid solution composition.

***(3) solubility*:** Solubility data have been collected and tabulated for a great number of compounds of interest [see Linke and Seidell (1965)]. Even if data are not available, they are relatively easy to determine for non-volatile electrolytes. The analysis to estimate $\gamma_\pm$ usually uses a solubility product (K_{sp}). From Eq. (12-2), K_{sp} can be written as an equilibrium constant for the reaction given.

$$K_{sp} = \frac{a_i^{\nu_i} a_j^{\nu_j}}{a_{ij,solid}} = \gamma_\pm^\nu \boldsymbol{m}_\pm^\nu \tag{12-26}$$

where $a_{ij,solid}$ is the activity of pure solid which is set to unity by convention. $\boldsymbol{m}_\pm$ is given by Eq. (12-12a) for single electrolytes and by Eq. (12-12b) for mixed electrolytes. K_{sp} is either known as a $f(T)$ or can be calculated from known or estimated ΔG_f^o values for the ionic species involved (see also Sections 8.8 and 13.3). Thus by measuring the solubility, $\boldsymbol{m}_\pm$ is determined directly, and $\gamma_\pm$ can be calculated from Eq. (12-26).

***(4) electrochemical potential (emf)*:** For isothermal electrochemical redox reactions occurring under nearly reversible conditions (at very low currents) in cells containing an anode and cathode, the chemical potential or Gibbs free energy change can be related to the measured electrical potential. At low currents, in the electrochemical fuel cell shown in Figure 12-3, the isothermal reversible electrical work of moving a charged species across a net potential or emf of the cell, $\exists$, is proportional to the magnitude of $\exists$. For a differential amount of charge transferred:

$$\delta W_{rev} = \exists d\underline{q} = \exists\left(\sum \nu_i z_i\right) F d\xi = n_e F \exists d\xi \tag{12-27}$$

where n_e = moles of electrons transferred in the reaction per mole of reactant, ξ = the extent of reaction [see Eq. (6-51)], and F = Faraday's constant = 96,485 coulombs/mol. Quite arbitrarily we have introduced a sign convention in Eq. (12-27) as well. This can be seen by referring to the cell diagram in Figure 12.3. Here the electrodes labelled 1 and 2 are at different potential levels Ψ_1 and Ψ_2 so the net cell potential or emf $\exists$ is defined by our convention as

$$\exists \equiv \Psi_2 - \Psi_1$$

In Section 5.8, we modified the Fundamental Equation to include a contribution to the work for electrical charge transport in the expression for $d\underline{U}$. By taking the second Legendre transform with respect to $\underline{S} \to T$ and $\underline{V} \to P$ conjugate variable changes, we obtain the Gibbs free energy $\underline{G}$ in differential form with $Pd\underline{V}$ and $\exists d\underline{q}$ work effects included

$$d\underline{G} = -\underline{S}dT + \underline{V}dP + \sum \mu_i dN_i + n_e F \exists d\xi = -\underline{S}dT + \underline{V}dP + \left[\sum \nu_i \mu_i + n_e F \exists\right] d\xi$$

which can be integrated from $\xi = 0$ to 1, noting that $d\underline{G} = 0$ for a system at equilibrium at constant T, P, and total mass. Thus, for an isobaric, isothermal, electrochemical cell operating at reversible equilibrium conditions:

$$\boxed{\Delta G_{rx} \equiv \sum \nu_i \mu_i = -n_e F \exists} \tag{12-28}$$

If the reactants and products involved in the cell reaction are in their standard states (that is 1 bar, 298 K (25°C), unit molality, etc.) then ΔG_{rx} becomes $\Delta G_{rx}^{\square}$ and $\exists$ becomes $\exists^{\square}$, the standard electrochemical potential:

$$\Delta G_{rx}^{\square} = -n_e F \exists^{\square} \tag{12-29}$$

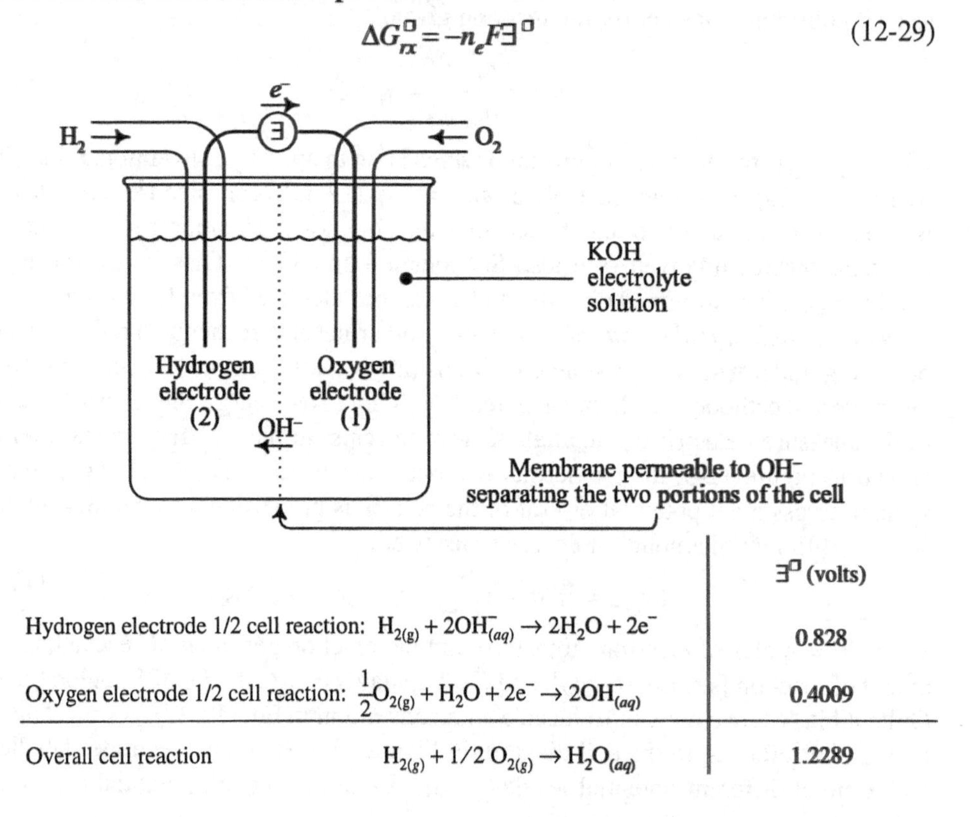

		$\exists^{\square}$ (volts)
Hydrogen electrode 1/2 cell reaction:	$H_{2(g)} + 2OH^-_{(aq)} \rightarrow 2H_2O + 2e^-$	0.828
Oxygen electrode 1/2 cell reaction:	$\frac{1}{2}O_{2(g)} + H_2O + 2e^- \rightarrow 2OH^-_{(aq)}$	0.4009
Overall cell reaction	$H_{2(g)} + 1/2\ O_{2(g)} \rightarrow H_2O_{(aq)}$	1.2289

Figure 12.3 Electrochemical hydrogen-oxygen fuel cell operating in an alkaline electrolyte such as KOH at 25°C, 1 bar with platinum electrodes.

With the definition of activity given in Eq. (12-6), we can compare ΔG_{rx} for the cell at any real condition to $\Delta G_{rx}^{\square}$ at standard state conditions using a modified form of the criteria for a reversible chemical reaction at equilibrium [Eq. (6-55)]. The modified criteria is actually given above in Eq. (12-28), where in the presence of a finite electric potential $\exists$, we see that ΔG_{rx} is not equal to zero. With $\mu_i = \mu_i^{\square} + RT\ln a_i$, the difference between ΔG_{rx} and $\Delta G_{rx}^{\square}$ is given by,

$$\Delta G_{rx} - \Delta G_{rx}^{\square} = \sum \nu_i(\mu_i - \mu_i^{\square}) = \sum RT\,\nu_i(\ln a_i)$$

By substituting Eqs. (12-28) and (12-29),

$$\exists = \exists^{\square} - \frac{RT}{n_e F} \ln\left(\prod (a_i)^{\nu_i}\right) \tag{12-30}$$

Equation (12-30) is commonly called the Nernst equation. It provides a direct means for determining how a measured cell voltage will vary as a function of the activities of the reactants and products involved in the cell reaction.

For a single ij electrolyte that dissociates according to reaction (12-2), Eq. (12-30) can be written as:

$$\ln\left(\frac{a_i^{\nu_i} a_j^{\nu_j}}{a_{ij,solid}}\right) = \frac{n_e F}{RT}(\exists^{\square} - \exists) \tag{12-31}$$

With $a_{ij,solid} = 1.0$ and Eq. (12-9b), Eq. (12-31) can be reconfigured to provide an expression for determining $\gamma_{\pm}$ in terms of known or measurable quantities:

$$\ln \gamma_{\pm} = \frac{n_e F}{\nu RT}(\exists^{\square} - \exists) - \ln \boldsymbol{m}_{\pm} \tag{12-32}$$

Things get more tedious algebraically for mixed electrolytes with common ions present but the general approach is the same. In order to actually use Eq. (12-32), both $\exists$ and $\exists^{\square}$ are needed. $\exists$ is measured at the conditions of interest. While at the infinite dilution condition ($\boldsymbol{m} \to 0$), $\gamma_{\pm} \to 1.0$ and $\exists^{\square}$ can be estimated by extrapolation of $\exists$ versus $\boldsymbol{m}$ data to $\boldsymbol{m} = 0$. To facilitate the extrapolation to zero molality, the theoretical Debye- Hückel model is often used as described in the next section. For many systems of interest in electrochemical applications, standard potentials can be obtained by combining values tabulated for 1/2 cell reactions. The $\exists^{\square}$ values cited in Table 12.2 have been extracted from various sources including W.M. Latimer (1952), *The Oxidation States of the Elements and Their Potentials in Aqueous Solution*, 2nd ed., Prentice-Hall, Englewood Cliffs, N.J.; Bard *et al.* (1985), *Standard Potentials in Aqueous Solution*, Marcel Dekker, New York; and Pitzer (1995), *Thermodynamics* (Table 19-3, p. 361), McGraw-Hill, New York.

Before we leave this section, it is appropriate to illustrate the range of experimental $\gamma_{\pm}$ values that are observed for typical strong electrolytes. Data for many common strong electrolytes in aqueous solutions at 25°C are widely available in the literature [see, for example, Harned and Owen (1958), Robinson and Stokes (1970) and Zemaitis (1986)]. A few representative examples are plotted as a function of molality in Figure 12.4 where it is obvious that both large positive ($\gamma_{\pm} > 1$) and negative ($\gamma_{\pm} < 1$) deviations exist over a wide range. The great challenge for modelers of electrolyte systems is to provide a systematic way of predicting and correlating such diverse behavior.

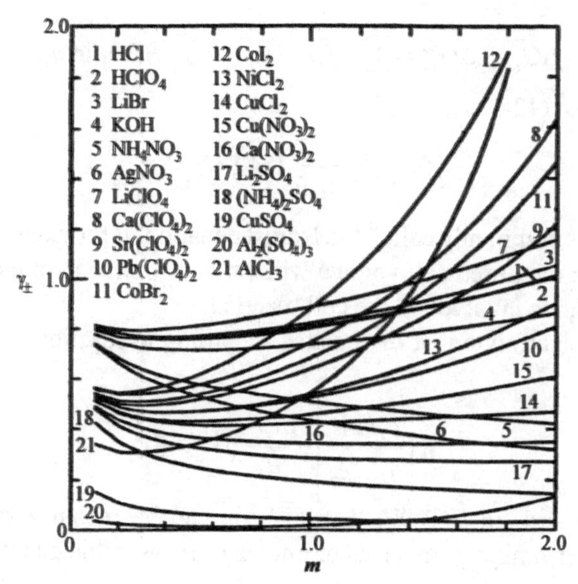

Figure 12.4 Experimental mean ionic activity coefficient $\gamma_{\pm}$ as a function of molality (***m***) for several strong electrolytes in aqueous solutions at 25°C [data from Robinson and Stokes (1970), figure adapted from Meissner and Tester (1972), with permission of the American Chemical Society].

12.3 Debye-Hückel Model (theoretical)

The theory proposed in 1923 by Debye and Hückel provides a rigorous anchor point for most engineering models of electrolyte solutions. Strictly speaking, the theoretical results of Debye-Hückel theory are exact only in the infinite dilution limit. The basic scheme they employed separates the non-idealities of electrolyte solutions into separate contributions that can be superimposed. This was an important step because it simplified the problem of accounting for long range coulombic forces from solvent-ion and ion-ion interactions which are fundamentally different than the dispersive van der Waals forces acting between neutral molecules. As discussed in Chapter 10, electrostatic forces can be as large or larger than van der Waals forces and they are effective over larger separation distances (see Table 10.1). These coulombic effects lead to large deviations from ideal behavior even in dilute solutions.

Debye-Hückel theory incorporates several key simplifications:

(1) the solvent is assumed to be a structureless continuum with a fixed permittivity (ε) and dielectric strength (D_s)

(2) an ion atmosphere concept was introduced to estimate the mean effect of a specified, spherically symmetric ion charge on all other ions

(3) this mean effect was approximated by the insertion of a single ion into a medium described by a continuous Poisson-Boltzmann distribution of charge density around the specified ion.

Table 12.2 Standard Electrode Potentials at 25°C

Electrode reaction	$\mathcal{E}^0$ (volts)
$Li^+ + e^- = Li$	−3.04
$K^+ + e^- = K$	−2.925
$Ca^{2+} + 2e^- = Ca$	−2.85
$Na^+ + e^- = Na$	−2.714
$H_2 + 2e^- = 2H^-$	−2.25
$Al^{3+} + 3e^- = Al$	−1.66
$Zn(CN)_4^{2-} + 2e^- = Zn + 4CN^-$	−1.26
$ZnO_2^{2-} + 2H_2O + 2e^- = Zn + 4OH^-$	−1.216
$Zn(NH_3)_4^{2+} + 2e^- = Zn + 4NH_3$	−1.03
$Sn(OH)_6^{2-} + 2e^- = HSnO_2^- + H_2O + 3OH^-$	−0.90
$Fe(OH)_2 + 2e^- = Fe + 2OH^-$	−0.877
$2H_2O + 2e^- = H_2 + 2OH^-$	−0.828
$Fe(OH)_3 + 3e^- = Fe + 3OH^-$	−0.77
$Zn^{2+} + 2e^- = Zn$	−0.763
$Ag_2S + 2e^- = 2Ag + S^{2-}$	−0.69
$Fe^{2+} + 2e^- = Fe$	−0.44
$Bi_2O_3 + 3H_2O + 6e^- = 2Bi + 6OH^-$	−0.44
$PbSO_4 + 2e^- = Pb + SO_4^{2-}$	−0.356
$Ag(CN)_2^- + e^- = Ag + 2CN^-$	−0.31
$Ni^{2+} + 2e^- = Ni$	−0.250
$AgI + e^- = Ag + I^-$	−0.151
$Sn^{2+} + 2e^- = Sn$	−0.136
$Pb^{2+} + 2e^- = Pb$	−0.126
$Cu(NH_3)_4^{2+} + 2e^- = Cu + 4NH_3$	−0.12
$Fe^{3+} + 3e^- = Fe$	−0.036
$2H^+ + 2e^- = H_2$	0.000
$AgBr + e^- = Ag + Br^-$	+0.095
$HgO(r) + H_2O + 2e^- = Hg + 2OH^-$	+0.098
$Sn^{4+} + 2e^- = Sn^{2+}$	+0.15
$AgCl + e^- = Ag + Cl^-$	+0.222
$Hg_2Cl_2 + 2e^- = 2Hg + 2Cl^-$	+0.2676
$Cu^{2+} + 2e^- = Cu$	+0.337
$Ag(NH_3)_2^+ + e^- = ag + 2NH_3$	+0.373
$I_2 + 2e^- = 2I^-$	+0.5355
$Hg_2SO_4 + 2e^- = 2Hg + SO_4^{2-}$	+0.6151
$Fe^{3+} + e^- = Fe^{2+}$	+0.771
$Ag^+ + e^- = Ag$	+0.7991
$Br_{2(l)} + 2e^- = 2B_r^-$	+1.066
$O_2 + 4H^+ + 4e^- = 2H_2O$	+1.229
$Cl_{2(g)} + 2e^- = 2Cl^-$	+1.3583
$PbO_2 + SO_4^{2-} + 4H^+ + 2e^- = PbSO_4 + 2H_2O$	+1.685
$O_3 + 2H^+ + 2e^- = O_2 + H_2O$	+2.076
$F_2 + 2e^- = 2F^-$	+2.87

Note: all reactive species are at unit activity ($a_i = 1$) in their standard states

The derivations used here to describe Debye-Hückel theory are largely phenomonological and are based in classical electrostatics. Alternate statistical mechanical based derivations are also available, see McQuarrie (1976) for example.

We begin by modeling a dilute electrolyte solution as a mixture of spherical ions of charges q_+ and q_- and uniform radius r_o immersed in a solvent of constant dielectric strength, D_s. The total reversible work under isobaric and isothermal conditions can be related to the change in Gibbs energy or chemical potential, ΔG or $\Delta\mu$, and can be separated into two parts associated with the charging process:

$$W_{rev,i} = W_i + W_{ij}$$

where

W_i = work required to charge a single isolated spherical ion of radius r_o in the pure solvent

W_{ij} = additional work required to charge the ion in the presence of all other ions.

We are most interested in evaluating W_{ij} as it scales with the interactions between ions and is directly related to the activity coefficient γ_i, appropriately defined with an infinite dilution referenced standard state for ion i.

$$W_{ij} = W_{rev,i} - W_i = \Delta\mu_{ij} = kT \ln \gamma_i \qquad (12\text{-}33)$$

where all terms in Eq. (12-33) are on a per ion basis.

For an isolated conducting sphere contained in an infinite continuum, the electrical work associated with the charging process from $q = 0$ to $q = q_i$ is given by integration of an equation analogous to Eq. (12-27) with $\exists$, the cell potential, replaced by Φ_{ii}, the electrostatic potential, given by Coulomb's law:

$$W_i = \int_0^{q_i} \Phi_{ii}\, dq_i = \int_0^{q_i} \frac{q_i}{4\pi\varepsilon r_o} dq_i = \frac{q_i^2}{8\pi\varepsilon r_o} = \frac{(ez_i)^2}{8\pi\varepsilon r_o} \qquad (12\text{-}34)$$

where ε = permittivity = $\varepsilon_o D_s$, ε_o = vacuum permittivity = $8.8541878 \times 10^{-12}$ [(coulomb)^2/Jm or Farad/m], D_s = static dielectric constant of solvent, q_i = charge on ion i = ez_i , and e = elementary electron charge = $1.6021773 \times 10^{-19}$ coulombs (C). (Note that to avoid ambiguities we will continue to use SI units rather than introduce esu units here.)

Now if we can calculate $W_{rev,i}$, then W_{ij} and γ_i can be determined using Eq. (12-33). Consider the following geometry of charges immersed in the solvent, ion i is inserted into a dielectric continuum containing n charged ions:

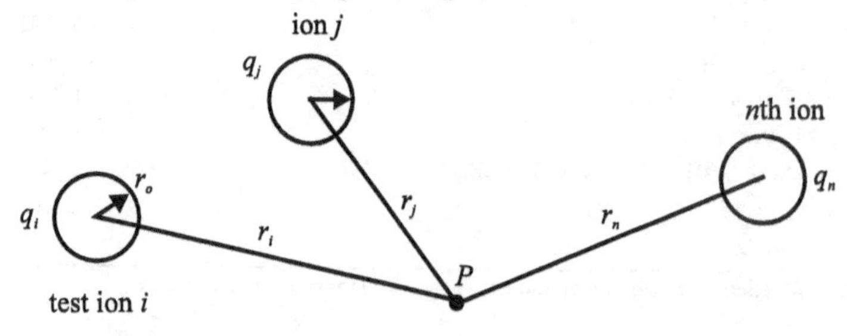

At P, the total electric potential $\Phi(r)$ can be related to the charge density ($\rho_\pm$ = total charge per unit volume) by using the Poisson equation from classical electrostatics. For a spherically symmetric system.

$$\nabla^2\Phi = \frac{1}{r^2}\frac{\partial}{\partial r}\left(r^2\frac{\partial\Phi}{\partial r}\right) = -\frac{\rho_\pm}{\varepsilon} \tag{12-35}$$

If $\rho_\pm$ is constant or is an explicit function of Φ or r, Eq. (12-35) can be integrated, and we can, in principle, estimate the Φ_i at the surface of ith ion. Up to now, everything done by Debye and Hückel is essentially rigorous. To achieve a tractable solution, more approximations are needed.

Approximation (1): The charge density is assumed to follow a Boltzmann distribution in Φ. At any point in the dielectric continuum,

$$\rho_\pm = C_+z_+F + C_-z_-F = F\sum_i C_iz_i \tag{12-36}$$

where C_+, C_-, and C_i (in general) are total ion concentrations in mol/m^3 at a particular value of Φ,

$$C_+ = \langle C_+^o\rangle \exp\left[-z_+ e\,\Phi/kT\right]$$

$$C_- = \langle C_-^o\rangle \exp\left[-z_- e\,\Phi/kT\right]$$

$$C_i = \langle C_i^o\rangle \exp\left[-z_i e\,\Phi/kT\right]$$

The bracketed $\langle C_{+,-,i}^o\rangle$ quantities refer to bulk or average ion concentrations. Where $\Phi = 0$ an electrically neutral region exists. However, with $\Phi = 0$, the distribution of ions is uniform, thus $\rho_\pm = 0$ as electroneutrality holds:

$$\text{for } \Phi = 0: \qquad \langle C_+^o\rangle z_+ + \langle C_-^o\rangle z_- = 0 \tag{12-37}$$

By substituting values of C_+ and C_- given above into Eq. (12-36), $\rho_\pm$ is

$$\rho_\pm = F\left[z_+\langle C_+^o\rangle \exp\left(-z_+ e\,\Phi/kT\right) + z_-\langle C_-^o\rangle \exp\left(-z_- e\,\Phi/kT\right)\right] \tag{12-38}$$

or in general,

$$\rho_\pm = F\sum_j z_j\langle C_j^o\rangle \exp\left(-z_j e\,\Phi/kT\right)$$

Approximation (2): A polynomial expansion for the exponentials in Eq. (12-38) is used to simplify the $\rho_\pm$ equation.

If $z_j e\,\Phi/kT < 1$ then $e^{-x} = 1 - x + \frac{x^2}{2!} - \frac{x^3}{3!} + \cdots$ converges and $\rho_\pm$ can be approximated by

$$\rho_\pm \approx F\left[\langle C_+^o\rangle z_+ + \langle C_-^o\rangle z_- - \frac{e\Phi}{kT}\left[\langle C_+^o\rangle z_+^2 + \langle C_-^o\rangle z_-^2\right] + \cdots\right] \tag{12-39}$$

With Eq. (12-37) (electroneutrality), the first two terms are zero, so with $e/k = F/R$ and retaining only the second-order terms in z_j^2:

$$\rho_{\pm} \approx -\frac{F^2\Phi}{RT}\left[<C_+^o> z_+^2 + <C_-^o> z_-^2\right] = -\frac{F^2\Phi}{RT}\sum_j^{all\ ions} <C_j^o> z_j^2 \tag{12-40}$$

Now we can return to the Poisson equation for Φ (Eq. (12-35)) with an expression for $\rho_{\pm}$ in terms of Φ and reformulate the equation to an ODE since $\Phi = f(r)$ only:

$$\frac{1}{r^2}\left(\frac{d}{dr}\left(r^2\frac{d\Phi}{dr}\right)\right) - \kappa^2\Phi = 0 \tag{12-41}$$

where

$$\kappa^2 \equiv \frac{F^2}{\varepsilon RT}\sum_j^{all\ ions} <C_j^o> z_j^2 \tag{12-42}$$

where we have generalized the equation for $\rho_{\pm}$ to include an arbitrary mixture of + and – ions of different types. κ has a physical interpretation that is forthcoming. The ODE of Eq. (12-41) is simplified by using the similarity transform $\zeta = \Phi r$:

$$\frac{d^2\zeta}{dr^2} - \kappa^2\zeta = 0$$

which has the general solution:

$$\Phi = \frac{B_1\exp(-\kappa r)}{r} + \frac{B_2\exp(+\kappa r)}{r} \tag{12-43}$$

where B_1 and B_2 are integration constants. Applying boundary conditions is the next step. As $r \to \infty$, $\Phi \to 0$ so $B_2 = 0$, thus

$$\Phi = \frac{B_1\exp(-\kappa r)}{r} \tag{12-44}$$

To evaluate B_1 we need an additional approximation.

Approximation (3): Expand **exp(–κr)** ***and retain only the first two terms.*** This results in a linearization of Φ from Eq. (12-44):

$$\Phi = \frac{B_1}{r}\left(1 - \kappa r + \frac{\kappa^2 r^2}{2!} - \cdots\right) \approx \frac{B_1}{r}(1 - \kappa r) \tag{12-45}$$

If the ion concentration is zero (infinitely-dilute state), then $\kappa \to 0$ and the potential at point P is due only to the inserted test ion i. In this case,

$$\Phi_{\kappa=0} = \frac{z_i e}{4\pi\varepsilon r}$$

But with $\kappa = 0$, Eq. (12-45) becomes

$$\Phi_{\kappa=0} = \frac{B_1}{r}$$

Thus,

$$B_1 = \frac{z_i e}{4\pi\varepsilon} \tag{12-46}$$

Now we incorporate this result back into Eq. (12-45) to obtain a reasonable approximation of $\Phi(r)$ in dilute solutions where Approximation 3 is reasonable:

$$\Phi \approx \frac{z_i e}{4\pi\varepsilon}\left(\frac{1}{r} - \kappa\right) \tag{12-47}$$

and the total reversible work for the charging process of the ith ion at $r = r_o$ in the presence of other ions in a dilute state is given by,

$$W_{rev,i} = \int_0^{q_i} \Phi dq_i = \int_0^{z_i} \frac{z_i e}{4\pi\varepsilon}\left(\frac{1}{r_o} - \kappa\right) e dz_i \tag{12-48}$$

To perform the integration, we assume that κ is constant at a small value corresponding to a dilute state. Effectively then, $W_{rev,i}$ is a measure of the total work of inserting and charging the central test ion i in the atmosphere created by the distribution of ions at concentrations $\langle C_i^o \rangle$. Strictly speaking, we should account for the change in κ as q_i increases, but as a first approximation we can neglect the effect which mathematically is equivalent to treating the test ion as a point charge ($r_o \to 0$). Integrating Eq. (12-48) with this approximation, we get

$$W_{rev,i} = \frac{(z_i e)^2}{8\pi\varepsilon r_o} - \frac{(z_i e)^2 \kappa}{8\pi\varepsilon} \tag{12-49}$$

where the first term corresponds to self charging of the test ion i in a dielectric medium (W_i) and the second to charging in the atmosphere (W_{ij}). We see this directly by using Eqs. (12-33) and (12-34) which allow us to express the activity coefficient γ_i for ion i as,

$$kT \ln \gamma_i = W_{ij} = W_{rev,i} - W_i = \frac{-(z_i e)^2 \kappa}{8\pi\varepsilon} \tag{12-50}$$

Because we must have both + and – ions, the mean ionic activity coefficient is introduced. For a single ij electrolyte, using + and – subscripts to represent i and j ions:

$$\gamma_\pm = (\gamma_+^{\nu_+} \gamma_-^{\nu_-})^{1/\nu} \qquad \nu = \nu_+ + \nu_- \tag{12-51}$$

Using Eqs. (12-50) and (12-51), we get

$$\nu \ln \gamma_\pm = \frac{-e^2 \kappa}{8\pi\varepsilon kT}(\nu_+ z_+^2 + \nu_- z_-^2) \tag{12-52}$$

With electroneutrality, $\nu_+ z_+ + \nu_- z_- = 0$. Thus, we can create an equivalent expression for $\nu_+ z_+^2 + \nu_- z_-^2$ by multiplying the equality for electroneutrality first by z_+ and then by z_- and adding them to get

$$\nu_+ z_+^2 + \nu_- z_-^2 = -(\nu_+ + \nu_-)(z_+ z_-) = -\nu(z_+ z_-) \tag{12-53}$$

With $e/k = F/R$, $kN_A = R$, and $\varepsilon = \varepsilon_o D_s$, and Eq. (12-53), Eq. (12-52) becomes

$$\ln \gamma_{\pm} = \frac{F^2 \kappa z_+ z_-}{8\pi \varepsilon_o D_s N_A RT} \tag{12-54}$$

The ionic strength I is introduced in κ as defined in Eq. (12-42):

$$\kappa = \left[\frac{F^2}{\varepsilon RT} \sum_i \langle C_i^o \rangle z_i^2 \right]^{1/2} = \left(\frac{2000\, F^2 \rho_s}{\varepsilon RT}\right)^{1/2} I^{1/2} \tag{12-55}$$

where we note that $\langle C_i^o \rangle$ has units of mol/m^3 so to convert to molarity C_i in mol/L we used

$$C_i = \langle C_i^o \rangle / 1000$$

In addition, in dilute aqueous solutions at 25°C, molarity and molality are approximately equal, $C_i \approx m_i$, and for any dilute solution $C_i = \rho_s m_i$, thus the ionic strength is given by:

$$I = \frac{1}{2} \sum_i m_i z_i^2 \approx \frac{1}{2000} \sum_i \frac{\langle C_i^o \rangle z_i^2}{\rho_s} \tag{12-56}$$

where ρ_s = solvent density in kg/L. Using Eq. (12-56), we easily derive the second form for κ given in Eq. (12-55). Converting Eq. (12-54) to a $\log_{10}$ basis and introducing the result from Eq. (12-55) with $\varepsilon = \varepsilon_o D_s$, we get the common final form of the ***Debye-Hückel limiting law***:

$$\log_{10} \gamma_{\pm} = \frac{z_+ z_-}{8\pi} \left(\frac{F^2}{\varepsilon_o D_s RT}\right)^{3/2} \left(\frac{(2000)^{1/2} \rho_s^{1/2} I^{1/2}}{2.303\, N_A}\right) \tag{12-57}$$

which with substitutions for R, N_A, F, and ε_o becomes:

$$\boxed{\log_{10} \gamma_{\pm} = \frac{1.8248 \times 10^6 \rho_s^{1/2}}{(D_s T)^{3/2}} (z_+ z_-)\, I^{1/2} = -A\, |z_+ z_-|\, I^{1/2}} \tag{12-58}$$

where the absolute value | | sign has been introduced and the constant 1.8248×10^6 carries the units of K$^{3/2}$ L$^{1/2}$ /mol$^{1/2}$. If the solvent is water at 25°C, $D_s = 78.54$, $\rho_s = 1.0$ kg/L, and $A = 0.510$, or

$$\log_{10} \gamma_{\pm} \text{ (25°C, aqueous)} = -0.510\, |z_+ z_-|\, I^{1/2}$$

The major effects of solute and solvent properties in the Debye-Hückel model can be summarized for dilute solution of strong electrolytes:

(i) square root dependence of $\log_{10} \gamma_{\pm}$ on ionic strength I
(ii) explicit dependence on the ion charge product $|z_+ z_-|$
(iii) dependence on solvent properties due only to the dielectric constant (D_s) and density (ρ_s).

Example 12.1

Is the behavior shown in Figure 12.4 consistent with the Debye-Hückel limiting law?

Solution

Equation (12-58) indicates that all the concentration dependence is in the $I^{1/2}$ term. Thus if the data are to be consistent with the Debye-Hückel law they must converge to a single curve if we replot the reduced activity coefficient $\Gamma_\pm$ as given in Eq. (12-14). For aqueous single electrolytes at 25°C,

$$\log_{10} \Gamma_\pm = \frac{1}{|z_+ z_-|} \log_{10} \gamma_\pm = -0.510\, I^{1/2} \tag{12-59}$$

We re-examine some data given in Figure 12.4 in the dilute region in the figure below. Here we see that Eq. (12-59) is followed and thus the data are consistent with the limiting Debye-Hückel law in the dilute limit ($I \leq 0.01$), as they must be.

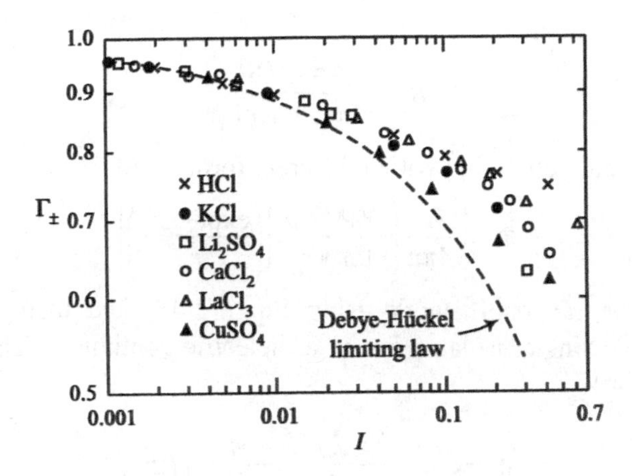

Example 12.2

Can Approximation (3) be eliminated to produce a more accurate result for predicting $\gamma_\pm$ using the Debye-Hückel model?

Solution

In order to evaluate the constant B_1 in Eq. (12-44), an alternate approach is needed. The total charge contained in the volume of interest is $q_i = z_i e$. This result must be given by integration of the charge density $\rho_\pm$ from the surface of the spherical ion at $r = r_o$ to $r = \infty$. In spherical coordinates, with angular symmetry assumed, this integration is given by

$$z_i e = \int_{r_o}^{\infty} 4\pi r^2 \rho_\pm dr \tag{12-60}$$

Now we can use this result to produce a modified form of the Debye-Hückel model.

Extension of the Debye-Hückel model. Combining Eqs. (12-40) and (12-44) and incorporating the result into Eq. (12-60) yields an alternative formulation of the Poisson-Boltzmann distribution:

$$z_i e = -\int_{r_o}^{\infty} 4\pi r^2 \left[\frac{F^2}{RT}\sum_i <C_i^o> z_i^2\right] B_1 \frac{\exp(-\kappa r)}{r}\, dr \tag{12-61}$$

With further simplification

$$z_i e = \frac{-4\pi B_1 F^2}{RT}\left[\sum_i <C_i^o> z_i^2\right]\int_{r_o}^{\infty} r \exp(-\kappa r) dr \tag{12-62}$$

With κ defined by Eq. (12-42),

$$z_i e = -4\pi\varepsilon B_1 \kappa^2 \left(\frac{r}{\kappa} + \frac{1}{\kappa^2}\right)\exp(-\kappa r)\Bigg|_{r_o}^{\infty} \tag{12-63}$$

Solving for B_1

$$B_1 = \frac{z_i e \exp(\kappa r_o)}{4\pi\varepsilon(1+\kappa r_o)} \tag{12-64}$$

Proceeding as we did before, but with a different form for Φ

$$\Phi = \left[\frac{z_i e}{4\pi\varepsilon}\left(\frac{\exp(\kappa r_o)}{1+\kappa r_o}\right)\left(\frac{\exp(-\kappa r)}{r}\right)\right] \tag{12-65}$$

we recalculate the reversible work using Eq. (12-48) and then subtract off the contribution of the single, isolated ion in a dielectric continuum. This results in the following form for W_{ij}

$$W_{ij} = -\frac{e^2 N_A \sum <C_i^o> z_i^2}{12\pi\varepsilon}\kappa\, g(\kappa r_o)$$

with $\quad g(\kappa r_o) = \dfrac{3}{(\kappa r_o)^3}\left[\ln(1+\kappa r_o) - \kappa r_o + \dfrac{(\kappa r_o)^2}{2}\right].$

We now use the formulation of Eq. (12-50) as a basis, remembering that κ is a function of $q_i = ez_i$, which can be scaled by inserting the term $q_i/z_i e$ into the expression for κ given in Eq. (12-55). Thus, $\ln \gamma_i$ is given by

$$kT \ln \gamma_i = \frac{(z_i e)^2}{8\pi\varepsilon}\frac{\kappa}{1+\kappa r_o} + \frac{\overline{V}_i kT}{24\pi N_A r_o^3}(\kappa r_o)^3 f(\kappa r_o) \tag{12-66}$$

where

$$f(\kappa r_o) = \frac{3}{(\kappa r_o)^3}\left[1 + \kappa r_o - \frac{1}{1 + \kappa r_o} - 2\ln(1 + \kappa r_o)\right]$$

The second term involving $\overline{V}_i$ in Eq. (12-66) appears as a result of the relationship between the isothermal, reversible electrical work for the ion interaction W_{ij} and ΔG^{EX}, the excess Gibbs free energy which is used to obtain $\ln \gamma_i$. In molecular SI units:

$$kT \ln \gamma_i = \frac{1}{N_A}\left(\frac{\partial \Delta \underline{G}^{EX}}{\partial N_i}\right)_{T,P,N_j[i]} = W_{ij}$$

Since $\underline{V}$ can change as N_i changes at constant T and P, the partial molar volume $\overline{V}_i$ of the solute ion also appears in the differentiation with respect to mole number N_i. The dependence of W_{ij} on $\underline{V}$ is contained in the term κ.

Invoking the modified electroneutrality condition [Eq. (12-53)] and following the approach used before to generate $\gamma_\pm$, with $e = F/N_A$ we get:

$$\ln \gamma_\pm = \frac{-F^2 |z_+ z_-|}{8\pi\varepsilon_o D_s N_A^2 kT}\left(\frac{\kappa}{1 + \kappa r_o}\right) + \frac{\overline{V}_{ij}(\kappa r_o)^3}{2\pi N_A r_o^3 \nu} f(\kappa r_o) \tag{12-67}$$

where $\overline{V}_{ij}$ is partial molar volume of the electrolyte ij. Fortunately for many applications even to reasonably high ionic strengths, the second term in Eq. (12-67) can be neglected. Thus, Eq. (12-67) is usually simplified to give the ***extended Debye-Hückel*** form:

$$\boxed{\log_{10} \gamma_\pm = \frac{-A |z_+ z_-| I^{1/2}}{1 + B r_o I^{1/2}}} \tag{12-68}$$

where

$$A = \frac{(2000)^{1/2} F^3 \rho_s^{1/2}}{2.303 (\varepsilon_o D_s RT)^{3/2} (8\pi N_A)} = \frac{1.8248\times10^6 \rho_s^{1/2}}{(D_s T)^{3/2}}\left(\frac{\text{kg}}{\text{mol}}\right)^{1/2} \quad (T \text{ in K})$$

$$B = \left(\frac{2000 F^2 \rho_s}{\varepsilon_o D_s RT}\right)^{1/2} = \frac{5.0292\times10^{11} \rho_s^{1/2}}{(D_s T)^{1/2}}\,\text{m}^{-1}\left(\frac{\text{kg}}{\text{mol}}\right)^{1/2}$$

Typically, the ionic radius r_o ranges from 2 to 6×10^{-10} m. For water at 25°C,

$$A = 0.510\left(\frac{\text{kg}}{\text{mol}}\right)^{1/2};\; B = 3.2865\times10^9\,\text{m}^{-1}\left(\frac{\text{kg}}{\text{mol}}\right)^{1/2};\; r_o = 3.5\times10^{-10}\,\text{m}$$

Thus $r_o B \approx 1\left(\frac{\text{kg}}{\text{mol}}\right)^{1/2}$ and from Eq. (12-68):

$$\log_{10} \gamma_\pm (25°\text{C, aqueous}) \approx \frac{-0.510 |z_+ z_-| I^{1/2}}{1 + I^{1/2}}$$

which for small values of I reduces to the Debye-Hückel limiting law given in Eq. (12-58).

Debye screening length. Now that we have developed the working equations for estimating $\gamma_\pm$ in dilute solutions, we can use the Debye-Hückel model to provide a semi-quantitative picture of solution structure on a molecular level. Intuitively, we would expect ions of opposite charge to cluster together, so that some level of local molecular ordering will result. In a global sense, each ion would, on average, be surrounded by a cloud containing a net excess of oppositely charged ions. This "ion atmosphere" has a characteristic radius of order $1/\kappa$—which has been termed the "Debye screening length". Let's now try to see if this concept is reasonable.

We can estimate the thickness of the ion atmosphere by looking at the magnitude of the charge density $\rho_\pm$ as a function of the distance from the surface of a reference ion. Using the approximate expression for $\rho_\pm$ developed for dilute electrolytes [Eq. (12-40)] along with the result for Φ given in Eq. (12-44) [with Eq. (12-46) for B_1], grouping terms yields

$$\rho_\pm = \frac{-z_i e\, \kappa^2}{4\pi} \frac{\exp(-\kappa r)}{r} \tag{12-69}$$

where $$\kappa = \left(\frac{2000\, e^2 N_A \rho_s I}{\varepsilon_o D_s kT} \right)^{1/2} \quad \text{and} \quad r_D \equiv \frac{1}{\kappa} = \text{Debye length}$$

κ is written in molecular-scale SI units and will have inverse length units of m^{-1} as $1/\kappa$ is the Debye length. In a spherical shell from r to $r + dr$, the amount of charge contained is

$$dq_i = \rho_\pm 4\pi r^2 dr = -z_i e\, \kappa^2\, r \exp(-\kappa r) dr$$

Integrating as we did in Example 12.2 but now from $r = 0$ to ∞

$$z_i e = z_i e \int_0^\infty \kappa^2 r \exp(-\kappa r)\, dr$$

so for proper normalization,

$$\int_0^\infty \kappa^2 r \exp(-\kappa r)\, dr = \int_0^\infty f(r)\, dr = 1.0 \tag{12-70}$$

The integrand $f(r)$ can be thought of as a normalized radial distribution function of charge density. Figure 12.5 plots $f(r)$ as a function of r where we see that a maximum in $f(r)$ occurs at the Debye length, $r_D = 1/\kappa$. As D_s decreases, κ increases--decreasing r_D, which indicates that ionic clustering is more intensely pair correlated or locally structured on a molecular level. A similar effect results if I increases. By increasing the electrolyte concentration, the Debye screening effect is less effective.

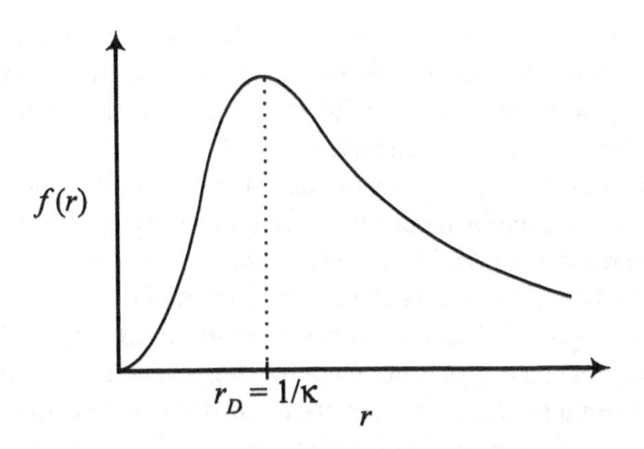

Figure 12.5 Debye-Hückel radial charge distribution function for strong electrolytes.

12.4 Beyond Debye-Hückel Theory

Modeling behavior at higher concentrations presents a great challenge for researchers. While the Debye-Hückel model is rigorous in the infinite dilute limit, and as such, provides an asymptotic framework that all models must quantitatively adhere to, it does not work well for ionic strengths above about 0.01 and it gets progressively worse as ionic strength increases to levels that are most relevant in practical engineering situations involving electrolytes. While $\gamma_{\pm}$ decreases with increasing concentration initially, many electrolytes show large positive deviations ($\gamma_{\pm} >> 1.0$) at finite concentrations while others show fairly strong negative deviations ($\gamma_{\pm} << 1.0$) (see Figure 12.4). Ideally we would like to have a robust theoretical model that predicts this range of behavior.

Countless investigators have wrestled with the problem of developing a suitable model to extend Debye-Hückel theory to finite concentrations. From Hückel's early work in the 1920's to the research contributions by MacInnes, Harned, Scatchard, and Guggenheim in the 30's and 40's, to the models and correlations proposed in the 60's, 70's and 80's by Bromley, Pitzer, Meissner, and Chen there is one common theme, theory can only go so far--experimental data are always needed to specify model parameters [see MacInnes (1947), Pitzer (1995), Robinson and Stokes (1970), and Harned and Owen (1958) for further discussion]. This situation is comparable to the requirements of fitting parameters to data that pervades the development of constitutive activity coefficient models for non-electrolytes discussed in Chapter 11.

While rigorous formulations based in statistical mechanics can be written for specifying ΔG^{EX} and $\gamma_{\pm}$ for solutions of strong electrolytes, they can only be quantitatively evaluated with assumed intermolecular potential functions that need both energetic and size parameters. Unfortunately, these potential parameters are often fit to data themselves. Although the physics of ion-ion and ion-molecule interactions can be captured with relatively simple potential functions, complexities arise in dense liquid solutions requiring additional assumptions, such as pairwise additivity, to produce tractable results. All usable models for electrolytes at finite concentrations are inherently semi-empirical to some extent. Nonetheless, as we saw in Chapter 11 for non-electrolytes, having a theoretical framework assists the development of practical constitutive property models for electrolytes as well.

A good example of using rigorous theory as a basis is found in Pitzer's development of his so-called ion interaction model for strong electrolytes. Pitzer begins with the osmotic virial expansion from McMillan-Mayer theory, he truncates the expansion and then uses empirical fitting to specify the coefficients and makes other adjustments to produce a workable model. Other approaches have also used varying degrees of theory and empiricism to produce engineering models. These include work by Meissner who employed a Corresponding States framework to correlate $\gamma_{\pm}$ to permit extrapolation to extremely high concentrations, by Chen and coworkers who successfully superimposed Pitzer's extended Debye-Hückel model for ion-ion interactions onto Renon and Prausnitz's NRTL local composition model for molecule-molecule interactions, and by Planche and coworkers who used the mean spherical approximation in an Ornstein-Zernike integral equation model to capture both ionic and non-ionic effects. Although it is not possible to provide detailed derivations in this text, we outline the important concepts and approaches that have been used to develop these models in the sections that follow.

12.5 Pitzer Ion Interaction Model

Pitzer used McMillan-Mayer theory as a starting point. Their basic working equation is a virial equation in concentration that produces the osmotic pressure, Π, of the solution. The framework is similar to what we presented in Section 11.6, but more specifically structured to match Debye-Hückel limiting behavior in dilute solutions [see Pitzer (1973, 1980, and 1995) for details].

A modified approach is used here to derive the form of the osmotic expansion using integral equation theory [Rasaiah and Friedman (1969)]. To begin the discussion, consider a first-order, restricted primitive model which assumes that the electrolyte can be represented as a mixture of hard spheres of radius ρ_o and charge $z_i e$ in a dielectric continuum consisting of the solvent. Assuming that the Poisson-Boltzmann model which was used to derive the Debye -Hückel model, still applies, then we can relate the resulting charge density $\rho_{\pm}$ to a normalized radial distribution function introduced in Section 10.7:

$$g_{ij}(r) = \frac{\rho_{\pm}}{z_i e} = \exp\left[\left(\frac{z_i z_j e^2}{4\pi\varepsilon_o D_s}\right)\left(\frac{\exp(\kappa r_o)}{1+\kappa r_o}\right)\left(\frac{\exp(-\kappa r)}{r}\right)\right] \tag{12-71}$$

where $g_{ij}(r)$ is the probability of finding an ion of type i near one of type j, with i and j interchangeable. Note that in Eq. (12-71) we have used the more exact form for $\rho_{\pm}$ given in Example 12.2.

Molecular simulation using the Monte Carlo (MC) method will also provide a direct estimate of $g_{ij}(r)$. To include the effects of a repulsive core (excluded volume) and coulombic attraction or repulsion, the following potential is used for the restricted primitive model:

$$\begin{aligned} &\textit{Hard core} \quad \Phi_{ij} = \infty \quad r < r_o \\ &\textit{Coulombic} \quad \Phi_{ij} = \frac{z_i z_j e^2}{4\pi\varepsilon_o D_s r} \quad r > r_o \end{aligned} \tag{12-72}$$

Even if you have some reservations about the approximations involved in generating Eq. (12-71), the fact that the predicted $g_{ij}(r)$ results are in agreement with the deterministic Monte Carlo simulation for a typical 1:1 electrolyte up to a concentration of about 1 molar is encouraging. This is particularly important because the MC simulation is self-consistent and "exact" for the specified potential, limited only by the inherent uncertainties of using a finite number of particle interactions to produce an ensemble averaged quantity. Thus, one is tempted to accept that the $g_{ij}(r)$ function given in Eq. (12-71) properly captures the important physics including hard core excluded volume and long range electrostatic effects.

With expressions for $g_{ij}(r)$ and $\Phi_{ij}(r)$ we can use the pressure equation from integral equation theory to obtain the osmotic coefficient ϕ^*_{MM} [see McQuarrie (1986), p 261 ff]:

$$\phi^*_{MM} \equiv \frac{\Pi}{ckT} = 1 - \frac{1}{6ckT} \sum_i \sum_j c_i c_j \int_0^\infty r \left(\frac{\partial \Phi_{ij}}{\partial r} \right) g_{ij}(r)\, 4\pi r^2\, dr \tag{12-73}$$

where c_i and c_j are concentrations of the dissociated ionic species and c is the total concentration of solute in units of molecules per unit volume. With $\Phi_{ij}(r)$ given by Eq. (12-72) and $g_{ij}(r)$ given by Eq. (12-71) we can, in principle, obtain ϕ^*_{MM} and ϕ^*_s [see Eq. (12-23b)]. However, there are some computational considerations, for example, the integrand goes to infinity at $r = r_o$. These problems have been addressed by Barker and Henderson (1976) who added a third term to Eq. (12-73) that effectively accounts for the kinetic effect of the hard core:

$$\frac{2\pi r_o^3}{3c} \sum_i \sum_j c_i c_j g_{ij}(r_o + \delta)$$

where $g_{ij}(r_o + \delta)$ is $g_{ij}(r)$ at a value of r differentially larger than $r = r_o$. To obtain a closed form solution, we must simplify the functional form for $g_{ij}(r)$. Following what we did with Debye-Hückel theory, two choices are evident using Eq. (12-71):

(1) linear approximation $\quad g_{ij}(r) \approx 1 - \zeta_{ij}$

(2) parabolic approximation $\quad g_{ij}(r) \approx 1 - \zeta_{ij} + \dfrac{\zeta_{ij}^2}{2}$

where

$$\zeta_{ij} \equiv \left(\frac{z_i z_j e^2}{4\pi\varepsilon_o D_s}\right)\left(\frac{\exp(\kappa r_o)}{1+\kappa r_o}\right)\left(\frac{\exp(-\kappa r)}{r}\right)$$

ζ_{ij} is the argument of the exp [ ] function in Eq. (12-71). Either approximation yields a result that is compatible with the limiting behavior given by Debye-Hückel theory at low I ($\kappa \rightarrow 0$). The parabolic approximation gives the following result for ϕ^*_{MM}

$$\phi^*_{MM} = 1 - \frac{\kappa^3}{24\pi c\,(1+\kappa r_o)} + c\left[\frac{2\pi\, r_o^3}{3} + \frac{\kappa^4 r_o}{48\pi c^2\,(1+\kappa r_o)^2}\right] \tag{12-74}$$

Equation (12-74) can now be used to estimate $\gamma_\pm$ by Gibbs-Duhem integration [Eq. (12-16)]:

$$\ln \gamma_\pm = -\,b\kappa\left[\frac{1}{3(1+\kappa r_o)} + \frac{2}{3\kappa r_o}\ln\,(1+\kappa r_o)\right]$$

or using an expansion for the ln ( ) term and combining terms:

$$\ln \gamma_\pm = -b\kappa\,(1 - 2/3\;\kappa r_o + 5/9\;\kappa^2\, r_o^2 - \cdots) \tag{12-75a}$$

where

$$b \equiv \frac{|z_i z_j| e^2}{8\pi\varepsilon_o D_s kT} \quad \text{and} \quad \kappa = \left[\left(\frac{2000\, e^2\, N_A\, \rho_s}{\varepsilon kT}\right) I\right]^{1/2}$$

An alternate expression for ln $\gamma_\pm$ comes from the charging process that we developed in our extended treatment of Debye-Hückel theory (see Example 12.2). By neglecting the second term in Eq. (12-67) and reformulating according to Eq. (12-75), we obtain:

$$\ln \gamma_\pm = \frac{-b\kappa}{1+\kappa r_o} = -b\kappa(1 - \kappa r_o + \kappa^2\, r_o^2 - \cdots) \tag{12-75b}$$

One sees that either Eq. (12-75a) or (12.75b) gives the same asymptotic behavior in an infinitely dilute solutions as κ goes to zero. Figure 12.6 compares results from this integral equation theory approach to those obtained by MC simulation using the same potential given in Eq. (12-72) with $r = r_o$ as the only parameter needed. The Debye-Hückel limiting law result is also plotted for comparison. Again the agreement between the molecular simulation and the integral equation theory result using an approximated $g_{ij}(r)$ is remarkable. In trying to apply the restricted primitive model to other electrolytes that exhibit a wider range of $\gamma_\pm$ versus I behavior, even with r_o taken as fully adjustable, larger errors appear.

Pitzer recognized that additional refinements were needed before one could produce a viable working equation for mixtures of electrolytes at higher concentrations. Using superposition of long-range coulombic effects and short-range non-ionic interactions in the conceptual framework proposed earlier by Guggenheim and others, Pitzer (1973) produced his ion interaction equations for electrolytes. His formulation extends the Debye-Hückel equation in a manner that is consistent with the McMillan-Mayer theoretical osmotic virial expansion. Non-idealities are given in a thermodynamically consistent fashion in terms of the excess Gibbs free energy with an added term to account for long-range electrostatic interactions and retaining only the second and third virial terms in the expansion:

$$\frac{\Delta \underline{G}^{EX}}{RT} = m_w f(I) + \frac{1}{m_w} \sum_i \sum_j \lambda_{ij}(I)\, n_i n_j + \frac{1}{m_w^2} \sum_i \sum_j \sum_k \Lambda_{ijk}\, n_i n_j n_k \qquad (12\text{-}76)$$

where

$f(I)$ = Debye-Hückel term for long-range electrostatic ion-ion effects, depends on temperature and solvent properties, ρ_s and D_s

m_w = mass of water (kg)

$\lambda_{ij}(I)$ = second osmotic virial coefficient that depends on ionic strength I

Λ_{ijk} = third osmotic virial coefficient, assumed constant

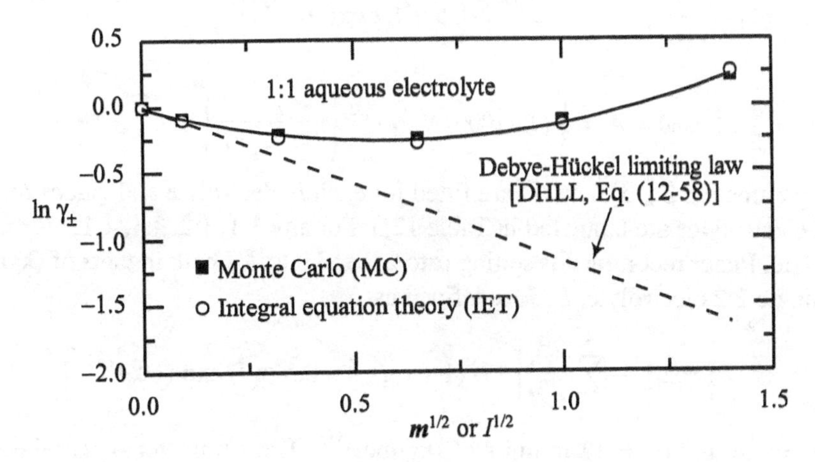

Figure 12.6

Mean ionic activity coefficient predictions from Andersen and Chandler (1971) [see also McQuarrie (1976), p 349 Table 15.4]

m	$(\ln \gamma_\pm)_{IET}$	$(\ln \gamma_\pm)_{MC}$	$(\ln \gamma_\pm)_{DHLL}$
0.00911	− 0.0957	− 0.0973	− 0.1121
0.10376	− 0.2273	− 0.2311	− 0.3783
0.42502	− 0.2587	− 0.2643	− 0.7656
1.0001	− 0.1263	− 0.1265	− 1.1744
1.9676	+0.2587	+0.2540	− 1.6472

For a single electrolyte, Eq. (12-76) is simplified considerably:

$$\frac{\Delta \underline{G}^{EX}}{RT} = m_w f(I) + \frac{2}{m_w} n_i n_j \left(B_{ij} + \frac{1}{m_w} n_i z_i C_{ij}\right) \tag{12-77}$$

$RT\ln \gamma_i = \overline{\Delta G_i^{EX}}$ is obtained by appropriate differentiation of Eq. (12-77) to give $\ln\gamma_i$ and with the usual definitions, $\gamma_\pm$ for the ***Pitzer ion interaction model*** is given by:

$$\boxed{\ln \gamma_\pm = |z_+ z_-| f^\gamma + m\left(\frac{2\,\nu_+\nu_-}{\nu}\right) B_\pm^\gamma + m^2\left(\frac{2(\nu_+\nu_-)^{3/2}}{\nu}\right) C_\pm^\gamma} \tag{12-78}$$

where

$$f^\gamma = \text{Debye-Hückel term} = -3\,A_\phi\left[\frac{I^{1/2}}{1+bI^{1/2}} + \frac{2}{b}\ln\left(1+bI^{1/2}\right)\right]$$

$$B_\pm^\gamma = 2\beta_o + \frac{2\beta_1}{\alpha^2 I}\left[1 - \left(1 + \alpha\, I^{1/2} - 0.5\,\alpha^2 I\right)\exp\left(-\alpha I^{1/2}\right)\right]$$

$$C_\pm^\gamma = \frac{3}{2}\,C_\pm^\phi \quad \text{and} \quad A_\phi = \frac{1}{3}\left(2\pi(1000)\,N_A\,\rho_s\right)^{1/2}\left(\frac{e^2}{4\pi\varepsilon_o D_s kT}\right)^{3/2}$$

The parameters β_o, β_1, and $C_\pm^\phi$ are fitted for each ij electrolyte and values for several typical electrolytes are tabulated in Table 12.3. For any 1:1, 1:2, 3:1, 4:1, . . . , i:1, or 1:j electrolyte, Pitzer recommends setting α to 2.0 and b to 1.2 both in units of $(\text{kg/mol})^{1/2}$. For a single 2:2 electrolyte, $B_\pm^\gamma$ is redefined as

$$B_\pm^\gamma = 2\beta_o + \sum_{i=1}^{2}\frac{2\beta_i}{\alpha_i^2 I}\left[1 - \left(1 + \alpha_i I^{1/2} - 0.5\,\alpha_i^2 I\right)\exp\left(-\alpha_i I^{1/2}\right)\right] \tag{12-79}$$

with $\alpha_1 = 1.4$ and $\alpha_2 = 12$ in units of $(\text{kg/mol})^{1/2}$. The parameter A_ϕ is related to the Debye-Hückel limiting law A parameter [see Eq. (12-58)]. The temperature dependence of A_ϕ for water is given in Table 12.4.

The extension of Pitzer's models to mixtures of electrolytes is straightforward but, nonetheless, mathematically tedious. In addition to the f^γ, $B_\pm^\gamma$, and $C_\pm^\gamma$ terms that appear in Eq. (12-78), other interaction terms Φ_{ij}^* and Ψ_{ijk}^* are retained from the truncated third-order virial expansion expressed in Eq. (12-76) [see Pitzer (1973) and (1995) for more details]. The physical feature they represent is analogous to the k_{ij} binary interaction parameter that was introduced in the mixture form of various $P\underline{V}TN$ EOS models given in Section 9.2. However, now the interactions are between pairs of common cations or anions, e.g., HCl-LiCl or NaCl-$NaNO_3$, or just between ion pairs without common ions, e.g., NaCl-KNO_3. In general, summing over all positive cation and anion interactions:

Table 12.3 Pitzer Ion Interaction Model Parameters for 25°C [from Pitzer (1995)]

Electrolyte	β_o	β_1	β_2	$C_{\pm}^{\phi}$
HCl	0.1775	0.2945	–	0.00080
LiCl	0.1494	0.3074	–	0.00359
NaCl	0.0765	0.2664	–	0.00127
KCl	0.0484	0.2122	–	– 0.00084
NH_4NO_3	– 0.0154	0.1120	–	– 0.00003
Na_2SO_4	0.01869	1.0994	–	0.005549
K_2SO_4	0.4995	0.7793	–	–
$FeCl_2$	0.3359	1.5323	–	– 0.00861
$NiCl_2$	0.3499	1.5300	–	– 0.00471
$CaCl_2$	0.3159	1.6140	–	– 0.00034
$MgSO_4$	0.2210	3.343	– 37.23	0.0250
$CuSO_4$	0.2340	2.527	– 48.33	0.0044
$NiSO_4$	0.1702	2.907	– 40.06	0.0366
$AlCl_3$	0.6993	5.8447	–	0.00273

Note: parameters only apply up to a maximum molality of 4-6 mol/kg

Table 12.4 Temperature Dependence of the Debye-Hückel A_ϕ Parameter for Aqueous Electrolytes [from Pitzer (1995)] (P = 1 bar for $T < 100°C$, $P=P_{vp}$ for $T > 100°C$)

T °C	A_ϕ $(kg/mol)^{1/2}$	T °C	A_ϕ $(kg/mol)^{1/2}$
0	0.3767	150	0.5295
25	0.3915	200	0.6228
50	0.4103	250	0.7535
100	0.4606	300	0.960
		350	1.44

$$\frac{\Delta \underline{G}^{EX}}{RT} = m_w f(I) + \frac{1}{m_w} \sum_c \sum_a m_c m_a \left[B_{ca} + \sum_c m_c z_c C_{ca} \right] + \sum_c \sum_{c'} m_c m_{c'} \left[2\,\Phi_{cc'} + \sum_a m_a \Psi^*_{cc'a} \right] + \sum_a \sum_{a'} m_a m_{a'} \left[2\,\Phi^*_{aa'} + \sum_c m_c \Psi^*_{aa'c} \right] \tag{12-80}$$

where B_{ca}, $\Phi^*_{cc'}$, $\Phi^*_{aa'}$, $\Psi^*_{cc'a}$, $\Psi^*_{aa'c}$ can be expressed in terms of the second and third osmotic virial coefficients, λ_{ij} and Λ_{ijk}. Ultimately, these coefficients are expressed as functions of I with constants fitted to data. The $\Phi^*_{cc'}$ and $\Phi^*_{aa'}$ parameters represent true binary interactions while the mixture parameters $\Psi^*_{cc'a}$ and $\Psi^*_{aa'c}$ represent ternary interactions between triplets of cations and anions. Frequently these ternary terms are fitted directly to data or simply omitted from the correlation. For many practical applications involving mixtures, individual ion rather than mean activity coefficients are preferred. These can be obtained by differentiating $\Delta \underline{G}^{EX}$ in Eq. (12-80), for example,

$$RT \ln \gamma_c = \left(\frac{\partial \Delta \underline{G}^{EX}}{\partial n_c} \right)_{T,\,P,\,n_i[c]}$$

If required, individual γ_i's can be appropriately combined to express $\gamma_\pm$ using Eq.(12-11).

Like many other activity coefficient models of non-ideal behavior of importance to chemical engineering operations, such as the Wilson, NRTL, and UNIQUAC models, the Pitzer ion interaction equations require pure component and binary parameters. Later in the chapter we will illustrate how well the Pitzer model does in concentrated electrolyte solutions. But before leaving this discussion of Pitzer's work we need to mention his extension of the Debye-Hückel theory for the long-range ion-ion interaction contribution. His recommended formulation is in terms a mole-fraction-based ionic strength I_x model for individual ion activity coefficients where

$$I_x = \frac{1}{2} \sum_i^{all\ ions} x_i z_i^2$$

The resulting equation for γ_j of a specific ionic component j of charge z_j is given for both infinite dilution and pure fused salt reference states in terms of a single adjustable parameter, ρ^*:

Infinite dilution

$$(\ln \gamma_j^{**})^{PDH} = -z_j^2 A_x \left[\frac{2}{\rho^*} \ln (1 + \rho^* I_x^{1/2}) + \frac{I_x^{1/2}(1 - 2\,I_x / z_j^2)}{(1 + \rho^* I_x^{1/2})} \right] \tag{12-81}$$

Pure fused salt

$$(\ln \gamma_j)^{PDH} = -z_j^2 A_x \left[\frac{2}{\rho^*} \ln \frac{(1+\rho^* I_x^{1/2})}{(1+\rho^* (I_x^o)^{1/2})} + \frac{I_x^{1/2}(1-2I_x/z_j^2)}{(1+\rho^* I_x^{1/2})} \right] \quad (12\text{-}82)$$

with

$$A_x = \frac{1}{3}\left(\frac{2\pi \times 10^6 N_A \rho_s}{m_s} \right)^{1/2} \left(\frac{e^2}{4\pi\varepsilon_o D_s kT} \right)^{3/2} \quad (12\text{-}83)$$

$$\rho^* = r_o \left(\frac{2 \times 10^6 e^2 N_A \rho_s}{m_s \varepsilon_o D_s kT} \right)^{1/2} = \text{closest approach parameter} \quad (12\text{-}84)$$

where r_o = hard core radius; I_x^o = ionic strength of pure fused salt (see definition of I_x on p 534); m_s = solvent molecular weight, g/mol; ρ_s = solvent density, kg/L.

Pitzer's recommendation for multicomponent systems where ρ^* varies with mole fraction, is to either take a fixed average value for ρ^* or to treat it as a fitted parameter rather than use the defining Eq. (12-84) above.

There have been several refinements of the Pitzer model that have utilized an integral equation approach for modeling electrolytes. Of these, many have used the so-called *mean spherical approximation* (MSA) in an Ornstein-Zernike integral equation format. The model by Planche and Renon (1981) is a good example of such an approach. The major difference with their model is that they attempt to capture short-range effects using a hard-sphere term with a binary contact interaction energy W_{ij} that is temperature dependent. This introduces three binary parameters, one of which is the hard core diameter σ_{ij} that is easily calculated as an average of pure component values, while the other two require fitting to actual data. The main advantages of the Planche and Renon model are that (1) it can be used for both mixtures of electrolytes and non-electrolytes, (2) only binary and pure component parameters are needed, (3) changes in dielectric constant as a function of composition are accounted for and (4) the model has been tested on over 30 systems with considerable success [see also Ball *et al.* (1985)].

12.6 Meissner Corresponding States Model

Just as we were able to generalize $P\underline{V}TN$ properties of gases and liquids using the Law of Corresponding States to express the compressibility Z as a function of reduced temperature and pressure (T_r, P_r), Meissner recognized a possible analogous route to generalizing the non-ideal $\gamma_\pm$ behavior of strong electrolytes. The experimentally observed variation of $\gamma_\pm$ with concentration or ionic strength is large as illustrated in Figure 12.4. Some electrolytes like HCl and LiCl show strong positive deviations ($\gamma_\pm > 1$) while others like $CuSO_4$ show negative deviations ($\gamma_\pm < 1$). Of course, all electrolytes must match the infinite dilute condition, $\gamma_\pm \to 1$ as I or $\boldsymbol{m} \to 0$, by approaching it along the Debye-Hückel line at low concentrations [see Eq. (12-58)].

This asymptotic behavior was discussed in Example 12.1, where it was immediately apparent that the reduced activity coefficient Γ_{ij}^{o} provides a limiting result that was independent of electrolyte type. For aqueous systems at 25°C, in the Debye-Hückel limit:

$$\log_{10} \Gamma_{ij}^{o} = -\,0.510\, I^{1/2} \tag{12-85}$$

with

$$\Gamma_{ij}^{o} \equiv (\gamma_{\pm})^{1/|z_+ z_-|}$$

The superscript (o) has been added to emphasize the fact that Γ_{ij}^{o} refers to a single ij electrolyte. Meissner and Tester (1972) explored the idea of using Γ_{ij}^{o} outside the Debye-Hückel region to correlate pure single electrolyte $\gamma_{\pm}$ data for over 120 different electrolytes.

With few exceptions, the plots of $\log_{10} \Gamma_{ij}^{o}$ versus I shown in Figure 12.7 formed a family of non-intersecting curves quite analogously to the form of a compressibility chart ($Z = f\,(T_r\,, P_r)$). Thus the chaotic situation of intersecting curves shown in Figure 12.4 can be converted to an ordered family of curves shown in Figure 12.7 by just scaling $\gamma_{\pm}$ by its charge factor $|z_+ z_-|$. This approach is obviously consistent with the Debye-Hückel limiting law, but what was surprising was the extension of orderly behavior to significantly higher concentrations even up to ionic strengths of 20 or more! Later refinements by Kusik and Meissner (1973, 1978) and Meissner (1980) produced the generalized correlation of $\Gamma_{ij}^{o} = f\,(I)$ shown on log-log coordinates in Figure 12.8.

This generalized correlation has many advantages. With only one data point for $\gamma_{\pm}$ at a particular I, one determines the proper Γ_{ij}^{o} curve for that electrolyte for extrapolation (or interpolation) to any other value of I (or $\boldsymbol{m}$), including those compositions that even exceed the saturation value (solubility) of that particular salt. This feature is particularly valuable for multicomponent systems. Each line plotted in Figure 12.8 can be defined by a single empirical parameter, q_{ij}^{o}, which is used in the following set of equations to correlate Γ_{ij}^{o} versus I for aqueous electrolytes at 25°C. These equations define the ***Meissner model***:

$$\begin{aligned}
&\Gamma_{ij}^{o} = [1 + B(1 + 0.1\, I)^{q_{ij}^{o}} - B]\, \Gamma_{ij}^{DH} \\
&\log_{10} \Gamma_{ij}^{DH} = \frac{-\,0.5107\, I^{1/2}}{1 + C I^{1/2}} \\
&B = 0.75 - 0.065\, q_{ij}^{o} \\
&C = 1 + 0.055\, q_{ij}^{o} \exp\,(-0.023\, I^{3})
\end{aligned} \tag{12-86}$$

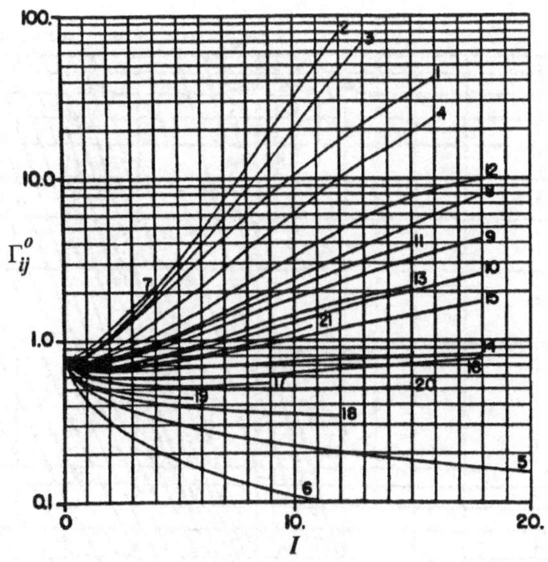

Figure 12.7 Reduced activity coefficient plotted as a function of ionic strength for the same set of electrolytes given in Figure 12.4 [reprinted with permission from Meissner and Tester (1972), American Chemical Society].

Example 12.3

Estimate $\gamma_{\pm}$ of NaCl using the Meissner model at 25°C from a dilute solution to a supersaturated solution at m = 20 mol NaCl/kg H_2O.

Solution

The q_{ij}^{o} value from Table 12.5 is 2.23. Eq. (12-86) can be used directly with $m = I$ for this 1:1 salt. Results are plotted in Figure 12.10 and are compared with experimental data up to saturation at about 6 molal. A sample calculation for $m = 6$ follows

$$B = 0.75 - 0.065(2.23) = 0.605$$

$$C = 1 + 0.055(2.23)\exp(-0.023\,(6)^3) = 1.0009$$

$$\log_{10}\Gamma_{ij}^{DH} = \frac{-0.5107\,(6)^{1/2}}{1 + 1.0009(6)^{1/2}} = -0.3624 \quad or \quad \Gamma_{ij}^{DH} = 0.4341$$

$$\Gamma_{ij}^{o} = \gamma_{\pm} = (1 + 0.605(1 + .1(6))^{2.23} - 0.605)\,0.4341$$

$$\gamma_{\pm} = 0.9206 \quad \text{versus} \quad 0.986 \text{ experimentally}$$

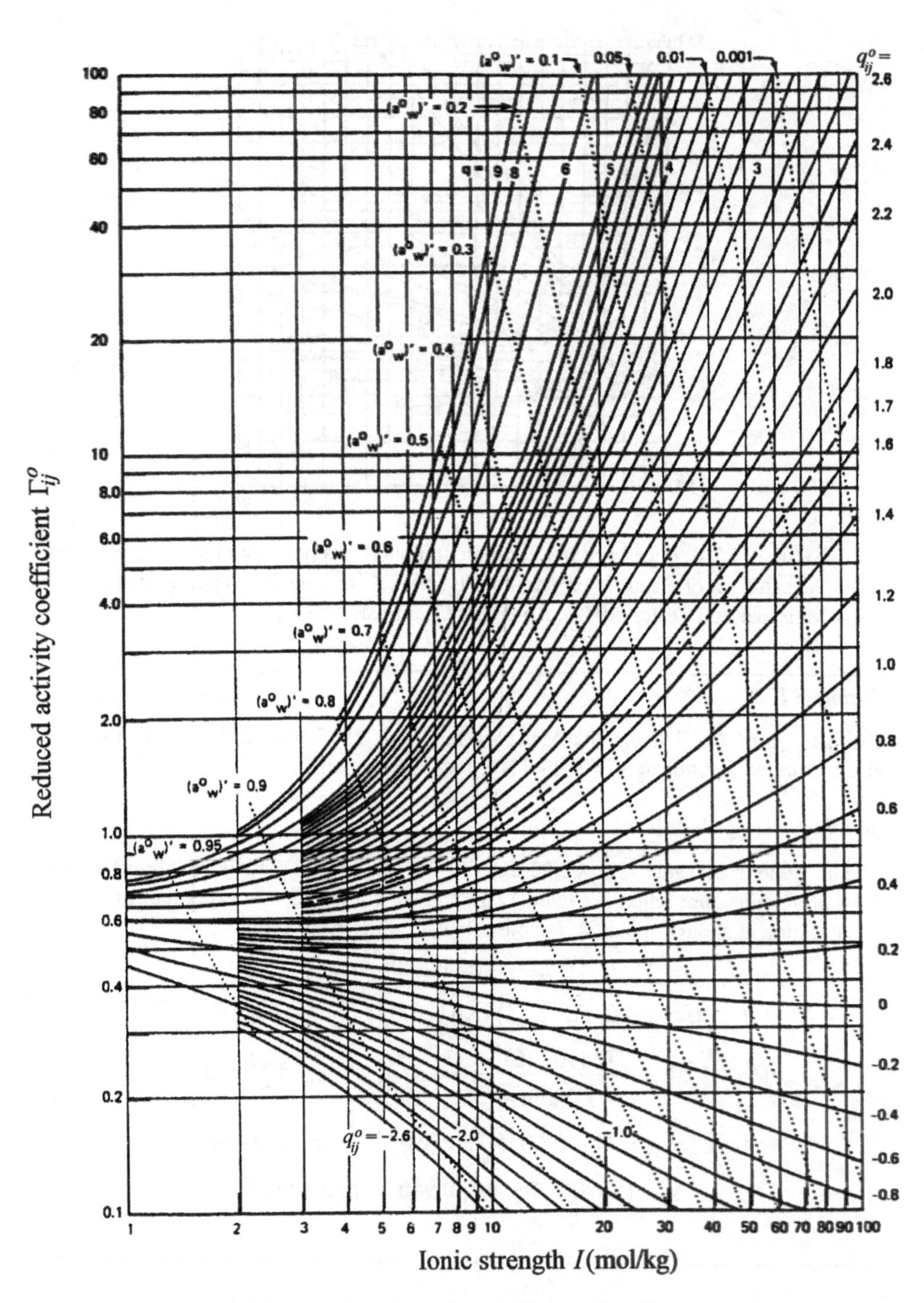

Figure 12.8 Meissner reduced activity coefficient chart for aqueous strong electrolytes [reprinted with permission from Meissner (1980), American Chemical Society].

Also shown on Figure 12.8 are dotted lines of constant water activity $(a_w^o)'$ assuming a singly charged basis for the ij electrolyte in solution (that is a 1:1 electrolyte is the basis). To convert any other electrolyte of type $|z_+:z_-|$ of the same q_{ij}^o value and ionic strength I, the following equation is used

$$\log_{10}(a_w^o) = 0.0156\, I\left(1 - \frac{1}{|z_+z_-|}\right) + \log_{10}(a_w^o)' \qquad (12\text{-}87)$$

Table 12.5 Meissner q_{ij}^o Parameters for Selected Aqueous Electrolytes at 25°C [abridged version of Table 1 from Kusik and Meissner (1978)]

Electrolyte	q_{ij}^o
HCl	6.69
LiCl	5.62
NaCl	2.23
KCl	0.92
NH_4NO_3	- 1.15
Na_2SO_4	- 0.19
K_2SO_4	- 0.25
$FeCl_2$	2.16
$NiCl_2$	2.33
$CaCl_2$	2.40
$MgSO_4$	0.15
$CuSO_4$	0.00
$NiSO_4$	0.025
$AlCl_3$	1.92
$Al_2(SO_4)_3$	0.36

Temperature variations for q_{ij}^o can also be treated:

$$q_{ij}^o(T) = q_{ij}^o(T_{ref})\left(1 - \frac{T - T_{ref}}{|z_+z_-|}\right) \qquad (12\text{-}88)$$

where T_{ref} is the reference temperature where $q_{ij}^o(T_{ref})$ is known.

Although empirically generated, Eq. (12-86) rigorously follows the Debye-Hückel relationship $\log_{10}\Gamma_{ij}^{DH}$ at small I. The fact that the scaling of $\gamma_\pm$ with $|z_+z_-|$ works so well at all finite ionic strengths does not follow trends projected by the semi-empirical models of Guggenheim, Bromley, Pitzer, and others where the $|z_+z_-|$ factor only appears in the long-range coulombic term. A key limitation of the Meissner model is that q_{ij}^o must be known. Kusik and Meissner (1978) and Zemaitis *et al.* (1986) tabulate q_{ij}^o values for over 120 electrolytes, while we list a few representative values in Table 12.5. Of course, one must keep in mind that a minimum of one experimental data point is needed to establish a q_{ij}^o value.

The extension of the Meissner model to multicomponent electrolyte mixtures involves a few key assumptions. Most important is the mixing rule used to generate Γ_{ij}^{mix} for an ij ion pair in a mixture containing like and unlike cations and anions. We selected the option suggested by Kusik and Meissner (1978), which expresses Γ_{ij}^{mix} as a charge and mole fraction weighted sum of single (pure) electrolyte Γ_{ij}^o values for all possible cation (i) and anion (j) pairs in the mixture. Following the nomenclature introduced in Section 12.1:

$$\log_{10}\Gamma_{ij}^{mix} = \frac{z_i}{z_i + z_j}\sum_{j'}^{even}\frac{(z_i + z_{j'})^2}{2z_i z_{j'}}X_{j'}^*\log_{10}\Gamma_{ij'}^o + \frac{z_j}{z_i + z_j}\sum_{i'}^{odd}\frac{(z_{i'} + z_j)^2}{2z_{i'}z_j}X_{i'}\log_{10}\Gamma_{i'j}^o \tag{12-89}$$

where $X_{j'}^* \equiv I_{j'}/I = m_{j'}z_{j'}^2/\sum_{total} m_i z_i^2$ and a similar set of terms for $X_{i'}^*$. Note that both z_i and z_j are taken as positive numbers.

In addition, the mixture q_{ij} values (q_{ij}^{mix}) are given by a sum over all possible single pure salt q_{ij}^o values weighted by their appropriate ionic strengths. Thus,

$$q_{ij}^{mix} = \sum_{j'}^{even} X_{j'}^* q_{ij'}^o + \sum_{i'}^{odd} X_{i'}^* q_{i'j}^o \tag{12-90}$$

Then q_{ij}^{mix} can be used with Eq. (12-86) or Figure 12.8 to calculate Γ_{ij}^{mix} directly. However, this approach and the direct use of Eq. (12-89) only give the same results for the specific case of a mixture consisting of similarly charged electrolyte types, e.g., a mixture of LiCl, HCl, and NaCl or $CaCl_2$ and $Ni(NO_3)_2$.

The water activity for the mixed electrolyte system (a_w^{mix}) can be estimated by direct integration of the Gibbs-Duhem relationship. This yields the exact result:

$$\log_{10}(a_w^{mix}) = \frac{m_w}{500}\sum_i\sum_j X_{ij}^*\left[\frac{I}{z_i z_j \ln 10} + \int_0^I I' d[\log_{10}\Gamma_{ij}^{mix}(I')]\right] \tag{12-91}$$

where m_w is the molecular weight of water (g/mol) and $X_{ij}^* = I_{ij}/I$ equals the ionic strength fraction of the component ion-pair ij. Various other closed-form approximations to Eq. (12-91) have been proposed by Kusik and Meissner (1978) but as pointed out by Resch (1995) several inconsistencies and limitations exist with their approximations. Thus direct integration is recommended whenever possible.

Meissner's model is a beautifully crafted example of the power of an empirical method for extending limited $\gamma_\pm$ data to extremely concentrated regimes at ionic strengths of 20 or more. It does suffer slightly from the fact that the working equations are written explicitly for $\gamma_{\pm}$, γ_{ij}, or Γ^{o}_{ij} rather than ΔG^{EX}.

12.7 Chen Local Composition Model

Chen and coworkers (1982) proposed a model for electrolyte solutions that combines Pitzer's extended Debye-Hückel equation for long-range ion-ion interactions with the NRTL model of Renon and Prausnitz (1968) for short-range interactions in a local composition framework. When applied to mixed solvent systems, the Born equation is used to convert reference conditions from infinite dilution in the mixed solvent system to infinite dilution in an aqueous solution.

***Assumptions*:** There are two fundamental assumptions in the Chen model

(1) like-ion repulsion Like ion repulsion effects are sufficiently large on a molecular level that the "local composition" around a specific ion can only consist of a neutral species (e.g., solvent) or an oppositely charge ion. Around a central neutral solvent both cations and anions can coexist in the local composition region. This situation is illustrated in Figure 12.9, and, as we will see, it places a mathematical restraint on the local mole fraction (x_{ij}).

(2) local electroneutrality The distribution of (+) cations and (–) anions around a neutral species is such that the net "local" ionic charge is zero.

The superposition of long and short range effects and the reference state transformation in ΔG^{EX} is given by:

$$\Delta G^{EX} = \Delta G^{EX,PDH} + \Delta G^{EX,Born} + \Delta G^{EX,LC} \tag{12-92}$$

The superscripts *PDH*, *Born* and *LC* refer to the Pitzer-Debye-Hückel, Born approximation correction, and local composition formulations, respectively. By differentiation of Eq.

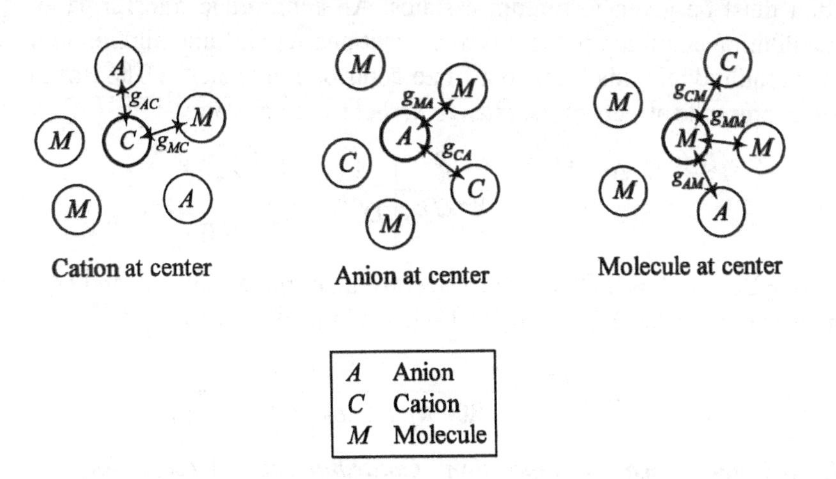

Figure 12.9 Three types of cells in the Chen model according to the like-ion repulsion and local electroneutrality assumptions [from Chen *et al.* (1982)].

(12-92) at constant T, P, $n_j[i]$, activity coefficients are generated for individual species i which can be an ion or neutral molecule. Thus, the ***Chen model*** is constituted as:

$$\boxed{\ln \gamma_i^{**} = \ln \gamma_i^{**,PDH} + \ln \gamma_i^{**,Born} + \ln \gamma_i^{**,LC}} \tag{12-93}$$

where the double asterisk (**) is inserted to emphasize that unsymmetric normalization is adopted with a *mole fraction* basis for both solvent and solute:

Component	*Limiting reference condition*	*(Standard state)*
for solvent (s)	$\gamma_s \to 1.0$ as $x_s \to 1.0$	(pure solvent)
for any solute ion (j)	$\gamma_j \to 1.0$ as $x_j \to 0$	(hypothetical pure solute with infinite dilution behavior)

Individual ion γ_i's can be combined using an equation like Eq. (12-11) to provide a mean ionic activity coefficient $\gamma_\pm$.

***PDH Long-Range Contribution*:** Here, the extended Debye-Hückel treatment suggested by Pitzer [Eq. (12-81)] for the infinite dilution reference condition is used to account for long-range coulombic forces. One notes that there is only one adjustable parameter in the extended Pitzer-Debye-Hückel equation--ρ^*, the closest approach parameter. Compositional dependence is given in terms of the mole fraction based ionic strength I_x. Solvent properties are included in the dielectric constant D_s and the molar density ρ_s/m_s.

***Born Approximation Contribution*:** In situations where mixed solvents exist in the presence of electrolytes, for example a mixture of isopropanol, water and NaCl, special attention must be given to reference states. An appropriate transformation from the infinite dilution condition in the mixed solvent phase to infinite dilution in the aqueous phase is required and this leads to a finite contribution to ΔG^{EX}. The Born equation is used to capture the effects of the change in dielectric constant:

$$\frac{\Delta G^{EX,Born}}{RT} = \frac{e^2}{8\pi kT\varepsilon_o}\left[\frac{1}{D_s^{mix}} - \frac{1}{D_s^w}\right]\frac{\sum_i x_i z_i^2}{100\, r_B} \tag{12-94}$$

where r_B = Born radius, D_s^{mix} = dielectric constant of mixed solvent, and D_s^w = dielectric constant of pure water. Equation (12-94) leads immediately to

$$\ln \gamma_i^{**,Born} = \frac{e^2 z_i^2}{800\,\pi\, kT\, \varepsilon_o\, r_B}\left[\frac{1}{D_s^{mix}} - \frac{1}{D_s^w}\right] \tag{12-95}$$

***Short-Range Local Composition Contribution*:** Except for a few small modifications to accommodate the ionic characteristics of electrolytes, the theoretical development and nomenclature introduced in Section 11.7.3 for the NRTL model were used by Chen *et al.* (1982) to formulate the short-range contribution to their model. The effective *local mole fractions* are defined in terms of *modified bulk mole fractions* and a Boltzmann distribution for energetic interactions:

for bulk mole fractions:

$$X_j = x_j z_j \text{ for ions } (j = c, a) \quad \text{and} \quad X_j = x_j \text{ for neutral molecules } (j = s)$$

for local mole fractions:

$$\frac{X_{ji}}{X_{ii}} = \frac{X_j}{X_i} G_{ji} = \frac{X_j}{X_i} \exp(-\alpha_{ji}\,\tau_{ji})$$

with both single (i–j) and paired species (i–j + i–i) interactions:

$\alpha_{ji} = \alpha_{ij}$ (non–randomness factor frequently set to 0.2)

$g_{ji} = g_{ij}$ (interaction energy used in $\tau_{ji} = (g_{ji} - g_{ii})/RT$)

With electroneutrality and other assumed symmetries:

$$\tau_{cs} = \tau_{as} \equiv \tau_{ca,\,s} \quad \text{and} \quad \tau_{sc,\,ac} = \tau_{sa,\,ca} \equiv \tau_{s,\,ca}$$

With the model's assumptions, the constraining equations for X_{ij} are:

$$X_{cs} + X_{as} + X_{ss} = 1 \quad \text{(central solvent molecule)}$$

$$X_{sc} + X_{ac} = 1 \quad \text{(central cation)} \quad \text{and} \quad X_{sa} + X_{ca} = 1 \quad \text{(central anion)} \tag{12-96}$$

$\Delta G^{EX,LC}$ is usually structured with symmetric normalization so in order to make it compatible with the PDH formulation, an infinite dilution reference condition is needed:

$$\Delta G^{EX,LC} = \Delta G^{EX,LC}_{symmetric} - x_c \ln \gamma_c^{\infty} - x_a \ln \gamma_a^{\infty} \tag{12-97}$$

where γ_c^{∞} and γ_a^{∞} are infinite dilution values of γ. Using Eq. (12-97), the unsymmetrically normalized γ_j^{**} contribution can be determined. For the cation (c), this is given by:

$$\frac{1}{z_c} \ln \gamma_c^{**,LC} = \frac{X_s^2\,\tau_{cs}\,G_{cs}}{(X_c G_{cs} + X_a G_{as} + X_s)^2} - \frac{X_a\,\tau_{sa}\,X_s\,G_{sa}}{(X_c + X_s\,G_{sa})^2} + \frac{X_s\,\tau_{sc}\,G_{sc}}{(X_a + X_s\,G_{sc})} - \tau_{sc} - G_{cs}\,\tau_{cs} \tag{12-98}$$

and a similar expression for the anion (a). The solvent (s) expression takes on a slightly different form [see Chen *et al.* (1982) and Chen and Evans (1986)].

The Chen model has been extended to mixtures of electrolytes and to mixed solvent systems [Chen and Evans (1986) and Mock *et al* (1986)]. The latest version of the model is available as an option in the ASPEN PLUS simulator and is described in their Release 9 *Physical Property Manual* [Aspen Technology (1994)].

An important feature of the Chen model is that non-idealities are fully represented with relatively few adjustable parameters, including one parameter ρ^* for long-range electrostatic effects. Frequently, ρ^* is set to a constant value, e.g., for aqueous systems $\rho^* \approx 14.9$, the Born radius r_B, and a maximum of three binary parameters α_{ij}, τ_{ij}, and

τ_{ji} result. Also, α_{ij} is commonly set to 0.2 for the local composition part of the model (see Table 12.6). The performance of the Chen model is characterized in the next section.

Table 12.6 Binary Parameters for the Chen Local Composition Model at 25°C Note that ρ^* was set to 14.9 and α_{ij}=0.2 [see Chen *et al.* (1982) for a complete tabulation].

Electrolyte Type	Compound	$\tau_{s,ca}$	$\tau_{ca,s}$
	HCl	10.089	-5.212
	LiCl	10.031	-5.154
1:1	NaCl	8.885	-4.549
	KCl	8.064	-4.107
	NH_4NO_3	7.359	-3.526
	Na_2SO_4	8.389	-4.539
1:2	K_2SO_4	9.247	-4.964
2:1	$FeCl_2$	9.118	-5.377
	$NiCl_2$	10.751	-6.013
	$CaCl_2$	11.396	-6.218
2:2	$MgSO_4$	11.346	-6.862
	$NiSO_4$	11.378	-6.837
3:1	$AlCl_3$	10.399	-6.255
3:2	$Al_2(SO_4)_3$	10.646	-7.116

12.8 Performance of Electrolyte Models in Engineering Practice

In Sections 12.3 through 12.7 we have discussed various approaches to modeling non-ideal effects in strong electrolyte solutions. Now we can evaluate just how well these models do in predicting the properties of electrolyte systems of interest to us. To simplify matters, we restrict our comparison to the activity coefficients of single strong electrolytes in aqueous solutions.

Keep in mind that, with the exception of the Debye-Hückel limiting law and several modest extensions of it, adjustable fitted parameters are required to produce a workable electrolyte model. In general, the greater the number of parameters the better the prediction. One can find many pure component and binary parameters for specific electrolytes and solvents tabulated in handbooks such as the one authored by Zemaitis *et al.* (1986) or they be found in a computerized database such as DIPPR. In some cases, actual data are used to regress suitable parameter values before the model is evaluated for that electrolyte.

Experimental activity coefficient data and model predictions are compared in Figure 12.10 for several strong aqueous electrolytes at 25°C, 1:1 type—NaCl and LiCl, 1:2 type—Na_2SO_4 and 2:1 type—$CaCl_2$. One immediately sees that the Debye-Hückel and Pitzer-Debye-Hückel (with $\rho^*= 14.9$) model predictions show large deviations at molalities above 0.05. In contrast, the Pitzer ion interaction model, the Meissner corresponding states model, and the Chen local composition model all do quite well even up to 6 molal. For different electrolytes or these electrolytes at different conditions such as higher temperatures or in the presence of other solvents, agreement between predictions and experiment is in general much poorer. For example, for NaCl at 300°C at a confining pressure slightly above its saturation vapor pressure, the predicted results for $\gamma_\pm$ from the Pitzer, Chen, and Meissner models for shown in Figure 12.11 are considerably poorer in general than the fit at 25°C in Figure 12.10. The Pitzer and Chen models are superior to the Meissner model partly because the temperature dependence of solvent and solute properties is somewhat oversimplified in the Meissner approach while the other two models have more temperature dependent parameters and a larger explicit temperature dependence in their mathematical form.

Example 12.4

In electrolyte mixture applications one of the common uses for $\gamma_\pm$ models is in the prediction of salt solubility in the presence of a common ion. For example, the solubility of NaCl in general decreases as HCl or KCl is added due to the presence of the common Cl^- ion. To solve this problem correctly $\gamma_{\pm,NaCl}$ must be known as a function of temperature and composition. Compare predicted solubilities for NaCl in HCl with the experimental data listed below for saturated aqueous solutions at 25°C reported by Linke and Seidell (1965).

ρ(solution) g/cm^3	C_{HCl} mol/L	C_{NaCl} mol/L
1.198	0	5.43
1.187	0.50	4.88
1.178	0.89	4.48
1.151	2.27	3.15
1.132	3.49	2.08
1.120	4.50	1.33
1.116	5.25	0.907
1.176	6.10	0.544
1.213	7.07	0.293
1.130	7.98	0.158
1.146	9.24	0.091
1.197	13.41	0.017

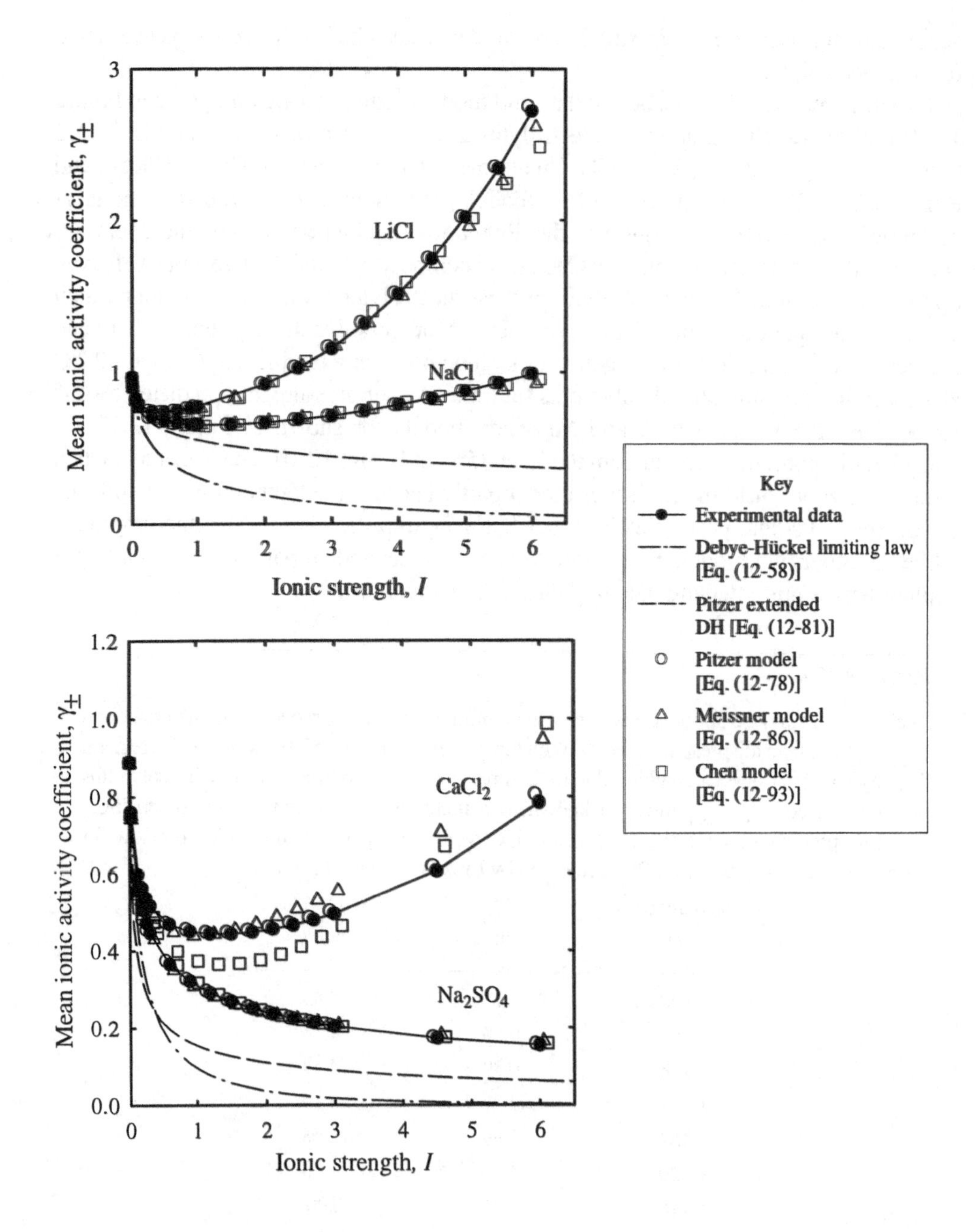

Figure 12.10 Experimental and Predicted mean ionic activity coefficients for NaCl and LiCl (a) and $CaCl_2$ and Na_2SO_4 (b) in water at 25°C. Model parameters are found in Tables 12.3 through 12.6. Regressed data are from Robinson and Stokes (1970).

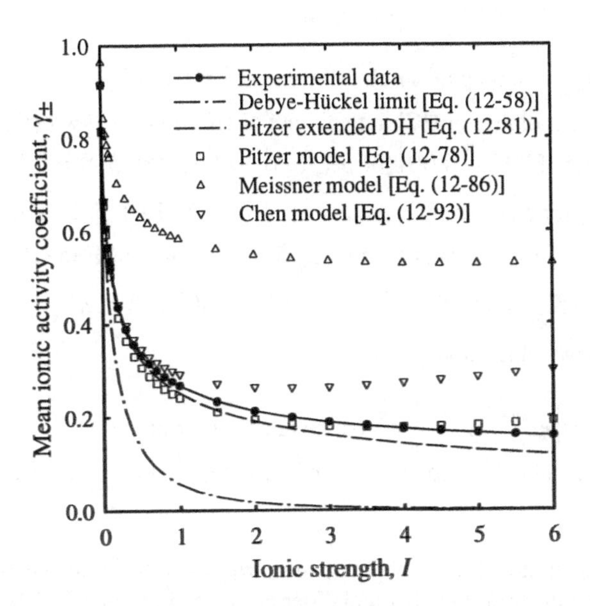

Figure 12.11 Experimental and predicted mean ionic activity coefficient for NaCl in water at 300°C. Model parameters are found in Tables 12.3 through 12.6. Data from Zemaitis *et al.* (1986).

Solution

A sample calculation for one concentration of HCl, $C_{HCl} = 4.5$ mol/L, will illustrate the approach. We begin by examining the solubility of pure NaCl in water to produce an estimate of the solubility product, K_{sp}, which represents the solubility equilibrium.

$$\text{for } C_{HCl} = 0 \quad C^{sat}_{NaCl} = 5.43 \text{ mol/L} \quad or \quad m^{sat}_{NaCl} = C^{sat}_{NaCl}/(\rho_s - 0.001\, C^{sat}_{NaCl}\, \mathcal{M}_{NaCl})$$

$$K_{sp} \equiv \frac{a^{\nu_+}_{Na^+}\, a^{\nu_-}_{Cl^-}}{a_{NaCl(s)}} = \gamma^2_{\pm} m^2_{\pm} = (\gamma_{\pm} m^{sat}_{NaCl})^2$$

where $\nu_+ = \nu_- = 1$ and $a_{NaCl(s)}$ is set to 1.0. With $\mathcal{M}_{NaCl} = 58.5$ g/mol and $\rho_s = 1.198$ g/cm^3:

$$m^{sat}_{NaCl} = \frac{5.43}{0.8803} = 6.168 \text{ mol/kg } H_2O$$

Using Figure 12.10(a), $\gamma_{\pm}$ for NaCl at 6.168 molal, 25°C is approximately 0.99. Alternatively, of course, we could use one of the models to estimate $\gamma_{\pm}$ at this molality for NaCl. Now K_{sp} becomes

$$K_{sp} = (6.168)^2\,(0.99)^2 = 37.2871$$

With no hydration or solvation products present, the activity of water does not affect K_{sp}, so from equilibrium thermodynamics $K_{sp} = f(T)$ only. Thus, as HCl is added, we need to estimate how $\gamma_{\pm}$ for NaCl changes at the new ionic strength and composition of the HCl-NaCl mixture. Then we can calculate $m^{sat}_{Na^+}$ at its new saturation condition in the presence of the common Cl^- ion. An iterative solution is required.

(1) Initialize $m_{Na^+}^{sat}$ at its pure solubility value = 6.168 mol/kg.

(2) Calculate the molality of HCl in a mixture of HCl and NaCl with the data $\rho_s = 1.12$ and

$$m_{HCl} = 4.5/[1.12 - 0.001((4.5)(36.5) + (1.33)(58.5))] = 5.126 \text{ mol/kg}$$

(3) Calculate $\gamma_\pm$ for NaCl using a suitable model for the HCl-NaCl mixture.

(4) Construct solubility product accounting for the common Cl^- ion

$$K_{sp} = 37.2871 = (\gamma_{\pm,\,NaCl}^2)\,(m_{Na^+}^{sat*})(5.126 + m_{Na^+}^{sat*})$$

(5) Solve for new value for $m_{Na^+}^{sat*}$

(6) Test to see if $\left| m_{Na^+}^{sat*} - m_{Na^+}^{sat} \right| < \varepsilon$ (some preset criterion)

If yes, the calculation is complete. If no, reset $m_{Na^+}^{sat}$ to $m_{Na^+}^{sat*}$, and return to step (3)

Using the Meissner model for $\gamma_\pm$ of the mixture, $m_{Na^+}^{sat} = 1.70$ compared to an experimental value of 1.5. Results for other HCl molalities are given in Figure 12.12.

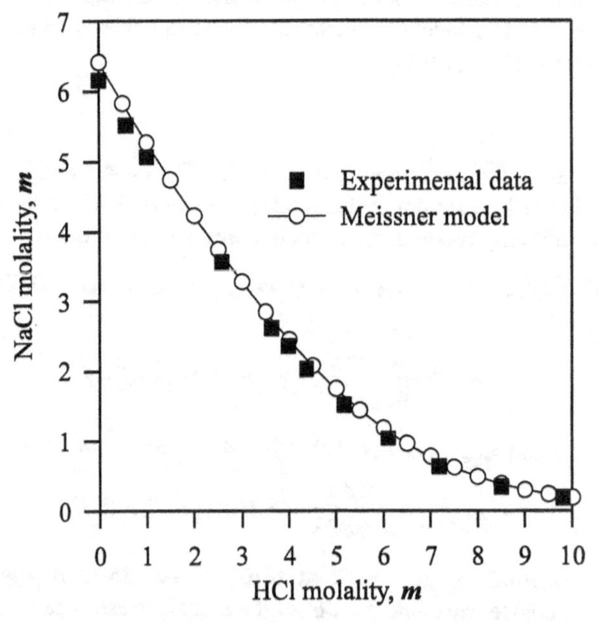

Figure 12.12 Solubility of NaCl in $HCl\text{-}H_2O$ at 25°C.

12.9 Modeling Multisolvent Mixed Electrolyte Systems

An important class of problems for chemical engineers involve vapor-liquid, solid-liquid, and liquid-liquid equilibria where salts, water, and organics are present.

For example, in distillation and extraction, the presence of salts can significantly alter the distribution of specific components between phases *via* a salting-out or salting-in mechanism. Of course, what is happening from the perspective of classical thermodynamics is that the activity coefficients are being altered by the presence of dissolved salt. In order to carry out process design or optimization, it is frequently necessary to estimate the magnitude of these effects. Naturally, this can be done experimentally, but thermodynamic models can provide both a systematic way to correlate and extend the data to assist the engineering analysis process.

Thermodynamic models must be particularly robust to capture the non-idealities that are present is such complex systems. Chen and Evans (1986), and Mock *et al.* (1986) have used the Chen local composition model for mixed solvent with considerable success to electrolyte concentrations of about 6 molal. A key element of their model is contained in the Born equation approximation term [Eq. (12-95)] which up to this point in our discussion was not required because we were dealing with electrolytes in a single solvent of defined dielectric strength. Conceptually, there may be some problems with the Born correction (for example see the discussion at the end of Section 12.1).

Zerres and Prausnitz (1994) have developed an alternative approach that avoids the Born equation and replaces the treatment of short-range effects with the NRTL local composition model approach with a simpler model. Their model superimposes three contributions, (1) chemical solvation effects using a chemical equilibrium formulation with one adjustable parameter, (2) physical non-ionic, short-range effects using an extended van Laar equation for ΔG^{EX} defined in terms of volume fractions (Φ^*) for salt-free binaries and the Pitzer truncated osmotic virial equation to account for the contribution due to ion-molecule interactions which introduces five binary parameters and one ternary parameter, and (3) coulombic long range effects due to ion-ion interactions using an extended Debye-Hückel model.

In addition, Zerres and Prausnitz give appropriate attention to proper conversion from a McMillan-Mayer to a Lewis and Randall framework for the long range contribution to $\gamma_{\pm}$. A linear mixing rule with volume fraction scaling is used to estimate the dielectric constant of the mixed solvent. While their model is too new to have been exhaustively tested, it does represent an viable alternative to the Chen model.

12.10 Summary and Recommendations

The key features of electrolyte modeling have now been discussed. We began with a discourse of the standard states and conventions used for electrolytes and how they differed from non-dissociated, non-electrolytes. Then we moved on to a detailed derivation of the Debye-Hückel model for single strong electrolytes where we saw how several approximations were used to produce a linearized form of the Poisson-Boltzmann distribution of charges that led to an estimate of the isothermal work associated with ion-ion interactions which allowed us to produce an expression for the mean ionic activity coefficient. The Debye-Hückel model results represent an important asymptotic anchor point that all theoretical and empirical models must fit in the dilute limit.

However, with the exception of the exact limiting law behavior of Debye-Hückel theory, for predictions at practical finite concentrations, one is ultimately forced to rely on semi-empirical models with parameters fitted to experimental data. Unfortunately the "truth is in the details" when it comes to sorting out what is fundamentally on firm theoretical ground and what is an exercise in curve fitting. This separation is not meant to degrade the contribution of the theoretical work on electrolytes as many advances have occurred since the seminal work of Debye and Hückel. In particular, the demonstrated agreement between predictions using the restricted primitive model and deterministic molecular simulations and the fact that the McMillan-Mayer theoretical osmotic virial approach leads to useful results with a minimal number of adjustable parameters are encouraging. Superimposing electrostatic coulombic effects onto a hard-sphere repulsive core provides a viable and realistic means to extend Debye-Hückel theory to account for both long-range coulombic and excluded volume effects. At the very least, the theoretical aspects of electrolyte behavior on a molecular level provide a framework for the semi-empirical models to build on.

Pitzer's ion interaction model captures the physics of solute-solute intermolecular effects within a McMillan-Mayer framework while maintaining the proper asymptotic behavior given by Debye-Hückel theory in the dilute limit. Meissner's model represents a creative extension of Corresponding States for electrolytes that provides an easily visualized way of extrapolating limited data to high concentrations and to mixtures. Chen's model incorporates Pitzer's extended Debye-Hückel equations for long-range coulombic effects into a local composition model structure that is based on the NRTL equation.

The models selected for detailed discussion are representative, but by no means inclusive of all approaches for predicting and correlating the behavior of electrolyte solutions. These and other engineering models do a reasonable job for single strong electrolytes in aqueous environments but they all encounter problems when mixtures of solvents and/or electrolytes are present or when strongly solvated effects exist or if there is only partial dissociation of the electrolyte. Unfortunately these situations are encountered frequently in practical applications, such as geochemical, biological and biochemical systems, metals undergoing electrochemical corrosion, and in materials processing where crystallization is used. To produce workable models for these situations, theory can only go so far and certain *ad hoc* features must be introduced. For example, the truncation of higher order effects or the empirical fitting of model parameters rather than obtaining them from *ab initio* calculations. In some instances, the magnitude of fitted parameters may go beyond what is theoretically possible if the model is to apply rigorously. This is, unfortunately, the state-of-the-art today.

As we have said in earlier discussions of models for non-ideal behavior, never underestimate the value of a few reliable data points. Real experimental data not only helps keep the modelers honest but it may be the only way to obtain a usable form for the constitutive model itself. As difficult as it is to predict the future, one thing is certain, given the importance of electrolytes to such a wide range of phenomena in physical and biological sciences, electrolyte research is bound to be vigorous in the decades ahead. With continued evolution of both experimental and computer methods, major advances in understanding electrolyte behavior are possible. Our strongest recommendation is to "stay tuned" to the literature to keep up with new developments.

References

Andersen, H.C. and D. Chandler (1971), "Mode expansion in equlibrium statistical mechanics, III. Optimized convergence and application to ionic solution theory," *J. Chem. Phys.*, **55**, p 1497-1504.

Aspen Technology (1994) *Physical Property Methods and Models,* ASPEN PLUS Reference Manual Release 9, Cambridge, MA, Appendices B and C.

Ball, F-X., H. Planche, W. Furst, and H. Renon (1985) "Representation of deviation from ideality in concentrated aqueous solutions of electrolytes using a mean spherical approximation model," *AIChE J.* **31**(8), p 1233-1240.

Bard, A.J., R. Parsons, and J. Jordan (1985) *Standard Potentials in Aqueous Solutions,* Marcel Dekker, New York, NY.

Barker, J.A. and D. Henderson (1976), "What is "liquid"? Understanding the states of matter," *Rev. Modern Physics* **48**, p 587-671.

Cardoso, M.I.E.De M and J.P. O'Connell (1987), "Activity coeffcients in mixed solvent electrolyte solutions," *Fluid Phase Equilibria* **33**, p 315-326.

Chen, C.C., H. I. Britt, J.F. Boston, and L.B. Evans (1982), "Local composition model for excess Gibbs energy of electrolyte solutions, Part 1: Single solvent, single completely dissociated electrolyte systems," *AIChE J.* **28**(4), p 588-596.

Chen, C.C. and L.B. Evans (1986), " A local composition model for the excess Gibbs energy of aqueous electrolyte systems," *AIChE J.* **32**(3), p 444-454.

Kusik C.L. and H.P. Meissner (1973), " Vapor pressures of water over aqueous solutions of strong electrolytes," *Ind. Eng. Chem. Proc. Des. Dev.* **12**(1), p 112-115.

Kusik C.L. and H.P. Meissner (1978), "Electrolyte activity coefficients in inorganic processing," in *Fundamental Aspects of Hydrometallurgical Processes,* AIChE Symposium Series, p 14-20.

Latimer, W.M. (1952) *The Oxidation States of the Elements and Their Potentials in Aqueous Solutions*, 2nd ed., Prentice-Hall, Englewood Cliffs, NJ.

Linke. W.F. and A. Seidell (1965) *Solubilities of Inorganic and Metal-Organic Compounds*, vol II, Amer. Chem. Soc., Washington, D.C., p 962.

Meissner, H.P. and J.W. Tester (1972), "Activity coefficients of strong electrolytes in aqueous solutions," *Ind. Eng. Chem. Proc. Des. Dev.* **11**(1), p 128-133.

Meissner, H. P. (1980), "Prediction of activity coefficients of strong electrolytes in aqueous solutions," in *Thermodynamics of Aqueous Systems with Industrial Applications*, Amer. Chem. Soc., Washington, D.C., p 495-511.

Mock, B., L.B. Evans, and C.C. Chen (1986), "Thermodynamic representation of phase equilibria in mixed-solvent electrolyte systems," *AIChE J.* **32**(10), p 1655-1664.

Pailthorpe, B.A, D.J. Mitchell, and B.W. Ninham (1984), "Ion-solvent interactions and the activity coefficients of real electrolyte solutions," *J. Chem. Soc., Faraday Trans.* 2,**80**, p 115-139.

Pitzer, K.S. (1973), "Thermodynamics of electrolytes. I. Theoretical basis and general equations," *J. Phys. Chem.* **77**(2), p 268-277.

Pitzer, K.S. (1980), "Electrolytes from dilute solutions to fused salts," *J. Amer. Chem. Soc.* **102**(9), p 2902-2906.

Planche, H. And H. Renon (1981), "Mean spherical approximation applied to a simple but nonprimitive model of interaction for electrolyte solutions and polar substances," *J. Phys. Chem.* **85**(25), p 3924-3929.

Rasaiah, J.C. and H.L. Friedman (1969), "Integral equation computations for aqueous 1-1 electrolytes. Accuracy of the method," *J. Chem. Phys.* **50**, p 3965-3976.

Resch, T.J. (1995), "*A Framework for the Modeling of Suspended Multicomponent Particulate Systems with Applications to Atmospheric Aerosols*," PhD Thesis, Chemical Engineering Department, Massachusetts Institute of Technology, Cambridge, MA.

Zerres, H. and J.M. Prausnitz (1994), "Thermodynamics of phase equilibria in aqueous-organic systems with salt," *AIChE J.*, **40** (4), p676-691.

Suggested readings:

Alberty, R.A. and R.J. Silbey (1992), *Physical Chemistry*, Wiley, New York, NY, Chapter 8, Sections 8.1-8.3,8.7-8.8, p 236-243, p 250-255. [Electrochemical cells]

Castellan, G.W. (1983), *Physical Chemistry*, 3rd ed., Benjamin-Cummings, Menlo Park, CA, Chapter 16, p 347-367. [Electrolyte activity and standard state conventions, Debye-Hückel theory]

Harned, H.S. and B.B. Owen (1958), *The Physical Chemistry of Electrolyte Solutions*, 3rd ed, Reinhold, New York, NY. [treatise on electrolytes with comprehensive listing of data]

Klotz, I.M. (1950), *Chemical Thermodynamics - Basic Theory and Methods*, Chapter 22, p 300-328. [excellent treatment of standard states and conventions]

MacInnes, D.A. (1947), *The Principles of Electrochemistry*, Reinhold, New York, NY, reprinted in 1961 by Dover, NY. [pioneering seminal work on electrochemistry]

McQuarrie, D.A. (1976), *Statistical Mechanics*, Harper and Row, New York, NY, Chapter 15, p 326-351.

Pitzer, K.S. (1995), *Thermodynamics*, 3rd edition, McGraw-Hill, New York, Chapters 15-17, p 254-322. [up to date revision of Lewis and Randall text—clear and comprehensive treatment of electrolytes]

Robinson, R.A. and R.H. Stokes (1970), *Electrolyte Solutions*, 2nd ed., Butterworth and Co., London. [classic treatment with comprehensive data listings]

Zemaitis, J.F.,Jr., D.M. Clark, M. Rafal, and N.C. Scrivner (1986), *Handbook of Aqueous Electrolytes*, Design Institute for Physical Property Data (DIPPR), Amer. Inst. Chem. Eng., New York, NY. [excellent reference book with comprehensive listing of data and bibliographical sources, and electrolyte model parameters for the Chen, Pitzer, Meissner and other models]

Problems

12.1. The activity coefficient is often used to express a quantitative deviation from the Lewis and Randall rule.

(a) Given: $\gamma_1 = \exp\,[5[1 + 2(1 - x_2)]x_2^2]$ for a 1-2 binary mixture at fixed T and P, sketch a plot of γ_1 versus x_2 over the interval from $x_2 = 0$ to 1. Does component 1 exhibit positive or negative deviations from the Lewis and Randall rule? What is the standard state for component 1? What is the value of γ_1 at infinite dilution ($x_1 \to 0$)?

(b) Develop an expression for $\gamma_2 = f\,[x_1, x_2]$ and sketch its behavior on the same plot as in part (a).

(c) Frequently in dealing with electrolytes, an unsymmetrical normalization is used to define a hypothetical standard state. This is consistent with the condition that $\gamma_\pm = 1.0$ at infinite dilution. How would the fugacity or chemical potential of the solvent be expected to behave in a strong electrolyte solution of NaCl in H_2O as the NaCl concentration decreases toward zero?

12.2. Describe how you would use electrochemical cell potential measurements to estimate the mean ionic activity coefficient $\gamma_\pm$ of NaCl as a function of concentration [Hint: you must first determine how to estimate $\exists^o$ and you can assume that the Debye-Hückel limiting law holds for $\gamma_\pm$ as molality (m) $\to$ 0].

12.3. Using the Meissner and Pitzer models (a) estimate the mean ionic activity coefficient of LiCl at 25°C for concentrations ranging from 0.1 molal at 10 molal. (b) Estimate the solubility of LiCl in a 10 molal solution of HCl? (c) Would you expect the solubility of LiCl to be higher or lower in a 10 molal Na_2SO_4 solution than in the 10 molal HCl solution in part (b). At 25°C, the solubility of pure LiCl is about 9.0 mol/L.

12.4. Derive Eq. (12-87) and show that it can be applied at any temperature. How are values of a_w^o obtained experimentally for salt solutions of NaCl, K_2SO_4, and $AlCl_3$? Can a similar procedure be used for HCl?

12.5. Dr. E.M. Faraday, a distinguished electrochemical engineering professor, has just returned from a year's sabbatical leave where she taught in the physical chemistry department of a prestigious Western University. In that department, she was asked to teach geochemistry students about the molality and molarity reference scales for electrolyte activity coefficients. In order not to offend her geochemist colleagues and upset their traditional approach, while still maintaining her rigorous thermodynamics demeanor, she asked her students to develop equations to convert from a molality scale to one based on mole fraction x_i. One astute student, known as "Big Wall Pete" for his rock climbing abilities, produced the equation, after several hours of hard work, that was identical to Eq. (12-13), but he lost his notes and Prof. E.M. F. wants to include it in her updated version of her class notes. Can you help her out?

12.6. Gabe and Matt Jones, distant cousins to Rocky and specialists in electrolyte modeling, are arguing about the fine points of standard states and electrolyte models and need your help. Gabe maintains that the Debye-Hückel limiting law (DHLL) must be followed for all electrolytes in the dilute limit because it is "exact" in that region. Matt claims that form of the $\gamma_\pm$ equation may yield different results. For example, he points out that Pitzer's model has two forms for $\gamma_\pm$, one given by Eq. (12-78) and the other by Eq. (12-81) while both forms have an infinite dilution reference condition. The first is on a molality basis while the second is a mole fraction basis. Matt says these are not equivalent to the DHLL which states that

$$\log_{10} \gamma_\pm = -0.510\, |z_+ z_-|\, I^{1/2}$$

12.7. RPI (Recombinant Processes, Inc.) Industries is exploring a possible acquisition of a new BioTech firm that claims it has unraveled the problems associated with cell membrane transport by developing a new concept for altering osmotic pressure. They claim that the presence of glucose in a normal saline solution will change the osmotic pressure by an amount that exceeds the ideal solution limit by more than a factor of two. Data are given below for the osmotic pressure of solutions of glucose and NaCl in aqueous solutions at 25°C. What is your opinion of their claims?

molality	glucose	NaCl
0.01	0.1904	0.3526
0.10	2.026	3.566
0.20	4.322	7.223
0.5	12.84	18.73
0.8	23.80	31.04

12.8. Estimate the Debye length κ^{-1} for LiCl, NaCl, and $CuCl_2$ in aqueous solutions at 25°C and 300°C from concentrations ranging from 0.1 to 10 molal. Plot and interpret your results.

12.9. Using Eq. (12-71) as a basis for $g_{ij}(r)$ can you develop a way of estimating the size of a solvent molecule and the most likely distance between ions as a function of concentration?

12.10. The $\Delta G^o_{298\text{ K}}$ for the following dissolution (solid-liquid) equilibrium reaction is 3500 cal/mol

$$K_2X_{(s)} = 2K^+_{(aq)} + X^{2-}_{(aq)}$$

Estimate the solubility of K_2X at 25°C if $\gamma_\pm$ has been measured to be 6.0 at $I = 10$ (mol/kg).

Estimating Physical Properties 13

Up to this point in Part II we have concentrated on describing a wide variety of constitutive models and conventions used to represent the equilibrium thermodynamic properties of pure components and mixtures. Central to this discussion has been developing a framework to correlate and predict the properties of real fluids that show varying degrees of deviation from ideal gas or ideal solution behavior. We emphasized the importance of having experimental data available for use in developing the semi-empirical extensions of theories. The main problem is that these theoretical models are incomplete or inadequate in their description of non-ideal behavior. To improve their utility for engineering applications, model parameters are fit to experimental data. The amount of data required could be small, but sufficient to determine the model's empirical parameters. For example, two data points, might be all that are required for a two-parameter model of liquid phase activity coefficients, such as the infinite dilution γs for a binary. Or if we want to use the compressibility model via Corresponding States for $Z = f(T_r, P_r)$ or the van der Waals or similar cubic EOS for specifying the $P\underline{V}TN$ properties of a pure fluid then two parameters, the critical constants T_c and P_c, are required. To use three-parameter $P\underline{V}TN$ models, like the Peng-Robinson cubic EOS, then an additional parameter, the Pitzer acentric factor ω, is needed. If more data are available then a regression can be used to specify the required parameters in some optimal fashion, for example, to minimize the sum of squares of differences between experimental and predicted property values.

13.1 Approaches for Property Prediction and Estimation

Chemical engineers are repeatedly faced with inadequate physical property data to specify model parameters required for design calculations or for process engineering analysis. Inevitably, there is never sufficient time or money to measure all the physical properties needed, particularly when complex mixtures of components are present. Thus, engineers must rely on their resourcefulness and creative skills to estimate the property information they need. And perhaps more importantly, they must be able to know how good their estimates are, so they can assess the uncertainties of their design or analysis. Appropriate sensitivity analysis to physical properties should always be carried out before final decisions are reached.

The range of thermodynamic physical property estimation is naturally quite broad. At one extreme, one may be able to locate the vast majority of data required in the open literature or in some computerized database, then only statistical regression, extrapolation and/or interpolation are required to estimate the properties you need. This really is property prediction by correlation. At the other extreme, no experimental data are available and one really must *estimate* the required properties. As there are many possible options open, which are continually evolving and being upgraded, engineers must gain the skills and experience needed to make proper choices. For instance, if you were to use one of the computerized process design and simulation software packages that exist on the open market today, you will be relying at least in part on the property estimation and modeling capabilities of that product to correctly generate accurate information. Consequently, you are relying on the skills and expertise of the staff that maintains the physical property modeling and predictive components of that software. The staff must not only be fixing problems or "bugs" but they also must be continually upgrading their software with the latest improvements and additions to property databases to keep them up-to-date and reliable.

While we cannot cover all aspects of property estimation in this chapter, we hope to point you in the right direction by providing the conceptual basis for some of the commonly employed estimation techniques that are in use today. For an in-depth review of property correlation and estimation methods readers should consult Reid *et al.* (1987) and Lyman *et al.* (1990).

The following examples will amplify the points made in these opening paragraphs.

Example 13.1

You are a new engineer on the staff of a company that produces speciality chemicals. One of the chemists has just invented a new synthesis procedure for a proprietary organic compound, Alpha-xyz, important to the pharmaceutical industry. The company wants to test market the compound soon and they want you to develop a process design that will produce 10,000 kg of 99.99% pure product. The chemist has been so busy with developing her synthesis of Alpha-xyz that all she knows about its physical properties are its normal boiling point, molecular weight, and molecular structure. The synthesis recipe involves several reaction steps and distillation separations with common solvents used throughout the process.

Solution

The options open to you include:

(1) Review the literature and computerized physical property databases to see what is known about the pure solvents used and hopefully something about pure Alpha-xyz itself as well its behavior in mixtures with the solvents.

(2) Go into the laboratory and make the required measurements of all properties needed to carryout the design.

(3) Measure only a few selected properties that cannot be estimated very accurately, like the activity coefficients of Alpha-xyz in dilute solutions of solvent B as a function of temperature.

(4) Estimate all physical properties needed using whatever methods are available.

Obviously there are alternatives among these options, but for now assume that only options (1) and (4) are feasible. In the first option, we need to know exactly what kind of data are needed to do the design calculations and where to find them. The answer to the first question has been covered in the earlier chapters of Part II for pure components and mixtures of gases, liquids, and even solids. The second question as to the location of information, is in and of itself a very challenging task as there are many sources of data ranging from published papers in the scientific literature, to handbooks, to compilations of data in written or electronic form. Data compilation, searching, and evaluation are discussed in Section 13.2.

In option (4) where everything is estimated, we need recipes or algorithms for making these estimates that will ensure some known level of confidence in the results. For example, with the molecular weight and structure known for Alpha-xyz, can one estimate such pure component parameters as the critical constants and the enthalpy and Gibbs free energy of formation? The answer is yes assuming that the structure of Alpha-xyz contains elements arranged in molecular or functional groups that are found in other common organic compounds. The methods for estimating properties based on molecular functional structure are called group contribution methods. In the main, they have been enormously successful at predicting the properties of pure materials as you will see from the discussion in Section 13.3. The conceptual basis for pure component molecular group contributions is illustrated in Figure 13.1. In addition, considerable success has also been achieved in the extension of the group contribution method to estimating the activity coefficients of components in mixtures. These extensions are covered in Section 13.4.

Being able to evaluate the derived thermodynamic properties, such as U, S, H, A, and G, is very important to chemical engineering because they permit both first and second law and phase and chemical equilibrium calculations to be performed for specific systems involving both pure components and mixtures. For example, to perform power cycle state point calculations to estimate turbine, compressor, or pump power levels, and heat exchanger or condenser duties, both ΔH and ΔS are needed over a range of real fluid conditions, including those well outside the ideal-gas region. Or to make a chemical equilibrium calculation, both ΔG^o_{rx} and ΔH^o_{rx} for the reaction under standard state conditions may be needed.

While most systems of interest to chemical engineers are mixtures, pure component properties play an important role. For example, we saw in Section 9.2 that relatively simple mixing or combining rules involving compositionally-weighted pure component parameters could be used to generate $P\underline{V}TN$ information about the mixture at specified conditions of temperature, pressure or density, and composition using an equation of state like the cubic Peng-Robinson EOS. Although a binary δ_{ij} interaction parameter is

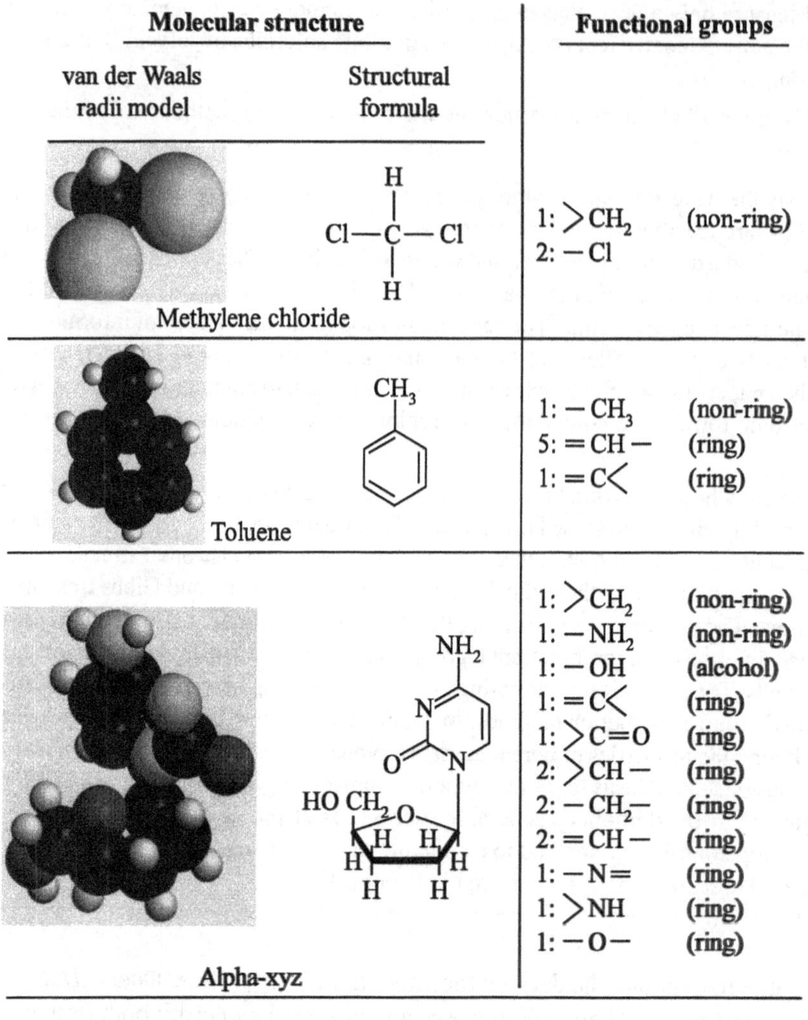

Molecular structure (van der Waals radii model; Structural formula)	Functional groups	
Methylene chloride	1: $>CH_2$	(non-ring)
	2: $-Cl$	
Toluene	1: $-CH_3$	(non-ring)
	5: $=CH-$	(ring)
	1: $=C<$	(ring)
Alpha-xyz	1: $>CH_2$	(non-ring)
	1: $-NH_2$	(non-ring)
	1: $-OH$	(alcohol)
	1: $=C<$	(ring)
	1: $>C=O$	(ring)
	2: $>CH-$	(ring)
	2: $-CH_2-$	(ring)
	2: $=CH-$	(ring)
	1: $-N=$	(ring)
	1: $>NH$	(ring)
	1: $-O-$	(ring)

Figure 13.1 Concept of molecular group contributions.

needed in many cases for accurate predictions, the critical constants and acentric factor of the pure components are always required. The following example shows how derived thermodynamic properties can be estimated for a pure gas or liquid.

Example 13.2

Describe how you would estimate *H, S, U, A,* and *G* for a pure compound in a stable fluid state over temperatures and pressure ranging from 25°C, 1 bar to 400°C, 300 bar.

Solution

Again there are a number of options open to us, including

(1) Search the literature for a tabulation of derived properties as a $f(T, P, \text{or } \rho)$ in the form of a chart or table. The steam tables or charts (Figures 8.11a and 8.11b) are good examples of these. Figures 8.12a and 8.12b provides another example for carbon dioxide.

(2) Find and utilize a computerized representation of (1). For example, the properties of pure water and steam have been correlated empirically with a multiparameter high precision EOS which is easily accessible electronically [Moore (1972) and Haar, Gallagher, and Kell (1984)].

(3) Calculate the derived properties needed by combining a series of isothermal departure functions (for real to ideal-gas state property changes) and ideal-gas state temperature variations. As we know from our discussions in Section 8.5, to do this we require a $P\underline{V}TN$ EOS suitable for non-ideal fluids to model density-dependent, configurational effects and an ideal-gas state heat capacity which represents intramolecular energy storage effects and is only a function of temperature, $C_p^o(T)$.

Strictly speaking we can calculate specific values of H, S, U, A, and G only relative to some prescribed reference state. Frequently, the reference state for enthalpies and Gibbs energies of formation, ΔH_f^o, and ΔG_f^o, is the pure compound in its stable state of aggregation at ambient conditions of 25°C (298 K), 1 atm (1.013 bar) (see Section 8.8). Once we calculate values of H and G relative to this reference state, then we can use the $P\underline{V}TN$ EOS to generate the other derived quantities. Consider $H_i(T, P)$ for a gaseous compound referenced to an ideal gas state at 298 K, 1 atm:

$$H_i(T, P) = \underbrace{\Delta H_{f,i}^o\,(298\text{ K}, 1.013\text{ bar})}_{\textit{ideal gas standard state}} + \underbrace{[H_i(T, P) - H_i^o(T, P^o)]}_{\substack{\textit{real to ideal gas}\\ \textit{departure function}}} + \underbrace{\int_{298\text{ K}}^{T} C_{p,i}^o(T)\, dT}_{\textit{heat capacity function}} \tag{13-1}$$

where the first term is the enthalpy for formation for pure i at 298 K in its ideal-gas standard state, the second term in [ ] is the isothermal departure function at T from P to $P^o = RT/\underline{V}$ (*ideal gas*), and the third term is the change of H in the ideal-gas state from 298 K to T. In systems where the stable phase is not an ideal gas at ambient conditions, the phase change (latent) enthalpy contributions must also be included in Eq. (13-1) and the heat capacity integral would need to use C_p values corresponding to the correct phases involved between 298 K and T. To evaluate the departure function contribution some type of *$P\underline{V}TN$ model or* EOS is needed. In today's environment, frequently a pressure-explicit cubic EOS is employed. If so, then pure component critical constants T_c and P_c, and possibly the Pitzer acentric factor ω, are required.

For chemical equilibrium calculations, one frequently needs to evaluate the thermodynamic equilibrium constant K_a for a reversible reaction (rx) (see Section 16.5).

In these situations we resort to the common definition of K_a in terms of the standard ΔG^o_{rx} for the reaction where all reactants and products are in their standard states at temperature T:

$$\ln K_a = -\Delta G^o_{rx}/RT = -\sum_i \frac{\nu_{i,rx}\,\Delta G^o_{f,i}(T)}{RT} \tag{13-2}$$

where $\nu_{i,rx}$ is the stoichiometric coefficient and $\Delta G^o_{f,i}(T)$ is the Gibbs energy of formation at T of reactant (or product) i involved in the reaction. Using the Gibbs-Helmholtz relationship, we can express K_a at any temperature as,

$$\ln K_a = \ln K_a(298\ K) + \int_{298\,\mathrm{K}}^{T} \frac{\Delta H^o_{rx}\,dT}{RT^2} \tag{13-3}$$

In order to complete the calculation both ΔG^o_{rx} and ΔH^o_{rx} need to be known as a function of temperature. They are normally expressed in terms of the standard Gibbs energy and enthalpy of formation at 298 K, 1.013 bar, and in the stable state of aggregation for all components participating in the reaction. If all reacting species are gases, then one can apply appropriate corrections for temperature using the ideal-gas state heat capacity, $C^o_p(T)$. In this case, ln K_a becomes:

$$\ln K_a(T) = \ln K_a(298\ \mathrm{K}) + \frac{1}{R}\int_{298\,\mathrm{K}}^{T}\left[\sum_i \nu_{i,rx}\left(\Delta H^o_{f,i} + \int_{298}^{T} C^o_{pi}\,dT\right)\right]\frac{dT}{T^2} \tag{13-4}$$

From these examples, we see the importance of establishing values for pure component properties.

13.2 Sources of Physical Property Data

Normally pure component and mixture properties are compiled and tabulated separately. The pure component and mixture properties of most interest in chemical engineering practice are listed below in Table 13.1. Under mixtures we have only listed *model parameters* (rather than properties) in Table 13.1 that are important to the calculation of volumetric $P\underline{V}TN$ mixture properties using a specific EOS model or for obtaining activity coefficients using a specific ΔG^{EX} model. Both single and multiphase systems of are interest. Multiphase properties are typically categorized in several important classes of phase equilibria such as vapor-liquid, liquid-liquid, and solid-fluid (see Chapters 15 and 17). Representative properties for single strong electrolytes and electrolyte mixtures are discussed in Chapter 12.

At the beginning of the 20th century compiled data were restricted to just a few publications, most prominently in the International Critical Tables, Landolt-Bornstein and Beilstein compilations. Over the last 20 years things have changed considerably, mostly because of computers and the creation of large databases that can be accessed easily, for example by electronic transmission or diskettes. Today, the sources of information are widely distributed and in multimedia form—tables, charts, and computerized lists and accessible databases are the most common. Many of these databases are well maintained and upgraded as new data become available. Table 13.2 lists some of the commonly used sources of pure component and mixture data in both published and electronic form. Table 13.3 gives examples of United States (US) research centers that focus on physical property measurements and data correlation. An excellent example of current trends in effective data management can be found in the process flowsheet simulators that are growing in use in the chemical engineering community. These systems must handle large databases and provide a variety of physical property modeling options to be competitive. Section 13.5 covers this application in more detail.

13.3 Group Contribution Methods for Estimating Pure Component Properties

This section deals with the methods used to estimate the pure component properties listed in Table 13.1. The molecular group contribution concept illustrated in Figure 13.1 suggests that a particular property of the unknown compound such as its critical temperature or pressure or its enthalpy of formation can be synthesized by adding together the contributions to that property that can be attributed to specific molecular functional groups. This seems intuitively reasonable as all macroscopic properties are related to the structure and interaction of the molecules comprising the system.

The most straightforward recipe for implementing a group contribution approach would be to assume that the superposition of effects from the functional groups is linear. To keep the process simple, normally contributions are added as a weighted sum of the number of groups of a certain type in the unknown compound times the contribution of that functional group to the property of interest. In order to implement the method for a specific property, we must select the structures of the molecular groups and then determine their contributions to that property. Although there are many ways to do this, the usual method is statistically based with some creative heuristic input to dissect the molecular structures of the compounds used in the data set to produce a rational and consistent set of substructures. Like everything else that we have done so far for modeling, non-ideal experimental data are needed to complete this task. In principle, the more data one has on different compounds the better the estimation of a particular property will be.

Table 13.1 Pure Component and Mixture Parameters and Properties

Pure component properties

Molecular weight or molar mass m_i

Normal boiling point T_b

Critical temperature T_c

Critical pressure P_c

Critical volume V_c

Critical compressibility Z_c

Pitzer acentric factor ω

$P\underline{V}TN$ volumetric properties

Liquid molar volume $V^{(l)}$

Ideal gas state heat capacities C_p^o

Standard enthalpy of formation ΔH_f^o

Standard Gibbs free energy of formation ΔG_f^o

Vapor pressure P_{vp}

Enthalpy of vaporization ΔH_{vap}

Mixture parameters

δ_{ij} binary interaction parameters for various $P\underline{V}TN$ EOSs, for example

- Peng-Robinson
- Redlich-Kwong-Soave

Binary interaction parameters for various ΔG^{EX}/activity coefficient models, for example

- Margules $A, B, C, \ldots$
- Wilson Λ_{ij}
- NRTL τ_{ij}, α_{ij}
- UNIQUAC τ_{ij}
- UNIFAC and ASOG parameters for molecular group interactions a_{ij}

UNIQUAC size parameters for molecular volume and area r_i and q_i

Table 13.2 Representative Compilations and Databases of Physical Properties

Source	Attributes
American Petroleum Institute *Technical Data Book—Petroleum Refining* 3rd ed. (1981)	Critical properties and other data
ASHRAE *Thermodynamic Properties of Refrigerants* (1969) and *Fundamentals* (1981)	Pure component thermodynamics properties, tables and charts for many refrigerants
CRC *Handbook of Chemistry and Physics* (1995 ff)	General pure component data source, published annually
DIPPR Design Institute for Physical Property Data	Database on pure components, mixtures, and electrolytes; computerized, maintained and well-documented
Benson, *Thermochemical Kinetics* (1968)	Thermochemical data and correlations
JANAF Thermochemical Data, Dow Chemical Company	Combustion and thermochemical data and correlations
DECHEMA Dortmund databank (see Gmehling *et al.* (1977))	VLE and LLE phase diagrams and activity coefficient data
International Critical Tables	General historical source of experimental data tabulations
ASME *Steam Tables* (1967 ff)	
NBS/NRC *Steam Tables* (1984) by Haar, Gallagher, and Kell	
Bondi, *Physical Properties of Molecular Crystals, Liquids, and Glasses* (1968)	Molecular sizes and other physical data
Beilstein and Landolt-Bornstein	General historical sources of thermodynamic property data
van Krevelen, *Properties of Polymers:Their Estimation and Correlation with Chemical Structure* (1976)	Extensive polymer data on pure component and mixtures
Natural Gas Processors Assn., *Engineering Data Book* (1981)	Data tabulations for natural gas components
ELDATA - *The International Electronic Journal of Physico-Chemical Data* (1995 *et al.*)	New electronic source of experimental data
Hougan, Watson and Ragatz, *Chemical Process Principles, part 2 Thermodynamics,*(1959)	Excellent treatment of the application of Corresponding States
Perry's *Chemical Engineers' Handbook*, R.H. Perry and D. Green (editors), 6th ed. (1984)	Reference book with pure component data including vapor pressure, critical constants, heat capacities, and many thermodynamic diagrams, revised periodically

Table 13.2 Representative Compilations and Databases of Physical Properties (cont.)

Source	Attributes
Reid, Prausnitz, and Poling, *The Properties of Gases and Liquids - Property Data Bank*, Appendix A, 4th ed. (1987)	Pure component properties for 618 compounds
Stull, Westrum, and Sinke, *The Chemical Thermodynamics of Organic Compounds* (1969)	Thermochemical and vapor pressure data

Table 13.3 Several US Research Centers Specializing in Thermodynamic Properties

CINDAS - Purdue University

DIPPR - American Institute of Chemical Engineers AIChE

Thermodynamics Research Center - Texas A&M University

Thermal Research Laboratory, Dow Chemical Company

Center for Applied Thermodynamics Studies - University of Idaho

National Center for Thermodynamic Data of Minerals, Reston VA

National Institute for Standards and Technology (NIST) Gaithersburg, MD

- Chemical Thermodynamics Data Center
- Crystal Data Center
- Alloy Phase Diagram Center
- Phase Diagrams for Ceramists Center

Although the group contribution approach has evolved considerably over the last 50 years, and many successful applications have appeared [for example see Reid *et al.* (1987)], they all retain the fundamental assumption of linearized group additivity. In all cases a considerable amount of *ad hoc* empiricism is used to select the data sets and to specify the functional groups that are needed to produce the quantitative estimates of each group's contribution to the property of interest. Think of the calculational tradeoffs between picking groups with more atoms arranged in different ways to capture increased complexities in structure versus picking simpler group structures to increase number of groups of a given type that will be found among the data. One needs to strike a proper balance between group complexity and number of groups to achieve optimal results. A good analogy would be to consider a strand of DNA. At one extreme, the DNA molecule can be viewed as a collection of atoms of C, N, H, O, and P, and at another extreme as an ordered sequence of four specific pyrimidine nitrogenous bases (cytosine, thymine, adenine, and guanine) attached to a phosphated polysaccharide backbone polymer. The correct substructure really depends on what property you are trying to estimate. If you're interested in the ideal-gas state heat capacity, then the number and type of individual bonds will be important. If you're interested in linking one's DNA to a blood sample taken at the scene of a crime then the exact ordering of the four nitrogenous bases in the double helix strand of DNA are the key.

Assuming that the molecular structure is known, then the molecular weight is immediately calculable from the constituent elements. Therefore, molecular weight is usually not an issue unless you are faced with a very complex mixture, dispersed polymer or an inherently complex natural substance like a liquid derived from coal processing. Under these circumstances you could possibly utilize psuedocritical properties based on a different averaging scheme [Reid *et al.* (1987) and Aspen Technology (1998)]. For a dispersed polymer system, a technique suggested by Prausnitz and coworkers (1984 - 1986) called "continuous thermodynamics" could be used where integral or ensemble averaging methods are employed over a known molecular weight distribution. In any case, a different type of estimation method is needed to predict properties for these kinds of systems.

Fortunately for most pure component properties, the substructure of groups is not that difficult to select. Group contribution methods have been applied successfully to the prediction of normal boiling points, all the critical constants, the acentric factor, the ideal gas state heat capacity, and the enthalpy and Gibbs free energy of formation of literally hundreds of compounds. In reference to these properties (and others), the reader should consult Reid *et al.* (1987) and Lyman *et al.* (1990) for detailed treatment. For the use of group contribution methods in the estimation of properties related to the distribution of organic chemicals in the environment (i.e., aqueous solubility, equilibrium phase partition coefficients, etc.) see Schwarzenbach *et al.* (1993). In the discussion that follows we will illustrate the power of group contribution techniques.

Critical Constants (T_c, P_c, V_c, Z_c) and Normal Boiling Point (T_b) The critical constants and normal boiling point of a pure component are important for specifying parameters for various EOS's or as scaling parameters for the Law of Corresponding States. Given the fact that careful experimental measurements of these properties are not a popular area for research today, there is a real need to be able to estimate the critical constants to within a few percent error. A number of group contribution options exist for estimating these, including those due to Ambrose, Fedors, and Joback [see Reid *et al.* (1987) for details]. By way of example, we first discuss Joback's modification of Lydersen's 1955 algorithm for predicting T_c, P_c, V_c, and Z_c. With a specified set of functional groups and contributions provided by regression of data from over 400 compounds, Table 13.4 was developed by Joback (1984). His combining equations for specific group additivity for the critical constants are as follows for component i :

$$T_c = \frac{T_b}{\left[0.584 + 0.965 \sum_j v_j^{(i)} \Delta_{T_c} - (\sum_j v_j^{(i)} \Delta_{T_{cj}})^2\right]} \quad \text{(in K)}$$

$$P_c = \frac{1}{\left[0.113 + 0.0032\, n_a - \sum_j v_j^{(i)} \Delta_{P_{cj}}\right]^2} \quad \text{(in bar)} \tag{13-5}$$

$$V_c = 17.5 + \sum_j v_j^{(i)} \Delta_{V_{cj}} \quad \text{(in cm}^3\text{/mol)}$$

where n_a is the total number of atoms in the molecule, $v_j^{(i)}$ is the number of groups of type j in compound i, and the sums run over all groups that are found in the compound of interest.

Most commonly, the normal boiling point T_b is known or can be easily measured. However there will be instances where it too must be estimated. Here the Joback method can be used but it provides only an approximate value of T_b (an average error of 12.9 K was observed for about 400 compounds examined by Joback). Using the values of Δ_{T_b} in Table 13.4 and the following equation, T_b is estimated as:

$$T_{b,i} = 198 + \sum_j v_j^{(i)} \Delta_{T_{bj}} \tag{13-6}$$

As an alternative, Klincewicz and Reid (1984) provide an empirical relationship that correlates T_b with T_c with an average error of only 1 to 2%:

$$T_{c,i} = 50.2 - 0.16m_i + T_{b,i} \tag{13-7}$$

Equation (13-7) could be used with the molecular weight m_i to estimate either $T_{b,i}$ or $T_{c,i}$. Now, we can proceed to examine a few examples of how to use the Joback method.

Example 13.3

Estimate the critical constants and boiling point of the compounds Alpha-xyz, toluene, and methylene chloride, given in Figure 13.1.

Solution

Let's begin by estimating properties for the two solvents toluene and methylene chloride since we can compare their predicted properties from the Joback's technique to values tabulated in Appendix G. After that we can look at the unknown compound Alpha-xyz.

The functional groups are given in Figure 13.1, so we start the process by using Table 13.4 to generate the contributions for each molecule.

Table 13.4 Joback Group Contributions for the Critical Constants and Normal Boiling Point [from Reid *et al.* (1987), p 15]

	Δ			
	T_c	P_c	V_c	T_b
non-ring increments:				
$-CH_3$	0.0141	-0.0012	65	23.58
$>CH_2$	0.0189	0	56	22.88
$>CH-$	0.0164	0.0020	41	21.74
$>C<$	0.0067	0.0043	27	18.25
$=CH_2$	0.0113	-0.0028	56	18.18
$=CH-$	0.0129	-0.0006	46	24.96
$=C<$	0.0117	0.0011	38	24.14
$=C=$	0.0026	0.0028	36	26.15
$\equiv CH$	0.0027	-0.0008	46	9.20
$\equiv C-$	0.0020	0.0016	37	27.38
Ring increments:				
$-CH_2-$	0.0100	0.0025	48	27.15
$>CH-$	0.0122	0.0004	38	21.78
$>C<$	0.0042	0.0061	27	21.32
$=CH-$	0.0082	0.0011	41	26.73
$=C<$	0.0143	0.0008	32	31.01
Halogen increments:				
$-F$	0.0111	-0.0057	27	-0.03
$-Cl$	0.0105	-0.0049	58	38.13
$-Br$	0.0133	0.0057	71	66.86
$-I$	0.0068	-0.0034	97	93.84
Oxygen increments:				
$-OH$ (alcohol)	0.0741	0.0112	28	92.88
$-OH$ (phenol)	0.0240	0.0184	-25	76.34
$-O-$ (non-ring)	0.0168	0.0015	18	22.42
$-O-$ (ring)	0.0098	0.0048	13	31.22
$>C=O$ (non-ring)	0.0380	0.0031	62	76.75
$>C=O$ (ring)	0.0284	0.0028	55	94.97
$O=CH-$ (aldehyde)	0.0379	0.0030	82	72.24
$-COOH$ (acid)	0.0791	0.0077	89	169.09
$-COO-$ (ester)	0.0481	0.0005	82	81.10
$=O$ (except as above)	0.0143	0.0101	36	-10.50
Nitrogen increments:				
$-NH_2$	0.0243	0.0109	38	73.23
$>NH$ (non-ring)	0.0295	0.0077	35	50.17
$>NH$ (ring)	0.0130	0.0114	29	52.82
$>N-$ (non-ring)	0.0169	0.0074	9	11.74
$-N=$ (non-ring)	0.0255	-0.0099	—	74.60
$-N=$ (ring)	0.0085	0.0076	34	57.55
$-CN$	0.0496	-0.0101	91	125.66
$-NO_2$	0.0437	0.0064	91	152.54
Sulfur increments:				
$-SH$	0.0031	0.0084	63	63.56
$-S-$ (non-ring)	0.0119	0.0049	54	68.78
$-S-$ (ring)	0.0019	0.0051	38	52.10

For toluene:

$$\begin{array}{llllll} & & \text{-}CH_3 & =CH\text{-} & =C< & \\ T_c: & \sum \nu_j^{(i)} \Delta_{Tcj} = & (1)(0.0141) & +5(0.0082) & +(1)(0.0143) & = 0.0694 \\ P_c: & \sum \nu_j^{(i)} \Delta_{Pcj} = & -(1)(0.0012) & +5(0.0011) & +(1)(0.0008) & = 0.0051 \\ V_c: & \sum \nu_j^{(i)} \Delta_{Vcj} = & (1)(65) & +5(41) & +(1)(32) & = 302 \\ T_b: & \sum \nu_j^{(i)} \Delta_{Tbj} = & (1)(23.58) & +5(26.73) & +(1)(31.01) & = 188.24 \end{array}$$

Now, we use Eq. (13-5) to obtain T_c, P_c, V_c and Eq. (13-6) for T_b.

$T_b = 198 + 188.24 = 386.24$ K versus 383.8 K (exp)

$$T_c = \frac{386.24}{0.6462} = 597.7 \text{ K versus } 591.8 \text{ K (exp)}$$

$$P_c = \frac{1}{(0.113 + 0.0032(15) - 0.0051)^2} = 41.1 \text{ bar versus } 41 \text{ bar (exp)}$$

$V_c = 17.5 + 302 = 319.5$ cm^3/mol versus 316 cm^3/mol (exp)

For methylene chloride:

$$\begin{array}{lllll} & & >CH_2 & \text{-}Cl & \\ T_c: & \sum \nu_j^{(i)} \Delta_{Tcj} = & 0.0189 & +2(0.0105) & = 0.0399 \\ P_c: & \sum \nu_j^{(i)} \Delta_{Pcj} = & 0 & +2(-0.0049) & = -0.0098 \\ V_c: & \sum \nu_j^{(i)} \Delta_{Vcj} = & 56 & +2(58) & = 172 \\ T_b: & \sum \nu_j^{(i)} \Delta_{Tbj} = & 22.88 & +2(38.13) & = 99.14 \end{array}$$

and

$T_b = 198 + 99.14 = 297.14$ K versus 313 K (exp)

$$T_c = \frac{297.14}{0.6209} = 478.55 \text{ K versus } 510 \text{ K (exp)}$$

$$P_c = \frac{1}{(0.113 + 0.0032(5) + 0.0098)^2} = 51.91 \text{ bar versus } 63 \text{ (exp)}$$

$V_c = 17.5 + 172 = 189.5\ \text{cm}^3/\text{mol}$ (no experimental value available)

The agreement between predicted and experimental values is excellent for toluene and reasonable for methylene chloride. Part of the problem for methylene chloride is that the T_b value is off and that propagates through to T_c. For example, if we use the experimental T_b value we get $T_c = 504$ K. In addition, the experimental values for T_c, P_c, and V_c themselves could be in error as they are more difficult to measure than T_b.

For Alpha-xyz, the estimation procedure is more tedious than for toluene or methylene chloride because there are more groups. To illustrate the application, let's estimate only V_c using Eq. (13-5) and Table 13.4.

Functional Groups	$v_j^{(i)}\ \Delta_{Vcj}$
2 (–CH_2–) ring	2(48)
2 (>CH–) ring	2(38)
1 (>C=O) ring	55
1 (>CH_2) non-ring	56
2(=CH–) ring	2(41)
1(=C<) ring	32
1(–OH) alcohol	28
1(–NH_2) non-ring	38
1(–N=) ring	34
1(>NH) ring	29
1(–O–) ring	13

$$\sum v_j^{(i)}\ \Delta_{Vcj} = 539$$

$V_c = 17.5 + 539 = 556.5\ \text{cm}^3/\text{mol}$

Pitzer Acentric Factor (ω) If vapor pressure data are available then ω can be obtained directly from its definition given in Section 8.2 [Eq. (8-35)]. For compound i:

$$\omega \equiv -\log_{10}\left[\frac{P_{vp}\ (\text{at } T_r = 0.7)}{P_c}\right] - 1.0 \qquad (13\text{-}8)$$

If you have experimental or estimated values of T_c, P_c, and T_b but no further information on the vapor pressure, then the Clapeyron equation (see Section 15.4) can be used to estimate ω as, for example,

$$\omega = \frac{3}{7}\left[\left(\frac{T_b/T_c}{(1 - T_b/T_c)}\right)\log_{10}\left(\frac{P_c}{1.013}\right)\right] - 1 \qquad (13\text{-}9)$$

where P_c is in bars and T_c and T_b in K. An alternative approach using Z_c gives,

$$\omega = \frac{0.291 - Z_c}{0.080} \tag{13-10}$$

Values of ω are also listed in Appendix G for several compounds.

PVTN Volumetric Properties In Chapter 8 we covered a number of constitutive equations of state (EOSs) that are commonly used to represent the volumetric properties of non-ideal fluids. These were generally either in cubic, pressure-explicit or virial polynomal form, and had parameters that were fit at the critical point and/or along the critical isotherm. In addition, we discussed the use of Corresponding States as a route to determining $P\underline{V}TN$ behavior. In all these cases, appropriate critical constants are needed either to evaluate constants or for proper scaling in reduced coordinates. Once we have values for T_c , P_c , V_c , and Z_c we usually can predict reasonably accurate volumetric properties for the fluid phases of a pure component using a suitable $P\underline{V}TN$ EOS. More accurate predictions require more specialized treatment such as a multiparameter, high-precision EOS like the HKL equation for water (Haar *et al.* (1984)). Alternatively, we may restrict the use of the EOS to low density gas-like conditions and use a separate correlating equation for liquid density or volume as discussed below. Thus, volumetric property estimation involves two steps, first the *choice of $P\underline{V}TN$ EOS model you want to use* and second *the estimation of parameters needed for the model.*

Liquid Molar Volume ($V^{(L)}$) Most $P\underline{V}TN$ EOSs have limited accuracy for predicting liquid densities unless they have many adjustable parameters specific for a particular substance or unless special attention has been paid to fitting density, as was the case with the addition of volume translation discussed in Section 8.3. As mentioned above, separate correlations are frequently used for saturated liquid density or molar volume. Again there are several options open to us, but most are connected to the approach suggested by Rackett (1970):

$$V_{sat}^{(L)} = \frac{RT_c}{P_c} Z_{RA}^{[1+(1-T_r)^{2/7}]} \tag{13-11}$$

where Z_{RA} is an empirical constant for each compound. If the volume is known at a particular temperature then Z_{RA} can be evaluated. Another convenient feature of the Rackett equation is that it has been successfully extended to complex liquid mixtures using pseudocritical property methods. Values of the Rackett parameter have been tabulated, but if they are not available then Corresponding States can be used with the acentric factor:

$$Z_{RA} = 0.29056 - 0.08775\,\omega \tag{13-12}$$

in a formulation suggested by Yamada and Gunn (1973).

Ideal-Gas State Heat Capacity (C_p^o) The calculation of C_p^o from theory was outlined in Section 8.4 where we saw that C_v^o or $C_p^o = C_v^o + R$ for an attenuated ideal-gas state contained translational, rotational, and vibrational contributions. After superimposing these contributions and making a few simplifying assumptions, an explicit temperature dependence for C_p^o results that in normal chemical engineering practice is captured by a polynomial expansion in T:

$$C_p^o = a + bT + cT^2 + dT^3 + \ldots \tag{13-13}$$

where a incorporates R from Eq. (8-82). As we noted in Chapter 9, the heat capacity of ideal gas *mixtures* in an attenuated state is simply given by,

$$C^o_{p\,mix} = \sum y_i\, C^o_{pi}$$

where y_i is the mole fraction of component i. Thus, once we estimate the pure component C^o_{pi}s then we only need to know the composition to get the mixture value.

Group contribution methods have been applied to the prediction of ideal-gas state heat capacities. The temperature effects are primarily associated with the activation of intramolecular vibrational motion as the translational and rotational modes of the molecule are normally all activated at temperatures of interest to chemical processing. Thus, intuitively one might expect that C_p^o could be correlated by group contributions as they are related directly to the number of bonds of particular types. Again we illustrate the application of group contributions using the Joback method. Other good techniques exist, such as those suggested by Benson and Yamada [see Reid *et al.* (1987) for details], but they are somewhat harder to use and they are comparable to Joback's in terms of accuracy—about 1 to 2% error is typical. Table 13.5 provides the group contributions for C^o_{pi} with the following combining equation for component i:

$$C_p^o = C^o_{pi} = \left(\sum_j v_j^{(i)}\,\Delta_{aj} - 37.93\right) + \left(\sum_j v_j^{(i)}\,\Delta_{bj} + 0.210\right)T + \left(\sum_j v_j^{(i)}\,\Delta_{cj} - 3.91\times 10^{-4}\right)T^2 + \left(\sum_j v_j^{(i)}\,\Delta_{dj} + 2.06\times 10^{-7}\right)T^3 \tag{13-14}$$

with T in K and C_p^o in J/mol K. The DIPPR database utilizes a different formulation for C_p^o based on the treatment proposed by Aly and Lee (1981):

$$C^o_{pi} = C_{1,i} + C_{2,i}\left[\frac{C_{3,i}/T}{\sinh\,(C_{3,i}/T)}\right]^2 + C_{4,i}\left[\frac{C_{5,i}/T}{\cosh\,(C_{5,i}/T)}\right]^2 \tag{13-15}$$

where $C_{1,i} - C_{5,i}$ are fitted constants. Both Eq. (13-14) and (13-15) are applicable from about 280 to 1100 K. By using sinh and cosh functions the DIPPR equation is more consistent with the mathematical structure of the vibrational and electronic contributions. As an exercise, prove that the DIPPR form is self-consistent, assuming the electronic contribution has a similar mathematical form to the vibrational contribution given in Eqs. (8-92) - (8-94) (see Problem 13.1).

Standard Enthalpy and Gibbs Free Energy of Formation **(ΔH_f^o and ΔG_f^o)** Again many options are available for estimating these properties. Methods by Benson, Yoneda, and Joback have been used widely with great success and are recommended. Benson's and Yoneda's algorithms are more involved and tedious because they attempt to capture second nearest neighbor effects [see Reid *et al.* (1987)]. However, with many fully automated property estimation systems in place this is less of a problem now. To illustrate the application of group contributions we again selected the Joback method partly because it is easier to use and it is reasonably accurate (to about 5 to 10 kJ/mol). The operating equations are:

$$\Delta H_f^o\ (298\text{ K}) = 68.29 + \sum_j \nu_j^{(i)}\ \Delta_{Hj}$$

$$\Delta G_f^o\ (298\text{ K}) = 53.88 + \sum_j \nu_j^{(i)}\ \Delta_{Gj} \qquad \text{(in kJ/mol)} \quad (13\text{-}16)$$

where Δ_{Hj} and Δ_{Gj} represent specific group contributions to enthalpy and Gibbs energy, respectively. Tabulated values for various groups of interest are given in Table 13.5. Values for both ΔH_f^o and ΔG_f^o at standard conditions of 298 K are given in Appendix G for selected compounds.

Example 13.4

Estimate the ideal gas state heat capacity function and the standard enthalpy and Gibbs free energy of formation for toluene and methylene chloride using the Joback method.

Solution

For toluene, contributions are obtained from Table 13.5 and summed over the molecular groups present.

$$\begin{array}{lllll} & -CH_3 & =CH- & =C< & \\ \sum \nu_j^{(i)}\ \Delta_{aj} = & 1.95E{+}1 & +\ 5(-2.14) & +\ (-8.25) & = 0.550 \\ \sum \nu_j^{(i)}\ \Delta_{bj} = & -8.08E{-}3 & +\ 5(5.74E{-}2) & +\ 1.01E{-}1 & = 3.799 \times 10^{-1} \\ \sum \nu_j^{(i)}\ \Delta_{cj} = & 1.53E{-}4 & +\ 5(-1.64E{-}6) & +\ (-1.42E{-}4) & = 2.80 \times 10^{-6} \\ \sum \nu_j^{(i)}\ \Delta_{dj} = & -9.67E{-}8 & +\ 5(-1.59E{-}8) & +\ 6.78E{-}8 & = -1.08 \times 10^{-7} \end{array}$$

Therefore, using Eq. (13-14),

$$C_{p_{toluene}}^o = -37.38 + 0.5899T - 3.88 \times 10^{-4} T^2 + 9.80 \times 10^{-8}\ T^3$$

Table 13.5 Joback Group Contributions for the Ideal Gas State Heat Capacity and Enthalpy and Gibbs free energy of formation [from Reid *et al.* (1987)]

	Δ Values					
	Δ_H kJ/mol	Δ_G kJ/mol	Δ_a	Δ_b J/mol K	Δ_c	Δ_d
Non-ring increments:						
$-CH_3$	-76.45	-43.96	1.95E+1	-8.08E-3	1.53E-4	-9.67E-8
$>CH_2$	-20.64	8.42	-9.09E-1	9.50E-2	-5.44E-5	1.19E-8
>CH–	29.89	58.36	-2.30E+1	2.04E-1	-2.65E-4	1.20E-7
>C<	82.23	116.02	-6.62E+1	4.27E-1	-6.41E-4	3.01E-7
$=CH_2$	-9.63	3.77	2.36E+1	-3.81E-2	1.72E-4	-1.03E-7
=CH–	37.97	48.53	-8.00	1.05E-1	-9.63E-5	3.56E-8
=C<	83.99	92.36	-2.81E+1	2.08E-1	-3.06E-4	1.46E-7
=C=	142.14	136.70	2.74E+1	-5.57E-2	1.01E-4	-5.02E-8
≡CH	79.30	77.71	2.45E+1	-2.71E-2	1.11E-4	-6.78E-8
≡C–	115.51	109.82	7.87	2.01E-2	-8.33E-6	1.39E-9
Ring increments:						
$-CH_2-$	-26.80	-3.68	-6.03	8.54E-2	-8.00E-6	-1.80E-8
>CH–	8.67	40.99	-2.05E+1	1.62E-1	-1.60E-4	6.24E-8
>C<	79.72	87.88	-9.09E+1	5.57E-1	-9.00E-4	4.69E-7
=CH–	2.09	11.30	-2.14	5.74E-2	-1.64E-6	-1.59E-8
=C<	46.43	54.05	-8.25	1.01E-1	-1.42E-4	6.78E-8
Halogen increments:						
–F	-251.92	-247.19	2.65E+1	-9.13E-2	1.91E-4	-1.03E-7
–Cl	-71.55	-64.31	3.33E+1	-9.63E-2	1.87E-4	-9.96E-8
–Br	-29.48	-38.06	2.86E+1	-6.49E-2	1.36E-4	-7.45E-8
–I	21.06	5.74	3.21E+1	-6.41E-2	1.26E-4	-6.87E-8
Oxygen increments:						
–OH (alcohol)	-208.04	-189.20	2.57E+1	-6.91E-2	1.77E-4	-9.88E-8
–OH (phenol)	-221.65	-197.37	-2.81	1.11E-1	-1.16E-4	4.94E-8
–O– (non-ring)	-132.22	-105.00	2.55E+1	-6.32E-2	1.11E-4	-5.48E-8
–O– (ring)	-138.16	-98.22	1.22E+1	-1.26E-2	6.03E-5	-3.86E-8
>C=O (non-ring)	-133.22	-120.50	6.45	6.70E-2	-3.57E-5	2.86E-9
>C=O (ring)	-164.50	-126.27	3.04E+1	-8.29E-2	2.36E-4	-1.31E-7
O=CH– (aldehyde)	-162.03	-143.48	3.09E+1	-3.36E-2	1.60E-4	-9.88E-8
–COOH (acid)	-426.72	-387.87	2.41E+1	4.27E-2	8.04E-5	-6.87E-8
–COO– (ester)	-337.92	-301.95	2.45E+1	4.02E-2	4.02E-5	-4.52E-8
=O (except as above)	-247.61	-250.83	6.82	1.96E-2	1.27E-5	-1.78E-8

Δ_a, Δ_b, Δ_c, Δ_d for C_p [Eq. (13-14)]

Δ_H for ΔH_f^o ; Δ_G for ΔG_f^o [Eq. (13-16)]

continued

Table 13.5 (continued) Δ Values

	Δ_H kJ/mol	Δ_G kJ/mol	Δ_a	Δ_b J/mol K	Δ_c	Δ_d
Nitrogen increments:						
$-NH_2$	-22.02	14.07	2.69E+1	-4.12E-2	1.64E-4	-9.76E-8
>NH (non-ring)	53.47	89.39	-1.21	7.62E-2	-4.86E-5	1.05E-8
>NH (ring)	31.65	75.61	1.18E+1	-2.30E-2	1.07E-4	-6.28E-8
>N– (non-ring)	123.34	163.16	-3.11E+1	2.27E-1	-3.20E-4	1.46E-7
–N= (non-ring)	23.61	—	—	—	—	—
–N= (ring)	55.52	79.93	8.83	-3.84E-3	4.35E-5	-2.60E-8
=NH	93.70	119.66	5.69	-4.12E-3	1.28E-4	-8.88E-8
–CN	88.43	89.22	3.65E+1	-7.33E-2	1.84E-4	-1.03E-7
$-NO_2$	-66.57	-16.83	2.59E+1	-3.74E-3	1.29E-4	-8.88E-8
Sulfur increments:						
–SH	-17.33	-22.99	3.53E+1	-7.58E-2	1.85E-4	-1.03E-7
–S– (non-ring)	41.87	33.12	1.96E+1	-5.61E-3	4.02E-5	-2.76E-8
–S– (ring)	39.10	27.76	1.67E+1	4.81E-3	2.77E-5	-2.11E-8

Again following a similar procedure for methylene chloride

$$>CH_2 \qquad -Cl$$

$$\sum v_j^{(i)} \Delta_{aj} = -9.09E{+}1 + 2(3.33E{+}1) = -2.43 \times 10^{-1}$$

$$\sum v_j^{(i)} \Delta_{bj} = 9.5E{-}2 + 2(-9.63E{-}2) = -9.76 \times 10^{-2}$$

$$\sum v_j^{(i)} \Delta_{cj} = -5.44E{-}5 + 2(1.87E{-}4) = 3.20 \times 10^{-4}$$

$$\sum v_j^{(i)} \Delta_{dj} = 1.19E{-}8 + 2(-9.96E{-}8) = -1.87 \times 10^{-7}$$

and

$$C^o_{p_{meth.\ chloride}} = -62.23 + 0.1124T - 7.10 \times 10^{-5}\, T^2 + 1.90 \times 10^{-8}\, T^3$$

For the enthalpy and Gibbs energy of formation, Eq. (13-16) along with Table 13.5 is used:

for toluene

$$\Delta H^o_f = 68.29 + \sum v_j^{(i)} \Delta_{Hj} = 68.29 - 19.57 = 48.72 \text{ versus 50 kJ/mol (exp)}$$

$$\Delta G^o_f = 53.88 + \sum v_j^{(i)} \Delta_{Gj} = 53.88 + 66.59 = 120.47 \text{ versus 122.1 kJ/mol (exp)}$$

for methylene chloride

$$\Delta H^o_f = 68.29 - 163.74 = -95.45 \text{ versus -95.46 kJ/mol (exp)}$$

$$\Delta G^o_f = 53.88 - 120.20 = -66.32 \text{ versus -68.91 kJ/mol (exp)}$$

In both cases the agreement is remarkable.

Vapor Pressure (P_{vp}) and Heat of Vaporization (ΔH_{vap}) These properties are of enormous importance to chemical engineering operations, particularly separations involving distillation and gas absorption. Although they are not normally predicted directly form group contributions, an indirect route is frequently followed in the absence of sufficient data to produce a correlation. For example, with measured or predicted values of T_c, P_c, and T_b one can approximate the vapor pressure - temperature behavior with an integrated form of the Clapeyron equation [see Section 15.4 and Eq. (15-63)]:

$$\ln P_{vp} = A - \frac{B}{T} \tag{13-17}$$

where the parameter B is related to heat of vaporization, $B = \Delta H_{vap} / R\, \Delta Z_{vap}$ and A is the constant of integration. The values of A and B depend on the particular substance. By using Eq. (13-17) at the critical point where $T = T_c$ and $P = P_c$ and at the normal boiling point at 1.013 bar (1 atm) we obtain an interpolating equation for the vapor pressure at any temperature between T_c and T_b.

$$\ln\left(\frac{P_{vp}}{P_c}\right) = h^*\left(1 - \frac{1}{T_r}\right) \tag{13-18}$$

with P_{vp} and P_c in bars,

$$h^* = T_{br}\left[\frac{\ln (P_c/1.013)}{1 - T_{br}}\right]$$

and the reduced temperatures, $T_{br} = T_b / T_c$ and $T_r = T/T_c$

This treatment assumes that $\Delta H_{vap} / \Delta Z_{vap}$ is a constant, independent of temperature, which is approximately correct for most pure substances except near their critical points. To achieve higher accuracy, the Antoine equation is frequently used where:

$$\ln P_{vp} = A - \frac{B}{T + C} \tag{13-19}$$

Normally, A, B, and C are fitted constants. Note that if $C = 0$, then the Clapeyron equation results. One of the attractive features of the Antoine equation is that the constants have been tabulated for a large number of substances [for example see Appendix G of this text and Appendix A of Reid *et al.* (1987)]. Also extended forms of the Antoine equation have been developed by Wagner and others for improved accuracy.

In the complete absence of vapor pressure data, an estimate of ΔH_{vap} can be produced using Corresponding States where the reduced form is given by:

$$\Delta H_{vap_r} \equiv \frac{\Delta H_{vap}}{RT_c} = f(T_r, \omega) \tag{13-20}$$

Various forms of $f(T_r, \omega)$ have been proposed. For example, the extension of Pitzer's model by Carrruth and Kobayashi (1972) which gives for $0.6 < T_r \leq 1.0$:

$$\frac{\Delta H_{vap}}{RT_c} = 7.08\,(1 - T_r)^{0.354} + 10.95\,\omega\,(1 - T_r)^{0.456} \tag{13-21}$$

or the Reidel equation can be used which again assumes that $\Delta H_{vap}/\Delta Z_{vap}$ is constant and gives,

$$\Delta H_{vap} = 1.093\,RT_c\left[T_{br}\,\frac{\ln P_c - 1.013}{0.930 - T_{br}}\right] \tag{13-22}$$

where P_c is expressed in bar units. We can also use the Antoine equation to estimate ΔH_{vap} with an estimated value for ΔZ_{vap}:

$$\Delta H_{vap} = RT_c\,\Delta Z_c\left(\frac{B}{T_c}\right)\left(\frac{T_r}{T_r + C/T_c}\right)^2 \quad (\text{at } T_r) \tag{13-23}$$

Before we leave the general topic of group contribution methods for pure components, one final word of caution is needed. As Mavrovouniotis (1990) is quick to point out, while group contribution techniques clearly represent the constituent molecular substructures that make up the molecule, in most situations they do not consider the detailed bond structure and connectivity of the groups within the molecule. This produces some problems and inaccuracies. For example, isomers frequently will produce the same value of an estimated property. Compounds with significant bond resonance where hybridization of conjugate forms occur are frequently not well represented and require special treatment and corrections to improve the accuracy of the group contribution method. In addition, ionic species have not yet been treated by group additivity approaches, again largely because of the delocalization and distribution of charge density by conjugation.

As alternatives to the traditional group contribution methods, Mavrovouniotis (1990) and Constantinou and Gani (1994) have introduced what they called the *ABC* approach, which is short for ***Atoms, Bonds, and Conjugates***, that attempts to more rigorously account for the details of the bond structure and its connectivity. They have achieved some success in being able to distinguish between isomer thermochemical properties, but their methods still require further development and testing with other properties.

This completes our discussion of pure component property estimation; now we can move on to mixtures.

13.4 Group Contribution Methods for Estimating Mixture Properties

While the estimation of pure component properties proceeded in a reasonably straightforward manner using group contribution methods with considerable success and accuracy, the situation with mixtures is very different. First of all, key effects that cause compositionally dependent non-ideal behavior as expressed in mixture fugacity or activity coefficients are primarily due to intermolecular interactions between unlike components of the mixture. This situation is inherently more difficult to deal with using group contributions.

For example, when a pressure explicit equation of state is employed, the magnitude of the binary interaction parameter δ_{ij} is extremely important in scaling the level of non-ideal behavior. There are no completely general estimation methods for δ_{ij}—most often values are established by fitting experimental binary data, for example from VLE or LLE measurements. Thus, one must rely on compilations of binary interaction coefficients or be willing to regress suitable values from existing data when employing mixture $P\underline{V}TN$ EOSs. Good examples of doing this are found in Chapter 15 where phase equilibrium applications are discussed. For instance, when cubic mixture EOSs are used to predict solubility in supercritical fluids the majority of the effect is controlled by the magnitude of δ_{ij} over a very small range, typically $0.01 < \delta_{ij} < 0.5$ (see Section 15.8).

For activity coefficient models things are similar with a few exceptions that employ a group contribution approach, such as the ASOG and UNIFAC methods. Thus if you are interested in using a mixture ΔG^{EX} model like the Wilson, NRTL, or UNIQUAC models described in Section 11.7, binary parameters as given in Tables 11.2 and 13.1 are needed. Normally, model parameters are fit to experimental data of some type. It might be specific VLE or LLE measurements over a range of temperatures and pressures or a solubility determination or a colligative property measurement such as osmotic pressure. In any case, a direct measurement of non-ideal behavior is the most direct and accurate route for estimating γ. Reid *et al.* (1987) and Prausnitz *et al.* (1986) discuss this subject in much greater depth and anyone really interested in estimating activity coefficients should consult these references.

As a "last resort", in the absence of any reliable data, activity coefficients can be estimated with group contributions techniques. The ASOG (Analytical Solution of Groups) method developed by Wilson and coworkers [Wilson and Deal (1962)] employs a predictive method to estimate the binary coefficients of the Wilson equation (Section 11.7.3) while the UNIFAC (UNIquac group-Functional Activity Coefficient) method developed by Prausnitz and coworkers [Fredenslund *et al.* (1975)] utilizes the UNIQUAC model (Section 11.7.4) as a starting point. Both of these methods follow the approach suggested by Irving Langmuir in the 1920's to extend the group contribution concept to mixtures. In this framework, the activity coefficient of each component is a sum over all possible binary interactions between the functional groups on that component and all other groups that comprise the components in the mixture of interest. For example, the activity coefficient of phenol in a solution of acetone would be partly due to a contribution from the hydroxyl group of phenol interacting with just the ketone group on acetone. Just as was the case for pure component group contributions, accurate correlations depend on the proper selection of the functional groups and on having a sufficiently large database of binary systems to regress each group's contribution to γ. Because the UNIFAC and ASOG methods are quite similar in their formulations, we restrict our discussion here to UNIFAC. Furthermore, because Reid *et al.* (1987), Smith

Van Ness and Abbott (1996) and Sandler (1989) present excellent summaries of the UNIFAC method, we will only briefly discuss the model here.

UNIFAC uses the combinatorial part of the UNIQUAC model ($\ln \gamma_i^C$)exactly as given in Table 11.2. Keep in mind that only pure component size parameters for area (q_i) and volume (r_i) are needed for this part. In UNIFAC, these are obtained by summing over the functional groups k that comprise component i:

$$q_i = \sum_k v_k^{(i)} Q_k \qquad r_i = \sum_k v_k^{(i)} R_k \tag{13-24}$$

where Q_k and R_k are the size parameters for functional group k. The residual part of UNIQUAC is replaced with,

$$\ln \gamma_i^R = \sum v_k^{(i)} [\ln \Gamma_k - \ln \Gamma_k^{(pure\ i)}] \tag{13-25}$$

where Γ_k is the contribution of functional group k to the residual part of the activity coefficient and $\Gamma_k^{(pure\ i)}$ is the contribution of group k in a hypothetical standard state consisting of pure i ($x_i = 0$) at the same T and P as the mixture to produce a symmetrically normalized system where $\gamma_i \to 1.0$ as $x_i \to 1.0$. Γ_k is evaluated using an equation similar to the residual formulation for UNIQUAC given in Table 11.2:

$$\ln \Gamma_k = Q_k \left(1 - \ln \sum_m \theta_m \tau_{mk} - \sum_m \left(\frac{\theta_m \tau_{km}}{\sum_n \theta_n \tau_{nm}} \right) \right) \tag{13-26}$$

where the sums run over the total number of groups to capture interactions between groups and for any functional group k:

$$\theta_k = \frac{X_k \frac{z}{2} Q_k}{\sum_m X_m \frac{z}{2} Q_m}$$

The parameter X_k is the mole fraction of group k in the mixture,

$$X_k = \frac{\sum_j v_k^{(j)} x_j}{\sum_j \sum_m v_m^{(j)} x_j} \tag{13-26}$$

The only missing part is the energetic binary interaction parameter between groups j *and* k which is given as,

$$\tau_{jk} = \exp\left[\frac{u_{jk} - u_{kk}}{RT}\right] = \exp\left[\frac{-a_{jk}}{T}\right] \tag{13-27}$$

Specific values for τ_{jk} or a_{jk} are determined by regression of a large quantity of binary activity coefficient data to obtain the best set of fitted parameters. Table 13.6 gives values of a_{jk} for interactions between six representative groups. Reid *et al.* (1987) give a fairly complete listing of interaction parameters. Modifications to UNIFAC appear periodically where the temperature dependence of the τ_{jk} term is altered to improve the prediction of γ_i. Updated databases of model parameters are maintained by a number of groups, including J.M. Prausnitz (1995), Dortmund, and Aspen Technology (1998).

In many situations the UNIFAC and ASOG models will give very reasonable predictions even when highly non-ideal conditions exist. In other cases, totally incorrect γs are estimated which can translate to significant errors in engineering computations. For example, in very dilute solutions, γ_i for the dilute components gets very large as it approaches γ_i^∞ at infinite dilution, thus even modest errors in γ_i can result in extremely large errors (of an order of magnitude or more) in the predicted separation of the dilute components in a multistage distillation or extraction. In these difficult systems, you must, to quote Reid, Prausnitz and Poling (1987, p. 380), "Face the facts, you cannot get something for nothing . . . If you want reliable results you will need some reliable experimental data." For the example given, a few measurements of dilute behavior would have provided a route to regress UNIQUAC binary parameters directly to produce a much more realistic estimate of γ_i.

13.5 Applications to Modern Process Analysis and Simulation

We now have a modeling framework for predicting and correlating the thermodynamic properties of real fluids that needs to be connected to the process of actually making estimates of the magnitude of non-ideal effects important to practical engineering calculations. Over the past several decades the chemical engineering community has been very resourceful in compiling data for pure components and mixtures and now one frequently has access to these compilations electronically. Furthermore, there are many techniques for estimating the properties of pure materials and mixtures that require only a minimal amount of information—typically molecular structure, molecular weight, and sometimes normal boiling point. Property estimation methods in practice today rely on empirical functional group contribution techniques convenient to use and very accurate for most pure component parameters.

For mixtures, the situation is less tractable—binary interaction parameters for $P\underline{V}TN$ EOS or $\Delta G^{EX}-\gamma$ models should come from experimental data to ensure that the most accurate property estimates result. In the absence of such data, first-order approximations can be made using group contribution techniques like UNIFAC or ASOG for γ_i or Corresponding States for $P\underline{V}TN$ volumetric properties such as the fugacity coefficient $\hat{\phi}_i$.

Most modern chemical process flowsheet simulators such as ASPEN PLUS™, ProSim™, or ChemStation™ have well maintained physical property models and databases that can generate results for very complex systems. However, in order to get accurate design and simulation estimates one must be careful about selecting the models and parameters to use. The default or standard set of property models is selected too often by innocent users who believe that the people who have developed the simulator must be "experts" on all the physical properties they need to carry out their simulation. In reality, the quality of any simulation depends in large part on the engineering skills of the user not the software provider to select and modify the physical property models and parameters as required.

Table 13.6 Sampling of UNIFAC Group-Group Interaction Parameters [from Reid *et al.* (1987), Tables 8-21 and 8-22]

group j	group k	1 $-CH_3$	2 $-CH_2$	3 $-CH$	5 $-OH$	21 $-CCl$	13 $-CH_2O$
		a_{jk} (in K)					
1 $-CH_3$		0	86.020	61.130	986.500	35.930	251.500
2 $-CH_2$		−35.360	0	38.810	524.100	204.600	214.500
3 $-CH$		−11.120	3.446	0	636.100	−18.810	32.140
5 $-OH$		156.400	457.000	89.600	0	75.620	28.060
21 $-CCl$		91.460	97.510	4.680	562.200	0	225.400
13 $-CH_2O$		83.360	26.510	52.130	237.700	301.100	0

Figure 13.2 illustrates the calculational path and user inputs into the algorithm used for estimating thermodynamic derived properties using an equation of state (EOS) approach for all fluid phases. Note that in this instance the user is supplying only the components and their molecular structure, no new experimental data or models are being provided. Thus the simulator must either have all the required data in its database or it must have the capability to estimate the unknown parameters needed.

The issues and concerns to keep in mind regarding the estimation of physical properties are:

(1) Never underestimate the value of a few *reliable* data points to assist in the estimation and validation of predicted physical property values.

(2) Never overestimate the quality or appropriateness of a particular property model. Whether it is theoretical or empirically-based, or something in between, is not as relevant as whether it works well for your specific application.

(3) Be careful not to use semi-empirical models with fitted parameters outside their range.

(4) Always be skeptical about computer generated results which often are so neatly formatted and packaged with high quality output (for example, color graphics and tabulations of results) that they might be assumed to be correct. You should as a general rule conduct sensitivity studies and run validation problems that either represent asymptotic limits where rigorous theories apply or are specific problems whose solutions are known.

Despite these limitations, you should remain very optimistic about the creative power that you have to estimate reasonable property values with only limited knowledge about the behavior of the components involved.

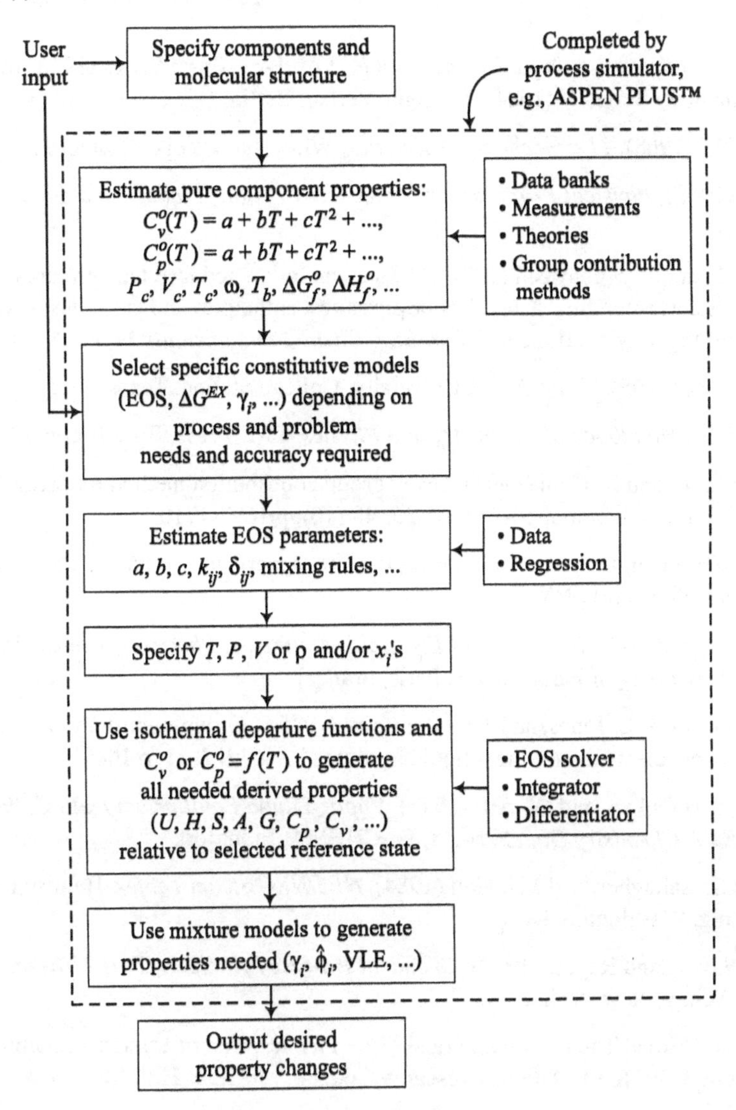

Figure 13.2 Typical flow chart for thermodynamic property calculations using an $PV\underline{T}N$ EOS and/or ΔG^{EX} approach.

References

Aly, F.A. and L.L. Lee (1981), "Self-consistent equations for calculating the ideal gas heat capacity, enthalpy, and entropy," *Fluid Phase Equilibria* **6**, p 169-179.

American Petroleum Institute in *Technical Data Book Petroleum Refining,* 3rd ed. Houston, Texas (1981).

ASHRAE, *Thermodynamic Properties of Refrigerants* (1969) and *Fundamentals*, American Society of Heating, Refrigeration, and Air Conditioning Engineers, Atlanta, Georgia (1993ff).

Beilstein's ***Handbuch der Organischen Chemie***, Vol. 1-27, Basic Series (1918) to Supplementary Series V (1984), Springer Verlag, Berlin.

Benson, S.W. (1968), *Thermochemical Kinetics,* Wiley, New York, Chapter 2.

Bondi, A. (1968), *Physical Properties of Molecular Crystals, Liquids, and Glasses,* Wiley, New York.

Carruth, G.F. and R. Kobayashi (1972), "Extension to low reduced temperatures of three-parameter corresponding states: Vapor pressures, enthalpies and entropies of vaporization, and liquid fugacity coefficients," *Ind. Eng. Chem. Fundamentals* **11**, p 509-517.

Chem Share™, (1995), Texas A&M University, College Station, Texas.

CRC (1995 ff), *Handbook of Chemistry and Physics, CRC Press*, Boca Raton, FL.

Constantinou, L. and R. Gani (1994), "New group contribution method for estimating properties of pure compounds," *AIChE J.* **40** (10), p 1697-1710.

DIPPR Design Institute for Physical Property Data, American Institute of Chemical Engineers, New York, NY.

ELDATA (1995ff)- *The International Electronic Journal of Physico-Chemical Data* [editorial offices, University of Paris, CNRS, Paris, France]

Fredenslund, Aa, R.L. Jones, and J.M. Prausnitz (1975), "Group-contribution estimation of activity coefficients in non-ideal liquid mixtures," *AIChE J.* **21**, p 1086.

Gmehling, J.,U.Onken, and W. Arlt (1977), *Vapor-Liquid Equilibrium Data Collection, DECHEMA Chemistry Data Series* 1, DECHEMA, Frankfort.

Haar, L., J.S. Gallagher, and J.H. Kell (1984), *NBS/NRC Steam Tables,* Hemisphere Publishing, Washington, DC.

Hougan, Watson and Ragatz (1959), *Chemical Process Principles, Part 2 Thermodynamics,* 2nd ed, Wiley, New York, NY.

International Critical Tables of Numerical Data, Physics, Chemistry and Technology (1926), Washburn, E.W. (ed.), National Research Council, McGraw-Hill, New York.

JANAF Thermochemical Data and Tables (1971), Thermal Research Laboratory, Dow Chemical Company, Midland, MI, 2nd ed., U.S. National Bureau of Standards, NSRDS-NB337 (updated paper and electronic versions available)

Joback, K.G. (1984), Master's thesis, Chemical Engineering Department, Massachusetts Institute of Technology, Cambridge, MA.

Keenan, J.H., F.G. Keyes, P.G. Hill, and J.G. Moore (1978), *Steam Tables,* Wiley, New York

Klincewicz, K.M. and R.C. Reid (1984), "Estimation of critical properties with group contribution methods," *AIChE J.* **30**, p 137-142.

Landolt-Bornstein, *Physikalisch-chemische tabellen* (1923) and *Zahlenwerte und Funktionen aus Naturwissenschaften und Technik*, Bands 1-4 (1950 - 1972), Springer Verlag, Berlin.

Mavrovouniotis, M.L. (1990), "Estimation of properties from conjugate forms of molecular structures: the ABC Approach," *Ind. Eng. Chem. Research* **29**, p 1943-1953

Moore, K.V. (1972), *ASTEM - A Collection of FORTRAN Subroutines to Evaluate the 1967 ASME* Equations of State for Water/Steam and Derivatives of these Equations, Aerojet Nuclear Co. Prepared for the US Atomic Energy Commission Idaho Falls, ID.

Natural Gas Processors Assn. (1981), *Engineering Data Book.*

Prausnitz, J.M. and others, various papers on "Continuous Thermodynamics": Kotterman, R.L., D. Dimitrelis, and J.M. Prausnitz (1984), *Ber. Bunsenges. Phys. Chem.* **88**, p 796-801; Kotterman, R.L. and J.M. Prausnitz (1985), *Ind. Eng. Chem. Process Design and Develoopment* **24**, p 434-443; Chou, G.F. and J.M. Prausnitz (1986), *Fluid Phase Equilibria* **30**, p 75-82; Kotterman, R.L., R. Bender, and J.M. Prausnitz (1985), *Ind. Eng. Chem. Process Design and Development* **24**, p 194-203.

Perry, R.H., D.W. Green, and J.O. Maloney (eds) (1984), *Perry's Chemical Engineers' Handbook*, 6th ed, McGraw-Hill, New York, NY, Section 3 Physical and Chemical Data.

Rackett, H.G. (1970), "Equation of state for saturated liquids," *J. Chem. Eng. Data* **15**, p 514-517.

Sandler, S. (1989), *Chemical and Engineering Thermodynamics*, 2nd ed., Wiley, New York, NY, p 340-345.

Schwarzenbach, R.P., P.M. Gschwend, and D.M. Imboden (1993), *Environmental Organic Chemistry*, Wiley, New York, NY.

Simulation Sciences, Inc. (1995), ProSim™, Fullerton, CA.

Smith, J.M., H.C. Van Ness, and M.M. Abbott (1996), *Introduction to Chemical Engineering Thermodynamics*, 5th ed., McGraw-Hill, New York, NY, Appendix B, p 740-746.

Stull, D.R., E.F. Westrum, and G.C. Sinke (1969), *The Chemical Thermodynamics of Organic Compounds*, Wiley, New York, NY.

van Krevelen, D.W. (1976), *Properties of Polymers:Their Estimation and Correlation with Chemical Structure*, 2nd ed., Elsevier, Amsterdam.

Wilson, G.M. and C.H. Deal (1962), "Activity coefficients and molecular structure," *Ind. Eng. Chem. Fundamentals* **1**, p 20-24.

Yamada, T. and R.P. Gunn (1973), "Saturated liquid molar volumes. The Rackett equation," *J. Chem. Eng. Data* **18**, p 234-236.

Suggested Readings

Aspen Technology (1998), *Aspen Plus® Version 10* and associated documentation (on-line and printed), 10 Canal Park, Cambridge, MA—detailed descriptions of computerized physical properties models and databases for property data and model parameters.

Lyman, W.J., W.F. Reehl, And D.H. Rosenblatt (eds.) (1990), *Handbook of Chemical Property Estimation Methods: Environmental Behavior of Organic Compounds*, American Chemical Society, Washington DC. [an environmentally oriented reference on methods for obtaining estimates of property values for organic chemicals—of particular importance to the material covered in this chapter are the following: (1) Activity coefficients—Chapter 11; (2) Boiling point—Chapter 12; (3) Heat of vaporization—Chapter 13; (4) Vapor pressure—Chapter 14; (5) Densities of vapors, liquids, and solids—Chapter 19; and (6) Heat capacity—Chapter 23].

Prausnitz, J.M., R.N. Lichtenthaler, E. Gomez de Azevedo (1986), *Molecular Thermodynamics of Fluid Phase Equilibria*, 2nd ed., Prentice-Hall, Englewood Cliffs, NJ [a comprehensive reference on the theory and empiricism behind models for condensed phase non-ideal behavior. For information on activity coefficient models - see Sections 6.10-6.15, p 226-268 and on techniques for estimating activity coefficients - see Appendix VI, p 544-551].

Reid, R.C. J.M. Prausnitz and B.E. Poling (1987) *The Properties of Gases and Liquids,* 4th ed. McGraw-Hill, New York, NY. [the complete reference on the prediction of physical properties—of particular importance to the material covered in this chapter are the following: (1) Pure components - Chapter 2; (2) Ideal gas heat capacities and enthalpy and Gibbs energy of formation - Chapter 6; (3) Vapor pressures and enthalpies of vaporization - Chapter 7; and (4) UNIFAC group contribution techniques for γ_i - Chapter 8, p 314-337].

Problems

13.1. Verify that the Aly-Lee formulation for C_p^o given in Eq. (13-15) is consistent with the molecular interpretation of C_p^o and C_v^o given in Chapter 8. Also for toluene and methylene chloride do you feel that both Eq. (13-14) and Eq. (13-15) will provide essentially the same accuracy for the predicted value of C_p^o at 300 K (see Example 13.4).

13.2. The experimental isobaric ideal gas state heat capacity for propylene is given by

$$C_p^o(\text{exp}) = 88.16 + 5.374T - 2.757 \times 10^{-3}T^2 + 5.239 \times 10^{-7}\, T^3 \text{ (J/kg K)}$$

and critical constants and acentric factors are

$$T_c = 365 \text{ K},\ P_c = 46.20 \text{ bar},\ \omega = 0.148$$

How do these values compare with estimated values using Joback's group contribution method?

13.3. Estimate the critical constants and vapor pressures of phenylene oxide.

13.4. The mixture properties of caffeine and carbon dioxide are important to certain decaffeination processes using supercritical fluid extraction. Unfortunately many of caffeine's properties are not known. Fluid phase equilibria calculations will require both vapor pressures and $P\underline{V}TN$ properties (see Problem 15.38).

Property	Caffeine	Carbon Dioxide
Structural formula	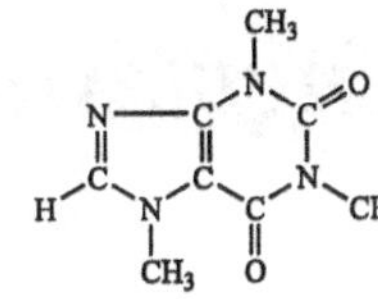	O=C=O
	1,3,7 Trimethylxanthine	CO_2
Molecular weight	194.2	44.01
T_c (in K)	—	304.1
P_c (in bar)	—	73.8
ω acentric factor	—	0.239
solid phase density (kg/m^3)	1230	—
melting point (in °C)	236.0	-55.6

Estimate the density of a mixture of a 5% mol caffeine in CO_2 mixture at 80 bar and 310 K. Also estimate the vapor pressure of liquid caffeine and compare it with the experimental data from Ebeling and Franck (1984) in *Ber. Bunsenges. Phys. Chem.* **88**, p 862-865 and from Bothe and Cammenga (1979) in *J. of Thermal Analysis* **16**, p 267-275. Solid and liquid caffeine vapor pressure data can be represented by the following correlations:

for solid caffeine ($T < 236°C$)

$\log_{10}(P \text{ in Pa}) = -5781/T + 15.031$ (from Bothe and Cammenga)

$\log_{10}(P \text{ in bar}) = -5411/T + 7.46 + \log_{10}(0.08314\, T)$ (from Ebeling and Franck)

for liquid caffeine ($T > 236°C$)

$\log_{10}(P \text{ in Pa}) = -3918/T + 11.143$ (from Bothe and Cammenga)

Part III Applications

Practical Heat Engines and Power Cycles 14

This chapter deals with the conversion of various forms of energy, including chemical, thermal, and hydraulic, into useful work or power. Although today these topics are more generic to the core curriculum of mechanical engineering, chemical engineers are frequently faced with making decisions regarding equipment design and selection or the specification of operating conditions to improve the energy conversion and utilization efficiency of a particular chemical process. Issues relating to cogeneration of electricity and process steam, cascaded refrigeration for sub-ambient temperature processing, and heat integration to reduce process steam and cooling water requirements are examples of such design choices. In addition, with growing interest in alternative non-CFC refrigerants and power production from lower temperature energy resources such as geothermal, solar thermal and process waste heat, selection of non-aqueous working fluids as prime movers in refrigeration, heat pump, or heat engine cycles have become an issue that chemical engineers are often asked to resolve. The methodologies developed in this chapter will assist you in carrying out these performance evaluations.

The conceptual framework for using the combined First and Second Laws of thermodynamics to analyze the feasibility and efficiency of various heat to work or power generation schemes was developed in Chapter 4. But at that stage of our development we were restricted to treat systems with idealized physical properties—including ideal gases, incompressible fluids, and isobaric processes at constant density or volume. Examples include the reversible adiabatic compression of an ideal gas and the isobaric cooling of a liquid under constant density conditions. For the ideal gas case, any PdV or VdP work contributions were easily calculated and for the incompressible or isobaric, constant density case, these work effects could be neglected. Now that we have introduced more comprehensive physical property models in Part II, we can examine how heat engines will perform using working fluids whose properties are non-ideal over at least a portion of the operating envelope of a specific power cycle. Of particular interest is when phase changes or critical points are encountered. A key question is whether these non-ideal working fluid properties can be exploited in such a way to improve performance and efficiency.

The analysis of energy conversion and utilization has been called *Second Law Analysis,* which on the surface may seem correct, but as we know from the development of Part I, both the First and Second Laws apply simultaneously. More appropriately, the central theme should be the maximization of work or power output by reducing irreversible losses, or stated another way, the minimization of parasitic losses that reduce the work producing potential. By reaffirming the connectivity between the First and Second Laws using a different framework, your appreciation for quantifying the irreversibilities of practical processes in terms of entropy generation should be enhanced.

We begin by expanding our analytical tools by introducing availability, lost work and exergy concepts to facilitate the analysis and to permit comparison of actual work generating capacities with maximum limits. The thermodynamic elements of pinch technology will be covered to illustrate an approach for optimization of heat exchanger network design to maximize efficiency by integrating hot and cold streams. Turbine, pump and compressor designs will also be considered, in part to illustrate the relevance of the thermodynamics of compressible fluid flow, and in part because they, like heat exchangers, are important components of power cycles. In the final section we will consider three specific heat to work conversion systems--Rankine, Brayton, and Otto cycles. Each will be discussed separately as they are important to many chemical engineering operations. In addition, they will provide an opportunity to illustrate how a process can be analyzed when real working fluids are employed.

14.1 Availability, Lost Work, and Exergy Concepts

The maximum possible work that can be produced by a given process occurs only at completely reversible conditions. This situation, of course, represents a hypothetical idealization. Irreversibilities exist in all *real* processes that lead to the net generation of entropy. Entropy increases may occur in the system itself or the surroundings or both—in all cases the total entropy change of the universe is positive. A key problem for many who study thermodynamics is to be able to fully appreciate the powerful influence that the First and Second Laws have on the performance of work producing or consuming processes. To enhance your understanding of these sometimes subtle and frequently misunderstood concepts, this chapter generalizes the evaluation of maximum work and provides a means for locating and quantifying the irreversibilities present in any process.

First, we consider a somewhat simplified, non-reacting open system that produces work, as shown in Figure 14.1. The process depicted operates at steady state with no kinetic (KE) or potential (PE) energy effects and with all boundary and internal energy transfers occurring reversibly and quasi-statically. The only heat interaction permitted is one that occurs between the system and the surroundings which in this case is a very large heat reservoir which remains at a constant temperature of T_o. Here heat is rejected to the environment at its so-called ambient "dead state" where $T = T_o$ and $P = P_o$. To keep the heat exchange process reversible, a Carnot heat engine operates between the system temperature T and T_o. All work quantities are stored or removed reversibly from a work reservoir (no friction or other dissipative effects are present). In order for mass flow into and out of the system and other processes within the system boundary to be carried out reversibly, all gradients of intensive properties must be kept infinitesimally small. If mixing of species occurs, it too must be done reversibly using an approach similar to what was introduced in Section 9.9. Figure 14.1 schematically illustrates the work, heat, and mass flows. At steady state the total system energy $\underline{E}$ and mass or moles N are constant. Thus,

$$d\underline{E} = 0 \qquad \text{and} \quad d\underline{E}/dt = 0$$

$$\delta n = \delta n_{in} = \delta n_{out} \quad \text{and} \quad \delta N/\delta t = 0$$

For the moment we assume that the total system volume $\underline{V}$ is time invariant as well. This will eliminate any secondary work effects due to the system volume expanding or contracting against the ambient environment at P_o. A steady state First Law balance around the process (primary system) yields:

$$d\underline{E} = 0 = \delta Q_s + \delta W_s + (H_{in} - H_{out})\,\delta n \tag{14-1}$$

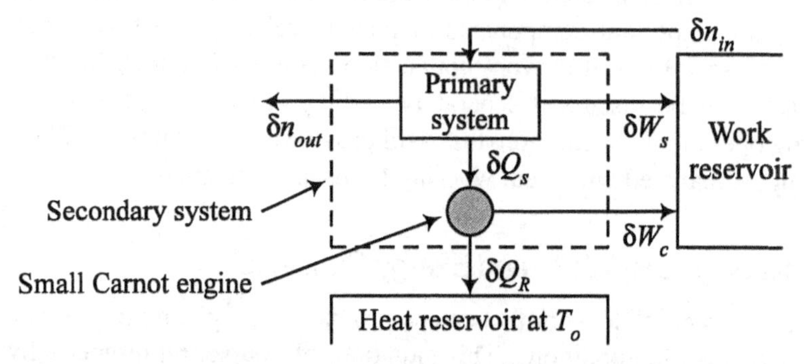

Figure 14.1 Steady state, non-reacting system. Directional sign conventions for work and heat flows correspond to a net work producing process.

Note that the direction of the heat or work interaction determines whether it is positive (entering) or negative (leaving). A steady state First Law balance around the Carnot heat engine gives:

$$d\underline{E} = 0 = -\delta Q_s + \delta Q_R + \delta W_c$$

Here we must change the sign of δQ_s to account for its directional change relative to the original primary system. Rearranging,

$$\delta Q_s = +\,\delta Q_R + \delta W_c \tag{14-2}$$

A steady state entropy balance for the composite secondary system—the process and heat engine yields:

$$d\underline{S} = 0 = \,+\,\delta Q_R/T_o + (S_{in} - S_{out})\delta n \tag{14-3}$$

Combining Eqs. (14-1) through (14-3) to eliminate δQ_s and δQ_R and solving for the total work interaction (Carnot + shaft work) gives:

$$\delta W_{total} = \delta W_c + \delta W_s = [(H_{out} - H_{in}) - T_o(S_{out} - S_{in})]\delta n \tag{14-4}$$

Because we are dealing with a reversible process, Eq. (14-4) yields the maximum work that could be produced (or the minimum work required) for the steady state process operating between the inlet and outlet conditions specified. The quantity in square brackets [ ] in Eq. (14-4) is equal to the maximum work per mole processed and is called ***the change in availability or exergy which has been given the symbol*** $\boldsymbol{\Delta B}$. Thus,

$$\boldsymbol{B} \equiv H - T_o S \quad \text{and} \quad \boldsymbol{\Delta B} \equiv \Delta H - T_o \Delta S = (H_{out} - H_{in}) - T_o (S_{out} - S_{in}) \tag{14-5}$$

and

$$(\delta W_{total})_{rev} = \delta W_{max} = \delta W_c + \delta W_s = \boldsymbol{\Delta B}\, \delta n \tag{14-6}$$

or by taking the time derivative we get the maximum power:

$$\dot{W}_{max} = \boldsymbol{\Delta B}\, (\delta n/\delta t) = \dot{n}\, \boldsymbol{\Delta B} \tag{14-7}$$

Clearly the availability or exergy ($\boldsymbol{B}$) is a state function in the strictest mathematical sense so that the maximum (or minimum) work associated with any steady state process is also independent of path.

Now we can proceed to a more general case where additional heat interactions at temperatures different than T_o and multiple mass flows are permitted and the system is no longer restricted to steady state operation. From the First Law open system equation in Chapter 3, the overall energy balance including KE and PE effects is given as,

$$d\underline{E} = d(\underline{U} + \frac{N\langle v\rangle^2}{2} + Ng\langle z\rangle)$$

$$d\underline{E} = \sum_i \delta Q_i + \delta Q_R + \sum_j \delta W_j + \sum_{in} (H_{in} + v_{in}^2/2 + gZ_{in})\, \delta n_{in} \tag{14-8}$$

$$- \sum_{out} (H_{out} + v_{out}^2/2 + gZ_{out})\, \delta n_{out}$$

In order to be able to complete the calculation we need to assume quasi-static operation so that Eq. (14-8) can be integrated or differentiated as a well-behaved continuous function. To simplify the terminology that will be carried forward, we define a new term, H^*, for all flowing streams following normal mechanical engineering practice [H^* has been labeled the "methalpy" by Kestin (1979)]:

$$H^* \equiv H + v^2/2 + gz \tag{14-9}$$

Recasting Eq. (14-8) in terms of transient derivatives:

$$\frac{d\underline{E}}{dt} = \sum_i \frac{\delta Q_i}{\delta t} + \frac{\delta Q_R}{\delta t} + \sum_j \frac{\delta W_j}{\delta t} + \sum_{in} H_{in}^* \frac{\delta n_{in}}{\delta t} - \sum_{out} H_{out}^* \frac{\delta n_{out}}{\delta t}$$

or

$$\frac{d\underline{E}}{dt} = \sum_i \dot{Q}_i + \dot{Q}_R + \sum_j \dot{W}_j + \sum_{in} H_{in}^* \dot{n}_{in} - \sum_{out} H_{out}^* \dot{n}_{out} \tag{14-10}$$

The reason for retaining the KE and PE terms is to keep the development general as our goal is to be able to analyze all processes of interest to us. For example, certain fluid flow situations in rotating turbomachinery will require use of the KE terms while hydroelectric power applications utilize the conversion of PE into KE and then to electricity using a liquid turbine.

An entropy balance for the general system becomes:

$$d\underline{S} = \sum_i \frac{\delta Q_i}{T_i} + \frac{\delta Q_R}{T_o} + \sum_{in} S_{in}\,\delta n_{in} - \sum_{out} S_{out}\,\delta n_{out} + \delta \underline{S}_{gen} \tag{14-11}$$

and

$$\frac{d\underline{S}}{dt} = \sum_i \frac{\dot{Q}_i}{T_i} + \frac{\dot{Q}_R}{T_o} + \sum_{in} S_{in}\,\dot{n}_{in} - \sum_{out} S_{out}\,\dot{n}_{out} + \dot{\underline{S}}_{gen} \tag{14-12}$$

where the terms involving entropy generation are *always positive for real processes* and *zero for reversible processes.* Equations (14-8) through (14-11) can be combined to eliminate all terms involving Q_R,

$$\delta W = \sum_j \delta W_j = +d(\underline{E} - T_o\underline{S}) - \sum_i \left(1 - \frac{T_o}{T_i}\right)\delta Q_i - \sum_{in} (H^* - T_oS)_{in}\,\delta n_{in}$$
$$+ \sum_{out} (H^* - T_oS)_{out}\,\delta n_{out} + T_o\delta\underline{S}_{gen} \tag{14-13}$$

and

$$\dot{W} = \sum_j \dot{W}_j = \frac{+d(\underline{E} - T_o\underline{S})}{dt} - \sum_i \left(1 - \frac{T_o}{T_i}\right)\dot{Q}_i - \sum (H^* - T_oS)_{in}\,\dot{n}_{in}$$
$$+ \sum_{out} (H^* - T_oS)_{out}\,\dot{n}_{out} + T_o\,\dot{\underline{S}}_{gen} \tag{14-14}$$

Equations (14-13) and (14-14) are completely general. We can immediately compare reversible operation, where $\dot{\underline{S}}_{gen}$ and $\delta\underline{S}_{gen} = 0$, and $W_{rev} = W_{max}$, with operation under real conditions where some work production potential is lost. Remember that for net work production $W < 0$, thus

$$\delta W = \delta W_{max} + T_o\delta\underline{S}_{gen} \tag{14-15}$$

The term $T_o\delta\underline{S}_{gen}$ is often called the "lost work". We emphasize that $\delta\underline{S}_{gen}$ is a *path dependent* term and always is ≥ 0, that is, *the lost work is always positive or zero*:

$$-(\delta W_{max} - \delta W) = |\,\delta W_{max} - \delta W\,| = T_o\delta\underline{S}_{gen} \geq 0 \tag{14-16}$$

Again the availability or exergy appears in a slightly modified form involving the methalpy H^*,

$$\boldsymbol{B}^* = H^* - T_oS \tag{14-17}$$

Once we remove $\underline{S}_{gen}$ from Eqs. (14-13) or (14-14), we generate the maximum work which is only dependent on the end states and the nature of the transient storage term and heat flux boundary conditions which are themselves defined over a reversible path with no dissipative or lost work effects occurring.

For steady state, reversible, adiabatic cases where KE and PE effects are negligible, the Q_i's are set to zero, and

$$\frac{d(\underline{E} - T_o\underline{S})}{dt} = 0 \ ; \qquad H^* \to H \quad \text{and} \quad W \to W_{max} = W_{rev}$$

and Eqs. (14-6) and (14-7) are recovered with

$$\boldsymbol{B} = H - T_oS \quad \text{and} \quad \Delta\boldsymbol{B} = (H - T_oS)_{out} - (H - T_oS)_{in}$$

$$\delta W_{max} = \delta n\, \Delta\boldsymbol{B} \quad \text{and} \quad \dot{W}_{max} = \dot{n}\, \Delta\boldsymbol{B} \tag{14-18}$$

In general, a loss in availability or exergy is associated with entropy generation. The magnitude of the loss is a direct measure of the degree of thermodynamic irreversibility occurring in the process. To illustrate this point, let's rewrite Eq. (14-18) into a fully transient form by defining *net availability or exergy* as the difference between the total work and the *PV* work due to expansion or contraction of the system volume $\underline{V}$ against the ambient restraining pressure P_o.

$$W_{net} = W - P_o\delta\underline{V} \quad \text{and} \quad \dot{W}_{net} = \dot{W} - P_o\delta\underline{V}/\delta t \tag{14-19}$$

By removing the atmospheric displacement work we obtain a more realistic measure of the useful or recoverable work in situations where the system expands. Thus from Eqs. (14-14) and (14-19), with the $P_o\delta\underline{V}$ taken quasi-statically as $P_od\underline{V}$, we get:

$$\dot{W}_{net} = \underbrace{\frac{+d(\underline{E} + P_o\underline{V} - T_o\underline{S})}{dt}}_{\textit{transient storage}} \underbrace{- \sum_i \left(1 - \frac{T_o}{T_i}\right)\dot{Q}_i}_{\textit{reversible Carnot heat transfer}}$$

$$\underbrace{- \sum_{in} (H^* - T_oS)_{in}\dot{n}_{in} + \sum_{out} (H^* - T_oS)_{out}\dot{n}_{out}}_{\textit{bulk flow availability}} \quad \underbrace{+ T_o\dot{\underline{S}}_{gen}}_{\textit{irreversible losses}} \tag{14-20}$$

In practice, when W_{net} or $\dot{W}_{net} < 0$ it will equal the shaft work W_s or power that is extracted from a heat engine or turbine (if > 0, they correspond to the work or power required by a heat pump or compressor). Now consider a few examples to illustrate the use of Eq. (14-20):

For a reversible, steady state process with all heat interactions at T_o, and no KE or PE effects:

$$(\dot{W}_{net})_{rev} = \dot{W}_{max} = \dot{n}\Delta\boldsymbol{B} \quad \text{as } \dot{\underline{S}}_{gen} \to 0 \tag{14-21a}$$

For a non-reversible, steady state, constant volume, adiabatic process:

$$\dot{W}_{net} = \dot{n}\,\Delta B^* + T_o\dot{S}_{gen} = \sum [(H^* - T_oS)_{out} - (H^* - T_oS)_{in}]\,\dot{n} + T_o\dot{S}_{gen} \tag{14-21b}$$

The conditions above apply to a great many idealized or real processes that require evaluation, including power cycles, pumps, compressors, and turbines operating continuously and heat exchangers if we neglect external heat losses. Consequently, we can estimate the irreversible loss of availability or exergy by simply calculating the actual work produced (or consumed) relative to reversible operation. This immediately leads us to operating definitions for efficiencies which are discussed in the next section. A very important characteristic of the availability or exergy function is that it is a true state function in a thermodynamic sense and thus is independent of the actual path of the process. As was the case in Chapters 3 and 4, this means that alternative paths can be constructed between the same initial and final states to facilitate calculation of the maximum work.

Example 14.1

In order to fully appreciate the value of the availability concept, show how it can be used to calculate the maximum electrical power that could be generated from hot fluid flowing at 75 kg/s from a geothermal well located in the high mountains of New Mexico. Because the reservoir system is in deep crystalline granite rock, the geothermal fluid is essentially pure water (< 500 ppm total dissolved solids). It is available at the wellhead in a pressurized state at 200 bar and 240°C. The average ambient air temperature at the well site is about 5°C, reflecting the high altitude location at 9000 ft (3 km) above sea level where the ambient pressure is 0.7 bar. To simplify your calculation you can assume that the density and heat capacity are constant at 1000 kg/m^3 and 4350 J/kg K, respectively. If you prefer more precise results, feel free to use the steam tables.

Solution

Using Eq. (14-18) for steady state, reversible, adiabatic operation:

$$\dot{W}_{max} = \dot{n}\,\Delta B = \dot{n}(\Delta H - T_o\Delta S)$$

with $\Delta H \equiv H_{out} - H_{in}$ and $\Delta S \equiv S_{out} - S_{in}$

Now we need to specify the incoming and outgoing conditions. Let's assume no losses in the temperature or pressure of the fluid between the wellhead and the power plant. Thus the incoming conditions are 240°C and 200 bar. The exiting condition from the power plant needs to be selected to result in maximum work which intuitively will occur at the ambient dead state conditions of T_o and P_o of 5°C and 0.7 bar. For now, we will neglect the $P_o d\underline{V}$ work term and the effect of pressure on H and S as the geothermal fluid is in a condensed liquid state. Thus,

$$\Delta H \approx \int_{T_{in}}^{T_o} <C_p>\, dT \quad \text{and} \quad \Delta S \approx \int_{T_{in}}^{T_o} \frac{<C_p>}{T}\, dT$$

and

$$\dot{W}_{max}^{app} = \dot{n} <C_p> \left[(T_o - T_{in}) - T_o \ln \frac{T_o}{T_{in}} \right]$$

$$\dot{W}_{max}^{app} = (75)(4350) \left[(5 - 240) - (273.15 + 5) \ln \frac{(273.15 + 5)}{(273.15 + 240)} \right]$$

$$\dot{W}_{max}^{app} = -21.09 \times 10^6 \text{ J/s} = -21.09 \text{ MW}$$

If we use the steam tables and include the effect of $P \rightarrow P_o$ then

$$\dot{W}_{max}^{st} = 75[(20.98 - 1040) - 278.15(.0761 - 2.6674)] \text{ kJ/s} = -22.37 \text{ MW}$$

So we see a relatively small effect of pressure on the maximum power. Effectively, the pressure effects should scale as

$$\dot{W}_{max}^{P \rightarrow P_o} \approx \dot{n} \int_{P_{in}}^{P_o} V dP \approx \dot{n} <V> (P_o - P_{in}) = -(75)\left(\frac{1}{915}\right)(200 - 0.7)10^5 \text{ J/s}$$

$$\dot{W}_{max}^{P \rightarrow P_o} \approx -1.63 \text{ MW}$$

For this particular example, we could have obtained the identical result if we constructed an infinite set of small Carnot heat engines operating over the temperature range from T_{in} to T_o for the geothermal fluid and with the heat sink kept at T_o. Using the Clausius statement of the Second Law for a reversible process:

$$\frac{\delta Q_H}{T_H} + \frac{\delta Q_C}{T_o} = 0$$

and from the First Law noting that the sign of δQ_H is positive and $\delta Q_C = \delta Q_R$ is negative relative to the set of Carnot engines

$$W_{max} = \int \delta Q_H - \int \delta Q_R = Q_H - Q_R$$

If we neglect the *PV* work effects as before, we obtain the same result as we did using availability.

As we might have expected using availability or exergy analysis does not introduce any new laws or fundamental postulates. Nonetheless, the framework of such analysis makes the concept and calculation of maximum work easier to implement. Other relevant examples are found in Bejan (1988), Kestin (1979), Kestin *et al.* (1980), Bett *et al.* (1975), and Milora and Tester (1976).

14.2 Carnot, Cycle, and Utilization Efficiencies

There are many ways to characterize the performance of a heat engine. Specific types of efficiencies commonly fall into two groups--those that relate the work produced to the heat supplied and those that relate the work or power produced to some measure of the maximum work producing potential of the energy source that is being converted. The ideal Carnot efficiency which is of the first type can be modified for real power cycles to reflect the same concept. Recall that

$$\eta_{carnot} \equiv -\delta W_{rev}/\delta Q_H = (T_H - T_C)/T_H$$

For practical heat engines we can define a cycle efficiency in a similar fashion to the Carnot efficiency as,

$$\eta_{cycle} \equiv \frac{\textit{net useful work or power produced}}{\textit{heat input to the engine}} = \frac{-\Sigma W_j}{Q_H} = \frac{-\dot{W}_{net}}{\dot{Q}_H} \tag{14-22}$$

where $\dot{Q}_H$ equals the heat transfer rate from the energy source to the heat engine and $\dot{W}_{net}$ is the net power produced (with our convention, $\dot{W}_{net} < 0$ and $\dot{Q}_H > 0$ relative to the heat engine). Equation (14-22) is normally restricted to steady state power cycle operations. The heat input Q_H comes from a variety of sources. Q_H may simply be a non-latent enthalpy change from the heat source as was the case in Example 14.1, or it may be the enthalpy of a combustion reaction associated with burning a chemical fuel such as coal, natural gas, or gasoline to produce carbon dioxide and water as oxidation products, or it may be the thermal energy required to maintain isothermal operation of an electrochemical fuel cell.

When the efficiency is defined in terms of the combustion enthalpy or heating value of a chemical fuel, then

$$\eta_{fuel} \equiv \frac{\dot{W}_{net}}{\dot{n}_{fuel}\Delta H_c} \tag{14-23}$$

Typically either a high heating value combustion enthalpy $\Delta H_{c_{(HHV)}}$ corresponding to $H_2O_{(l)}$ and $CO_{2(g)}$ as products or a low heating value $\Delta H_{c_{(LHV)}}$ corresponding to $H_2O_{(g)}$ and $CO_{2(g)}$ as products is used. There are a few other versions of Carnot-type efficiencies that are important in heat engine analysis, see Heywood (1988) for examples of ones relevant to internal combustion engines.

An equally valuable efficiency used to rate heat engine performance is called the thermodynamic utilization efficiency η_u which compares the net work or power produced to the maximum possible work or power that could be produced under fully reversible, quasi-static conditions with $\underline{S}_{gen} = 0$,

$$\eta_u \equiv \frac{W_{net}}{W_{max}} \text{ or } \frac{\dot{W}_{net}}{\dot{W}_{max}} \text{ (heat engines)} \quad \eta_u^* \equiv \frac{W_{min}}{W_{net}} \text{ or } \frac{\dot{W}_{min}}{\dot{W}_{net}} \text{ (heat pumps)} \tag{14-24}$$

With the availability defined in Section 14.1, we can relate η_u to $\Delta \boldsymbol{B}$ *for a steady state power cycle* using heat engines:

$$\eta_u \equiv \frac{\dot{W}_{net}}{\dot{n}\Delta \boldsymbol{B}} \tag{14-25}$$

where we have implicitly excluded the atmospheric $P_o d\underline{V}$ displacement work since it really does not contribute to the net useful power.

The Carnot, cycle and utilization efficiencies are related. For a power cycle system with an isothermal hot reservoir at T_H and a constant ambient temperature sink at T_o,

$$\eta_{cycle} = \eta_u \, \eta_{carnot} = \eta_u \left(\frac{T_H - T_o}{T_H} \right) \tag{14-26}$$

as $T_H \to \infty$ or $T_o \to 0\text{ K}$ then $\eta_{carnot} \to 1.0$. As the heat engine's performance approaches its theoretical reversible limit then $\eta_u \to 1.0$ as well. In order to have high absolute cycle efficiencies both η_u and η_{carnot} need to be high.

For low-temperature energy sources such as geothermal and non-concentrated solar thermal, T_H is generally less than 300°C and may be as low as 25°C for ocean thermal energy. Ambient temperatures range typically between approximately 0°C and 35°C. With such source and sink temperature levels, these low-temperature resources yield limiting Carnot efficiencies ranging from about 8% to 50%. In contrast, for fossil-fuel-fired power plants, maximum heat source temperatures range from 1000°C to 2000°C which increases Carnot limiting efficiencies to about 75% to 85%. Naturally any real power plant will operate at lower efficiency, but what is remarkable is that the utilization efficiency for both low and high temperature plants frequently are of comparable magnitudes ranging from about 50% to 65%.

Figure 14.2 illustrates an important feature of the availability function. We have plotted ΔB defined relative to a dead state at T_o as a function of T along with typical values for four associated efficiencies, η_{cycle}, η_{COP}, η_u^*, and η_u. For these situations, above T_o we produce power with a heat engine cycle, while below T_o we are employing a heat pump/refrigeration cycle that requires work where a coefficient of performance (COP), which relates the heating or cooling rate delivered to the power expended, is more appropriate than a cycle efficiency to characterize performance. Several important trends are evident in Figure 14.2. For $T > T_o$, η_{cycle} increases as both η_u and η_{Carnot} increase. As T approaches T_o from below, the ideal η_{COP} gets arbitrarily large:

$$\eta_{COP} = \frac{\dot{Q}_C}{\dot{W}_{min}} = f\left(\frac{T_C}{T_o - T_C}\right) \quad \text{or} \quad \frac{\dot{Q}_H}{\dot{W}_{min}} = f\left(\frac{T_H}{T_H - T_o}\right)$$

$$\eta_{COP} \to \infty \quad \text{as} \quad T_H \text{ or } T_C \to T_o$$

As discussed earlier both η_u and η_u^* vary less with temperature with each reaching a maximum value at a particular temperature depending on the specific properties of the working fluid selected.

Losses in utilization efficiency can always be related to irreversibilities that manifest themselves in entropy generation. From our earlier analysis, this results in a decrease in the work producing potential of the system. The availability or exergy can be used to keep track of these changes. In real power cycles losses are caused by several phenomena, including

(1) heat exchange across finite temperature differences

(2) heat losses to the ambient environment

(3) frictional pressure losses during fluid flow

(4) fluid expansion or contraction across valves or nozzles

(5) mixing of two or more components.

In cases (1) through (5) we can estimate the magnitude of the loss by calculating the decrease in availability that occurs. For example, between any two states in a steady state process:

$$\Delta \boldsymbol{B}_{12} = \boldsymbol{B}_2 - \boldsymbol{B}_1 = \dot{W}_{max}/\dot{n}$$

Thus the total availability change can be used to scale the actual loss between states. For example, if some work is recovered then a utilization efficiency defined for that step in the overall process can be calculated to show the magnitude of the loss.

$$\dot{W}_{loss} = \dot{W}_{max} - \dot{W}_{net} = (1 - \eta_u)_{12}\, \dot{W}_{max} = \dot{n}\,(1 - \eta_u)_{12} \Delta \boldsymbol{B}_{12}$$

If no work is recovered, as in a conventional heat exchanger, then $\eta_u = 0$ and $\Delta \boldsymbol{B}_{12}$ becomes a direct measure of the loss in work producing potential. Alternatively, we could have defined an efficiency for the heat exchange process as a ratio of exiting to entering availability:

$$\eta_{Hex} \equiv \frac{\Delta \dot{\underline{B}}_{out}}{\Delta \dot{\underline{B}}_{in}} = \frac{\sum\limits_{out} \Delta \dot{\underline{B}}}{\sum\limits_{in} \Delta \dot{\underline{B}}} \tag{14-27}$$

with all streams referenced to a common dead state at T_o and P_o. For cases where multiple heat and work interactions occur, such as in a multistep chemical process, we can still use availability or exergy analysis to produce a so-called 2nd law efficiency

$$\eta_{2^{nd}\,Law} = \frac{\underbrace{\sum\limits_{out} \Delta \dot{\underline{B}}_{out}}_{\text{exiting heat flow}} + \underbrace{\sum\limits_{i} \dot{W}_{out,\,i}}_{\text{exiting power}}}{\underbrace{\sum\limits_{in} \Delta \dot{\underline{B}}_{in}}_{\text{entering heat flow}} + \underbrace{\sum\limits_{j} \dot{W}_{in,\,j}}_{\text{entering power}}} \tag{14-28}$$

Practically speaking we are treating the work/power steps as pure exergy and the heat transfer steps are scaled by their work equivalents. In engineering design, many tradeoffs exist that can be evaluated using an availability/ exergy approach. For example, we will utilize this method to analyze power cycle performance in Section 14.5. For further discussion of this topic see Bejan (1988).

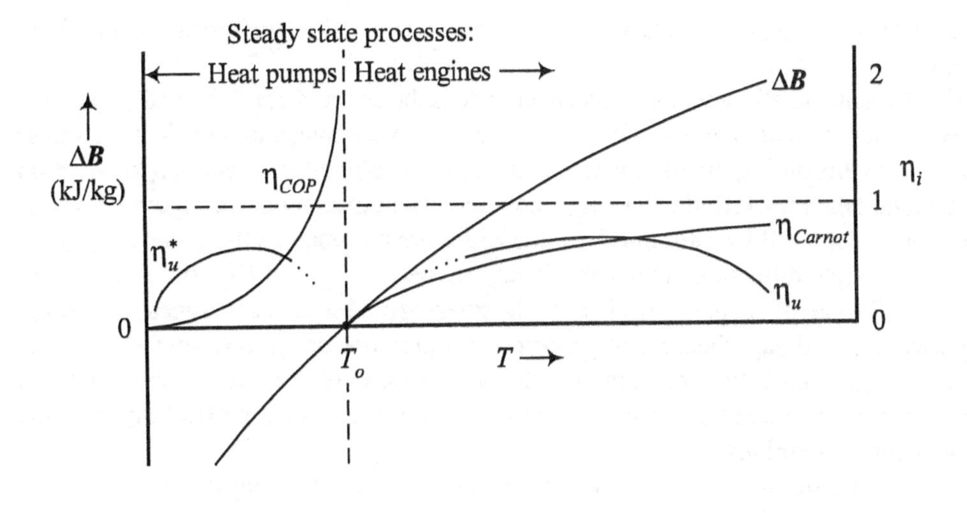

$$\Delta B \equiv (H - H_o) - T_o (S - S_o) \quad \text{per kg of working fluid}$$

Figure 14.2 Changes in efficiencies and in the availability or exergy function, ΔB, referenced to an ambient dead state at T_o, as a function of heat source temperature.

14.3 Heat Integration and Pinch Technology

A major concern in chemical processing is heat exchange. The operations needed to heat or cool process streams require a supply of external fuel or power which can represent a significant fraction of the total operating costs. Furthermore heat exchanger equipment costs can represent a large portion of the total capital investment. Early process designs viewed these requirements as a utility function that was external to the chemical plant where separate steam, chilled water, and/or refrigeration plants would provide all flows of hot and cold fluids needed for heat exchange. Now, partly to conserve energy and materials resources, heat integration methods have been developed to aid designers. Two specific problems are addressed in determining the optimal arrangement for an integrated heat exchange network (HEN). The first has to do with distributing internal and external resources given the required temperatures and heat loads for all streams and the second with actually specifying the exchangers within the network. Thermodynamics is primarily concerned with the first problem.

Cascading and transfer of heat between process streams is carried out as much as possible to minimize external energy requirements. Consider for instance, albeit on a much smaller scale, regenerative feedwater heating in steam power cycles and the recuperators used to preheat streams in gas turbine cycles. These are simple examples of heat integration involving only two process streams. Methods of treating multiple process streams in a systematic way have also been developed. For example, Linnhoff and associates (1982, 1983) have produced a new approach called "pinch technology" that aids the process of heat integration and specification of the HEN. Sama (1993) has taken a different approach that is more directly connected to traditional Second Law Analysis.

Fundamental to all methods of heat exchange is the concept of a "pinched" condition where the temperatures of opposing streams within an exchanger approach one another. We can see this on Figure 14.3, where temperature is plotted as a function the fraction of the total heat transferred for a single countercurrent exchanger. In Figure 14.3 (line (a)) constant mass flow rates and heat capacities are assumed so the T *versus* Q/Q_{total} lines are straight reflecting the linear relationship between $\underline{H}$ and T for both streams. In Figure 14.3 (line (b)), at pressures $P < P_c$, the entering cold stream undergoes preheating, vaporization, and superheating. At supercritical pressure conditions with $P > P_c$, no phase change occurs during heating but the heat capacity of the cold stream varies and can result in significant curvature to the T *versus* Q/Q_{total} line, particularly near the critical point conditions.

Pinch points occur in each case where the temperatures of the counterflowing streams differ by some minimal amount. A limiting case where $\Delta T_{pinch} = 0$ would occur in Figure 14.3 (line (a)) if $(\dot{m}C_p)_H/(\dot{m}C_p)_C > 1$ or where the heat exchanger surface area is very large. Of course, large area heat exchangers and pumping fluids at high flow rates are costly options. Thus in engineering practice a compromise is struck to keep ΔT_{pinch} within a fairly limited range, 5 to 20°C is typical.

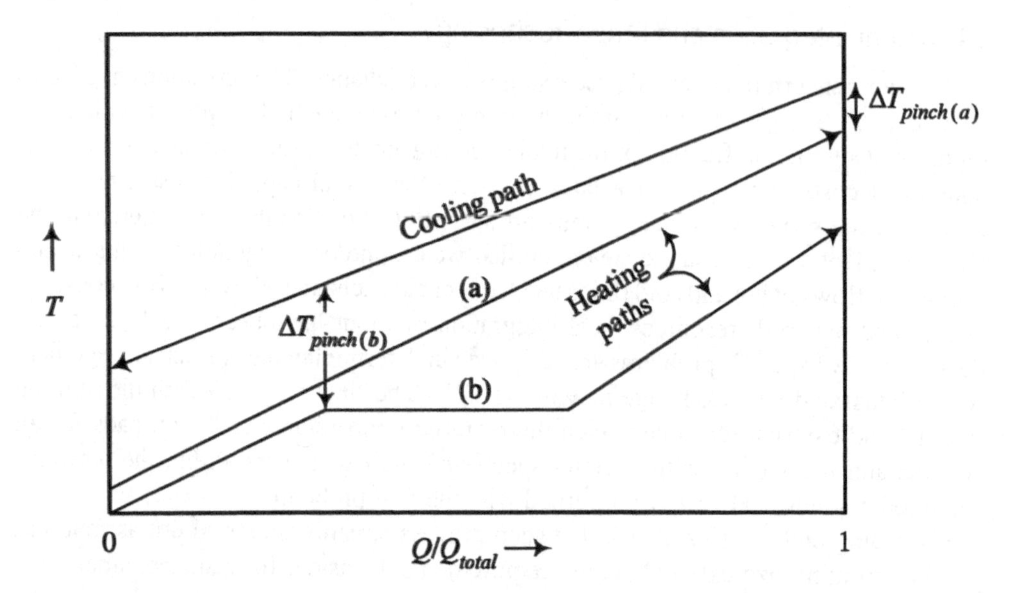

Figure 14.3 Temperature-heat transfer diagram for a single step countercurrent heat exchange process [(a) constant heat-capacity heating with no phase change; (b) heating with preheat, boiling, and superheat].

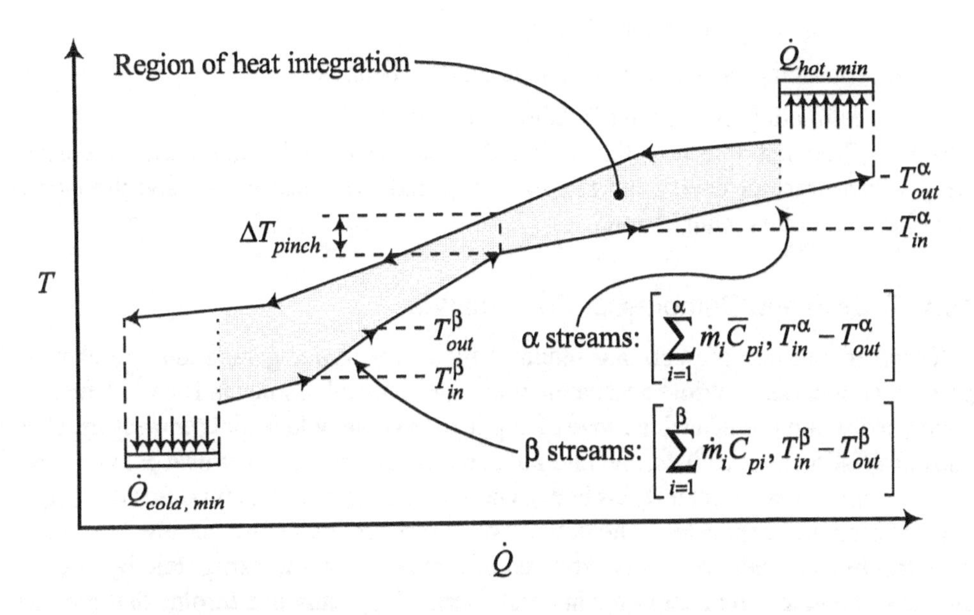

Figure 14.4 Composite temperature-heat transfer diagram for a multistep process to illustrate Pinch Technology.

Knowledge of the thermodynamic properties $[H = f(T, P)]$ of each fluid is required to develop the *T-Q* plot. In addition, one must specify ΔT_{pinch} as a design parameter. For a specified $\dot{m}C_p$ for each stream, the required heat exchanger surface area will depend on ΔT_{pinch} and the general shape of the *T-Q* lines for each fluid via the traditional design equation:

$$Area = \int_{in}^{out} \frac{\delta \dot{Q}}{\langle U_o \rangle \, \Delta T} \qquad (14\text{-}29)$$

where $\langle U_o \rangle$ is the overall heat transfer coefficient, $\Delta T = T_H - T_C$, and $\delta \dot{Q}$ is the differential amount of heat transferred between the hot and cold streams. While estimating the magnitude of $\langle U_o \rangle$ is outside of the realm of thermodynamics the *T-Q* relationship depends directly on the effect of temperature and pressure on enthalpy which requires data or a suitable $PVTN$ equation of state and an ideal gas state heat capacity.

As many streams of different composition, temperature and pressure are involved in a typical chemical process, we need to normalize and scale all the heating and cooling requirements in a simple systematic way in order that they can be represented on a single *T-Q* plot. This is the essence of Linnhoff's pinch technology methodology which facilitates the exploration of heat integration options. Figure 14.4 illustrates a composite *T-Q* plot for a multistream process whose $\dot{m}C_p$ products for all streams are constants. With this composite diagram one can establish heuristic rules for optimizing the design of the heat exchanger network.

To sum up, thermodynamics is important to this process in three areas:

(1) accurate specification of properties

(2) use of the First Law to calculate the *T*-*Q* behavior

(3) estimating the loss in availability that occurs.

Ultimately the main factors that affect the total heat exchanger network cost are ΔT_{pinch}, heat transfer coefficient-pressure drop tradeoffs, heat losses, and the number and placement of exchanger units.

14.4 Turbine and Compressor Performance

Turbines and compressors are required for many power generation and chemical process applications. While aeronautical and mechanical engineers focus on fuel, air, water, and steam use in a wide range of applications that include air compressors, steam and combustion gas turbines, and internal combustion engines, chemical engineers deal more commonly with other fluids both as pure components and in mixtures that require compression and expansion. The devices discussed in this section all involve rotating turbomachinery where power is extracted or emplaced on a working fluid by effecting an exchange of kinetic energy for internal energy. Typically in a turbine that produces net power, fluid expansions through nozzles are used to generate kinetic energy that is delivered to rotating turbine buckets to convert a part of it into rotating shaft work. The process is reversed for compressors.

While the details of the designs for accomplishing these tasks are left for aeronautical and mechanical engineers, the key thermodynamic principles of compressible fluid flow and gas dynamics are important to chemical engineers, particularly in the selection of appropriate working fluids and operating conditions.

Consider, for example, a single stage of a multistage, axial turboexpander illustrated in Figure 14.5. Here fluid is accelerated to high velocities, approaching sonic, in a fixed set of converging nozzles (the stator). The momentum of this accelerated flow is captured by a train of turbine buckets (the rotor) attached to a rotating shaft. Each stator-rotor combination represents a single turbine stage. In order to fully appreciate the thermodynamic consequences of such a design, we must revisit the First Law for open systems and make the connection between the kinetic energy and enthalpic contributions in compressible flow in convergent nozzles.

Let's begin with a simple converging nozzle defined by the σ-system boundary shown below:

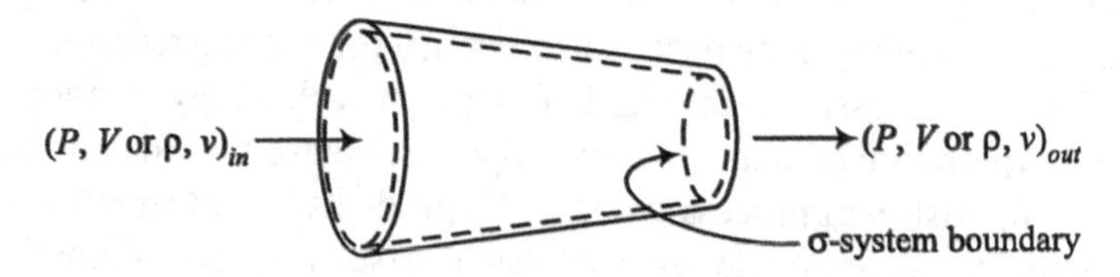

At any position along the length of the nozzle, let the cross-sectional area be $\underline{a}$. We apply the First Law assuming steady state, no shaft work, no PE effects, and ideal gas behavior

$$d\underline{E} = d\underline{U} + d(KE) = \delta Q + (H_{in} - H_{out})\,\delta m + (KE_{in} - KE_{out})\,\delta m = 0 \quad (14\text{-}30)$$

Note we are using a mass basis for the system with $\delta m = \delta m_{in} = \delta m_{out}$. If we assume that flow in the nozzle is reversible and adiabatic (or isentropic), $\delta Q = 0$ and without dissipative effects we can write.

$$\Delta H|_S \equiv H_{out} - H_{in} = -(KE_{out} - KE_{in}) = -\frac{1}{2}(v_{out}^2 - v_{in}^2) \quad (14\text{-}31)$$

Conceptually, we can visualize how a portion of the gas enthalpy is converted to rotating shaft work in a turbine stage. As shown in Figure 14.5, fluid is accelerated through the nozzles of the stator converting some of its enthalpy to kinetic energy. The high velocity flow exiting the nozzle impinges on a particular turbine rotor blade in its path and causes it to move--thus contributing to the shaft work. In the process of doing this the gas decelerates losing KE. The net effect is to remove $\Delta H|_S$ given in Eq. (14-31) from the fluid and convert most of it to shaft work. The exact amount of work produced relative to $\Delta H|_S$ is determined by the stage efficiency.

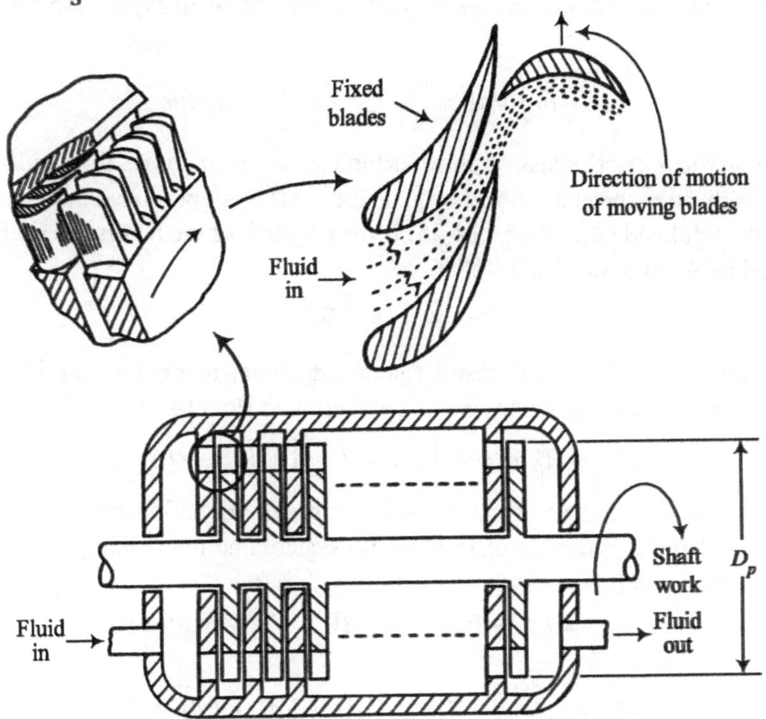

Figure 14.5 Basic turbine design for a multistage, stator/rotor axial flow machine [adapted from Bett, Rowlinson, and Saville (©) (1975) with the permission of the Athlone Press, 1996].

$$\eta_{t,\,i} \equiv \frac{\Delta H|_{actual}}{\Delta H|_{S}} \tag{14-32}$$

Since we are treating the process of fluid acceleration through the stator nozzle as a reversible, adiabatic or isentropic expansion we are implicitly assuming that $\eta_{t,\,i}$ is close to 1.0. The mass flow rate under steady state conditions comes directly from the continuity equation

$$\dot{m} = \frac{\delta m}{\delta t} = \rho \underline{a} v = \frac{\underline{a} v}{V} \qquad \text{(in kg/s)} \tag{14-33}$$

where v = fluid velocity (m/s), V = specific volume (m^3/kg), and ρ = density (kg/m^3). From laws of mechanics, a force balance is constructed where net force on fluid element = $d(mv)/dt$ = time rate of change of momentum or equivalently

$$-\underline{a}dP = \dot{m}dv$$

now with Eq. (14-33),

$$-VdP = vdv \tag{14-34}$$

By multiplying both sides by $\dot{m}$ and integrating, we obtain an expression for the change of KE.

$$-\dot{m}\int VdP = \frac{\dot{m}}{2}\left[v_{out}^2 - v_{in}^2\right] = (\Delta KE)\dot{m} \tag{14-35}$$

Thus, for an ideal, frictionless impulse turbine as depicted above, the fluid's ΔKE gain as it passes through the stator nozzle equals the shaft work produced in the turbine rotor because the left hand side of Eq. (14-35) is the negative of the reversible shaft work rate as derived in Section 4.8 [Eq. (4-68)]:

$$\dot{W}_S = \dot{m}\int VdP \tag{14-36}$$

Returning to the First Law result for an adiabatic, reversible nozzle contraction, Eq. (14-31) can be multiplied by $\dot{m}$ and equated with Eq. (14-35) to give:

$$(H_{out} - H_{in})\dot{m} = \dot{m}\int dH = +\dot{m}\int VdP = -(\Delta KE)\dot{m} \tag{14-37}$$

An equivalent result also comes from the Fundamental Equation and the definition of H as $U + PV$. Using a unit mass of fluid as the system and recognizing that a reversible, adiabatic expansion process is isentropic:

$$dU = TdS - PdV = -PdV \qquad \text{(with } dS = 0\text{)}$$

and

$$dH = dU + d(PV) = -PdV + VdP + PdV = VdP$$

Now by dividing through by $\dot{m}$ and using Eq. (14-35):

$$\int VdP = -\left(\frac{v_{out}^2}{2} - \frac{v_{in}^2}{2}\right) \tag{14-38}$$

For an adiabatic, reversible expansion of an ideal gas:

$$PV^{\kappa} = \text{const.}$$

where $\kappa = C_p/C_v$. Thus the integral of Eq. (14-38) can be evaluated:

$$\int_{P_{in}}^{P} VdP = \int_{P_{in}}^{P} \left(P_{in}V_{in}^{\kappa}\right)^{1/\kappa} \frac{dP}{P^{1/\kappa}} = -\frac{\kappa}{\kappa - 1} P_{in}V_{in}\left(1 - (P/P_{in})^{(\kappa-1)/\kappa}\right) \tag{14-39}$$

If $v_{in} = 0$ (incoming fluid at rest), Eq. (14-38) can now be rewritten as:

$$v_{out}^2 = \frac{2\kappa}{\kappa - 1} P_{in}V_{in}\left[1 - \left(\frac{P}{P_{in}}\right)^{(\kappa-1)/\kappa}\right] \tag{14-40}$$

We can now return to Eq. (14-33) to express the exiting mass flow rate per unit area $(\dot{m}/\underline{a})_{out}$. By combining this result with Eq. (14-40) and with PV^{κ} = constant:

$$\left(\frac{\dot{m}}{\underline{a}}\right)_{out}^2 = \left(\frac{v_{out}}{V}\right)^2 = \left[\frac{v_{out}}{V_{in}}\left(\frac{P}{P_{in}}\right)^{1/\kappa}\right]^2 = \frac{2\kappa}{\kappa - 1}\frac{P_{in}}{V_{in}}\left[\left(\frac{P}{P_{in}}\right)^{2/\kappa} - \left(\frac{P}{P_{in}}\right)^{(\kappa+1)/\kappa}\right] \tag{14-41}$$

One notes that $\dot{m}/\underline{a} = f(P/P_{in})$ goes through a maximum at a critical pressure ratio r_c

$$r_c \equiv (P/P_{in})_{critical} = [2/(\kappa + 1)]^{(\kappa/(\kappa-1))} \tag{14-42}$$

where v_{out} at r_c corresponds to the speed of sound. Therefore, the maximum mass flow rate in a converging nozzle occurs when the velocity reaches sonic conditions $v_c = (\kappa RT/m)^{1/2}$.

By substituting r_c into the $[\dot{m}/\underline{a}]^2$ expression of Eq. (14-41):

$$\left(\frac{\dot{m}}{\underline{a}}\right)_{max}^2 = \frac{P_{in}}{V_{in}}\kappa\left[\frac{2}{\kappa + 1}\right]^{(\kappa+1)/(\kappa-1)} \tag{14-43}$$

By using,

$$V_{in} = \frac{RT_{in}}{mP_{in}} \tag{14-44}$$

$$\dot{m} = \underline{a}P_{in}\left[\frac{\kappa m}{RT_{in}}\left[\frac{2}{\kappa + 1}\right]^{(\kappa+1)/(\kappa-1)}\right]^{1/2} \tag{14-45}$$

where m = molecular weight (kg/mol). Frequently a constant C_a is inserted in Eq. (14-45) to account for irreversible flow effects, including friction and turbulence dissipation. For example, in Problem 3.11 C_a = 0.6. Nearly ideal nozzles with carefully designed hydrodynamic surfaces can approach reversible, adiabatic (isentropic) conditions with C_a approaching 0.90 or higher.

Now that we have carried out the basic thermodynamic analysis of fluid flow through the stator nozzle, how do we use these results to aid us in designing the turbine? For rotating turbomachines, including axial and radial flow turbines and compressors, there are three key parameters that need to be set: (1) the allowable pressure drop per stage, (2) the enthalpy change per stage, and (3) the blade rotational speed and stage diameter. In turbine design, the conditions for the last stage (at lowest pressure) are crucial. Fortunately, because of the typical low pressures of order 1 bar or less, gas densities are low and ideal gas behavior is approximated. Thus our preceding theoretical development now can be used for last stage conditions with confidence.

(1) Allowable pressure drop, r_i: Equation (14-42) gives the critical pressure ratio for sonic or choked conditions in the nozzle of stage i as a function κ, the heat capacity ratio. Over a range of C_p/C_v's, Table 14.1 shows that r_c varies from 0.487 to 0.6065 as we go from simple diatomic gases to very complex molecules.

Table 14.1 Stage Pressure Ratios for Sonic, Choked Flow of Ideal Gases as a Function of Heat Capacity Ratio $\kappa = C_p/C_v$

κ	r_c
1.67	0.487
1.40	0.528
1.10	0.584
1.01	0.604
1.00	$0.607 = e^{-1/2}$

$r_i \equiv P_i/P_{i-1}$

*i*th stage

$P_{i-1} \rightarrow$ $\rightarrow P_i$

(2) Stage enthalpy drop: Under isentropic conditions, the change in enthalpy across the stage or the enthalpy drop is given by:

$$\Delta H|_{S,i} = H_{i-1} - H_i = -\int_{P_{i-1}}^{P_i} V dP = \frac{\kappa}{\kappa - 1} \frac{RT_i}{m}\left[(r_i)^{-\frac{\kappa-1}{\kappa}} - 1\right] \quad (14\text{-}46)$$

where RT_i/m has replaced $P_{in}V_{in}$ in Eq. (14-39).

(3) Rotational speed and stage diameter: Dimensional analysis has shown that for geometrically similar turbines, individual stage efficiency is a function of four dimensionless groups.

$$\eta_t = \frac{\dot{W}_{net}}{\dot{m}\Delta H|_{S,i}} = f[Re, Ma, n_s, d_s] \quad (14\text{-}47)$$

The Reynolds number (Re) is scaled to the rotational blade tip speed ($\dot{N}_r D_p$) and the blade "pitch" diameter (D_p) which is the maximum diameter of the outer turbine rotor hub (see Figure 14.5). The Mach number (Ma) is associated with the nozzle spouting velocity relative to the speed of sound. Balje (1982, 1981) proposed two groups that he called the specific speed (n_s) and specific diameter (d_s) which are unambiguously defined in SI units as follows:

$$n_s \equiv \frac{\dot{N}_r \left(\dot{\underline{V}}_i\right)^{1/2}}{(\Delta H|_{S,i})^{3/4}} \qquad d_s \equiv \frac{D_p \; (\Delta H|_{S,i})^{1/4}}{\left(\dot{\underline{V}}_i\right)^{1/2}} \tag{14-48}$$

Other groups could be used as well [see Milora and Tester (1976)] but the results are identical. Note that $\dot{\underline{V}}_i$ is the volumetric flow rate at the exit to stage i in m^3/s, and $\Delta H|_{S,i}$ is the isentropic stage enthalpy drop given by Eq. (14-46) in joules per kilogram (J/kg). $\dot{N}_r$ is the rotational speed in revolutions per second (rps) and D_p is in meters (m). For most practical applications using radial or axial turbines, highly turbulent, subsonic flow exists ($Re \geq 10^6$ and $Ma \leq 1$) and Re and Ma effects can be safely neglected. Balje demonstrated that certain values of n_s and d_s yielded optimal performance with maximum values of $\eta_{t,i}$ of 0.85 or more:

$$n_s^{opt} = 0.124 \qquad d_s^{opt} = 2.379$$

Thus, we can now inter-relate $\dot{N}_r$, D_p, $\dot{\underline{V}}_i$, and $\Delta H|_{S,i}$ to produce design charts like Figure 14.6. Here rotational speeds were selected to be compatible with 60 cycle/sec AC power generation. It is common practice to keep r_i somewhere between 0.7 and 0.95 of r_c.

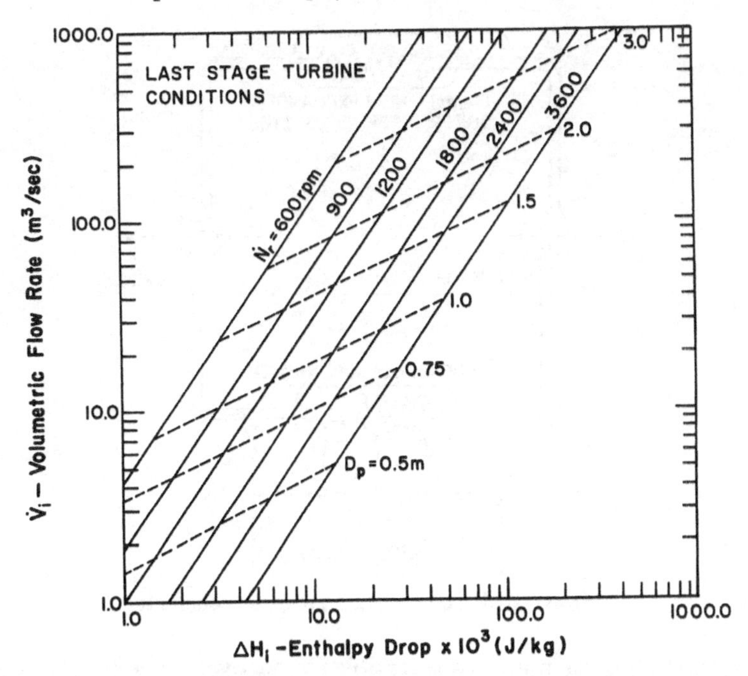

Figure 14.6 Stage volumetric flow rate as a function of stage enthalpy drop and rotational speed for axial flow turbines with maximum efficiencies ($\eta_t \approx 0.85$) [from Milora and Tester (1976)].

14.5 Power Cycle Analysis

Having discussed specific components and availability/exergy concepts we can now look at the performance of the entire power cycle. Three specific cycles, Rankine, Brayton, and Otto, have been selected for discussion here, partly to illustrate the analysis methodology and partly because of their prominence in today's heat to power conversion technology. A brief section on vapor compression refrigeration cycles is also included for completeness.

Rankine Cycles: Figure 14.7 schematically represents a typical Rankine Cycle where a pressurized liquid working fluid is vaporized in the boiler, expanded to a lower pressure in a turbine to extract work, liquefied in a condenser, and repressurized using a feed pump. In all four steps, irreversibilities occur. In certain operating regions one effect may dominate its control of overall performance in terms of influencing the values of η_u or η_{cycle}. We will examine the nature of these irreversibilities shortly, but to put things in proper context again we resort to the reversible Carnot cycle first introduced in Chapter 4.

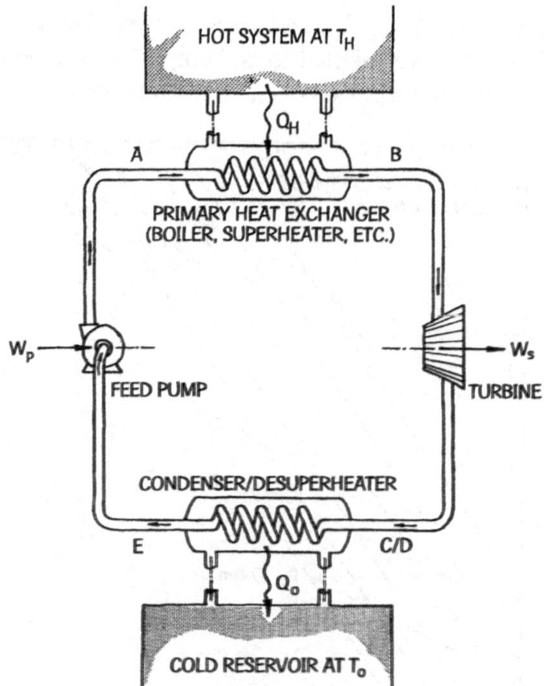

Figure 14.7 General Rankine cycle schematic.

An ideal Rankine cycle mimics a reversible Carnot cycle operating between two heat reservoirs where heat is extracted and rejected isothermally and work is produced or expended isentropically by expansion or compression. In Figure 14.8(a) this behavior is represented by the perfect rectangle *ABCDE* on a *T-S* diagram. The Carnot cycle yields maximum work as it represents the maximum conversion of heat into work via the Carnot efficiency:

$$\eta_{cycle} = \frac{-\delta W_{rev}}{\delta Q_H} = \frac{T_H - T_o}{T_H} \tag{14-46}$$

(a) Carnot with constant T_H and T_C (b) Carnot with variable T_H and constant T_C

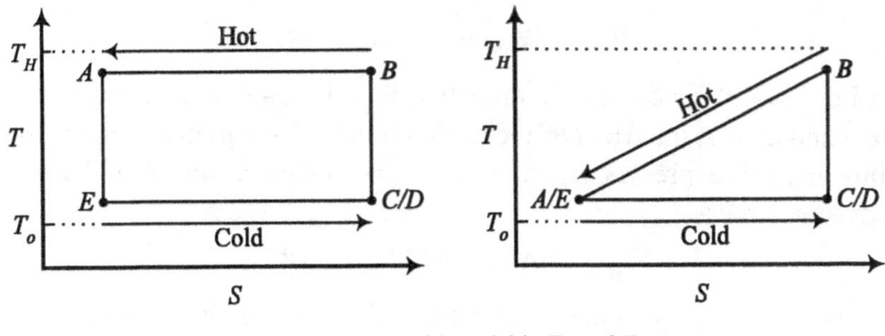

(c) Brayton with variable T_H and T_C

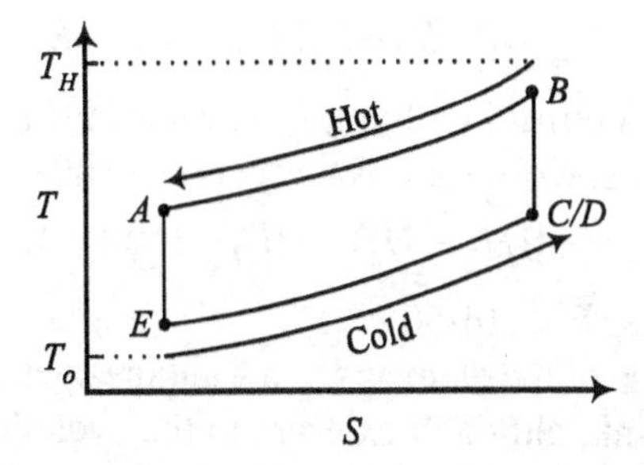

Figure 14.8 Temperature-entropy diagrams for ideal cycles. Labeled points correspond to Figure 14.7. Note that points C and D coincide on each diagram because we are not distinguishing between desuperheating and condensation in this idealization.

The only caveat needed at this point is that a small but finite ΔT is needed between the hot reservoir at T_H and the cycle at $T_B = T_A$ and between the cold reservoir at T_o and the cycle at $T_E = T_D$, to keep heat exchanger areas finite--so there will be some loss of availability or exergy associated with the heat exchange process. But since we are talking about an ideal Rankine cycle at this point in our discussion anyway--let's set the ΔT's to zero for now. Because entropy is an exact differential and the closed loop path $ABCDEA$ consists of two isothermal and two isentropic steps:

$$\oint dS = 0 = (S_B - S_A) - (S_D - S_E) \tag{14-47}$$

Thus along the isothermal paths at T_o and T_H the entropy changes are identical:

$$S_B - S_A = S_D - S_E \tag{14-48}$$

Over the isothermal paths ($A \rightarrow B$ and $D \rightarrow E$), the change in S is given by

$$S_B - S_A = \frac{\dot{Q}_H}{\dot{n}\, T_H} \quad \text{and} \quad S_D - S_E = \frac{\dot{Q}_o}{\dot{n} T_o} \tag{14-49}$$

where $\dot{Q}_H = \dot{n}(H_B - H_A)$ and $\dot{Q}_o = \dot{n}(H_D - H_E)$ and the net power produced (taken as a negative quantity) is the difference between the turbine and pump work rates at steady state:

$$\dot{W}_{net} = \dot{W}_t - \dot{W}_p = -(\dot{Q}_H - \dot{Q}_o) \tag{14-50}$$

Using Eq. (14-49) we can easily show that W_{net} is equal to the product of the mass flow rate $\dot{n}$ and area of the *ABCDE* rectangle on the *T-S* diagram of Figure 14.8(a). An availability analysis expresses *the maximum work as the net sum of all* $\Delta \boldsymbol{B}$ *changes* for this steady state process:

$$\dot{W}_{max} = \dot{n}\Delta\boldsymbol{B} = \dot{n}\,[\Delta\boldsymbol{B}_{hot} + \Delta\boldsymbol{B}_{cold}] \tag{14-51}$$

Since we have implicitly assumed that the cold reservoir is at ambient dead state conditions with $T_o = T_{cold}$, we can proceed directly to,

$$\dot{W}_{max} = \dot{n}\,[H_B - H_A - T_o\,(S_B - S_A)\,] - \dot{n}\,[H_D - H_E - T_o\,(S_D - S_E)\,] \tag{14-52}$$

note that the two lost work terms involving $T_o\Delta S$ cancel using Eq. (14-48)--as expected, there should not be any lost work for this perfect reversible cycle. Thus,

$$\dot{W}_{max} = \dot{n}\,[(H_B - H_A) - (H_D - H_E)\,] = \dot{W}_t - \dot{W}_p \tag{14-53}$$

which is the same result as Eq. (14-50).

Now let's explore some of the elements of a Rankine cycle in more detail. The nature of the heat source and sink, although external to the cycle itself, can have a profound effect on the choice of working fluids and operating conditions. So far in our analysis we have assumed that the hot and cold reservoirs are isothermal. For them to remain isothermal means they have to be very large relative to the amount of heat extracted. In reality, because energy costs money, neither reservoir can remain isothermal. If energy is extracted from the source and rejected to the sink, then their temperatures will change. Figures 14.8(b) and 14.8(c) show two possibilities.

In order to achieve the ideal Rankine cycle of Figure 14.8(a), the working fluid would have to vaporize isothermally at T_H and condense isothermally at T_o. For pure component fluids, we recognize that will happen only in a liquid-vapor coexistence region where latent heat can be added or removed isothermally. For this situation the pressures at T_H and T_o would correspond to the saturation vapor pressures at those temperatures. In Figure 14.8(b) only the heat source varies in temperature while the heat sink remains at T_o. In this case the perfect cycle is a triangle on *T-S* coordinates. In Figure 14.8(c) both heat source and heat sink temperatures vary, and now the ideal cycle becomes a parallelogram. Although these last two cases are not Carnot cycles in the strictest sense they are nonetheless "Carnotized" to achieve maximum work output given the specification of the heat source and sink. The cycle in Figure 14.8(c) is actually an ideal

Brayton cycle which is comparable to a gas turbine engine cycle where all heat transfer occurs without any condensation or vaporization.

While, in principle, the utilization efficiency η_u approaches unity in the ideal limit, in practice, the performance we require will depend on economic considerations. In general, designers are balancing the cost of the heat supply and rejection streams against the cost of converting the energy to useful power. If η_u is very small then the resource is being used poorly but the power plant cost might be lower, at the other extreme at high η_u, energy supply costs would be minimized but the plant costs would be proportionally higher. In real Rankine cycles, η_u typically is about 40 to 65% as a result of problems internal to the Rankine cycle such as turbine and pump efficiencies less than 100% and irreversibilities introduced by the presence of finite temperature and pressure differences required for heat exchange and to circulate fluids with frictional losses.

For a variable temperature heat source (Figure 14.8), a Rankine cycle operating at sub-critical pressures results in a very non-ideal heating path relative to the smooth cooling path of the heat source. This results in large exergy losses due to the higher, on average, temperature difference between streams in the primary heat exchanger as shown in Figure 14.9(a). The operating conditions of the Rankine cycle can be adapted to come closer to an ideal Carnotized cycle. One approach is to heat the working fluid at supercritical conditions to obtain a better match between the cooling path of the heat source and the heating path of the working fluid. Figure 14.9(b) illustrates the idea. In the example that follows this approach is explored for the conversion of a low-temperature heat source to electricity using a non-aqueous working fluid.

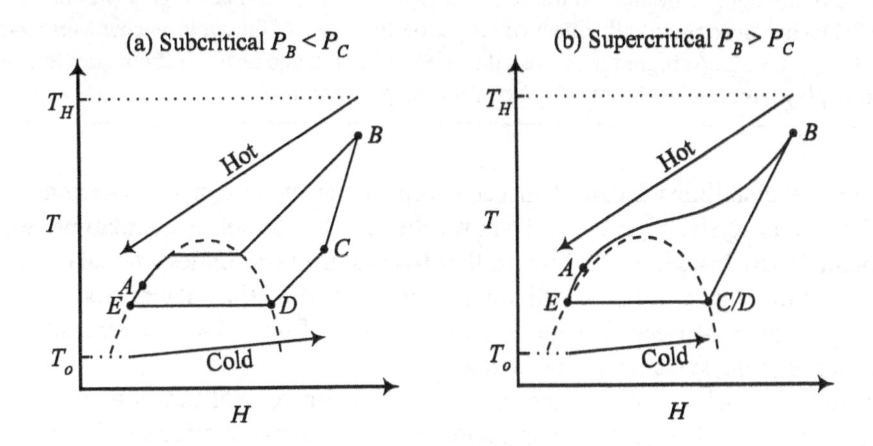

Figure 14.9 Subcritical and supercritical Rankine cycles.

Example 14.2

What is the optimum turbine inlet pressure for maximum utilization of a geothermal energy source in a Rankine cycle that employs R-115 as a working fluid? The geothermal source consists of a 150°C liquid water stream flowing at 100 kg/s with an average heat capacity of 4300 J/kg K and kept at a sufficiently high pressure to avoid any flashing. At the geothermal site, a 3°C average ambient air temperature exists. R-115 (C_2ClF_5) has a critical temperature of 80°C and pressure of 31.57 bar and is therefore suitable for both subcritical and supercritical operation. Other operating parameters for the system are: $\eta_t = 0.85$, $\eta_p = 0.80$, $T_B = 135°C$ for R-115 at point *B* in Figure 14.7, and $\Delta T_{pinch} = 10°C$ for both hot and cold streams.

Solution

In order to make cycle performance evaluations derived properties such as *H* and *S*, must be known at state points *A, B, C, D*, and *E*. To do this we either use a thermodynamic chart or table or its electronic equivalent. If these are not available, we can estimate the derived properties using a suitable $P\underline{V}TN$ EOS to calculate departure functions and an ideal gas state heat capacity and possibly other empirical correlations. In this example, the Martin-Hou EOS (see Table 8.1) and a standard polynomial expansion for C_p^o in *T* were used for gas phase properties along with empirical correlations for vapor pressure, liquid density, and enthalpy of vaporization [see Milora and Tester (1976) for details].

The results for four different turbine inlet pressures are shown in Figure 14.10. As we go from sub- to supercritical operating pressures both η_u and η_{cycle} increase corresponding to improvements in the heating and cooling paths. This reduces the loss of availability for these steps by decreasing the average temperature difference. Notice the large amount of sensible heat rejection (desuperheating) and the proportionally larger ΔT's in the primary heating step at a reduced pressure of 0.87. The pinched condition occurs in the primary heat exchanger at the point where R-115 vaporizes. At a supercritical pressure ($P_r = 1.24$) there is still some curvature to the R-115 *T-H* heating line, and at a higher pressure of $P_r = 2.54$ it becomes essentially linear where the performance of the cycle reaches a maximum with $\eta_u = 63.2\%$. At higher pressures, the effects of pump and turbine efficiencies less than unity play a more significant role in controlling performance.

The irreversibilities incurred in each step of the cycle can be determined using availability analysis. Figure 14.11 shows the results of such a calculation (see also Problem 14.16). In addition to showing the effects of irreversibilities internal to the cycle, the availability loss associated with reinjecting geothermal fluid at temperatures greater than T_o are also included. For the analysis shown in Figure 14.11, frictional pressure drops through the system have been neglected.

Where η_u reaches a maximum, the sum of all irreversibilities is a minimum. As pressure increases from sub- to supercritical we see the major effects of improved heat transfer. Above the optimum operating pressure, the effects of the turbine and pump irreversibilities, which scale with the magnitude of ΔP or ΔH in each device, begin to take over control as the heat transfer processes are essentially optimized and further pressure increases do not improve heat exchanger performance. In addition, at these higher operating pressures, lower turbine efficiencies result when the expansion enters the liquid-vapor dome.

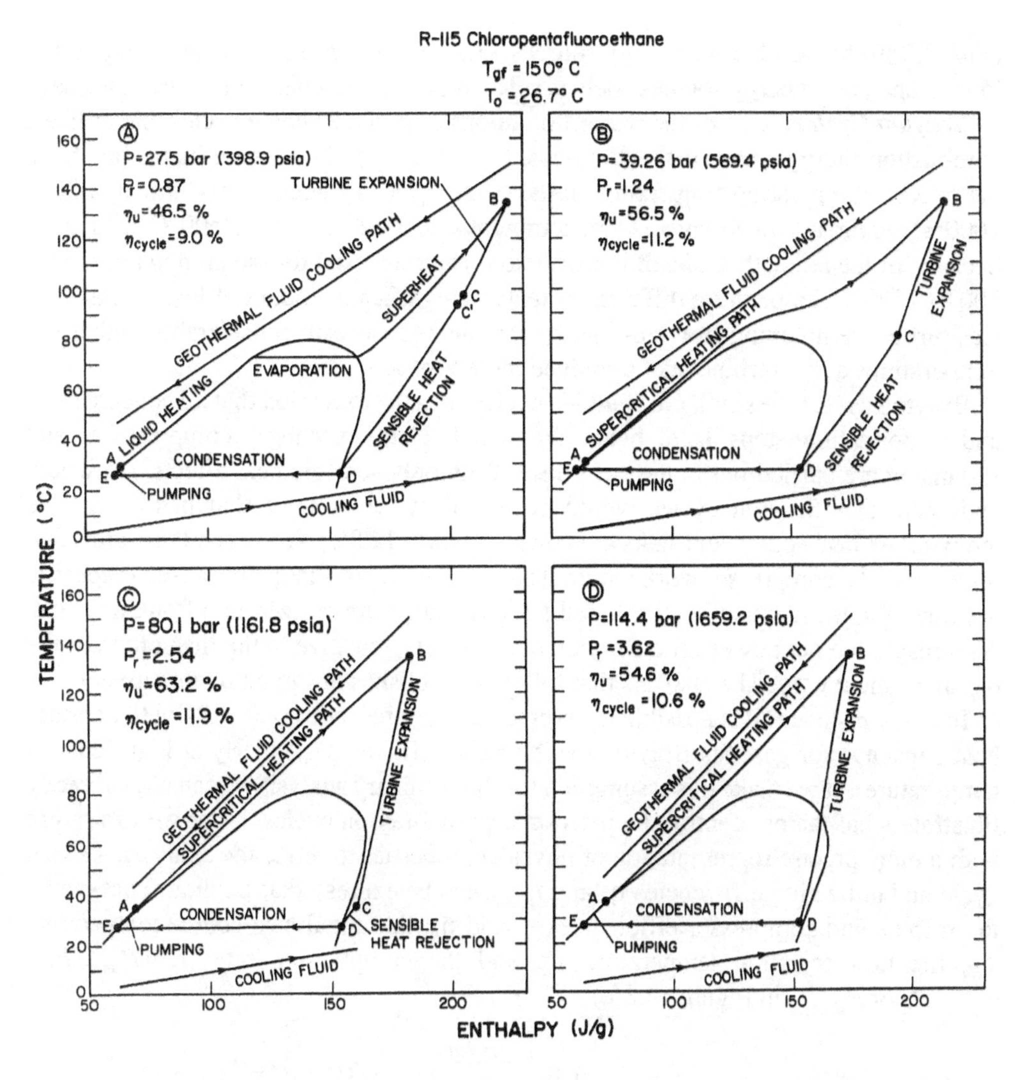

Figure 14.10 Effect of turbine inlet pressure on rankine cycle performance, temperature-enthalpy diagrams (*T-H*) for sub- and supercritical R-115 cycles [from Milora and Tester (1976)].

One would expect that each resource temperature will yield a different value of turbine inlet pressure for optimal performance of a given cycle type and working fluid. The specific output plot of Figure 14.12 for different propane Rankine cycles operating between 190 and 275°C illustrates this point. As the heat source temperature increases, the optimum pressure increases from 100 to 240 bar. Furthermore, each working fluid will achieve its *absolute maximum* performance at a particular resource temperature where its thermodynamic properties best match the energy source and heat rejection conditions. These points are discussed in detail by Milora and Tester (1976) and Kestin *et al.* (1980) for geothermal energy sources, but they can be applied in general to other low-temperature energy sources such as solar, waste process heat, and ocean thermal.

***Brayton Cycles*:** One of the major limitations of steam Rankine cycles that utilize a combustion energy source is that the flame temperature of the burning fuel is far above the maximum operating temperature of the steam cycle. Typically fuels burn at 1000 to 2000°C while the maximum steam temperatures used are about 500°C, primarily because of the strength limits of construction materials used for the components. This 500 to 1500°C temperature difference creates a significant amount of lost availability. Gas turbines employing Brayton-type cycles can operate with considerably higher gas temperatures at the turbine inlet to reduce these losses.

Brayton cycles are similar to Rankine cycles with the exception that all condensation and vaporization steps have been eliminated. Heat exchange, compression, and expansion are carried out in the gas phase. With only sensible heat effects, we would anticipate that Brayton cycles would be excellent for systems that had a variable temperature heat source and sinks as shown in Figure 14.8(c). However, there are a few problems. The compressor work required in Brayton cycles is larger because the specific volume of a gas is much higher than the liquid that is pressurized in a Rankine cycle. As a result, the net power for a Brayton cycle is very sensitive to the magnitude of the pump efficiency η_p. The analysis that follows establishes this point quantitatively.

In many instances it is possible to assume ideal gas behavior ($PV = RT$ with constant heat capacity) for gaseous Brayton cycle working fluids, particularly at high reduced temperatures. If we make this assumption for a preliminary analysis we can more directly illustrate what factors control the performance of Brayton cycles. Later we can return with a more precise representation of physical properties to refine the analysis. The net cycle and utilization efficiencies η_u and η_{cycle} can be expressed as explicit functions of the turbine and compressor efficiencies η_t and η_p, the initial heat source temperature T_H, the heat rejection temperature T_o, and the compression ratio $r_c = P_{max}/P_{min} = P_A/P_E$ or P_B/P_C in Figure 14.8(c):

$$\eta_{cycle} = \frac{\dot{W}_t - \dot{W}_p}{\dot{Q}_{AB}} = \frac{\eta_p \eta_t \left[\dfrac{T_{wf,\,out}}{T_o}\right](1 - 1/\alpha) - (\alpha - 1)}{\eta_p \left(\dfrac{T_{wf,\,out}}{T_o} - 1\right) - (\alpha - 1)} \tag{14-54}$$

$$\eta_u = \frac{\eta_{cycle} \dot{Q}_{AB}}{\dot{n}_H \Delta B_H} = \frac{\eta_{cycle} \dot{Q}_{AB}}{<C_p> \dot{n} \left[T_B - T_A - T_o \ln(T_B/T_A)\right]} \tag{14-55}$$

where $\alpha \equiv$ specific pressure ratio $= (P_{max}/P_{min})^{(\kappa-1)/\kappa}$, $T_{wf,\,out} = T_B$ on Figure 14.8(c), and $\kappa \equiv C_p/C_v$.

As η_t and η_p approach unity, η_{cycle} approaches the ideal Brayton efficiency for a fully Carnotized cycle. As the compression ratio increases to its maximum limit of

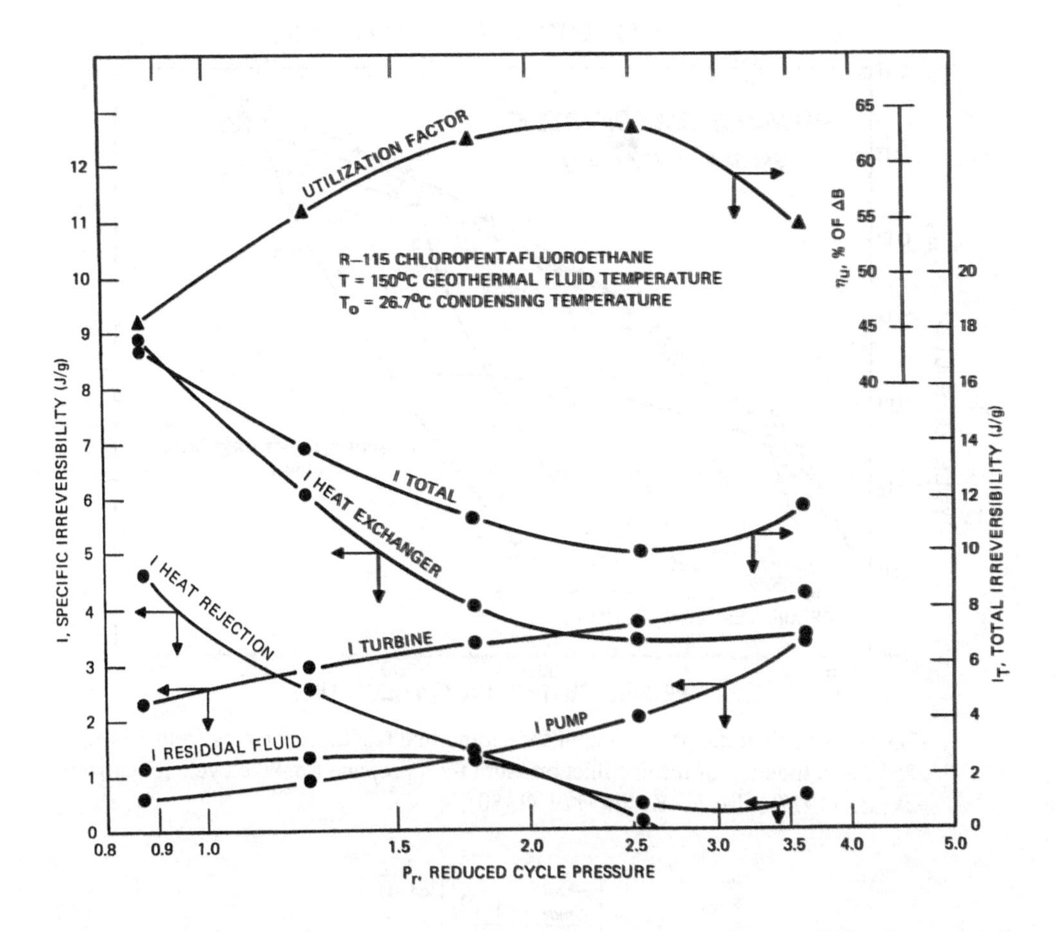

Figure 14.11 Specific component irreversibility and utilization efficiency η_u as a function of reduced cycle pressure [see Figure 14.10 for other cycle parameters; adapted from Milora and Tester (1976)].

$$r_c)_{max} = \left[\frac{T_{wf}^{out}}{T_o}\right]^{\frac{\kappa}{\kappa-1}} \tag{14-56}$$

η_{cycle} is maximized but η_u goes to zero, the implication is that the amount of heat extracted from the hot fluid also goes to zero. In normal engineering practice with combustion heat sources, the gas inlet temperature to the turbine is limited by the strength of the materials used in the turbine blades. With the primary heat exchanger gas outlet temperature fixed at $T_{max,\,wf}$ and the minimum heat rejection temperature set at T_o, Eq. (14-54) can be differentiated with respect to α to determine the optimum compression ratio:

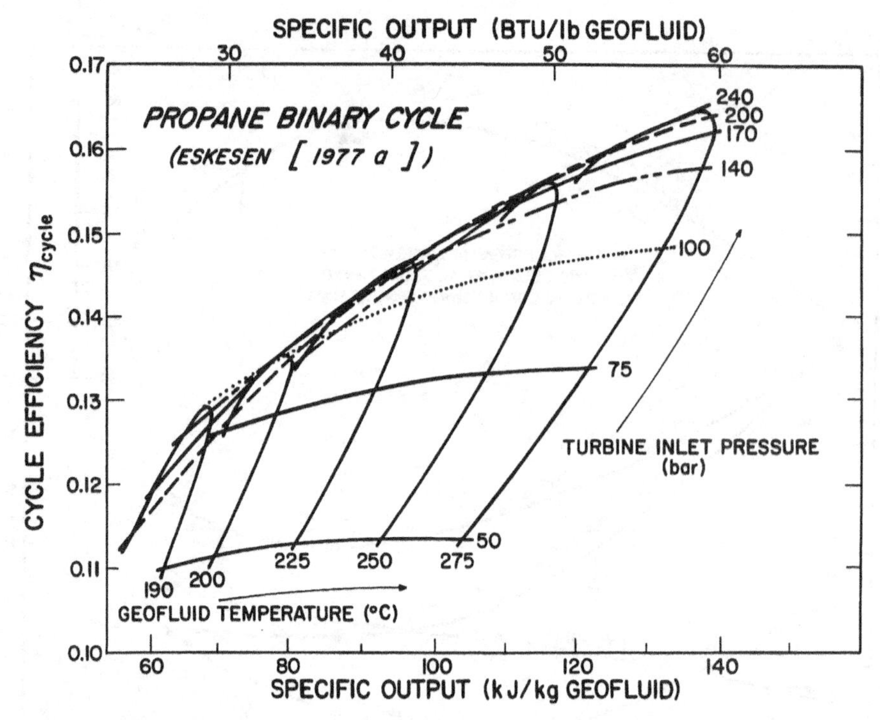

Figure 14.12 Specific power output for geothermal resources varying from 190°C to 275°C as a function of turbine inlet pressure for a propane Rankine cycle [data from Eskesen (1977); also see Kestin *et al.* (1980)].

$$r_c)_{opt} = \left[\frac{T^{out}_{max,\,wf}\,\eta_p\eta_t}{T_o}\right]^{\frac{\kappa}{2(\kappa-1)}} \tag{14-57}$$

In order to maintain high overall cycle and utilization efficiencies, η_t and η_p should be at least 85% or better. With today's technology it is now possible to achieve 90% for these component efficiencies, thus high temperature gas turbines employing Brayton type cycles are becoming popular as they are now more fuel efficient.

In a typical gas turbine power plant, air is compressed to high pressure where it enters the combustion chamber and a liquid or gaseous hydrocarbon fuel is injected. Heat is added to the system in the form of the exothermic energy of the oxidation reactions of the fuel. This process can be modeled as an isobaric addition of heat $\dot{Q}_{AB}$ equal approximately to the rate of fuel addition $\dot{n}_{fuel}$ times the enthalpy of combustion ΔH_c. The combustion product gases, primarily water, carbon dioxide and nitrogen, are expanded to produce power. Ideally, both the compression and expansion steps are carried out isentropically. Finally, similar to any Brayton cycle, sensible heat is rejected to the environment as the hot gases from the turbine exhaust are mixed with the atmosphere at T_o and P_o. Unavoidably, a significant irreversibility results from this form of heat rejection. If the compression and expansion steps are reversible, then the overall cycle efficiency is only a function of the pressure ratio P_{max}/P_{min}:

$$\eta_{cycle}\Big|_S = 1 - \left(\frac{P_{min}}{P_{max}}\right)^{\frac{\kappa-1}{\kappa}} \tag{14-58}$$

With η_p and η_t both 1.0, η_{cycle} is given by Eq. (14-54). There are a number of measures that can be taken to improve the performance of gas turbine systems: for example, regenerative heat exchange where hot gases are used to preheat incoming air and intercooling between compression stages. Another direction taken lately has been to develop higher temperature materials for turbine blades and nozzles or to actively cool turbine components so that higher injection temperatures ($T_{max,wf}$) can be used.

***Combined cycles*:** To reduce the availability loss from the heat that is rejected above T_o in a gas turbine system, a hybrid concept is used that combines the best features of a gas turbine cycle with a steam Rankine cycle. These have been labeled combined or integrated power cycles. Brayton-type cycles perform better at higher temperatures and steam cycles can take advantage of lower heat rejection temperatures by adjusting the steam condensing pressure to produce nearly ideal isothermal heat rejection conditions relatively close to T_o. Thus if they could be combined as shown in Figure 14.13 with the gas turbine cycle interacting with the hot combusting fuel and the steam cycle receiving its heat from the heat rejection step of the gas turbine cycle as it is cooled before releasing the exhaust gas to the environment, a much lower rate of availability loss would occur. This cascading situation reduces the availability loss in two major ways. First it decreases the average temperature difference between hot and cold streams. Working fluid temperatures on the hot side of the gas turbine cycle are closer on average to the fuel combustion temperatures and on the cold side steam condensing temperatures are closer to the ambient temperature. Second, some of the sensible heat from the gas turbine exhaust is also used to preheat the water before vaporization in the steam cycle. All of these factors move the combined hybrid cycle's performance closer to that of an ideal Carnot cycle. With proper consideration for advances in turbomachinery performance and the evolution of advanced high temperature materials, one would anticipate that the upper limit for combustion-based, combined gas-turbine-steam cycle heat to power efficiencies is about 60%. As of 1997, combined efficiencies of over 55% have been achieved.

Another approach for improving overall cycle performance heat sources at temperatures of 200°C or less is to use so-called bottoming cycles employing low temperature boiling fluids whose densities near T_o would be much higher than steam under similar conditions, thus turbine sizes can be kept small. Still another novel approach is to use working fluids that are mixtures rather than pure components. The Kalina cycle is an excellent example of such a system where a mixture of water and ammonia has been proposed for moderate to low temperature applications [see Kalina (1983), Bejan (1988), and Tester (1982)]. In this case the properties of ammonia are particularly well suited to low-temperature applications.

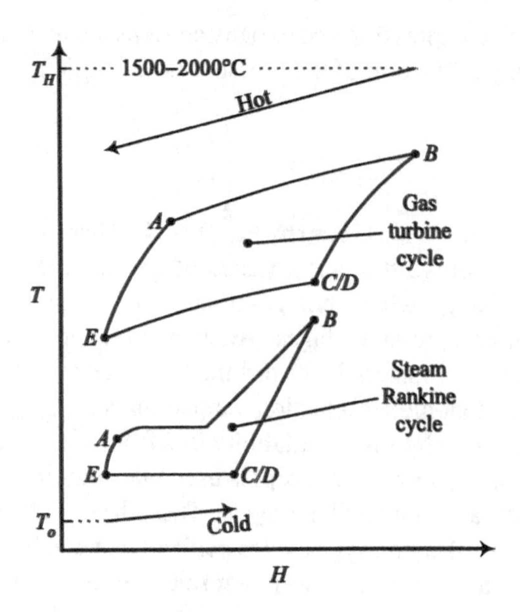

Figure 14.13 Combined gas turbine (Brayton-type) and steam Rankine cycles.

***Otto cycle*:** Internal combustion engines provide the power for most of the automobiles in use today. From a thermodynamics perspective, the heat to work conversion process used is a practical modification of the basic Otto cycle given in Figures 14.14(a) and 14.14(b). The intake stroke of the engine brings fresh combustion air and fuel into the cylinder along a constant pressure path to point D. Here the mixture is adiabatically compressed to point A, where the mixture is ignited using a high voltage pulse across the spark plug where it compresses at constant volume to point B primarily due to the increase in temperature. Work is extracted by expansion to point C and then the exhaust valve is opened to remove the combusted mixture again at constant volume to point D. The four steps of an actual spark ignition engine cycle are then (1) intake, (2) compression and ignition, (3) expansion, and (4) exhaust. Strictly speaking, this is not a power cycle in the sense that the system is open as mass flows into and out of the engine. But for steady state operation, the four-step process is repeated over many identical cycles that trace out the pressure-volume trajectory shown in Figure 14.14(c). Although the Otto cycle's P-V trace only approximates the actual four-stroke cycle, it is useful to carry out a preliminary thermodynamic analysis of the Otto cycle as it will provide a convenient and accurate way to qualitatively explain what controls the performance of real internal combustion engines. In a similar manner to a combusting gas turbine, the primary heat input to the system is a direct result of combustion and the fuel conversion efficiency is the appropriate scaling parameter for performance. If the combustion efficiency approaches 100% as is typically the case when heat losses are small, then $\dot{Q}_{AB} = \dot{n}_{fuel}\Delta H_c$ and the cycle and fuel efficiencies are equivalent.

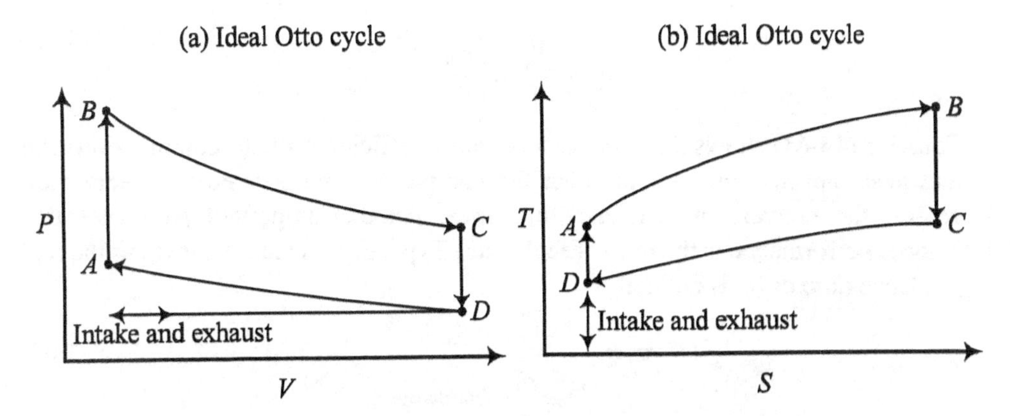

(c) Actual internal combustion engine (spark ignition)

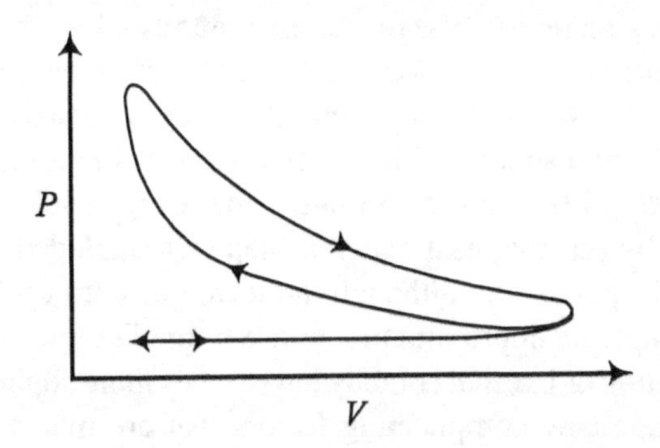

Figure 14.14 Otto and internal combustion cycles.

$$\eta_{cycle} = \frac{\dot{W}_{net}}{\dot{Q}_{AB}} = \eta_{fuel} = \frac{\dot{W}_{net}}{\dot{n}_{fuel}\Delta H_c} \tag{14-59}$$

The work steps can be approximated by a reversible, adiabatic or isentropic expansion and compression of an ideal gas. In this case the choice of the heat capacity ratio $\kappa = C_p/C_v$ must reflect the mixture composition. The analysis of these isentropic steps is actually done using the adiabatic closed system form of the First Law with $dU = -PdV$. Now we can rewrite the fuel efficiency as:

$$\eta_{fuel} = \frac{\dot{n}\left[(U_B - U_C) - (U_A - U_D)\right]}{\dot{n}_{fuel}\Delta H_c} \tag{14-60}$$

but with PV^κ equal to a constant for the isentropic steps, Eq. (14-60) becomes:

$$\eta_{fuel} = 1 - [1/r_c^*]^{\kappa-1} \tag{14-61}$$

with the compression ratio r_c^* defined somewhat differently than before:

$$r_c^* \equiv \frac{V_C}{V_B} = \left(\frac{P_B}{P_C}\right)^{1/\kappa} \tag{14-62}$$

Equation (14-61) shows the direct dependence of efficiency on the compression ratio r_c^* and heat capacity ratio κ. As either the compression ratio or heat capacity ratio increases, the conversion efficiency increases. Another important parameter that influences performance is the ratio of fuel to air. Typically a quantity known as the fuel equivalence ratio or ϕ_f is defined:

$$\phi_f \equiv \frac{(\dot{n}_{fuel}/\dot{n}_{air})}{(\dot{n}_{fuel}/\dot{n}_{air})_{stoichiometric}} \tag{14-63}$$

In general, as the fuel to air mixture becomes leaner and substoichiometric ($\phi_f < 1.0$), the efficiency increases. The results are plotted on Figure 14.15 for both ideal gas and fuel-gas mixture Otto cycles along with actual performance data from operating engines. We see that the effective utilization efficiency η_u equals 0.35/0.46 or about 75%--this is an η_u of a real spark ignition engine relative to an ideal Otto cycle for a fuel-air mixture with $\phi_f = 1.0$. A more accurate value for η_u would be somewhat lower if the availability lost by sensible heat rejection step were included.

As Heywood (1988) points out, although the ideal gas Otto cycle shows important trends, it is far too simple an approximation to use to predict real engine performance. For quantitative modeling of internal combustion engines more sophisticated models are required to capture the many complicating factors that are missed by the Otto cycle analysis. In particular, the behavior of the combustion process is grossly oversimplified. In addition, the effects of fuel composition, heat transfer between the hot combustion products and the cylinder wall, and oil leakage can have significant effects on performance.

14.6 Summary

The central theme of this chapter has been developing approaches to modeling the performance of practical heat engines that are used for power generation. This area is certainly one of the earliest applications of the First and Second Laws of classical thermodynamics and as such serves to reinforce and enhance our appreciation for the generality of these important postulates even when used to describe the behavior of non-ideal working fluids. Although the concepts are well-established and at the foundation of Thermodynamics, their application to evaluating alternative energy technologies is of great interest. In addition, concern about the use of energy worldwide has led to a renewed interest in industrial energy efficiency and conservation. This has resulted in the development of new tools and methodologies to assess the thermodynamic performance of complex, multistep processes, such as pinch technology for heat integration of a chemical plant and availability/exergy analysis of processes used to manufacture energy intensive products. To illustrate the factors that can alter performance we chose three power cycles to analyze--

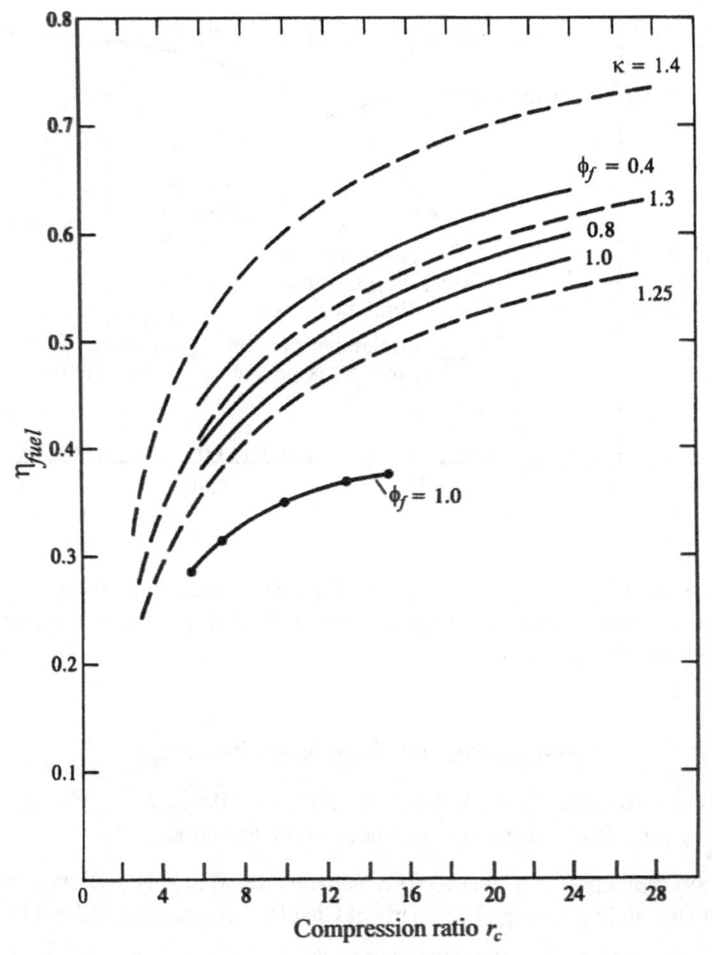

Figure 14.15 Internal combustion engine fuel efficiency as a function of compression ratio for ϕ_f= 1.0. Solid lines correspond to fuel-air Otto cycle conditions for various ϕ_f's and dotted lines correspond to ideal gas Otto cycles for various κ's [from Heywood (1988)].

Rankine, Brayton, and Otto cycles were selected because of their relevance to engineering design issues that face chemical engineers. Individual components such as heat exchange and turbine and compressor performance were discussed first followed by a treatment of the entire cycle. Figure 14.16 summaries the range of cycle efficiencies observed for these and other power conversion systems. One can see that considerable enlightenment has resulted in the development of these energy technologies with engineers striving to improve performance towards its Second Law--Carnot limit.

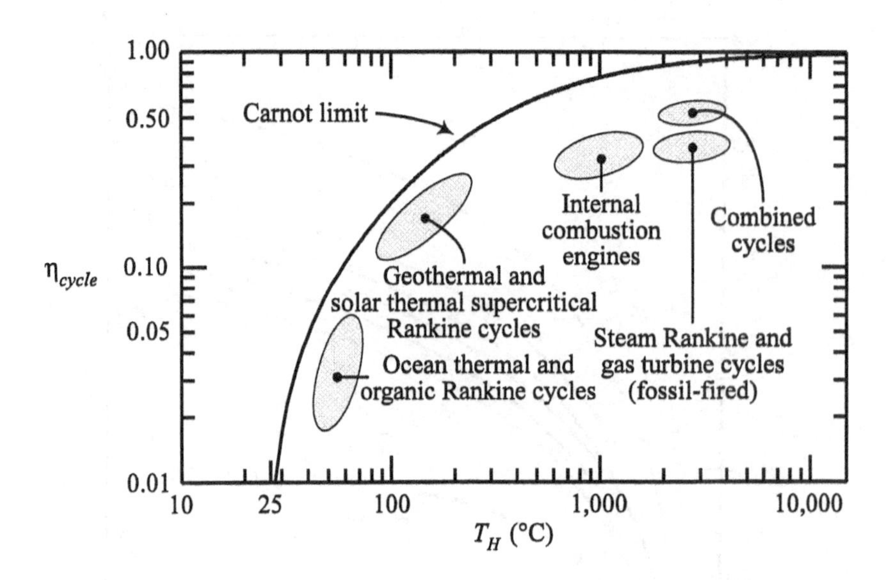

Figure 14.16 Typical cycle efficiencies for heat to work conversion as a function of heat source temperature. Heat rejection is assumed to occur at average ambient conditions of 25°C and 1 bar.

References and Suggested Readings

Balje, O.E. (1981), *Turbomachines*, Wiley, New York and (1962), *J.Eng. Power, ASME Trans.* **84** (1), p 83. [turbine and compressor performance criteria]

Bejan, A. (1988), *Advanced Engineering Thermodynamics*, Wiley, New York, Chapter 3 [on exergy and availability concepts] and parts of Chapter 5 [on exergy analysis of systems]

Bett, K.E., J.S. Rowlinson, and G. Saville (1975), *Thermodynamics for Chemical Engineers,* MIT Press, Cambridge, MA. [treats turbomachinery and flow problems in power cycles]

Eskesen, J.H. (1977), "Study of practical cycles for geothermal power plants," General Electric Company report COO-26194 UC-66, Contract No. EY-76-C-02-219, April. [power cycle analysis, case studies]

Gyftopoulos, E.P. and G.P. Berretta (1991), *Thermodynamics: Foundations and Applications*, MacMillan, New York. [excellent mechanical engineering perspective on exergy and availability concepts]

Heywood, J.B. (1988), *Internal Combustion Engine Fundamentals*, McGraw-Hill, New York, Chapter 2, sections 2.5, 2.8, and 2.10 and Chapter 3, sections 3.5.5 and 3.6.2 [on efficiencies] and Chapter 5 [on modeling internal combustion engine cycles]

Kalina, A.I. (1983), "Combined cycle and waste heat recovery power systems based on a novel thermodynamic energy cycle utilizing low-temperature heat for power generation," *Amer. Soc. of Mech. Eng.* paper 83-JPGC-GT-3, p 1-5. [hybrid power cycle]

Kestin, J. (1979), *A Course in Thermodynamics*, revised printing of Vol. 1, Hemisphere Publishing, Washington, DC, p 40.

Kestin, J. *et al.* (eds.) (1980), "Available work in geothermal energy,"in *A Sourcebook on the Production of Electricity from Geothermal Energy,*" DOE/RA/28320-2, US Department of Energy, US Government Printing Office, Washington, DC., Chapter 3 and "Availability, the concept and associated terminology," *Energy* **5**, p 679-692. [on concepts and operating equations for availability]

Linnhoff, B. (1993), "New concepts in thermodynamics for better chemical process design," *Proc. Royal Soc. London* **386** (1790), p 1-33 and Linnhoff, B. *et al.* (1982), "A user guide on process integration for the efficient use of energy," *Inst. of Chemical Engineers,* Rugby, UK. [pinch technology]

Milora, S. L. and J.W. Tester (1976), *Geothermal Energy as a Source of Electric Power: Thermodynamic and Economic Design Criteria*, MIT Press, Cambridge, MA, Chapter 3, and Chapter 4 [on geothermal power cycles], and Chapter 5. [on turbine and pump design criteria]

Sama, D.A. (1993), "The use of the second law of thermodynamics in the design of heat exchangers, heat exchanger networks, and processes," *Proc. Inter. Conf. on Energy Systems and Ecology ENSEC '93*, Cracow, Poland (July 5-9, 1993). [alternatives to pinch technology]

Tester, J.W. (1982), "Energy conversion and economic issues for geothermal energy," Chapter 14 in the *Handbook of Geothermal Energy,* Gulf Publishing Co., Houston, TX, p 471-586. [power cycle analysis]

Problems

14.1. You have been asked to design and size a high pressure feed pump for a supercritical Rankine cycle application using a proprietary working fluid formulated by MITY Industries. Their fluid is called Kryptonol because of its high temperature stability and other "indestructible" properties. MITY Industries will only tell us that Kryptonol is a highly "non-ideal" complex organic molecule consisting of H, C, O, Cl, and F atoms whose stoichiometry and structure is known and whose critical properties are: $T_c = 300$ K, $P_c = 95$ bar, and $V_c = 10^{-3}$ m^3/kg.

Your pump should pressurize a flow of 100 kg/s of Kryptonol from 1 to 100 bar at an initial temperature of 315 K. Clearly state how you would estimate the required thermophysical properties as well as how you would obtain power requirements and unit sizes (including number of stages and stage diameter). You can utilize the results of Balje's analysis where $\eta_p = f\,[n_s, d_s, Ma, Re]$ as was done for turbine design. Explain all steps, simplifications, and assumptions that you use--you do not have to calculate actual numbers!

14.2. As an introduction to the use of non-aqueous working fluids in geothermal power cycles, you have been asked to provide preliminary design parameters for the primary heat exchanger of a supercritical Rankine cycle. Hot geofluid enters the exchanger at 150°C and 100 bar at a flow rate of 1870 kg/s. Heat is transferred to a counter-currently flowing stream of a CFC R-115 (monochloropentafluoroethane, C_2ClF_5) which enters at 31°C and 39.26 bar. The R-115 stream is vaporized at supercritical conditions ($P_r = 1.24$) and exits the exchanger at 135°C.

You can assume that the pressure drop on either the R-115 or geofluid side of the exchanger is negligible. In addition, the heat capacity (C_p) for the geofluid can be initially assumed to be constant at 4182 J/kg K. The *PVT* properties of R-115 can be represented by the Peng-Robinson equation with the following critical constants and acentric factor (ω):

$$T_c = 353.1 \text{ K},\ P_c = 31.57 \text{ bar},\ V_c = 1.6310\times10^{-3} \text{ m}^3\text{/kg},$$
$$Z_c = 0.271, \text{ and } \omega = 0.253$$

The reduced ideal-gas-state heat capacity of R-115 is given by:

$$C_p^*/R = a + bT_r + cT_r^2$$

with $a = 2.4760$, $b = 14.0108$, and $c = -3.6790$

(a) Plot a temperature-heat transferred (T-Q/Q_T) diagram for the exchanger showing the temperature profiles for both streams if the minimum pinch temperature difference is set at 10°C. Q/Q_T in this case is the fraction of heat exchanged so it will vary from 0 to 1 along the length of the exchanger. What is the mass flow rate of R-115 and what is the temperature of the geofluid as it exits the exchanger?

(b) If the overall heat transfer coefficient for the exchanger is U_o, develop an expression to estimate the required heat transfer area. Do not calculate numbers!

(c) If the heat capacity of the geofluid can no longer be assumed constant, i.e., $C_{p,\,geofluid} = f[T, P]$, describe how you would modify your analysis to answer parts (a) and (b).

14.3. Suppose that you had at your disposal liquefied natural gas (LNG) and high pressure pure oxygen available at 150 bar, 25°C. The LNG can be assumed to be pure methane (CH_4) at its normal boiling point (1 bar, 111K).

(a) Devise a multi-step process to obtain the maximum power with a 10 mol/s LNG feed combined with 40 mol/s of oxygen gas.

(b) Estimate the maximum power. Prevailing ambient conditions at the site are 25°C, 1 bar. State all assumptions made.

Data provided in Appendix G may be used.

14.4. A proprietary fluid formulated by MITY Industries has been proposed for use in a Brayton power cycle application. MITY will only tell us that their fluid is an organic molecule, containing H, C, F, and Cl atoms, whose stoichiometry, molecular structure, and critical properties are known. You have been asked to design a rotary staged compressor to pressurize a flow of 100 kg/sec from 10 to 100 bar. Compression in each stage is adiabatic followed by interstage cooling between stages to maintain isothermal conditions at 400 K. Although not much is known about the properties of this fluid, its *PVT* properties are expected to be very non-ideal under these operating conditions. A high precision equation of state of modified Lee-Kesler form has been proposed by the manufacturers for your use:

$$Z/Z_c = P_rV_r/T_r = 1 + B/V_r + C/V_r^2 + D\exp[-E/V_r]$$

$$P_r = P/P_c; \; T_r = T/T_c; \; V_r = V/V_c$$

$$T_c = 370 \text{ K}; \; P_c = 50 \text{ bar}; \; V_c = 2 \times 10^{-3} \, m^3/\text{kg}; \; Z_c = 0.267;$$

$$m = 86 \text{ g/mol (molecular weight)}$$

where B, C, D, E are specified empirical constants whose values are known.

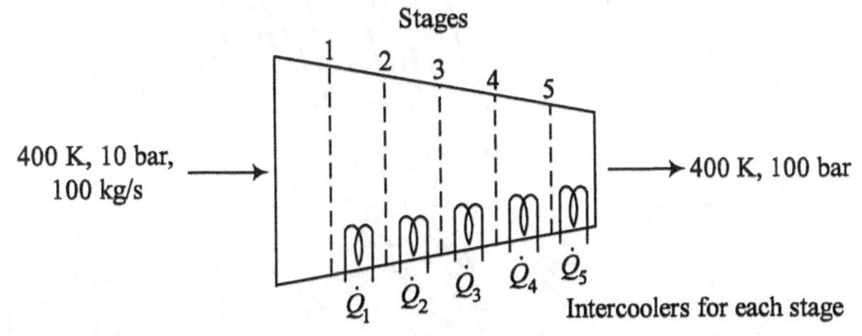

Clearly state how you would estimate other thermodynamic properties that might be required in your analysis (Techniques described in Chapter 13 may be used if desired). Describe how you would obtain design parameters for your compressor including power and interstage cooling levels required, number of compressor stages and maximum inlet stage diameter. You can utilize the results of Balje's analysis where the compressor efficiency $\eta_p = f[n_s, d_s, Ma, Re]$, and assume that the compressor is to be operated at the highest efficiency that is practically possible [a Balje plot of specific diameter versus specific speed is given in Figure P14.4; see Eqs. (14-47) and (14-48)]. For hydrodynamic reasons, the stage pressure ratio must be at or below 1.585. Explain all steps, simplifications, and additional assumptions that you make--you do not have to calculate actual numbers! If you propose computer computations, describe the algorithm you intend to use.

14.5. For the following set of conditions you are to estimate and optimize the performance of a Rankine cycle designed for conversion of geothermal heat into electricity. The cycle has been simplified somewhat to aid your calculations.

Primary heat exchanger:

- inlet geofluid - 14 MPa, 200°C, 1000 kg/s
- outlet working fluid - 185°C
- minimum approach temperature difference = 10°C
- no pressure drop

Turbo-generator:

- dry stage efficiency = 85% with 1% penalty for each 1% moisture
- outlet (exhaust) pressure = vapor pressure at 26.7°C

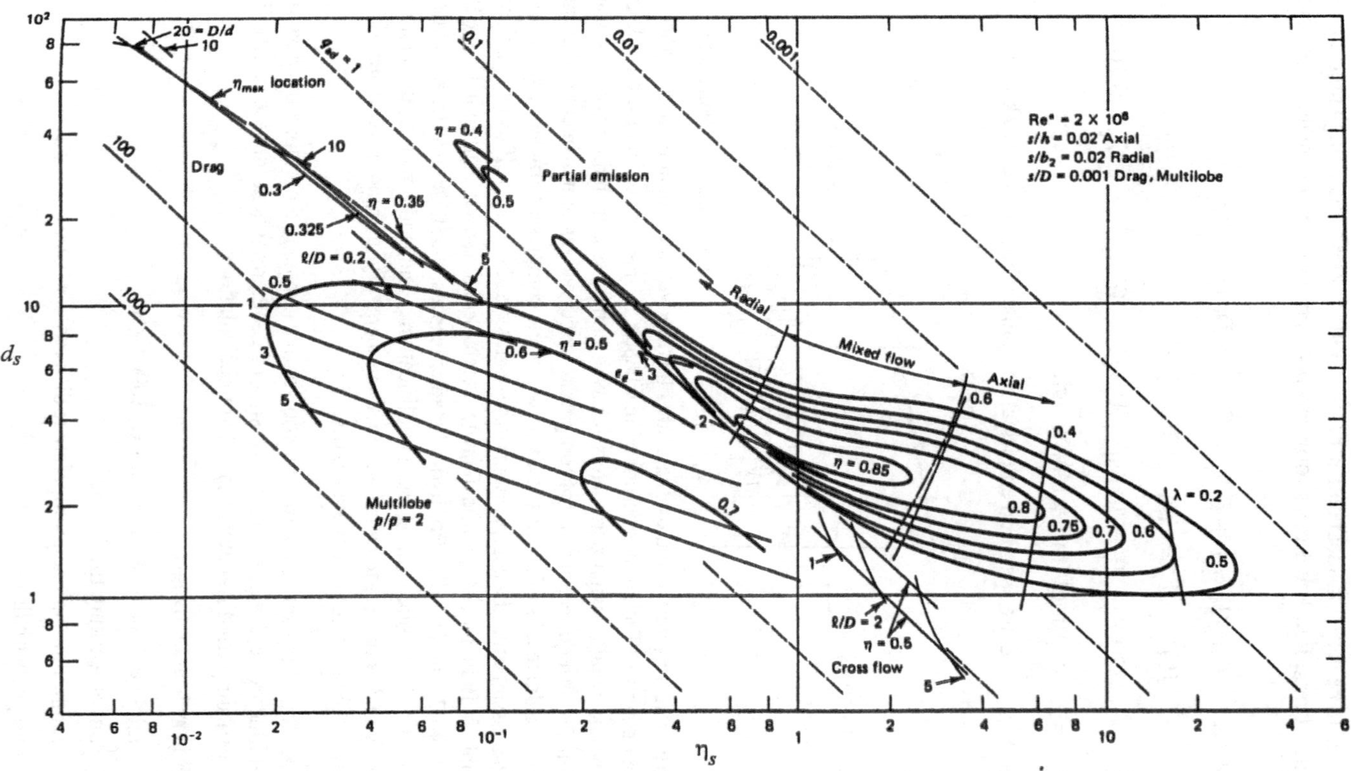

Figure P14.4 A $n_s d_s$ Baljé diagram for single stage compressors, note that n_s is based on $\dot{N}_r$ in radians per second in this figure (1 revolution = 2 π radians) [from Baljé (1981) *Turbomachines,* reprinted by permission of John Wiley and Sons, Inc.].

Desuperheater/Condenser:

- outlet temperature = 26.7°C
- no pressure drop
- excess cooling water available at 26.7°C

Feed Pump:

- overall efficiency = 80%
- compress along saturated liquid curve $V = V^{sat}$

Thermodynamic Properties:

- *PVT* - use one of the cubic equations of state given in Table 8.1.
- $C_p^*(T) = A + BT + CT^2 + DT^3 + \ldots$ -- where *A*, *B*, *C*, *D*, etc. are obtained from data or estimated from group contribution methods (see Section 13.5)

(a) Using R-114 ($C_2Cl_2F_4$) as the working fluid compare your result for an optimal η_u with $\eta_u = 0.70$ obtained by Milora and Tester (1976, Figure 9, p. 45). What power production level can be expected?

(b) Using the attached pressure-enthalpy chart for R-114, calculate η_u for the optimum operating conditions determined in part (a).

(c) Since the production of R-114 has been phased out due to its damaging effect on the Earth's ozone layer, select a non-aqueous fluid of your choice and, with the same format of representing its thermodynamic properties as for R-114 in part (a), determine the optimum η_u for a 200°C geofluid.

(d) Also determine optimal η_u's for the fluid selected in part (c) at geofluid temperatures of 100, 125, 150, 175, 225, and 250°C. You can assume that the working fluid leaves the primary heat exchanger at a temperature 15°C below the inlet geofluid temperature.

14.6. In hot dry rock geothermal energy extraction, thermal energy is removed from rock contained in deep, hot underground reservoirs. Typically, pressurized water is pumped down one well and convectively circulated through a region of fractured rock and heated by conduction through the fracture walls into the solid rock. The hot water is then pumped to the surface in a second well where it can be used as a heat source for a Rankine power cycle to produce electricity. A thermodynamic model of such a reservoir system, assuming no heat recharge from the surrounding rock, is illustrated by the temperature-entropy diagram shown in Figure P14.6. T_H represents the temperature of the rock mass at any time during the heat extraction process and T_o is the ambient sink temperature (assumed constant). The rock is initially at point 1 where the temperature is T_1. The lowest temperature for economical heat extraction is assumed to be T_2 at point 2.

(a) Develop an expression to estimate the amount of work that could be produced from such a hot dry rock system. State all assumptions and indicate on a *T-S* diagram what this amount of work is equivalent to.

(b) If T_1 is equal to 327°C as specified by existing natural conditions in a particular hot dry rock reservoir at 6 km depth and if T_o = 15°C, develop an expression for the thermal efficiency defined as the ratio of the work produced in part (a) divided by

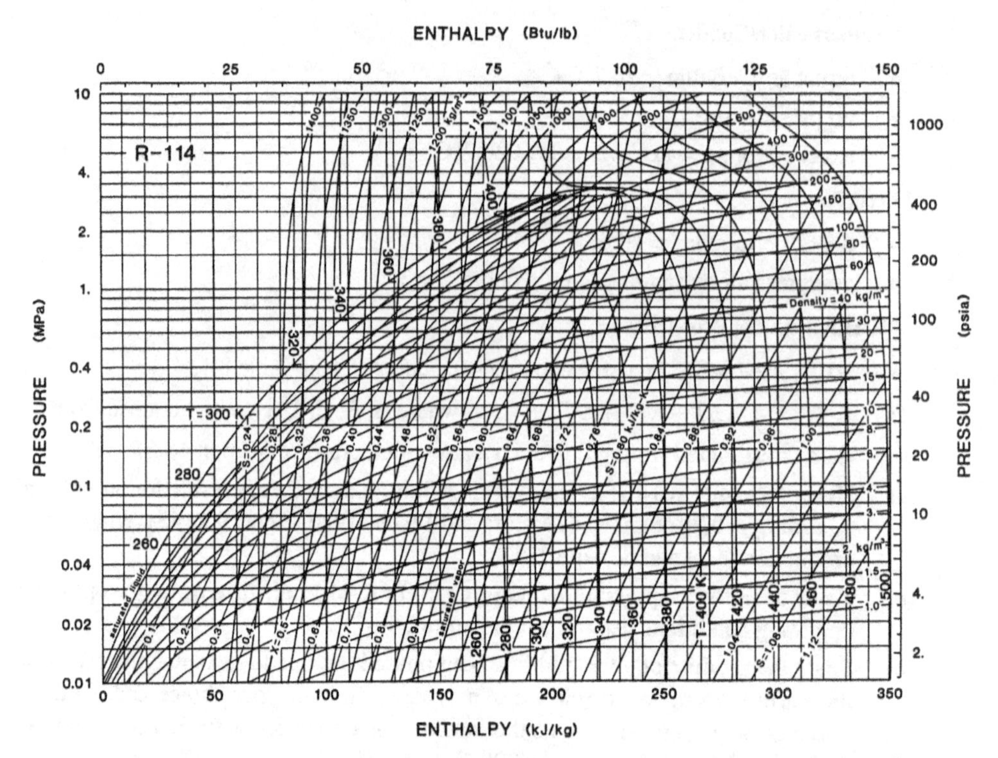

Figure P14.5 Pressure-enthalpy chart for R-114 (from 1993 ASHRAE Handbook, Fundamentals, used with permission of ASHRAE, Atlanta, Georgia).

the enthalpy change of the reservoir. Sketch how this efficiency will vary as a T_2 decreases from 327°C to 15°C.

(c) Once the hot water reaches the surface it can be used to generate electric power, for example, by using it as a primary heat source in a Rankine cycle employing an organic working fluid. Selection of a suitable working fluid for a geothermal resource that operates over a temperature range from T_1 to T_2 over its 10 year lifetime presents a problem in that no single compound can perform optimally over the entire temperature range. A suggested approach is to use a binary mixture where each component has different boiling points and critical properties. A proprietary mixture consisting of compounds A and B has been proposed for such a system. The molecular structure and molecular weight are all known for each as are critical constants (T_c, V_c, P_c, Z_c) and normal boiling points. Describe how you would estimate the maximum power output of a practical Rankine cycle employing a given A-B mixture composition and a specific rock temperature somewhere between T_1 and T_2. You should explain how various state points in the Rankine cycle will be evaluated, for example, at the inlet and outlet points for the heat exchanger, turbine, feed pump, and condenser-desuperheater. Heat rejection can be assumed to be to air at 15°C (T_o). You can assume that the Redlich-Kwong equation of state can be used to model the $P\underline{V}TN$ properties of the mixture. State all additional assumptions made and describe how you will estimate all physical properties needed. Only the data given below are available for your use.

Working Fluid Properties

	Compound A C_4H_9Cl	Compound B C_3H_7F
T_c (K)	520	390
P_c (bar)	36	43
V_c (cm^3/g)	4.58	4.5
Z_c	0.283	0.277
Molecular weight	92	78
Boiling point at 1 bar (°C)	35	10

Rock Properties

C_p (J/kg K)	1000
k (W/m K)	2.8
ρ (kg/m^3)	2700

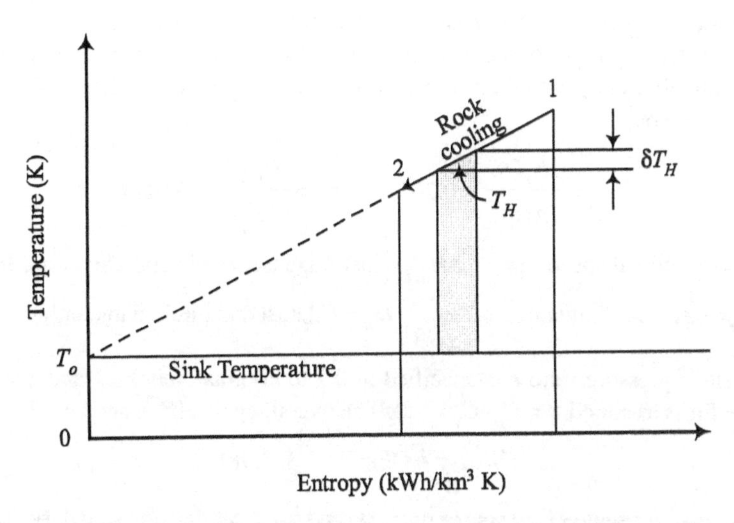

Figure P14.6 Temperature-entropy plot for rock reservoir system.

14.7. A mixture of 20 mol % NH_3 and 80 mol % H_2O is available at a high pressure (P_A) and high temperature (T_A). You have been asked to estimate the work recovery rate possible if the mixture were expanded adiabatically to a lower pressure P_B in a specially designed non-condensing, multistage, corrosion-resistant turbine. In addition, your client wants to know how much such a turbine might cost; thus you will need to estimate the exhaust end area, where

$$\text{Area} = \frac{\pi}{4} D_p^2 \quad \text{where } D_p = \text{the pitch or turbine wheel outside diameter}$$

Make your estimate of the area assuming that you want to operate this turboexpander at the highest efficiency possible at Balje-determined optimum parameters (n_s, d_s). The

$\underline{PV}TN$ properties of the mixture can be adequately described with a cubic EOS. Ideal-gas-state heat capacities for pure NH_3 and H_2O are given by standard correlations such as given in Appendix G.

(a) Describe how you would estimate the power produced from a 1 mole/s stream of the mixture. Do not attempt to calculate numbers but describe your solution procedure using a flow sheet or stepwise algorithm. Give important equations when relevant.

(b) Describe how you would estimate D_p for this particular mixture assuming that the exhaust pressure conditions at P_B are low enough to assume ideal gas behavior. Also at the exhaust conditions, you should note that no liquid is present.

14.8. Develop an expression for the total last stage turbine exhaust area a^* (number of turbines times area per exhaust end) required for an idealized Rankine cycle having a power input P in J/s. Assume that all heat is rejected from the cycle at a constant condensing temperature T_o and that the exhaust pressure is low enough to assume ideal gas behavior for the saturated vapor of the working fluid. Also assume that each turbine operates with optimum efficiency such that: (in SI units) π_1 and π_2 are equivalent to the Balje' n_s and d_s parameters.

$$\pi_1 = \frac{\dot{N}_r D_P}{(\Delta H_i)^{1/2}} = 0.265 \quad ; \quad \pi_2 = \frac{\dot{N}_r^2 \underline{\dot{V}}_i}{(\Delta H_i)^{3/2}} = 0.0124$$

$\dot{N}_r$ = rotational speed, rps $\quad \Delta H_{S,i}$ = last stage isentropic enthalpy drop, J/kg

$\underline{\dot{V}}_i$ = last stage flow rate, m^3/sec $\quad D_p$ = exhaust end pitch diameter, m

The stage pressure ratio r is specified at 0.7 to eliminate choking and a suitable mean value for is assumed $\kappa = C_p/C_v = 1.2$. The overall cycle efficiency is:

$$\eta_{cycle} = P/\dot{q}_{HE} = P/(P + \dot{q}_C)$$

where $\dot{q}_{HE}$ = primary heat source duty, J/s and $\dot{q}_C$ = condensing step duty--heat rejection rate, J/s

(a) Express your answer for a^* as a function of the following cycle operation and working fluid property variables:

$$a^* = n_e \pi D_p^2/4 = f\,[\eta_{cycle}, P, T_o, m, V_g^{sat}, \Delta H_{vap}]$$

where:

V_g^{sat} = specific volume of saturated vapor at T_o, m^3/kg

ΔH_{vap} = molar enthalpy of vaporization at T_o, J/kg

m = molecular weight, g/mol

n_e = number of exhaust ends

(b) At a given power level, how do the exhaust area requirements for NH_3 and water compare at $T_o = 26.7°C$ (80°F) where:

	H_2O	NH_3
V_g^{sat}	39 m^3/kg	0.122 m^3/kg
ΔH_{vap}	2438 kJ/kg	1161 kJ/kg

14.9. (New York Journal American - January 1983) "As incongruous as it may seem, there is a plan afoot to use waters of the Arctic Ocean as a source of power to generate electricity. And while the idea is a long way from being perfected, the theory is almost too simple to be true." An engineer at the University of Alaska in Fairbanks, Rocky Jones Byrd, and his cousin Larry have come up with the proposal, which takes advantage of the temperature difference between the cold arctic air at -93°C and the ocean water beneath the polar cap at -3°C. Power would be generated in a Rankine cycle employing a non-aqueous, low-boiling point working fluid consisting of pure methane. According to Rocky and Larry, the conversion method, which has been tested elsewhere, would produce no waste, pose no radiation hazard, would entail no fuel cost, and would cause minimal pollution overall.

(a) Assuming that a flow rate of 100 kg/s of sea water is available, what is the maximum amount of work that could be produced? For sea water, you can assume $C_p = 4200$ J/kg K and $\rho = 1000$ kg/m^3.

(b) Qualitatively sketch out a temperature-heat transferred (T-Q or T-H) diagram showing state points for a supercritical CH_4 cycle, the sea water cool-down path, and the air heat-up path. Figure P14.9 is a methane P-H diagram.

(c) Assuming that the supercritical cycle is selected such that a turbine inlet pressure is 6 MPa (870 psia) and its temperature is -13°C, what is the work produced for each kg expanded through the turbine with an 85% efficiency with respect to reversible adiabatic operation ($\eta_t = 0.85$) to an exhaust pressure of 3.28 MPa?

14.10. MITY Industries has come up with a unique proposal to recover power during peak load periods using a geothermally heated hot gas storage concept. Their design is shown in Figure P14.10. Basically, air initially at 300 K is compressed during off-peak hours and charged to a deep storage cavity where it is heated rapidly as it contacts hot rock. During peak periods, the process is reversed to provide hot, compressed gas to drive a turbine/generator to produce electricity with depressurized air exhausted to the environment. In their design, the turbine and compressor are actually the same device just run in reverse. The efficiency of the turbine-compressor is constant and equal to η as a fraction of ideal adiabatic power level during expansion or compression. You can assume that air is an ideal gas with $C_p = 30$ J/mol K and that the rock surrounding the cavity is impermeable to air penetration.

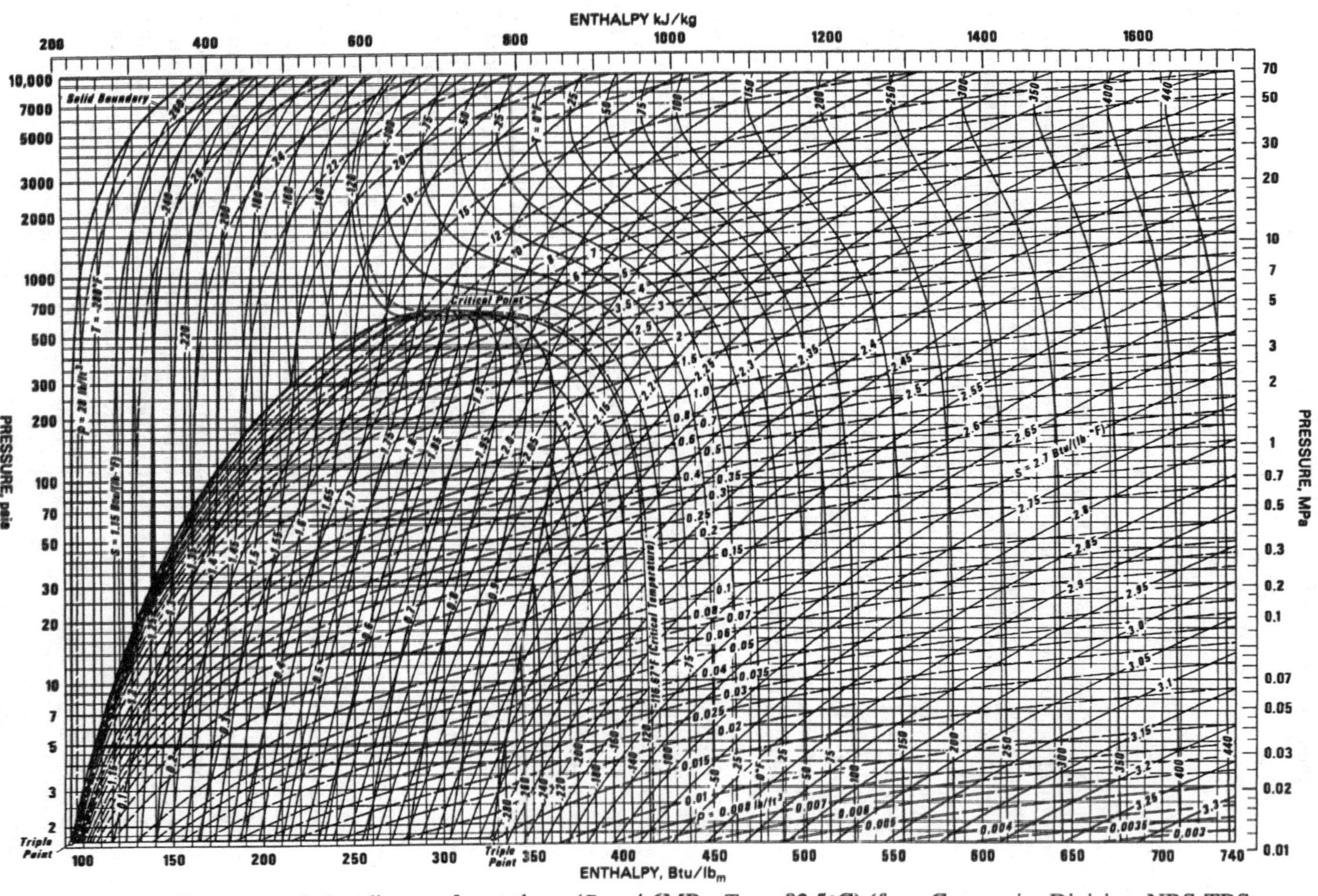

Figure P14.9 Pressure-enthalpy diagram for methane ($P_C = 4.6$MPa, $T_C = -82.5$°C) (from Cryogenics Division, NBS-TBS, Boulder, Colorado, Chart 2029, April 1, 1977).

(a) For an initial rock temperature of 600 K develop an expression to estimate the net work that could be produced for a 24-hr cycle that reaches a maximum storage pressure of P_{max}.

(b) For the same amount of heat exchange between the rock in the cavity and the gas during the compression stage as occurred in part (a), show how you would estimate the net absolute maximum work possible.

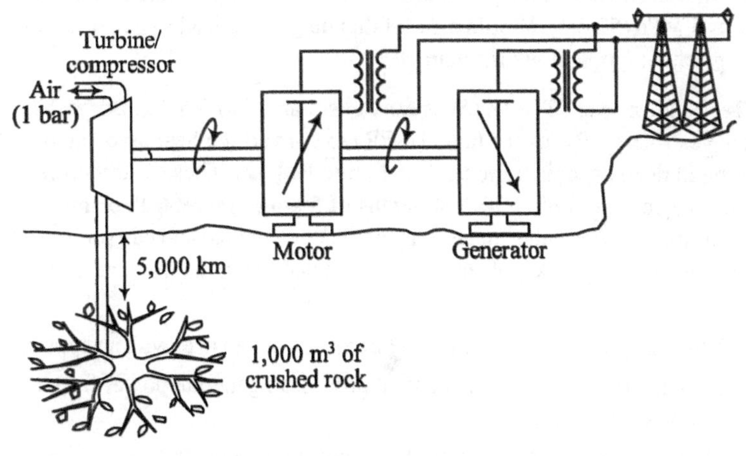

Figure P14.10

14.11. Consider a counter-current shell and tube heat exchanger operating continuously to cool a complex polymer solution whose density is constant at 800 kg/m^3 but whose heat capacity C_p is known to vary over the temperature range of interest as:

$$C_p = C_p^o + aT^2 \quad (T \text{ in K})$$

$$\text{where } C_p^o = 2000 \text{ J/kg K and } a = 2 \times 10^{-2} \text{ J/kgK}^3$$

The polymer solution enters the exchanger at 10 kg/s on the shell side at 127°C and must be cooled to 27°C before further processing in the plant. Cooling water is available at 2°C but must not be heated to more than 27°C to avoid overloading the chilled-water unit.

Ultimately, heat from the water chiller is rejected to the environment at 0°C. Over a temperature range from 0°C to 27°C, water can be assumed to have a constant heat capacity of 4000 J/kg K. The external surfaces of the heat exchanger are well insulated.

(a) Sketch a temperature-enthalpy plot to show the temperature profiles for the water and polymer streams. What is the minimum mass flow of water?

(b) Carry out a thermodynamic analysis of the exchanger to estimate the irreversible loss of work producing potential. You can assume negligible pressure drops on the shell and tube sides.

(c) Under the specified mass flow and inlet temperature conditions of part (a), show how you would calculate the maximum amount of work that could be obtained from an appropriately designed process involving heat interactions only with the cold and hot streams. For this idealized case, the outlet temperatures of both the water and polymer streams can vary to provide maximum work output. They are no longer constrained by original process specifications. Does the maximum amount of work change if the exchanger is operated co-currently rather than counter-currently? Explain your answer.

(d) Measurements on the polymer side of the exchanger indicate that the pressure drop is 5 MPa (50 bar). Would a ΔP of this magnitude lead to a measurable loss of work producing potential? Explain.

14.12. Cal Tech Industries (CTI), The Western Subsidiary of MITY Industries, has proposed and Under Sea Energy Recovery unit (USER) to convert the heat associated with volcanoes erupting in deep oceanic trenches (see Figure P14.12). These "submarine" volcanoes are known to eject hot, molten lava at depths of 5 km or more. CTI claims that their USER is the "ultimate" ocean-thermal energy conversion device and can produce 0.34 kW-hr of electricity for every 1 kg of lava that is ejected from the volcano at a temperature of 1730°C.

(a) What do you think of their claim? Back up your answer with an appropriate analysis.

(b) Sketch a flow sheet for a practical process to generate power from hot lava using a steam Rankine cycle.

(c) Describe where the irreversibilities are in your design.

(d) Estimate the actual work producing potential of your design.

Rock Properties	Ocean Properties
$\rho_{solid} = \rho_{liquid} = 2500$ kg/m^3	Initial temperature = 0°C
$C_{pliquid} = C_{psolid} = 1000$ J/kg K	$\rho = 1000$ kg/m^3
$\Delta H_{fusion} = 100000$ J/kg $\Delta H_{vap} = 10^6$ J/kg	$C_p = 4200$ J/kg K
rock melting temperature = 1000°C	
rock boiling temperature = 2000°C	
(decomposition occurs above 2000°C)	

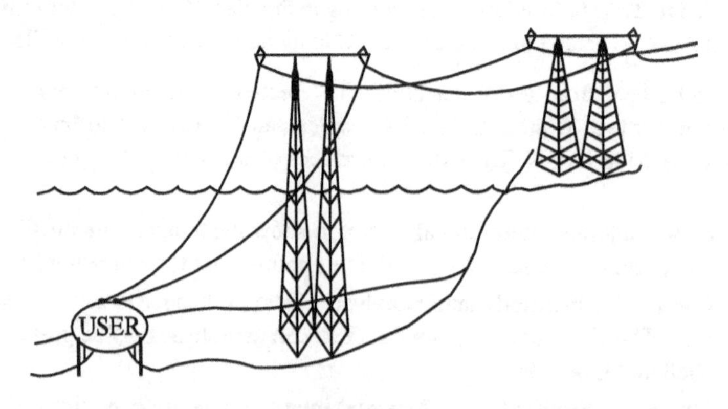

Figure P14.12

14.13. Solar energy has been considered to drive refrigeration and air conditioning systems. In one concept, a Rankine cycle would use collected solar radiation as its heat source. The turbine power produced would then be used to directly drive the compressor on a refrigeration unit. Both the solar Rankine cycle and the refrigeration cycle would use a chlorofluorocarbon R-114 ($C_2Cl_2F_4$) as a working fluid. A schematic of the coupled cycle is shown below. The fluid in the solar-heated loop is a liquid mixture of propylene glycol and water (50% by wt). One of the major problems with solar energy collection for this concept is that heat losses from the collector increase as the average temperature of the collector increases, while operation at the highest possible temperature is desirable to maximize the efficiency of the Rankine cycle in converting heat from the solar source to work.

You have been asked to analyze the performance of this system to select an optimal operating temperature for the collectors. You can assume that the collector heat loss is given by the following approximate expression:

$$\dot{q}_{loss} = k_1 (\overline{T} - T_o) + k_2 (\overline{T}^4 - T_o^4)$$

to account for convective and radiative effects where k_1 and k_2 are empirically determined constants. $\overline{T}$ is the average collector temperature and T_o is the ambient air temperature (both $\overline{T}$ and T_o are in K). You can assume an average $T_o = 27°C = 300$ K. Over a collector temperature range from 100°C to 300°C you can also assume that the Rankine cycle performance expressed in terms of a utilization efficiency η_u is given by:

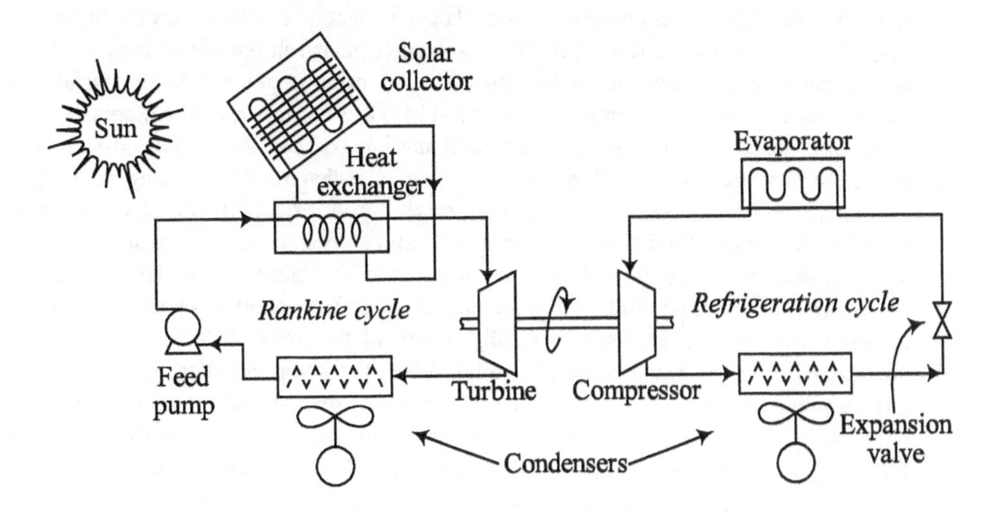

Figure P14.13 Coupled cycle schematic.

$$\eta_u = W/W_{max} = 0.1 + 0.6\left(\frac{\overline{T} - 300}{300}\right) \quad (\overline{T} \text{ in K})$$

Given these constraints, describe how you would estimate $\overline{T}$ for maximum work output from the Rankine cycle. State any assumptions made and show all necessary equations. Just describe your solution procedure. You do not have to calculate a specific numerical answer for $\overline{T}$.

14.14. An alternative technique employing a "gravity-head" Rankine cycle for converting geothermal heat into electricity has been proposed which, according to the inventor, Hugh Matthews, "significantly improves the efficiency of conversion over the competition." You are to analyze the proposed design and compare it with a more conventional Rankine cycle.

For the conventional Rankine cycle (see Figure P14.14(a)), you can assume that hot geothermal brine is pumped to the surface using a downhole pump operated by a 580 kW submerged electric motor. Heat is then extracted from the hot brine to vaporize an organic working fluid (R-114, $C_2Cl_2F_4$) in a countercurrent heat exchanger. The vaporized organic working fluid expands through a turbine-generator, is condensed using air and cooling water and is pressurized to the maximum cycle operating pressure in a feed pump before it enters the exchanger to complete the closed loop.

The "Gravity-Head" or GH cycle is similar to a conventional Rankine cycle in many respects but, as shown in Figure P14.14(b), it differs in two important areas: the first is the method with which heat is removed from the geothermal brine and added to the working fluid (again R-114), and the second is the process for increasing the pressure of R-114 to the turbine inlet pressure. In the GH cycle, heat is removed from the brine in a 2100 ft. (640 m) vertical downhole heat exchanger placed in the upper section of the production hole as shown in Figure P14.14(b). As brine is pumped to the surface, it transfers heat to a countercurrent stream of R-114. The pressure of R-114 increases with increasing depth as it flows down the well until it reaches the turbine inlet of the turbine-driven brine pump. Consequently, heat is added to the working fluid at a continually increasing pressure due to this "gravity-head" effect. The density difference between the organic fluid flowing downward and the organic vapor flowing upward in the insulated riser pipe is sufficient to maintain a self-pumped circulation, thereby eliminating the working fluid feed pump entirely. Geothermal brine, on the other hand, is pumped to the surface using a small amount of the available energy of the hot, pressurized R-114 to rotate a turbine which drives the pump impeller. This downhole turbine-pump unit or TPU should be more efficient than the electrically-driven one required for the conventional plant. The wet condition of the R-114 vapor at the top of the riser pipe necessitates separation into two phases before expansion in two separate turbines, one designed for liquid and the other for vapor. The exhaust streams from the turbines alre combined as they enter the condensers as shown in Figure P14.14(b). The power obtained from the liquid turbine is sufficient to operate the reinjection pump with direct drive from the turbine.

The inventor of the GH process claims that the combination of reduced parasitic pumping for both working fluid and brine and a superior scheme for transferring heat from the brine to R-114 will result in a substantial improvement of performance over the conventional Rankine cycle with an electrically-driven downhole brine pump. His comprehensive comparison results are given in Tables P14.1 and P14.2 and Figures P14.14(c) through P14.14(f). For optimum operating conditions a supercritical ($P > P_c$) turbine inlet pressure is required for both cycles. Pressure-enthalpy plots (Figures P14.14(c) and P14.14(d)) are given as are temperature-heat transfer diagrams (Figures P14.14(e) and P14.14(f)) for each cycle. The inventor claims that the GH cycle will produce 4800 kW while the conventional Rankine cycle would produce only 3940 kW with the same brine flow rate and inlet brine temperature.

(a) Calculate the maximum power that could be produced from the geothermal brine.

(b) Is the inventor's claim that the R-114 feed pump can be eliminated legitimate? Does the GH cycle violate any fundamental concepts or laws?

(c) When the inventor says that the heat exchange scheme between brine and R-114 is superior for the GH process as compared to a conventional Rankine process, he might be suggesting that less heat exchanger area is required for the GH process. Develop an expression for estimating the heat transfer area required for each cycle to decide whether the inventor's claim is true.

(d) Assuming that the GH cycle works, do you agree with the inventor's claim that the power output of the GH cycle will be 22% higher than the conventional Rankine cycle (4800 kW vs. 3940 kW)? Explain your answer.

Conditions Common to Both Cycles

Geothermal Brine

- Temperature at inlet to downhole pump = 360°F (180°C)
- Mass flow rate = 154 lb/s = 70 kg/s
- Heat capacity C_p = 1 Btu/lb °F = 4.186 kJ/kg K

Ambient Air

- Temperature 68°F (20°C)

R-114 Working Fluid (see Figure P14.5 for *P-H* diagram)

- Turbine inlet temperature = 340°F (170°C)
- Turbine inlet pressure = 900 psia (6.21 MPa)
  (Note: for GH cycle, this pressure corresponds to turbine pump inlet pressure)
- P_c = 475 psia (3.26 MPa)
- T_c = 295°F (419 K or 146°C)
- V_c = 0.029 ft^3/lb (0.0017 m^3/kg)

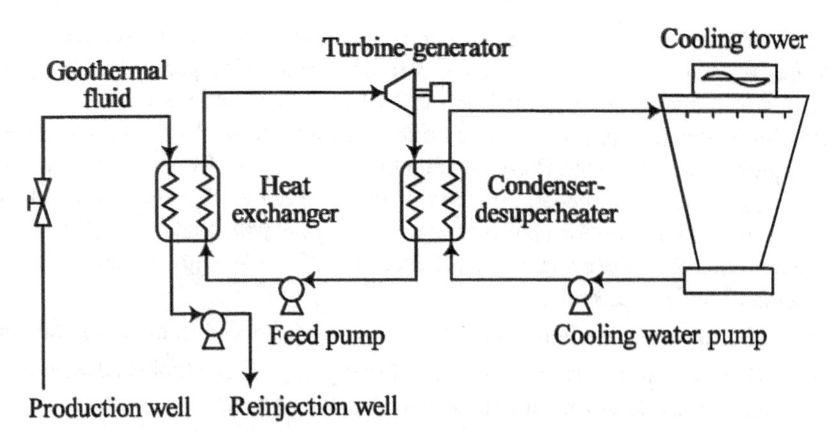

Figure P14.14(a) Conventional Rankine cycle schematic.

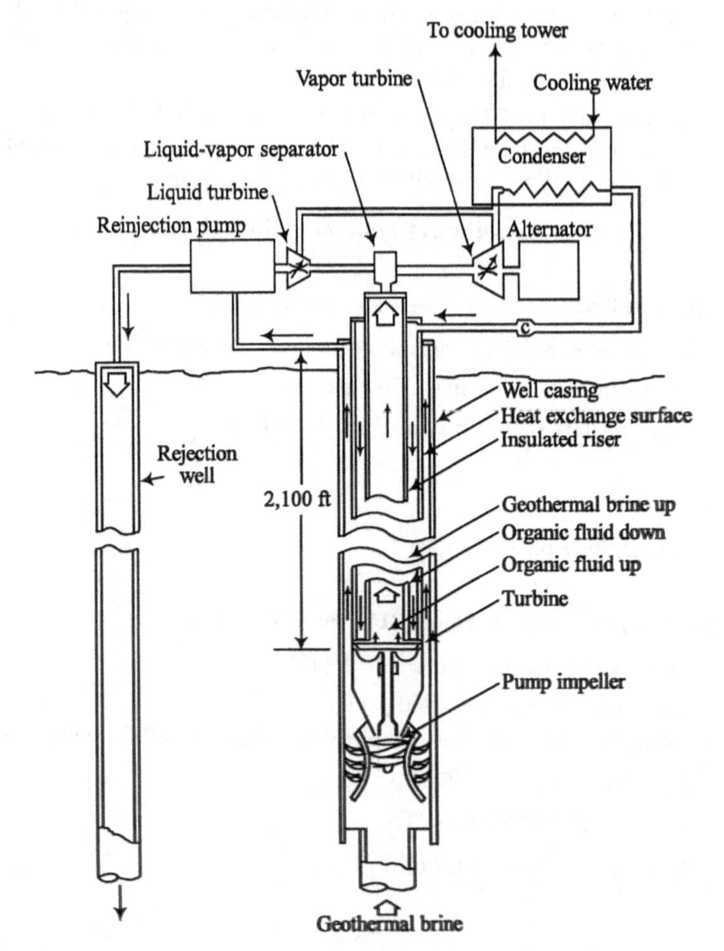

Figure P14.14(b) Gravity head conversion process schematic.

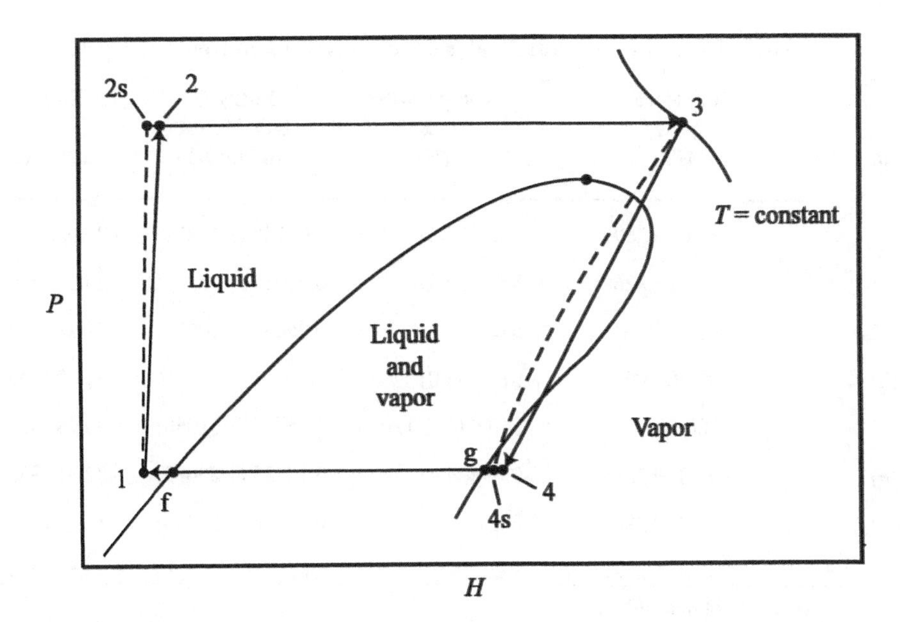

Figure P14.14(c) Simplified pressure-enthalpy (*P-H*) diagram for R-114 conventional Rankine cycle showing state points given in Table P14.1.

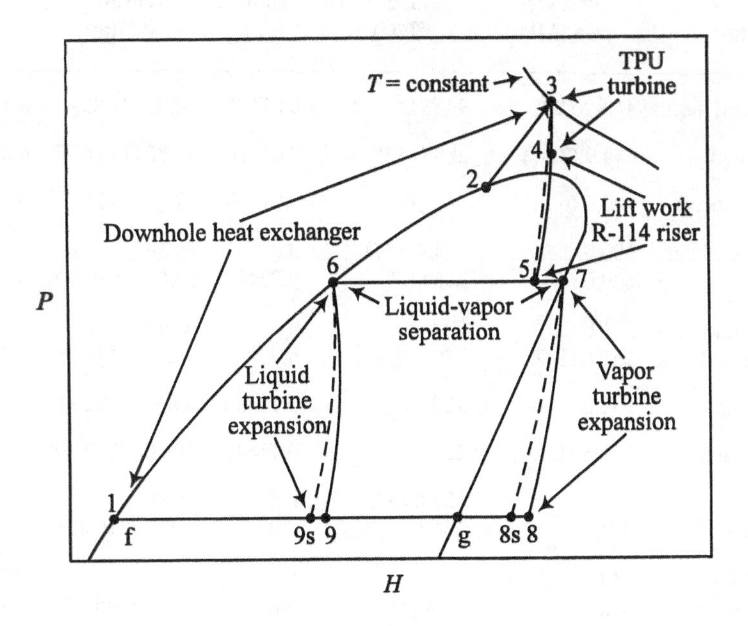

Figure P14.14(d) Simplified pressure-enthalpy (*P-H*) diagram for R-114 showing gravity head cycle and state points given in Table P14.2.

Table P14.1 State-Point Analysis for Conventional Rankine Cycle

State	Pressure psia (MPa)	Temperature °F (°C)	Entropy BTU/lbm R (kJ/kg K)	Enthalpy BTU/lbm (kJ/kg)
f (f)	39.0 (0.269)	90.0 (32.2)	0.06188 (1.10824)	29.90 (231.41)
1 (1)	39.01 (0.269)	85.0 (29.4)	0.06039 (1.1020)	29.35 (230.12)
2s (2s)	920 (6.34)	n.d.	0.06039 (1.1020)	31.28 (234.62)
2 (2)	920 (6.34)	91.4 (33.0)	n.d.	31.70 (235.59)
3 (3)	900 (6.21)	340.0 (171.1)	0.15795 (1.51050)	96.78 (386.96)
g (g)	41.8 (0.288)	93.9 (34.4)	0.15735 (1.50695)	82.31 (353.31)
4s (4s)	41.8 (0.288)	n.d.	0.15796 (1.51050)	82.78 (354.41)
4 (4)	41.8 (0.288)	n.d.	n.d.	84.88 (359.29)

n.d. = not determined
s = isentropic ideal expansion or compression

Table P14.2 State-Point Analysis for Gravity Head Power Cycle

State	Pressure psia (MPa)	Temperature °F (°C)	Volume ft^3/lb_m	Enthalpy BTU/lb_m (kJ/kg)	Entropy BTU/lb_m °R (kJ/kg K)
1 or f (sat'd liq.)	41.8 (0.29)	94 (34)	0.01126	30.89 (71.85)	n.d.
2 (sat'd liq.)	444.9 (3.07)	288 (142)	0.02016	85.76 (199.48)	n.d.
3 (supercritical)	900.0 (6.21)	340 (171)	0.01939	97.53 (226.85)	0.15808 (0.6618)
4s (supercritical)	600.0	314.4(157)	n.d.	96.38	0.15808
4	600.0 (4.1)	318 (159)	0.02410	96.67 (224.85)	0.15900 (0.6657)
5s (17% liq.)	231.6	229	n.d.	93.44	0.15900
5 (17% liq.)	231.6 (1.60)	229 (109)	n.d.	93.73 (218.02)	n.d.
6 (sat'd liq.)	231.6	229	0.01434	65.60 (152.59)	0.11791 (0.4937)
7 (sat'd vapor)	231.6 (1.60)	229	0.11840	100.00 (232.60)	0.16743 (0.7010)
8s (vapor)	41.8	128.0 (53)	n.d.	89.00	0.16743
8 (vapor)	41.8 (0.29)	136.0 (58)	n.d.	90.65 (210.85)	0.17000 (0.7118)
9s (43% liq.)	41.8	94.4	n.d.	61.20	0.11791
9 (40% liquid)	41.8 (0.29)	94.4 (35)	n.d.	62.52 (145.42)	0.12000 (0.5024)
8 & 9(vapor)	41.8	110.0 (43)	n.d.	85.87 (199.73)	0.16200 (0.6783)

n.d. = not determined
s = isentropic ideal expansion or compression

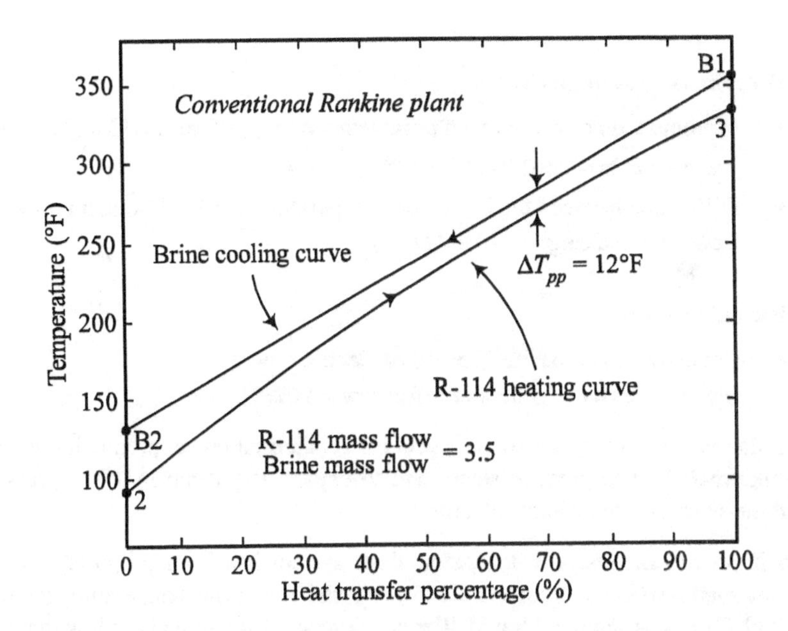

Figure P14.14(e) Temperature-heat transferred diagram for the conventional plant. State points 2 and 3 correspond to values for R-114 in Table P14.1 while B1 and B2 correspond to inlet and outlet geothermal.

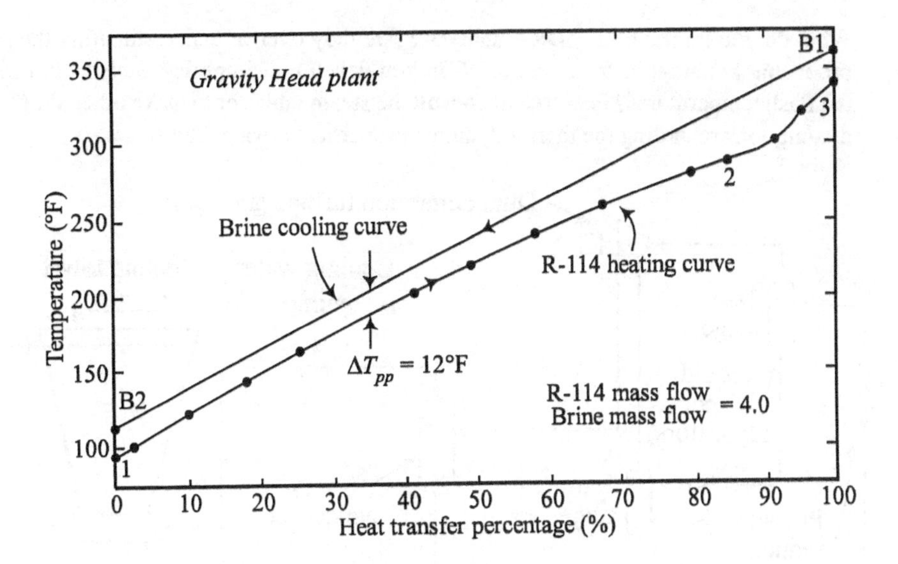

Figure P14.14(f) Temperature-heat transferred diagram for the downhole heat exchange process in the Gravity Head system. State points 1, 2, and 3 correspond to values for R-114 in Table P14.2 while B1 and B2 correspond to inlet and outlet geothermal brine conditions respectively.

Heat Exchanger Design Criteria

- Minimum approach or pinch point temperature differences [ΔT_{pp}] between geothermal brine and R-114 = 12°F (6.7°C)
- Difference between the brine inlet temperature and R-114 outlet temperature from heat exchanger = 20°F (11°C)

Parasitic Power

- Condenser fans require 375 kW of electric power
  Electric alternator/generator efficiency = 96%

14.15. An alternative to using a Rankine cycle for geothermal power production is to flash the geothermal fluid to produce steam and to expand the steam in a low-pressure steam turbine-generator to generate electricity.

To improve efficiency, a dual-stage flash system has been proposed for a specific geothermal resource in Southern California, where the brine temperature at the wellhead is 230°C (T_{gf}) as saturated liquid. The salt content of the brine is so low that pure water properties can be used. An analysis conducted by HDRA (Hot Dry Rock Associates) claims that optimum conditions for the 2-stage flash are reached with maximum power output when the stage 1 flash temperature is 167°C and stage 2 is 107°C. The ambient dead state temperature of 37.8°C (T_o) is indicative of the hot desert locale.

What do you think of the HDRA analysis? Are they over or underestimating the power producing potential of the resource? Can you develop a simplified analysis to estimate the flash temperatures? Feel free to consult the steam tables or to make other simplifying assumptions regarding the thermodynamic properties of water and steam.

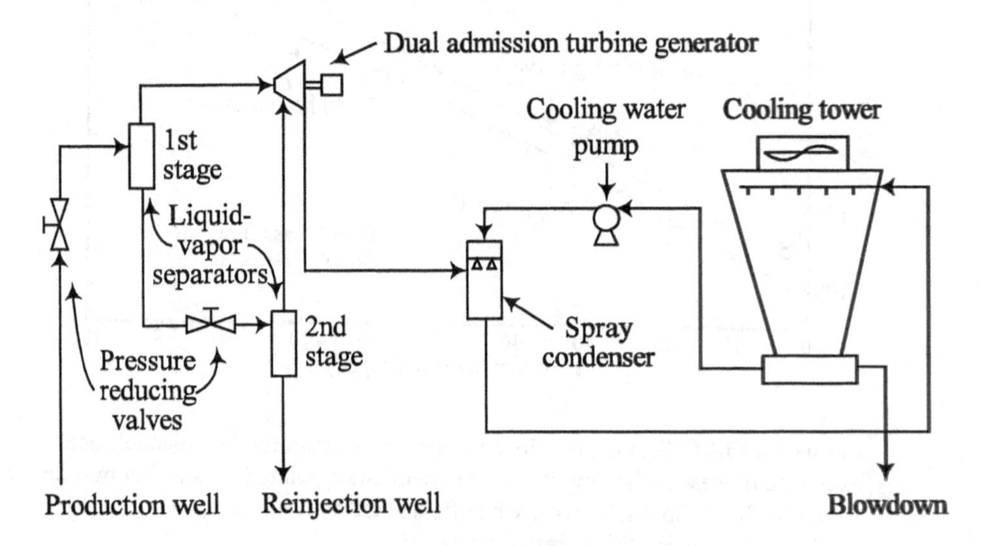

Figure P14.15(a) Dual flash system.

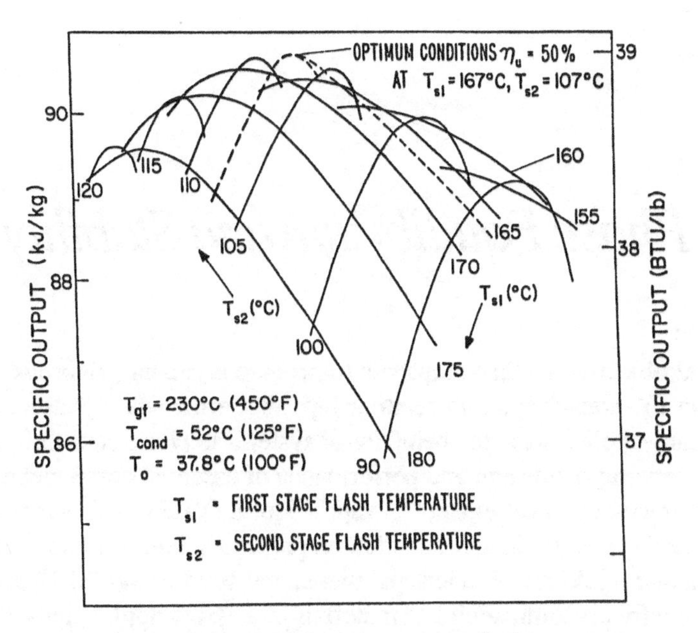

Figure P14.15(b) Specific output versus flash temperature.

14.16. In developing the results shown in Figure 14.11, the loss in work producing potential was calculated for each step in the R-115 cycle (see also Figure 14.9):

(1) feed pumping from E to A
(2) primary heat exchange from A to B
(3) turbine expansion from B to C
(4) desuperheating (if needed) from C to D
(5) condensing from D to E

In addition, the lost availability by reinjecting geothermal fluid at $T > T_o$ was determined.

(a) Derive general expressions for the specific irreversibility I for each step above where

$$I \equiv \Delta \boldsymbol{B}_{loss} \text{ and } \dot{W}_{lost} = \dot{n}\Delta \boldsymbol{B}_{loss} = \dot{n}I$$

(b) Check estimated values of I at a reduced turbine inlet pressure of 2.54 and compare with results plotted in Figure 14.11.

(c) Describe conditions where I goes to zero for the turbine and feed pump. How close can one approach these conditions, practically speaking?

Phase Equilibrium and Stability 15

The central objective of this chapter is to provide a general, rigorous approach for understanding the constraints and relationships that arise when phase coexistence is present. In many applications, the behavior of systems in phase equilibrium is a critical factor in determining the design and performance of essential components in chemical engineering processes. For example, vapor-liquid (VLE) and vapor-liquid-liquid (VLLE) equilibria are at the heart of distillation processes; liquid-liquid equilibria (LLE) for extraction and liquid membrane separations; and solid-liquid (SLE) and solid-solid (SSE) equilibria for crystallization and materials processing applications. Proper design of equipment requires characterization of equilibrium phase behavior as a function of temperature and pressure. Typically, this involves establishing how various components partition themselves among the phases present.

In addition to these equilibrium properties, behavior at phase stability limits is important in engineering practice as well. For example, predicting whether additional gas, liquid or solid phases will nucleate under metastable conditions has applications to designing materials and to ensuring the safety of operations that handle metastable fluids. Furthermore, a wide range of phenomena associated with behavior near the critical points of fluids are being exploited to improve separations as well as chemical reaction kinetics. The equilibrium criteria always form the starting point for further developments in phase equilibrium. This will become apparent in the treatment in this chapter.

The criteria for the coexistence of phases in equilibrium were developed in Section 6.5, where we demonstrated that the temperatures, pressures, and component chemical potentials were equal in all phases. These criteria are valid even if chemical reactions occur in one or more phases. They may not, however, necessarily be valid if there are any constraints to the flow of mass or energy between phases. Thus, it is often stated that such criteria only apply in the general sense for *simple*, multiphase systems. Example 15.1, discussed later, illustrates a *nonsimple* system.

In order to facilitate the mathematical treatments of phase equilibria that follow in Sections 15.3 – 15.6, we begin by introducing the Gibbs phase rule and then proceed to describing important applications using equilibrium phase diagrams for single and multicomponent systems. This approach is designed to review material that is normally presented in a first course in chemical engineering thermodynamics as a stepping stone to a more advanced treatment. In Section 15.7, practical applications involving phase stability will be discussed.

This chapter only provides a framework for tackling phase equilibria problems which represent a large body of chemical engineering research. Interested readers should pursue more detailed monographs by Rowlinson (1969), Prausnitz *et al.* (1986) and Walas (1985) for further study.

15.1 The Phase Rule

In Section 2.7 it was stated in Postulate I that $(n+2)$ *independently variable* properties characterize completely the stable equilibrium state of a simple system. We admitted that there may be cases in which properties were not independently variable, although we have not yet met such cases (with the trivial exception of excluding one mole fraction from a set of n in a mixture).

We are now interested in extending our treatment to multiphase systems to determine the minimum set of variables to describe completely such systems both in extent and intensity. We will find even here that a general result may only be found for simple, composite systems. For such a system with π phases let us describe each phase separately and then relate all properties by invoking the criteria of phase equilibrium. If we were to apply Postulate I to each phase separately, we could choose any $(n+2)$ properties of each phase, provided that each set includes no more than $(n+1)$ intensive variables (see Section 5.2).

For each phase, a particularly convenient set of $(n+2)$ properties is

$$T^{(s)}, P^{(s)}, x_1^{(s)}, \ldots, x_{n-1}^{(s)}, N^{(s)} \tag{15-1}$$

where superscript (s) is used as a dummy index to denote a phase. For a composite system containing π phases we have a set such as Eq. (15-1) for each phase, or $\pi(n+2)$ properties. To determine which of these $\pi(n+2)$ properties are not independent, we must apply the criteria of phase equilibria, which for this system are

$$T^{\alpha} = T^{\beta} = \cdots = T^{(s)} = \cdots = T^{\pi} \tag{15-2}$$

$$P^{\alpha} = P^{\beta} = \cdots = P^{(s)} = \cdots = P^{\pi} \tag{15-3}$$

$$\mu_j^{\alpha} = \mu_j^{\beta} = \cdots = \mu_j^{(s)} = \cdots = \mu_j^{\pi} \quad (j = 1, \ldots, n) \tag{15-4}$$

There are $(\pi-1)$ equalities in each of Eqs. (15-2) and (15-3) and $n(\pi-1)$ in Eq. (15-4) for a total of $(n+2)(\pi-1)$. If we transform this set of $(n+2)(\pi-1)$ equalities into relations containing only the properties in the $\pi(n+2)$ set of the kind given in Eq. (15-1), the $(n+2)(\pi-1)$ equalities can be used as restraining equations to determine the relationships between the $\pi(n+2)$ properties of the coexisting phases.

Let us carry out this transformation. We see that the set of variables in Eq. (15-1) is particularly convenient because we need transform only the chemical potentials. Specifically, we can expand these by relations of the kind that we have used in Chapter 8:

$$\mu_j^{(s)} = g_j^{(s)}(T^{(s)}, P^{(s)}, x_1^{(s)}, \ldots, x_{n-1}^{(s)}) \tag{15-5}$$

Thus, each of the equalities of Eq. (15-4) takes the form

$$g_j^{\alpha}(T^{\alpha}, P^{\alpha}, x_1^{\alpha}, \ldots, x_{n-1}^{\alpha}) = g_j^{\beta}(T^{\beta}, P^{\beta}, x_1^{\beta}, \ldots, x_{n-1}^{\beta}) \tag{15-6}$$

and we have $n(\pi-1)$ such relations. The $(n+2)(\pi-1)$ restraining equations of Eqs. (15-2), (15-3), and (15-6) involve only $\pi(n+1)$ *intensive* variables of the $\pi(n+2)$ properties in Eq. (15-1). Clearly, then, the extensive variables included in the set of Eq. (15-1) (i.e., $N^{\alpha}, N^{\beta}, \ldots, N^{(s)}, \ldots, N^{\pi}$) are not related by the criteria of phase equilibria and hence are independently variable.

Of the remaining $\pi(n+1)$ intensive variables of the set given by Eq. (15-1), we have $(n+2)(\pi-1)$ restraining equations so that the number of independently variable intensive properties is $\pi(n+1) - (n+2)(\pi-1) = (n+2-\pi)$. Thus, we conclude that to describe the composite simple system completely, there are $(n+2-\pi)$ independently variable intensive properties and π independently variable extensive properties for the total of $(n+2)$ variables required by Postulate I.

Although we have derived this result using specific sets of properties, the result is of general validity and not restricted to any particular set of intensive and extensive variables. For example, one could readily show that instead of the set of Eq. (15-1), we could have started with the set $T^{(s)}, P^{(s)}, \overline{S}_1^{(s)}, \ldots, \overline{S}_{n-1}^{(s)}, S^{(s)}$. In place of the expansion of Eq. (15-5), we would then have to use $\mu_j^{(s)} = g_j^{(s)}(T^{(s)}, P^{(s)}, \overline{S}_1^{(s)}, \ldots, \overline{S}_{n-1}^{(s)})$. Additional transformations are required if the chosen set contains intensive variables other than $T^{(s)}$ and $P^{(s)}$, but the final result is unchanged.

The general conclusion, which is one of the most important results of thermodynamics of multiphase systems, can be stated as follows: *For a composite simple system containing π phases and n components in which chemical reactions do not occur, there are $(n+2-\pi)$ independently variable intensive properties, and therefore at least π extensive properties must be included in the set of n+2 properties necessary to describe the composite system completely.*

This result was first expressed by J. Willard Gibbs in 1875 and is commonly referred to as the *Gibbs phase rule*. In most texts the number of *independently variable intensive* properties is referred to as the variance or degrees of freedom, denoted here by $\mathcal{F}$, and the phase rule is written as

$$\mathcal{F} = n + 2 - \pi \tag{15-7}$$

Let's consider a binary system, where

$$\mathcal{F} = 4 - \pi \tag{15-8}$$

. When only a single phase is present, $\mathcal{F} = 3$, and we must specify three intensive properties such as T, P, and x_i to describe completely all other intensive properties. We have dealt with such systems in Chapter 9.

Although the phase rule appears to be very simple, the application in some cases leads to results that are difficult to interpret. For example, in many binary liquid-vapor systems, dew and bubble points are monotonic functions of composition. There is no reason to expect this always to be the case. In fact, we frequently encounter systems in which the pressure, for an isothermal section, or the temperature, for an isobaric section, attains a maximum or minimum value. These systems are called *azeotropic mixtures* and a minimum-boiling type is illustrated in Figure 15.1. The azeotrope is that state in which the concentrations in the liquid and vapor phases become identical. Thus, for this state the relative volatility ($\alpha_{AB} \equiv (y_A/x_A)/(y_B/x_B)$) becomes unity and separation processes such as simple distillation become impossible.

As shown in Figure 15.1 in the isothermal section at $T = T_z$, the pressure at the azeotrope is noted as P_z. The isobaric section is drawn for $P = P_z$ and again the azeotropic temperature is T_z. The azeotropic concentrations are drawn to be the same in both figures.

The azeotropic mixture illustrates one of the limitations of the Gibbs phase rule. The variance of a two-phase binary is 2. There are, however, cases in which we cannot necessarily describe completely the intensive properties of the phases in a mixture simply by specifying the temperature and pressure. For example, in Figure 15.1(a), for the section at $T = T_z$, choose a pressure P' less than P_z. At T_z, P', there are *two* equally valid sets of coexisting liquid and vapor compositions: The Gibbs phase rule apparently breaks down. The problem is that we are expecting too much. As seen in the next section, the relation we obtain for $P = f(x)_T$ can be multivalued.

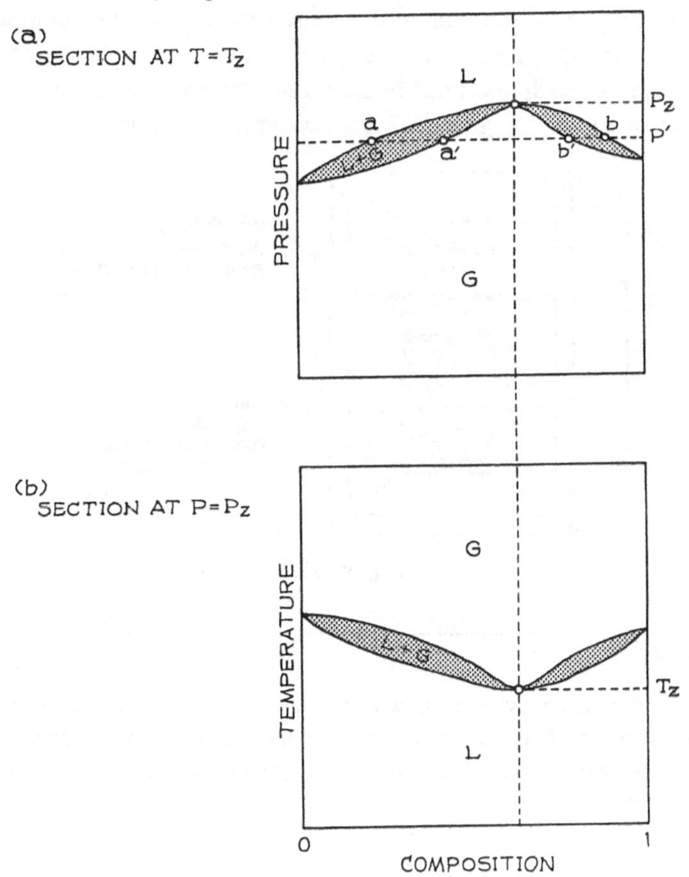

Figure 15.1 Isothermal and isobaric sections for a minimum boiling azeotropic system.

The apparent difficulty in interpreting the Gibbs phase rule is avoided if we reconsider the discussion earlier in this section. We stated that there are ***certain*** sets of intensive properties which, when selected, will completely define the system. But this statement did not imply that we could choose *any* set. Thus, for the case of an azeotropic system, we may not be able to use the pressure-temperature set. On the other hand, if we specify temperature and liquid composition, we always define a unique state for which pressure and vapor composition are determined. (Chapter 17 provides an alternative treatment of azeotropic phenomena as a particular type of indifferent state phase behavior, see Sections 17.3 and 17.6 for further discussion).

It is also interesting to note that to specify the system completely (i.e., extensive as well as intensive properties), we could use pressure and temperature provided that we also include N_A and N_B in the set of $(n+2)$ properties required by Postulate I. In this case, the average mole fractions of both components are known, and hence we can make an unambiguous choice regarding the branch of the azeotrope that is the correct one (e.g., in Figure 15.1(a), a–a' or b–b').

Before concluding the discussion of the phase rule, let us note that it does not apply to composite systems that are not simple systems. For nonsimple systems, no general rule can be developed; each case must be analyzed separately with respect to the internal constraints that are present. Figure 15.2 and Example 15.1 illustrate this point.

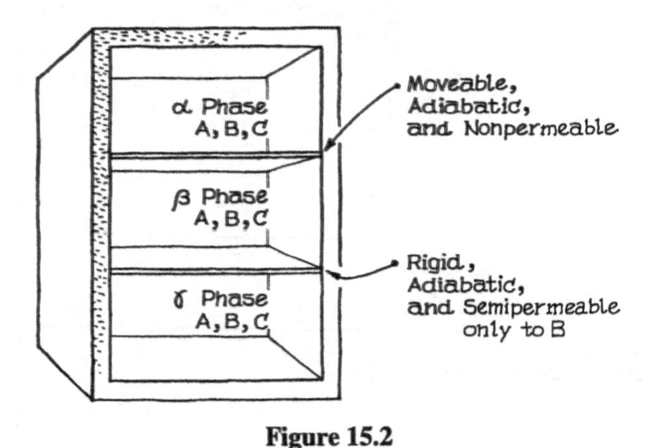

Figure 15.2

Example 15.1

The isolated composite system shown in Figure 15.2 contains three phases with components A, B, and C in each phase. Determine the minimum number of properties necessary to describe completely the composite system and suggest how these properties should be chosen.

Solution

Because the composite system is not a simple system, Postulate I does not apply. To determine the number of independently variable properties, we proceed by choosing $n+2$

or five properties for each phase and use the particular set of equilibrium criteria that applies to the given internal restraints. Let us choose the following $\pi(n+2)$ or 15 properties:

$$T^\alpha, T^\beta, T^\gamma, P^\alpha, P^\beta, P^\gamma, x_A^\alpha, x_A^\beta, x_A^\gamma, x_B^\alpha, x_B^\beta, x_B^\gamma, N^\alpha, N^\beta, N^\gamma$$

The criteria of equilibrium can be shown to be (see Sections 6.4 and 6.5)

$$T^\beta = T^\gamma \; ; \; P^\alpha = P^\beta \; ; \; \mu_B^\beta = \mu_B^\gamma$$

These three restraining equations allow us to eliminate three variables from the set of 15. Nevertheless, we clearly cannot eliminate any arbitrary three. We can, however, eliminate one from each of the following three sets:

$$1. \; (T^\beta, T^\gamma) \quad 2. \; (P^\alpha, P^\beta) \quad 3. \; (T^\beta, T^\gamma, P^\beta, P^\gamma, x_A^\beta, x_A^\gamma, x_B^\beta, x_B^\gamma)$$

Thus, one satisfactory set of independently variable properties is $T^\alpha, P^\alpha, P^\gamma, x_A^\alpha, x_A^\beta, x_A^\gamma, x_B^\alpha, x_B^\beta, x_B^\gamma, N^\alpha, N^\beta, N^\gamma$. The required number of 12 may be compared to the value of five if the system were a simple one.

Numerous cases are encountered in phase equilibria in which we find it difficult to measure the low concentration of a component in one of the coexisting phases (e.g., for an aqueous solution of an inorganic salt in equilibrium with water vapor, the concentration of salt in the vapor phase is usually immeasurably small). In such cases in which there are no artificial walls separating the phases, we could either view the system as a simple one in which the concentration of insoluble component is very small, or we could treat it as if the phases were separated by a movable, diathermal barrier that is impermeable to the insoluble component. Either interpretation leads to the same practical result, provided that we are not interested in determining the concentration of the component in the phase in which its concentration is very small. In the first case (simple system), we would not use the restraining equation of equality of chemical potentials of the insoluble component because we are not interested in its concentration in one of the phases; in the second case (impermeable barrier), we do not have an equality of chemical potentials, and we assume that the concentration of the insoluble component is zero in one phase. Thus, from the pragmatic viewpoint, the two cases are equivalent.

15.2 Phase Diagrams

Later in this chapter we develop a number of useful equations that relate properties in different phases that are in equilibrium. These relationships can be represented graphically by property *phase diagrams*. The term phase diagram has many connotations, but we will almost exclusively limit ourselves to those diagrams that employ the primitive variables temperature, pressure, volume, and composition.

Pure component. For a pure component, composition is invariant and the pressure-temperature diagram is quite simple as illustrated in Figure 15.3. Each curve represents a boundary between single-phase domains. For states existing *on* any of these curves, the two bordering phases coexist. Thus, the curves are the *P*-*T* relationship that is predicted by the phase rule for monovariant systems. Figure 15.3 is typical of all pure substances, except that the slope of the liquid-solid equilibrium curve may be negative instead of positive, as drawn. As will become apparent in Section 15.3, a positive slope is consistent with a material that becomes more dense upon solidification, but a negative slope would mean that the solid phase is less dense (e.g., ice-water). Only a very few materials show this latter behavior.

Although not shown on Figure 15.3, there may be a number of different solid phases each of which is the stable phase in some pressure-temperature domain.

The point shown as *R* is called the ***triple point***, at which the three phases, gas, liquid, and solid, all coexist. From the phase rule, the variance at the triple point is zero for a pure substance; hence, it is a singular condition. The curve separating the gas and solid regions is the ***sublimation*** curve, the curve between liquid and gas is the ***vaporization*** curve, and the curve between liquid and solid is the ***freezing*** (or melting) curve. Other triple points are possible in pure component systems. For example, two distinct solid phases and a liquid melt or three distinct solid phases could exist at triple points. A common example of a triple point where three solid phases coexist is encountered at high pressure in the pure water system. In all cases, however, three monovariant *P*-*T* lines intersect at a unique point to establish the coexistence of the three phases.

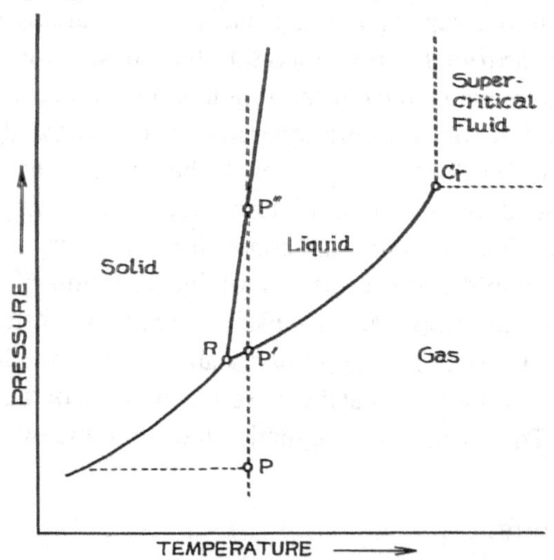

Figure 15.3 Single-component pressure-temperature phase diagram.

Following the freezing or sublimation curves away from the triple point, we see that there is no definite point of termination (except, of course, $P \to 0$, $T \to 0$). However, for the vaporization curve, there is a temperature above which no liquid phase exists. This is shown as *Cr* in Figure 15.3 and is readily recognized as the critical point. The properties of the gas and liquid phases become identical at *Cr*. The naming of the domains above the critical point vary, but usually if $T < T_c$ and $P > P_c$, we say that the material is in a *compressed liquid-like* state. For $P > P_c$ and $T > T_c$, we define the material to be in a *supercritical fluid* state and if $P < P_c$ and $T > T_c$, we are in the *gas* domain.

Since Figure 15.3 conveniently allows us to denote states of a system, we could show values of other properties at such states by adding an additional coordinate perpendicular to the plane of the paper. We illustrate this with the specific volume in Figure 15.4 in which an isometric view of the *PVT* space model is shown. To relate Figures 15.3 and 15.4, we also show the pressure-temperature phase diagram of the material in Figure 15.4. The *P-T* gas state of Figure 15.3 is shown as point A on the isometric. The vapor-liquid state at P' is shown as points B and the solid-liquid state at P'' becomes points D. For the latter two states, two specific volumes are shown; each corresponds to a different equilibrium phase. Straight lines connecting points B or D project as points on the pressure-temperature plane. These lines are called *tie-lines* because they relate the properties of the phases that are in equilibrium at the same temperature and pressure.

All equilibrium states of the material must lie on the solid (S), liquid (L), or gas (G) surfaces or on the curves of intersection between these surfaces. A state in the *G-S* domain, for example, point E, is in reality a mixture of gas and solid with properties of each phase as shown on the phase boundary curves at E^S and E^G. A state not on any surface shown is not an equilibrium state.

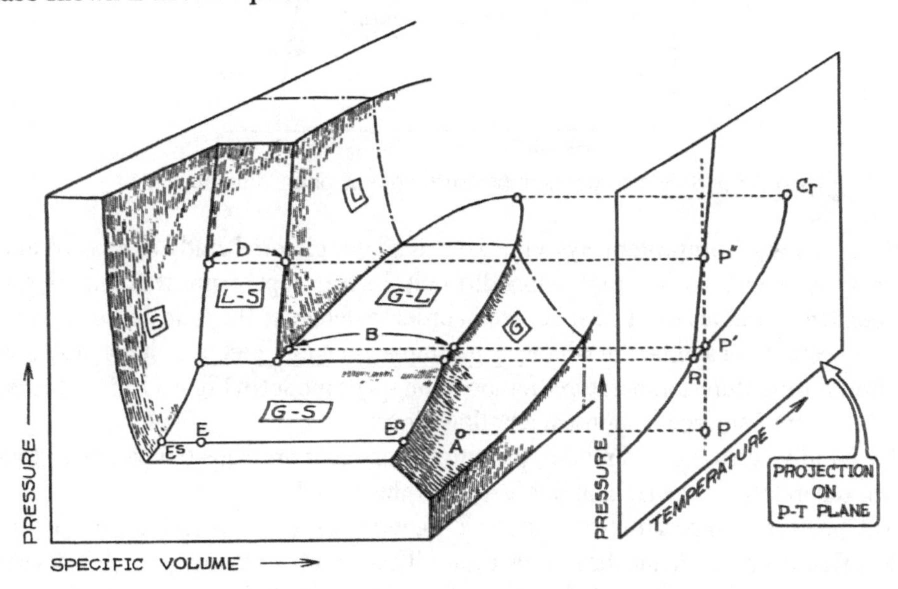

Figure 15.4 *PVT* diagram for a pure material.

The dot-dashed curve shown in Figure 15.4 represents the trace of an intersection with a plane T = constant. This trace is shown to illustrate the more customary way to show a three-dimensional diagram as that of Figure 15.4. Suppose one *projected* the space model onto a *P-V* plane and included a number of T = constant traces. Then a figure such as 15.5 would result. (The solid-liquid volume change is greatly exaggerated for clarity; usually such volume changes are negligible compared to the volume change in going from a liquid to a gas.) Other projections besides the *P-V* section shown in Figure 15.5 could be drawn using different sets of intensive variables; some other kinds are illustrated in the figures included with the problems of Chapter 6. These two-dimensional projections of three-dimensional space models are very convenient to use in tracing property value changes during various processes.

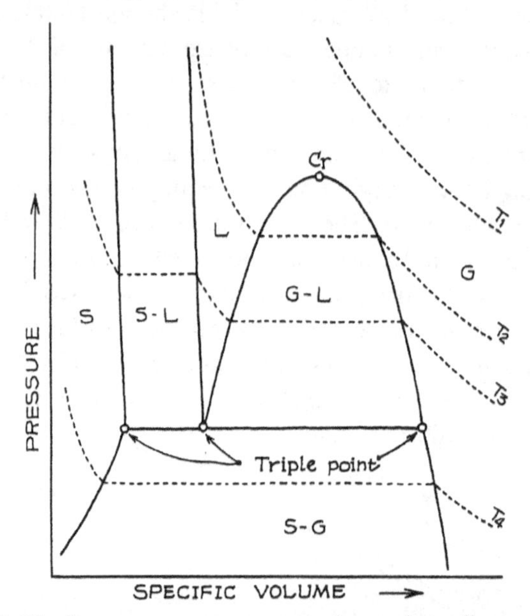

Figure 15.5 Single component pressure-volume phase diagram.

Binary, Two-Component System. To extend our pictorial study of phase diagrams to binary systems is considerably more difficult since, except on phase boundaries, three independent variables must be specified in order to delineate the state of the system. For the simplest case of a binary in which each component is completely soluble in all phases, we show the pressure-temperature-composition space model in Figure 15.6.[1] This model is really a set of surfaces that form a hollow figure.

To aid in visualizing this model, the pure component pressure-temperature plots are shown separately on the right and left sides of the model.

Any pressure-temperature-composition point occurs: (a) on a surface, (b) within the hollow figure, or (c) outside the hollow figure. There is a *G-L* surface, which is not shown in Figure 15.6 because it lies below the *L-G* surface. The hollow region between them terminates at the critical points curve, Cr_A–Cr_B. Any point in PTx_i space outside the hollow region defined by the coexistence surfaces corresponds to a single-phase state which is all gas, all liquid, or all solid, depending on the PTx_i values chosen. If the point

1 This diagram and many others shown in this chapter were based on those shown in the set of articles by C.E. Wales (1963) as well as in the monographs by Ricci (1966) and Findlay *et al.* (1951).

falls on a surface, it represents a state that is saturated and in equilibrium with at least one other phase.

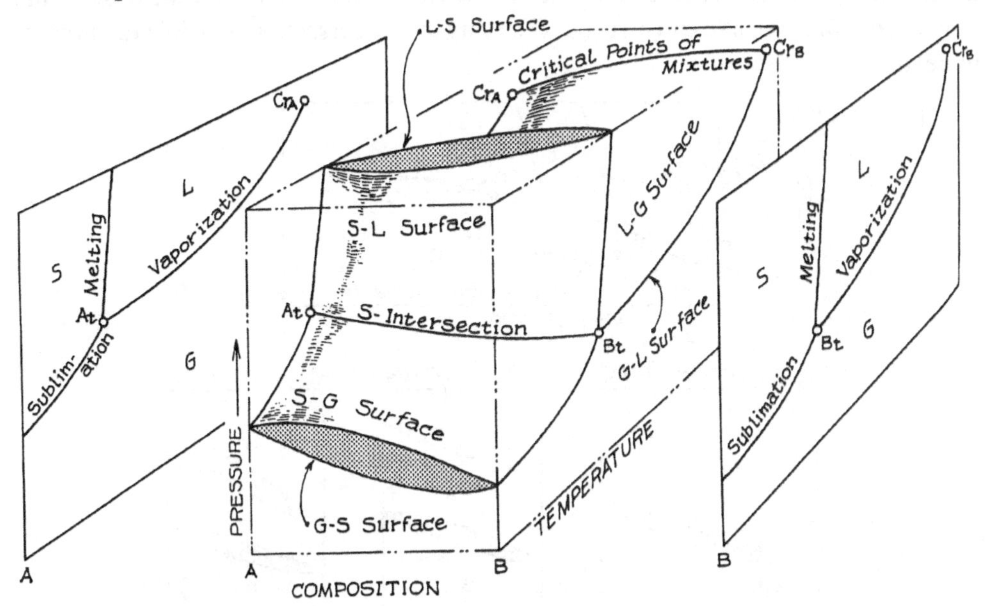

Figure 15.6 PTx_i space model of an A-B binary system in which both components are miscible in all proportions.

Finally, points that fall within a hollow region represent a mixture of two (or more) phases in equilibrium, each of which would exist separately on a surface and which would be on a tie-line drawn through the point in question. Thus, single phases do not exist inside the hollow region. Note also that the liquid-vapor, solid-vapor, and solid-liquid coexistence regions all show continuous solubility or miscibility from pure A to pure B. In the solid phase region this assumes the existence of a continuous solid solution.

To obtain a clearer picture of the shape of the space model, it is convenient to show isothermal and isobaric sections. This is readily accomplished by projecting all intersections onto a common pressure-temperature plane. Such projections are given in Figures 15.7a and 15.8a. Then at the various pressures shown in the former, and at the different temperatures shown in the latter, isobaric and isothermal sections are drawn in Figures 15.7b and 15.8b. That is, the set of curves separating two single-phase regions in Figures 15.7b and 15.8b are the loci of tie-lines bordering the hollow region of Figure 15.6. On each of these sections the pure component triple points (A_t and B_t), as well as the critical points (Cr_A and Cr_B), are shown for reference.

On the space model of Figure 15.6 there are six different surfaces and three curves of intersection. For the solid, there is the surface that represents solid solution states in equilibrium with gas (*S-G*) and with liquid (*S-L*). Similarly, there are (*L-G*), (*L-S*), (*G-L*), and (*G-S*) surfaces. Of this set of six, only the first two and a portion of the third are visible in Figure 15.6. For the intersections there is one between (*S-G*) and (*S-L*), another between (*L-S*) and (*L-G*), and a third between (*G-S*) and (*G-L*). These are termed the *S*, *L*, and *G* intersections, respectively. They represent the loci of states for the triple points of mixtures; all begin and end at A_T and B_T. Only the *S* intersection is visible in the space model drawn.

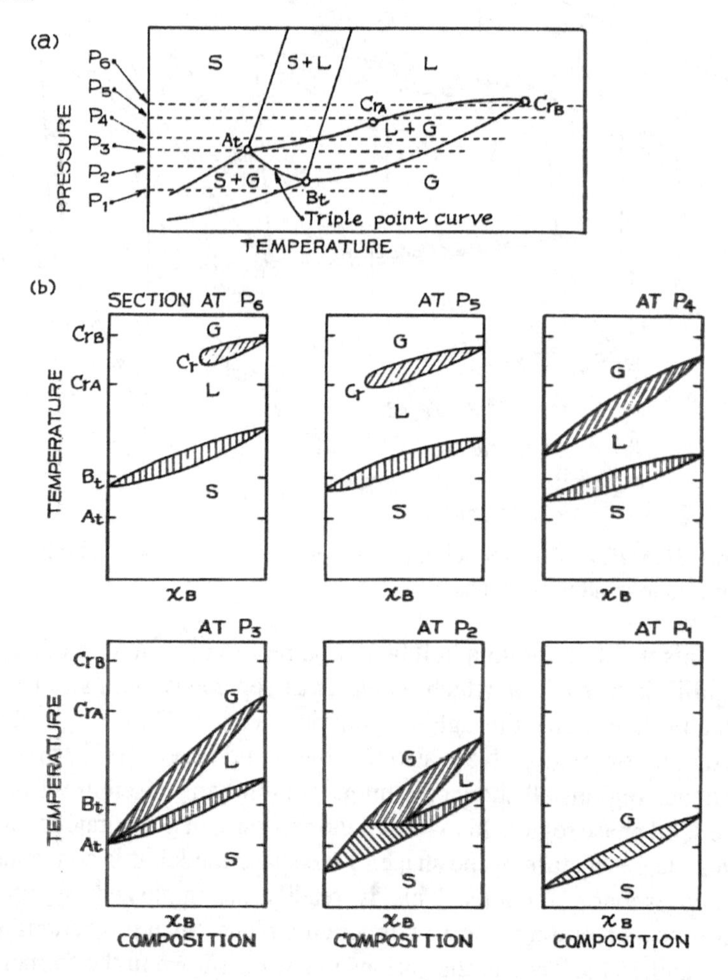

Figure 15.7 Isobaric sections of the PTx_i space model of Figure 15.6.

All intersections project on a common triple point curve in the pressure-temperature plane as shown in Figures 15.7a and 15.8a. This must be so since, for three phases of a binary system to be in equilibrium, the variance is unity; specification of *either* the pressure or temperature defines the state of the system completely.

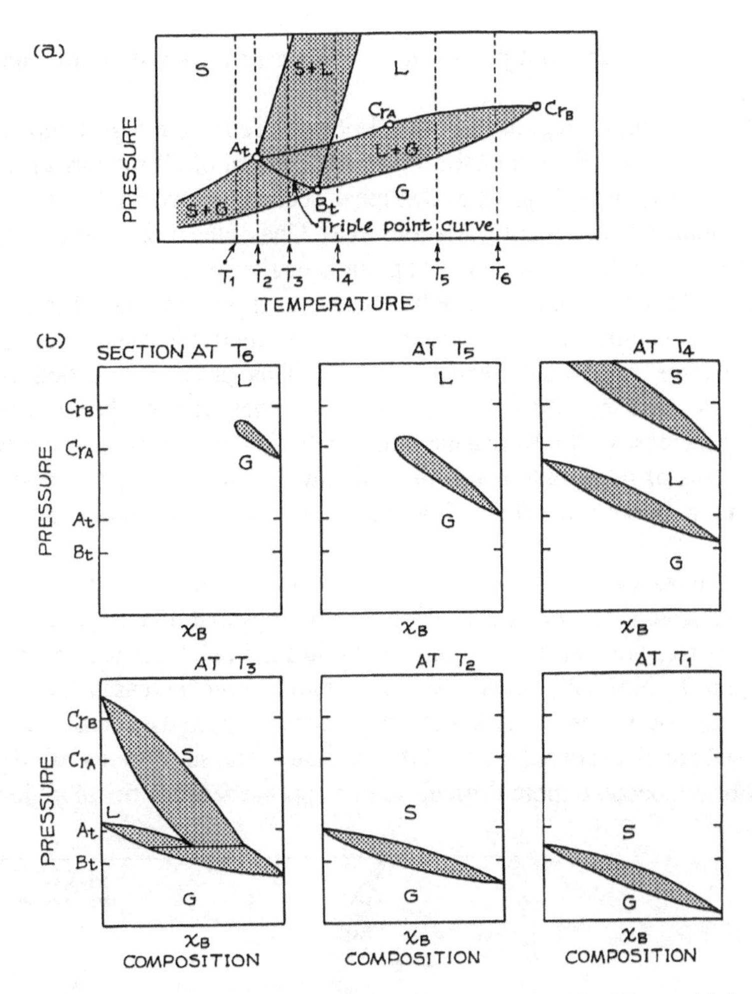

Figure 15.8 Isothermal sections of the PTx_i space model of Figure 15.6.

The triple point region may be shown in another manner by employing Figures 15.7a or 15.7b along with projections of the pressure-composition and temperature-composition planes. In other words, view Figure 15.6 from the side, end, and top. These three views are given in Figure 15.9. At some chosen temperature, T, draw a vertical line through the top and side views. For this equilibrium state, the composition of the three phases may be immediately found as x^S, x^L, and x^G. The equilibrium pressure is read from the side view at the intersection of the isotherm with the triple point curve. The composition values may also be found in the left-end view at this equilibrium pressure.

Still another way to visualize the space model is to section it at constant composition as indicated in Figure 15.10. The hollow nature of the figure is clearly seen. In Figure 15.11 the pressure-temperature projection of the section has been drawn. Such constant composition sections are known as *isopleths*. No tie-lines can be drawn on isoplethal diagrams because two or more phases do not necessarily coexist at any one common composition.

The hollow, gas-liquid region in Figure 15.6 is closed at the high-temperature, end on a single space curve. Such a curve represents the loci of all mixture critical points and it connects Cr_A and Cr_B. The isoplethal section shown in Figure 15.11 is a particularly simple one and many variations exist. One of the more interesting is shown in Figure 15.12. Here the mixture critical point is marked at Cr_m.

An isothermal process may be visualized to occur from *a* to *b* (see Figure 15.12) in which the system is initially similar to a gas but in a supercritical state, then becomes a gas-liquid mixture, and finally ends as a gas. This process is called ***retrograde condensation*** because initially as one enters the two-phase region, the amount of liquid relative to gas increases with decreasing pressure. Following the isobar *d* to *c*, we would find another type of retrograde behavior in that we would have in succession, liquid, liquid-gas, and a supercritical fluid above both the critical temperature and pressure of the mixture.

For temperatures and pressures between the critical points of the pure components, isothermal and isobaric sections consist of closed loops that do not traverse the entire composition range. Several of these are shown in Figures 15.7b and 15.8b at P_5 and P_6 and at T_5 and T_6. Normally, as the pressure or temperature increases, the liquid-vapor lens becomes thinner and there is a smaller difference between the composition of liquid and vapor. The practical consequence of this change is that separations based on vapor-liquid equilibrium become more difficult as one approaches the critical region.

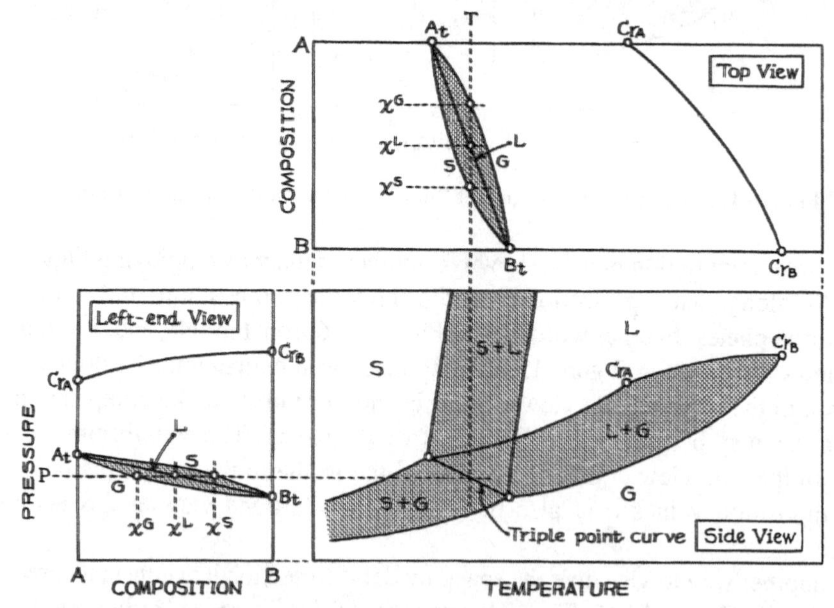

Figure 15.9 Triple point and critical point intersections for the space model of Figure 15.6 of an A-B binary.

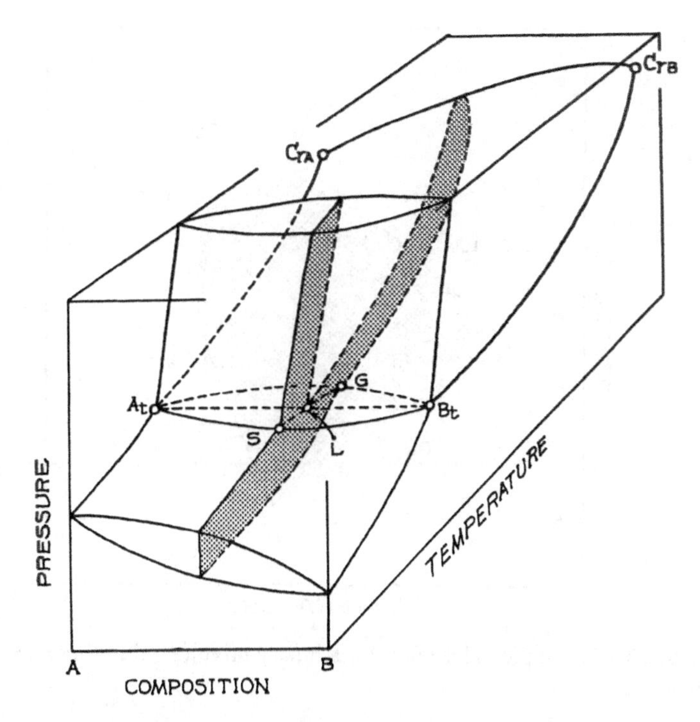

Figure 15.10 Constant composition (isoplethal) section of the A-B binary PTx_i space model of Figure 15.6 (shown as striped).

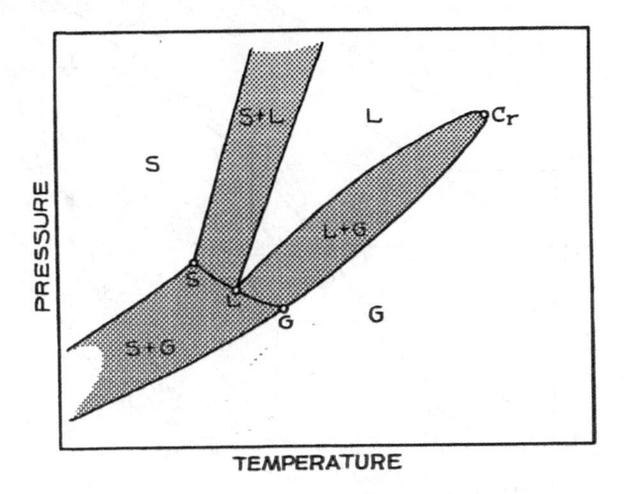

Figure 15.11 Pressure-temperature projection of the isoplethal section shown in Figure 15.10.

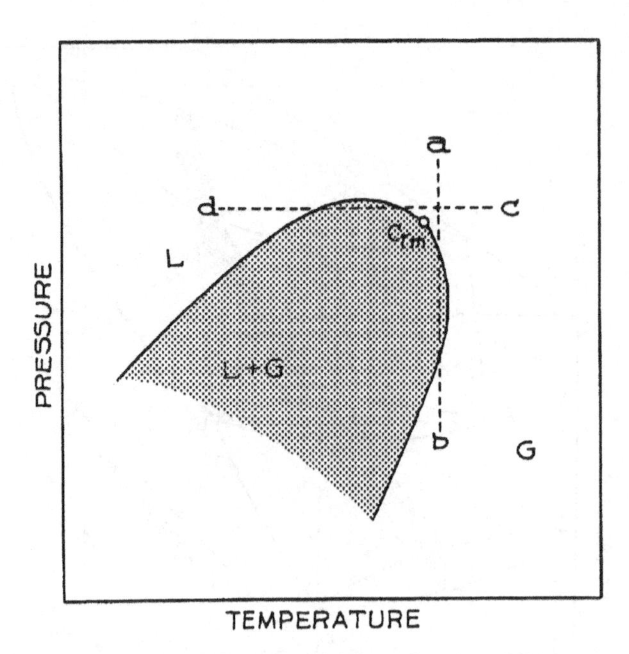

Figure 15.12 Isoplethal section for a binary mixture exhibiting retrograde phenomena.

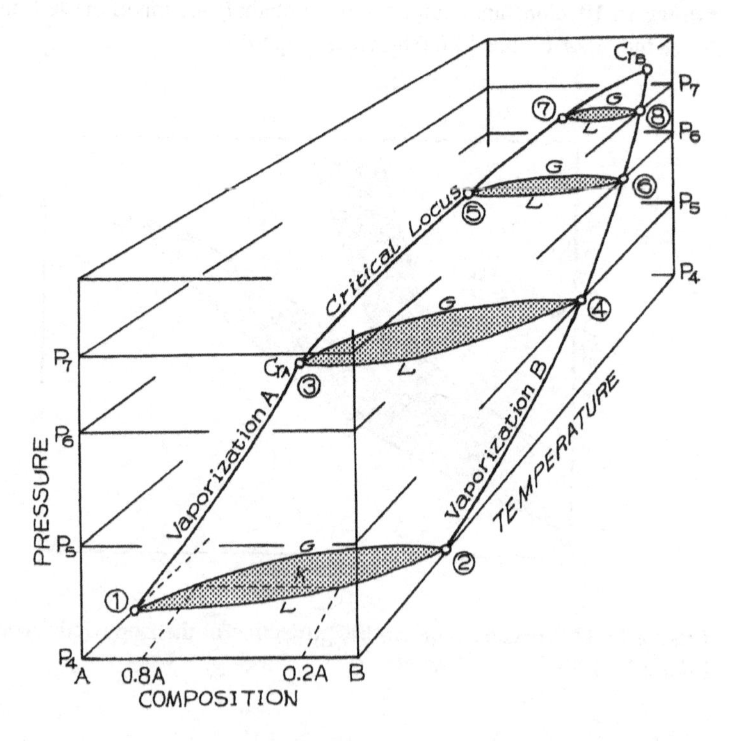

Figure 15.13 Isobaric sections of the liquid-gas region for an A-B binary.

Another point of interest involves relationships between the composition of equilibrium phases. For example, if we examine only the liquid-vapor portion of the space model as shown in Figure 15.13 and make isobaric sections at P_4 to P_8, we could then plot the composition of the vapor against that of the liquid at the different pressures shown. This has been done in Figure 15.14. Each set of compositions was taken from the termini of tie-lines in the isobaric sections. One tie-line point (K) is shown at P_4 on Figure 15.13; this tie-line provides a single vapor-liquid equilibrium datum point as indicated in Figure 15.14. Above the critical pressure of A, the isobars on the x-y plot are foreshortened.

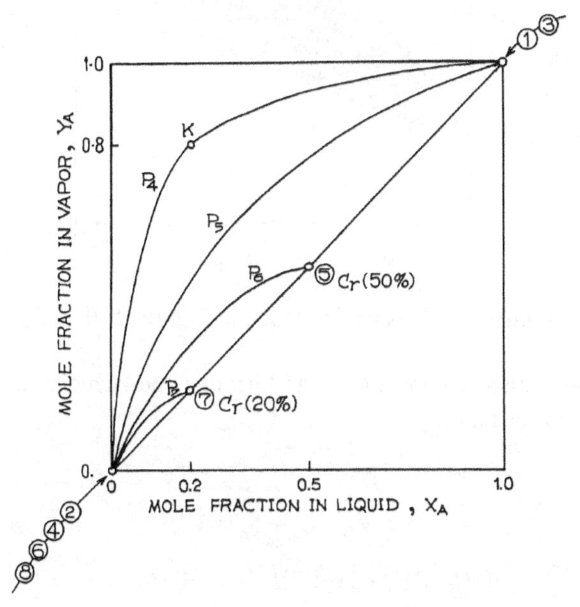

Figure 15.14 An x-y diagram for the A-B binary system of Figure 15.13.

Example 15.2

Given a system at P_2 in Figure 15.7b and as shown in Figure 15.15, trace and describe the equilibrium states as one cools a mixture originally at point 1.

Solution

At point 1, the system is a gas. Cooling to point 2, the first liquid begins to form. This occurs at the dew point temperature. The first liquid appears with a composition appropriate to point 2^L. Cooling further until the system temperature is equal to 3, we now have a gas-liquid mixture with a vapor phase at 3^G and a liquid at 3^L. From a simple mass balance, it can be shown that the ratio: moles of gas to moles of liquid is in the same proportion as the lengths of the tie-lines $(\overline{3-3^L})/(\overline{3-3^G})$. At 4, we would find a gas phase at 4^G and a liquid at 4^L. At this temperature we have a triple point and we could have three phases in equilibrium (i.e., 4^G, 4^L, and 4^S); thus, the relative amounts of the three phases cannot be determined except within certain limits. To see this, assume that we had one mole of the

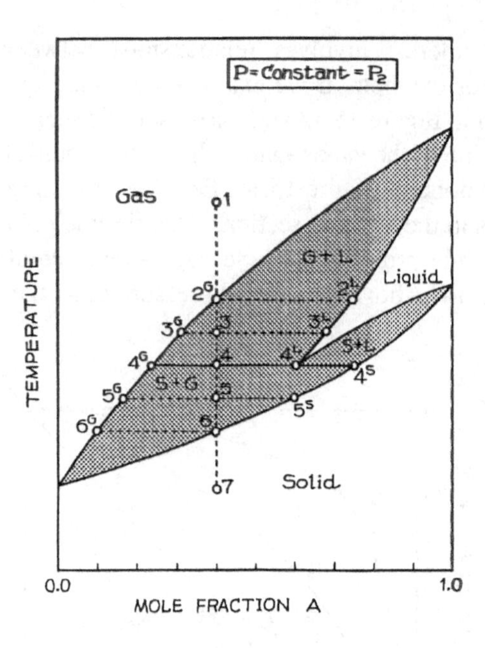

Figure 15.15 Isobaric section of Figure 15.7b at P_2.

mixture originally with a composition x_0. At 4, let there be V_4 moles of vapor, L_4 moles of liquid, and S_4 moles of solid.

Then,

$$V_4 + L_4 + S_4 = 1$$

$$[x(4^G)]V_4 + [x(4^L)]L_4 + [x(4^S)]S_4 = x_0$$

With only these two equations, the values of V_4, L_4, and S_4 cannot be found.
If we cool to point 5, we find a solid and a gas phase in equilibrium at 5^S and 5^G. The gas-to-solid mole ratio is $(\overline{5\text{–}5^S})/(\overline{5\text{–}5^G})$. Finally at 6, we are at a ***bubble point*** (though usually bubble points refer to liquid states with incipient vaporization). Below 6, the mixture is a solid.

As we discussed earlier, azeotropic mixtures introduce a complication in that the dew and bubble point curves are multivalued. Figure 15.1 illustrates this for a minimum boiling azeotrope where identical bubble and dew points are found at two different compositions to the right and left of the azeotropic composition where $y_i = x_i$.

If we were to visualize the space model for an azeotropic system, the loci of azeotropes would be a curve that would result from making a seam down the gas-liquid hollow lens. Azeotropic compositions usually change with temperature or pressure variations so that one may often bypass the azeotropic limitation in distillation by operating columns at different pressures.

Another phenomena commonly observed in binary systems is the *miscibility gap*. In some mixtures there exist regions in which a homogeneous phase separates into two distinct phases that are immiscible. This may occur in the liquid, solid, or, less frequently, in the gas phase. Each of the two phases so formed may themselves be mixtures, in which case the two components are said to be *partially miscible*. Alternatively, the phases may consist of pure components and in this case the two components are said to be *immiscible*.

Let us consider a liquid phase with a miscibility gap and in which the components are partially miscible. We make sections as shown in Figures 15.7b at P_4 and 15.8b at T_4, but as modified on Figure 15.16 to show a closed-loop gap in the liquid phase. Within the closed loop, two liquid phases exist; the composition of the two liquids would be those at the termini of tie-lines.

The temperatures and pressures labeled K_{TU}, K_{TL}, K_{PU}, and K_{PL} are called the upper and lower critical solution (or consolute) temperatures and pressures respectively. Above an upper consolute limit or below a lower one, the two components are completely miscible and form only a single liquid phase.

Many systems only exhibit either an upper or lower consolute limit. These cases would be expected when the miscibility gap intersects a two-phase lens. For example, again using a liquid miscibility gap as an illustration, we would find only a lower consolute temperature when the miscibility gap intersects the liquid-gas lens and only an upper consolute temperature when the gap overextends the solid-gas lens. Carrying this reasoning further, when a miscibility gap is relatively large, we may not find either an upper or lower consolute limit.

For systems in which the miscibility gap does extend into another two-phase region, the isothermal and isobaric sections take on added complexity. In Figure 15.17 we show a case in which the gap of Figure 15.16 extends into the liquid-gas lens. In this example, only a lower consolute temperature and an upper consolute pressure exist.

Should an azeotropic mixture also show a partial miscibility gap, the isothermal and isobaric sections might appear as in Figure 15.18. At the pseudoazeotropic point the gas is now in equilibrium with two different liquid phases. Also, consider an isothermal compression of a gas at 1. Upon reaching 2, a liquid of composition $2'$ forms. Increasing the pressure then results in the formation of additional liquid until 3 is reached. At this pressure we have a triple point with two liquid phases at $3'$ and $3''$ and a gas at $3'''$. Further compression leads to the complete condensation of the vapor and there remains a system containing only two liquid phases. Increasing the pressure further produces a change in the relative amounts and compositions of the two liquid phases until at 4 there exists only a single liquid.

For those systems with wide miscibility gaps, the concentrations of the lean and rich phases approach zero and unity, respectively. This occurs only rarely with liquid-phase gaps but is relatively common for solid-phase gaps.

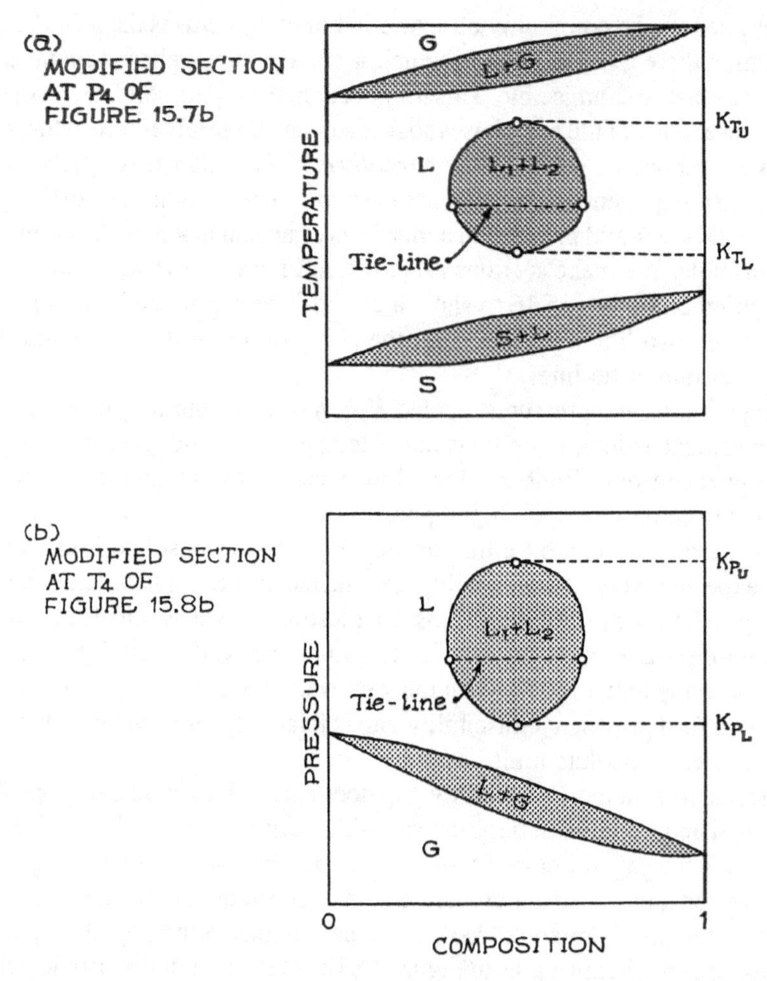

Figure 15.16 Isobaric and isothermal sections with a closed miscibility gap.

Up to now we have dealt with completely miscible (consolute) vapor-liquid and vapor-solid system as shown in Figure 15.6 and only liquid-liquid systems that show a miscibility gap as shown in Figure 15.16. Another important class of problems involves solid-liquid and solid-liquid-vapor systems where the solid phase (or phases) is only partially miscible. This limited solubility can extend from essentially pure solid to solid solutions with high concentrations of the second component. In some binary systems, azeotropic-type behavior is seen in the solid-liquid coexistence region where a maximum or minimum melting point can exist in an otherwise continuously miscible solid solution.

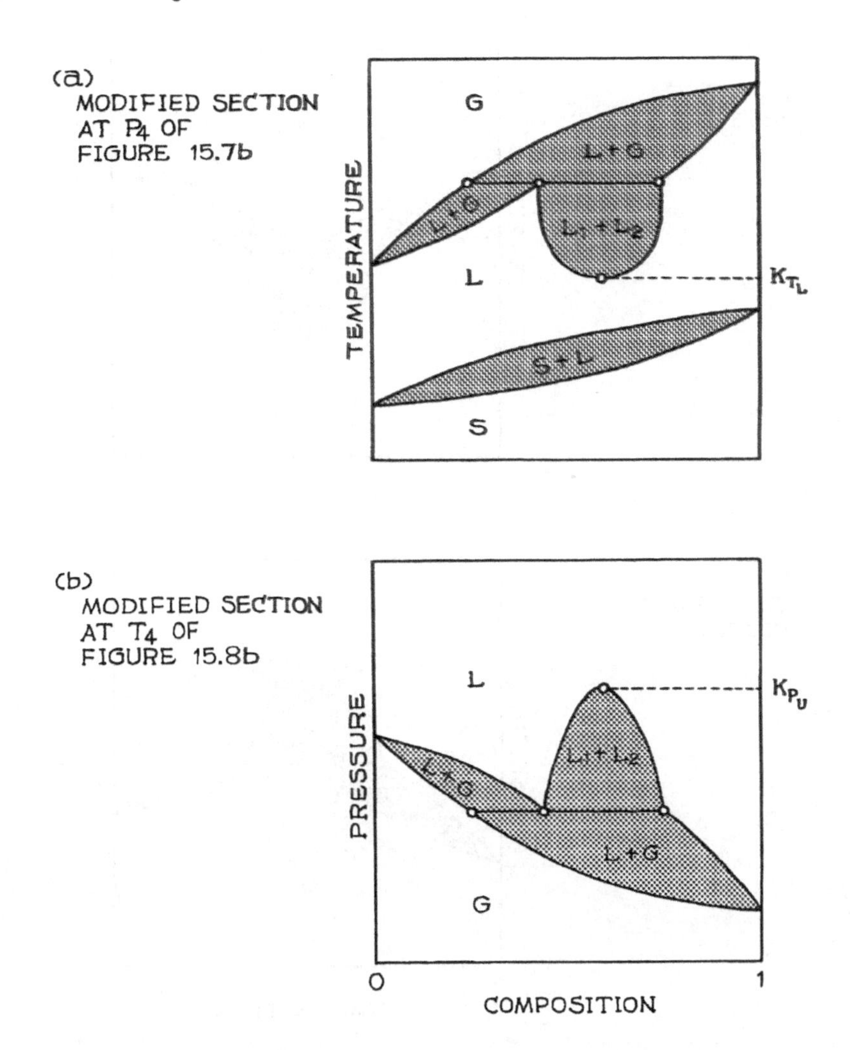

Figure 15.17 Isobaric and isothermal sections with a partial miscibility gap.

Figure 15.19 is a PTx_i plot for a binary (A-B) system that shows complete phase separation in the solid phases present, which are represented as pure A and pure B. The corresponding P-T projection for such a system is given in Figure 15.20. Here a new type of invariant condition appears, with $n = 2$ and $\pi = 4$, $\mathcal{F} = 0$ occurring at a so-called ***quadruple point*** shown as E on Figure 15.20. In a similar fashion to what we did for the completely miscible binary system of Figure 15.6, we can prepare isobaric, T–x_i slices at different pressures. In this case, however, the *eutectic* temperatures and compositions along the line C_2–E_3 must be known in order to produce quantitative diagrams. Two important examples of T–x_i are given in Figure 15.21 (Iron-carbon system at 1 bar) and in Figure 15.22 (sodium chloride-water system at 250 bar). In Figure 15.21 an extended region of carbon solid solubility is shown in the γ-Fe phase.

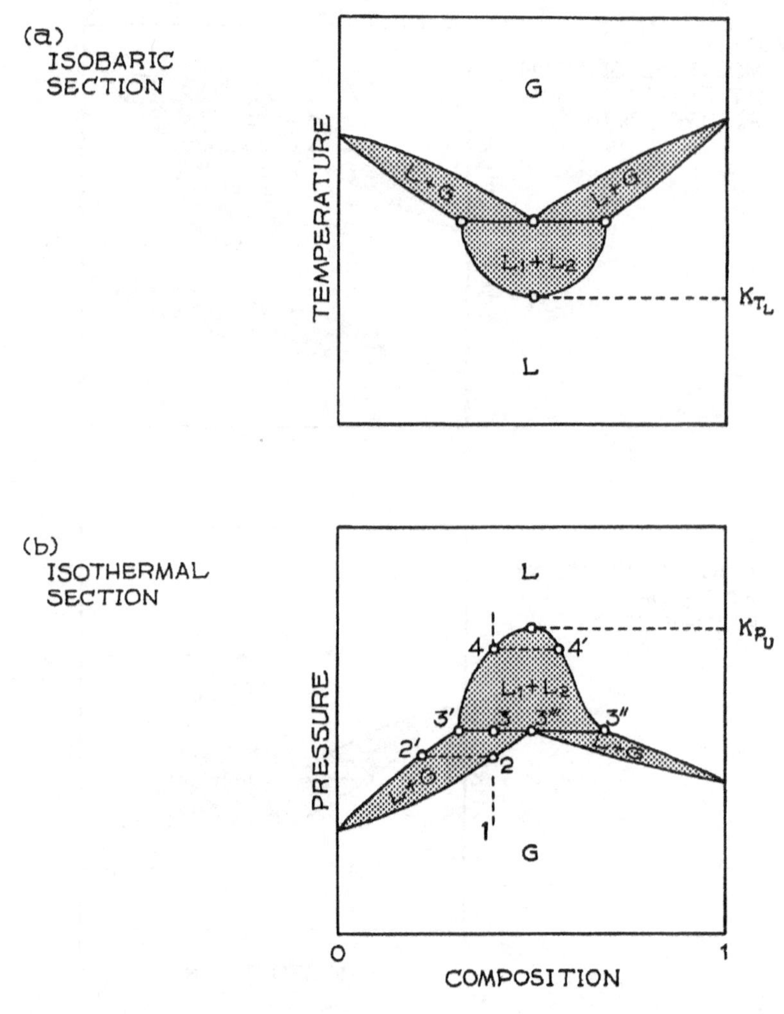

Figure 15.18 Isobaric and isothermal sections of an azeotropic system with a partial miscibility gap.

Ternary systems. For multiphase systems involving more than two components, graphical representation becomes more complex. For example, in a ternary system, composition could be plotted on a triangular diagram and the temperature or pressure included in such a way that a triangular prism is formed. Isothermal or isobaric sections can then be made.

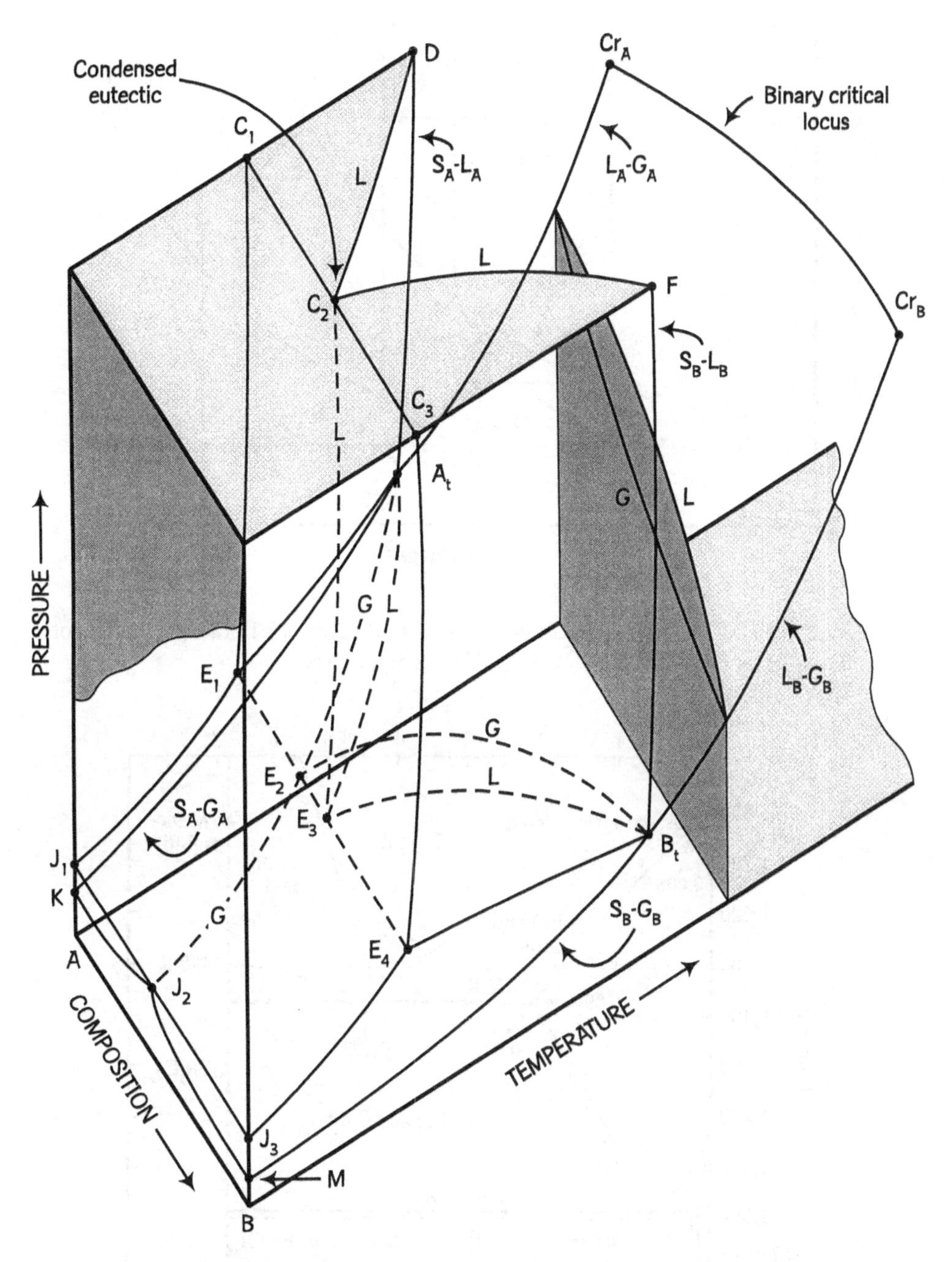

Figure 15.19 A PTx_i space diagram for a binary (A-B) system consolute as liquids and immiscible as solids. Both components are appreciably volatile over the whole temperature range. There are no solid solutions, no solid compounds and no azeotropes (adapted from notes provided by Prof. R. Thorpe, Cornell University, Ithaca, NY).

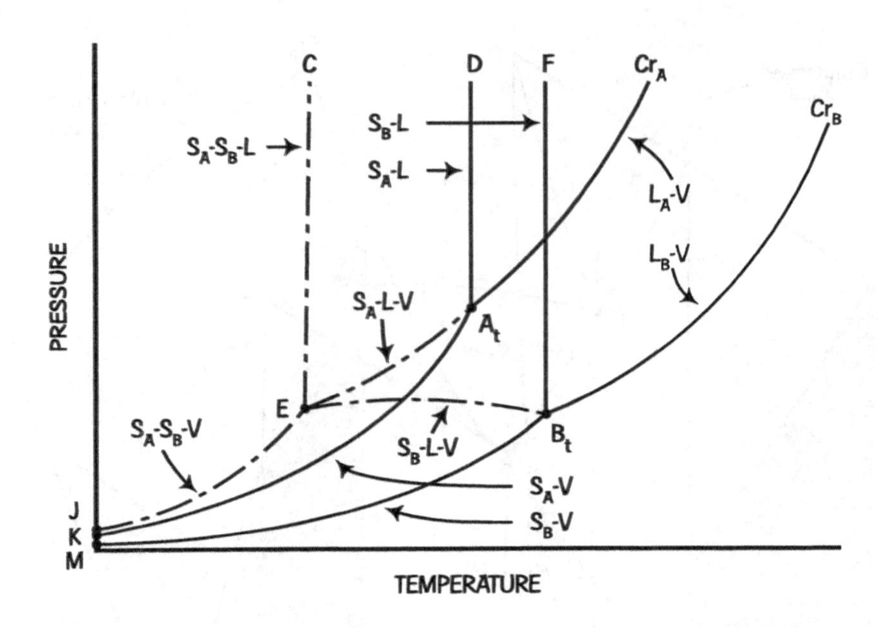

Figure 15.20 Pressure-temperature projection of Figure 15.19 (adapted from notes provided by Prof. R. Thorpe, Cornell University, Ithaca, NY).

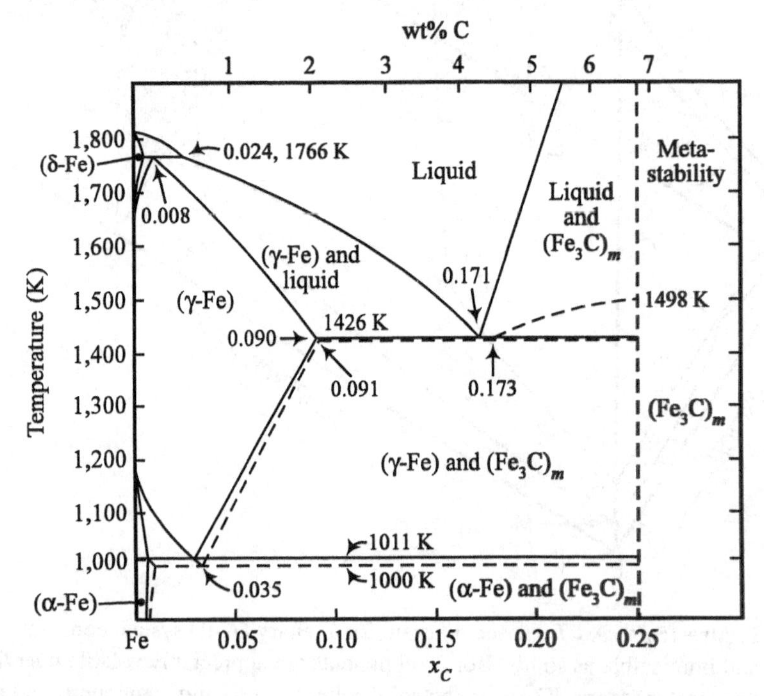

Figure 15.21 Iron-carbon system at 1 bar. Note that cementite $(Fe_3C)_m$ is a metastable phase [adapted from Lupis (1983) based on data from Hultgren *et al.* (1973), used with permission].

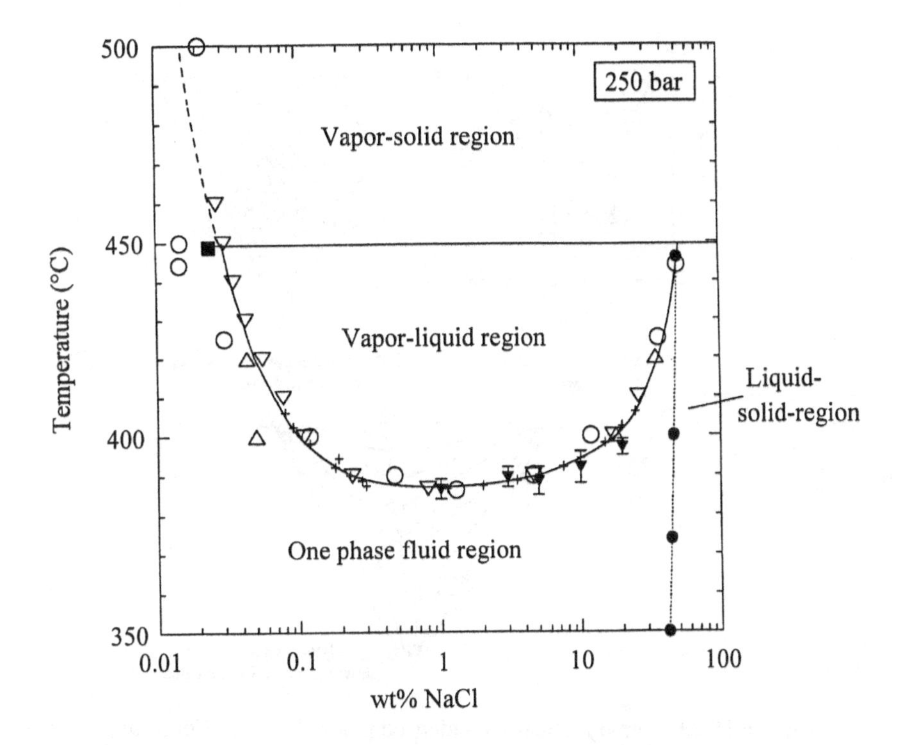

—— Compilation of Bischoff and Pitzer (1989)
○ Sourirajan and Kennedy (1962)
■ Bischoff *et al.* (1986)
● Estimated from Linke (1958)
▼ Armellini and Tester (1991)
△ Parisod and Plattner (1981)
- - - Prediction of Pitzer and Pabalan (1986)
▽ Olander and Liander (1950)
\+ Khaibullin and Borisov (1966)

Figure 15.22 A temperature-composition NaCl-H_2O phase diagram at 250 bar [from Armellini and Tester (1991)].

An example is given in Figure 15.23 for an *A*-*B*-*C* ternary that exhibits complete miscibility in the liquid phase and complete immiscibility in three solid phases (pure *A*, pure *B*, and pure *C*). Three binary eutectics (e_1, e_2, and e_3) make up the three outer faces of the prism and a ternary eutectic at *E* is determined by the intersection of three binary crystallization paths from e_1 to *E*, e_2 to *E*, and e_3 to *E*. In addition to the triangular prism approach it is also possible to represent some features of phase behavior on a pseudo-binary diagram as shown in Figure 15.24. Here NaCl-Na_2SO_4-H_2O system equilibrium data are plotted on a water-free (T–x_i) basis for various pressures. Since NaCl and Na_2SO_4 are non-volatile components relative to water this is reasonable. Note that the binary eutectic system exists when $P_{H_2O} = 0$ and that the eutonic line represents a monovariant condition where four phases coexist (solid NaCl, solid Na_2SO_4, liquid, and vapor). One can also represent these equilibria by a series of isothermal slices at a fixed pressure shown in Figure 15.25.

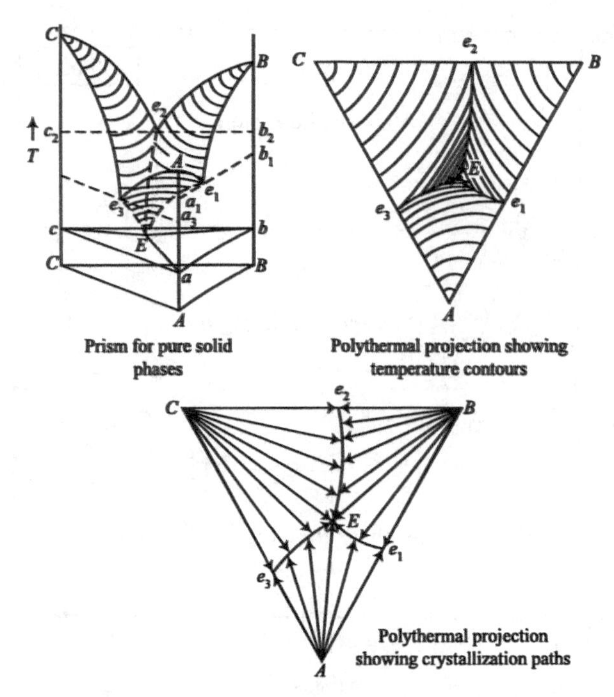

Figure 15.23 Ternary isobaric condensed phase diagram. Consolute in the liquid, and completely immiscible in the solid phases of A, B, and C [from Ricci (1966), *The Phase Rule and Heterogeneous Equilibrium*, Dover, New York].

Supercritical fluid systems. The final category of systems we will consider in this section involves supercritical solvents that can exhibit anomalous phase behavior including high solubilities for solutes that would be insoluble under normal, ambient conditions. In Figure 15.26, two P-T projections are shown for any A-B binary. In case (a), where both A and B components are volatile, normal eutectic behavior with a continuous monovariant boundary between the triple points of A and B and the quadruple point E and a continuous binary locus of critical points is observed. Typically this behavior occurs when A and B have comparable volatilities. In case (b), where component B is non-volatile, the 3-phase (S_B, L, F, (or V)) line is discontinuous caused by its intersection with the critical point locus. Two *end points* appear, a lower critical end point or LCEP and an upper critical end point or UCEP.

The subsequent solubility and phase behavior of a non-volatile solute in an otherwise supercritical solvent can be quite unusual. For example, Figure 15.27 gives a calculated phase diagram for the naphthalene-carbon dioxide (CO_2) system. The critical point of pure CO_2 is $T_c = 304.1$ K and $P_c = 73.8$ bar. We note that the solubility of naphthalene given as the mole fraction x_N on the y-axis (logarithmic scale) increases almost four-orders of magnitude for a 50 K increase in temperature from 300 to 350 K along isobars in the range of 65 to 90 bar.

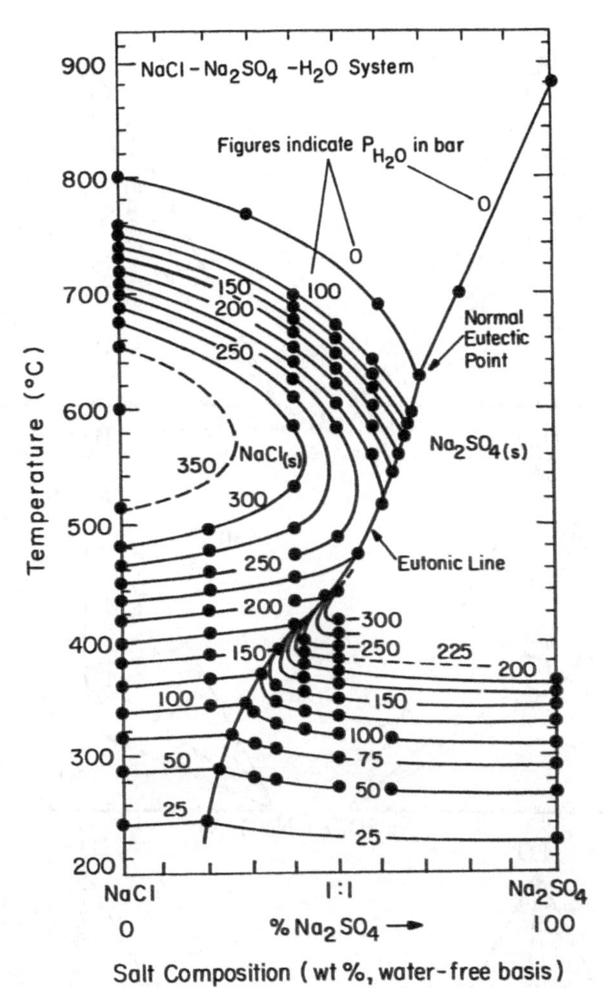

Figure 15.24 Projections of saturated solution isobars in the $NaCl$-Na_2SO_4-H_2O system [from Armellini, Hong, and Tester (1994) adapted using data from Ravich *et al.* (1953)].

In Section 15.2 we have introduced only a few simple examples of phase diagrams. For other examples and a more detailed discussion, the interested reader should consult primary references such as the monographs by Ricci (1966), Prince (1966), Rhines (1956), and Findlay *et al.* (1951).

15.3 The Differential Approach for Phase Equilibrium Relationships

We shall develop two approaches for evaluating the analytical relationships between properties of coexisting phases: the differential approach, described in this section, and the integral approach, treated in Sections 15.5 and 15.6. Although the integral approach is more commonly used, the differential approach is conceptually more elucidating. It is for this reason that the latter is developed here in some detail.

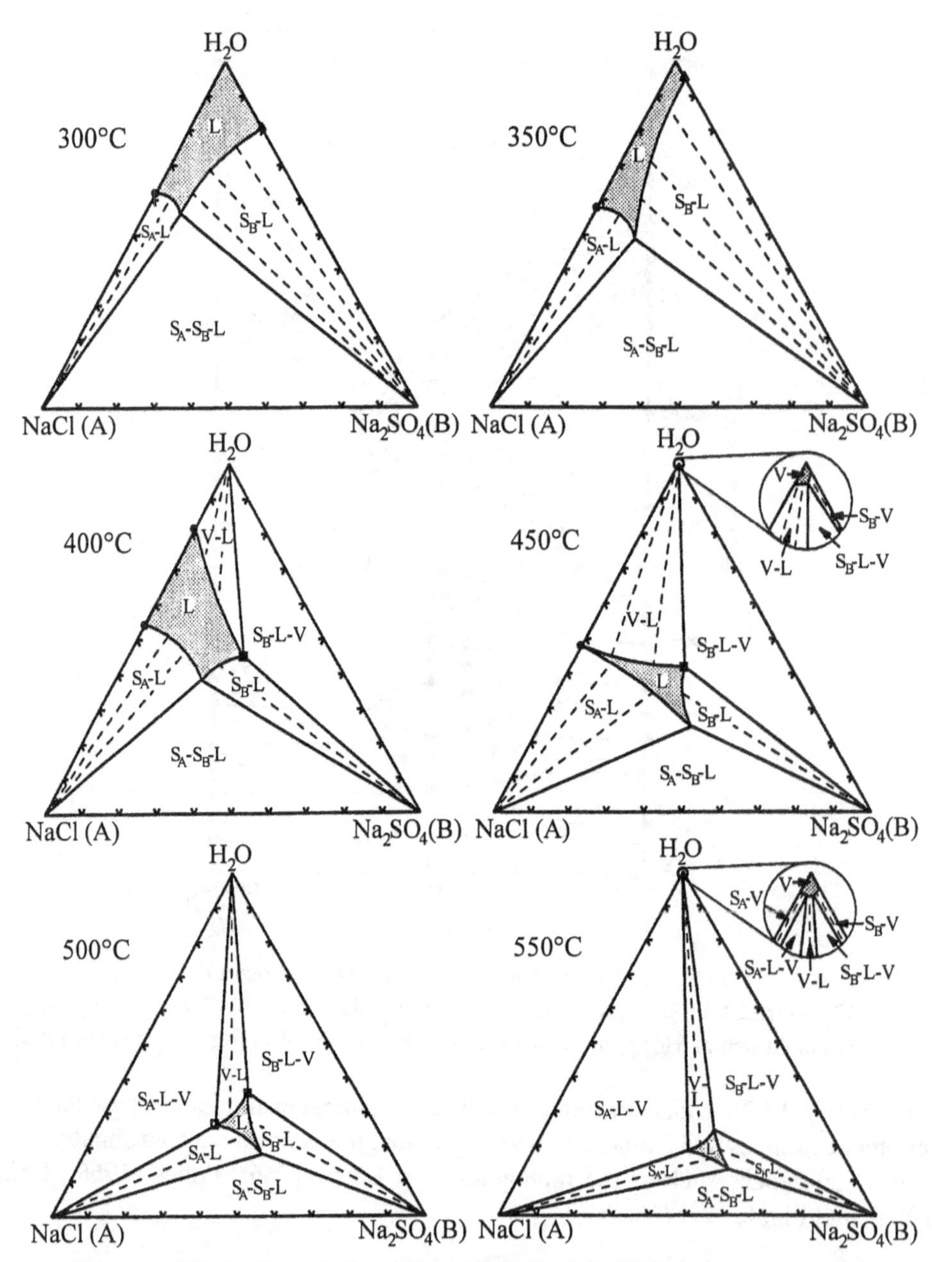

Figure 15.25 Isothermal slices of the $NaCl$-Na_2SO_4-H_2O ternary system phase diagram at 250 bar [from Armellini, Tester, and Hong (1994)].

We first treat a simple binary system of components A and B with two coexisting phases α and β in which each component is present in each phase. The reasoning is then extended to three coexisting phases of a binary and to the general case of multicomponent, multiphase simple systems.

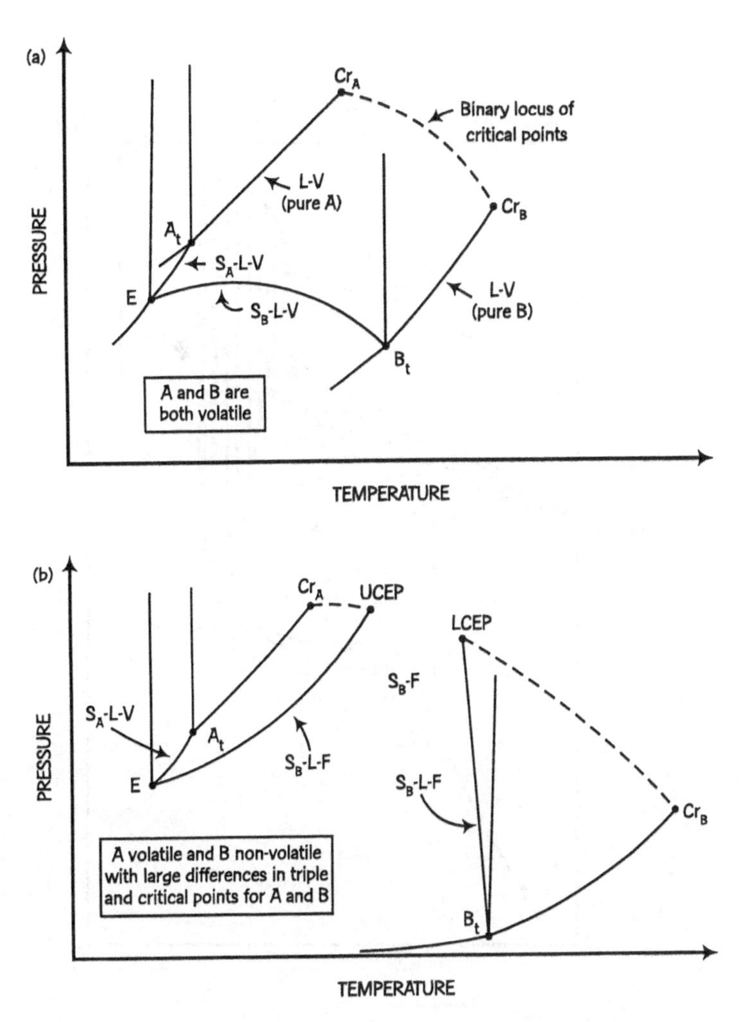

Figure 15.26 Pressure-temperature projections for an *A-B* binary.

To develop analytical relationships, we use the theoretical basis described in Section 15.1 with one modification: For the criteria of equilibrium between coexisting phases, instead of employing chemical potentials [as in Eq. (15-4)], we use the equivalent criteria of equality of the fugacity of each component in each phase.[2] Thus, the criteria of equilibrium for a binary mixture coexisting as α and β phases are

---

2 Note that whereas equality of component fugacities is completely equivalent to equality of chemical potentials, equality of component activities is not necessarily equivalent. The discrepancy exists because different reference states of *unit activity* are sometimes chosen for a component in different phases. For fugacities, the reference state is always chosen as an ideal gas at the system temperature.

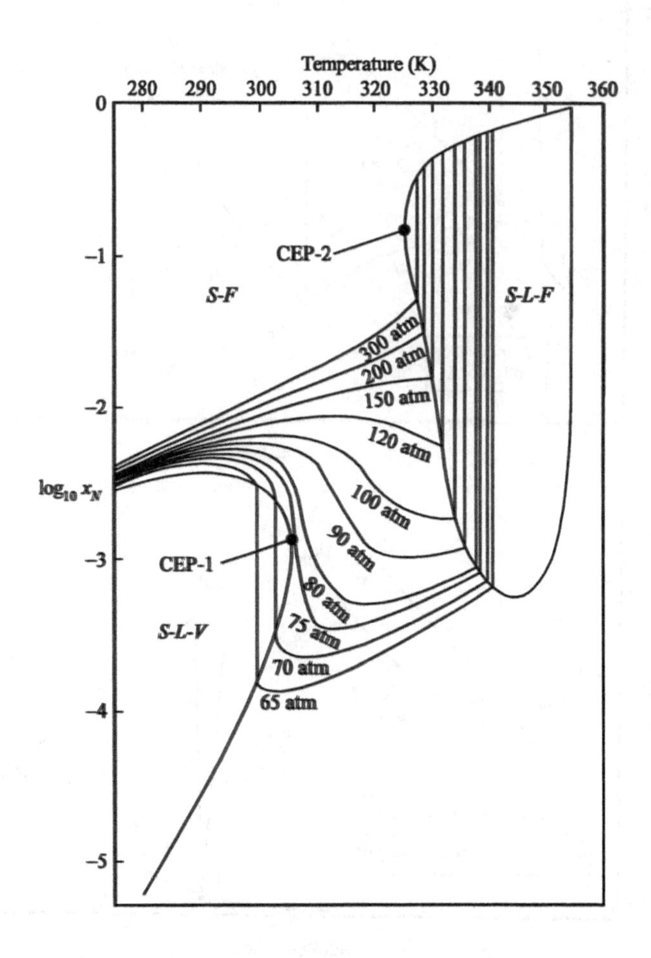

Figure 15.27 Naphthalene-carbon dioxide solubility map calculated from the Peng-Robinson EOS with δ_{12}=0.11. Two critical end points (CEP-1 and CEP-2) are shown between two-phase [solid(*S*)-fluid(*F*)] and three-phase [*S-L-V* and *S-L-F*] regions [from Hong, G., ScD Thesis, Department of Chemical Engineering, MIT, Cambridge, MA (1981)].

$$T^{\alpha} = T^{\beta} \tag{15-9}$$

$$P^{\alpha} = P^{\beta} \tag{15-10}$$

$$\hat{f}_A^{\alpha} = \hat{f}_A^{\beta} \quad or \quad \ln \hat{f}_A^{\alpha} = \ln \hat{f}_A^{\beta} \tag{15-11}$$

and

$$\hat{f}_B^{\alpha} = \hat{f}_B^{\beta} \quad or \quad \ln \hat{f}_B^{\alpha} = \ln \hat{f}_B^{\beta} \tag{15-12}$$

To obtain differential equations involving P, T, and phase compositions, we take the total derivatives of Eq. set (15-9) through (15-12) and expand all fugacities in terms of T, P, and the pertinent phase compositions. In view of Eqs. (15-9) and (15-10), no phase designation need be made for T and P.

For example, suppose that we choose T, P, and x_A as the $(n+1)$ variables in each phase. Thus, for component A, we have

$$d \ln \hat{f}_A^\alpha = \left(\frac{\partial \ln \hat{f}_A^\alpha}{\partial T}\right)_{P,x^\alpha} dT + \left(\frac{\partial \ln \hat{f}_A^\alpha}{\partial P}\right)_{T,x^\alpha} dP + \left(\frac{\partial \ln \hat{f}_A^\alpha}{\partial x_A^\alpha}\right)_{T,P} dx_A^\alpha$$

or

$$d \ln \hat{f}_A^\alpha = -\frac{\overline{H}_A^\alpha - H_A^*}{RT^2} dT + \frac{\overline{V}_A^\alpha}{RT} dP + \left(\frac{\partial \ln \hat{f}_A^\alpha}{\partial x_A^\alpha}\right)_{T,P} dx_A^\alpha \tag{15-13}$$

$$d \ln \hat{f}_A^\beta = \left(\frac{\partial \ln \hat{f}_A^\beta}{\partial T}\right)_{P,x^\beta} dT + \left(\frac{\partial \ln \hat{f}_A^\beta}{\partial P}\right)_{T,x^\beta} dP + \left(\frac{\partial \ln \hat{f}_A^\beta}{\partial x_A^\beta}\right)_{T,P} dx_A^\beta$$

or

$$d \ln \hat{f}_A^\beta = -\frac{\overline{H}_A^\beta - H_A^*}{RT^2} dT + \frac{\overline{V}_A^\beta}{RT} dP + \left(\frac{\partial \ln \hat{f}_A^\beta}{\partial x_A^\beta}\right)_{T,P} dx_A^\beta \tag{15-14}$$

where H_A^* refers to the enthalpy of A in an ideal-gas state at the temperature of the system. Note that the expansions of Eqs. (15-13) and (15-14) are written separately for each phase. As discussed in Chapters 5 and 9, these expansions are always valid for each phase, regardless of whether or not the phase is in equilibrium with other phases.

The fugacity expansions for component B are more convenient if we use the same set of intensive variables. Thus, the equations corresponding to Eqs. (15-13) and (15-14) for component B are

$$d \ln \hat{f}_B^\alpha = -\frac{\overline{H}_B^\alpha - H_B^*}{RT^2} dT + \frac{\overline{V}_B^\alpha}{RT} dP + \left(\frac{\partial \ln \hat{f}_B^\alpha}{\partial x_A^\alpha}\right)_{T,P} dx_A^\alpha \tag{15-15}$$

$$d \ln \hat{f}_B^\beta = -\frac{\overline{H}_B^\beta - H_B^*}{RT^2} dT + \frac{\overline{V}_B^\beta}{RT} dP + \left(\frac{\partial \ln \hat{f}_B^\beta}{\partial x_A^\beta}\right)_{T,P} dx_A^\beta \tag{15-16}$$

Equating differentials of $\ln \hat{f}_i$, we get

$$-\frac{\overline{H}_A^\alpha - \overline{H}_A^\beta}{RT^2}dT + \frac{\overline{V}_A^\alpha - \overline{V}_A^\beta}{RT}dP + \left(\frac{\partial \ln \hat{f}_A^\alpha}{\partial x_A^\alpha}\right)_{T,P} dx_A^\alpha - \left(\frac{\partial \ln \hat{f}_A^\beta}{\partial x_A^\beta}\right)_{T,P} dx_A^\beta = 0 \quad (15\text{-}17)$$

and

$$-\frac{\overline{H}_B^\alpha - \overline{H}_B^\beta}{RT^2}dT + \frac{\overline{V}_B^\alpha - \overline{V}_B^\beta}{RT}dP + \left(\frac{\partial \ln \hat{f}_B^\alpha}{\partial x_A^\alpha}\right)_{T,P} dx_A^\alpha - \left(\frac{\partial \ln \hat{f}_B^\beta}{\partial x_A^\beta}\right)_{T,P} dx_A^\beta = 0 \quad (15\text{-}18)$$

where we must remember that $\overline{H}_i^\alpha$ and $\overline{V}_i^\alpha$ are evaluated at x_A^α, T, and P, whereas $\overline{H}_i^\beta$ and $\overline{V}_i^\beta$ are at x_A^β, T, and P. Equations (15-17) and (15-18) are two differential equations that must be satisfied simultaneously. We can solve by eliminating any one of the differentials in T, P, x_A^α, and x_A^β, thereby obtaining one differential equation in three variables. Integration of that equation yields the desired result.

Thus, the relationship $P = f(T, x_A^\beta)$ can be obtained by eliminating dx_A^α simultaneously from Eqs. (15-17) and (15-18). This final result may be simplified by noting that the coefficients of the two dx_A^α terms and the two dx_A^β are related by the Gibbs-Duhem equation for component fugacities, that is, for a binary system of phase π,

$$x_A^\pi \left(\frac{\partial \ln \hat{f}_A^\pi}{\partial x_A^\pi}\right)_{T,P} + x_B^\pi \left(\frac{\partial \ln \hat{f}_B^\pi}{\partial x_A^\pi}\right)_{T,P} = 0 \quad (15\text{-}19)$$

Thus, multiplying Eq. (15-17) by x_A^α and multiplying Eq. (15-18) by x_B^α, adding the two equations, and applying Eq. (15-19) to each phase, the result simplifies to

$$-\frac{x_A^\alpha(\overline{H}_A^\alpha - \overline{H}_A^\beta) + x_B^\alpha(\overline{H}_B^\alpha - \overline{H}_B^\beta)}{RT^2}dT + \frac{x_A^\alpha(\overline{V}_A^\alpha - \overline{V}_A^\beta) + x_B^\alpha(\overline{V}_B^\alpha - \overline{V}_B^\beta)}{RT}dP$$

$$- \left(x_A^\alpha - \frac{x_A^\beta x_B^\alpha}{x_B^\beta}\right)\left(\frac{\partial \ln \hat{f}_A^\beta}{\partial x_A^\beta}\right)_{T,P} dx_A^\beta = 0 \quad (15\text{-}20)$$

Solving explicitly for dP,

$$dP = \frac{1}{T}\frac{x_A^\alpha(\overline{H}_A^\alpha - \overline{H}_A^\beta) + x_B^\alpha(\overline{H}_B^\alpha - \overline{H}_B^\beta)}{x_A^\alpha(\overline{V}_A^\alpha - \overline{V}_A^\beta) + x_B^\alpha(\overline{V}_B^\alpha - \overline{V}_B^\beta)}dT$$

$$+ \frac{RT(x_A^\alpha - x_A^\beta x_B^\alpha / x_B^\beta)(\partial \ln \hat{f}_A^\beta / \partial x_A^\beta)_{T,P}}{x_A^\alpha(\overline{V}_A^\alpha - \overline{V}_A^\beta) + x_B^\alpha(\overline{V}_B^\alpha - \overline{V}_B^\beta)} dx_A^\beta \quad (15\text{-}21)$$

Since P is a property and since T and x_A^β form an independent set for P under conditions of two-phase equilibrium, Eq. (15-21) must be an exact differential equation. Therefore, it follows immediately that

$$\left(\frac{\partial P}{\partial T}\right)_{x_A^\beta,[\alpha-\beta]} = \frac{1}{T}\frac{x_A^\alpha(\overline{H}_A^\alpha - \overline{H}_A^\beta) + x_B^\alpha(\overline{H}_B^\alpha - \overline{H}_B^\beta)}{x_A^\alpha(\overline{V}_A^\alpha - \overline{V}_A^\beta) + x_B^\alpha(\overline{V}_B^\alpha - \overline{V}_B^\beta)} \tag{15-22}$$

and

$$\left(\frac{\partial P}{\partial x_A^\beta}\right)_{T,[\alpha-\beta]} = \frac{RT(x_A^\alpha - x_A^\beta x_B^\alpha/x_B^\beta)(\partial \ln \hat{f}_A^\beta/\partial x_A^\beta)_{T,P}}{x_A^\alpha(\overline{V}_A^\alpha - \overline{V}_A^\beta) + x_B^\alpha(\overline{V}_B^\alpha - \overline{V}_B^\beta)} \tag{15-23}$$

where the subscript $[\alpha-\beta]$ denotes conditions under which phases α and β coexist at equilibrium. Note that the partial derivatives in the left-hand side of Eqs. (15-22) and (15-23) do not imply that x_A^α is constant; x_A^α will in fact change as T is varied under conditions of α–β phase equilibrium and constant x_A^β. Similarly, x_A^α will change as x_A^β is varied with phase equilibrium and constant T. (Recall that for this case the variance is 2, so that x_A^α can be expressed as a function of T and x_A^β.) Using the *xyz*-1 rule (Section 5.3),

$$\left(\frac{\partial T}{\partial x_A^\beta}\right)_{P,[\alpha-\beta]}\left(\frac{\partial P}{\partial T}\right)_{x_A^\beta,[\alpha-\beta]}\left(\frac{\partial x_A^\beta}{\partial P}\right)_{T,[\alpha-\beta]} = -1 \tag{15-24}$$

and it follows that

$$\left(\frac{\partial T}{\partial x_A^\beta}\right)_{P,[\alpha-\beta]} = -RT^2\frac{(x_A^\alpha - x_A^\beta x_B^\alpha/x_B^\beta)(\partial \ln \hat{f}_A^\beta/\partial x_A^\beta)_{T,P}}{x_A^\alpha(\overline{H}_A^\alpha - \overline{H}_A^\beta) + x_B^\alpha(\overline{H}_B^\alpha - \overline{H}_B^\beta)} \tag{15-25}$$

Example 15.3

Prove that for a minimum-boiling azeotrope at constant temperature the pressure maximizes at the azeotropic concentration.

Solution

An azeotrope occurs when $x_A^V = x_A^L$, at which point the term $(x_A^V - x_A^L x_B^V/x_B^L)$ in Eq. (15-25) vanishes. By inspection of Eq. (15-23), it is clear that $(\partial P/\partial x_A^L)_{T,[L-V]}$ also vanishes at the azeotropic concentration. To prove that a temperature minimum corresponds to a pressure maximum, all we need show is that $(\partial T/\partial x_A^L)_{P,[L-V]}$ and $(\partial P/\partial x_A^L)_{T,[L-V]}$ have opposite signs. The ratio of these two derivatives is $-(\partial P/\partial T)_{x_A^L,[L-V]}$ by Eq. (15-24). Equation (15-22) shows that with $\alpha = V$ and $\beta = L$, the enthalpy and volume changes are always positive. Therefore, $(\partial P/\partial T)_{x_A^L,[L-V]}$ is always positive and, hence, $(\partial T/\partial x_A^L)_{P,[L-V]}$ and $(\partial P/\partial x_A^L)_{T,[L-V]}$ always have opposite signs.

In the procedure described above we evaluated the partial derivatives involving T, P, and x_A^β by eliminating the dx_A^α terms in solving Eqs. (15-17) and (15-18). Similarly, we could have worked with T, P, and x_A^α by eliminating dx_A^β [the results would be identical to Eqs. (15-22), (15-23), and (15-25) with x_A^α replacing x_A^β], or T, x_A^α, and x_A^β by eliminating dP, or P, x_A^α, and x_A^β by eliminating dT. For example, to obtain the partial derivatives in T, x_A^α, and x_A^β, eliminating dP from Eqs. (15-17) and (15-18) yields

$$\frac{1}{T}\left[\frac{\overline{H}_A^\alpha - \overline{H}_A^\beta}{\overline{V}_A^\alpha - \overline{V}_A^\beta} - \frac{\overline{H}_B^\alpha - \overline{H}_B^\beta}{\overline{V}_B^\alpha - \overline{V}_B^\beta}\right] dT - RT\left[\frac{(\partial \ln \hat{f}_A^\alpha/\partial x_A^\alpha)_{T,P}}{\overline{V}_A^\alpha - \overline{V}_A^\beta} - \frac{(\partial \ln \hat{f}_B^\alpha/\partial x_A^\alpha)_{T,P}}{\overline{V}_B^\alpha - \overline{V}_B^\beta}\right] dx_A^\alpha$$
$$+RT\left[\frac{(\partial \ln \hat{f}_A^\beta/\partial x_A^\beta)_{T,P}}{\overline{V}_A^\alpha - \overline{V}_A^\beta} - \frac{(\partial \ln \hat{f}_B^\beta/\partial x_A^\beta)_{T,P}}{\overline{V}_B^\alpha - \overline{V}_B^\beta}\right] dx_A^\beta = 0 \tag{15-26}$$

Simplifying Eq. (15-26) by applying the Duhem equation for fugacity for each phase, we obtain

$$-\frac{(\overline{H}_A^\alpha - \overline{H}_A^\beta)(\overline{V}_B^\alpha - \overline{V}_B^\beta) - (\overline{H}_B^\alpha - \overline{H}_B^\beta)(\overline{V}_A^\alpha - \overline{V}_A^\beta)}{RT^2}\, dT$$
$$+\frac{x_B^\alpha(\overline{V}_B^\alpha - \overline{V}_B^\beta) + x_A^\alpha(\overline{V}_A^\alpha - \overline{V}_A^\beta)}{x_B^\alpha}\left(\frac{\partial \ln \hat{f}_A^\alpha}{\partial x_A^\alpha}\right)_{T,P} dx_A^\alpha \tag{15-27}$$
$$-\frac{x_B^\beta(\overline{V}_B^\alpha - \overline{V}_B^\beta) + x_A^\beta(\overline{V}_A^\alpha - \overline{V}_A^\beta)}{x_B^\beta}\left(\frac{\partial \ln \hat{f}_A^\beta}{\partial x_A^\beta}\right)_{T,P} dx_A^\beta = 0$$

From Eq. (15-27), it follows that

$$\left(\frac{\partial x_A^\beta}{\partial x_A^\alpha}\right)_{T,[\alpha-\beta]} = \frac{x_B^\alpha(\overline{V}_B^\alpha - \overline{V}_B^\beta) + x_A^\alpha(\overline{V}_A^\alpha - \overline{V}_A^\beta)}{x_B^\beta(\overline{V}_B^\alpha - \overline{V}_B^\beta) + x_A^\beta(\overline{V}_A^\alpha - \overline{V}_A^\beta)}\left(\frac{x_B^\beta}{x_B^\alpha}\right)\frac{(\partial \ln \hat{f}_A^\alpha/\partial x_A^\alpha)_{T,P}}{(\partial \ln \hat{f}_A^\beta/\partial x_A^\beta)_{T,P}} \tag{15-28}$$

$$\left(\frac{\partial T}{\partial x_A^\alpha}\right)_{x_A^\beta,[\alpha-\beta]} = \frac{x_B^\alpha(\overline{V}_B^\alpha - \overline{V}_B^\beta) + x_A^\alpha(\overline{V}_A^\alpha - \overline{V}_A^\beta)}{(\overline{H}_A^\alpha - \overline{H}_A^\beta)(\overline{V}_B^\alpha - \overline{V}_B^\beta) - (\overline{H}_B^\alpha - \overline{H}_B^\beta)(\overline{V}_A^\alpha - \overline{V}_A^\beta)} \times \frac{RT^2}{x_B^\alpha}\left(\frac{\partial \ln \hat{f}_A^\alpha}{\partial x_A^\alpha}\right)_{T,P} \tag{15-29}$$

where $(\partial T/\partial x_A^\beta)_{x_A^\alpha,[\alpha-\beta]}$ is given by Eq. (15-29) with x_A^α and x_A^β interchanged.

From the partial derivatives obtained from Eqs. (15-22), (15-23), (15-25), (15-28), and (15-29), we could construct a variety of isoplethal, isothermal, or isobaric cross sections. To evaluate an isopleth (e.g., $P = f(T)$ at constant x_A^β), we must integrate $(\partial P/\partial T)_{x_A^\beta,[\alpha-\beta]}$. To carry out the integration of Eq. (15-22), we must express the right-hand side as a function of P, T, and x_A^β, where x_A^β is held constant during integration. From physical property data of individual phases, we can express $\overline{H}_i^\alpha$, $\overline{V}_i^\alpha$ and $\overline{H}_i^\beta$, $\overline{V}_i^\beta$ as a function of P, T, x_A^α and P, T, x_A^β, respectively. However, x_A^α is not a constant during the integration since it varies with T. Therefore, in addition to physical property data, we must know x_A^α as a function of T and x_A^β under a $\alpha - \beta$ phase equilibrium. This relationship can be obtained by integration of Eq. (15-28), and, therefore, P must be expressed as a function of x_A^α, x_A^β, and T. This functionality we originally sought by integration of Eq. (15-22). Thus, to obtain a rigorous solution to an isopleth requires simultaneous solution of Eqs. (15-22) and (15-28). Although the procedure is complicated for the rigorous case, in practice it can usually be simplified considerably by using judicious approximations. One example is given below.

Example 15.4

A binary system of components A and B coexists in liquid-vapor equilibrium at 448 K and at low pressure. The vapor phase can be considered an ideal mixture of ideal gases; the liquid-phase activity coefficients can be approximated by the van Laar equations:

$$\ln \gamma_A = \frac{A_{12}}{[1 + (A_{12}/A_{21})(x_A/x_B)]^2} \tag{15-30}$$

and

$$\ln \gamma_B = \frac{A_{21}}{[1 + (A_{21}/A_{12})(x_B/x_A)]^2} \tag{15-31}$$

where A_{12} and A_{21} are functions of temperature, and γ_A and γ_B can be considered independent of pressure.

(a) Determine the y-x relationship at a constant temperature of 448 K, and indicate how a $(P–x)_T$ diagram would be constructed.

(b) If isothermal $P–x$ data were available instead and activity coefficients were not known, would it be possible to construct the $y–x$ relationship? If so, indicate the procedure one would follow. Note that for a liquid phase obeying the van Laar equations, the following limiting law can be shown to pertain:

$$\lim_{x_A \to 0} \frac{y_A}{x_A} = \left(\frac{P_{vp_A}}{P_{vp_B}}\right) \exp(A_{12}) \tag{15-32}$$

where P_{vp_A} and P_{vp_B} are the vapor pressures of pure A and B at the temperature of the system. For the system in question at 448 K, $P_{vp_A} = 5.65$ bar and $P_{vp_B} = 8.98$ bar.

Solution

(a) The $y–x$ relationships at constant T can be determined from Eq. (15-28) if we let α = L = liquid and $\beta = V$ = vapor. Thus,

$$\left(\frac{\partial y_A}{\partial x_A}\right)_{T,[\alpha-\beta]} = \frac{x_A(\overline{V}_A^V - \overline{V}_A^L) + x_B(\overline{V}_B^V - \overline{V}_B^L)}{y_A(\overline{V}_A^V - \overline{V}_A^L) + y_B(\overline{V}_B^V - \overline{V}_B^L)}\left(\frac{y_B}{x_B}\right)\frac{(\partial \ln \hat{f}_A^L/\partial x_A)_{T,P}}{(\partial \ln \hat{f}_A^V/\partial y_A)_{T,P}} \tag{15-33}$$

For an ideal vapor mixture of ideal gases, the following simplifications are applicable:

$$\overline{V}_i^V = V_i^V = \frac{RT}{P} \quad (i = A \text{ or } B) \tag{15-34}$$

$$\left(\frac{\partial \ln \hat{f}_A^V}{\partial y_A}\right)_{T,P} = \frac{1}{y_A} \tag{15-35}$$

For conditions far removed from the critical point, there is an additional simplification:

$$\overline{V}_i^V = \frac{RT}{P} >> \overline{V}_i^L \tag{15-36}$$

Expressing $\hat{f}_A$ in the liquid phase using an activity coefficient, $\hat{f}_A = \gamma_A x_A f_A^{pure}$, then

$$\left(\frac{\partial \ln \hat{f}_A^L}{\partial x_A}\right)_{T,P} = \frac{1}{x_A} + \left(\frac{\partial \ln \gamma_A}{\partial x_A}\right)_{T,P} \tag{15-37}$$

Substituting Eqs. (15-34) through (15-37) into Eq. (15-33) and simplifying, we obtain

$$\left(\frac{\partial y_A}{\partial x_A}\right)_{T,[\alpha-\beta]} = \frac{y_A y_B}{x_A x_B}\left[1 + x_A\left(\frac{\partial \ln \gamma_A}{\partial x_A}\right)_{T,P}\right] \tag{15-38}$$

rearranging and integrating at constant T

$$\int \frac{dy_A}{y_A(1-y_A)} = \int \frac{dx_A}{x_A(1-x_A)} + \int \frac{1}{1-x_A}\left(\frac{\partial \ln \gamma_A}{\partial x_A}\right)_{T,P} dx_A \tag{15-39}$$

Equation (15-39) can be integrated directly by evaluating the activity coefficient partial derivative from Eq. (15-30). It is, however, more convenient to rearrange terms prior to integration. Since

$$\frac{1}{1-x_A} = \frac{1}{x_B} = \frac{1-x_B+x_B}{x_B} = \frac{x_A}{x_B} + 1$$

$$\frac{1}{1-x_A}\left(\frac{\partial \ln \gamma_A}{\partial x_A}\right)_{T,P} = \frac{x_A}{x_B}\left(\frac{\partial \ln \gamma_A}{\partial x_A}\right)_{T,P} + \left(\frac{\partial \ln \gamma_A}{\partial x_A}\right)_{T,P} \tag{15-40}$$

From the Duhem expression for γ,

$$\frac{x_A}{x_B}\left(\frac{\partial \ln \gamma_A}{\partial x_A}\right)_{T,P} = -\left(\frac{\partial \ln \gamma_B}{\partial x_A}\right)_{T,P}$$

Thus,

$$\int \frac{1}{1-x_A}\left(\frac{\partial \ln \gamma_A}{\partial x_A}\right)_{T,P} dx_A = \int\left[\frac{\partial \ln (\gamma_A/\gamma_B)}{\partial x_A}\right]_{T,P} dx_A \tag{15-41}$$

Substituting Eq. (15-41) into Eq. (15-39), and using the indefinite integral method, we obtain

$$\ln\frac{y_A}{1-y_A} = \ln\frac{x_A}{1-x_A} + \ln\frac{\gamma_A}{\gamma_B} + C \tag{15-42}$$

The constant of integration can be evaluated using the limiting condition given in the problem statement. That is, as $x_A \to 0$ and $y_A \to 0$, $\gamma_A \to \exp(A_{12})$, $\gamma_B \to 1$; thus, from Eqs. (15-32) and (15-42),

$$\ln\left(\lim_{x_A\to 0}\frac{y_A}{x_A}\right) = \lim_{x_A\to 0}\left(\ln\frac{y_A}{x_A}\right) = \ln\frac{\exp(A_{12})P_{vp_A}}{P_{vp_B}} = A_{12} + C \tag{15-43}$$

and

$$C = \ln\frac{P_{vp_A}}{P_{vp_B}} \tag{15-44}$$

so

$$\frac{y_A}{1-y_A} = \frac{x_A}{1-x_A}\left(\frac{P_{vp_A}}{P_{vp_B}}\right)\frac{\gamma_A}{\gamma_B} \quad (T\text{ constant}) \tag{15-45}$$

Substituting Eqs. (15-30) and (15-31) into Eq. (15-45) and simplifying, we obtain the desired result:

$$\frac{y_A}{1-y_A} = g(x_A) = \frac{P_{vp_A}}{P_{vp_B}}\left(\frac{x_A}{x_B}\right)\exp\left[\frac{A_{12}A_{21}(A_{21}x_B^2 - A_{12}x_A^2)}{(A_{12}x_A + A_{21}x_B)^2}\right] \quad (T\text{ constant}) \tag{15-46}$$

or

$$y_A = \frac{g(x_A)}{1+g(x_A)} \quad (T\text{ constant}) \tag{15-47}$$

To construct a P–x diagram at constant T, we must integrate Eq. (15-23). Substituting the simplifying assumptions given above into Eq. (15-23) (with β = liquid), we have

$$\left(\frac{\partial P}{\partial x_A}\right)_{T,[\alpha-\beta]} = \frac{RT(y_A - x_A y_B/x_B)[(1/x_A) + (\partial \ln \gamma_A/\partial x_A)_{T,P}]}{RT/P} \qquad (15\text{-}48)$$

or

$$\left(\frac{\partial \ln P}{\partial x_A}\right)_{T,[\alpha-\beta]} = \left(y_A - \frac{x_A y_B}{x_B}\right)\left\{\frac{1}{x_A} + x_B\left[\frac{\partial \ln (\gamma_A/\gamma_B)}{\partial x_A}\right]_{T,P}\right\} \qquad (15\text{-}49)$$

Thus, with T constant:

$$\int d\ln P = \int \frac{1}{1-x_A}\left[\frac{g(x_A)}{1+g(x_A)} - x_A\right]\left\{\frac{1}{x_A} + x_B\left[\frac{\partial \ln (\gamma_A/\gamma_B)}{\partial x_A}\right]_{T,P}\right\} dx_A \qquad (15\text{-}50)$$

The integration is complex; the final result is much easier to obtain by the integral approach described in Section 15.5. The point to note is that simplifications resulting from the facts that one phase is a vapor (i.e., $\underline{V}^V >> \underline{V}^L$) and that the vapor is ideal results in a procedure which, although complex, is manageable. Furthermore, we have seen that in the frequently occurring cases in which these assumptions are valid, knowledge of the condensed phase activity coefficient is all that is required to generate $(y\text{–}x)_T$ and $(P\text{–}x)_T$ diagrams.

(b) If we did not have an expression for the activity coefficient, and if we did have isothermal P–x data, we should be able to reverse the procedure above to construct the y–x relationship. Clearly, if we assumed that the liquid phase behaved as a van Laar liquid, all terms in the right-hand side of Eq. (15-50) could be expressed as functions of x_A and the van Laar constants, A_{12} and A_{21}. Thus, these constants could be evaluated by curve-fitting Eq. (15-50) to the experimental P–x data, and these constants could then be used in Eq. (15-46) to obtain the y–x relationship. Alternatively, we could use other applicable activity coefficient-liquid composition correlations and, in an analogous procedure, evaluate the coefficients in the expansion from the P–x data. Similarly, if the vapor phase were not ideal, an appropriate real-gas equation of state would have to be used for the vapor properties. Thus, it is possible to obtain y–x equilibrium information from the relatively simple measurement of equilibrium pressure without recourse to measurement of vapor phase composition.

We have seen in the discussion of Eqs. (15-13) through (15-18) that in order to generate isothermal, isobaric, and isoplethal cross sections, we must have knowledge of the temperature, pressure, and concentration dependency of the fugacities (or chemical potentials). If, however, we did not have knowledge of all property relations, we could develop some diagrams from knowledge of other cross sectional diagrams. The minimum number of such diagrams needed to specify all others is equivalent to determining the minimum number of independent partial derivatives involving temperature, pressure, and concentration.

For a binary system involving two phases, we require the minimum number of independent partials involving the variables T, P, y_A, and x_A. In Section 5.3 we faced a similar problem for the four variables S, T, V, and P. There we found that all but four partials could be eliminated as independent by mathematical manipulation, and one of these four could be eliminated by thermodynamic reasoning (e.g., the Maxwell reciprocity relation was obtained from the fact that these variables satisfied the Fundamental Equation, $U = f(S, V)$). Since the reciprocity condition does not apply here, it follows that there are four independent partials involving T, P, y_A, and x_A, but that no more than two can be chosen from any one set of the combinations of (T, P, x_A^α), (T, P, x_A^β), $(T, x_A^\alpha, x_A^\beta)$, and $(P, x_A^\alpha, x_A^\beta)$. Thus, any cross section could be obtained, for example, for a vapor-liquid binary system, from T–x, T–y, P–x, and P–y diagrams. Alternatively, in the general case in which simplifying assumptions are not applicable, we cannot obtain a y–x diagram from P–x data alone.

The results obtained for the two-phase binary system can be generalized to include additional phases and components. For a binary system involving the three phases α, β, and γ, the criteria of equilibrium become

$$T^\alpha = T^\beta = T^\gamma \tag{15-51}$$

$$P^\alpha = P^\beta = P^\gamma \tag{15-52}$$

$$\ln \hat{f}_A^\alpha = \ln \hat{f}_A^\beta = \ln \hat{f}_A^\gamma \tag{15-53}$$

$$\ln \hat{f}_B^\alpha = \ln \hat{f}_B^\beta = \ln \hat{f}_B^\gamma \tag{15-54}$$

These criteria are identical to solving simultaneously the two cases of α–β and β–γ phase equilibria. For the α–β equilibrium case, the relationships developed previously (namely, Eqs. (15-17), (15-18), (15-21), and (15-27)) are still valid. For the β–γ equilibrium, we would obtain identical equations with α replaced by γ. We would then solve the two sets simultaneously.

For example, by eliminating x_A^α from Eqs. (15-17) and (15-18), we obtained Eq. (15-21), which we can write as

$$dP = \left(\frac{\partial P}{\partial T}\right)_{x_A^\beta,[\alpha-\beta]} dT + \left(\frac{\partial P}{\partial x_A^\beta}\right)_{T,[\alpha-\beta]} dx_A^\beta \tag{15-55}$$

where the partials are given by Eqs. (15-22) and (15-23). The analogous relation for the β–γ equilibrium in which we eliminate x_A^γ would then be

$$dP = \left(\frac{\partial P}{\partial T}\right)_{x_A^\beta,[\beta-\gamma]} dT + \left(\frac{\partial P}{\partial x_A^\beta}\right)_{T,[\beta-\gamma]} dx_A^\beta \tag{15-56}$$

From Eq. (15-55) and (15-56), we could solve simultaneously to obtain P as a function of T by eliminating dx_A^β, P as a function of x_A^β by eliminating T, or T as a function of x_A^β by eliminating P. For example, let us eliminate x_A^β. Thus,

$$\left(\frac{\partial x_A^\beta}{\partial P}\right)_{T,[\beta-\gamma]}\left[dP-\left(\frac{\partial P}{\partial T}\right)_{x_A^\beta,[\beta-\gamma]}dT\right]=\left(\frac{\partial x_A^\beta}{\partial P}\right)_{T,[\alpha-\beta]}\left[dP-\left(\frac{\partial P}{\partial T}\right)_{x_A^\beta,[\alpha-\beta]}dT\right] \quad (15\text{-}57)$$

or

$$dP=\frac{(\partial P/\partial T)_{x_A^\beta,[\alpha-\beta]}(\partial x_A^\beta/\partial P)_{T,[\alpha-\beta]}-(\partial P/\partial T)_{x_A^\beta,[\beta-\gamma]}(\partial x_A^\beta/\partial P)_{T,[\beta-\gamma]}}{(\partial x_A^\beta/\partial P)_{T,[\alpha-\beta]}-(\partial x_A^\beta/\partial P)_{T,[\beta-\gamma]}}dT$$

or

$$dP=-\left[\frac{(\partial x_A^\beta/\partial T)_{P,[\alpha-\beta]}-(\partial x_A^\beta/\partial T)_{P,[\beta-\gamma]}}{(\partial x_A^\beta/\partial P)_{T,[\alpha-\beta]}-(\partial x_A^\beta/\partial P)_{T,[\beta-\gamma]}}\right]dT \quad (15\text{-}58)$$

The bracket in Eq. (15-58) is clearly $(\partial P/\partial T)_{[\alpha-\beta-\gamma]}$, which can be written as $(\partial P/\partial T)$ for the monovariant system of three phases in equilibrium. Equation (15-58) represents, in effect, a locus of triple points. Note that for each set of T and P that satisfies Eq. (15-58), x_A^β will vary. We could have obtained this variation of x_A^β with T, for example, by eliminating dP from Eqs. (15-55) and (15-56). Other variables, such as x_A^α and x_A^γ, could be expressed as functions of T, P, or x_A^β in an analogous manner.

In the general case of n components distributed between π phases, we obtain, *for each component*, $(\pi - 1)$ equations of the form

$$-\frac{\overline{H}_i^\alpha-\overline{H}_i^\beta}{RT^2}dT+\frac{\overline{V}_i^\alpha-\overline{V}_i^\beta}{RT}dP+\sum_{j\neq k}^{n}\left[\left(\frac{\partial \ln \hat{f}_i^\alpha}{\partial x_j^\alpha}\right)_{T,P,x_i^\alpha[x_j^\alpha,x_k^\alpha]}dx_j^\alpha-\left(\frac{\partial \ln \hat{f}_i^\beta}{\partial x_j^\beta}\right)_{T,P,x_i^\beta[x_j^\beta,x_k^\beta]}dx_j^\beta\right]=0 \quad (15\text{-}59)$$

Thus, we have $n(\pi - 1)$ equations of this form relating the $((n-1)(\pi)+2)$ variables involving T, P, and x. Solving these equations simultaneously, we can eliminate $(n(\pi-1)-1)$ variables, resulting in a differential equation involving $(n+3-\pi)$ variables. From this equation, the differential of any one variable is expressed as a function of the remaining $(n+2-\pi)$ variables.[3]

---

3 Note that the Gibbs-Duhem equations (one for each phase for a total of π) can be used to simplify the $n(\pi$-1) equations of the form of Eq. (15-59) but not to reduce the number of variables. That is, the Duhem equations reduce the number of coefficients of the dx_j terms and therefore reduce the amount of physical property data required in integrations to obtain the final result.

15.4 Pressure-Temperature Relations

In phase equilibria applications, the total system vapor pressure is often a strong function of temperature. Also it turns out that the derivative dP/dT is related to the enthalpy (or entropy) and volume changes during a phase transformation. The ability to correlate and predict vapor pressure as a function of temperature or to derive enthalpies changes of certain phase reactions is of considerable interest to chemical engineers. We shall illustrate a general approach by examining a few simple, but common systems, all of which involve a vapor phase in addition to one or more condensed phases.

The simplest case encountered is a pure liquid in equilibrium with its vapor. At equilibrium, we have $f^V = f^L$ in addition to temperature and pressure equalities. Thus,

$$d \ln f^V = d \ln f^L \tag{15-60}$$

Since the fugacity of a pure material is a function only of T and P, expanding we obtain

$$\left(\frac{\partial \ln f^V}{\partial T}\right)_P dT + \left(\frac{\partial \ln f^V}{\partial P}\right)_T dP = \left(\frac{\partial \ln f^L}{\partial T}\right)_P dT + \left(\frac{\partial \ln f^L}{\partial P}\right)_T dP \tag{15-61}$$

Substituting Eqs. (9-112) and (9-114) into Eq. (15-61) and collecting terms,

$$\left(\frac{dP}{dT}\right)_{[L-V]} = \frac{H^V - H^L}{T(V^V - V^L)} = \frac{\Delta H_{vap}}{T\,\Delta V_{vap}} \tag{15-62}$$

Expressing $\Delta V^{vap} = (RT/P)(Z^V - Z^L) = (RT/P)\,\Delta Z_{vap}$, then

$$\left[\frac{d \ln P}{d(1/T)}\right]_{[L-V]} = \frac{-\Delta H_{vap}}{R\,\Delta Z_{vap}} \tag{15-63}$$

Equation (15-63) is called the Clausius-Clapeyron equation when $\Delta Z_{vap} \cong 1.0$. The ratio $(\Delta H_{vap}/\Delta Z_{vap})$ is a weak but essentially linear function of temperature except near the critical point. Often, at low pressures, the assumption is made that over a nominal temperature range, the ratio is constant and thus Eq. (15-63) is readily integrated. When such an assumption is not valid or when high accuracy is desired, other integration techniques are available (see Sections 13.2 and 13.3).

Next, let us consider the case in which we have a nonvolatile solute, such as an inorganic salt, dissolved in a volatile solvent such as water. The vapor above the solution is then essentially pure water. Since the system is divariant, the pressure-temperature relation for such a system is not unique unless we place a further restriction on the system. Two kinds of restricted systems will be illustrated.

Let the salt concentration in the liquid be constant as the system temperature (or pressure) is varied. For the volatile water component (denoted by subscript w),

$$d \ln f_w^V = d \ln \hat{f}_w^L \tag{15-64}$$

Expanding in terms of T and P for pure vapor and in terms of T, P, and x_w for the liquid mixture, we obtain noting that $dx_w = 0$,

$$\left(\frac{\partial \ln f_w^V}{\partial T}\right)_P dT + \left(\frac{\partial \ln f_w^V}{\partial P}\right)_T dP = \left(\frac{\partial \ln \hat{f}_w^L}{\partial T}\right)_{P,x} dT + \left(\frac{\partial \ln \hat{f}_w^L}{\partial P}\right)_{T,x} dP \qquad (15\text{-}65)$$

Substituting for the partial derivatives of fugacity one finds that

$$\left(\frac{\partial P}{\partial T}\right)_{x,[L-V]} = \frac{H_w^V - \overline{H}_w^L}{T(V_w^V - \overline{V}_w^L)} \qquad (15\text{-}66)$$

The numerator represents the enthalpy change in vaporizing 1 mole of water from the solution at constant composition. The quantity H_w^V is not the enthalpy of saturated water vapor since, at a given T, the system pressure does not correspond to the equilibrium vapor pressure for pure water. The pressure correction, however, is ordinarily small and usually neglected. In fact, the numerator is often expanded by adding and subtracting H_w^L, the enthalpy of pure *liquid* water at the system temperature and pressure. Then

$$H_w^V - \overline{H}_w^L = (H_w^V - H_w^L) + (H_w^L - \overline{H}_w^L) \qquad (15\text{-}67)$$

When pressure corrections to the enthalpy of pure water are neglected, the first term in Eq. (15-67) is the enthalpy of vaporization of pure water at the system temperature, $\Delta H_{vap,\ pure\ w}$; the second term is then the partial molar enthalpy of mixing, $\overline{\Delta H_w^L}$.

Then, if the additional assumptions that $V_w^V >> \overline{V}_w^L$ and $V_w^V = RT/P$ are invoked,

$$\left[\frac{\partial \ln P}{\partial(1/T)}\right]_{x,[L-V]} = \frac{-(\Delta H_{vap,\ pure\ w} - \overline{\Delta H_w^L})}{R} \qquad (15\text{-}68)$$

In theory, then, from P-T data at constant liquid composition, one could determine partial molar enthalpies of mixing. Both $\overline{\Delta H_w^L}$ and $\overline{\Delta H_s^L}$ can be found, the former from Eq. (15-68) and the latter from $\overline{\Delta H_w^L}$ and the Gibbs-Duhem equation. Nevertheless, neat as this approach appears, it is unfortunately not a very useful one. Even with very accurate $(P\text{–}T)_x$ data, differentiation leads to some error and since, usually $\overline{\Delta H_w^L} < \Delta H_{vap,\ pure\ w}$, it is difficult to extract accurate values of $\overline{\Delta H_w^L}$ from the data. Direct calorimetric determination of enthalpies of mixing is usually the preferred way to measure this property.

Instead of keeping the salt solution concentration constant, one could impose the restriction that the solution is at all times saturated with the nonvolatile solute (i.e., there is always some undissolved salt present). In this case since $\pi = 3$ and $n = 2$, we have a univariant system.

Following the same general treatment, we have a liquid mixture, a pure water vapor phase, and a pure salt solid phase. Thus,

$$d \ln f_w^V = d \ln \hat{f}_w^L \qquad (15\text{-}69)$$

and

$$d \ln f_s^S = d \ln \hat{f}_s^L \qquad (15\text{-}70)$$

Expanding the fugacity in the pure phases in terms of T and P and the liquid phase in terms of T, P, and x_w, and substituting for the fugacity partials with respect to T and P, we obtain

$$-\frac{H_w^V - \overline{H}_w^L}{RT^2}\,dT + \frac{V_w^V - \overline{V}_w^L}{RT}\,dP = \left(\frac{\partial \ln \hat{f}_w^L}{\partial x_w}\right)_{T,P} dx_w \tag{15-71}$$

and

$$-\frac{H_s^S - \overline{H}_s^L}{RT^2}\,dT + \frac{V_s^S - \overline{V}_s^L}{RT}\,dP = \left(\frac{\partial \ln \hat{f}_s^L}{\partial x_w}\right)_{T,P} dx_w \tag{15-72}$$

To determine the P-T relation, we must solve Eqs. (15-71) and (15-72) simultaneously for dx_w and then use the Gibbs-Duhem relation to eliminate one of the fugacity partials with respect to concentration. A simple procedure is to multiply Eqs. (15-71) and (15-72), respectively, by x_w and x_s. Upon adding the resultant equations, we see that terms in dx_w drop out by virtue of the Gibbs-Duhem equation. Thus,

$$\left[\frac{\partial P}{\partial T}\right]_{[S\text{-}L\text{-}V]} = \frac{x_w(H_w^V - \overline{H}_w^L) + x_s(H_s^S - \overline{H}_s^L)}{T\,[x_w(V_w^V - \overline{V}_w^L) + x_s(V_s^S - \overline{V}_s^L)]} \tag{15-73}$$

To compare this case to Eq. (15-68), assume that the predominant term in the denominator is $x_w V_w^V$ and that $V_w^V = RT/P$; then

$$\left[\frac{\partial \ln P}{\partial (1/T)}\right]_{[S\text{-}L\text{-}V]} = -\frac{x_w(H_w^V - \overline{H}_w^L) + x_s(H_s^S - \overline{H}_s^L)}{Rx_w} \tag{15-74}$$

The term $\partial \ln P/\partial(1/T)$ is again related to an enthalpy change due to evaporating x_w moles of water while simultaneously crystallizing x_s moles of salt. The division by x_w simply yields the result as the enthalpy change per mole of water evaporated.

We might carry Eq. (15-74) one step further. The enthalpy of mixing of water and salt is often of interest. This ΔH_{mix} is defined as

$$\Delta H_{mix} = x_w(\overline{H}_w^L - H_w^L) + x_s(\overline{H}_s^L - H_s^S) \tag{15-75}$$

where H_w^L is the enthalpy of the pure water at the system T and P. If one adds and subtracts $x_w H_w^L$ to the right-hand side of Eq. (15-74), then

$$\left[\frac{d \ln P}{d(1/T)}\right]_{[S\text{-}L\text{-}V]} = \frac{\Delta H_{mix}}{Rx_w} - \frac{\Delta H_{vap,\,pure\,w}}{R} \tag{15-76}$$

Thus, from P-T data for saturated solutions and the heat of vaporization of the volatile component, one can determine, at least approximately, heats of mixing. An alternative approach to this problem is found in Section 17.4.

Now let us consider a binary liquid-vapor system in which both components exist in each phase. For this divariant system, we shall use the additional restriction of constant liquid composition.

The pressure-temperature relation in this case follows directly from Eq. (15-22), where we denote the liquid phase by β and vapor phase by α. Thus,

$$\left(\frac{\partial P}{\partial T}\right)_{x,[L-V]} = \frac{1}{T}\frac{y_A(\overline{H}_A^V - \overline{H}_A^L) + y_B(\overline{H}_B^V - \overline{H}_B^L)}{y_A(\overline{V}_A^V - \overline{V}_A^L) + y_B(\overline{V}_B^V - \overline{V}_B^L)} \tag{15-77}$$

Assuming that $\overline{V}_A^V >> \overline{V}_A^L$, $\overline{V}_B^V >> \overline{V}_B^L$, and noting that

$$y_A\overline{V}_A^V + y_B\overline{V}_B^V = V^V = \frac{RT}{P} \tag{15-78}$$

$$y_A\overline{H}_A^V + y_B\overline{H}_B^V = H^V \tag{15-79}$$

then

$$\left[\frac{\partial \ln P}{\partial(1/T)}\right]_{x,[L-V]} = -\frac{H^V - y_A\overline{H}_A^L - y_B\overline{H}_B^L}{R} \tag{15-80}$$

As expected, the slope of $\ln P$ versus $1/T$ corresponds to an enthalpy change. This enthalpy change is, however, somewhat unusual. The term H^V represents the enthalpy of the saturated vapor mixture at y_A, y_B. The terms $\overline{H}_A^L$ and $\overline{H}_B^L$ represent the partial molar enthalpies of the saturated liquid at x_A and x_B. To visualize the situation, consider Figure 15.28. The enthalpy of the saturated vapor and liquid of a binary mixture of A and B is shown as a function of liquid and vapor mole fraction. Point C represents the enthalpy of a saturated vapor at y_A. This vapor is in equilibrium with a liquid of composition x_A with enthalpy H^L (point D). The diagram is drawn for either constant T or P. Consider a tangent to the lower curve at D. The tangent intersects the left-hand ordinate (pure B) at E and the right-hand ordinate (pure A) at F. Let us consider the significance of these points E and F.

If Figure 15.28 is drawn at constant temperature, then

$$dH^L = \left(\frac{\partial H^L}{\partial P}\right)_{T,x_A} dP + \left(\frac{\partial H^L}{\partial x_A}\right)_{P,T} dx_A \tag{15-81}$$

Dividing Eq. (15-81) by dx_A and placing the restriction of saturated solution yields

$$\left(\frac{\partial H^L}{\partial x_A}\right)_{T,[L-V]} = \left(\frac{\partial H^L}{\partial P}\right)_{T,x_A}\left(\frac{\partial P}{\partial x_A}\right)_{T,[L-V]} + \left(\frac{\partial H^L}{\partial x_A}\right)_{P,T} \tag{15-82}$$

but

$$\overline{H}_A^L = H^L + x_B\left(\frac{\partial H^L}{\partial x_A}\right)_{T,P} \tag{15-83}$$

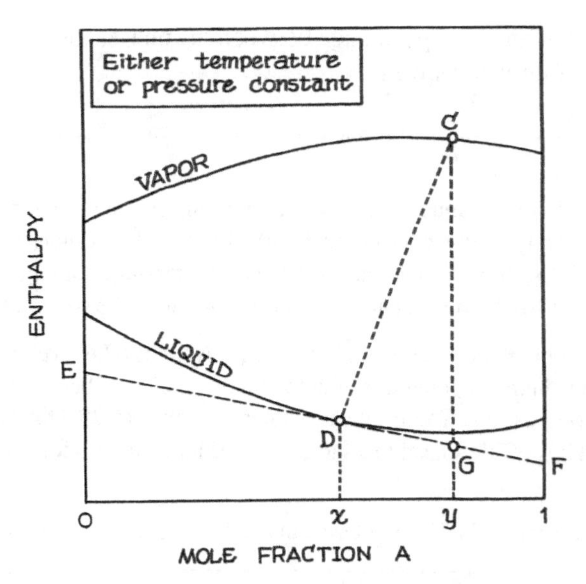

Figure 15.28 Enthalpy-concentration diagram.

so by multiplying Eq. (15-82) by x_B and subtracting Eq. (15-83) from the result, we obtain

$$\overline{H}_A^L = H^L + x_B\left(\frac{\partial H^L}{\partial x_A}\right)_{T,[L-V]} - x_B\left(\frac{\partial H^L}{\partial P}\right)_{T,x_A}\left(\frac{\partial P}{\partial x_A}\right)_{T,[L-V]} \tag{15-84}$$

A similar relation may be written for $\overline{H}_B^L$.

The $\overline{H}_A^L$ and $\overline{H}_B^L$ terms are those required for Eq. (15-80). The first two terms on the right-hand side of Eq. (15-84) yield point F on Figure 15.28. (If the equation were written for $\overline{H}_B^L$, the same terms would represent point E). The last term on the right-hand side is a term that corrects for pressure changes in the system with composition; it is evaluated at point D. Ordinarily, it is small and usually neglected. If so, then the numerator in Eq. (15-80) is clearly represented by the vertical distance $\overline{CG}$; the physical significance of this enthalpy term is then the change in enthalpy from a saturated vapor at C to a subcooled liquid at G with the same composition.

Similar reasoning yields, for a constant-pressure case,

$$\overline{H}_A^L = H^L + x_B\left(\frac{\partial H^L}{\partial x_A}\right)_{P,[L-V]} - x_B C_p^L\left(\frac{\partial T}{\partial x_A}\right)_{P,[L-V]} \tag{15-85}$$

Since the last term on the right-hand side usually is not negligible, point G will have to be adjusted.

Many other cases requiring the interpretation of the P-T relations of liquid-vapor mixtures could be treated. The few discussed here point out the typical approach; all yielded, to a first approximation, an equation relating $\ln P$ to $1/T$ with the slope of the curve proporational to an enthalpy change between the phases. Section 17.4 provides an alternative general approach as well.

Example 15.5

A new process calls for an aqueous feed of ammonium nitrate at high pressure at 25°C. The pressure currently being considered is 10 kilobars. To complete our design, the solubility of NH_4NO_3 in the feedstream is desired. A literature search of this system has provided the data shown below. What is your best estimate of the solubility at 10 kilobars?

Data: At 25°C and 1 bar, the solubility of NH_4NO_3 in water is 67.63 wt% salt. At a pressure of about 1 bar, the chemical potentials of both components have been determined at 25°C as a function of composition. These data are shown in Figure 15.29. Note that the chemical potential of NH_4NO_3 is referenced to a saturated solution, whereas for water the reference is pure water.

Volumetric data for the NH_4NO_3-H_2O system at 25°C are given in Figure 15.30.

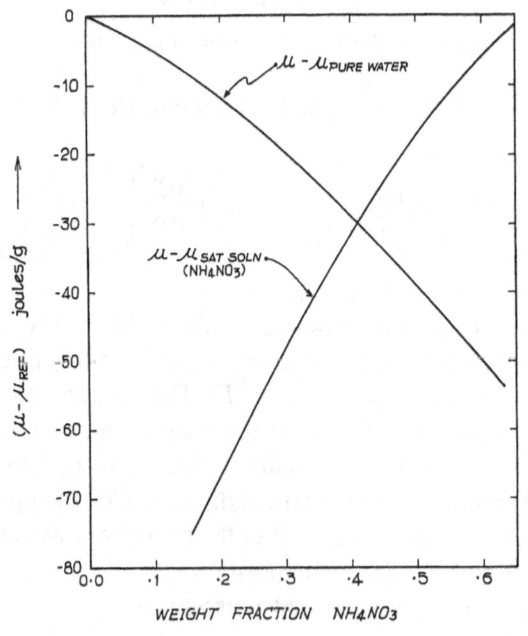

Figure 15.29 Chemical potentials for water and NH_4NO_3 in solution at 25°C, 1 bar.

Solution

Let x be the solubility (weight fraction) of NH_4NO_3 in the high-pressure solution. Now visualize an isothermal cyclic process beginning with a solution at x, at 1 bar. The change in chemical potential of NH_4NO_3 from this solution to the solid phase at 1 bar can be found from Figure 15.29 once x is known:

$$\mu_{NH_4NO_3}(1 \text{ bar, solid}) - \mu_{NH_4NO_3}(1 \text{ bar, liquid at } x) = -f(x)$$

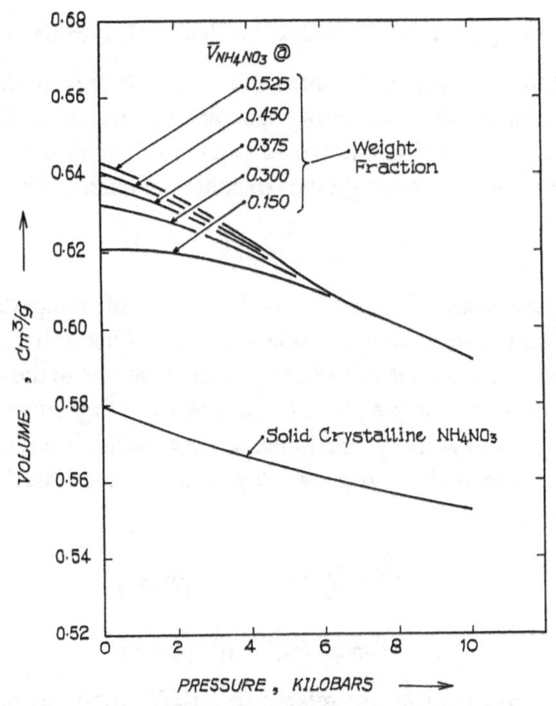

Figure 15.30 Partial molal volume and pure volumes for NH_4NO_3 at 25°C.

Next, compress the pure solid to 10 kbar.

$$\mu_{NH_4NO_3}(10 \text{ kbar, solid}) - \mu_{NH_4NO_3}(1 \text{ bar, solid}) = \int_{1 \text{ bar}}^{10 \text{ kbar}} V^S_{NH_4NO_3}\, dP$$

This solid is in equilibrium with liquid at x, i.e.,

$$\mu_{NH_4NO_3}(10 \text{ kbar, liquid at } x) - \mu_{NH_4NO_3}(10 \text{ kbar, solid}) = 0$$

Finally expand this liquid back to 1 bar,

$$\mu_{NH_4NO_3}(1 \text{ bar, liquid at } x) - \mu_{NH_4NO_3}(10 \text{ kbar, liquid at } x) = \int_{10 \text{ kbar}}^{1 \text{ bar}} \overline{V}^L_{NH_4NO_3}\, dP$$

Adding, we obtain

$$-f(x) = \int_{1 \text{ bar}}^{10 \text{ kbar}} [\overline{V}^L(x) - V^S]\, dP$$

By trial and error, with Figure 15.29 to obtain $f(x)$ and Figure 15.30 to obtain $\overline{V}^L(x)$ and V^S as a function of pressure, x is found to be 0.295. Thus, at 10 kbar the solubility of NH_4NO_3 is only about 29.5 wt%.

15.5 The Integral Approach to Phase Equilibrium Relationships

The integral approach proceeds from the same starting point as the differential approach, namely, equating component fugacities in each coexisting phase. Instead of differentiating these relationships, however, we treat them directly in the integral form. In general, the procedure is to equate the fugacities of each component

$$\hat{f}_i^{\alpha} = \hat{f}_i^{\beta} \tag{15-86}$$

and then to expand these fugacities as functions of temperature, pressure, and composition in each phase. Two approaches are used. One utilizes an equation of state to determine component fugacities in *both* phases, whereas in the other, an equation of state is employed only for the gas phase (if present) and fugacities in condensed phases are correlated in terms of activity coefficients. Both methods are described below.

Equation-of-state approach. Component fugacities are expressed in terms of a fugacity coefficient as given in Eq. (9-120).

$$\begin{aligned} \hat{f}_i^{\alpha} &= \hat{\phi}_i^{\alpha} P x_i^{\alpha} \quad (\alpha\text{–phase}) \\ \hat{f}_i^{\beta} &= \hat{\phi}_i^{\beta} P x_i^{\beta} \quad (\beta\text{–phase}) \end{aligned} \tag{15-87}$$

and the fugacity coefficients $\hat{\phi}_i$ are related to $PV\underline{T}N$ properties of the mixture by Eq. (9-142). To illustrate these relations for a case where α = vapor and β = liquid, one obtains

$$\frac{y_i}{x_i} = \frac{\hat{\phi}_i^L}{\hat{\phi}_i^V} \tag{15-88}$$

where $\hat{\phi}_i^V$ is the fugacity coefficient of component i in the vapor phase and is a function of T, P, and vapor composition $(y_1, \ldots, y_i, \ldots, y_{n-1})$.

In Eq. (9-144) we show a relation for $\hat{\phi}_i$ as determined from the Peng-Robinson equation of state [Eq. (9-26)]. For a n-component system, the phase rule requires that n-intensive variables be specified. Suppose that these were (n-1) liquid-phase mole fractions and the system temperature. There are n equations of the form of Eq. (15-88) to calculate the unknown (n-1) vapor-phase mole fractions and the system pressure. (Note that in Eq. (9-144) that compressibility factors are also present. These would be determined separately for both phases, at the system temperature and pressure, at the appropriate phase composition.)

Clearly, the solution of the n equations in Eq. (15-88) requires machine computation as the expressions for $\hat{\phi}_i$ are highly nonlinear. The success of this method depends, however, on the choice of an appropriate mixture equation of state. Few have been shown to be of much practical value to calculate phase equilibria. The Benedict-Webb-Rubin EOS (or one of its many modifications) has been used for a number of years to determine phase equilibrium behavior of mixtures of light hydrocarbons. More recently, phase behavior has been correlated with two-constant modifications of the Redlich-Kwong EOS such as suggested by Soave or by Peng and Robinson [see also Sections 8.3 (Tables 8.1 and 8.2)]. For systems that are essentially nonpolar, these correlations have often been quite successful. In Figure 15.31 we show experimental and calculated pressures for the isobutane-carbon dioxide system at 311 K where the Peng-Robinson equation of EOS determines phase volumes and fugacity coefficients. The sensitivity of the computed pressures to the isobutane-carbon dioxide interaction parameter [see Eq. (9-29)] is well illustrated.

Another application of the equation-of-state approach lies in computing solubilities of relatively nonvolatile solids in supercritical fluids. The term *supercritical fluid* in this instance indicates the solvent phase is at a temperature and pressure above the critical-state values of the pure solvent. Returning to Eq. (15-86), let α be the supercritical phase and β be the solid phase, with i the solute. Then $\hat{f}_i^{\alpha}$ is determined from an equation of state as shown in Eq. (15-87). $\hat{f}_i^{\beta}$ is now the fugacity of pure solid solute at the temperature and pressure of the system (assuming no solvent gas dissolves in the solid). Therefore,

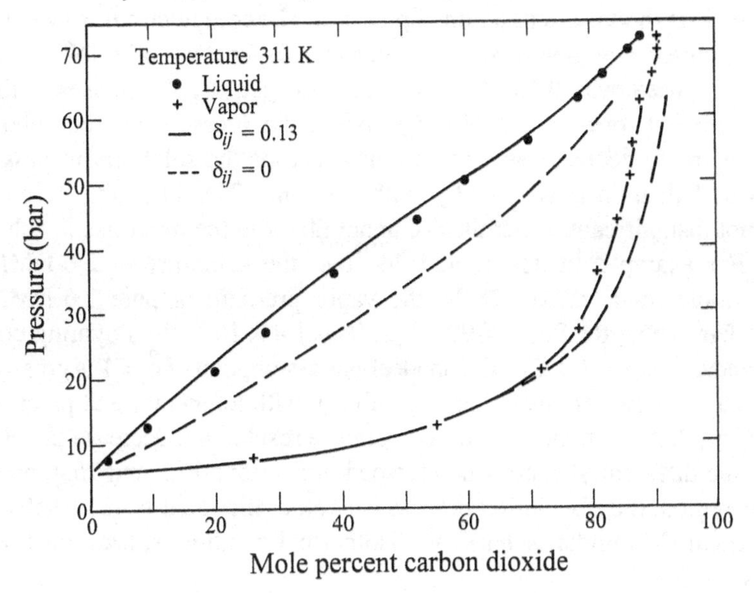

Figure 15.31 Estimation of the phase behavior of isobutane-carbon dioxide system using the Peng-Robinson equation of state [adapted from D.-Y. Peng and D.B. Robinson, *Ind. Eng. Chem. Fundam.*, **15**, 59 (1976)].

$$\hat{f}_i^{\alpha} = \hat{\phi}_i^{\alpha}(P, T, y_i)Py_i = f_i^{\beta}(P, T)$$
$$= P_{vp,i}\phi_i(P_{vp,i}, T)\exp\left(\int_{P_{vp,i}}^{P} \frac{V_i^{\beta}}{RT}\,dP\right) \qquad (15\text{-}89)$$

The fugacity coefficient for vapor above pure solid i at its vapor pressure, ϕ_i, is normally very close to unity. Assuming that the solid is incompressible ($V_i^\beta \neq f(P)$), Eq. (15-89) can then be written as

$$y_i = \left(\frac{P_{vp,i}}{P}\right)\frac{1}{\hat{\phi}_i^\alpha}\left\{\exp\left[\frac{V_i^\beta}{RT}(P - P_{vp,i})\right]\right\} \tag{15-90}$$

Thus, the mole fraction of the solute in the supercritical fluid is given as a product of three terms. The first is called the *ideal solubility* and would simply indicate that, at constant temperature, y_i is proportional to P^{-1}. The third term is the *Poynting correction* and measures the effect of pressure on the solid fugacity. This term increases with pressure. The second term reflects the non-ideality of the fluid-phase mixture of solvent and solute. $\hat{\phi}_i^\alpha$ is close to unity at low pressures, but it can become very small (to give large values of y_i) at high pressures. To illustrate Eq. (15-90), consider Figure 15.32. In this figure the solubility of solid 2,6-dimethylnaphthalene (2,6-DMN) in supercritical carbon dioxide is shown as a function of pressure. The computed curves were made with the Peng-Robinson equation of state to determine $\hat{\phi}_i$ for the 2,6-DMN. The interaction parameter, δ_{ij}, used was 0.10. At low pressures, $y_{2,6\text{–DMN}}$ decreases with pressure, reflecting the first term in Eq. (15-90). When the pressure exceeds about 20 bar, $\hat{\phi}_{2,6\text{–DMN}}$ begins to decrease and, in so doing, causes the solubility to pass through a minimum and then increase rapidly with pressure. The Poynting correction term, although not insignificant, is not the dominant factor in the rapid rise in solubility with pressure. For example, at 318 K and 244 bar, the solubility of 2,6-DMN is about 6.3×10^{-3} mole fraction. At 318 K, the vapor pressure of pure 2,6-DMN is about 3.5×10^{-5} bar. Thus, for Eq. (15-90), $P_{vp,i}/P \sim 1.4 \times 10^{-7}$, the Poynting correction is about 4.2 and $\hat{\phi}_i^{-1}$ is 10,700. Clearly, an ideal-gas assumption of $\hat{\phi}_i = 1$ is grossly incorrect.

Figure 15.32 shows an inversion in solubility with temperature at pressures around 80 to 100 bar, that is, an increase in temperature results in a decrease in solubility. By the use of the differential approach described in Sections 15.3 and 15.4, one can show that in this region, the dissolution process is exothermic. That is, to dissolve 2,6-DMN in supercritical CO_2 under isobaric and isothermal conditions, heat must be removed from the system.

We have indicated only a few examples of the equation-of-state approach for determining phase equilibrium behavior. As noted, the success of this method hinges on having an accurate mixture equations of state for all the phases present. Progress in this area will most certainly lead to a wider use of this technique, even to the computation of phase behavior in liquid-liquid systems.

Activity coefficient approach for condensed phases. The component fugacity is given as shown in Eq. (9-189),

$$\hat{f}_i^\sigma = \hat{f}_i^+ \gamma_i^+ x_i^\sigma \tag{15-91}$$

where $\hat{f}_i^+$ is the fugacity of i in a reference state defined by $T^+, P^+, x_1^+, \ldots, x_{n-1}^+$ and a state of aggregation which may be different from that of the mixture. Since $\hat{f}_i^\sigma$ and x_i^σ are properties of the actual mixture in phase (σ), it follows that γ_i^+, the activity coefficient, is a function of the actual mixture properties as well as the reference-state properties, that is,

$$\gamma_i^+ = g_i(T, P, x_1^\sigma, \ldots, x_{n-1}^\sigma, T^+, P^+, x_1^+, \ldots, x_{n-1}^+) \tag{15-92}$$

Although we have the option of defining reference states in a very general manner, in actual fact, only a few are ever used. In every case T^+ is defined to be equal to the system temperature T and in *almost* all cases, the composition in the reference state is chosen to be the pure component at T, P^+, and in the same condensed state (σ) as the mixture.

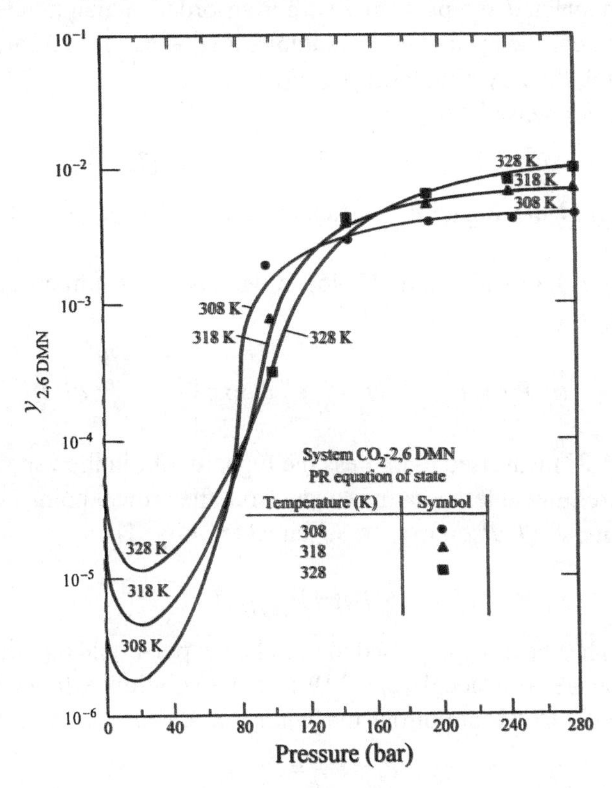

Figure 15.32 Solubility of 2,6-dimethylnaphthalene in supercritical carbon dioxide [adapted from R.T. Kurnik, S.J. Holla, and R.C. Reid (1981), *J. Chem. Eng. Data*, **26**, p 47].

Equation (15-91) then becomes

$$\hat{f}_i^\sigma = f_i^\sigma(T, P^+)\, \gamma_i(T, P^+, x_i^\sigma, \ldots, x_{n-1}^\sigma)\, x_i^\sigma \tag{15-93}$$

For systems where the temperature does not exceed the critical temperature of any component, P^+ is usually set equal to the system pressure P. It is then convenient to relate $f_i^\sigma(T, P)$ to the fugacity of pure i at its vapor pressure, $P_{vp,i}$

$$\ln\left(\frac{f^{\sigma}_{i,P}}{f^{\sigma}_{i,P_{vp,i}}}\right) = \int_{P_{vp,i}}^{P}\left(\frac{\partial \ln f_i}{\partial P}\right)_T dP = \frac{1}{RT}\int_{P_{vp,i}}^{P} V_i^{\sigma}\, dP \tag{15-94}$$

where the volume in the integral is the molar volume of pure i at T in the same condensed state as the mixture.

$$f^{\sigma}_{i,P} = f^{\sigma}_{i,P_{vp,i}} \exp\left(\int_{P_{vp,i}}^{P} \frac{V_i^{\sigma}}{RT}\, dP\right) \tag{15-95}$$

The exponential term in Eq. (15-95) is a *generalized Poynting correction factor*. It can usually be neglected for pressures within an order of magnitude of $P_{vp,i}$ and at temperatures not near the critical. (For example, if $(P - P_{vp,i}) = 10$ bar, $T = 400$ K, and $V_i^{\sigma} = 100$ cm^3/mol, the Poynting term is 1.031.)

Also, for pure condensed i at T, $P_{vp,i}$,

$$f_i^{\sigma}(T, P_{vp,i}) = f_i^{V}(T, P_{vp,i}) = P_{vp,i}\, \phi_i^{V}(T, P_{vp,i}) \tag{15-96}$$

where $\phi_i^{V}(T, P_{vp,i})$ is the fugacity coefficient of pure i vapor [see Eqs. (9-121) and (9-127)].

With Eqs. (15-91), (15-95), and (15-96), if we have a system consisting of a vapor and liquid phase,

$$\hat{\phi}_i^{V} P y_i = P_{vp,i}\, \phi_i^{V}(T, P_{vp})\gamma_i^{\sigma} x_i^{\sigma} \exp\left(\int_{P_{vp,i}}^{P} \frac{V_i^{\sigma}}{RT}\, dP\right) \tag{15-97}$$

Note that Eq. (15-87) was used to express the fugacity of i in the vapor phase.

At low pressures and at temperatures away from the critical point, both the Poynting correction term and $\phi_i^{V}(T, P_{vp,i})$ may be set equal to unity. Then

$$\hat{\phi}_i^{V} P y_i = P_{vp,i}\gamma_i^{\sigma} x_i^{\sigma} \tag{15-98}$$

Equation (15-98) has been widely used to correlate vapor-liquid equilibrium data. Note that if the gas phase were ideal ($\hat{\phi}_i = 1.0$) and the solution exhibited ideal behavior ($\gamma_i^{\sigma} = 1.0$), then Eq. (15-98) simplifies to *Raoult's law*:

$$P_i = P y_i = P_{vp,i} x_i^{\sigma} \tag{15-99}$$

Frequently, one encounters simulations at modest pressures where the gas phase can be treated as an ideal gas mixture ($\hat{\phi}_i^{V} = 1.0$). In these cases, a *modified form of Raoult's Law* where all non-idealities are assumed to reside in the liquid mixture phase. Hence,

$$P_i = P y_i = \gamma_i^{L} x_i^{L} P_{vp,i} \tag{15-100}$$

For *liquid-liquid equilibria*, Eqs. (15-86) and (15-87) become

$$\gamma_i^{\alpha} x_i^{\alpha} = \gamma_i^{\beta} x_i^{\beta} \tag{15-101}$$

as the reference state has been chosen to be the same for each condensed phase.

In Eq. (15-98) or (15-101), the condensed phase activity coefficients, besides being a function of composition, vary with the system temperature and pressure. The pressure effect is often neglected at low pressures resulting in a wide variety of constitutive property models of the form:

$$\gamma^{\sigma} = g(T, x_1^{\sigma}, \ldots, x_{n-1}^{\sigma})$$

As noted in Chapters 9 and 11, a solution model for the condensed phase is developed from theory and the examination of experimental data. The excess Gibbs energy of the mixture is obtained, and the component activity coefficients found from Eq. (9-185). An example was given in Section 9.8 to illustrate this method using the very simple relation $\underline{\Delta G}^{EX} = NCRTx_1x_2$ for a binary of 1 and 2.

Other activity coefficient models are discussed in Section 11.7 and summarized in Table 11.1. Note in Table 11.1 that the Margules and van Laar models assume that the constants in the ΔG^{EX} function are temperature independent so, for the activity coefficient expressions, the linear temperature dependence enters simply from Eq. (9-185). This proportionality is only very approximate. For the Wilson, NRTL, and UNIQUAC models, the parameters are assumed to be functions of temperature.

For most accurate correlations of vapor-liquid equilibrium, the Wilson, NRTL, or UNIQUAC models are now employed. The last is particularly useful since a significant amount of work has been expended to develop an activity coefficient estimation scheme (UNIFAC) to allow UNIQUAC to be used on systems where no or very few experimental data exist (see Section 13.4 for further discussion).

Example 15.6

At atmospheric pressure, the binary system of *n*-butanol-water exhibits a minimum-boiling azeotrope and partial miscibility in the liquid phase. *Txy* data at 1.022 bar are as follows:

Vapor and Liquid Mole Fractions of *n*-Butanol in Equilibrium with Water at 1.022 bar

T (°C)	y	x
100	0.0	0.0
95.8	0.150	0.008
95.4	0.161	0.009
92.8	0.237	0.019
92.8	0.240	0.020
92.7	0.246	0.098
92.7	0.246	0.099
92.7	0.246	0.247
93.0	0.250	0.454
93.0	0.247	0.450
96.3	0.334	0.697
94.0	0.276	0.583
96.6	0.340	0.709
100.8	0.444	0.819
106.4	0.598	0.903
106.8	0.612	0.908
110.9	0.747	0.950
117.5	1.000	1.000

Source: T.E. Smith and R.F. Bonner (1949), "Vapor-liquid equilibrium still for partially miscible liquids," *Ind. Eng. Chem.*, **41**, p 2867.

(a) Using only the azeotropic point, estimate the *Txy* equilibrium properties for this system. Use the van Laar activity coefficient correlation in Table 11.1.

(b) Estimate the compositions of the two liquid phases that occur in this immiscible system.

Solution

(a) From the table of data provided, the best estimate of the azeotropic point is $T \sim 92.7°\text{C} = 365.9$ K, $y_{n\text{-butanol}} = x_{n\text{-butanol}} = 0.246$.

At this low pressure, the vapor phase is essentially ideal ($\hat{\phi}_i = 1.0$), and Eq. (15-98) may be used to estimate the activity coefficients. At 365.9 K, $P_{vp,\, n\text{-butanol}}$ is about 0.378 bar and $P_{vp,\, \text{water}} = 0.774$ bar. Thus,

$$\gamma_{n\text{-butanol}} = \frac{(1.022)(0.246)}{(0.378)(0.246)} = 2.71 \quad \text{and} \quad \gamma_{\text{water}} = \frac{(1.022)(0.754)}{(0.774)(0.754)} = 1.32$$

The van Laar activity coefficient expressions in Table 11.1 may be algebraically rearranged to solve for the constants A and B:

$$\frac{A}{R} = T \ln \gamma_1 \left(1 + \frac{x_2 \ln \gamma_2}{x_1 \ln \gamma_1}\right)^2 \quad \text{and} \quad \frac{B}{R} = T \ln \gamma_2 \left(1 + \frac{x_1 \ln \gamma_1}{x_2 \ln \gamma_2}\right)^2$$

With n-butanol as component 1 and water as 2,

$$\frac{A}{R} = (365.9)(\ln 2.71)\left[1 + \frac{(0.754)(\ln 1.32)}{(0.246)(\ln 2.71)}\right]^2 = 1253 \text{ K}$$

Similarly, $\frac{B}{R} = 479 \text{ K}$

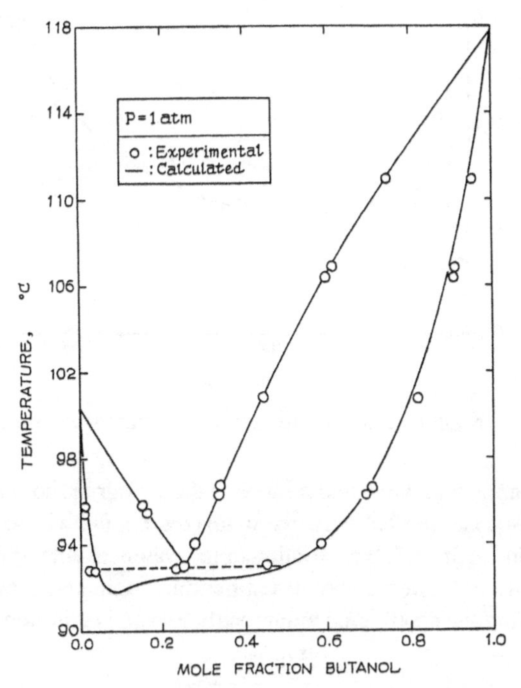

Figure 15.33 *Txy* diagram for n-butanol-water system.

With these parameters,

$$\ln \gamma_1 = \frac{1253/T}{[1 + 2.62(x_1/x_2)]^2} \quad \text{and} \quad \ln \gamma_2 = \frac{479/T}{[1 + 0.382(x_2/x_1)]^2}$$

If Eq. (15-98) is written for both n-butanol and water and added to eliminate vapor- phase compositions, with $\hat{\phi}_i = 1.0$,

$$P = x_1 P_{vp,1}(T)\, \gamma_1(T, x_1) + x_2 P_{vp,2}(T)\, \gamma_2(T, x_1)$$

P is set at 1.02 bar. For a given value of x_1 and the activity coefficient correlations, one iterates to find the system temperature. Vapor compositions are then found from Eq. (15-98) with vapor pressures for each component at the boiling temperature. Accurate vapor pressures are necessary. Experimental data are compared with computed values in Figures 15.33 and 15.34. The agreement is quite good except for the liquid phase at low n-butanol concentrations. The temperature correction to the activity-coefficient correlations is very approximate ($T \ln \gamma_i = \text{constant}$), but is probably satisfactory over the small temperature range involved.

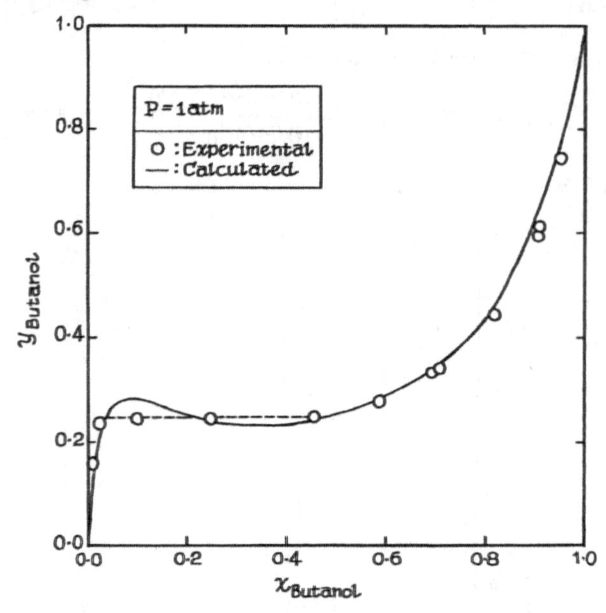

Figure 15.34 x-y diagram for n-butanol-water.

(b) The computed maxima and minima in the x-y curve should alert one to the fact that a liquid-phase split has occurred in this region. For a binary system to be stable, as shown in Figure 7.7, the variation of the chemical potential with composition must be positive (under constant-temperature and constant-pressure constraints). Neglecting the small temperature variations, this criterion reduces to

$$\frac{d \ln (\gamma_i x_i)}{dx_i} > 0 \tag{15-102}$$

or

$$\frac{d \ln a_1}{dx_1} > 0, \quad \frac{d \ln a_2}{dx_2} > 0 \tag{15-103}$$

where a_i is the activity.

Calculated values of a_1 and a_2 are shown in Figure 15.35. Since the stability criteria are violated on the spinodal curve, we know that the binodal curve lies outside the spinodal curve. Thus, by inspection of Figure 15.35, we see that the binodal (equilibrium) mole fractions of n-butanol are *less* than about 0.09 and greater than about 0.34. To determine the concentrations in the two immiscible liquid phases, the equilibrium criterion, Eq. (15-101), is used for both components. A trial-and-error solution yields n-butanol mole fractions of about 0.04 and 0.50 in the two phases. Experimental values are about 0.02 and 0.34. The calculated compositions were used to draw the dashed lines on Figures 15.33 and 15.34.

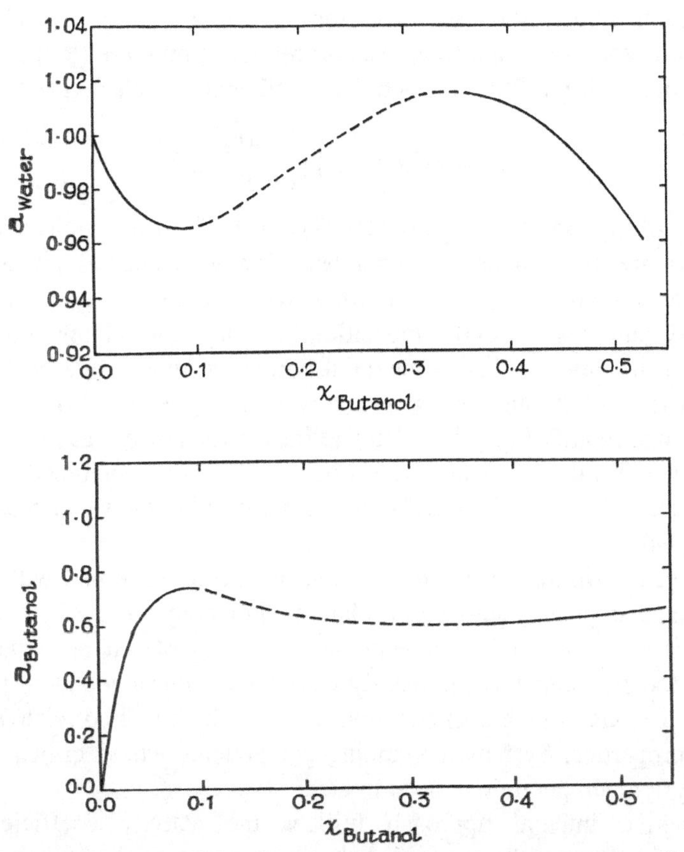

Figure 15.35 Activities in the *n*-butanol-water system at one atmosphere.

In theory, if the activity-coefficient correlations contained an accurate temperature function, this procedure could be used to determine both liquid-phase compositions as a function of temperature — even to predict consolute temperatures. In general, however, the van Laar form with $T \ln \gamma =$ constant is not satisfactory for this purpose.

When the system is at high pressure (but still with the temperature below the critical temperature of any component), Eq. (15-97) is still applicable. The Poynting correction is not necessarily negligible and ϕ_i^V and $\hat{\phi}_i^V$ cannot ordinarily be set equal to unity. The more difficult problem in applying Eq. (15-97) lies in the activity-coefficient term. We saw in Table 11.1 that an approximate temperature correction was included, but no pressure correction. At low pressures, even for isothermal cases where pressure can vary, the activity coefficient is a weak function of pressure and such corrections are neglected. This may not be true at high pressures.

One way to treat this problem is to define a constant-pressure activity coefficient to be used in Eq. (15-97). Presumably, any of the activity coefficient correlations in Table 11.1 could then be used.

For illustration, let us still retain the reference state for the fugacity to be the system pressure (which can vary). Choose a *constant* reference pressure (RP) for the activity coefficient. Since, in Eq. (15-97) the activity coefficient is evaluated at P, then

$$\gamma_i(P) = \gamma_i(RP) \exp\left(\int_{RP}^{P} \frac{\overline{\Delta V}_i^{\alpha}}{RT}\, dP\right) \tag{15-104}$$

Substitution of Eq. (15-104) into Eq. (15-97) now leads to an activity coefficient that is only temperature and composition dependent. Of course, one has now to evaluate an exponential term involving partial volumes which themselves are a function of temperature, system pressure, and composition. Although correlations could (and have) been developed to estimate this exponential correction, none have been particularly successful.[4] It is worth noting, however, that when this pressure correction is made to γ_i, it is common to modify Eq. (15-94) and define the reference fugacity at the value of RP chosen for γ_i. While the exponential terms may then be combined, one loses the desirable feature of Eq. (15-95) that $f^{\sigma}_{i,P_{vp}}$ is easily related to the vapor pressure of pure i as in Eq. (15-96).

In summary, the discussion in this section has been concerned with the integral approach to express phase equilibrium relations. For nonpolar systems, especially at elevated pressures, more and more emphasis is being placed on equation-of-state approaches. Modern computers can readily solve the nonlinear equations [Eq. (15-88)] provided that one has available an equation of state which will correlate *mixture PVT* composition properties. Systems containing components whose critical temperature exceeds the system temperature may be treated.

The alternative integral approach utilizes the activity coefficient and the reference-state fugacity for the condensed phases(s). At relatively low pressures and for systems with no supercritical components, this approach has been widely and successfully used even for very polar mixtures. Reference states are normally defined as the pure component in the same condensed state as the mixture. Many activity coefficient-composition correlations are available (see Section 11.7). Frequently, the temperature correction in these correlations is only approximate, and no pressure correction is present. These limitations are normally of little importance except in cases where the pressure is high and/or the temperature is near the critical point for the mixture. Under the latter conditions, you can develop the thermodynamics to yield, for example, pressure-independent activity coefficients, but you must then include other terms involving volumetric properties of the condensed phase. Rarely are the necessary data to determine such correction factors available.

---

4 I. Wichterle [(1978), *Fluid Phase Eq.*, **2**, p 143] discusses many of the proposed values of RP, running from 0 to 1000 bar.

15.6 Equilibrium in Systems with Supercritical Components

In the activity-coefficient approach of the previous section, it was assumed that the system temperature did not exceed the critical temperature of any component. This allowed us to define reference fugacity states as *pure* components in the condensed state. We could have selected reference fugacity states other than for the pure material, but there was no advantage to be gained by such a selection.

When we have a condensed phase with one or more components that cannot exist in the same condensed phase when pure and at the system temperature, this usually implies that the system temperature exceeds the critical temperature of the components. Carbon dioxide in a liquid hydrocarbon mixture at a temperature higher than 304 K (the critical temperature of CO_2) would be an example of a *supercritical* component. In Section 9.8 we considered such cases briefly for binary systems. We expand this treatment and extend it to multicomponent systems in this section.

Reviewing briefly the development in Section 9.8, refer to Figure 9.8. Component fugacities are plotted as a function of composition for a binary liquid system of solute (A) and solvent (B) at some temperature. Component fugacities are shown only over a limited concentration range since it was assumed that the solute could not exist as a liquid when in the pure state.

The reference fugacity for the solvent was defined (as in Section 15.5) as the pure liquid at the system temperature and pressure. For the solute, a hypothetical reference fugacity state was defined by drawing a line tangent to the f_A curve as $x_A \to 0$. The intersection of this tangent with the $x_A = 1.0$ ordinate yielded the reference fugacity state for the solute. This value, f_A^{**}, was called the infinite-dilution reference (or standard) state. When we introduce a reference fugacity state such as f_A^{**} (with the usual reference state for the solvent), we say that the system is *unsymmetrically normalized*. In Section 15.5 only symmetrically normalized systems were discussed. Activity coefficients for the unsymmetrical case are defined in Eqs. (9-197) and (9-198).

We wish to stress that f_A^{**} has no physical significance, whereas f_B is related to a well-defined liquid state (i.e., pure B at T and P). f_A^{**} must be obtained from experimental data or by an estimation derived from some solution theory. An alternative type of reference state for the solute might also be proposed. Suppose that we arbitrarily extrapolate the $\hat{f}_A$ versus x_A curve to $x_A = 1.0$. This would locate a reference state, f_A, which is also fictitious. Again, f_A is related to no physical state of the solute and must be determined from experimental data. But with its introduction, we now have a symmetrically normalized system with γ_A defined in the same way as γ_B [see Eq. (9-197)], that is,

$$\gamma_A = \frac{\hat{f}_A}{x_A f_A} \tag{15-105}$$

Since $\hat{f}_A$ and x_A do not depend on the choice of the reference state, we can divide Eq. (15-105) by Eq. (9-198) to obtain

$$\frac{\gamma_A}{\gamma_A^{**}} = \frac{f_A^{**}}{f_A} = \gamma_A^{\infty} = \text{constant} \tag{15-106}$$

Since γ_A^{**} approaches unity as $x_A \to 0$, we can define the constant γ_A^{∞} as

$$\gamma_A^{\infty} = \lim_{x_A \to 0} \gamma_A \tag{15-107}$$

The infinite-dilution reference fugacity f_A^{**} can then be simply related to f_A and γ_A^{∞} as

$$f_A^{**} = f_A \gamma_A^{\infty} \tag{15-108}$$

or

$$\ln f_A^{**} = \ln f_A + \ln \gamma_A^{\infty} \tag{15-109}$$

and, with Eq. (15-106),

$$\ln \gamma_A - \ln \gamma_A^{**} = \ln \gamma_A^{\infty} \tag{15-110}$$

Example 15.7

Assume that we are interested in a liquid binary mixture of A and B with A as the supercritical component. Using the van Laar correlation in Table 11.1, show how one may relate activity coefficients of A to composition using both symmetrical and unsymmetrical normalization. How is f_A related to f_A^{**}?

Solution

In Table 11.1 we let component 1 be A and component 2 be B. Also, to avoid confusion, let the van Laar constants be C and D rather than A and B. Then, since Table 11.1 is developed for the symmetrically normalized case,

$$RT \ln \gamma_A = C\left(1 + \frac{Cx_A}{Dx_B}\right)^{-2} \quad \text{and} \quad RT \ln \gamma_B = D\left(1 + \frac{Dx_B}{Cx_A}\right)^{-2}$$

Note that even though component A is supercritical, the definition of γ_A is given by Eq. (15-105) (i.e., the reference fugacity f_A is implied). γ_B is given by Eq. (9-197). Then, with Eq. (15-107),

$$\ln \gamma_A^{\infty} = \frac{C}{RT}$$

so with Eqs. (15-109) and (15-107),

$$\ln f_A^{**} = \ln f_A + \frac{C}{RT} \quad \text{and} \quad \ln \gamma_A^{**} = \ln \gamma_A - \frac{C}{RT} \tag{15-111}$$

and further simplification yields,

$$\ln \gamma_A^{**} = \frac{C}{RT}\left[\left(1 + \frac{Cx_A}{Dx_B}\right)^{-2} - 1\right] \tag{15-112}$$

Clearly, $\gamma_A^{**} \to 1$ as $x_A \to 0$.

The object of Example 15.7 and the preceding development was to illustrate that, for binary liquid systems involving a supercritical component, one could employ either the symmetrical or unsymmetrical normalized activity coefficient. The two reference-state fugacities for the supercritical component are related by Eq. (15-109) and the resulting activity coefficients by Eq. (15-110). In both instances, the reference fugacities (f_A or f_A^{**}) are fictitious pure-component properties and numerical values must be found from experimental data. There is no advantage to be gained from working with one or the other for binary systems. For multicomponent cases, however, the treatment using f_A may have conceptual advantages.[5]

We consider a ternary liquid mixture containing A, B, and C with A the supercritical component. The reference-state fugacities are f_A, f_B, and f_C. The later two refer to the pure liquid states of B and C at T and P. The first is a fictitious reference state; for a binary system it can be defined in an unambiguous manner, but for multicomponent systems, it will be shown to vary with the *relative* composition of B and C. (Use of f_A^{**} would not improve the situation.)

Following the treatment of Van Ness and Abbott (see footnote 5), we make use of the *mixture* fugacity defined earlier in Eqs. (9-153) and (9-154), where we noted that the appropriate for the mixture fugacity was

$$\frac{\partial}{\partial N_i}(N \ln f_m)_{T,\,P,\,N_j[i]} = \ln \frac{\hat{f}_i}{x_i} \tag{15-113}$$

For reasons that will become apparent later, we define the reference fugacity for the supercritical component, f_A, in terms of f_{AB} and f_{AC}, where these refer to the reference fugacities of A in A, B and A, C binaries with symmetrical normalization at the same T, P.

$$\ln f_A = x'_B \ln f_{AB} + x'_C \ln f_{AC} \tag{15-114}$$

with

$$x'_B \equiv \frac{x_B}{x_B + x_C}, \qquad x'_C \equiv \frac{x_C}{x_B + x_C} \tag{15-115}$$

---

5 H.C. Van Ness and M.M. Abbott [(1979), *AIChE J.*, **25**, p. 645] have presented a detailed analysis of both supercritical reference fugacities. Our treatment follows their development quite closely.

Therefore, the value for f_A in the ternary is related to binary properties and the relative ratios of components B and C.

We now define a function $(\Delta \ln f_m)$ as

$$\Delta \ln f_m = \ln f_m - x_A \ln f_A - x_B \ln f_B - x_C \ln f_C \tag{15-116}$$

The function $(\Delta \ln f_m)$ can be shown to be equivalent to $\Delta G^{EX}/RT$ for a mixture with the reference states as defined for f_A, f_B, and f_C. We multiply Eq. (15-116) by the total moles in the system, N, and then perform three partial differentiations with respect to N_A, N_B, and N_C. Using Eq. (15-113), one obtains:

$$\frac{\partial}{\partial N_A}(N \Delta \ln f_m)_{N_B, N_C} = \ln \frac{\hat{f}_A}{x_A} - \ln f_A \tag{15-117}$$

$$\frac{\partial}{\partial N_B}(N \Delta \ln f_m)_{N_A, N_C} = \ln \frac{\hat{f}_B}{x_B} - \ln f_B - N_A \frac{\partial}{\partial N_B}(\ln f_A) \tag{15-118}$$

$$\frac{\partial}{\partial N_C}(N \Delta \ln f_m)_{N_A, N_B} = \ln \frac{\hat{f}_C}{x_C} - \ln f_C - N_A \frac{\partial}{\partial N_C}(\ln f_A) \tag{15-119}$$

In the differentiation in Eq. (15-117), $\ln f_A$ is not a function of N_A at constant N_B and N_C, by virtue of Eq. (15-114). With the definition of γ_i as

$$\gamma_i = \frac{\hat{f}_i}{x_i f_i} \tag{15-120}$$

We also define

$$N_A \frac{\partial}{\partial N_B}(\ln f_A) = \ln \gamma_B^R \tag{15-121}$$

$$N_A \frac{\partial}{\partial N_C}(\ln f_A) = \ln \gamma_C^R \tag{15-122}$$

and substitute Eqs. (15-120) through (15-122) into Eqs. (15-117) through (15-119):

$$\frac{\partial}{\partial N_A}(N \Delta \ln f_m)_{N_B, N_C} = \ln \gamma_A \tag{15-123}$$

$$\frac{\partial}{\partial N_B}(N \Delta \ln f_m)_{N_A, N_C} = \ln \frac{\gamma_B}{\gamma_B^R} \tag{15-124}$$

$$\frac{\partial}{\partial N_C}(N \Delta \ln f_m)_{N_A, N_B} = \ln \frac{\gamma_C}{\gamma_C^R} \tag{15-125}$$

Then, since $N \Delta \ln f_m$ has all the properties of a $\Delta \underline{B}$ function as discussed in Chapter 9, we multiply each term in Eqs. (15-123) through (15-125) by the appropriate mole fraction and sum. The result is

$$\Delta \ln f_m = x_A \ln \gamma_A + x_B \ln \frac{\gamma_B}{\gamma_B^R} + x_C \ln \frac{\gamma_C}{\gamma_C^R} \tag{15-126}$$

The terms γ_B^R and γ_C^R may be evaluated from their definitions [Eqs. (15-121) and (15-122)] with Eq. (15-114), the results are

$$\ln \gamma_B^R = \frac{x'_C x_A}{x_B + x_C} \ln \frac{f_{AB}}{f_{AC}} \tag{15-127}$$

$$\ln \gamma_{CB}^R = \frac{x'_B x_A}{x_B + x_C} \ln \frac{f_{AC}}{f_{AB}} \tag{15-128}$$

where it can be shown that

$$x_B \ln \gamma_B^R + x_C \ln \gamma_C^R = 0 \tag{15-129}$$

So, Eq. (15-126) becomes

$$\Delta \ln f_m = x_A \ln \gamma_A + x_B \ln \gamma_B + x_C \ln \gamma_C \tag{15-130}$$

with

$$\Delta \ln f_m = \frac{\Delta \underline{G}^{EX}}{RT} \tag{15-131}$$

The partial properties of $(N \Delta \ln f_m)$ or $\Delta \underline{G}^{EX}/RT$ are given by Eqs. (15-123) through (15-125). As a simple example, assume that for the ternary mixture

$$\frac{\Delta \underline{G}^{EX}}{RT} = \frac{\Delta \underline{G}_{AB}^{EX}}{RT} + \frac{\Delta \underline{G}_{AC}^{EX}}{RT} + \frac{\Delta \underline{G}_{BC}^{EX}}{RT} \tag{15-132}$$

with

$$\frac{\Delta \underline{G}_{ij}^{EX}}{RT} = C_{ij} x_i x_j \tag{15-133}$$

Then, for the ternary,

$$\frac{\Delta \underline{G}^{EX}}{RT} = \Delta \ln f_m = C_{AB} x_A x_B + C_{AC} x_A x_C + C_{BC} x_B x_C \tag{15-134}$$

and

$$\frac{\partial}{\partial N_A}\left(\frac{\Delta G^{EX}}{RT}\right)_{N_B, N_C} = C_{AB} x_B + C_{AC} x_C - \frac{\Delta \underline{G}^{EX}}{RT} \tag{15-135}$$

$$\frac{\partial}{\partial N_B}\left(\frac{\Delta \underline{G}^{EX}}{RT}\right)_{N_A,\, N_C} = C_{BC}x_C + C_{AB}\,x_A - \frac{\Delta G^{EX}}{RT} \tag{15-136}$$

$$\frac{\partial}{\partial N_C}\left(\frac{\Delta \underline{G}^{EX}}{RT}\right)_{N_B,\, N_C} = C_{BC}x_B + C_{AC}\,x_A - \frac{\Delta G^{EX}}{RT} \tag{15-137}$$

Finally, with Eqs. (15-123) through (15-125) with Eq. (15-131),

$$\ln \gamma_A = C_{AB}\,x_B + C_{AC}x_C - \frac{\Delta G^{EX}}{RT} \tag{15-138}$$

$$\ln \gamma_B = C_{BC}\,x_C + C_{AB}\,x_A - \frac{\Delta G^{EX}}{RT} + \frac{x_C' x_A}{x_B + x_C}\ln\frac{f_{AB}}{f_{AC}} \tag{15-139}$$

$$\ln \gamma_C = C_{BC}\,x_B + C_{AC}\,x_A - \frac{\Delta G^{EX}}{RT} + \frac{x_B' x_A}{x_B + x_C}\ln\frac{f_{AC}}{f_{AB}} \tag{15-140}$$

Equations (15-138) through (15-140) allow one to calculate the ternary activity coefficients once the binary parameters C_{AB}, C_{AC}, and C_{BC}, and reference fugacities f_{AB} and f_{AC} have been determined. The component fugacities of each component in the ternary are then expressed as

$$\hat{f}_i = x_i \gamma_i f_i \tag{15-141}$$

The extension of the ternary treatment to multicomponent systems is available. No comment was made concerning the pressure correction of f_A, if such should be necessary. No satisfactory methods have yet been developed, and the case is similar to the subcritical cases discussed in Section 15.5.

15.7 Phase Stability Applications

In Section 7.2 the stability criteria for pure, binary, and ternary systems were introduced and several simple examples were described. In this section, we elaborate on two key applications that are important to chemical engineering. The first relates to the onset of phase splitting as miscibility limits are reached and nucleation of a second phase initiates while the second deals with phenomena occurring near critical points.

Miscibility limits and phase nucleation. Here we focus on pure component and binary systems to ease the nomenclature and algebraic burden. The extension to ternary and higher-order multicomponent systems is straightforward. For pure systems, any of the criteria given in Table 7.1 can be used. When pressure-explicit EOS models are employed, derivative forms for the Helmholtz free energy are most convenient as seen in Example 7.2 for a van der Waals fluid and Example 15.8 for a fluid modeled with the Peng-Robinson EOS. For binary systems, in principle, any of the criteria cited in Table 7.2 can be used, but in practice, the forms involving the Gibbs free energy are most frequently used with a ΔG^{EX} model--particularly for cases involving liquid-liquid phase splitting. If an accurate $P\underline{V}TN$ EOS is available, then Helmholtz free energy derivatives employing the $\pounds_i$ determinant are used (see Section 7.2.2).

As we saw in Section 15.2, liquid-liquid miscibility gaps are frequently encountered. As illustrated in Figure 15.36, T_c is commonly referred to as an upper critical solution temperature (UCST). If a minimum rather than a maximum critical temperature was observed, then it is called a lower critical solution temperature (LCST). Occasionally, both a UCST and an LCST are observed (see Figure 15.16). A key objective of thermodynamics analysis in this area is to be able to predict the conditions (T, P, x_i) where phase separation exists. With a robust ΔG^{EX} model for the binary such a calculation is possible; one simply finds the points where ΔG_{mix} as a function of x_i or N_i has multiple points of tangency as shown in Figure 15.36 for various temperatures and pressures. Limits of metastable behavior are defined by a change in sign of the derivative. For phase α,

$$\Delta G^{\alpha}_{mix} = \sum_{i=1}^{2} x_i\,[\overline{G}^{\alpha}_i - G^{\alpha}_i] = \sum_{i=1}^{2} x_i\,[\mu^{\alpha}_i - G^{\alpha}_i] \qquad (15\text{-}142)$$

or

$$\Delta G^{\alpha}_{mix} = x_1\,\overline{\Delta G}^{\alpha}_1 + x_2\,\overline{\Delta G}^{\alpha}_2 \qquad (15\text{-}143)$$

At a point of double tangency shown on Figure 15.36, $\overline{\Delta G}^{\alpha}_i$ is the same for both phases L_1 and L_2. Consequently, this is the same as equating the chemical potentials given in Eq. (15-142) because,

$$\overline{\Delta G}^{L_1}_i = \overline{G}^{L_1}_i - G_i = \overline{G}^{L_2}_i - G_i = \overline{\Delta G}^{L_2}_i = \overline{\Delta G}_i \qquad (15\text{-}144)$$

where the pure component reference state G_i is at the same T and P and state of aggregation as pure liquid i. Also for a two-phase system at some average, mixed composition x_1 where $x^{L_1}_1 < x_1 < x^{L_2}_1$ and:

$$\Delta G_{mix} = \Delta G^{L_2}_{mix}\,\theta + \Delta G^{L_1}_{mix}\,(1-\theta) \qquad (15\text{-}145)$$

where

$$\theta = \text{moles of } L_2 \text{ phase/(total moles of } L_1 + L_2 \text{ phases)}$$

$$\Delta G_{mix} = [x^{L_2}_1\,\theta + x^{L_1}_1\,(1-\theta)]\,\overline{\Delta G}_1 + [x^{L_2}_2\,\theta + x^{L_1}_2\,(1-\theta)]\,\overline{\Delta G}_2 \qquad (15\text{-}146)$$

For a stable system with n=2, $m = n + 2 = 4$, and

$$y^{(m-2)}_{(m-1)(m-1)} = y^{(2)}_{33} \geq 0 \quad (= 0 \text{ at limit of stability}) \qquad (15\text{-}147)$$

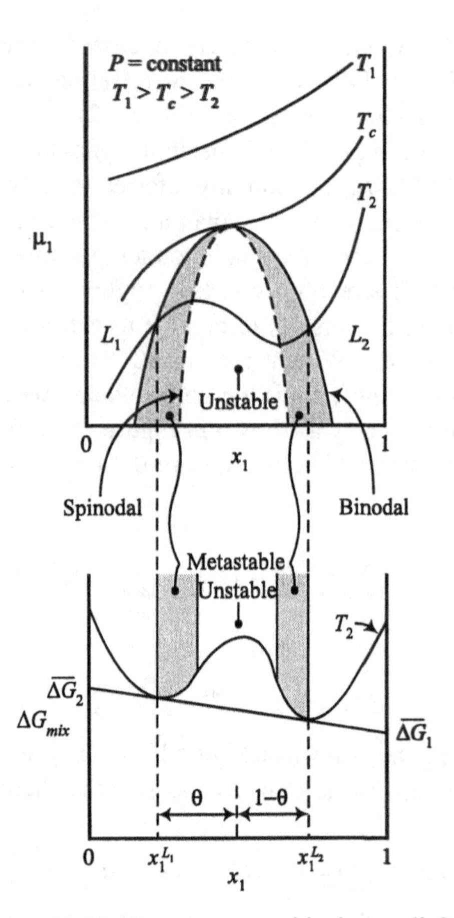

Figure 15.36 Binary system with phase splitting.

or equivalently with $y^{(0)} = \underline{U} = U(\underline{S}, \underline{V}, N_1, N_2)$ and normal variable ordering:

$$\underline{G}_{N_1N_1} \geq 0 \ ;$$

$$\left(\frac{\partial^2 \underline{G}}{\partial N_1^2}\right)_{T,P,N_2} \geq 0 \ ; \ \left(\frac{\partial \mu_1}{\partial N_1}\right)_{T,P,N_2} \geq 0 \ \ or \ \left(\frac{\partial \mu_1}{\partial x_1}\right)_{T,P} \geq 0 \tag{15-148}$$

An alternative expression can be developed using Eq. (15-143):

$$\left(\frac{\partial \Delta G_{mix}}{\partial x_1}\right)_{T,P} = (\overline{G}_1 - G_1) - (\overline{G}_2 - G_2) + x_1\left(\frac{\partial \overline{G}_1}{\partial x_1}\right)_{T,P} + x_2\left(\frac{\partial \overline{G}_2}{\partial x_1}\right)_{T,P} \tag{15-149}$$

where the last two terms on the RHS sum to zero from the Gibbs-Duhem equation at constant temperature and pressure. Now, by rewriting Eq. (15-149) as:

$$\left(\frac{\partial \Delta G_{mix}}{\partial x_1}\right)_{T,P} = \mu_1 - G_1 - (\mu_2 - G_2) \tag{15-150}$$

we can express the second derivative as:

$$\left(\frac{\partial^2 \Delta G_{mix}}{\partial x_1^2}\right)_{T,P} = \left(\frac{\partial \mu_1}{\partial x_1}\right)_{T,P} - \left(\frac{\partial \mu_2}{\partial x_1}\right)_{T,P} \tag{15-151}$$

where the pure component derivatives vanish. By using the Gibbs-Duhem equation, Eq. (15-151) can be simplified to:

$$\left(\frac{\partial^2 \Delta G_{mix}}{\partial x_1^2}\right)_{T,P} = \left(\frac{\partial \mu_1}{\partial x_1}\right)_{T,P} \left(1 + \frac{x_1}{x_2}\right) \tag{15-152}$$

At a stability limit corresponding to the spinodal,

$$\left(\frac{\partial \mu_1}{\partial x_1}\right)_{T,P} = 0 \quad and \quad \left(\frac{\partial^2 \Delta G_{mix}}{\partial x_1^2}\right)_{T,P} = 0 \tag{15-153}$$

Geometrically, the spinodal is defined as a point of inflection on the ΔG_{mix} versus x_1 plot (Figure 15.36) and the binodal as a point of double tangency. This approach can also be applied to phase separation in solid-solid systems involving limited solubility in solid solutions as illustrated in Figures 15.21 and 15.37.

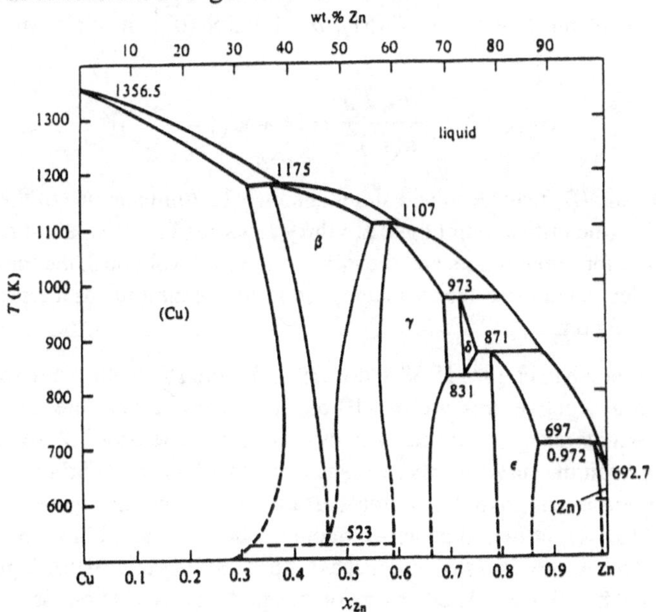

Figure 15.37 Phase diagram of the copper-zinc system [from Lupis (1983) based on data from Hultgren *et al.* (1973), used with permission].

Example 15.8

Determine the limits of stability for both liquid *and* vapor n-hexane over a wide range of pressures.

Solution

We can use any of the criteria given in Table 7.1. If, however, we select an equation of state such as the Peng-Robinson [Eq. (8-46)] to represent both the vapor and liquid properties of n-hexane, the first criterion employing A_{VV} is the most convenient.

The Peng-Robinson equation of state is

$$P = \frac{RT}{V-b} - \frac{a(\omega, T_r)}{V(V+b)+b(V-b)} \tag{A}$$

At the limit of stability, $A_{VV} = 0$ or $(\partial P/\partial \underline{V})_{T,N} = 0$. With N constant,

$$\left(\frac{\partial P}{\partial V}\right)_T = 0 = \frac{-RT}{(V-b)^2} + \frac{2a(\omega, T_r)\,(V+b)}{[V(V+b)+b(V-b)]^2} \tag{B}$$

The parameters for the Peng-Robinson equation are defined in Eqs. (8-47) through (8-51). For n-hexane: $T_c = 507.4$ K, $P_c = 29.7$ bar, and $\omega = 0.296$. Thus, with Eq. (8-48), $a(T_c) = 2.739$ J m^3/mol^2; with Eq. (8-51), $b = 1.105 \times 10^{-4}$ m^3/mol; with Eq. (8-50), $\kappa = 0.807$; and

$$\alpha(\omega, T_r) = \frac{a(\omega, T_r)}{a(T_c)} = [1 + \kappa\,(1 - T_r^{1/2})]^2$$

Equations (A) and (B) may be solved simultaneously to eliminate one of the variable set (P, V, T). Below the critical point there are three roots for V. The smallest represents the limit of stability for liquid phase, and the largest yields the volume at the limit of stability for the subcooled vapor. The intermediate value is not meaningful, as it represents a state in the unstable region.

The results are shown in Figures 15.38 and 15.39. In Figure 15.38, the liquid spinodal curve is drawn from a negative pressure of –10 bar to a pressure near the critical pressure. (Negative pressures simply indicate that the liquid is in tension.) Also shown is the saturated liquid volume curve. In this case, since the liquid is in equilibrium with the vapor, negative pressure is not possible. A single isotherm is drawn in Figure 15.38. For this temperature (467 K), at pressures above about 16 bar, the liquid hexane is subcooled. Below about 16 bar, the liquid is superheated. The limit of stability for liquid hexane at 467 K is about 0.8 to 0.9 bar. Another way of interpreting this last point is to visualize the heating of liquid hexane under a pressure of 1 bar. The normal boiling point is reached at about 342 K, but if nucleation from surfaces can be suppressed, the results shown above indicate that it is possible to heat this liquid hexane to 467 K before spontaneous nucleation occurs. Note that the 467 K isotherm has a slope of zero when it intersects the spinodal curve. This is necessary to match the $\underline{A}_{VV} = 0$ criterion at the limit of stability.

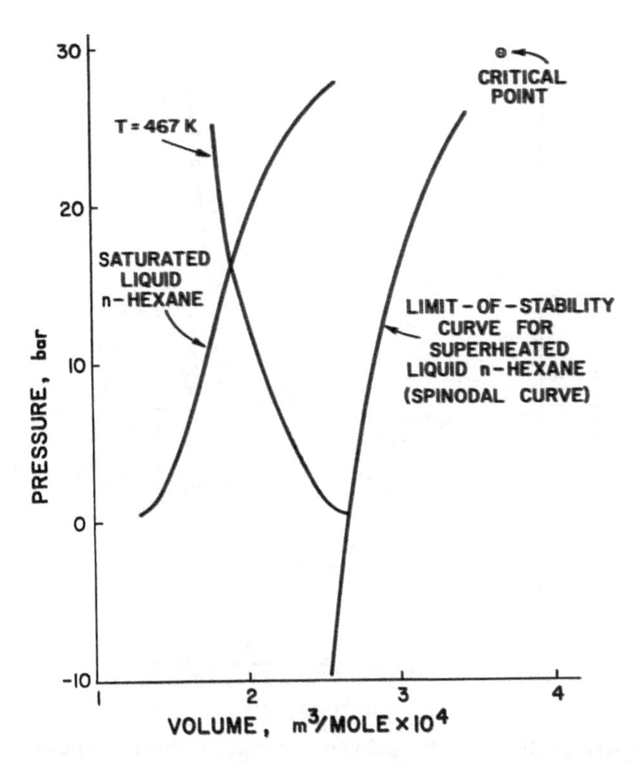

Figure 15.38 Estimated limit of stability for liquid *n*-hexane.

In Figure 15.39, the spinodal curve is compared against the saturated vapor pressure curve. Both intersect at the critical point. The spinodal curve is nearly linear and extends into the negative-pressure region. Also shown in Figure 15.39 are experimental data for the limit of stability of *n*-hexane as measured by Skripov and Ermakov [(1964), *Zh. Fiz. Khim.*, **38**, 396]. The agreement is quite good considering that the Peng-Robinson equation of state does not predict liquid-phase properties with high accuracy.

For the vapor spinodal, refer to Figure 15.40. Here the limit of stability for subcooled *n*-hexane is shown to the left of the curve designating saturated vapor. Between these two curves we have a metastable region with subcooled vapor. Also plotted in this graph is the 467 K isotherm. At pressures below about 16 bar, states on this isotherm are in the superheated *vapor* region. Above 16 bar, we have subcooled vapor. The limit of stability for this isotherm is about 20 bar.

Matching Figures 15.38 and 15.40, one would find the spinodal curves to be continuous and to pass through the critical point.

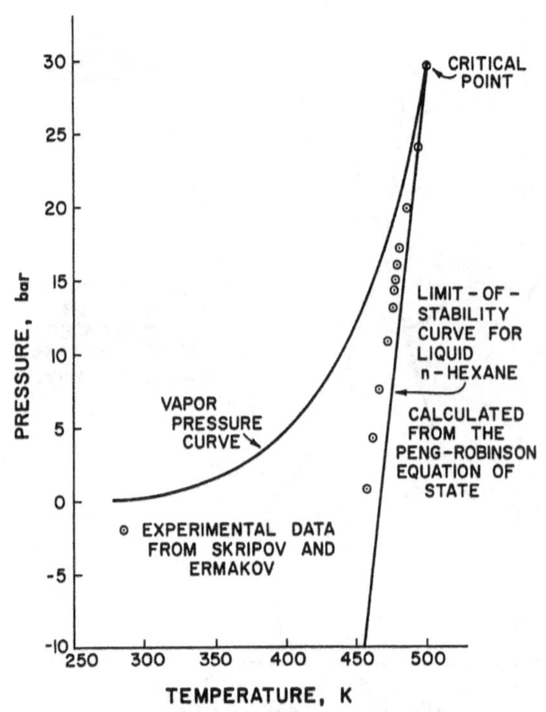

Figure 15.39 Spinodal and vapor pressure curve for n-hexane.

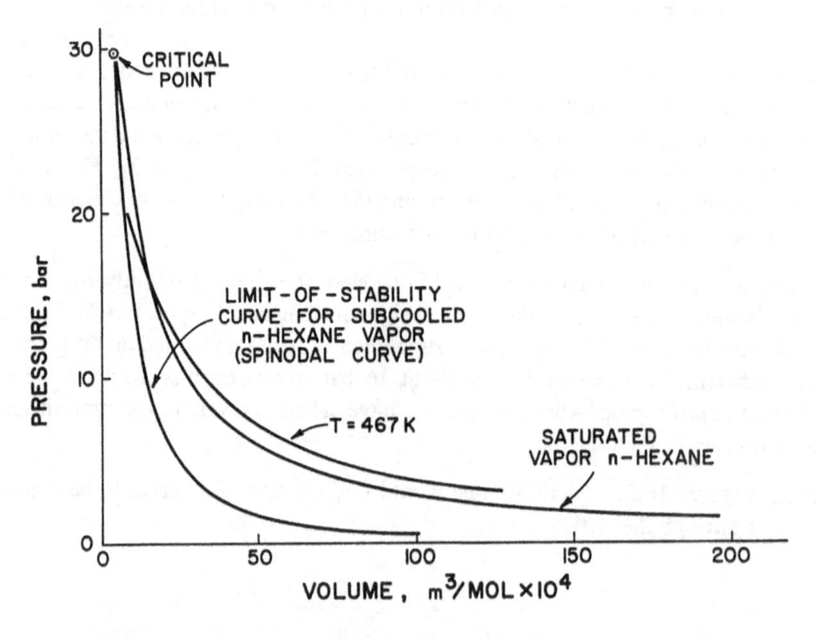

Figure 15.40 Estimated limit-of-stability for vapor n-hexane.

Critical phenomena and spinodal decomposition. Behavior near critical points can be exploited in chemicals and materials processing. For example, the high levels of solubility that frequently occur in supercritical fluids for otherwise insoluble components, see Figures 15.27 and 15.32 for example, have been effectively utilized for separation and purification processes in systems involving polymers, metals, and ceramics.

By entering an intrinsically unstable region rapidly by abrupt changes in pressure (e.g., by expansion) or in temperature (e.g., by quenching) immediate nucleation of a second phase occurs. Unstable regions inside of spinodal curves can be reached most easily near critical points, as depicted in Figure 15.36 for a binary system exhibiting either a UCST or an LCST. As we enter such regions a unique mechanism of phase separation and growth occurs that is characterized by the evolution of small-scale compositional fluctuations that grow in time. This phenomena is known as *spinodal decomposition* and is in direct contrast to the normal nucleation and growth mechanism. Nucleation and growth is more likely to occur in the metastable region between the binodal and spinodal curves. These two mechanisms are conceptualized in Figure 15.41 for a binary system [see also Cahn (1968), Kingery *et al.* (1976), Kwei and Wang (1978), and Hashimoto (1986) for further discussion].

In these applications, we would like to be able to define the spinodal region near any critical points as this would allow us to estimate the depth of penetration into the intrinsically unstable region. As before, to do this we couple a suitable constitutive property model (ΔG^{EX} or a $P\underline{V}TN$ EOS) that captures the non-idealities with the phase stability criteria. Figure 15.42 shows the results of such a calculation where a reduced EOS model developed by Sanchez and Lacombe was used for the polyethylene-pentane system by Kiran (1994) to estimate both binodal and spinodal curves. In addition, Kiran is currently pursuing dynamic light scattering methods that could characterize the time scales and concentration fluctuations to verify the profiles shown in Figure 15.42.

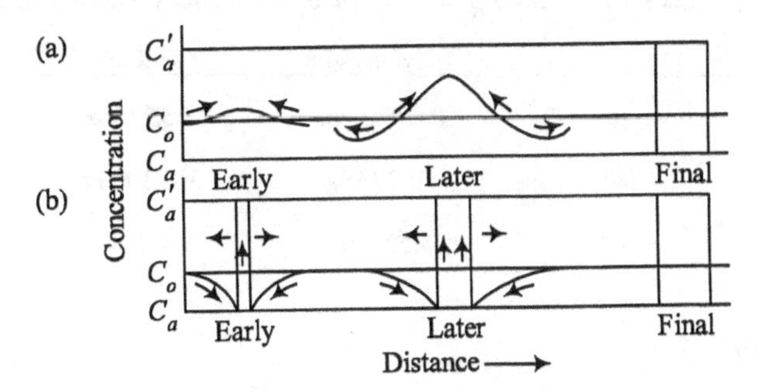

Figure 15.41 Schematic evaluation of concentration profiles for (a) spinodal decomposition and (b) nucleation and growth [adapted from J.W. Cahn (1968)].

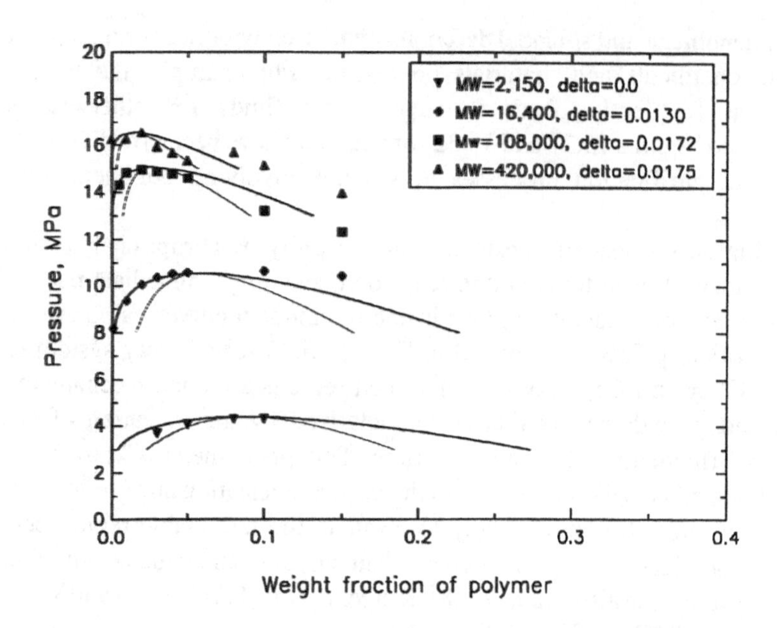

Figure 15.42 Equilibrium and spinodal envelope predictions for the polyethylene-pentane system from Kiran (1994). Experimental data shown as points for four different molecular weight fractions for polyethylene. Lines represent predictions of the Sanchez-Lacombe model, delta (δ_{ij}) is the binary interaction parameter used with standard mixing rules [reprinted by permission of Kluwer Academic Publishers].

The reduced form of the Sanchez-Lacombe EOS model is:

$$\rho_R^2 + P_R + T_R\,[\ln(1 - \rho_{R)} + (1 - 1/r)\rho_R] = 0$$

where $\rho_R \equiv \rho/\rho^*$, $P_R \equiv P/P^*$, and $T_R \equiv T/T^*$. For the binary shown in Figure 15.42

	polyethylene	pentane
ρ^*	0.895 g/cm^3	0.755 g/cm^3
P^*	359 MPa	310.1 MPa
T^*	521 K	441 K

15.8 Summary

This chapter illustrates one of the most important applications of classical thermodynamics--the characterization, correlation, and prediction of the phase behavior of pure components and mixtures. The Gibbs phase rule was shown to be a useful means of determining the intrinsic degrees of freedom of any multiphase, multicomponent system at equilibrium. Likewise, phase diagrams in $PTVx_i$ coordinates illustrated many attributes of phase behavior including phase partitioning, stable coexistence regions, metastability regions, and critical points.

The mathematical criteria for phase equilibrium, that is constant T, P, and μ_i (or $\hat{f}_i$) for all phases, were linked to constitutive EOS or γ_i property models of non-ideal behavior using both a differential and an integral approach. Key results include characterization of vapor pressure-temperature relationships for monovariant systems, isobaric T–x_i, and isothermal P–x_i phase coexistence boundary prediction for VLE, LLE, SLE as well as supercritical fluid applications. Phase stability criteria were also coupled to constitutive property models to estimate spinodal stability limits.

References and Suggested Readings

Armellini, F.J. and J.W. Tester (1991), "Experimental methods for studying salt nucleation and growth from supercritical water," *J. Supercritical Fluids*, **4**, p 254-264.

Armellini, F.J., G. Hong, and J.W. Tester (1994), "Precipitation of sodium chloride and sodium sulfate in water from sub- to supercritical conditions: 150° to 550°C, 100 to 300 bar," *J. Supercritical Fluids*, **7**, p 147-158.

Cahn, J.W. (1968), "Spinodal decomposition," *Trans. Met. Soc. AIME*, **242**, p 166-180.

Case, L.O. (1939), *Elements of the Phase Rule*, Edwards Letter Shop, Ann Arbor, MI.

Findlay, A., A.N. Campbell, and N.O. Smith (1951), *The Phase Rule*, Dover Publications, Inc., New York.

Hashimoto, T. (1986), "Structure formation in polymer mixtures by spinodal decomposition," Ch. 6 in *Current Topics in Polymer Science*, Vol. II, Oltenbrite, *et al.* (eds.), Hauser, New York.

Hong, G. (1981), "Binary phase diagrams from a cubic equation of state," PhD Thesis, Chemical Engineering Department, Mass. Inst. of Technology, Cambridge, MA. 532 pp.

Hultgren, R., P.D. Desai, D.T. Hawkins, M. Gleiser, and K.K. Kelley (1973), *Selected Values of the Thermodynamic Properties of Binary Alloys*, ASM, Metals Park, Ohio.

Kingery, W.D., H.K. Bowen, and D.R. Uhlmann (1976), *Introduction to Ceramics*, 2nd. ed., Wiley, New York, Ch. 8.

Kiran, E. (1994), "Polymer formation, modifications and processing in or with supercritical fluids," in *Supercritical Fluids*, Kluwer, The Netherlands, p 541-588.

Kwei, T.K. and T.T. Wang (1978), "Phase separation behavior of polymer-polymer mixtures," in *Polymer Blends*, Vol. 1, Paul, D.R. and S. Newman (eds.), Academic Press, New York.

Lupis, C.H.P. (1983), *Chemical Thermodynamics of Materials*, Prentice Hall PTR, Englewood Cliffs, NJ, Figure 19, p 217 and Figure 21, p 218.

Prausnitz, J.M., R.N. Lichtenthaler, and E. Gomes de Azevedo (1986), *Molecular Thermodynamics of Fluid-Phase Equilibria*, 2nd ed., Prentice Hall, Englewood Cliffs, New Jersey.

Prince, A. (1966), *Alloy Phase Equilibria*, Elsevier, New York.

Ravich, M.I. et al. (1953), "Solubility and vapor pressure of saturated solutions in the NaCl-Na_2-SO_4-H_2O system at high temperatures," *Akad. Nauk.* SSSR **22**, p 240-254.

Ricci, J.E. (1966), *The Phase Rule and Heterogeneous Equilibrium*, Dover Publications, New York.

Rhines, F.N. (1956), *Phase Diagrams in Metallurgy*, McGraw-Hill, New York, 340 p

Rowlinson, J.S. (1969), *Liquids and Liquid Mixtures*, 2nd ed., Butterworth, London.

Walas, S.N. (1985), *Phase Equilibria in Chemical Engineering*, Butterworth, Boston.

Wales, C.E. (1963), *Chemical Engineering*, May 27, p 120; June 24, p 111; July 22, p 141; Aug. 19, p 167; and Sept. 16, p 187.

Zernike, J. (1957), *Chemical Phase Theory*, Gregory Lounz, New York.

Problems

15.1. If a small-diameter wire is passed over a block of ice and weights are attached to each end of the wire, the wire apparently cuts through the ice but leaves no trace of the path (Figure P15.1). This phenomenon is often called *regelation*.

Many physics texts explain this phenomenon by the fact that the high pressure exerted by the wire lowers the ice freezing point and thus the wire "melts through." Similar reasoning is invoked to explain the occurrence of a water lubrication film beneath ice skates. What is your opinion of this explanation?

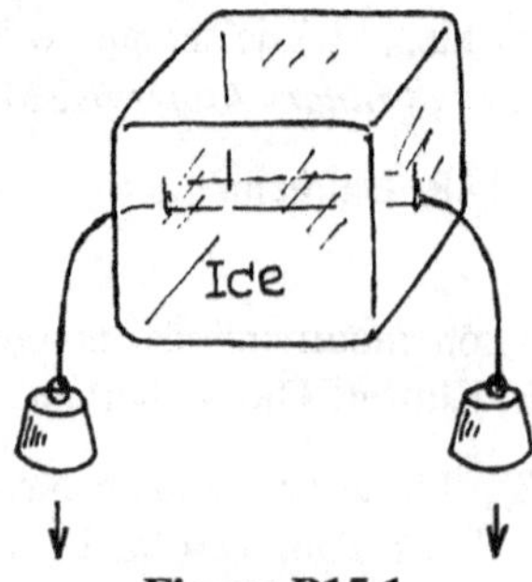

Figure P15.1

15.2. For a pure component, a first-order transition is a change in which the first and all higher derivatives of the chemical potential, μ, have a discontinuity at the point of change. A second-order transition is one in which only the second and higher derivatives of μ with respect to P and T have a discontinuity. Solid methane, for instance, shows, at a particular temperature, a discontinuous C_p because of the onset of free rotation of the molecules.

In other words, in the first-order transition, the chemical potential is continuous at the point of change, but it has a discontinuous first derivative. In a second-order transition, it is the specific entropy and the specific volume that are continuous as at the point of change and yet have discontinuous first derivatives. Derive the analogs to the Clapeyron equation for a second-order phase transition.

15.3. A saturated solution of naphthalene in chlorobenzene is prepared at 293 K and 1 bar. The liquid phase is decanted and the liquid mole fraction of naphthalene is determined to be 25.6%. The solid-free solution is placed in a piston-cylinder and compressed isothermally to 500 bar. Estimate the mole fraction of naphthalene in the liquid under these conditions.

Data: For solutions of naphthalene in chlorobenzene, the specific liquid volume, cm^3/mol, is independent of pressure and is given as

$$V = 102 + 26x, \quad x = \text{mole fraction naphthalene}$$

Ideal liquid solution properties may be assumed.

	Naphthalene	Chlorobenzene
Melting point (K)	353.3	228.0
Molar volume (cm^3/mol)	115 (solid)	102 (liquid)
Latent heat of fusion (kJ/mol)	18.46	—

15.4. In a binary solution of two components, the eutectic point is the lowest freezing point of the mixture. It is less than the freezing point of either pure component.

Assume that the liquid phase forms an ideal solution and that all solid phases are pure components (i.e., no mixed crystals form). The vapor phase forms an ideal-gas mixture. Some data that may be of use are given below. Neglect any pressure effects. Assume that ΔH values do not vary with temperature.

	Nitrogen	Oxygen
Freezing point (K)	63.3	54.4
$\Delta H_{vaporization}$ (J/mol)	6000	7490
$\Delta H_{sublimation}$ (J/mol)	6720	7940
ΔH_{fusion} (J/mol)	721	447

Estimate the eutectic point (i.e., composition and temperature) for a liquid-air mixture (O_2 and N_2).

15.5. It is late fall and you suddenly remember that you have not put antifreeze in the radiator of your car. All service stations nearby are closed and it is predicted that the temperature will drop to −10°C tonight. In desperation you visit the chemical engineering laboratories to borrow some suitable chemical to mix with water to make a noncorrosive antifreeze. There are formamide, urea, methanol, ethanol, glycerol, sucrose, and other such chemicals: which one to use, and how much? Then you dimly remember that in physical chemistry someone told you that in dilute solutions there was the same freezing-point depression per mole of solute for all such materials. Make any and all assumptions necessary to estimate rapidly the value of this constant ($\Delta T/N$). ΔT is the freezing-point depression in °C when N moles of solute are dissolved in 1 kg of water. $\Delta H_{f,H_2O} = 335$ kJ/kg.

The next day, when you have time to reflect on your action the preceding evening, you note that all commercial antifreezes are aqueous solutions. Why is this so? Do all aqueous *solutions* have a freezing point below that of pure water? Is it possible to have an aqueous solution with a freezing point higher than pure water?

15.6. It is a fact that occasionally during rapid loading of liquid oxygen and hydrogen into missile tanks, one or the other of the tanks has imploded, with catastrophic results (see Figure P15.6). Such tanks are constructed to withstand an internal pressure somewhat above atmospheric but will collapse if the external pressure significantly exceeds the internal pressure.

During the initial part of the loading cycle, the cryogen may be all gas, a mixture of gas and liquid, or all liquid. The cryogen is pumped through a side port and has intimate contact with the gas already in the tank. Assume ideal gases.

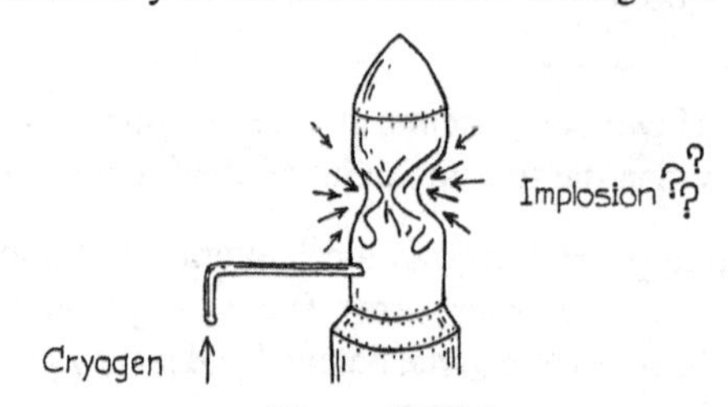

Figure P15.6

Data: Tank volume: 300 m^3 and $C_{p,\text{ vapor}} = 24.2(H_2)$, $29.3(O_2)$, J/mol K

	Vapor pressure (bar)	Saturation temperature (K)	Heat of vaporization (kJ/kg)
O_2	2.55	100	200
H_2	2.55	24	458

(a) Demonstrate that a pressure decrease cannot occur if only gas flows into the tank.

(b) Derive a general relation to relate the fractional change in tank pressure and temperature with the fractional increase in the mass of gas in the tank for the case in which you think the pressure drops the maximum amount. The feed is all liquid, saturated at the pressure in the transfer line, and when this liquid enters, it is immediately vaporized by contact with gas present in the tank. Clearly state any assumptions made.

(c) If the feed were a mixture of liquid and gas, saturated at the transfer line pressure, derive a relation to determine the critical quality of the feed. The critical quality is defined as that fraction of vapor above which no pressure drop is possible irrespective of any rate process.

(d) Determine the fractional change in pressure for both hydrogen and oxygen for part (b) when the mass of gas in the tank has doubled. The liquid in the transfer line is saturated at 2.55 bar and the initial tank gas pressure and temperature are 1 bar and 278 K, respectively. Assume that the initial gas in the oxygen tank is oxygen and that the initial gas in the hydrogen tank is hydrogen.

(e) For the same transfer line pressure and initial tank conditions as in part (d), determine the critical quality for a saturated mixture of gas and liquid cryogen using the relation derived in part (c). Do the calculation for both hydrogen and oxygen.

(f) As an engineer, what recommendations could you make to minimize implosion hazards during loading?

15.7. The partial pressures of water over sodium nitrate solutions are shown below at various temperatures and compositions.

Concentration (g $NaNO_3$/100 g H_2O)	Partial pressure of water (N/m^2)		
	0°C	25°C	50°C
0	610	3,168	12,340
10	589	3,057	11,890
20	571	2,952	11,480
30	553	2,852	11,080
40	539	2,758	10,610
50	524	2,672	10,330
60	511	2,589	9,990
70	497	2,510	9,670
73	493[a]		
80		2,438	9,370
90		2,369	9,080
92		2,356[a]	
100			8,810
110			8,560

[a]The solution is saturated.

In addition, the following data are available for saturated solutions of $NaNO_3$.

T (°C)	Partial pressure of water (N/m^2)	Concentration of $NaNO_3$ (g/100 g H_2O)
0	493	73.0
20	1,760	88.2
25	2,356	92.0
40	5,210	104.8
60	13,150	124.0
80	28,810	148.0
100	56,260	176.0
120	99,730	210.6

Determine the enthalpy change for crystallization at 25°C.

$$\Delta H_{cry} = x_{H_2O}(H^L - \overline{H}^L)_{H_2O} + x_{NaNO_3}(H^S - \overline{H}^L)_{NaNO_3}$$

Also calculate $(H^S - \overline{H}^L)_{NaNO_3}$ at 25°C, saturated solution.

15.8. A calorimeter has recently been constructed for NASA to determine the thermal conductivities of various superinsulations. In essence, this calorimeter is a thick-walled copper sphere with a radius of 0.506 m; the superinsulation is placed on the outside of the sphere. The insulation is 1.27 cm thick and consists of many layers of 6 μm aluminized Mylar separated by thin spacers made of glass-fiber paper, silk netting, or other such materials. The tests are carried out in high-vacuum environmental chambers. The outer layer of insulation is maintained at 294 K. The calorimeter is filled with liquid hydrogen.

Heat transfer through the insulation is manifested by the boiling of the liquid hydrogen, and the mass flow of vapor is measured as a function of time. The vapor vent line may be considered adiabatic (actually, it has a separate liquid hydrogen guard to prevent axial conduction) and is of sufficient cross-sectional area that the pressure in the calorimeter is equal to the prevailing atmospheric pressure.

In any given test, the calorimeter is filled with liquid hydrogen and the liquid level held about constant until the calorimeter heat leak reaches a steady-state value. At this time the approximate head of liquid hydrogen is measured and the test started. During a test the flow rate of vented hydrogen vapor is measured at various times and the barometric pressure recorded. The assumption that the copper shell and both hydrogen phases are at the same temperature is believed to be quite good; also, the liquid and vapor phases are always in equilibrium with each other at the prevailing atmospheric pressure existing at the exit end of the vent line.

Some real test data are shown below.

Test XX-3
Mass of Liquid Hydrogen in Tank at Start of Test = 38.4 kg

Time (h)	Mass flow rate of vented hydrogen vapor [g/h (±0.2)]	Barometric pressure (bar)
0	40.8	1.016
2	41.3	1.011
4	42.2	1.007
6	44.5	0.989[a]
8	43.5	0.991
10	44.9	0.987
12	43.1	1.000
14	39.5	1.012
16	39.9	1.016
18	40.8	1.017
20	41.3	1.012

[a]A severe thunderstorm and squall occurred between the hours of about 6 and 10, hence the drop in barometric pressure.

Saturation Properties of Hydrogen

		Enthalpy (kJ/kg)	
T (K)	P (bar)	Liquid	Vapor
20.0	0.932	200	657
20.4	1.013	205	659
21	1.206	210	662

After an appropriate thermodynamic analysis, determine the effective thermal conductivity of the insulation.

15.9. The following data have been reported for the vapor-liquid equilibrium between ethyl alcohol and n-hexane at 1.013 bar.

T (°C)	Mole fraction alcohol in liquid	Mole fraction alcohol in vapor
76.0	0.990	0.905
73.2	0.980	0.807
67.4	0.940	0.635
65.9	0.920	0.580
61.8	0.848	0.468
59.4	0.755	0.395
58.7	0.667	0.370
58.35	0.548	0.360
58.1	0.412	0.350
58.0	0.330	0.340
58.25	0.275	0.330
58.45	0.235	0.325
59.15	0.102	0.290
60.2	0.045	0.255
63.5	0.010	0.160
66.7	0.006	0.065

The vapor pressures of ethyl alcohol and n-hexane are cited below.

T (°C)	$P_{vp,\text{ alcohol}}$ (bar)	T (°C)	$P_{vp,\ n\text{-hexane}}$ (bar)
50.0	0.2933	-53.9	1.33×10^{-3}
52.0	0.3233	-34.5	6.66×10^{-3}
56.0	0.3891	-25.0	1.33×10^{-2}
58.0	0.4266	-14.1	2.67×10^{-2}
60.0	0.4670	-2.3	5.33×10^{-2}
62.0	0.5108	5.4	8.00×10^{-2}
64.0	0.5578	15.8	1.33×10^{-1}
66.0	0.6085	31.6	2.67×10^{-1}
68.0	0.6629	49.6	5.33×10^{-1}
70.0	0.7215	68.7	1.013
72.0	0.7844	93.0	2.027
74.0	0.8519	131.7	5.066
76.0	0.9241	166.6	10.13
78.4	1.0133		

(a) What does thermodynamics tell us about the validity of these data?

(b) What can one conclude about the heat of solution and the entropy of solution?

15.10. Suppose that one were interested in determining xyP relations for an *isothermal* system of components 1 and 2 at low pressures (where the vapor phase forms an ideal-gas mixture). In one type of experiment, no vapor composition measurements can be made directly but P–x data can be obtained over the entire range of liquid compositions.

(a) Show clearly how vapor compositions and liquid-phase activity coefficients may then be determined.

(b) Repeat part (a) assuming that pressure-vapor composition data were available and γ, x values were to be calculated.

(c) In a given binary vapor-liquid equilibrium system at constant temperature, the total pressure is graphed as a function of liquid mole fraction. A tangent is drawn to the total pressure curve as the mole fraction of one component becomes vanishingly small (Figure P15.10). What limits are placed on the slope of this tangent?

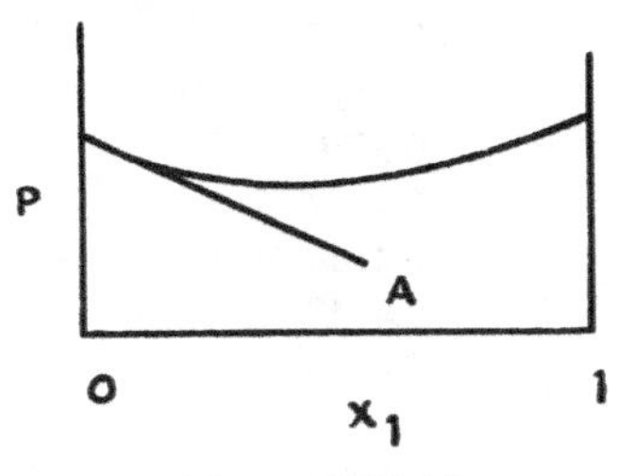

Figure P15.10

15.11. Vapor-liquid equilibrium data are shown below for the binary system isooctane-perfluoroheptane at 303, 323, and 343 K.

Data and allowable assumptions:

The vapor phase in equilibrium with the liquid may be considered an ideal-gas mixture. Specific volumes of liquids are negligible compared to vapor specific volumes. The enthalpies of vaporization of pure isooctane and perfluoroheptane at 323 K are 34.33 and 35.31 kJ/mol, respectively.

(a) A saturated liquid stream containing 50 mole% isooctane is flowing in a pipeline at 323 K. What is the minimum isothermal, isobaric work required to separate this stream into two liquid fractions that contain 90 and 5 mole% isooctane, respectively? Diagram your proposed process scheme.

(b) A batch vessel at 323 K contains 100 moles of liquid with 50 mole% isooctane. Devise a scheme to boil this liquid in such a manner that the pressure remains constant at the original value; the temperature is also to remain constant at 323 K. Calculate the heat interaction for the process ending with only vapor in the vessel.

Vapor-Liquid Equilibrium Data
for Isooctane-Perfluoroheptane ($T = 323$ K)

Mole fraction isooctane in liquid	Mole fraction isooctane in vapor	Pressure (bar)
0	0	0.3156
0.0560	0.1455	0.3632
0.1680	0.2940	0.4052
0.2555	0.3495	0.4176
0.2850	0.3615	0.4182
0.3760	0.3910	0.4204
0.4270	0.3997	0.4200
0.6450	0.4240	0.4146
0.7990	0.4540	0.4025
0.8258	0.4755	0.3915
0.8718	0.5015	0.3789
0.9085	0.5435	0.3612
0.9870	0.8790	0.2331
1	1	0.1955

Vapor-Liquid Equilibrium Data
for Isooctane-Perfluoroheptane ($T = 303$ K)

Mole fraction isooctane in liquid	Mole fraction isooctane in vapor	Pressure (bar)
0	0	0.1308
0.0380	0.1345	0.1505
0.0712	0.2180	0.1620
0.0990	0.2614	0.1688
0.1514	0.3198	0.1765
0.2058	0.3595	0.1812
0.3060	0.3869	0.1845
0.3348	0.3935	0.1848
0.3910	0.4080	0.1848
0.4298	0.4081	0.1847
0.5505	0.4165	0.1842
0.8448	0.4555	0.1762
0.9195	0.5040	0.1612
0.9550	0.5705	0.1403
0.9881	0.6805	0.1007
1	1	0.0833

Vapor-Liquid Equilibrium Data
for Isooctane-Perfluoroheptane ($T = 343$ K)

Mole fraction isooctane in liquid	Mole fraction isooctane in vapor	Pressure (bar)
0	0	0.6715
0.0785	0.1540	0.7711
0.1600	0.2570	0.8190
0.3190	0.3565	0.8515
0.3992	0.3810	0.8532
0.3995	0.3825	0.8521
0.6938	0.4471	0.8242
0.8020	0.4870	0.7905
0.8815	0.5590	0.7300
0.9540	0.7155	0.5945
0.9825	0.8918	0.4789
1	1	0.4082

15.12. With interest in the utilization of solar and geothermal energy, some research is focused on developing techniques to "store" this energy. One concept involves the melting of inorganic salts, as energy may then be recovered at a later time by allowing the fused salts to solidify. Most pure inorganic salts, however, melt at relatively high temperatures. Salt mixtures usually melt at temperatures below those of the pure components. The minimum melting temperature for a mixture is termed the eutectic point. (Assume that, in the case of interest, solid solutions do not form.)

We would like to estimate this eutectic point as accurately as possible to aid us in evaluating the concept of fused-salt thermal-energy storage.

(a) Consider the system LiCl-KCl. What would you estimate to be the atmospheric-pressure eutectic temperature and composition? Assume that the liquid phase of molten LiCl-KCl forms an ideal solution and liquid enthalpies are not a strong function of temperature. Some data for this system are shown below.

Salt	Melting point (K)	Enthalpy of fusion (kJ/mol)	Entropy of fusion (J/mol K)
LiCl	883	19.93	22.57
KCl	1043	26.54	25.46

(b) The experimentally determined value of the eutectic point is $x_{LiCl} \approx 0.59$ at $T \approx 618$ K. As can be seen from the ideal solution result from part (a), the prediction assuming ideal solutions is quite poor. What would be your estimate of the eutectic temperature for the LiCl-KCl system if regular solution behavior were assumed and the heat of mixing were expressed as

$$\Delta H = x_{LiCl}x_{KCl}(-17{,}580 - 377x_{LiCl})$$

with ΔH in J/mol and x is the mole fraction? Assume that ΔH is not a strong function of temperature.

15.13. The *International Critical Tables*, Vol. III, p. 313, lists the boiling points for mixtures of acetaldehyde and ethyl alcohol at various pressures. The data are summarized below for mixtures containing 80 mole% ethyl alcohol in the liquid.

From these data determine the molar enthalpy of vaporization of ethyl alcohol at 320.7 K, from a liquid mixture containing 80% ethyl alcohol. [That is, what is $(\overline{H}^V - \overline{H}^L)$ alcohol?]

T (K)	P (N/m2)	Mole fraction ethyl alcohol in vapor
331.3	9.319×10^4	0.318
320.7	5.306×10^4	0.385
299.9	1.027×10^4	0.330

Note: The vapor phase may be considered to be an ideal-gas mixture, but the liquid phase is a non-ideal solution.

15.14. In preparing to make a simulation study of an evaporation process to remove water from concentrated sucrose solutions, we need to know, as accurately as possible, the enthalpy change in vaporizing water from these concentrated solutions. We do have some new, unpublished experimental data that show the boiling point of the solution at several pressures and compositions (see below).

Boiling Temperatures for Sucrose[a]-Water Solutions

$P = 1.013$ bar		$P = 0.8781$ bar		$P = 0.7350$ bar	
Brix[b]	T (°C)	Brix[b]	T (°C)	Brix[b]	T (°C)
0	100	0	96.1	0	91.5
75.0	107.5	75.0	103.5	75.0	99.5
77.4	109	77.4	105.5	77.4	101.8
78.7	110	78.7	106		
80.0	110.5	80.0	107	80.0	103
81.3	112	81.3	108.5	81.3	104
82.7	113	82.7	109.8	82.7	105
84.2	114.5	84.2	111	84.2	106.5
85.8	116.5	85.8	113	85.8	108
87.2	118.5	87.2	115	87.2	110.5
88.8	122	88.8	118	88.8	112.5
90.5	127	90.5	122	90.5	116
92.3	130.5	92.3	125	92.3	121
94	138	94	131	94	128
		95	135	95	131

[a]Sucrose may be considered as pure (no invert) with the empirical formula $C_{12}H_{22}O_{11}$.
[b]Brix = weight percent sucrose.

Making *as few assumptions* as possible, calculate $H_W^V - \overline{H}_W^L$ for a 90 Brix solution at a pressure of 1.013 bar. Clearly show all steps in your calculations and justify any assumptions made. Discuss your result if you feel it is unreasonable.

15.15. We reproduce below a section on anesthesia from W.J. Moore [(1972), *Physical Chemistry*, 4th ed. Prentice-Hall, Englewood Cliffs, NJ, p 241].

Mechanism of Anesthesia

One of the most fascinating unsolved problems in medical physiology is the mechanism by which gases produce anesthesia and narcosis. Many anesthetics, such as krypton and xenon, are apparently inert chemically; in fact, it would appear that all gases produce an anesthetic effect at high enough pressures. Cousteau, in *The Silent World*, gives a memorable account of the nitrogen narcosis experienced at great depths, *l'ivresse des grandes profondeurs*, which has claimed the life of more than one diver.

[The accompanying table] summarizes some data on mice, which were tested by a criterion based on their righting reflex. Animals in a test chamber were rocked off their feet; if a mouse did not replace all four feet on the floor within 10 s, the anesthetic was awarded a "knockout." The results indicate that even helium produces narcosis at high pressures.

Early attempts to understand the causes of anesthesia were made by Meyer (1899) and Overton (1901), who found a good correlation between the solubility of a gas in a lipid (olive oil) and its narcotic efficacy. Since nerve cell membranes are composed mainly of lipids and proteins, it was suggested that the anesthetic molecules dissolve in the membranes and block the process of nerve conduction in some way as yet unknown. The activity of dissolved anesthetic necessary to produce anesthesia is in the range 0.02 to 0.05, so that the anesthetic can certainly alter the properties of the membrane to some considerable extent.

Best Estimates of Anesthetic Pressures for Mice (Righting Reflex)

Key number	Gas	Pressure (atm)	Key number	Gas	Pressure (atm)
1	He	190	10	C_2H_4	1.1
2	Ne	>110	11	C_2H_2	0.85
3	Ar	24	12	Cyclo-C_3H_6	0.11
4	Kr	3.9	13	CF_4	19
5	Xe	1.1	14	SF_6	7.0
6	H_2	85	15	CF_2Cl_2	0.4
7	N_2	35	16	$CHCl_3$	0.008
8	N_2O	1.5	17	Halothane	0.017
9	CH_4	5.9	18	Ether	0.032

Estimate the anesthetic pressure for carbon tetrachloride. Clarify and justify (where possible) any assumptions made. Note that the value of X in the accompanying figure from Moore refers to the solubility of the gas in olive oil when the partial pressure of the gas is 1 bar.

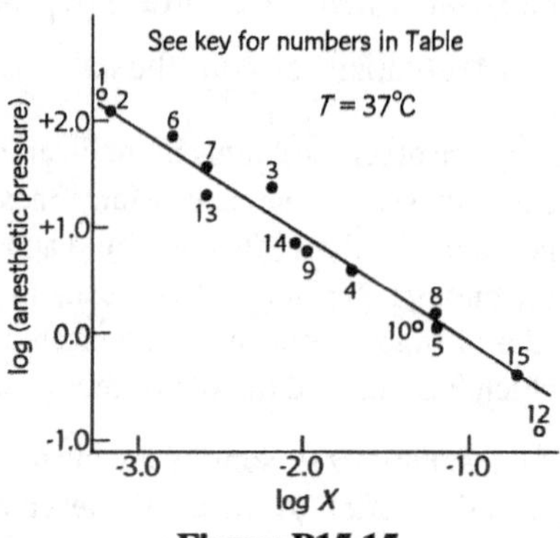

Figure P15.15

Property data for carbon tetrachloride and for the compounds listed in the table are as follows:

Gas	Critical pressure (bar)	Normal boiling point (K)	Critical temperature (K)
He	2.27	4.21	5.19
Ne	27.6	27.0	44.4
Ar	48.7	87.3	150.8
Kr	55.0	119.8	209.4
Xe	58.4	165.0	289.7
H_2	13.0	20.4	33.2
N_2	33.9	77.4	126.2
N_2O	72.4	184.7	309.6
CH_4	46.0	111.7	190.6
C_2H_4	50.4	169.4	282.4
C_2H_2	61.4	189.2	308.3
Cyclopropane	54.9	240.2	397.8
CF_4	37.4	145.2	227.6
SF_6	37.6	209.3	318.7
CF_2Cl_2	41.2	243.4	385.0
$CHCl_3$	54.7	334.3	536.4
Halothane[a]	—	—	—
Ether	36.4	307.7	466.7
CCl_4	45.6	349.7	556.4

[a] $C_2HBrClF_3$.

15.16. Some isobaric vapor-liquid equilibrium data at 1.013 bar for the system acetone(A)-water(W) are as follows:

T(°C)	Mole fraction acetone: Liquid	Mole fraction acetone: Vapor	Differential enthalpy of mixing for the saturated liquid (J/mol): $\overline{\Delta H}_A$	Differential enthalpy of mixing for the saturated liquid (J/mol): $\overline{\Delta H}_W$
100	0	0	5296.3	0
84.75	0.02	0.4451	—	—
75.13	0.05	0.6340	3715.8	43.1
68.19	0.10	0.7384	2357.2	154.1
65.02	0.15	0.7813	—	—
63.39	0.20	0.8047	2357.2	481.5
61.45	0.30	0.8295	-564.4	817.7
60.39	0.40	0.8426	-990.2	1038.3
59.91	0.50	0.8518	-1030.0	1063.4
59.55	0.60	0.8634	-845.7	837.4
58.79	0.70	0.8791	-566.5	86.7
58.07	0.80	0.9017	-286.8	-548.5
57.07	0.90	0.9371	-78.7	-1737.5
54.14	1.00	1.0000	0	-3274.1

In addition to these data, the vapor pressures of the pure components are as follows:

T(°C)	$P_{vp,\text{ acetone}}$ (bar)	$P_{vp,\text{ water}}$ (bar)
15	0.1933	0.0171
20	0.2437	0.0233
30	0.3769	0.0424
40	0.5633	0.0737
50	0.8171	0.1233
60	1.154	0.1992
70	1.591	0.3116
80	2.148	0.4734
90	2.843	0.7010
100	3.697	1.013

(a) Assuming the vapor phase to be an ideal-gas mixture, determine the entropy change in mixing pure liquid water and liquid acetone, at 1.013 bar, to form a 40 mole% acetone solution. The pure components and final solution are at 60.39°C.

(b) Calculate the total entropy change for a process in which 0.6 mole of liquid water and 0.4 mole of liquid acetone, both at 60.39°C, are separately vaporized, expanded reversibly and isothermally to a pressure so that they may be added to a vapor mixture at 1.013 bar (with 84.26 mole% acetone), reversibly across semipermeable membranes. One mole of this vapor is then condensed to a liquid with a composition of 40 mole% acetone at 60.39°C. Compare your result with the relation determined in part (a).

(c) Comment on the consistency of the data.

15.17. W.L. Jolley and Joel Hildebrand studied the solubility and entropy of solution of gases in nonpolar solvents. They have plotted the equilibrium solubility of various inert gases in benzene at 298 K when there is a partial pressure of 1.013 bar of the inert gas over the solution. x is the equilibrium solubility (mole fraction) of the gas in benzene, $\overline{S}$ is the partial molar entropy of the gas in solution, and S(gas) is the entropy of the pure gas at 298 K and 1.013 bar (see Figure P15.17).

(a) Which gases become more soluble with an increase in temperature at temperatures greater than 298 K? Prove your reasoning and state clearly any simplifying assumptions that you make in the proof.

(b) For some gases, there is an entropy *increase* when the pure gas dissolves in benzene to form a solution. How might you explain this fact?

(c) Estimate ΔH (vaporization) of pure benzene at 298 K and 1.013 bar. The vapor pressure of pure benzene at this temperature is 0.126 bar.

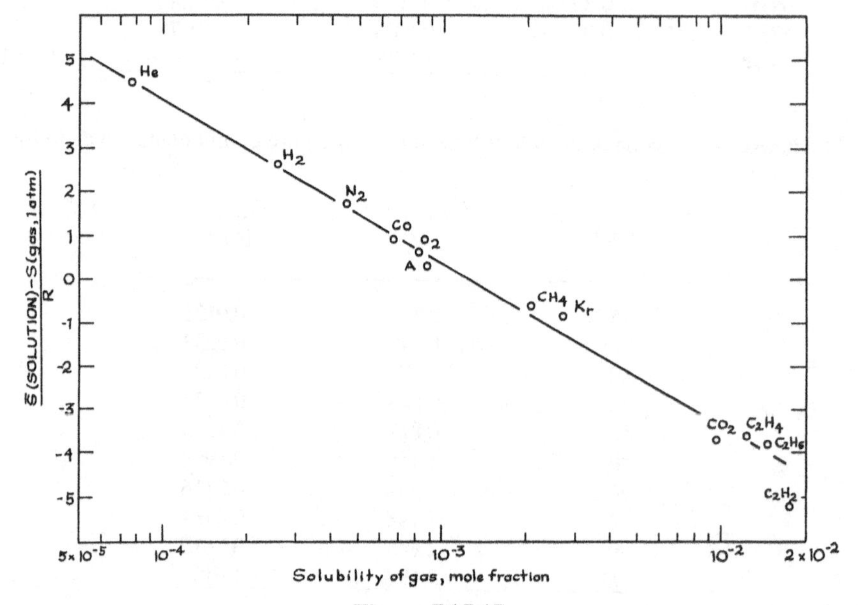

Figure P15.17

15.18. Use a liquid solution model or your own creative imagination to devise a two-constant arithmetic expression that you think will properly express ΔG^{EX} as a function of composition for a nonideal binary liquid mixture. Remember that $\Delta \underline{G}^{EX}$ is proportional to the total number of moles and should go to zero as x_A and $x_B = 0$.

Using this proposed equation, determine algebraic (constant temperature, constant pressure) expressions for the activity coefficients as a function of composition and estimate the x-y curve for ethanol and water at 1 atm (1.013 bar) using only the azeotropic point.

$$x_{\text{alcohol}} = y_{\text{alcohol}} = 0.8943 \text{ at } 78.17°\text{C}$$

The vapor pressures of the pure components as a function of temperature are as follows:

	Vapor pressure (bar)	
T (°C)	Ethanol	Water
76	0.9239	0.4013
78	0.9999	0.4360
80	1.083	0.4733
82	1.169	0.5120
84	1.267	0.5546
86	1.368	0.6000
88	1.469	0.6493
90	1.583	0.6999
92	1.707	0.7546
94	1.831	0.8133
96	1.964	0.8759
98	2.108	0.9426
100	2.257	1.013

You may wish to test your calculated x-y results with the following isobaric liquid-vapor equilibrium data.

T (°C)	x Mole fraction ethanol, liquid	y Mole fraction ethanol, vapor
95.7	0.0190	0.1700
90.0	0.0600	0.3560
86.4	0.1000	0.4400
84.3	0.1600	0.5040
83.3	0.2000	0.5285
82.3	0.2600	0.5570
81.8	0.3000	0.5725
81.2	0.3600	0.5965
80.7	0.4000	0.6125
80.2	0.4600	0.6365
79.8	0.5000	0.6520
79.4	0.5600	0.6775
79.13	0.6000	0.6965
—	0.6600	0.7290
78.6	0.7000	0.7525
—	0.7600	0.7905
78.3	0.8000	0.8175
—	0.8600	0.8640
78.17	0.8943	0.8943

Source: Data from J.S. Carey, Sc.D. thesis, M.I.T., 1929.

15.19. The values of ΔG^{EX} and ΔH^{EX} for a particular binary liquid mixture are shown in Figure P15.19 as a function of the mole fraction of component A. One can describe these curves by: $\Delta G^{EX} = \eta\, \Delta H^{EX}$. It was also found for T, P constant that $\overline{\Delta G_A^{EX}} = \eta_A\, \overline{\Delta H_A^{EX}}$ and $\overline{\Delta G_B^{EX}} = \eta_B\, \overline{\Delta H_B^{EX}}$. The constants η, η_A, and η_B are not functions of temperature.

(a) Derive a relationship to show how the activity coefficient varies with temperature for component A if the pressure and composition were fixed.

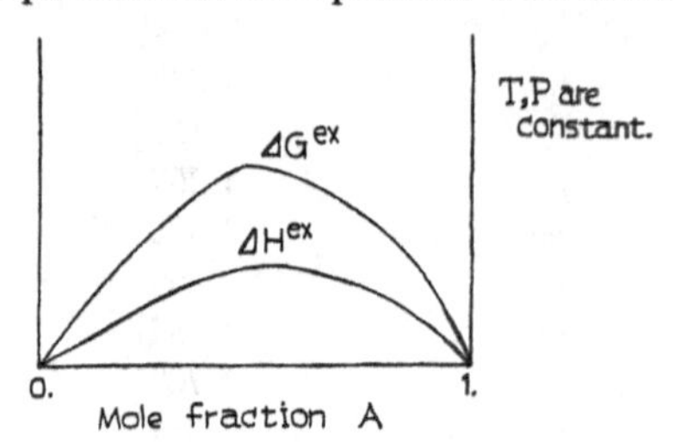

Figure P15.19

(b) Using the expression derived in part (a), keeping the pressure and composition constant, and supposing that $\gamma_A = 1.2$ at 300 K, what is γ_A at 400 K if $\eta = 2$?

(c) Discuss the cases $\eta_A > 1$, $\eta_A = 1$, and $0 < \eta_A < 1$ from the standpoint of $\overline{\Delta S_A^{EX}}$ and indicate what type of liquid solution one might expect in each of these cases. Assume that $\gamma_A > 1$.

15.20 It is claimed that a low-temperature liquid mixture may be made by mixing ice and pure sulfuric acid in a Dewar flask. If the ice and acid were originally at 0°C, what would be the final temperature and liquid composition when 1 kg of acid is poured over 4 kg of ice?

Data:

(1) The heat capacity of sulfuric acid solutions may be approximated by

$$C_p = 4.186 - 3.349x \text{ kJ/kg soln-K}$$

where x is the weight fraction H_2SO_4. Assume that the heat capacity does not vary with temperature. The heat capacity of water ice is 2.09 kJ/kg K.

(2) Freezing point of sulfuric acid solutions

Wt% H_2SO_4	T (°C)
0	0
10	-5
20	-13
30	-35
40	-62
50	-35

(3) The enthalpy change in the dilution of 1 kg H_2SO_4 with water (at 0°C) to form a solution with x weight fraction H_2SO_4 is approximated as

$$\Delta H = 255.8x - 732.6 \text{ kJ/kg acid} \qquad (0.01 < x < 0.3)$$

(4) Vapor pressure of pure ice

T (°C)	P (N/m^2)
0	611
-10	260
-20	104
-30	38.1
-40	12.9

(5) Partial pressure of water over sulfuric acid solutions

$$\log_{10} p_W = A - \frac{B}{T} \qquad (p_w, \text{N/m}^2; T, \text{K})$$

Wt% H_2SO_4	A	B
0	11.071	2260
10	11.050	2259
20	11.050	2268
30	10.989	2271
40	10.969	2299
50	10.957	2357

15.21. A well-insulated tank is vented to the atmosphere. Initially, it contains a hot aqueous solution of sodium hydroxide at 90°C, with a concentration of 45 wt% caustic. Connected to the tank is a steam header with saturated steam at 1 bar, 100°C [Figure P15.21(a)].

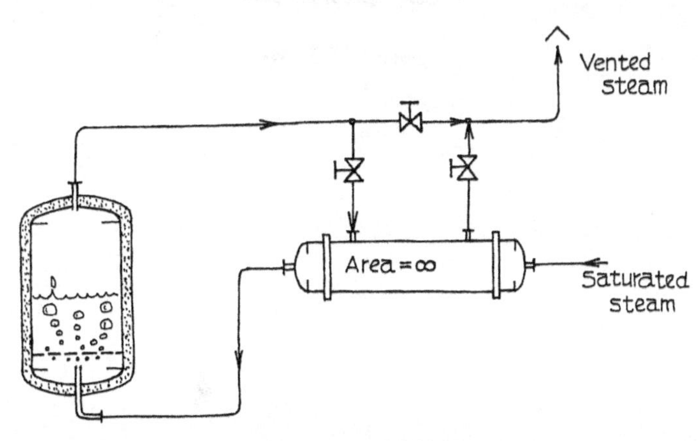

Figure P15.21(a)

Assume that the vented steam is in equilibrium with the solution at all times and neglect any sensible heat effects for the vessel walls or inert gases over the solution. The enthalpy of water vapor at 1 bar and 100°C is 2675 kJ/kg. The heat capacity of water vapor at 1 bar is about 1.91 kJ/kg K.

(a) If the steam line were opened and steam allowed to bubble slowly into the tank, plot the resulting solution temperature as a function of sodium hydroxide concentration.

(b) Plot a curve showing the kilograms of steam condensed and vented as a function of the sodium hydroxide concentration.

(c) If the vent line from the tank were arranged [as shown in Figure P15.21(a)] so that the vented gas was passed through an "infinite-area" heat exchanger countercurrent to the entering steam, how would the answers in parts (a) and (b) be affected? Data are given in Figures P15.21(b) and (c).

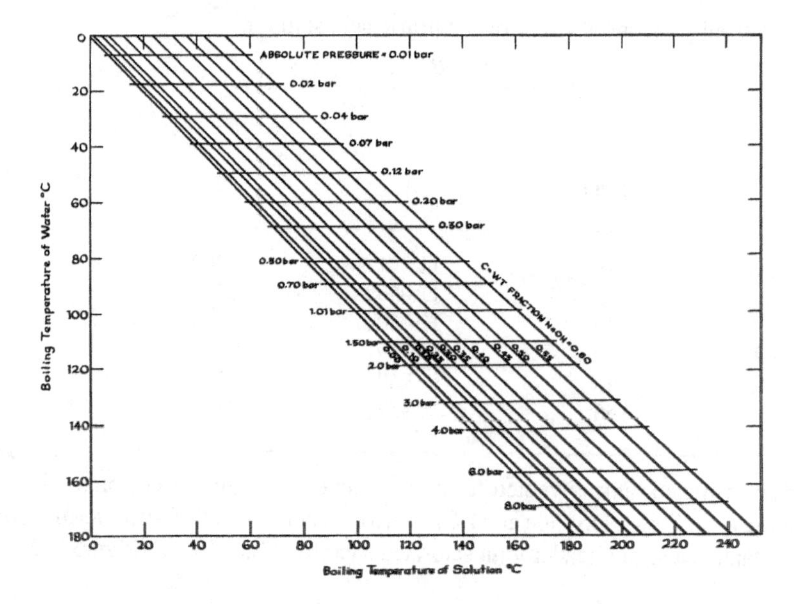

Figure P15.21(b)

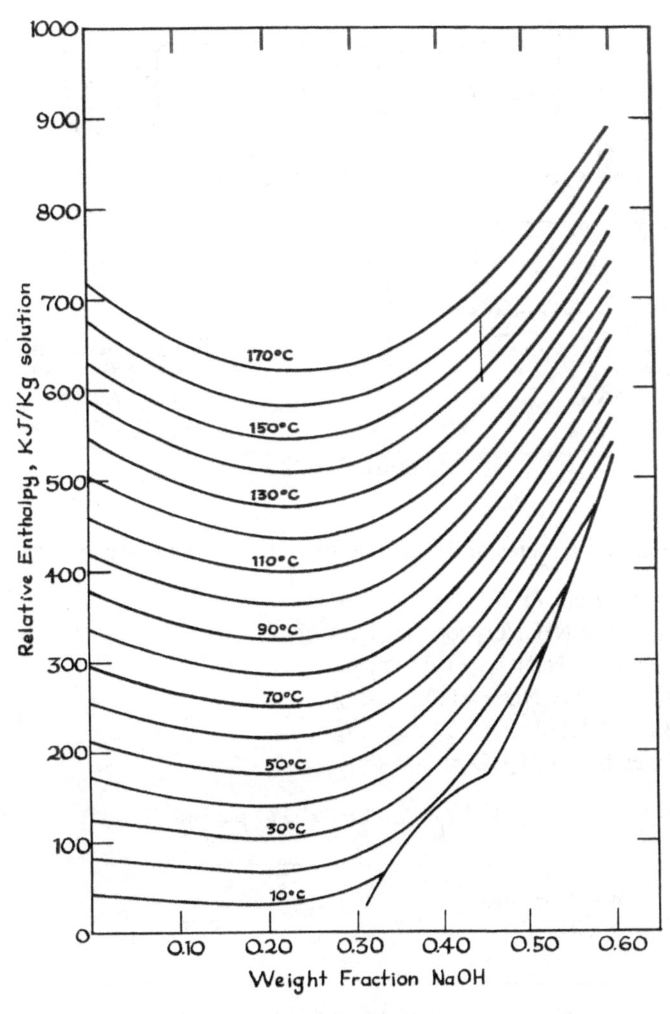

Figure P15.21(c)

15.22. An aqueous solution of 15 wt% NH_4NO_3 is to be concentrated to 30 wt% NH_4NO_3 by reverse osmosis. The feed solution is pressurized to 1520 bar and passed through a cell containing a membrane permeable to water but not to NH_4NO_3. Based on results of laboratory tests, a commercial process design had been formulated. The proposed process is illustrated in Figure P15.22 together with the proposed operating conditions. There is a significant pressure drop in the high-pressure chamber since a high fluid velocity is necessary to minimize the mass-transfer resistance in the bulk fluid phase.

Data: At 1 bar and 298 K the chemical potentials of both species have been determined as a function of composition. These data are shown in Figure 15.29. For the chemical potential, ammonium nitrate is referred to a saturated solution, while the reference for H_2O is pure water.

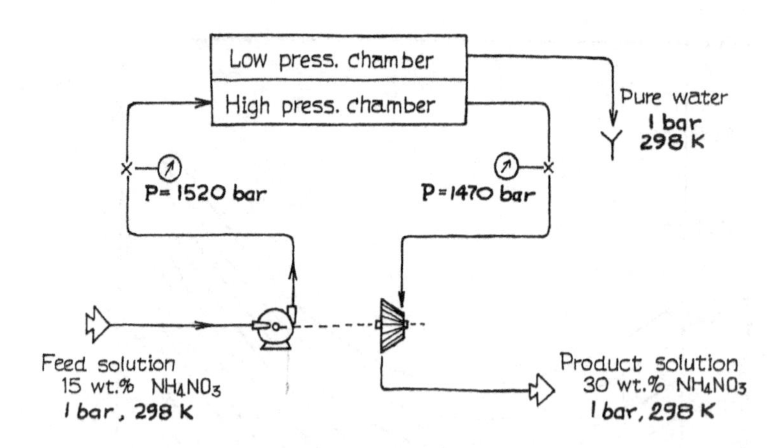

Figure P15.22

At 298 K and for pressures of 0.1 to 2000 bar, the following assumptions may be made:

$\underline{V}$(pure water) = 1 cm^3/g
$\overline{V}_W$(independent of NH_4NO_3 conc.) = 1 cm^3/g
$\underline{V}$(pure crystalline NH_4NO_3) = 0.58 cm^3/g
$\overline{V}_{NH_4NO_3}$(15 wt% NH_4NO_3) = 0.622 cm^3/g
$\overline{V}_{NH_4NO_3}$(30 wt% NH_4NO_3) = 0.632 cm^3/g
Molecular weight of NH_4NO_3 = 80.04

T = 298 K, P = 1 bar

Weight fraction NH_4NO_3	$P_i/P_{vp,\ water}$	$P_i/P_{vp,\ NH_4NO_3}$
0	1.0	0
0.10	0.9601	0.0483
0.15	0.9405	0.0750
0.20	0.9215	0.1120
0.25	0.8990	0.1565
0.30	0.8746	0.2115

At 298 K, $P_{vp,\text{ water}} = 0.0316$ bar, $P_{vp,\ NH_4NO_3\ pure} = 10^{-5}$ bar. Assume ideal gases.

(a) Describe in detail a reversible process that will accomplish the same overall separation as that described (i.e., will accept the same feed solution and produce the same product solutions). You are free to use any kind of any finite number of pumps, membranes, heat exchangers, or any other device dictated by your creativity. Present a flow sheet summarizing your process; indicate the operating conditions (i.e., pressure and composition) at each point in the process.

(b) For your reversible process, calculate the work requirement.

(c) The net work requirement for the reverse osmosis process shown in Figure P15.22 is 86 J/g of feed solution. What is the overall efficiency of this process?

15.23. Comment on the following often-quoted rule: "If the solution of a solute in a solvent is endothermic, the solubility of the solute increases with an increase in temperature; the converse is also true."

Prepare a case in detail to show if this rule is true; if there are any cases where it is false, describe clearly the conditions under which the violation occurs. Your analysis should include, but not be limited to, the special case of an ideal liquid solution with negligible pressure effects.

Illustrate your analysis with the case of lithium chloride dissolving into water. For this system some thermodynamic data are shown data below. Other data sources may also be used if desired.

Solubility of LiCl in H_2O

Temperature (°C)	Moles LiCl / kg H_2O	Equilibrium solid phase
0	15.3	$LiCl \cdot 2H_2O$
10	17.3	
12.5	17.8	$LiCl \cdot 2H_2O/LiCl \cdot H_2O$
20	18.8	$LiCl \cdot H_2O$
40	21.2	
60	24.0	
80	27.1	
98	30.7	$LiCl \cdot H_2O/LiCl$
100	30.8	LiCl

Source: *International Critical Tables*, Vol. IV ((1930), McGraw-Hill, New York, p. 233).

Heat of Formation at 298.2 K

Formula	State	ΔH_f (kJ/mol)
LiCl	Crystal	-409.05
$LiCl{\cdot}H_2O$	Crystal	-713.0
$LiCl{\cdot}2H_2O$	Crystal	-1013.6
$LiCl_{(aq)}$	$3H_2O$	-429.65
	$4H_2O$	-434.30
	$5H_2O$	-437.232
	$8H_2O$	-440.912
	$10H_2O$	-441.933
	$15H_2O$	-443.164
	$20H_2O$	-443.746
	$25H_2O$	-444.069
	$50H_2O$	-444.726
	$100H_2O$	-445.120
	$200H_2O$	-445.392
	$400H_2O$	-445.601
	∞H_2O	-446.217
H_2O	liquid	-286.030

Source: U.S. Bureau of Standards Circular 500, p. 433.

15.24. A supercritical fluid separation process involves the extraction of solid solutes into a gas which is somewhat above its critical temperature and pressure.

(a) Assume that the solid solute is in equilibrium with the supercritical gas at some temperature T and pressure P. No gas dissolves in the solid solute, nor is the specific volume of the solid a strong function of pressure. Derive an expression to allow one to estimate the mole fraction of the solute in the supercritical gas phase. Assume that one has an equation of state to express the system pressure as a function of the volume, temperature, and composition of the vaporized solute-supercritical gas phase.

(b) Using the relationship found in part (a), write a computer program and use it to estimate the equilibrium mole fraction of phenanthrene in supercritical carbon dioxide at 313 K at pressures ranging from 1 to 500 bar, in 10-bar increments. Use the Peng-Robinson equation of state to describe the *PVTy* relationship for the gas phase. This equation is given in Eqs. (9-26) through (9-30) and Table 8.1 (see also Example 9.7). Data for CO_2 and phenanthrene are given below:

	Phenanthrene		CO_2
Molecular weight	178		44
T_c (K)	878		304.2
P_c (bar)	29.0		73.76
Acentric factor, ω	0.44		0.225
Solid volume, V^S (cm^3/mol)	181.9		
Normal boiling point, T_b (K)	612.6		
ΔH_{vap} at T_b (kJ/mol)	55.68		
Interaction parameter, δ_{ij}		0.10	

(c) Using the thermodynamic relationship derived in part (a), at 313 K, what is the *minimum* solubility of phenanthrene in CO_2? Is there a *maximum* solubility?

(d) Derive an expression to allow an estimation of the heat of solution of solid phenanthrene into supercritical CO_2. What experimental data would you require to determine this heat of solution at various temperatures and pressures?

15.25. A 0.02832 m^3 chamber that is well insulated contains initially 13.6 kg of dry-ice flakes in equilibrium with CO_2 vapor. The pressure is maintained at 1.013 bar by a safety relief valve. A steady stream of propane at 1.013 bar and 273.2 K is injected into the bottom of the chamber and passes up through the dry ice bed.

Data: The acentric factors for CO_2 and propane are 0.225 and 0.152, respectively. Propane has a critical temperature of 369.8 K and a critical pressure of 42.4 bar. If you need vapor-phase properties of the superheated vapor phase, use the Peng-Robinson equation of state (see Chapters 8 and 9).

Assuming that any gases leaving the chamber are in equilibrium with any solid and/or liquid phase present, plot the temperature and mole fraction of propane in the vapor as a function of the amount of propane injected up until the time at which all dry-ice flakes disappear.

Thermodynamic Properties of Saturated Propane

T (K)	P (bar)	V (liquid) (m^3/kg $\times 10^3$)	V (vapor) (m^3/kg)	H (liquid) (kJ/kg)	H (vapor) (kJ/kg)	S (liquid) (kJ/kg K)	S (vapor) (kJ/kg K)
211.0	0.390	1.654	1.011	378.2	823.4	3.682	5.791
216.5	0.516	1.673	0.780	389.8	830.4	3.738	5.774
222.1	0.674	1.687	0.610	401.7	837.4	3.793	5.753
227.6	0.869	1.706	0.483	413.6	843.9	3.847	5.737
233.2	1.103	1.725	0.385	425.7	850.6	3.900	5.723
238.8	1.391	1.744	0.313	438.2	857.4	3.953	5.711
244.3	1.727	1.764	0.253	450.8	864.1	4.006	5.698
249.9	2.134	1.785	0.208	463.8	870.9	4.057	5.687
255.4	2.607	1.806	0.171	476.8	877.4	4.108	5.675
261.0	3.161	1.829	0.144	490.1	883.9	4.158	5.665
266.5	3.792	1.854	0.120	503.8	889.9	4.208	5.656
272.1	4.530	1.880	0.100	517.1	895.7	4.257	5.648
277.6	5.364	1.907	0.0830	530.1	901.3	4.305	5.641
283.2	6.309	1.936	0.0712	543.8	906.9	4.353	5.634
288.8	7.370	1.966	0.0614	557.3	912.3	4.401	5.627
294.3	8.570	2.003	0.0533	571.5	917.4	4.448	5.622
300.0	9.901	2.041	0.0465	585.9	922.0	4.495	5.616

Note: It may be assumed that, should propane and CO_2 form a liquid solution, the solution will exhibit ideal-solution behavior. State clearly and justify any additional assumptions that you make.

Thermodynamic Properties of Saturated Carbon Dioxide[a]

T (K)	P (bar)	V (condensed phase) (m^3/kg $\times 10^4$)	V (vapor) (m^3/kg $\times 10^4$)	H (condensed phase) (kJ/kg $\times 10^{-2}$)	H (vapor) (kJ/kg)	S (condensed phase) (kJ/kg K)	S (vapor) (kJ/kg K)
177.6	0.219	6.293	15,180	−2.826	3.005	2.539	5.823
188.8	0.614	6.355	5,730	−2.698	3.070	2.609	5.709
199.9	1.53	6.442	2,375	−2.561	3.124	2.681	5.526
205.4	2.343	6.492	1,570	−2.482	3.142	2.721	5.457
211.0	3.506	6.542	1,060	−2.384	3.156	2.766	5.393
216.5	5.159	6.611	725	−2.279	3.161	2.815	5.328
216.6	5.178	6.611	722	−2.277	3.161	2.816	5.327
216.6	5.178	8.490	722	−0.319	3.161	3.720	5.327
222.2	6.529	8.640	579	−0.214	3.177	3.767	5.295
227.6	8.150	8.796	468	−0.109	3.191	3.814	5.264
233.2	10.05	8.971	382	0	3.205	3.859	5.235
238.8	12.26	9.152	314	0.105	3.215	3.904	5.207
244.3	14.82	9.351	260	0.218	3.222	3.948	5.180
249.9	17.74	9.564	217	0.323	3.226	3.991	5.151
255.4	21.06	9.801	181	0.437	3.231	4.034	5.128
261.0	24.83	10.08	152	0.558	3.226	4.080	5.103
266.5	29.08	10.38	128	0.684	3.217	4.127	5.077
272.1	33.84	10.73	108	0.823	3.205	4.177	5.052
277.6	39.15	11.16	90.1	0.970	3.180	4.225	5.022
283.2	45.06	11.66	75.2	1.126	3.140	4.278	4.989
288.8	51.61	12.30	62.1	1.291	3.073	4.335	4.951
294.3	58.84	13.18	50.2	1.482	2.966	4.396	4.909
299.9	66.79	14.79	37.9	1.719	2.761	4.477	4.838
304.2	73.82	21.45	21.45	2.256	2.256	4.647	4.647

[a]Properties at temperatures below 216.6 K refer to solid as the condensed phase.

15.26 The stability criteria, Eq. (9-43), are often used to relate constants in an equation of state to critical constants. To illustrate, the well-known two-constant equation of state of Redlich-Kwong [Eq. (7-60)],

$$P = \frac{RT}{V-b} - \frac{a}{T^{1/2}V(V+b)}$$

employs a and b as constants for a pure material. To nondimensionalize these constants, we define new parameters, Ω_a and Ω_b:

$$a = \frac{\Omega_a R^2 T_c^{5/2}}{P_c} \quad \text{and} \quad b = \frac{\Omega_b R T_c}{P_c}$$

The terms Ω_a and Ω_b are dimensionless and can be evaluated from the stability criteria. Show that for this particular equation of state they are pure numbers and equal to

$$\Omega_a = [9(2^{1/3} - 1)]^{-1} \text{ and } \Omega_b = \frac{2^{1/3} - 1}{3}$$

15.27. Calculate the temperature at the limit of stability for a liquid mixture of 95% ethane and 5% *n*-butane at 1 bar. Use the Peng-Robinson equation of state and assume that the binary interaction parameter for ethane-*n*-butane is zero.

	Ethane	*n-Butane*
T_c (K)	305.4	425.2
P_c (bar)	48.8	38.0
ω	0.098	0.193

15.28. For a binary mixture of *B* and *C*, show that immiscibility occurs if

$$x_B\left(\frac{\partial \ln \gamma_B}{\partial x_B}\right)_{T,P} + 1 = 0$$

where γ_B is the activity coefficient of component *B*.

15.29. Rocky Jones claims that the Margules equation can be used for LLE problems even at the consolute point or limit of miscibility of a binary mixture. What do you think? Explain your answer, provide equations and suitable mathematical criteria. State and justify all assumptions made. The Margules equation can be expressed in ΔG^{EX} form as:

$$\Delta G^{EX} = RT\, x_1 x_2 (Ax_1 + Bx_2)$$

where *A* and *B* are binary parameters, independent of composition. Can *A* and *B* take on any value? How would you determine them experimentally?

15.30. Given that the activity coefficients for a binary liquid mixture are

$$RT \ln \gamma_1 = Cx_2^2 \quad \text{and} \quad RT \ln \gamma_2 = C(1 - x_2)^2$$

where *C* is a constant,

(a) Derive the appropriate mathematical criteria that must be satisfied for two stable liquid phases to exist.

(b) Express temperature as a function of composition along the boundary between metastable and unstable states in the two-liquid-phase region.

(c) Estimate the composition and temperature at the critical point. Above this critical temperature only one stable liquid phase could exist for any composition.

15.31. Below its triple point of –56.2°C, the vapor pressure of solid carbon dioxide is given by:

$$\ln P_{vp} = -3115/T + 16.00$$

where P_{vp} is in bars and T in K. The latent heat of fusion for pure CO_2 is 1990 cal/mol.

(a) Estimate the vapor pressure of liquid CO_2 at 0°C. State any and all assumptions used in making your estimate.

(b) Some investigators have argued that the presence of small amounts of low volatility hydrocarbons in the liquid phase lowers the CO_2 vapor pressure below its value for the pure state at the same temperature. Furthermore, they contend that for contaminant levels of approximately 1 mole % or less the decrease in vapor pressure will be linearly proportional to the mole fraction of contaminant. Develop an expression describing phase equilibrium in a 2-component system (CO_2 plus contaminant) to show under what conditions their assertion might be correct.

15.32. Several reputable culinary experts have argued that adding salt to a boiling pot of water that will be used to cook pasta decreases the cooking time because of a substantial increase in the temperature of boiling. You are to analyze this hypothesis to see if it is reasonable from a thermodynamic standpoint to establish a relationship between boiling temperature and amount of salt added. Clearly state all assumptions and simplifications.

$$\Delta H_{vap} \text{ of } H_2O = 2400 \text{ kJ/kg} \quad \text{and} \quad \Delta H_{fus} \text{ of } H_2O = 335 \text{ kJ/kg}$$

15.33. Often in dealing with polymer systems, the solution properties of a monodisperse polymer dissolved in a low-molecular-weight solvent are required. One example deals with the anomalous concentration dependence of the ratio of the osmotic pressure to polymer concentration (Π/C). This behavior is explained qualitatively using the Flory-Huggins model for $\Delta \underline{G}$. Another experimental observation is the appearance of a "cloud point" at a particular pressure, temperature, and polymer concentration which signifies the appearance of a second liquid phase. This phase-splitting behavior also should be predictable from the Flory-Huggins model with $\Delta \underline{G}$ given as:

$$\Delta \underline{G}/RT = N_1 \ln \Phi_1^* + N_2 \ln \Phi_2^* + \chi \Phi_1^* \Phi_2^* (N_1 + rN_2)$$

$$\Phi_1^* = N_1/(N_1 + rN_2) \quad \text{and} \quad \Phi_2^* = rN_2/(N_1 + rN_2)$$

where r = chain length (# of monomer units) and χ = Flory parameter

Furthermore, Flory and Schultz observed in 1952 that as the molecular weight (or chain length) of the polymer increased, the maximum temperature of the 2-phase region increased and the volume fraction at this maximum temperature decreased as shown in Figure P15.33(a) below. This maximum temperature point can be treated as a true critical point or upper consolute point, corresponding to the intersection of the binodal (or coexistence) and the spinodal curves.

(a) Show that:

$$\overline{G}_1 - G_1 = \mu_1 - \mu^o(\text{pure}) = RT\left[\ln(1 - \Phi_2^*) + (1 - 1/r)\,\Phi_2^* + \chi \Phi_2^{*2}\right]$$

(b) Using the appropriate stability criteria, derive two mathematical equations that are satisfied at this critical point. These equations should involve the volume fraction

of the polymer Φ_2^*, the chain length r, and the Flory parameter χ.
Hint: $(\partial\mu_i/\partial x_i)_{T,P} = 0$ is equivalent to $(\partial\mu_i/\partial\Phi_i^*)_{T,P} = 0$

(c) By satisfying these equations simultaneously derive general expressions for the polymer volume fraction Φ_{2c}^* and Flory parameter χ_c at this critical point. Show that as the polymer chain length increases to ∞, $\Phi_{2c}^* \rightarrow 1/(r)^{1/2}$ and $\chi_c \rightarrow 1/2$.

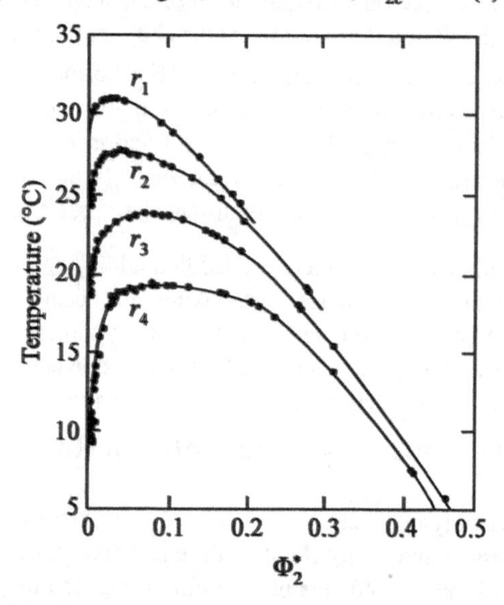

Figure P15.33(a) Cloud point curves of polystyrene dissolved in cyclohexane. Each curve represents a different chain length r for a monodisperse polymer. $r_1 > r_2 > r_3 > r_4$.

(d) If the Flory parameter χ_c at the critical point can be approximated as $\chi_c = w/RT_c$ with w = constant, is the behavior shown in Figure P15.33(b) below consistent with the results of part (c) above? Explain.

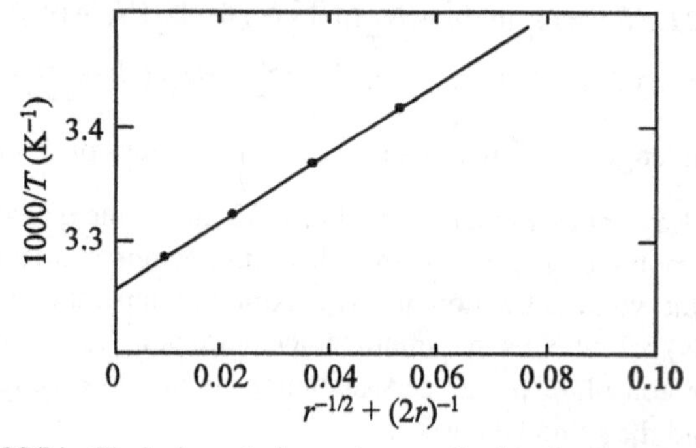

Figure P15.33(b) Chain length dependence of critical temperature T_c.

15.34. One separation process with many potential applications in chemical engineering involves the use of water in its supercritical state as a selective solvent. Although it is well known that ionic solids such as NaCl are readily soluble in H_2O at normal temperatures and pressures, their solubility drops off markedly under supercritical conditions. For NaCl a decrease from about 40 wt % at 70°C, 0.1 MPa to 0.004% by weight at 350°C, 25 MPa is observed. In contrast, organic compounds such as benzene (C_6H_6) that are essentially insoluble in H_2O at normal temperatures and pressures become very soluble in supercritical water.

(a) Investigators have proposed that these solubility phenomena are caused by non-ideal effects. Describe how you would estimate the solubility of C_6H_6 in H_2O as a function of T and P for temperatures ranging from 25 to 500°C, and pressures ranging from 0.1 to 30 MPa. Describe any and all property models needed, state assumptions used and thermodynamic data required, particularly those regarding *PVT* properties. Do no attempt to calculate numbers, just give your approach.

(b) According to the phase diagram in Figure 15.22, starting with a liquid solution containing 0.3 wt % NaCl at 25.0 MPa, and isobarically heating above the vapor-liquid equilibrium line ($T > T_{sat} = 390°C$) results in more liquid forming. This assumes that nucleation of liquid droplets is not a problem. However, if nucleation doesn't occur readily, it seems possible that a "maximum degree of superheat" above T_{sat} will be reached at the spinodal curve. Furthermore, behavior at 390°C, 1% NaCl seems to resemble a critical point. Develop a thermodynamic analysis to consider the validity (or non-validity) of these observations.

(c) Given a two-liquid phase feed of 0.5% C_6H_6 dispersed in an aqueous brine containing 10 wt % NaCl at 25°C and 25.0 MPa, design a process to separate H_2O and C_6H_6 as a dense vapor phase from a concentrated liquid solution of NaCl and H_2O. Sketch your process and relate its operating parameters to data given on the phase diagram of Figure 15.22.

(d) Discuss any important tradeoffs and practical considerations relevant to your design.

$$\text{for } H_2O: \; P_c = 22.1 \text{ MPa}, \quad T_c = 374°\text{C, and } \; V_c = 3.11 \times 10^{-3} \text{ m}^3\text{/kg}$$

15.35. The importance of carbon dioxide as a "greenhouse gas" has recently been in the news. Oceanographers and atmospheric scientists are attempting to model the movement of CO_2 in the earth's ecosystem. An important removal mechanism occurs when CO_2 is absorbed in the oceans and converted to calcium carbonate ($CaCO_3$) or limestone by certain marine organisms. Estimates of the solubility of CO_2 in ocean water as a function of temperature and depth are needed by scientists at MIT's Woods Hole Oceanographic Institution who are modeling CO_2 uptake by various marine organisms. Describe how you would estimate the solubility of CO_2 in ocean water over a temperature range from 0°C to 30°C and ocean depths ranging from 0 to 5000 m. The ambient air composition is, by volume, approximately: N_2 (79%), O_2 (20%), H_2O (1%) and CO_2 (0.035%). Ocean water can be approximated as a 3.5 wt % NaCl solution. You can also assume that the Peng-Robinson EOS can adequately model the $PV\underline{T}N$ behavior of air. Clearly state all additional assumptions made in your analysis.

15.36. Describe how you would estimate the temperature-composition behavior of a mixture of ortho- (o) and para- (p) cresol at 1 bar. Describe also how you would estimate the eutectic composition. Available data are listed below. State any assumptions made in your analysis regarding the behavior of *o*- and *p*-cresol.

Property	*o*-cresol	*p*-cresol
melting point, °C	30.94	34.80
triple point, °C	30.50	33.60
saturated liquid density (g/cm^3)	1.027	1.018
refractive index	1.5361	1.5312

Vapor pressure data P_{vp} in mm Hg, T in K

liquid-vapor $\ln P_{vp}^{[L-V]} =$ $15.91 - 3305/(T - 108)$—*o*-cresol
$16.20 - 3479/(T - 111)$—*p*-cresol

solid-vapor $\ln P_{vp}^{[S-V]} =$ $22.32 - 4536/(T - 109)$—*o*-cresol
$22.42 - 4690/(T - 112)$—*p*-cresol

15.37. The utilization of natural gas reserves with high carbon dioxide (CO_2) content poses an enrichment problem that requires separation. In addition, the proposed use of CO_2 for enhanced oil recovery creates further need to characterize the vapor-liquid equilibrium properties of CH_4-CO_2 mixtures. Cryogenic distillation has been proposed at low temperatures (<–50°C) and pressures below 4 MPa (600 psia) where the formation of solid CO_2 is possible.

Figure P15.37(a) [Holmes *et al.* (May 1982) ***Hydrocarbon Processing***, p 131] and Figure P15.37(b) [Donnelly and Katz (1954), *Ind. Eng. Chem.*, Vol. 46(3), p 511] show the *PTx* behavior of the CH_4-CO_2 system. You have been asked to develop a computational model for predicting the VLE behavior between the 3-phase (solid CO_2-liquid-vapor) line and the binary critical locus.

You can assume that the Peng-Robinson $PV\underline{T}N$ equation of state applies to the binary with the following critical constants and other data:

Property	CO_2	CH_4
P_c	7.38 MPa	4.60 MPa
V_c	93.9 cm^3/mol	99.2 cm^3/mol
T_c	304.1 K	190.4 K
Z_c	0.274	0.288
ω	0.239	0.011
triple pt.	–58°C, 0.55 MPa	–182°C, 0.012 MPa

A good starting value for the binary interaction parameter δ_{ij} is 0.09. In addition, the vapor pressure of pure solid CO_2 is available from Perry's 6th edition as:

P_{vp} (mm Hg)	T (°C)
1	−134.3
5	−124.4
10	−119.5
20	−114.4
40	−108.6
60	−104.8
100	−100.2
200	− 93.0
400	− 85.7
760	− 78.2

The density of solid CO_2 can be taken as 1540.26 kg/m^3. The ideal gas state heat capacity in J/mol K for CO_2 and CH_4 are as follows:

$$CO_2:\ C_p^o = 19.80 + 0.07344\,T - 5.602\times10^{-5}\,T^2 + 1.715\times10^{-8}\,T^3$$
$$CH_4:\ C_p^o = 19.25 + 0.05213\,T + 1.197\times10^{-5}\,T^2 - 1.132\times10^{-8}\,T^3$$

You are to predict the binary vapor-liquid envelope between the 3-phase and critical loci for the following overall compositions of 12, 29.5, 45.7, and 82% by mol CH_4 as shown in Figure P15.37(b). In addition, you should show how your model works for predicting the vapor pressure curves for pure CO_2 and CH_4 from their critical points to their triple points. In preparing your solution you will need to develop suitable mathematical criteria that are applicable along the binary critical locus. You may want to consult the paper by Peng and Robinson [(1977), *AIChE J.*, **23**(2), p 137] to help you with this development.

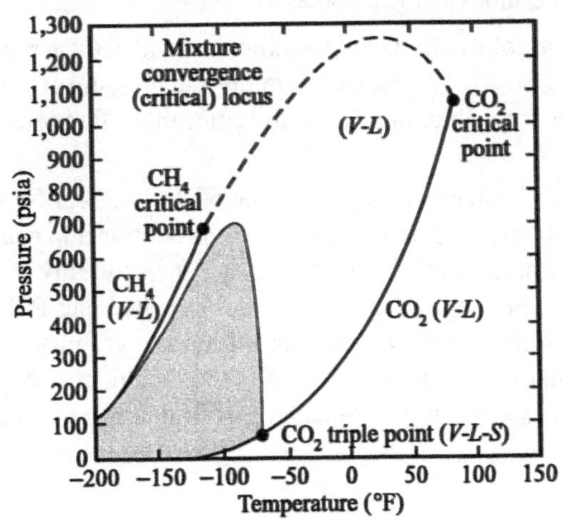

Figure P15.37(a) Phase diagram CH_4-CO_2 binary [Holmes (1982)].

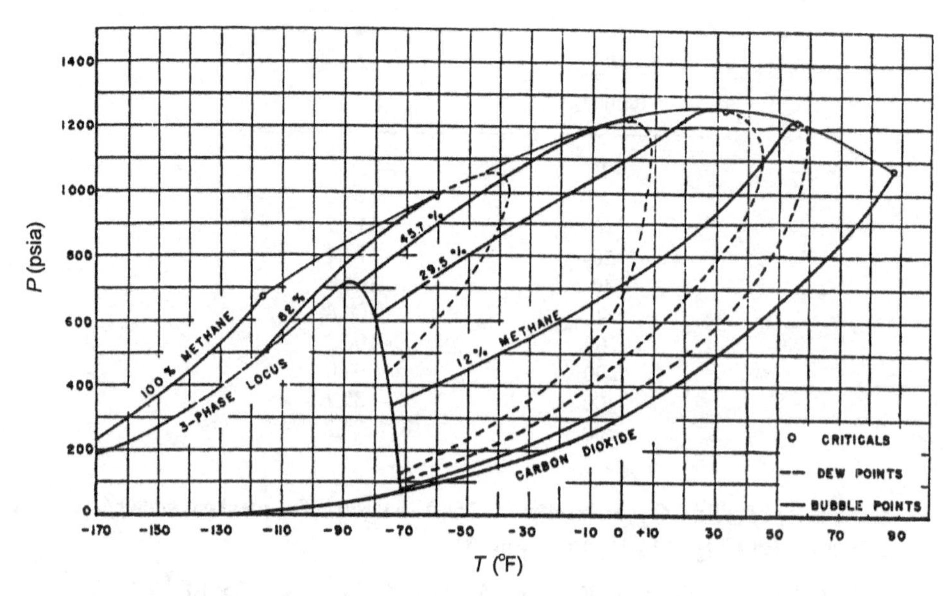

Figure P15.37(b) Pressure-temperature projections for the CH_4-CO_2 binary [Donnelly and Katz (1954)].

15.38. According to McHugh and Krukonis [(1986) *Supercritical Fluid Extraction*, Butterworths, London], coffee decaffeination with carbon dioxide has been an active area of process development resulting in many US and foreign patents. For example, the extraction of neat caffeine into supercritical CO_2 is used in several commercial processes, with current worldwide production levels well in excess of 60,000,000 lbs per year. In this problem you are to analyze several fundamental aspects of the supercritical extraction process from a thermodynamic perspective.

(a) Assume that solid caffeine is in equilibrium with CO_2 at pressures and temperatures somewhat above the critical temperature and pressure of pure CO_2. Derive a general expression for estimating the equilibrium mole fraction of caffeine dissolved in supercritical CO_2.

(b) Experimental solubility data along the 60°C isotherm for caffeine in supercritical CO_2 are shown in Figure P15.38. Using the relationship you derived in part (a) write a computer program and use it to estimate the solubility of caffeine in CO_2 at 60°C over a pressure range from 1 to 350 bar. Use the Peng-Robinson cubic equation of state to describe the $PVTy_i$ relationship for the gas phase. Other property data for caffeine and CO_2 are listed below. You should treat δ_{12}, the binary interaction parameter, as adjustable to obtain a best fit to the experimental solubility data.

Property	Caffeine	Carbon dioxide
Structural formula		O=C=O
	1,3,7 Trimethylxanthine	CO_2
Molecular weight	194.2	44.01
T_c (in K)	*	304.1
P_c (in bar)	*	73.8
ω acentric factor	*	0.239
solid phase density (kg/m^3)	1230	—
melting point (in °C)	236.0	–55.6

*Since the critical constants are not available for caffeine you will need to estimate values in order to use the Peng-Robinson equation of state. There are a number of techniques available but for the purposes of this problem, we suggest that you should use Joback's modification of the Lydersen group contribution method. This method was developed at MIT by Kevin Joback and is discussed in Section 13.3.

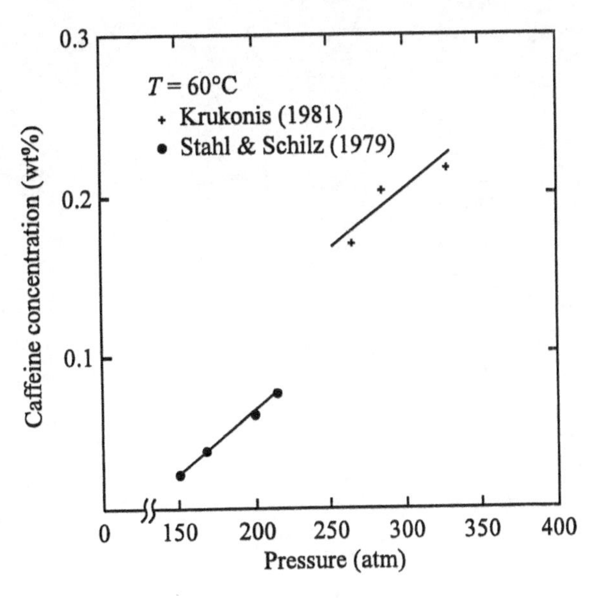

Figure P15.38 Caffeine solubility data at 60°C.

Once you have estimated the critical constants for caffeine you can then estimate the acentric factor ω using the vapor pressure data from Ebeling and Franck (1984) in *Ber. Bunsenges. Phys. Chem.* **88**, p 862-865 and from Bothe and Cammenga (1979) in *J. of Thermal Analysis*, **16**, p 267-275. Solid and liquid caffeine vapor pressure data can be represented by the following correlations:

for solid caffeine ($T < 236°C$)

$\log_{10}(P_{vp} \text{ in Pa}) = -5781/T + 15.031$ (from Bothe and Cammenga)

$\log_{10}(P_{vp} \text{ in bar}) = -5411/T + 7.46 + \log_{10}(0.08314\, T)$ (from Ebeling and Franck)

for liquid caffeine ($T > 236°C$)

$\log_{10}(P_{vp} \text{ in Pa}) = -3918/T + 11.143$ (from Bothe and Cammenga)

Chemical Equilibria 16

In Chapter 6 the criteria of chemical equilibrium were developed. We had chosen an isolated system with π phases and n components. No internal barriers were present. We allowed only a single chemical reaction to occur among i components ($i \leq n$) with specified stoichiometry. At equilibrium it was shown that the temperatures and pressures were equal in all phases, as were the chemical potentials of all components. The introduction of the single chemical reaction led to Eq. (6-55), which indicated that the sum of products of the chemical potential of each component multiplied by its stoichiometric multiplier in the assumed reaction was zero:

$$\sum_{i=1}^{n} \nu_i \mu_i = 0 \tag{6-55}$$

In the present chapter we analyze in greater depth the concepts behind chemical equilibria and employ the results to calculate compositions of equilibrium systems. There have been a number of published papers that relate to our objective. In most instances, however, the primary aim of the authors was to describe computational algorithms to obtain numerical results in complex systems. Although we note some of the algorithms, the primary intent of the chapter is on fundamental concepts, and the interested reader is referred to the literature for computing methods.

16.1 Problem Formulation and General Approach

In essence we wish to obtain thermodynamic relationships applicable to a multicomponent system in chemical equilibrium. The system is defined as stated in Postulate I by two independently variable properties plus the masses of all components initially present. In the equilibrium state we also need to specify two constraints so that the moles (or masses) of all chemical entities may be determined. In the development in Chapter 6, we chose the system to be isolated so that the constraints were constant *total* energy and volume (as well as mass). The equilibrium criterion was then given in Eq. (6-39). With the phase equilibrium relationships for T and P [Eqs. (6-44) and (6-45)] and the energy and volume restrictions [Eqs. (6-40) and (6-41)], Eq. (6-39) may be written as

$$\delta \underline{S}\Big|_{\underline{U},\underline{V}} = \sum_{s=1}^{\pi} \sum_{j=1}^{n} \underline{S}_{N_j}^{(s)} \, \delta N_j^{(s)} = 0 \tag{16-1}$$

With $\underline{S}_{N_j}^{(s)} = -\mu_j^{(s)}/T$, Eq. (16-1) becomes

$$\sum_{s=1}^{\pi} \sum_{j=1}^{n} \mu_j^{(s)} \, \delta N_j^{(s)} = 0 \tag{16-2a}$$

Note that the mole balance given in Eq. (6-42) was not used since chemical reactions can occur; we will have to develop an alternative relation to allow the conservation of atoms rather than moles of specific components. Also, to this point the equality of chemical potentials in the different phases was not invoked, since it will be convenient in our treatment to allow this fact to appear as a consequence of the analysis.

Let us stop briefly and examine Eq. (16-2a). The variations $\delta N_j^{(s)}$ could arise due to a transfer of δN_j between phases or as a result of one or more chemical reactions that involve N_j. Irrespective of the origin of the variations, within the limitation of constant total $\underline{U}$ and $\underline{V}$, the sum in Eq. (16-2a) must be zero for an equilibrium state. If we had not chosen a system at constant total $\underline{U}$ and $\underline{V}$, but instead, a system at constant T and P, then the equilibrium criterion would have been Eq. (6-32) (i.e., the Gibbs energy is minimized). Expanding $\underline{G}(T, P, N_j^{(s)})$ with T, P constant, one again obtains Eq. (16-2a). Thus, the same equilibrium criterion is obtained and virtual variations of $\delta N_j^{(s)}$ must now be conceived bearing the constant T, P restrictions in mind. Equation (16-2a) would also have been obtained if the system were held at constant volume and temperature [here, the Helmholtz energy would be minimized and Eq. (6-29) would be used]. Other constraints could be used, but as long as no restrictions are placed on any $\mu_j^{(s)}$ or $\delta N_j^{(s)}$ (as by, for example, a semipermeable membrane), Eq. (16-2a) is obtained as the basic equilibrium criterion.

There is one additional important comment relevant to the understanding and use of Eq. (16-2a). The component summation is designated as $j = 1, \ldots, n$. The choice of the components present in the equilibrium mixture must be made by the investigator on some rational basis. Clearly, the components in the j vector are not limited to those that might have been used to synthesize the system initially. Usually, but not always, the j vector will contain the original components in addition to others which may appear as a result of some chemical transformation.

Consider some simple examples. If H_2 and O_2 were the only species charged to the original system, what should we include in our j vector at equilibrium? Obviously, no component that contained elements other than H or O could be specified (i.e., no nuclear transformations are considered). H_2O would be an obvious choice of another component if evidence were available indicating that it could form under the actual state of the system. Should the H radical be included--or H_2O_2--or other components? If water is to be present, is it in the vapor phase, the solid, the liquid, or all three? Questions such as these are vital to the solution of any chemical equilibrium problem, but they are often not easily answered. Generally, one can include all the possible components desired--in as many phases as desired--and then force the computational scheme to show that actual equilibrium compositions of most are negligibly small (or zero). This general inclusion principle also applies to selecting components which exist in more than one phase. The disadvantage of including many components lies in greater computational complexity. But if a possible component is not included, thermodynamics provides no mechanism for its introduction.

The important concept to keep in mind when dealing with chemical equilibria problems is embodied in Postulate I. If we specify the initial charge of both reacting and non-reacting (inert) components along with two independent properties such as T and P, the final equilibrium state is completely specified. This includes a complete description of the phases present and their composition. In most chemical engineering applications, we will fix T and P and want to estimate the equilibrium composition. In these situations the basic problem reduces to determining the minimum of the total Gibbs free energy $\underline{G}$, which can be expressed as the following for a system consisting of π phases and n components:

$$\underline{G} = \sum_{s=1}^{\pi} \sum_{j=1}^{n} \overline{G}_j^{(s)} N_j^{(s)} = \sum_{s=1}^{\pi} \sum_{j=1}^{n} \mu_j^{(s)} N_j^{(s)}$$

subject to the constraints of constant T, P, and total mass, M. The initial charge of components (N_i^o where i=1, . . . , n) must also be known which permits one to apply an elemental atom balance as a conservation equation. The minimization of $\underline{G}$ for the total system is equivalent to satisfying a set of reaction constraining equations given in Eq. (6-55):

$$\sum_{j=1}^{n} \nu_{j,r} \mu_j^{(s)} = 0 \quad \text{for all reactions } r$$

If we have appropriate constitutive models for $\overline{G}_j$ or μ_j as a function of T, P, and x_i, then we can proceed directly to determine the equilibrium condition as a constrained non-linear minimization of $\underline{G}$ (see Example 16.5). Without specifying any particular reaction (or set of reactions) for the n-component system, a non-stoichiometric formulation results

$$\delta \underline{G}\Big|_{T,P,M} = \sum_{s=1}^{\pi} \sum_{j=1}^{n} \mu_j^{(s)} \delta N_j^{(s)} = \sum_{j=1}^{n'} \mu_j \delta N_j = 0 \qquad \text{(16-2b)}$$

where in the last summation we have assumed that $\mu_j^{(1)} = \ldots = \mu_j^{(s)} = \ldots = \mu_j^{(\pi)}$ (phase equilibrium) and set $n \leq n' \leq n\pi$.

Alternatively, we can reformulate the problem by incorporating the stoichiometry of all reactions using the extent of reaction variable, ξ_r, for each reaction r that was introduced in Chapter 6 [see Eq. (6-51)]. Now, with $d\xi_r \equiv \delta N_{j,r}/\nu_{j,r}$:

$$\delta \underline{G}\Big|_{T,P,M} = \sum_{j=1}^{n'} (\nu_{j,r}\,\mu_j)\,\delta\xi_r = 0 \tag{16-2c}$$

for all reactions r. Here $\nu_{j,r}$ is the stoichiometric coefficient for component j in reaction r. If a particular component does not participate in the reaction, then $\nu_{j,r} = 0$ while $\nu_{j,r} > 0$ (positive) for products and $\nu_{j,r} < 0$ (negative) for reactants by convention. Clearly to carry out the minimization of $\underline{G}$ properly, one must correctly specify the independent components and/or independent chemical reactions!

In summary, to treat systems in chemical equilibrium, we specify the overall system by the total number of atoms (or the mass) of each element present, and we place two additional independent restraints (e.g., T and P) as required by Postulate I. We can then use Eqs. (16-2) in a variety of equivalent forms to compute the moles of all selected components in all phases.

16.2 Conservation of Atoms

So far, we have stressed that there is a selection process to specify the components present in the equilibrium system. These components are energetically favorable aggregates of atoms which comprise the system. Let us assume that the total number of elements present is ℓ and that each component present can be represented as $\underset{\sim}{W}_j$. Following Smith (1980), we consider identical chemical species in different phases as distinct components. Thus $H_2O_{(g)}$ and $H_2O_{(l)}$ would be represented by two different $\underset{\sim}{W}_j$'s (the formula vector).

A matrix $\underset{\approx}{\mathcal{D}}$ is formed with the columns as formula vectors, $\underset{\sim}{W}_j (j = 1, \ldots, n')$ and rows as elements $(i = 1, \ldots, \ell)$. $\underset{\approx}{\mathcal{D}}$ represents a ℓ-dimensional vector space with n' components.

Thus,

j components ⟶

$$\underset{\sim}{W}_j \equiv \begin{bmatrix} d_{1j} \\ d_{2j} \\ \cdot \\ \cdot \\ \cdot \\ d_{\ell j} \end{bmatrix} \quad \text{and} \quad \underset{\approx}{\mathcal{D}} \equiv \begin{bmatrix} d_{11} & \cdots & d_{1n'} \\ \vdots & d_{ij} & \vdots \\ d_{\ell 1} & \cdots & d_{\ell n'} \end{bmatrix} = [d_{ij}] \tag{16-3a}$$

i elements/atoms ⟵

The element d_{ij} then represents the number of atoms i in component j. If b_i is the total number of atoms of i (divided by Avogadro's number) present in the system, the conservation of atoms leads to

$$\sum_{j=1}^{n'} d_{ij} N_j - b_i = 0, \quad i = 1, \ldots, \ell \tag{16-3b}$$

Since we are treating the same chemical species in different phases as distinct entities, the value of n' is set as $n \le n' \le n\pi$, where n is the number of distinct chemical entities [as in Eq. (16-2b)] and π is the number of phases.

As Zeleznik and Gordon (1968) point out, if ionization occurs, Eq. (16-3b) should also include a conservation of charges. This is done by assuming charge is the $(\ell + 1)$st element, letting $b_{\ell+1} = 0$ and then $d_{\ell+1,j}$ is the charge on the j component.

The elemental atom balance given by Eq. (16-3b) can also be vectorized to:

$$\underset{\approx}{\mathcal{D}}\underset{\sim}{N} = \underset{\sim}{b} \tag{16-3c}$$

where

$$\underset{\sim}{N} \equiv \begin{bmatrix} N_1 \\ \cdot \\ \cdot \\ \cdot \\ N_{n'} \end{bmatrix} \quad \text{and} \quad \underset{\sim}{b} \equiv \begin{bmatrix} b_1 \\ \cdot \\ \cdot \\ \cdot \\ b_\ell \end{bmatrix}$$

In addition to Eq. (16-3), there is a nonnegativity constraint,

$$N_j \ge 0 \tag{16-4}$$

16.3 Nonstoichiometric Formulation

In this section we combine Eqs. (16-2) and (16-3) to provide a set of equations from which N_j may be calculated. The technique is termed *nonstoichiometric* because no specific chemical reactions are introduced.

We first write Eq. (16-2b) with the chosen formula vector set $j = 1, \ldots, n'$ as

$$\sum_{j=1}^{n'} \mu_j \, \delta N_j = 0 \qquad (T, P, \text{mass fixed}) \tag{16-5}$$

We then multiply each atom balance equation in Eq. (16-3b) by a Lagrange multiplier η_i and, after taking variations, we combine with Eq. (16-5). The result is

$$\sum_{j=1}^{n'} \left(\mu_j + \sum_{i=1}^{\ell} \eta_i d_{ij} \right) \delta N_j + \sum_{i=1}^{\ell} \left(\sum_{j=1}^{n'} d_{ij} N_j - b_i \right) \delta \eta_i = 0 \tag{16-6}$$

Treating the variations of N_j and η_i as independent, we recover Eq. (16-3b) and, also,

$$\mu_j + \sum_{i=1}^{\ell} \eta_i d_{ij} = 0, \quad j = 1, \ldots, n' \tag{16-7}$$

Equations (16-3b) and (16-7) comprise the necessary relations to solve for the n' values of N_j and the ℓ values of η_i. That is, the unknowns total $n' + \ell$, while the equations in Eqs. (16-3b) and (16-7) also total $n' + \ell$. Also, since η_i and d_{ij} are independent of the phase(s) that contain components with the same structural atomic formula, Eq. (16-7) implies that for these chemically identical components, μ_j is the same in all phases.

Example 16.1

Assume that we have a given system in chemical equilibrium at some specified T and P. The elements comprising the system are C, H, and O. We have assumed that the only components present in the equilibrium mixture are CO_2, H_2, H_2O, and C. We also have reason to believe that water is present only in the gas phase, but inconsequential amounts of carbon appear in the gas. We then have four components: (1) $CO_{2(g)}$; (2) $H_{2(g)}$; (3) $H_2O_{(g)}$; (4) $C_{(s)}$. Develop the nonstoichiometric relations to allow calculation of N_{CO_2}, N_{H_2}, N_{H_2O}, and N_C at equilibrium, assuming that b_C, b_H, and b_O are, respectively, the total quantities of C, O, and H present.

Solution

We first develop the matrix $\underset{\approx}{\mathcal{D}}$.

$$\begin{array}{c|cccc|} & CO_2 & H_2 & C & H_2O \\ \hline C & 1 & 0 & 1 & 0 \\ H & 0 & 2 & 0 & 2 \\ O & 2 & 0 & 0 & 1 \end{array} = \underset{\approx}{\mathcal{D}}$$

Note that the columns denote the components ($j = 1, 2, 3, 4$), whereas the rows include the elements ($i = 1, 2, 3$). A matrix element, say d_{24}, would then represent the atoms of H in a molecule of H_2O. With Eqs. (16-3) and (16-7),

$$N_{CO_2} + N_C = b_C \qquad (\alpha)$$

$$2N_{H_2} + 2N_{H_2O} = b_H \qquad (\beta)$$

$$2N_{CO_2} + N_{H_2O} = b_O \qquad (\gamma)$$

$$\mu_{CO_2} + \eta_C + 2\eta_O = 0 \qquad (A)$$

$$\mu_{H_2} + 2\eta_H = 0 \qquad (B)$$

$$\mu_C + \eta_C = 0 \qquad (C)$$

$$\mu_{H_2O} + 2\eta_H + \eta_O = 0 \qquad (D)$$

We can express the chemical potential of any component j as

$$\mu_j = G_j^o + RT \ln \frac{\hat{f}_j}{f_j^o} \qquad (16\text{-}8)$$

where Eq. (9-107) was used twice, once to specify μ_j in the reacting mixture and once for μ_j in some arbitrarily defined pure standard state at the system temperature T where

$$\mu_j^o = G_j^o$$

and the fugacity of pure j in this standard state is f_j^o. For a gas phase, Eq. (9-120) may be used to express $\hat{f}_j$ in terms of pressure, mole fraction, and fugacity coefficient, for example,

$$\hat{f}_j = \hat{\phi}_j(y_j, T, P)Py_j \tag{16-9}$$

If a liquid solution were present, Eq. (16-9) may be used if $\hat{\phi}_j$ (liquid) can be determined from a suitable $PV\underline{T}N$ EOS. Alternatively, Eq. (9-158) can be introduced to employ activities. If the reference state for j is the pure material, Eq. (9-164) is applicable:

$$\hat{f}_j = \gamma_j(x_j, T, P)f_j(T, P)x_j \tag{16-10}$$

where γ_j would be evaluated from a ΔG^{EX} model. With either Eq. (16-9) or (16-10) substituted into Eq. (16-8), μ_j becomes a function of mole fractions (as well as T and P and the reference-state conditions). Thus, the set of seven equations (A) through (D), (α) through (γ) for N_{CO_2}, H_{H_2}, N_C, N_{H_2O}, η_C, η_H, η_O may be solved.

The astute reader of Example 16.1 will probably have noted that the seven equations developed can be greatly simplified and the solution expedited if (B) is multiplied by 2, (C) by –1, and (D) by –2, and then (A) through (D) added. Then all η_i terms cancel and the set of equations is immediately reduced from seven to four, with the unknowns being the mole numbers of CO_2, H_2, C, and H_2O.

Suppose, however, that the problem discussed in Example 16.1 is made more realistic by adding CO as a component in the equilibrium mixture. Equations (A) through (D) would be unaffected, but $\underline{\underline{\mathcal{D}}}$ would have an additional column of CO with the column elements 1, 0, 1. Thus, a new relation is necessary:

$$\mu_{CO} + \eta_C + \eta_O = 0 \tag{E}$$

and (α) and (γ) would have to be modified to include the moles of CO. There are now eight equations in eight unknowns. One still suspects that there is some convenient scheme to collapse the set of equations (A) through (E) to eliminate η_C, η_H, and η_O. The systematic method that will accomplish this elimination is, in fact, embodied in the *stoichiometric* formulation discussed in Section 16.4.

Example 16.2

Suppose that in Example 16.1 we had reason to believe that water existed in the equilibrium mixture both in the gas and liquid states. Also, we suspect that significant quantities of CO_2 dissolve in the water, but we neglect any solubility of H_2 (and carbon) in the aqueous phase. How do the nonstoichiometric relations change?

Solution

Following the format of Example 16.1, since $n' = 6$, $\underset{\approx}{\mathcal{D}}$ is expanded to include two columns for CO_2 [$CO_{2(g)}$ and $CO_{2(l)}$] and two columns for water [$H_2O_{(g)}$ and $H_2O_{(l)}$]. The elements in each of these column sets are, of course, identical. When Eq. (16-7) is applied, we find that we have two equations for CO_2 [for $\mu_{CO_{2(g)}}$ and $\mu_{CO_{2(l)}}$] and, similarly for H_2O. Combining these, we obtain

$$\mu_{CO_{2(g)}} = \mu_{CO_{2(l)}} \quad \text{and} \quad \mu_{H_2O_{(g)}} = \mu_{H_2O_{(l)}}$$

as expected for this specified condition of liquid-vapor phase equilibrium. There are three equations from Eq. (16-3b), containing as variables $N_{CO_{2(g)}}$, $N_{CO_{2(l)}}$, N_C, $N_{H_2O_{(g)}}$, $N_{H_2O_{(l)}}$, and N_{H_2}, and six more with the three Lagrange multipliers, η_C, η_H, and η_O, totaling nine. The solution is obtained from the set of four equations from Eq. (16-7), three equations from Eq. (16-3b), and the two equalities of chemical potential of CO_2 and H_2O since $\mu = f(T, P, \text{composition})$.

16.4 Stoichiometric Formulation

Before developing this formulation, we should return briefly to Eq. (16-3) and modify it slightly to make it more general. The index on i was set to include $i = 1, \ldots, \ell$. Actually, ℓ should be replaced by the integer that yields the rank of $\underset{\approx}{\mathcal{D}}$. In matrix mathematics, the rank of any matrix is the smallest non-vanishing determinant formed by reducing the number of rows or columns of the matrix. In other words, if

$$\det[\underset{\approx}{\mathcal{D}}] = |\underset{\approx}{\mathcal{D}}| = 0$$

then you must reduce the number of rows or columns until

$$\det[\text{reduced } \underset{\approx}{\mathcal{D}}] \neq 0$$

In most cases the rank will equal the number of chemical elements, ℓ, but as shown in Example 16.3, there are exceptions.

Example 16.3

Assume that an equilibrium mixture contains CO, H_2 and methanol. Write the nonstoichiometric relations.

Solution

$$\begin{array}{c|ccc|} & CO & H_2 & CH_3OH \\ C & 1 & 0 & 1 \\ H & 0 & 2 & 4 \\ O & 1 & 0 & 1 \end{array} = \underset{\approx}{\mathcal{D}}$$

The three equations from Eq. (16-7) are

$$\begin{array}{lllll} \mu_{CO} & +\eta_C & & +\eta_O & =0 \\ \mu_{H_2} & & +2\eta_H & & =0 \\ \mu_{CH_3OH} & +\eta_C & +4\eta_H & +\eta_O & =0 \end{array}$$

But, with Eq. (16-3)

$$\begin{array}{llll} N_{CO} & & +N_{CH_3OH} & =b_C \\ & 2N_{H_2} & +4N_{CH_3OH} & =b_H \\ N_{CO} & & +N_{CH_3OH} & =b_O \end{array}$$

We see that the C and O balances are identical. Thus, the moles of CO and CH_3OH are not independent. The rank of $\underset{\approx}{\mathcal{D}}$ is 2, whereas $\ell = 3$. Therefore, in Eq. (16-3), i should equal 1 and 2.

With this minor, but important change in Eq. (16-3), we now introduce a variation of the formulation which effectively eliminates the Lagrange multipliers η_i. As Zeleznik and Gordon (1968) point out, these undetermined multipliers originated since we desired to treat each δN_j as independently variable. Another approach is to use the concept of ***independent reactions***, in which the δN_j for each reaction is not independent. If we had no atom restrictions, the number of conceivable reactions one could formulate would be n'. But with the restrictions imposed by Eq. (16-3), this number is reduced to $n' - C$, where C is the rank of $\underset{\approx}{\mathcal{D}}$. We write each of these *independent reactions* as

$$\sum_{j=1}^{n'} \mathbf{W}_j \nu_{j,r} = 0 \quad or \quad \underset{\sim}{\mathbf{W}}_j \underset{\sim}{\nu}_r \quad \text{for} \quad r = 1, \ldots, n' - C \tag{16-11}$$

where $\underset{\sim}{\mathbf{W}}_j$ is the formula vector for component j and $\underset{\sim}{\nu}_r$ is the column vector of stoichiometric coefficients $\nu_{j,r}$. Note that we have adopted a vector subscript notation for $\nu_{j,r}$, the stoichiometric coefficient for component j in reaction r. The conservation of elements requires that

$$\sum_{j=1}^{n'} d_{ij} \nu_{j,r} = 0 \quad \text{for} \quad r = 1, \ldots, n' - C \tag{16-12}$$

Thus by employing the $\underset{\approx}{\mathcal{D}}$ matrix, we obtain

$$\underset{\approx}{\mathcal{D}} \underset{\sim}{\nu}_r = 0 \tag{16-13}$$

If we include all independent $(n' - C)$ reactions, the column vector $\underset{\sim}{\nu}_r$ can be expanded to a full, two-dimensional matrix $\underset{\approx}{\nu}$ with

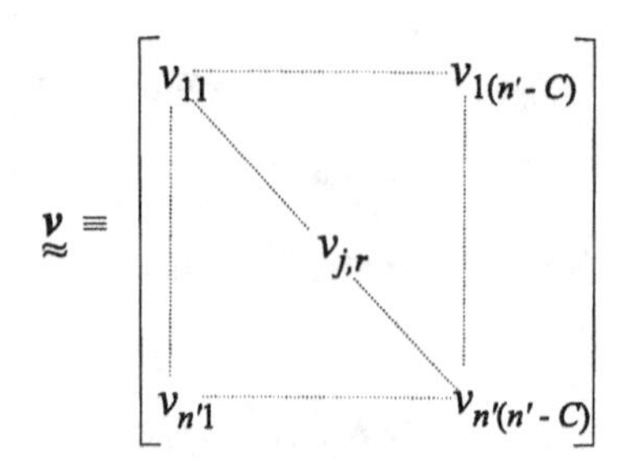

such that

$$\underset{\approx}{\mathcal{D}}\,\underset{\approx}{\nu} = 0$$

Example 16.4

For the CO_2, H_2, C, H_2O, CO example described in Example 16.1 and in the following text, specify the number of independent reactions and suggest appropriate stoichiometric coefficient vectors.

Solution

Rewriting $\underset{\approx}{\mathcal{D}}$ for this case,

$$\begin{array}{c|ccccc|l} & CO_2 & H_2 & C & H_2O & CO & \\ C & 1 & 0 & 1 & 0 & 1 & \\ H & 0 & 2 & 0 & 2 & 0 & = \underset{\approx}{\mathcal{D}} \\ O & 2 & 0 & 0 & 1 & 1 & \end{array}$$

When $\underset{\approx}{\mathcal{D}}$ is reduced to echelon form by Gauss-Jordan elimination, it can be shown the rank C of $\underset{\approx}{\mathcal{D}}$ is 3. With $n' = 5$, the number of independent reactions is $5 - 3 = 2$. Equation (16-11) or (16-12) states that

$$\begin{bmatrix} d_{11} & d_{12} & d_{13} & d_{14} & d_{15} \\ d_{21} & d_{22} & d_{23} & d_{24} & d_{25} \\ d_{31} & d_{32} & d_{33} & d_{34} & d_{35} \end{bmatrix} \times \begin{bmatrix} \nu_1 \\ \nu_2 \\ \nu_3 \\ \nu_4 \\ \nu_5 \end{bmatrix} = \begin{bmatrix} 0 \\ 0 \\ 0 \end{bmatrix}$$

With the given elements in $\underset{\approx}{\mathcal{D}}$,

$$\begin{array}{llllll} \nu_1 & & +\nu_3 & & +\nu_5 & = 0 \\ & +2\nu_2 & & +2\nu_4 & & = 0 \\ 2\nu_1 & & & +\nu_4 & +\nu_5 & = 0 \end{array}$$

Eliminating, for example, ν_3, ν_4, and ν_5.

$$\nu_3 = \nu_1 - \nu_2$$
$$\nu_4 = -\nu_2$$
$$\nu_5 = \nu_2 - 2\nu_1$$

If $\nu_1 = 1$ and $\nu_2 = 0$, then $\nu_3 = 1$, $\nu_4 = 0$, $\nu_5 = -2$, one independent reaction would be

$$CO_2 + C - 2CO = 0$$

For the second reaction, let $\nu_1 = 0$ and $\nu_2 = 1$; then $\nu_3 = -1$, $\nu_4 = -1$, and $\nu_5 = 1$. Then

$$H_2 - C - H_2O + CO = 0$$

Thus, there are two stoichiometric column vectors, [1, 0, 1, 0, –2] and [0, 1, –1, –1, 1] for the two reactions. These are two possible independent reactions. Other methods of eliminating ν_j values or in choosing different values of ν_1, ν_2 may lead to different reactions, but these can always be obtained by linear combinations of the two shown above.

Equation (16-12) [or (16-13)] is a very important relation. We can use it directly to provide an equilibrium formulation with no Lagrange multipliers as follows. In Eq. (16-7), for a particular reaction r, multiply by $\nu_{j,r}$ and sum

$$\sum_{j=1}^{n'} \nu_{j,r}\mu_j + \sum_{j=1}^{n'} \nu_{j,r} \sum_{i=1}^{\ell} \eta_i d_{ij} = 0 \qquad (16\text{-}14)$$

Inverting the two summations in the second term, and using Eq. (16-12), we obtain

$$\sum_{j=1}^{n'} \nu_{j,r}\mu_j = 0, \qquad r = 1, \ldots, n' - C \qquad (16\text{-}15)$$

In this stoichiometric formulation we have n' variables (N_j). There are $n' - C$ equations of the form of Eq. (16-15) and C atom constraint balances from Eq. (16-3). (As noted earlier, in the majority of all situations one may replace C by ℓ, the number of chemical elements.) Whereas in this method we have eliminated the Lagrange multipliers, it is now necessary to specify the appropriate sets of stoichiometric multipliers for all independent reactions. In complex systems, this can sometimes lead to difficulties.

It is emphasized that the independent reactions introduced in the stoichiometric formulation would not normally have any connection with the actual chemical reaction path. They were developed as an artifice to relate δN, for each reaction, to changes in a single variable by the stoichiometric coefficient vector. That is, for each independent reaction,

$$\delta N_{j,r} = \nu_{j,r}\, d\xi_r, \qquad r = 1, \ldots, n' - C \qquad (16\text{-}16)$$

The variable ξ_r is called the *extent of reaction* for reaction r.

$$\delta N_j = \sum_{r=1}^{n'-C} \delta N_{j,r} = \sum_{r=1}^{n'-C} v_{j,r}\, d\xi_r \tag{16-17}$$

Integrating Eq. (16-17) yields

$$N_j = N_{j(initial)} + \sum_{r=1}^{n'-C} v_{j,r}\, \xi_r, \quad j = 1, \ldots, n' \tag{16-18}$$

N_j represents the moles of j in the equilibrium mixture and equals the moles of N_j present initially as modified by chemical reactions. Since μ_j is a function of N_j at a given P and T, then Eq. (16-18) may be used with Eq. (16-15) to provide $n' - C$ relations between the $n' - C$ extents of reaction. For example, assume that we have only a gas phase which behaves as an ideal gas mixture; then Eq. (16-8) becomes

$$\mu_j = G_j^o + RT \ln y_j \tag{16-19}$$

or

$$\mu_j = G_j^o + RT \ln \frac{N_j}{N} \tag{16-20}$$

In arriving at Eq. (16-19) from Eq. (16-8), we made several assumptions: (1) the fugacity of j in the standard state is given as P since an ideal gas is chosen; and (2) the fugacity coefficient of j in the mixture is unity—this is a consequence of the ideal-gas assumption.

In Eq. (16-20) we substituted for the mole fraction of j the mole ratio of N_j to the total moles,

$$N = \sum_{j=1}^{n'} N_j \tag{16-21}$$

In Eq. (16-21), the total moles N would include all reactants and products *and* any inert components (where $v_{j,r} = 0$). If desired, the inert components may be separated from the sum in Eq. (16-21) and carried along as an additive constant. If we now combine Eqs. (16-15), (16-18), and (16-20), we obtain

$$\sum_{j=1}^{n'} v_{j,r} \left[G_j^o + RT \ln \frac{N_{j(initial)} + \sum_{r=1}^{n'-C} v_{j,r}\, \xi_r}{N} \right] = 0, \quad r = 1, \ldots, n' - C \tag{16-22}$$

Solutions of Eqs. (16-22) are described in review papers by Smith (1980) and Gautam and Seider (1979).

16.5 Equilibrium Constants

For cases wherein there are many phases or many components or several independent reactions, numerical computation is necessary to estimate chemical equilibrium conditions. In simple cases, where hand calculation is possible, the approaches discussed in Sections 16.3 and 16.4 may be modified. Return to Eq. (16-15) and use Eq. (16-8) for μ_j. Assume for the moment that there is only a single independent reaction. Then

$$\sum_{j=1}^{n'} v_j \left(G_j^o + RT \ln \frac{\hat{f}_j}{f_j^o} \right) = 0$$

or

$$\frac{-\left(\sum_{j=1}^{n'} v_j G_j^o \right)}{RT} = \ln \left[\prod_{j=1}^{n'} \left(\frac{\hat{f}_j}{f_j^o} \right)^{v_j} \right] \tag{16-23}$$

The *standard Gibbs energy for the reaction* is written

$$\Delta G_{rx}^o = \sum_{j=1}^{n'} v_j G_j^o \tag{16-24}$$

with all components in their respective standard states denoted by the superscript (^{o}). The *equilibrium constant* for the reaction is defined as

$$K_a \equiv \prod_{j=1}^{n'} \left(\frac{\hat{f}_j}{f_j^o} \right)^{v_j} \tag{16-25}$$

Therefore, Eq. (16-23) becomes

$$K_a = \exp\left(\frac{-\Delta G_{rx}^o}{RT} \right) \tag{16-26}$$

It is clear that the same procedure could be followed if there were more than one independent reaction. Equations (16-23) through (16-26) would then be written for each reaction. The appropriate stoichiometric coefficient vector $v_{j,r}$ could be found as before, and:

$$\Delta G_{rx,r}^o = \sum_{j=1}^{n'} v_{j,r} G_j^o$$

$$\prod_{j=1}^{n'} \left(\frac{\hat{f}_j}{f_j^o} \right)^{v_{j,r}} = K_{a,r} , \qquad \text{for} \quad r = 1, \ldots, n' - C \tag{16-27}$$

$$K_{a,r} = \exp\left(\frac{-\Delta G^o_{rx,r}}{RT}\right)$$

Note that G_j^o, f_j^o, and $\hat{f}_j$ are independent of which reaction is considered. For the presentation here we have assumed that the standard states are pure components at the same temperature T but at some fixed pressure like 1 bar or 1 atm.

The introduction of equilibrium constants into the stoichiometric formulations is usually only of value when there are one or two independent reactions. For more complex systems, the general methods described earlier are used. However, it is of value to explore further the concept of a single independent reaction employing Eqs. (16-24) through (16-26), as the concepts developed are applicable in the general case even when no equilibrium constant is formally used.

We continue to treat the case of a system in chemical equilibrium with a single independent reaction. Also, we assume that the stoichiometric vector has been determined. Thus, of the n' components in the π phases, the stoichiometric vector element of some may be zero; these are termed *inert components*, as they do not enter into the equilibrium calculations.

When computing ΔG^o_{rx} in Eq. (16-24) or μ_j in Eq. (16-8), values of the standard Gibbs energies for the reacting components must be specified. This standard state must be chosen at the temperature of the equilibrium system, but the pressure, composition, and state (gas, liquid, solid) may be selected for convenience. It is clear from Eq. (16-26) that K_a depends only on the properties of the reacting components in their assumed standard states. Within the expression for K_a [see Eq. (16-25)], there also appear the fugacities of the components in the same standard state.

Thus, for any numerical value of K_a, there is an implied set of standard states chosen for the components. One normally denotes this by writing the independent reaction with a shorthand notation to specify the standard states. For example, if the reaction were written

$$\nu_1 \mathbf{W}_{1(g)} + \nu_2 \mathbf{W}_{2(l)} + \nu_3 \mathbf{W}_{3(s)} = 0 \tag{16-28}$$

the designation in the parentheses following the formula vector $\mathbf{W}_j$ indicates the standard state chosen. The symbol (g) after $\mathbf{W}_1$ means that the standard state of component 1 is a pure gas, usually assumed to be in an ideal-gas-state at 1 bar (often abbreviated as "unit fugacity") at the system temperature. The symbol (l) following $\mathbf{W}_2$ normally means that the standard state for component 2 is a pure liquid at its vapor pressure at the system temperature. In some instances, however, the same symbol would indicate that the standard state for 2 is a pure liquid at the system pressure (or at 1 bar) at the system temperature. Since Gibbs energies or fugacities are relatively insensitive to pressure—except near the critical point—the selection of a standard state pressure is often not important. In a similar way, (s) indicates a solid in its most stable state at T. The defining pressure is usually, but not always, the vapor pressure.

To convert Eq. (16-25) to a function of concentrations, the mixture fugacity, $\hat{f}_j$, is expanded as a function of concentration by using Eq. (16-9) or (16-10) if j is present in a vapor phase or in a condensed phase, respectively. If a component is present in two or more phases, either fugacity expansion may be used since $\hat{f}_j$ is equal in all phases at equilibrium.

Should all species taking part in the reaction be present as vapors, substitution of Eq. (16-9) into Eq. (16-25) yields

$$K_a = \left(\prod_{j=1}^{n'} \hat{\phi}_j^{\nu_j}\right)\left(\prod_{j=1}^{n'} y_j^{\nu_j}\right) P^{\nu} = K_{\phi} K_y P^{\nu} \tag{16-29}$$

where $\nu = \Sigma_{j=1}^{n} \nu_j$, and the standard-state fugacities, f_j^o, have been taken as unit fugacity. For gases this standard state is equivalent to an ideal-gas state at 1 atm or 1 bar pressure. If the vapor mixture forms an ideal solution, then (see Section 9.7)

$$\hat{\phi}_j = \frac{f_j}{P} \quad \text{(ideal solution)} \tag{16-30}$$

where f_j is the fugacity of pure vapor j at the temperature and pressure of the mixture. Substituting Eq. (16-30) into Eq. (16-29) yields

$$K_a = K_{f/P} K_y P^{\nu} \quad \text{(ideal vapor solution)} \tag{16-31}$$

For liquid or solid mixtures, with Eq. (16-10),

$$\frac{\hat{f}_j}{f_j^o} = \frac{f_j}{f_j^o} \gamma_j x_j \tag{16-32}$$

If the standard-state pressure is taken as P_{vp_j} or 1 bar and the system pressure is P, then a Poynting correction is used

$$RT \ln \frac{f_j}{f_j^o} = \int_{(P_{vp_j} \text{ or } 1 \text{ bar})}^{P} V_j \, dP \tag{16-33}$$

Unless P is very large, the integral in Eq. (16-33) is normally small; if it were neglected, then substitution of Eq. (16-32) into Eq. (16-25) yields

$$K_a = \left(\prod_{j=1}^{n'} \gamma_j^{\nu_j}\right)\left(\prod_{j=1}^{n'} x_j^{\nu_j}\right) = K_{\gamma} K_x \tag{16-34}$$

(condensed phase; pressure correction neglected)

We recall that K_a is dependent on the standard state. The selection of the standard state is made in a pragmatic sense; that is, it is usually determined by the available data. For a few simple reactions, K_a has been expressed as a function of temperature, as shown later in Figure 16.1. When such data are unavailable, K_a can be determined from ΔG_{rx}^o by using Eq. (16-24) or from ΔH_{rx}^o and ΔS_{rx}^o by Eq. (16-35).

$$\Delta G_{rx}^o = \Delta H_{rx}^o - T \Delta S_{rx}^o \tag{16-35}$$

Since it is impractical to list ΔG^o_{rx} (or ΔH^o_{rx}, ΔS^o_{rx}) for every reaction, tables are available for a large number of compounds showing the Gibbs energy and enthalpy of formation of the species from the elements. In these tables the function ΔG^o_{rx} becomes, for each species, ΔG^o_f and similarly, ΔH^o_{rx} becomes ΔH^o_f. To obtain ΔG^o_{rx} and ΔH^o_{rx}, at the temperature of interest,

$$\Delta G^o_{rx} = \sum_{j=1}^{n'} v_j \, \Delta G^o_{f_j} \tag{16-36}$$

and

$$\Delta H^o_{rx} = \sum_{j=1}^{n'} v_j \, \Delta H^o_{f_j} \tag{16-37}$$

where, again, $v_j = 0$ for those species that do not participate in the reaction. For elements as introduced in Section 8.8, by convention,[1]

$$\Delta H^o_f = \Delta G^o_f = 0 \qquad \text{(at all temperatures for elements)} \tag{16-38}$$

Several tabulations form an excellent reference source for values of ΔH^o_f and ΔG^o_f over a wide range of temperatures (see also Appendix G). These compilations also show values of C^o_p, the heat capacity in the ideal-gas standard state [see for example Appendix G and Reid *et al.* (1987), Stull *et al.* (1969), and Stull and Prophet (1971)].

If ΔG^o_{rx} and ΔH^o_{rx} are available at one temperature, ΔG^o_{rx} can be found at any other temperature as follows:

$$\frac{d(\Delta G^o_{rx}/T)}{dT} = \frac{1}{T}\frac{d\,\Delta G^o_{rx}}{dT} - \frac{\Delta G^o_{rx}}{T^2} = -\frac{\Delta S^o_{rx}}{T} - \frac{\Delta G^o_{rx}}{T^2} = -\frac{\Delta H^o_{rx}}{T^2} \tag{16-39}$$

Equation (16-39) may be integrated if ΔH^o_{rx} is known at one temperature T_1 since

$$\Delta H^o_{rx,T} = \Delta H^o_{rx,T_1} + \int_{T_1}^{T} \Delta C^o_p \, dT \tag{16-40}$$

where

$$\Delta C^o_p = \sum_{j=1}^{n'} v_j C^o_{p_j} \tag{16-41}$$

Before completing this discussion, it is appropriate to describe briefly the Third Law of Thermodynamics. Experimental evidence indicates that the entropy change in a chemical reaction becomes negligible as the absolute temperature approaches zero; thus,

---

1 For gaseous elements such as oxygen, nitrogen, etc., there is no problem with this convention. For elements that are solids at the temperature of interest, a clear statement of the crystal form is also necessary. For example, the standard state for carbon is based on graphite. Should other forms of carbon be present in a system, the ΔH^o_f and ΔG^o_f for such forms are not zero.

$$\Delta S^o_{rx}(T=0\ \text{K}) = \Delta S^o_{rx,0} = 0 \tag{16-42}$$

With the additional stipulation that the entropy state of elements is zero at $T = 0$ K, it follows that $\Delta S^o_{rx,0} = 0$ for all materials; or, as more often stated, $S^o_{0j} = 0$. With this base, it is possible to refer to an absolute entropy that can be calculated by integrating with actual heat capacity data through the solid phase (at very low temperatures), the liquid phase, and into the vapor phase. Entropy changes that result from phase transitions are also included [see for example Eq. (11-15)]. As discussed in Chapter 8, the determination of such absolute entropies obviously requires considerable data[2] and will not be considered further in this text.

In many instances, the equilibrium constant of a reaction is known at one temperature and an extrapolation to other temperatures is necessary. From Eqs. (16-26) and (16-39),

$$\frac{d\ln K_a}{dT} = -\frac{1}{R}\frac{d(\Delta G^o_{rx}/T)}{dT} = \frac{\Delta H^o_{rx}}{RT^2} \tag{16-43}$$

Often ΔH^o_{rx} does not vary appreciably with temperature and, therefore, $\ln K_a$ is nearly linear in $1/T$. A number of reaction equilibrium constants have been plotted in this fashion on Figure 16.1. From the slope one can find a temperature-average value of ΔH^o_{rx}, that is,

$$-\Delta H^o_{rx} = \frac{R\ln(K_{T_2}/K_{T_1})}{1/T_2 - 1/T_1} \tag{16-44}$$

Positive slopes on Figure 16.1 then correspond to reactions with $\Delta H^o_{rx} < 0$ (i.e., exothermic reactions).

In summary, to determine the equilibrium constant of a reaction:

1. If the standard states of all components are chosen as pure materials in an ideal-gas unit fugacity state:
   (a) Use Figure 16.1, if applicable.
   (b) Determine the enthalpy and Gibbs energy of formation for all components (see Section 8.8) and use Eqs. (16-36) and (16-37).
   (c) If other sources are utilized that allow the Gibbs energy and enthalpy of formation to be determined at only a single temperature, Eq. (16-43) may be integrated to find ΔG^o_{rx} and $\ln K_a$ at other temperatures with ΔH^o either assumed constant or expressed as a function of temperature, as in Eq. (16-40).

   For such standard states, f_j^o in Eq. (16-25) is then set at 1 bar and $\hat{f}_j$ must then also be expressed in units of bar.

2. If one or more of the components is chosen with a standard state that differs from the pure-component ideal gas-unit fugacity state, f_j^o in Eq. (16-25) must be the true fugacity

---

2 For example, since as $T \to 0$, $C_{p(s)} \to 0$, special caution must be used to integrate at low temperatures. All phase transformations (first and higher order) must also be included.

of j in this chosen state, and to obtain ΔG^o_{rx}, the chemical potential of j in the same state must be used.

(a) Some tabulations list the Gibbs energy of formation of species in states that are different from ideal gas-unit fugacity states, and these may be used.

(b) If the Gibbs energy of formation of the species is known in any reference state, it may be converted to the desired state by devising a reversible process and calculating the change in Gibbs energy. For example, suppose that ΔG^o_f were available for an ideal gas-unit fugacity state but that the desired standard state for this material was a liquid, pure at a pressure P that is greater than P_{vp} at the same temperature. We could correct ΔG^o_f as follows:

$$\Delta G^o_f\,(\text{liquid}, P) = \Delta G^o_f\,(\text{ideal gas-unit fugacity}) + \int_{1\ \text{bar}}^{P_{vp}} V^V\, dP + \int_{P_{vp}}^{P} V^L\, dP$$

The first integral represents the change in Gibbs energy in an isothermal variation from 1 bar to the vapor pressure (also in bar). There is no Gibbs energy change in condensation at P_{vp}. The third term reflects the Gibbs energy change of the liquid as the pressure changes from P_{vp} to the system pressure.

3. If no values of the Gibbs energy of formation can be located, they may be approximated by group-contribution methods. [See Sections 13.3 and 13.4 and Reid *et al.* (1987)].

In most of the cases discussed above, the standard state was chosen at a fixed pressure; thus, neither K_a nor ΔG^o_{rx} was a function of pressure. If, however, the standard state for any of the reactants or products were to be chosen at the system pressure, P, then both K_a and ΔG^o_{rx} become functions of this pressure. From Eq. (16-24),

$$\left(\frac{\partial \Delta G^o_{rx}}{\partial P}\right)_T = \sum_{j=1}^{n'} \nu_j \left(\frac{\partial G^o_j}{\partial P}\right)_T = -RT\left(\frac{\partial \ln K_a}{\partial P}\right)_T \tag{16-45}$$

If G^o_j is not a function of P, this derivative vanishes. If G^o_j is a function of P, then $(\partial G^o_j/\partial P)_T = V^o_j$, the molar volume of j in the chosen standard state. It is usually more convenient to use a ΔG^o_{rx} (or K_a) that is independent of pressure. In this regard, referring to the liquid-phase example shown in 2(b) above, we see that if the standard state were chosen as the pure liquid at its vapor pressure, ΔG^o_f would be pressure independent.

Example 16.5

As noted in Section 16.1, a closed system at constant temperature and pressure is in equilibrium when it attains a minimum Gibbs energy. Consider a system initially charged with 1 mole of pure I_2, which is maintained at 800°C and 1 bar, in which the following dissociation reaction occurs:

$$I_{2(g)} = 2I_{(g)}$$

$$\Delta H^o_{rx} = 156.6\ \text{kJ/mol} \quad \text{and} \quad \Delta S^o_{rx} = 108.4\ \text{J/mol K}$$

The standard states are pure vapors at unit fugacity and the gas mixture is ideal.

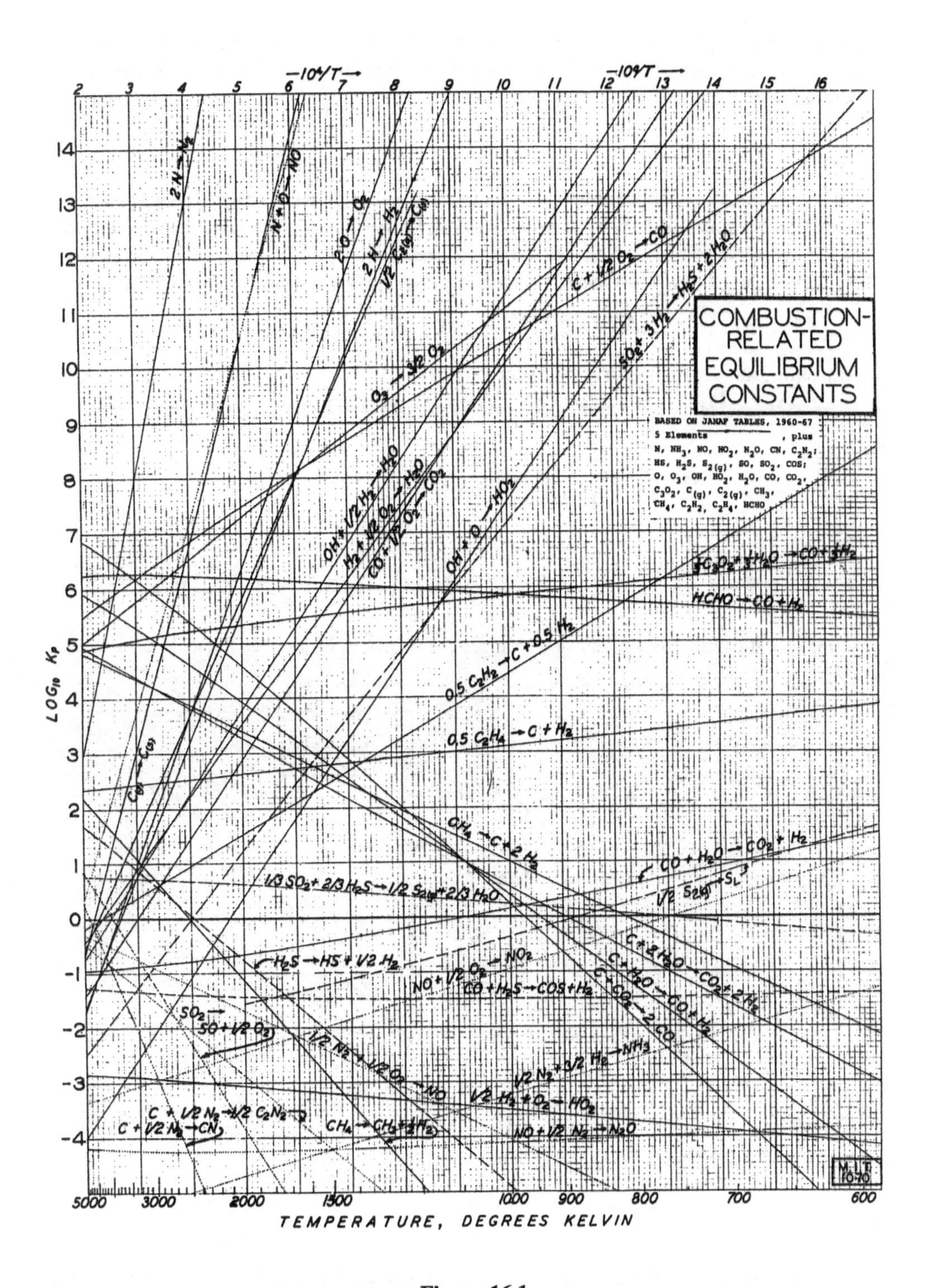
COMBUSTION-RELATED EQUILIBRIUM CONSTANTS
BASED ON JANAF TABLES, 1960-67
TEMPERATURE, DEGREES KELVIN
LOG10 Kp
10⁴/T

Figure 16.1

Determine the equilibrium composition by first calculating the Gibbs energy of the mixture as a function of the moles of I_2 dissociated [denoted by ξ, see Eq. (16-18)] and then determine the minimum in $\underline{G}$ with respect to ξ. Note that $100\,\xi$ represents the % conversion.

Solution

Let us form a mixture containing $(1-\xi)$ moles of I_2 and 2ξ moles of I by a two-step process. First select ξ moles of I_2 in its standard state and assume that it decomposes to 2ξ moles of I in its standard state. Then let us mix the remaining $(1-\xi)$ I_2 moles with the newly formed 2ξ moles of I. The changes in enthalpy and entropy for these two steps are:

$$\Delta\underline{H}_1 = \xi\,\Delta H^o_{rx} \quad \text{and} \quad \Delta\underline{S}_1 = \xi\,\Delta S^o_{rx}$$

Asssuming an ideal gas mixture for step 2:

$$\Delta\underline{H}_2 = \Delta\underline{H}_{mix} = 0 \quad \text{and} \quad \Delta\underline{S}_2 = \Delta\underline{S}_{mix} = -R\,[(1-\xi)\ln y_{I_2} + (2\xi)\ln y_I]$$

$$\Delta\underline{S}_2 = -R\left[(1-\xi)\ln\frac{1-\xi}{1+\xi} + (2\xi)\ln 2\frac{\xi}{1+\xi}\right]$$

Since

$$\Delta\underline{G} = \Delta\underline{H} - T\,\Delta\underline{S} = \xi\,\Delta H^o_{rx} - T\xi\,\Delta S^o_{rx} + RT\left[(1-\xi)\ln\frac{1-\xi}{1+\xi} + (2\xi)\ln 2\frac{\xi}{1+\xi}\right]$$

Values for the enthalpy and entropy changes are shown in Figures 16.2 and 16.3.

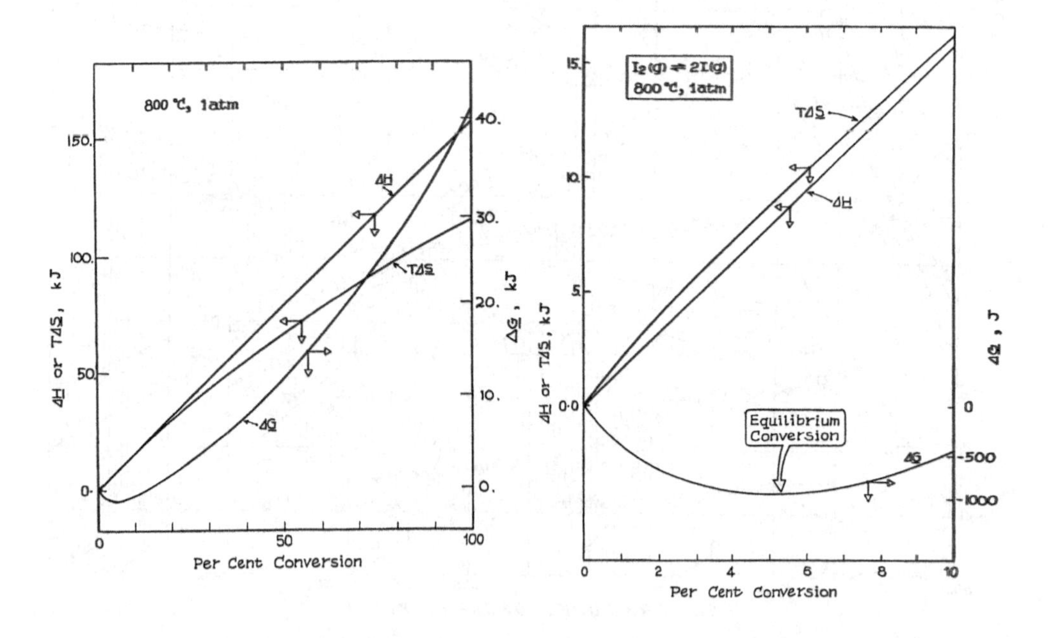

Figure 16.2 ΔH, $T\Delta S$, and ΔG for the reaction $I_{2(g)} \rightarrow 2I_{(g)}$.

Figure 16.3 Expanded view of Figure 16.2

$\Delta \underline{G}$ attains a minimum value when ξ is about 0.052 (i.e., when some 5.2% of the original iodine has decomposed). Returning to the $\Delta \underline{G}$ expression derived above to find the minimum value of $\Delta \underline{G}$ as ξ is varied, we can also differentiate with respect to ξ and set the result equal to zero; that is,

$$0 = (\Delta H^o_{rx} - T\,\Delta S^o_{rx}) + RT \ln \frac{4\xi^2}{1-\xi^2} = \Delta G^o_{rx} + RT \ln \frac{y_I^2}{y_{I_2}}$$

This is, of course, the same result that we would obtain from Eqs. (16-26) and (16-29) using unit-fugacity standard states and assuming ideal gases and 1 bar pressure (i.e., $K_\phi = 1, P^\nu = 1$), with $\Delta G^o_{rx} = \Delta H^o_{rx} - T\,\Delta S^o_{rx} = 156{,}000 - (1{,}073)(108.4) = 40.3$ kJ/mol I_2. The important point of this simple example is to note that it was the entropy change in mixing that led to the minimization of $\Delta \underline{G}$. Without it, in this case, the equilibrium mixture would have contained only pure iodine as I_2.

With only a slight increase in complexity, gas-phase nonidealities may be included. For such a case,

$$\Delta \underline{H} = \xi\,\Delta H^o_{rx} + \Delta \underline{H}^{EX} \quad \text{and} \quad \Delta \underline{S} = \xi\,\Delta S^o_{rx} + \Delta \underline{S}^{ID} + \Delta \underline{S}^{EX}$$

$$\Delta \underline{H}^{EX} - T\,\Delta \underline{S}^{EX} = \Delta \underline{G}^{EX} = 2\xi RT \ln \gamma_I + (1-\xi) RT \ln \gamma_{I_2}$$

Then,

$$\frac{d\,\Delta G}{d\xi} = 0 = \Delta G^o_{rx} + RT \ln \frac{\gamma_I^2}{\gamma_{I_2}} + RT \ln \frac{y_I^2}{y_{I_2}}$$

or

$$\Delta G^o_{rx} = -RT \ln \frac{(y_I \gamma_I)^2}{y_{I_2}\gamma_{I_2}} = -RT \ln K_a$$

since $y_j \gamma_j = \hat{f}_j$. To proceed further in this case requires data or a model for the variation of γ_I and γ_{I_2} (or $\Delta \underline{H}^{EX}$, $\Delta \underline{S}^{EX}$) with composition.

Although ΔH^o_{rx} is commonly referred to as the standard enthalpy of reaction, it is, strictly speaking, the enthalpy change between products and reactants in their respective standard states. ΔH^o_{rx} is not the enthalpy change when the reactants and products are present in an equilibrium mixture. For an equilibrium mixture, the enthalpy change for the reaction is

$$\Delta H_{rx} = \sum_{j=1}^{n'} \nu_j \overline{H}_j \qquad (16\text{-}46)$$

where $\overline{H}_j$ is the partial molar enthalpy of component j at the system temperature, pressure, and in the equilibrium mixture. For ideal-gas mixtures, there are no enthalpy changes on mixing or pressure effects on enthalpies, so that, in this case, $\Delta H^o_{rx} = \Delta H_{rx}$.

16.6 The Phase Rule for Chemically Reacting Systems

We have seen in Section 16.3 that the criteria of phase equilibrium are valid even when chemical reactions occur within the system. In addition, we find one more restraining equation in the form of Eq. (16-15) or (16-25) for each independent chemical reaction.

Since the criteria of phase equilibrium were used previously to develop the phase rule, Eq. (15-7), we need only modify the rule to account for chemical equilibrium. That is, the variance should be reduced by one for each independent chemical reaction:

$$\mathcal{F} = n + 2 - \pi - r \tag{16-47}$$

where r is the number of independent chemical reactions and n the number of chemically distinct species. The phase rule does not imply that any set of $\mathcal{F}$ intensive variables can be used to describe completely all other intensive variables. Consider, for example, a two-phase system containing five species in which one isomerization reaction occurs between species 1 and 2. For this system, $\mathcal{F} = 5 + 2 - 2 - 1 = 4$. If one of the phases present were an ideal vapor phase, the chemical equilibrium expression,

$$K_a = K_y = \frac{y_1}{y_2} \tag{16-48}$$

can be interpreted as a function of y_1, y_2, and T. Thus, only two of these three intensive variables are independent, hence a set of four intensive variables, necessary to satisfy the phase rule, must contain at least two variables other than y_1, y_2, and T.[3]

Since Postulate I is valid for systems in which chemical reactions occur, n+2 independently variable properties are still required to specify the intensity and extent of the system. Thus, it follows that a minimum of $\pi + r$ extensive variables are necessary in addition to the maximum of $n + 2 - \pi - r$ intensive variables. As illustrated in the following problem, there are special cases in which some intensive variables are set by specifying certain extensive variables. In such cases, the dependent intensive variables cannot be used in the set of $n + 2$ that are necessary to describe completely the system.

---

3 If the phase containing the reactant species was not ideal, Eq. (16-48) would contain a K_ϕ term which, in theory, is a function of all y_j, and hence the restriction on y_1, y_2, and T would appear to be removed. If, however, the nonideality was not large, we might find that astronomical pressures are required to satisfy a set of y_1, y_2, and T which was far removed from that necessary to satisfy Eq. (16-48).

Example 16.6

For each of the following cases, determine a set of $n+2$ independent variables which include $n+2-\pi-r$ intensive variables. Consider also the special case in which only the reactants are initially charged, and these are fed in the ratio of their stoichiometric coefficients. Assume that the vapor phase is ideal in each case.

(a) $A_{(g)} + B_{(g)} = C_{(s)}$ (A and B are not soluble in C)

(b) $A_{(g)} + B_{(g)} = C_{(g)} + D_{(g)}$

Solution

(a) Since $n=3$, $\pi=2$, and $r=1$, $F=n+2-\pi-r=3+2-2-1=2$. Possible intensive variables are T, P, y_A (species C is present as a pure solid). The equilibrium relationship is

$$K_a = \frac{\hat{f}_C^s/f_C^o}{(\hat{f}_A^g/f_A^o)(\hat{f}_B^g/f_B^o)} = g(T) \tag{16-49}$$

Choose f_C^o as the fugacity of pure solid C at T and the vapor pressure of $C_{(s)}$ at T and neglect any effect of pressure on the fugacity of the pure solid. Then, assuming unit activity for the pure C solid phase (s), $\hat{f}_C^s/f_C^o = 1$. Choosing the standard states of gaseous species A and B as unit fugacity, $\hat{f}_A^g/f_A^o = y_A P$ and $\hat{f}_B^g/f_B^o = y_B P$. Thus, Eq. (16-49) becomes

$$K_a = \frac{1}{(y_A y_B P^2)} \tag{16-50}$$

Thus, any two of the three intensive variables can be used in conjunction with Eq. (16-50) to determine the third intensive variable.

To determine the extent of the system, we must specify $\pi + r = 3$ extensive variables. Let us choose these as N_{A_0}, N_{B_0}, and N_{C_0}. Defining ξ as the extent of reaction [see Eq. (16-18)], Eq. (16-50) can be written as

$$y_A y_B = \left(\frac{N_{A_0} - \xi}{N_{A_0} + N_{B_0} - 2\xi}\right)\left(\frac{N_{B_0} - \xi}{N_{A_0} + N_{B_0} - 2\xi}\right) = \frac{1}{K_a P^2} \tag{16-51}$$

If T and P were the two intensive variables specified, Eq. (16-51) could be used to determine ξ and, thence, y_A and y_B. Thus, T, P, N_{A_0}, N_{B_0}, and N_{C_0} would be an independent set of $n+2$ variables.

For the special case in which A and B are fed in the stoichiometric ratio of 1:1, $N_{A_0} = N_{B_0}$, and Eq. (16-51) reduces to

$$\frac{1}{4} = \frac{1}{K_a P^2} \tag{16-52}$$

Clearly, both T and P could not be considered independent. Only one can be selected as independent since Eq. (16-52) can then be used to calculate the other. Since $\nu_A = \nu_B$, y_A and y_B are fixed at 0.5 by the stoichiometric feed condition. Thus, the mole fraction is an intensive variable which is set by the stoichiometric feed condition. Only one other intensive variable can be set independently for this special case, and that intensive variable must be either T or P. A set of $(n + 2)$ variables would then have to include three extensive variables in addition to T or P (e.g., T, y_A, N_{B_0}, N_{C_0}, and $\underline{V}$).

(b) Since $n = 4$, $\pi = 1$, and $r = 1$, $\mathcal{F} = n + 2 - \pi - r = 4 + 2 - 1 - 1 = 4$. One set of intensive variables is T, P, y_A, y_B, and y_C. Choosing the standard-state fugacities as unity, the equilibrium relationship becomes

$$K_a = \frac{y_C y_D}{y_A y_B} \tag{16-53}$$

Thus, any three variables from the set T, y_A, y_B, y_C are independent, and three of these, in addition to P, can be used to specify the intensity of the system. Let us assume that we are given T, P, y_A, y_B. To define the extent of the system, we require $\pi + r = 2$ extensive variables. Let us choose these as N_{A_0} and N_{B_0}.

Given T, P, y_A, y_B, N_{A_0}, N_{B_0}, we could first calculate K_a from T and then solve for y_C using Eq. (16-53) in the form

$$\frac{y_C(1 - y_A - y_B - y_C)}{y_A y_B} = K_a \tag{16-54}$$

Knowing y_C, we can solve next for y_D by difference. The total moles initially fed, N_0, and the extent of reaction at equilibrium, ξ, can be determined by solving simultaneously the two equations for y_A and y_B:

$$y_A = \frac{N_{A_0} - \xi}{N_0}$$

$$y_B = \frac{N_{B_0} - \xi}{N_0}$$

The initial moles of C and D can then be found from the following equations:

$$y_C = \frac{N_{C_0} + \xi}{N_0} \quad \text{or} \quad N_{C_0} = y_C N_0 - \xi$$

and

$$N_{D_0} = N_0 - N_{A_0} - N_{B_0} - N_{C_0}$$

For the special case in which A and B are fed in the stoichiometric ratio of 1:1, $N_{A_0} = N_{B_0}$, and $N_{C_0} = N_{D_0} = 0$, and Eq. (16-54) reduces to

$$\frac{(0.5 - y_A)^2}{y_A^2} = K_a \tag{16-55}$$

Thus, y_A and T cannot be considered as independent. We are free to choose y_A and P or T and P, but not y_A, T, and P. A set of $n + 2$ variables would then have to include four extensive variables in addition to y_A and P or T and P. Such a set could be N_{A0}, N_{B0}, N_{C0}, N_{D0}, y_A, P.

In Example 16.6, the special cases of stoichiometric feeds are sometimes cited as examples of cases that do not obey the phase rule, as stated in Eq. (16-47). If we recall that the phase rule states that there are certain sets of $\mathcal{F}$ intensive variables that describe the intensity of the system, the special cases do not disobey the rule. For example, in case (a) of Example 16.6, when we specify stoichiometric feed conditions, y_A is set and must be included as one of the two degrees of freedom allowed by the phase rule. Similarly, in case (b), the two variables y_B and y_C are set by specifying y_A, P, and the stoichiometric feed condition, and hence y_B and y_C must be included as two of the four degrees allowed by the phase rule.

It should be noted that for a single-phase system, specifying the moles charged plus T and P is always sufficient to determine $\mathcal{F}$ intensive variables. In this case, we need only determine the r values of ξ_r (one for each independent reaction) to calculate the $(n-1)$ equilibrium compositions. Given T and P, there is one equilibrium relationship for each reaction, or r equations in the r unknown values of ξ.

A final word of caution is in order. When the phase rule is applied to a given system, it is assumed that we know the number of species and phases present and the number of reactions that occur. We can, however, always charge materials to a "black box" and control the temperature and pressure; nature will then decide for us which reactions occur and which species and phases are present at equilibrium. If we try to specify T and P and also n, π, and r, we run the risk of overspecifying the system. In such cases, one or more phases and species will disappear by one or more reactions going to completion.

For example, if we charged $CaCO_{3(s)}$, $CaO_{(s)}$, and $CO_{2(g)}$ to a vessel and set T and P, it is unlikely that the three phases will coexist in chemical equilibrium. In this case, $\mathcal{F}=n+2-\pi-r=3+2-3-1=1$. Thus, we can control the temperature or the pressure, but not both, if we are to have three phases present at equilibrium. If we set the temperature at 898°C, the equilibrium pressure of CO_2 is 1 atm. If we set the pressure higher than 1 atm, all the CO_2 will disappear by reaction with CaO; if we set the pressure lower than 1 atm, all the $CaCO_3$ will decompose to CO_2 and CaO. In either case, $\mathcal{F}=2+2-2-0=2$, which is in agreement with the fact that we have set T and P independently. It is important to recognize that reactions in a heterogeneous system can go to completion, whereas reactions in a homogeneous phase can never be 100% complete.

Example 16.7

Barium sulfide is made by reducing barium sulfate-rich ore with coke in a rotary kiln. It has been proposed that the reduction be carried out by contacting the ore with CO in a fluidized bed. At the design temperature, the equilibrium constant for the reduction reaction

$$BaSO_{4(s)} + 4CO_{(g)} = BaS_{(s)} + 4CO_{2(g)} \tag{1}$$

was determined experimentally by equilibrating a mixture of CO and CO_2 with a mixture of BaS and $BaSO_4$ solids. The applicability of the experimental results to the industrial process has been questioned for the following reason: the barium sulfate ore contains appreciable amounts of Fe_2O_3 and Fe_3O_4. Since the reaction

$$3Fe_2O_{3(s)} + CO_{(g)} = 2Fe_3O_{4(s)} + CO_{2(g)} \tag{2}$$

is known to proceed at the temperature under consideration, it is suggested that the gas phase in equilibrium with $BaSO_4$-BaS will be different from that in equilibrium with the ore ($BaSO_4$-BaS-Fe_2O_3-Fe_3O_4).

(a) How many intensive variables can be set independently for an equilibrium system containing $BaSO_{4(s)}$, $BaS_{(s)}$, $Fe_2O_{3(s)}$, $Fe_3O_{4(s)}$, $CO_{(g)}$, $CO_{2(g)}$? Assume that no solid solutions are formed.

(b) If the following initial system was maintained at a fixed temperature and 1 bar pressure, what would be the final equilibrium composition?

1 mole ore (70 mole% $BaSO_4$, 5 mole% Fe_2O_3, 25 mol % Fe_3O_4)

0.5 mole CO (no CO_2 initially present)

Assume that at this temperature $K_{a_1} = 10^8$, $K_{a_2} = 20$, and that the vapor phase is an ideal-gas mixture.

Solution

(a) There are six species, five phases, and two independent reactions; thus, $\mathcal{F} = 1$. In this particular case, however, the two equilibria cannot be satisfied simultaneously because

$$\frac{y_{CO_2}}{y_{CO}} = (K_{a_1})^{1/4} = K_{a_2}$$

Since $(K_{a_1})^{1/4}$ is 100 and K_{a_2} is 20, at equilibrium, one of the solids must disappear, leaving five species, four phases, and only one reaction. Thus, $\mathcal{F} = 2$.

(b) Since the temperature and pressure are set, we must determine which solid phase disappears. The initial (y_{CO_2}/y_{CO}) ratio is zero; as Fe_2O_3 and $BaSO_4$ are reduced, (y_{CO_2}/y_{CO}) increases. When this ratio reaches 20, reaction (2) is satisfied, but reaction (1) is not. Thus, the ratio will lie between 20 and 100 while Fe_3O_4 is oxidized to Fe_2O_3 and $BaSO_4$ is reduced to BaS. The cycle will continue until Fe_3O_4 or $BaSO_4$ is depleted. The first to vanish will be the limiting reactant for the overall reaction

$$BaSO_{4(s)} + 8Fe_3O_{4(s)} = BaS_{(s)} + 12Fe_2O_{3(s)} \qquad [(1) + (2)]$$

Since N_{j_0}/ν_i is 0.7 for $BaSO_4$ and 0.031 for Fe_3O_4, the limiting reactant is Fe_3O_4, and it will disappear. Thus, at equilibrium, we will have 0.05 + (0.25)(3)/(2) = 0.425 mole of Fe_2O_3. From K_{a_1},

$$\frac{y_{CO_2}}{y_{CO}} = \frac{1 - y_{CO}}{y_{CO}} = 100$$

or $y_{CO} = 0.01, y_{CO_2} = 0.99$. Thus, $N_{CO} = (0.01)(0.5) = 0.005$ and $N_{CO_2} = 0.495$. Originally, by an oxygen balance,

$$4N^o_{BaSO_4} + 3N^o_{Fe_2O_3} + 4N^o_{Fe_3O_4} + N^o_{CO} = (4)(0.7) + (3)(0.05) + (4)(0.25) + 0.5 = 4.45$$

where the superscript (o) refers to the moles fed. At equilibrium,

$$4.45 = 4N_{BaSO_4} + 3N_{Fe_2O_3} + N_{CO} + 2N_{CO_2} = 4N_{BaSO_4} + (3)(0.425) + 0.005 + (2)(0.495)$$

or

$$N_{BaSO_4} = 0.545 \text{ mole} \quad \text{and} \quad N_{BaS} = 0.7 - 0.545 = 0.155 \text{ mole}$$

Since we have found that reactions (1) and (2) cannot be satisfied simultaneously, we might question if they are truly independent. If we apply the criteria of Section 16.4 to the system of $BaSO_{4(s)}$, $BaS_{(s)}$, $Fe_2O_{3(s)}$, $Fe_3O_{4(s)}$, $CO_{(g)}$, $CO_{2(g)}$, we would indeed find that there are two independent reactions. Applying the phase rule to this system, we find that it is univariant. Thus, if we specify the pressure, we can determine the temperature for which $(K_{a_1})^{1/4} = K_{a_2}$. Alternatively, if we specify only the temperature, we should be able to find a pressure and gas-phase composition such that

$$K_{a_1} = \frac{a_{BaS}(a_{CO_2})^4}{a_{BaSO_4}(a_{CO})^4} \quad \text{and} \quad K_{a_2} = \frac{(a_{Fe_3O_4})^2 a_{CO_2}}{(a_{Fe_2O_3})^3 a_{CO}}$$

Since activities of the solids are weak functions of pressure, it may be necessary to go to extremely high or low pressures before these expressions are simultaneously satisfied.

16.7 Effect of Chemical Equilibrium on Thermodynamic Properties

In the nonstoichiometric and stoichiometric formulations for systems in chemical equilibrium, we implicitly assumed that the system was at a known temperature and pressure. In many real problems, different properties may be specified to define the equilibrium state. For example, suppose that we had a tubular, adiabatic reactor into which reactants are passed at a steady flow. For the exit mixture, the total enthalpy must be the same as the entering material, that is,

$$\underline{H}(T, P, N_1, \ldots, N_n)_{leaving} = \underline{H}(T, P, N_1, \ldots, N_n)_{entering} \tag{16-56}$$

By specifying chemical equilibrium and one other variable at the reactor exit (e.g., the pressure), the problem is then defined.

In other cases there may be a relationship between the initial (or entering) and final entropy state, as, for example, in an expansion turbine.

Often, but not always, it is convenient to estimate the final temperature and pressure and then iterate to obtain appropriate values to conform to given enthalpy or entropy balances. This technique is illustrated for a very simple case in Example 16.8.

Example 16.8

The system NO_2-N_2O_4 is a particularly good one to study the effects of chemical reactions on thermodynamic properties. The stoichiometry is simple, the enthalpy of dimerization of the NO_2 is large, and the kinetics are so rapid that the system may always be assumed to be in chemical equilibrium.

If the reaction is written as

$$N_2O_{4(g)} = 2NO_{2(g)}$$

with unit fugacity standard states, then

$$\Delta G^o_{rx} = 5.733 \times 10^4 - 176.7T, \quad \text{J/mol } N_2O_4$$

The heat capacity of the mixture of N_2O_4 and NO_2, at constant composition (i.e., with no chemical reaction) is essentially independent of composition and is given as

$$C_p(\text{no reaction}) = 4.18 \times 10^{-4}T - 0.7264, \quad \text{J/g K}$$

Suppose that we choose a base of 1 mole (92 g) of N_2O_4 initially. Then, with Eq. (16-18), $N_{N_2O_4} = 1 - \xi$, $N_{NO_2} = 2\xi$, $N = 1 + \xi$. Let us assume that NO_2-N_2O_4 systems form an ideal-gas mixture. Then $\hat{f}_{N_2O_4} = y_{N_2O_4}P = (1-\xi)P/(1+\xi)$, $\hat{f}_{NO_2} = y_{NO_2}P = 2\xi P/(1+\xi)$. Since $f^o_{N_2O_4} = f^o_{NO_2} = 1$ bar [that is, ideal-gas standard states at 1 bar, equivalent to unit fugacity], then, with Eqs. (16-25) and (16-26),

$$K_a = \frac{4\xi^2 P}{1-\xi^2} = \exp\left(\frac{-\Delta G^o_{rx}}{RT}\right) \quad \text{or} \quad \xi = \left(\frac{K_a}{K_a + 4P}\right)^{1/2} \qquad (16\text{-}57)$$

For example, if T = 400 K and P = 10 bar, $\Delta G^o_{rx} = -1.335 \times 10^4$ J/mol N_2O_4, $K_a = \exp[-(-1.335 \times 10^4)/8.314 \times 400] = 55.39$ and $\xi = [55.39/(55.39 + 4 \times 10)]^{1/2} = 0.762$. Under these conditions, the mole fraction NO_2 = (2)(0.762)/(1 + 0.762) = 0.865.

(a) Derive expressions to calculate $\Delta \underline{H}$ and $\Delta \underline{S}$ for a mixture of NO_2 and N_2O_4 between T_1, P_1, and T_2, P_2. Assume a base of 1 mole of N_2O_4 (or 92 g of mixture).

(b) Suppose that one had a N_2O_4-NO_2 mixture at 400 K and 10 bar, and this mixture was expanded across an insulated valve to 1 bar. Estimate the exit temperature.

(c) A N_2O_4-NO_2 mixture at 400 K and 10 bar is expanded across a reversible, adiabatic turbine to 1 bar. How much work is produced per mole of NO_2 (92 g of mixture)?

(d) Calculate the effective heat capacity $(\partial \underline{H}/\partial T)_P$ of a 92 g gas mixture of N_2O_4 and NO_2 in chemical equilibrium between 290 K and 370 K at 1 bar.

Solution

(a) At a given T and P, the total enthalpy of the N_2O_4-NO_2 gas mixture may be written as

$$\underline{H} = \overline{H}_{N_2O_4} N_{N_2O_4} + \overline{H}_{NO_2} N_{NO_2}$$

The partial molar enthalpy of, say, $\overline{H}_{N_2O_4}$ at T, P, and y_i can be expressed as follows:

$$\overline{H}_{N_2O_4} = \overline{\Delta H}^{EX}_{N_2O_4} + \overline{\Delta H}^{ID}_{N_2O_4} + \int_{P^o}^{P} \left(\frac{\partial H_{N_2O_4}}{\partial P}\right)_T dP \Bigg|_{N_2O_4} + H^o_{N_2O_4} \tag{16-58}$$

where P^o is the standard-state pressure, 1 bar. This expression can be greatly simplified since $\overline{\Delta H}^{ID}_{N_2O_4} = 0$ and, if the mixture is ideal, $\overline{\Delta H}^{EX}_{N_2O_4} = 0$. Also, if the effect of pressure on the enthalpy of pure N_2O_4 between the system pressure P and the standard-state pressure (1 bar) is negligible, then

$$\overline{H}_{N_2O_4} \approx H^o_{N_2O_4}(T, P^o)$$

$\overline{H}_{NO_2}$ can be obtained in a similar manner. With $N_{N_2O_4} = 1 - \xi$ and $N_{NO_2} = 2\xi$,

$$\underline{H} = \xi(2H^o_{NO_2} - H^o_{N_2O_4}) + H^o_{N_2O_4}$$

Then, since $(2H^o_{NO_2} - H^o_{N_2O_4}) = \Delta H^o_r$,

$$\underline{H} = \xi \Delta H^o_{rx} + H^o_{N_2O_4}$$

The enthalpy difference between two states T_1, P_1 (with $\xi = \xi_1$) and T_2, P_2 ($\xi = \xi_2$), is, assuming that $\Delta H^o_{rx} \neq f(T)$,

$$\underline{H}_2 - \underline{H}_1 = \Delta H^o_{rx}(\xi_2 - \xi_1) + \int_{T_1}^{T_2} C^o_{pN_2O_4}\, dT \tag{16-59}$$

Note that by assuming that ΔH^o_{rx} is independent of temperature, we are saying that:

$$2C^o_{pNO_2} = C^o_{pN_2O_4}$$

A similar formulation to Eq. (16-59) may be obtained for $\underline{S}$. Here, however, even for an ideal-gas mixture, $\overline{\Delta S}^{ID}_{N_2O_4}$ is not zero, but

$$\overline{\Delta S}^{ID}_{N_2O_4} = -R \ln y_{N_2O_4} = -R \ln \frac{1-\xi}{1+\xi}$$

Also, for N_2O_4 treated as an ideal gas,

$$\int_{P^o}^{P} \left(\frac{\partial S}{\partial P}\right)_T dP = -R \ln \frac{P}{P^o}$$

Then, with $P^o = 1$ bar:

$$\underline{S} = \overline{S}_{N_2O_4} N_{N_2O_4} + \overline{S}_{NO_2} N_{NO_2} = \overline{S}_{N_2O_4}(1-\xi) + \overline{S}_{NO_2}(2\xi)$$

$$= \left[\overline{\Delta S}^{EX}_{N_2O_4} - R \ln \left(\frac{1-\xi}{1+\xi}\right) - R \ln P + S^o_{N_2O_4}(T, P^o)\right](1-\xi)$$

$$+ \left[\overline{\Delta S}^{EX}_{NO_2} - R \ln \left(\frac{2\xi}{1+\xi}\right) - R \ln P + S^o_{NO_2}(T, P^o)\right] 2\xi$$

Assuming that $\overline{\Delta S}^{EX}_{N_2O_4} = \overline{\Delta S}^{EX}_{NO_2} = 0$,

$$\underline{S} = -R\xi \ln\left(\frac{4\xi^2 P}{1-\xi^2}\right) + \xi\,\Delta S^o_{rx} - R\ln\frac{1-\xi}{1+\xi} - R\ln P + S^o_{N_2O_4}(T, P^o)$$

The first two terms may be simplified to give

$$\underline{S} = \frac{\xi\,\Delta H^o_{rx}}{T} - R\ln\left[\frac{P(1-\xi)}{1+\xi}\right] + S^o_{N_2O_4}(T, P^o)$$

where ΔH^o_{rx} is the standard enthalpy change of reaction. Then,

$$\underline{S}_2 - \underline{S}_1 = \Delta H^o_{rx}\left(\frac{\xi_2}{T_2} - \frac{\xi_1}{T_1}\right) - R\ln\left[\frac{P_2}{P_1}\left(\frac{1-\xi_2}{1+\xi_2}\right)\frac{1+\xi_1}{1-\xi_1}\right] + \int_{T_1}^{T_2} C^o_{pN_2O_4}\frac{dT}{T} \qquad (16\text{-}60)$$

(b) We use Eq. (16-59). ΔH^o_{rx} is estimated from $\ln K_a = -\Delta G^o_{rx}/RT = -(5.733\times10^4/RT) + (176.7/R)$ and $[d\ln K_a/d(1/T)] = -\Delta H^o_{rx}/R = -5.733\times10^4/R$ or $\Delta H^o_{rx} = 5.733\times10^4$ J/mol N_2O_4 reacted. Also, at the start of this example, ξ was calculated at 400 K, 10 bar to be 0.762. Then, across a cracked, insulated valve, $\Delta\underline{H} = 0$:

$$0 = 5.733\times10^4\,(\xi_2 - 0.762) + 92\left[\int_{400}^{T_2}(4.18\times10^{-4}T + 0.7264)\,dT\right]$$

Since $\xi_2 = f(T_2, P_2)$ from Eq. (16-57), then, by iteration, for $P_2 = 1$ bar, $T_2 = 360$ K, and $\xi_2 = 0.819$

In passing through the valve, N_2O_4 reacted to form more NO_2, and the temperature dropped since the reaction is endothermic. If no reaction had occurred, the exit temperature would have been the same as that at the inlet since ideal-gas behavior was assumed.

(c) Across the reversible adiabatic turbine $\Delta\underline{S} = 0$ and we will use Eqs. (16-60) and (16-57) to determine T_2 and ξ_2. With $T_1 = 400$ K, $P_1 = 10$ bar, $\xi_1 = 0.762$, and $\Delta H^o_{rx} = 5.733\times10^4$ J/mol N_2O_4, P_2 is set to 1 bar and Eq. (16-60) becomes,

$$0 = 5.733\times10^4\left(\frac{\xi_2}{T_2} - \frac{0.762}{400}\right) - 8.314\ln\left[\frac{1}{10}\left(\frac{1-\xi_2}{1+\xi_2}\right)\frac{1.762}{0.238}\right]$$
$$+ 92\int_{400}^{T_2}(4.18\times10^{-4}T + 0.7264)\frac{dT}{T}$$

Then by iteration, $T_2 = 340.5$ K and $\xi_2 = 0.636$.

In this case, while passing through the turbine, NO_2 reacted to form more N_2O_4 even as the system pressure dropped. With these T_1, P_1, ξ_1, T_2, P_2, ξ_2 values and Eq. (16-59),

$$\Delta\underline{H} = 5.733\times10^4(0.636 - 0.762) + 92\int_{400}^{340.5}(4.18\times10^{-4}T + 0.7264)\,dT$$

$$= -1.12\times10^4\ \text{J/mol } N_2O_4$$

Thus, for each 92 g of mixture fed to the reversible turbine, 1.12×10^4 J of work could be obtained. If no reaction had occurred, $T_2 = 315.5$ K and $\Delta \underline{H} = -6.8 \times 10^3$ J/mol N_2O_4. Thus, even though the outlet temperature is smaller in the "no-reaction" case, less work is obtained per mole (92 g) of N_2O_4-NO_2 mixture than when rapid chemical equilibrium was considered.

(d) In this case we can use Eq. (16-59); differentiating with respect to T, we obtain

$$\left(\frac{\partial \underline{H}}{\partial T}\right)_P = \Delta H^o_{rx}\left(\frac{d\xi}{dT}\right)_P + C^o_{pN_2O_4}$$

But since $K_a = 4\xi^2 P/(1 - \xi^2)$ and $d \ln K_a/d(1/T) = -\Delta H^o_{rx}/R$, then

$$\left(\frac{d\xi}{dT}\right)_P = \frac{\xi(1 - \xi^2)\Delta H^o_{rx}}{2RT^2} \tag{16-61}$$

so

$$(C_p)_{rx\ mix} = \left(\frac{\partial \underline{H}}{\partial T}\right)_P = \frac{\xi(1 - \xi^2)(\Delta H^o_{rx})^2}{2RT^2} + C^o_{pN_2O_4} \tag{16-62}$$

Basis: 1 mole or 92 g of N_2O_4 charged initially which is equivalent to 92 g of mixture

T (K)	$C^o_{pN_2O_4}$ (J/mol N_2O_4 K)	$\frac{\xi(1-\xi^2)(\Delta H^o_r)^2}{2RT^2}$ (J/mol N_2O_4 K)	$\left(\frac{\partial \underline{H}}{\partial T}\right)_P$ (J/mol N_2O_4 K)
290	78.0	158	236
300	78.4	433	511
310	78.8	549	628
320	79.1	645	724
330	79.5	686	766
340	79.9	649	729
350	80.3	545	625
360	80.7	411	492
370	81.1	287	368

It can be seen from this example that the heat capacity of a reacting gas mixture can be significantly affected by the chemical reaction in those temperature regions where ξ is a strong function of temperature.

In the more general case, where there may be many independent reactions, the simple techniques developed in Example 16.8 would not be practicable, and a sensitivity matrix is developed in conjunction with the computer algorithm [Smith (1969)]. Such a matrix is a convenient method to determine the effects on the equilibrium solution by changing some input parameter—or the specified path. Le Châtelier's principle, discussed in Section 16.8, addresses the same question of the system sensitivity, although usually only single reaction systems are treated.

16.8 Le Châtelier's Principle in Chemical Equilibria

If a system in thermal, mechanical, and chemical equilibrium were perturbed from the equilibrium state, changes would usually occur within the system to reestablish a new equilibrium state. For example, if the system temperature were suddenly increased at constant pressure and extent of reaction, the system would no longer be in chemical equilibrium. If the system were then isolated from the environment, the reaction would occur in such a way as to decrease the temperature (i.e., the system would adjust itself to reduce the effect of the initial perturbation). This simple example can be easily proved from Eq. (16-43), irrespective of whether the reaction is exothermic or endothermic. Almost all equilibrium systems behave in this manner for all kinds of perturbations, and the general rule that defines such behavior is called the Le Châtelier principle. One of the simplest statements of this principle is given by Maass and Steacie (1939): "If an attempt is made to change the pressure, temperature, or concentration of a system in equilibrium, then the equilibrium will shift, if possible, in such a manner as to diminish the magnitude of the alteration in the factor that was varied."

A system that follows Le Châtelier's principle is said to moderate with respect to the perturbation. It is interesting to examine the behavior of a chemically reacting system in light of this principle. The simple method proposed by Katz (1961) will be used first. For those interested, other treatments are available in Prigogine and Defay (1954) and in Callen (1985).

To develop the Katz technique, the mathematics are considerably simplified if only ideal-gas mixtures are considered and standard states are limited to those corresponding to a pure ideal gas at 1 bar. A brief discussion is presented later to indicate the extension to other cases.

Consider the ***general reaction of an ideal gas mixture***; $aA_{(g)} + bB_{(g)} = cC_{(g)} + dD_{(g)}$ and define the following ratios:

$$Q_p \equiv \frac{p_C^c\, p_D^d}{p_A^a\, p_B^b} \tag{16-63}$$

$$Q_N \equiv \frac{N_C^c\, N_D^d}{N_A^a\, N_B^b} = \frac{Q_p N^\nu}{P^\nu} \tag{16-64}$$

$$Q_y \equiv \frac{y_C^c\, y_D^d}{y_A^a\, y_B^b} = \frac{Q_N}{N^\nu} = \frac{Q_p}{P^\nu} \tag{16-65}$$

where $\nu = c + d - a - b$. At equilibrium with $K_\phi = 1.0$, $K_a = K_p = Q_p$, $K_y = Q_y$, etc., but the Q definitions are applicable whether or not equilibrium has been established.

If the reaction proceeds to the right, Q_N increases. We can easily prove this by taking the logarithm of Eq. (16-64) and differentiating with respect to the extent of reaction, ξ. The result is

$$\frac{dQ_N}{d\xi} = Q_N\left(\frac{c^2}{N_C} + \frac{d^2}{N_D} + \frac{a^2}{N_A} + \frac{b^2}{N_B}\right) \tag{16-66}$$

Since $dQ_N/d\xi$ is clearly positive, Q_N increases as the reaction proceeds to form more C and D. One may also show that Q_y increases with ξ by a similar technique.

The key to the entire treatment is to consider the system in chemical equilibrium; if a perturbation should occur, the equilibrium constants are unaffected (except for temperature changes), but the values of Q may change. The subsequent increase or decrease in Q_N or Q_y indicates the direction of the shift in the reaction. Several examples are treated below. *All cases are isothermal.*

(1) Addition of an inert gas at constant pressure

$$K_p = \text{constant} = \frac{Q_N P^\nu}{N^\nu}$$

P is a constant and as N increases, Q_N will then increase if $\nu > 0$ and decrease if $\nu < 0$. Thus, the reaction occurs to form more C and D if $\nu > 0$ and less C and D if $\nu < 0$.

(2) Addition of an inert gas at constant volume

$$K_p = \text{constant} = \frac{Q_N P^\nu}{N^\nu} = Q_N\left(\frac{RT}{\underline{V}}\right)^\nu$$

Since T and $\underline{V}$ are constant, Q_N does not change, and there is no change in the moles of reactants and products present at equilibrium.

(3) Variation of system pressure

$$K_p = \text{constant} = Q_y P^\nu$$

For an increase in P, the reaction shifts to the right if $\nu < 0$ and to the left if $\nu > 0$. Similar, but opposite, conclusions hold if P is decreased.

(4) Addition of a reactant or product at constant volume

$$K_p = \text{constant} = Q_N\left(\frac{RT}{\underline{V}}\right)^\nu$$

If C or D were added, Q_N increases initially, but since all other terms are constant, the reaction must take place to the left to decrease Q_N to its original value. Similarly, the reaction occurs to the right if A or B were to be added.

(5) Addition of a reactant or product at constant pressure

This case requires a little more consideration. Since $Q_p = Q_y P^\nu$ and P is constant, a change in Q_p indicates a corresponding change in Q_y.

If the addition of one of the components leads to an increase in the ratio denoted by Q_p, then the reaction must occur in such a way to lower Q_p to its original value since, at equilibrium, $K_p = Q_p$. Now,

$$Q_p = \frac{P^\nu Q_N}{N^\nu}$$

If component j is added, with all the moles of other components constant:

$$\left(\frac{\partial Q_p}{\partial N_j}\right)_{T,\,P,\,N_i[j]} = \frac{P^\nu}{N^{2\nu}}\left[N^\nu\left(\frac{\partial Q_N}{\partial N_j}\right)_{T,\,P,\,N_i[j]} - (Q_N)\left(\frac{\partial N^\nu}{\partial N_j}\right)_{T,\,P,\,N_i[j]}\right]$$

But

$$Q_N = \prod_{j=1}^{k} N_j^{\nu_j}; \qquad \left(\frac{\partial Q_N}{\partial N_j}\right)_{T,P,N_i[j]} = \frac{\nu_j Q_N}{N_j}; \qquad \text{and} \qquad \left(\frac{\partial N^\nu}{\partial N_j}\right)_{T,P,N_i[j]} = \nu N^{(\nu-1)}$$

Substituting, we obtain

$$\left(\frac{\partial \ln Q_p}{\partial N_j}\right)_{T,P,N_i[j]} = \frac{1}{N_j}(\nu_j - y_j\nu) \tag{16-67}$$

From Eq. (16-67), Q_p increases with the addition of N_j provided that $(\nu_j - y_j\nu) > 0$. In this case the reaction then occurs to decrease Q_p to its original value. For example, consider the reaction,

$$2NH_3 = N_2 + 3H_2 \quad (\nu = 2)$$

For ammonia, the term $[-2 - y_{NH_3}(2)]$ is certainly never greater than zero and Q_p decreases with addition of NH_3. Reaction then occurs to Q_p (i.e., the amounts of N_2 and H_2 will increase because of ammonia decomposition). For hydrogen addition, it is easily shown that the reaction will occur to form more ammonia.

For nitrogen, $\nu_{N_2} - y_{N_2}\ \nu = 1 - 2y_{N_2}$. This term is greater than zero if $y_{N_2} < 0.5$, but less than zero if $y_{N_2} > 0.5$. For the case $y_{N_2} < 0.5$, Q_p increases with the addition of nitrogen, so that the reaction will occur to reduce Q_p (i.e., N_2 and H_2 react to form more NH_3). If, however, $y_{N_2} > 0.5$, Q_p decreases with nitrogen addition and additional NH_3 must decompose to reëstablish Q_p to its original value. Such behavior as noted in the last case is often referred to as a case in which the system does not "moderate" with respect to the variation. It is interesting to note that, although the ammonia system may not moderate with respect to the addition of nitrogen (i.e., react in such a way that the effect of the nitrogen addition is decreased), the final concentration of nitrogen, expressed, say, as mol/liter, is still less than the original nitrogen concentration before addition. Thus, to this extent, the system moderates even with respect to nitrogen.

Another example involves the methanol synthesis reaction:

$$CO_{(g)} + 2H_{2(g)} = CH_3OH_{(g)}$$

	Initial moles	Moles in an equilibrium state
CO	A	$A - x$
H_2	B	$B - 2x$
CH_3OH	0	x
		$A + B - 2x$

$$K_p = \frac{p_{CH_3OH}}{p_{CO}p_{H_2}^2} = \frac{y_{CH_3OH}}{P^2 y_{CO} y_{H_2}^2} = \frac{(x)(A + B - 2x)^2}{(A - x)(B - 2x)^2 P^2}$$

At about 500°C, $K_p \approx 10^{-6}$; unless the pressure is high, the amount of methanol formed is small. To illustrate the effect of adding H_2 or CO, assume a low pressure of 1 bar. The equilibrium constant expression may then be approximated as

$$\frac{x}{K_p} \sim \frac{AB^2}{(A + B)^2}$$

Consider two cases; in the first, start with 1 mole of CO and allow chemical equilibrium to be attained with various amounts of hydrogen. As shown in Figure 16.4, the amount of CH_3OH present at equilibrium continually increases with an increase in hydrogen. This is what is predicted, since

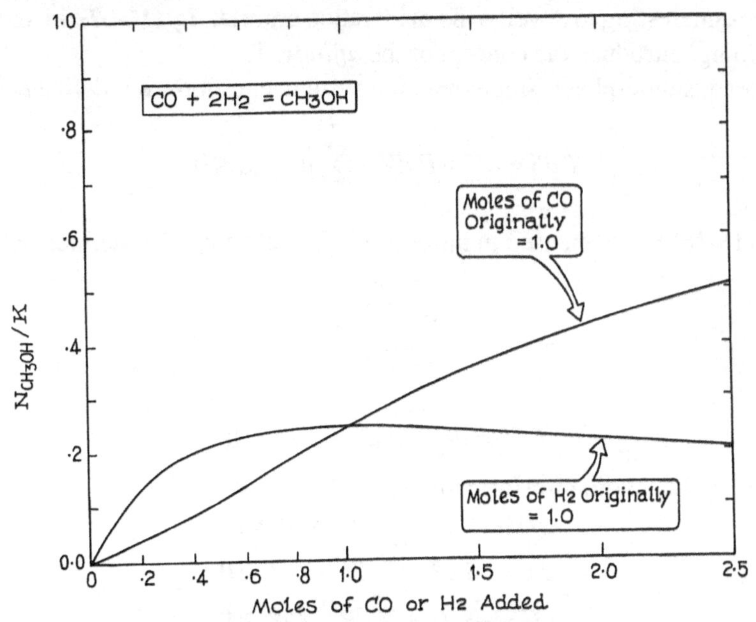

Figure 16.4 Variation of methanol produced with the addition of CO or H_2.

$$\nu_{H_2} - y_{H_2}\nu = -2 - (-2)y_{H_2} < 0$$

for all values of y_{H_2} up to unity. From Eq. (16-67), Q_p decreases, and the reaction occurs to raise Q_p to its original value. For the second case, start with 1 mole of H_2 and add different quantities of CO. The value of x/K_p increases as CO is first added, but then decreases if more than 1 mole is added. Again we note that

$$\nu_{CO} - y_{CO}\nu = -1 - (-2)y_{CO} = 2y_{CO} - 1$$

If $y_{CO} < 0.5$, from Eq. (16-67), Q_p decreases with CO addition, and the subsequent reaction occurs to form more CH_3OH and bring Q_p to its original level. If $y_{CO} > 0.5$ the opposite effect occurs, and the addition of CO leads to a decrease in the yield of CH_3OH. The maximum yield of CH_3OH occurs when 1 mole of CO has been added to 1 mole of H_2.

In the more general sense, moderation may be examined from the point of view of starting with a system in equilibrium. The system is perturbed by changing one of the variables. We are interested in following the behavior of the system back to an equilibrium state (which may be different from the original state). If, in the change from the nonequilibrium state to the equilibrium state, there is a corresponding change in the perturbing variable opposite to that which originally caused the upset, we speak of the system as moderating with respect to that variable. We have already seen, for example, in the ammonia-hydrogen-nitrogen equilibria, if the system were perturbed at constant temperature and pressure by adding hydrogen, the system will react such that more ammonia is formed. This change reduces the amount of hydrogen, and the system moderates. Many other examples could be cited. Instead of considering them in detail, however, it is interesting to develop the moderation criteria, Eq. (16-67), in another way and, in so doing, introduce the concept of the *affinity*, Y.

For a closed, single-phase, single-reaction system, equilibrium is defined as

$$T\,d\underline{S} = d\underline{U} + P\,d\underline{V} - \sum_{j=1}^{n} \mu_j \nu_j\, d\xi \le 0 \tag{16-68}$$

where Eq. (16-16) has been used to introduce ξ. The affinity, Y, is defined as

$$Y \equiv -\sum_{j=1}^{n} \mu_j \nu_j \tag{16-69}$$

Thus,

$$T\,d\underline{S} = d\underline{U} + P\,d\underline{V} + Y\,d\xi \le 0 \tag{16-70}$$

Similarly, by Legendre transformations,

$$\begin{aligned} d\underline{H} &= T\,d\underline{S} + \underline{V}\,dP - Y\,d\xi \ge 0 \\ d\underline{A} &= -\underline{S}\,dT - P\,d\underline{V} - Y\,d\xi \ge 0 \\ d\underline{G} &= -\underline{S}\,dT + \underline{V}\,dP - Y\,d\xi \ge 0 \end{aligned} \tag{16-71}$$

Thus, the criterion for chemical equilibria, Eq. (16-15), now may be written as

$$Y = 0 \tag{16-72}$$

Of more immediate interest is the fact that if a system were not in equilibrium, then, regardless of what sets of variables (P, T), $(\underline{U}, \underline{V})$, $(\underline{S}, \underline{V})$, or $(T, \underline{V})$ are kept constant, the reaction must occur so that

$$Y\, d\xi > 0 \quad \text{(real process)} \tag{16-73}$$

or, for simultaneous reactions,

$$\sum_{m=1}^{r} Y^{(m)}\, d\xi^{(m)} > 0 \tag{16-74}$$

Presumably, some reaction could take place, so that viewed singly, Eq. (16-73) is violated, but for the entire system Eq. (16-74) still holds. Considering only a single reaction, a generalized reaction rate R may be defined as

$$R = \frac{d\xi}{dt} \tag{16-75}$$

so that for any actual chemical reaction,

$$YR > 0 \tag{16-76}$$

Returning to the original problem, we start with a system in chemical equilibria and perturb it by adding a small amount of one of the reactive components. We are interested in the behavior of the perturbed system as it approaches the new equilibrium state. We hold the pressure and temperature constant. Let i be the added component. The change in affinity is

$$Y \sim Y_0 + \left(\frac{\partial Y}{\partial N_i}\right)_{T,\, P,\, N_j[i]} \delta N_i \tag{16-77}$$

for small values of δN_i. Equation (16-72) requires that $Y_0 = 0$ because the derivative is evaluated at the original equilibrium state. Then from Eq. (16-76) and (16-77),

$$\left[\left(\frac{\partial Y}{\partial N_i}\right)_{T,P,N_j[i]} \delta N_i\right] \left(\frac{1}{\nu_i}\frac{dN_i}{dt}\right) > 0 \tag{16-78}$$

where δN_i is the amount of i added (positive in this case) and dN_i/dt is the response of the system to the addition. If $dN_i/dt < 0$, the system moderates with respect to an addition of N_i. For dN_i/dt to be negative, however, and still satisfy the inequality given by Eq. (16-78),

$$\frac{1}{\nu_i}\left(\frac{\partial Y}{\partial N_i}\right)_{T,P,N_j[i]} < 0 \quad \text{(for moderation)} \tag{16-79}$$

To explore the consequences of Eq. (16-79), from Eq. (16-69)

$$\frac{1}{\nu_i}\left(\frac{\partial Y}{\partial N_i}\right)_{T,P,N_j[i]} = -\frac{1}{\nu_i}\sum_{j=1}^{n}\nu_j\left(\frac{\partial \mu_j}{\partial N_i}\right)_{T,P,N_j[i]} = -\frac{1}{\nu_i}\sum_{j\neq i}\nu_j\left(\frac{\partial \mu_j}{\partial N_i}\right)_{T,P,N_j[i]} - \left(\frac{\partial \mu_i}{\partial N_i}\right)_{T,P,N_j[i]}$$

But,

$$\left(\frac{\partial \mu_i}{\partial N_i}\right)_{T,P,N_j[i]} = -\frac{1}{N_i}\sum_{j\neq i}N_j\left(\frac{\partial \mu_j}{\partial N_i}\right)_{T,P,N_j[i]}$$

so

$$\frac{1}{\nu_i}\left(\frac{\partial Y}{\partial N_i}\right)_{T,P,N_j[i]} = -\frac{1}{\nu_i}\sum_{j=1}^{n}N_j\left(\frac{\nu_j}{N_j}-\frac{\nu_i}{N_i}\right)\left(\frac{\partial \mu_j}{\partial N_i}\right)_{T,P,N_j[i]} < 0 \quad \text{(for moderation)} \tag{16-80}$$

The right-hand side of Eq. (16-80) must be less than 0 for the system to moderate with additional N_i. Note that this expression may be greatly simplified for the case of ideal gases. Here,

$$\left(\frac{\partial \mu_j}{\partial N_i}\right)_{T,P,N_j[i]} = RT\left(\frac{\partial \ln y_j}{\partial N_i}\right)_{T,P,N_j[i]} \tag{16-81}$$

Equation (16-80) then simplifies to

$$\frac{1}{\nu_i}\left(\frac{\partial Y}{\partial N_i}\right)_{T,P,N_j[i]} = \frac{RT}{N_i}\left(\frac{y_i\nu}{\nu_i} - 1\right) \tag{16-82}$$

Thus, for moderation,

$$\frac{y_i\nu}{\nu_i} - 1 < 0 \quad \text{or} \quad \frac{y_i\nu - \nu_i}{\nu_i} < 0 \quad \text{(to moderate)} \tag{16-83}$$

Equation (16-83) is a more compact form than Eq. (16-67), which requires subsequent steps to determine changes in Q_p.

This technique, in which the affinity concept is used to study the effects of system upsets, is very powerful and not limited to ideal gases, although the final results may be of such a form as to hinder immediate physical visualization of the consequences [i.e., Eq. (16-80)].

Example 16.9

Apply Eq. (16-83) to the decomposition reaction

$$NH_2COONH_{4(s)} = 2NH_{3(g)} + CO_{2(g)}$$

Solution

The amount of solid present is unimportant. Thus,

$$\nu_{CO_2} = 1, \quad \nu_{NH_3} = 2, \quad \nu = 3$$

For CO_2 addition, to have the reaction proceed in order to form solid,

$$\frac{(3)(y_{CO_2}) - 1}{1} < 0, \quad y_{CO_2} < \tfrac{1}{3}$$

Similarly, to form more solid by ammonia addition,

$$\frac{(3)(y_{NH_3}) - 2}{2} < 0, \quad y_{NH_3} < \tfrac{2}{3}$$

16.9 Summary

For a system held at constant temperature (T) and pressure (P) the condition of chemical equilibrium corresponds to a minimum in the Gibbs free energy (G) subject to the additional constraint that

$$\sum_{j=1}^{n} \nu_j \mu_j = 0$$

for each chemical reaction that occurs. In this chapter two completely equivalent approaches were discussed that represent chemical equilibrium. The first used a nonstoichiometric formulation where only components needed to be specified and reaction stoichiometries were not required. And the second incorporated a set of assumed reactions and components, tested for their independence, and finally used them to establish equilibrium phase compositions.

Equilibrium constants (K_a) were shown to be useful for the analysis of simple systems with one or two significant reactions. They provided a means for estimating temperature and pressure effects as well as a technique for using standard Gibbs energies and enthalpies of formation to estimate the magnitude of K_a. The effects of chemical equilibrium on thermophysical properties can be large in regions where K_a changes rapidly as a function of T or P.

The Gibbs phase rule was easily modified to account for the presence of equilibrium reactions by reducing the number of degrees of freedom by one for each independent equilibrium reaction present. And finally, a general treatment of Le Châtelier's principle was introduced to quantitatively formulate how an applied stress (of temperature, pressure, or composition) shifts the equilibrium to a new equilibrium state.

References

Brewer, Bromley, Gilles, and Lofgren (1949), *The Transuranic Elements*, NNES, Div. IV, Volume 14B, McGraw-Hill, New York, p 861.

Callen, H.B. (1985), *Thermodynamics and an Introduction to Thermostatistics*, 2nd ed. Wiley, New York.

Cantor, and Schimmel (1980), *Biophysical Chemistry*, Freeman, New York, p 283-285.

Gautam R. and W.R. Seider (1979), "Computation of phase and chemical equilibrium," *AIChE J.*, 25, p 991.

Gilliland, E.R., R.C. Gunness, and V.O. Bowles (1936), "Free energy of ethylene hydration," *Ind. Eng. Chem.*, **28**, p 370.

Katz, L. (1961), "A systematic way to avoid Le Châtelier's principle in chemical reactions," *J. Chem. Ed.*, **38**, p 375.

Maass, O. and E.W.R. Steacie (1939), *Introduction to the Principles of Physical Chemistry*, Wiley, New York.

Prigogine, I. and R. Defay, *Chemical Thermodynamics*, trans., D.H. Everett, Longmans, and Green, New York. (1954), Chap. 17.

Reid, R.C., J.M. Prausnitz and B.E. Poling (1987), *Properties of Gases and Liquids*, 4th ed. McGraw-Hill, New York, Chap. 7.

Smith, W.R. (1969), "The effects of changes in problem parameters on chemical equilibrium calculations," *Can. J. Chem. Eng.*, **47**, p 95.

Smith, W.R. (1980), "The computation of chemical equilibria in complex systems," *Ind. Eng. Chem. Fundam.*, **19**, p 1.

Stull, D.R., E.F. Westrum, Jr., and G.C. Sinke (1969), *The Chemical Thermodynamics of Organic Compounds*, Wiley, New York.

Stull, D.R. and H. Prophet (1971), *JANAF Thermochemical Tables*, 2nd ed., NSRDS-NBS 37.

Whillier (1977), "Recovery of energy from water going down mine shafts," *J. So. African Inst. Min. Met.*, p 183.

Zeleznik, F.J. and S. Gordon (1968), "Calculation of complex chemical equilibria," *Ind. Eng. Chem.*, **60**(6), p 27.

Problems

16.1. A reaction occurs between A and B to form C as follows:

$$A_{(l)} + 3B_{(g)} \leftrightarrow C_{(l)}$$

where the gas standard state denotes unit fugacity, and the liquid standard states are pure liquid at the vapor pressure corresponding to the reaction temperature.

(a) Set up an expression for the chemical equilibrium constant in terms of the compositions, fugacities of pure components, and activity coefficients. Assume that there are two phases (liquid and gas) at equilibrium and that both phases are ideal solutions. Clearly define all quantities and describe briefly how they may be determined from theory or experiment.

(b) If the chemical reaction equilibrium constant were written so that all standard states were gas at unit fugacity, how would this constant be related to the K calculated in part (a)?

(c) Which of the values of K calculated above would vary more if the system pressure were doubled? Why?

(d) From the phase rule, how many independent variants are there for this system in chemical and phase equilibrium?

16.2. We are interested in studying the reduction of carbon dioxide with hydrogen to form carbon and water over an iron catalyst. This reaction may be of value in future long-range manned NASA missions.

Although the feed gases are pure CO_2 and H_2, the reaction mixture is known to contain, in addition, CO, CH_4, H_2O, and C (solid). The catalyst, although elemental iron initially, may be partially oxidized to FeO or reacted with carbon to form FeC at the reaction conditions.

Data:

	ΔG_f^o (298 K) (kJ/mol)	ΔH_f^o (298 K) (kJ/mol)	Vapor pressure (298 K)
CO_2	–394.6	–393.8	Sublimes
$H_{2(g)}$	0	0	Above critical temperature
C(graphite)	0	0	Not known
$H_2O_{(g)}$	–228.7	–242.0	3,160 N/m^2 (liquid)
C(diamond)	2.87	2.02×10^{-3}	Not known

(a) The temperature and pressure are set and the mole fraction of CO_2 in the $CO_2 \cdot H_2$ feed is varied to study the equilibrium yield. How many independent reactions are there present for this system?

(b) One of the suspected reactions is

$$CO_{2(g)} + 2H_{2(g)} = C_{(s)} + 2H_2O$$

What is the value of K for this reaction at 298 K? Feel free to use some or all of the data given above.

(c) Should the reaction in part (b) have been written so that the standard state for water, instead of being a vapor at unit fugacity, was pure liquid at its vapor pressure at 298 K, how would the equilibrium constant be affected? (That is, calculate the new value.) Would the yield of water increase or decrease at equilibrium?

16.3. Two engineers are discussing a particular reaction A = B + 2D, and there seems to be a dispute as to whether the yield of products would increase or decrease with temperature. Both agree that an increase in temperature produces less favorable yields if the reaction were exothermic.

JoAnna D. has written the reaction as

$$A = B_{(g)} + 2D_{(g)}$$

whereas Brian P. indicates

$$A = B_{(l)} + 2D_{(g)}$$

At 373 K, heats of formation are as follows:

	ΔH_f°(kJ/mol at 373 K)
A	-146.5
$B_{(g)}$	-41.9
$D_{(g)}$	-46.1

At 373 K, the enthalpy of vaporization of B is 29.3 kJ/mol.

Assuming that these heats of formation are relatively insensitive to temperature and that the gas phase is ideal, would an increase in temperature give a larger or smaller yield in the reactor for a feed of 1 mole of A? Discuss.

16.4. We are in the process of designing a sulfuric acid concentrator, and we need accurate vapor-liquid equilibrium data for the H_2SO_4-H_2O system. We have located an article which reports the partial pressures of H_2SO_4 and H_2O as a function of liquid composition at constant temperature. Since the article is rather old, one of our engineers suggested that we determine the consistency of the data by applying the Gibbs-Duhem equation. He used the following equation:

$$N_A d \ln p_A + N_w d \ln p_w = \frac{\underline{V}}{RT} dP \tag{A}$$

where N_A and N_W are the total moles (liquid plus vapor) of H_2SO_4 and H_2O, respectively; p is partial pressure, P is total pressure, and $\underline{V}$ is the total volume (liquid plus vapor).

The calculations from the data do not provide consistency, as determined by Eq. (A). However, another one of our engineers claims that Eq. (A) is not applicable because sulfuric acid decomposes in the vapor phase according to the reaction

$$H_2SO_{4(g)} = H_2O_{(g)} + SO_3$$

She claims that the proper consistency test is

$$N_A d \ln p_A + N_w d \ln p_w + N_T d \ln p_T = \frac{V}{RT} dP \tag{B}$$

where subscripts A, w, and T refer to H_2SO_4, H_2O, and SO_3, respectively.

(a) Given that the decomposition reaction occurs, is either one or both of Eqs. (A) and (B) a valid test of consistency? Clearly explain your reasoning.

(b) At the temperature in question, we have calculated a value of K_p $(= p_w p_T / p_A)$ from Gibbs energies of formation. Although the article does not clearly state how the partial pressures were obtained, we assume that they measured the total pressures and the mole fraction of sulfur-containing species in the vapor phase (i.e., $y_A + y_T$). Assuming that we know p_A $[= (y_A + y_T)P]$ and P as functions of liquid composition at constant T, and assuming that K_p is known, derive an expression to test the consistency of the data.

16.5. Zinc oxide is reduced by roasting it with carbon in a closed retort in which the gaseous and liquid products of reaction may be continuously removed. Air is carefully excluded from the retort so that the reaction is free to proceed under its own equilibrium pressure. The actual process involves two reactions:

$$ZnO_{(s)} + CO_{(g)} \leftrightarrow Zn_{(g)} + CO_{2(g)} \tag{1}$$

$$C_{(s)} + CO_{2(g)} \leftrightarrow 2CO_{(g)} \tag{2}$$

Zinc produced in the reaction may be removed from the retort as either a gas or a liquid or both, depending on the operating temperature and pressure.

Data:

Temperature (K)	$\Delta G^o_{1,T}$ (kJ/mol)	$\Delta G^o_{2,T}$ (kJ/mol)
1100	58.54	-24.99
1200	47.01	-42.53
1300	35.22	-59.94
1400	23.76	-77.26
1500	12.27	-94.47

Material	Melting Point (K)	Boiling Point (K)	Density (g/cm^3)	Molecular weight
Zn	692	1180	7.2	65.38
ZnO	2248	Unknown	5.6	81.37
C	3823	5100	2.3	12.01

(a) What does the phase rule indicate about the variance of this system under conditions where liquid zinc is produced? Under conditions where it is not produced? *Explain carefully.*

(b) Calculate the equilibrium reaction pressure as a function of the roasting temperature.

(c) Determine the temperature at which the operation must be conducted in order that the products may be withdrawn at a pressure of 1.0 bar.

16.6. Assume that the equations shown below represent chemical reactions. Compute the value of ΔG^o_{298} for each. The standard states are as follows: I_2 gas, unit fugacity; I_2 liquid, pure liquid, 1 bar; I_2 solid, pure solid, 1 bar. The densities of liquid and solid iodine may be taken to be 1.03 g/cm^3, and the gas iodine to be an ideal gas.

(a) $I_{2(s)} \leftrightarrow I_2$

(b) $I_{2(s)} \leftrightarrow I_{2(l)}$

(c) $I_{2(l)} \leftrightarrow I_2$

Some vapor-pressure data for iodine are as follows:

T (K)	P_{vp} (N/m^2)	T (K)	P_{vp} (N/m^2)
223	4.9×10^{-3}	333	5.75×10^{2}
233	2.5×10^{-2}	343	1.096×10^{3}
243	1.07×10^{-1}	353	2.01×10^{3}
253	4.0×10^{-1}	363	3.57×10^{3}
263	1.32	373	6.07×10^{3}
273	3.99	387.4[a]	1.20×10^{4}
283	1.077×10^{1}	393	1.48×10^{4}
293	2.69×10^{1}	403	2.09×10^{4}
303	6.28×10^{1}	423	3.92×10^{4}
313	1.37×10^{2}	433	5.25×10^{4}
323	2.88×10^{2}	457.6[b]	1.013×10^{5}

[a]Melting point.
[b]Normal boiling point.

16.7. Michael K. Jones, a close relative of Rocky and Rochelle and avid inventor, claims to be able to produce diamonds from β-graphite at room temperature by a process involving the application of 37 kbar pressure. In view of the data shown below, are his claims to be taken seriously?

$$\text{Specific gravity, } \beta\text{-graphite} = 2.26$$
$$\text{Specific gravity, diamond} = 3.51$$
$$C_{\beta\text{-graphite}} = C_{\text{diamond}} \quad \Delta G^o_{298} = 2870 \text{ J/g-atom}$$

Both solids are incompressible and no solid solutions are formed.

16.8. Gilliland *et al.* (1936) investigated the hydration of ethylene by placing ethylene, water, and sulfuric acid in an agitated pressure vessel kept at constant temperature by a condensing vapor. After analyzing the liquid phase for ethanol until constant composition indicated that equilibrium had been attained, a sample of the vapor phase was withdrawn and analyzed. One run showed the following data:

$$T = 527 \text{ K}, \quad P = 267.7 \text{ bar}$$

	Mole fraction	
	Liquid phase	Gas phase
Ethyl alcohol	0.084	0.075
Ethylene	0.0197	0.250
H_2O	0.881	0.675
H_2SO_4	0.015	—

(a) Estimate ΔG^o at 527 K for this reaction.

(b) The values reported in the literature for absolute entropies, S^o, and heats of formation are as follows:

	S^o(298 K) (J/mol K)	ΔH^o_f(298 K) (kJ/mol)
C_2H_5OH	278.0	-237.0
$C_2H_{4(g)}$	220.6	+52.3
$H_2O_{(g)}$	188.8	-241.14

Compute $\Delta G^o_{527\text{ K}}$ form these data.

16.9. In a laboratory investigation a high-pressure gas reaction $A \leftrightarrow 2B$ is being studied in a flow reactor at 473 K and 100 bar. At the end of the reactor, the gases are in chemical equilibrium, and their composition is desired.

Unfortunately, to make any analytical measurements on this system, it is necessary to bleed off a small side stream through a low-pressure conductivity cell operating at 1 bar. It is found that when the side stream passes through the sampling valve, the temperature drops to 373 K, and the conductivity cell gives compositions of $y_A = 0.5$ and $y_B = 0.5$.

Data and allowable assumptions:

Heat of the reaction, $\Delta H = 29.3$ kJ/mol of A reacting, independent of temperature.

Heat capacity: 29.3 J/mol K for B independent of temperature. The gas mixture is ideal at all pressures, temperatures, and compositions. Assume no heat transfer in the sampling line or across the sampling valve.

(a) Calculate the composition of the gas stream before the sampling valve.

(b) Are the gases in chemical equilibrium after the sampling valve? (Show definite proof for your answer.)

16.10. Brewer, Bromley, Gilles, and Lofgren (*The Transuranic Elements*, NNES, Div. IV, Volume 14B, McGraw-Hill, 1949, p. 861) discuss some interesting experiments with plutonium chloride ($PuCl_3$).

Small vials of solid $PuCl_3$ crystals were heated to 973 K and a mixture of argon and chlorine gas passed through them at 8.0×10^4 Pa. The exit gases were cooled to 298 K in a trap that contained a fine filter. Solid $PuCl_3$ was found on the filter after an experiment.

Four experiments were made. In each, 2,000 cm^3 of an argon-chlorine gas mixture was passed through the $PuCl_3$. (The 2,000 cm^3 was measured at 8.00×10^4 Pa and 298 K.) The reported data are as follows:

Mole fraction in the sweep gas		
Chlorine	Argon	Quantity of PuC13 found in trap (mg)
0	1.0	0
0.0625	0.9375	32
0.25	0.75	73
1.0	0	143

There seems to be a disagreement concerning the mechanism of transport of the $PuCl_3$. One group simply states that although $PuCl_3$ is not particularly volatile, it is vaporized in the sweep gas at 973 K and trapped at the low temperature outside the vial. The other group suggests that a chemical reaction is involved and proposes that

$$PuCl_{3(s)} + \tfrac{1}{2}Cl_{2(g)} = PuCl_4$$

occurs at 973 K and that the $PuCl_4$ vapor is carried into the trap, where it decomposes to $PuCl_3$ and Cl_2. Skepticism has greeted the latter explanation since $PuCl_4$ has never been observed. The atomic weight of plutonium is 239; for chlorine, 35.5, and for argon, 40.

What opinions do you have concerning these two theories? Support your statement with quantitative calculations.

16.11. We do not seem to be able to locate any data for the Gibbs energy of reaction, nor ΔH^o for a reaction shown as

$$D_{(g)} = A_{(g)} + B_{(g)}$$

However, one of our research students has conducted a few experiments as described below. A constant-volume pressure vessel was immersed in a constant-temperature bath. The vessel was first evacuated and then filled with pure gaseous D at 298 K and 1 bar. After heating the closed vessel to 473 K, reaction occurred to form A and B. When no further pressure change was noted, the gauge read 3 bar. Further heating to 523 K caused a pressure increase to 3.30 bar.

With these data, can one calculate $\Delta G^o_{523\text{ K}}$ and ΔH^o? If so, provide numerical values. If not, describe what additional data are required. You may assume that the vapor phase forms an ideal-gas mixture and that ΔH^o is not a function of temperature.

16.12. A very efficient plant has two gas streams, the properties of which are tabulated below. The management would like to extract the maximum work possible from the streams before discarding the products in a collection vessel at 300 K and 1 bar. Rocky Jones, has suggested a black box, which he claims will serve the purpose (see Figure P16.12).

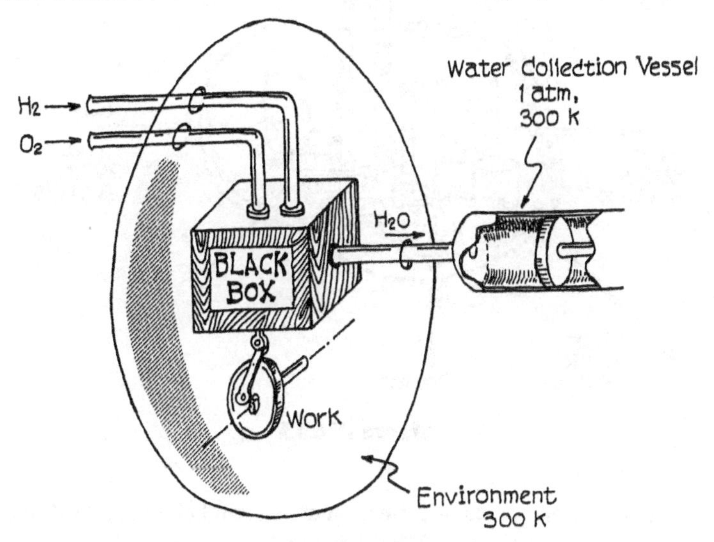

Figure P16.12

	Pure oxygen	Pure hydrogen
Temperature (K)	300	600
Pressure (bar)	2	3
Flow rate (mole/s)	1	2
C_p (J/mol K)	29.3	29.3

Data: $\Delta G^o_{f\,H_2O} = -228.8$ kJ/mol; $\Delta H^o_{f\,H_2O} = -241.9$ kJ/mol at 300 K with products and reactants in the standard states of pure gases, unit fugacity. The vapor pressure of liquid water at 300 K is 0.036 bar.

What is the maximum power obtainable from Rocky's box?

16.13. In our sulfuric acid plant we have a process stream of pure SO_2 gas at 2 bar and 600 K flowing at a rate of 6.4 kg/s. This is to be mixed with a pure oxygen stream at 293 K and 1 bar. The oxygen flow rate is 1.6 kg/s. The SO_2 and O_2 are reacted for form SO_3, which is then separated from any unreacted SO_2 and O_2 (which are recycled) and cooled to 293 K before flowing to the absorbers.

One of our new employees has indicated that we could start with the same reactants and end with the same product SO_3 but, at the same time, obtain significant "free" work from the process. His ideas are a bit sketchy, but I illustrate them in Figure P16.13. He indicates that all heat sinks and sources in the environment are at 293 K.

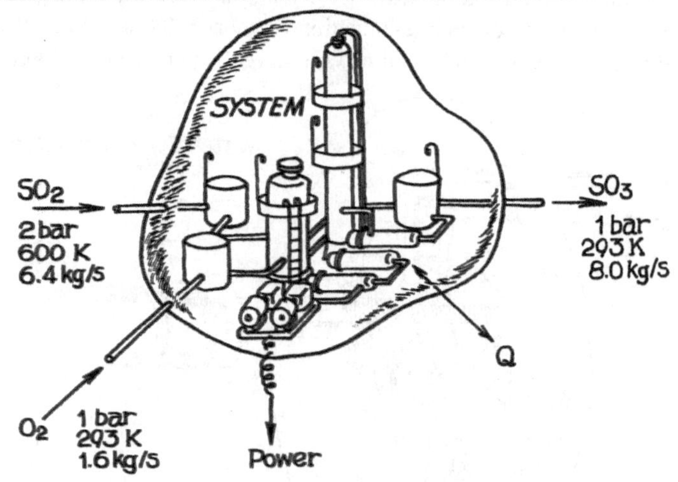

Figure P16.13

Some Gibbs energy and heats of formation are shown below at 293 K and 600 K. Also assume that $C_p\,(SO_2\ and\ SO_3) = 41.9$ J/mol K and $C_p\,(O_2) = 29.3$ J/mol K.

	$\Delta G^o_{f,293}$ (kJ/mol)	$\Delta H^o_{f,293}$ (kJ/mol)	$\Delta G^o_{f,600}$ (kJ/mol)	$\Delta H^o_{f,600}$ (kJ/mol)
SO_2	-300.2	-297.1	-302.2	-301.0
O_2	0	0	0	0
SO_3	-412.4	-436.7	-428.3	-483.2

Could you tell me how much power we could produce from his process? Make a clear sketch of your proposed scheme.

16.14. Suppose that you have at your disposal 1 mole of liquefied natural gas (LNG) and a cylinder containing 4 moles of high-pressure oxygen gas.

Data:

Assume that the LNG is pure methane, saturated liquid, at 1 bar pressure.
The saturation temperature is 111 K.
The oxygen cylinder is at 150 bar and 298 K.

The environment is the atmosphere at constant temperature and pressure of 298 K and 1 bar. The composition of the atmosphere, on a dry basis, in mole fraction, is

O_2: 0.20
CO_2: 0.00025

with the remainder as nitrogen and inert gases.
The relative humidity of the atmosphere is 100%.
The Gibbs energies of formation and enthalpies of formation of a few compounds are as follows: ($T = 298$ K)

	ΔH_f^o (J/mol)	ΔG_f^o (J/mol)
CO	-110,577	-137,334
$CO_{2(g)}$	-393,702	-394,572
$CH_{4(g)}$	-74,883	-50,818
$H_2O_{(g)}$	-241,951	-228,702
$H_2O_{(l)}$	-285,975	-237,304
$C_{(s)}$	0	0
$O_{2(g)}$	0	0

where (g) refers to a standard state of unit fugacity (bar) and (l) refers to a standard state of liquid, 298 K.The heat capacity, C_p, of methane gas is 34.25 J/mol K and for oxygen 29.30 J/mol K; both may be assumed independent of temperature. The heat of vaporization of methane at 111 K (1 bar) is 8,183 J/mol. The vapor pressure of water at 298 K is 2330 Pa.

Devise (and describe) a series of processes and calculate the *maximum* work you could obtain from these starting materials.

16.15. It has been suggested that liquid air can serve as the ultimate pollution-free automotive fuel. The liquid would be used as the working fluid in an open-cycle Rankine engine that only has heat interactions with the atmosphere.

The cycle is shown schematically in Figure P11.15. Liquid air in a vented, insulated fuel-storage tank (a) is pumped into a high-pressure vaporizer (b), where it is heated by atmospheric air to ambient temperature (298.2 K). The resulting high-pressure gas is then fed to a multistage expansion engine (c) that drives the vehicle. The engine obviously operates without a condenser and exhausts to the atmosphere (d). The air would probably be reheated between expansion stages by the use of suitable reheat coils (e).

You are invited to comment on the theoretical, as well as practical feasibility of the scheme. A useful result would be to estimate the energy cost per kilojoule delivered to the wheels of a small "city"-type automobile, by a liquid-air-fueled engine and compare this cost with those for gasoline and lead-acid battery operation. The essential question of whether this engine, in drawing heat only from the atmosphere, violates the Second

Law, should also be discussed. Other thermodynamic data for liquid air and gasoline can be obtained from standard handbooks.

Efficiency: The efficiency relative to a reversible adiabatic process for high-pressure air motors that are currently commercially available is about 85%.

On the same basis, the efficiency of an automobile (new) gasoline engine is about 12%.

Costs:
Liquid air: ≈ \$0.22/kg
Gasoline: ≈ \$400/m^3
Lead-acid battery: Stores 22.05 W-h/kg
Replacement cost = \$6.6/kg
Lifetime = 300 charge cycles
Charge efficiency: 1.33 W-h charge to store 1.0 W-h
Cost of charging: 0.27/kW-h

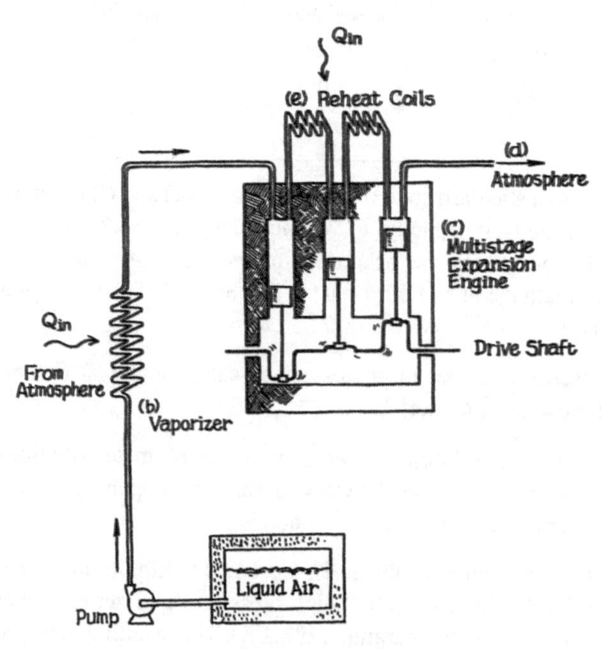

Figure P16.15

16.16. Whillier (*J. So. African Inst. Min. Met.*, April 1977, p. 183) discusses the problem of supplying cooling water to deep gold mines.

As normally carried out, a water-chilling refrigeration plant is located on the surface to cool water before it is circulated down into the mine (see Figure P16.16(a)). Water leaves this plant at point 1 and flows down to control valve 2. Assume that the flow is adiabatic and frictionless.

At 2, the water is reduced in pressure and flows to an underground storage tank 3, from which it is circulated through the mine and back to the surface. The bottom of the mine where the workers are located and where tank 3 is placed may be assumed to be at the same pressure as at the mine surface—1 bar.

The deepest gold mine in South Africa is about 3.4 km. For this mine, if the water temperature at point 1 is 10°C, what is the temperature in tank 3?

Whillier has suggested that control valve 2 be replaced by a power-producing turbine. If such a turbine were installed and operated at 70% efficiency, what would the temperature in tank 3 be for this case? (Assume that the turbine operates in an adiabatic manner.)

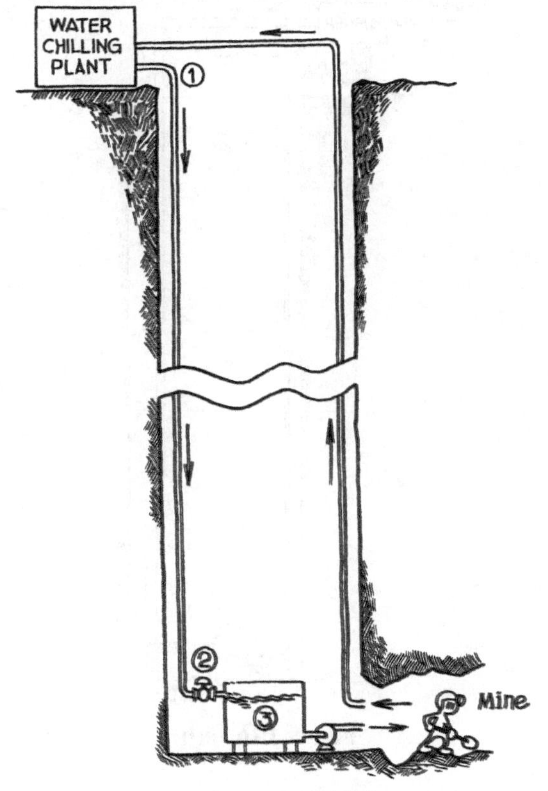

Figure P16.16(a)

Data:
$g = 9.8$ m/s^2
C_P(water, liquid) = 4.19 kJ/kg K
ρ(water, liquid) = 1000 kg/m^3 (assume constant)
Neglect heat leaks to piping and tanks

At the time I was writing this problem, I discussed it with one of my closest friends, R.M.S. Potter. He was only mildly interested in such a prosaic situation, but he did return in a short while with a startling idea related to the problem. It seems, as usual, that he has found the solution to our energy problems. What he proposes is really quite simple. He will drill a deep hole in the earth using his Rock Melting Subterrene, and at the bottom he would place a power-producing turbine. He would allow water to flow down a pipe to the bottom of the hole and pass through the turbine. Then, he would use the power from the turbine to drive a generator to produce electric power that would be used to electrolyze the water that passed through the turbine. The hydrogen and oxygen produced would fill the hole (except for the water pipe) and, at the surface, these gases would be burned to produce power and form liquid water to be recirculated.

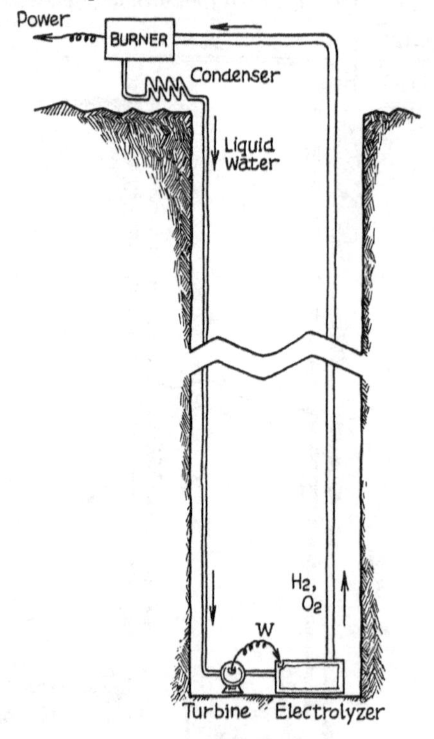

Figure P16.16(b)

Could you let me have your valued judgment on the feasibility of this process? (*Note*: For water taken as a gas in its standard state

$$\Delta H^o_{f298} = -1.59 \times 10^7 \text{ J/kg} \quad \text{and} \quad \Delta G^o_{f298} = -1.32 \times 10^7 \text{ J/kg}$$

16.17. (a) For a general reaction at a given temperature and pressure,

$$aA + bB = cC + dD$$

Suppose that one began initially with pure A and B and desired to vary the initial mole ratio (A/B) to maximize the concentration of C and D in the equilibrium mixture. Show for the case of an ideal-gas mixture that the desired initial ratio $(N_{A_0}/N_{B_0}) = a/b$.

(b) What is the maximum conversion of N_2 to ammonia obtainable in a Haber ammonia unit operating at 300 bar and 773 K when stoichiometric quantities of hydrogen and nitrogen are used?

Use as a basis 1 mole of nitrogen:

$$\tfrac{1}{2}N_{2(g)} + \tfrac{3}{2}H_{2(g)} = NH_{3(g)}$$

$$\Delta G^o(\text{J/mol NH}_3) = 51.54\,T \log T - 51.50T + \frac{1.058}{10^3}T^2 - \frac{3.54}{10^6}T^3 - 38{,}390$$

(c) Ammonia at 3.45 bar and 373 K is passed through an adiabatic reversible turbine to an exit pressure of 1 bar. Assuming complete reversibility and that chemical equilibrium exists at all times, what is the maximum work that can be obtained?

NH_3 Properties

P (bar)	*T* (K)	*S* (J/gK)	*H* (J/g)
3.45	373	6.202	1686.4
1.01	282	6.202	1493.5

Source: National Bureau of Standards Circular 142.

16.18. One mole of ethylene and 1 mole of benzene are fed to a constant-volume batch reactor and heated to 600 K. On the addition of a Friedel-Crafts catalyst, an equilibrium mixture of ethylbenezene, benzene, and ethylene is formed:

$$C_6H_{6(g)} + C_2H_{4(g)} \leftrightarrow C_6H_5C_2H_{5(g)}$$

The pressure in the reactor before the addition of the catalyst (i.e., before any reaction has occurred) is 2 bar.

Data (in kJ/mol):

$C_6H_{6(l)}$: $\Delta G^o_{f298} = 113.5$, $\Delta H^o_{f298} = 52.21$
C_2H_4: $G^o_{f298} = 68.20$, $\Delta H^o_{f298} = 52.34$
$C_6H_5C_2H_5$: $G^o_{f298} = 130.75$, $\Delta H^o_{f298} = 29.81$

Physical Properties

	Boiling point (°C)	ΔH_{vap} at 1 bar (kJ/mol)	Average C_p (J/mol K)
C_6H_6	80.1	30.77	113
C_2H_4	-103.71	13.52	54
$C_6H_5C_2H_5$	136.19	36.01	167

Calculate the total heat removed by the cooling fluid in a heat exchanger used to maintain the reactor temperature constant at 600 K as the reaction proceeds to equilibrium.

16.19. Hydrogen molecules exist in ortho and para forms. The difference results from the fact the nuclear spins of the individual atoms in a molecule may be either parallel or antiparallel. The equilibrium ratio between the forms is determined almost completely by the temperature. At room temperature and above, the ratio is constant at 25% para and 75% ortho; at liquid hydrogen temperatures (20 K) the ortho form is present in negligible amounts. Equilibrium ratios at other temperatures are shown in the accompanying table. The actual ratios for any hydrogen-gas system are, however, strongly influenced by the rates of transformation. For example, at room temperature and above, the homogeneous rates are reasonably large, but a catalyst is necessary to achieve any appreciable transformation at 20 K.

From an engineering point of view, gaseous hydrogen when cooled and liquefied converts to the lower-energy form (para) in an exothermic reaction, and as cooling and liquefaction rates are fast compared to the conversion rate, freshly liquefied hydrogen differs only slightly from the 25%-*p*, 75%-*o* composition of the feed gas. Transformation to the para form then occurs at low temperatures, and this requires additional low-temperature refrigeration.

A prospective client is proposing to use stored liquid hydrogen (which is available and fully converted to the para form) to cool a stream of hot helium. The hydrogen gas is to be exhausted around 420 K and fed to another part of the system. Although the size of the heat exchanger is not particularly important, one would like to cool the helium stream as much as possible with a given hydrogen flow rate. The client feels that the residence time of the hydrogen in the system is so short that the exit will still exist in a pure para form. It is felt, however, that if a good catalyst could be developed, the endothermic transformation from $p \rightarrow o$ could be carried out in the heat exchanger and this additional heat effect used to cool the helium to a lower temperature.

At present he wants us to share in the cost to develop a catalyst that will yield an equilibrium gas mixture [of $H_2(o)$ and $H_2(p)$] at all positions in the heat exchanger.

Write a concise memorandum indicating your opinion of the scheme and recommending our course of action.

Temperature (K)	Equilibrium fraction *para*-hydrogen	Temperature (K)	Equilibrium fraction *para*-hydrogen
< 10	1.0	90	0.42882
10	0.999999	100	0.38620
20	0.99821	120	0.32959
30	0.97021	150	0.28603
40	0.88727	200	0.25974
50	0.77054	250	0.25264
60	0.65569	300	0.25072
70	0.55991	> 300	0.25
80	0.48537		

16.20. A rigid, well-insulated gas storage tank of 0.3-m^3 capacity is filled originally with hydrogen gas at 1 bar and 30 K. Connected to the tank is a large high-pressure manifold containing hydrogen gas at 100 bar and 40 K. The valve is opened and hydrogen is allowed to flow into the tank until the gas inside attains a temperature of 50 K. Assume ideal gases and that the ortho-para equilibrium shift is infinitely rapid (see Problem 16.19 for data on equilibrium composition). Heat-capacity data for pure *para*-hydrogen indicate that over the temperature range of interest here, $C_P/R \approx 2.50$. The tank walls and hardware are assumed to have zero heat capacity. Assume that the contents of the tank are well mixed at all times.

(a) What is the pressure when the temperature of the gas inside the tank reaches 50 K?

(b) What would the temperature be if the manifold and tank pressure were increased without limit?

(c) What is the entropy change of the universe of this process?

16.21. Freshly liquefied hydrogen that has not been catalyzed consists of a 3:1 ortho-para mixture. On standing, there is a slow shift of the mixture toward the equilibrium concentration, a fact that complicates the problem of storing the liquid for any length of time.

Assume that 1 mole of hydrogen at room temperature and 1 bar is quickly liquefied (in order not to affect the ortho/para ratio) and then allowed to reach an equilibrium state. From the data in Problem 16.19 showing the equilibrium fraction of *para*-hydrogen as a function of temperature, this should occur when the concentration of para is about 99.8%. During this change the system is chosen to be adiabatic and always vented at 1 bar.

Data: Heat of vaporization of hydrogen is assumed independent of ortho/para ratio and equals 454 kJ/kg at 20.4 K, the boiling temperature at 1 bar. There is no appreciable difference between the vapor- and liquid-phase concentration of the two components (i.e., the relative volatility of *ortho*- to *para*-hydrogen is unity over the entire liquid range).

How many moles of liquid hydrogen remain after the final equilibrium state is reached?

16.22. Gaseous nitrogen peroxide consists of a mixture of NO_2 and N_2O_4, and chemical equilibrium between these components is rapidly established. It has been suggested that this gas mixture be employed as a heat transfer medium. To evaluate this proposal, the heat capacity of the equilibrium mixture must be determined as a function of temperature.

Ideal gases may be assumed for this temperature range at 1 bar. $\Delta G^o = 57{,}330 - 176.72T(\text{K})$ for the reaction $N_2O_4 = 2NO_2$. Unit fugacities are assumed for the standard states, and ΔG^o is in J/mol. C_p for the frozen equilibrium mixture may be approximated as:

$$C_p(\text{frozen equilibrium}) = 4.19 \times 10^{-4} T\,(\text{K}) + 0.726 \quad \text{in J/g K}$$

Calculate and plot the effective heat capacity $(\partial H/\partial T)_P$ for the equilibrium mixture between 290 and 370 K at 1 bar.

16.23. The system N_2O_4-NO_2 remains in chemical equilibrium under essentially all conditions (i.e., if a step change in pressure or temperature is imposed, it takes only about 0.1 to 0.2 μs for the system to react and attain equilibrium under the new conditions).

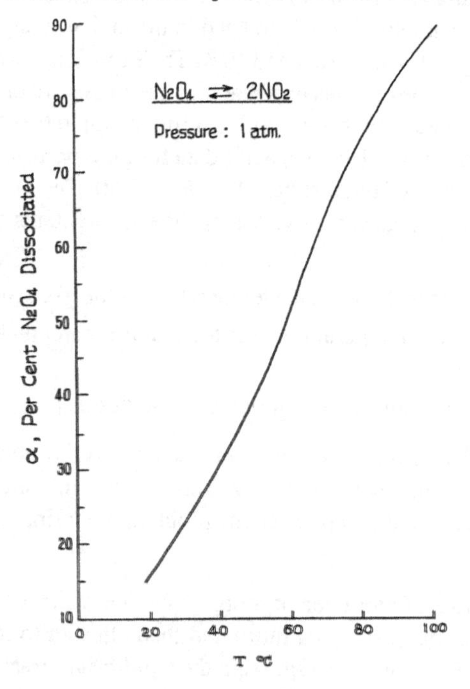

Figure P16.23

In a portion of a plant design, we are faced with estimating the change in temperature of a high-pressure N_2O_4-NO_2 gas mixture as it is throttled across a valve. The valve is well insulated, and the flow rate is steady. The upstream temperatures and pressures are 353 K and 5 bar. The upstream mole fractions are $y_{NO_2} = 0.637$ and $y_{N_2O_4} = 0.363$. The downstream pressure is 1 bar.

Available data: The standard Gibbs energy of the reaction and the *frozen* heat capacity are given in Problem 16.22. The molecular weights of N_2O_4 and NO_2 are 92 and 46, respectively. Although not strictly true, the gas mixture may be assumed to be ideal (i.e., the reaction $P\underline{V} = NRT$ is applicable and there is no effect of pressure on the enthalpy of the pure components). The fraction N_2O_4 dissociated to NO_2 is shown in Figure P16.23.

What is the downstream temperature and composition?

16.24. As noted in Problem 16.23, the chemical kinetics are such that gas mixtures of NO_2-N_2O_4 are always in chemical equilibrium.

Suppose that a mixture of these gases is expanded from a pressure and temperature, P_1, T_1 to a lower pressure P_2 in an adiabatic, reversible turbine.

(a) Write a differential equation that expresses how the pressure, temperature, and fraction N_2O_4 dissociated vary during the expansion. (Assume a basis of 1 mole N_2O_4 and let α be the fraction dissociated.)

(b) Show clearly how the adiabatic, reversible work may be calculated.

(c) Repeat parts 9(a) and (b) if the turbine were isothermal (at T_1) and reversible. Assume ideal gas mixtures.

16.25. The thermal conductivity of reacting gas mixtures is often found to be much larger than would be expected from molecular considerations. If a temperature gradient exists in the gas, then, in different temperature regions, the concentration of reactive species may be different; this concentration gradient causes a diffusion flux that adds to the normal thermal conduction heat flux since there is a transport of energy by molecular diffusion.

A convenient system to study this phenomenon utilizes nitrogen dioxide. The rate of the reaction

$$2NO_2 \leftrightarrow N_2O_4$$

is very rapid in both directions and, for most studies, the mixture may always be assumed to be in chemical equilibrium (see Problem 16.23).

Data:

Binary Diffusion Coefficients (at 1 bar)

T (K)	D (cm^2/s)
300	0.06532
350	0.08843
400	0.1147
500	0.1757
600	0.2467

Thermal Conductivity if No Reaction ("Frozen" Conductivity)

T(K)	W/m K $\times 10^2$
294	1.3
316	1.6
327	1.7
350	2.0
372	2.2
394	2.4

Source: "Product Bulletin, Nitrogen Tetroxide," Allied Chemical, Nitrogen Division, New York.

Assume that $\Delta H^o_{rx} = -58.07$ kJ/mol N_2O_2 is independent of temperature.

Derive an expression that shows the additional contribution to thermal conductivity because of the diffusion flux for this system. Plot this difference as a function of temperature. Also plot the temperature gradient between two plates 1 cm apart if the top plate is at 400 K and the bottom plate at 300 K and there is a mixture of NO_2 and N_2O_4 between the plates. The pressure is 1 bar. What is the heat flux between the plates?

16.26. One mole of H_2 and 1 mole of I_2 are placed in a constant-volume container at 473 K, the total pressure being 1 bar. Since a catalyst is present, the following reaction is always in equilibrium:

$$\tfrac{1}{2}H_{2(g)} + \tfrac{1}{2}I_{2(s)} = HI$$

$$\text{with } \Delta G^o_{rx,\,298\text{ K}} = 1323\text{ J} \quad \text{and} \quad \Delta H^o_{rx,\,298\text{ K}} = 24.74\text{ kJ}$$

The vapor pressure of HI from 222.4 K (melting point) to 237.7 K (normal boiling point) is

$$\log_{10} P_{vp} = 9.755 - \frac{1{,}129}{T}$$

where P_{vp} is in N/m^2 and T is in K. The vapor pressure of solid HI is 80.40 N/m^2 at 184.3 K. The vapor pressures of solid and liquid iodine are given in Problem 16.6.

Make a plot of the total pressure in the container as a function of the temperature as the temperature in the box is reduced from 473 K to 223 K. Assume that I_2 is insoluble in liquid HI. Also, plot the partial pressure of each component as a function of temperature.

16.27. When one is studying the properties of a "pure" substance in vapor-liquid equilibria, many simple equations, such as the Clausius-Clapeyron equation, may be derived. A little thought about this kind of analysis reveals, however, that the situation is more complex. For example, take liquid HF. It undoubtedly exists in rather complex equilibria between monomeric HF and polymers of the form $(HF)_x$; also vapor HF is known to consist of a chemical equilibrium mixture of HF and $(HF)_6$ with traces of other polymers. Other examples might include acetic acid, alcohols, or the classic N_2O_4-NO_2 case wherein the liquid and vapor phases contain both N_2O_4 and NO_2 in phase and chemical equilibrium. Finally, one might even cite the case of water; in the liquid phase we are reasonably certain that the molecules are not completely monomeric in nature; in the vapor, there have been theories advanced to allow for the presence of $(H_2O)_x$. With these thoughts in mind, the question is again raised about results obtained from analyses that consider only the monomer.

To demonstrate your ability to handle problems of this sort, consider a situation in which there is an equilibrium of the form

$$nA \leftrightarrow B$$

This chemical equilibrium relation holds for both liquid and vapor phases; also, phase equilibria is attained. Examples of this relation might be:

B	A	n
N_2O_4	NO_2	2
$(HF)_6$	HF	6
$(CH_3COOH)_2$	CH_3COOH	2

Choose a base of 1 gram-formula weight of A in each phase to simplify the analysis. Let α^g be the number of moles of monomer reacting in the vapor phase to form α^g/n moles of polymer. Thus, in the vapor phase there are $(1-\alpha^g)$ moles of monomer and α^g/n moles of polymer. Similarly, let α^L be the comparable parameter in the liquid phase.

(a) Suppose that you could measure experimentally the vapor pressure of this two-phase A-B system as a function of temperature. What would the slope $(dP/dT)_{saturation}$ represent? Make a rigorous analysis and express your results in the usual thermodynamic nomenclature, such as partial molar quantities, α^g, α^L, etc. Also describe in words what your answer means.

(b) If the vapor pressure data were plotted as ln P against $1/T$, the "apparent latent heat" can be calculated as

$$\Delta H_{apparent} = -R\frac{d \ln P}{d(1/T)}$$

How would $\Delta H_{apparent}$ be related to the enthalpy term found in part (a)? In these calculations assume that liquid volumes are negligible compared to vapor volumes and that the vapor phase behaves as an ideal-gas mixture.

(c) For the case $6HF \leftrightarrow (HF)_6$, vapor-pressure data, plotted as ln P versus $1/T$ gives an apparent ΔH of 24.70 kJ/20.1 g HF at 273.2 K [R.L. Jarry and Wallace Davis, Jr., "The Vapor Pressure, Association, and Heat of Vaporization of Hydrogen Fluoride," *J. Phys. Chem.*, **57**, 600 (1953)]. Also, a calorimetric enthalpy of vaporization gives 7.03 kJ/20.1 g HF at the same temperature [Karl Fredenhagen, "Physikalischchemische Messungen am Fluorwasserstoff," *Z. Anorg. Chem.*, **210**, 210 (1933)]. Estimate the mole fractions of HF and $(HF)_6$ in the vapor phase at 273.2 K.

16.28. It is well known that vapors of very polar gases are often quite nonideal even at low pressures. One reason for this behavior is that polymerization undoubtedly occurs and the effective number of moles is less than the apparent value based on the mass of gas and the molecular weight of the monomer.

Suppose that we attempt to account for the "apparent" nonideality by allowing a monomer W to undergo the following dimerization:

$$2W \leftrightarrow W_2$$

We further assume that:

1. No higher polymerization reactions occur.
2. W and W_2 behave as ideal gases in the usual sense and their mixture is an ideal gas mixture.

3. The mixture is always in chemical equilibrium with

$$K \equiv \frac{p_{w_2}}{p_w^2}$$

Let α be the fraction W which associates.

(a) Derive an equation to relate the apparent compressibility factor Z to α, where $Z \equiv P\underline{V}/NRT$. N is the mass of W in volume $\underline{V}$ divided by the molecular weight of monomer.

(b) Derive an equation relating the apparent compressibility factor to the system temperature and pressure if $\Delta G^o = \Delta H^o - T\,\Delta S^o$ and unit fugacity standard states are employed. (Assume that ΔH^o and ΔS^o are not functions of temperature.)

(c) Let W be monomeric acetic acid. For acetic acid:

$$\Delta H^o_{rx} = -58.62 \text{ kJ/mol of dimer}$$

$$\Delta S^o_{rx} = -138.2 \text{ J/mol K}$$

At 391 K and 1 bar, what fraction of the acetic acid would you predict would be in the form of dimers? What would be the apparent compressibility factor?

16.29. Some recent experiments have been carried out to determine the equilibrium partial pressure of oxygen over molten potassium oxides. The data were taken in the following way. Samples of pure KO_2 were placed in a MgO boat in an evacuated tube. The tube was inserted into an oven at a sufficiently high temperature that the oxide melted. The pressure of the evolved oxygen was measured after equilibrium was attained. From this pressure measurement, the tube volume, and the oven temperature, the moles of oxygen evolved could be ascertained. From this value and the original sample weight, the atomic O/K ratio of the oxide liquid could be calculated. Next, a known amount of oxygen was bled out of the system, and the system allowed to come to equilibrium.

Again the oxygen pressure was measured, and the liquid O/K ratio calculated. The data indicated that at any given temperature level oxygen partial pressure depends only on the liquid O/K ratio. The data are as follows:

	Partial pressure of oxygen (bar)		
O/K ratio	773 K	873 K	923 K
1.0	0.018	0.071	0.13
1.1	0.10	0.13	0.21
1.2	0.26	0.26	0.31
1.3	0.44	0.41	0.44
1.4	0.62	0.56	0.61

(a) Are these data consistent with the phase rule? Demonstrate. There is essentially nothing known about the structure of the liquid phase. It is black, probably has a high electrical conductivity, and contains some or all of the following species: K^+, O_2^-, K_2O, K_2O_2, KO_2.

(b) Estimate accurately the heat evolved or absorbed if a reaction occurs so that the liquid absorbs oxygen isothermally at 873 K as it changes from an O/K ratio of 1.0 to 1.4. Express your answer on a basis of 1 g-atom of potassium in the liquid.

(c) A "simple" picture of the liquid shows it to be an "ideal" mixture of liquid K_2O, K_2O_2, and KO_2. Demonstrate how you would calculate equilibrium constants for the reactions given below, using only the p_{O_2} and O/K values measured experimentally.

$$2KO_{2(l)} \rightarrow K_2O_{2(l)} + O_2 \quad (1)$$

$$2KO_{2(l)} \rightarrow K_2O_{(l)} + \tfrac{3}{2}O_2 \quad (2)$$

(d) How could you test the hypothesis in part (c) to see whether it is reasonable?

(e) Many thermodynamicists would describe the system as follows:

$$\tfrac{1}{2}O_{2(g)} \leftrightarrow [O_{(l)}]$$

where [O(l)] represents the oxygen in the liquid phase.

$$K = \frac{[O_{(l)}]}{p_{O_2}^{1/2}}$$

By defining an activity coefficient such that

$$[O_{(l)}] = (\gamma)\ (\text{conc. } O_2 \text{ in liquid})$$

with $\gamma \rightarrow 1.0$ as concentration $\rightarrow 0$. Show how the activity coefficient and equilibrium constant may be determined from the data at any given temperature and composition. Assume for simplicity that you have at your disposal the partial pressures of oxygen over the entire range of O/K, ratios from 0 to 2.

16.30. In recovery furnaces where kraft paper liquor is burned, evidence has been accumulating that the sulfur emission is far less than might be expected from the sulfur present initially in the liquor. This result, of course, is very desirable, but to optimize furnace operation, some ideas regarding the mechanism of such a sulfur trapping would be helpful.

The liquor is a very complex mixture of carbonaceous material and a number of inorganic salts (both solids and liquids) through which air is blow to evaporate water and to burn the carbon (and other combustibles). Sulfur in the original liquor ends up as various salts (e.g., Na_2S), a few ppm are found in the flue gas as SO_2 and SO_3. Surprisingly, a significant amount of S also appears as Na_2SO_4 in the stack electrostatic precipitator.

Several of the current theories hypothesize a general mechanism which allows a decomposition of sodium carbonate in the furnace. Ideas differ as to the reaction products, but the net result seems to be the formation of reactive sodium species in the furnace gases. These sodium species then can trap SO_2/SO_3 as sulfites or sulfates and end as precipitates in the flue.

Prepare a brief memorandum (with suitable appendices for documentation) to indicate your opinion of the sodium carbonate decomposition mechanism. To allow a more closely defined and simpler case, let us assume that we have, initially, an evacuated, cold bed of Na_2CO_3 and carbon. We plan to heat this bed to temperatures typical of those reported for kraft liquor furnaces (i.e., between 1,000 and 1,400 K). Then we will measure the pressure and gas composition. To aid in the design of our experiment, estimate what pressures and gas compositions you would expect in this temperature range. The JANAF thermochemical tables are a good source of data.

16.31. (a) Following are some interesting articles clipped from The New York Times.[4] Please read them carefully and prepare a well-written, brief memorandum which we can submit to our Director of Research. She needs this analysis to advise our Board of Directors whether to purchase an interest in the invention.

(b) Mr. Rocky Jones has also approached us to suggest methods to produce hydrogen from water. As usual, he is very indefinite, but I was able to obtain at least one specific proposal that is described in Figure P16.31(a).

In our West Bacon plant, we have a situation where we are now venting (wastefully!) 3 kg of steam each and every second. This steam is piped from the [*Deleted, proprietary data*] reactor at 1,000 K and 39.2 bar. Mr. Jones proposes to run this steam into his unit and vent saturated steam at 1 bar. A portion of this saturated steam effluent is recycled to the unit and hydrogen and oxygen gases are (somehow) produced in separate streams.

Please analyze this "Jones box" and let me have your considered opinion as to its operability and the amount of hydrogen we might hope to obtain each hour.

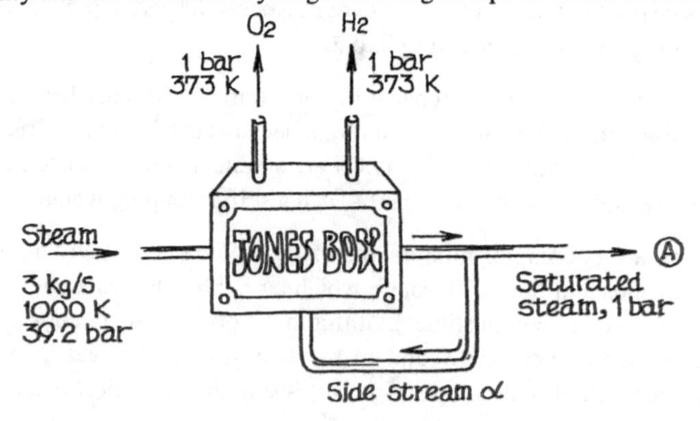

Figure P16.31(a) Box appears to be adiabatic and rigid.

---

4 © 1977 by the New York Times Company, reprinted by permission.

Experts Dispute Contention That Device Can Extract Cheap Energy From Water

By ROBERT LINDSEY
Special to The New York Times

LOS ANGELES, March 28—A machine that its developers contend can extract cheap energy from water has been tested here and has touched off a flurry of speculation on Wall Street, a Federal investigation of possible stock manipulation, and virtually unanimous skepticism from scientists who say the machine cannot do what its inventors say it can.

Its inventors say the device separates tap water, in a continuous, self-sustaining reaction with virtually no outside energy, into its two component parts—oxygen and hydrogen, a fuel that can be used to heat homes or power automobiles without pollution.

Scientists who were questioned say such a machine is theoretically impossible, but developers insist recent tests have shown its feasibility. Rumors of the invention have caused wild trading in the stock of one company associated with the device, as well as an investigation by the Securities and Exchange Commission.

Last week, officials of two commercial testing laboratories here said that, in preliminary reports, they had examined and tested the device and that, as the inventors contend, it produced combustible hydrogen as well as oxygen from water, for periods of 19 and 30 minutes.

Yesterday, a reporter watched as water, apparently from the municipal supply system of a suburban community here, was piped into the prototype device, a stainless steel box somewhat larger than a big trunk. A few knobs were turned, and a jet of flaming gas shot out of a tube and kept burning until the machine was turned off about 15 minutes later.

It was impossible to determine if flammable gas or other energy sources had been concealed in the box, or whether the demonstration was a hoax in any other way. However, Sam Leach, the machine's 61-year-old inventor, said that, except for electricity necessary to initiate the process, the energy from the reaction had come from the water itself.

Specialists Skeptical

"The water is being split into hydrogen and oxygen," he asserted. "The reaction is self-sustaining."

When the few details known about the process were explained to a half dozen of the country's leading specialists in hydrogen energy technology, everyone said the device could not operate as the inventors contended. Several called it a fraud.

The dream of liberating the energy of hydrogen in every drop of water has excited scientists for at least a century and has accelerated in the last three years. A discovery of a cheap hydrogen-oxygen separation system would obviously have enormous impact on world economics, industry, and the balance of power.

It has long been known that hydrogen and oxygen can be split by applying great amounts of energy. High school chemistry students observed the process in electrolysis, in which an electric current passes through water containing a salt or alkali and separates oxygen and hydrogen.

Nuclear reactors have also been used to do the job. Every method requires the use of far more energy to obtain hydrogen than that derived from it.

Mr. Leach, a well-to-do, reclusive Southern California inventor who says he has more than 70 patents, accused other researchers of "tunnel vision" in failing to apply generally known principles to the problem that he says he has solved.

"As described to me, the system violates the principles of thermodynamics," said Dr. Bernard M. Abraham of the Argonne National Laboratory, Argonne, Ill. "You can't get something for nothing; there's no way they can do it."

"You can't get anymore energy out of it than you put into it," said Dr. R. H. Wentorf Jr. a physical chemist in the General Electric Company's research and development center at Schenectady, N.Y. "I'd be very cautious about it; some trick is being used."

Scoffers Were Expected

"Sure, they've got a magic machine—they put something in it called 'dreamium,'" scoffed a scientist at the California Institute of Technology.

"We expect them to say we can't do it," said Morris Mirkin, who founded and then sold the national Budget Rent-a-Car Company, and who with his family owns rights to the technology for all applications except housing. "But they're wrong, and we'll prove it."

The researchers who were interviewed said there were several ways to stage a demonstration that seemed to prove water was being separated into a combustible fuel—for instance, concealing a hidden source of energy within the machine, perhaps a hidden battery or electric lines, or using one of several kinds of metals that could be liberated of inherent hydrogen and oxygen in a relatively short time until exhausted.

However, a spokesman for the two laboratories that checked the machine, the Smith-Emery Company of Los Angeles and the Approved Engineering Test Laboratories of Encino, Calif., said the machine had been disassembled and reassembled in front of their observers and no source of hidden energy had been found.

"I could find no evidence of hanky-panky," said Gordon Walker, who directed the test for Smith-Emery.

Rumors that a machine had been invented that might provide an almost inexhaustible amount of cheap energy have circulated in the financial community since last summer. They have caused extensive speculation and volatile ups and downs in the price of stock of the Presley Companies, a Newport Beach, Calif., homebuilding concern that has been hit hard by the national slump in construction.

The value of its stock has soared from about $2 last summer to 20¾ on the American Stock Exchange, until the Securities and Exchange Commission suspended trading on Thursday until April 3.

According to government informants, the commission is investigating dealing in this stock, alleged failure to disclose certain information about the project, and other possible violations of Federal securities laws. An S.E.C. investigator told a reporter today, "You'll be hearing more about this from us."

According to Mr. Leach, hydrogen and oxygen are separated in the machine by a process that uses two stainless steel chambers. Each contains a granular metal called a "reactant," not otherwise indentified. The reactant acts much like a sponge to absorb oxygen.

Started by Electricity

To start the process, a small electric charge is said to be applied to heat the reactant and vaporize a flow of water into steam. The steam is said to pass over the heated reactant in one chamber, and the reactant grabs, or "sequesters," the oxygen while letting hydrogen pass through for any intended use.

At some point, the reactant "sponge" becomes filled with oxygen, Mr. Leach said, and it has to be emptied for the next cycle.

This is done, he continued, by heating the chamber in such a way that the oxygen is removed, and the reactant is regenerated. The hydrogen-producing reaction is said to take place at about 850 to 900 degrees Fahrenheit, and the oxygen-releasing reaction at 1,000 to 1,200 degrees.

According to Mr. Leach, the energy used to remove the oxygen in the second step is heat generated during the oxygen-steam reaction in the first step in the adjacent chamber. This heat is also used to provide additional steam, he said.

The two-step process—first removing the oxygen, then exhausting it from the saturated reactor—is said to continue indefinitely, as long as more water is added, according to Mr. Leach.

He said he had run his prototype machine, the fourth since the development began, a maximum of two hours, and contended that the limit was not in the fundamental process but in the equipment. Despite the doubts of other scientists, who say they do not see where he could obtain energy for the oxygen-purging stage, Mr. Leach said no additional energy was needed, and that the metal reactant was not consumed, eroded or damaged in the process.

Details not Explained

Mr. Leach will not disclose the specific developments he learned in order to make the reaction work. However, an aide said that patents on the procedure and equipment had been applied for in 86 countries and that at least one would probably disclose details about May 1.

"It's an exothermic reaction that is self-sustaining," Mr. Leach said. "The scientists [who doubt the development] are well aware of it. I did the mathematical modeling, and after two years, before I had worked on any piece of equipment, I knew the inventing was over.

"These are all commonly known; what I've done is bring some things together that other people hadn't thought about doing," he said. "The important thing is we provided an environment for these reactions to occur, the timing and so forth. I know why they're saying it can't work; I can't fault them for that, but they're missing something; thermodynamicists follow certain things blindly, like tunnel vision."

"There'll be additional tests," Mr. Leach said, "to certify that we are in fact splitting water, but I have no doubt at all that we have done it."

Split of Water Elements for Fuel Draws Interest, Along With Scorn

By ROBERT LINDSEY
Special to The New York Times

LOS ANGELES, May 5—Despite continuing skepticism over its effectiveness, a controversial process whose inventor claims can separate water into its component parts—hydrogen and oxygen—at low cost has drawn the interest of at least two major corporations.

The Bechtel Corporation of San Francisco began negotiating with Sam Leslie Leach, the inventor of the process, late last winter, and has indicated that it was interested in purchasing a worldwide exclusive license for the process. Today, the Aluminum Company of America in Pittsburgh acknowledged that it had expressed interest in the process, although no contract negotiations nor any testing of the technique have taken place.

An invention capable of separating the hydrogen from water at low cost, theoretically at least, would have substantial importance in helping the world meet energy needs. Pure hydrogen can be used as a fuel much as gasoline or natural gas.

Inventors have tried to split water for more than a century, much as alchemists sought to turn stone into gold. But every process tested in the past has required more energy to separate the hydrogen and oxygen than could be derived from the process itself. Scientists have derided assertions, including those of Mr. Leach, saying that what he claims to have invented amounts to a "perpetual motion machine" that violates certain basic laws of physics.

Assertion First Made Last Year

Mr. Leach, a well-to-do 62-year-old Californian who has a number of patents in optics and other fields, first made the assertion that he had developed an economic water-splitting process last year. Since then, he said he had repeatedly asked the United States Energy Research and Development Administration to evaluate his process, but he says the Government has not taken him seriously.

No information has been made public to indicate that Mr. Leach has made the technical breakthrough that he claims. Neither Bechtel nor Alcoa has tested the Leach machine, but the inventor has discussed some elements of the scientific principles involved with research specialists from both companies and they apparently have taken his presentations seriously enough to enter into detailed consideration of the process.

Bechtel is a large, diversified construction and engineering concern heavily involved in a wide variety of energy projects. Alcoa is the nation's largest producer in the energy-intensive aluminum industry.

A Bechtel spokesman said today the company would have no comment on the company's interest in the machine.

Technical Experts Sent by Others

A number of other major corporations and representatives of at least two foreign governments, Israel and Brazil, have sent technical experts to a suburban laboratory near here to witness demonstrations of the prototype machine in which tap water is introduced at one end and oxygen and hydrogen—the hydrogen in the form of a flame—is exhausted.

Although Mr. Leach's claims have brought the disapproval and derision of scientists, they have perplexed industry, the Securities and Exchange Commission and Wall Street. After word leaked out about the machine last year, shares of the Presley Companies, a California home builder that has a license for one application of the process, skyrocketed.

Subsequently, the S.E.C. investigated and concluded that, based on his patent applications, there was no evidence to support Mr. Leach's assertions of the breakthrough.

Mr. Leach's response was that in his early 10 applications he had outlined only portions of the technology and that he has since submitted additional ground.

He has hinted obliquely that the process is related to what he says is the discovery of an overlooked source of energy in certain materials that, when processed in a certain way, provides the energy needed to sustain the water-splitting operation.

Data: Assume ideal gases:

C_P(water vapor) ~ 35 J/mol K

ΔH^o_f 373 K(water vapor) ~ -2.42×10^5 J/mol

ΔG^o_f 373 K(water vapor) ~ -2.254×10^5 J/mol

ΔH_g 373 K ~ 40,000 J/mol

(c) You know I have a high opinion of Rocky, but I wonder if he hasn't neglected a real opportunity in his simple Jones box. It would seem to me that we could go even further! Could we take his exit steam (saturated vapor, 1 bar) and condense it in another "box" to saturated liquid water at 1 bar and produce even more H_2? For example, starting from stream *A* in Figure P16.31(a), we would have the result shown in Figure P16.31(b). Please comment on my extension.

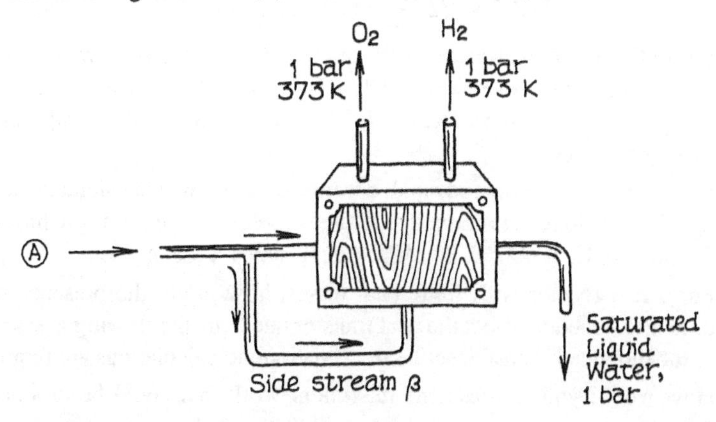

Figure 16.31(b) Again the box (?) appears to be adiabatic and rigid.

16.32. Given the following Dissociation reaction:

$$(NH_4)_2CO_{3(s)} = 2NH_{3(g)} + CO_{2(g)} + H_2O_{(g)} \quad \Delta H^o = -40 \text{ kJ/mol}$$

How many degrees of freedom exist if 1 mole of $(NH_4)_2CO_3$ is charged to the system? Is it possible to obtain a dissociation vapor pressure curve for this system? Develop an expression for the equilibrium constant in terms of standard state fugacities, total pressure, and degree of conversion α. How will the system respond to an increase in *P*? How will it respond to an increase in *T*?

16.33. The following phase reaction

$$B_{(s)} + A_{(g \text{ or } l)} = AB_{(s)}$$

can be treated as either a 3-component system with one reaction or as a 2-component system with compound formation. You can assume that both solid phases are pure and that A is present either in a liquid or vapor binary mixture of A and B. Sketch the *P-T* behavior around the quadruple point (1 bar, 150°C) where 4 phases are present (solid A, vapor, liquid, and solid AB). Your sketch should go from the triple point for pure A at 10 bar and 200°C to the congruent melting point of the solid compound AB at 20 bar, 250°C. Indicate phases in equilibrium for monovariant lines and invariant points. A eutectic mixture of solid A and solid AB is known to exist at approximately 150°C over a wide range of pressures above 1 bar. For the system, solid B, vapor, and solid AB, devise an expression for dP/dT and show what this reduces to if A is much more volatile than B at low pressures.

16.34. Although data are sparse, several studies have suggested that chlorinated hydrocarbons like CH_3Cl can be efficiently oxidized to CO_2, H_2O, and HCl at very high pressures (greater than 25 MPa) and temperatures above 400°C. The reaction is known to be exothermic with the following stoichiometry:

$$CH_3Cl + 3/2\,O_2 = CO_2 + H_2O + HCl$$

(a) Under the highly non-ideal conditions that exist, describe how you would estimate the equilibrium conversion of CO_2, H_2O, and HCl starting with a known feed composition of O_2 and CH_3Cl at 400°C, 30 MPa. State all assumptions and data or models required.

(b) If the kinetics are assumed rapid, describe how you would calculate the temperature rise that would occur in a well-insulated tubular reactor for a stoichiometric feed of O_2 and CH_3Cl at 400°C, 30 MPa. Again, state all assumptions.

(c) For a real reactor with finite heat losses, how might the presence of the above oxidation reaction affect the heat transfer rate from the flowing gases in the reactor to the tube walls? Just describe the situation; no calculations are required.

(d) How would you estimate the maximum work that could be produced from the product gas mixture produced in part (b)? The ambient temperature is 300 K.

16.35. Rocky Jones has just come up with a new invention to process an "unconventional" source of natural gas. We at MITY Industries would like you to evaluate his design before securing venture capital for commercialization. His device is known as the "RJ Supercritical Geopressured Oxidizer/Converter" and is intended to generate *copious* amounts of electric power from the geopressurized aqueous brines found in deep aquifers off the Gulf Coast of Texas. These brines are claimed to be saturated with CH_4 and available at 177°C and 1000 bar. The annual-averaged ambient temperature and air pressure in this part of Texas is 27°C, 1.0001 bar, respectively.

(a) Even if Rocky's process works, there is some question as to just how much methane is contained in the geopressured brine at 177°C, 1000 bar. Describe how you would estimate the amount of dissolved CH_4 contained in the brine under saturated conditions. As a first approximation, you can assume that the dissolved salt concentration is low enough to be neglected. State your assumptions and clearly describe how you will evaluate any thermodynamic properties needed.

(b) Describe and sketch an idealized process that would yield the maximum possible amount of work per kg of brine—entering at 177°C 1000 bar. Not much is known about Rocky's device as it is highly proprietary, but we do know that the exiting stream only contains CO_2, H_2O and some dissolved salts at 27°C and 1 bar. This indicates that all the CH_4 has been oxidized via the following reaction:

$$CH_4 + 2O_2 \rightarrow CO_2 + 2H_2O$$

O_2 presumably enters the process at 1 bar, 27°C.

(c) Describe how you would estimate the work output of your process per kg of brine. State what property information and/or correlations would be required to make a quantitative evaluation.

16.36. Two young thermodynamicists, Kermit Kelvin and Henrietta Helmholtz, were arguing over whether a neutral aqueous solution always has a pH of 7.0. Henrietta claims that as the temperature increases from say 25°C to 200°C, the pH decreases to about 5.5 for a neutral solution. Kermit disagrees, maintaining that the pH remains at 7.0 for a neutral solution regardless of the temperature. Develop an appropriate thermodynamic analysis to decide who is correct. State and justify all assumptions used. The following information is provided:

$$pH = -\log_{10}[a_{H^+}], \quad a_{H^+} = \text{activity of } H^+ \text{ ion}$$

$$H_2O_{(aq)} = H^+_{(aq)} + OH^-_{(aq)}$$

where the (aq) implies an aqueous liquid phase and the equilibrium constant for the above reaction is $K_a = 10^{-14}$ at 25°C and 10^{-12} at 120°C.

16.37. *Amphiphiles* are molecules which consist of a *hydrophilic* region which is soluble in water and a *hydrophobic* region which is not. When amphiphiles are placed in water they can achieve segregation of their hydrophobic regions from water by aggregating into colloidal aggregates called *micelles*. The resulting solution is called a *micellar* solution.

$$n \text{ monomers} = n\text{-type micelle} \qquad (1)$$

Consider a micellar solution at temperature T and pressure P composed of N_w water molecules and N_A amphiphiles which are distributed into N_1 monomers in chemical equilibrium with N_n n-type micelles. Typically $n = 50$.

The micellar solution can be described by the following Gibbs free energy model:

$$\underline{G} = N_w\mu_w^o(T, P) + N_1\mu_1^o(T, P) + N_n\mu_n^o(T, P) +$$
$$k_BT(N_w \ln (X_w) + N_1 \ln (X_1) + N_n \ln (X_n)), \qquad (2)$$

where k_B is the Boltzmann constant, $\mu_i^o(T, P)$ with $i = w, 1, n$ are standard chemical potentials, and $X_i = N_i/(N_w + N_1 + N_n)$ with $i = w, 1, n$ are mole fractions.

(a) Derive expressions for the chemical potentials of the water, monomers and n-type micelles.

(b) What is the condition of chemical equilibrium for the chemical reaction in Eq. (1)? Express your results in terms of chemical potentials.

(c) Use the result in (b) to derive a formula relating X_n to X_1. What is the physical meaning of the various terms appearing in that formula?

(d) For a given mole fraction of amphiphiles, $X_A = N_A/(N_w + N_1 + N_n)$, derive an implicit nonlinear equation which relates X_1 to X_A. Do not attempt to solve this equation.

(e) To analyze qualitatively the properties of the equation derived in (d), it is not necessary to actually solve for X_1 as a function of X_A. Instead, define a characteristic mole fraction:

$$X^* = \exp\left\{\frac{(\mu_n^o - n\mu_1^o)}{(n-1)} k_B T\right\}$$

and normalize all mole fractions in the equation by X^*. You will then obtain an implicit nonlinear equation relating (X_1/X^*) to (X_A/X^*). Analyze this equation in the limits (1) $X_A << X^*$, (2) $X_A = X^*$ and (3) $X_A >> X^*$ and draw schematically the behavior of X_1 and X_n versus X_A. What is the physical meaning of X^*?

16.38. Chemotrophic microorganisms obtain energy for growth through a redox process where oxidation of organic compounds leads to the synthesis of an "energized" compound called ATP (adenosine-5′-triphosphate) from ADP (adenosine-5′-diphosphate) and inorganic phosphate, P_i. ATP, in turn, is used to provide energy for biosynthesis. Thus, when evaluating free energy transduction in living systems, one can consider ATP to represent a "quantum" of available energy. The value of this quantum is the minimum free energy required to synthesize one mole of ATP; this value is approximately -10.5 kcal/mol (-44 kJ/mol) under quasi-reversible conditions. As a consequence, the thermodynamic efficiency (η) of energy transduction is dependent on the cell's ability to couple mechanistically ATP energization with biosynthesis.

In the lactic acid ($C_3H_6O_3$) fermentation, glucose ($C_6H_{12}O_6$) is converted through a complex series of reactions to pyruvate ($C_3H_4O_3$) which leads to the generation of ATP; these coupled reactions can be depicted as:

$$\text{glucose} + 2\,\text{NAD}^+ \longrightarrow 2\,\text{pyruvate} + 4\,\text{H}^+ + 2\,\text{NADH}$$

$$n\,\text{ADP} + n\,P_i \curvearrowright n\,\text{ATP}$$

The generation of 2 moles of NADH (a reduced form of nicotinamide adenine dinucleonitide, NAD^+) is coupled with an electron-accepting partial process. Because the ADP to ATP transfer process is quantized, n must take on integer values. From mechanistic and thermodynamic considerations, the amount of energy transfer during the production of lactic acid from glucose cannot exceed that required by the ADP to ATP energization. The ability of an organism to grow depends on the subsequent coupling of an ATP to ADP transfer process to biosynthesis of cell matter. In this process, ATP releases its "energy" according to the simplified net reaction:

$$\text{glucose} \longrightarrow \text{cell mass}$$

$$m\,\text{ATP} \curvearrowright m\,\text{ADP} + m\,P_i$$

Again m must be an integer.

During the lactic acid fermentation, 2 moles of glucose (a neutral molecule) are consumed for every mole of lactic acid that is accumulated. The cell must maintain the intracellular pH at 7 in spite of the fact that it is producing lactic acid. It can do this by expending energy to maintain a pH gradient across the cell membrane. Protonated lactic acid is freely diffusible across the cell membrane, but lactate is not. The pK_a of lactic acid is about 4.

(a) Calculate the standard Gibbs free-energy change for conversion of glucose in lactic acid fermentation. Please state all assumptions.

(b) A thermodynamic efficiency (η) for microbial growth can be defined as the ratio of the work producing potential of ATP produced from ADP by the energizing process to the work consumed by converting glucose to lactic acid. Calculate the maximum and expected values for the lactic acid fermentation. From your knowledge of typical thermodynamic efficiencies for machines and combustion devices, state the value of η that is most "reasonable."

(c) Realizing that under non-equilibrium physiological conditions, the concentration of substrate and product are not in their standard state values of 1 mol/liter and vary with time and that the temperature is typically 37°C and not 25°C, estimate the range of actual ΔG values for lactic acid fermentation. A typical initial glucose concentration is 50 g/liter and most of it can all be converted to lactic acid in cells. State all assumptions.

Substrates	Products	ΔG° (25°C) kcal/mol	ΔH° (25°C) kcal/mol
$2NAD^+$ + glucose	2 pyruvate + $4H^+$ + 2NADH	-26.8	+1
pyruvate + H_2	lactate	-10.3	+2

Note: All standard states are ideal solution of solutes in water at 25°C and concentrations of 1 mole/liter.

16.39. Geochemists have suggested that ratios of certain ion concentrations in natural water can be used to estimate *in situ* reservoir temperatures. Common examples of reversible geochemical reactions are those involving cation exchange among K^+, Na^+, and Ca^{+2} ions that are common elemental constituents to many minerals. For example,

$$K^+_{(aq)} + (Na_xCa_y)\text{ plagioclase}_{(s1)} + 4y\ SiO_{2(s2)} = \text{K-feldspar}_{(s3)} + xNa^+_{(aq)} + yCa^{+2}_{(aq)}$$

where the phases are

(aq) - aqueous liquid phase
($s1$) - solid plagioclase feldspar [$Na_xAlSi_3O_8 \cdot Ca_yAl_2Si_2O_8$]
($s2$) - solid quartz [pure SiO_2]
($s3$) - solid potassium (K) feldspar [$KAlSi_3O_8$]

Experimental data collected from natural hot springs, aquifers and oil reservoirs, as plotted in the figure below, show that the quantity

$$\log_{10}\left[[Na^+]/[K^+]\right] + \beta \log_{10}\left[\sqrt{[Ca^{+2}]}/[Na^+]\right]$$

scales linearly with $1/T$. Consequently, measuring the K^+, Na^+, and Ca^{+2} ion concentrations may provide an empirical means of estimating the *in situ* fluid temperatures within reservoirs, a so-called geothermometer.

(a) Is the observed behavior reasonable from a thermodynamic standpoint? State any and all assumptions required to support your arguments.

(b) The dissolved SiO_2 concentration in the aqueous phase can also be measured along with K^+, Na^+, and Ca^{+2} concentrations. Describe how you would estimate the ΔH of crystallization for the following phase reaction:

$$SiO_{2(aq)} = SiO_{2(s2)}$$

from the correlations shown in Figure P16.39 below combined with corresponding values for $SiO_{2(aq)}$ for each fluid sample. State all assumptions used.

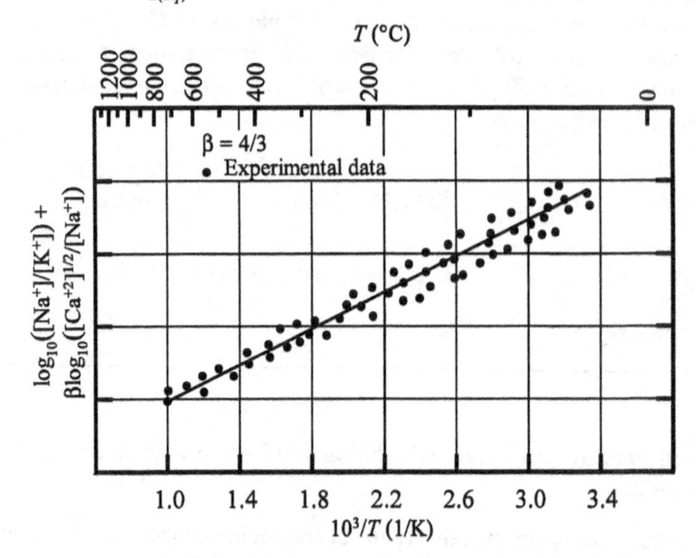

Figure P16.39 Na-K-Ca geothermometer data correlations.

16.40. In the commercial production of ammonia, a high-pressure, high-temperature gas phase synthesis process is used. N_2 and H_2 are reacted in the presence of a catalyst at 327°C and at pressures ranging from 50 to 300 bar:

$$N_{2(g)} + 3H_{2(g)} = 2NH_{3(g)}$$

The overall conversion is known to be limited by equilibrium.

(a) What is the magnitude of the equilibrium constant at 327°C? (Some data are provided below).

(b) Starting with a stoichiometric feed of 1 part N_2 to 3 parts H_2, *describe* how you would calculate the conversion as a function of pressure from 50 to 300 bar at a reaction temperature of 327°C. You can assume that the $P\underline{V}TN$ properties of the vapor mixture can be adequately modeled with the Peng-Robinson EOS. State and justify all assumptions and indicate what data and/or physical property information would be required to produce a quantitative answer.

(c) Would you expect the molar heat capacity at constant pressure (C_p) to vary as the temperature is raised from 327°C? Explain your answer.

$$\Delta G^o_{(298\text{ K})} = -16{,}450 \text{ J/mol NH}_3 \text{ and } \Delta H^o_{(298\text{ K})} = -46{,}110 \text{ J/mol NH}_3$$

Standard states: gases at unit fugacity, $T = 298$ K, $P = 1$ bar

16.41. The following discussion is found on p. 283-285 of Cantor and Schimmel's 1980 textbook on *Biophysical Chemistry.*

"Before considering data on free energy of transfer, it is important to consider the distinction between *unitary* and *cratic* contributions to free energy" (see Gurney, 1953; Kauzmann, 1959). The reason for this consideration is that we must be sure all data are considered on the same basis. to illustrate the difficulty encountered, consider the reaction

$$A + B = AB \tag{1}$$

with an equilibrium constant defined as

$$K = (AB)/(A)(B) \tag{2}$$

and a standard free energy change of

$$\Delta G^o = -RT \ln K \tag{3}$$

If we choose concentration units of mole/liter, then ΔG^o corresponds to the change in free energy when A and B, each at a standard state of 1 mole/liter, are combined to give AB at 1 mole/liter. However, the numerical value of K (and consequently that of ΔG^o) will be different if a different standard state (or a different set of concentration units) is employed. This dependence of the free energy change on the choice of the standard state is, of course, true only for reactions in which the number of reactant species is not equal to the number of product species; in these cases, the magnitude of the equilibrium constant depends on the choice of standard state, unlike the situation for a unimolecular reaction.

"The dependence of ΔG^o on the standard state is a purely statistical effect that arises from the expression for the partial molal entropy; consider the partial molal entropy S_A of A:

$$S_A = S'_A - R \ln x_A \tag{4}$$

where x_A is the mole fraction of A. (We are considering dilute solutions, and therefore we are ignoring activity-coefficient effects.) The term $-R \ln x_A$ is called the *cratic* contribution to the entropy; it is a purely statistical term arising from the mixing of A with solvent molecules. This contribution is independent of the chemical nature of A. The

S'_A term is called the *unitary* contribution; it reflects the characteristics of A itself and its interaction with solvent. The standard entropy change for the reaction of Equation (1) becomes

$$\Delta S^o = S'_{AB} - S'_A - S'_B + R \ln x_M = \Delta S' + R \ln x_M \tag{5}$$

where x_M is the mole fraction of a molecule when its concentration is 1 mole/liter. Because the water concentration is 55.6 molar in dilute aqueous solution, the cratic contribution to the entropy change is about $-R \ln 55.6 \approx -8$ cal/mol K, this corresponds to a cratic contribution to the free energy of about -2.4 kcal/mole. If a different standard state and different concentration units were chosen, a different contribution would result. The unitary change in entropy, ΔS^o_u, thus is obtained from

$$\Delta S^o_u = \Delta S^o - R \ln x_M \tag{6}$$

Comment on the author's contentions regarding the dependence of ΔG^o and equilibrium constant (K) on standard states when $\nu = \Sigma \nu_i$ is zero or not. Is it purely a "statistical effect?"

Also, what is your interpretation of the "*cratic*" and "*unitary*" contributions to the standard entropy change for the reaction Eq. (1) ΔS^o?

16.42. Several prominent Soviet and U.S. scientists have conjectured for some time that vast amounts of methane (CH_4) lay trapped beneath the surface of the earth in solid hydrate form. An important discovery in 1934 demonstrated that water and methane under pressure can form a solid phase that exists well above the normal freezing point of pure water. This solid phase has been described as a non-stoichiometric clathrate compound where gaseous methane molecules are trapped in a lattice structure consisting of water molecules. With all vacant lattice cages filled, the reaction to form the so-called methane gas hydrates is given by:

$$CH_{4(g)} + 5\tfrac{3}{4} H_2O_{(l)} = CH_4 \cdot 5\tfrac{3}{4} H_2O_{(s)} \tag{1}$$

ΔH^o for this phase reaction is approximately constant at -14.5 kcal/gmole CH_4 with pure $CH_{4(g)}$, pure $H_2O_{(l)}$ and pure solid hydrate at T and P of the mixture taken as the standard states.

As a consultant to the Underground/Undersea Division of the MITY Corporation, you have been asked to establish whether the *in situ* conditions existing on the North Slope of Alaska near Prudhoe Bay are such that stable hydrates may exist. Phase equilibrium data in the form of a projected P-T diagram for the CH_4-H_2O system are shown in Figure P16.42 below.

(a) Referring to the data shown on the P-T plot, below 0°C (273 K) the phases known to exist are solid hydrate, ice, and a vapor rich in methane. Above 0°C, ice disappears and a water-rich liquid phase appears along with solid hydrate and a methane-rich vapor. Is this behavior consistent with the phase rule? If so, locate the quadruple point and sketch the other monovariant P-T lines that intersect at the quadruple point and indicate what phases would be in equilibrium. In addition, since the *P-T* lines shown are projected from the water-rich composition region, one monovariant line should intersect with the pure H_2O *P-T* phase diagram. Indicate this intersection on your sketch. The triple point for pure H_2O occurs at 0.0098°C, 600 Pa.

(b) Given that the North Slope's geothermal temperature gradient (i.e., how fast temperature increases with increasing depth) is 0.03°C/m and that the ambient surface temperature in that region is approximately -12°C, establish whether methane hydrate can form at all and if it can, give the depth range possible for stable hydrates to exist. You may assume that the rock formation is sufficiently permeable to support a hydrostatic pressure gradient of 9.7 kPa/m and that the permafrost region extends to a depth of 400 m where the temperature is 0°C.

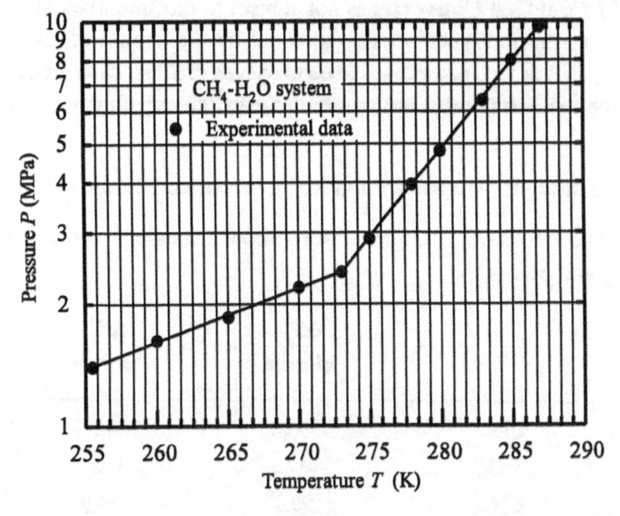

Figure P16.42 Pressure-temperature projection for the methane-water binary system.

16.43. A method for removing CO_2 has been suggested. Air at 1 atm containing about 1% volume CO_2 is bubbled through a saturated solution of lime (CaO) in water at 25°C. Estimate the maximum amount of CO_2 that could be removed by this method. State and justify all assumptions made. Also discuss how the phase rule would be applied to this system. Some data are listed below:

Substance	ΔG^o_{f298} cal/mol
$CO_{2(g)}$	-94,260
$H_2O_{(l)}$	-56,700
CO_3^{2-} (standard state at 1 molal)	-126,400
Solubility of $Ca(OH)_2$	0.0211×10^{-3} mol/m^3
$K_w = [H^+][OH^-]$ (ion product of water)	10^{-14}
$K_{sp} = [Ca^{+2}][CO_3^{-2}]$ (solubility product of $CaCO_3$)	0.87×10^{-8}

16.44. MITY Industries needs your help again. Rocky and Rochelle Jones have proposed a new device that they say even outdoes their other work producing machines. They claim that they can convert a stream of natural gas (pure CH_4) directly to electricity by reacting it with pure oxygen in a proprietary device they call the **R^2J Fuel Cell Converter**.

Although we know very little about how their system works, we have measured the off gas and found that it consists only of CO_2 and H_2O. The whole system seems to operate isothermally at ambient conditions (298 K) as well. They also claim that their device is even more efficient than the best Rankine cycle that would utilize the 1,000°C temperatures produced by rapidly burning the CH_4 stream in pure O_2. As such, they state that the **R^2J Fuel Cell Converter** is not subject to the "mundane limitations imposed by the 2nd law of thermodynamics as expressed by the Carnot efficiency." The efficiency attainable by this device is reported to be in excess of 90%. Although no power production rates relative to heating rates are reported, the efficiency claimed by the inventors seems high.

Please set us straight. Are Rocky and Rochelle correct? Some data are provided below.

Enthalpies and Gibbs free energies of formation at standard state conditions of 298 K and 1 atm are given below.

Species	ΔG_f^o kJ/mole	ΔH_f^o kJ/mole
$CO_{2(g)}$	-394.4	-393.5
$CO_{(g)}$	-137.2	-110.5
$CH_{4(g)}$	-50.8	-74.9
$H_2O_{(l)}$	-237.1	-285.6
$H_2O_{(g)}$	-228.6	-241.8
$OH^-_{(aq)}$	-138.7	1,536.2
$H^+_{(aq)}$	1,517.0	-1,43.6

16.45. During the continuous casting of steel, and particularly aluminum-killed steel, it is desirable to prevent oxidation of liquid metal being cast. Constituents of the melt, such as iron, aluminum, manganese, chromium, titanium and the like, are converted to oxide when exposed to air. These oxides are sources of non-metallic inclusions in the solidified metal and can lead to serious defects. At the same time, the loss of these constituents through oxidative processes changes the composition of the resultant alloy. To prevent such oxidation of the molten metal, various fluxes are added to the molten steel in the tundish during the continuous casting of steel. These fluxes, frequently utilized in powder form to facilitate melting, tend to float on top of the tundish and are carried along it inside of the mold, forming a protective layer of liquid flux over the surface of the molten steel in the mold, as it solidifies. In this manner they control the heat transfer from the steel to the mold wall, tending to render it uniform and preventing the "bridging" or joining of steel to copper. The flux thus lubricates the passage of the forming solidified steel shell through oxidation of the steel by insulating it from the atmosphere, and insulates the top of the strand in the mold. A flux will normally be added to the undish at a rate of about 1 kg per 1,000 kg of steel, varying from about 0.75 kg to about 1.25 kg per 1,000 kg depending on the internal surface area of the mold. A typical flux has the following ingredients (in wt%): SiO_2 – 36.1%; CaF_2 – 43.9%; Na_2O – 15.1%; B_2O_3 – 2.6%; K_2O – 2.2%; Al_2O_3 – 0.1%.

The steel industry has been mystified for years as to the cause of extensive and expensive deterioration of continuous casting equipment such as rollers, structure members, hydraulic lines and other components which constrain and manipulate the forming strand after it emerges from the mold. Damage from such deterioration is very expensive not only in parts and labor to replace or repair them, but also in "down-time" in which entire facilities are suspended from production activities.

Past research has accumulated overwhelming evidence that the cause of metal loss is not physical or entirely physical (i.e., erosion caused by the cooling water). Recent studies strongly suggest that the flux, or a component in the flux, is responsible for the high degree of iron (Fe) metal loss.

(a) Identify all independent reactions which could lead to metal loss assuming the components of the reactions are CaF_2, CaO, H_2O, HF, O_2, FeO, and Fe_2O_3.

(b) What temperature corresponds to a $\Delta G^o(T) \approx 0$ for the corrosive reaction involving the formation of HF identified in part (a)?

(c) Determine the maximum K_a for the corrosive reaction from part (b) in the available process temperature range shown in Figure P16.45. What is the maximum generation of corrosive agent capable of being produced if we assume 1 kg of flux per 1,000 kg of steel with a production rate of 227,000 kg of steel per hour?

(d) Knowing that the once-through direct-contact cooling water flow rates are approximately 200 liters/sec and that the cooling water has a pH of about 8.2, design an experiment that verifies whether the possible reactions determined in part (a) do in fact occur. If your experiment requires sampling of materials or fluids, state specifically where the sample points should be located (a graphical depiction is preferred). Where would you expect to see the greatest amount of deterioration; the least deterioration?

	CaF_2	HF	FeO	Fe_2O_3	CaO	$H_2O_{(l)}$
ΔG^o	-1173.496	-274.646	-251.429	-743.523	-603.501	-237.141
ΔH^o	-1225.912	-272.546	-272.044	-825.503	-635.089	-285.830
m	78.08	20.01	71.85	159.7	56.08	18.01

Note: ΔG^o and ΔH^o are in kJ/mol, m is the molecular weight.

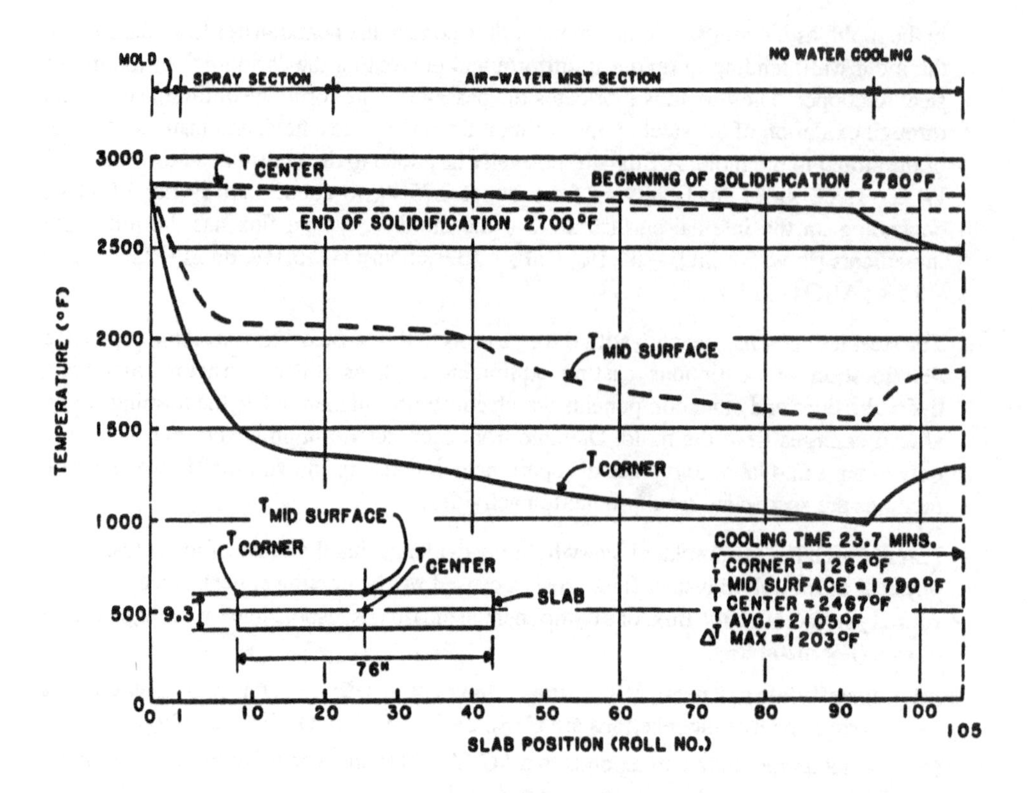

Figure P16.45 Temperature as a function of steel slab position.

Charge efficiency: 1.33 W-h charge to store 1.0 W-h
Cost of charging: 0.27/kW-h

Generalized Treatment of Phase and Chemical Equilibria 17

Many applications of thermodynamics in chemical engineering require knowledge about the equilibrium behavior of multicomponent, multiphase systems. For example, how does the vapor pressure change with temperature over a saturated salt solution or how does the azeotropic composition and boiling temperature vary with the total pressure in a three-component system? Another set of important examples involves prediction of colligative property behavior such as freezing point and osmotic pressure changes as a second component is added.

This chapter extends the development of phase and chemical reaction equilibria of Chapters 15 and 16. The primary objective here is to provide a coherent, completely general treatment of the inter-relationships that exist among temperature (T), pressure (P), and composition (x_i) variables. A matrix/determinant format is introduced to provide a framework for a comprehensive and robust description of TPx_i variations in invariant and monovariant, multiphase equilibrium systems with or without chemical reactions occurring.

17.1 Phase Rule Constrained Parameter Variability

In general, the Gibbs phase rule tells us a great deal about how we expect a system at equilibrium to behave. The intensive degrees of freedom, $\mathcal{F}$, is given by Eq. (16-47)

$$\mathcal{F} = n + 2 - \pi - r \tag{17-1}$$

where n = number of components, π = number of phases, and r = number of chemical reactions. Strictly speaking, an additional term should be subtracted from the right-hand side of Eq. (17-1) to account for additional criteria such as a stoichiometric ratio, an isobaric or isothermal process constraint, the critical point criteria, a selective membrane, and electroneutrality. Unless stated otherwise, these additional constraints will not be considered in the general treatment that follows. Thus, a system will be *invariant* if $\mathcal{F}=0$ or if the number of phases is

$$\pi = n + 2 - r \tag{17-2}$$

For a pure, non-reacting system with $n = 1$ and $r = 0$, when $\pi = 3$ the system is invariant. This corresponds to a triple point for the system. For a binary, non-reacting system ($n = 2$, $r = 0$), four phases are required for invariance at a quadruple point. *Monovariant* behavior will prevail if $\mathcal{F} = 1$, or if the number of phases is

$$\pi = n + 1 - r \tag{17-3}$$

Monovariant systems commonly observed experimentally include variations of vapor pressure as a function of temperature along a multiphase coexistence curve. For example, for single component systems two-phase regions are monovariant; for two component systems three-phase regions, and so forth. The approach of using Eq. (17-1) to determine $\mathcal{F}$ can be carried out for any number of components, phases, and reactions. The phase rule is completely general, providing the phase assemblage is at equilibrium and that other constraints such as stoichiometric ratios do not exist.

17.2 Matrix/Determinant Formalism Applied to Generalized Gibbs-Duhem and Reaction Equilibrium Expressions

As discussed in Section 6.5, the general Gibbs criteria of entropy maximization ($\delta \underline{S} = 0$) at equilibrium for an assemblage of π phases and n components can be used in conjunction with the method of Lagrange multipliers to develop specific equilibrium criteria for a simple system with no external force fields or internal constraints:

$$\delta \underline{S} = \sum_{s=2}^{\pi} \left(\frac{1}{T^{(s)}} - \frac{1}{T^{(1)}} \right) \delta \underline{U}^{(s)} + \sum_{s=2}^{\pi} \left(\frac{P^{(s)}}{T^{(s)}} - \frac{P^{(1)}}{T^{(1)}} \right) \delta \underline{V}^{(s)} - \sum_{s=2}^{\pi} \sum_{j=1}^{n} \left(\frac{\mu_j^{(s)}}{T^{(s)}} - \frac{\mu_j^{(1)}}{T^{(1)}} \right) \delta N_j^{(s)} \tag{17-4}$$

Because $\delta \underline{S}$ variations must be zero for any possible independent variation in $\underline{U}^{(s)}$, $\underline{V}^{(s)}$ or $N_j^{(s)}$, it follows that

$$T^{(1)} = T^{(2)} = \ldots = T^{(s)} = \ldots = T^{(\pi)} = T \tag{17-5a}$$

$$P^{(1)} = P^{(2)} = \ldots = P^{(s)} = \ldots = P^{(\pi)} = P \tag{17-5b}$$

$$\mu_j^{(1)} = \mu_j^{(2)} = \ldots = \mu_j^{(s)} = \ldots = \mu_j^{(\pi)} = \mu_j, \quad j = 1, \ldots, n \tag{17-5c}$$

Thus, in general, the superscripts denoting a particular phase can be omitted for T, P, and μ_j for a simple, multiphase system at equilibrium.

For the moment, let's restrict our attention to a multicomponent phase assemblage at equilibrium where no chemical reactions occur and where there are no external force fields or passive constraints (for example, membranes) to energy or mass transfer between phases. One can easily see that specification of T, P, and $\mu_1, \ldots, \mu_n$ or of $n+2$ intensive variables completely fixes the system. According to the phase rule, if the number of phases π equals the number of components n, $\mathcal{F} = 2$ and the system is bivariant, but if T and P are also specified, $\mathcal{F} = 0$ and the system becomes invariant. By starting with the Fundamental Equation for internal energy $\underline{U}$ as the basis function $y^{(0)}$:

$$\underline{U} = y^{(0)} = f(\underline{S}, \underline{V}, N_1, \ldots, N_n) = T\underline{S} - P\underline{V} + \sum_{i=1}^{n} \mu_i N_i \tag{17-6}$$

with

$$d\underline{U} = dy^{(0)} = Td\underline{S} - Pd\underline{V} + \sum_{i=1}^{n} \mu_i dN_i$$

we can develop a completely equivalent mathematical form by taking the total Legendre transformation in n+2 coordinates to produce $y^{(n+2)}$:

$$y^{(n+2)} = \underline{U} - T\underline{S} + P\underline{V} - \sum_{i=1}^{n} \mu_i N_i = 0 \tag{17-7}$$

Because $y^{(n+2)} = 0 =$ constant, the total differential $dy^{(n+2)}$ also equals zero and can be expanded as a function of T, P, and μ_i using Eqs. (17.6) and (17.7):

$$dy^{(n+2)} = -\underline{S}dT + \underline{V}dP - \sum_{i=1}^{n} N_i d\mu_i = 0 \tag{17-8}$$

which is of course the Gibbs-Duhem equation. Equation (17-8) can be written independently for each phase (s) so that a set of π Gibbs-Duhem equations results for the assemblage at equilibrium where the constraint equations (Eqs. (17-5a), (17-5b), (17-5c) are followed:

$$\begin{aligned}
-\underline{S}^{(1)}\, dT + \underline{V}^{(1)}\, dP - \sum_{j=1}^{n} N_j^{(1)}\, d\mu_j &= 0 \\
\vdots \qquad\quad \vdots \qquad\quad &\vdots \\
-\underline{S}^{(s)}\, dT + \underline{V}^{(s)}\, dP - \sum_{j=1}^{n} N_j^{(s)}\, d\mu_j &= 0 \\
\vdots \qquad\quad \vdots \qquad\quad &\vdots \\
-\underline{S}^{(\pi)}\, dT + \underline{V}^{(\pi)}\, dP - \sum_{j=1}^{n} N_j^{(\pi)}\, d\mu_j &= 0
\end{aligned} \tag{17-9}$$

In addition, if r chemical reactions occur they too must adhere to separate criteria of chemical equilibrium where for a reaction k involving n components

$$\sum_{j=1}^{n} \nu_{j,k}\, \mu_j = \sum_{j=1}^{n} \nu_j^{(k)}\, \mu_j = 0 \quad \text{or} \quad \sum_{j=1}^{n} \nu_j^{(k)}\, d\mu_j = 0 \quad \text{for} \quad k = 1, \ldots, r \tag{17-10}$$

Each reaction k is represented with a set of stoichiometric coefficients $\nu_{j,k} = \nu_j^{(k)}$ [Note that we have modified the notation for $\nu_j^{(k)}$ to emphasize that the reaction (k) equilibrium constraint is on the same level as the phase (s) equilibrium constraint]. The stoichiometry of each reaction k with W_j components is

$$\sum_{j=1}^{n} \nu_j^{(k)} W_j = 0 \quad \text{for} \quad k = 1, \ldots, r \tag{17-11}$$

where $\nu_j^{(k)}$ is positive for a product and negative for a reactant by convention.

Equations (17-9) and (17-10) can be combined to provide a complete set of equations that constrain the system and must be satisfied at equilibrium. This equation set can be written in vector/matrix form for a system of π phases and r reactions as,

$$-\begin{pmatrix} \underline{S}^{(1)} \\ \vdots \\ \underline{S}^{(\pi)} \\ 0^{(1)} \\ \vdots \\ 0^{(r)} \end{pmatrix} (dT) + \begin{pmatrix} \underline{V}^{(1)} \\ \vdots \\ \underline{V}^{(\pi)} \\ 0^{(1)} \\ \vdots \\ 0^{(r)} \end{pmatrix} (dP) - \begin{pmatrix} N_1^{(1)} & \cdots & N_j^{(1)} & \cdots & N_n^{(1)} \\ \vdots & & \vdots & & \vdots \\ N_1^{(\pi)} & \cdots & N_j^{(\pi)} & \cdots & N_n^{(\pi)} \\ \nu_i^{(1)} & & \nu_j^{(1)} & & \nu_n^{(1)} \\ \vdots & & \vdots & & \vdots \\ \nu_1^{(r)} & \cdots & \nu_j^{(r)} & \cdots & \nu_n^{(r)} \end{pmatrix} \begin{pmatrix} d\mu_1 \\ \vdots \\ d\mu_j \\ \vdots \\ d\mu_n \end{pmatrix} = 0 \tag{17-12a}$$

or in shorthand vector notation

$$-\underset{\sim}{S}^{(s)}\, dT + \underset{\sim}{V}^{(s)}\, dP - \underset{\approx}{N}_j^{(s)}\, d\underset{\sim}{\mu}_j = 0 \tag{17-12b}$$

Equations (17-12a, b) are completely general and can be used to interpret behavior in equilibrium systems in the absence of effects due to external fields and body forces.

17.3 Invariant Systems ($\mathcal{F}=0$)

The most common invariant condition encountered is at the triple point of a single component, non-reacting system where three phases coexist. From the phase rule with n=1 and r=0, $\mathcal{F}$=0 when π=3. For example, for water at 0.0076°C and 611.3 Pa, ice, liquid, and vapor are in equilibrium. Other triple points in pure systems can occur where three solid phases or two solid phases and a vapor or liquid phase coexist. If we rewrite Eq. (17-12) for the case of a pure component system at its triple point

$$-\begin{bmatrix} \underline{S}^{(1)} \\ \underline{S}^{(2)} \\ \underline{S}^{(3)} \end{bmatrix} dT + \begin{bmatrix} \underline{V}^{(1)} \\ \underline{V}^{(2)} \\ \underline{V}^{(3)} \end{bmatrix} dP - \begin{bmatrix} N_1^{(1)} \\ N_1^{(2)} \\ N_1^{(3)} \end{bmatrix} d\mu_1 = 0 \tag{17-13a}$$

We can put Eq. (17-13a) in intensive form by dividing by $N_1^{(s)}$ for each phase (s):

$$-\begin{bmatrix} S^{(1)} \\ S^{(2)} \\ S^{(3)} \end{bmatrix} dT + \begin{bmatrix} V^{(1)} \\ V^{(2)} \\ V^{(3)} \end{bmatrix} dP - \begin{bmatrix} 1 \\ 1 \\ 1 \end{bmatrix} d\mu_1 = 0 \tag{17-13b}$$

Using the method of determinants and Cramer's Rule (see Appendix D) we can easily show that the conditions where Equation set (17-13b) is satisfied corresponds to the trivial solution where $dT = dP = d\mu_1 = 0$. Thus the state of the system is fixed at the triple point (tp) with $T = T_{tp}$, $P = P_{tp}$, and $\mu_1 = \mu_{tp}$.

This approach can be completely generalized to a multicomponent system. When the number of reactions r plus the number of phases π is two more than the number of components, $\mathcal{F}$ is always zero and the system is invariant. In this situation, the number of variables (n+2) in the Equation set (17-12a,b) is equal to the number of equations ($\pi + r$). Again using Cramer's rule, $dT = dP = d\mu_1 = d\mu_2 = \ldots = d\mu_n = 0$ (providing the det $\underset{\approx}{N_j^{(s)}}$ is not zero) and the system is completely specified at an invariant point.

Other invariant systems are possible when additional constraints are present. For example, consider a closed, non-reacting, multicomponent composite system consisting of π phases at equilibrium. Mass conservation can be expressed as

$$N_j = \sum_{s=1}^{\pi} N_j^{(s)} \quad \text{for all } j = 1, \ldots, n \tag{17-14}$$

According to the phase rule, an invariant system ($\mathcal{F}=0$) will exist when T and P are fixed and $\pi = n$. Thus, Eq. (17-12a) becomes

$$-\begin{pmatrix} N^{(1)} & \cdots & N_j^{(1)} & \cdots & N_n^{(1)} \\ \vdots & & \vdots & & \vdots \\ N_1^{(s)} & \cdots & N_j^{(s)} & \cdots & N_n^{(s)} \\ \vdots & & \vdots & & \vdots \\ N_1^{(\pi)} & \cdots & N_j^{(\pi)} & \cdots & N_n^{(\pi)} \end{pmatrix} \begin{pmatrix} d\mu_1 \\ \vdots \\ d\mu_j \\ \vdots \\ d\mu_n \end{pmatrix} = 0 \tag{17-15}$$

or

$$\underset{\approx}{N_j^{(s)}}\, \underset{\sim}{d\mu_j} = 0 \tag{17-16}$$

where $\underset{\approx}{N_j}$ is an $n \times n$ or $\pi \times \pi$ square matrix. Equation (17-15) or (17-16) is the matrix form of the Gibbs-Duhem equation applying to π phases with T and P fixed. If the determinant of $\underset{\approx}{N_j^{(s)}}$ is not zero, then $\underset{\sim}{d\mu_j} = 0$ and the system is completely fixed as

$$d\mu_1 = d\mu_2 = \ldots = d\mu_j = \ldots d\mu_n = 0 \tag{17-17}$$

This will be the normal case that is consistent with the phase rule.

A second case is also of interest to consider, that is when the determinant of $\underset{\approx}{N_j^{(s)}} = 0$. In this situation, the system is not fixed since the $d\mu_i$'s are not necessarily zero. Mathematically,

$$\det \underset{\approx}{N}_j^{(s)} = |\underset{\approx}{N}_j^{(s)}| = 0 \tag{17-18}$$

when:

(1) a constant ratio exists between any two columns or rows,

(2) any two columns or rows are equal, or

(3) any row or column is made up entirely of zeros.

In general, $|\underset{\approx}{N}_j^{(s)}| = 0$ when one row of the column can be expressed as a linear combination of the other rows or columns.

One important point concerns the initial constraint that the total number of moles of each component are fixed since the system is closed [Eq. (17-14)]. This restricts the *overall* composition (Z_j) of each component j in the mixed phase system:

$$Z_j \equiv \frac{N_j}{\sum_{j=1}^{n} N_j} \tag{17-19}$$

By using Eq. (17-14) and the fact that the total number of moles, N_{total}, in the system is $\sum_{j=1}^{n} N_j$:

$$Z_j = \frac{\sum_{s=1}^{\pi} N_j^{(s)}}{N_{total}}, \quad j = 1, \ldots, n \tag{17-20}$$

In summary, the thermodynamic equilibrium condition for any closed system is specifically limited to one unique state when the number of components (n) plus two equals the number of phases (π) plus the number of reactions (r), providing the determinant of the mole number matrix is not zero, that is $\mathcal{F}=0$ when $n+2=\pi+r$ if:

$$|\underset{\approx}{N}_j^{(s)}| \neq 0 \tag{17-21}$$

This is essentially a restatement of the Gibbs phase rule, however, it provides one important qualification regarding indeterminancy for the condition given Eq. (17-21). When $|\underset{\approx}{N}_j^{(s)}| = 0$, we cannot say the system is invariant. Under these conditions, the rank of the $\underset{\approx}{N}_j^{(s)}$ matrix is less than n. This is equivalent to saying that the number of *independent* components is also less than n. In general, the $\underset{\approx}{N}_j^{(s)}$ matrix is $n \times (\pi + r)$ and its rank will be determined by striking out whole rows and columns to form the largest square array that has a non-zero determinant.

If $|N_j^{(s)}| = 0$ for the case of a system where $n = \pi$ and T and P are fixed [Eq. (17-15)], then the number of independent components is less than the number of phases present and the system is over-constrained. An azeotrope in a binary mixture provides an excellent model of this behavior and is presented in Example 17.1.

The general conditions for invariance are consistent with the first postulate and can be viewed as a restatement of Duhem's theorem [Prigogine and Defay (1954), p. 188]. If Eq. (17-21) is followed, temperature and pressure must be independently variable, thus an $n = \pi$ system invariant when the total masses of each component are specified along with the pressure and temperature.

The phase rule as written in Eq. (17-1) is expressed in terms of the system's intensive degrees of freedom. Strictly speaking, the overall extensive state of the system, as stated mathematically for $j = 1, \ldots, n$ components in Eq. (17-20), must be fixed in order for the phase rule to provide an unambiguous analysis.

Example 17.1

Consider the ethanol (A) - benzene (B) binary, liquid vapor system that forms a minimum boiling azeotrope at a constant total pressure $P = 750$ mm Hg. Data are plotted in the form of a T-x,y diagram shown in Figure 17.1.

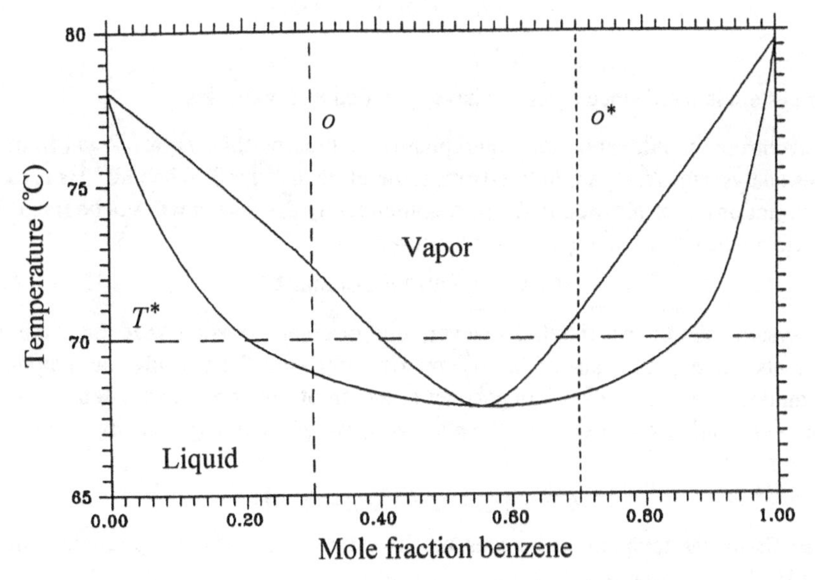

Figure 17.1 Binary isobaric phase diagram for the ethanol-benzene system at 750 mm Hg pressure [data taken from the International Critical Tables (1928)].

By first inspection, it might appear that we have violated the phase rule; however, noting the closed system restriction given by Eq. (17-14), we can find a solution to our dilemma. At T^*, there are two allowable isobaric states (o) and (o^*) shown above in Figure 17.1.

However, for given molar amounts of A and B, present, Z_j is fixed and the system can only exist in one state.

Using the matrix formulation previously developed, several interesting points are brought out. For the A-B binary:

$$\left[N_j^{(s)}\right] = \underset{\approx}{N_j^{(s)}} = \begin{bmatrix} N_A^V & N_B^V \\ N_A^L & N_B^L \end{bmatrix} \tag{17-22}$$

where the superscripts: V = vapor phase and L = liquid phase. Note that when we are in the liquid phase, $\pi = 1$ and $n = 2$, there is *no* vapor phase present at equilibrium and

$$N_A^V = N_B^V = 0 \tag{17-23}$$

Therefore $|\underset{\approx}{N_j^{(s)}}| = 0$ and as stated before, variation of μ_i's are allowed and the system is not fixed just by specifying T and P. For a non-reacting system $r = 0$ but with T and P fixed this adds two additional constraints; thus, the phase rule gives one degree of freedom:

$$\mathcal{F} = n + 2 - \pi - r - 2 = 1 \quad (T \text{ and } P \text{ constant}) \tag{17-24}$$

Now if we move up to the bubble point on the liquid-vapor equilibrium line, n=2 and $\pi = 2$ since both liquid and vapor phases are present, and in general $|\underset{\approx}{N_j^{(s)}}| \neq 0$. Therefore, in order to satisfy Eq. (17-16)

$$d\underset{\sim}{\mu_j} = d\mu_j \ \text{(for all } j) = 0 \tag{17-25}$$

and the system is fixed since again we have specified n+2 variables.

At the azeotropic point, vapor and liquid phases are present, thus $n = \pi = 2$ and from our previous analysis the $N_j^{(s)}$'s are in fixed ratios; therefore, $|\underset{\approx}{N_j^{(s)}}| = 0$. When $|\underset{\approx}{N_j^{(s)}}|$ is zero, we expect variations to be allowed in the μ_i's, consequently the system will not be fixed. The phase rule on the other hand tells us for $n = \pi = 2$ that

$$\mathcal{F} = 0 \quad (T \text{ and } P \text{ constant}) \tag{17-26}$$

so the system should be fixed. However, our original assertion that there are two components is wrong. The rank of the $\underset{\approx}{N_j^{(s)}}$ matrix is unity and there is only one component. The components are no longer independent at the azeotropic point in the sense that one stoichiometric ratio exists to reduce the *number of independent components to one*. With $n = 1$ and $\pi = 2$:

$$\mathcal{F} = n + 2 - \pi = 1 + 2 - 2 = 1 \tag{17-27}$$

so either fixing the temperature or pressure fixes the system at the azeotropic point, as expected.

Prigogine and Defay (1954) refer to the general condition where the determinant of the mole number matrix vanishes ($|\underset{\approx}{N_j^{(s)}}| = 0$) as an indifferent state. In Section 17.6 we will show how this indifferent condition can be applied to prove the Gibbs-Konovalow theorems which describe the behavior of pressure and temperature derivatives with respect to composition. For a two-component system, one possible indifferent state corresponds to an azeotrope at constant pressure, which is equivalent to saying that the azeotropic point represents an extremum in the temperature-composition curve as depicted in the *T-x,y* diagram of Figure 17.1. It is also interesting to note that, at least in principle, a multicomponent (n>2) azeotrope is also possible.

17.4 Monovariant Systems $\mathcal{F}$=1: Pressure-Temperature Variations

If $\mathcal{F} = 1 = n + 2 - \pi - r$, then

$$\pi + r = n + 1 \tag{17-28}$$

For a monovariant system of n components, π phases and r reactions we would expect, by analogy with a single component system as described in Section 15.4, that the derivative dP/dT could be expressed in a general Clapeyron form as,

$$\left.\frac{dP}{dT}\right|_{\mathcal{F}=1} = \frac{|\Delta \underset{\approx}{H}|}{T\,|\Delta \underset{\approx}{V}|} \quad \text{or equivalently} \quad \frac{|\Delta \underset{\approx}{S}|}{|\Delta \underset{\approx}{V}|} \tag{17-29}$$

where, $|\Delta \underset{\approx}{H}|$, $|\Delta \underset{\approx}{V}|$ and $|\Delta \underset{\approx}{S}|$ are appropriately defined determinants. The equivalent entropy form was first derived by Gibbs in 1876 and has been later used by Prigogine and Defay (1954) and Gilmont (1959). However, a rigorous derivation of the more useful form involving enthalpy has not appeared in the literature.

In order to specify the system we need to write a Gibbs-Duhem equation for each phase (s) and a constraint for each chemical reaction (r). We can write the Gibbs-Duhem relationship using a format with chemical potential or fugacity for each phase (s), following the approach introduced Section 17.2.

In $\boldsymbol{\mu_1}$ ***(chemical potential) format:*** The extensive form of the Gibbs-Duhem equation comes directly from the $(n + 2)^{th}$ Legendre transform of

$$y^{(0)} = \underline{U} = f(\underline{S}, \underline{V}, N_1, \ldots, N_n) = f(z_1, \ldots, z_{n+2})$$

which for phase (s) is

$$dy^{(n+2)} = -\sum_{i-1}^{n+2} z_i d\xi_i = -\underline{S}^{(s)}\, dT + \underline{V}^{(s)}\, dP - \sum_{i=1}^{n} N_i^{(s)}\, d\mu_i = 0$$

By dividing by $N^{(s)} = \sum_{i=1} N_i^{(s)}$ to get mole fractions, the intensive form results:

$$\frac{dy^{(n+2)}}{N^{(s)}} = -S^{(s)}\, dT + V^{(s)}\, dP - \sum_{i=1}^{n} x_i^{(s)}\, d\mu_i = 0 \tag{17-30}$$

Or alternatively in $\hat{f}_i$ (fugacity) format: Here, energy representation is not good because of the T term in $RT \ln \hat{f}_i$, so we use entropy representation for the Fundamental Equation:

$$y^{(0)} = \underline{S} = f(\underline{U}, \underline{V}, N_1, \ldots, N_n)$$

z_i	ξ_i
$\underline{U}$	$1/T$
$\underline{V}$	P/T
N_1	$-\mu_1/T$
$\vdots$	$\vdots$
N_n	$-\mu_n/T$

$$dy^{(0)} = d\underline{S} = \left[\frac{1}{T}\right] d\underline{U} + \left[\frac{P}{T}\right] d\underline{V} - \sum_{i=1}^{n} \left[\frac{\mu_i}{T}\right] dN_i$$

$$dy^{(n+2)} = -\sum_{i=1}^{n+2} z_i d\xi_i = 0$$

for phase (s)

$$dy^{(n+2)} = -\underline{U}^{(s)}\, d\left[\frac{1}{T}\right] - \underline{V}^{(s)}\, d\left[\frac{P}{T}\right] + \sum_{i=1}^{n} N_i^{(s)}\, d\left[\frac{\mu_i}{T}\right] = 0$$

$$\left[\frac{\underline{U}^{(s)}}{T^2} + \frac{P\underline{V}^{(s)}}{T^2}\right] dT - \left[\frac{\underline{V}^{(s)}}{T}\right] dP + \sum_{i=1}^{n} N_i^{(s)}\, d\left[\frac{\mu_i}{T}\right] = 0$$

or

$$\left[\frac{\underline{H}^{(s)}}{T^2}\right] dT - \left[\frac{\underline{V}^{(s)}}{T}\right] dP + \sum_{i=1}^{n} N_i^{(s)}\, d\left[\frac{\mu_i}{T}\right] = 0 \tag{17-31}$$

Again by dividing by $N^{(s)}$ to get mole fractions,

$$\left[\frac{H^{(s)}}{T^2}\right] dT - \left[\frac{V^{(s)}}{T}\right] dP + \sum_{i=1}^{n} x_i^{(s)}\, d\left[\frac{\mu_i}{T}\right] = 0 \tag{17-32}$$

Using the definition of fugacity as $\mu_i = RT \ln \hat{f}_i + \lambda_i(T)$ and the Gibbs-Helmholtz relation to modify the third term:

$$d\left[\frac{\mu_i}{T}\right] = R\, d \ln \hat{f}_i + d\left[\frac{\lambda_i(T)}{T}\right] = R\, d \ln \hat{f}_i - \frac{H_i^o}{T^2}\, dT \tag{17-33}$$

where H_i^o is the reference ideal-gas state enthalpy. Equation (17-32) can be rewritten as:

$$\left[\frac{H^{(s)} - \sum_{i=1}^{n} x_i^{(s)} H_i^o}{RT^2}\right] dT - \frac{V^{(s)}}{RT} dP + \sum_{i=1}^{n} x_i^{(s)}\, d \ln \hat{f}_i = 0 \tag{17-34}$$

Equations (17-30), (17-32), and (17-34) are equivalent Gibbs-Duhem expressions, good for any phase in the assemblage of π phases.

For any chemical reaction, the constraining equilibrium criteria was stated earlier in Eq. (17-10) and for the kth reaction can be written as:

$$\nu_1^{(k)}\,d\mu_1 + \nu_2^{(k)}\,d\mu_2 + \ldots + \nu_n^{(k)}\,d\mu_n = 0$$

Thus, for r reactions at a given temperature T

$$\sum_{i=1}^{n} \nu_i^{(k)}\,d\left[\frac{\mu_i}{T}\right] = 0 \qquad \text{for } k = 1, \ldots, r \tag{17-35}$$

Now we can combine the π equations for the phase equilibria with the r equations for the reaction equilibria. First, let's use Eq. (17-32) for the phase equilibria and Eq. (17-35) for the reaction equilibria. This provides the following set of $\pi + r$ linear equations:

$$\begin{aligned}
\frac{H^{(1)}}{T^2}dT - \frac{V^{(1)}}{T}\,dP + \sum_{i=1}^{n} x_i^{(1)}\,d\left[\frac{\mu_i}{T}\right] &= 0\\
\vdots \qquad\quad \vdots \qquad\qquad \vdots \qquad\qquad &\ \ \vdots\\
\frac{H^{(\pi)}}{T^2}dT - \frac{V^{(\pi)}}{T}\,dP + \sum_{i=1}^{n} x_i^{(\pi)}\,d\left[\frac{\mu_i}{T}\right] &= 0\\
\sum_{i=1}^{n} \nu_i^{(1)}\,d\left[\frac{\mu_i}{T}\right] &= 0\\
\vdots \qquad\qquad &\ \ \vdots\\
\sum_{i=1}^{n} \nu_i^{(r)}\,d\left[\frac{\mu_i}{T}\right] &= 0
\end{aligned} \tag{17-36}$$

Equation set (17-36) is underdefined in that the number of unknowns (n+2) is greater than the number of equations ($\pi + r$) by 1 from Eq. (17-28). However, for this monovariant system we can solve for the derivative dP/dT using Cramer's rule (see Appendix D):

$$\left.\frac{dP}{dT}\right]_{\mathcal{F}=1} = \frac{|\Delta \underset{\approx}{H}|}{T\,|\Delta \underset{\approx}{V}|} \tag{17-37}$$

where the determinants are defined by:

$$|\Delta \underset{\approx}{H}| \equiv \begin{vmatrix} H^{(1)} & x_1^{(1)} & x_2^{(1)} & \cdots & x_n^{(1)} \\ \vdots & \vdots & \vdots & & \\ H^{(\pi)} & x_1^{(\pi)} & x_2^{(\pi)} & \cdots & x_n^{(\pi)} \\ 0 & \nu_1^{(1)} & \nu_2^{(1)} & \cdots & \nu_n^{(1)} \\ \vdots & \vdots & \vdots & & \\ 0 & \nu_1^{(r)} & \nu_2^{(r)} & \cdots & \nu_n^{(r)} \end{vmatrix} \tag{17-38}$$

$$|\Delta \underset{\approx}{V}| \equiv \begin{vmatrix} V^{(1)} & x_1^{(1)} & x_2^{(1)} & \cdots & x_n^{(1)} \\ \vdots & \vdots & \vdots & & \\ V^{(\pi)} & x_1^{(\pi)} & x_2^{(\pi)} & \cdots & x_n^{(\pi)} \\ 0 & \nu_1^{(1)} & \nu_2^{(1)} & \cdots & \nu_n^{(1)} \\ \vdots & \vdots & \vdots & & \\ 0 & \nu_1^{(r)} & \nu_2^{(r)} & \cdots & \nu_n^{(r)} \end{vmatrix} \tag{17-39}$$

By using the properties of determinants, $|\Delta \underset{\approx}{H}|$ and $|\Delta \underset{\approx}{V}|$ can usually be simplified greatly.

As an alternative approach to the one above, we could have substituted fugacity and used Eq. (17-34) for the phase equilibria and Eqs. (17-35) and (17-33) for the reaction equilibria. Using this approach, for an assemblage of π phases and r reactions:

$$\begin{aligned}
&\left[\frac{H^{(1)} - \sum_{i=1}^{n} x_i^{(1)} H_i^o}{RT^2}\right] dT - \frac{V^{(1)}}{RT}\, dP + \sum_{i=1}^{n} x_i^{(1)}\, d\ln \hat{f}_i = 0 \\
&\qquad \vdots \qquad\qquad\qquad \vdots \qquad\qquad \vdots \\
&\left[\frac{H^{(\pi)} - \sum_{i=1}^{n} x_i^{(\pi)} H_i^o}{RT^2}\right] dT - \frac{V^{(\pi)}}{RT}\, dP + \sum_{i=1}^{n} x_i^{(\pi)}\, d\ln \hat{f}_i = 0 \\
&-\frac{\sum_{i=1}^{n} \nu_i^{(1)} H_i^o}{RT^2}\, dT \;+\; 0 \;+\; \sum_{i=1}^{n} \nu_i^{(1)}\, d\ln \hat{f}_i = 0 \\
&\qquad \vdots \qquad\qquad\qquad \vdots \qquad\qquad \vdots \\
&-\frac{\sum_{i=1}^{n} \nu_i^{(r)} H_i^o}{RT^2}\, dT \;+\; 0 \;+\; \sum_{i=1}^{n} \nu_i^{(r)}\, d\ln \hat{f}_i = 0
\end{aligned} \tag{17-40}$$

Equation (17-40) is equivalent to a set of $(\pi + r)$ equations in $(n+1)$ unknowns for which $\mathcal{F}$=1, hence the system is monovariant. As was done before, the dP/dT derivative can also be solved for using Cramer's rule:

$$\left.\frac{dP}{dT}\right]_{\mathcal{F}=1} = \frac{|\Delta \underset{\approx}{H}^*|}{|\Delta \underset{\approx}{V}^*|} \tag{17-41}$$

where

$$|\Delta \underset{\approx}{H}^*| \equiv \begin{vmatrix} \left[H^{(1)} - \Sigma\, x_i^{(1)} H_i^0\right]/RT^2 & x_1^{(1)} & \cdots & x_n^{(1)} \\ \vdots & \vdots & & \vdots \\ \left[H^{(\pi)} - \Sigma\, x_i^{(\pi)} H_i^0\right]/RT^2 & x_1^{(\pi)} & \cdots & x_n^{(\pi)} \\ -\Sigma\, \nu_i^{(1)} H_i^0/RT^2 & \nu_1^{(1)} & \cdots & \nu_n^{(1)} \\ \vdots & \vdots & & \vdots \\ -\Sigma\, \nu_i^{(r)} H_i^0/RT^2 & \nu_1^{(r)} & \cdots & \nu_n^{(r)} \end{vmatrix} \tag{17-42}$$

and

$$|\Delta \underset{\approx}{V}^*| \equiv \begin{vmatrix} V^{(1)}/RT & x_1^{(1)} & \cdots & x_n^{(1)} \\ \vdots & \vdots & & \vdots \\ V^{(\pi)}/RT & x_1^{(\pi)} & \cdots & x_n^{(\pi)} \\ 0 & \nu_1^{(1)} & \cdots & \nu_n^{(1)} \\ \vdots & \vdots & & \vdots \\ 0 & \nu_1^{(r)} & \cdots & \nu_n^{(r)} \end{vmatrix} \tag{17-43}$$

Note that the $\Delta \underset{\approx}{H}^*$ and $\Delta \underset{\approx}{V}^*$ matrices are square because $(\pi + r) = (n + 1)$. The $|\Delta \underset{\approx}{H}^*|$ and $|\Delta \underset{\approx}{V}^*|$ determinants can be simplified using the properties of determinants. As a first step, $|\Delta \underset{\approx}{V}^*|$ can have the $1/RT$ factor taken outside as a multiplier:

$$|\Delta \underset{\approx}{V}^*| \equiv \frac{1}{RT} \begin{vmatrix} V^{(1)} & x_1^{(1)} & \cdots & x_n^{(1)} \\ \vdots & \vdots & & \vdots \\ V^{(\pi)} & x_1^{(\pi)} & \cdots & x_n^{(\pi)} \\ 0 & \nu_1^{(1)} & \cdots & \nu_n^{(1)} \\ \vdots & \vdots & & \vdots \\ 0 & \nu_1^{(r)} & \cdots & \nu_n^{(r)} \end{vmatrix} \tag{17-44}$$

In addition, the $|\Delta \underset{\approx}{H}^*|$ determinant can be expanded into two determinants with the $1/RT^2$ factor removed:

$$|\underset{\approx}{\Delta H^*}| \equiv \frac{1}{RT^2}\begin{vmatrix} H^{(1)} & x_1^{(1)} & \dots & x_n^{(1)} \\ \vdots & \vdots & & \vdots \\ H^{(\pi)} & x_1^{(\pi)} & \dots & x_n^{(\pi)} \\ 0 & \nu_1^{(1)} & \dots & \nu_n^{(1)} \\ \vdots & \vdots & & \vdots \\ 0 & \nu_1^{(r)} & \dots & \nu_n^{(r)} \end{vmatrix} + \frac{1}{RT^2}\begin{vmatrix} -\Sigma x_i^{(1)} H_i^o & x_1^{(1)} & \dots & x_n^{(1)} \\ \vdots & \vdots & & \vdots \\ -\Sigma x_i^{(\pi)} H_i^o & x_1^{(\pi)} & \dots & x_n^{(\pi)} \\ -\Sigma \nu_i^{(1)} H_i^o & \nu_1^{(1)} & \dots & \nu_n^{(1)} \\ \vdots & \vdots & & \vdots \\ -\Sigma \nu_i^{(r)} H_i^o & \nu_1^{(r)} & \dots & \nu_n^{(r)} \end{vmatrix} \tag{17-45}$$

because each element in the first column of the second determinant is a linear combination of the other elements in the row we can show that the second determinant is zero. For example, consider a two-component, monovariant system with $\pi = 2$, $r = 1$, the second determinant in Eq. (17-45) is equal to

$$-\frac{1}{RT^2}\begin{vmatrix} \left[x_1^{(1)} H_1^o + x_2^{(1)} H_2^o\right] & x_1^{(1)} & x_2^{(1)} \\ \left[x_1^{(2)} H_1^o + x_2^{(2)} H_2^o\right] & x_1^{(2)} & x_2^{(2)} \\ \left[\nu_1^{(1)} H_1^o + \nu_2^{(1)} H_2^o\right] & \nu_1^{(1)} & \nu_2^{(1)} \end{vmatrix} \tag{17-46}$$

which can be expanded to

$$H_1^o \begin{vmatrix} x_1^{(1)} & x_1^{(1)} & x_2^{(1)} \\ x_1^{(2)} & x_1^{(2)} & x_2^{(2)} \\ \nu_1^{(1)} & \nu_1^{(1)} & \nu_2^{(1)} \end{vmatrix} + H_2^o \begin{vmatrix} x_2^{(1)} & x_1^{(1)} & x_2^{(1)} \\ x_2^{(2)} & x_1^{(2)} & x_2^{(2)} \\ \nu_2^{(1)} & \nu_1^{(1)} & \nu_2^{(1)} \end{vmatrix} \tag{17-47}$$

both determinants are equal to zero because two columns are identical in each. Thus,

$$H_1^o\,|0| + H_2^o\,|0| = 0 \tag{17-48}$$

By combining all results from Eqs. (17-45) and (17-48) we have shown in general that the determinants for $|\underset{\approx}{\Delta H^*}|$ and $|\underset{\approx}{\Delta V^*}|$ are identical to the ones given in Eqs. (17-38) and (17-39) with

$$|\underset{\approx}{\Delta H^*}| = |\underset{\approx}{\Delta H}|/RT^2 \quad \text{and} \quad |\underset{\approx}{\Delta V^*}| = |\underset{\approx}{\Delta V}|/RT \tag{17-49}$$

Thus the expression for dP/dT is the same as that obtained by the approach using chemical potentials in entropy representation (Eqs. (17-35) and (17-37)).

Example 17.2

Consider the case of a three-component, three-phase system with one reaction

$$\nu_1 C_1^{(1)} + \nu_2 C_2^{(2)} = \nu_3 C_3^{(3)} \tag{17-50}$$

where superscripts (1), (2) and (3) refer to three distinct phases and components j=1, 2, and 3 can exist, in principle, in all phases.

Solution

From the phase rule we get a monovariant system:

$$\mathcal{F} = n + 2 - \pi - r = 3 + 2 - 3 - 1 = 1$$

In setting up Equation set (17-36) for 3 phases and 1 reaction, we get

$$\begin{aligned}
&\frac{H^{(1)}}{T^2}\,dT - \frac{V^{(1)}}{T}\,dP + x_1^{(1)}\,d\left[\frac{\mu_1}{T}\right] + x_2^{(1)}\,d\left[\frac{\mu_2}{T}\right] + x_3^{(1)}\,d\left[\frac{\mu_3}{T}\right] = 0\\
&\frac{H^{(2)}}{T^2}\,dT - \frac{V^{(2)}}{T}\,dP + x_1^{(2)}\,d\left[\frac{\mu_1}{T}\right] + x_2^{(2)}\,d\left[\frac{\mu_2}{T}\right] + x_3^{(2)}\,d\left[\frac{\mu_3}{T}\right] = 0\\
&\frac{H^{(3)}}{T^2}\,dT - \frac{V^{(3)}}{T}\,dP + x_1^{(3)}\,d\left[\frac{\mu_1}{T}\right] + x_2^{(3)}\,d\left[\frac{\mu_2}{T}\right] + x_3^{(3)}\,d\left[\frac{\mu_3}{T}\right] = 0\\
&\qquad\qquad + \nu_1\,d\left[\frac{\mu_1}{T}\right] + \nu_2\,d\left[\frac{\mu_2}{T}\right] + \nu_3\,d\left[\frac{\mu_3}{T}\right] = 0
\end{aligned} \tag{17-51}$$

which can be simplified immediately to eliminate $d[\mu_3/T]$ to give a monovariant system of three equations in four unknowns. Thus,

$$\begin{aligned}
&\frac{H^{(1)}}{T^2}\,dT - \frac{V^{(1)}}{T}\,dP + \left[x_1^{(1)} - \frac{\nu_1}{\nu_3}x_3^{(1)}\right]d\left[\frac{\mu_1}{T}\right] + \left[x_2^{(1)} - \frac{\nu_2}{\nu_3}x_3^{(1)}\right]d\left[\frac{\mu_2}{T}\right] = 0\\
&\frac{H^{(2)}}{T^2}\,dT - \frac{V^{(2)}}{T}\,dP + \left[x_1^{(2)} - \frac{\nu_1}{\nu_3}x_3^{(2)}\right]d\left[\frac{\mu_1}{T}\right] + \left[x_2^{(2)} - \frac{\nu_2}{\nu_3}x_3^{(2)}\right]d\left[\frac{\mu_2}{T}\right] = 0\\
&\frac{H^{(3)}}{T^2}\,dT - \frac{V^{(3)}}{T}\,dP + \left[x_1^{(3)} - \frac{\nu_1}{\nu_3}x_3^{(3)}\right]d\left[\frac{\mu_1}{T}\right] + \left[x_2^{(3)} - \frac{\nu_2}{\nu_3}x_3^{(3)}\right]d\left[\frac{\mu_2}{T}\right] = 0
\end{aligned} \tag{17-52}$$

Solving for dP/dT using Cramer's rule, we get for $\pi = 3$, $n = 3$, and $r = 1$:

$$\left.\frac{dP}{dT}\right|_{\mathcal{F}=1} = \frac{\begin{vmatrix} H^{(1)} & \left[x_1^{(1)} - \frac{\nu_1}{\nu_3}x_3^{(1)}\right] & \left[x_2^{(1)} - \frac{\nu_2}{\nu_3}x_3^{(1)}\right] \\ H^{(2)} & \left[x_1^{(2)} - \frac{\nu_1}{\nu_3}x_3^{(2)}\right] & \left[x_2^{(2)} - \frac{\nu_2}{\nu_3}x_3^{(2)}\right] \\ H^{(3)} & \left[x_1^{(3)} - \frac{\nu_1}{\nu_3}x_3^{(3)}\right] & \left[x_2^{(3)} - \frac{\nu_2}{\nu_3}x_3^{(3)}\right] \end{vmatrix}}{T\begin{vmatrix} V^{(1)} & \left[x_1^{(1)} - \frac{\nu_1}{\nu_3}x_3^{(1)}\right] & \left[x_2^{(1)} - \frac{\nu_2}{\nu_3}x_3^{(1)}\right] \\ V^{(2)} & \left[x_1^{(2)} - \frac{\nu_1}{\nu_3}x_3^{(2)}\right] & \left[x_2^{(2)} - \frac{\nu_2}{\nu_3}x_3^{(2)}\right] \\ V^{(3)} & \left[x_1^{(3)} - \frac{\nu_1}{\nu_3}x_3^{(3)}\right] & \left[x_2^{(3)} - \frac{\nu_2}{\nu_3}x_3^{(3)}\right] \end{vmatrix}} \tag{17-53}$$

Now consider a specific case for $\pi = 3$, $n = 3$, and $r = 1$ such as the chloroform-water clathrate equilibrium

$$CHCl_{3(g)} + 17\ H_2O_{(L_1)} = CHCl_3 \cdot 17\ H_2O_{(H)} \tag{17-54}$$

where (g) = gas, (L_1) = liquid rich in H_2O, and (H) = solid hydrate. For practical purposes with $1 \equiv H_2O$, $2 \equiv CHCl_3$, $3 \equiv$ hydrate = $CHCl_3 \cdot 17\ H_2O$

$$x_3^{(g)} = x_1^{(g)} \approx 0 \qquad x_2^{(g)} \approx 1.0 \text{ (pure } CHCl_3 \text{ in gas)}$$
$$x_3^{(L_1)} = x_2^{(L_1)} \approx 0 \qquad x_1^{(L_1)} \approx 1.0 \text{ (pure water in liquid)}$$
$$x_2^{(H)} = x_1^{(H)} = 0 \qquad x_3^{(H)} \approx 1.0 \text{ (pure solid hydrate)}$$

For the reaction given in Eq. (17-54) above, $\nu_1 = -17$, $\nu_2 = -1$, and $\nu_3 = +1$. Thus, from Eq. (17-53)

$$\frac{dP}{dT} = \frac{\begin{vmatrix} H^{(g)} & 0 & 1 \\ H^{(L_1)} & 1 & 0 \\ H^{(H)} & 17 & 1 \end{vmatrix}}{T\begin{vmatrix} V^{(g)} & 0 & 1 \\ V^{(L_1)} & 1 & 0 \\ V^{(H)} & 17 & 1 \end{vmatrix}} = \frac{H^{(g)} + 17H^{(L_1)} - H^{(H)}}{T\,[V^{(g)} + 17V^{(L_1)} - V^{(H)}]} \tag{17-55}$$

Furthermore, if $V^{(g)} >> V^{(L_1)}$ or $V^{(H)}$ and the pressure P is low enough to consider the vapor phase to be ideal, then $V^{(g)} \approx RT/P$ and an equation analogous to the Clausius-Clapeyron equation results:

$$\frac{dP}{dT} = -\frac{\Delta H^o_{rx}}{RT^2/P} \quad \text{or} \quad \left.\frac{d\ln P}{d(1/T)}\right|_{g-L_1-H} = +\frac{\Delta H^o_{rx}}{R} \tag{17-56}$$

where $\Delta H^o_{rx} \equiv H^{(H)} - 17\, H^{(L_1)} - H^{(g)}$ which is equivalent to a standard state enthalpy change for the phase reaction (17-54).

Example 17.3

Derive an expression for dP/dT for a three-phase, two-component system consisting of solid salt (S), water vapor (V) and salt-water liquid (L). Assume salt is non-volatile, therefore with $w \equiv$ water and $s \equiv$ salt: $x_w^{(V)} \approx 1.0$, $x_s^{(S)} = 1.0$, and $x_w^{(L)} + x_s^{(L)} = 1.0$

Solution

$\mathcal{F} = n + 2 - \pi - r = 2 + 2 - 3 - 0 \leftarrow$ [no reactions ($r = 0$)] $= 1$, and the system is monovariant. Eqs. (17-37) through (17-39) reduce to:

$$\left.\frac{dP}{dT}\right|_{S-L-V} = \frac{\begin{vmatrix} H^{(V)} & 1 & 0 \\ H^{(L)} & x_w^{(L)} & x_s^{(L)} \\ H^{(S)} & 0 & 1 \end{vmatrix}}{T\begin{vmatrix} V^{(V)} & 1 & 0 \\ V^{(L)} & x_w^{(L)} & x_s^{(L)} \\ V^{(S)} & 0 & 1 \end{vmatrix}} \tag{17-57a}$$

which can be expanded by cofactors to give:

$$\left.\frac{dP}{dT}\right|_{S-L-V} = \frac{H^{(V)}\begin{vmatrix} x_w^{(L)} & x_s^{(L)} \\ 0 & 1 \end{vmatrix} - 1\begin{vmatrix} H^{(L)} & x_s^{(L)} \\ H_s^{(S)} & 1 \end{vmatrix} + 0\begin{vmatrix} H^{(L)} & x_w^{(L)} \\ H^{(S)} & 0 \end{vmatrix}}{T\left[V^{(V)}\begin{vmatrix} x_w^{(L)} & x_s^{(L)} \\ 0 & 1 \end{vmatrix} - 1\begin{vmatrix} V^{(L)} & x_s^{(L)} \\ V^{(S)} & 1 \end{vmatrix} + 0\begin{vmatrix} V^{(L)} & x_w^{(L)} \\ V^{(S)} & 0 \end{vmatrix}\right]} \tag{17-57b}$$

which in turn can be evaluated as:

$$\left.\frac{dP}{dT}\right|_{S-L-V} = \frac{x_w^{(L)}\, H^{(V)} - H^{(L)} + x_s^{(L)}\, H^{(S)}}{T\left[x_w^{(L)}\, V^{(V)} - V^{(L)} + x_s^{(L)}\, V^{(S)}\right]} \tag{17-58}$$

But,

$$H^{(L)} = x_w^{(L)}\, \overline{H}_w^{(L)} + x_s^{(L)}\, \overline{H}_s^{(L)} \quad \text{and} \quad V^{(L)} = x_w^{(L)}\, \overline{V}_w^{(L)} + x_s^{(L)}\, \overline{V}_s^{(L)} \tag{17-59}$$

Therefore, Eq. (17-58) can be rewritten as:

$$\left.\frac{dP}{dT}\right|_{S-L-V} = \frac{x_w^{(L)}\left[H^{(V)} - \overline{H}_w^{(L)}\right] + x_s^{(L)}\left[H_s^{(S)} - \overline{H}_s^{(L)}\right]}{T\left[x_w^{(L)}\left[V^{(V)} - \overline{V}_w^{(L)}\right] + x_s^{(L)}\left[V_s^{(S)} - \overline{V}_s^{(L)}\right]\right]} \tag{17-60}$$

which is the same as Eq. (15-73).

17.5 Monovariant Systems: Temperature-Composition Variations

In the previous section, we discussed pressure-temperature variations for monovariant systems, where a generalized equation for dP/dT was developed in terms of determinants involving enthalpy and volume cast in a Clapeyron equation format. Of equal importance in equilibrium systems is the temperature-composition behavior of isobaric systems. This leads directly to solubility behavior and a description of condensed phase diagrams involving eutectics, solid solutions, and other types of solid phases. In addition, the approach can be used to describe coexistence behavior for vapor-liquid systems involving azeotropes.

Although we will only develop isobaric relationships in this section, the equations can easily be converted to isothermal, pressure-composition variations. Furthermore, to reduce the complexity of the matrices and determinants, we have limited our treatment to non-reacting ($r = 0$), multiphase, multicomponent systems. Again the extension is straightforward. For an isobaric, non-reacting ($r = 0$) system where $\pi = n$:

$$\mathcal{F} = n + 2 - \pi - r - 1 \text{ (constant } P) = 1 \tag{17-61}$$

Thus, the system is monovariant and following from Eq. (17-36), we can write:

$$\begin{array}{c} \dfrac{H^{(1)}}{T^2}dT - \dfrac{V^{(1)}}{T}dP + \displaystyle\sum_{i=1}^{n} x_i^{(1)}\, d\left[\frac{\mu_i}{T}\right] = 0 \\ \vdots \qquad \vdots \qquad \vdots \qquad \vdots \\ \dfrac{H^{(\pi)}}{T^2}dT - \dfrac{V^{(\pi)}}{T}dP + \displaystyle\sum_{i=1}^{n} x_i^{(\pi)}\, d\left[\frac{\mu_i}{T}\right] = 0 \end{array} \tag{17-62}$$

We begin by examining variations of μ_1/T and T with $dP = 0$, noting that we have selected component i=1, completely arbitrarily, as the reference component. Using Cramer's rule to simplify Eq. (17-62) (see Appendix D).

$$\frac{1}{T^2}\begin{vmatrix} H^{(1)} & x_2^{(1)} & \ldots & x_n^{(1)} \\ \vdots & \vdots & & \vdots \\ H^{(\pi)} & x_2^{(\pi)} & \ldots & x_n^{(\pi)} \end{vmatrix} dT + \begin{vmatrix} x_1^{(1)} & x_2^{(1)} & \ldots & x_n^{(1)} \\ \vdots & \vdots & & \vdots \\ x_1^{(\pi)} & x_2^{(\pi)} & \ldots & x_n^{(\pi)} \end{vmatrix} d\left[\frac{\mu_1}{T}\right] = 0 \tag{17-63}$$

or in shorthand notation:

$$\frac{1}{T^2}\,|\Delta \underset{\approx}{H}^{\infty}|\,dT + |\underset{\approx}{x}_i^{(s)}|\,d\left[\frac{\mu_1}{T}\right] = 0. \tag{17-64}$$

Note that $|\Delta \underset{\approx}{H}^{\infty}|$ in Eqs. (17-63) and (17-64) is different than $|\Delta \underset{\approx}{H}|$ defined earlier in Eq. (17-38). In order for Eq. (17-64) to be practical, we need to transform from $d[\mu_1/T]$ to $\{T, x_i\}$ coordinates. This can be done by expanding $d[\mu_1/T]$ as a function of T and x_i (i=1,...,n) at constant pressure for an arbitrary phase β in the assemblage ($\alpha,\ldots,\beta,\ldots,\pi$). Strictly speaking, we do not need a superscript β for μ_1 because the chemical potentials are equal in all phases, but in practice, we will need some type of constitutive property model for phase β to allow evaluation of the derivatives.

$$d\left[\frac{\mu_1}{T}\right] = d\left[\frac{\mu_1^\beta}{T}\right] = \left[\frac{\partial(\mu_1^\beta/T)}{\partial T}\right]_{P,x_i^\beta} dT + \sum_{i=1}^{n}\left[\frac{\partial\,(\mu_1^\beta/T)}{\partial x_i^\beta}\right]_{T,\,P,x_j^\beta[i,\,k]} dx_i^\beta \tag{17-65}$$

By applying the Gibbs-Helmholtz equation for the temperature derivative of μ_1^β/T and simplifying the term inside the summation:

$$d\left[\frac{\mu_1}{T}\right] = -\frac{\overline{H}_1^\beta}{T^2}\,dT + \sum_{i=1}^{n}\frac{1}{T}\left[\frac{\partial\mu_1^\beta}{\partial x_i^\beta}\right]_{T,P,x_j^\beta[i,k]} dx_i^\beta \tag{17-66}$$

Using Eq. (17-66), Eq. (17-64) can now be rewritten as:

$$\frac{1}{T^2}\,|\Delta \underset{\approx}{H}^{\infty}|\,dT + |\underset{\approx}{x}_i^{(s)}|\left[-\frac{\overline{H}_1^\beta}{T^2}\,dT + \sum_{i=1}^{n}\frac{1}{T}\left[\frac{\partial\mu_1^\beta}{\partial x_i^\beta}\right]_{T,P,x_j^\beta[i,k]} dx_i^\beta\right] = 0 \tag{17-67}$$

Since we are dealing with a monovariant equilibrium system, we need only examine temperature variations with respect to a single mole fraction parameter. If we choose x_1^β arbitrarily, then the summation in Eq. (17-67) reduces to a single term, as all other compositional variation terms dx_i^β ($i = 2, \ldots, n$) are constrained by the multiphase equilibrium:

$$\frac{1}{T^2}\left[|\Delta \underset{\approx}{H}^{\infty}| - \overline{H}_1^\beta\,|\underset{\approx}{x}_i^{(s)}|\right]dT + \frac{1}{T}\,|\underset{\approx}{x}_i^{(s)}|\left[\frac{\partial\mu_1^\beta}{\partial x_1^\beta}\right]_{T,P} dx_1^\beta = 0 \tag{17-68}$$

or

$$\left[\frac{\partial T}{\partial x_1^\beta}\right]_{P,[\mathcal{F}=1]} = \frac{-T\,|\underset{\approx}{x}_i^{(s)}|\left(\partial\mu_1^\beta/\partial x_1^\beta\right)_{T,P}}{\left[|\Delta \underset{\approx}{H}^{\infty}| - \overline{H}_1^\beta\,|\underset{\approx}{x}_i^{(s)}|\right]} \tag{17-69}$$

In Eqs. (17-68) and (17-69), the partial derivative $(\partial\mu_1/\partial x_1^\beta)$ is only restricted to constant T and P variations as the $[\alpha,\ldots,\pi]$ phase assemblage at equilibrium already constrains all other compositional variations. One can also introduce fugacity at this point, using $\mu_i^\beta = RT \ln \hat{f}_i^\beta + \lambda_i(T)$. Thus,

$$\left[\frac{\partial \mu_1^\beta}{\partial x_1^\beta}\right]_{T,P} = RT\left[\frac{\partial \ln \hat{f}_1^\beta}{\partial x_1^\beta}\right]_{T,P} \tag{17-70}$$

and Eq. (17-69) can be rewritten as:

$$\left[\frac{\partial T}{\partial x_1^\beta}\right]_{P,[\mathcal{F}=1]} = \frac{-RT^2\,|\underline{\underline{x}}_i^{(s)}|\left[\dfrac{\partial \ln \hat{f}_1^\beta}{\partial x_1^\beta}\right]_{T,P}}{\left[|\Delta \underline{\underline{H}}^{\infty}| - \overline{H}_1^\beta\,|\underline{\underline{x}}_i^{(s)}|\right]} \tag{17-71}$$

Clearly, Eq. (17-69) or (17-71) can be generalized for any component $i = 1, 2, \ldots, n$ and used to describe the temperature-composition behavior of any π-phase monovariant system. If $n=2$, the algebraic complexity is kept to a minimum. This provides an excellent illustrative example commonly found in many thermodynamic texts (for example, see Eq. (15-25) in this text and Eq. (18.47) in Prigogine and Defay (1954)). For a two-component, two-phase, non-reacting isobaric system:

$$\pi = n = 2 \text{ (}\alpha \text{ and } \beta \text{ phases) and } \mathcal{F} = n + 2 - \pi - r - 1 = 2 + 2 - 2 - 0 - 1 = 1 \tag{17-72}$$

$$|\underline{\underline{x}}_i^{(s)}| = \begin{vmatrix} x_1^\alpha & x_2^\alpha \\ x_1^\beta & x_2^\beta \end{vmatrix} = x_1^\alpha x_2^\beta - x_2^\alpha x_1^\beta = \left[x_1^\alpha - \frac{x_2^\alpha x_1^\beta}{x_2^\beta}\right] x_2^\beta \tag{17-73}$$

$$|\Delta \underline{\underline{H}}^{\infty}| = \begin{vmatrix} H^\alpha & x_2^\alpha \\ H^\beta & x_2^\beta \end{vmatrix} = H^\alpha x_2^\beta - H^\beta x_2^\alpha \tag{17-74}$$

Using $H^\alpha = x_1^\alpha \overline{H}_1^\alpha + x_2^\alpha \overline{H}_2^\alpha$ and $H^\beta = x_1^\beta \overline{H}_1^\beta + x_2^\beta \overline{H}_2^\beta$, Eq. (17-74) becomes,

$$|\Delta \underline{\underline{H}}^{\infty}| = \left(x_1^\alpha \overline{H}_1^\alpha + x_2^\alpha \overline{H}_2^\alpha\right) x_2^\beta - \left(x_1^\beta \overline{H}_1^\beta + x_2^\beta \overline{H}_2^\beta\right) x_2^\alpha \tag{17-75}$$

and

$$\overline{H}_1^\beta\,|\underline{\underline{x}}_i^{(s)}| = \begin{vmatrix} \overline{H}_1^\beta x_1^\alpha & x_2^\alpha \\ \overline{H}_1^\beta x_1^\beta & x_2^\beta \end{vmatrix} = x_1^\alpha x_2^\beta \overline{H}_1^\beta - x_2^\alpha x_1^\beta \overline{H}_1^\beta \tag{17-76}$$

Substituting Eqs. (17-73), (17-75), and (17-76) into Eq. (17-69) and combining terms gives,

$$\left[\frac{\partial T}{\partial x_1^\beta}\right]_{[\alpha-\beta],P} = \frac{-T\left[x_1^\alpha - \dfrac{x_2^\alpha x_1^\beta}{x_2^\beta}\right]\left[\dfrac{\partial \mu_1^\beta}{\partial x_1^\beta}\right]_{T,P}}{x_1^\alpha\left[\overline{H}_1^\alpha - \overline{H}_1^\beta\right] + x_2^\alpha\left[\overline{H}_2^\alpha - \overline{H}_2^\beta\right]} \tag{17-77}$$

which is the same as Eq. (15-25) once $RT(\partial \ln \hat{f}_1^\beta/\partial x_1^\beta)_{T,P}$ is substituted for $(\partial \mu_1^\beta/\partial x_1^\beta)_{T,P}$.

Example 17.4

At 1 bar, a continuous solid solution is found in the copper-nickel binary system. Experimental evidence suggests that the freezing point of the solution increases as nickel is added even down to the infinite dilution limit ($x_{Ni} \to 0$). Develop a relationship to estimate how the freezing point of a copper-nickel solution varies as the mole fraction of nickel increases from 0. Clearly state all assumptions regarding how you are modeling the solid and liquid phases. Melting points and heats of fusion are listed below for pure copper and nickel.

	T_{melt}	ΔH_{fusion}
	(°C)	(J/mol)
Cu	1083	13,012
Ni	1455	17,573

Solution

Because $n = 2$, $\pi = 2$, $r = 0$, and the system is isobaric

$$\mathcal{F} = n + 2 - \pi - r = 2 + 2 - 2 - 0 - 1 \text{ (constant } P\text{)} = 1$$

so the system is monovariant and Eq. (17-77) can be used directly with component subscripts for Cu ≡ 1 and for Ni ≡ 2; and superscripts for phases $\alpha \equiv$ solid solution and $\beta \equiv$ liquid.

At this point we need a model for liquid solution so that $(\partial \mu_1^\beta/\partial x_1^\beta)_{T,P}$ can be evaluated. In general,

$$\mu_1^\beta = \mu_1^o + RT \ln\left(\gamma_1^\beta x_1^\beta\right) \tag{17-78}$$

so

$$\left[\frac{\partial \mu_1^\beta}{\partial x_1^\beta}\right]_{T,P} = RT\left[\frac{\partial \ln\left(\gamma_1^\beta x_1^\beta\right)}{\partial x_1^\beta}\right]_{T,P} \tag{17-79}$$

so we need a ΔG^{EX} model to get $RT \ln \gamma_1^\beta = \overline{\Delta G}_1^{EX,\beta}$. As $x_2^\beta = x_{Ni}^L \to 0$, $\gamma_1^\beta \to 1.0$ and ideal solution behavior exists. For this limiting condition

$$\left[\frac{\partial \mu_1^{\beta}}{\partial x_1^{\beta}}\right]_{T,P,x_2^{\beta}\to 0} = RT\left[\frac{1}{x_1^{\beta}}\right]_{T,P} \tag{17-80}$$

and

$$\overline{H}_1^{\alpha} = H_1^{\alpha}\ ,\ \ \overline{H}_1^{\beta} = H_1^{\beta}\ ,\ \ \overline{H}_2^{\alpha} = H_2^{\alpha}\ ,\ \ \text{and}\ \overline{H}_2^{\beta} = H_2^{\beta}$$

Thus, this ideal solution behavior can be substituted into Eq. (17-77) to give:

$$\left[\frac{\partial T}{\partial x_1^{\beta}}\right]_{[\alpha-\beta],P} = \frac{-RT^2\left[x_1^{\alpha} - \dfrac{x_2^{\alpha}x_1^{\beta}}{x_2^{\beta}}\right]\left[\dfrac{1}{x_1^{\beta}}\right]_{T,P}}{x_1^{\alpha}\left[H_1^{\alpha} - H_1^{\beta}\right] + x_2^{\alpha}\left[H_2^{\alpha} - H_2^{\beta}\right]} \tag{17-81}$$

which as $x_1^{\beta} \to 1.0$ and $x_1^{\alpha} \to 1.0$ becomes:

$$\lim_{x_2^{\beta}\to 0}\left[\frac{\partial T}{\partial x_1^{\beta}}\right]_{[\alpha-\beta],P} = \frac{-RT^2\left[K_x - 1\right]}{\Delta H_{fus,1}} \tag{17-82}$$

where $\Delta H_{fus,1} \equiv H_1^{\beta} - H_1^{\alpha}$ and K_x a distribution coefficient defined as

$$K_x \equiv \frac{x_2^{\alpha}}{x_2^{\beta}} = \frac{x_{\text{Ni}}^{ss}}{x_{\text{Ni}}^{L}} \tag{17-83}$$

For this system, which is nearly ideal, K_x can be assumed constant. Thus, by separating variables in Eq. (17-82) and integrating

$$\int_{T_{f,1}}^{T} \frac{dT}{T^2} \cong -\frac{R}{\Delta H_{fus,1}}\left[K_x - 1\right]\left[\int_{x_1^{\beta}=1}^{x_1^{\beta}} dx_1^{\beta}\right] \tag{17-84}$$

and with $x_2^{\beta} = 1 - x_1^{\beta}$, and $dx_1^{\beta} = -dx_2^{\beta}$, Eq. (17-84) becomes:

$$-\left[\frac{1}{T} - \frac{1}{T_{f,1}}\right] \cong -\frac{R}{\Delta H_{fus,1}}\left[1 - K_x\right]\, x_2^{\beta} \tag{17-85}$$

As $x_2^{\beta} \to 0$, T is close to $T_{f,1}$, so Eq. (17-85) can be approximated as,

$$T_{f,1} - T \cong \frac{RT_{f,1}^2}{\Delta H_{fus,1}}\left[1 - K_x\right]\left[x_2^{\beta}\right] \tag{17-86}$$

or

$$T_{f,\text{Cu}} - T \cong \frac{RT_{f,\text{Cu}}^2}{\Delta H_{fus,\text{Cu}}}\left[1 - \frac{x_{\text{Ni}}^{ss}}{x_{\text{Ni}}^{L}}\right]\left[x_{\text{Ni}}^{L}\right] \tag{17-87}$$

so if nickel (2) is more soluble in the solid solution phase than in the liquid phase, (K_x> 1) and $T > T_f$ then the freezing point of the mixture increases as nickel is added! If nickel is less soluble in the solid solution ($K_x < 1$)and $T < T_f$ then the freezing point decreases as nickel is added. The T-x diagram for the Cu-Ni binary system is sketched in Figure 17.2. Note that $K_x > 1$, thus as nickel is added the freezing point increases.

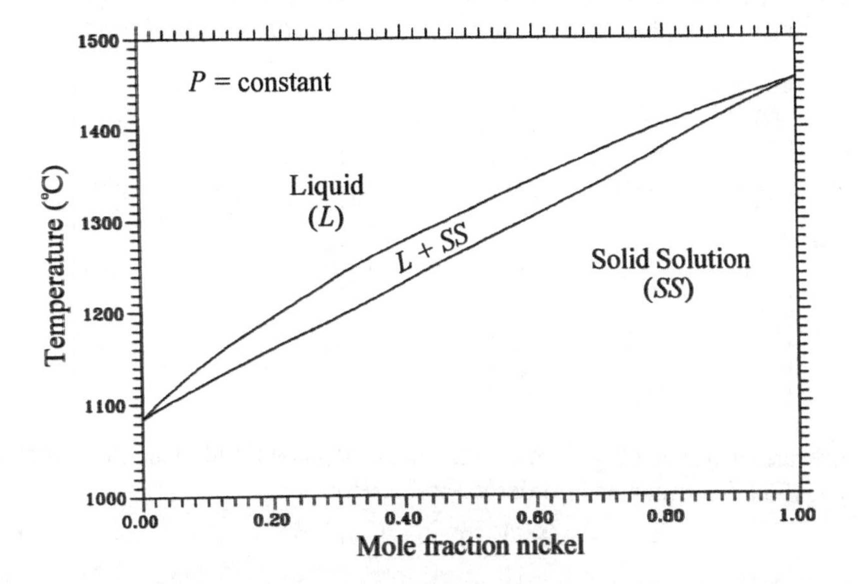

Figure 17.2 Binary isobaric phase diagram for the copper-nickel system at 1 atm pressure. [data from the Metals Handbook (1973)].

In general, if K_x is greater than one, the freezing point will increase as the second component is added, regardless of the magnitude of the activity coefficient (γ_i) provided the denominator and numerator of Eq. (17-77) have the same sign. From stability arguments, $(\partial\mu_1^\beta/\partial x_1^\beta)_{T,P}$ is always greater than 0. In addition, if α is selected as the solid phase, then the denominator will most likely be negative at the limit of pure component (2), consequently from Eq. (17-77)

$$\left[\frac{\partial T}{\partial x_1^\beta}\right]_{[\alpha-\beta],P} > 0 \quad \text{if} \quad K_x \equiv \frac{x_2^\alpha}{x_2^\beta} > 1 \quad (\text{as } x_2 \to 1)$$

This result can also be generalized to any two-component, two-phase system. For example, a non-azeotropic, binary, vapor-liquid system will show a boiling point elevation at one compositional extreme as the second, less volatile component is added and a boiling point depression as the more volatile component is added.

17.6 Indifferent States and Azeotropic Behavior $\mathcal{F} \geq 2$

In this section we consider a general approach to proving that the Gibbs-Konovalow theorems are followed if a multicomponent, multiphase system is in an indifferent state. As defined by Prigogine and Defay (1954), an indifferent state corresponds to a certain condition whereby "the equilibrium state is 'indifferent' to the mass of the phases present." In polyvariant systems, the necessary and sufficient conditions for indifference are found when all determinants of $\pi + r$ order of the $\underset{\approx}{N}_i^{(s)}$ or $\underset{\approx}{x}_i^{(s)}$ matrix are zero.

Recall that for a π-phase system of n components with r chemical reactions:

$$\underset{\approx}{N}_j^{(s)} \equiv \begin{bmatrix} N_1^{(1)} & \cdots & N_j^{(1)} & \cdots & N_n^{(1)} \\ \vdots & & \vdots & & \vdots \\ N_1^{(\pi)} & \cdots & N_j^{(\pi)} & \cdots & N_n^{(\pi)} \\ \vdots & & \vdots & & \vdots \\ v_1^{(1)} & \cdots & v_j^{(1)} & \cdots & v_n^{(1)} \\ \vdots & & \vdots & & \vdots \\ v_1^{(r)} & \cdots & v_j^{(r)} & \cdots & v_n^{(r)} \end{bmatrix} \qquad \underset{\approx}{x}_j^{(s)} \equiv \begin{bmatrix} x_1^{(1)} & \cdots & x_j^{(1)} & \cdots & x_n^{(1)} \\ \vdots & & \vdots & & \vdots \\ x_1^{(\pi)} & \cdots & x_j^{(\pi)} & \cdots & x_n^{(\pi)} \\ \vdots & & \vdots & & \vdots \\ v_1^{(1)} & \cdots & v_j^{(1)} & \cdots & v_n^{(1)} \\ \vdots & & \vdots & & \vdots \\ v_1^{(r)} & \cdots & v_j^{(r)} & \cdots & v_n^{(r)} \end{bmatrix} \tag{17-88}$$

A polyvariant system ($\mathcal{F} \geq 2$), of n components, π phases, and r reactions, implies that $n \geq \pi + r$. If all determinants of order $\pi + r$ are zero:

$$|\underset{\approx}{N}_j^{(s)}| = 0 \quad \text{or} \quad |\underset{\approx}{x}_j^{(s)}| = 0 \tag{17-89}$$

then the system is in an indifferent state. It should be remembered that all invariant ($\mathcal{F} = 0$) and monovariant ($\mathcal{F} = 1$) systems are indifferent because we are always dealing with a set of Equations (17-36) where the number of equations is less than or equal to the number of unknowns. Prigogine and Defay (1954) discuss the mathematical implications of Eq. (17-89) and show that there are ($\mathcal{F} - 1$) conditions of indifference.

If we return to our original formulation of the combined Gibbs-Duhem and reaction equilibrium constraints in matrix form given in Eq. (17-12b), we immediately see that several other relationships appear:

$$-\underset{\sim}{S}^{(s)}\, dT + \underset{\sim}{V}^{(s)}\, dP - \underset{\approx}{N}_i^{(s)}\, d\underset{\sim}{\mu}_j = 0 \tag{17-12b}$$

In an indifferent state $|\underset{\approx}{N}_j^{(s)}| = 0$, so if the system is isobaric, an extremum in temperature will exist ($dT = 0$) or conversely for an isothermal system, an extremum in pressure will exist ($dP = 0$). This is a general proof of the first and second theorems due to Gibbs and Konovalow.

Example 17.5

In Section 3, we introduced the concept of an indifferent state for a non-reacting system at a liquid-vapor azeotropic point. In that case, $n = \pi = 2$, so $\mathcal{F} = 2 + 2 - 2 - 0 = 2$ and there is just $\mathcal{F}$-1 or one condition of indifference.

$$\left| x_j^{(s)} \right| = \begin{vmatrix} y_1 & y_2 \\ x_1 & x_2 \end{vmatrix}$$

where the y's refer to vapor phase mole fractions and the x's to liquid phase mole fractions in the conventional sense. Because $x_i = y_i$ at the azeotrope,

$$\left| x_j^{(s)} \right| = 0$$

It is also true that, for an isobaric system as depicted in Figure 17.1, the azeotropic point corresponds to an extremum in temperature--in this case a minimum boiling point.

Example 17.6

Another important class of indifferent states occurs in three-component, three-phase systems with no reactions. In this case, $\mathcal{F} = 3 + 2 - 3 - 0 = 2$ and again there is only one condition of indifference when:

$$\left| x_j^{(s)} \right| = \begin{vmatrix} x_1^{\alpha} & x_2^{\alpha} & x_3^{\alpha} \\ x_1^{\beta} & x_2^{\beta} & x_3^{\beta} \\ x_1^{\gamma} & x_2^{\gamma} & x_3^{\gamma} \end{vmatrix} = 0$$

In order for $|x_j^{(s)}| = 0$, either $x_i^{\alpha} = x_i^{\beta}$ or $x_i^{\alpha} = x_i^{\gamma}$ or $x_i^{\gamma} = x_i^{\beta}$ or some linear relationship exists between the phase compositions. This could occur as a three-component, ternary azeotrope which are known to exist.

Example 17.7

When chemical reactions are present, indifferent states can also appear. This is usually the case when stoichiometric ratios exist. For example, starting with pure solid $(NH_4)_2CO_3$ in a vessel and heating, three vapor components could form whose compositions are in fixed ratios:

$$(NH_4)_2CO_{3(s)} = 2\ NH_{3(g)} + CO_{2(g)} + H_2O_{(g)}$$

In this system, $n = 4$, $\pi = 2$, $r = 1$ and $\mathcal{F} = 4 + 2 - 2 - 1 = 3$, so two conditions of indifference are possible. In this system, the composition matrix $\underset{\approx}{x}_j^{(s)}$ is given by:

$$\underset{\approx}{x}_j^{(s)} = \begin{bmatrix} y_1^g & y_2^g & y_3^g & y_4^g \\ x_1^s & x_2^s & x_3^s & x_4^s \\ 2 & 1 & 1 & -1 \end{bmatrix} \qquad \begin{aligned} 1 &\equiv NH_3 \\ 2 &\equiv CO_2 \\ 3 &\equiv H_2O \\ 4 &\equiv (NH_4)_2CO_3 \end{aligned}$$

But we can assume that the solid phase is pure $(NH_4)_2CO_3$ and that there is no vapor species of undissociated carbonate, thus

$$|\underset{\approx j}{x}^{(s)}| = \begin{vmatrix} \frac{1}{2} & \frac{1}{4} & \frac{1}{4} & 0 \\ 0 & 0 & 0 & 1 \\ 2 & 1 & 1 & -1 \end{vmatrix}$$

By expanding with cofactors, there are three independent determinants of order $(\pi + r)$ or (3×3) of the $\underset{\approx j}{x}^{(s)}$ matrix that must be zero. They are

$$\begin{vmatrix} \frac{1}{2} & \frac{1}{4} & \frac{1}{4} \\ 0 & 0 & 0 \\ 2 & 1 & 1 \end{vmatrix} = 0 \qquad \begin{vmatrix} \frac{1}{4} & \frac{1}{4} & 0 \\ 0 & 0 & 1 \\ 1 & 1 & -1 \end{vmatrix} = 0 \qquad \begin{vmatrix} \frac{1}{2} & \frac{1}{4} & 0 \\ 0 & 0 & 1 \\ 2 & 1 & -1 \end{vmatrix} = 0 \qquad (17\text{-}98)$$

For another case that is simpler, consider the dissociation of NH_4Cl into two gaseous components, NH_3 and HCl:

$$NH_4Cl_{(s)} = HCl_{(g)} + NH_{3(g)}$$

Now there are two phases and three components and one reaction, so $\mathcal{F} = 3 + 2 - 2 - 1 = 2$, so there is only one condition of indifference. Again, it is easy to show that $|\underset{\approx j}{x}^{(s)}| = 0$.

$$|\underset{\approx j}{x}^{(s)}| = \begin{vmatrix} \frac{1}{2} & \frac{1}{2} & 0 \\ 0 & 0 & 1 \\ 1 & 1 & -1 \end{vmatrix} = 0 \qquad \begin{matrix} 1 \equiv HCl \\ 2 \equiv NH_3 \\ 3 \equiv NH_4Cl \end{matrix}$$

17.7 Summary

The objective of this chapter was to develop a coherent, generalized approach to treat temperature, pressure, and compositional variations in any multicomponent, multiphase system at equilibrium. Gibbs-Duhem relations for each phase and chemical equilibrium constraints for each reaction are coupled using a straightforward methodology for expressing variations in T, P, and N_i (or x_i) that can be applied generally to any system in phase and chemical equilibria. With a matrix/determinant format, machine computations utilizing linear algebra methods are relatively easy to implement even when a large number of components, reactions, and phases are present.

References

Gibbs, J.W. (1976), "On the Equilibrium of Heterogeneous Substances," in *Transactions of the Connecticut Academy* 3, 108-248 (October 1875 - May 1876) and 343-524 (May 1877 - July 1878) appearing in *The Collected Works of J. W. Gibbs, Volume 1, Thermodynamics* Yale University Press, New Haven, CT, 1957.

Gilmont, R. (1959), *Thermodynamic Principles for Chemical Engineers*, Prentice Hall, Englewood Cliffs, NJ.

International Critical Tables of Numerical Data, Physics, Chemistry and Technology (1928), Volume III, first edition, McGraw-Hill, New York, NY, 1928, p 313.

Metals Handbook (American Society for Metals, Metals Park, OH, 8th ed., Volume 8, 1973).

Prigogine, I. and R. Defay (1954), *Chemical Thermodynamics*, translated by D. H. Everett, Longmans and Green, London.

Problems

17.1. Several processes involving distillation, liquid-liquid extraction, and/or crystallization have been proposed to separate a ternary mixture of methanol (MeOH), water, and NaCl into its pure components. The mixture is initially at 25°C which is also the temperature of the surroundings. You can assume that all non-idealities can be represented as a sum of three binary interactions using a two-suffix Margules format for each interaction such that

$$\Delta G^{EX}/RT = Ax_1x_2 + Bx_1x_3 + Cx_2x_3$$

where x_i = mole fraction of component i in the ternary mixture and A, B, and C are binary parameters obtained from the following sets of binary-only data at 25°C:

$A = 3.0$ for $[H_2O, MeOH] = [1,2]$ binary

$B = 2.0$ for $[H_2O, NaCl] = [1,3]$ binary

$C = 1.0$ for $[MeOH, NaCl] = [2,3]$ binary

Initially the liquid mixture composition is as follows: $x_1 = 0.4$, $x_2 = 0.5$, and $x_3 = 0.1$

(a) As a basis for comparison, you are to devise an idealized process to estimate the minimum work required to separate the initial mixture into its pure components. Describe and sketch the process and calculate W_{min}.

(b) Estimate the activity coefficients of MeOH and water at 25°C at their initial compositions in the mixture.

(c) Estimate the vapor composition and total vapor pressure if the ternary liquid mixture at its initial composition is vaporized at 25°C. State and justify any assumptions made. Pure component vapor pressures at 25°C are:

$P_{vp\ MeOH} = 0.18$ bar
$P_{vp\ H_2O} = 0.03$ bar
$P_{vp\ NaCl} = 0.00001$ bar

(d) In one proposed separation scheme, water and MeOH are evaporated at 1 bar batch-wise from the mixture. Is it possible to achieve a reasonable separation of MeOH and water during the evaporation by collecting fractions of overhead product as a function of time? At some point, solid salt appears. How many degrees of freedom exist at this point? Sketch a temperature-composition (T–x_1) diagram for this process to describe how liquid compositions behave as evaporation proceeds after solid salt appears.

(e) Following the proposed batch-wise evaporation step, the saturated liquid salt solution would be cooled isobarically until ice formed. Describe how you would calculate the temperature where solid ice could first appear. How many degrees of freedom exist at this point? Clearly state what assumptions you are making and what data are required.

17.2. An important problem to the geothermal well-drilling industry is corrosion of steel casing in deep wellbores. Oxygen is entrained at the surface as drilling fluids are recirculated and pumped down the well. Dissolved O_2 concentrations need to be maintained at minimum levels to keep corrosion rates down. Using your knowledge of phase equilibria at high pressures and temperatures, describe how you would estimate the solubility of O_2 in hot geothermal fluid at 380°C in a wellbore at a depth of 2 km. State and justify any and all assumptions made. Densities, critical point data and acentric factors for H_2O and O_2 are as follows:

Property	H_2O	O_2
P_c	220 bar	50 bar
T_c	374°C	–118°C
ρ_c	430 kg/m^3	323 kg/m^3
ω	0.34	0.02
ρ (1 bar, 25°C)	1000 kg/m^3	—

17.3. Several chlorinated hydrocarbons are known to form a special class of compounds with water called gas hydrates or clathrates. These compounds consist of a cage-like lattice structure formed by the water component with the second component (so-called guest molecules) occupying the cages in the lattice. A typical phase reaction for the chloroform ($CHCl_3$)-water (H_2O) hydrate forming system is given in Eq. (17-54). Other phases are also possible as shown in the attached *P-T* projection (Figure P17-3): I = pure solid ice and L_2 - $CHCl_3$ organic-rich liquid phase.

(a) Perform a phase rule analysis on this system. How many components are there, what are they, and are there any restrictions which could reduce the number of degrees of freedom? Are the data given in Figure P17.3(a) thermodynamically consistent? What phases are in equilibrium at the quadruple points A and B? Is it possible for the reaction given in Eq. (17-54) to go to completion? Or to not occur at all? At what temperatures and pressures are these true?

Using the data of Figure P17.3, describe what phases are stable and where transitions occur in isobaric heating at 100 mm Hg starting with solid hydrate in $CHCl_3$-rich liquid (L_2) at –4°C.

(b) What stability criteria would you apply to estimate whether the L_1–L_2–G three-phase (heterogeneous azeotrope) system has a critical end point? What phases would you expect to disappear at temperatures and pressures above this critical point? Given an appropriate pressure-explicit equation of state to describe the *PVT* behavior of the vapor and liquid phases, describe how you would determine the conditions of the critical end point.

(c) What conditions and/or assumptions must be met to have an approximately linear relationship between $\log_{10}P$ and $1/T$ for the three-phase monovariant L_1-H-G system as shown in Figure P17.3(b)? Explain how you would use the results of your analysis to estimate the standard enthalpy of reaction (ΔH^o_{rx}) for the phase reaction given in Eq. (17-54).

(d) One possible way of avoiding solid clathrate formation is to add another component to the mixture of $CHCl_3$ and water that would lower the "freezing point" of pure hydrate. This additional component should also *not* form a solid hydrate itself. Hexane was found to be a suitable component. Experiments were conducted as follows: a three-phase mixture consisting of a water-rich liquid (L_1), a $CHCl_3$-rich organic liquid (L_2), and a solid hydrate ($CHCl_3 \cdot 17\,H_2O$) was prepared with a known amount of hexane added. The entire system was allowed to come to equilibrium at constant pressure (1 bar) and the temperature was determined as a function of the mole fraction of hexane (water-free basis, x_{hexane}).

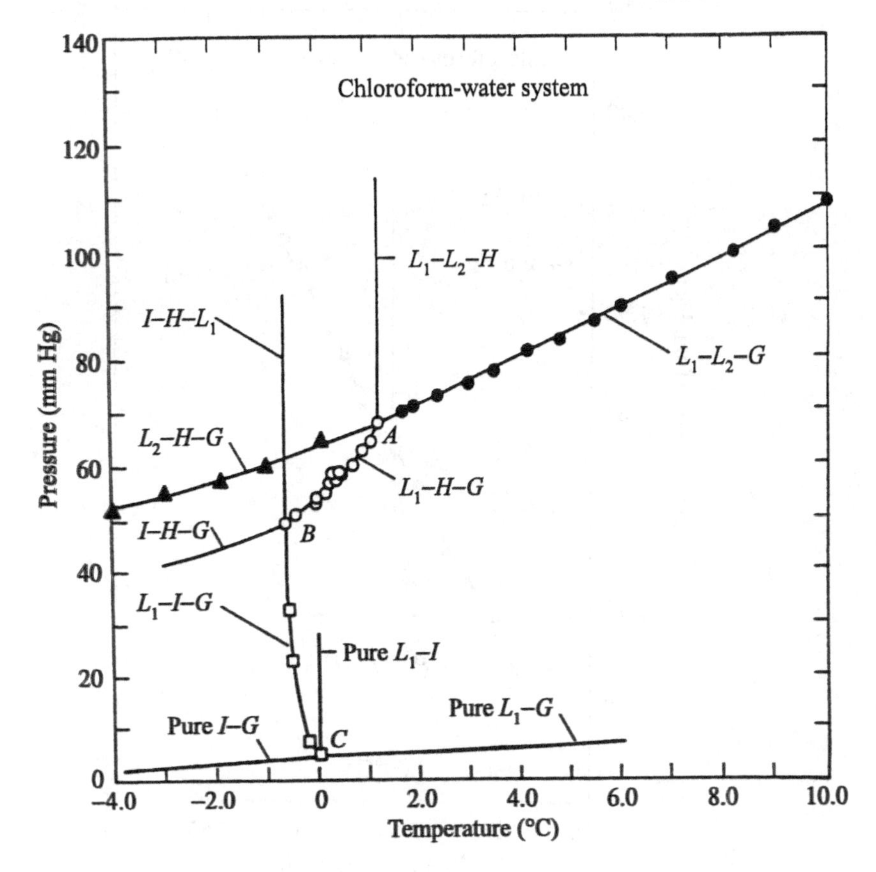

Figure P17.3(a) Pressure-temperature projection for the chloroform-water system.

If ΔT is defined as the difference between the freezing point with and without hexane present, ΔT varied linearly with x_{hexane} as x_{hexane} approached zero:

$$\Delta T = T(1 \text{ bar}, x_{\text{hexane}} = 0) - T(1 \text{ bar}, x_{\text{hexane}}) \approx 5.6x_{hexane} \tag{1}$$

Assuming ideal solution behavior for hexane in L_2 and water in L_1 and that the solid hydrate phase (H) is pure, perform an appropriate thermodynamic analysis to show that the limiting behavior of Eq. (1) is correct. You can assume that the standard enthalpy of reaction is:

$$\Delta H^o_{rx} \equiv H^o_H - 17\, H^o_{\text{H}_2\text{O}_{(\text{L}_1)}} - H^o_{\text{CHCl}_{3(\text{L}_2)}} = -108 \text{ kJ/mole} = -26 \text{ kcal/mole}$$

where the superscript (o) refers to pure components at T and 1 bar pressure in the states of aggregation indicated.

You can also assume that the solubilities of $CHCl_3$ and hexane in liquid water are negligible and the solubility of water in either liquid hexane or $CHCl_3$ is negligible. Hexane and $CHCl_3$ on the other hand are mutually soluble.

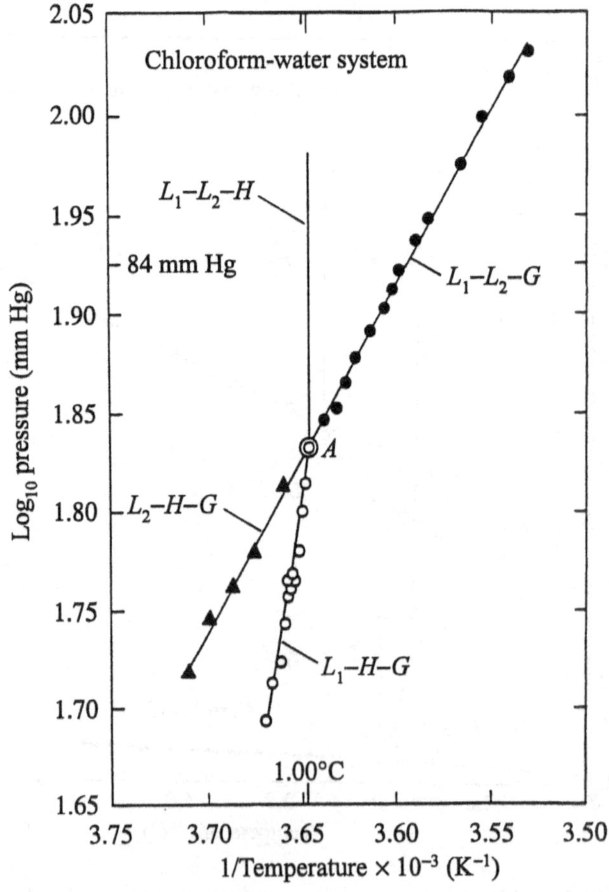

Figure P17.3(b) Log pressure-reciprocal temperature plot for the chloroform-water system.

17.4. Many two-component systems exhibit limited mutual solubility in the liquid phase and frequently result in phase splitting to form two liquid phases. In one particular application, where two liquid phases are known to exist between water and methylene chloride (CH_2Cl_2), a flash separation has been proposed to transfer a portion of the most volatile component (CH_2Cl_2) into a vapor phase. In the flash separator, a two-liquid phase, water-methylene chloride mixture at temperature T_i and pressure P_i is passed through an insulated valve to reduce the pressure, causing a vapor phase to appear.

(a) Such three-phase systems (L_1-L_2-V) are frequently referred to as "heterogeneous azeotropes" because the compositions of all phases and the temperatures are presumably fixed. Is this consistent with the phase rule?

(b) Assuming that the state of the system can be specified as in part (a) and if T_i and P_i are known, describe how you would estimate the temperature and composition of both liquid phases and the vapor phase for various values of the flash pressure P. For most applications of interest T_i will be less than 35°C with total vapor pressures never larger than 2 atm. Both liquid phases can be described by ΔG^{EX} functions given as

for water-rich liquid phase: $\Delta G^{EX} = f[T, x_w]$

for methylene chloride-rich liquid phase: $\Delta G^{EX} = g[T, x_c]$

where the subscripts, w = water and c = methylene chloride are used.

(c) Experimentally it has been observed that the total vapor pressure P_{vp} over this 3-phase, 2-component system can be specified as a function of temperature T of the following form:

$$\ln P_{vp} = A/T + B$$

where A and B are empirical constants. Perform an appropriate thermodynamic analysis to verify that the form of the equation is theoretically correct. What is the term A approximately equal to if water and methylene chloride are essentially mutually insoluble as liquids and the vapor pressure of pure methylene chloride is more than 10 times greater than pure water at the same temperature?

(d) Starting with a liquid-liquid-vapor phase mixture, as the pressure is reduced, the temperature eventually decreases to a thermal arrest point where a solid phase (S_1) appears as a fourth phase. Describe this process on a projected P-T phase diagram to illustrate that the arrest point corresponds to a quadruple point for this system. Label the other monovariant lines that will intersect this point. Under what conditions can the temperature be reduced below the quadruple point by decreasing the pressure?

17.5. The azeotropic point in a binary mixture of vapor and liquid phases results in special properties. For example, at the azeotropic point the relative volatility is unity because the vapor and liquid compositions are identical. In addition, a plot of the logarithm of the azeotropic pressure (P_{az}) versus $1/T$ in K^{-1} is typically linear for many real systems. Is this behavior reasonable? For the situation where a liquid mixture at its azeotropic composition is vaporized at fixed pressure, how would you estimate the heat required to vaporize 1 mole of the liquid?

17.6. For many two-component systems whose components are very different chemically, it is frequently found that two essentially immiscible liquid phases are formed. When the total vapor pressure over such a two-liquid phase system is measured, one commonly finds that the logarithm of the vapor pressure (P) varies linearly with the reciprocal of the absolute temperature ($1/T$). Explain this behavior using the Gibbs phase rule with appropriate thermodynamic arguments. Determine an expression for P if the liquid phases are completely immiscible.

17.7. While ice skating on the Charles River, Rochelle and Rocky Jones got into a heated discussion as to how their skates worked. Both agreed that a small amount of ice melts under the blade to reduce friction but they disagreed as to what caused the ice to melt. Rochelle, being the better student in Thermodynamics (Subject 10.40), claimed that melting was caused by lowering the freezing point of water under the locally high pressure exerted by the skate blades. Rocky, who was a top Transport (Subject 10.50) student, maintained that melting was caused by heat generated by the friction between the blade and the ice. Describe how you would analyze this problem from a thermodynamics perspective. What data or physical properties would you need to reach a quantitative conclusion regarding Rochelle's hypothesis? Feel free to use outside sources of data to substantiate your assertions.

17.8. Consider the solidification of a pure liquid in a long vertical cylinder closed at the bottom and open at the top to an atmospheric pressure of 1 bar. In the first experiment, the cylinder is partially filled with liquid and cooled to –5°C. Under these conditions, the fluid solidifies below a particular level, with liquid remaining above this level. If the temperature is further lowered to –6°C, the solid-liquid interface moves upward by 100 cm. Estimate the density of the solid phase. Note that the pressure at the original position of the solid-liquid interface remains constant. For this substance, $\Delta H_{fus} = 8$ kJ/kg and $\rho_{liquid} = 1000$ kg/m^2.

17.9. The concentration buildup and movement of carbon dioxide in the environment is of increasing concern because of its role as a greenhouse gas and the possibility of global warming. Scientists claim that in the deep ocean, CO_2 may be "immobilized" or stored in a solid phase:

$$CO_2 \cdot 5\ 3/4\ H_2O_{(H)}$$

which is a so-called water clathrate or gas hydrate (H).

Understanding the stability of the CO_2 solid hydrate phase as a function of temperature and pressure is an important first step. Phase equilibrium data are presented in the form of P-T projections for the CO_2-H_2O binary system in Figure P17.9. Univariant loci are shown as are divariant fields.

(a) For the CO_2-water binary system, specify the maximum pressure and temperature at which solid hydrate could appear in the presence of a liquid phase rich in water and a gas phase rich in CO_2.

(b) Analyze the P-T behavior along the three-phase locus (H-$H_2O_{(l)}$-$CO_{2(g)}$) where H = solid hydrate, $H_2O_{(l)}$ = water-rich liquid phase and $CO_{2(g)}$ = CO_2-rich gas phase.

Is the linear behavior of ln P versus T shown in Figure P17.9 reasonable? Explain using an appropriate thermodynamic analysis. For the small temperature range given, a variation in $1/T$ can be approximately linearized as a variation in T.

(c) How would you modify your analysis in part (b) to account for the presence of dissolved salts in sea water? You can model the ocean as a 3.5 wt% NaCl solution that is fully dissociated. With no further data available, describe how you would predict where the three-phase locus would lie. State all additional assumptions made.

(d) One concept being pursued for sequestering CO_2 would utilize ocean depths of 1000 m or more, where the temperature ranges from 1 to 2°C. In this case, liquid CO_2 would be released at depth. Would solid hydrate form? If so, what factors would influence the rate of formation?

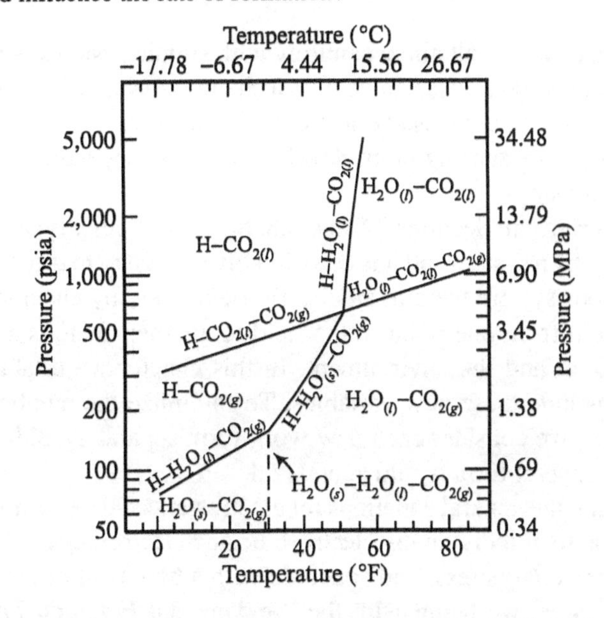

Figure P17.9 Pressure-temperature projection for the CO_2-H_2O binary system.

17.10. Consider the isobaric behavior at 1 atm for the binary condensed system: [copper (Cu) - gold (Au)]. The melting point of pure Cu is 1084°C and of pure Au is 1064°C. A continuous solid solution is known to exist for all compositions ($0 < x_{Au} < 1$) and at $x_{Au} = 0.565$, T = 889°C the liquid and solid phases have the same composition.

(a) Sketch a temperature-composition (T–x_i) diagram at 1 atm.

(b) Develop a suitable thermodynamic analysis to quantitatively determine the monovariant lines and invariant points in your T–x_i diagram at 1 atm.

(c) Show that the behavior observed at $x_{Au} = 0.565$ is consistent with the conditions of an indifferent state.

Systems under Stress, in Electromagnetic or Potential Fields 18

To this point we have dealt almost entirely with simple systems in which work done on or by the system could be associated with a change in system boundaries (i.e., $P\, d\underline{V}$). With these limitations we expressed $\underline{U}$ as a function of ($\underline{S}$, $\underline{V}$, $N_1,..., N_n$) in a Fundamental Equation and from this starting point developed useful Legendre transforms, various equilibrium relationships, etc.

As discussed briefly in Sections 3.8 and 5.8, there exist other types of work that may also be done on or by a system and it is occasionally necessary to take these into account. For example, if our system were in an electric field, then, by changing the field or by moving the system from one point in the field to another, there is a *work interaction* between the system and the environment. In this chapter we establish Fundamental Equations that include these new variables. To minimize the number of terms in the derived equations, we consider each new work form separately, although it should be obvious that the general form would contain all.

We first develop the general equations for electrostatic and electromagnetic work and then in succession treat individually electrostatic and electromagnetic systems, systems under (one-dimensional) stress, and conclude with a treatment of potential and kinetic energies. In all cases, we begin with the Fundamental Equation for internal energy expressed in differential form and include the new work terms of interest:

$$d\underline{U} = T\, d\underline{S} - P\, d\underline{V} + \sum_j\ [\delta W_{rev}]_j + \sum_i \mu_i\, dN_i \tag{18-1}$$

18.1 Electromagnetic Work

In the theory of electrodynamics, Maxwell's equations occupy the same preeminent position reserved for Newton's laws in classical mechanics (i.e., a vast amount of empirical evidence accumulated over the past century has led scientists to believe that all macroscopic electromagnetic phenomena are governed by these equations). They are adopted here as the basis of an expression for electromagnetic work. Following the treatment of Stratton (1941), the final result is a relation for the work *done on the system by external sources to establish the field*:

$$W = \int_{\underline{V}} \left[\int_0^{\mathbf{D}} \mathbf{E} \cdot d\mathbf{D} + \int_0^{\mathbf{B}} \mathbf{H} \cdot d\mathbf{B} \right] d\underline{V} \tag{18-2}$$

where: $\mathbf{E}$ = electric field strength, V/m or kg m/s^{3}A
$\mathbf{D}$ = electric displacement, A·s/m^2 or coulombs/m^2 (C/m^2)
$\mathbf{H}$ = magnetic field strength, A/m
$\mathbf{B}$ = magnetic induction, V·s/m^2, tesla or kg/s^{2}A (note that 10^4 gauss = 1 tesla)

The integral is taken over the entire system volume subject to the field, $\underline{V}$. For the work to be given by Eq. (18-2), the electric and magnetic fields must vanish at the boundaries of $\underline{V}$. Various special cases of interest arise that permit simplification of Eq. (18-2). For example, the field and displacement (or induction) vectors can be aligned in parallel which simplifies the scalar dot product. If the total system volume $\underline{V}$ includes the free space as well as the volume of the material contained in the system, one frequently sets $\underline{V} = \underline{V}_s$ = constant. However, as was the case in considering only $P\underline{V}$ work, you must be careful about the fact that $\underline{V}$, in general, can be a function of T, P, $\mathbf{E}$ (or $\mathbf{D}$), and $\mathbf{H}$ (or $\mathbf{B}$). The last two sets of variables refer to problems in electrostatics and magnetostatics, respectively. Taken collectively they represent a general class of electromagnetic problems. In many situations simplified constitutive laws exist to relate electromagnetic properties. A constant-volume, linear dielectric with invariant properties contained between two parallel charged plates is a good example of such a simplified system.

The electric vectors $\mathbf{E}$ and $\mathbf{D}$ are related by the permittivity, ε:

$$\mathbf{D} = \varepsilon \mathbf{E} \tag{18-3}$$

where in free space $\varepsilon \rightarrow \varepsilon_o = 8.854 \times 10^{-12}$ (A·s/V·m) or farad (F)/m. For real substances, ε is a function of the material, temperature, pressure, and electric field strength.

In a similar manner $\mathbf{B}$ is related to $\mathbf{H}$ by the permeability, $\mu_{\mathbf{M}}$:

$$\mathbf{B} = \mu_{\mathbf{M}} \mathbf{H} \tag{18-4}$$

where in free space $\mu_{\mathbf{M}} = \mu_{\mathbf{M}}^o = 4\pi \times 10^{-7}$ (V·s/A·m) or henry (H)/m. For real substances, the magnetic permeability, $\mu_{\mathbf{M}}$, is a function of the material, temperature, pressure, and magnetic field strength.

The constants ε_o and $\mu_{\mathbf{M}}^o$ are related to the velocity of electromagnetic waves in free space by Eq. (18-5):

$$c = (\varepsilon_o \mu_{\mathbf{M}}^o)^{-1/2} = 2.9979 \times 10^8 \text{ m/s} \tag{18-5}$$

where c is the speed of light. In general, ε and $\mu_{\mathbf{M}}$ are symmetric tensors of rank 2[1], but as used here, we consider only the cases in which $\mathbf{D}$ and $\mathbf{E}$ are parallel, as are $\mathbf{B}$ and $\mathbf{H}$. Thus, ε and $\mu_{\mathbf{M}}$ can be visualized as scalar multipliers. This simplification limits the treatment of electric systems to simple geometries (e.g., to parallel-plate condensers or long uniformly-wound solenoids), but the principal concepts and results are not greatly affected.

---

[1] For example, $\mathbf{D_x} = \varepsilon_{xx}\mathbf{E_x} + \varepsilon_{xy}\mathbf{E_y} + \varepsilon_{xz}\mathbf{E_z}$, etc.

Several other parameters are commonly used to describe electric systems. For example, the static dielectric constant, D_s, and electric and magnetic susceptibilities, χ_e and χ_m, are defined as:

$$D_s \equiv \frac{\varepsilon}{\varepsilon_o} \tag{18-6}$$

$$\chi_e \equiv \frac{\varepsilon}{\varepsilon_o} - 1 \text{ and } \chi_m \equiv \frac{\mu_{\mathbf{M}}}{\mu_{\mathbf{M}}^o} - 1 \tag{18-7}$$

It is also convenient to define two other quantities. We shall refer to **P** as the *electric polarization* and **M** as the *magnetization* and define them as:

$$\mathbf{P} \equiv \mathbf{D} - \varepsilon_o\mathbf{E} = \varepsilon_o(D_s - 1)\mathbf{E} \tag{18-8}$$

$$\mathbf{M} \equiv \frac{\mathbf{B}}{\mu_{\mathbf{M}}^o} - \mathbf{H} = \chi_m\mathbf{H} \tag{18-9}$$

It is obvious that, as defined, **P** and **M** vanish in free space.

Materials that have constant electrical permittivity (ε) and magnetic permeability ($\mu_{\mathbf{M}}$) independent of electric and magnetic field strength are called *linear* dielectrics and paramagnetics.

A critical issue in treating electromagnetic effects are the units used to describe values for the properties **H**, **B**, **D**, and **E** and for ε, ε_o, $\mu_{\mathbf{M}}$, and $\mu_{\mathbf{M}}^o$. In SI units ε and ε_o are expressed in farads/m or A^2s^4/m^3 kg and $\mu_{\mathbf{M}}$ and $\mu_{\mathbf{M}}^o$ are in henrys/m or m kg/A^2s^2. Thus, using the two constitutive property laws given in Eqs. (18-3) and (18-4) for situations involving linear isotropic materials in time-steady fields with **H** and **B** and/or **E** and **D** aligned, the scalar dot products in Eq. (18-2) can be simplified to evaluate the differential work contributions due to electric polarization given by

$$\delta W_{\mathbf{E}} = \mathbf{E} \cdot d\mathbf{D} = \varepsilon \mathbf{E} d\mathbf{E} \tag{18-10a}$$

and due to magnetic polarization

$$\delta W_{\mathbf{M}} = \mathbf{H} \cdot d\mathbf{B} = \mu_{\mathbf{M}} \mathbf{H} d\mathbf{H} \tag{18-10b}$$

Both δW's have units of J/m^3, so they really represent the work per unit volume as is evident from the form of Eq. (18-2). This presents a problem when reconstructing the Fundamental Equation to include electromagnetic work effects. In our normal energy formulation given in Section 5.1, $\underline{U}$ is an extensive quantity, first order in mass, thus $W_{\mathbf{E}}$ and $W_{\mathbf{M}}$ need to be expressed in extensive form as well. Rosensweig (1989) carefully developed an approach to correctly transform these work quantities to proper extensive form. In general, the extensive differential work is given as the dot product of an intensive force vector and an extensive differential displacement vector. In electromagnetic systems the displacements **D** and **B** are converted to extensive form by multiplying them by the total system volume, $\underline{V}$. Thus,

$$\underline{\mathbf{D}} \equiv \mathbf{D}\underline{V} \quad \text{and} \quad \underline{\mathbf{B}} \equiv \mathbf{B}\underline{V} \tag{18-11}$$

Now we can construct the Fundamental Equation for an n-component, open, variable volume system with time steady electromagnetic work. First in differential form:

$$d\underline{U} = Td\underline{S} - Pd\underline{V} + \mathbf{E}d(\mathbf{D}\underline{V}) + \mathbf{H}d(\mathbf{B}\underline{V}) + \sum_{i=1}^{n} \mu_i dN_i \tag{18-12}$$

Equation (18-12) can be Euler integrated to give $\underline{U}$:

$$\underline{U} = T\underline{S} - P\underline{V} + \mathbf{E}(\mathbf{D}\underline{V}) + \mathbf{H}(\mathbf{B}\underline{V}) + \sum_{i=1}^{n} \mu_i N_i \tag{18-13}$$

Or, in general, using Eq. (18-11):

$$\underline{U} = U(\underline{S}, \underline{V}, \underline{\mathbf{B}}, \underline{\mathbf{D}}, N_i) \tag{18-14}$$

Now we can recast Eqs. (18-8) and (18-9) as

$$\mathbf{B} = \mu_{\mathbf{M}^o}(\mathbf{H} + \mathbf{M}) \quad \text{and} \quad \mathbf{D} = \varepsilon_o \mathbf{E} + \mathbf{P} \tag{18-15}$$

Thus, Eq. (18-12) can be rewritten as

$$d\underline{U} = Td\underline{S} - Pd\underline{V} + \mathbf{E}d((\varepsilon_o\mathbf{E} + \mathbf{P})\underline{V}) + \mathbf{H}d(\mu_{\mathbf{M}^o}(\mathbf{H} + \mathbf{M})\underline{V}) + \sum_i \mu_i dN_i \tag{18-16}$$

and Eq. (18-13) becomes

$$\underline{U} = T\underline{S} - P\underline{V} + \mathbf{E}(\varepsilon_o\mathbf{E} + \mathbf{P})\underline{V} + \mathbf{H}\mu_{\mathbf{M}^o}(\mathbf{H} + \mathbf{M})\underline{V} + \sum_i \mu_i N_i \tag{18-17}$$

We can now define a modified pressure P^* and internal energy $\underline{U}^*$ which will simplify the algebra:

$$P^* \equiv P - \frac{\mu_{\mathbf{M}^o}\mathbf{H}^2}{2} - \frac{\varepsilon_o\mathbf{E}^2}{2} \tag{18-18}$$

$$\underline{U}^* \equiv \underline{U} - \frac{\mu_{\mathbf{M}^o}\mathbf{H}^2\underline{V}}{2} - \frac{\varepsilon_o\mathbf{E}^2\underline{V}}{2} \tag{18-19}$$

Effectively, $\underline{U}^*$ is the excess internal energy over that produced by the **H** and **E** fields in a free space (vacuum) volume $\underline{V}$. With these modifications, we can reconstruct the Fundamental Equation for an open, n-component system that excludes the space occupied by the mass of the system [Rosensweig (1989)]:

$$d\underline{U}^* = Td\underline{S} - P^*d\underline{V} + \mathbf{E}d(\mathbf{P}\underline{V}) + \mu_{\mathbf{M}^o}\mathbf{H}d(\mathbf{M}\underline{V}) + \sum_i \mu_i dN_i \tag{18-20}$$

18.2 Electrostatic Systems

In purely electrostatic systems we can set **H** and **M** to zero and re-examine Eqs. (18-16) and (18-20). Under these conditions:

$$d\underline{U} = Td\underline{S} - Pd\underline{V} + \mathbf{E}d((\varepsilon_o\mathbf{E} + \mathbf{P})\underline{V}) + \sum_i \mu_i dN_i \tag{18-21}$$

and

$$d\underline{U}^* = Td\underline{S} - P^*d\underline{V} + \mathbf{E}d(\mathbf{P}\,\underline{V}) + \sum_i \mu_i dN_i \tag{18-22}$$

with $\underline{U}^* = \underline{U} - \frac{\varepsilon_o\mathbf{E}^2V}{2}$ and $P^* = P - \frac{\varepsilon_o\mathbf{E}^2}{2}$ in this case from Eqs. (18-18) and (18-19).

The term $\varepsilon_o\mathbf{E}^2\,\underline{V}/2$ does not depend on the properties of the material comprising the system. We emphasize that $\mathbf{P}$ is the component of the electric polarization vector parallel to the field $\mathbf{E}$.

Equation (18-22) is the Fundamental Equation that we were seeking. It is immediately obvious that a number of Legendre transforms may be written in order to obtain independent variable sets other than $\underline{U}^*(\underline{S}, \underline{V}, \mathbf{P}, N_1, \ldots, N_n)$. Also, since the Fundamental Equation is homogeneous, the Euler form is

$$\underline{U}^* = T\underline{S} - P^*\underline{V} + \mathbf{E}\mathbf{P}\underline{V} + \sum_i \mu_i N_i \tag{18-23}$$

and the Gibbs-Duhem relation becomes

$$\underline{S}\,dT - \underline{V}\,dP^* + \mathbf{P}\underline{V}d\mathbf{E} + \sum_i N_i\,d\mu_i = 0 \tag{18-24}$$

With the exception of the new electrostatic terms, these equations are identical to those derived earlier. The chemical potential now, however, becomes for component k

$$\mu_k = \overline{U}_k^* - T\overline{S}_k + P^*\overline{V}_k - \mathbf{E}\left(\frac{\partial(\mathbf{P}\underline{V})}{\partial N_k}\right)_{T,\,P,\,N_j[k]} \tag{18-25}$$

$$= \overline{U}_k^* - T\overline{S}_k + P^*\overline{V}_k - \mathbf{E}(\overline{\mathbf{P}\underline{V}})_k \tag{18-26}$$

To maintain our previous contention that μ_k is a partial molar Gibbs energy, we see that we should define the Gibbs energy as the transform

$$\underline{G}(T, P, \mathbf{E}, N_j) = \underline{U}^* - T\underline{S} + P^*\underline{V} - \mathbf{E}\mathbf{P}\underline{V} \tag{18-27}$$

It is important to recognize that this function is no different from the Gibbs energy used previously in this book. The designation of the variable $\mathbf{E}$ was not noted earlier since we implicitly assumed that our systems were either not dielectrics ($D_s = 1.0$), the field was zero, or there was no interaction between an imposed field and our system. This same reasoning holds for all other transforms and, in addition, to the equilibrium and stability criteria developed earlier.

For dielectric materials in an electric field, the concept of the pressure is somewhat different from the same material in the absence of a field. Pressures within a body are modified by the presence of the field and to apply the Fundamental Equation, one should consider that the pressure term refers to that value acting on the boundaries of the system.

Solids, however, often present a complex problem in the theory of elasticity. A detailed discussion of this case is given by Landau and Lifshitz (1960).

Several examples are presented below to illustrate the application of the Fundamental Equation including an electrostatic term.

Example 18.1

For a dielectric material present in an electrostatic field, the variation in the system volume with changes in field strength at constant temperature, pressure, and mole numbers is called *electrostriction*. Estimate the fractional change in volume of hydrogen gas at 293 K and 20.2 MN/m^2 when the electric field on the system is increased from 0 to 10^6 V/m. Experimental data show that for hydrogen under these conditions a plot of ln P versus the dielectric constant yields essentially a straight line with a slope of 24.8; that is, at 293 K and for fields between 0 and 10^6 V/m,

$$\left(\frac{\partial \ln P}{\partial D_s}\right)_{T,\mathbf{E},N} = 24.8$$

Solution

We first wish to obtain the partial derivative $(\partial \underline{V}/\partial \mathbf{E})_{T,P,N}$. The third Legendre transform of Eq. (18-23) with $y^{(0)} = \underline{U}^*$ is

$$y^{(3)} = f(T, P, \mathbf{E}, N) = \underline{U} - T\underline{S} + P\underline{V} - \underline{V}\,\mathbf{E}\mathbf{P}$$

$$dy^{(3)} = -\underline{S}\,dT + \underline{V}\,dP - \underline{V}\,\mathbf{P}d\mathbf{E} + \sum_i \mu_i\,dN_i$$

Assuming that the field is constant and taking $\mathbf{P}$ as the component of $\mathbf{P}$ parallel to $\mathbf{E}$, and using Eq. (18-8) lead to:

$$dy^{(3)} = -\underline{S}\,dT + \underline{V}\,dP - \underline{V}\varepsilon_o(D_s - 1)\,\mathbf{E}\,d\mathbf{E} + \sum_i \mu_i\,dN_i$$

and therefore

$$\left(\frac{\partial \underline{V}}{\partial \mathbf{E}}\right)_{T,P,N} = -\frac{\partial}{\partial P}[\underline{V}\varepsilon_o(D_s - 1)\mathbf{E}]_{T,\mathbf{E},N} \approx -\underline{V}\varepsilon_o\mathbf{E}\left(\frac{\partial D_s}{\partial P}\right)_{T,\mathbf{E},N} \approx \frac{-\underline{V}\varepsilon\mathbf{E}}{P}\left(\frac{\partial \ln P}{\partial D_s}\right)^{-1}_{T,\mathbf{E},N}$$

where we have assumed that $\underline{V} \neq f(P)$. Separating variables and integrating, we obtain

$$\ln\left(\frac{\underline{V}_{\mathbf{E}}}{\underline{V}_{\mathbf{E}=0}}\right) \approx -\frac{\varepsilon_o}{P(24.8)(2)}\mathbf{E}^2$$

ε_o is a constant equal to 8.85×10^{-12} As/V m. P is 20.2 MN/m^2, and $\mathbf{E} = 10^6$ V/m.

$$\frac{\underline{V}_{\mathbf{E}}}{\underline{V}_{\mathbf{E}=0}} \approx \exp\left[\frac{-8.85 \times 10^{-12}(10^6)^2}{(20.2)(10^6)(2)(24.8)}\right] \approx \exp(-9 \times 10^{-9}) = 0.9999^+$$

In this case, the contraction in volume is extremely small. Normally, for macroscopic systems, electrostriction produces negligible changes in volume, although for molecular-scale nanometer sized systems local fields can be very large and there can be significant effects. For example, applied to ions in solution, electrostriction is often invoked to account qualitatively for the decrease in total volume when ionic solutions are mixed. Furthermore, there are several practical devices that employ electrostrictive effects using strong electric fields but where only small displacements are required. For example, special air blowers for electronic equipment and positioning probes in scanning tunneling microscopes (STM) have been designed to use electrostriction effects to achieve high levels of performance.

Example 18.2

Indicate how the entropy of a dielectric changes as an electric field is applied at constant temperature, pressure, and mole numbers.

Solution

If we use the same Legendre transform developed in Example 18.1.

$$\left(\frac{\partial \underline{S}}{\partial \mathbf{E}}\right)_{T,P,N} = \frac{\partial}{\partial T}[\underline{V}\varepsilon_o(D_s - 1)\mathbf{E}]_{P,\mathbf{E},N} = \varepsilon_o \mathbf{E}\underline{V}\left(\frac{\partial D_s}{\partial T}\right)_{P,\mathbf{E},N}$$

For most materials, ε, and therefore D_s, decrease with an increase in temperature; thus, the entropy decreases as the material is polarized. Perhaps this result could have been anticipated from a consideration of the fact that, in an electric field, the molecules would tend to align their dipoles with the field and the overall randomness of the system would decrease. At constant $\underline{V}$ and N, during polarization there also would be a heat interaction of $T\,d\underline{S}$, and since $d\underline{S} < 0$, heat would be evolved.

Example 18.3

Estimate the work required to establish a field when a dielectric is present within a parallel-plate capacitor of constant volume and the field is increased from 0 to $\mathbf{E}$.

Solution

This work is given by integrating Eq. (18-10a) and multiplying by $\underline{V}$. With Eq. (18-6),

$$W_e = \frac{\underline{V}\varepsilon_o D_s}{2}\mathbf{E}^2$$

To estimate W_e, choose a basis of 1 kg-mol. For a field of 10^6 V/m, selecting as examples, air, methyl alcohol, fused silica, water, and ferroelectric titanate:

Material	Conditions	D_s	Volume, $\underline{V}$ (m^3/kg-mol)	Work (J/kg-mol)	Approximate breakdown (V/m)
Air	273 K, 1 bar	1.0006	22.4	100	3.0×10^6
Methyl alcohol (liquid)	293 K	31.2	0.0402	5.5	—
Fused silica	293 K	4	0.027	0.48	10^7
Water (liquid)	298 K, 1 bar	80	0.018	6.425	—
Ferroelectric titanate	298 K	2100	0.025	234.2	2.12×10^6

Compared to the value of RT at 300 K [$\sim 2.5\times 10^6$ J/kg-mol], these work terms are indeed insignificant; more energy could be stored at higher field strengths, but the value of 10^6 V/m used in the example is close to the breakdown strength of most materials.

Example 18.4

A flat plate capacitor is placed inside a constant volume system containing a dielectric material. As the capacitor is charged, show how the chemical potential and concentration vary with field strength.

Solution

Let us first find the derivative

$$\left(\frac{\partial \mu_i}{\partial \mathbf{E}}\right)_{T,\,\underline{V},\,N}$$

For this derivative we desire the Second Legendre transform of $\underline{U}^* = y^{(0)}$, with Eq. (8-22)

$$y^{(2)} = f(T, \mathbf{E}, \underline{V}, N)$$

$$y^{(2)} = \underline{U}^* - T\underline{S} - \underline{V}\mathbf{EP}$$

$$dy^{(2)} = -\underline{S}\,dT - \underline{V}\mathbf{P}\,d\mathbf{E} - \mathbf{PE}d\underline{V} - P^* d\underline{V} + \sum_i \mu_i\,dN_i$$

and

$$\left(\frac{\partial \mu_i}{\partial \mathbf{E}}\right)_{T,\,\underline{V},\,N} = -\underline{V}\left(\frac{\partial \mathbf{P}}{\partial N_i}\right)_{T,\,\underline{V},\,\mathbf{E},\,N_j[i]} = -\underline{V}\varepsilon_o\mathbf{E}\left(\frac{\partial D_s}{\partial N_i}\right)_{T,\,\underline{V},\,\mathbf{E},\,N_j[i]}$$

where Eq. (18-8) has been used. Defining the variation of D_s with N_i as D_s' and assuming that it is independent of field strength; then by integrating, we get

$$\mu_i(\mathbf{E}) - \mu_i\,(\mathbf{E}=0) = -\frac{\underline{V}\,\varepsilon_o\mathbf{E}^2 D_s'}{2}$$

This equation indicates the variation of μ_i with $\mathbf{E}$. As suggested by Guggenheim (1967), the value of $\mu_i(\mathbf{E}=0)$ can be related to a standard-state value at the same temperature by

$$\mu_i = \mu_i^o + RT\ln\frac{\hat{f}_i}{\hat{f}_i^o} \qquad (\mathbf{E}=0)$$

where the superscript represents some arbitrarily chosen standard state at T. Also, for simplicity, let us assume that the material is an ideal gas:

$$\hat{f}_i = y_i P = P_i = \frac{N_i RT}{\underline{V}}$$

Substituting yields

$$\mu_i(\mathbf{E}) = \mu_i^o + RT\ln\frac{N_i RT}{\hat{f}_i^o\underline{V}} - \frac{\underline{V}\,\varepsilon_o\mathbf{E}^2 D_s'}{2}$$

The dielectric constant can be expressed by the Debye equation as

$$\frac{D_s - 1}{D_s + 2} = \frac{NN_A}{3\underline{V}}\left(\alpha + \frac{d_m^2}{3\varepsilon_o kT}\right)$$

where D_s is the dielectric constant, α the molecular polarizability, d_m the dipole moment, and N_A is Avogadro's number. For ideal gases $D_s \sim 1$; thus (for a pure gas),

$$D_s' = \left(\frac{\partial D_s}{\partial N}\right)_{T,\underline{V}} = \frac{N_A}{\underline{V}}\left(\alpha + \frac{d_m^2}{3\varepsilon_o kT}\right)$$

Then,

$$\mu_i(\mathbf{E}) = \mu_i^o + RT\ln\frac{CRT}{\hat{f}_i^o} - \frac{\varepsilon_o N_A\mathbf{E}^2(\alpha + d_m^2/3\varepsilon_o kT)}{2}$$

where C is the concentration. Assume now that this gas in the field $\mathbf{E}$ is in equilibrium with more pure gas outside the field at temperature T. For the external gas, $\mathbf{E}=0$ and we could again write a similar expression for μ_i:

$$\mu_i^{outside} = \mu_i^o + RT\ln\frac{C^{outside}RT}{\hat{f}_i}$$

At equilibrium, since $\mu_i^{outside} = \mu_i^{inside}$,

$$RT \ln \frac{C^{inside}}{C^{outside}} = \frac{\varepsilon_o N_A \mathbf{E}^2(\alpha + d_m^2/3\varepsilon_o kT)}{2}$$

Suppose that our system consisted of HCl. $\alpha \approx 2.6 \times 10^{-30}$ m^3/molecule and $d_m \approx 1$ debye $= 0.3335 \times 10^{-29}$ C m/molecule. k is 1.38×10^{-23} J/molecule K and let $T = 300$ K. For the other terms, $\varepsilon_o = 8.854 \times 10^{-12}$As/Vm, $N_A = 6.022 \times 10^{23}$ molecules/mol, and assume that $\mathbf{E} = 10^7$ V/m. Then,

$$RT \ln \frac{C^{inside}}{C^{outside}} = 8.854 \times 10^{-12} \; (6.022 \times 10^{23})(10^{14})$$

$$\times \left[2.6 \times 10^{-30} + [(0.3335)^2(10^{58})(8.854 \times 10^{-12})(1.38)(10^{-23})(300)(3)]^{-1}\right] 0.5$$

$$= (53.3)(10^{25})(2.6 + 101)(10^{-30})(0.5) = 2.8 \times 10^{-2} \text{ J/mol}$$

$$\frac{C^{inside}}{C^{outside}} = \exp\left[\frac{2.8 \times 10^{-2}}{(8.314)(300)}\right] = \exp\,(1.1 \times 10^{-5})$$

The enhancement is obviously very small. Only at much higher field strengths would one be able to show an appreciable difference between the concentrations inside and outside the field.

It is perhaps clear from the examples shown above that in most cases electric fields affect the thermodynamic properties of a system very little. To obtain significant effects, extremely large field strengths are required, but before they may be attained, breakdown normally occurs. Nevertheless, in the immediate vicinity of ions in a solution, very high field strengths do exist and significantly affect certain molecular properties. For further discussion of electrostatic effects, see Section 12.3 on Debye-Huckel theory.

18.3 Magnetic Systems

We can develop the thermodynamics of systems in magnetic fields in much the same way as we did for systems in electrostatic fields. In this case. however, we must deal with the magnetic induction, **B**, the magnetic field strength, **H**, and the magnetization (or magnetic moment per unit volume or magnetic polarization vector), **M**. They are related as shown in Eq. (18-9). **M** is zero unless there is material from the system of interest in the field.

As before, the work done by the system is given by Eq. (18-10b), and the Fundamental Equation is obtained from Eq. (18-16) by setting **E** to 0. Thus,

$$d\underline{U} = T\,d\underline{S} - P\,d\underline{V} + \mathbf{H}\,d(\mu_{\mathbf{M}^o}(\mathbf{H} + \mathbf{M})\underline{V}) + \sum_i \mu_i\,dN_i \tag{18-28}$$

The free space contribution in the magnetic work term is normally separated and combined with $\underline{U}$ to define a modified internal energy $\underline{U}^*$ and pressure P^* using Eqs. (18-18) and (18-19):

$$\underline{U}^* = \underline{U} - \frac{\mu_{\mathbf{M}^o}\mathbf{H}^2\underline{V}}{2} \tag{18-29}$$

$$P^* = P - \frac{\mu_{\mathbf{M}^o}\mathbf{H}^2}{2}$$

so that

$$d\underline{U}^* = T\,d\underline{S} - P^* d\underline{V} + \mu_{\mathbf{M}^o}\mathbf{H}\,d(\mathbf{M}\underline{V}) + \sum_i \mu_i\,dN_i \tag{18-30}$$

This Fundamental Equation is analogous to Eq. (18-22) for electrostatic systems and as such, the Euler form, the Gibbs-Duhem, and various Legendre transforms are readily obtained. To illustrate the use of such forms, see the examples that follow.

Example 18.5

Holding the temperature, pressure, and mole numbers constant, how is the entropy of a material affected by changes in the magnetic field strength?

Solution

We shall want to find a Legendre transform for the variables (T, P^*, $\mathbf{H}$, N). Thus,

$$y^{(3)} = \underline{U}^* - T\underline{S} + P^*\underline{V} - \mu_{\mathbf{M}^o}\underline{V}\mathbf{H}\mathbf{M}$$

$$dy^{(3)} = -\underline{S}\,dT + \underline{V}\,dP^* - \mu_{\mathbf{M}^o}\,\underline{V}\,\mathbf{M}\,d\mathbf{H} + \sum_i \mu_i\,dN_i$$

and

$$\left(\frac{\partial \underline{S}}{\partial \mathbf{H}}\right)_{T,P,N} = \mu_{\mathbf{M}^o}\,\underline{V}\left(\frac{\partial \mathbf{M}}{\partial T}\right)_{P,\mathbf{H},N} = \mu_{\mathbf{M}^o}\,\underline{V}\,\mathbf{M}\,\delta_I$$

where δ_I is equal to $(\partial \ln \mathbf{M}/\partial T)$ and is called the thermal magnetization coefficient. Values of δ_I are not often known. For water, Camp and Johnson (1967) report that $\delta_I \approx 5 \times 10^{-4}$ K^{-1}. For water, $\mu_{\mathbf{M}} \approx 1.3 \times 10^{-6}$ Vs/m A and $\chi_m = -9.06 \times 10^{-6}$. Then, for a value of $\mathbf{B}$ = 1000 gauss = 0.1 weber/m^2 = 0.1 Vs/m^2 = 0.1 tesla from Eq. (18-4),

$$\mathbf{H} = \frac{\mathbf{B}}{\mu_{\mathbf{M}}} = \frac{0.1}{1.3 \times 10^{-6}} = 7.7 \times 10^4 \text{ A/m}$$

$$\mathbf{M} = \chi_m \mathbf{H} = (-9.06 \times 10^{-6})7.7 \times 10^4 \approx -0.7 \text{ A/m}$$

Let $\underline{V} = 1$ m^3, so that

$$\mathbf{M}\underline{V} = -0.7 \text{ Am}^2$$

Then

$$\left(\frac{\partial S}{\partial \mathbf{H}}\right)_{T,P,N} = (4\pi \times 10^{-7})(-0.7)(5 \times 10^{-4}) \approx -3 \times 10^{-10} \text{ J/(A/m) K}$$

For 1 m^3 of water, with an increase in **B** from 0 to 1000 gauss, the entropy decreases, but very slightly.

Example 18.6

For a pure component, at constant pressure, show how the boiling point varies with the magnetic field strength, **H**.

Solution

For phase equilibrium, the temperature and chemical potential are the same in the vapor and liquid. Without a magnetic field effect, the pressures would also be equal, but with polarization effects present this is not the case. Rosensweig (1989) has worked out a number of relevant cases. In this problem, we assume a 2-layer system of vapor and liquid at equilibrium with **B** oriented perpendicular to the layers. The Gibbs-Duhem equation that is obtained from our expression for $d\underline{U}^*$ in Eq. (18-30) provides a starting point, for a single pure component j.

$$\underline{S}dT - \underline{V}dP^* + \mu_{\mathbf{M}^o}\, \mathbf{M}\underline{V}\, d\mathbf{H} + N_j d\mu_j = 0$$

and we can express the chemical potential of pure component j in terms of

$$d\mu_j = -\overline{S}_j dT + \overline{V}_j dP^* - \mu_{\mathbf{M}^o}\, \overline{(\mathbf{M}V)}_j d\mathbf{H} \tag{18-31}$$

where the overbar refers to modified partial quantities. For example, in this case

$$\overline{(\mathbf{M}\underline{V})}_j = \left(\frac{\partial(\mathbf{M}\underline{V})}{\partial(N_j m_j)}\right)_{T,P^*,N_i[j]} = \frac{(\mathbf{M}_j\underline{V}_j)}{N_j m_j} = \frac{(\mathbf{M}_j V_j)}{m_j}$$

for pure compnent j where m_j is the molecular weight of j. Thus, we are using a mass rather than mole basis to define $\overline{S}_j$, $\overline{V}_j$, and $\overline{(\mathbf{M}V)}_j$. At equilibrium we can show that

$$T^{(V)} = T^{(L)}$$

$$\mu_j^{(V)} = \mu_j^{(L)} \quad or \quad d\mu_j^{(V)} = d\mu_j^{(L)}$$

$$P^{(V)} - (\mathbf{HB})^{(V)} = P^{(L)} - (\mathbf{HB})^{(L)}$$

by following an approach similar to that used to establish phase equilibrium criteria in the absence of magnetic fields given in Section 6.5. It is interesting to note that the usual mechanical equilibrium condition of uniform pressure across phase boundaries has been replaced by a new criteria that says P varies between phases depending on the magnitude of the product of **HB**.

With **B** oriented perpendicular to the layered vapor and liquid phases, the magnetic induction is continuous across the interface:

$$B^{(V)} = B^{(L)} = B_{ext}$$

For a constant external environmental pressure P_{ext}, we can express the modified pressure, in general, as

$$P^* = P - \frac{\mu_{\mathbf{M}}^o \mathbf{H}^2}{2}$$

For convenience, we can define an equivalent external magnetic field strength referenced to free space as

$$\mathbf{B}_{ext} \equiv \mu_{\mathbf{M}}^o \mathbf{H}_{ext}$$

so that

$$P_{ext}^* = P_{ext} - \frac{\mu_{\mathbf{M}}^o \mathbf{H}_{ext}^2}{2} = \text{constant}$$

and $dP_{ext}^* = 0$. Using Eq. (18-31) and the equality of chemical potentials above

$$\left(\frac{\partial T}{\partial \mathbf{H}_{ext}}\right)_{P_{ext}^*,\, L-V} = \frac{-\left(\overline{\mathbf{M}_j V_j}^{(V)} - \overline{\mathbf{M}_j V_j}^{(L)}\right)\mu_{\mathbf{M}}^o}{m_j(\overline{S}_j^{(V)} - \overline{S}_j^{(L)})}$$

but since $\mathbf{M}_j^{(V)} = \chi_j^{(V)}\mathbf{H}_{ext}$, $\mathbf{M}_j^{(L)} = \chi_j^{(L)}\mathbf{H}_{ext}$ and

$$\overline{S}_j^{(V)} - \overline{S}_j^{(L)} = S_j^{(V)} - S_j^{(L)} = \Delta S_{vap} = \frac{\Delta H_{vap}}{T}$$

$$\left(\frac{\partial T}{\partial \mathbf{H}_{ext}}\right)_{P_{ext}^*,\, L-V} = \frac{-\mu_{\mathbf{M}}^o \mathbf{H}_{ext}\, T((\chi_j V_j)^{(V)} - (\chi_j V_j)^{(L)})}{m_j\, \Delta H_{vap}}$$

If we assume that $(\chi_j V_j)^{(V)}$, $(\chi_j V_j)^{(L)}$ and ΔH_{vap} are relatively constant over the range of $\mathbf{H}$ from 0 to $\mathbf{H}_{ext}$, we can separate variables and integrate.

$$\ln\left[\frac{T(at\ \mathbf{H}_{ext})}{T(at\ \mathbf{H}_{ext}=0)}\right] = \ln\left[\frac{T_{\mathbf{H}}}{T_o}\right] = \frac{-\mu_{\mathbf{M}}^o \mathbf{H}_{ext}^2}{2\,\Delta H_{vap}}\left[\frac{(\chi_j V_j)^{(V)}}{m_j} - \frac{(\chi_j V_j)^{(L)}}{m_j}\right]$$

For oxygen we can illustrate the magnitude of the effect of $\mathbf{H}_{ext}$ on the boiling point at 1 bar external pressure. Using data provided by Rosensweig (1989, p 407).

$T_o = 90$ K at 1 bar, $\mathbf{H} = 0$

$\Delta H_{vap} = 6640$ J/mol $= 212.5$ kJ/kg

$\frac{1}{m_j}(\chi_j V_j)^{(V)} = 4.345 \times 10^{-6}\ \text{m}^3/\text{kg}$

$\frac{1}{m_j}(\chi_j V_j)^{(L)} = 2.92 \times 10^{-6}\ \text{m}^3/\text{kg}$

With $\mu_{\mathbf{M}}^o = 4\pi \times 10^{-7}$ Vs/Am, then

$$\ln \frac{T_{\mathbf{H}}}{T_o} = -4.2134 \times 10^{-18} \, (\mathbf{H}_{ext})^2$$

where $\mathbf{H}_{ext}$ is in A/m. In 1995, the world's strongest continuous magnetic field ($\mathbf{B}_{ext}$) is about 37 tesla (or 370,000 gauss) which is over 500,000 times the earth's natural field of about 0.7 gauss. Given this, let's examine a range of 1 to 50 tesla for $\mathbf{B}_{ext}$. With $\mathbf{H}_{ext} = \mathbf{B}_{ext}/\mu_{\mathbf{M}}^o$, we get

$\mathbf{B}_{ext}$ tesla	$\mathbf{H}_{ext}$ A/m	$\ln \frac{T_{\mathbf{H}}}{T_o}$	$\Delta T = T_{\mathbf{H}} - T_o$ K
1	7.95×10^5	-2.663×10^{-6}	-2.4×10^{-4}
37	2.944×10^7	-3.651×10^{-3}	-0.328
50	3.9789×10^7	-6.671×10^{-3}	-0.598

So we see that the effect is still not very large even with extremely strong magnetic fields! See also Meachin and Biddulph (1978) who show theoretically and experimentally that there is essentially no effect of magnetic fields up to 10^7 A/m on the vapor pressures of oxygen, nitrogen, and argon. A similar relationship can be derived for estimating the effect of electric fields on boiling point [see Lyon (1961) and Sharma (1971)].

It is clear from the examples shown that electromagnetic fields affect thermodynamic properties only when field strengths are extremely high for most materials. But for certain rare earth elements, magnetostriciton effects can be larger by one to two orders of magnitude. Materials containing these rare earth components have been used in a number of important applications, including echo ranging and ultrasonic sound generation. Furthermore, high density magnetic recording now uses pickup heads based on a large magnetostriction effect.

Another extremely important application employs so-called ferrofluids to provide high performance rotating shaft seals and active cooling in special environments. Ferrofluids contain nanometer sized (< 10 *nm*) magnetite (Fe_3O_4) particles that are stabilized by a surfactant in a liquid medium. Pressure in a magnetic ferrofluid increases by up to one bar per stage. In commercial rotary shaft seals this magnetic-field effect allows a ferrofluid to reach the levels of dimensional tolerance required for high speed computer hard drive applications where small space clearances are needed to exclude contamination and low friction levels are needed to keep power consumption levels within bounds. Ferrofluids are also currently used to cool loudspeaker devices.

In addition, the magnetocaloric effect is utilized for cooling below 1 K [see, for example, Giauque *et al.* (1927, 1933) and Debye (1926)]. Referring to Example 18.5, we showed that

$$\left(\frac{\partial \underline{S}}{\partial \mathbf{H}}\right)_{T,P,N} = \mu_{\mathbf{M}^o}\frac{V}{-}\left(\frac{\partial \mathbf{M}}{\partial T}\right)_{P,\mathbf{H},N} \tag{18-32}$$

From Eq. (18-9), $\mathbf{M} = \chi_m \mathbf{H}$ and at constant N, Eq. (18-32) becomes

$$\left(\frac{\partial S}{\partial (\mathbf{H}^2)}\right)_{T,P,N} = \mu_{\mathbf{M}^o}\frac{V}{2}\left(\frac{\partial \chi_m}{\partial T}\right)_{P,\mathbf{H},N}$$

For many paramagnetic materials at low temperatures the relation between χ_m and T is given by Curie's Law:

$$\chi_m = \frac{C}{T}$$

where C is a constant. Therefore,

$$\left(\frac{\partial S}{\partial (\mathbf{H}^2)}\right)_{T,P,N} = -\frac{\mu_{\mathbf{M}^o} C V}{2T^2}$$

At low temperatures, the entropy will decrease considerably when the material is isothermally magnetized. The next step is to isolate thermally the sample and remove it from the field. A temperature drop occurs on demagnetization as the interested reader can verify by determining the sign of $(\partial T/\partial \mathbf{H})_{\underline{S},P,N}$.

In certain chemical reacting systems, measurable magnetic field effects have been observed. For example, emulsion polymerization studies reported by Turro (1983) indicated that fields of 0.1 tesla have an effect on polymerization rates.

18.4 Thermodynamics of Systems under Stress

It is rare for chemical engineers to become involved with thermodynamic analyses of systems that are in tension or compression. Significant problems arising in this area usually fall heir to applied mechanicists, who are interested in the behavior of materials under static or dynamic loads. The materials involved are solid instead of fluid since, for the latter, the consequences of the imposition of a force can be treated by pressure-volume terms introduced earlier in the book. In fact, solids are almost always considered to be elastic (i.e., although they may be stressed and deformed, the total volume is invariant). In the thermodynamics of such systems, the system volume is then assumed constant and no $P\underline{V}$ terms are used.

In the discussion of stressed systems, we refer to the strain resulting from an applied stress. These terms are readily visualized in a qualitative way since it is common experience to apply a force to, for example, rubber and expect a change in the dimensions of the specimen. A careful definition of terms is, however, necessary before any quantitative relations can be meaningful. Such definitions will be our first task, although the ultimate objective is to formulate a Fundamental Equation for stressed systems.

When we use the term "stress," we obviously imply a force/area. Stress is, therefore, a vector quantity with both a magnitude and direction. Similarly, for strain we describe the movement of an element of the system in a particular direction. Consider Figure 18.1. A bar of an elastic material is fixed to a rigid plate and subjected to a tensile force, F_x. The sides are unconstrained. Because of F_x, the bar is elongated by Δx and is in tension. The stress in this case, $F_x/\underline{a}$, produces a strain, $\Delta x/x$. Also, as shown, there are induced strains $\Delta y/y$ and $\Delta z/z$ as well as induced stresses in the y and z directions. Thus, the situation is complicated even in this relatively simple example. For any body under stress, we must consider the strain in the three coordinate directions. Also, one should allow for *shear* (i.e., a stress so directed that there is a rotation or a twisting of the system).

It is not within the scope of this book to develop in detail three-dimensional elastic theory as applications rarely occur in chemical engineering. We will limit our treatment to a one-dimensional case in which there is but one applied stress or force. Thus we return to Figure 18.1. Here the work done on the system is $F_x\Delta x$. The fact that there is a finite Δy or Δz is immaterial since we are only interested in force-distance interactions operating over the system boundaries and F_x is the only force to fit this criterion. In the general case with both linear movement and rotation we should have to consider that the work done on the system consisted of six terms. Three terms would involve the linear movement in the three rectangular coordinates due to stresses in these directions. There would, however, be three other terms to account for motion in, for example, the y and z direction from an x-directed stress. There are not nine terms because of the conservation of angular momentum (i.e., the x-movement due to a y-stress is equal to the y-movement due to an x-stress). See Callen (1960) for further discussion.

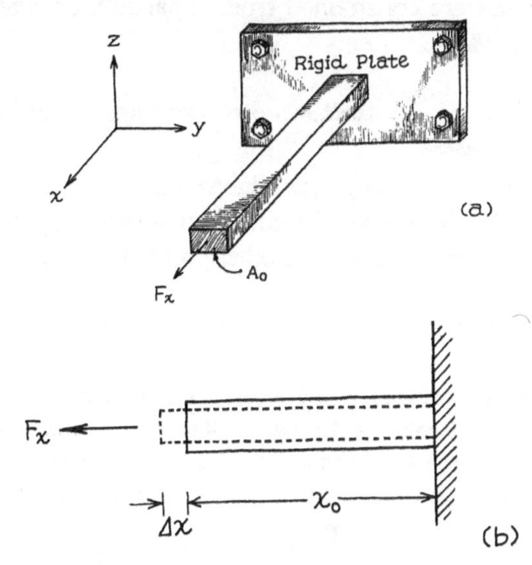

Figure 18.1 Extension under stress.

As used in a thermodynamics study, the designation of work as simply $F_x\,dx$ is inconvenient since the terms do not blend neatly with other extensive properties in a Fundamental Equation. Let us, therefore, multiply and divide by the total system volume, $\underline{V}$,

$$\delta W = \frac{\underline{V}}{\underline{V}} F_x \, dx = \frac{\underline{V} F_x \, dx}{\underline{a} x} = \underline{V} \sigma_x \, d\Omega \tag{18-33}$$

where we have defined the stress as $F_x/\underline{a} \equiv \sigma_x$ and the differential linear strain as $d\Omega = dx/x$ as the fractional extension (or contraction) in the same direction as σ_x. This work is then an extensive property. The Fundamental Equation in this case (with $\underline{V}$ = constant) is then:

$$d\underline{U} = T \, d\underline{S} + \sigma_x \underline{V} \, d\Omega + \sum_i \mu_i \, dN_i \tag{18-34}$$

Once we have reached this point, the Euler relation, the Gibbs-Duhem equation, and many Legendre transforms can be written immediately. The Euler form is, for example,

$$\underline{U} = T\underline{S} + \sigma_x \underline{V}\Omega + \sum_i \mu_i N_i \tag{18-35}$$

Various coefficients may also be defined with the σ_x–Ω notation. When a system is acted upon by a single force (as in Figure 18.1), the stress-strain ratio is called *Young's modulus, Y*:

$$Y \equiv \left(\frac{\partial \sigma_x}{\partial \Omega}\right)_T \tag{18-36}$$

If, however, the sides are constrained from expanding or contracting in the other directions, then the derivative

$$\left(\frac{\partial \sigma_x}{\partial \Omega}\right)_{T,\ restrained} \equiv \textit{isothermal elastic stiffness coefficient} \tag{18-37}$$

Many other solid properties can be introduced, including heat capacities at constant length (or stress), etc. We illustrate one application of thermodynamics to such systems in the example below and provide several other problems at the end of the chapter.

Example 18.7

An adiabatic, elastic deformation of a body is normally accompanied by a change in temperature. This phenomenon is known as the *thermoelastic effect*. Estimate the magnitude of this effect for a specimen of Armco iron that is stressed from 0 to 103 MPa at 773 K. Under these conditions, the coefficient of linear thermal expansion is

$$\alpha_{\sigma_x} = \frac{1}{L_o}\left(\frac{\partial L_o}{\partial T}\right)_{\sigma_x} = 16.8 \times 10^{-6} \text{ K}^{-1}$$

and the heat capacity at constant stress is

$$C_{\sigma_x} = T\left(\frac{\partial S}{\partial T}\right)_{\sigma_x} \approx 9.1 \text{ cal/mol K}$$

Solution

We desire the partial derivative $(\partial T/\partial \sigma_x)_{\underline{S}, N}$. Thus, we employ a Legendre transform, $y^{(1)} = f(\sigma_x, \underline{S}, N)$. From Eq. (18-34).

$$dy^{(1)} = -\Omega \underline{V}\, d\sigma_x + T\, d\underline{S} + \sum_i \mu_i dN_i$$

$$\left(\frac{\partial T}{\partial \sigma_x}\right)_{\underline{S}, N} = -\underline{V}\left(\frac{\partial \Omega}{\partial \underline{S}}\right)_{\sigma_x, N} = \frac{\underline{V}(\partial \Omega/\partial T)_{\sigma_x, N}}{(\partial \underline{S}/\partial T)_{\sigma_x, N}}$$

From the definition of Ω,

$$\alpha_{\sigma_x} = \left(\frac{\partial \Omega}{\partial T}\right)_{\sigma_x, N}$$

Thus,

$$\int_{T_1}^{T_2} \frac{dT}{T} = -\int_{\sigma_{x_1}}^{\sigma_{x_2}} \frac{\alpha_{\sigma_x}}{C_{\sigma_x} N/\underline{V}}\, d\sigma_x$$

The term $(C_{\sigma_x} N/\underline{V})$ is simply the heat capacity at constant stress on a unit volume basis. Assuming that both it and α_{σ_x} are not strong functions of σ_x (see below), then

$$\ln \frac{T_2}{T_1} = -\frac{\alpha_{\sigma_x}(\Delta \sigma_x)}{C_{\sigma_x} N/\underline{V}}$$

For small differences between T_2 and T_1,

$$\ln \frac{T_2}{T_1} \approx \frac{T_2 - T_1}{T} = \frac{\Delta T}{T}$$

thus

$$\Delta T \approx -\frac{\alpha_{\sigma_x} T(\Delta \sigma_x)}{C_{\sigma_x} N/\underline{V}}$$

α_{σ_x} has been given a $16.8 \times 10^6\ K^{-1}$.

$$\frac{C_{\sigma_x} N}{\underline{V}} = \frac{(9.1)(4.19)}{(55.8)}\, 7.6 \times 10^6 = 5.18\ MJ/m^3$$

where the atomic weight was chosen as 55.8 and the density as 7.6 g/cm^3. Thus,

$$\Delta T = -\frac{16.8 \times 10^{-6}(103)(773)}{5.18} = -0.26\ K$$

The expected temperature drop would be about 0.26 K for this adiabatic stress. This value is in excellent agreement with the value of 0.25 K reported by Rocca and Bever (1950).

These same authors discuss the variation of $\alpha_{\sigma_x}/(C_{\sigma_x}N/\underline{V})$ with σ_x and show that for stresses of the order of 100 MPa, this group varies only 2 to 3%.

It is also interesting to note that a *temperature drop must occur upon stretching* if $\alpha_{\sigma_x} > 0$. For some substances in which there is a contraction upon heating, the ***thermoelastic effect*** will be positive (i.e., the material will heat when adiabatically stretched).

18.5 Systems in Body-Force Fields or under Acceleration Forces

We have now treated several systems in which there were work terms of other than the $Pd\underline{V}$ type. Once we expressed this work (in an extensive manner), we simply included it in the general Fundamental Equation and from there proceeded to derive the desired Legendre transforms and partial derivatives. Implicit in this treatment was the supposition that the internal energy $\underline{U}$ was a function of variables other than $\underline{S}$, $\underline{V}$, $N_1,..., N_n$ (e.g., for systems in electrostatic fields $\underline{U} = \underline{U}(\underline{S}, \underline{V}, \mathbf{P}, N_1,..., N_n)$, where $\mathbf{P}$ is the component of the electric polarization parallel to the electric field $\mathbf{E}$).

It was then a simple task to develop any desired partial derivative from the appropriate Fundamental Equation. We also indicated in Section 18.3 that it was logical to expand the definitions of $\underline{H}$, $\underline{A}$, $\underline{G}$, etc., to include new system variables. Some authors, however, retain the earlier definitions for these well-known properties and define new properties if electrostatic, electromagnetic, etc., terms are to be included. Either approach is quite satisfactory, although the duality is often confusing. It is preferable to avoid definitions if possible and only consider the necessary independent variables. A Legendre transform is then employed to obtain the desired new Fundamental Equation.

Except for learning new notations introduced in Sections 18.2 through 18.4, there are no new principles introduced. When, however, one wishes to study systems in potential (body-force) fields or as acted upon to produce acceleration (or deceleration), then a somewhat different approach is taken as introduced earlier in Section 3.8. As we will show below, the work can readily be expressed in terms appropriate for these systems, but it is not immediately obvious how to use these terms to develop the Fundamental Equation. We must carefully consider the question of whether the internal energy of a system depends on the position of the system in a potential field or on its velocity. Although we could, if we desired, extend the definition of $\underline{U}$ to include these variables, the results would run counter to our intuitive desire to preserve the character of $\underline{U}$ to be independent of position and velocity as much as possible. True, if we have a nonhomogeneous electromagnetic field, $\underline{U}$ is already a function of position, but we shall overlook such inconsistencies.

Although this is a text on classical thermodynamics, and little or no note has been taken of the constituent molecules in a system, we must admit that the desire to eliminate potential and acceleration fields from the definition of $\underline{U}$ relates to the fact that we would like to feel that those variables that affect $\underline{U}$ also modify some molecular properties. Perhaps the molecular velocity or spacing, or rotational frequency, etc., might be affected. These properties cannot, however, be related neatly to the overall system velocity or, for example, to the system's position in a potential field. These points are discussed further in Chapter 10.

The net result of these admittedly intuitive feelings is to arrive at the decision to express the total energy of a system as a sum of the familiar internal energy and other energies that may be associated with potential or acceleration fields as we did earlier in Chapter 3 (see Eq. (3-61));

$$\text{total system energy} = (\text{internal} + \text{potential} + \text{kinetic})\ \text{energy}$$

or

$$\underline{E} = \underline{U} + \underline{PE} + \underline{KE} \tag{18-38}$$

The $\underline{PE}$ term is found by considering the work effects necessary to move a system in a force field. If a system is moved a distance $d\mathbf{L}$ against or with a collinear force $\mathbf{F}$, then

$$d(\underline{PE}) = +\,\delta W = -\,\mathbf{F}\cdot d\mathbf{L} = -\,m\,\mathbf{a}\cdot d\mathbf{L} = m\,d\Phi \tag{18-39}$$

where the potential Φ is defined as

$$d\Phi \equiv -\,\mathbf{a}\cdot d\mathbf{L} \tag{18-40}$$

The potential energy of the system is then

$$\underline{PE} = m\Phi \tag{18-41}$$

since the mass m was considered constant. To illustrate, consider a system in a gravitational field with an acceleration $\mathbf{g}$. Measure $\mathbf{L}$ in the up direction and $\mathbf{g}$ in the down direction. Then, since $\mathbf{L}$ and $\mathbf{g}$ are collinear vectors but with opposite directions,

$$d\Phi = g\,dL \tag{18-42}$$

$$\underline{PE} = m\int d\Phi = m\Phi = m\int g\,dL \tag{18-43}$$

No limits have been placed on the integration since one may arbitrarily define the potential energy to be zero at some reference value of L and then relate all other values to this base state.

As another example, consider a solid-body rotation. Let $\mathbf{L} = \mathbf{r} =$ the radius measured *outward* from the center of rotation. Here the centrifugal acceleration is $\omega^2\mathbf{r}$ and is also directed radially *outward*. Then

$$d\Phi = -\,\omega^2\mathbf{r}\cdot d\mathbf{r} \tag{18-44}$$

and

$$\underline{PE} = m\int d\Phi = m\Phi = -\,m\int \omega^2 r\,dr \tag{18-45}$$

Example 18.8

What is the potential of a synchronous satellite at a distance r from the center of the earth and at an altitude where $g = g_r$?

Solution

From the definitions above, $d\Phi = (g_r - \omega^2 r)dr$

For synchronous operation, $d\Phi = 0$ or $g_r = \omega^2 r$

The kinetic energy of a system is, of course,

$$\underline{\mathrm{KE}} = \tfrac{1}{2} m v^2 \tag{18-46}$$

and thus

$$\underline{E} = \underline{U} + \underline{\mathrm{PE}} + \underline{\mathrm{KE}} = \underline{U} + m\Phi + \tfrac{1}{2} m v^2 \tag{18-47}$$

To express the differential form for $\underline{E}$, and still employ the customary expansion for $\underline{U}$, we see immediately that there must be terms on the right-hand side involving m, Φ, and v. If we use Eq. (18-1) with no electromagnetic or stress/strain related work terms, for simplicity, with the molecular weight of component j being m_j,

$$d\underline{E} = d\underline{U} + d(\underline{\mathrm{PE}}) + d(\underline{\mathrm{KE}})$$

or

$$d\underline{E} = T\, d\underline{S} - P\, d\underline{V} + m\, d\Phi + m\, d\frac{v^2}{2} + \sum_j \left(\mu_j + m_j\Phi + \frac{m_j v^2}{2}\right) dN_j \tag{18-48}$$

Let us now examine these terms more carefully. The $m\, d\Phi$ and $m\, d(v^2/2)$ terms indicate work interactions between the system and the environment in undergoing a change in Φ or v. For example, let the process be the fall of the system from a higher to a lower elevation in a gravitational field and, in so falling, a weight is raised in the environment. Now $d\Phi = g\, dL < 0$ (since L is decreasing) and thus the work done by the system is negative as indicated by the rise of the weight in the surroundings.

Within the summation we also find terms involving Φ and v. The interpretation is simple. As the mole numbers change because of mass addition or loss, there is introduced or removed potential and kinetic energy from the system. Of course, since Eq. (18-48) represents a quasi-static process, all such interchange must be such that the specific potential and kinetic energies of the mass entering and leaving the system must be equal to that present in the system. This is rather obvious since only one Φ or v is noted. The analogy here is between the Fundamental Equation $\underline{U} = \underline{U}(\underline{S}, \underline{V}, N)$ and the open-system, combined First and Second Laws. We first write a completely general open-system First Law relation for this case by inspection:

$$d\underline{E} = d\left(\underline{U} + \Phi m + \frac{m v^2}{2}\right) = \delta Q_\sigma + \delta W_\sigma + \left[\sum_{j=1}^{n} \left(H_j + \Phi\, m_j + \frac{m_j v^2}{2}\right) \delta n_j\right]_{in} - \left[\sum_{j=1}^{n} \left(H_j + \Phi\, m_j + \frac{m_j v^2}{2}\right) \delta n_j\right]_{out} \tag{18-49}$$

$$m = \sum_{j=1}^{n} m_j N_j \quad \text{and} \quad \underline{U} = NU$$

where the terms N, N_j, and n_j (in or out) represent moles and m is the total system mass. Then it is obvious that without the potential and kinetic energy terms, Eq. (18-49) reduces to the general open-system First Law expression, Eq. (3-60). With $\Phi = gz$ we get the equivalent of Eq. (3-63) on a mass basis. To show the connection between Eqs. (18-48) and (18-49) in another way, assume that the interchange of mass, heat, and work with the environment is reversible,

$$H_{j\,in} = H_{j\,out} = H_j \text{ (system)}$$

$$S_{j\,in} = S_{j\,out} = S_j \text{ (system)}$$

In addition, we are also assuming that the interchange of potential, kinetic, and internal energy *within the system* is completely reversible as well. In other words, there are no hydrodynamic dissipation effects even when gradients of velocity (v) and Φ exist inside the σ-boundary of the system. All of this is equivalent to saying that there is no entropy generation within the system.

Then we can develop an equivalent combined First and Second Law expression by incorporating an entropy balance for this reversible system (see Section 4.7):

$$d\underline{S} = \frac{\delta Q_r}{T} + \left[\sum_{j=1}^{n} S_j\, \delta n_j\right]_{in} - \left[\sum_{j=1}^{n} S_j\, \delta n_j\right]_{out} \tag{18-50}$$

and

$$\delta W_\sigma = -P\,d\underline{V} + m\,d\Phi + mv\,dv \tag{18-51}$$

$$d\underline{E} = T\,d\underline{S} - P\,d\underline{V} + m\,d\Phi + mv\,dv + \left[\sum_{j=1}^{n}\left(\mu_j + m_j\,\Phi + \frac{m_j v^2}{2}\right)\delta n_j\right]_{in}$$

$$- \left[\sum_{j=1}^{n}\left(\mu_j + m_j\,\Phi + \frac{m_j v^2}{2}\right)\delta n_j\right]_{out} \tag{18-52}$$

For the completely reversible case, all intensive properties (T, P, μ_j) are constant, Φ and v are defined locally within the system as well as at the boundaries at particular coordinates (x, y, and z), and from a mole balance $(\delta n_j)_{in} - (\delta n_j)_{out} = dN_j$, thus by substitution, Eq. (18-48) is obtained.

Applying Euler's theorem to Eq. (18-48) and noting that Φ and v are invariant with changes in mass, we obtain

$$\underline{E} = T\underline{S} - P\underline{V} + \sum_{j=1}^{n}\left(\mu_j + m_j\Phi + \frac{m_j v^2}{2}\right)N_j$$

or

$$\underline{E} = \underline{U} + \sum_{j=1}^{n} \left(m_j \Phi + \frac{m_j v^2}{2} \right) N_j \tag{18-53}$$

Since $\underline{E} = \underline{E}(\underline{S}, \underline{V}, \Phi, v, N_j)$, it is now possible to employ Legendre transforms as developed in Chapter 5 to obtain other fundamental representations and relationships between partial derivatives.

Example 18.9

Obtain the transforms (a) $f(T, P, \Phi, v, N_1, \ldots, N_n)$ and (b) $f(T, P, m, v, N_1, \ldots, N_n)$ and show how the pressure varies with potential at constant T and velocity in a closed system.

Solution

(a) $f(T, P, \Phi, v, N_1, \ldots, N_n)$

$$y^{(2)} = \underline{E} - T\underline{S} + P\underline{V}$$

$$dy^{(2)} = -\underline{S}\, dT + \underline{V}\, dP + m\, d\Phi + mv dv + \sum_{j=1}^{n} \left(\mu_j + m_j \Phi + \frac{m_j v^2}{2} \right) dN_j$$

(b) $f(T, P, m, v, N_1, \ldots, N_n)$

$$y^{(3)} = \underline{E} - T\underline{S} + P\underline{V} - m\Phi$$

$$dy^{(3)} = -\underline{S}\, dT + \underline{V}\, dP - \Phi\, dm + mv dv + \sum_{j=1}^{n} \left(\mu_j + m_j \Phi + \frac{m_j v^2}{2} \right) dN_j$$

From this last relation,

$$\left(\frac{\partial P}{\partial \Phi} \right)_{T, v, N, m} = -\left(\frac{\partial m}{\partial \underline{V}} \right)_{T, P, v, N} = -\rho$$

where ρ is the mass density. This relation is, of course, the usual equation to determine a hydrostatic pressure when $\Phi = gz$.

The final point of interest in this section involves the criteria of equilibrium in body-force fields. We develop such criteria only for a system in a potential energy field because no useful criteria are found for systems with kinetic energy (i.e., it is readily shown that systems with the lowest velocities are most stable).

For a system in a potential energy field, one criterion often given is that

$$dP = -\rho\, d\Phi \tag{18-54}$$

but this was already derived in Example 18.9. Furthermore, it may easily be shown that, at equilibrium, there can be no gradients of temperature for a system in a potential field.

The more interesting criteria result from an examination of the variation of chemical potential with Φ.

Consider Figure 18.2. We show a system at equilibrium in a potential field. That is, there is a measurable difference in Φ between points A and B. (For example, one might consider the mixture to be in a vertical pipe in a gravitational field with A and B representing two heights.) The temperature of the system is uniform although the pressure varies with Φ [see Eq. (18-54)]. At A and B we have semipermeable membranes that allow component j to pass freely into side tube C. At A the chemical potential of j is equal on both sides of the membrane; a similar statement may be made at B. Thus,

$$\mu_{j,mix}^{A} = \mu_{j,pure}^{A}$$

$$\mu_{j,mix}^{B} = \mu_{j,pure}^{B}$$

Subtracting yields

$$\mu_{j,\,mix}^{A} - \mu_{j,mix}^{B} = \mu_{j,\,pure}^{A} - \mu_{j,\,pure}^{B} \tag{18-55}$$

But in the side tube, where only pure j is present,

$$\mu_{j,\,pure}^{A} - \mu_{j,\,pure}^{B} = \int_{P^B}^{P^A} V_j\,dP = \int_{\Phi^B}^{\Phi^A} (-V_j\,\rho_j)d\Phi = -m_j(\Phi^A - \Phi^B) \tag{18-56}$$

Combining Eqs. (18-55) and (18-56) and noting that A and B could have been chosen at random,

$$\mu_{j,\,mix}^{A} + m_j\Phi^A = \mu_{j,\,mix}^{B} + m_j\Phi^B = \text{constant} \tag{18-57}$$

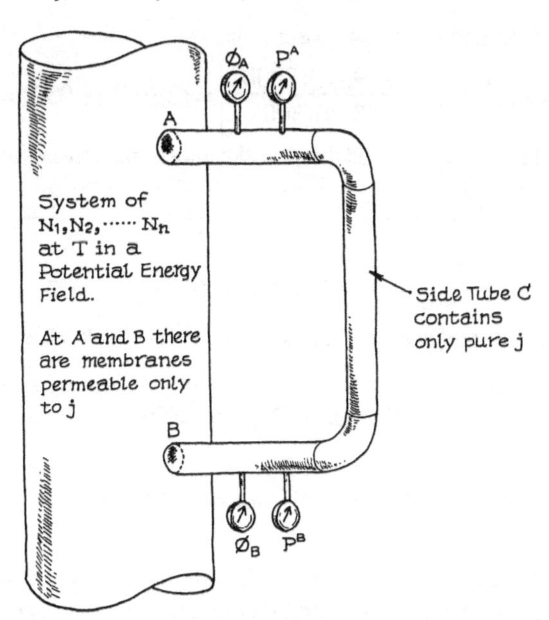

Figure 18.2 The effect of potential fields on chemical potentials.

Equation (18-57) provides the desired equilibrium criteria. It shows that the chemical potential is not a constant in a potential energy field but rather the sum of chemical potential and the product of the molecular weight times the potential is constant.

Example 18.10

A deep, narrow well hole has been capped and left undisturbed for years. It is believed that the temperature of the gas does not vary significantly with depth and that convection currents are negligible. The pressure at the well top is 2 bar, the temperature 300 K, and the gas composition is 70 mole % helium and 30 mole % methane. The hole is 2 km deep. What is the helium mole fraction at the bottom? Assume ideal gases.

Solution

With z measured down from the surface and $\Phi = 0$ on the surface, $\Phi = -gz$. Then from Eq. (18-56)

$$\mu_{He,\,b} - \mu_{He,\,t} = -m_{He}(\Phi_b - \Phi_t) = m_{He} g z_b$$

where the subscripts b and t represent bottom and top conditions. $m_{He} = 4$ kg/kg-mol, $g = 9.81$ m/s^2, and $z_b = 2000$ m. Also, for ideal gases,

$$\mu_{He,\,b} - \mu_{He,\,t} = RT \ln \frac{\hat{f}_{He,\,b}}{\hat{f}_{He,\,t}} = RT \ln \frac{P_{He,\,b}}{P_{He,\,t}}$$

Thus, with $T = 300$ K, and $R = 8314$ J/kg-mol K,

$$P_{He,\,b} = P_{He,\,t} \exp\left[\frac{(4)(9.81)(2000)}{(8314)(300)}\right] = (2)(0.7)(1.032) = 1.44 \text{ bar}$$

In a similar manner, $P_{CH_4,\,b} = 0.68$ bar, so the mole fraction helium at the bottom is 1.44/(1.44 + 0.68) = 0.68.

References

Callen, H.B. (1960), *Thermodynamics*, 1st ed., Wiley, New York, Chap. 13.

Camp F.W. and E.F. Johnson (1967), "The effect of strong magnetic fields on chemical engineering systems," MATT-67, Plasma Physics Laboratory, Princeton University, Princeton, NJ.

Debye, P. (1926), "Einige Bemerkungen zur Magnetisierung bei tiefer Temperatur," *Ann. Phys.*, **81**, p 1154.

Giauque, W.F. (1927), "Paramagnetism and the Third Law of thermodynamics. Interpretation of the low-temperature magnetic susceptibility of gadolinium sulfate," *J. Am. Chem. Soc.*, **49**, p 1870.

Giauque, W.F. and D.P. MacDougall (1933), "Attainment of temperatures below 1° absolute by demagnetization of $Gd_2(SO_4)_3 \cdot 8H_2O$," *Phys. Rev.*, **43**, p 768.

Guggenheim, E.A. (1967), *Thermodynamics*, North-Holland, Amsterdam, Chap. 10, Section 10.09, p 336-337.

Landau, L.P. and E.M. Lifshitz (1960), "Electrodynamics of continuous media," Vol. 8, *Course of Theoretical Physics*, translated by G.B. Sykes and G.S. Bell Pergamon Press, New York.

Lyon, R.K. (1961), "Effect of strong electrical fields on the boiling points of some alcohols," *Nature*, **192**, p 1285.

Meachin, A.J. and M.W. Biddulph (1978), "The effect of high magnetic fields on the vapor pressures of nitrogen, oxygen, and argon," *Cryogenics*, p 29.

Rocca R. and M.B. Bever (1950), "The thermoelastic effect," *Trans. AIME*, **88** *Journal of Metals*, (Feb. 1950), p 327.

Rosensweig, R.E. (1989), "Thermodynamics of electromagnetism," Chapter 13, in *Thermodynamics: an Advanced Textbook for Chemical Engineers*, by G. Astarita Plenum Press, New York, p 365-438.

Sharma, R.C. (1971), "Effects of electric field on the boiling points of liquids," *J. Appl. Phys.*, **42**, p 1234.

Stratton, J.A. (1941), *Electromagnetic Theory*, McGraw-Hill, NY.

Turro, N.J. (1983), "Micellar systems as "supercages" for reactions that germinate radical pairs: magnetic effects, *J. Amer. Chem. Society*, **105** (7), p 1861-1868.

Problems

18.1. We have a system α confined between the plates of a condenser in which a strong electric field is present. The material in α is a pure substance. At the boundaries of the condenser, the α system is contiguous with a system β made of the same pure material; the electric field in β is zero. The system β is itself in contact with a reservoir R_P through an adiabatic, frictionless piston; reservoir R_P is large and the pressure within is constant. System β is also in contact with a large reservoir R_T through a rigid, diathermal gate. The temperature in R_T is constant. What are the equilibrium criteria for the α–β system?

18.2. For a mixture of reacting gases, show how the chemical equilibrium constant varies with the electric field strength.

18.3. Present a similar analysis as requested in Problem 18.2 but for the magnetic field strength.

18.4. What is the work required to establish a magnetic field of 10,000 gauss in a long solenoid that contains 1 kg-mol of air at standard conditions? Assume that $\mu_{air} = \mu_o$.

18.5. A sutdent is carrying out an experiment in which he desires to hang a balance pan a definite distance above a tabletop (see Figure 18.5). To accomplish this, he decides to use a rubber band that he has conveniently found in his pocket. He readily obtains the desired separation distance by adding mass to the pan in order to stretch the rubber band the correct amount. the student has, however, just come in from the outside and when used, the rubber band is at the temperature outside the laboratory. He works rapidly and when the pan and weights are hung on the rubber band, it quickly stretches to a length of 30 cm.

Figure P18.5

Continuing the experiment, as the rubber band approaches room temperature, the student notices with irritation that the pan is rising. To counteract this, he gradually adds weights to the pan to maintain the same stretched length (30 cm). A total of 14 g has to be added until the rubber band is at room temperature and no more movement occurs.

Is it summer or winter outside?

Assume that the original extension of the rubber band is adiabatic and reversible. Initially, the rubber band was 7.5 cm long and had an (unstretched) cross section measuring 1.5 x 1.5 mm. Rubber does not crystallize when stretched and data for this particular rubber band show that $(\partial\sigma_x/\partial T)_L = 4410$ N/m^2 K, and $T(\partial S/\partial T)_L = 1900$ J/kg K. The rubber band has a specific gravity of 0.95 and the room temperature is 25°C.

18.6. Crackpot Inventions, Inc. is currently designing an elastic-rod Carnot engine (see Figure P18.6). In this device, work will be produced by heat transfer from a high-temperature (T_A) to a low-temperature (T_B) bath by the extension and relaxation of an elastic rod. This device will operate in a four-step cycle:

(1) Isothermally stretching the elastic rod from L_1 to L_2 while transferring Q_B to the low-temperature bath.

(2) Adiabatically and reversibly stretching the rod from L_2, T_B to L_3 to T_A.

(3) Isothermally relaxing the rod from L_3 to L_4, while absorbing Q_A from the high-temperature bath.

(4) Adiabatically and reversibly relaxing the rod to its initial length.

As it now stands, the device is fitted for a 25.4-cm rod that is capable of stretching to 30.5 cm. Crackpot wants to know what temperature they should use in their high-temperature bath so that their device can produce at least 34 J of work per cycle while using the room-temperature environment as the cold bath.

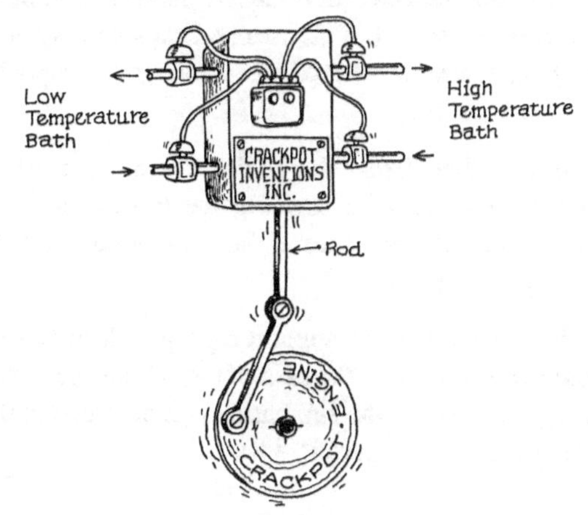

Figure P18.6

Their elastic rod has the following characteristics:

Size: 25.4 by 0.25 by 0.25 cm (unstretched)
$(\partial\sigma_x/\partial T)_L = 7.0$ MPa/K
$C_L = T(\partial S/\partial T)_L = 3.2$ J/g K
Specific gravity: 1.3

Room temperature is 297 K. What must be the temperature of the high- temperature bath? What is the efficiency of the engine?

18.7. A manned spacecraft is to be the payload of a large, multistage rocket. At liftoff, the rocket will accelerate vertically with a constant acceleration. Only later will a trajectory be programmed (Figure P18.7).

Figure P18.7

Prior to launch, the manned capsule is filled with 20% O_2 and 80% He at 290 K and 1 bar and is sealed until the high-acceleration launch phase is completed. During this same period, any heat transfer to or within the manned capsule may be neglected (i.e., it will remain isothermal), and assume that there is no gas composition variation due to respiration.

It is necessary to ensure that during the launch acceleration period the oxygen partial pressure never drops below 0.17 bar. Assuming that the capsule is 3 m high and of constant cross section, what is the total pressure at the leading edge (top) of the capsule at the maximum tolerable acceleration?

18.8. We have just received a proposal to support a project designed to desalinate water for submerged submarines. Attached to the skin of the submarine would be a membrane that is semipermeable to water. When the submarine is submerged to the correct depth, water would pass into the submarine (Figure P18.8).

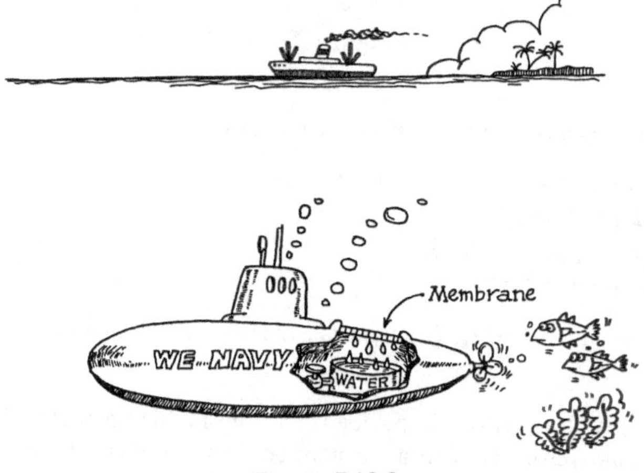

Figure P18.8

We assume that the seawater density and temperature remain constant with depth and are 1024 kg/m^3 and 290 K, respectively. The submarine interior is at 1 bar and has a humidity equivalent to saturation at 290 K (1767 Pa). The partial pressure over the ocean at the surface is 1671 Pa. The salt concentration in seawater is approximately 35,000 ppm.

At what depth should the submarine cruise to make the system work?

18.9. An equimolar mixture of He^3 and He^4 is pumped into a small-diameter, 30.5-m-long, well-insulated vertical tube. The initial pressure is 1 bar and the temperature 5 K. The gas mixture is ideal and no convection currents are present.

Derive a relation to express the equilibrium composition as a function of height and determine the fraction of He^4 and the total pressure at the bottom of the tube. How would your results change if the tube length were allowed to increase without limit?

18.10. Preliminary reports indicate the atmosphere of Saturn consists primarily of methane and hydrogen (see Figure P18.10). At the outer boundary, skimpy data show:

$T \approx -150°C$
$P \approx 10^{-3}$ bar
50% CH_4 and 50% H_2

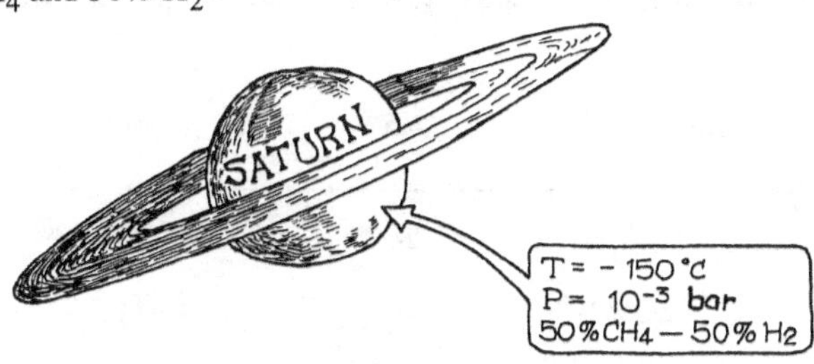

Figure P18.10

From other observations, it is believed that the depth of the atmosphere is about 25,000 km. With a constant gravitational acceleration equal to 1.1 g_{earth}, describe the atmospheric concentration and pressure variations with altitude and point out any interesting conclusions.

18.11. To separate the uranium isotopes $^{235}UF_6$ and $^{238}UF_6$ by centrifugation, one is limited by the mechanical strength of the material on the periphery. Recent reports, however, have indicated that some new alloys will withstand tangential velocities of 400 m/s at 400 K. Estimate a separation factor α_{8-5}:

$$\alpha_{8-5} = \frac{y_8^r y_5^0}{y_5^r y_8^0}$$

where the subscripts indicate the isotope and the superscripts r and 0 refer to the rim and center. Compare this α with the value of 1.0043 as determined from a single gaseous diffusion separation stage.

If the pressure at the center were 1 bar, what is your best estimate of the rim pressure? Assume ideal gases. Comment.

18.12. A centrifuge 30 cm in radius rotating at 15,000 rpm contains a mixture of benzene and toluene at 340 K. After equilibrium is reached, the mixture at the center is sampled and found to be a vapor with 6700 Pa partial pressures of both benzene and toluene.

Plot the concentration and total pressure profile as a function of radius. Assume that the gas is ideal and that any liquid phase present forms an ideal solution. At 340 K, the vapor pressures of the benzene and toluene are 0.624 and 0.234 bar, respectively.

18.13. In some of the new high-thrust rockets, extreme acceleration fields are expected. Combustion experts are already analyzing problems expected to be encountered in such an environment. You are asked to aid in the program to estimate the equilibrium constant for the reaction

$$CO + \tfrac{1}{2}O_2 = CO_2$$

at 1000 K and 1000 times normal gravity. At 1 g, $\log_{10} K_{1000} = 10.3$. Other data are shown below:

	$\Delta H_{f,298}$ (J/mol)	A	$B \times 10^3$	$C \times 10^7$
O_2	0	26.2	11.50	-32.2.
CO_2	-3.93×10^5	28.7	35.73	-103.6
CO	-1.11×10^5	26.2	8.76	-19.2

C_p (J/mol K) = $A + BT/2 + CT^2/3$, where T is in °C.

18.14. Tritium separation from deuterium and hydrogen is not easily accomplished. Yet there are now suggestions that it can be more readily carried out in ultra-centriguges. Estimate what increase in tritium concentration you might expect at the rim draw-off point if the feed is at atmospheric pressure and is comprised of 5% T_2, 50% D_2, and 45% H_2. The unit is to operate isothermally at 43 K with a rim velocity of 400 m/s. Do you visualize any practical problems?

18.15. Randy Wahoo, a graduate student adept in the application of thermodynamics, was overheard disputing the reasoning of a pragmatic meterologist, Phil M. Strip. They were discussing the pressure and composition profile in the earth's atmosphere. Both agreed that as a basis, they would choose the earth's surface, where

$N_2/O_2 = 4.0$
Air is saturated with water ($P_{H_2O} = 840$ Pa at 278 K)
$T = 278$ K
$P = 1$ bar

Randy would prefer to model the atmosphere as quiescent and isothermal in an equilibrium state. Using this model, estimate the N_2/O_2 ratio, the partial pressure of water, and the total pressure at an altitude of 1 km. (Neglect CO_2.)

Phil claims that this model is sheer nonsense. His own model involves visualizing a unit mass of air at sea level expanding polytropically as it is moved up above the surface of the earth. By polytropic expansion, he means that the air expands by a relation PV^n = constant. He has selected n to be 1.2. No compositional changes are allowed. Again calculate the pressure, the water partial pressure, and the temperature at an altitude of 1 km. Compare and contrast both models.

18.16. We present a variation on a scheme that has surfaced several times in the literature. (See, for example, *Scientific American*, Dec. 1971, p 100 and Apr. 1972, p 110, as well as *Science*, June 2, 1972, p 1011 and Dec. 15, 1972, p 1199.)

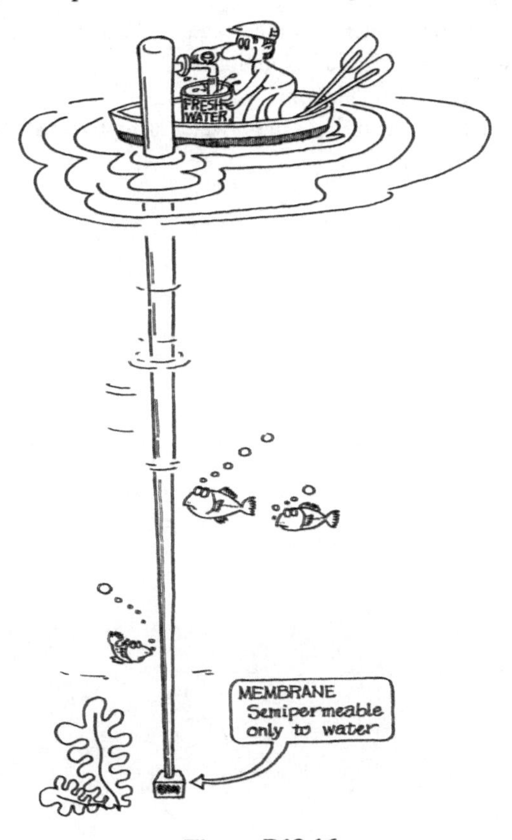

Figure P18.16

We suggest in Figure P18.16 a simple scheme to reclaim fresh water from the ocean. A long pipe is lowered into the ocean and on the bottom of this pipe we install a membrane permeable to water. As we all know, the density of seawater exceeds that of fresh water; therefore, at any depth, the pressure in the ocean would exceed that which would be exerted by a comparable column of fresh water. If we make the pipe of sufficient length, the pressure difference at the bottom would exceed the osmotic pressure and water from the sea would flow into the pipe.

Comment on the feasibility of this process. Assume that the ocean is an equilibrium system.

How would your evaluation change if we took a more realistic ocean and assume that currents mix the ocean to such a degree that there is neither concentration nor temperature variation with depth?

Data:

Assume that the ocean temperature = 10°C
P_{vp} for water at 10°C = 1230 Pa.
The partial pressure of water over the ocean, at the surface, is 1207 Pa
(this value assumes a 3.5% NaCl solution).
Density of seawater = 1.0255 g/cm^3
The partial molar volume of H_2O in seawater is approximately equal to the pure molar volume of H_2O.

Thermodynamics of Surfaces 19

In the absence of gravity or other body forces, free liquids assume a spherical shape. Intermolecular attractive forces operate to pull the molecules together; this attraction results in the formation of spheres. A convenient way to characterize this behavior is to assume that the surface is in tension (i.e., each element of the surface layer experiences tensile forces from neighboring elements); the net result is a surface somewhat similar to that in an elastic balloon.

The analogy to the balloon may be carried further. We know that there exists a pressure difference between the inside and outside of a balloon because of these tensile forces. Similarly, in a bubble or drop of liquid, we will show that the internal pressure also exceeds that outside the drop. To demonstrate this, we will again use the concept of surface tension forces.

In this chapter we first examine surfaces from the standpoint of work interactions (i.e., we derive an expression to allow us to determine the work necessary to vary both the *volume and area* of a surface layer). Following this, we explore the effects of curvature on small liquid drops and crystal nuclei. In the remainder of the chapter we develop the important thermodynamic expressions for surface layers and also apply the criteria of equilibrium to the nucleation of new phases. The extensive subject matter of surfaces is examined only from a classical thermodynamic point of view; interesting and very important phenomena such as wetting, surface activity, and adhesion must be left for the reader to enjoy in specialized texts, such as Adamson (1982).

19.1 Surface Tension

To illustrate more clearly the mechanical analog of surface forces, consider a plane interface between phases α and β. The element shown in Figure 19.1 measures x units wide, y units deep, and has a thickness τ such that the top and bottom of the element are located in the homogeneous phases α and β, respectively. Each side of this parallelpiped is subjected to a normal, compressive pressure P that is equal at all points.

This equality results from the criteria of phase equilibria derived for *plane* surfaces in Chapter 6. Also, in the two directions perpendicular to the thickness, τ, there is the tension force σ *per unit length* (i.e., the tensile force on the front and back faces would be σx; on each side face, σy). In SI units, σ is expressed in Newtons per meter (N/m).

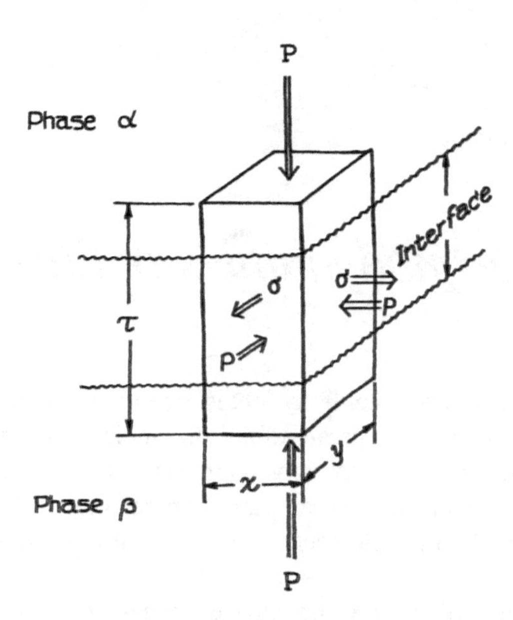

Figure 19.1 Element in plane interface.

If we choose the parallelpiped as our system, the work done by the system on the surroundings when the volume is increased by $dx\,dy\,d\tau$ is negative and given as

$$-\delta W = (Pxy)\,d\tau + (P\tau y - \sigma y)\,dx + (P\tau x - \sigma x)dy$$
$$-\delta W = P(xy\,d\tau + \tau y\,dx + \tau x\,dy) - \sigma(y\,dx + x\,dy)$$

or with $\underline{V} = xy\tau$ and $\underline{a} = xy$

$$\delta W = -P\,d\underline{V} + \sigma\,d\underline{a} \tag{19-1}$$

In this case, the work term given by Eq. (19-1) includes both volume deformation ($P\,d\underline{V}$) and surface deformation ($\sigma\,d\underline{a}$) contributions.

19.2 Equilibrium Considerations

Consider the equilibrium system shown in Figure 19.2. It is composed of two distinct parts: a bulk phase α and a small fragment of a different phase β. Phase α is maintained at constant temperature and pressure by contact with the large isobaric and isothermal reservoirs R_P and R_T.

Except for the presence of the β phase, the situation is identical to that shown in Figure 6.5 and as described as case (c) in Section 6.3. There, with no β phase, it was shown that the system attained an equilibrium state when the total Gibbs energy was a minimum.

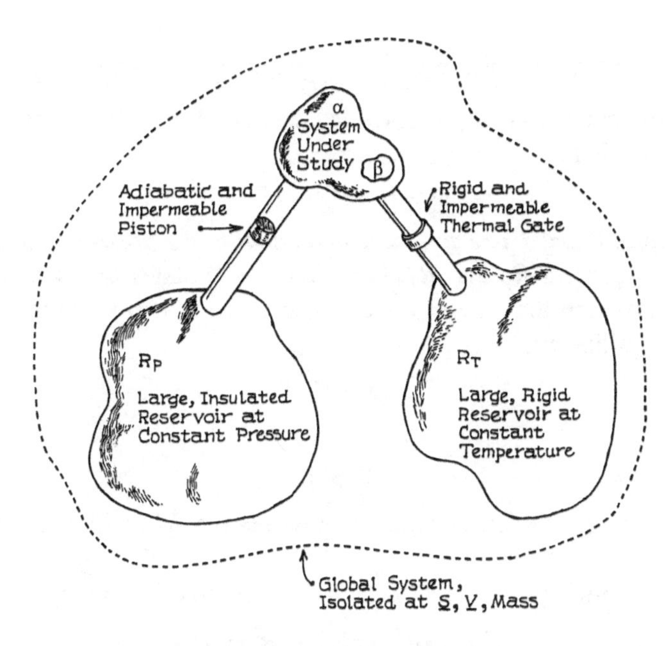

Figure 19.2 System interactions with constant pressure or temperature reservoirs.

The present case is, however, somewhat different because the small element of phase β does not necessarily have the same intensive properties as phase α even though both phases are in equilibrium. In addition, we must consider energy contributions for the surface phase (σ) between α and β.

If the same nomenclature as in Section 6.3 is used, the criterion for *stable* equilibrium states is

$$\Delta \underline{U}^{\Sigma} = \Delta \underline{U} + \Delta \underline{U}^{R_P} + \Delta \underline{U}^{R_T} > 0$$

where the superscript Σ represents the global system and terms that are not superscripted refer to the α – β system. As shown before, for variations in volume or entropy,

$$\Delta \underline{U}^{R_P} = -P\,\Delta \underline{V}^{R_P} = P\,\Delta \underline{V} \qquad (19\text{-}2)$$

$$\Delta \underline{U}^{R_T} = T\,\Delta \underline{S}^{R_T} = -T\,\Delta \underline{S} \qquad (19\text{-}3)$$

where the global conservation of total volume and entropy has been employed. The P and T values in Eqs. (19-2) and (19-3) are those that characterize the continuous phase α but not necessarily phase β or σ. Also,

$$\Delta \underline{U} = \Delta \underline{U}^{\alpha} + \Delta \underline{U}^{\beta} + \Delta \underline{U}^{\sigma} \qquad (19\text{-}4)$$

where β and σ represent the β and surface phases, respectively. If Eqs. (19-2), (19-3), and (19-4) are substituted into Eq. (6-16) with the expansions

$$\Delta \underline{V} = \Delta \underline{V}^{\alpha} + \Delta \underline{V}^{\beta} + \Delta \underline{V}^{\sigma} \qquad (19\text{-}5)$$

$$\Delta \underline{S} = \Delta \underline{S}^{\alpha} + \Delta \underline{S}^{\beta} + \Delta \underline{S}^{\sigma} \qquad (19\text{-}6)$$

then,

$$(\Delta \underline{U}^{\alpha} + P\,\Delta \underline{V}^{\alpha} - T\,\Delta \underline{S}^{\alpha}) + (\Delta \underline{U}^{\beta} + P\,\Delta \underline{V}^{\beta} - T\,\Delta \underline{S}^{\beta}) + (\Delta \underline{U}^{\sigma} + P\,\Delta \underline{V}^{\sigma} - T\,\Delta \underline{S}^{\sigma}) > 0 \quad (19\text{–}7)$$

In a fashion analogous to that presented in Chapter 14, let's introduce a term called the availability, $\underline{B}$, where

$$\underline{B} \equiv \underline{U} + P\underline{V} - T\underline{S} \quad (19\text{-}8)$$

In this definition, P and T *are constants and equal to the pressure and temperature in the external reservoirs* R_P and R_T. Equation (19-7) states that availability is the thermodynamic function that attains a maximum or minimum value in the $\alpha - \beta - \sigma$ system; then, at equilibrium,

$$d\underline{B} = 0 \quad (19\text{-}9)$$

and for stability

$$d^m\underline{B} > 0 \quad (19\text{-}10)$$

where $d^m\underline{B}$ is the lowest order, nonvanishing derivative. We will utilize the stability criterion later in Section 19.7.

At equilibrium, with Eqs. (19-7), (19-8), and (19-9), in differentials

$$d\underline{B} = d\underline{B}^{\alpha} + d\underline{B}^{\beta} + d\underline{B}^{\sigma} = 0 \quad (19\text{-}11)$$

To determine the specific equilibrium criteria for this case, $\underline{U}^{\alpha}$, $\underline{U}^{\beta}$, and $\underline{U}^{\sigma}$ are expanded using the Fundamental Equations in energy representation,

$$d\underline{U}^{\alpha} = T\,d\underline{S}^{\alpha} - P\,d\underline{V}^{\alpha} + \sum_{j=1}^{n} \mu_j^{\alpha}\,dN_j^{\alpha} \quad (19\text{-}12)$$

$$d\underline{U}^{\beta} = T^{\beta}\,d\underline{S}^{\beta} - P^{\beta}\,d\underline{V}^{\beta} + \sum_{j=1}^{n} \mu_j^{\beta}\,dN_j^{\beta} \quad (19\text{-}13)$$

$$d\underline{U}^{\sigma} = T^{\sigma}\,d\underline{S}^{\sigma} - P^{\sigma}\,d\underline{V}^{\sigma} + \sigma\,d\underline{a} + \sum_{j=1}^{n} \mu_j^{\sigma}\,dN_j^{\sigma} \quad (19\text{-}14)$$

With Eqs. (19-11) through (19-14) subject to the constraints of $\underline{V}^{\alpha} + \underline{V}^{\beta} + \underline{V}^{\sigma}$ = constant and $\underline{S}^{\alpha} + \underline{S}^{\beta} + \underline{S}^{\sigma}$ = constant, and with the understanding that $T = T^{\alpha}$ and $P = P^{\alpha}$ are constants and are determined by the reservoir temperatures and pressures. In this situation, we are effectively saying that the bulk continuous phase (α) is kept at constant T and P. With these assumptions, we get,

$$(T^{\beta} - T)\,d\underline{S}^{\beta} + (T^{\sigma} - T)\,d\underline{S}^{\sigma} + \sum_{j=1}^{n} \mu_j^{\alpha}\,dN_j^{\alpha} + \sum_{j=1}^{n} \mu_j^{\beta}\,dN_j^{\beta}$$
$$+ \sum_{j=1}^{n} \mu_j^{\sigma}\,dN_j^{\sigma} - (P^{\beta} - P)\,d\underline{V}^{\beta} - (P^{\sigma} - P)\,d\underline{V}^{\sigma} + \sigma\,d\underline{a} = 0 \quad (19\text{-}15)$$

Before examining Eq. (19-15), the conservation of mass restraint must be included. For each component, this is equivalent to saying that

$$dN_j = dN_j^{\alpha} + dN_j^{\beta} + dN_j^{\sigma} = 0 \tag{19-16}$$

If Eq. (19-16) is multiplied by $-\mu_j^{\alpha}$ (chosen as a Lagrange multiplier) and added to Eq. (19-15), each variation is independent and the coefficients must be zero. Therefore,

$$T^{\beta} = T = T^{\alpha} \tag{19-17}$$

$$T^{\sigma} = T = T^{\alpha} \tag{19-18}$$

$$\mu_j^{\beta} = \mu_j^{\alpha} \tag{19-19}$$

$$\mu_j^{\sigma} = \mu_j^{\alpha} \tag{19-20}$$

and

$$(P^{\beta} - P)\, d\underline{V}^{\beta} + (P^{\sigma} - P)\, d\underline{V}^{\sigma} - \sigma\, d\underline{a} = 0 \tag{19-21}$$

The first four equalities are as we expected: the temperatures and chemical potentials of each component are equal throughout the system. In this regard, our results do not differ from what we found earlier for phase equilibrium criteria. Equation (19-21) is, however, a different result. It expresses a condition of mechanical equilibrium where now there is a connection between $\underline{V}^{\beta}$, $\underline{V}^{\sigma}$, and $\underline{a}$ that depends on the geometry of the β phase.

First, let us consider the term $(P^{\sigma} - P)\, d\underline{V}^{\sigma}$. We have some latitude in defining the extent of the surface phase and, depending on our choice, we could have different values of P^{σ} and $d\underline{V}^{\sigma}$. If we were to define our surface phase boundary so that the surface had properties similar to phase α, then $P^{\sigma} \approx P^{\alpha} = P$ and $(P^{\sigma} - P)\, d\underline{V}^{\sigma}$ is essentially zero.

Usually, however, the surface layer is associated with the discontinuous phase (in this case, β), so the $(P^{\sigma} - P)\, d\underline{V}^{\sigma}$ term is incorporated into the $(P^{\beta} - P)\, d\underline{V}^{\beta}$ term. Equation (19-21) then reduces to

$$P^{\beta} - P = P^{\beta} - P^{\alpha} = \sigma \frac{d\underline{a}}{d\underline{V}^{\beta}} \tag{19-22}$$

The final relation expresses the pressure difference across a curved interface in an equilibrium system.

Example 19.1

What would the pressure difference $(P^{\beta} - P^{\alpha})$ be for the following cases?

(a) If phase β were a sphere of radius r.

(b) If phase β were assumed to be a solid with the shape of a cylinder or pancake disk of radius r and some fixed height h. (h is some multiple of the lattice spacing.) This case is often used in two-dimensional nucleation theories involving crystallization on a flat surface.

(c) The phase boundary between the α and β phases consists of arbitrarily curved surface that is completely characterized by two radii of curvature C_1 and C_2 as shown in Figure 19.3.

Solution

(a) Let the radius of the sphere of phase β be r. Then, $\underline{a} = 4\pi r^2$, $\underline{V}^\beta = (4/3)\pi r^3$ and then, from Eq. (19-22),

$$P^\beta - P^\alpha = \sigma \frac{d\underline{a}}{d\underline{V}^\beta} = \frac{2\sigma}{r} \tag{19-23}$$

(b) For the assumed pancake shape, only the area of the curved surface is of interest. Thus $\underline{a} = 2\pi rh$ and $\underline{V}^\beta = \pi r^2 h$.

$$P^\beta - P^\alpha = \sigma \frac{d\underline{a}}{d\underline{V}^\beta} = \frac{\sigma}{r} \tag{19-24}$$

(c) In Figure 19.3 a small element of the surface is shown. The area is xy. If we

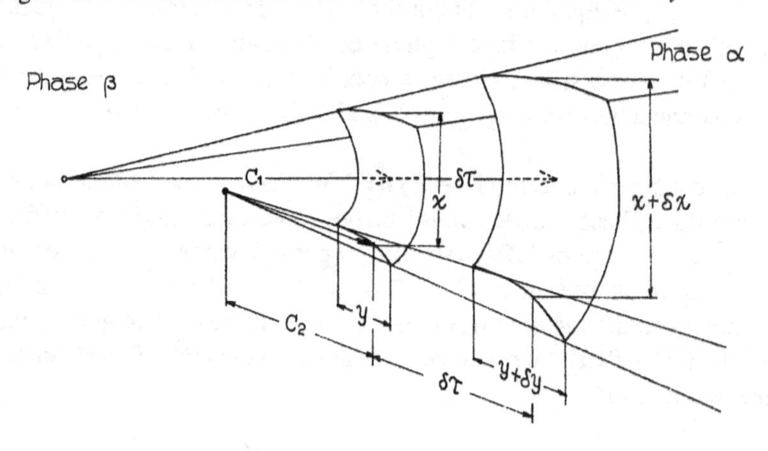

Figure 19.3

extend the element by $\delta\tau$ and thereby increase C_1 to $C_1 + \delta\tau$ and C_2 to $C_2 + \delta\tau$, then the increase in area and volume of phase β becomes

$$\delta\underline{a} = (y + \delta y)(x + \delta x) - xy = y\,\delta x + x\,\delta y$$

$$\delta\underline{V}^\beta = xy\,\delta\tau$$

But, by similar triangles,

$$\frac{x + \delta x}{C_1 + \delta\tau} = \frac{x}{C_1} \quad \text{and} \quad \frac{y + \delta y}{C_2 + \delta\tau} = \frac{y}{C_2}$$

Then,

$$\delta x = \frac{x}{C_1}\,\delta\tau \quad \text{and} \quad \delta y = \frac{y}{C_2}\,\delta\tau$$

Substituting the δx and δy terms into the expression for $\delta \underline{a}$ and using Eq. (19-22),

$$P^\beta - P^\alpha = \sigma\left(\frac{d\underline{a}}{d\underline{V}^\beta}\right) = \frac{(xy\delta\tau/C_1 + xy\delta\tau/C_2)\sigma}{xy\delta\tau}$$

$$P^\beta - P^\alpha = \sigma\left(\frac{1}{C_1} + \frac{1}{C_2}\right) \tag{19-25}$$

Eq. (19-25) is the general form of the Young-Laplace equation [see also Adamson (1982), p 6-8]. For a sphere, $C_1 = C_2 = r$, and Eq. (19-25) reduces to Eq. (19-23); for a plane surface, $C_1 = C_2 = \infty$ and $P^\beta = P^\alpha$. The latter case yields the pressure equality we found earlier when we studied phase equilibrium for planar surfaces (see Section 6.5).

19.3 Effects of Pressure Differences across Curved Interfaces

For our equilibrium system containing a large bulk phase α and fragments of a different phase β, we have shown that the pressure in phase β exceeds that in α. If β is spherical, the pressure difference is given by Eq. (19-23). If β has a different curvature, we can always revert to the general case as shown by Eq. (19-25). These relations show how the pressure difference depends on the size and shape of the fragment of phase β.

To estimate the pressure inside the small fragments of phase β, it is convenient to relate P^β to the true equilibrium pressure in the system when a phase with the same *composition* and *temperature* as phase α is in equilibrium with a *planar* surface of phase β. Several examples are given below to illustrate the methodology.

Phase β: small spherical liquid drops; Phase α: vapor

Let us begin with the case where a *pure* component vapor at P^α and T^α is in equilibrium with small liquid drops with radii r. The pressure inside the drop is P^β but, by Eq. (19-17), the drop temperature $T^\beta = T^\alpha$. The chemical potential equality between the liquid drop and vapor phase is expressed in terms of fugacity (see Section 9.7) with $T = T^\alpha = T^\beta$:

$$f^\beta(P^\beta, T) = f^\alpha(P^\alpha, T) \tag{19-26}$$

The fugacity of the vapor (α) phase may be written in terms of its fugacity coefficient ϕ^v:

$$f^\alpha(P^\alpha, T) = \phi^v(P^\alpha, T)\, P^\alpha \tag{19-27}$$

For the fugacity of the liquid drop (β) phase, we can relate the fugacity at P^β to the planar equilibrium pressure P_{vp} by the Poynting correction [see Eq. (15.94)]

$$f^\beta(P^\beta, T) = f^\beta(P_{vp}, T)\exp\left[\left(\frac{1}{RT}\right)\int_{P_{vp}}^{P^\beta} V^\beta\, dP\right] \tag{19-28}$$

The fugacity of the liquid drop at P_{vp} and T is equal to the fugacity of the pure vapor at P_{vp}, and T, so, in an analogous fashion to Eq. (19-27),

$$f^{\beta}(P_{vp}, T) = f^{v}(P_{vp}, T) = \phi^{v}(P_{vp}, T)P_{vp} \tag{19-29}$$

Combining Eqs. (19-26) through (19-29), we obtain

$$\phi^{v}(P_{vp}, T)P_{vp} \exp\left[\left(\frac{1}{RT}\right)\int_{P_{vp}}^{P^{\beta}} V^{\beta}\,dP\right] = \phi^{v}(P^{\alpha}, T)P^{\alpha} \tag{19-30}$$

Equation (19-30) relates the pressures P^{β} and P^{α} to P_{vp} and other properties of the system. If we further assume that the pure vapor behaves as an ideal gas at P_{vp} and P^{α}, both fugacity coefficients are unity. Also, assume that the liquid molar volume, V^{β}, is not a strong function of pressure. Then, Eq. (19-30) simplifies to

$$P^{\alpha} = P_{vp} \exp\left[\frac{V^{\beta}(P^{\beta} - P_{vp})}{RT}\right] \tag{19-31}$$

From Example 19.1, using the Young-Laplace equation, $P^{\beta} = P^{\alpha} + 2\sigma/r$, so

$$P^{\alpha} = P_{vp} \exp\left[\frac{V^{\beta}(P^{\alpha} - P_{vp} + 2\sigma/r)}{RT}\right] \tag{19-32}$$

In most cases $(P^{\alpha} - P_{vp}) << 2\sigma/r$, so Eq. (19-32) becomes

$$P^{\alpha} = P_{vp} \exp\left(\frac{2\sigma V^{\beta}}{r RT}\right) \tag{19-33}$$

This is the Kelvin equation, to illustrate its use, suppose that our α phase were pure water vapor at P^{α} and 313 K. At this temperature, the vapor pressure of water is 7370 Pa. That is, if the water vapor were in equilibrium with a *planar liquid* water surface, the pressure would be 7370 Pa. However, if the water vapor were in equilibrium with small spherical water droplets, the pressure would exceed 7370 Pa. For instance, if the droplet radii were 10 nm, then $P^{\alpha} = 8070$ Pa. If droplet radii could be made as small as 1 nm, $P^{\alpha} \approx 20{,}000$ Pa. In using Eq. (19-33) to make such estimates, we have assumed that the surface tension of water is independent of drop size. This assumption is questionable for very small drops.

Example 19.2

If spherical drops of water (of the same size) were in equilibrium with water vapor at 1.1 bar and 373 K, estimate the droplet radii and the pressure within the drops. At 373 K, the surface tension of water is 0.0589 N/m and the molar volume is 1.87×10^{-5} m^3/mol. The vapor pressure of water at 373 K is 1.013 bar.

Solution

With Eq. (19-33),

$$P^{\alpha} = 1.1 = 1.013 \exp\left[\frac{(2)(0.0589)(1.87\times 10^{-5})}{(8.314)(373)\,r}\right]$$

with $r = 8.6 \times 10^{-9}$ m = 8.6 nm,

$$P^{\beta} - P^{\alpha} = \frac{2\sigma}{r} = \frac{(2)(0.0589)}{8.6 \times 10^{-9}} = 1.37 \times 10^{7} \text{ Pa}$$

and

$$P^{\beta} = 1.1 \times 10^{5} + 1.37 \times 10^{7} = 1.38 \times 10^{7} \text{ Pa}$$

Within such drops, the *calculated* pressure is very large. It is this high pressure that increases the fugacity of the liquid water so that a higher vapor phase fugacity results at equilibrium.

A similar treatment is possible for the case where we have a vapor-phase *mixture* at T^{α} and P^{α} in equilibrium with small liquid drops. Let x_j^{β} represent liquid mole fractions within the droplet and y_j^{α} be the vapor mole fractions in phase α. The analogs for Eqs. (19-26) through (19-30) with $T = T^{\alpha} = T^{\beta}$ are:

$$\hat{f}_j^{\beta}(P^{\beta}, T, x_j^{\beta}) = \hat{f}_j^{\alpha}(P^{\alpha}, T, y_j^{\alpha}) \tag{19-34}$$

$$\hat{f}_j^{\alpha}(P^{\alpha}, T, y_j^{\alpha}) = \hat{\phi}_j^{v}(P^{\alpha}, T, y_j^{\alpha})P^{\alpha}y_j^{\alpha} \tag{19-35}$$

$$\hat{f}_j^{\beta}(P^{\beta}, T, x_j^{\beta}) = \phi_j^{v}(P_{vpj}, T)P_{vpj}\gamma_j^{\beta}x_j^{\beta} \exp\left[\left(\frac{1}{RT}\right)\int_{P_{vpj}}^{P^{\beta}} V_j^{\beta}\, dP\right] \tag{19-36}$$

In Eq. (19-35), $\hat{\phi}_j^{v}$ is the fugacity coefficient of component j in the vapor. γ_j^{β} is the activity coefficient of j in the liquid drop phase and depends on P^{β} and x_j^{β}. $\phi_j^{v}(P_{vpj}, T)$ is the fugacity coefficient of pure vapor j at T and P_{vpj}. Substituting Eqs. (19-35) and (19-36) into Eq. (19-34) leads to a rigorous but unwieldy expression. To simplify matters, assume the vapor phase is an ideal-gas mixture ($\hat{\phi}_j^{v} = 1.0$), the liquid drop is an ideal solution ($\gamma_j^{\beta} = 1.0$), the pure vapor of j at P_{vpj} is ideal ($\phi_j^{v} = 1.0$), and that $V_j^{\beta} \neq f(P)$. Then

$$P^{\alpha}y_j^{\alpha} = P_{vpj}x_j^{\beta} \exp\left[\frac{V_j^{\beta}(P^{\beta} - P_{vpj})}{RT}\right] \tag{19-37}$$

This equation may be compared to Eq. (19-31) for a pure component under similar conditions.

There are n relations of the form of Eq. (19-37) and, as shown in Example 19.1, one equation showing how P^{α} and P^{β} depend on drop size (r). These are sufficient to solve for the $(n - 1)$ unknown mole fractions in the β phase as well as for P^{α} and P^{β}. Note that the exponential term in Eq. (19-37) can be considered as a factor that enhances the vapor pressure of each component in the liquid. Since molar volumes and vapor pressures differ from one component to another, the equilibrium liquid drop composition will differ from that of a planar liquid surface in equilibrium with the same vapor.

Phase β: spherical vapor bubble; Phase α: liquid

Let us again start with a pure-component case. The α phase is now liquid at P^α, while the β phase consists of vapor bubbles of radii r. (The temperatures in both phases are equal.) Following the same technique illustrated above,

$$f^\beta(P^\beta) = f^\alpha(P^\alpha) \tag{19-38}$$

$$\phi(P^\beta)P^\beta = \phi(P_{vp})P_{vp} \exp\left[\left(\frac{1}{RT}\right)\int_{P_{vp}}^{P^\alpha} V^\alpha \, dP\right] \tag{19-39}$$

Assuming ideal gases and $V^\alpha \neq f(P)$,

$$P^\beta = P_{vp} \exp\left[\frac{V^\alpha(P^\alpha - P_{vp})}{RT}\right] \tag{19-40}$$

Equation (19-40) is identical to Eq. (19-31) with the α–β superscripts interchanged. The interpretation is, however, different, as illustrated in Example 19.3.

Example 19.3

We have a bulk liquid phase of pure ethane at 1 bar pressure and at a temperature of 270 K. This liquid is in equilibrium with small vapor bubbles of radii r. Calculate r and, also, the pressure within the bubbles.

At 270 K the vapor pressure of ethane is 22.1 bar, the liquid molar volume is 7.38×10^{-5} m^3/mol, and the surface tension is 3.5×10^{-3} N/m.

Solution

In this case we have liquid ethane, which is highly superheated. (From Problem 7.3 we see that the ethane is in a state very close to the limit of stability from a thermodynamic point of view.) If the liquid ethane were in equilibrium with a planar vapor interface, the pressure would be 22.1 bar.

With Eq. (19-40),

$$P^\beta = (22.1 \times 10^5) \exp\left[\frac{(7.38 \times 10^{-5})(10^5 - 22.1 \times 10^5)}{(8.314)(270)}\right] = 20.6 \times 10^5 \text{ Pa}$$

with

$$P^\beta - P^\alpha = \frac{2\sigma}{r} = (20.6 \times 10^5 - 10^5) = \frac{(2)(3.5 \times 10^{-3})}{r}$$

$$r = 3.6 \times 10^{-9} \text{ m} = 3.6 \text{ nm}$$

As we noted in the liquid drop-vapor case, the pressure inside the vapor bubble of the β phase is significantly higher than the bulk liquid α phase, 20.6 bar versus 1 bar. In contrast to the liquid droplet case, however, the vapor pressure was slightly reduced over its planar value of 22.1 bar.

As we shall see later, calculation of the equilibrium bubble (or drop) size is an important step in formulating a theory of nucleation of phase β from a bulk α phase. The equilibrium bubble pressure and size for a superheated liquid mixture can be determined by a similar approach.

Phase β: solid spherical particles; Phase α: liquid

For this situation, consider a pure liquid in equilibrium with finely divided solid of the same material. If the solid phase were large and planar, the temperature would correspond to the freezing point of the liquid. The fact that the solid is composed of small fragments leads to a prediction that the freezing point will be depressed. If we assume that the solid fragments are spherical in shape with radii r,

$$\Delta T = T_r - T_m \approx \frac{-2V^{(s)}\sigma T_m}{r\,\Delta H_{fus}} \tag{19-41}$$

where

T_r = freezing point of the liquid in equilibrium with the finely divided solid (K)
T_m = normal freezing point of the liquid in equilibrium with a planar solid (K)
$V^{(s)}$ = molar volume of the solid T_m and at the liquid pressure (m^3/mol)
σ = solid-liquid interfacial tension (N/m)
ΔH_{fus} = enthalpy of fusion (J/mol)

The derivation of Eq. (19-41) is left as a problem at the end of the chapter. Actual calculations of ΔT are difficult as few values of σ are known. In fact, the measurement of ΔT for a known value of r has been used to infer values of σ.

Phase β: solid spherical particles; Phase α: liquid solution

In this case one is interested in exploring whether the solubility of a solute in a solution is modified if the solution is in equilibrium with small solid fragments rather than with large (planar) solute crystals. To formulate the problem, suppose that we have a binary solution of 1 and 2 at some pressure and temperature P^α, T^α. This solution is in equilibrium with small fragments of a pure solute phase (component 1) at T^α and P^β. Then

$$f_1^\beta(P^\beta) = \hat{f}_1^\alpha(P^\alpha, x_1^\alpha) \tag{19-42}$$

Next, assume that a liquid solution of 1 and 2 at the same temperature and pressure (P^α) were in equilibrium with a planar solute surface. In this case the solute concentration in the liquid is the "equilibrium" solubility $x_{1,eq}^\alpha$

$$f_1^\beta(P^\alpha) = \hat{f}_1^\alpha(P^\alpha, x_{1,eq}^\alpha) \tag{19-43}$$

Note that in the planar case, the solid pressure is P^{α}. Taking logarithms of Eqs. (19-42) and (19-43) and subtracting, then, with Eq. (8-158), we obtain

$$\ln \frac{a_1}{a_{1,\,eq}} = \frac{V^{(s)}}{RT}(P^{\beta} - P^{\alpha}) \tag{19-44}$$

where a_1 is the activity of the solute in the solution that is in contact with the small solid solute particles and $a_{1,\,eq}$ is the solute activity in a solution in contact with a large planar surface (the normal equilibrium solubility case). Approximating the ratio of activities by the ratio of concentrations and using $P^{\beta} - P^{\alpha} = 2\sigma/r$ yields

$$C_1 = C_{1,\,eq} \exp\left(\frac{V^{(s)} 2\sigma}{rRT}\right) \tag{19-45}$$

Since the exponential term is greater than unity, the solubility C_1 exceeds the equilibrium solubility $C_{1,\,eq}$. If the solid ionizes during dissolution, the exponential argument should be divided by i, the increase in the number of particles during dissolution and electrolytic dissociation.

Equation (19-45) may be criticized from many points of view. Small particles are probably not spherical, and a departure from this simple shape can considerably affect C/C_{eq} values [see, for example, Jones (1913)]. Also, the correspondence between concentration and activity may not be acceptable for highly soluble compounds. Finally, and most important, it is probably not reasonable to associate σ with a solid-liquid interfacial tension, since, for small aggregates of only a few molecules, this interfacial tension loses any meaning. Harbury (1946) suggests that a fictitious interfacial tension, σ', be used where, from the few data available, σ' is much less than σ.

A modification of Eq. (19-45) has been used in the development of two-dimensional nucleation and growth theories, [see Ohara and Reid (1973) for further discussion], where it is assumed that crystal growth is controlled by the rate of formation of small pancake-shaped nuclei on a large planar crystallite surface. For such a solid shape, as shown in Example 19.1, $P^{\beta} - P^{\alpha} = \sigma/r$, so, when Eq. (19-45) is derived, the factor of 2 has been dropped from the exponential term. Then, if the solution supersaturation for solute 1 is defined as

$$\mathrm{S} \equiv \frac{C_1 - C_{1,\,eq}}{C_{1,\,eq}} \tag{19-46}$$

the equilibrium pancake radius is

$$r = \frac{\sigma V^{(s)}}{RT \ln (1 + \mathrm{S})} \tag{19-47}$$

A definite size of a two-dimensional nucleus is then associated with a given supersaturation of the solution.

19.4 Pure-Component Relations

The Fundamental Equations for any two phases α and β, separated by a surface phase σ, have already been given in Eqs. (19-12) through (19-14). Let us assume that the interface between α and β is not highly curved so that $P^\alpha = P^\beta = P^\sigma$ and begin our development by considering a closed system in which there is but a single nonreacting component j present. We have also shown that, at equilibrium, $T^\alpha = T^\beta = T^\sigma = T$ and $\mu_j^\alpha = \mu_j^\beta = \mu_j^\sigma = \mu_j$. The total Legendre transform of Eqs. (19-12) through (19-14) yields Gibbs-Duhem equations [see Eq. (5-105)] for the α and β phases,

$$-S^\alpha\,dT + V^\alpha\,dP - d\mu_j = 0 \tag{19-48}$$

$$-S^\beta\,dT + V^\beta\,dP - d\mu_j = 0 \tag{19-49}$$

and for the *surface phase*, in extensive form,

$$-\underline{S}^\sigma\,dT + \underline{V}^\sigma\,dP - \underline{a}\,d\sigma - N_j^\sigma\,d\mu_j = 0 \tag{19-50}$$

We divide Eq. (19-50) by $\underline{a}$ to obtain the intensive form:

$$-S^\sigma\,dT + \tau\,dP - d\sigma - \Gamma\,d\mu_j = 0 \tag{19-51}$$

In Eq. (19-51), $S^\sigma = \underline{S}^\sigma/\underline{a}$ is the entropy per unit area. Similarly, $\tau = \underline{V}^\sigma/\underline{a}$ is the thickness of the surface and $\Gamma \equiv N_j^\sigma/\underline{a}$ is the surface concentration, moles of j per unit area.

Eliminating dP and $d\mu_j$ from Eqs. (19-48), (19-49), and (19-51),

$$-\frac{d\sigma}{dT} = (S^\sigma - \Gamma S^\beta) - \frac{(\tau - \Gamma V^\beta)(S^\alpha - S^\beta)}{V^\alpha - V^\beta} \tag{19-52}$$

Equation (19-52) is symmetrical in α and β and terms containing these superscripts may be interchanged. The Euler form of Eq. (19-14) may be written as

$$\sigma = H^\sigma - TS^\sigma - \Gamma\mu_j \tag{19-53}$$

If Eq. (19-52) is multiplied by T and the result added to Eq. (19-53) and then the Euler form of Eq. (19-13), as multiplied by Γ, is used to eliminate the $\Gamma\mu$ term, we get:

$$\sigma - T\frac{d\sigma}{dT} = (H^\sigma - \Gamma H^\beta) - \frac{(\tau - \Gamma V^\beta)(H^\alpha - H^\beta)}{V^\alpha - V^\beta} \tag{19-54}$$

where $H^\sigma \equiv \underline{H}^\sigma/\underline{a} = (\underline{U}^\sigma + P\underline{V}^\sigma)/\underline{a}$.

Equations (19-52) and (19-54) relate the surface tension and surface tension-temperature gradient to thermodynamic properties such as entropy, enthalpy, and volume. It is not immediately obvious, however, that these relations are invariant with respect to the variations in thickness, τ, of the surface. Suppose for instance, that τ is extended by a slight amount into phase β, then

$$\frac{dS^\sigma}{d\tau} = \frac{S^\beta}{V^\beta}$$

$$\frac{d\Gamma}{d\tau} = \frac{1}{V^\beta}$$

Differentiating Eq. (19-52) with respect to a variation in τ into phase β, the right-hand side becomes zero. Thus, the relation is not dependent on this particular variation in τ. Similar results are obtained if τ is increased into phase α.

This invariance leads, of course, to an indefiniteness about terms such as S^σ, H^σ, τ, and Γ. As noted above, for a given τ, there is a given S^σ. If τ is increased in phase β, per unit area, the entropy will increase by $S^\beta(d\tau/V^\beta)$. This new S^σ will depend on the value of S^β to a degree proportionate to the increase in τ. We are accustomed to visualizing the properties of the surface layer to be more representative of a less mobile phase or of a discontinuous phase in a two-liquid system. For example, in a liquid-gas system, we often imagine the surface phase to possess nearly liquid-like properties. (Note that this is always possible by choosing the position of the surface layer in a manner to exclude any appreciable gas phase.) By choices of this nature, we may simplify Eqs. (19-52) and (19-54) to more common forms. For example, in a gas–liquid system, if phase β were liquid and phase α gas, then τ/Γ is comparable to $V^\beta = V^L$, which is much less than $(V^\alpha - V^L) = (V^V - V^L)$. Equations (19-52) and (19-54) then become

$$-\frac{d\sigma}{dT} = S^\sigma - \Gamma S^L \tag{19-55}$$

$$\sigma - T\frac{d\sigma}{dT} = H^\sigma - \Gamma H^L \tag{19-56}$$

$$\sigma = G^\sigma - \Gamma G^L \tag{19-57}$$

Even in these simplified forms, S^σ, H^σ, and Γ are still indefinite and depend on τ, although neither $(S^\sigma - \Gamma S^L)$ nor $(H^\sigma - \Gamma H^L)$ depend on τ.

Surface tension decreases with increasing temperature, and a convenient estimation equation for nonpolar materials is given by Brock and Bird (1955):

$$\sigma = P_c^{2/3} T_c^{1/3}\, Q(1 - T_r)^{11/9} \tag{19-58}$$

where P_c is in bar, T_c in K, and σ in N/m.

$$Q = 1.207 \times 10^{-4}\left(1 + \frac{T_{b_r} \ln P_c}{1 - T_{b_r}}\right) - 2.81 \times 10^{-4} \tag{19-59}$$

and $T_{b_r} = T_b/T_c$.

Other estimation techniques have been suggested [Reid, Prausnitz, and Poling (1987)], and Scriven and Davis with their colleagues Carey and Boniorno (1976, 1978) have employed mean field theory to calculate surface tensions.

Since $d\sigma/dT$ is negative, $(S^\sigma - \Gamma S^L)$ is positive. This quantity represents the difference between the entropy (per unit area) in the surface film and the entropy the same surface would have if it had the properties of the bulk liquid phase. In a similar manner, $\sigma - T(d\sigma/dT)$ may be visualized as the enthalpy difference between the surface film and bulk liquid. With Eq. (19-57), the surface tension itself may be considered to be the difference in Gibbs surface energy and the bulk liquid Gibbs energy expressed on a unit area basis. This isothermal Gibbs energy change is often defined to be the reversible work required to bring material from the bulk liquid to form a unit area of surface.

Before leaving this section on pure component surface thermodynamics, let's review the developments presented thus far. In the initial sections of this chapter, from mechanical considerations only, the concept of surface tension was developed. These same arguments led to some interesting conclusions such as the pressure difference existing across curved interfaces, the enhanced vapor pressure over small convex liquid or solid surfaces, etc. The application of thermodynamics in this development was minimal although it was proved that in any equilibrium multiphase system, regardless of curvature, the temperature and chemical potentials were equal in all phases.

From these concepts, the surface tension and its temperature gradient were related to fundamental thermodynamic properties of the surface. The utility of this latter development may be questioned. The surface layer emphasized in the foregoing paragraphs is held by some to be completely fictitious and devoid of true physical significance; its depth is indefinite, it is a composite of the properties of the bounding phases, and the absolute values of S^σ, H^σ, etc. are dictated solely by the exact position and thickness of the layer--as defined by the user--not by nature.

19.5 Multicomponent Relations

The thermodynamics of surface layers in multicomponent systems is a straight-forward extension from those developed in Section 19.4 for single components, although the algebra is considerably more complex. The relations employed are shown below; as before, the surface-layer equations are expressed in terms of a unit area of surface (i.e., $S^\sigma = \underline{S}^\sigma/\underline{a}$; $\tau = \underline{V}^\sigma/\underline{a}$; $\Gamma_j = N_j^\sigma/\underline{a}$). The Gibbs-Duhem relation, Eq. (19-51), becomes

$$-d\sigma = S^\sigma\, dT - \tau\, dP + \sum_{k=1}^{n} \Gamma_k\, d\mu_k \tag{19-60}$$

For both phases α and β one may express the chemical potential of any component j as, for example, in phase β,

$$d\mu_j = -\overline{S}_j^\beta\, dT + \overline{V}_j^\beta\, dP + \sum_{k \neq i} \left(\frac{\partial \mu_j}{\partial x_k^\beta} \right)_{T,\, P,\, x[i,\, k]} dx_k^\beta \tag{19-61}$$

At *constant composition*, in the β and α phases, with Eqs. (19-60) and (19-61),

$$-d\sigma = \left(S^\sigma - \sum_{k=1}^{n} \Gamma_k \overline{S}_k^\beta \right) dT - \left(\tau - \sum_{k=1}^{n} \Gamma_k \overline{V}_k^\beta \right) dP \tag{19-62}$$

and from equations of the form of Eq. (19-61) for any component k, with all $dx = 0$,

$$dP = \frac{\overline{S}_k^\alpha - \overline{S}_k^\beta}{\overline{V}_k^\alpha - \overline{V}_k^\beta} dT \qquad (\text{all } x^\beta \text{ constant}) \tag{19-63}$$

Equation (19-62) may be substituted into Eq. (19-63) to eliminate all pressure terms. Applying Eq. (19-62) to a vapor-liquid system where the vapor phase is α and the liquid phase is β, $\tau/\Sigma\Gamma_k$ is comparable to the liquid-phase volume $V^L = \Sigma_{k=1}^n x_k^L \overline{V}_k^L$. Assuming ideal solutions where $\overline{V}_k^L = V_k^L$,

$$\tau - \sum_{k=1}^n \Gamma_k \overline{V}_k^L \approx \tau - V^L \sum_{k=1}^n \Gamma_k \approx \sum_{k=1}^n \Gamma_k \left(\frac{\tau}{\sum_i \Gamma_i} - V_k^L \right) \approx 0 \tag{19-64}$$

Thus, in this case,

$$-\left(\frac{\partial \sigma}{\partial T}\right)_L = S^\sigma - \sum_{k=1}^n \Gamma_k \overline{S}_k^L \quad (\text{constant composition in } L \text{ phase}) \tag{19-65}$$

Equation (19-64) is a good approximation to eliminate the "dP" term since the small residue $\Sigma \Gamma_i \overline{V}_i^L$ must still be divided [as from Eq. (19-63)] by a large number, $(\overline{V}_k^V - \overline{V}_k^L) \approx \Delta V_{vaporization}$, to reduce further the magnitude of the multiplier of the "dP" term. The form of Eq. (19-65) is identical to Eq. (19-55) and the discussion following the latter equation is also applicable to Eq. (19-65). That is, the equation is invariant with respect to the position or thickness of the surface layer as long as each bounding surface of the interfacial layer is in a different phase.

There is, however, a limitation to either Eq. (19-61) or (19-62). That is, the composition of all components in the liquid phase must be held constant. This restriction originates from the elimination of the $d\mu_j$ terms in Eq. (19-60) with Eq. (19-61) as simplified by forcing the terms containing $(\partial\mu_j/\partial x_k^L)dx_k^L$ to be zero. There is no thermodynamic inconsistency here since from the phase rule there are n independent variables and if $(n-1)$ compositions in the liquid phase are fixed, then σ can be expressed as a function of temperature only. Similar statements may be made if the derivation of Eq. (19-62) were made by using a relation of the form of Eq. (19-61) but for the α phase. The final resulting equation (19-62) or (19-65) does have this composition restriction, and as such, the general applicability is reduced. As in the case of many other multicomponent thermodynamic relations, other composition restrictions may be imposed and different results obtained.

19.6 Surface Tension—Composition Relationships

Euler integration of Eq. (19-14) allows one to define a Gibbs energy for the surface phase.

$$\underline{G}^{\sigma} = \underline{U}^{\sigma} - T\underline{S}^{\sigma} + P\underline{V}^{\sigma} = \sigma\underline{a} + \sum_{j=1}^{n} \mu_j N_j^{\sigma} \tag{19-66}$$

Let us restrict our presentation to a binary system of 1 and 2. Then, if Eq. (19-66) is divided by the total moles in the surface layer, $N^{\sigma} = N_1^{\sigma} + N_2^{\sigma}$,

$$G^{\sigma} = \frac{\underline{G}^{\sigma}}{N^{\sigma}} = \sigma\Omega + \mu_1 x_1^{\sigma} + \mu_2 x_2^{\sigma} \tag{19-67}$$

where G^{σ} is the Gibbs energy per mole in the surface layer, σ is the surface tension of the mixture, and x_1^{σ}, x_2^{σ} are the mole fractions of 1 and 2 in this layer. Ω is the area per mole and is often written

$$\Omega = x_1^{\sigma}\overline{\Omega}_1 + x_2^{\sigma}\overline{\Omega}_2 \tag{19-68}$$

$\overline{\Omega}_1$ and $\overline{\Omega}_2$ are partial molar areas for components 1 and 2. Substituting Eq. (19-68) into Eq. (19-67), we obtain

$$G^{\sigma} = x_1^{\sigma}(\mu_1 + \sigma\overline{\Omega}_1) + x_2^{\sigma}(\mu_2 + \sigma\overline{\Omega}_2) \tag{19-69}$$

Next, we can express the *surface* Gibbs energies for *pure* components 1 and 2 at the same temperature as

$$G_1^{\sigma} = \mu_1^o + \sigma_1\Omega_1 \quad \text{and} \quad G_2^{\sigma} = \mu_2^o + \sigma_2\Omega_2 \tag{19-70}$$

In Eq. (19-70), μ_1^o and μ_2^o refer to chemical potentials of pure 1 and 2, σ_1 and σ_2 are the pure-component surface tensions, and Ω_1 and Ω_2 are the pure-component surface areas per mole of material.

With Eqs. (19-69) and (19-70), we can form a mixing function,

$$\begin{aligned}\Delta G^{\sigma} = G^{\sigma} - x_1^{\sigma}G_1^{\sigma} - x_2^{\sigma}G_2^{\sigma} &= x_1^{\sigma}[(\mu_1 + \sigma\overline{\Omega}_1) - (\mu_1^o + \sigma_1\Omega_1)] \\ &+ x_2^{\sigma}[(\mu_2 + \sigma\overline{\Omega}_2) - (\mu_2^o + \sigma_2\Omega_2)]\end{aligned} \tag{19-71}$$

If we then write the analogous equation for the Gibbs energy mixing function for the bulk liquid phase,

$$\Delta G = x_1(\mu_1 - \mu_1^o) + x_2(\mu_2 - \mu_2^o) = x_1 RT \ln a_1 + x_2 RT \ln a_2 \tag{19-72}$$

Comparing Eqs. (19-71) and (19-72) suggests that it might be convenient to define "surface chemical potentials" as follows:

$$\zeta_1 \equiv \mu_1 + \sigma\overline{\Omega}_1 \quad \text{and} \quad \zeta_2 \equiv \mu_2 + \sigma\overline{\Omega}_2$$

and for pure component reference states, (19-73)

$$\zeta_1^o \equiv \mu_1^o + \sigma_1\Omega_1 \quad \text{and} \quad \zeta_2^o \equiv \mu_2^o + \sigma_2\Omega_2$$

Thus,

$$\begin{aligned}\zeta_1 - \zeta_1^o &= RT \ln a_1^{\sigma} = RT \ln \gamma_1^{\sigma} x_1^{\sigma} \\ \zeta_2 - \zeta_2^o &= RT \ln a_2^{\sigma} = RT \ln \gamma_2^{\sigma} x_2^{\sigma}\end{aligned} \tag{19-74}$$

Equation (19-74) then allows us to introduce surface activities and activity coefficients in terms of surface chemical potentials. Using Eqs. (19-73) and (19-74) for component 1, and noting that $\mu_1 - \mu_1^o = RT \ln a_1 = RT \ln \gamma_1 x_1$, we get

$$\zeta_1 - \zeta_1^o = (\mu_1 - \mu_1^o) + (\sigma\overline{\Omega}_1 - \sigma_1\Omega_1) = RT \ln \gamma_1^\sigma x_1^\sigma \tag{19-75}$$
$$= RT \ln \gamma_1 x_1 + (\sigma\overline{\Omega}_1 - \sigma_1\Omega_1)$$

Then

$$x_1^\sigma = \frac{x_1\gamma_1}{\gamma_1^\sigma} \exp\left[\frac{\sigma\overline{\Omega}_1 - \sigma_1\Omega_1}{RT}\right] \tag{19-76}$$

In Eq. (19-76), the surface composition of component 1 is related to the bulk composition (x_1) and activity coefficient (γ_1). The surface properties involved are γ_1^σ, σ, σ_1, $\overline{\Omega}_1$, and Ω_1. Writing the comparable equation for component 2 and adding,

$$x_1^\sigma + x_2^\sigma = 1 = \frac{x_1\gamma_1}{\gamma_1^\sigma} \exp\left[\frac{\sigma\overline{\Omega}_1 - \sigma_1\Omega_1}{RT}\right] + \frac{x_2\gamma_2}{\gamma_2^\sigma} \exp\left[\frac{\sigma\overline{\Omega}_2 - \sigma_2\Omega_2}{RT}\right] \tag{19-77}$$

Before discussing the use of Eq. (19-77), let us simplify it to illustrate an important point. Assume ideal solutions in both the bulk and surface layers and let $\overline{\Omega}_1 = \overline{\Omega}_2 = \Omega_1 = \Omega_2 = \Omega$; then, after rearrangement and expansion, we get

$$\sigma = x_1\sigma_1 + x_2\sigma_2 - \frac{\Omega}{2RT} x_1 x_2 (\sigma_1 - \sigma_2)^2 \tag{19-78}$$

With an estimate of Ω, the mixture surface tension may then be determined. Also, one should note that Eq. (19-78) predicts that σ is smaller than the mole fraction (bulk liquid) average of the pure-component surface tensions.

Returning to Eq. (19-76) or (19-77), these were used by Sprow and Prausnitz (1966) to estimate surface compositions and mixture surface tensions for simple liquid mixtures. In this case the surface was modeled as a regular solution to obtain surface activity coefficients. Also, surface areas were correlated with liquid volumes raised to the two-thirds power.

A different approach to estimate surface compositions uses the Gibbs-Duhem equation for the surface layer [Eq. (19-50)] [see also Shih and Chen (1968) and Guggenheim (1967)]. For a binary of 1 and 2 at constant temperature and pressure,

$$\underline{a}\, d\sigma + N_1^\sigma\, d\mu_1 + N_2^\sigma\, d\mu_2 = 0 \tag{19-79}$$

Dividing by N^σ, the moles in the surface layer, we obtain

$$\Omega\, d\sigma + x_1^\sigma\, d\mu_1 + x_2^\sigma\, d\mu_2 = 0 \tag{19-80}$$

where $\Omega = \underline{a}/N^\sigma$ and $x_1^\sigma = N_1^\sigma/N^\sigma$. From the Gibbs-Duhem equation written for the bulk liquid, at constant temperature and pressure,

$$x_1\, d\mu_1 + x_2\, d\mu_2 = 0 \tag{19-81}$$

Eliminating $d\mu_2$ between Eqs. (19-80) and (19-81) and making use of Eq. (19-68) yields

$$x_1^\sigma = \frac{x_1 - x_2\overline{\Omega}_2(d\sigma/d\mu_1)}{1 + x_2(\overline{\Omega}_1 - \overline{\Omega}_2)(d\sigma/d\mu_1)} \tag{19-82}$$

As in Eq. (19-76), partial areas $\overline{\Omega}_1$ and $\overline{\Omega}_2$ are necessary to estimate surface compositions when employing Eq. (19-82). The derivative $(d\sigma/d\mu_1)$ could be written as

$$\frac{d\sigma}{d\mu_1} = \frac{d\sigma/dx_1}{d\mu_1/dx_1} = \frac{d\sigma/dx_1}{RT\, d\ln(\gamma_1 x_1)/dx_1} \tag{19-83}$$

and evaluated from data showing how the mixture surface tension and bulk-liquid activity coefficients vary with composition.

Example 19.4

Estimate the ethanol surface mole fraction in a 20 mole % ethanol-water solution at 298 K. Some mixture data are given below for $T = 298$ K

Mole fraction ethanol	Surface tension (N/m × 10^3)	Ethanol activity coefficient	Liquid molar volume (cm^3/mol)
1.0	22.0	1.0	58.7
0.9	22.6	1.0	54.3
0.8	23.2	1.02	49.9
0.7	23.85	1.06	45.4
0.6	24.6	1.13	41.0
0.5	25.4	1.25	37.7
0.4	26.35	1.45	33.1
0.3	27.6	1.76	29.1
0.2	29.7	2.27	25.4
0.1	36.6	3.02	21.6
0.04	47.9	3.42	19.5
0	72.2	—	18.05

Solution

Equation (19-82) is used. To estimate the partial molar areas for ethanol and water, the relation given by Sprow and Prausnitz (1966) is used:

$$\overline{\Omega}_j = \left(\frac{\overline{V}_j^\sigma}{N_A}\right)^{2/3} \times N_A$$

where $\overline{V}_j^\sigma$ is the partial molar volume of component j in the surface layer. We will assume that this value is the same as in the bulk phase and, for ethanol mole fractions greater than about 0.4, the molar volume data show that $\bar{V}_{EtOH}$ = 58.7 cm^3/mol and $\bar{V}_{H_2O}$ = 15.8 cm^3/mol. Thus,

$$\Omega_{EtOH} = \left(\frac{58.7}{6.022 \times 10^{23}}\right)^{2/3} (6.022 \times 10^{23})$$

$$= 1.28 \times 10^9 \text{ cm}^2/\text{mol} = 1.28 \times 10^5 \text{ m}^2/\text{mol}$$

$$\Omega_{H_2O} = 5.3 \times 10^8 \text{ cm}^2/\text{mol} = 5.3 \times 10^4 \text{ m}^2/\text{mol}$$

To determine $(d\sigma/d\mu_{EtOH})$, this derivative is written as $(1/RT)[d\sigma/d \ln (\gamma x)_{EtOH}]$ and σ plotted versus $\ln (\gamma x)_{EtOH}$. The slope when $x_{EtOH} = 0.2$ is -1.6×10^{-2} N/m^2. Therefore,

$$\frac{d\sigma}{d\mu_{EtOH}} = \frac{-1.6 \times 10^{-2}}{(8.314)(298)} = -6.4 \times 10^{-6} \text{ mol/m}^2$$

With Eq. (19-82),

$$x^{\sigma}_{EtOH} = \frac{0.2 - (0.8)(5.3 \times 10^4)(-6.4 \times 10^{-6})}{1 + (0.8)(1.28 \times 10^5 - 5.3 \times 10^4)(-6.4 \times 10^{-6})} = 0.77$$

It is interesting to note that if Eq. (19-76) is used to estimate x^{σ}_{EtOH}, and γ^{σ}_{EtOH} is equated to γ_{EtOH} in the data table at $x_{EtOH} \sim 0.77$, then $\gamma^{\sigma}_{EtOH} = 1.04$.

$$x^{\sigma}_{EtOH} = \frac{(0.2)(2.27)}{1.04} \exp\left[\frac{(29.7 - 22.0)(10^{-3})(1.28 \times 10^5)}{(8.314)(298)}\right] = 0.65$$

The agreement between the two methods to estimate x^{σ}_{EtOH} is not particularly good. In Eq. (19-76), due to the exponential term, the calculated value of x^{σ}_{EtOH} is quite sensitive to the surface tensions and areas used in the computation.

19.7 Nucleation

Nucleation refers to the birth process of a new phase. It obviously has many important ramifications in science and engineering.

If a homogeneous, particle-free, pure vapor is cooled until the pressure is equal to the vapor pressure, thermodynamics would indicate that a liquid (or solid) phase should form. The actual facts show that this frequently does not occur. In such a case, the temperature must be decreased below the normal dew-point or saturation temperature before the second phase appears. Such undercooling is, however, not found if the gas contains many "dust-like" particles or if the gas were "seeded" with a small quantity of the liquid (or solid) phase when the dew point was reached. Similar statements apply to the precipitation of solids from liquids or the superheating of liquids.

The reason for these phenomena lies in the fact, developed in Section 19.3, that small "nuclei" of the new phase have higher vapor pressures (or more accurately, higher chemical potentials) than those of a bulk, planar phase. These nuclei will only be stable when present in a supersaturated mother phase.

The equilibrium and stability criteria that are applicable to the formation of a new phase have already been given in Eqs. (19-9) and (19-10); that is, at equilibrium, the availability function, $\underline{B}$, is either a minimum for stable equilibrium or a maximum for unstable equilibrium.

Let us assume that we have initially a β-free phase, α, which is always at T and P as dictated by isothermal and isobaric reservoirs (see Figure 19.2). We shall allow the formation of a β phase but at the same time insist that phase α remain at T and P. The moles in the α–β system are constant. We wish to determine the change in availability during this process. From the definition of $\underline{B}$ in Eq. (19-8),

$$\Delta\underline{B} = \underline{B} - \underline{B}_{int} \tag{19-84}$$

where $\underline{B}_{int}$ is the initial total availability before the β phase forms. Thus,

$$\begin{aligned}\Delta\underline{B} &= (\underline{U}^{\alpha} + P\underline{V}^{\alpha} - T\underline{S}^{\alpha}) + (\underline{U}^{\beta} + P\underline{V}^{\beta} - T\underline{S}^{\beta}) \\ &\quad + (\underline{U}^{\sigma} + P\underline{V}^{\sigma} - T\underline{S}^{\sigma}) - (\underline{U}^{\alpha} + P\underline{V}^{\alpha} - T\underline{S}^{\alpha})_{int}\end{aligned} \tag{19-85}$$

Now

$$N^{\alpha}_{int} = N^{\alpha} + N^{\beta} + N^{\sigma} \tag{19-86}$$

and

$$\begin{aligned}U^{\alpha} &= TS^{\alpha} - PV^{\alpha} + \mu^{\alpha} \\ U^{\beta} &= TS^{\beta} - P^{\beta}V^{\beta} + \mu^{\beta} \\ U^{\sigma} &= T\frac{\underline{S}^{\sigma}}{N^{\sigma}} - P^{\sigma}\frac{\underline{V}^{\sigma}}{N^{\sigma}} + \sigma\frac{\underline{a}}{N^{\sigma}} + \mu^{\sigma}\end{aligned} \tag{19-87}$$

so that substitution of Eqs. (19-86) and (19-87) into Eq. (19-85) yields

$$\begin{aligned}\Delta\underline{B} &= N^{\beta}[(\mu^{\beta} - \mu^{\alpha}) + (P - P^{\beta})V^{\beta}] + \sigma\underline{a} \\ &\quad + N^{\sigma}[(\mu^{\sigma} - \mu^{\alpha}) + (P - P^{\sigma})V^{\sigma}]\end{aligned} \tag{19-88}$$

Equation (19-88) is, of course, very similar to Eq. (19-15); the principal difference is that in the former we assumed that the temperatures were everywhere equal in the final system. A finite, rather than infinitesimal change was also proposed.

If we proceed one step further and assume that the final state is an equilibrium state, $\Delta\underline{B}$ equals the change in availability when a homogeneous phase α changes to an α–β system with small fragments of the β phase in equilibrium with the residual α phase. Also, at this terminal state, as we have shown in Eqs. (19-19) and (19-20), there is equality of chemical potentials throughout the system, so that Eq. (19-88) simplifies to

$$\Delta\underline{B} = N^{\beta}(P - P^{\beta})V^{\beta} + \sigma\underline{a} \tag{19-89}$$

where we have also neglected the very small contribution due to $N^{\sigma}(P - P^{\sigma})V^{\sigma}$. If phase β is a sphere, then from Eq. (19-23) with $\underline{a} = 4\pi r^2$ and $N^{\beta} = (4/3)\pi r^3/V^{\beta}$,

$$\Delta\underline{B} = \frac{4}{3}\pi\sigma r^2 = \frac{\frac{16}{3}\pi\sigma^3}{(P^\beta - P)^2} \tag{19-90}$$

Expressions for $(P^\beta - P)$ that were derived in Section 19.3 may be used to determine $\Delta\underline{B}$ in terms of P^α and other system variables.

Relative to the question of stability, we note from Eq. (19-88) that

$$\Delta\underline{B} = f(N^\beta, r)$$

since $\underline{V}^\beta$ and $\underline{a}$ depend on r. If we expand the availability in a Taylor series around the state of equilibrium, we get

$$\underline{B} - \underline{B}_{eq} = B_r\,\delta r + B_N\,\delta N^\beta + B_{rr}\,\delta r^2 + 2B_{rN}\,\delta r\,\delta N^\beta + B_{NN}\,(\delta N^\beta)^2 + \ldots \tag{19-91}$$

where we have adopted a shorthand notation for partial derivatives [e.g., $B_r = (\partial\underline{B}/\partial r)_{N^\beta}$, $B_{NN} = (\partial^2\underline{B}/(\partial N^\beta)^2)_r$, etc.]. The first-order terms are zero at equilibrium; that is, from Eq. (19-88), neglecting the surface phase since it contributes negligibly, and noting that μ^α, P, and σ are constants, we obtain

$$B_r = N^\beta\left(\frac{\partial\mu^\beta}{\partial r}\right)_{N^\beta} - \frac{4}{3}\pi r^3\left(\frac{\partial P^\beta}{\partial r}\right)_{N^\beta} + (P - P^\beta)(4\pi r^2) + 8\pi\sigma r \tag{19-92}$$

The first two terms cancel since

$$\left(\frac{\partial\mu^\beta}{\partial r}\right)_{N^\beta} = V^\beta\left(\frac{\partial P^\beta}{\partial r}\right)_{N^\beta} = \frac{\frac{4}{3}\pi r^3}{N^\beta}\left(\frac{\partial P^\beta}{\partial r}\right)_{N^\beta} \tag{19-93}$$

Thus,

$$B_r = (P - P^\beta)(4\pi r^2) + 8\pi\sigma r \tag{19-94}$$

and, as expected, *at equilibrium*, using Eq. (19-23), we have

$$B_r = 0$$

Next, we will show that B_N is also zero at equilibrium. Beginning again with Eq. (19-88) with a spherical nucleus assumed ($N^\beta V^\beta = 4/3\;\pi r^3$), we get

$$B_N = (\mu^\beta - \mu^\alpha) + N^\beta\left(\frac{\partial\mu^\beta}{\partial N^\beta}\right)_r - \frac{4}{3}\pi r^3\left(\frac{\partial P^\beta}{\partial N^\beta}\right)_r \tag{19-95}$$

but

$$\left(\frac{\partial\mu^\beta}{\partial N^\beta}\right)_r = V^\beta\left(\frac{\partial P^\beta}{\partial N^\beta}\right)_r = \frac{\frac{4}{3}\pi r^3}{N^\beta}\left(\frac{\partial P^\beta}{\partial N^\beta}\right)_r \tag{19-96}$$

Therefore,

$$B_N = \mu^\beta - \mu^\alpha = 0 \tag{19-97}$$

by Eq. (19-19). Next, considering the second-order terms, they may be rearranged by forming a sum of squares:

$$\underline{B} - \underline{B}_{eq} = B_{NN}\left(\delta N^\beta + \frac{B_{rN}}{B_{NN}}\delta r\right)^2 + \frac{(B_{NN}B_{rr} - B_{rN}^2)(\delta r)^2}{B_{NN}} \tag{19-98}$$

We need to develop expressions for B_{NN}, B_{rN}, and B_{rr}.

B_{NN} **:** Beginning with Eq. (19-97) and differentiating with respect to N^β, with μ^α a constant,

$$B_{NN} = \left(\frac{\partial \mu^\beta}{\partial N^\beta}\right)_r = V^\beta\left(\frac{\partial P^\beta}{\partial N^\beta}\right)_r \tag{19-99}$$

To obtain a more convenient expression for B_{NN}, since phase β is at constant T,

$$P^\beta = f(N^\beta, \underline{V}^\beta)$$

Applying Euler's theorem yields

$$0 = \underline{V}^\beta\left(\frac{\partial P^\beta}{\partial \underline{V}^\beta}\right)_{N^\beta} + N^\beta\left(\frac{\partial P^\beta}{\partial N^\beta}\right)_{\underline{V}^\beta} \tag{19-100}$$

Let

$$\kappa_T = -\left(\frac{\partial \ln \underline{V}}{\partial P}\right)_{T,N} \tag{19-101}$$

Then, with Eqs. (19-100), (19-101), and $\underline{V}^\beta = \frac{4}{3}\pi r^3$,

$$B_{NN} = \frac{\frac{4}{3}\pi r^3}{\kappa_T (N^\beta)^2} \tag{19-102}$$

B_{rr}**:** Differentiating Eq. (19-94) with respect to r, and using Eq. (19-23), we get

$$B_{rr} = -8\pi\sigma - \frac{16\pi\sigma^2}{(P^\beta - P)^2}\left(\frac{\partial P^\beta}{\partial r}\right)_{N^\beta} \tag{19-103}$$

But

$$\left(\frac{\partial P^\beta}{\partial r}\right)_{N^\beta} = \left(\frac{\partial P^\beta}{\partial \underline{V}^\beta}\right)_{N^\beta}\left(\frac{\partial \underline{V}^\beta}{\partial r}\right)_{N^\beta} = -\frac{3}{\kappa_T r} \tag{19-104}$$

So

$$B_{rr} = \frac{24\pi\sigma}{(P^\beta - P)\kappa_T} - 8\pi\sigma = \frac{12\pi r}{\kappa_T} - 8\pi\sigma \tag{19-105}$$

B_{rN} **:** Beginning with Eq. (19-94) and differentiating with respect to N^β, we have

$$B_{rN} = -4\pi r^2 \left(\frac{\partial P^\beta}{\partial N^\beta}\right)_r = -\frac{4\pi r^2}{\kappa_T N^\beta} \tag{19-106}$$

where Eqs. (19-100) and (19-101) have also been used.

Clearly, B_{NN} is positive since $\kappa_T > 0$. Thus, the stability of our system is determined solely by the coefficient $(B_{NN} B_{rr} - B_{rN}^2)$ in Eq. (19-98).

$$B_{NN} B_{rr} - B_{rN}^2 = \frac{\frac{4}{3}\pi r^3}{\kappa_T (N^\beta)^2}\left(\frac{12\pi r}{\kappa_T} - 8\pi\sigma\right) - \frac{16\pi^2 r^4}{\kappa_T^2 (N^\beta)^2} = \frac{(-8\pi\sigma)(\frac{4}{3}\pi r^3)}{\kappa_T (N^\beta)^2} \tag{19-107}$$

Also,

$$\frac{B_{NN} B_{rr} - B_{rN}^2}{B_{NN}} = -8\pi\sigma < 0 \tag{19-108}$$

The interpretation of this result is that a phase β embryo, in equilibrium with a mother phase α, is in a state of *unstable* equilibrium.

In this examination, the formation of a new phase β from a mother phase α actually proceeds by increasing the availability of a system—a progression that is counter to our intuition and to usual trends in thermodynamics. Normally, systems seek a state of lowest availability. In a molecular sense, by fluctuations in density, small fragments of a new phase are formed from a mother phase α which is in a metastable condition. All these new fragments are unstable and disappear until a fragment of a critical size is formed. Then, and only then, can the fragment grow to form a bulk new phase β.

The change in availability to form an equilibrium-sized sphere of phase β [Eq. (19-90)] is often employed in nucleation theory, wherein it is assumed that the rate of nucleation is proportional to $\exp(-\Delta B/RT)$. An interesting study of the superheating of liquids which demonstrates the utility of the availability concept is summarized in the work of Apfel (1971).

References and Suggested Readings

Adamson, A.W. (1982), *Physical Chemistry of Surfaces*, 4th ed., Wiley-Interscience, NY. [Chapters I and II provide complementary material from a Physical Chemist's perspective]

Apfel, R.E. (1970), "Vapor cavity formation in liquids," *Tech. Memo.* **62**, Acoustics Research Laboratory, Harvard University, Cambridge, MA.; see also Apfel, R.E. (1971), *J. Chem. Phys.*, **54**, p 62.

Brock, J.R. and Bird, R.B. (1955), "Surface tension and the principle of corresponding states," *AIChE J.*, **1**, p 174-177.

Carey, B.S., Scriven, L.E., and Davis, H.T. (1978), "Semiempirical theory of surface tensions of pure and normal alkanes and alcohols," *AIChE J.*, **24**, p. 1076-1080; and Bongiorno, V., Scriven, L.E., and Davis, H.T. (1976), "Molecular theory of fluid interfaces," *J. Coll. Int. Sci.*, **57**, p 462-475.

Guggenheim, E.A. (1967), *Thermodynamics*, 8th ed., North Holland, Amsterdam, p 159-169, 213-217, 237-239. [excellent review of surface thermodynamics]

Harbury, L. (1946), "Solubility and melting point as functions of particle size," *J. Phys. Chem.*, **50**, p 190.

Jones, W.J. (1913), "Über die Beziehung zwischen geometrischer Form und Dampfdruck, Löslichkeit, und Formenstabilität," *Ann. Phys.*, **41**, p 441.

Ohara, M. and Reid, R.C. (1973), *Modelling Crystal Growth Rates from Solution*, Prentice-Hall, Englewood Cliffs, NJ., Chap. 2.

Reid, R.C., Prausnitz, J.M., and Poling, B.E. (1987), *Properties of Gases and Liquids*, 4th ed. McGraw-Hill, NY., Chapter 12, p 632-655.

Shih, Y-P. and Chen, S-A. (1968), "A note on the thermodynamics of surface tensions of binary solution," *AIChE J.*, 14, p 973.

Sprow, F.B. and Prausnitz, J.M. (1966), "Surface tensions of simple liquid mixtures," *Trans. Faraday Soc.*, **62**, p 1105.

Problems

19.1. In a distillation tray there is considerable spray of very fine liquid droplets. Since small droplets have higher vapor pressures than plane liquid-vapor interfaces, we would like to know whether or not the small spray mists have significantly different relative volatilities than those of plane interfaces. Take, for example, ethanol-water at 298 K. Using the data given below, estimate the relative volatility of alcohol-to-water for 0.1-μm-diameter drops and compare this to the volatility for plane surfaces.

x (mole % ethanol)	P_{water} (Pa)	$P_{ethanol}$ (Pa)
0	3170	0.0
10	2890	2370
20	2720	3570
30	2590	4160
40	2450	4560
50	2310	4920
60	2110	5350
70	1770	5850
80	1330	6440
90	730	7110
100	0.0	7870

(a) Liquid composition, 50 mole % water.

(b) Planar vapor-liquid equilibrium at 298 K.

(c) Surface tension of 50 mole % ethanol-water solution at 298 K = 0.0254 N/m.

(d) Densities of liquid at 298 K are given in Example 19.4.

(e) Assume that the surface tension is independent of drop size.

(f) Assume that the partial molar volumes of both components in the liquid are independent of pressure.

19.2. Estimate the composition of the vapor in small vapor bubbles in equilibrium with a superheated liquid mixture of 95 mole % ethane and 5 mole % *n*-butane at 270 K and 1 bar pressure. Also, what is the pressure inside the vapor bubble and equilibrium bubble radius? Assume that the liquid mixture forms an ideal solution and that the vapor in the bubble is an ideal-gas mixture. Other data are shown below at $T = 270$ K and $P_{(liquid)} =$ 1 bar.

	Ethane	*n*-Butane
$T_r = T/T_c$	0.884	0.635
P_{vp}, Pa	2.21×10^6	9.12×10^4
$V_{(liquid)}$ (m^3/mol)	7.38×10^{-5}	9.68×10^{-5}

The surface tension of the mixture is estimated to be

$$\sigma = (3.54x_E + 15.3x_B - 5.02\, x_E x_B)\,(10^{-3})\ \text{J/m}^2$$

where x_E is the mole fraction ethane and x_B the mole fraction *n*-butane.

19.3. In the expression for surface work, Eq. (19-1), the pressure-volume term is given as $P\,d\underline{V}$. If one were dealing with a curved surface, the internal pressure differs from the external pressure. Which pressure should one use in Eq. (19-1)—or does it make any difference? Why?

19.4. Derive Eq. (19-41) and state any assumptions made.

19.5. What is the general equation showing the approximate slope of a plot of ln P (equilibrium) versus $1/T$ for small droplets of liquid? For water at 293 K where $\sigma = 0.07275$ N/m, $\Delta H_{vap} = 4.19 \times 10^4$ J/mol, $V^L = 18\ cm^3$/mol; for $r = 10^{-7}$ cm, what is the slope in K? How does the heat of vaporization vary with drop size?

19.6. A Ph.D. thesis at Syracuse University carried out by Dr. Fernandez has studied the growth rates of water-ice thin platelets in a subcooled water-salt solution. It was postulated that the platelets grew by extension of scalloped edges of the crystal perpendicular to the basal plane. His physical model studied two-dimensional growth of a single scallop as shown (see Figure P19.6).

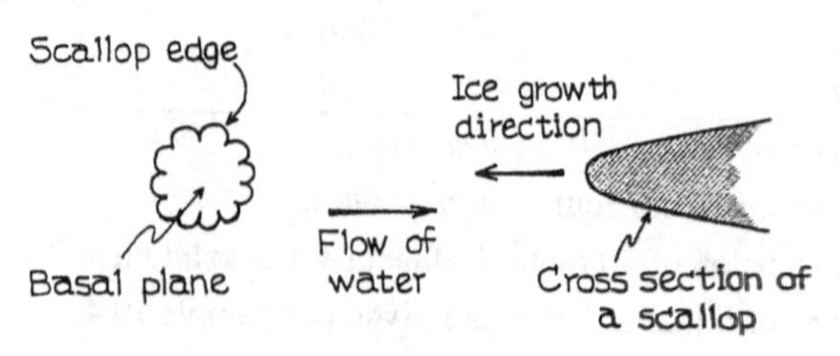

Figure P19.6

The geometric shape of a scallop is believed to be nearly parabolic, but the curvature near the tip is almost constant; thus, it is postulated that Eq. (19-41) may be used to estimate the tip surface temperature. In this case, the temperature at the ice tip T_e is given by

$$T_e = T_m - \frac{2\sigma T_m V^{(s)}}{r\Delta H_{fus}}$$

where σ is the interfacial tension between solution and pure water-ice, T_m is the freezing point of pure water, 273.2 K, $V^{(s)}$ is the molar volume of ice, ΔH_{fus} is the heat of fusion of ice, and r is the tip curvature. Assume no heat conduction through the ice.

(a) Derive this relationship for a salt solution containing x weight percent salt. What approximations are necessary?

Assuming that the equation given above is applicable, let the growth process be controlled by the rate of heat transfer away from the tip; that is,

$$q = h(T_e - T_\infty) \quad \text{and} \quad q = \frac{R\Delta H_{fus}}{V^{(s)}}$$

where q is the heat flux, h the heat transfer coefficient, T_∞ the bulk temperature in the salt bath, and R the linear growth rate of the tip.

The heat transfer coefficient for flow normal to a parabolic cylinder may be approximated as

$$\frac{hr}{k} = A\left(\frac{vr}{\eta}\right)^{1/2}$$

where r is the tip radius, k and η the thermal conductivity and kinematic viscosity of liquid water, respectively, v the impinging liquid velocity, and A is a constant.

(b) From these relations and the concept of a steady-state value of R during growth, determine how R depends on the experimental variable $\Delta T \equiv T_m - T_\infty$. For example, does R double if ΔT is doubled?

19.7. The surface tension of very dilute aqueous solutions of butanol has been measured at 298 K and reported by Harkins and Wampler [*J. Am. Chem. Soc.*, **53** (1931), 850]. These authors also report the activities of dilute solutions as a function of molality as follows:

Molality (m)	Activity[a]	Surface tension ($N/m \times 10^3$)
0.00329	0.00328	72.80
0.00658	0.00654	72.26
0.01320	0.01304	70.82
0.0264	0.02581	68.00
0.0536	0.05184	63.14
0.1050	0.09892	56.31
0.2110	0.19277	48.08
0.4330	0.37961	38.87
0.8540	0.71189	29.87

[a]Standard state, $a = m$ as $m \to 0$.

Estimate the mole fraction butanol in the interface at a bulk molality of 0.1050. If you wish, you may assume that, on the average, one molecule of butanol occupies about 0.27 nm^2 and one molecule of water, 0.07 nm^2.

19.8. A spherical bubble is enclosed by a thin ethanol film; the bubble is 1 mm in diameter and the film is 1×10^{-4} mm thick. The bubble is in equilibrium with surrounding air, which is at 1 bar and 298 K.

Data:

For ethanol, the surface tension is believed to be insensitive to pressure variations but is related to temperature as

$$\sigma = 0.021\left(1 - \frac{T - 298}{250}\right) \qquad (T \text{ in K}, \sigma \text{ in N/m})$$

The surface heat capacity is a constant:

$$C_a = T\left(\frac{\partial S}{\partial T}\right)_{\underline{a}} = 2.72 \text{ J/g K}$$

The density of ethanol is assumed independent of temperature or pressure and equals 0.79 g/cm^3 and $\Delta H_{vap,\,ethanol} = 930$ J/g at 298 K.

(a) Evaluate the surface energy of the ethanol film, $(\partial \underline{U}^\sigma / \partial \underline{a})_{P,T}$, J/m^2.

(b) A small tube is inserted into the bubble and air (at 298 K) blown in so that it is expanded rapidly to a diameter of 3 mm. Neglect any heat or mass transfer from the air to the liquid film and assume that the temperature of the film is uniform throughout. What is the temperature in this film after the expansion?

Appendixes

A Summary of the Postulates

I. For closed simple systems with given internal restraints, there exist stable equilibrium states that can be characterized completely by two independently variable properties in addition to the masses of the particular chemical species initially charged.

II. In processes for which there is no net effect on the environment, all systems (simple and composite) with given internal restraints will change in such a way that they approach one and only one stable equilibrium state for each simple subsystem. In the limiting condition, the entire system is said to be at equilibrium.

III. For any states, (1) and (2), in which a closed system is at equilibrium, the change of state represented by (1) → (2) and/or the reverse change (2) → (1) can occur by at least one adiabatic process, and the adiabatic work interaction between this system and its surroundings is determined uniquely by specifying the end states (1) and (2).

IV. If the sets of systems *A*, *B* and *A*, *C* each have no heat interaction when connected across nonadiabatic walls, then there will be no heat interaction if systems *B* and *C* are also so connected.

B Mathematical Relations of Functions of State

Let B be any property, primitive or derived, of a system and let x, y, and z be independently variable properties of a single-component system. [The results can readily be generalized to $(n+2)$ indepenent variables for an n-component system.] Since B is a function of state, a function f exists such that

$$B = f(x, y, z) \tag{B-1}$$

The function $f(x, y, z)$ is usually specified to within an arbitrary constant because derived properties are usually defined in terms of measured *differences* between two states. If the function f is known, differences in the value of B between two stable equilibrium states can be calculated as

$$\Delta B = B_2 - B_1 = f(x_2, y_2, z_2) - f(x_1, y_1, z_1) \tag{B-2}$$

In many cases, we may not know $f(x, y, z)$ explicitly, but we may have the differential form of Eq. (B-1):

$$dB = \left(\frac{\partial f}{\partial x}\right)_{y,z} dx + \left(\frac{\partial f}{\partial y}\right)_{x,z} dy + \left(\frac{\partial f}{\partial z}\right)_{x,y} dz \tag{B-3}$$

If all three partial derivatives are known, $f(x, y, z)$ can be evaluated to within an arbitrary constant by the method of indefinite integrals. Integrating first with respect to x, we obtain

$$f(x, y, z) = \int \left(\frac{\partial f}{\partial x}\right)_{y, z} dx + g(y, z) \tag{B-4}$$

where y and z are held constant in the integration and g is a function of y and z only. If we differentiate Eq. (B-4) with respect to y at constant x and z, and if we equate the result to the known function $(\partial f/\partial y)_{x, z}$, we get

$$\left(\frac{\partial g}{\partial y}\right)_z = \left(\frac{\partial f}{\partial y}\right)_{x, z} - \frac{\partial}{\partial y}\left[\int \left(\frac{\partial f}{\partial x}\right)_{y, z} dx\right]_{x, z} \tag{B-5}$$

Integrating with respect to y while holding x and z constant, we obtain

$$g(y, z) = \int \left(\frac{\partial f}{\partial y}\right)_{x, z} dy - \int \frac{\partial}{\partial y}\left[\int \left(\frac{\partial f}{\partial x}\right)_{y, z} dx\right]_{x, z} dy + g'(z) \tag{B-6}$$

where g' is a function of z only. Substituting Eq. (B-6) into Eq. (B-4) yields

$$f(x, y, z) = \int \left(\frac{\partial f}{\partial x}\right)_{y, z} dx + \int \left(\frac{\partial f}{\partial y}\right)_{x, z} dy - \int \frac{\partial}{\partial y}\left[\int \left(\frac{\partial f}{\partial x}\right)_{y, z} dx\right]_{x, z} dy + g'(z) \tag{B-7}$$

The function $g'(z)$ can be evaluated to within an arbitrary constant by repeating the procedure.

For functions of more than two variables, the method of indefinite integrals is somewhat laborious, and it is only worth the effort if an analytical solution of $f(x, y, z)$ is desired.

In general, we are interested in evaluating numerical differences in B, and a somewhat simpler solution can usually be obtained by integrating Eq. (B-3) over a specific path. Since B is a state function, the value of ΔB will be independent of the path chosen for integration; hence, any convenient path will suffice. One such path is to proceed from x_1, y_1, z_1 to x_2, y_1, z_1 to x_2, y_2, z_1 and then to the final state x_2, y_2, z_2. In this case it can be readily shown that

$$\Delta B = \int_{x_1}^{x_2} \left(\frac{\partial f}{\partial x}\right)_{y_1, z_1} dx + \int_{y_1}^{y_2} \left(\frac{\partial f}{\partial y}\right)_{x_2, z_1} dy + \int_{z_1}^{z_2} \left(\frac{\partial f}{\partial z}\right)_{x_2, y_2} dz \tag{B-8}$$

Finally, if we are faced with the problem of evaluating the difference in a function ϕ given a differential equation of the type

$$d\phi = M\,dx + N\,dy + Q\,dz \tag{B-9}$$

we can use any of the methods described above if it can be shown that ϕ is a state function of the variables, x, y, and z. It can be shown that the necessary and sufficient requirement for this condition is that each of the following equations be satisfied:

$$\left(\frac{\partial M}{\partial y}\right)_{x, z} = \left(\frac{\partial N}{\partial x}\right)_{y, z} \quad \left(\frac{\partial M}{\partial z}\right)_{x, y} = \left(\frac{\partial Q}{\partial x}\right)_{y, z} \quad \left(\frac{\partial N}{\partial z}\right)_{x, y} = \left(\frac{\partial Q}{\partial y}\right)_{x, z} \tag{B-10}$$

Alternatively, if any of equation set (B-10) is not satisfied, $d\phi$ is not an exact differential (e.g., either ϕ is not a state function or ϕ is a state function of variables other than x, y, and z). In this case $\int d\phi$ is called a *line integral* because the value of $\Delta\phi$ will depend on the specific path used for integration.

Some important thermodynamic variables are not state functions or properties (i.e., work and heat interactions). Differentials of these functions are denoted by using a δ in place of the d of the differential operator as in $\delta\phi$. Such functions are sometimes referred to as *Pfaffians*.

C Derivation of Euler's Theorem

Consider a function $f(a, b, x, y)$ which is homogeneous to the degree $h > 0$ in x and y and to degree 0 in a and b. By definition, if the variables x and y are each multiplied by a factor k, the value of $f(a, b, kx, ky)$ will be increased by a factor of k^h. Thus, for any value of k, we have

$$f(a, b, X, Y) = k^h f(a, b, x, y) \tag{C-1}$$

where

$$X = kx \quad \text{and} \quad Y = ky$$

Equating the total differentials of Eq. (C-1) and treating k as a variable since Eq. (C-1) is valid for all k, we obtain

$$\begin{aligned} &\frac{\partial}{\partial a}[f(a, b, X, Y)]_{b, X, Y}\,da + \frac{\partial}{\partial b}[f(a, b, X, Y)]_{a, X, Y}\,db \\ &\quad + \frac{\partial}{\partial X}[f(a, b, X, Y)]_{a, b, Y}\,dX + \frac{\partial}{\partial Y}[f(a, b, X, Y)]_{a, b, X}\,dY \\ &= (k^h)\frac{\partial}{\partial a}[f(a, b, x, y)]_{b, x, y}\,da + (k^h)\frac{\partial}{\partial b}[f(a, b, x, y)]_{a, x, y}\,db \\ &\quad + (k^h)\frac{\partial}{\partial x}[f(a, b, x, y)]_{a, b, y}\,dx + (k^h)\frac{\partial}{\partial y}[f(a, b, x, y)]_{a, b, x}\,dy \\ &\quad + (hk^{h-1})[f(a, b, x, y)]\,dk \end{aligned} \tag{C-2}$$

but

$$dX = k\,dx + x\,dk \quad \text{and} \quad dY = k\,dy + y\,dk \tag{C-3}$$

Substituting Eq. (C-3) into Eq. (C-2), and collecting terms, we obtain

$$\begin{aligned} &\left\{\frac{\partial}{\partial a}[f(a, b, X, Y)] - (k^h)\frac{\partial}{\partial a}[f(a, b, x, y)]\right\} da \\ &\quad + \left\{\frac{\partial}{\partial b}[f(a, b, X, Y)] - (k^h)\frac{\partial}{\partial b}[f(a, b, x, y)]\right\} db \\ &\quad + \left\{(k)\frac{\partial}{\partial X}[f(a, b, X, Y)] - (k^h)\frac{\partial}{\partial x}[f(a, b, x, y)]\right\} dx \\ &\quad + \left\{(k)\frac{\partial}{\partial Y}[f(a, b, X, Y)] - (k^h)\frac{\partial}{\partial y}[f(a, b, x, y)]\right\} dy \\ &\quad + \left\{(x)\frac{\partial}{\partial X}[f(a, b, X, Y)] + (y)\frac{\partial}{\partial Y}[f(a, b, X, Y)] - hk^{h-1}f(a, b, x, y)\right\} dk = 0 \end{aligned} \tag{C-4}$$

Since a, b, x, y, and k are independent, Eq. (C-4) is valid only if the coefficients of da, db, dx, dy, and dk are each zero. Thus,

$$\frac{\partial}{\partial a}[f(a, b, X, Y)] = (k^h)\frac{\partial}{\partial a}[f(a, b, x, y)] \tag{C-5}$$

$$\frac{\partial}{\partial b}[f(a, b, X, Y)] = (k^h)\frac{\partial}{\partial b}[f(a, b, x, y)] \tag{C-6}$$

$$\frac{\partial}{\partial X}[f(a,b,X,Y)] = (k^{h-1})\frac{\partial}{\partial x}[f(a,b,x,y)] \tag{C-7}$$

$$\frac{\partial}{\partial Y}[f(a,b,X,Y)] = (k^{h-1})\frac{\partial}{\partial y}[f(a,b,x,y)] \tag{C-8}$$

$$(x)\frac{\partial}{\partial X}[f(a,b,X,Y)] + (y)\frac{\partial}{\partial Y}[f(a,b,X,Y)] = (hk^{h-1})[f(a,b,x,y)] \tag{C-9}$$

Substituting Eqs. (C-7) and (C-8) into Eq. (C-9), we obtain

$$(x)\frac{\partial}{\partial x}[f(a,b,x,y)] + (y)\frac{\partial}{\partial y}[f(a,b,x,y)] = h[f(a,b,x,y)] \tag{C-10}$$

Equation (C-10) is a form of Euler's theorem that applies specifically to a function of four variables where only two (x and y) are homogeneous to a degree greater than zero. Note that Eq. (C-10) contains terms only in those variables for which f is homogeneous to degree h.

Applications. The thermodynamic functions of interest to us are special cases of homogeneous functions. In particular, these functions are either homogeneous to the first degree in mass (extensive) with $h = 1.0$ or homogeneous to the zeroth degree in mass (intensive) with $h = 0$. The arbitrary multiplier, k, will always be equal to the mass or moles of the system, N (or $1/N$ as the case may be).

Energy:

$$\underline{U} = f(\underline{S}, \underline{V}, N) \tag{C-11}$$

Since $\underline{U}$ is first order in mass (or moles), and since $\underline{S}$, $\underline{V}$, N are all proportional to mass, we have

$$\underline{U}(k\underline{S}, k\underline{V}, kN) = k\,\underline{U}(\underline{S}, \underline{V}, N)$$

Therefore, $x = \underline{S}$; $y = \underline{V}$; and $z = N$ so that Eq. (C-10) leads to

$$\underline{U} = \left(\frac{\partial \underline{U}}{\partial \underline{S}}\right)_{\underline{V},N}\underline{S} + \left(\frac{\partial \underline{U}}{\partial \underline{V}}\right)_{\underline{S},N}\underline{V} + \left(\frac{\partial \underline{U}}{\partial N}\right)_{\underline{S},\underline{V}}N = T\underline{S} - P\underline{V} + \mu N \tag{C-12}$$

Enthalpy:

$$\underline{H} = f(\underline{S}, P, N) \tag{C-13}$$

If we multiply the mass (or moles) N by k, we will increase $\underline{S}$ and N by a factor of k, but P will remain unchanged. That is,

$$\underline{H}(k\underline{S}, P, kN) = k\,\underline{H}(\underline{S}, P, N)$$

Thus, $a = P$; $x = \underline{S}$; and $y = N$ so that from Eq. (C-10),

$$\underline{H} = \left(\frac{\partial \underline{H}}{\partial \underline{S}}\right)_{P,N}\underline{S} + \left(\frac{\partial \underline{H}}{\partial N}\right)_{\underline{S},P}N = T\underline{S} + \mu N \tag{C-14}$$

If we consider the general case of a function $\underline{B}$ in $(n+2)$ variables of which m are homogeneous to the first degree in mass (or moles) N and $(n+2-m)$ are homogeneous to the zeroth degree in mass, then

$$\underline{B} = NB = \sum_{i=1}^{m}\left(\frac{\partial \underline{B}}{\partial \underline{Z}_i}\right)\underline{Z}_i \tag{C-15}$$

where $\underline{Z}_i$ is extensive for $i = 1, \ldots, m$ (e.g., $\underline{S}$, $\underline{V}$, or N).

D Cramer's Rule and Determinant Properties

For a set of n equations in n unknowns $(z_1,..., z_n)$, we can write

$$
\begin{aligned}
a_{11}z_1 + a_{12}z_2 + \cdots + a_{1n}z_n &= b_1 \\
a_{21}z_1 + a_{22}z_2 + \cdots + a_{2n}z_n &= b_2 \\
\vdots \\
a_{n1}z_1 + a_{n2}z_2 + \cdots + a_{nn}z_n &= b_n
\end{aligned} \tag{D-1}
$$

or in vector form

$$[\underset{\approx}{a}]\,[\underset{\sim}{z}] = [\underset{\sim}{b}] \tag{D-2}$$

Then Cramer's rule follows if det $[\underset{\approx}{a}] \neq 0$ with:

$$z_j = \frac{\det\,[\underset{\approx}{a}^{[j]},\ \underset{\sim}{b}]}{\det\,[\underset{\approx}{a}]} = \frac{|\underset{\approx}{a}^{[j]},\ \underset{\sim}{b}|}{|\underset{\approx}{a}|} \tag{D-3}$$

where $|\underset{\approx}{a}^{[j]},\ \underset{\sim}{b}|$ is the determinant formed by replacing the jth coloumn of $|\underset{\approx}{a}|$ by the column vector consisting of n elements of $|\underset{\sim}{b}|$, or

$$|\underset{\approx}{a}^{[j]}, \underset{\sim}{b}| \equiv \begin{vmatrix} a_{11} & \cdots & a_{1(j-1)} & b_1 & a_{1(j+1)} & \cdots & a_{1n} \\ a_{21} & \cdots & a_{2(j-1)} & b_2 & a_{2(j+1)} & \cdots & a_{2n} \\ \vdots & & \vdots & \vdots & \vdots & & \vdots \\ a_{n1} & \cdots & a_{n(j-1)} & b_n & a_{n(j+1)} & \cdots & a_{nn} \end{vmatrix} \tag{D-4}$$

Application of Eq. (D-3) to invariant systems with n equations in n unknowns is straightforward; see for example Eqs. (17-13a) and (17-18).

For monovariant systems, the situation is complicated because the number of equations is one less than the number of unknowns. Cramer's rule can still be applied but some modifications are necessary. Consider the following example. Starting with Gibbs-Duhem equation set (17-36) for π phases with no chemical reactions ($r = 0$). In this case, $\pi = n + 1$ because $\mathcal{F} = 1 = n + 2 - \pi$. The total number of unknowns (dT, dP, $d[\mu_1/T]$, $d[\mu_2/T]$,..., $d[\mu_n/T]$) is equal to $(n + 2)$ and the number of equations is π. Thus, the equation set is under-specified. To show how we can solve for one variable in terms of another using Cramer's rule, transform equation set (17-36) by moving all terms in dT to the right-hand side. Thus,

$$
\begin{aligned}
-\frac{V^{(1)}}{T}\,dP + \sum_{i=1}^{n} x_i^{(1)}\,d[\mu_i/T] &= \frac{-H^{(1)}}{T^2}\,dT \\
\vdots \qquad\qquad \vdots \qquad\qquad &\vdots \\
-\frac{V^{(\pi)}}{T}\,dP + \sum_{i=1}^{n} x_i^{(\pi)}\,d[\mu_i/T] &= \frac{-H^{(\pi)}}{T^2}\,dT
\end{aligned}
\tag{D-5}
$$

Now we have a set of $\pi = n + 1$ equations in $(n + 1)$ unknowns, where Cramer's rule can be applied directly. In this system:

$$
[\underset{\approx}{a}] = \begin{bmatrix} -\dfrac{V^{(1)}}{T} & x_1^{(1)} & x_2^{(1)} & \cdots & x_n^{(1)} \\ \cdot & \cdot & \cdot & \cdot & \cdot \\ \cdot & \cdot & \cdot & \cdot & \cdot \\ \cdot & \cdot & \cdot & \cdot & \cdot \\ -\dfrac{V^{(\pi)}}{T} & x_1^{(\pi)} & x_2^{(\pi)} & \cdots & x_n^{(\pi)} \end{bmatrix} \qquad [\underset{\sim}{b}] = \begin{bmatrix} -\dfrac{H^{(1)}}{T^2}\,dT \\ \cdot \\ \cdot \\ \cdot \\ -\dfrac{H^{(\pi)}}{T^2}\,dT \end{bmatrix} = \begin{bmatrix} -\dfrac{H^{(1)}}{T^2} \\ \cdot \\ \cdot \\ \cdot \\ -\dfrac{H^{(\pi)}}{T^2} \end{bmatrix} dT \tag{D-6}
$$

If we are interested in the solution for $dP = z_1$, then we can now apply Eq. (D-3) to get

$$
dP = z_1 = \frac{|\underset{\approx}{a}^{[1]}, \underset{\sim}{b}|}{|\underset{\approx}{a}|} \tag{D-7}
$$

or rewritten in Clapeyron form as

$$
\frac{dP}{dT} = \frac{\begin{vmatrix} -H^{(1)}/T^2 & x_1^{(1)} & \cdots & x_n^{(1)} \\ \cdot & \cdot & \cdot & \cdot \\ \cdot & \cdot & \cdot & \cdot \\ \cdot & \cdot & \cdot & \cdot \\ -H^{(\pi)}/T^2 & x_1^{(\pi)} & \cdots & x_n^{(\pi)} \end{vmatrix}}{\begin{vmatrix} -V^{(1)}/T & x_1^{(1)} & \cdots & x_n^{(1)} \\ \cdot & \cdot & \cdot & \cdot \\ \cdot & \cdot & \cdot & \cdot \\ \cdot & \cdot & \cdot & \cdot \\ -V^{(\pi)}/T & x_1^{(\pi)} & \cdots & x_n^{(\pi)} \end{vmatrix}} = \frac{|\Delta \underset{\approx}{H}|}{T\,|\Delta \underset{\approx}{V}|} \tag{D-8}
$$

Equation (D-8) is identical to Eq. (17-37).

E Generalized Cubic EOS Solver

Many *PVTN* equations of state are pressure-explicit and cubic in volume. For example, the Peng-Robinson EOS given in Eq. (8-46) can be rearranged as a cubic equation:

$$V^3+\left(b-\frac{RT}{P}\right)V^2+\left(\frac{a(\omega,T_r)}{P}-3b^2-\frac{2bRT}{P}\right)V+\left(b^3+\frac{b^2RT}{P}-\frac{a(\omega,T_r)b}{P}\right)=0 \qquad \text{(E-1)}$$

where the parameters $a(\omega, T_r)$ and b are as defined in Eqs. (8-47) through (8-51). Substituting $Z=PV/RT$ into Eq. (E-1) yields:

$$Z^3-(1-B)Z^2+(A-3B^2-2B)Z-(AB-B^2-B^3)=0 \qquad \text{(E-2)}$$

where

$$A\equiv\frac{a(\omega,T_r)P}{R^2T^2}=\frac{0.4572\,\alpha(\omega,T_r)P_r}{T_r^2} \qquad \text{(E-3)}$$

$$B\equiv\frac{bP}{RT}=\frac{0.0778P_r}{T_r} \qquad \text{(E-4)}$$

The objective frequently is to solve for V or Z with T and P specified.

***Analytical Solution*:** For any cubic equation of the form given below there will be three roots, x_1, x_2, and x_3:

$$x^3+p_1x^2+p_2x+p_3=0 \qquad \text{(E-5)}$$

Let $$Q\equiv\frac{3p_2-p_1^2}{9};\quad R\equiv\frac{9p_1p_2-27p_3-2p_1^3}{54};\quad D^*\equiv Q^3+R^2 \qquad \text{(E-6)}$$

D^* is called the discriminant and its value determines the domain of the roots to Eq. (E-5): (1) If $D^*<0$, all roots are real and unequal (2) if $D^*=0$, all roots are real and at least two are equal and (3) if $D^*>0$, only one root is real and two are imaginary.

For $D^*>0$, the real root is given by

$$x_1=S+T-\frac{1}{3}p_1 \qquad \text{(E-7)}$$

and the two imaginary roots ($i\equiv\sqrt{-1}$)

$$x_{2,3}=-\frac{1}{2}(S+T)-\frac{1}{3}p_1\pm\frac{1}{2}i\sqrt{3}\,(S-T) \qquad \text{(E-8)}$$

where

$$S=(R+\sqrt{D^*})^{1/3}\quad\text{and}\quad T=(R-\sqrt{D^*})^{1/3}$$

If $D^*=0$, $S=T$ and the imaginary components disappear--so 3 real roots result. For $D^*<0$, the roots are given by:

$$x_1 = 2\sqrt{-Q}\cos\left(\frac{\theta}{3}\right) - \frac{1}{3}p_1\ ;\quad x_2 = 2\sqrt{-Q}\cos\left(\frac{\theta}{3} + 120^\circ\right) - \frac{1}{3}p_1\ ;$$
$$x_3 = 2\sqrt{-Q}\cos\left(\frac{\theta}{3} + 240^\circ\right) - \frac{1}{3}p_1 \qquad \text{(E-9)}$$

where $\cos\theta = R/\sqrt{-Q^3}$

Example E.1

Calculate the equilibrium liquid and vapor molar volume of CO_2 at 35.64 bar and 274 K using the Peng-Robinson EOS. For CO_2 use $P_c = 73.76$ bar, $T_c = 304.2$ K, and $\omega = 0.225$.

Solution

Use the compressibility form of the EOS [Eq. (E-2)] with the parameters defined in Eqs. (8-49), (8-50), (E-3), and (E-4) and the properties cited above:

$$\kappa(\omega) = 0.70795\ ;\quad \alpha(\omega,T_r) = 1.07342\ ;\quad A = 0.29227\ ;\quad B = 0.04173$$

Thus, Eq. (E-2) becomes:

$$Z^3 - 0.95827\,Z^2 + 0.20359\,Z - 0.01038 = 0 \qquad \text{(E-10)}$$

Using Eq. (E-6) to calculate Q, R, and D^*,

$$Q = -\,0.03417\ ;\quad R = 0.00527\ ;\quad D^* = -1.21 \times 10^{-5}$$

With $D^* < 0$, there are three real roots, as expected since P and T are below their critical values. Using Eq. (E-9):

$$\cos\theta = 0.83434 \quad \text{and} \quad \theta = 33.4^\circ$$

Thus,

$$Z_1 = 0.68215\ ;\quad Z_2 = 0.07614\ ;\quad \text{and } Z_3 = 0.19998 \qquad \text{(E-11)}$$

In the vapor-liquid region, only the largest (vapor) and smallest (liquid) roots are meaningful because the middle root is in a region of intrinsic instability. If the system is in equilibrium, then $P = P_{vp} = f(T)$ and we are on the vapor-liquid coexistence curve.

At any other condition, no particular conclusions can be made regarding relationships between P and T from the roots, other than smallest value of Z (or V) corresponds to liquid and the largest value to vapor.

To evaluate V_g and V_ℓ we use the defining equation for Z

$$V_g = \frac{Z_1RT}{P} = \frac{(0.68215)(8.314)(274)}{(35.638 \times 10^5)} = 4.36 \times 10^{-4}\ \text{m}^3/\text{mol or } 436\ \text{cm}^3/\text{mol}$$

and

$$V_\ell = \frac{Z_2RT}{P} = 4.87 \times 10^{-5}\ \text{m}^3/\text{mol or } 48.7\ \text{cm}^3/\text{mol}$$

The experimental values for CO_2 are $V_g = 438.9$ cm^3/mol and $V_\ell = 47.7$ cm^3/mol.

***Numerical approximation*:** An alternative method for the solution of cubic equations is to use the general technique of successive substitutions, which can be applied to any nonlinear equation in one variable. As the name suggests, this method employs an iterative scheme starting with an initial guess to converge upon a solution. To use this method, the general nonlinear equation to be solved:

$$f(x) = 0 \tag{E-12}$$

must be transformed to the following form:

$$x = F(x) \tag{E-13}$$

The iterative process then consists of choosing a value x_j and using it to calculate a new value x_{j+1} using Eq. (E-13) as follows:

$$x_{j+1} = F(x_j) \tag{E-14}$$

after which the calculated x_{j+1} becomes the new value for x_j in the next iteration. Iterations continue until $x_{j+1} = x_j$, which then is the solution.

A variation of the method of successive substitutions which results in accelerated convergence is due to Wegstein. In this version, x_{j+1} is not calculated directly from $F(x_j)$ but instead from an intermediate value y_{j+1} and a parameter q as shown in Eqs. (E-15) and (E-16):

$$y_{j+1} = F(x_j) \tag{E-15}$$

$$x_{j+1} = q\, x_j + (1 - q)\, y_{j+1} \tag{E-16}$$

The value of q is arbitrary, but its range is related to the nature of the substitution process. (see Perry *et al.*, 1984). For the first iteration step q is set to zero and in subsequent iterations according to the recursive formula:

$$q_j = \frac{F(x_j) - F(x_{j-1})}{(F(x_j) - F(x_{j-1})) - (x_j - x_{j-1})} \tag{E-17}$$

Initial guesses for the roots also are arbitrary, but some helpful guidelines for cubic EOSs are available to ensure rapid convergence. For example, a good initial value for the liquid volume is $V_0 = b$ or $Z_0 = bP/RT$ and for the vapor volume is $V_0 = RT/P$ or $Z_0 = 1$ corresponding to an ideal gas.

Even with a convergent solution, we still need to check for thermodynamic consistency. With a cubic pressure-explicit EOS, it is most convenient to use the $A_{VV} > 0$ stability criteria for a pure component. If both roots satisfy the criteria, then the larger one is the vapor volume and the smaller one is the liquid volume. The third root would be at an intermediate volume but it would be unstable with $A_{VV} < 0$. For the Peng-Robinson EOS,

$$A_{VV} = -\left(\frac{\partial P}{\partial V}\right)_T = \frac{-RT}{(V-b)^2} + \frac{a(\omega, T_r)(2V + 2b)}{[V(V+b) + b(V-b)]^2} \tag{E-18}$$

so we can quickly test to see if $A_{VV} > 0$ for given values of $V(T, P)$. As expected, only the stable liquid and vapor values of V in Example E.1 satisfy the stability criteria. Now we can test the Wegstein method for convergence and accuracy.

Example E.2

Repeat Example E.1 using Wegstein's method.

Solution

Rearranging Eq. (E-10) into the form of Eq. (E-13) we get

$$Z = 0.05098 - 4.91183\,Z^3 + 4.70686\,Z^2 = F(Z) \tag{E-19}$$

For the first iteration (j = 0) to determine the liquid volume, the initial guess for Z_0 is calculated from $Z_0 = bP/RT$, where $b = 0.0778\,RT_c/P_c$ (for the Peng- Robinson EOS) = 2.67×10^{-5} m^3/mol. Thus,

$$Z_0 = \frac{(2.67 \times 10^{-5})(35.638)}{(8.314)(274)} = 0.04173$$

Equation (E-15) gives y_1 using $F(Z)$ in Eq. (E-19):

$$y_1 = f(Z_0) = 0.05882$$

with q = 0 for the first iteration, $Z_1 = y_1$ as well. For the second iteration (j = 1), $y_2 = F(Z_1) = 0.06626$ and $q = 0.77201$ using Eq. (E-17). Then from Eq. (E-16),

$$Z_2 = (-0.77201)(0.05882) + (1-(-0.77201))(0.06626) = 0.07201$$

The results of six iterations are summarized in the table below.

j	Z_j	$y_{j+1} = F(Z_j)$	q	Z_{j+1}
0	0.04173	0.05882	0	0.05882
1	0.05882	0.06626	-0.77201	0.07201
2	0.07201	0.07355	-1.23483	0.07546
3	0.07546	0.07567	-1.59066	0.07601
4	0.07601	0.07602	-1.69117	0.07603
5	0.07603	0.07603	-1.70607	0.07603
6	0.07603	0.07603	-1.70663	0.07603

The converged value of Z is 0.07603 or V = 48.6 cm^3/mol. From our previous discussion, we know this corresponds to the liquid volume as it is the smallest root and satisfies the stability criteria ($A_{VV} > 0$).

To estimate the vapor volume we start with an initial choice of $Z_0 = 1$ and proceed as we did for the liquid phase. Again six iterations are required for convergence, but the value of $Z = 0.07603$ is the same as determined for the liquid phase starting with $Z_0 = bP/RT$. Thus we need a different starting value for Z. If we try $Z_0 = 0.9$ which is more realistic for these values of T and P, Z converges to 0.20014 after six iterations. Unfortunately, this value of Z lies in an unstable region ($A_{VV} < 0$). Choosing a value of $Z_0 = 0.8$, we converge to Z = 0.68210 corresponding to V = 436 cm^3/mol after six iterations. Based on our earlier analysis, this is stable vapor phase volume.

Other successive substitution methods that can be used to solve cubic EOSs include the Newton-Raphson method, Broyden's method, and the method of False Position [see Perry *et al.* (1984) and Walas (1985) for further discussion].

Computer-based solutions: Computer-based numerical algorithms can be written or computer-based algebraic software packages can be used to determine the roots of EOSs. Algorithms are typically written in a structured format employing FORTRAN 77, C, or Microsoft® BASIC and may use the EOS-solving strategies presented in this appendix. The reader is advised to explore Press's (1992) *Numerical Recipes in C* (general coding) and Walas's (1985) Appendix C (EOS coding) for additional information regarding this EOS-root-determining approach.

Software packages can also be used, though care is required since these packages may report values which are not always accurate. In algebraic packages such as Maple®, Mathematica®, and Matlab®, polynomials (up to cubic in nature) are solved rigorously, and polynomials of greater order will be numerically solved with the program's computational engine. Spreadsheet applications such as Microsoft Excel® and Lotus 1-2-3® are also capable of determining EOS roots once the user is familiar with operations, internal numerical routines, and macro programming of the application.

References

Perry, R.H., D.W. Green, and J.O. Maloney (eds.) (1984), *Perry's Chemical Engineers' Handbook*, 6th ed., McGraw-Hill, New York, NY.

Press, W.H., S.A. Peukolsky, W.T. Vetterling, and B.P. Flannery (1992), *Numerical Recipes in C*, 2nd ed., Cambridge University Press, Cambridge, UK.

Walas, S.M. (1985), *Phase Equilibria in Chemical Engineering*, Butterworth, Boston, Appendix C.

F *General Mixture Relationships for Extensive and Intensive Properties*

Any extensive variable, $\underline{B}$, can be expanded in differential form as a function of $(n + 2)$ other variables and, for single-phase, simple systems there are no restrictions on the choice of variables. In Table F.1, columns (b) and (c), we illustrate two variable sets that are commonly employed for mixtures. For reference, the corresponding relations for pure materials are shown in column (a).

In column (b), the independent variables are n mole numbers plus two other properties. In column (c), we have $(n - 1)$ mole fractions, the total number of moles, and two other properties. In this latter set, we have eliminated one mole fraction (i.e., x_n) so that any of the other $(n - 1)$ mole fractions can be varied independently. Thus, when $\underline{B}$ is expressed as a partial derivative with respect to x_i, x_n must vary so that $dx_n = -dx_i$.

Equation (F-2a-c) represents chain-rule expansions of $\underline{B}$ in terms of the independent variables. One should note that the subscript N_i in column (b) denotes that *all* mole numbers are held constant. In column (c) the subscript N indicates that the total moles are constant. Of course, the combined x and N subscripts in column (c) are equivalent to the N_i in column (b). Also, subscripts such as $N_j[i]$ denote that all N_j are held constant except N_i. Similarly, the subscript $x[i, n]$ means that all x_j are constant except x_i and x_n.

Equation (F-2a-c) may be integrated by using Euler's theorem as shown in Appendix C. If Y_1 and Y_2 are extensive, Eq. (F-3a-c) is obtained; if they are intensive, the equations simplify to the form of Eq. (F-4a-c).

If Y_1 and Y_2 are are intensive variables, the form of Eq. (F-4b) indicates that the extensive property $\underline{B}$ may be expressed simply as a weighted average of the partial derivatives $(\partial \underline{B}/\partial N_i)_{Y_1, Y_2, N_j[i]}$. In such cases it has been found that T and P form a convenient set of Y_1 and Y_2, and we define the derivative as a *partial molar property*, $\overline{B}_i$:

$$\overline{B}_i \equiv \left(\frac{\partial \underline{B}}{\partial N_i}\right)_{T, P, N_j[i]} \tag{F-1}$$

Note that partial molar properties are intensive and depend on the temperature, pressure, and composition of the system. The corresponding property for a pure material is, of course, the specific property, B.

With the definition of $\overline{B}_i$, Eqs. (F-2b) and (F-4b) are transformed to Eqs. (F-7b) and (F-6b), as shown in Table F.1. Although not shown, many of the derivatives in column (c) can also be rewritten to employ partial molar properties. For example, let us express the derivative $(\partial \underline{B}/\partial x_i)_{T, P, x[i, n], N}$ of Eq. (F-7c) in terms of partial molar quantities. Starting with Eq. (F-7b), divide all differentials by dx_j and impose upon these derivatives the restraints that T, P, all x_i except x_j and x_n (i.e., $x[j, n]$), and N be constant. The first two terms on the right-hand side vanish; thus,

$$\left(\frac{\partial \underline{B}}{\partial x_j}\right)_{T, P, x[j, n], N} = \sum_{i=1}^{n} \overline{B}_i \left(\frac{\partial N_i}{\partial x_j}\right)_{T, P, x[j, n], N} \tag{F-2}$$

Since

$$N_i = x_i N \tag{F-3}$$

$$\left(\frac{\partial N_i}{\partial x_j}\right)_{T, P, x[j, n], N} = \begin{Bmatrix} 0 & i \neq j, n \\ 1 & i = j \\ -1 & i = n \end{Bmatrix} N \tag{F-4}$$

so that Eq. (F-2) becomes

$$\left(\frac{\partial \underline{B}}{\partial x_j}\right)_{T, P, x[j, n], N} = N(\overline{B}_j - \overline{B}_n) \tag{F-5}$$

The general relations for an intensive property are given in Table F.2; columns (b) and (c) refer to a mixture and column (a) to a pure material. Note that when Y_1 and Y_2 are intensive variables, the sets in columns (a) and (c) involve $(n + 1)$ intensive and one extensive variable. Since any intensive variable can be expressed as a function of any other $(n + 1)$ intensive properties, it follows that $(\partial B/\partial N)$ in Eqs. (F-16a) and (F-16c) must vanish.

Equations (F-18b) and (F-18c) are the differential forms of the two most commonly used sets for expressing an intensive property of a mixture. They contain the terms $(\partial B/\partial N_i)_{T, P, N_j[i]}$ and $(\partial B/\partial x_i)_{T, P, x[i, n]}$, each of which can be expressed in terms of partial molar quantities. Since $(\partial B/\partial N_i)_{T, P, N_j[i]}$ already contains the partial molar set of properties T, P, $N_1, \ldots, N_n$, the transformation follows directly from the substitution of $B = \underline{B}/N$:

$$\left(\frac{\partial B}{\partial N_i}\right)_{T,P,N_j[i]} = \left(\frac{\partial \underline{B}/N}{\partial N_i}\right)_{T,P,N_j[i]} = \frac{\overline{B}_i}{N} - \frac{\underline{B}}{N^2}\left(\frac{\partial N}{\partial N_i}\right)_{T,P,N_j[i]} = \frac{1}{N}(\overline{B}_i - B) \qquad \text{(F-6)}$$

In a similar manner, $(\partial B/\partial x_i)_{T,P,x[i,n]}$ can be related to $(\partial \underline{B}/\partial x_i)_{T,P,x[i,n]}$:

$$\left(\frac{\partial B}{\partial x_i}\right)_{T,P,x[i,n]} = \left(\frac{\partial \underline{B}/N}{\partial x_i}\right)_{T,P,x[i,n]} = \frac{1}{N}\left(\frac{\partial \underline{B}}{\partial x_i}\right)_{T,P,x[i,n]} = \overline{B}_i - \overline{B}_n \qquad \text{(F-7)}$$

Note that N is held constant during differentiation because $(\partial/\partial x_i)_{T,P,x[i,n]}$ implies that N is constant. With Eqs. (F-6) and (F-7), Eqs. (F-18b) and (F-18c) of Table F.2 become

$$dB = \left(\frac{\partial B}{\partial T}\right)_{P,N_i} dT + \left(\frac{\partial B}{\partial P}\right)_{T,N_i} dP + \frac{1}{N}\sum_{i=1}^{n} \overline{B}_i\, dN_i - \frac{B}{N}\, dN \qquad \text{(F-8)}$$

$$dB = \left(\frac{\partial B}{\partial T}\right)_{P,x} dT + \left(\frac{\partial B}{\partial P}\right)_{T,x} dP + \sum_{i=1}^{n-1} (\overline{B}_i - \overline{B}_n)\, dx_i \qquad \text{(F-9)}$$

Expanding the sum in Eq. (F-9), we obtain

$$dB = \left(\frac{\partial B}{\partial T}\right)_{P,x} dT + \left(\frac{\partial B}{\partial P}\right)_{T,x} dP + \sum_{i=1}^{n} \overline{B}_i\, dx_i \qquad \text{(F-10)}$$

Example F.1

Relate the partial derivative $(\partial/\partial x_i)_{T,P,x[i,k]}$ to $(\partial/\partial x_i)_{T,P,x[i,n]}$.

Solution

From Eq. (F-2),

$$\left(\frac{\partial B}{\partial x_i}\right)_{T,P,x[i,n]} = \overline{B}_i - \overline{B}_n \quad \text{and} \quad \left(\frac{\partial B}{\partial x_i}\right)_{T,P,x[i,k]} = \overline{B}_i - \overline{B}_k$$

Then

$$\left(\frac{\partial B}{\partial x_i}\right)_{T,P,x[i,k]} - \left(\frac{\partial B}{\partial x_i}\right)_{T,P,x[i,n]} = \overline{B}_n - \overline{B}_k = \left(\frac{\partial B}{\partial x_n}\right)_{T,P,x[n,k]}$$

In Table F.2, note that we could have expressed B as a function of $(n + 1)$ intensive variables plus one extensive variable, e.g., $(T, P, x_1, \ldots, x_{n-1}, N)$ or $(T, P, x_1, \ldots, x_{n-1}, \underline{Y})$, where $\underline{Y}$ is extensive. In differential form, then

$$dB = \left(\frac{\partial B}{\partial T}\right)_{P,x,\underline{Y}} dT + \left(\frac{\partial B}{\partial P}\right)_{T,x,\underline{Y}} dP + \sum_{i=1}^{n-1}\left(\frac{\partial B}{\partial x_i}\right)_{T,P,x[i,n],\underline{Y}} dx_i + \left(\frac{\partial B}{\partial \underline{Y}}\right)_{T,P,x} d\underline{Y}$$

Applying Euler's theorem, $(\partial B/\partial \underline{Y})_{T,P,x} = 0$, regardless of what $\underline{Y}$ is chosen. It then follows that all derivatives such as $(\partial B/\partial T)_{P,x,\underline{Y}}$ can be written as $(\partial B/\partial T)_{P,x}$ where it is then implied that some extensive variable, $\underline{Y}$, is held constant.

Table F.1
General Extensive Property, $\underline{B}$

(a) Pure material Independent variables Y_1, Y_2, N	(b) Mixture Independent variables $Y_1, Y_2, N_1, \ldots, N_n$	(c) Mixture Independent variables $Y_1, Y_2, x_1, \ldots, x_{n-1}, N$
$\underline{B} = f(Y_1, Y_2, N)$ (F-1a)	$\underline{B} = f(Y_1, Y_2, N_1, \ldots, N_n)$ (F-1b)	$\underline{B} = f(Y_1, Y_2, x_1, \ldots, x_{n-1}, N)$ (F-1c)
$d\underline{B} = \left(\frac{\partial \underline{B}}{\partial Y_1}\right)_{Y_2, N} dY_1 + \left(\frac{\partial \underline{B}}{\partial Y_2}\right)_{Y_1, N} dY_2 + \left(\frac{\partial \underline{B}}{\partial N}\right)_{Y_1, Y_2} dN$ (F-2a)	$d\underline{B} = \left(\frac{\partial \underline{B}}{\partial Y_1}\right)_{Y_2, N_i} dY_1 + \left(\frac{\partial \underline{B}}{\partial Y_2}\right)_{Y_1, N_i} dY_2 + \sum_{i=1}^{n} \left(\frac{\partial \underline{B}}{\partial N_i}\right)_{Y_1, Y_2, N_j[i]} dN_i$ (F-2b)	$d\underline{B} = \left(\frac{\partial \underline{B}}{\partial Y_1}\right)_{Y_2, x, N} dY_1 + \left(\frac{\partial \underline{B}}{\partial Y_2}\right)_{Y_1, x, N} dY_2 + \sum_{i=1}^{n-1} \left(\frac{\partial \underline{B}}{\partial x_i}\right)_{Y_1, Y_2, x[i, n], N} dx_i + \left(\frac{\partial \underline{B}}{\partial N}\right)_{Y_1, Y_2, x} dN$ (F-2c)
Integration by Euler's theorem Assuming Y_1 and Y_2 are extensive properties		
$\underline{B} = \left(\frac{\partial \underline{B}}{\partial Y_1}\right)_{Y_2, N} Y_1 + \left(\frac{\partial \underline{B}}{\partial Y_2}\right)_{Y_1, N} Y_2 + \left(\frac{\partial \underline{B}}{\partial N}\right)_{Y_1, Y_2} N$ (F-3a)	$\underline{B} = \left(\frac{\partial \underline{B}}{\partial Y_1}\right)_{Y_2, N_i} Y_1 + \left(\frac{\partial \underline{B}}{\partial Y_2}\right)_{Y_1, N_i} Y_2 + \sum_{i=1}^{n} \left(\frac{\partial \underline{B}}{\partial N_i}\right)_{Y_1, Y_2 N_j[i]} N_i$ (F-3b)	$\underline{B} = \left(\frac{\partial \underline{B}}{\partial Y_1}\right)_{Y_2, x, N} Y_1 + \left(\frac{\partial \underline{B}}{\partial Y_2}\right)_{Y_1, x, N} Y_2 + \left(\frac{\partial \underline{B}}{\partial N}\right)_{Y_1, Y_2, x} N$ (F-3c)
Simplifications if Y_1 and Y_2 are intensive properties		
$\underline{B} = \left(\frac{\partial \underline{B}}{\partial N}\right)_{Y_1, Y_2} N$ (F-4a)	$\underline{B} = \sum_{i=1}^{n} \left(\frac{\partial \underline{B}}{\partial N_1}\right)_{Y_1, Y_2, N_j[i]} N_i$ (F-4b)	$\underline{B} = \left(\frac{\partial \underline{B}}{\partial N}\right)_{Y_1, Y_2, x} N$ (F-4c)
$\left(\frac{\partial \underline{B}}{\partial N}\right)_{Y_1, Y_2} = \frac{\underline{B}}{N} = B$ (F-5a)	$\left(\frac{\partial \underline{B}}{\partial N_i}\right)_{Y_1, Y_2 N_j[i]} \neq \frac{\underline{B}}{N_i}$ (F-5b)	$\left(\frac{\partial \underline{B}}{\partial N}\right)_{Y_1, Y_2, x} = \frac{\underline{B}}{N} = B$ (F-3c)
With $Y_1 = T$ and $Y_2 = P$		
$\underline{B} = \left(\frac{\partial \underline{B}}{\partial N}\right)_{T, P} N = BN$ (F-6a)	$\underline{B} = \sum_{i=1}^{n} \left(\frac{\partial \underline{B}}{\partial N_1}\right)_{T, P, N_j[i]} N_i = \sum_{i=1}^{n} \overline{B}_i N_i$ (F-6b)	$\underline{B} = \left(\frac{\partial \underline{B}}{\partial N_i}\right)_{T, P, x} N = BN$ (F-6c)
Eq. (F-2a) becomes $d\underline{B} = \left(\frac{\partial \underline{B}}{\partial T}\right)_{P, N} dT + \left(\frac{\partial \underline{B}}{\partial P}\right)_{T, N} dP + B\, dN$ (F-7a)	Eq. (F-2b) becomes $d\underline{B} = \left(\frac{\partial \underline{B}}{\partial T}\right)_{P, N_i} dT + \left(\frac{\partial \underline{B}}{\partial P}\right)_{T, N_i} dP + \sum_{i=1}^{n} \overline{B}_i\, dN_i$ (F-7b)	Eq. (F-2c) becomes $d\underline{B} = \left(\frac{\partial \underline{B}}{\partial T}\right)_{P, x, N} dT + \left(\frac{\partial \underline{B}}{\partial P}\right)_{T, x, N} dP + \sum_{i=1}^{n-1} \left(\frac{\partial \underline{B}}{\partial x_i}\right)_{T, P, x[i, n], N} dx_i + B\, dN$ (F-7c)

Table F.2
General Intensive Property, B

(a) Pure material Independent variables Y_1, Y_2, N	(b) Mixture Independent variables $Y_1, Y_2, N_1, \ldots, N_n$	(c) Mixture Independent variables $Y_1, Y_2, x_1, \ldots, x_{n-1}, N$
$B = f(Y_1, Y_2, N)$ (F-13a)	$B = f(Y_1, Y_2, N_1, \ldots, N_n)$ (F-13b)	$B = f(Y_1, Y_2, x_1, \ldots, x_{n-1}, N)$ (F-13c)
$dB = \left(\frac{\partial B}{\partial Y_1}\right)_{Y_2, N} dY_1 + \left(\frac{\partial B}{\partial Y_2}\right)_{Y_1, N} dY_2 + \left(\frac{\partial B}{\partial N}\right)_{Y_1, Y_2} dN$ (F-14a)	$dB = \left(\frac{\partial B}{\partial Y_1}\right)_{Y_2, N_i} dY_1 + \left(\frac{\partial B}{\partial Y_2}\right)_{Y_1, N_1} dY_2 + \sum_{i=1}^{n} \left(\frac{\partial B}{\partial N_i}\right)_{Y_1, Y_2, N_j[i]} dN_i$ (F-14b)	$dB = \left(\frac{\partial B}{\partial Y_1}\right)_{Y_2, x, N_i} dY_1 + \left(\frac{\partial B}{\partial Y_2}\right)_{Y_1, x, N_1} dY_2 + \sum_{i=1}^{n-1} \left(\frac{\partial B}{\partial x_i}\right)_{Y_1, Y_2, x[i, n], N} dx_i + \left(\frac{\partial B}{\partial N}\right)_{Y_1, Y_2, x} dN$ (F14c)
Integration by Euler's theorem Assuming Y_1 and Y_2 are extensive properties		
$0 = \left(\frac{\partial B}{\partial Y_1}\right)_{Y_2, N} Y_1 + \left(\frac{\partial B}{\partial Y_2}\right)_{Y_1, N} Y_2 + \left(\frac{\partial B}{\partial N}\right)_{Y_1, Y_2} N$ (F-15a)	$0 = \left(\frac{\partial B}{\partial Y_1}\right)_{Y_2, N} Y_1 + \left(\frac{\partial B}{\partial Y_2}\right)_{Y_1, N_i} Y_2 + \sum_{i=1}^{n} \left(\frac{\partial B}{\partial N_i}\right)_{Y_1, Y_2, N_j[i]} N_i$ (F-15b)	$0 = \left(\frac{\partial B}{\partial Y_1}\right)_{Y_2, x, N} Y_1 + \left(\frac{\partial B}{\partial Y_2}\right)_{Y_1, x, N} Y_2 + \left(\frac{\partial B}{\partial N}\right)_{Y_1, Y_2, x} N$ (F-15c)
Simplifications if Y_1 and Y_2 are intensive properties		
$0 = \left(\frac{\partial B}{\partial N}\right)_{Y_1, Y_2}$ (F-16a)	$0 = \sum_{i=1}^{n} \left(\frac{\partial B}{\partial N_i}\right)_{Y_1, Y_2, N_j[i]} N_i$ (F-4b) $\left[but \left(\frac{\partial B}{\partial N_i}\right)_{Y_1, Y_2, N_j[i]} \neq 0 \right]$ (F-16b)	$0 = \left(\frac{\partial B}{\partial N}\right)_{Y_1, Y_2, x}$ (F-16a)
With $Y_1 = T$ and $Y_2 = P$		
$0 = \left(\frac{\partial B}{\partial N}\right)_{T, P}$ (F-17a) Eq. (F-14a) becomes $dB = \left(\frac{\partial B}{\partial T}\right)_{P} dT + \left(\frac{\partial B}{\partial P}\right)_{T} dP$ (F-18a)	$0 = \sum_{i=1}^{n} \left(\frac{\partial B}{\partial N_i}\right)_{T, P, N_j[i]} N_i$ (F-17b) Eq. (F-14b) becomes $dB = \left(\frac{\partial B}{\partial T}\right)_{P, N_i} dT + \left(\frac{\partial B}{\partial P}\right)_{T, N_i} dP + \sum_{i=1}^{n} \left(\frac{\partial B}{\partial N_i}\right)_{T, P, N_j[i]} dN$ (F-18b)	$0 = \left(\frac{\partial B}{\partial N}\right)_{T, P, x}$ (F-17c) Eq. (F-14c) becomes $dB = \left(\frac{\partial B}{\partial T}\right)_{P, x} dT + \left(\frac{\partial B}{\partial P}\right)_{T, x} dP + \sum_{i=1}^{n-1} \left(\frac{\partial B}{\partial x_i}\right)_{T, P, x[i, n]} dx_i$ (F-18c)

G Pure Component Property Data

The following tables contain pure component property data for 22 liquids and gases. Most of the data were taken from Reid *et al.* (1987) and from the latest version of *Physical Property Methods and Models* in ASPEN PLUS™ (1995). Values for the constants used in the Antoine equation for pure component vapor pressure were obtained from Reid *et al.* (1977).

In the table of ideal-gas-state heat capacity data, the powers of ten which appear next to the correlation constants a, b, c, and d, represent values by which the actual constants have been multiplied, e.g., the specific heat of argon $C^o_{p,\mathrm{Ar}} = 20.8$ J/mol K. Caution is advised as data correlations are valid only over limited temperature ranges. For the heat capacity correlations, we recommend a range of 298 K to 1500 K. For the Antoine correlation, a temperature range is supplied. Tabulated are the following:

m_i : molecular weight (g/mol)
T_c : critical temperature (K)
P_c : critical pressure (bar)
V_c : critical volume (cm^3/mol)
Z_c : critical compressibility, P_cV_c/RT_c
ω : Pitzer's accentric factor
d_m : dipole moment (debyes, D)

C^o_p : ideal-gas-state heat capacity (J/mol K)
a, b, c, d : specific heat constants
P_{vp} : vapor pressure (mm Hg)
A, B, C : Antoine constants
ΔH^o_f (298.15 K) : Standard ideal-gas-state enthalpy of formation at 1 atm (kJ/mol)
ΔG^o_f (298.15 K) : Standard ideal-gas-state Gibbs free energy of formation at 1 atm (kJ/mol)

$$C^o_p = a + bT + cT^2 + dT^3 \quad \text{J/mol K} \; (T \text{ in K})$$

$$\ln P_{vp} = A - \frac{B}{T+C} \quad P_{vp} \text{ in mm Hg } (T \text{ in K})$$

References

Reid, R.C., J.M. Prausnitz, and B.E. Poling (1987), *The Properties of Gases and Liquids*, 4th edition, McGraw-Hill, New York.

Reid, R.C., J.M. Prausnitz, and T.K. Sherwood (1977), *The Properties of Gases and Liquids*, 3rd edition, McGraw-Hill, New York.

ASPEN PLUS™ Release 9 (1995), Volume 2: *Physical Property Methods and Models*, Volume 3: *Physical Property Data*.

Compound	m_i (g/mol)	T_c (K)	P_c (bar)	V_c (cm^3/mol)	Z_c	ω	d_m (D)	ΔH_f^o (298.15 K) kJ/mol	ΔG_f^o (298.15 K) kJ/mol
Ar	39.948	150.8	48.7	74.9	0.291	0.001	0.0	0	0
C_2H_5OH	46.069	513.9	61.4	167.1	0.240	0.644	1.7	–235.0	–168.4
C_2H_6	30.070	305.4	48.8	148.3	0.285	0.099	0.0	–84.74	–32.95
C_3H_8	44.094	369.8	42.5	203.0	0.281	0.153	0.0	–103.9	–234.9
C_6H_5OH	94.113	694.2	61.3	229.0	0.240	0.438	1.6	–96.42	–32.90
C_6H_6	78.114	562.2	48.9	259.0	0.271	0.212	0.0	82.98	129.7
CH_2Cl_2	84.933	510.0	63.0	—	—	0.199	1.8	–95.46	–68.91
$(CH_3)_2CO$	58.080	508.1	47.0	209.0	0.232	0.304	2.9	–217.7	–153.2
CH_3COOH	60.052	592.7	57.9	171.0	0.201	0.447	1.3	–435.1	–376.9
CH_3OH	32.042	512.6	80.9	118.0	0.224	0.556	1.7	–201.3	–162.6
CH_4	16.043	190.4	46.0	99.2	0.288	0.011	0.0	–74.90	–50.87
$CHCl_3$	119.378	536.4	53.7	238.9	0.293	0.218	1.1	–101.3	–685.8
CO	28.010	132.9	35.0	93.2	0.295	0.066	0.1	–110.6	–137.4
CO_2	44.01	304.2	73.8	93.9	0.274	0.225**	0.0	–393.8	–394.6
H_2	2.016	33.0	12.9	64.3	0.303	–0.216	0.0	0	0
H_2O	18.015	647.3	221.2	57.1	0.235	0.344	1.8	–242.0	–228.8
N_2	28.013	126.2	33.9	89.8	0.290	0.039	0.0	0	0
N_2O	44.013	309.6	72.4	97.4	0.274	0.165	0.2	81.60	103.7
NH_3	17.031	405.5	113.5	72.5	0.244	0.250	1.5	–45.72	–16.16
NO	30.006	180	64.8	57.7	0.250	0.588	0.2	90.43	86.75
O_2	31.999	154.6	50.5	73.4	0.288	0.021	0.0	0	0
SO_3	80.058	491.0	82.1	127.3	0.256	0.481	0.0	–396.0	–371.3

** $\omega = 0.239$ if metastable liquid-phase vapor pressure is used at $T_r = 0.7$ where solid CO_2 exists (see data in Problem 15.25)

Compound	$a \times 10^{-1}$	$b \times 10^{2}$	$c \times 10^{4}$	$d \times 10^{8}$	A	B	C	T range (K)
Ar	2.080	—	—	—	15.2330	700.51	–5.84	81 — 94
C_2H_5OH	0.9014	21.41	–0.8390	0.1373	18.9119	3803.98	–41.68	270 — 369
C_2H_6	0.5409	17.81	–0.6938	0.8713	15.6637	1511.42	–17.16	130 — 199
C_3H_8	–0.4224	30.63	–1.586	3.215	15.7260	1872.46	–25.16	164 — 249
C_6H_5OH	–3.584	59.83	–4.827	15.27	16.4279	3490.89	–98.59	345 — 481
C_6H_6	–3.392	47.39	–3.017	7.130	15.9008	2788.51	–52.36	280 — 377
CH_2Cl_2	1.295	16.23	–1.302	4.208	16.3029	2622.44	–41.70	229 — 332
$(CH_3)_2CO$	0.6301	26.06	–1.253	2.038	16.6513	2940.46	–35.93	241 — 350
CH_3COOH	0.4840	25.49	–1.753	4.949	16.8080	3405.57	–56.34	290 — 430
CH_3OH	2.115	7.092	0.2587	–2.852	18.5875	3626.55	–34.29	257 — 364
CH_4	1.925	5.213	0.1197	–1.132	15.2243	897.84	–7.16	93 — 120
$CHCl_3$	2.400	18.93	–1.841	6.657	15.9732	2696.79	–46.16	260 — 370
CO	3.087	–1.285	0.2789	–1.272	14.3686	530.22	–13.15	63 — 108
CO_2	1.980	7.344	–0.5602	1.715	22.5898	3103.39	–0.16	154 — 204
H_2	2.714	0.9274	–0.1381	0.7645	13.6333	164.90	3.19	14 — 25
H_2O	3.224	0.1924	0.1055	–0.3596	18.3036	3816.44	–46.13	284 — 441
N_2	3.115	–1.357	0.2680	–1.168	14.9542	588.72	–6.60	54 — 90
N_2O	2.162	7.281	–0.5778	1.830	16.1271	1506.49	–25.99	144 — 200
NH_3	2.731	2.383	0.1707	–1.185	16.9481	2132.50	–32.98	179 — 261
NO	2.935	–0.09378	0.09747	–0.4187	20.1314	1572.52	–4.88	95 — 140
O_2	2.811	-3.680×10^{-4}	0.1746	–1.065	15.4075	734.55	–6.45	63 — 100
SO_3	1.921	13.74	–1.176	3.700	20.8403	3995.70	–36.66	290 — 332

Index

D

E

P

Q

R

S

T

U

V

W

Y

Z

H Conversion Factors and Physical Constants

Length

1 m = 3.28084 ft = 39.3701 in
1 ft = 12 in = 0.3048 m = 30.48 cm
1 Å = 10^{-10} m = 10^{-8} cm = 0.1 nm
1 mile (mi) = 5280 ft = 1609 m

Mass (m)

1 kg = 2.20462 lb_m = 35.2739 oz
1 metric tonne = 1000 kg

Volume (V)

1 m^3 = 35.3147 ft^3 = 264.17 gal
= 1000 L
1 L = 0.03531 ft^3 = 61.02 in^3
1 ft^3 = 0.02832 m^3 = 7.477 gal
1 gal = 3.785 L

Density (ρ)

1 g/cm^3 = 1000 kg/m^3
= 62.428 lb_m/ft^3

Dipole moment (d_m)

1 D = 3.336×10^{-30} C m

Force (F)

1 N = 1 kg m/s^2
= 10^5 dyne
= 0.224809 lb_f

Time (t)

1 min = 60 s
1 hr = 3600 s
1 day = 86,400 s
1 yr = 365.24 day = 3.156×10^7 s

Pressure (P)

1 bar = 10^5 N/m^2 (Pa)
= 10^6 dyne/cm^2
= 0.986923 atm
= 14.5038 psia
= 750.061 torr
1 atm = 1.01325 bar
= 760 mm Hg at 0°C (torr)
= 10.333 m H_2O at 4°C
= 14.697 lb_f/in^2 (psia)

Energy (E)

1 J = 1 N m = 10^7 erg
1 ft lb_f = 1.356 J
1 cal = 4.186 J
1 Btu = 1055 J = 252 cal = 778 ft lb_f
1 eV = 1.602×10^{-19} J
1 kWh = 3.600×10^6 J

Power

1 W = 1 J/s
1 hp = 746 W = 550 ft lb_f s^{-1}
1 Btu/hr = 0.293 W

Surface tension (σ)

1 dyne/cm = 10^{-3} N/m

Velocity (v)

1 m/s = 3.281 ft/s
1 mi/min = 60 mi/hr = 88 ft/s
1 km/hr = 0.2778 m/s = 0.6214 mi/hr
1 mi/hr = 1.467 ft/s = 0.4470 m/s